SUMMARIES—WHERE to LOOK for HELP

PREREQUISITES

Your algebra, trigonometry, or precalculus textbooks, if you kept them, are good places to look because you are familiar with the notation and presentation. If you do not have these books for reference, we have provided the necessary review in *Student Mathematics Handbook*. Also see

Properties of absolute value, see Table 1.1, p. 3
Exact values of trigonometric functions, see Table 1.2, page 9
Directory of curves, see Table 1.3, page 27
Definition of inverse trigonometric functions, see Table 1.4, p. 38

TECHNOLOGY

See *Technology Manuals* for *Maple*, MATLAB, and *Mathematica*. We have also found the TI-92 to be of great value in working many calculus problems. Inexpensive software includes Derive (which is included with the TI-92) and Converge for graphical representation of many calculus concepts.

DIFFERENTIATION AND GRAPHING

Definition, p. 100
Derivative formulas, see back endsheet
Holes, poles, jumps, and continuity, see Figure 2.18, p. 70
First derivative and second derivative tests, p. 205 and 209
Graphing strategy, see Table 4.1, p. 226
Optimization procedure, p. 238

INTEGRATION

Definition, p. 274 (indefinite), p. 293 (definite)
Summation formulas, p. 287
Fundamental Theorem of Calculus, p. 303 (first), p. 306 (second)
Integration formulas, see back endsheet
Short integration table, see Appendix D
Integration table, see *Student Mathematics Handbook*
Volumes of revolution, see Table 6.1, p. 371
Integration strategy, see Table 7.2, pp. 457–458
Modeling applications involving integration, see Table 10.3, pp. 689–690

INFINITE SERIES

Guidelines for determining convergence of series, Table 8.1, pp. 540–541
Power series for elementary functions, Table 8.2, p. 563

ANALYTIC GEOMETRY VECTORS, AND THREE DIMENSIONS

Summary of volumes of revolution, Table 6.1, p. 371
Directory of polar-form curves, Table 6.2, p. 379
Catalog of quadric surfaces, Table 9.1, p. 625
Data on planetary orbits, Table 10.1, p. 657
Curvature formulas, Table 10.2, p. 671
Summary of velocity, acceleration, and curvature, Table 10.4, p. 690
Conversion formulas for coordinate systems, Section 12.7

FUNCTIONS OF SEVERAL ARIABLES

Comparison of important integral theorems, see Chapters 11–13 Cumulative Review, pp. 000–000
Properties of vector fields, see Chapter 13 Review, pp. 000-000

DIFFERENTIAL EQUATIONS

Introduction to differential equations, see Section 5.6, p. 316
Summary of growth models, see Table 7.3, p. 466
First-order differential equations, Section 7.6, p. 461
Summary of strategies for solving first-order differential equations, Table 14.1, p. 000
Second-order linear differential equations, Section 14.3, p. 000

MISCELLANEOUS FORMULAS

SPECIAL TRIGONOMETRIC LIMITS: $\displaystyle\lim_{h\to 0}\frac{\sin h}{h}=1$ $\qquad\qquad\displaystyle\lim_{h\to 0}\frac{\cos h - 1}{h}=0$

APPROXIMATION INTEGRATION METHODS: Let $\Delta x = \dfrac{b-a}{n}$ and, for the kth subinterval, $x_k = a + k\Delta x$; also suppose f is continuous throughout the interval $[a, b]$.

Rectangular Rule: (right endpoints) $\displaystyle\int_a^b f(x)\,dx \approx [f(a + \Delta x) + f(a + 2\Delta x) + \cdots + f(a + n\Delta x)]\Delta x$

Trapezoidal Rule: $\displaystyle\int_a^b f(x)\,dx \approx \tfrac{1}{2}[f(x_0) + 2f(x_1) + 2f(x_2) + \cdots + 2f(x_{n-1}) + f(x_n)]\Delta x$

Simpson Rule: $\displaystyle\int_a^b f(x)\,dx \approx \tfrac{1}{3}[f(x_0) + 4f(x_1) + 2f(x_2) + \cdots + 4f(x_{n-1}) + f(x_n)]\Delta x$

ARC LENGTH: Let f be a function whose derivative f' is continuous on the interval $[a, b]$. Then the **arc length**, s, of the graph of $y = f(x)$ between $x = a$ and $x = b$ is given by the integral

$$s = \int_a^b \sqrt{1 + [f'(x)]^2}\,dx$$

If C is a parametrically described curve $x = x(t)$, $y = y(t)$ that is simple for $a < t < b$, then

$$s = \int_a^b \sqrt{\left(\frac{dx}{dt}\right)^2 + \left(\frac{dy}{dt}\right)^2}\,dt$$

SERIES Suppose there is an open interval I containing c throughout which the function f and all its derivatives exist.

Taylor Series of f: $f(x) = f(c) + \dfrac{f'(c)}{1!}(x - c) + \dfrac{f''(c)}{2!}(x - c)^2 + \dfrac{f'''(c)}{3!}(x - c)^3 + \cdots$

Maclaurin Series of f: $f(x) = f(0) + \dfrac{f'(0)}{1!}x + \dfrac{f''(0)}{2!}x^2 + \dfrac{f'''(0)}{3!}x^3 + \cdots$

FIRST-ORDER LINEAR D.E. The general solution of the first-order linear differential equation

$$\frac{dy}{dx} + P(x)y = Q(x)$$

is given by

$$y = \frac{1}{I(x)}\left[\int Q(x)I(x)\,dx + C\right]$$

where $I(x) = e^{\int P(x)\,dx}$ and C is an arbitrary constant.

VECTOR-VALUED FUNCTIONS Del operator: $\nabla = \dfrac{\partial}{\partial x}\mathbf{i} + \dfrac{\partial}{\partial y}\mathbf{j} + \dfrac{\partial}{\partial z}\mathbf{k}$

Gradient: $\nabla f = \dfrac{\partial f}{\partial x}\mathbf{i} + \dfrac{\partial f}{\partial y}\mathbf{j} + \dfrac{\partial f}{\partial z}\mathbf{k} = f_x\mathbf{i} + f_y\mathbf{j} + f_z\mathbf{k}$

Laplacian: $\nabla^2 f = \dfrac{\partial^2 f}{\partial x^2} + \dfrac{\partial^2 f}{\partial y^2} + \dfrac{\partial^2 f}{\partial z^2} = f_{xx} + f_{yy} + f_{zz}$

Derivatives of a vector field $\mathbf{F} = u\mathbf{i} + v\mathbf{j} + w\mathbf{k}$:

$$\operatorname{div}\mathbf{F} = \frac{\partial u}{\partial x} + \frac{\partial v}{\partial y} + \frac{\partial w}{\partial z} = \nabla \cdot \mathbf{F} \qquad \textit{This is a scalar derivative.}$$

$$\operatorname{curl}\mathbf{F} = \begin{vmatrix} \mathbf{i} & \mathbf{j} & \mathbf{k} \\ \dfrac{\partial}{\partial x} & \dfrac{\partial}{\partial y} & \dfrac{\partial}{\partial z} \\ u & v & w \end{vmatrix} = \nabla \times \mathbf{F} \qquad \textit{This is a vector derivative.}$$

SINGLE VARIABLE
CALCULUS

THIRD EDITION

Monty J. Strauss
Texas Tech University

Gerald L. Bradley
Claremont McKenna College

Karl J. Smith
Santa Rosa Junior College

Prentice Hall, Upper Saddle River, New Jersey 07458

Library of Congress Cataloging in Publication Data

Strauss, Monty J.
  Single variable calculus / Monty J. Strauss, Gerald L. Bradley, Karl J. Smith–3rd ed.
  p. cm.
  Rev. ed. of: Single variable calculus / Gerald L. Bradley, 2nd ed. c1999
  Includes index.
  ISBN 0-13-041550-2

  1. Calculus.   I. Title: Calculus. II. Bradley, Gerald L., Single variable calculus.
  III. Smith, Karl J. IV. Title.

QA303.2 S774 2002   CIP
515--dc21   20021036031

Acquisition Editor: *George Lobell*
Editor in Chief: *Sally Yagan*
Production Editor: *Lynn Savino Wendel*
Vice President/Director of Production and Manufacturing: *David W. Riccardi*
Senior Managing Editor: *Linda Mihatov Behrens*
Executive Managing Editor: *Kathleen Schiaparelli*
Assistant Managing Editor: *Bayani Mendoza de Leon*
Manufacturing Buyer: *Alan Fischer*
Manufacturing Manager: *Trudy Pisciotti*
Marketing Manager: *Angela Battle*
Assistant Editor of Media: *Vince Jansen*
Marketing Assistant: *Rachel Beckman*
Development Editor: *Irene Nunes*
Editor in Chief, Development: *Carol Trueheart*
Art Director: *Heather Scott*
Assistant to Art Director: *John Christiana*
Creative Director: *Carole Anson*
Director of Creative Services: *Paul Belfanti*
Managing Editor, Audio/Video Assets: *Grace Hazeldine*
Art Editor: *Thomas Benfatti*
Assistant Managing Editor, Math Media Production: *John Matthews*
Interior Designer: *Anne Flanagan*
Editorial Assistant/Supplements Editor: *Melanie Van Benthuysen*
Photo Researcher: *Karen Pugliano*
Photo Editor: *Beth Boyd*
Cover Designer: *Anne Flanagan*
Cover Photo: Inside the Louvre Pyramid (which is shown on back cover) looking upward.
Cover Photo Credits: *Antionio Martinelli*
Historical Quest Portraits: *Steven S. Nau*
Art Studio: Artworks
            Senior Manager: *Patty Burns*
            Production Manager: *Ronda Whitson*
            Manager, Production Technologies: *Matt Haas*
            Project Coordinator: *Jessica Einsig*
            Illustrator: *Mark Landis*

©2002, 1999, 1995 by Prentice-Hall, Inc.
Upper Saddle River, New Jersey 07458

Printed in the United States of America

10 9 8 7 6 5 4 3 2 1

ISBN 0-13-041550-2

Pearson Education Ltd.
Pearson Education Australia PTY, Limited
Pearson Education Singapore, Pte. Ltd
Pearson Education North Asia Ltd.
Pearson Education Canada, Ltd.
Pearson Educación de Mexico, S.A. de C.V.
Pearson Education—Japan
Pearson Education Malaysia, Pte. Ltd.

Contents

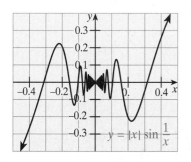
$y = |x| \sin \dfrac{1}{x}$

3 Differentiation   97

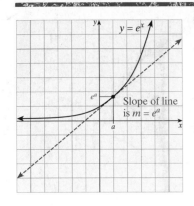

4 Additional Applications of the Derivative   183

5 Integration   271

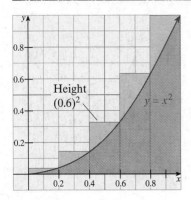

6 Additional Applications of the Integral   355

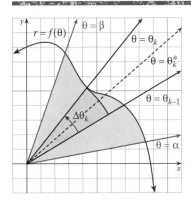

7 Methods of Integration   425

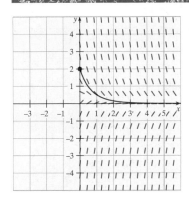

8 Infinite Series   493

*The shaded chapters 11–14 appear only in *Multivariable Calculus*. Chapters 9 and 10 appear in both *Single Variable Calculus* and *Multivariable Calculus*.

12 Multiple Integration   777

Tangent plane

$\mathbf{R}_v\Delta v$

S_k

S

$\mathbf{R}_u\Delta u$

Δy

D_k

Δx

13 Vector Analysis   859

14 Introduction to Differential Equations   935

Exact solution

(x_0, y_0)

$hf(x_0, y_0)$

h

x_0

$x_0 + h$

Appendices

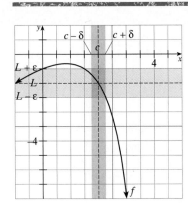

Index   A-71

Preface

This text was developed to blend the best aspects of calculus reform with the reasonable goals and methodology of traditional calculus. It achieves this middle ground by providing sound development, stimulating problems, and well-developed pedagogy within a framework of a traditional structure. "Think, then do," is a fair summary of our approach.

New to this Edition

The acceptance and response from our first two editions has been most gratifying. For the third edition, we wanted to take a good book and make it even better. If you are familiar with the previous editions, the first thing you will notice is that we have added a new coauthor, Monty J. Strauss. His added expertise, and his attention to accuracy and detail, as well as his many years of experience teaching calculus, have added a new dimension to our exposition. Here is what is new in the third edition:

Organization

- In this edition we introduce e^x and $\ln x$ in Chapter 2 after we have defined the notion of a limit. This is beneficial because it allows the number e to be properly defined using limits. We also assume a knowledge of the conic sections and their graphs. A free *Student Mathematics Handbook* is available that provides review and reference material on these transcendental functions.
- l'Hôpital's Rule is now covered earlier in Chapter 4. This placement allows instructors to explore more interesting applications like curve sketching.
- A new section covering applications to business, economics, and the life sciences has been added to Chapter 6 on Applications of the Integral. This new material is designed to help students see how calculus relates to and is used in other disciplines.
- The chapter on polar coordinates and parametric forms has been distributed to other chapters in the book. The polar coordinate system and graphing in polar forms is in Chapter 6 in the context of the integration topic of finding areas. Parametric representation of curves now appears in the book where it is first needed, in Chapter 9.
- Modeling continues as a major theme in this edition. Modeling is now introduced in Section 3.4, and then appears in almost every section of the book. These applications are designated MODELING PROBLEMS. Some authors use the words "Modeling Problem" to refer to any applied problem. In the third edition of *Calculus*, we make a distinction between *modeling problems* and *application problems* by defining a modeling problem as follows. A **modeling problem** is a problem that requires that the reader make some assumptions about the real world in order to derive or come up

with the necessary mathematical formula or mathematical information to answer the question. These problems also include real-world examples of modeling by citing the source of the book or journal that shows the modeling process.

Problem Sets

- We have added a new major category of problems, called **counterexample problem**. A *counterexample* is an example that disproves a proposition or theorem, and in mathematics we are often faced with a proposition that is true or false, and our task is to prove the proposition true or to find a counterexample to disprove the proposition. In the third edition of *Calculus*, we attempt to build the student's ability with this type of situation to mean that the student must either find justification that the proposition is true or else find a counterexample. We believe this new form of problem to be important for preparing the student for future work in not only advanced mathematics courses, but also for analytically oriented courses.
- **Exploration Problems** explore concepts which may prove true or false and provide opportunities for innovative thinking.
- **Interpretation Problems** require exposition that requires a line of thinking that is not directly covered in the textbook.

Supplements

- **Interactive CD** (free with every new copy). The new CD-ROM is designed to enhance students' computational and conceptual understanding of calculus. This CD-ROM is not an add-on of extra material to the text but rather an incredibly useful expansion of the text. See the media supplement portion of the Walk-Through for a complete description of each CD.
- **TestGen-EQ.** This easy to use test generator contains all of the questions from the printed Test Item File.
- **Prentice Hall Online Homework Grader.** For more details, see the PH Homework Grader section in the Walk-Through.

Hallmark Features

Some of the distinguishing characteristics of the earlier editions are continued with this edition:

- It is possible to begin the course with either Chapter 1 or Chapter 2 (where the calculus topics begin).
- We believe that students *learn* mathematics by *doing* mathematics. Therefore, the **problems and applications** are perhaps the most important feature of any calculus book. You will find that the problems in this book extend from routine practice to challenging. The problem sets are divided into *A* Problems (routine), *B* Problems (requiring independent thought), and *C* Problems (theory problems). You will find the scope and depth of the problems in this book to be extraordinary while engineering and physics examples and problems play a prominent role, we include applications from a wide variety of fields, such as biology, economics, ecology, psychology, and sociology. In addition, the chapter summaries provide not only topical review, but also many miscellaneous exercises. Although the chapter reviews are typical of examinations, the miscellaneous problems are not presented as graded problems, but rather as a random list of problems loosely tied to the ideas of that chapter. In addition, there are cumulative reviews located at natural subdivision points in the text: Chapters 1–5, Chapters 6–8, Chapters 1–10, and Chapters 11–13. For a full description of each type of problem available, see pages xviii through xxi of Walk-Through.
- We understand that students often struggle with prerequisite material. Further, it is often frustrating for instructors to have to reteach material from previous courses. As

a result, we have created a unique **Student Mathematics Handbook**. This handbook functions as a **"Just-in-Time" Precalculus Review** that provides precalculus drill/review material, a catalog of curves, analytic geometry, and integral tables. Students are guided through the text to this handbook by a SMH symbol located in the text margin. This guide is entirely author-written and offered FREE with every new copy of the text.

(SMH)

- We have taken the introduction of **differential equations** seriously. Students in many allied disciplines need to use differential equations early in their studies and consequently cannot wait for a postcalculus course. We introduce differential equations in a natural and reasonable way. Slope fields are introduced as a geometric view of antidifferentiation in Section 5.1, and then are used to introduce a graphical solution to differential equations in Section 5.6. We consider separable differential equations in Chapter 5 and first-order linear equations in Chapter 7, and demonstrate the use of both modeling a variety of applied situations. Exact and homogeneous differential equations appear in Chapter 14, along with an introduction to second-order linear equations. The "early and often" approach to differential equations is intended to illustrate their value in continuous modeling and to provide a solid foundation for further study.

- **Visualization** is used to help students develop better intuition. Much of this visualization appears in the wide margins to accompany the text. Also, since tough calculus problems are often tough geometry (and algebra) problems, this emphasis on graphs will help students' problem-solving skills. Additional graphs are related to the student problems, including answer art.

- We have included dozens of "TECHNOLOGY NOTES" devoted to the use of technology. We strive to keep such references "platform neutral" because specific calculators and computer programs frequently change and are better considered in separate technology manuals. These references are designed to give insight into how technological advances can be used to help understand calculus. Problems requiring a graphing calculator or software and computer also appear in the exercises. On the other hand, problems that are not specially designated may still use technology (for example, to solve a higher-degree equation). Several technology manuals are also available at a discount price. See Instructor/Student Supplement section in the Walk-Through for details.

- **Guest essays** provide alternate viewpoints. The questions that follow are called MATHEMATICAL ESSAYS and are included to encourage individual writing assignments and mathematical exposition. We believe that students will benefit from individual writing and research in mathematics. Another pedagogical feature is the **"What this says:"** box in which we rephrase mathematical ideas in everyday language. In the problem sets we encourage students to summarize procedures and processes or to describe a mathematical result in everyday terms.

- **Group research projects**, each of which appears at the end of a chapter and involves intriguing questions whose mathematical content is tied loosely to the chapter just concluded. These projects have been developed and class-tested by their individual authors, to whom we are greatly indebted. Note that the complexity of these projects increases as we progress through the book and the mathematical maturity of the student is developed.

- We continue to utilize the **humanness** of mathematics. History is not presented as additional material to learn. Rather we have placed history into *problems* that lead the reader from the development of a concept to actually participating in the discovery process. The problems are designated as Historical Quest problems. The problems are not designed to be "add-on or challenge problems," but rather to become an integral part of the assignment. The level of difficulty of Quest problems ranges from easy to difficult. An extensive selection of biographies of noted mathematicians can be found on the internet site accompanying this text (www.prenhall.com/strauss)

Accuracy and Error Checking

Because of careful checking and proofing by several people besides each of the three authors, the authors and publisher believe this book to be substantially error free. For any errors remaining, the authors would be grateful if they were sent to Monty J. Strauss at **M.Strauss@ttu.edu** or Monty J. Strauss, Department of Mathematics and Statistics, Texas Tech University, Lubbock, TX 79409-1042.

Acknowledgments

The writing and publishing of a calculus book is a tremendous undertaking. We take this responsibility very seriously because a calculus book is instrumental in transmitting knowledge from one generation to the next. We would like to thank the many people who helped us in the preparation of this book. First, we thank our editor George Lobell, who led us masterfully through the development and publication of this book. We sincerely appreciate Henri Feiner, who not only worked all of the problems but also read and critiqued each word of the manuscript, and Carol Williams who transcribed our manuscript into TEX. We would like to thank Dennis Kletzing for his work in transforming the manuscript into finished pages. Finally, we would like to thank Lynn Savino Wendel, who led us through the production process.

Of primary concern is the accuracy of the book. We had the assistance of many: Henri Feiner, who read the entire manuscript and offered us many valuable suggestions; and Nancy and Mary Toscano, who were meticulous in their checking of our manuscript. Thanks also to the accuracy checkers of the previous editions, Jerry Alexanderson, Mike Ecker, Ken Sydel, Diana Gerardi, Kurt Norlin, Terri Bittner, Nancy Marsh, and Mary Toscano. We would also like to thank the following readers of the text for the many suggestions for improvement:

REVIEWERS OF THE THIRD EDITION:

Gregory Adams, Bucknell University

Robert Bakula, Ohio State University

J. Caggiano, Arkansas State Univery

James T. Campbell, University of Memphis

Lin Dearing, Clemson University

Stan Dick, George Mason University

Tevian Dray, Oregon State University

Michael W. Ecker, Pennsylvania State University, Wilkes-Barre Campus

Anda Gadidov, Gannon University

Ruth Gornet, Texas Tech University

Julia Hassett, Oakton Community College

Isom H. Herron, Rensselaer Polytechnic Institute

Michael G. Hilgers, University of Missouri–Rolla

Jason P. Huffman, Jacksonville State University

James E. Jamison, University of Memphis

Jeuel G. LaTorre, Clemson University

Ira Wayne Lewis, Texas Tech University

Maura B. Mast, University of Massachusetts–Boston

Mark Naber, Monroe County Community College

Chris Peterson, Colorado State University

Siew-Ching Pye, California State Polytechnic University–Pomona

Joe Rody, Arizona State University

Yongwu Rong, The George Washington University
Eric Rowley, Utah State University
John E. Santomas, Villanova University
Carl Seaquist, Texas Tech University
Dennis Wacker, Saint Louis University
W. Thurmon Whitley, University of New Haven
Teri Woodington, Colorado School of Mines
Cathleen M. Zucco-Teveloff, Trinity College

REVIEWERS OF THE FIRST TWO EDITIONS:

Gerald Alexanderson, Santa Clara University
David Arterburn, New Mexico Tech
Neil Berger, University of Illinois at Chicago
Michael L. Berry, West Virginia Wesleyan College
Linda A. Bolte, Eastern Washington University
Brian Borchers, New Mexico Tech
Barbara H. Briggs, Tennessee Technical University
Robert Broschat, South Dakota State University
Robert D. Brown, University of Kansas
Dan Chiddix, Ricks College
Philip Crooke, Vanderbilt University
Ken Dunn, Dalhousie University
Michael W. Ecker, Pennsylvania State University, Wilkes-Barre Campus
John H. Ellison, Grove City College
Mark Farris, Midwestern State University
Sally Fieschbeck, Rochester Institute of Technology
William P. Francis, Michigan Technological University
Stuart Goldenberg, California Polytechnic State University, San Luis Obispo Campus
Harvey Greenwald, California Polytechnic San Luis Obispo
Richard Hitt, University of South Alabama
Joel W. Irish, University of Southern Maine
Roger Jay, Tomball College
John H. Jenkins, Embry Riddle Aeronautical University
Clement T. Jeske, University of Wisconsin–Platteville
Kathy Kepner, Paducah Community College
Daniel King, Sarah Lawrence College
Lawrence Kratz, Idaho State University
Don Leftwich, Oklahoma Christian University
Sam Lessing, Northeast Missouri University
Estela S. Llinas, University of Pittsburgh at Greensburg
Pauline Lowman, Western Kentucky University
Ching Lu, Southern Illinois University at Edwardsville
William E. Mastrocola, Colgate University
Philip W. McCartney, Northern Kentucky University
E. D. McCune, Stephen F. Austin State University
John C. Michels, Chemeketa Community College

Judith Ann Miller, Delta College

Ann Morlet, Cleveland State University

Dena Jo Perkins, Oklahoma Christian University

Pamela B. Pierce, College of Wooster

Judith Reeves, California Polytechnic State Universty, Pomona Campus

Jim Roznowski, Delta College

Peter Salamon, San Diego State University

Connie Schrock, Emporia State University

Tatiana Shubin, San Jose State University

Jo Smith, Ketering University

Anita E. Solow, DePauw University

Lowell Stultz, Kalamazoo Valley Community College

Tingxiu Wang, Oakton Community College

Monty J. Strauss
Gerald L. Bradley
Karl J. Smith

SINGLE VARIABLE
CALCULUS

A Guide to Using this Text

HALLMARK FEATURES

Unique Student Mathematics Handbook ▶

(Included on the Student CD-ROM as well as in an optional print supplement)

This "Just-in-Time Precalculus Review" provides precalculus drill/review material, a catalogue of curves, analytic geometry, and integral tables. Students are guided from the text to this handbook by a **SMH** symbol located in the text margin. This guide is entirely text-author written. See the description of the Student CD-ROM for other features.

page 159

CD-ROM included free in both the Instructor and Student Editions.

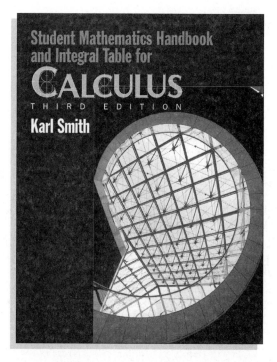

May be shrinkwrapped with any version of the text.

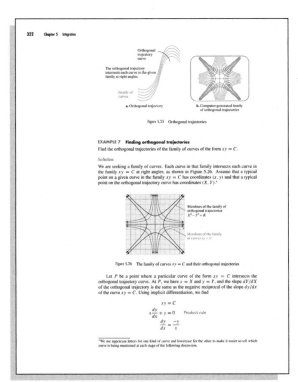

page 322

◀ Emphasis on visualization

Over 1900 graphs and pictures are featured throughout the text. Designed to supplement the text prose, the art program helps student to develop mathematical intuition and problem-solving skills.

Effective use of technology ▶

Graphing calculators and computer algebra systems are carefully integrated throughout the text. The "Technology Notes" feature gives students insight into how technological advances can be used to help understand calculus. Instructors can also encourage students to use technology to complete exercise sets. Laboratory manuals are available at a discount when packaged with the text. See supplements section of a complete list of available technology.

page 124

PROBLEM SETS

"What Does This Say?" Problems ▶

These problems ask students to respond to questions or to rephrase sentences in their own words. Designed to help students develop the essential skill of communicating mathematical ideas.

4.4 Curve Sketching with Asymptotes: Limits Involving Infinity   **227**

4.4 PROBLEM SET

Ⓐ 1. **WHAT DOES THIS SAY?** Outline a method for curve sketching.

2. **WHAT DOES THIS SAY?** What are critical numbers? Discuss the importance of critical numbers in curve sketching.

45. The *ideal speed* v for a banked curve on a highway is modeled by the equation

$$v^2 = gr \tan \theta$$

page 227

Interpretation Problems ▶

These problems require exposition that requires a line of thinking that is not directly covered in the textbook. Encourages students to think critically about calculus concepts.

58. **Interpretation Problem** Evaluate $\lim\limits_{x \to 0}\left[x^2 - \dfrac{\cos x}{1,000,000,000}\right]$. Explain why a calculator solution may lead to an incorrect conclusion about the limit.

page 69

21. $\lim\limits_{x \to 0^+} \dfrac{\sin x}{\sqrt{x}}$

22. $\lim\limits_{x \to 0^+} \dfrac{1 - \cos \sqrt{x}}{x}$

23. $\lim\limits_{x \to 0} \dfrac{\sin 2x}{x}$

24. $\lim\limits_{x \to 0} \dfrac{\sin 4x}{9x}$

25. $\lim\limits_{t \to 0} \dfrac{\tan 5t}{\tan 2t}$

26. $\lim\limits_{x \to 0} \dfrac{\cot 3x}{\cot x}$

27. $\lim\limits_{x \to 0} \dfrac{1 - \cos x}{\sin x}$

28. $\lim\limits_{x \to 0} \dfrac{\sin^2 x}{2x}$

29. $\lim\limits_{x \to 0} \dfrac{\sin^2 x}{x^2}$

30. $\lim\limits_{x \to 0} \dfrac{x^2 \cos 2x}{1 - \cos x}$

31. $\lim\limits_{x \to 0} \dfrac{\sec x - 1}{x \sec x}$

32. $\lim\limits_{x \to \pi/4} \dfrac{1 - \tan x}{\sin x - \cos x}$

page 69

◀ **Computational Problems**

Graded exercise sets that are progressively more challenging and grouped into three categories. 'A' problems are routine, 'B' problems require independent thought, and 'C' problems focus on theory. Makes assigning exercises easier and allows students to gain confidence as they progress through the exercises.

20. Modeling Problem Two towns A and B are 12.0 mi apart and are located 5.0 and 3.0 mi, respectively, from a long, straight highway, as shown in Figure 4.62.

Figure 4.62   Building the shortest road

A construction company has a contract to build a road from A to the highway and then to B. Analyze a model to determine the length (to the nearest tenth of a mile) of the *shortest* road that meets these requirements.

page 246

Counterexample Problems ▶

These problems help students develop the ability to construct counterexamples. Students are asked to formulate an example that satisfies certain conditions.

◀ **Modeling Problems**

Designed to help students see the applications of calculus to multiple disciplines and real-world situations.

67. Counterexample Problem When the line segment $y = x - 2$ between $x = 1$ and $x = 3$ is revolved about the x-axis, it generates part of a cone. The formula obtained in Section 6.4 says that the surface area of the cone is

$$\int_1^3 2\pi (x - 2)\sqrt{2}\, dx$$

but this integral is equal to 0. What is wrong? What is the actual surface area?

page 421

37. Exploration Problem Suppose that $f(t)$ is continuous for all t and that for any number x it is known that the average value of f on $[-1, x]$ is

$$A(x) = \sin x$$

Use this information to deduce the identity of f.

page 333

◀ **Exploration Problems**

These problems explore concepts which may prove true or false and provide opportunities for innovative thinking.

PROBLEM SETS

Spy Problems ▶

A student favorite, these problems which run throughout the text like a movie serial, trace the heroic events of an international spy. His survival requires the students' help by successfully answering calculus questions.

---

54. Spy problem Just as the Spy is about to catch up with Scélérat (Problem 70 of Chapter 10 Supplementary Problems), the snow gives way and he falls into a cavern. He staggers to his feet and removes his skis. Why is it so warm? Good grief—the cave is a large roasting oven! Fortunately, he is wearing his heat-detector ring, which indicates the direction of greatest temperature decrease. Suppose the bunker is coordinatized so that the temperature at each point (x, y) on the floor of the bunker is given by

$$T(x, y) = 3(x - 6)^2 + 1.5(y - 1)^2 + 41$$

degrees Fahrenheit, where x and y are in feet. The Spy begins at the point $(1, 5)$ and stumbles across the room at the rate of 4 ft/min, always moving in the direction of maximum temperature decrease. But he can last no more than 2 minutes under these conditions! Assuming that there is an escape hole at the point where the temperature is minimal, does he make it or is the Spy toast at last?

page 748

---

81. HISTORICAL QUEST

SRINIVASA RAMANUJAN
1887–1920

Ramanujan was that rarest of mathematicians, an instinctive genius with virtually no formal training. Like his Hindu predecessor, Bhaskara (see the Historical Quest in Problem 55 of Section 4.4), Ramanujan had an uncanny instinct for numerical "truth" and conceived of his results in much the same way that a sculptor "sees" a statue in a raw block of stone. In his short life, he initiated new ways of thinking about number theory and made conjectures that are the subject of mathematical inquiry to this day.

A story related by his friend and mentor, the eminent British number theorist G. H. Hardy (1877–1947), serves to illustrate the resonance between Ramanujan's mind and the concept of number. Ramanujan was ill and Hardy came to visit him in the hospital. At a loss for how to begin the conversation, Hardy idly remarked that he had arrived in a taxi with the "dull" number 1729. Ramanujan immediately grew excited and exclaimed, "No, Hardy, it is a very interesting number, for it is the smallest integer that can be expressed as the sum of cubes in two different ways." ($1729 = 1^3 + 12^3 = 9^3 + 10^3$.)

In a letter from Ramanujan to G. H. Hardy at Cambridge University,* Ramanujan stated

$$1 - 5\left(\frac{1}{2}\right)^3 + 9\left(\frac{1 \cdot 3}{2 \cdot 4}\right)^3 - 13\left(\frac{1 \cdot 3 \cdot 5}{2 \cdot 4 \cdot 6}\right)^3 + \cdots = \frac{2}{\pi}$$

Hardy spent a great deal of time wondering how this sum could be equal to $2/\pi$. ∎

page 568

◀ Historical Quest Problems

These problems help students to place calculus in a historical context, and further encourages them to take part in the mathematical discovery process.

---

60. Journal Problem *Crux,* problem by John A. Winterink* ∎
Prove the validity of the following simple method for finding the center of a conic: For the central conic,

$$\phi(x, y) = ax^2 + 2hxy + by^2 + 2gx + 2fy + c = 0$$

$ab - h^2 \neq 0$, show that the center is the intersection of the lines $\partial\phi/\partial x = 0$ and $\partial\phi/\partial y = 0$.

page 719

◀ Journal Problems

Journal problems are reprinted from leading mathematics journals. Citations are provided if students wish to locate additional information. A great resource for further exploration.

END-OF-CHAPTER PROBLEMS

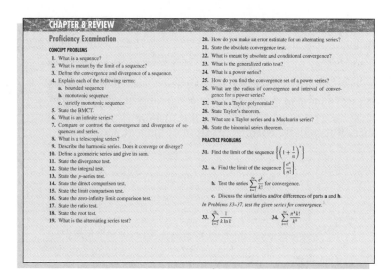

page 566

◀ Concept Problems

Concept problems are designed to develop students' conceptual understanding of calculus by learning how to think and draw conclusions about calculus topics.

◀ Practice Problems

Practice problems are specifically designed to help students develop their computational skills.

▼ Putnam Examination Problems

These challenging problems are provided to give insight into the types of problems that are asked in mathematical competitions, like the Mathematical Association of America's annual competition.

97. Putnam Examination Problem Let f be a real-valued function having partial derivatives defined for $x^2 + y^2 < 1$ that satisfies $|f(x, y)| \leq 1$. Show that there exists a point (x_0, y_0) in the interior of the unit circle such that
$$[f_x(x_0, y_0)]^2 + [f_y(x_0, y_0)]^2 \leq 16.$$

98. Putnam Examination Problem Find the smallest volume bounded by the coordinate planes and a tangent plane to the ellipsoid
$$\frac{x^2}{a^2} + \frac{y^2}{b^2} + \frac{z^2}{c^2} = 1$$

99. Putnam Examination Problem Find the shortest distance between the plane $Ax + By + Cz + 1 = 0$ and the ellipsoid
$$\frac{x^2}{a^2} + \frac{y^2}{b^2} + \frac{z^2}{c^2} \leq 1$$

page 774

Supplementary Problems

Describe the domain of each function given in Problems 1–4.

1. $f(x, y) = \sqrt{16 - x^2 - y^2}$ **2.** $f(x, y) = \dfrac{x^2 - y^2}{x - y}$

3. $f(x, y) = \sin^{-1} x + \cos^{-1} y$ **4.** $f(x, y) = e^{x+y} \tan^{-1}\left(\dfrac{y}{x}\right)$

Find the partial derivatives f_x and f_y for the functions defined in Problems 5–10.

5. $f(x, y) = \dfrac{x^2 - y^2}{x + y}$ **6.** $f(x, y) = x^3 e^{3y/(2x)}$

7. $f(x, y) = x^2 y + \sin \dfrac{y}{x}$ **8.** $f(x, y) = \ln\left(\dfrac{xy}{x + 2y}\right)$

9. $f(x, y) = 2x^3 y + 3xy^2 + \dfrac{y}{x}$ **10.** $f(x, y) = xye^{xy}$

For each function given in Problems 11–15, describe the level curve or level surface $f = c$ for the given values of the constant c.

11. $f(x, y) = x^2 - y; c = 2, c = -2$

12. $f(x, y) = 6x + 2y; c = 0, c = 1, c = 2$

13. $z = f(x, y) = \begin{cases} \sqrt{x^2 + y^2} & \text{if } z \geq 0 \\ |y| & \text{if } z < 0 \end{cases}$
$c = 0, c = 1, c = -1$

14. $f(x, y, z) = x^2 + y^2 + z^2; c = 16, c = 0, c = -25$

15. $f(x, y, z) = x^2 + \dfrac{y^2}{2} + \dfrac{z^2}{9}; c = 1, c = 2$

Evaluate the limits in Problems 16 and 17, assuming they exist.

16. $\displaystyle\lim_{(x,y)\to(1,1)} \dfrac{xy}{x^2 + y^2}$ **17.** $\displaystyle\lim_{(x,y)\to(0,0)} \dfrac{x + ye^{-x}}{1 + x^2}$

page 771

▲ Supplemental Problems

Extensive sets of problems (usually about 100) occur at the end of each chapter. These problems are based on randomly selected topics from the complete chapter. Also helps students to check their comprehension of the entire chapter.

MEDIA SUPPLEMENTS

◀ **New Interactive Student CD-ROM**

(free with every new copy)

The new student CD-ROM is designed to enhance students' computational and conceptual understanding of calculus. This CD-ROM is not an add-on of extra material to the text but rather an incredibly useful expansion of the text and includes:

Live Examples

The heart of this CD is the Live Examples powered by Live Math. Nearly every geometric text example is animated. The traditional text environment is enhanced and brought to life with animations and questions exploring what-if scenarios. Algebraic solutions are paired with geometric solutions. There are hundreds of Live Examples to accompany the complete text.

True/False Study Questions

There are 10 T/F questions per text section. These questions focus on the core ideas of calculus. These questions are specifically designed to encourage students to read (not just thumb through the text to find examples matching the questions).

Homework Starters

Each text section includes teaching hints for 3–5 homework problems. These hints often include geometric animations. There are more than 150 Homework Starters.

Unique Student Mathematics Handbook

A unified and complete treatment of prerequisite material, easily referenced and keyed to the textbook. Comes free on the Student CD-ROM. Also available in print format.

MEDIA SUPPLEMENTS

Companion Website ▼

www.prenhall.com/strauss

 This website contains much of what is on the CD-ROM along with other activities better suited to the Internet and includes:

Live Examples

The heart of this website lies in the Live Examples powered by Live Math. Nearly every geometric text example is animated. The traditional text environment is now enhanced and brought to life with animations and questions exploring what-if scenarios. Algebraic solutions are paired with geometric solutions. There are hundreds of Live Examples.

Homework Starters

Each text section includes teaching hints for 3–5 homework problems. These hints often include geometric animations. There are more than 150 Homework Starters.

True/False Study Questions

There are 10 T/F questions per text section. These questions focus on the core ideas of calculus. These questions are specifically designed to encourage students to read (not just thumb through the text to find examples matching the questions).

Links and Destinations

There are roughly five to ten links per text section to other Internet based sites with calculus oriented or related material. For example, one fascinating link is the enormous History Archive Site in St. Andrews, Scotland, which is loaded with hundreds, if not thousands, of biographies and other fascinating historical material.

MEDIA SUPPLEMENTS

Prentice Hall Online Homework Grader ▼

There are over 2000 questions from this calculus text made available in algorithmic form that allows for professors to distribute, collect, grade, and record assignments.

Assessment and Homework Management

This web-based system is designed to help you teach and help your students learn. This system was developed to automate the production and delivery of web-based assignments ranging from both homework and self-paced practice to secure, proctored testing.

For instructors, the system provides a more natural transition from traditional paper-based assignments to time-saving, automatically graded online homework, that improves on the limitations of current course management systems. Students prefer this system because it provides immediate, individualized performance feedback and the opportunity to generate unlimited practice through algorithmic study sessions.

What can you do with the Homework Grader? You can...

- Easily create self-scoring, web-delivered tests and homework assignments from your own time-tested problem files.
- Use the 2000 textbook problems already loaded for use.
- Edit, modify, or expand existing question banks (whether publisher-provided or homegrown).
- Control the rules and policies surrounding your assignments, including time limitations, ability to retake assignments, feedback content, results reporting, low-stakes or high-stakes forms, and formats.
- Publish your assignments to your own class-specific website, where your students will log in to complete the work.

For further details go to www.prenhall.com/math

TEXT OPTIONS

Single Variable Calculus, 3e

Includes Chapters 1-10 of the three semester full text. This version is intended for a two-semester course in single variable calculus.

SUPPLEMENTS FOR SINGLE VARIABLE CALCULUS

INSTRUCTOR

Instructor's Edition (0-13-035303-5)

The Instructor's Edition includes the answers to all odd-numbered exercises and detailed Teaching Notes.

TestGen-EQ (0-13-092018-5)

This easy to use test generator contains all of the questions from the Printed Test Bank.

Printed Test Bank (0-13-062028-9)

The test bank offers a full complement of additional problems that correlate to exercises presented in the text. Computerized versions of the the test bank are also available.

Instructor's Solutions Manual (0-13-062026-2)

Includes solutions to all the exercises in the text.

STUDENT

Student Survival and Solutions Manual (0-13-067245-9)

Contains carefully worked out solutions to all odd-numbered exercises in the text. Contains helpful hints to bridge the gap between the textbook and a working knowledge of calculus. These hints anticipate the types of errors and difficulties a student may have while taking the course.

Student Mathematics Handbook (0-13-092021-5)

A unified and complete treatment of prerequisite material, easily referenced and keyed to the textbook. Comes free on the Student CD-ROM. Also available in print format.

Single Variable Text and Student Mathematics Handbook
(0-13-074449-2)

Please see Supplement section of this Guide for more complete descriptions.

TEXT OPTIONS

Multivariable Calculus, 3e

Includes Chapters 9-14 of the three-semester full text. This version is intended for a one-semester course in multivariable calculus.

SUPPLEMENTS FOR MULTIVARIABLE CALCULUS

INSTRUCTOR

TestGen-EQ (0-13-092018-5)

This easy to use test generator contains all of the questions from the Printed Test Bank.

Printed Test Bank (0-13-062028-9)

The test bank offers a full complement of additional problems that correlate to exercises presented in the text. Computerized versions of the the test bank are also available.

Instructor's Solutions Manual (0-13-062026-2)

Includes solutions to all the exercises in the text.

STUDENT

Student Survival and Solutions Manual (0-13-067245-9)

Contains carefully worked out solutions to all odd-numbered exercises in the text. Contains helpful hints to bridge the gap between the textbook and a working knowledge of calculus. These hints anticipate the types of errors and difficulties a student may have while taking the course.

Student Mathematics Handbook (0-13-092021-5)

A unified and complete treatment of prerequisite material, easily referenced and keyed to the textbook. Comes free on the Student CD-ROM. Also available in print format.

Multivariable Text and Student Mathematics Handbook
(0-13-074440-9)

Please see Supplement section of this Guide for more complete descriptions.

Functions and Graphs

PREVIEW

This chapter uses several topics from algebra and trigonometry that are essential for the study of calculus. The necessary prerequisites for this course are reviewed in a companion book, *Student Mathematics Handbook*, and are briefly discussed in this chapter. This handbook is provided free with new copies of this textbook.

The concept of function is an especially important prerequisite for calculus. Even if you wish to skip most of this initial chapter, you should review the basic notions of functions and inverse functions in Sections 1.3 and 1.4.

PERSPECTIVE

Although modern science requires the use of many different skills and procedures, calculus is the primary mathematical tool for dealing with change. Sir Isaac Newton, one of the discoverers of calculus, once remarked that to accomplish his results, he "stood on the shoulders of giants." Indeed, calculus was not born in a moment of divine inspiration but developed gradually, as a variety of apparently different ideas and methods merged into a coherent pattern. The purpose of this initial chapter is to lay the foundation for the development of calculus.

1.1   Preliminaries

IN THIS SECTION   distance on a number line, absolute value, distance in the plane, trigonometry, solving trigonometric equations

This section provides a quick review of some fundamental concepts and techniques from precalculus mathematics. If you have recently had a precalculus course, you may skip over this material.

Algebra, geometry, and trigonometry are important ingredients of calculus. Even though we will review many ideas from algebra, geometry, and trigonometry, we will not be able to develop every idea from these courses before we use it in calculus. For example, the law of cosines from trigonometry may be needed to solve a problem in a section that never mentions trigonometry in the exposition. For this reason, we have made available a separate reference manual, *Student Mathematics Handbook*, which includes the background material you will need for this course. We suggest that you keep it close at hand. References to this handbook are indicated by the logo **SMH**

Every mathematical book that is worth reading must be read "backwards and forwards," if I may use the expression. I would modify Lagrange's advice a little and say, "Go on, but often return to strengthen your faith." When you come on a hard and dreary passage, pass it over; come back to it after you have seen its importance or found its importance or found the need for it further on.

——George Chrystal,
Algebra, Part 2 (Edinburgh, 1889)

DISTANCE ON A NUMBER LINE

You are probably familiar with the set of **real numbers** and several of its subsets, including the counting or natural numbers, the integers, the rational numbers, and the irrational numbers.

The real numbers can be visualized most easily by using a **one-dimensional coordinate system** called a **real number line**, as shown in Figure 1.1.

SMH The real numbers, *along with these subsets, are found in Chapter 2 of the Student Mathematics Handbook.*

Figure 1.1   Real number line

Notice that a number a is less than a number b if it is to the left of b on a real number line, as shown in Figure 1.2. Similar definitions can be given for $a > b$, $a \leq b$, and $a \geq b$.

Figure 1.2   Geometric definition of *less than*

The location of the number 0 is chosen arbitrarily, and a unit distance is picked (meters, feet, inches, ...). Numbers are ordered on the real number line according to the following order properties.

Order Properties

For all real numbers a, b, and c

   Trichotomy law: Exactly one of the following is true:

$$a < b, \quad a > b, \quad \text{or} \quad a = b.$$

   Transitive law of inequality: If $a < b$ and $b < c$, then $a < c$.

   Additive law of inequality: If $a < c$ and $b < d$, then $a + b < c + d$.

   Multiplicative law of inequality: If $a < b$, then

$$ac < bc \text{ if } c > 0 \quad \text{and} \quad ac > bc \text{ if } c < 0.$$

ABSOLUTE VALUE

Absolute Value

The **absolute value** of a real number a, denoted by $|a|$, is

$$|a| = \begin{cases} a & \text{if } a \geq 0 \\ -a & \text{if } a < 0 \end{cases}$$

The number x is located $|x|$ units away from 0 — to the right if $x > 0$ and to the left if $x < 0$.

Absolute value is used to describe the distance between points on a number line.

Distance Between Two Points on a Number Line

The **distance** between the numbers x_1 and x_2 on a number line is

$$|x_2 - x_1|$$

WARNING Note that
$|x_2 - x_1| = |x_1 - x_2|$

For example, the distance between 2 and -3 is $|2 - (-3)| = 5$ units, and between -2 and -3 is $|-2 - (-3)| = 1$ unit.

Several properties of absolute value that you will need in this course are summarized in Table 1.1.

TABLE 1.1 Properties of Absolute Value

Let a and b be any real numbers.

Property	Comment
1. $|a| \geq 0$	**1.** Absolute value is nonnegative.
2. $|-a| = |a|$	**2.** The absolute value of a number and the absolute value of its opposite are equal.
3. $|a|^2 = a^2$	**3.** If an absolute value is squared, the absolute value signs can be dropped because both squares are nonnegative.
4. $|ab| = |a|\,|b|$	**4.** The absolute value of a product is the product of the absolute values.
5. $\left\|\dfrac{a}{b}\right\| = \dfrac{|a|}{|b|}, \quad b \neq 0$	**5.** The absolute value of a quotient is the quotient of the absolute values.
6. $-|a| \leq a \leq |a|$	**6.** This property is true because $|a|$ is either a or $-a$.
7. Let $b \geq 0$; $|a| = b$ if and only if $a = \pm b$	**7.** This property is useful in solving absolute value equations. See Example 1.
8. Let $b > 0$; $|a| < b$ if and only if $-b < a < b$	**8. and 9.** These are the main properties used in solving absolute value inequalities. See Example 2.
9. Let $b > 0$; $|a| > b$ if and only if $a > b$ or $a < -b$	
10. $|a + b| \leq |a| + |b|$	**10.** This property is called the **triangle inequality**. It is used in both theory and numerical computations involving inequalities.

WARNING "p if and only if q" is used to mean that both a statement and its converse are true. That is,

if p, then q, and

if q, then p.

For example, property 8 has two parts:

 (i) If $|a| < b$, then $-b < a < b$;
 (ii) If $-b < a < b$, then $|a| < b$.

Property 7 is sometimes stated for any a or b as $|a| = |b|$ if and only if $a = \pm b$. Since $|b| = \pm b$, it follows that this property is equivalent to property 7. Also, properties 8 and 9 are true for $\leq$ and $\geq$ inequalities. Specifically, if $b \geq 0$, then

$$|a| \leq b \quad \text{if and only if} \quad -b \leq a \leq b$$

and

$$|a| \geq b \quad \text{if and only if} \quad a \geq b \text{ or } a \leq -b$$

A convenient notation for representing intervals on a number line is called **interval notation** and is summarized in the accompanying table. Note that a solid dot (•) at an endpoint of an interval indicates that it is included in the interval, whereas an open dot (○) indicates that it is excluded. An interval is **bounded** if both of its endpoints are real numbers. A bounded interval is **open** if it includes neither of its endpoints, **half-open** if it includes only one endpoint, and **closed** if it includes both endpoints. The symbol "∞" (pronounced *infinity*) is used for intervals that are not limited in one direction or another. In particular, $(-\infty, \infty)$ denotes the entire real number line.

 If this notation is new to you, please check the *Handbook* (Section 2.7) for further examples. We will use this interval notation to write the solutions of absolute value equations and absolute value inequality problems.

Name of Interval	Inequality Notation	Interval Notation	Graph
Closed interval	$a \leq x \leq b$	$[a, b]$	
	$a \leq x$	$[a, \infty)$	
	$x \leq b$	$(-\infty, b]$	
Open interval	$a < x < b$	(a, b)	
	$a < x$	(a, ∞)	
	$x < b$	$(-\infty, b)$	
Half-open interval	$a < x \leq b$	$(a, b]$	
	$a \leq x < b$	$[a, b)$	
Real number line	All real numbers	$(-\infty, \infty)$	

Absolute Value Equations   Absolute value property 7 allows us to solve absolute value equations easily. For this reason property 7 is called the **absolute value equation property**.

EXAMPLE 1   Solving an equation with an absolute value on one side
Solve $|2x - 6| = x$.

Solution
If $2x - 6 \geq 0$, then $|2x - 6| = 2x - 6$ so that we solve

$$2x - 6 = x \quad \text{or} \quad x = 6 \qquad \text{Property 7}$$

If $2x - 6 < 0$, then $|2x - 6| = -(2x - 6)$ so that we solve

$$-(2x - 6) = x$$
$$-3x = -6$$
$$x = 2$$

The solutions are $x = 6$ and $x = 2$.   ■

The absolute value expression $|x - a|$ can be interpreted as the distance between x and a on a number line. An equation of the form

$$|x - a| = b$$

is satisfied by two values of x that are a given distance b from a when represented on a number line. For example, $|x - 5| = 3$ states that x is 3 units from 5 on a number line. Thus, x is either 2 or 8.

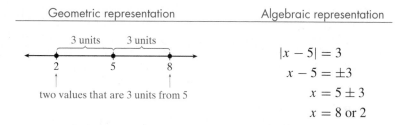

Absolute Value Inequalities   Because $|x - 5| = 3$ states that the distance from x to 5 is 3 units, the inequality $|x - 5| < 3$ states that the distance from x to 5 is less than 3 units, whereas $|x - 5| > 3$ states that the distance from x to 5 is greater than 3 units.

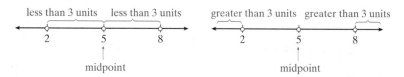

EXAMPLE 2   Solving an absolute value inequality

Solve $|2x - 3| \leq 4$.

Solution

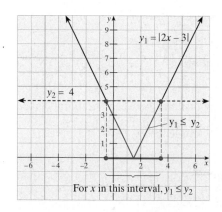

For x in this interval, $y_1 \leq y_2$

Figure 1.3 Solving an inequality by graphing

Algebraic solution:

$$-4 \leq 2x - 3 \leq 4 \qquad \textit{Property 8}$$
$$-4 + 3 \leq 2x - 3 + 3 \leq 4 + 3$$
$$-1 \leq 2x \leq 7$$
$$-\frac{1}{2} \leq \frac{2x}{2} \leq \frac{7}{2}$$
$$-\frac{1}{2} \leq x \leq \frac{7}{2}$$

The solution is the interval $[-\frac{1}{2}, \frac{7}{2}]$.

Geometric solution: Graph $y_1 = |2x - 3|$ and $y_2 = 4$, as shown in Figure 1.3. Because we are looking for $|2x - 3| \leq 4$, we note those x-values on the real number line for which the graph of y_1 is below the graph of y_2. We see that the interval is $[-0.5, 3.5]$ or $[-\frac{1}{2}, \frac{7}{2}]$.   ■

When absolute value is applied to measurement, it is called **tolerance.** Tolerance is an allowable deviation from a standard. For example, a cement bag whose weight w lb is "90 lb plus or minus 2 lb" might be described as having a weight given by $|w - 90| \le 2$. When considered as a tolerance, the expression $|x - a| \le b$ may be interpreted as x being compared to a with an **absolute error** of measurement of b units. Consider the following example.

EXAMPLE 3   Absolute value as a tolerance

Suppose you purchase a 90-lb bag of cement. It will not weigh exactly 90 lb. The material must be measured, and the measurement is approximate. Some bags will weigh as much as 2 lb over 90 lb, and some will weigh as much as 2 lb under 90 lb. If so, the bag could weigh as much as 92 lb or as little as 88 lb. State this as an absolute value inequality.

Solution

Let $w = $ weight of the bag of cement in pounds. Then

$$90 - 2 \le w \le 90 + 2$$
$$-2 \le w - 90 \le 2$$

Equivalently, $|w - 90| \le 2$.   ■

DISTANCE IN THE PLANE

Absolute value is used to find the distance between two points on a number line. To find the distance between two points in a coordinate plane, we use the *distance formula*, which is derived by using the Pythagorean theorem.

THEOREM 1.1   Distance between two points in the plane

The distance d between the points $P_1(x_1, y_1)$ and $P_2(x_2, y_2)$ in the plane is given by

$$d = \sqrt{(\Delta x)^2 + (\Delta y)^2} = \sqrt{(x_2 - x_1)^2 + (y_2 - y_1)^2}$$

where Δx (read "delta x") is the **horizontal change $x_2 - x_1$** (sometimes called *run*) and Δy (read "delta y") is the **vertical change $y_2 - y_1$** (sometimes called *rise*).

Proof   Using the two points, form a right triangle by drawing lines through the given points parallel to the coordinate axes, as shown in Figure 1.4.
    The length of the horizontal side of the triangle is $|x_2 - x_1| = |\Delta x|$, and the length of the vertical side is $|y_2 - y_1| = |\Delta y|$. Then

$$d^2 = |\Delta x|^2 + |\Delta y|^2 \qquad \text{Pythagorean theorem}$$
$$d^2 = (\Delta x)^2 + (\Delta y)^2 \qquad \text{Absolute value property 3}$$
$$d = \sqrt{(\Delta x)^2 + (\Delta y)^2} \qquad \text{Solve for } d \qquad \square$$

Figure 1.4 Distance formula

Midpoint formula   Related to the formula for the distance between two points is the formula for finding the midpoint of a line segment, as shown in Figure 1.4.

Midpoint Formula

The **midpoint,** M, of the segment with endpoints $P_1(x_1, y_1)$ and $P_2(x_2, y_2)$ has coordinates

$$M\left(\frac{x_1 + x_2}{2}, \frac{y_1 + y_2}{2}\right)$$

Notice that the coordinates of the midpoint of a segment are found by averaging the first and second components of the coordinates of the endpoints, respectively. You are asked to derive this formula in Problem 74.

Relationship between an equation and a graph **Analytic geometry** is that branch of geometry that ties together the geometric concept of position with an algebraic representation—namely, coordinates. For example, you remember from algebra that a line can be represented by an equation. Precisely what does this mean? Can we make such a statement that is true for any curve, not just for lines? We answer in the affirmative with the following definition.

Graph of an Equation

> The **graph of an equation** in two variables x and y is the collection of all points $P(x, y)$ whose coordinates (x, y) satisfy the equation.

There are two frequently asked questions in analytic geometry:

1. Given a graph (a geometrical representation), find the corresponding equation.
2. Given an equation (an algebraic representation), find the corresponding graph.

In Example 4, we use the distance formula to derive the equation of a circle. This means that if x and y are numbers that satisfy the equation, then the point (x, y) must lie on the circle. Conversely, the coordinates of any point on the circle will satisfy the equation.

EXAMPLE 4   Using the distance formula to derive an equation of a graph

Find the equation of a circle with center (h, k) and radius r.

Solution

Let (x, y) be any point on a circle. Recall that a circle is the set of all points in the plane a given distance from a given point. The given point (h, k) is the center and the given distance is the radius r.

$$r = \text{DISTANCE FROM } (h, k) \text{ TO } (x, y)$$
$$r = \sqrt{(x - h)^2 + (y - k)^2}$$
$$r^2 = (x - h)^2 + (y - k)^2, \quad \text{or} \quad (x - h)^2 + (y - k)^2 = r^2 \qquad \blacksquare$$

A **unit circle** is a circle of radius 1 and center at the origin, so its equation is $x^2 + y^2 = 1$.

EXAMPLE 5   Finding the equation of a circle

Find the equation of the circle with center $(3, -5)$ that passes through the point $(1, 8)$.

Solution

See Figure 1.5. The radius is the distance from the center to the given point:

$$r = \sqrt{(1 - 3)^2 + [8 - (-5)]^2}$$
$$= \sqrt{4 + 169}$$
$$= \sqrt{173}$$

Thus, the equation of the circle is

$$(x - 3)^2 + [y - (-5)]^2 = (\sqrt{173})^2$$
$$(x - 3)^2 + (y + 5)^2 = 173 \qquad \blacksquare$$

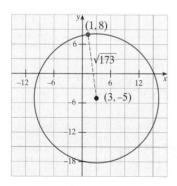

Figure 1.5 Circle with center $(3, -5)$ passing through $(1, 8)$

EXAMPLE 6   Graphing a circle given its equation

Sketch the graph of the circle whose equation is

$$4x^2 + 4y^2 - 4x + 8y - 5 = 0$$

Solution

We need to convert this equation into standard form. To do this we use a process called **completing the square** (see Section 2.6 of the *Student Mathematics Handbook*).

$$4x^2 + 4y^2 - 4x + 8y - 5 = 0$$

$$x^2 + y^2 - x + 2y = \tfrac{5}{4} \qquad \text{Coefficients of squared terms should be 1.}$$

$$(x^2 - x \quad) + (y^2 + 2y \quad) = \tfrac{5}{4} \qquad \text{Associate } x\text{-terms and } y\text{-terms.}$$

$$\left[x^2 - x + (-\tfrac{1}{2})^2\right] + (y^2 + 2y + 1^2) = \tfrac{5}{4} + \tfrac{1}{4} + 1 \qquad \text{Complete the squares by adding } \tfrac{1}{4} \text{ and 1 to both sides.}$$

$$(x - \tfrac{1}{2})^2 + (y + 1)^2 = \tfrac{10}{4} = \tfrac{5}{2}$$

This is a circle with center at $(\tfrac{1}{2}, -1)$ and radius $\sqrt{\tfrac{5}{2}}$. The graph is shown in Figure 1.6. ∎

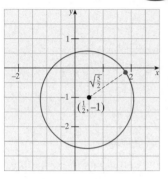

Figure 1.6 Sketch of
$4x^2 + 4y^2 - 4x + 8y - 5 = 0$

TRIGONOMETRY

One of the prerequisites for this book is trigonometry. If you need some review, consult a text on trigonometry or Chapter 4 of the *Student Mathematics Handbook* that accompanies this book. A summary of useful trigonometric formulas is found in Appendix E.

Angles are commonly measured in degrees and radians. A **degree** is defined to be $\tfrac{1}{360}$ revolution and a **radian** $\tfrac{1}{2\pi}$ revolution. Thus, to convert between degree and radian measure use the following formula:

$$\frac{\theta \text{ measured in degrees}}{360} = \frac{\theta \text{ measured in radians}}{2\pi}$$

WARNING In calculus, angles are usually measured in terms of radians rather than degrees since radian measure often leads to simpler formulas. You may assume that when we write expressions such as $\sin x$, $\cos x$, and $\tan x$, the angle x is in radians unless otherwise specified by a degree symbol.

EXAMPLE 7   Converting degree measure to radian measure

Convert $255°$ to radian measure.

Solution
$$\frac{255}{360} = \frac{\theta}{2\pi}$$

$$\theta = \left(\frac{\pi}{180}\right)(255) \approx 4.450589593 \qquad \blacksquare$$

EXAMPLE 8   Converting radian measure to degree measure

Express 1 radian in terms of degrees.

Solution
$$\frac{\theta}{360} = \frac{1}{2\pi}$$

$$\theta = \left(\frac{180}{\pi}\right)(1)$$

$$\approx 57.29577951° \qquad \blacksquare$$

SOLVING TRIGONOMETRIC EQUATIONS

There will be many times in calculus when you will need to solve a trigonometric equation. As you may remember from trigonometry, solving a trigonometric equation is equivalent to evaluating an inverse trigonometric relation. Inverse functions will be introduced in Section 1.4; for now, we will solve trigonometric equations whose solutions involve the values in Table 1.2 (called a **table of exact values**). These exact values from trigonometry are reviewed in the *Student Mathematics Handbook*.

(SMH)

TABLE 1.2   Exact Trigonometric Values

Function \ Angle θ	0	$\dfrac{\pi}{6}$	$\dfrac{\pi}{4}$	$\dfrac{\pi}{3}$	$\dfrac{\pi}{2}$
$\cos\theta$	1	$\dfrac{\sqrt{3}}{2}$	$\dfrac{\sqrt{2}}{2}$	$\dfrac{1}{2}$	0
$\sin\theta$	0	$\dfrac{1}{2}$	$\dfrac{\sqrt{2}}{2}$	$\dfrac{\sqrt{3}}{2}$	1
$\tan\theta$	0	$\dfrac{\sqrt{3}}{3}$	1	$\sqrt{3}$	undefined

It is customary to use the values from Table 1.2 whenever possible. The approximate calculator values will be given only when necessary.

EXAMPLE 9   Evaluating trigonometric functions

Evaluate $\cos\frac{\pi}{3}$; $\sin\frac{5\pi}{6}$; $\tan(-\frac{5\pi}{4})$; $\sec 1.2$; $\csc(-4.5)$; and $\cot 180°$.

Solution

If you use a calculator, make certain it is in the proper mode (radian or degree).

$\cos\frac{\pi}{3} = \frac{1}{2}$	Exact value; Quadrant I
$\sin\frac{5\pi}{6} = \frac{1}{2}$	Exact value; Quadrant II
$\tan\left(-\frac{5\pi}{4}\right) = -1$	Exact value; Quadrant II
$\sec 1.2 \approx 2.759703601$	Approximate calculator value
$\csc(-4.5) \approx 1.022986384$	Approximate calculator value
$\cot 180°$ is not defined.	▪

EXAMPLE 10   Solving a trigonometric equation by factoring

Solve $2\cos\theta\sin\theta = \sin\theta$ on $[0, 2\pi)$.

Solution
$$2\cos\theta\sin\theta - \sin\theta = 0$$
$$\sin\theta(2\cos\theta - 1) = 0$$

$\sin\theta = 0 \qquad 2\cos\theta - 1 = 0$

$\theta = 0, \pi \qquad\quad \cos\theta = \frac{1}{2}$

$\qquad\qquad\qquad\qquad \theta = \frac{\pi}{3}, \frac{5\pi}{3}$

WARNING   Do not divide both sides by $\sin\theta$, because you might lose a solution. Notice that if $\theta = 0$ or π, then $\sin\theta = 0$. You cannot divide by 0.   ▪

EXAMPLE 11   Solving a trigonometric equation using identities

Solve $\sin x + \sqrt{3}\cos x = 1$ on $[0, 2\pi)$.

Solution

For an algebraic solution,

$$\sqrt{3}\cos x = 1 - \sin x$$
$$3\cos^2 x = 1 - 2\sin x + \sin^2 x \qquad \text{Square both sides.}$$
$$3(1 - \sin^2 x) = 1 - 2\sin x + \sin^2 x$$
$$2 + 2\sin x - 4\sin^2 x = 0$$
$$2\sin^2 x - \sin x - 1 = 0$$
$$(2\sin x + 1)(\sin x - 1) = 0$$
$$\sin x = -\tfrac{1}{2} \qquad \sin x = 1$$
$$x = \tfrac{7\pi}{6}, \tfrac{11\pi}{6} \qquad x = \tfrac{\pi}{2}$$

However, since we squared both sides, we need to check for extraneous roots by substituting into the *original equation*, since squaring sometimes introduces extraneous solutions (that is, solutions that do not satisfy the given equation). Checking, we see that $x = \tfrac{7\pi}{6}$ is extraneous, and the solution is $x = \tfrac{\pi}{2}, \tfrac{11\pi}{6}$.

It is often worthwhile to check by finding a geometric solution. We graph the left and right sides of the equation on the same axes:

$$y_1 = \sin x + \sqrt{3}\cos x$$
$$y_2 = 1$$

The graphs are shown in Figure 1.7, and for a solution we look to the intersection points.

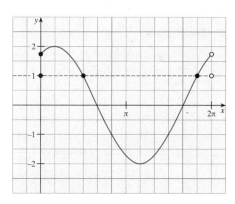

Figure 1.7   There are two intersection points, at $x = \tfrac{\pi}{2}$ and $x = \tfrac{11\pi}{6}$.

1.1   PROBLEM SET

A **1.** Fill in the missing parts in the table.

Inequality Notation	Interval Notation
$-3 < x < 4$	**a.**
b.	$[3, 5]$
c.	$[-2, 1)$
$2 < x \le 7$	**d.**

2. Fill in the missing parts in the table.

Inequality Notation	Interval Notation
a.	$(-\infty, -2)$
b.	$[\tfrac{\pi}{4}, \sqrt{2}]$
$x > -3$	**c.**
$-1 \le x \le 5$	**d.**

3. Represent each of the following on a number line.

    a. $-1 \le x < 4$ **b.** $-1 \le x \le 3$

    c. $(0, 2)$ **d.** $[-2, 1)$

4. Represent each of the following on a number line.

    a. $x > 2$ **b.** $x \le 4$

    c. $(-\infty, 2) \cup (2, \infty)$ **d.** $-2 < x \le 3$ or $x \ge 5$

In Problems 5–6, plot the given points P and Q on a Cartesian plane, find the distance between them, and find the coordinates of the midpoint of the line segment $\overline{PQ}$.

5. a. $P(2, 3), Q(-2, 5)$ **b.** $P(-2, 3), Q(4, 1)$

6. a. $P(-5, 3), Q(-5, -7)$ **b.** $P(-4, 3), Q(3, -4)$

(SMH) *Solve each equation in Problems 7–24. Assume that a, b, and c are known constants.*

7. $x^2 - x = 0$ **8.** $2y^2 + y - 3 = 0$

9. $y^2 - 5y + 3 = 17$ **10.** $x^2 + 5x + a = 0$

11. $3x^2 - bx = c$ **12.** $4x^2 + 20x + 25 = 0$

13. $|2x + 4| = 16$ **14.** $|5y + 2| = 12$

15. $|3 - 2w| = 7$ **16.** $|5 - 3t| = 14$

17. $|3x + 1| = -4$ **18.** $|1 - 5x| = -2$

19. $\sin x = -\frac{1}{2}$ on $[0, 2\pi)$

20. $(\sin x)(\cos x) = 0$ on $[0, 2\pi)$

21. $(2\cos x + \sqrt{2})(2\cos x - 1) = 0$ on $[0, 2\pi)$

22. $(3\tan x + \sqrt{3})(3\tan x - \sqrt{3}) = 0$ on $[0, 2\pi)$

23. $\cot x + \sqrt{3} = \csc x$ on $[0, 2\pi)$

24. $\sec^2 x - 1 = \sqrt{3}\tan x$ on $[0, 2\pi)$

Solve each inequality in Problems 25–34, and give your answer using interval notation.

25. $3x + 7 < 2$ **26.** $5(3 - x) > 3x - 1$

27. $-5 < 3x < 0$ **28.** $-3 < y - 5 \le 2$

29. $3 \le -y < 8$ **30.** $-5 \le 3 - 2x < 18$

31. $t^2 - 2t \le 3$ **32.** $s^2 + 3s - 4 > 0$

33. $|x - 8| \le 0.001$ **34.** $|x - 5| < 0.01$

In Problems 35–38, find an equation of the circle with given center C and radius r.

35. $C(-1, 2); r = 3$ **36.** $C(3, 0); r = 2$

37. $C(0, 1.5); r = 0.25$ **38.** $C(-1, -5); r = 4.1$

In Problems 39–42, find the centers and radii of the circles and then graph.

39. $x^2 - 2x + y^2 + 2y + 1 = 0$

40. $4x^2 + 4y^2 + 4y - 15 = 0$

41. $x^2 + y^2 + 2x - 10y + 25 = 0$

42. $2x^2 + 2y^2 + 2x - 6y - 9 = 0$

(SMH) *Use the sum and difference formulas from trigonometry to find the exact values of the expressions in Problems 43–46. Check by finding a calculator approximation.*

43. $\sin\left(-\frac{\pi}{12}\right)$ **44.** $\cos\frac{7\pi}{12}$

45. $\tan\frac{\pi}{12}$ **46.** $\sin 165°$

B **47. WHAT DOES THIS SAY?*** Describe a process for solving a quadratic equation.

---

*Many problems in this book are labeled **What Does This Say?** Following the question will be a question for you to answer in your own words, or a statement for you to rephrase in your own words. These problems are intended to be similar to the "What This Says" boxes.

48. WHAT DOES THIS SAY? Describe a process for solving absolute value equations.

49. WHAT DOES THIS SAY? Describe a process for solving absolute value inequalities.

50. WHAT DOES THIS SAY? Describe a process for solving trigonometric equations.

(SMH) *Specify the period for each graph in Problems 51–56. Also graph each curve.*

51. a. $y = \sin x$ **b.** $y = \cos x$ **c.** $y = \tan x$

52. a. $y = 2\sin 2\pi x$ **b.** $y = 3\cos 3\pi x$ **c.** $y = 4\tan\left(\frac{\pi x}{5}\right)$

53. $y = \tan(2x - \frac{\pi}{2})$ **54.** $y = 2\cos(3x + 2\pi) - 2$

55. $y = 4\sin(\frac{1}{2}x + 2) - 1$ **56.** $y = \tan(\frac{1}{2}x + \frac{\pi}{3})$

57. The current I (in amperes) in a certain circuit (for some convenient unit of time) generates the following set of data points:

Time	Current	Time	Current
0	−60.00000	10	30.00000
1	−58.68886	11	40.14784
2	−54.81273	12	48.54102
3	−48.54102	13	54.81273
4	−40.14784	14	58.68886
5	−30.00000	15	60.00000
6	−18.54102	16	58.68886
7	−6.27171	17	54.81273
8	6.27171	18	48.54102
9	18.54102	19	40.14784
		20	30.00000

Plot the data points and draw a smooth curve passing through these points. Determine possible values of A, B, C, and D so that the graph of these data is approximated by the equation $y - A = B\sin C(x - D)$.

58. Suppose that a point P on a water wheel with a 30-ft radius is d units from the waterline. If the water wheel turns at 6 revolutions per minute, the height of the point P above the waterline is given by the following set of data points:

Time	Height, d	Time	Height, d
0	−1.000	10	−1.000
1	4.729	11	4.729
2	19.729	12	19.729
3	38.271	13	38.270
4	53.271	14	53.270
5	59.000	15	59.000
6	53.271	16	53.271
7	38.271	17	38.271
8	19.729	18	19.730
9	4.729	19	4.730
		20	−1.000

Plot the data points and draw a smooth curve passing through these points. Determine possible values of A, B, C, and D so that the graph of these data is approximated by the equation

$$y - A = B\cos C(x - D)$$

Determine a possible equation of a curve that is generated by this water wheel.

59. The sun and moon tide curves are shown here.* During a new moon, the sun and moon tidal bulges are centered at the same longitude, so their effects are added to produce maximum high tides and minimum low tides. This produces maximum tidal range (the distance between the low and high tides).

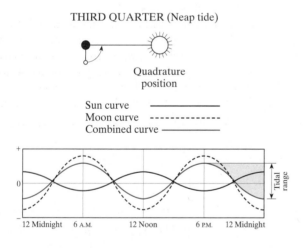

NEW MOON (Spring tide)

Write possible equations for the sun curve, the moon curve, and the combined curve. Assume the tidal range is 10 ft and the period is 12 hours.

60. The sun and moon tide curves are shown in the figure. During the third quarter (neap tide), the sun and moon tidal bulges are at right angles to each other. Thus, the sun tide reduces the effect of the moon tide.

THIRD QUARTER (Neap tide)

Write possible equations for the sun curve, the moon curve, and the combined curve. Assume the tidal range is 4 ft and the period is 12 hours.

61. A person is swimming at a depth of d units beneath the surface of the water. Because the light is refracted by the water, a viewer standing above the waterline sees the swimmer at an apparent depth of s units. (See Figure 1.8.)

Figure 1.8 Refracted light when viewing an underwater swimmer

In physics,[†] it is shown that if the person is viewed from an angle of incidence θ, then

$$s = \frac{3d\cos\theta}{\sqrt{7 + 9\cos^2\theta}}$$

a. If $d = 5.0$ meters and $\theta = 37°$, what is the apparent depth of the swimmer?

b. If the actual depth is $d = 5.0$ meters, what angle of incidence yields an apparent depth of $s = 2.5$ meters?

62. Journal problem   *The Mathematics Student Journal* by Murray Klamkin[‡] ■ Determine all the roots of the quartic equation $x^4 - 4x = 1$.

63. HISTORICAL QUEST   The division of one revolution into 360 equal parts (called degrees) is no doubt due to the sexagesimal (base 60) numeration system used by the Babylonians. Several explanations have been put forward to account for the choice of this number. (For example, see Howard Eves's *In Mathematical Circles*, Boston: Prindle, Weber & Schmidt, 1969.) One possible explanation is put forth by Otto Neugebauer, scholar and authority on early Babylonian mathematics and astronomy. In early Sumerian times, there existed a Babylonian mile, equal to about seven of our miles. Sometime in the first millennium B.C., when Babylonian astronomy reached the stage in which systematic records of the stars were kept, the Babylonian mile was adapted for measuring spans of time. Since a complete day was found to be equal to 12 time-miles, and one complete day is equivalent to one revolution of the sky, a complete circuit was divided into 12 equal parts. Then, for convenience, the Babylonian mile was subdivided into 30 equal parts. One complete circuit therefore has (12)(30) = 360 equal parts. ■

Show that the radius of a circle can be applied exactly six times to its circumference as a chord.

---

*From *Introductory Oceanography*, 5th ed., by H. V. Thurman, p. 253. Reprinted with permission of Merrill, an imprint of Macmillan Publishing Company. Copyright ©1988 Merrill Publishing Company, Columbus, Ohio.

[†]R. A. Serway, *Physics*, 3rd ed., Philadelphia: Saunders, 1992, p. 1007.

[‡]Most mathematics journals have problem sections that solicit interesting problems and solutions for publication. From time to time, we will reprint a problem from a mathematics journal. If you have difficulty solving a journal problem, you may wish to use a library to find the problem and solution as printed in the journal. The title of the journal is included as part of the problem, and we will generally give you a reference as a footnote. This problem is found in the named journal in Volume 28 (1980), issue 3, p. 2.

64. HISTORICAL QUEST The numeration system we use (base 10) evolved over a long period of time. It is often called the *Hindu-Arabic* system because its origins can be traced back to the Hindus in Bactria (now Afghanistan). Later, in A.D. 700, India was invaded by the Arabs who used and modified the Hindu numeration system, and, in turn, introduced it to Western civilization. The Hindu Brahmagupta stated the rules for operations with positive and negative numbers in the seventh century A.D. There are some indications that the Chinese had some knowledge of negative numbers as early as 200 B.C. On the other hand, the Western mathematician Girolamo Cardan (1501–1576) was calling numbers such as (–1) absurd as late as 1545. ∎

Write a paper on the history of the real number system.

C **65. Exploration Problem** If $ax + b = 0$, what effect does changing the sign of b have on the solution ($a \neq 0$; keep a fixed)?

66. Exploration Problem If $ax^2 + bx + c = 0$, what effect does changing c have on the solution ($a \neq 0$, keep a and b fixed)?

67. Exploration Problem If $\sin ax = b$, what effect does changing a have on the solution ($a \neq 0$; keep b fixed)?

68. If $c \geq 0$, show that $|x| \leq c$ if and only if $-c \leq x \leq c$.

69. Show that $-|x| \leq x \leq |x|$ for any number x.

70. Prove that $|a| = |b|$ if and only if $a = b$ or $a = -b$.

71. Prove that if $|a| < b$ and $b > 0$, then $-b < a < b$.

72. Prove the triangle inequality:

$$|x + y| \leq |x| + |y|$$

73. Show that $||x| - |y|| \leq |x - y|$ for all x and y.

74. Derive the **midpoint formula**

$$M\left(\frac{x_1 + x_2}{2}, \frac{y_1 + y_2}{2}\right)$$

for the midpoint of a segment with endpoints $P(x_1, y_1)$ and $Q(x_2, y_2)$.

1.2 Lines in the Plane

IN THIS SECTION slope of a line, forms for the equation of a line, parallel and perpendicular lines

SLOPE OF A LINE

A distinguishing feature of a line is the fact that its *inclination* with respect to the horizontal is constant. It is common practice to specify inclination by means of a concept called *slope*. A carpenter might describe a roof line that rises 1 ft for every 3 ft of horizontal "run" as having a slope or pitch of 1 to 3.

Let Δx and Δy represent, as before, the amount of change in the variables x and y, respectively. Then a nonvertical line L that rises (or falls) Δy units (measured from bottom to top) for every Δx units of run (measured from left to right) is said to have a *slope* of $m = \Delta y/\Delta x$. (If Δy is negative, then the "rise" is actually a fall; and if Δx is negative, then the run is actually right to left.) In particular, if $P(x_1, y_1)$ and $Q(x_2, y_2)$ are two distinct points on L, then the changes in the variables x and y are given by $\Delta x = x_2 - x_1$ and $\Delta y = y_2 - y_1$, and the slope of L is

$$m = \frac{\Delta y}{\Delta x} = \frac{y_2 - y_1}{x_2 - x_1} \quad \text{for } \Delta x \neq 0 \qquad \textit{See Figure 1.9.}$$

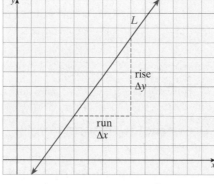

a. The slope of L is $m = \dfrac{\Delta y}{\Delta x}$.

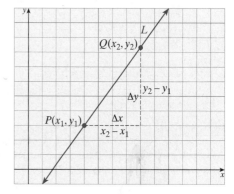

b. The slope is given by $m = \dfrac{y_2 - y_1}{x_2 - x_1}$.

Figure 1.9 The slope of a line

Slope of a Line

> A nonvertical line that contains the points $P(x_1, y_1)$ and $Q(x_2, y_2)$ has **slope**
>
> $$m = \frac{\Delta y}{\Delta x} = \frac{y_2 - y_1}{x_2 - x_1}$$

We say that a line with slope m is *rising* (when viewed from left to right) if $m > 0$, *falling* if $m < 0$, and *horizontal* if $m = 0$.

There is a useful trigonometric formulation of slope. The **angle of inclination** of a line L is defined to be the nonnegative angle ϕ ($0 \le \phi < \pi$) formed between L and the positive x-axis.

Angle of Inclination

WARNING Sometimes we say a vertical line has **infinite slope.**

> The **angle of inclination** of a line L is the angle ϕ ($0 \le \phi < \pi$) between L and the positive x-axis. Then the **slope** of L with inclination ϕ is
>
> $$m = \tan \phi$$

We see that the line L is *rising* if $0 < \phi < \frac{\pi}{2}$ and is *falling* if $\frac{\pi}{2} < \phi < \pi$. The line is *horizontal* if $\phi = 0$ and is *vertical* if $\phi = \frac{\pi}{2}$. Notice that if $\phi = \frac{\pi}{2}$, $\tan \phi$ is not defined; therefore the slope, m, is undefined for a vertical line.

To derive the trigonometric representation for slope we need to find the slope of the line through $P(x_1, y_1)$ and $Q(x_2, y_2)$, where ϕ is the angle of inclination. From the definition of the tangent, we have

$$\tan \phi = \frac{y_2 - y_1}{x_2 - x_1} = \frac{\Delta y}{\Delta x} = m \quad \text{See Figure 1.10.}$$

Lines with various slopes are shown in Figure 1.11.

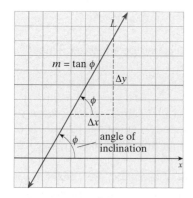

Figure 1.10 Trigonometric form for the slope of a line

Positive slope;
line rises.

Negative slope;
line falls.

Zero slope ($\Delta y = 0$);
line is horizontal.

Slope is undefined;
($\Delta x = 0$) line is vertical.

Figure 1.11   Examples of slope

FORMS FOR THE EQUATION OF A LINE

In algebra you studied several forms of the equation of a line. The derivations of some of these are reviewed in the problems. Here is a summary of the forms most frequently used in calculus.

Forms of Linear Equations

Standard Form:	$Ax + By + C = 0$	A, B, C constants
		(A and B not both 0)
Slope-intercept Form:	$y = mx + b$	Slope m, y-intercept $(0, b)$
Point-slope Form:	$y - k = m(x - h)$	Slope m, through point (h, k)
Two-intercept Form:	$\dfrac{x}{a} + \dfrac{y}{b} = 1$	Intercepts $(a, 0)$ and $(0, b)$
Horizontal line:	$y = k$	Slope 0
Vertical line:	$x = h$	Slope undefined

EXAMPLE 1 Deriving the two-intercept form of the equation of a line

Derive the equation of the line with intercepts $(a, 0)$ and $(0, b)$, $a \neq 0$, $b \neq 0$.

Solution

The slope of the equation passing through the given points is

$$m = \frac{b - 0}{0 - a} = -\frac{b}{a}$$

Use the point-slope form with $h = 0$, $k = b$. (You can use either of the given points.)

$$y - b = -\frac{b}{a}(x - 0)$$
$$ay - ab = -bx$$
$$bx + ay = ab$$
$$\frac{x}{a} + \frac{y}{b} = 1 \qquad \text{Divide both sides by } ab. \quad \blacksquare$$

Two quantities x and y that satisfy a linear equation $Ax + By + C = 0$ (A and B not both 0) are said to be *linearly related*. This terminology is illustrated in Example 2.

EXAMPLE 2 Linearly related variables

When a weight is attached to a helical spring, it causes the spring to lengthen. According to Hooke's law, the length d of the spring is linearly related to the weight w.* If $d = 4$ cm when $w = 3$ g and $d = 6$ cm when $w = 6$ g, what is the original length of the spring, and what weight will cause the spring to lengthen to 5 cm?

Solution

Because d is linearly related to w, we know that points (w, d) lie on a line, and the given information tells us that two such points are $(3, 4)$ and $(6, 6)$ as shown in Figure 1.12. We first find the slope of the line and then use the point-slope form to derive its equation.

$$m = \frac{6 - 4}{6 - 3} = \frac{2}{3}$$

Next, substitute into the point-slope form with $h = 3$ and $k = 4$:

$$d - 4 = \tfrac{2}{3}(w - 3)$$
$$d = \tfrac{2}{3}w + 2$$

The original length of the spring is found for $w = 0$:

$$d = \tfrac{2}{3}(0) + 2 = 2$$

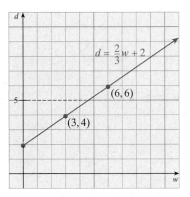

Figure 1.12 The length d of the spring is linearly related to the weight w of the attached object.

---

*Hooke's law is useful for small displacements, but for larger displacements, it may not be a good model.

The original length was 2 cm. To find the weight that corresponds to $d = 5$, we solve the equation

$$5 = \tfrac{2}{3}w + 2$$

$$\tfrac{9}{2} = w$$

Therefore, the weight that corresponds to a length of 5 cm is 4.5 g.   ■

PARALLEL AND PERPENDICULAR LINES

It is often useful to know whether two given lines are either parallel or perpendicular. A vertical line can be parallel only to other vertical lines and perpendicular only to horizontal lines. Cases involving nonvertical lines may be handled by the criteria given in the following box.

Slope Criteria for Parallel and Perpendicular Lines

If L_1 and L_2 are nonvertical lines with slopes m_1 and m_2, then

L_1 and L_2 are **parallel** if and only if $m_1 = m_2$;

L_1 and L_2 are **perpendicular** if and only if $m_1 m_2 = -1$, or $m_1 = -\dfrac{1}{m_2}$.

➡ **What This Says**   Nonvertical and nonhorizontal lines are parallel if and only if their slopes are equal. They are perpendicular if and only if their slopes are negative reciprocals of each other.

The key ideas behind these two slope criteria are displayed in Figure 1.13.

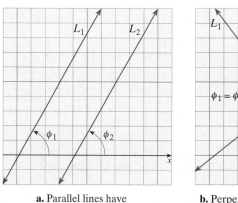
a. Parallel lines have equal slope: $m_1 = m_2$.

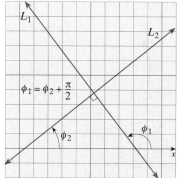
b. Perpendicular lines have negative reciprocal slopes: $m_1 m_2 = -1$.

Figure 1.13

EXAMPLE 3   Finding equations for parallel and perpendicular lines

Let L be the line $3x + 2y = 5$.

a. Find an equation of the line that is parallel to L and passes through $P(4, 7)$.
b. Find an equation of the line that is perpendicular to L and passes through $P(4, 7)$.

Solution

By rewriting the equation of L as $y = -\frac{3}{2}x + \frac{5}{2}$, we see that the slope of L is $m = -\frac{3}{2}$.

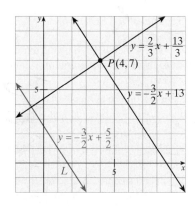

Figure 1.14 Graph of lines parallel and perpendicular to the given line $3x + 2y = 5$

a. Any line that is parallel to L must also have slope $m_1 = -\frac{3}{2}$. The required line contains the point $P(4, 7)$. Use the point-slope form to find the equation and write your answer in standard form:

$$y - 7 = -\frac{3}{2}(x - 4)$$
$$2y - 14 = -3x + 12$$
$$3x + 2y - 26 = 0$$

b. Any line perpendicular to L must have slope $m_2 = \frac{2}{3}$ (negative reciprocal of the slope of L). Once again, the required line contains the point $P(4, 7)$, and we find

$$y - 7 = \frac{2}{3}(x - 4)$$
$$3y - 21 = 2x - 8$$
$$2x - 3y + 13 = 0$$

These lines are shown in Figure 1.14. ∎

1.2 PROBLEM SET

Ⓐ **1. WHAT DOES THIS SAY?** Outline a procedure for graphing a linear equation.

In Problems 2–15, find the equation in standard form for the line that satisfies the given requirements.

2. passing through $(1, 4)$ and $(3, 6)$
3. passing through $(-1, 7)$ and $(-2, 9)$
4. horizontal line through $(-2, -5)$
5. passing through the point $(1, \frac{1}{2})$ with slope 0
6. slope 2 and y-intercept $(0, 5)$
7. vertical line through $(-2, -5)$
8. slope -3, x-intercept $(5, 0)$
9. x-intercept $(7, 0)$ and y-intercept $(0, -8)$
10. x-intercept $(4.5, 0)$ and y-intercept $(0, -5.4)$
11. passing through $(-1, 8)$ parallel to $3x + y = 7$
12. passing through $(4, 5)$ parallel to the line passing through $(2, 1)$ and $(5, 9)$
13. passing through $(3, -2)$ perpendicular to $4x - 3y + 2 = 0$
14. passing through $(-1, 6)$ perpendicular to the line through the origin with slope 0.5
15. perpendicular to the line whose equation is $x - 4y + 5 = 0$ where it intersects the line whose equation is $2x + 3y - 1 = 0$

In Problems 16–29, find, if possible, the slope, the y-intercept, and the x-intercept of the line whose equation is given. Sketch the graph of each equation.

16. $y = \frac{2}{3}x - 8$
17. $y = -\frac{5}{7}x + 3$
18. $y - 4 = 4.001(x - 2)$
19. $y - 9 = 6.001(x - 3)$
20. $5x + 3y - 15 = 0$
21. $3x + 5y + 15 = 0$
22. $2x - 3y - 2,550 = 0$
23. $6x - 10y - 3 = 0$
24. $\frac{x}{4} + \frac{y}{6} = 1$
25. $\frac{x}{2} - \frac{y}{3} = 1$

26. $y = 2x$
27. $x = 5y$
28. $y - 5 = 0$
29. $x + 3 = 0$

Ⓑ **30.** Find an equation for a vertical line(s) L such that a region bounded by L, the x-axis, and the line $2y - 3x = 6$ has area 3.

31. Find an equation for a horizontal line(s) M such that the region bounded by M, the y-axis, and the line $2y - 3x = 6$ has an area of 3.

32. Find the equation of the perpendicular line passing through the midpoint of the line segment connecting $(-3, 7)$ and $(4, -1)$.

33. Three vertices of a parallelogram are $A(1, 3)$, $C(4, 11)$, and $B(3, -2)$. If A and B lie on the same side, what is the fourth vertex?

Use a graphing calculator in Problems 34–45 to solve the systems of equations graphically, and approximate the coordinates of the intersection point. Some calculators have an INTERSECT feature. Check your owner's manual. Then solve the system algebraically and compare the results.

34. $\begin{cases} 2x - 3y = -8 \\ x + y = 6 \end{cases}$

35. $\begin{cases} 2x - 3y = -8 \\ 4x - 6y = 0 \end{cases}$

36. $\begin{cases} 2x - 3y = -8 \\ y = \frac{2}{3}x + \frac{8}{3} \end{cases}$

37. $\begin{cases} 3x - 4y = 16 \\ -x + 2y = -6 \end{cases}$

38. $\begin{cases} y = 3x + 1 \\ x - 2y = 8 \end{cases}$

39. $\begin{cases} 2x + 3y = 12 \\ -4x + 6y = 18 \end{cases}$

40. $\begin{cases} 2x + 3y = 15 \\ y = \frac{2}{3}x - 20 \end{cases}$

41. $\begin{cases} x + y = 12 \\ 0.6y = 0.5(40) \end{cases}$

42. $\begin{cases} x + 3y = \cos^2 \frac{\pi}{3} \\ x + y = -\sin^2 \frac{\pi}{3} \end{cases}$

43. $\begin{cases} x^2 + y^2 = 4 \\ x - y = 0 \end{cases}$

44. $\begin{cases} x^2 + y^2 = 10 \\ (x-3)^2 + y^2 = 9 \end{cases}$ **45.** $\begin{cases} x^2 + y^2 = 15 \\ (x+4)^2 + y^2 = 16 \end{cases}$

46. A life insurance table indicates that a woman who is now A years old can expect to live E years longer. Suppose that A and E are linearly related and that $E = 50$ when $A = 24$ and $E = 20$ when $A = 60$.

   a. At what age may a woman expect to live 30 years longer?

   b. What is the life expectancy of a newborn female child?

   c. At what age is the life expectancy zero?

47. On the Fahrenheit temperature scale, water freezes at $32°$ and boils at $212°$; the corresponding temperatures on the Celsius scale are $0°$ and $100°$. Given that the Fahrenheit and Celsius temperatures are linearly related, first find numbers r and s so that $F = rC + s$, and then answer these questions.

   a. Mercury freezes at $-39°C$. What is the corresponding Fahrenheit temperature?

   b. For what value of C is $F = 0$?

   c. What temperature is the same in both scales?

48. The average SAT mathematics scores of incoming students at an eastern liberal arts college have been declining in recent years. In 1996, the average SAT score was 575; in 2001, it was 545. Assuming the average SAT score varies linearly with time, answer these questions.

   a. Express the average SAT score in terms of time measured from 1996.

   b. If the trend continues, what will the average SAT score of incoming students be in 2014?

   c. When will the average SAT score be 455?

*__49.__ **Spy Problem**   An internationally famous spy has escaped from the headquarters of a diamond smuggling ring in the tiny Mediterranean country of Azusa. Our hero, driving a stolen milk truck at 72 km/hr, has a 40-minute head start on his pursuers, who are chasing him in a Ferrari going 168 km/hr. The distance from the smugglers' headquarters in Azusa to freedom in the neighboring country of Duarte is 83.8 km. Is this the end of the Spy, or does he live to return in Chapter 2?

50. A certain car rental agency charges $40 per day with 100 free miles plus 34¢ per mile after the first 100 miles. First express the cost of renting a car from this agency for one day in terms of the number of miles driven. Then draw the graph and use it to check your answers to these questions.

   a. How much does it cost to rent a car for a 1-day trip of 50 mi?

   b. How many miles were driven if the daily rental cost was $92.36?

51. A manufacturer buys $200,000 worth of machinery that depreciates linearly so that its trade-in value after 10 years will be $10,000. Express the value of the machinery as a function of its age and draw the graph. What is the value of the machinery after 4 years?

52. Show that if the point $P(x, y)$ is equidistant from $A(1, 3)$ and $B(-1, 2)$, its coordinates must satisfy the equation $4x + 2y - 5 = 0$. Sketch the graph of this equation.

53. Let $P_1(2, 6)$, $P_2(-1, 3)$, $P_3(0, -2)$, and $P_4(a, b)$ be points in the plane that are located so that $P_1 P_2 P_3 P_4$ is a parallelogram.

   a. There are three possible choices for P_4. One is $A(3, 1)$. What are the others, which we will call B and C?

   b. The centroid of a triangle is the point where its three medians intersect. Find the centroid of $\triangle ABC$ and of $\triangle P_1 P_2 P_3$. Do you notice anything interesting?

54. **Exploration Problem**   Ethyl alcohol is metabolized by the human body at a constant rate (independent of concentration). Suppose the rate is 10 mL per hour.

   a. Express the time t (in hours) required to metabolize the effects of drinking ethyl alcohol in terms of the amount A of ethyl alcohol consumed.

   b. How much time is required to eliminate the effects of a liter of beer containing 3% ethyl alcohol?

   c. Discuss how the function in part **a** can be used to determine a reasonable "cutoff" value for the amount of ethyl alcohol A that each individual may be served at a party.

55. Since the beginning of the month, a local reservoir has been losing water at a constant rate (that is, the amount of water in the reservoir is a linear function of time). On the 12th of the month, the reservoir held 200 million gallons of water; on the 21st, it held only 164 million gallons. How much water was in the reservoir on the 8th of the month?

56. To encourage motorists to form car pools, the transit authority in a certain metropolitan area has been offering a special reduced rate at toll bridges for vehicles containing four or more persons. When the program began 30 days ago, 157 vehicles qualified for the reduced rate during the morning rush hour. Since then, the number of vehicles qualifying has increased at a constant rate (that is, the number is a linear function of time), and 247 vehicles qualified today. If the trend continues, how many vehicles will qualify during the morning rush hour 14 days from now?

57. **Exploration Problem**   The value of a certain rare book doubles every 10 years. The book was originally worth $3.

   a. How much is the book worth when it is 30 years old? When it is 40 years old?

   b. Is the relationship between the value of the book and its age linear? Explain.

58. HISTORICAL QUEST   The region between the Tigris and Euphrates Rivers (present day Iraq) is rightly known as the Cradle of Civilization. During the so-called Babylonian period (roughly 2000–600 B.C.),[†] important mathematical ideas began to germinate in the region, including positional notation for numeration. Unlike their Egyptian contemporaries who usually wrote on fragile papyrus, Babylonian mathematicians recorded their ideas on clay tablets. One of these tablets, in the Yale Collection, shows a system equivalent to

$$\begin{cases} xy = 600 \\ (x+y)^2 - 150(x-y)^2 = 100 \end{cases} \blacksquare$$

Find a positive solution $(x > 0, y > 0)$ for this system correct to the nearest tenth.

---

*Take note: The Spy will accompany us on our journey through this book, if he can survive (which, of course, will require your help).

[†]For an interesting discussion of Mesopotamian mathematics, see *A History of Mathematics*, 2nd ed., by Carl B. Boyer, revised by Uta C. Merzbach, John Wiley and Sons, Inc., New York, 1968, pp. 26-47.

59. HISTORICAL QUEST The Louvre Tablet from the Babylonian civilization is dated about 1500 B.C. It shows a system equivalent to

$$\begin{cases} xy = 1 \\ x + y = a \end{cases} \blacksquare$$

Solve this system for x and y in terms of a.

60. Show that, in general, a line passing through $P(h, k)$ with slope m has the equation

$$y - k = m(x - h)$$

61. If $A(x_1, y_1)$ and $B(x_2, y_2)$ with $x_1 \neq x_2$ are two points on the graph of the line $y = mx + b$, show that

$$m = \frac{y_2 - y_1}{x_2 - x_1}$$

Use this fact to show that the graph of $y = mx + b$ is a line with slope m and then show that the line has y-intercept $(0, b)$.

62. Show that the distance s from the point (x_0, y_0) to the line $Ax + By + C = 0$ is given by the formula

$$s = \frac{|Ax_0 + By_0 + C|}{\sqrt{A^2 + B^2}}$$

63. Let L_1 and L_2 have slopes m_1 and m_2, respectively, and let ϕ be the angle between L_1 and L_2, as shown in Figure 1.15. Show that $\tan \phi = \dfrac{m_2 - m_1}{1 + m_1 m_2}$.

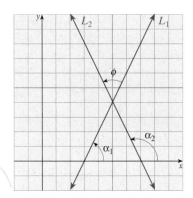

Figure 1.15 Angle ϕ between given lines

64. Journal Problem *Ontario Secondary School Mathematics Bulletin*[*] ■ Show that there is just one line in the family $y - 8 = m(x + 1)$ that is five units from the point $(2, 4)$.

65. The *centroid* of a triangle with vertices $A(x_1, y_1)$, $B(x_2, y_2)$, $C(x_3, y_3)$ is the point where its three medians intersect. It can be shown that the centroid $P(x_0, y_0)$ is located 2/3 the distance from each vertex to the midpoint of the opposite side. Use this fact to show that

$$x_0 = \frac{x_1 + x_2 + x_3}{3} \quad \text{and} \quad y_0 = \frac{y_1 + y_2 + y_3}{3}$$

*Volume 18 (1982), issue 2, p. 7.

1.3   Functions and Graphs

IN THIS SECTION definition of a function, functional notation, domain of a function, composition of functions, graph of a function, classification of functions

Scientists, economists, and other researchers study relationships between quantities. For example, an engineer may need to know how the illumination from a light source on an object is related to the distance between the object and the source; a biologist may wish to investigate how the population of a bacterial colony varies with time in the presence of a toxin; an economist may wish to determine the relationship between consumer demand for a certain commodity and its market price. The mathematical study of such relationships involves the concept of a *function*.

DEFINITION OF A FUNCTION

Function

> A **function** f is a rule that assigns to each element x of a set X a unique element y of a set Y. The element y is called the **image** of x under f and is denoted by $f(x)$ (read as "f of x"). The set X is called the **domain** of f, and the set of all images of elements of X is called the **range** of the *function*.

A function whose *name* is f can be thought of as the set of ordered pairs (x, y) for which each member x of the domain is associated with exactly one member $y = f(x)$ of the range. The function can also be regarded as a rule that assigns a unique "output"

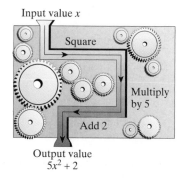

Input value x

Square

Multiply by 5

Add 2

Output value $5x^2 + 2$

Figure 1.16 A function machine

in the set Y to each "input" from the set X, just as the function "machine" shown in Figure 1.16 generates an output of $5x^2 + 2$ for each input x.

> ➡ **What This Says**   To be called a function, the rule must have the property that it assigns to each legitimate input one and only one output. For instance, -2 is not a legitimate input for the function that computes the square root because we are working in the set of real numbers.

A visual representation of a function is shown in Figure 1.17.

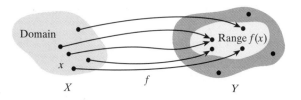

Domain

Range $f(x)$

x

X f Y

Figure 1.17   A function as a mapping

Note that it is quite possible for two different elements in the domain X to map into the same element in the range, and that it is possible for Y to include elements not in the range of f. If, however, the range of f does consist of all of Y, then f is said to map X **onto** Y. Furthermore, if each element in the range is the image of one and only one element in the domain, then f is said to be a **one-to-one function.** A function f is said to be **bounded** on $[a, b]$ if there exists a number B so that $|f(x)| \leq B$ for all x in $[a, b]$.

Most of our work will be with real-valued functions of a real variable; that is, functions whose domain and range are both sets of real numbers.*

FUNCTIONAL NOTATION

Functions can be represented in various ways, but usually they are specified by using a mathematical formula. It is traditional to let x denote the input and y the corresponding output, and to write an equation relating x and y. The letters x and y that appear in such an equation are called **variables.** Because the value of the variable y is determined by that of the variable x, we call y the **dependent variable** and x the **independent variable**.

In this book, when we define a function by an expression such as $f(x) = x^2 + 4x + 5$, we mean the function f is the set of all ordered pairs (x, y) satisfying the equation $y = x^2 + 4x + 5$. To **evaluate** a function f means to find the value of f for some particular value in the domain. For example, to evaluate f at $x = 2$ is to find $f(2)$.

EXAMPLE 1   Using functional notation

Suppose $f(x) = 2x^2 - x$. Find $f(-1)$, $f(0)$, $f(2)$, $f(\pi)$, $f(x + h)$, and

$$\frac{f(x + h) - f(x)}{h},$$

where x and h are real numbers and $h \neq 0$.

---

*Most functions that appear in this book belong to a very special class of **elementary functions**, defined by Joseph Liouville (1809–1882). Certain functions that appear in physics and higher mathematics are not elementary functions.

Solution

In this case, the defined function f tells us to subtract the independent variable x from twice its square. Thus, we have

$$f(-1) = 2(-1)^2 - (-1) = 3$$
$$f(0) = 2(0)^2 - (0) = 0$$
$$f(2) = 2(2)^2 - 2 = 6$$
$$f(\pi) = 2\pi^2 - \pi$$

To find $f(x + h)$ we begin by writing the formula for f in more neutral terms, say as

$$f(\square) = 2(\square)^2 - (\square)$$

Then we insert the expression $x + h$ inside each box, obtaining

$$f(\boxed{x + h}) = 2(\boxed{x + h})^2 - (\boxed{x + h})$$
$$= 2(x^2 + 2xh + h^2) - (x + h)$$
$$= 2x^2 + 4xh + 2h^2 - x - h$$

Finally, if $h \neq 0$,

$$\frac{f(x + h) - f(x)}{h} = \frac{[2x^2 + 4xh + 2h^2 - x - h] - [2x^2 - x]}{h}$$
$$= \frac{4xh + 2h^2 - h}{h}$$
$$= 4x + 2h - 1 \qquad \blacksquare$$

Note: The expression $\dfrac{f(x + h) - f(x)}{h}$ is called a **difference quotient** and is used in Chapter 3 to compute the *derivative*.

Certain functions are defined differently on different parts of their domain and are thus more naturally expressed in terms of more than one formula. We refer to such functions as **piecewise-defined functions**.

EXAMPLE 2   Evaluating a piecewise-defined function

If

$$f(x) = \begin{cases} x \sin x & \text{if } x < 2 \\ 3x^2 + 1 & \text{if } x \geq 2 \end{cases}$$

find $f(-0.5)$, $f(\frac{\pi}{2})$, and $f(2)$.

Solution

To find $f(-0.5)$, we use the first line of the formula because $-0.5 < 2$:

$$f(-0.5) = -0.5 \sin(-0.5) = 0.5 \sin 0.5 \quad \text{This is the exact value.}$$
$$\approx 0.2397 \qquad\qquad\qquad \text{This is the approximate value.}$$

To find $f(\frac{\pi}{2})$, we use the first line of the formula because $\frac{\pi}{2} \approx 1.57 < 2$:

$$f(\tfrac{\pi}{2}) = \tfrac{\pi}{2} \sin \tfrac{\pi}{2} = \tfrac{\pi}{2} \quad \text{This is the exact value.}$$

Finally, because $2 \geq 2$, we use the second line of the formula to find $f(2)$.

$$f(2) = 3(2)^2 + 1 = 13 \qquad \blacksquare$$

Functional notation can be used in a wide variety of applied problems, as shown by Example 3 and again in the problem set.

EXAMPLE 3   Applying functional notation

It is known that an object dropped from a height in a vacuum will fall a distance of s ft in t seconds according to the formula

$$s(t) = 16t^2, \qquad t \geq 0$$

a. How far will the object fall in the first second? In the *next* 2 seconds?
b. How far will it fall during the time interval $t = 1$ to $t = 1 + h$ seconds?
c. What is the average rate of change of distance (in feet per second) during the time $t = 1$ sec to $t = 3$ sec?
d. What is the average rate of change of distance during the time $t = x$ seconds to $t = x + h$ seconds?

Solution

a. $s(1) = 16(1)^2 = 16$. In the first second the object will fall 16 ft. In the next two seconds the object will fall

$$s(1 + 2) - s(1) = s(3) - s(1) = 16(3)^2 - 16(1)^2 = 128$$

The object will fall 128 ft in the next 2 sec.

b. $s(1 + h) - s(1) = 16(1 + h)^2 - 16(1)^2$
$$= 16 + 32h + 16h^2 - 16 = 32h + 16h^2$$

c. AVERAGE RATE $= \dfrac{\text{CHANGE IN DISTANCE}}{\text{CHANGE IN TIME}} = \dfrac{s(3) - s(1)}{3 - 1} = \dfrac{128}{2} = 64.$

The average rate of change is 64 ft/sec.

d. $\dfrac{s(x + h) - s(x)}{(x + h) - x} = \dfrac{s(x + h) - s(x)}{h}$ *Does this look familiar? See Example 1.*

$= \dfrac{16(x + h)^2 - 16x^2}{h} = \dfrac{16x^2 + 32xh + 16h^2 - 16x^2}{h} = 32x + 16h$ ■

DOMAIN OF A FUNCTION

In this book, unless otherwise specified, the domain of a function is the set of real numbers for which the function is defined. We call this the **domain convention** and it will be used throughout this text. If a function f is **undefined** at x, it means that x is not in the domain of f. The most frequent exclusions from the domain are those values that cause division by 0 or negative values under a square root. In applications, the domain is often specified by the context. For example, if x is the number of people on an elevator, the context requires that negative numbers and nonintegers be excluded from the domain; therefore, x must be an integer such that $0 \leq x \leq c$, where c is the maximum capacity of the elevator.

EXAMPLE 4   Domain of a function

Find the domain for the given functions.

a. $f(x) = 2x - 1$ **b.** $g(x) = 2x - 1, x \neq -3$

c. $h(x) = \dfrac{(2x - 1)(x + 3)}{x + 3}$ **d.** $F(x) = \sqrt{x + 2}$ **e.** $G(x) = \dfrac{4}{5 - \cos x}$

Solution

a. All real numbers; $(-\infty, \infty)$.
b. All real numbers except -3. This can be written as $(-\infty, -3) \cup (-3, \infty)$.
c. Because the expression is meaningful for all $x \neq -3$, the domain is all real numbers except -3.

d. F has meaning if and only if $x + 2$ is nonnegative; therefore, the domain is $x \geq -2$, or $[-2, \infty)$.

e. G is defined whenever $5 - \cos x \neq 0$. This imposes no restriction on x since $|\cos x| \leq 1$. Thus, the domain of G is the set of all real numbers; this can be written $(-\infty, \infty)$. ∎

Equality of Functions

Two functions f and g are **equal** if and only if both

1. f and g have the same domain.
2. $f(x) = g(x)$ for all x in the domain.

EXAMPLE 5   Equality of functions

Consider again the functions from Example 4:

$$f(x) = 2x - 1 \qquad g(x) = 2x - 1 \text{ for } x \neq -3 \qquad h(x) = \frac{(2x - 1)(x + 3)}{x + 3}$$

Does $h(x)$ equal $f(x)$ or $g(x)$?

Solution

A common mistake is to "reduce" the function h to obtain the function f:

$$\text{WRONG:} \quad h(x) = \frac{(2x - 1)(x + 3)}{x + 3} = 2x - 1 = f(x)$$

This is wrong because the functions f and h have different domains. In particular, -3 is a valid input for f, but it is not a valid input for h. We say that the function h has a **hole** at $x = -3$. In other words, this WRONG calculation is valid only if $x \neq -3$, so the correct answer is

$$\text{RIGHT:} \quad h(x) = \frac{(2x - 1)(x + 3)}{x + 3} = 2x - 1, \, x \neq -3$$

Therefore, $h(x) = g(x)$. ∎

COMPOSITION OF FUNCTIONS

There are many situations in which a quantity is given as a function of one variable that, in turn, can be written as a function of a second variable. Suppose, for example, that your job is to ship x packages of a product via Federal Express to a variety of addresses. Let x be the number of packages to ship, and let $f(x)$ be the weight of the x objects and $g(w)$ be the cost of shipping a package of weight w. Then

Weight is a function $f(x)$ of the number of objects x.

Cost is a function $g[f(x)]$ of the weight.

So we have expressed cost as a function of the number of packages. This process of evaluating a function of a function is known as *functional composition*.

Composition of Functions

The **composite function** $f \circ g$ is defined by

$$(f \circ g)(x) = f[g(x)]$$

for each x in the domain of g for which $g(x)$ is in the domain of f.

➡ **What This Says** To visualize how functional composition works, think of $f \circ g$ in terms of an "assembly line" in which g and f are arranged in series, with output $g(x)$ becoming the input of f, as illustrated in Figure 1.18.

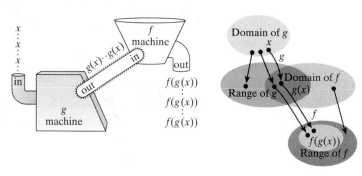

Figure 1.18 Composition of functions

EXAMPLE 6 **Finding the composition of functions**

If $f(x) = 3x + 5$ and $g(x) = \sqrt{x}$, find the composite functions $f \circ g$ and $g \circ f$.

Solution

The function $f \circ g$ is defined by $f[g(x)]$.

$$(f \circ g)(x) = f[g(x)] = f(\sqrt{x}) = 3\sqrt{x} + 5$$

The function $g \circ f$ is defined by $g[f(x)]$.

$$(g \circ f)(x) = g[f(x)] = g(3x + 5) = \sqrt{3x + 5}$$

That is, $f \circ g$ is not, in general, the same as $g \circ f$. ∎

WARNING Example 6 illustrates that functional composition is not commutative.

EXAMPLE 7 **An application of composite functions**

Air pollution is a problem for many metropolitan areas. Suppose a study conducted in a certain city indicates that when the population is p hundred thousand people, then the average daily level of carbon monoxide in the air is given by the formula

$$L(p) = 0.70\sqrt{p^2 + 3}$$

parts per million (ppm). A second study predicts that t years from now, the population will be $p(t) = 1 + 0.02t^3$ hundred thousand people. Assuming these formulas are correct, what level of air pollution should be expected in 4 years?

Solution

The level of pollution is $L(p) = 0.70\sqrt{p^2 + 3}$, where $p(t) = 1 + 0.02t^3$. Thus, the pollution level at time t is given by the composite function

$$(L \circ p)(t) = L[p(t)] = L(1 + 0.02t^3) = 0.70\sqrt{(1 + 0.02t^3)^2 + 3}$$

In particular, when $t = 4$, we have

$$(L \circ p)(4) = 0.70\sqrt{[1 + 0.02(4)^3]^2 + 3} \approx 2.00 \text{ ppm}$$ ∎

In calculus, it is frequently necessary to express a function as the composite of two simpler functions. Here is an example that illustrates how this can be done.

EXAMPLE 8   Expressing a given function as a composite of two functions

Express each of the following functions as the composite of two functions u and g so that $f(x) = g[u(x)]$.

a. $f(x) = (x^2 + 5x + 1)^5$ **b.** $f(x) = \cos^3 x$
c. $f(x) = \sin x^3$ **d.** $f(x) = \sqrt{5x^2 - x}$

Solution

There are often many ways to express $f(x)$ as a composite $g[u(x)]$. Perhaps the most natural is to choose u to represent the "inner" portion of f and g as the "outer" portion. Such choices are indicated in the following table.

Given Function $f(x) = g[u(x)]$	Inner Function $u(x)$	Outer Function $g(u)$
a. $f(x) = (x^2 + 5x + 1)^5$	$u(x) = x^2 + 5x + 1$	$g(u) = u^5$
b. $f(x) = \cos^3 x$	$u(x) = \cos x$	$g(u) = u^3$
c. $f(x) = \sin x^3$	$u(x) = x^3$	$g(u) = \sin u$
d. $f(x) = \sqrt{5x^2 - x}$	$u(x) = 5x^2 - x$	$g(u) = \sqrt{u}$

GRAPH OF A FUNCTION

Graphs have visual impact. They also reveal information that may not be evident from verbal or algebraic descriptions. Two graphs depicting practical relationships are shown in Figure 1.19.

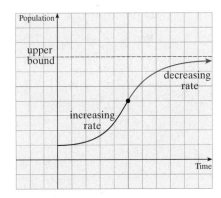

a. A production function
This graph describes the variation in total industrial production in a certain country over a five year time span. The fact that the graph has a peak suggests that production is greatest at the corresponding time.

b. Bounded population growth
This graph represents the growth of a population when environmental factors impose an upper bound on the possible size of the population. It indicates that the rate of population growth increases at first and then decreases as the size of the population gets closer and closer to the upper bound.

Figure 1.19   Two graphs with practical interpretations

To represent a function $y = f(x)$ geometrically as a graph, it is traditional to use a Cartesian coordinate system on which units for the independent variable x are marked on the horizontal axis and units for the dependent variable y on the vertical axis.

Graph of a Function

> The **graph** of a function f consists of points whose coordinates (x, y) satisfy $y = f(x)$, for all x in the domain of f.

In Chapter 4, we will discuss efficient techniques involving calculus that you can use to draw accurate graphs of functions. In algebra you began sketching lines by plotting points, but you quickly discovered that this is not a very efficient way to draw more complicated graphs, especially without the aid of a graphing calculator or computer. Table 1.3 includes a few common graphs you have probably encountered in previous courses. We will assume that you are familiar with their general shape and know how to sketch each of them either by hand or with the assistance of your graphing calculator. There are, however, certain functions, such as $f(x) = \sin \frac{1}{x}$, that are hard to graph even with a calculator or computer. That said, be assured that most functions that appear in this text or in practical applications can be graphed.

Vertical Line Test   By definition of a function, for a given x in the domain there is only one number y in the range. Geometrically, this means that any vertical line $x = a$ crosses the graph of a function at most once. This observation leads to the following useful criterion.

> ### The Vertical Line Test
>
> A curve in the plane is the graph of a function if and only if it intersects no vertical line more than once.

Look at Figure 1.20 for examples of the vertical line test.

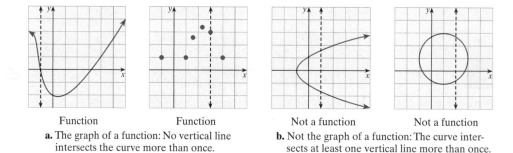

| Function | Function | Not a function | Not a function |

a. The graph of a function: No vertical line intersects the curve more than once.   **b.** Not the graph of a function: The curve intersects at least one vertical line more than once.

Figure 1.20   The vertical line test

Intercepts   The points where a graph intersects the coordinate axes are called *intercepts*. Here is a definition.

Intercepts

> If the number zero is in the domain of f and $f(0) = b$, then the point $(0, b)$ is called the **y-intercept** of the graph of f. If a is a real number in the domain of f such that $f(a) = 0$, then $(a, 0)$ is an **x-intercept** of f.

> ➡ **What This Says**   To find the x-intercepts, set y equal to 0 and solve for x. To find the y-intercept, set x equal to 0 and solve for y.

TABLE 1.3 Directory of Curves

Identity Function $y = x$	Standard Quadratic Function $y = x^2$	Standard Cubic Function $y = x^3$

Absolute Value Function $y = |x|$	Square Root Function $y = \sqrt{x}$	Cube Root Function $y = \sqrt[3]{x}$

Standard Reciprocal Function $y = \dfrac{1}{x}$	Standard Reciprocal Squared Function $y = \dfrac{1}{x^2}$	Standard Square Root Reciprocal Function $y = \dfrac{1}{\sqrt{x}}$

Cosine Function $y = \cos x$	Sine Function $y = \sin x$	Tangent Function $y = \tan x$

Secant Function $y = \sec x$	Cosecant Function $y = \csc x$	Cotangent Function $y = \cot x$

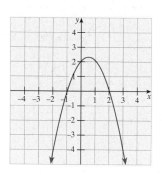

Figure 1.21 Graph of $f(x) = -x^2 + x + 2$ using the intercepts

Symmetry

EXAMPLE 9 Finding intercepts

Find all intercepts of the graph of the function $f(x) = -x^2 + x + 2$.

Solution

The y-intercept is $(0, f(0)) = (0, 2)$. To find the x-intercepts, solve the equation $f(x) = 0$. Factoring, we find that

$$-x^2 + x + 2 = 0$$
$$x^2 - x - 2 = 0$$
$$(x + 1)(x - 2) = 0$$
$$x = -1 \text{ or } x = 2$$

Thus, the x-intercepts are $(-1, 0)$ and $(2, 0)$, and the y-intercept is $(0, 2)$. The graph of f is shown in Figure 1.21. ∎

Symmetry There are two kinds of symmetry that help in graphing a function, as shown in Figure 1.22 and defined in the following box.

The graph of $y = f(x)$ is **symmetric with respect to the y-axis** if whenever $P(x, y)$ is a point on the graph, so is the point $(-x, y)$ that is the mirror image of P about the y-axis. Thus, y-axis symmetry occurs if and only if $f(-x) = f(x)$ for all x in the domain of f. A function with this property is called an **even function**.

The graph of $y = f(x)$ is **symmetric with respect to the origin** if whenever $P(x, y)$ is on the graph, so is $(-x, -y)$, the mirror image of P about the origin. Symmetry with respect to the origin occurs when $f(-x) = -f(x)$ for all x. A function that satisfies this condition is called an **odd function**.

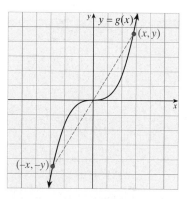

a. Graph of an even function f. Symmetry with respect to the y-axis: $f(-x) = f(x)$

b. Graph of an odd function g. Symmetry with respect to the origin: $g(-x) = -g(x)$

Figure 1.22 Graphs of even and odd functions

WARNING The graph of a nonzero function cannot be symmetric with respect to the x-axis. In other words, if the curve is symmetric with respect to the x-axis, such as the circle, then it will not be a function.

There are many functions that are neither odd nor even. You may wonder why we have said nothing about symmetry with respect to the x-axis, but such symmetry would require $f(x) = -f(x)$, which is precluded by the vertical line test (do you see why?).

EXAMPLE 10   Even and odd functions

Classify the given functions as even, odd, or neither.

a. $f(x) = x^2$ **b.** $g(x) = x^3$ **c.** $h(x) = x^2 + 5x$

Solution

a. $f(x) = x^2$ is *even*, because

$$f(-x) = (-x)^2 = x^2 = f(x)$$

b. $g(x) = x^3$ is *odd*, because

$$g(-x) = (-x)^3 = -x^3 = -g(x)$$

c. $h(x) = x^2 + 5x$ is *neither* even nor odd because

$$h(-x) = (-x)^2 + 5(-x) = x^2 - 5x$$

Note that $h(-x) \neq h(x)$ and $h(-x) \neq -h(x)$.

The graphs of these three functions are shown in Figure 1.23. Note the symmetry in the first two graphs and the lack of symmetry with respect to the axis or the origin in the third.

a. $f(x) = x^2$

b. $g(x) = x^3$

c. $h(x) = x^2 + 5x$

Figure 1.23   Three graphs illustrating the concept of symmetry   ■

CLASSIFICATION OF FUNCTIONS

We will now describe some of the common types of functions used in this text.

Polynomial Function

A **polynomial function** is a function of the form

$$f(x) = a_n x^n + a_{n-1} x^{n-1} + \cdots + a_2 x^2 + a_1 x + a_0$$

where n is a nonnegative integer and $a_n, \ldots, a_2, a_1, a_0$ are constants. If $a_n \neq 0$, the integer n is called the **degree** of the polynomial. The constant a_n is called the **leading coefficient** and the constant a_0 is called the **constant term** of the polynomial function. In particular,

A **constant function** is zero degree: $f(x) = a$
   Example: $f(x) = 5$

A **linear function** is first degree: $f(x) = ax + b$
   Example: $f(x) = 2x - \sqrt{2}$

A **quadratic function** is second degree: $f(x) = ax^2 + bx + c$
   Example: $f(x) = 3x^2 + 5x - \frac{1}{2}$

> A **cubic function** is third degree: $f(x) = ax^3 + bx^2 + cx + d$
> Example: $f(x) = \sqrt{2}x^3 - \pi x$
>
> A **quartic function** is fourth degree: $f(x) = ax^4 + bx^3 + cx^2 + dx + e$
> Example: $f(x) = 3x^4 - \frac{\sqrt{2}}{2}$

Rational Function

> A **rational function** is the quotient of two polynomial functions, $p(x)$ and $d(x)$, which can be written in the form
>
> $$f(x) = \frac{p(x)}{d(x)}, \qquad d(x) \neq 0$$
>
> Examples: $f(x) = x^{-1};$ $f(x) = \dfrac{x-5}{x^2 + 2x - 3};$ $f(x) = x^{-3} + \sqrt{2}\,x$

When we write $d(x) \neq 0$, we mean that all values c for which $d(c) = 0$ are excluded from the domain of d.

If r is any nonzero real number, the function $f(x) = x^r$ is called a **power function** with exponent r. You should be familiar with the following cases:

Integer powers ($r = n$, a positive integer): $f(x) = x^n = \underbrace{x \cdot x \cdot \ldots \cdot x}_{n \text{ factors}}$

Example: $f(x) = x^6$

Reciprocal powers (r is a negative integer): $f(x) = x^{-n} = \dfrac{1}{x^n}$ for $x \neq 0$

Example: $f(x) = x^{-4}$

Roots ($r = \frac{m}{n}$ is a nonzero rational number): $f(x) = x^{m/n} = \sqrt[n]{x^m} = (\sqrt[n]{x})^m$ for all x if n odd, for $x \geq 0$ if n even, $n \neq 0$ ($\frac{m}{n}$ is reduced)

Examples: $f(x) = x^{3/4};$ $f(x) = \sqrt[5]{x^2}$

Power functions can also have irrational exponents (such as $\sqrt{2}$ or π), but such functions must be defined in a special way (see Section 2.4).

A function is called **algebraic** if it can be constructed using a finite number of algebraic operations (such as adding, subtracting, multiplying, dividing, or taking roots) starting with polynomials. Functions that are not algebraic are called **transcendental**. The following are transcendental functions:

SMH

Trigonometric functions are the functions sine, cosine, tangent, secant, cosecant, and cotangent. The basic forms of these functions are reviewed in Chapter 4 of the *Student Mathematics Handbook*. You can also review these functions by consulting a trigonometry or precalculus textbook.

Exponential functions are functions of the form $f(x) = b^x$, where b is a positive constant, $b \neq 1$. We will introduce these functions in Section 2.4.

Logarithmic functions are functions of the form $f(x) = \log_b x$, where b is a positive constant, $b \neq 1$. We will also study these functions in Section 2.4.

1.3   PROBLEM SET

In Problems 1–12, find the domain of f and compute the indicated values or state that the corresponding x-value is not in the domain. Tell whether any of the indicated values are zeros of the function, that is, values of x that cause the functional value to be 0.

A   **1.** $f(x) = 2x + 3$; $f(-2)$, $f(1)$, $f(0)$

2. $f(x) = -x^2 + 2x + 3$; $f(0)$, $f(1)$, $f(-2)$

3. $f(x) = 3x^2 + 5x - 2$; $f(1)$, $f(0)$, $f(-2)$

4. $f(x) = x + \dfrac{1}{x}$; $f(-1)$, $f(1)$, $f(2)$

5. $f(x) = \dfrac{(x+3)(x-2)}{x+3}$; $f(2)$, $f(0)$, $f(-3)$

6. $f(x) = (2x-1)^{-3/2}$; $f(1)$, $f(\frac{1}{2})$, $f(13)$

7. $f(x) = \sqrt{x^2 + 2x}$; $f(-1)$, $f(\frac{1}{2})$, $f(1)$

8. $f(x) = \sqrt{x^2 + 5x + 6}$; $f(0)$, $f(1)$, $f(-2)$

9. $f(x) = \sin(1 - 2x)$; $f(-1)$, $f(\frac{1}{2})$, $f(1)$

10. $f(x) = \sin x - \cos x$; $f(0)$, $f(-\frac{\pi}{2})$, $f(\pi)$

11. $f(x) = \begin{cases} -2x + 4 & \text{if } x \le 1 \\ x + 1 & \text{if } x > 1 \end{cases}$; $f(3)$, $f(1)$, $f(0)$

12. $f(x) = \begin{cases} 3 & \text{if } x < -5 \\ x + 1 & \text{if } -5 \le x \le 5 \\ \sqrt{x} & \text{if } x > 5 \end{cases}$; $f(-6)$, $f(-5)$, $f(16)$

In Problems 13–20, evaluate the difference quotient $\dfrac{f(x+h) - f(x)}{h}$ for the given function.

13. $f(x) = 9x + 3$

14. $f(x) = 5 - 2x$ — $2x + 5$

15. $f(x) = 5x^2$

16. $f(x) = 3x^2 + 2x$

17. $f(x) = |x|$ if $x < -1$ and $0 < h < 1$

18. $f(x) = |x|$ if $x > 1$ and $0 < h < 1$

19. $f(x) = \dfrac{1}{x}$

20. $f(x) = \dfrac{x+1}{x-1}$

State whether the functions f and g in Problems 21–26 are equal.

21. $f(x) = \dfrac{2x^2 + x}{x}$; $g(x) = 2x + 1$

22. $f(x) = \dfrac{2x^2 + x}{x}$; $g(x) = 2x + 1$, $x \ne 0$

23. $f(x) = \dfrac{2x^2 - x - 6}{x - 2}$; $g(x) = 2x + 3$, $x \ne 2$

24. $f(x) = \dfrac{3x^2 - 7x - 6}{x - 3}$; $g(x) = 3x + 2$, $x \ne 3$

25. $f(x) = \dfrac{3x^2 - 5x - 2}{x - 2}$; $g(x) = 3x + 1$

26. $f(x) = \dfrac{(3x+1)(x-2)}{x-2}$, $x \ne 6$;

$g(x) = \dfrac{(3x+1)(x-6)}{x-6}$, $x \ne 2$

Classify the functions defined in Problems 27–34 as even, odd, or neither.

27. $f_1(x) = x^2 + 1$

28. $f_2(x) = \sqrt{x^2}$

29. $f_3(x) = \dfrac{1}{3x^3 - 4}$

30. $f_4(x) = x^3 + x$

31. $f_5(x) = \dfrac{1}{(x^3 + 3)^2}$

32. $f_6(x) = \dfrac{1}{(x^3 + x)^2}$

33. $f_7(x) = |x|$

34. $f_8(x) = |x + x^3|$

In Problems 35–40, find the composite functions $f \circ g$ and $g \circ f$.

35. $f(x) = x^2 + 1$ and $g(x) = 2x$

36. $f(x) = \sin x$ and $g(x) = 1 - x^2$

37. $f(t) = \sqrt{t}$ and $g(t) = t^2$

38. $f(u) = \dfrac{u - 1}{u + 1}$ and $g(u) = \dfrac{u + 1}{1 - u}$

39. $f(x) = \sin x$ and $g(x) = 2x + 3$

40. $f(x) = \dfrac{1}{x}$ and $g(x) = \tan x$

In Problems 41–50, express f as the composition of two functions u and g such that $f(x) = g(u)$, where u is a function of x.

41. $f(x) = (2x^2 - 1)^4$

42. $f(x) = (x^2 + 1)^3$

43. $f(x) = |2x + 3|$

44. $f(x) = \sqrt{5x - 1}$

45. $f(x) = \tan^2 x$

46. $f(x) = \tan x^2$

47. $f(x) = \sin \sqrt{x}$

48. $f(x) = \sqrt{\sin x}$

49. $f(x) = \sin\left(\dfrac{x + 1}{2 - x}\right)$

50. $f(x) = \tan\left(\dfrac{2x}{1 - x}\right)$

51. If point A in Figure 1.24 has coordinates $(2, f(2))$, what are the coordinates of P and Q?

Problem 51   **Problem 52**

Figure 1.24

52. If point B in Figure 1.24 has coordinates $(3, g(3))$, what are the coordinates of R and S?

Find the x-intercepts, if any, for the functions given in Problems 53–60.

53. $f(x) = 3x^2 - 5x - 2$

54. $f(x) = 6x^2 + 5x - 6$

55. $f(x) = (x - 15)(2x + 25)(3x - 65)(4x + 1)$

56. $f(x) = (x^2 - 10)(x^2 - 12)(x^2 - 20)$

57. $f(x) = 5x^3 - 3x^2 + 2x$

58. $f(x) = x^4 - 41x^2 + 400$

59. $f(x) = \dfrac{x^2 - 1}{x^2 + 2}$

60. $f(x) = \dfrac{x(x^2 - 3)}{x^2 + 5}$

B **61.** Suppose the total cost (in dollars) of manufacturing q units of a certain commodity is given by

$$C(q) = q^3 - 30q^2 + 400q + 500$$

a. Compute the cost of manufacturing 20 units.

b. Compute the cost of manufacturing the 20th unit.

62. An efficiency study of the morning shift at a certain factory indicates that an average worker who arrives on the job at 8:00 A.M. will have assembled

$$f(x) = -x^3 + 6x + 15x^2$$

CD players x hours later $(0 \le x \le 8)$.

a. How many players will such a worker have assembled by 10:00 A.M.?

b. How many players will such a worker assemble between 9:00 A.M. and 10:00 A.M.?

63. In physics, a light source of luminous intensity K candles is said to have *illuminance* $I = K/s^2$ on a flat surface s ft away. Suppose a small, unshaded lamp of luminous intensity 30 candles is connected to a rope that allows it to be raised and lowered between the floor and the top of a 10-ft-high ceiling. Assume that the lamp is being raised and lowered in such a way that at time t (in min) it is $s = 6t - t^2$ ft above the floor.

a. Express the illuminance on the floor as a composite function of t for $0 < t < 6$.

b. What is the illuminance when $t = 1$? When $t = 4$?

64. Biologists have found that the speed of blood in an artery is a function of the distance of the blood from the artery's central axis. According to *Poiseuille's law*, the speed (cm³/sec) of blood that is r cm from the central axis of an artery is given by the function

$$S(r) = C \cdot (R^2 - r^2)$$

where C is a constant and R is the radius of the artery.* Suppose that for a certain artery, $C = 1.76 \times 10^5$ cm/sec and $R = 1.2 \times 10^{-2}$ cm. (See Figure 1.25.)

Figure 1.25   Artery

a. Compute the speed of the blood at the central axis of this artery.

b. Compute the speed of the blood midway between the artery's wall and central axis.

---
*The law and the unit *poise*, a unit of viscosity, are both named for the French physician Jean Louis Poiseuille (1799–1869).

65. At a certain factory, the total cost of manufacturing q units during the daily production run is $C(q) = q^2 + q + 900$ dollars. On a typical workday, the number of units manufactured during the first t hours of a production run can be modeled by the function $q(t) = 25t$.

a. Express the total manufacturing cost as a function of t.

b. How much will have been spent on production by the end of the third hour?

c. When will the total manufacturing cost reach $11,000?

66. A ball is thrown directly upward from the edge of a cliff in such a way that t seconds later, it is

$$s = -16t^2 + 96t + 144$$

feet above the ground at the base of the cliff. Sketch the graph of this equation (making the t-axis the horizontal axis) and then answer these questions:

a. How high is the cliff?

b. When (to the nearest tenth of a second) does the ball hit the ground at the base of the cliff?

c. Estimate the time it takes for the ball to reach its maximum height. What is the maximum height?

67. Charles's law for gases states that if the pressure remains constant, then

$$V(T) = V_0 \cdot \left(1 + \frac{T}{273}\right)$$

where V is the volume (in.³), V_0 is the initial volume (in.³), and T is the temperature (in degrees Celsius).

a. Sketch the graph of $V(T)$ for $V_0 = 100$ and $T \ge -273$.

b. What is the temperature needed for the volume to double?

68. To study the rate at which animals learn, a psychology student performed an experiment in which a rat was sent repeatedly through a laboratory maze. Suppose that the time (in minutes) required for the rat to traverse the maze on the nth trial was approximately

$$f(n) = 3 + \frac{12}{n}$$

a. What is the domain of the function $f(x) = 3 + \frac{12}{x}$?

b. For what values of n does $f(n)$ have meaning in the context of the psychology experiment?

c. How long did it take the rat to traverse the maze on the third trial?

d. On which trial did the rat first traverse the maze in 4 minutes or less?

e. According to the function f, what will happen to the time required for the rat to traverse the maze as the number of trials increases? Will the rat ever be able to traverse the maze in less than 3 minutes?

69. The trajectory of a cannonball shot from the origin with initial velocity v and angle of inclination α (measured from level ground), is given by the equation

$$y = mx - 16v^{-2}(1 + m^2)x^2$$

where $m = \tan \alpha$. For this problem assume the initial velocity is $v = 200$ ft/s and that the angle of inclination is $42°$.

a. Using a graphing utility of a calculator, determine the maximum height reached by the cannonball.

b. Estimate the point where the cannonball will hit the ground.

c. Which curve in Table 1.3 shows the basic shape of this graph?

70. Draw the path of a cannonball (see Problem 69) for an angle of inclination of $47°$. Determine the angle α that will maximize the distance the cannonball will travel.

71. It is estimated that t years from now, the population of a certain suburban community will be

$$P(t) = 20 - \frac{6}{t+1}$$

thousand people.

a. What will the population of the community be nine years from now?

b. By how much will the population increase during the ninth year?

c. What will happen to the size of the population in the "long run"?

C **72. Journal problem:** *The Mathematics Student Journal.** ■ Given that $f(11) = 11$ and

$$f(x+3) = \frac{f(x) - 1}{f(x) + 1}$$

for all x, find $f(2000)$.

---

*Volume 28 (1980), issue 3, p. 2. Note that the journal problem requests $f(1979)$, which, no doubt, was related to the publication date. We have taken the liberty of updating the requested value.

73. HISTORICAL QUEST One of the best known mathematical theorems is the **Pythagorean theorem**, named after the Greek philosopher Pythagoras. Very little is known about the life of Pythagoras, but we do know he was born on the island of Samos. He founded a secret brotherhood called the Pythagoreans that continued for at least 100 years after Pythagoras was murdered for political reasons. Even though the cult was called a "brotherhood," it did admit women. According to Lynn Osen in *Women in Mathematics,*[†] the order was carried on by his wife and daughters after his death. In fact, women were probably more welcome in the centers of learning in ancient Greece than in any other age from that time until now. ■

PYTHAGORAS
ca. 500 B.C.

State and prove the Pythagorean theorem.

[†]MIT Press, Cambridge MA, 1975.

1.4 Inverse Functions; Inverse Trigonometric Functions

IN THIS SECTION inverse functions, criteria for the existence of an inverse f^{-1}, graph of f^{-1}, inverse trigonometric functions, inverse trigonometric identities

INVERSE FUNCTIONS

WARNING The symbol f^{-1} means the inverse of f and does not mean reciprocal, $1/f$.

For a given function f, we write $y_0 = f(x_0)$ to indicate that f maps the number x_0 in its domain into the corresponding number y_0 in its range. If f has an inverse f^{-1}, it is the function that reverses the effect of f in the sense that

$$f^{-1}(y_0) = x_0$$

For example, if

$$f(x) = 2x - 3, \text{ then } f(0) = -3, \ f(1) = -1, \ f(2) = 1,$$

and the inverse f^{-1} reverses f so that

$$
\begin{array}{lll}
f(0) = -3 & \mathbf{f^{-1}(-3) = 0} & \text{that is, } f^{-1}[f(0)] = 0 \\
f(1) = -1 & \mathbf{f^{-1}(-1) = 1} & \text{that is, } f^{-1}[f(1)] = 1 \\
f(2) = 1 & \mathbf{f^{-1}(1) = 2} & \text{that is, } f^{-1}[f(2)] = 2
\end{array}
$$

In the case where the inverse of a function is itself a function, we have the following definition.

Inverse Function

Let f be a function with domain D and range R. Then the function f^{-1} with domain R and range D is the **inverse of f** if

$$f^{-1}[f(x)] = x \qquad \text{for all } x \text{ in } D$$

and

$$f[f^{-1}(y)] = y \qquad \text{for all } y \text{ in } R$$

➡ **What This Says**   Suppose we consider a function defined by a set of ordered pairs (x, y) where $y = f(x)$. The image of x is y, as shown in Figure 1.26.

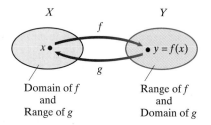

Figure 1.26   Inverse functions f and g

If y is a member of the domain of the function $g = f^{-1}$, then $g(y) = x$. This means that f matches each element of x to exactly one y, and g matches those same elements of y back to the original values of x. If you think of a function as a set of ordered pairs (x, y), the inverse of f is the set of ordered pairs with the components (y, x).

The inverse function f^{-1} reverses the effect of f. This relationship can be illustrated by function "machines," as shown in Figure 1.27. Note that the value input into the first "f machine" is x and the output $f(x)$ from this machine is the input into a second "f^{-1} machine" whose output is the original input value, x.

Figure 1.27   Inverse machines

The language of this definition suggests that there is only one inverse function of f. Indeed, it can be shown (see Problem 54) that if f has an inverse, then the inverse is unique.

EXAMPLE 1   Inverse of a given function defined as a set of ordered pairs

Let $f = \{(0, 3), (1, 5), (3, 9), (5, 13)\}$; find f^{-1}, if it exists.

Solution

The inverse simply reverses the ordered pairs:

$$f^{-1} = \{(3, 0), (5, 1), (9, 3), (13, 5)\}$$ ∎

EXAMPLE 2   Inverse of a given function defined by an equation

Let $f(x) = 2x - 3$; find f^{-1}, if it exists.

Solution

To find f^{-1}, let $y = f(x)$, then interchange the x and y variables, and finally solve for y.

$$\text{Given function:} \quad y = 2x - 3 \qquad \text{Then} \quad x = 2y - 3$$
$$2y = x + 3$$
$$\text{Inverse:} \quad y = \tfrac{1}{2}(x + 3)$$

Thus, we represent the inverse function as $f^{-1}(x) = \tfrac{1}{2}(x + 3)$. To verify that these functions are inverses of each other, we note that

$$f[f^{-1}(x)] = f\left[\tfrac{1}{2}(x+3)\right] = 2\left[\tfrac{1}{2}(x+3)\right] - 3 = x + 3 - 3 = x$$

and

$$f^{-1}[f(x)] = f^{-1}(2x - 3) = \tfrac{1}{2}[(2x - 3) + 3] = \tfrac{1}{2}(2x) = x$$

for all x.   ■

CRITERIA FOR EXISTENCE OF AN INVERSE f^{-1}

The inverse of a function may not exist. For example, both the functions

$$f = \{(0, 0), (1, 1), (-1, 1), (2, 4), (-2, 4)\} \quad \text{and} \quad g(x) = x^2$$

do not have inverses because if we attempt to find the inverses, we obtain relations that are not functions. In the first case, we find

$$\text{Possible inverse of } f: \{(0, 0), (1, 1), (1, -1), (4, 2), (4, -2)\}$$

This is not a function because not every member of the domain is associated with a single member in the range: $(1, 1)$ and $(1, -1)$, for example.

In the second case, if we interchange the x and y in the equation for the function g where $y = x^2$ and then solve for y, we find

$$x = y^2 \quad \text{or} \quad y = \pm\sqrt{x} \quad \text{for } x \geq 0$$

But this is not a function of x, because for any positive value of x, there are two corresponding values of y, namely, $\sqrt{x}$ and $-\sqrt{x}$.

A function f will have an inverse f^{-1} on the interval I when there is exactly one number in the domain associated with each number in the range. That is, f^{-1} exists if $f(x_1)$ and $f(x_2)$ are equal only when $x_1 = x_2$. A function with this property is said to be **one-to-one**. This is equivalent to the graphical criterion, called the *horizontal line test*, shown in Figure 1.28.

a. A function that has an inverse

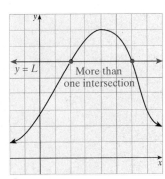

b. A function that does not have an inverse

Figure 1.28   Horizontal line test

> **Horizontal Line Test**
>
> A function f has an inverse if and only if no horizontal line intersects the graph of $y = f(x)$ at more than one point.

A function is said to be **strictly increasing** on an interval I if its graph is always rising on I, and **strictly decreasing** on I if the graph always falls on I. It is called **strictly monotonic** on I if it is either strictly increasing or strictly decreasing throughout that interval. A strictly monotonic function must be one-to-one and hence must have an inverse. For example, if f is strictly increasing on the interval I, we know that

$$x_1 > x_2 \quad \text{implies} \quad f(x_1) > f(x_2)$$

so there is no way to have $f(x_1) = f(x_2)$ unless $x_1 = x_2$. This observation is formalized in the following theorem.

THEOREM 1.2   A strictly monotonic function has an inverse

Let f be a function that is strictly monotonic on an interval I. Then f^{-1} exists and is monotonic on I (increasing if f is increasing and decreasing if f is decreasing).

Proof   We have already commented on why f^{-1} exists. To show that f^{-1} is strictly increasing whenever f is increasing, let y_1 and y_2 be numbers in the range of f, with $y_2 > y_1$. We will show that $f^{-1}(y_2) > f^{-1}(y_1)$. Because y_1, y_2 are in the range of f, there exist numbers x_1, x_2 in the domain I such that $y_1 = f(x_1)$ and $y_2 = f(x_2)$. Because $y_2 > y_1$, it follows that $f(x_2) > f(x_1)$, and because f is strictly increasing, we must have $x_2 > x_1$. Thus, $f^{-1}(y_2) > f^{-1}(y_1)$, and f^{-1} is strictly increasing. Similarly, if f is strictly decreasing, then so is f^{-1}. (The details are left for the reader.)   ❑

GRAPH OF f^{-1}

The graphs of f and its inverse f^{-1} are closely related. In particular, if (a, b) is a point on the graph of f, then $b = f(a)$ and $a = f^{-1}(b)$, so (b, a) is on the graph of f^{-1}. It can be shown that (a, b) and (b, a) are reflections of one another in the line $y = x$. (See Figure 1.29.) These observations yield the following procedure for sketching the graph of an inverse function.

Figure 1.29 The graphs of f and f^{-1} are reflections in the line $y = x$

> **Procedure for Obtaining the Graph of f^{-1}**
>
> If f^{-1} exists, its graph may be obtained by reflecting the graph of f in the line $y = x$.

INVERSE TRIGONOMETRIC FUNCTIONS

The trigonometric functions are not one-to-one, so their inverses do not exist. However, if we restrict the domains of the trigonometric functions, then the inverses exist on those domains. In trigonometry you probably distinguished between the sine curve with unrestricted domain and the sine curve with restricted domain by writing $y = \sin x$ and $y = \operatorname{Sin} x$, respectively. In calculus, it is not customary to make such a distinction by using a capital letter.

Let us consider the sine function first. We know that the sine function is strictly increasing on the closed interval $[-\frac{\pi}{2}, \frac{\pi}{2}]$, and if we restrict $\sin x$ to this interval, it does have an inverse, as shown in Figure 1.30.

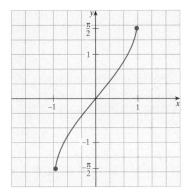

a. The graph of $y = \sin^{-1}x$ is obtained by reflecting the part of the sine on $\left[-\frac{\pi}{2}, \frac{\pi}{2}\right]$ about $y = x$.

b. The graph of the inverse sine function, $y = \sin^{-1}x$.

Figure 1.30   Inverse sine function

Inverse Sine Function

$$y = \sin^{-1} x \quad \text{if and only if} \quad x = \sin y \quad \text{and} \quad -\frac{\pi}{2} \le y \le \frac{\pi}{2}$$

The function $\sin^{-1} x$ is sometimes written $\arcsin x$.

WARNING   The function $\sin^{-1} x$ is NOT the reciprocal of $\sin x$. To denote the reciprocal, write $(\sin x)^{-1}$. In other words, $\sin^{-1} x \ne \dfrac{1}{\sin x}$, whereas $(\sin x)^{-1} = \dfrac{1}{\sin x}$.

Inverses of the other five trigonometric functions may be constructed in a similar manner. For example, by restricting $\tan x$ to the open interval $(-\frac{\pi}{2}, \frac{\pi}{2})$ where it is one-to-one, we can define the inverse tangent function as follows.

Inverse Tangent Function

$$y = \tan^{-1} x \quad \text{if and only if} \quad x = \tan y \quad \text{and} \quad -\frac{\pi}{2} < y < \frac{\pi}{2}$$

The function $\tan^{-1} x$ is sometimes written $\arctan x$.

The graph of $y = \tan^{-1} x$ is shown in Figure 1.31.

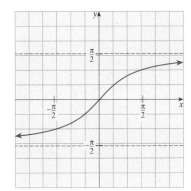

a. The graph of $y = \tan^{-1}x$ is obtained by reflecting the part of the tangent graph on $(-\frac{\pi}{2}, \frac{\pi}{2})$ about the line $y = x$.

b. The graph of $y = \tan^{-1}x$

Figure 1.31   Graph of the inverse tangent function

Definitions and graphs of four other fundamental inverse trigonometric functions are given in Table 1.4 and Figure 1.32, respectively.

TABLE 1.4 Definition of Inverse Trigonometric Functions

Inverse Function	Domain	Range
$y = \sin^{-1} x$	$-1 \le x \le 1$	$-\frac{\pi}{2} \le y \le \frac{\pi}{2}$ Quadrants I and IV
$y = \cos^{-1} x$	$-1 \le x \le 1$	$0 \le y \le \pi$ Quadrants I and II
$y = \tan^{-1} x$	$-\infty < x < +\infty$	$-\frac{\pi}{2} < y < \frac{\pi}{2}$ Quadrants I and IV
$y = \csc^{-1} x$	$x \ge 1$ or $x \le -1$	$-\frac{\pi}{2} \le y \le \frac{\pi}{2},\ y \ne 0$ Quadrants I and IV
$y = \sec^{-1} x$	$x \ge 1$ or $x \le -1$	$0 \le y \le \pi,\ y \ne \frac{\pi}{2}$ Quadrants I and II
$y = \cot^{-1} x$	$-\infty < x < +\infty$	$0 < y < \pi$ Quadrants I and II

a. $y = \cos^{-1} x$
$-1 \le x \le 1$
$0 \le y \le \pi$

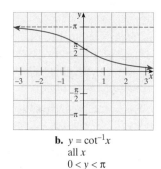
b. $y = \cot^{-1} x$
all x
$0 < y < \pi$

c. $y = \sec^{-1} x$
$|x| \ge 1$
$0 \le y \le \pi, y \ne \frac{\pi}{2}$

d. $y = \csc^{-1} x$
$|x| \ge 1$
$-\frac{\pi}{2} \le y \le \frac{\pi}{2}, y \ne 0$

Figure 1.32 Graphs of four inverse trigonometric functions

EXAMPLE 3 **Evaluating inverse trigonometric functions**

Evaluate the given functions.

a. $\sin^{-1}(\frac{-\sqrt{2}}{2})$ **b.** $\sin^{-1} 0.21$ **c.** $\cos^{-1} 0$ **d.** $\tan^{-1}(\frac{1}{\sqrt{3}})$

Solution

a. $\sin^{-1}(\frac{-\sqrt{2}}{2}) = -\frac{\pi}{4}$ Think: $x = \frac{-\sqrt{2}}{2}$ is negative, so y is in Quadrant IV; the reference angle is the angle whose sine is $\frac{\sqrt{2}}{2}$; it is $\frac{\pi}{4}$, so in Quadrant IV the angle is $-\frac{\pi}{4}$.

b. $\sin^{-1} 0.21 \approx 0.2115750$ By calculator; be sure to use radian mode and inverse sine (not reciprocal).

c. $\cos^{-1} 0 = \frac{\pi}{2}$ Memorized exact value.

d. $\tan^{-1}(\frac{1}{\sqrt{3}}) = \frac{\pi}{6}$ Think: $x = \frac{1}{\sqrt{3}}$ is positive, so y is in Quadrant I; the reference angle is the same as the value of the inverse tangent in Quadrant I. ∎

INVERSE TRIGONOMETRIC IDENTITIES

The definition of inverse functions yields four formulas, which we call the inversion formulas for sine and tangent.

WARNING The inversion formulas for $\sin^{-1}$ and $\tan^{-1}$ are valid only on the specified domains.

Inversion Formulas

$$\sin(\sin^{-1} x) = x \qquad \text{for } -1 \leq x \leq 1$$

$$\sin^{-1}(\sin y) = y \qquad \text{for } -\tfrac{\pi}{2} \leq y \leq \tfrac{\pi}{2}$$

$$\tan(\tan^{-1} x) = x \qquad \text{for all } x$$

$$\tan^{-1}(\tan y) = y \qquad \text{for } -\tfrac{\pi}{2} < y < \tfrac{\pi}{2}$$

EXAMPLE 4 Inversion formula for x inside and outside domain

Evaluate the given functions.

a. $\sin(\sin^{-1} 0.5)$ **b.** $\sin(\sin^{-1} 2)$ **c.** $\sin^{-1}(\sin 0.5)$ **d.** $\sin^{-1}(\sin 2)$

Solution

a. $\sin(\sin^{-1} 0.5) = 0.5$, because $-1 \leq 0.5 \leq 1$.
b. $\sin(\sin^{-1} 2)$ does not exist, because 2 is not between -1 and 1.
c. $\sin^{-1}(\sin 0.5) = 0.5$, because $-\tfrac{\pi}{2} \leq 0.5 \leq \tfrac{\pi}{2}$.
d. $\sin^{-1}(\sin 2) = 1.1415927$, by calculator.
   For exact values, notice that

$$\sin 2 = \sin(\pi - 2) \quad (\text{and } -\tfrac{\pi}{2} \leq \pi - 2 \leq \tfrac{\pi}{2})$$

so that we have $\sin^{-1}(\sin 2) = \sin^{-1}[\sin(\pi - 2)] = \pi - 2$. ∎

Some trigonometric identities correspond to inverse trigonometric identities, but others do not. For example,

$$\sin(-x) = -\sin x \quad \text{and} \quad \cos(-x) = \cos x$$

It is true that

$$\sin^{-1}(-x) = -\sin^{-1}(x)$$

but in general,

$$\cos^{-1}(-x) \neq \cos^{-1} x$$

(For a counterexample, try $x = 1$: $\cos^{-1}(-1) = \pi$; and $\cos^{-1} 1 = 0$.)

EXAMPLE 5 Proving inverse trigonometric identities

For $-1 \leq x \leq 1$, show that

a. $\sin^{-1}(-x) = -\sin^{-1} x$ **b.** $\cos(\sin^{-1} x) = \sqrt{1 - x^2}$

Solution

a. Let

$$y = \sin^{-1}(-x)$$

$\sin y = -x$	Definition of inverse sine
$-\sin y = x$	
$\sin(-y) = x$	Opposite angle identity
$-y = \sin^{-1} x$	Definition of inverse sine
$y = -\sin^{-1} x$	

Thus, $\sin^{-1}(-x) = -\sin^{-1} x$.

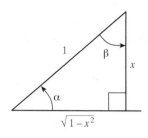

Figure 1.33 A reference triangle

b. Let $\alpha = \sin^{-1} x$, so $\sin \alpha = x$, where $-\frac{\pi}{2} \leq \alpha \leq \frac{\pi}{2}$. Here we consider the case where $0 \leq \alpha \leq \frac{\pi}{2}$, and the other case where $-\frac{\pi}{2} \leq \alpha < 0$ is handled similarly. Construct a right triangle with an acute angle α and hypotenuse 1, as shown in Figure 1.33. We call this triangle a **reference triangle**.

The side opposite α is x (since $\sin \alpha = x$), and by the Pythagorean theorem, the adjacent side is $\sqrt{1 - x^2}$. Thus, we have

$$\cos(\sin^{-1} x) = \cos \alpha$$
$$= \frac{\pm\sqrt{1 - x^2}}{1}$$
$$= \sqrt{1 - x^2}$$

Substitute $\cos \alpha = x$ and choose the positive value for the radical because α is in Quadrant I. ∎

Reference triangles, such as the one shown in Figure 1.33, are extremely useful devices for obtaining inverse trigonometric identities. For instance, let α and β be angles of a right triangle with hypotenuse 1. If the side opposite α is x (so that $\sin \alpha = x$), then $\alpha = \sin^{-1} x$ and $\beta = \cos^{-1} x$ so that

$$\sin^{-1} x + \cos^{-1} x = \alpha + \beta = \frac{\pi}{2} \quad \text{for } 0 \leq x \leq 1$$

since the acute angles of a right triangle must sum to $\pi/2$. The same reasoning can also be used to show that

$$\tan^{-1} x + \cot^{-1} x = \frac{\pi}{2}$$

and

$$\sec^{-1} x + \csc^{-1} x = \frac{\pi}{2}$$

Identities involving inverse trigonometric functions have a variety of uses. For instance, most calculators have keys for evaluating $\sin^{-1} x$, $\cos^{-1} x$, and $\tan^{-1} x$, but what about the other three inverse trigonometric functions? The answer is given by the following theorem, which allows us to compute $\sec^{-1} x$, $\csc^{-1} x$, and $\cot^{-1} x$ using reciprocal identities involving the three inverse trigonometric functions the calculator does have.

THEOREM 1.3   Reciprocal identities for inverse trigonometric functions

$$\sec^{-1} x = \cos^{-1} \frac{1}{x} \quad \text{if } |x| \geq 1$$
$$\csc^{-1} x = \sin^{-1} \frac{1}{x} \quad \text{if } |x| \geq 1$$
$$\cot^{-1} x = \frac{\pi}{2} - \tan^{-1} x$$

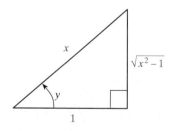

Figure 1.34 Reference triangle

Proof   We will prove that $\sec^{-1} x = \cos^{-1} \frac{1}{x}$ and leave the other two parts as an exercise. Let $y = \sec^{-1} x$. Then $\sec y = x$ and by using the reference triangle in Figure 1.34, we see that $\cos y = \frac{1}{x}$ so that

$$y = \sec^{-1} x = \cos^{-1} \frac{1}{x}$$

Note that the reference triangle is valid only if $|x| \geq 1$. ❑

EXAMPLE 6   Evaluating inverse reciprocal functions

Evaluate the given inverse functions using the inverse identities and a calculator.

a. $\sec^{-1}(-3)$ b. $\csc^{-1} 7.5$ c. $\cot^{-1} 2.4747$

Solution

a. $\sec^{-1}(-3) = \cos^{-1}(-\frac{1}{3}) \approx 1.910633236$

b. $\csc^{-1} 7.5 = \sin^{-1}(\frac{1}{7.5}) \approx 0.1337315894$

c. $\cot^{-1} 2.4747 = \frac{\pi}{2} - \tan^{-1} 2.4747 \approx 0.3840267299$ ∎

1.4   PROBLEM SET

A 1. **Exploration Problem**   Discuss the restrictions on the domain and range in the definition of the inverse trigonometric functions.

2. **Exploration Problem**   Discuss the use of reference triangles with respect to the inverse trigonometric functions.

Determine which pairs of functions defined by the equations in Problems 3–8 are inverses of each other.

3. $f(x) = 5x + 3$; $g(x) = \dfrac{x - 3}{5}$

4. $f(x) = \frac{2}{3}x + 2$; $g(x) = \frac{3}{2}x + 3$

5. $f(x) = \frac{4}{5}x + 4$; $g(x) = \frac{5}{4}x + 3$

6. $f(x) = \dfrac{1}{x}, x \neq 0$; $g(x) = \dfrac{1}{x}, x \neq 0$

7. $f(x) = x^2, x < 0$; $g(x) = \sqrt{x}, x > 0$

8. $f(x) = x^2, x \geq 0$; $g(x) = \sqrt{x}, x \geq 0$

Find the inverse (if it exists) of each function given in Problems 9–18.

9. $f = \{(4, 5), (6, 3), (7, 1), (2, 4)\}$

10. $g = \{(3, 9), (-3, 9), (4, 16), (-4, 16)\}$

11. $f(x) = 2x + 3$ 12. $g(x) = -3x + 2$

13. $f(x) = x^2 - 5, x \geq 0$ 14. $g(x) = x^2 - 5, x < 0$

15. $F(x) = \sqrt{x} + 5$ 16. $G(x) = 10 - \sqrt{x}$

17. $h(x) = \dfrac{2x - 6}{3x + 3}$ 18. $h(x) = \dfrac{2x + 1}{x}$

Give the exact values for functions in Problems 19–30.

19. a. $\cos^{-1} \frac{1}{2}$ b. $\sin^{-1}(-\frac{\sqrt{3}}{2})$

20. a. $\sin^{-1}(-\frac{1}{2})$ b. $\cos^{-1}(-\frac{1}{2})$

21. a. $\tan^{-1}(-1)$ b. $\cot^{-1}(-\sqrt{3})$

22. a. $\sec^{-1}(-\sqrt{2})$ b. $\csc^{-1}(-\sqrt{2})$

23. a. $\sin^{-1}(-\frac{\sqrt{3}}{2})$ b. $\sec^{-1}(-1)$

24. a. $\sec^{-1}(\frac{2}{\sqrt{3}})$ b. $\cot^{-1}(-1)$

25. $\cos(\sin^{-1} \frac{1}{2})$ 26. $\sin(\cos^{-1} \frac{1}{\sqrt{2}})$

27. $\cot(\tan^{-1} \frac{1}{3})$ 28. $\tan(\sin^{-1} \frac{1}{3})$

29. $\cos(\sin^{-1} \frac{1}{5} + 2\cos^{-1} \frac{1}{5})$

SMH   *Hint: Use the addition law for $\cos(\alpha + \beta)$.*

30. $\sin(\sin^{-1} \frac{1}{5} + \cos^{-1} \frac{1}{4})$

SMH   *Hint: Use the addition law for $\sin(\alpha + \beta)$.*

31. Suppose that α is an acute angle of a right triangle where

$$\sin \alpha = \frac{s^2 - t^2}{s^2 + t^2} \quad (s > t > 0)$$

Show that $\alpha = \tan^{-1}\left(\dfrac{s^2 - t^2}{2st}\right)$.

32. If $\sin \alpha + \cos \alpha = s$ and $\sin \alpha - \cos \alpha = t$, show that

$$\alpha = \tan^{-1}\left(\frac{s + t}{s - t}\right)$$

B *Sketch the graph of f in Problems 33–38 and then use the horizontal line test to determine whether f has an inverse. If f^{-1} exists, sketch its graph.*

33. $f(x) = x^2$, for all x 34. $f(x) = x^2, x \leq 0$

35. $f(x) = \sqrt{1 - x^2}$, on $(-1, 1)$

36. $f(x) = x(x - 1)(x - 2)$, on $[1, 2]$

37. $f(x) = \cos x$, on $[0, \pi]$

38. $f(x) = \tan x$, on $(-\frac{\pi}{2}, \frac{\pi}{2})$

Simplify each expression in Problems 39–44.

39. $\sin(2 \tan^{-1} x)$ 40. $\tan(2 \tan^{-1} x)$

41. $\tan(\cos^{-1} x)$ 42. $\cos(2 \sin^{-1} x)$

43. $\sin(\sin^{-1} x + \cos^{-1} x)$ 44. $\cos 2(\sin^{-1} x + \cos^{-1} x)$

45. Use reference triangles, if necessary, to justify each of the following identities.

a. $\cot^{-1} x = \frac{\pi}{2} - \tan^{-1} x$ for all x

b. $\csc^{-1} x = \sin^{-1}\left(\dfrac{1}{x}\right)$ for all $|x| \geq 1$

46. Use the identities in Theorem 1.3 to evaluate each function rounded to four decimal places.

a. $\cot^{-1} 0.67$ b. $\sec^{-1} 1.34$

c. $\csc^{-1} 2.59$ d. $\cot^{-1}(-1.54)$

47. Use the identities in Theorem 1.3 to evaluate each function rounded to four decimal places.

a. $\cot^{-1} 1.5$ b. $\cot^{-1}(-1.5)$

c. $\sec^{-1}(-1.7)$ d. $\csc^{-1}(-1.84)$

48. A painting 3 ft high is hung on a wall in such a way that its lower edge is 7 ft above the floor. An observer whose eyes are 5 ft above the floor stands x ft away from the wall. Express the angle θ subtended by the painting as a function of x.

49. To determine the height of a building (see Figure 1.35), select a point P and find the angle of elevation to be α. Then move out a distance of x units (on a level plane) to point Q and find that the angle of elevation is now β. Find the height h of the building, as a function of x.

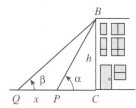

Figure 1.35   Determining height

○ 50. a. Find $\tan^{-1} 1 + \tan^{-1} 2 + \tan^{-1} 3$ using a calculator. Make a conjecture about the exact value. Prove your conjecture using reference triangles.

    b. Prove the conjecture from part **a** using trigonometric identities. *Hint*: Find $\tan(\tan^{-1} 1 + \tan^{-1} 2 + \tan^{-1} 3)$.

51. Prove the conjecture from Problem 50a using right triangles as follows. You may use the figure shown in Figure 1.36; assume that $\triangle ABC$, $\triangle ABD$, and $\triangle DEF$ are all right triangles with lengths of sides as shown in the figure.

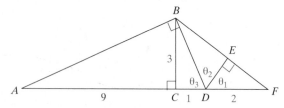

Figure 1.36 $\theta_1 + \theta_2 + \theta_3 = \pi$

52. We proved the identity

$$\csc^{-1} x = \sin^{-1}\left(\frac{1}{x}\right) \text{ for all } |x| \geq 1$$

By examining the graph of $y = \csc^{-1} x$, conjecture an identity of the form

$$\csc^{-1}(-x) = A + B \csc^{-1} x \text{ for } x > 1$$

Prove this identity, and then use it to evaluate $\csc^{-1}(-9.38)$.

53. Counterexample Problem   In Theorem 1.3 we proved the identity

$$\sec^{-1} x = \cos^{-1}\left(\frac{1}{x}\right) \text{ for all } x \geq 1$$

The graph of $y = \sec^{-1} x$ suggests that

$$\sec^{-1}(-x) = \pi + \sec^{-1} x \text{ for } x > 1$$

Either prove this identity or find a counterexample.

54. Show that if f^{-1} exists, it is unique. *Hint*: If g_1 and g_2 both satisfy

$$g_1[f(x)] = x = f[g_1(x)]$$
$$g_2[f(x)] = x = f[g_2(x)]$$

then $g_1(x) = g_2(x)$.

55. Suppose that $\triangle ABC$ is *not* a right triangle but has an obtuse angle γ. Draw $\overline{BD}$ perpendicular to $\overline{AC}$, forming right triangles $\triangle ABD$ and $\triangle BDC$ (with right angles at D). Show that

$$\frac{\sin \alpha}{a} = \frac{\sin \gamma}{c}$$

CHAPTER 1 REVIEW

Proficiency Examination

CONCEPT PROBLEMS

1. Characterize the following sets of numbers: natural numbers ($\mathbb{N}$), whole numbers ($\mathbb{W}$), integers ($\mathbb{Z}$), rational numbers ($\mathbb{Q}$), irrational numbers ($\mathbb{Q}'$), and real numbers ($\mathbb{R}$).

2. Define absolute value.

3. State the triangle inequality.

4. State the distance formula for points $P(x_1, y_1)$ and $Q(x_2, y_2)$.

5. Define slope in terms of angle of inclination.

6. List the following forms of the equation of a line:

    a. standard form

    b. slope-intercept form

    c. point-slope form

    d. horizontal line

    e. vertical line

7. State the slope criteria for parallel and for perpendicular lines.

8. Define function.

9. Define the composition of functions.

10. What is meant by the graph of a function?

11. Draw a quick sketch of an example of each of the following functions.

    a. identity function

    b. standard quadratic function

    c. standard cubic function

    d. absolute value function

    e. cube root function

    f. standard reciprocal function

    g. standard reciprocal squared function

    h. cosine function

    i. sine function

    j. tangent function

    k. secant function

    l. cosecant function

    m. cotangent function

    n. inverse cosine function

o. inverse sine function

p. inverse tangent function

q. inverse secant function

r. inverse cosecant function

s. inverse cotangent function

12. What is a polynomial function?

13. What is a rational function?

14. a. Define an inverse function.

    b. What is the procedure for graphing the inverse of a given function?

15. What is the horizontal line test?

16. State the inversion formulas for sine and tangent.

17. State the reciprocal inverse trigonometric identities.

PRACTICE PROBLEMS

18. Find an equation for the lines satisfying the given conditions:

    a. through $(-\frac{1}{2}, 5)$ with slope $m = -\frac{3}{4}$

    b. through $(-3, 5)$ and $(7, 2)$

    c. with x-intercept $(4, 0)$ and y-intercept $(0, -\frac{3}{7})$

    d. through $(-\frac{1}{2}, 5)$ and parallel to the line $2x + 5y - 11 = 0$

    e. the perpendicular bisector of the line segment joining $P(-3, 7)$ and $Q(5, 1)$

Sketch the graph of each of the equations in Problems 19–26.

19. $3x + 2y - 12 = 0$ **20.** $y - 3 = |x + 1|$

21. $y - 3 = -2(x - 1)^2$ **22.** $y = x^2 - 4x - 10$

23. $y = 2\cos(x - 1)$ **24.** $y + 1 = \tan(2x + 3)$

25. $y = \sin^{-1}(2x)$ **26.** $y = \tan^{-1} x^2$

27. If $f(x) = \dfrac{1}{x+1}$, what value(s) of x satisfy

$$f\left(\frac{1}{x+1}\right) = f\left(\frac{2x+1}{2x+4}\right)?$$

28. If $f(x) = \sqrt{\dfrac{x}{x-1}}$ and $g(x) = \dfrac{\sqrt{x}}{\sqrt{x-1}}$, does $f = g$? Why or why not?

29. If $f(x) = \sin x$ and $g(x) = \sqrt{1 - x^2}$, find the composite functions $f \circ g$ and $g \circ f$.

30. An open box with a square base is to be built for $96. The sides of the box will cost $3/ft² and the base will cost $8/ft². Express the volume of the box as a function of the length of its base.

Supplementary Problems*

In Problems 1–3, find the perimeter and area of the given figure.

1. the right triangle with vertices $(-1, 3)$, $(-1, 8)$, and $(11, 8)$

2. the triangle bounded by the lines $y = 5$, $3y - 4x = 11$, and $12x + 5y = 25$

3. the trapezoid with vertices $A(-3, 0)$, $B(5, 0)$, $C(2, 8)$, and $D(0, 8)$

In Problems 4–7, find an equation for the indicated line or circle.

4. the vertical line through the point where the line $y = 2x - 7$ intersects the parabola $y = x^2 + 6x - 3$

5. the circle that is tangent to the x-axis and is centered at $(5, 4)$

6. the circle with center on the y-axis that passes through the origin and is tangent to the line $3x + 4y - 40 = 0$

7. the line through the two points where the circles $x^2 + y^2 - 5x + 7y = 3$ and $x^2 + y^2 + 4y = 0$ intersect

8. Find constants A and B so that $\tan\left(x + \dfrac{\pi}{3}\right) = \dfrac{A + \tan x}{1 + B\tan x}$.

9. Find constants A and B so that $\sin^3 x = A\sin 3x + B\sin x$.

10. In a triangle, the perpendicular segment drawn from a given vertex to the opposite side is called the *altitude* on that side. Consider the triangle with vertices $(-2, 1)$, $(5, 6)$, and $(3, -2)$. Find an equation for each line containing an altitude of this triangle.

11. Show that the three lines found in Problem 10 intersect at the same point. Find the coordinates of this point. Is this the same point where the three medians meet?

12. Let $f(x) = x^2 + 5x - 9$. For what values of x is it true that $f(2x) = f(3x)$?

13. If an object is shot upward from the ground with an initial velocity of 256 ft/sec, its distance in feet above the ground at the end of t seconds is given by $s(t) = 256t - 16t^2$ (neglecting air resistance). What is the highest point for this projectile?

14. It is estimated that t years from now, the population of a certain suburban community will be

$$P(t) = \frac{11t + 12}{2t + 3}$$

thousand people. What is the current population of the community? What will the population be in 6 years? When will there be 5,000 people in the community?

15. Evaluate each of the given numbers (calculator approximations).

    a. $\tan^{-1} 2$ **b.** $\cot^{-1} 2$ **c.** $\sec^{-1}(-3.1)$

16. Evaluate each of the given numbers (exact values).

    a. $\sin(\cos^{-1} \frac{\sqrt{5}}{4})$ **b.** $\sin(2\tan^{-1} 3)$

    c. $\sin\left(\cos^{-1} \frac{3}{5} + \sin^{-1} \frac{5}{13}\right)$

17. Solve $\sqrt{x} = \cos^{-1} 0.317 + \sin^{-1} 0.317$

In Problems 18–21, find f^{-1} if it exists.

18. $f(x) = 2x^3 - 7$ **19.** $f(x) = \sqrt[7]{2x + 1}, x \geq -\frac{1}{2}$

20. $f(x) = \sqrt{\sin x}, 0 < x < \frac{\pi}{2}$ **21.** $f(x) = \dfrac{x+5}{x-7}, x \neq 7$

22. Show that for any constant $a \neq 1$, the function $f(x) = \dfrac{x+a}{x-1}$ is its own inverse.

23. Let $f(x) = \dfrac{ax+b}{cx+d}$, for constants $a, b, c,$ and d. Find $f^{-1}(x)$ in terms of $a, b, c,$ and d.

24. Find $f^{-1}(x)$ if $f(x) = \dfrac{x+1}{x-1}$. What is the domain of f^{-1}?

25. First show that $\tan^{-1} x + \tan^{-1} y = \tan^{-1}\left(\dfrac{x+y}{1-xy}\right)$ for $xy \neq 1$ whenever $-\dfrac{\pi}{2} < \tan^{-1}\left(\dfrac{x+y}{1-xy}\right) < \dfrac{\pi}{2}$. Then establish the following equations.

    a. $\tan^{-1} \frac{1}{2} + \tan^{-1} \frac{1}{3} = \frac{\pi}{4}$ **b.** $2\tan^{-1} \frac{1}{3} + \tan^{-1} \frac{1}{7} = \frac{\pi}{4}$

    c. $4\tan^{-1} \frac{1}{5} - \tan^{-1} \frac{1}{239} = \frac{\pi}{4}$

Note: The identity in part **c** will be used in Chapter 8 to estimate the value of π.

*The supplementary problems are presented in a somewhat random order, not necessarily in order of difficulty.

26. Counterexample Problem Each of the following equations may be either true or false. In each case, either show that the equation is generally true or find a counterexample.

a. $\tan^{-1} x = \dfrac{\sin^{-1} x}{\cos^{-1} x}$ b. $\tan^{-1} x = \dfrac{1}{\tan x}$

c. $\cot^{-1} x = \frac{\pi}{2} - \tan^{-1} x$ d. $\cos(\sin^{-1} x) = \sqrt{1 - x^2}$

e. $\sec^{-1}(\frac{1}{x}) = \cos^{-1} x$

27. Many materials, such as brick, steel, aluminum, and concrete, expand with increases in temperature. This is why spaces are placed between the cement slabs in sidewalks. Suppose you have a 100-ft length of material securely fastened at both ends, and assume that the buckle is linear. (It is not, but this assumption will serve as a worthwhile approximation.) If the height of the buckle is x ft and the percentage of swelling is y, then x and y are related as shown in Figure 1.37. Find the amount of buckling (to the nearest inch) for the following materials:

a. brick; $y = 0.03$ [This means $(0.03\%)(100 \text{ ft}) = 0.03$ ft, which is y in Figure 1.37.]

b. steel; $y = 0.06$

c. aluminum; $y = 0.12$ d. concrete; $y = 0.05$

Figure 1.37   Buckling of a given material

28. A ball has been dropped from the top of a building. Its height (in feet) after t seconds is given by the function

$$h(t) = -16t^2 + 256$$

a. How high will the ball be after 2 sec?

b. How far will the ball travel during the third second?

c. How tall is the building?

d. When will the ball hit the ground?

29. Suppose the number of worker-hours required to distribute new telephone books to x percent of the households in a certain rural community is given by the function

$$f(x) = \frac{600x}{300 - x}$$

a. What is the domain of the function f?

b. For what values of x does $f(x)$ have a practical interpretation in this context?

c. How many worker-hours were required to distribute new telephone books to the first 50% of the households?

d. How many worker-hours were required to distribute new telephone books to the entire community?

e. What percentage of the households in the community had received new telephone books by the time 150 worker-hours had been expended?

30. Find the area of each of the following plane figures:

a. the circle with $P(0, 0)$ and $Q(2, 3)$ endpoints of a diameter

b. the trapezoid with vertices $A(0, 0)$, $B(4, 0)$, $C(1, 3)$, and $D(2, 3)$

31. Find the volume and the surface area of each of the following solid figures:

a. a sphere with radius 4

b. a rectangular parallelepiped (box) with sides of length 2, 3, and 5

c. a right circular cylinder (including top and bottom) with height 4 and radius 2

d. an inverted cone with height 5 and top radius 3 (lateral surface only)

32. Consider the triangle with vertices $A(-1, 4)$, $B(3, 2)$, and $C(3, -6)$. Determine the midpoints M_1 and M_2 of sides $\overline{AB}$ and $\overline{AC}$, respectively, and show that the line segment $\overline{M_1 M_2}$ is parallel to side $\overline{BC}$ with half its length.

33. Generalize the procedure of Problem 32 to show that the line segment joining the midpoints of any two sides of a given triangle is parallel to the third side and has half its length.

34. Let $ABCD$ be a quadrilateral in the plane, and let P, Q, R, and S be the midpoints of sides $\overline{AB}$, $\overline{BC}$, $\overline{CD}$, and $\overline{DA}$, as shown in Figure 1.38. Show that $PQRS$ is a parallelogram.

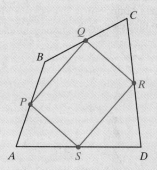

Figure 1.38   Problem 34

35. Find a constant c that guarantees that the graph of the equation

$$x^2 + xy + cy = 4$$

will have a y-intercept of $(0, -5)$. What are the x-intercepts of the graph?

36. A bus charter company offers a travel club the following arrangements: If no more than 100 people go on a certain tour, the cost will be $500 per person, but the cost per person will be reduced by $4 for each person in excess of 100 who takes the tour.

a. Express the total revenue R obtained by the charter company as a function of the number of people who go on the tour.

b. Sketch the graph of R. Estimate the number of people that results in the greatest total revenue for the charter company.

37. A mural 7 feet high is hung on a wall in such a way that its lower edge is 5 feet higher than the eye of an observer standing 12 feet from the wall. (See Figure 1.39.) Find the angle θ subtended by the mural at the observer's eye.

Figure 1.39   Problem 37

38. In Figure 1.40, ship A is at point P at noon and sails due east at 9 km/hr. Ship B arrives at point P at 1:00 P.M. and sails at 7 km/hr along a course that forms an angle of $60°$ with the course of ship A. Find a formula for the distance $s(t)$ separating the ships t hours after noon $(t \geq 1)$. Approximately how far apart (to the nearest kilometer) are the ships at 4:00 P.M.?

Figure 1.40   Problem 38

39. A manufacturer estimates that when the price for each unit is p dollars, the profit will be $N = -p^2 + 14p - 48$ thousand dollars. Sketch the graph of the profit function and answer these questions.

  a. For what values of p is this a profitable operation? (That is, when is $N > 0$?)

  b. What price results in maximum profit? What is the maximum profit?

40. Two jets bound for Los Angeles leave New York 30 minutes apart. The first travels 550 mph, and the second flies at 650 mph. How long will it take the second plane to pass the first?

41. To raise money, a service club has been collecting used bottles that it plans to deliver to a local glass company for recycling. Since the project began 8 days ago, the club has collected 2,400 pounds of glass, for which the glass company currently offers 15¢ per pound. However, since bottles are accumulating faster than they can be recycled, the company plans to reduce the price it pays by 1¢ per pound each day until the price reaches 0¢ fifteen days from now. Assuming that the club can continue to collect bottles at the same rate and that transportation costs make more than one trip to the glass company unfeasible, express the club's revenue from its recycling project as a function of the number of additional days the project runs. Draw the graph and estimate when the club should conclude the project and deliver the bottles in order to maximize its revenue.

42. An open box with a square base is to be built for $48. The sides of the box will cost $3/ft^2 and the base will cost $4/ft^2. Express the volume of the box as a function of the length of its base.

43. A closed box with a square base is to have a volume of 250 cubic feet. The material for the top and bottom of the box costs $2/ft^2, and the material for the sides costs $1/ft^2. Express the construction cost of the box as a function of the length of its base.

44. The famous author John Uptight must decide between two publishers who are vying for the rights to his new book, *Zen and the Art of Taxidermy*. Publisher A offers royalties of 1% of net proceeds on the first 30,000 copies sold and 3.5% on all copies in excess of that figure, and expects to net $2 on each copy sold. Publisher B will pay no royalties on the first 4,000 copies sold but will pay 2% on the net proceeds of all copies sold in excess of 4,000 copies, and expects to net $3 on each copy sold.

  a. Express the revenue John should expect if he signs with Publisher A (as a function P_A of the number of books sold, x). Likewise, find the revenue function P_B associated with Publisher B.

  b. Sketch the graphs of $P_A(x)$ and $P_B(x)$ on the same coordinate axes.

  c. For what value(s) of x are the two offers equivalent?

  d. With whom should he sign if he expects to sell 5,000 copies? 100,000 copies? 200,000 copies?

  e. State a simple criterion for determining which publisher he should choose if he expects to sell N copies.

*45. Putnam Examination Problem   If f and g are real-valued functions of one real variable, show that there exist numbers x, y such that

$$0 \leq x \leq 1,\ 0 \leq y \leq 1,\ \text{and}\ |xy - f(x) - g(y)| \geq \tfrac{1}{4}$$

*46. Putnam Examination Problem   Consider a polynomial $f(x)$ with real coefficients having the property $f(g(x)) = g(f(x))$ for every polynomial $g(x)$ with real coefficients. What is $f(x)$?

---

*The Putnam Examination is a national annual examination given under the auspices of the Mathematical Association of America and is designed to recognize mathematically talented college and university students. We include Putnam examination problems, which should be considered optional, to encourage students to consider taking the examination. Putnam problems used by permission of the Mathematical Association of America. Solutions to Putnam Examinations generally appear in *The American Mathematical Monthly,* approximately one year after the examination.

Calculus Was Inevitable

By John L. Troutman, Professor of Mathematics at Syracuse University

The invention of calculus is now credited jointly to Isaac Newton (1642–1727) and Gottfried W. Leibniz (1646–1716), when they produced their separate publications. However (around 1685) there was a bitter controversy throughout Europe as to whose work had been done first. Part of the explanation for this is the fact that each had done his actual work earlier (Newton in 1669; Leibniz, a little later); some can be attributed to the rivalry between scientists in England who championed Newton and those in Europe who supported Leibniz; but much simply reflects how ripe the intellectual climate was for the blooming of calculus. Indeed, even if neither of these great mathematicians had existed, it seems almost certain that the principles of calculus—including the fundamental theorem (introduced in Chapter 5)—would have been announced by the end of the seventeenth century.

Emergence of calculus was effectively demanded by the philosophical spirit of the times. Natural philosophers had long believed that the universe was constructed according to understandable mathematical principles, although they disagreed about just what these principles were and how they might be formulated. For example, the Pythagorean school (ca. 600 B.C.) maintained that everything consisted of (whole) numbers and their ratios; hence their consternation upon discovering that different entities such as $\sqrt{2}$ could be constructed. Next, the early astronomers announced that heavenly bodies move in circular orbits around the earth as center, and later that the earth itself must be a perfect sphere to reflect the divine hand of its Creator. Both of these assertions are now known to be false, and by 1612, Kepler had already explained why. Galileo (1564–1642) announced that the distance traveled by a heavy body falling from rest is proportional to the square of the elapsed time, and Fermat asserted (in 1657) that light moves along those paths that minimize the time of travel (we will discuss this principle in Chapter 4). The question was whether such laws could be formulated and justified mathematically, and what kind of mathematics would be appropriate to describe these phenomena.

For already from antiquity there was the warning of Zeno (ca. 450 B.C.), in the form of paradoxes, against unwise speculation about phenomena whose analysis involved infinite processes. In particular, he "proved" that motion was impossible if time consisted of individual "instants," and conversely, that covering a given distance was impossible if length was capable of infinite subdivision (see Chapter 8). Thus, although seventeenth-century scientists and philosophers might propose such principles, it was evident that there must be underlying subtleties in the mathematics required to support them (that, in fact, required another two centuries for satisfactory clarification).

To understand better how the mathematics developed, we must examine some of the previous attempts to solve the twin problems of classical origin that motivated the emergence of the calculus—those of finding the tangent line to a given planar geometrical curve and finding the area under the curve. Newton himself acknowledged that his greater vision resulted from his "standing on the shoulders of giants." Who were these giants, and what had they contributed? First, there were the efforts of the Greek mathematicians, principally Eudoxus, Euclid, and Archimedes, who had originated the geometric concepts of tangency and area between 400 B.C. and 200 B.C., together with examples of each, such as the construction of tangent lines to a circle and the area under a parabolic curve. The Hindu and Arabic mathematicians had extended the number system and the formal language of algebra with certain of its laws by 1300 A.D., but it was not until Newton's own era that the methods of algebra and geometry were combined satisfactorily in the analytic geometry of René Descartes (1596–1650) to produce the recognition that a geometrical curve could be regarded as the locus of points whose coordinates satisfied an algebraic equation. This provided potential numerical exactitude to geometric constructions as well as the possibility of giving geometric proofs for limiting algebraic arguments. Since Euclidian geometry was then generally regarded as the only reliable mathematics, it was the latter direction that was most frequently taken. And it was this direction that was taken in the seventeenth century by de Roberval, Fermat, Cavalieri, Huygens, Wallis, and others, not the least of whom was Newton's own teacher at Cambridge, Isaac Barrow (1630–1677). These giants, as Newton called them, obtained equations for tangent lines to, and the (correct) areas under, polynomial curves with equations $y = x^n$ for $n = 1, 2, 3, \ldots, 9$, and certain other geometrically defined curves such as the spiral and the cycloid.

In his analysis of tangency, Barrow incorporated the approximating infinitesimal triangle with sides Δx, Δy, Δs that is now standard in expositions of calculus, and Cavalieri attempted to "count" an indefinite number of parallel equidistant lines to obtain areas. It was known that in specific cases these problems were related, and that they were equivalent, respectively, to the kinematic problems of characterizing velocity and distance traveled during a motion, problems which directly confronted the paradoxes of Zeno.

All that remained was for some mathematicians to sense the generality underlying these specific constructions and to devise a usable notation for presenting the results. This was accomplished essentially independently by Newton (who, justly mistrusting the required limiting arguments, suppressed his own contributions until he could validate them geometrically) and by the only slightly less cautious Leibniz. However, as we have argued, by this time (about 1670), it was almost inevitable that someone should do so.

What calculus has provided is a mathematical language that, by means of the derivative, can describe the rates of change used to characterize various physical processes (such as velocity) and, by means of the integral, can show how macroscopic entities (such as area or distance) can emerge from properly assembled microscopic elements. Moreover, the fundamental theorem, which states that these are inverse operations, supplies an exact method for passing between these types of description. Finally, the ability to relate the results of limiting arguments by simple algebraic formulas permits the correct use of calculus while retaining skepticism regarding its foundations. This has enabled applications to go forward while mathematicians have sought an appropriate axiomatic basis.

Our present technological age attests to the success of this endeavor, and to the value of calculus.

Mathematical Essays

Use a library or references other than this textbook to research the information necessary to answer the questions in Problems 1–10.

1. HISTORICAL QUEST   Sir Isaac Newton was one of the greatest mathematicians of all time. He was a genius of the highest order but was often absent-minded. One story about Newton is that, when he was a boy, he was sent to cut a hole in the bottom of the barn door for the cats to go in and out. He cut two holes, a large one for the cat and a small one for the kittens.

ISAAC NEWTON
1642–1727

   Newton considered himself a theologian rather than a mathematician or a physicist. He spent years searching for clues about the end of the world and the geography of hell. One of Newton's quotations about himself is, "I seem to have been only like a boy playing on the seashore and diverting myself in now and then finding a smoother pebble or prettier shell than ordinary, whilst the great ocean of truth lay all undiscovered before me." ■

   Write an essay about Isaac Newton and his discovery of calculus. This essay should be at least 500 words.

2. HISTORICAL QUEST   At the age of 14, Gottfried Leibniz attempted to reform Aristotelian logic. He wanted to create a general method of reasoning by calculation. At the age of 20, he applied for his doctorate at the university in Leipzig and was refused (because, officials said, he was too young). He received his doctorate the next year at the University of Altdorf, where he made

GOTTFRIED LEIBNIZ
1646–1716

such a favorable impression that he was offered a professorship, which he declined, saying he had very different things in view. Leibniz went on to invent calculus, but not without a bitter controversy developing between Leibniz and Newton. Most historians agree that the bitterness over who invented calculus materially affected the history of mathematics. J. S. Mill characterized Leibniz by saying, "It would be difficult to name a man more remarkable for the greatness and universality of his intellectual powers than Leibniz." ■

   Write a 500-word essay about Gottfried Leibniz and his discovery of calculus.

3. HISTORICAL QUEST   Write a 500-word essay about the controversy surrounding the discovery of calculus by Newton and Leibniz.

4. The Greek mathematicians mentioned in this guest essay include Eudoxus, Euclid, and Archimedes. Write a short paper about contributions they made that might have been used by Newton.

5. What is the definition of elementary functions as given by Joseph Liouville?

6. The guest essay mentions the contributions of de Roberval, Fermat, Cavalieri, Huygens, Wallis, and Barrow toward the invention of calculus. Write a short paper about these contributions.

7. In this guest essay, Troutman argues that the invention of calculus was inevitable, and even if Newton and Leibniz had not invented it, someone else would have. Write a 500-word essay either defending or refuting this thesis.

8. HISTORICAL QUEST   Sophie Germain was one of the first women to publish original mathematical research in number theory. In her time, women were not admitted to first-rate universities and were not, for the most part, taken seriously, so she wrote at first under the pseudonym LeBlanc. The situation is not too different from that portrayed by Barbra Streisand in the movie *Yentl*. Even though Germain's most important research was in number theory, she was awarded the prize of the French Academy for a paper entitled "Memoir on the Vibrations of Elastic Plates." ■

SOPHIE GERMAIN
1776–1831

   As we progress through this book we will profile many mathematicians in the history of mathematics, and you will notice that most of them are white males. Why? Write a paper on the history of women mathematicians and their achievements. Your paper should include a list of many prominent women mathematicians and their primary contributions. It should also include a lengthy profile of at least one woman mathematician.

9. HISTORICAL QUEST   The Navajo are a Native American people who, despite considerable interchange and assimilation with the surrounding dominant culture, maintain a world view that remains vital and distinctive. The Navajo believe in a dynamic universe. Rather than consisting of objects and situations, the universe is made up of processes. Central to our Western mode of thought is the idea that things are separable entities that can be subdivided into smaller discrete units. For us, things that change through time do so by going from one specific state to another specific state. While we believe time to be continuous, we often even break it into discrete units or freeze it and talk about an instant or point in time. Among the Navajo, where the focus is on process, change is ever present; interrelationship and motion are of primary significance. These incorporate and subsume space and time.

   There are, in every culture, groups or individuals who think more about some ideas than do others. For other cultures, we know about the ideas of some professional groups or some ideas of the culture at large. We know little, however, about the mathematical thoughts of individuals in those cultures who are specially inclined toward mathematical ideas. In Western culture, on the other hand, we focus on, and record much about, those special individuals while including little about everyone else. Realization of this difference should make us particularly wary of any comparisons across cultures. Even more important, it should encourage finding out more about the ideas of mathematically oriented innovators in other cultures and, simultaneously, encourage expanding the scope of Western history to recognize and include mathematical ideas held by different groups within our culture or by our culture as a whole.* ■

   Write a paper discussing this quotation.

10. Book Report   "Ethnomathematics, as it is being addressed here, has the goal of broadening the history of mathematics to one that has a multicultural, global perspective." Read the book *Ethnomathematics* by Marcia Ascher (Pacific Grove: Brooks/Cole, 1991), and prepare a book report.

---

*From *Ethnomathematics* by Marcia Ascher, pp. 128–129 and 188–189.

2

Limits and Continuity

PREVIEW

Calculus is the mathematics of motion and change, while algebra, geometry, and trigonometry are more static in nature. The development of calculus in the seventeenth century by Newton, Leibniz, and others grew out of attempts by these and earlier mathematicians to answer certain fundamental questions about dynamic real-world situations. These investigations led to two fundamental procedures, *differentiation* and *integration*, which can be formulated in terms of a concept called the *limit*. We will introduce the limit in this chapter and study its basic properties. In addition, we will continue the investigation of functions begun in Chapter 1 by exploring the concept of functional continuity, and using limits to introduce exponential and logarithmic functions.

PERSPECTIVE

The limit is a mathematical tool for studying the tendency of a function as its independent variable approaches some value. For instance, in the guest essay at the end of Chapter 1, John Troutman mentions Zeno's paradoxes, which involve infinite processes. One such paradox can be stated as follows:

> Archilles fires an arrow, but it can never reach its target. Why? Because if the arrow takes T seconds to travel half the distance to the target, it will take $T/2$ seconds to finish half the rest, and so on for each half. No matter how close the arrow is to the target, it will take finite time to travel half the distance that remains, so the arrow will never reach its destination.

Common sense tells that the arrow will strike home in $2T$ seconds, so where is the error in reasoning? If we measure the total time of the arrow's flight by the "half, then half of that, and so on," approach of the paradox, it will take

$$T + \frac{T}{2} + \frac{T}{4} + \cdots$$

seconds for the arrow to strike the target; that is, an infinite sum of finite time intervals. The Greeks of Zeno's time assumed that such a sum would have to be infinite. However, the limit process can be used to show that this particular sum is not only finite, but equals $2T$, as common sense would suggest. We will study infinite sums in Chapter 8, but first it is necessary to introduce and explore a different kind of limit, the *limit of a function*, and that is the goal of this chapter.

In a very real sense, the concept of limit is the threshold to modern mathematics. You are about to cross that threshold, and beyond lies the fascinating world of calculus.

2.1   The Limit of a Function

IN THIS SECTION   informal computation of limits, one-sided limits, limits that do not exist, formal definition of a limit

The goal of this section is to define and explore what is meant by the *limit of a function.* We will begin with an informal discussion of the limit concept, emphasizing graphical and numerical computations. Then in Section 2.2, we will develop algebraic techniques for computing limits more systematically.

The development of the limit concept was a major breakthrough in the history of mathematics, and you should not be surprised or disappointed if certain aspects of this concept seem difficult to comprehend or apply. Be patient and this crucial concept will soon become part of your mathematical toolkit.

INFORMAL COMPUTATION OF LIMITS

The limit of a function f is a tool for investigating the behavior of $f(x)$ as x gets closer and closer to a particular number c. That is, the limit concerns the *tendency* of $f(x)$ for x near c rather than the *value* of f at $x = c$. The following example illustrates how such a tendency can be used to compute velocity.

EXAMPLE 1   Computing velocity as a limit

A freely falling body experiencing no air resistance falls $s(t) = 16t^2$ feet in t seconds. Express the body's velocity at time $t = 2$ as a limit.

Solution

What we want is the *instantaneous velocity* after 2 seconds, which can be thought of as the "speedometer reading" of the falling body at $t = 2$. To approximate the instantaneous velocity, we compute the average velocity of the body over smaller and smaller time intervals, either ending with or beginning with $t = 2$. For instance, over the time interval $1.9 \leq t \leq 2$, the average velocity, $\overline{v}(t)$, is

$$\overline{v}(t) = \frac{\text{DISTANCE TRAVELED}}{\text{ELAPSED TIME}} \qquad \textit{Where distance is in ft and time is in seconds}$$

$$= \frac{s(2) - s(1.9)}{2 - 1.9}$$

$$= \frac{16(2)^2 - 16(1.9)^2}{0.1}$$

$$= 62.4 \text{ ft/s}$$

Similar computations of average velocity over short time intervals ending or beginning with time $t = 2$ are contained in the following table:

Time interval	$1.8 \leq t \leq 2$	$1.9 \leq t \leq 2$	$1.99 \leq t \leq 2$	$1.999 \leq t \leq 2$	$2 \leq t \leq 2.0001$	$2 \leq t \leq 2.001$	$2 \leq t \leq 2.01$
Interval length (sec)	0.2	0.1	0.01	0.001	0.0001	0.001	0.01
Average velocity (ft/s)	60.8	62.4	63.84	63.98	64.0016	64.016	64.16

Examining the bottom row of this table, we see that the average velocity seems to be approaching the value 64 as the time intervals become smaller and smaller. Thus, it is reasonable to expect the velocity at the instant when $t = 2$ to be 64 ft/s.

In symbols, the average velocity of the falling body over the time interval $2 \leq t \leq 2 + h$ is given by

$$\frac{s(2 + h) - s(2)}{(2 + h) - 2} = \frac{16(2 + h)^2 - 16(2)^2}{h}$$

We say that the average velocity has a limiting value of 64 as the length h of the time interval tends to zero (gets smaller and smaller), and we denote this tendency by writing

$$\lim_{h \to 0} \frac{16(2 + h)^2 - 16(2)^2}{h} = 64$$ ∎

Here is a general, though informal, definition of the limit of a function.

Limit of a Function (Informal Definition)

The notation

$$\lim_{x \to c} f(x) = L$$

is read "the limit of $f(x)$ as x approaches c is L" and means that the functional values $f(x)$ can be made arbitrarily close to a unique number L by choosing x sufficiently close to c (but not equal to c).

➧ **What This Says**   If $f(x)$ becomes arbitrarily close to a single number L as x approaches c from either side, then we say that L is the limit of $f(x)$ as x approaches c. It means that the functional values of f "home in" on the number L as x gets closer and closer to c. The values of x can approach c from either side, but $x = c$ itself is excluded.

Notation: Sometimes we will find it convenient to represent the limit statement

$$\lim_{x \to c} f(x) = L$$

by writing $f(x) \to L$ as $x \to c$.

This informal definition of limit cannot be used in proofs until we give precise meaning to terms such as "arbitrarily close to L" and "sufficiently close to c." This will be done at the end of this section. For now, we will use this informal definition to explore a few basic features of the limiting process.

EXAMPLE 2   Informal computation of a limit

Use a table to guess the value of

$$L = \lim_{x \to -2} \frac{x^2 + x - 2}{x + 2}$$

Solution

It would be nice if we could simply substitute $x = -2$ into the formula for

$$f(x) = \frac{x^2 + x - 2}{x + 2}$$

but note that $f(-2)$ is not defined. This is of no consequence since the limit process is concerned only when x *approaches* -2 and has nothing to do with the value of f at $x = -2$. We form a table of values of $f(x)$ for x near -2:

x	-2.3	-2.1	-2.05	-2.001	-2	-1.9997	-1.995
$f(x)$	-3.3	-3.1	-3.05	-3.001	undefined	-2.9997	-2.995

The numbers on the bottom row of this table suggest that $f(x) \to -3$ as $x \to -2$; that is,

$$L = \lim_{x \to -2} \frac{x^2 + x - 2}{x + 2} = -3$$

The graph of f is shown in Figure 2.1. Notice that the graph is a line with a "hole" at the point $(-2, -3)$.   ∎

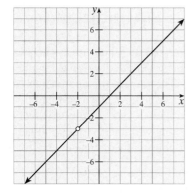

Figure 2.1 Graph of $f(x) = \dfrac{x^2 + x - 2}{x + 2}$

EXAMPLE 3 **Finding limits of trigonometric functions**

Evaluate $\lim\limits_{x \to 0} \sin x$ and $\lim\limits_{x \to 0} \cos x$.

Solution

We can evaluate these limits by table.

x	1	0.5	0.1	0.01	−0.01	−0.1	−0.5
$\sin x$	0.84	0.48	0.0998	0.0099998	−0.0099998	−0.0998	−0.48
$\cos x$	0.54	0.88	0.9950	0.9999500	0.99995	0.9950	0.88

The pattern of numbers in the table suggests that

$$\lim_{x \to 0} \sin x = 0 = \sin 0 \qquad \text{and} \qquad \lim_{x \to 0} \cos x = 1 = \cos 0 \qquad \blacksquare$$

EXAMPLE 4 **A computational dilemma**

Use a table to guess the value of

$$L = \lim_{x \to 0} \frac{2\sqrt{x + 1} - x - 2}{x^2}$$

Solution

Form a table of values of

$$f(x) = \frac{2\sqrt{x + 1} - x - 2}{x^2}$$

for x near 0:

x	−0.5	−0.1	−0.01	−0.001	0	0.001	0.005
$f(x)$	−0.3431	−0.2633	−0.2513	−0.2501	undefined	−0.2499	−0.2494

The numbers on the bottom line suggest that the limit is $L = -0.25$. $\qquad \blacksquare$

In Example 4, note that if you decide to try $x = 0.0000001$ just to make sure you have taken numbers close enough to 0, you may find that the calculator gives the value 0. Does this mean the limit is actually 0 instead of −0.25? No, the problem is that the numerator $2\sqrt{x + 1} - x - 2$ is so close to 0 when $x = 0.0000001$ that the calculator gives a false value for f.

The point is this: When you use your calculator to estimate limits, always be aware that this method can introduce errors. In the next section, we will develop an algebraic approach for computing limits that is more reliable. The three basic approaches (examining appropriate data by using a calculator and/or a table, drawing a graph to find a limit, or using algebraic rules) each has its virtues and faults. As you study this section and the next, you need to learn all three approaches, along with their virtues and faults, and this knowledge will serve you well for the rest of your mathematical course work.

ONE-SIDED LIMITS

Sometimes we will be interested in the limiting behavior of a function from only one side; that is, the limit as x approaches a number c from the left or the analogous limit from the right.

One-Sided Limits

Right-hand limit We write $\lim\limits_{x \to c^+} f(x) = L$ if we can make the number $f(x)$ as close to L as we please by choosing x sufficiently close to c on a small interval (c, b) *immediately to the right of c.*

Left-hand limit We write $\lim\limits_{x \to c^-} f(x) = L$ if we can make the number $f(x)$ as close to L as we please by choosing x sufficiently close to c on a small interval (a, c) *immediately to the left of c.*

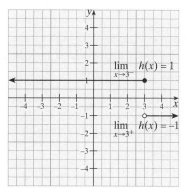

Figure 2.2 Graph of

$$h(x) = \begin{cases} 1 & \text{if } x \le 3 \\ -1 & \text{if } x > 3, \end{cases}$$

a graph illustrating one-sided limits

For instance, note that the graph in Figure 2.2 is "broken" at $x = 3$. The limit as x approaches 3 from the left is 1, while the limit from the right at the same point is -1.

Observe that the "two-sided" limit, $\lim\limits_{x \to 3} h(x)$, defined earlier in this section, does not exist since $h(x)$ does not tend toward a single limiting value L as x approaches $c = 3$ from either side. It should be clear that, in general, a two-sided limit cannot exist if the corresponding pair of one-sided limits are different. Conversely, it can be shown that if the two one-sided limits of a given function f as $x \to c^-$ and $x \to c^+$ both exist and are equal, then the two-sided limit $\lim\limits_{x \to c} f(x)$ must also exist. These observations are so important that we state them in the form of a theorem.

THEOREM 2.1 One-sided limit theorem

The two-sided limit $\lim\limits_{x \to c} f(x)$ exists if and only if the two one-sided limits $\lim\limits_{x \to c^-} f(x) = L$ and $\lim\limits_{x \to c^+} f(x)$ both exist and are equal. Furthermore, if $\lim\limits_{x \to c^-} f(x) = L = \lim\limits_{x \to c^+} f(x)$, then $\lim\limits_{x \to c} f(x) = L$. ❑

One-sided and two-sided limits are illustrated in Figure 2.3.

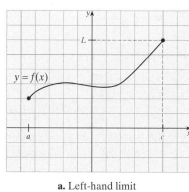

a. Left-hand limit
$\lim\limits_{x \to c^-} f(x) = L$

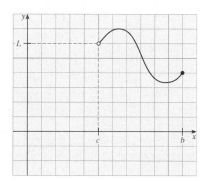

b. Right-hand limit
$\lim\limits_{x \to c^+} f(x) = L$

c. Two-sided limit
$\lim\limits_{x \to c} f(x) = L$

Figure 2.3 We say that $\lim\limits_{x \to c} f(x) = L$ if and only if $\lim\limits_{x \to c^-} f(x) = \lim\limits_{x \to c^+} f(x) = L$.

EXAMPLE 5 Estimating limits by graphing

Given the functions defined by the graphs in Figure 2.4, find the requested limits by inspection, if they exist.

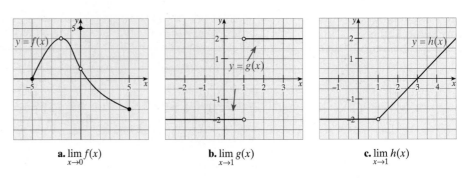

a. $\lim\limits_{x \to 0} f(x)$ **b.** $\lim\limits_{x \to 1} g(x)$ **c.** $\lim\limits_{x \to 1} h(x)$

Figure 2.4   Limits from a graph

Solution

a. Take a good look at the given graph; notice the open circles on the graph at $x = 0$ and $x = -2$ and also notice that $f(0) = 5$. To find $\lim\limits_{x \to 0} f(x)$ we need to look at both the left-hand and right-hand limits. Look at Figure 2.4a to find

$$\lim_{x \to 0^-} f(x) = 1 \quad \text{and} \quad \lim_{x \to 0^+} f(x) = 1$$

so $\lim\limits_{x \to 0} f(x)$ exists and $\lim\limits_{x \to 0} f(x) = 1$. Notice here that *the value of the limit as $x \to 0$ is not the same as the value of the function at $x = 0$.*

b. Look at Figure 2.4b to find

$$\lim_{x \to 1^-} g(x) = -2 \quad \text{and} \quad \lim_{x \to 1^+} g(x) = 2$$

so the *limit as $x \to 1$ does not exist.*

c. Look at Figure 2.4c to find

$$\lim_{x \to 1^-} h(x) = -2 \quad \text{and} \quad \lim_{x \to 1^+} h(x) = -2$$

so $\lim\limits_{x \to 1} h(x) = -2$.   ■

EXAMPLE 6   Evaluating a trigonometric limit using a table

Evaluate $\lim\limits_{x \to 0} \dfrac{\sin x}{x}$.

Solution

$f(x) = \dfrac{\sin x}{x}$ is an even function because

$$f(-x) = \frac{\sin(-x)}{-x} = \frac{-\sin x}{-x} = \frac{\sin x}{x} = f(x)$$

This means that we need to find only the right-hand limit at 0 because the limiting behavior from the left will be the same as that from the right. Consider the following table.

x	0.1 ----▶	0.05 ----▶	0.01 ----▶	0.001 ----▶	0
$f(x)$	0.998334	0.999583	0.9999833	0.999999833	Undefined

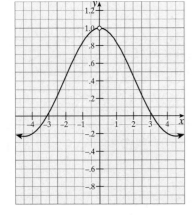

Figure 2.5 Graph of $f(x) = \dfrac{\sin x}{x}$

The table suggests that $\lim\limits_{x \to 0^+} \dfrac{\sin x}{x} = 1$ and hence that $\lim\limits_{x \to 0} \dfrac{\sin x}{x} = 1$. We will revisit this limit in Section 2.2. The graph of $f(x) = \dfrac{\sin x}{x}$ is shown in Figure 2.5.   ■

LIMITS THAT DO NOT EXIST

It may happen that a function f does not have a (finite) limit as $x \to c$. When $\lim\limits_{x \to c} f(x)$ fails to exist, the functional values $f(x)$ are said to **diverge** as $x \to c$.

EXAMPLE 7   A function that diverges

Evaluate $\lim\limits_{x \to 0} \dfrac{1}{x^2}$.

Solution

As $x \to 0$, the corresponding functional values of $f(x) = \dfrac{1}{x^2}$ grow arbitrarily large, as indicated in the following table.

	x approaches 0 from the left; $x \to 0^-$				x approaches 0 from the right; $x \to 0^+$		
x	-0.1 --▶	-0.05 --▶	-0.001 --▶	0	◀-- 0.001	◀-- 0.005	◀-- 0.01
$f(x) = \dfrac{1}{x^2}$	100	400	1×10^6	undefined	1×10^6	4×10^4	1×10^4

The graph of f is shown in Figure 2.6.

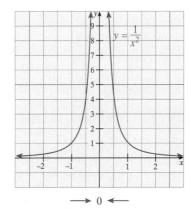

$$\longrightarrow \ 0 \ \longleftarrow$$

Figure 2.6 $\lim\limits_{x \to 0} \dfrac{1}{x^2}$ does not exist, and the graph illustrates that f rises without bound.

Geometrically, the graph of $y = f(x)$ rises without bound as $x \to 0$. Thus, $\lim\limits_{x \to 0} \dfrac{1}{x^2}$ does not exist.   ■

Infinite Limits

A function f that increases or decreases without bound as x approaches c is said to **tend to infinity** (∞) at c. We indicate this behavior by writing

$$\lim_{x \to c} f(x) = +\infty \qquad \text{if } f \text{ increases without bound}$$

and by

$$\lim_{x \to c} f(x) = -\infty \qquad \text{if } f \text{ decreases without bound}$$

Using this notation, we can rewrite the answer to Example 7 as

$$\lim_{x \to 0} \frac{1}{x^2} = +\infty$$

EXAMPLE 8   A function that diverges by oscillation

Evaluate $\lim_{x \to 0} \sin \dfrac{1}{x}$.

Solution

Note this is not the same as $\lim_{x \to 0} \dfrac{\sin x}{x}$. The values of $f(x) = \sin \dfrac{1}{x}$ oscillate infinitely often between 1 and -1 as x approaches 0. For example, $f(x) = 1$ for $x = \frac{2}{\pi}, \frac{2}{5\pi}, \frac{2}{9\pi}, \ldots$, and $f(x) = -1$ for $x = \frac{2}{3\pi}, \frac{2}{7\pi}, \frac{2}{11\pi}, \ldots$. The graph of $f(x)$ is shown in Figure 2.7.

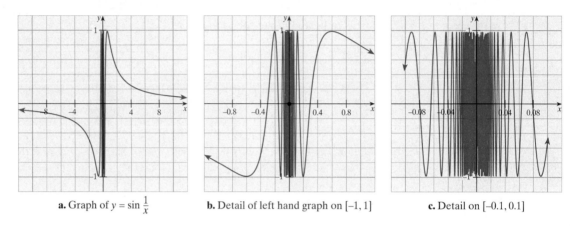

a. Graph of $y = \sin \frac{1}{x}$ **b.** Detail of left hand graph on $[-1, 1]$ **c.** Detail on $[-0.1, 0.1]$

Figure 2.7 $\sin \frac{1}{x}$ diverges by oscillation as $x \to 0$.

Because the values of $f(x)$ do not approach a unique number L as $x \to 0$, the limit does not exist. This kind of limiting behavior is called **divergence by oscillation**. ∎

In the next section, we will introduce some properties of limits that will help us evaluate limits efficiently. In the following problem set remember that the emphasis is on an intuitive understanding of limits, including their evaluation by graphing and by table.

FORMAL DEFINITION OF A LIMIT

Our informal definition of the limit provides valuable intuition and allows you to develop a working knowledge of this fundamental concept. For theoretical work, however, the intuitive definition will not suffice, because it gives no precise, quantifiable meaning to the terms "arbitrarily close to L" and "sufficiently close to c." In the nineteenth century, leading mathematicians, including Augustin-Louis Cauchy (1789–1857) and Karl Weierstrass (1815–1897), sought to put calculus on a sound logical foundation by giving precise definitions for the foundational ideas of calculus. The following definition, derived from the work of Cauchy and Weierstrass, gives precision to the limit notion.

**Limit of a Function
(Formal definition)**

> The limit statement
> $$\lim_{x \to c} f(x) = L$$
> means that for each number $\epsilon > 0$, there corresponds a number $\delta > 0$ with the property that
> $$|f(x) - L| < \epsilon \quad \text{whenever} \quad 0 < |x - c| < \delta$$

We show this definition graphically in Figure 2.8.

For each ε > 0
This forms an interval around L on the y-axis.

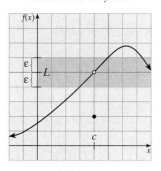

there is a δ > 0
This forms an interval around c on the x-axis.

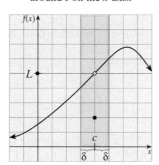

such that **if $0 < |x - c| < \delta$,**
This says that if x is in the δ-interval on the x-axis...

then $|f(x) - L| < \varepsilon$.
...then $f(x)$ is in the ε-interval on the y-axis.

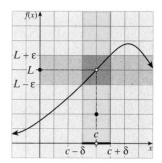

Figure 2.8 Formal definition of limit: $\lim\limits_{x \to c} f(x) = L$

Because the Greek letters ϵ (epsilon) and δ (delta) are traditionally used in this context, the formal definition of limit is sometimes called the **epsilon-delta** definition of the limit. The goal of this section is to show how this formal definition embodies our intuitive understanding of the limit process and how it can be used rigorously to establish a variety of results.

Do not be discouraged if this material seems difficult—it is. Probably your best course of action is to read this section carefully and examine the details of a few examples closely. Then, using the examples as models, try some of the exercises. This material often takes several attempts, but if you persevere, you should come away with an appreciation of the epsilon-delta process—and with it, a better understanding of calculus.

Behind the formal language is a fairly straightforward idea. In particular, to establish a specific limit, say $\lim\limits_{x \to c} f(x) = L$, a number $\epsilon > 0$ is chosen first to establish a desired degree of proximity to L, and then a number $\delta > 0$ is found that determines how close x must be to c to ensure that $f(x)$ is within ϵ units of L.

The situation is summarized in Figure 2.9, which shows a function that satisfies the conditions of the definition. Notice that whenever x is within δ units of c (but not equal to c), the point $(x, f(x))$ on the graph of f must lie in the rectangle (shaded region) formed by the intersection of the horizontal band of width 2ϵ centered at L and the vertical band of width 2δ centered at c. The smaller the ϵ-interval around the proposed limit L, generally the smaller the δ-interval will need to be for $f(x)$ to lie in the ϵ-interval. If such a δ can be found no matter how small ϵ is, then $f(x)$ and L are arbitrarily close, so L must be the limit. The following examples illustrate epsilon-delta proofs, one in which the limit exists and one in which it does not.

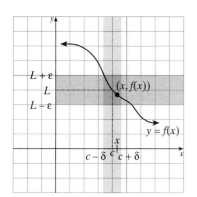

Figure 2.9 The epsilon-delta definition of limit

EXAMPLE 9   An epsilon-delta proof of a limit statement

Show that $\lim\limits_{x \to 2} (4x - 3) = 5$.

Solution

From the graph of $f(x) = 4x - 3$ (see Figure 2.10), we guess that the limit as $x \to 2$ is 5. The object of this example is to *prove* that the limit is 5. We have

$$|f(x) - L| = |4x - 3 - 5|$$
$$= |4x - 8|$$
$$= \underbrace{4|x - 2|}$$

This must be less than ϵ whenever $|x - 2| < \delta$.

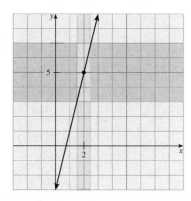

Figure 2.10 $\lim\limits_{x \to 2} (4x - 3) = 5$

For a given $\epsilon > 0$, choose $\delta = \dfrac{\epsilon}{4}$; then

$$|f(x) - L| = 4\,|x - 2| < 4\delta = 4\left(\frac{\epsilon}{4}\right) = \epsilon \qquad\blacksquare$$

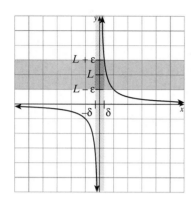

Figure 2.11 $\displaystyle\lim_{x\to 0}\frac{1}{x}$

EXAMPLE 10 An epsilon-delta proof that a limit does not exist

Show that $\displaystyle\lim_{x\to 0}\frac{1}{x}$ does not exist.

Solution

Let $f(x) = \dfrac{1}{x}$ and L be any number. Suppose that $\displaystyle\lim_{x\to 0} f(x) = L$. Look at the graph of f, as shown in Figure 2.11. It would seem that no matter what value of ϵ is chosen, it would be impossible to find a corresponding δ. Consider the absolute value expression required by the definition of limit: If

$$|f(x) - L| < \epsilon, \quad\text{or, for this example,}\quad \left|\frac{1}{x} - L\right| < \epsilon$$

then

$$-\epsilon < \frac{1}{x} - L < \epsilon \qquad \text{\textit{Property of absolute value}}$$
$$\text{\textit{(Table 1.1, p. 3)}}$$

and

$$L - \epsilon < \frac{1}{x} < L + \epsilon$$

If $\epsilon = 1$ (not a particularly small ϵ), then

$$\left|\frac{1}{x}\right| < |L| + 1$$
$$|x| > \frac{1}{|L| + 1}$$

If $\epsilon = 1$ (not a particularly small ϵ), then

$$\left|\frac{1}{x}\right| < |L| + 1$$
$$|x| > \frac{1}{|L| + 1}$$

In general, given any $\epsilon > 0$, then no matter how $\delta > 0$ is chosen, there will always be numbers x in the interval $0 < |x - 0| < \delta$ such that $\dfrac{1}{|x|} > L + \epsilon$. Since L was chosen arbitrarily, it follows that the limit does not exist. $\qquad\blacksquare$

Appendix B gives many of the important proofs in calculus, and if you look there, you will see many of them are given in ϵ-δ form. In the following problem set, remember that the emphasis is on an intuitive understanding of limits, including their evaluation by graphing and by table.

2.1   PROBLEM SET

A *Given the functions defined by the graphs in Figure 2.12, find the limits in Problems 1–6.*

Graph of *f*

Graph of *g*

Graph of *t*

Figure 2.12   Graphs of the functions *f*, *g*, and *t*

1. a. $\lim\limits_{x\to 3} f(x)$ **b.** $\lim\limits_{x\to 2} f(x)$ **c.** $\lim\limits_{x\to 0} f(x)$

2. a. $\lim\limits_{x\to -3} g(x)$ **b.** $\lim\limits_{x\to -1} g(x)$ **c.** $\lim\limits_{x\to 4^-} g(x)$

3. a. $\lim\limits_{x\to 4} t(x)$ **b.** $\lim\limits_{x\to -4} t(x)$ **c.** $\lim\limits_{x\to -5^+} t(x)$

4. a. $\lim\limits_{x\to 2^-} f(x)$ **b.** $\lim\limits_{x\to 2^+} f(x)$ **c.** $\lim\limits_{x\to 2} f(x)$

5. a. $\lim\limits_{x\to 3^-} g(x)$ **b.** $\lim\limits_{x\to 3^+} g(x)$ **c.** $\lim\limits_{x\to 3} g(x)$

6. a. $\lim\limits_{x\to 2^-} t(x)$ **b.** $\lim\limits_{x\to 2^+} t(x)$ **c.** $\lim\limits_{x\to 2} t(x)$

Find the limits by filling in the appropriate values in the tables in Problems 7–9.

7. $\lim\limits_{x\to 5^-} f(x)$, where $f(x) = 4x - 5$

$x \to 5^-$

x	2	3	4	4.5	4.9	4.99
$f(x)$	3					

$f(x) \to$?

8. $\lim\limits_{x\to 2^-} g(x)$, where $g(x) = \dfrac{x^3 - 8}{x^2 + 2x + 4}$

$x \to 2^-$

x	1	1.5	1.9	1.99	1.999	1.9999
$g(x)$	-1					

$g(x) \to$?

9. $\lim\limits_{x\to 2} h(x)$, where $h(x) = \dfrac{3x^2 - 2x - 8}{x - 2}$

	$x \to 2^-$				$2^+ \leftarrow x$			
x	1	1.9	1.99	1.999	2.001	2.1	2.5	3
$h(x)$	7							

$h(x) \to$? $\leftarrow h(x)$

10. Find $\lim\limits_{x\to 0} \dfrac{\tan 2x}{\tan 3x}$ using the following procedure based on the fact that $f(x) = \tan x$ is an odd function.

If $f(x) = \dfrac{\tan 2x}{\tan 3x}$, then

$$f(-x) = \frac{\tan(-2x)}{\tan(-3x)} = \frac{-\tan 2x}{-\tan 3x} = f(x).$$

Thus, we simply need to check for $x \to 0^+$. Find the limit by completing the following table.

$x \to 0^+$

x	1	0.5	0.1	0.01	0.001
$f(x)$	15.33				

$f(x) \to$?

Describe each illustration in Problems 11–16 using a limit statement.

11.

12.

13.

14.

15.

16.
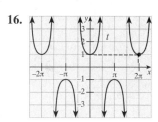

B **17. WHAT DOES THIS SAY?*** Explain a process for finding a limit.

Evaluate the limits in Problems 18–44 to two decimal places by graphing or by using a table of values, and check your answer using a calculator. If the limit does not exist, explain why.

18. $\lim\limits_{x\to 0^+} x^4$ **19.** $\lim\limits_{x\to 0^+} \cos x$

20. $\lim\limits_{x\to 2^-} (x^2 - 4)$ **21.** $\lim\limits_{x\to 3^-} (x^2 - 4)$

22. $\lim\limits_{x\to 1^+} \dfrac{1}{x - 3}$ **23.** $\lim\limits_{x\to -3^+} \dfrac{1}{x - 3}$

24. $\lim\limits_{x\to 3} \dfrac{1}{x - 3}$ **25.** $\lim\limits_{x\to \pi/2} \tan x$

---

*Many problems in this book are labeled What Does This Say? Following the question will be a question for you to answer in your own words, or a statement for you to rephrase in your own words. These problems are intended to be similar to the "What This Says" boxes.

26. a. $\lim\limits_{x\to 0}\dfrac{\cos x}{x}$

   b. $\lim\limits_{x\to \pi}\dfrac{\cos x}{x}$

27. a. $\lim\limits_{x\to 0}\dfrac{1-\cos x}{x}$

   b. $\lim\limits_{x\to \pi}\dfrac{1-\cos x}{x}$

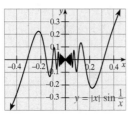

28. a. $\lim\limits_{x\to 0.4}|x|\sin\dfrac{1}{x}$

   b. $\lim\limits_{x\to 0}|x|\sin\dfrac{1}{x}$

29. $\lim\limits_{x\to 3}\dfrac{x^2+3x-10}{x-2}$

30. $\lim\limits_{x\to 3}\dfrac{x^2+3x-10}{x-3}$

31. $\lim\limits_{x\to \pi/2}\dfrac{2x-\pi}{\cos x}$

32. $\lim\limits_{x\to 1}\dfrac{\sin\frac{\pi}{x}}{x-1}$

33. $\lim\limits_{x\to 9}\dfrac{\sqrt{x}-3}{x-9}$

34. $\lim\limits_{x\to 9}\dfrac{\sqrt{x}-3}{x-3}$

35. $\lim\limits_{x\to 2}\dfrac{\sqrt{x+2}-2}{x-2}$

36. $\lim\limits_{x\to 1}\dfrac{\sqrt[3]{x}-1}{\sqrt{x}-1}$

37. $\lim\limits_{x\to 3^+}\dfrac{\sqrt{x-3}+x}{3-x}$

38. $\lim\limits_{x\to 4^+}\dfrac{\frac{1}{\sqrt{x}}-\frac{1}{2}}{x-4}$

39. $\lim\limits_{x\to 0}\dfrac{\sin 2x}{x}$

40. $\lim\limits_{x\to 0}\dfrac{\sin 3x}{x}$

41. $\lim\limits_{x\to 0}\dfrac{1-\frac{1}{x+1}}{x}$

42. $\lim\limits_{x\to 1}\dfrac{1-\frac{1}{x}}{x-1}$

43. $\lim\limits_{x\to 0}\cos\dfrac{1}{x}$

44. $\lim\limits_{x\to 0}\tan\dfrac{1}{x}$

45. A ball is thrown directly upward from the edge of a cliff and travels in such a way that t seconds later, its height (in feet) above the ground at the base of the cliff is

$$s(t)=-16t^2+40t+24$$

a. Compute the limit

$$v(t)=\lim\limits_{x\to t}\dfrac{s(x)-s(t)}{x-t}$$

to find the instantaneous velocity of the ball at time t.

b. What is the ball's initial velocity?

c. When does the ball hit the ground, and what is its impact velocity at time $t=0$?

d. When does the ball have velocity 0? What physical interpretation should be given to this time?

46. Tom and Sue are driving along a straight, level road in a car whose speedometer needle is broken but that has a trip odometer that can measure the distance traveled from an arbitrary starting point in tenths of a mile. At 2:50 P.M., Tom says he would like to know how fast they are traveling at 3:00 P.M., so Sue takes down the odometer readings listed in the table, makes a few calculations, and announces the desired velocity. What is her result?

time t	2:50	2:55	2:59	3:00	3:01	3:03	3:06
odometer reading	33.9	38.2	41.5	42.4	43.2	44.9	47.4

In Problems 47–48, estimate the limits by plotting points or by using tables.

47. $\lim\limits_{x\to 13}\dfrac{x^3-9x^2-45x-91}{x-13}$

48. $\lim\limits_{x\to 13}\dfrac{x^3-9x^2-39x-86}{x-13}$

49. HISTORICAL QUEST   By the second half of the 1700s, it was generally accepted that without logical underpinnings, calculus would be limited. Augustin-Louis Cauchy developed an acceptable theory of limits, and in doing so removed much doubt about the logical validity of calculus. Cauchy is described by the historian Howard Eves not only as a first-rate mathematician with tremendous mathematical productivity but also as a lawyer (he practiced law for 14 years), a mountain climber, and a painter (he worked in watercolors). Among other characteristics that distinguished him from his contemporaries, he advocated respect for the environment.

AUGUSTIN–LOUIS
CAUCHY
1789–1857

Cauchy wrote a treatise on integrals in 1814 that is considered a classic, and in 1816 his paper on wave propagation in liquids won a prize from the French Academy. It has been said that with his work the modern era of analysis began. In all, he wrote over 700 papers, which are, today, considered no less than brilliant. ∎

Cauchy did not formulate the ϵ-δ definition of limit that we use today, but formulated instead a purely arithmetical definition. Consult some history of mathematics books to find a translation of Cauchy's definition, which appeared in his monumental treatise, *Cours d'Analyse de l'Ecole Royale Polytechnique* (1821). As part of your research, find when and where the ϵ-δ definition of limit was first used.

⊙ *In Problems 50–55, use the formal definition of the limit to prove or disprove the given limit statement.*

50. $\lim\limits_{x\to 2}(x+3)=5$

51. $\lim\limits_{t\to 0}(3t-1)=0$

52. $\lim\limits_{x\to -2}(3x+7)=1$

53. $\lim\limits_{x\to 1}(2x-5)=-3$

54. $\lim\limits_{x\to 2}(x^2+2)=6$

55. $\lim\limits_{x\to 2}\dfrac{1}{x}=\dfrac{1}{2}$

56. The tabular approach is a convenient device for discussing limits informally, but if it is not used very carefully, it can be misleading. For example, for $x\neq 0$, let

$$f(x)=\sin\dfrac{1}{x}$$

a. Construct a table showing the values of $f(x)$ for $x = \frac{-2}{\pi}, \frac{-2}{9\pi}, \frac{-2}{13\pi}, \frac{2}{19\pi}, \frac{2}{7\pi}, \frac{2}{3\pi}$. Based on this table, what would you say about $\lim\limits_{x \to 0} f(x)$?

b. Construct a second table, this time showing the values of

$f(x)$ for $x = \frac{-1}{2\pi}, \frac{-1}{11\pi}, \frac{-1}{20\pi}, \frac{1}{50\pi}, \frac{1}{30\pi}, \frac{1}{5\pi}$. Now what would you say about $\lim\limits_{x \to 0} f(x)$?

c. Based on the results in parts **a** and **b**, what do you conclude about $\lim\limits_{x \to 0} \sin \frac{1}{x}$?

2.2   Algebraic Computation of Limits

IN THIS SECTION   computations with limits, using algebra to find limits, two special trigonometric limits, limits of piecewise-defined functions

COMPUTATIONS WITH LIMITS

In Section 2.1, we observed that finding limits by tables or by graphing is risky. In this section, we will explore a more exact way of computing limits that is based on the following properties.

Basic Properties and Rules for Limits

For any real number c, suppose the functions f and g both have limits at $x = c$.

Constant rule $\lim\limits_{x \to c} k = k$ for any constant k

Limit of x rule $\lim\limits_{x \to c} x = c$

Multiple rule $\lim\limits_{x \to c} [kf(x)] = k \lim\limits_{x \to c} f(x)$ for any constant k

The limit of a constant times a function is the constant times the limit of the function.

Sum rule $\lim\limits_{x \to c} [f(x) + g(x)] = \lim\limits_{x \to c} f(x) + \lim\limits_{x \to c} g(x)$

The limit of a sum is the sum of the limits.

Difference rule $\lim\limits_{x \to c} [f(x) - g(x)] = \lim\limits_{x \to c} f(x) - \lim\limits_{x \to c} g(x)$

The limit of a difference is the difference of the limits.

Product rule $\lim\limits_{x \to c} [f(x)g(x)] = \left[\lim\limits_{x \to c} f(x)\right]\left[\lim\limits_{x \to c} g(x)\right]$

The limit of a product is the product of the limits.

Quotient rule $\lim\limits_{x \to c} \dfrac{f(x)}{g(x)} = \dfrac{\lim\limits_{x \to c} f(x)}{\lim\limits_{x \to c} g(x)}$ if $\lim\limits_{x \to c} g(x) \neq 0$

The limit of a quotient is the quotient of the limits, as long as the limit of the denominator is not zero.

Power rule $\lim\limits_{x \to c} [f(x)]^n = \left[\lim\limits_{x \to c} f(x)\right]^n$ n is a rational number and the limit on the right exists.

The limit of a power is the power of the limit.

---

a. Limit of a constant: $\lim\limits_{x \to c} k = k$

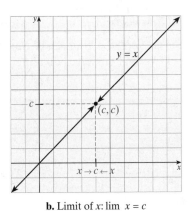

b. Limit of x: $\lim\limits_{x \to c} x = c$

Figure 2.13 Two basic limits

It is fairly easy to justify graphically the rules for the limit of a constant and the limit of x, as shown in Figure 2.13. All of these properties of limits (including the two basic limits shown in Figure 2.13) can be proved using the formal definition of limit.

EXAMPLE 1   Finding the limit of a polynomial function

Evaluate $\lim_{x \to 2}(2x^5 - 9x^3 + 3x^2 - 11)$.

Solution

$$\lim_{x \to 2}(2x^5 - 9x^3 + 3x^2 - 11) = \lim_{x \to 2}(2x^5) - \lim_{x \to 2}(9x^3) + \lim_{x \to 2}(3x^2) - \lim_{x \to 2}(11)$$

Sum and difference rules

$$= 2\left[\lim_{x \to 2} x^5\right] - 9\left[\lim_{x \to 2} x^3\right] + 3\left[\lim_{x \to 2} x^2\right] - 11$$

Multiple and constant rules

$$= 2\left[\lim_{x \to 2} x\right]^5 - 9\left[\lim_{x \to 2} x\right]^3 + 3\left[\lim_{x \to 2} x\right]^2 - 11$$

Power rule

$$= 2(2)^5 - 9(2)^3 + 3(2)^2 - 11 = -7$$

Limit of x rule ∎

Comment: If you consider Example 1 carefully, it is easy to see that if P is any polynomial, then the limit at $x = c$ can be found by substituting $x = c$ into the formula for $P(x)$.

Limit of a Polynomial Function

If P is a polynomial function, then

$$\lim_{x \to c} P(x) = P(c)$$

EXAMPLE 2   Finding the limit of a rational function

Evaluate $\lim_{z \to -1} \dfrac{z^3 - 3z + 7}{5z^2 + 9z + 6}$.

Solution

$$\lim_{z \to -1} \frac{z^3 - 3z + 7}{5z^2 + 9z + 6} = \frac{\lim_{z \to -1}(z^3 - 3z + 7)}{\lim_{z \to -1}(5z^2 + 9z + 6)} \qquad \text{Quotient rule}$$

$$= \frac{(-1)^3 - 3(-1) + 7}{5(-1)^2 + 9(-1) + 6}$$

$$= \frac{9}{2} \qquad \begin{array}{l}\text{Both numerator and denominator}\\ \text{are polynomial functions.}\end{array} \quad ∎$$

WARNING   You must be careful as to when you write the word "limit" and when you do not; pay particular attention to this when looking at the examples in this section.

Notice that if the denominator of the rational function is not zero, the limit can be found by substitution. If the limit of the denominator is zero, we find ourselves dividing by zero, an illegal operation.

Limit of a Rational Function

If Q is a rational function defined by $Q(x) = \dfrac{P(x)}{D(x)}$, then

$$\lim_{x \to c} Q(x) = \frac{P(c)}{D(c)}$$

provided $\lim_{x \to c} D(x) \neq 0$.

EXAMPLE 3   Finding the limit of a power (or root) function

Evaluate $\lim\limits_{x \to -2} \sqrt[3]{x^2 - 3x - 2}$.

Solution $\lim\limits_{x \to -2} \sqrt[3]{x^2 - 3x - 2} = \lim\limits_{x \to -2} (x^2 - 3x - 2)^{1/3}$

$$= \left[\lim\limits_{x \to -2} (x^2 - 3x - 2) \right]^{1/3} \qquad \text{Power rule}$$

$$= [(-2)^2 - 3(-2) - 2]^{1/3} = 8^{1/3} = 2 \qquad \blacksquare$$

Once again, for values of the function for which $f(c)$ is defined, the limit can be found by substitution.

In the previous section we used a table to find that $\lim\limits_{x \to 0} \sin x = 0$ and $\lim\limits_{x \to 0} \cos x = 1$. In the following example we use this information, along with the properties of limits, to find other trigonometric limits.

EXAMPLE 4   Finding trigonometric limits algebraically

Given that $\lim\limits_{x \to 0} \sin x = 0$ and $\lim\limits_{x \to 0} \cos x = 1$, evaluate:

a. $\lim\limits_{x \to 0} \sin^2 x$ **b.** $\lim\limits_{x \to 0} (1 - \cos x)$

Solution

a. $\lim\limits_{x \to 0} \sin^2 x = \left[\lim\limits_{x \to 0} \sin x \right]^2 \qquad \text{Power rule}$

$$= 0^2 \qquad \lim\limits_{x \to 0} \sin x = 0$$

$$= 0$$

b. $\lim\limits_{x \to 0} (1 - \cos x) = \lim\limits_{x \to 0} 1 - \lim\limits_{x \to 0} \cos x \qquad \text{Difference rule}$

$$= 1 - 1 \qquad \text{Constant rule and} \lim\limits_{x \to 0} \cos x = 1$$

$$= 0 \qquad \blacksquare$$

The following theorem states that we can find limits of trigonometric functions by direct substitution, as long as the number that x is approaching is in the domain of the given function. Proofs of several parts are outlined in the problem set, and in fact an examination of the graphs of the functions indicates why they are true.

THEOREM 2.2   Limits of trigonometric functions

If c is any number in the domain of the given function, then

$$\lim\limits_{x \to c} \cos x = \cos c \qquad \lim\limits_{x \to c} \sin x = \sin c \qquad \lim\limits_{x \to c} \tan x = \tan c$$

$$\lim\limits_{x \to c} \sec x = \sec c \qquad \lim\limits_{x \to c} \csc x = \csc c \qquad \lim\limits_{x \to c} \cot x = \cot c$$

Proof   We will show that $\lim\limits_{x \to c} \sin x = \sin c$. The other limit formulas may be proved in a similar fashion (see Problems 63–64). Let $h = x - c$. Then $x = h + c$, so $x \to c$ as $h \to 0$. Thus,

$$\lim\limits_{x \to c} \sin x = \lim\limits_{h \to 0} \sin(h + c)$$

Using the trigonometric identity

$$\sin(A + B) = \sin A \cos B + \cos A \sin B$$

and the limit formulas for sums and products, we find that

$$\lim_{x \to c} \sin x = \lim_{h \to 0} \sin(h + c)$$

$$= \lim_{h \to 0} [\sin h \cos c + \cos h \sin c]$$

$$= \lim_{h \to 0} \sin h \cdot \lim_{h \to 0} \cos c + \lim_{h \to 0} \cos h \cdot \lim_{h \to 0} \sin c$$

$$= 0 \cdot \cos c + 1 \cdot \sin c \qquad \lim_{h \to 0} \sin h = 0 \text{ and } \lim_{h \to 0} \cos h = 1$$

$$= \sin c$$

Note that $\sin c$ and $\cos c$ do not change as $h \to 0$ because these are constants with respect to h. ❑

EXAMPLE 5   Finding limits of trigonometric functions

Evaluate the following limits:

a. $\displaystyle \lim_{x \to 1} (x^2 \cos \pi x)$ **b.** $\displaystyle \lim_{x \to 0} \frac{x}{\cos x}$

Solution

a. $\displaystyle \lim_{x \to 1} (x^2 \cos \pi x) = \left[\lim_{x \to 1} x \right]^2 \cdot \left[\lim_{x \to 1} \cos \pi x \right] = 1^2 \cos \pi = -1$

b. $\displaystyle \lim_{x \to 0} \frac{x}{\cos x} = \frac{\displaystyle \lim_{x \to 0} x}{\displaystyle \lim_{x \to 0} \cos x} = \frac{0}{1} = 0$ ■

USING ALGEBRA TO FIND LIMITS

Sometimes the limit of $f(x)$ as $x \to c$ *cannot* be evaluated by direct substitution. In such a case, we look for another function that agrees with f for all values of x *except at the troublesome value* $x = c$. We illustrate this procedure with some examples.

EXAMPLE 6   Evaluating a limit using fractional reduction

Evaluate $\displaystyle \lim_{x \to 2} \frac{x^2 + x - 6}{x - 2}$.

Solution

If you try substitution on this limit, you will obtain

$$\text{If } x = 2, \text{ then } x^2 + x - 6 = 0.$$
$$\downarrow$$
$$\lim_{x \to 2} \frac{x^2 + x - 6}{x - 2} = \frac{0}{0}$$
$$\uparrow$$
$$\text{If } x = 2, \text{ then } x - 2 = 0.$$

The form $\frac{0}{0}$ is called an **indeterminate form** because the value of the limit cannot be determined without further analysis.

   If the expression is a rational expression, the next step is to try to simplify the function by factoring and checking to see if the reduced form is a polynomial.

$$\lim_{x \to 2} \frac{x^2 + x - 6}{x - 2} = \lim_{x \to 2} \frac{(x + 3)(x - 2)}{x - 2} = \lim_{x \to 2} (x + 3)$$

This simplification is valid only if $x \neq 2$. Now complete the evaluation of the reduced function by direct substitution. This is not a problem, because the limit is concerned with values *as x approaches* 2, not the value where $x = 2$. In the preceding expression

$$\lim_{x \to 2} \frac{x^2 + x - 6}{x - 2} = \lim_{x \to 2} (x + 3) = 5$$ ■

Another algebraic technique for finding limits is to rationalize either the numerator or the denominator to obtain an algebraic form that is not indeterminate.

EXAMPLE 7 **Evaluating a limit by rationalizing**

Evaluate $\lim\limits_{x \to 4} \dfrac{\sqrt{x}-2}{x-4}$.

Solution

Once again, notice that both the numerator and denominator of this rational expression are 0 when $x = 4$, so we cannot evaluate the limit by direct substitution. Instead, rationalize the numerator:

$$\lim_{x \to 4} \frac{\sqrt{x}-2}{x-4} = \lim_{x \to 4} \frac{\sqrt{x}-2}{x-4} \cdot \frac{\sqrt{x}+2}{\sqrt{x}+2} \qquad \text{Multiply by 1.}$$

$$= \lim_{x \to 4} \frac{x-4}{(x-4)(\sqrt{x}+2)}$$

$$= \lim_{x \to 4} \frac{1}{\sqrt{x}+2}$$

$$= \frac{1}{\sqrt{4}+2} = \frac{1}{4} \qquad \blacksquare$$

WARNING This method will work only if the resulting numerator allows the fraction to be simplified. Pay close attention to this example because it illustrates a procedure we will use over and over.

LIMITS OF PIECEWISE-DEFINED FUNCTIONS

In Section 1.3 we discussed *piecewise-defined functions*. To evaluate $\lim\limits_{x \to c} f(x)$ where the domain of f is divided into pieces, we first look to see whether c is a value separating two of the pieces. If so, we need to consider one-sided limits, as illustrated by the following examples.

EXAMPLE 8 **Limit of a piecewise-defined function**

Find $\lim\limits_{x \to 0} f(x)$, where $f(x) = \begin{cases} x+5 & \text{if } x > 0 \\ x & \text{if } x < 0 \end{cases}$.

Solution

Notice that $f(0)$ is not defined, and that it is necessary to consider left- and right-hand limits because $f(x)$ is defined differently on each side of $x = 0$. We have

$$\lim_{x \to 0^-} f(x) = \lim_{x \to 0^-} x \qquad f(x) = x \text{ to the left of } 0.$$
$$= 0$$
$$\lim_{x \to 0^+} f(x) = \lim_{x \to 0^+} (x+5) \qquad f(x) = x+5 \text{ to the right of } 0.$$
$$= 5$$

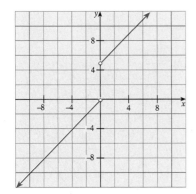

Figure 2.14 Graph of f

Because the left- and right-hand limits are not the same, we conclude that $\lim\limits_{x \to 0} f(x)$ does not exist. If we look at the graph, as shown in Figure 2.14, it is easy to see that the left- and right-hand limits are not the same. $\blacksquare$

EXAMPLE 9 **Limit of a piecewise-defined function**

Find $\lim\limits_{x \to 0} g(x)$, where $g(x) = \begin{cases} x+1 & \text{if } x > 0 \\ x^2+1 & \text{if } x < 0 \end{cases}$.

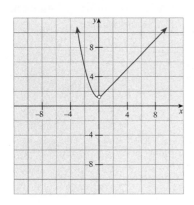

Figure 2.15 Graph of g

Solution
$$\lim_{x \to 0^-} g(x) = \lim_{x \to 0^-} (x^2 + 1) = 1$$
$$\lim_{x \to 0^+} g(x) = \lim_{x \to 0^+} (x + 1) = 1$$

Because the left- and right-hand limits are equal, we conclude that $\lim_{x \to 0} g(x) = 1$. This time, by looking at the graph in Figure 2.15, we see that the left- and right-hand limits are the same (even though the function is not defined at $x = 0$). Compare this graph with the graph in Example 8. ∎

TWO SPECIAL TRIGONOMETRIC LIMITS

The following theorem contains two additional limits that play an important role in calculus. We evaluated the first of these limits by table and graphically in Example 6 of Section 2.1.

THEOREM 2.3   Special limits involving sine and cosine

$$\lim_{h \to 0} \frac{\sin h}{h} = 1 \qquad \lim_{h \to 0} \frac{\cos h - 1}{h} = 0$$

Proof   A geometric argument justifying these rules is given at the end of this section. ❏

For now, let us concentrate on illustrating how these rules can be used.

EXAMPLE 10   Evaluation of trigonometric and inverse trigonometric limits

Find each of the following limits.

a. $\lim_{x \to 0} \dfrac{\sin 3x}{5x}$ **b.** $\lim_{x \to 0} \dfrac{\sin^{-1} x}{x}$ **c.** $\lim_{h \to 0} \dfrac{\cos h - 1}{h^2}$

Solution

a. We prepare the limit for evaluation by Theorem 2.3 by writing

$$\frac{\sin 3x}{5x} = \frac{3}{5}\left(\frac{\sin 3x}{3x}\right) \qquad \textbf{WARNING} \longrightarrow \quad \sin 3x \neq 3\sin x$$

Since $3x \to 0$ as $x \to 0$, we can set $h = 3x$ in Theorem 2.3 to obtain

$$\lim_{x \to 0} \frac{\sin 3x}{5x} = \lim_{x \to 0} \frac{3}{5}\left(\frac{\sin 3x}{3x}\right) = \frac{3}{5}\lim_{h \to 0}\frac{\sin h}{h} = \frac{3}{5}(1) = \frac{3}{5}$$

b. Let $u = \sin^{-1} x$, so $\sin u = x$. Thus $u \to 0$ as $x \to 0$, and

$$\lim_{x \to 0} \frac{\sin^{-1} x}{x} = \lim_{u \to 0} \frac{u}{\sin u} = 1$$

c. We begin by writing

$$\frac{\cos h - 1}{h^2} = \frac{(\cos h - 1)(\cos h + 1)}{h^2(\cos h + 1)}$$
$$= \frac{\cos^2 h - 1}{h^2(\cos h + 1)}$$
$$= \frac{-\sin^2 h}{h^2(\cos h + 1)} \qquad \text{Since } \sin^2 h + \cos^2 h = 1$$

Thus, we have

$$\lim_{h\to 0}\frac{\cos h - 1}{h^2} = \lim_{h\to 0}\frac{-\sin^2 h}{h^2(\cos h + 1)}$$

$$= -\left[\lim_{h\to 0}\frac{\sin^2 h}{h^2}\right]\left[\lim_{h\to 0}\frac{1}{\cos h + 1}\right]$$

$$= -\left[\lim_{h\to 0}\frac{\sin h}{h}\right]^2\left[\lim_{h\to 0}\frac{1}{\cos h + 1}\right]$$

$$= -(1)^2\left(\frac{1}{1+1}\right)$$

$$= -\frac{1}{2}$$ ∎

We conclude this section with a justification of Theorem 2.3. Our argument is based on the following useful property that is proved in Appendix A.

Squeeze Rule*

If $g(x) \le f(x) \le h(x)$ on an open interval containing c, and if

$$\lim_{x\to c} g(x) = \lim_{x\to c} h(x) = L, \text{ then } \lim_{x\to c} f(x) = L$$

➡ **What This Says**   If a function can be squeezed between two functions whose limits at a particular point c have the same value L, then that function must also have limit L at $x = c$.

With this squeeze rule in our toolkit, we are now ready to justify Theorem 2.3.

Proof   Proof that $\displaystyle\lim_{h\to 0}\frac{\sin h}{h} = 1$.

To prove the sine limit theorem requires some principles that are not entirely obvious. However, we can demonstrate its plausibility by considering Figure 2.16, in which AOC is a sector of a circle of radius 1 with angle h measured in radians. The line segments $\overline{AD}$ and $\overline{BC}$ are drawn perpendicular to segment $\overline{OC}$.

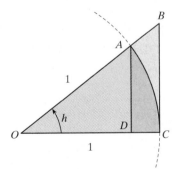

Assume $0 < h < \frac{\pi}{2}$; that is, h is in Quadrant I.

$|\widehat{AC}| = h$

$|\overline{AD}| = \sin h$

$|\overline{BC}| = \tan h = \dfrac{\sin h}{\cos h}$

$|\overline{OD}| = \cos h$

Figure 2.16   Trigonometric relationships in the proof that $\displaystyle\lim_{h\to 0}\frac{\sin h}{h} = 1$

---
*This rule is sometimes known as the **sandwich rule**.

Now, compare the area of the sector AOC with those of $\triangle AOD$ and $\triangle BOC$. In particular, since the area of the circular sector of radius r and central angle θ is $\frac{1}{2}r^2\theta$ (see the *Student Mathematics Handbook*), sector AOC must have area

$$\tfrac{1}{2}(1)^2 h = \tfrac{1}{2}h$$

We also find that $\triangle AOD$ has area

$$\tfrac{1}{2}\left|\overline{OD}\right|\left|\overline{AD}\right| = \tfrac{1}{2}\cos h \sin h,$$

and $\triangle BOC$ has area

$$\tfrac{1}{2}\left|\overline{BC}\right|\left|\overline{OC}\right| = \tfrac{1}{2}\left|\overline{BC}\right|(1) = \frac{1}{2}\frac{\sin h}{\cos h}$$

By comparing areas (see Figure 2.16), we have

AREA OF $\triangle AOD$		AREA OF SECTOR AOC		AREA OF $\triangle BOC$	
$\frac{1}{2}\cos h \sin h$	$\leq$	$\frac{1}{2}h$	$\leq$	$\dfrac{\sin h}{2\cos h}$	
$\cos h$	$\leq$	$\dfrac{h}{\sin h}$	$\leq$	$\dfrac{1}{\cos h}$	Divide by $\frac{1}{2}\sin h$.
$\dfrac{1}{\cos h}$	$\geq$	$\dfrac{\sin h}{h}$	$\geq$	$\cos h$	Take reciprocals (reverse order).
$\cos h$	$\leq$	$\dfrac{\sin h}{h}$	$\leq$	$\dfrac{1}{\cos h}$	Interchange left and right sides (reverse order again).

This same inequality holds in the interval $\left(-\frac{\pi}{2}, 0\right)$, which is Quadrant IV for the angle h. This can be shown by using the trigonometric identities $\cos(-h) = \cos h$ and $\sin(-h) = -\sin h$. Finally, we take the limit of all parts as $h \to 0$ to find

$$\lim_{h \to 0} \cos h \leq \lim_{h \to 0} \frac{\sin h}{h} \leq \lim_{h \to 0} \frac{1}{\cos h}$$

By Theorem 2.2, $\lim_{h \to 0} \cos h = \cos 0 = 1$. Thus,

$$1 \leq \lim_{h \to 0} \frac{\sin h}{h} \leq \frac{1}{1}$$

From the squeeze rule, we conclude that $\lim_{h \to 0} \dfrac{\sin h}{h} = 1$.
You are asked to prove the second part of this theorem in Problem 59.   ❏

2.2   PROBLEM SET

A *In Problems 1–32, evaluate each limit.*

1. $\lim\limits_{x \to -2} (x^2 + 3x - 7)$

2. $\lim\limits_{t \to 0}(t^3 - 5t^2 + 4)$

3. $\lim\limits_{x \to 3}(x + 5)(2x - 7)$

4. $\lim\limits_{x \to 0} \dfrac{\sin^{-1} x}{x - 1}$

5. $\lim\limits_{z \to 1} \dfrac{z^2 + z - 3}{z + 1}$

6. $\lim\limits_{x \to 3} \dfrac{x^2 + 3x - 10}{3x^2 + 5x - 7}$

7. $\lim\limits_{x \to \pi/3} \sec x$

8. $\lim\limits_{x \to \pi/4} \dfrac{1 + \tan x}{\csc x + 2}$

9. $\lim\limits_{x \to 1/3} \dfrac{x \sin \pi x}{1 + \cos \pi x}$

10. $\lim\limits_{x \to 6} \dfrac{\tan(\pi/x)}{x - 1}$

11. $\lim\limits_{u \to -2} \dfrac{4 - u^2}{2 + u}$

12. $\lim\limits_{x \to 2} \dfrac{x^2 - 4x + 4}{x^2 - x - 2}$

13. $\lim\limits_{x\to 1}\dfrac{\frac{1}{x}-1}{x-1}$

14. $\lim\limits_{x\to 0}\dfrac{(x+1)^2-1}{x}$

15. $\lim\limits_{x\to 1}\left(\dfrac{x^2-3x+2}{x^2+x-2}\right)^2$

16. $\lim\limits_{x\to 3}\sqrt{\dfrac{x^2-2x-3}{x-3}}$

17. $\lim\limits_{x\to 1}\dfrac{\sqrt{x}-1}{x-1}$

18. $\lim\limits_{y\to 2}\dfrac{\sqrt{y+2}-2}{y-2}$

19. $\lim\limits_{x\to 0}\dfrac{\sqrt{x+1}-1}{x}$

20. $\lim\limits_{x\to 0}\dfrac{\sqrt{x^2+4}-2}{x}$

21. $\lim\limits_{x\to 0^+}\dfrac{\sin x}{\sqrt{x}}$

22. $\lim\limits_{x\to 0^+}\dfrac{1-\cos\sqrt{x}}{x}$

23. $\lim\limits_{x\to 0}\dfrac{\sin 2x}{x}$

24. $\lim\limits_{x\to 0}\dfrac{\sin 4x}{9x}$

25. $\lim\limits_{t\to 0}\dfrac{\tan 5t}{\tan 2t}$

26. $\lim\limits_{x\to 0}\dfrac{\cot 3x}{\cot x}$

27. $\lim\limits_{x\to 0}\dfrac{1-\cos x}{\sin x}$

28. $\lim\limits_{x\to 0}\dfrac{\sin^2 x}{2x}$

29. $\lim\limits_{x\to 0}\dfrac{\sin^2 x}{x^2}$

30. $\lim\limits_{x\to 0}\dfrac{x^2\cos 2x}{1-\cos x}$

31. $\lim\limits_{x\to 0}\dfrac{\sec x-1}{x\sec x}$

32. $\lim\limits_{x\to \pi/4}\dfrac{1-\tan x}{\sin x-\cos x}$

33. WHAT DOES THIS SAY? How do you find the limit of a polynomial function?

34. WHAT DOES THIS SAY? How do you find the limit of a rational function?

35. WHAT DOES THIS SAY? How do you find

$$\lim\limits_{x\to 0}\dfrac{\sin ax}{x}$$

for $a\neq 0$?

B *In Problems 36–43, compute the one-sided limit or use one-sided limits to find the given limit, if it exists.*

36. $\lim\limits_{x\to 2^-}(x^2-2x)$

37. $\lim\limits_{x\to 1^+}\dfrac{\sqrt{x-1}+x}{1-2x}$

38. $\lim\limits_{x\to 2}|x-2|$

39. $\lim\limits_{x\to 3}|3-x|$

40. $\lim\limits_{x\to 0}\dfrac{|x|}{x}$

41. $\lim\limits_{x\to -2}\dfrac{|x+2|}{x+2}$

42. $\lim\limits_{x\to 2}f(x)$, where $f(x)=\begin{cases}3-2x & \text{if } x\leq 2\\ x^2-5 & \text{if } x>2\end{cases}$

43. $\lim\limits_{s\to 1}g(s)$, where $g(s)=\begin{cases}\dfrac{s^2-s}{s-1} & \text{if } s>1\\ \sqrt{1-s} & \text{if } s\leq 1\end{cases}$

WHAT DOES THIS SAY? *In Problems 44–51, explain why the given limit does not exist.*

44. $\lim\limits_{x\to 1}\dfrac{1}{x-1}$

45. $\lim\limits_{x\to 2^+}\dfrac{1}{\sqrt{x-2}}$

46. $\lim\limits_{t\to 2}\dfrac{t^2-4}{t^2-4t+4}$

47. $\lim\limits_{x\to 3}\dfrac{x^2+4x+3}{x-3}$

48. $\lim\limits_{x\to 1}f(x)$, where $f(x)=\begin{cases}2 & \text{if } x\geq 1\\ -5 & \text{if } x<1\end{cases}$

49. $\lim\limits_{t\to -1}g(t)$, where $g(t)=\begin{cases}2t+1 & \text{if } t\geq -1\\ 5t^2 & \text{if } t<-1\end{cases}$

50. $\lim\limits_{x\to \pi/2}\tan x$

51. $\lim\limits_{x\to 1}\csc \pi x$

In Problems 52–57, either evaluate the limit or explain why it does not exist.

52. $\lim\limits_{x\to 1}\dfrac{\frac{1}{x}-1}{\sqrt{x}-1}$

53. $\lim\limits_{x\to 0}\left(\dfrac{1}{x}-\dfrac{1}{x^2}\right)$

54. $\lim\limits_{x\to 5}f(x)$, where $f(x)=\begin{cases}x+3 & \text{if } x\neq 5\\ 4 & \text{if } x=5\end{cases}$

55. $\lim\limits_{x\to 2}g(t)$, where $g(t)=\begin{cases}t^2 & \text{if } -1\leq t<2\\ 3t-2 & \text{if } t\geq 2\end{cases}$

56. $\lim\limits_{x\to 2}f(x)$, where $f(x)=\begin{cases}2(x+1) & \text{if } x<3\\ 4 & \text{if } x=3\\ x^2-1 & \text{if } x>3\end{cases}$

57. $\lim\limits_{x\to 3}f(x)$, where $f(x)=\begin{cases}2(x+1) & \text{if } x<3\\ 4 & \text{if } x=3\\ x^2-1 & \text{if } x>3\end{cases}$

58. Interpretation Problem Evaluate $\lim\limits_{x\to 0}\left[x^2-\dfrac{\cos x}{1,000,000,000}\right]$. Explain why a calculator solution may lead to an incorrect conclusion about the limit.

C 59. Prove the second part of Theorem 2.3: $\lim\limits_{h\to 0}\dfrac{\cos h-1}{h}=0$.

60. Counterexample Problem Give an example for which neither $\lim\limits_{x\to c}f(x)$ nor $\lim\limits_{x\to c}g(x)$ exists, but $\lim\limits_{x\to c}[f(x)+g(x)]$ does exist.

61. Let $f(x)=\dfrac{1}{x^2}$ with $x\neq 0$, and let L be any fixed positive integer. Show that

$$f(x)>100L \text{ if } |x|<\dfrac{1}{10\sqrt{L}}$$

What does this imply about $\lim\limits_{x\to 0}f(x)$?

62. Use the sum rule to show that if $\lim\limits_{x\to c}[f(x)+g(x)]$ and $\lim\limits_{x\to c}f(x)$ both exist, then so does $\lim\limits_{x\to c}g(x)$.

63. Show that $\lim\limits_{x\to x_0}\cos x=\cos x_0$. *Hint:* You will need to use the trigonometric identity $\cos(A+B)=\cos A\cos B-\sin A\sin B$.

64. Show that $\lim\limits_{x\to x_0}\tan x=\tan x_0$ whenever $\cos x_0\neq 0$.

2.3 Continuity

IN THIS SECTION intuitive notion of continuity, definition of continuity, continuity theorems, continuity on an interval, the intermediate value theorem

INTUITIVE NOTION OF CONTINUITY

The idea of *continuity* may be thought of informally as the quality of having parts that are in immediate connection with one another, as shown in Figure 2.17. The idea evolved from the vague or intuitive notion of a curve "without breaks or jumps" to a rigorous definition first given toward the end of the nineteenth century (see Historical Quest Problem 45).

We begin with a discussion of *continuity at a point*. It may seem strange to talk about continuity *at a point*, but it should seem natural to talk about a curve being "discontinuous at a point." A few such discontinuities are illustrated by Figure 2.18.

$\lim\limits_{x \to c} f(x)$ exists and is equal to $f(c)$.

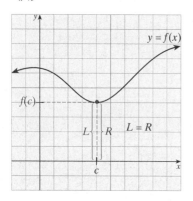

Figure 2.17 A function continuous at a point $x = c$

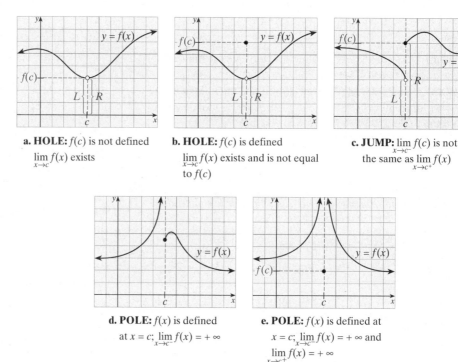

a. HOLE: $f(c)$ is not defined
$\lim\limits_{x \to c} f(x)$ exists

b. HOLE: $f(c)$ is defined
$\lim\limits_{x \to c} f(x)$ exists and is not equal to $f(c)$

c. JUMP: $\lim\limits_{x \to c^-} f(c)$ is not the same as $\lim\limits_{x \to c^+} f(x)$

d. POLE: $f(x)$ is defined at $x = c$; $\lim\limits_{x \to c^-} f(x) = +\infty$

e. POLE: $f(x)$ is defined at $x = c$; $\lim\limits_{x \to c^-} f(x) = +\infty$ and $\lim\limits_{x \to c^+} f(x) = +\infty$

Figure 2.18 Types of discontinuity: holes, poles, and jumps

DEFINITION OF CONTINUITY

Let us consider the conditions that must be satisfied for a function f to be continuous at a point c. First, $f(c)$ must be defined or we have a "hole" in the graph, as shown in Figure 2.18a. (An open dot indicates an excluded point.) In Figure 2.18b, there is a "hole" even though $f(c)$ is defined. If $\lim\limits_{x \to c} f(x)$ has one value as $x \to c^-$ and another as $x \to c^+$, then $\lim\limits_{x \to c} f(x)$ does not exist and there will be a "jump" in the graph of f, as shown in Figure 2.18c. Finally, if one or both of the one-sided limits at c are infinite $(+\infty$ or $-\infty)$, there will be a "pole" at $x = c$, as shown in Figures 2.18d and e.

Continuity of a Function at a Point

A function f is **continuous at a point $x = c$** if the following three conditions are satisfied:

1. $f(c)$ is defined;
2. $\lim\limits_{x \to c} f(x)$ exists;
3. $\lim\limits_{x \to c} f(x) = f(c)$.

A function that is not continuous at c is said to have a **discontinuity** at that point.

→ **What This Says**   The third condition $\lim\limits_{x \to c} f(x) = f(c)$ summarizes the idea behind continuity. It says that if x is close to c, then $f(x)$ must be close to $f(c)$.

If f is continuous at $x = c$, the difference between $f(x)$ and $f(c)$ is small whenever x is close to c because $\lim\limits_{x \to c} f(x) = f(c)$. Geometrically, this means that the points $(x, f(x))$ on the graph of f converge to the point $(c, f(c))$ as $x \to c$, and this is what guarantees that the graph is unbroken at $(c, f(c))$ with no "gap" or "hole," as shown in Figure 2.19.

We should note that not every continuous function can be graphed (for example, see Example 2e later in this section), so you cannot test continuity by graphing, even though graphing can help for many functions.

If f is continuous at c, the points $(x, f(x))$ converge to $(c, f(c))$ as $x \to c$.

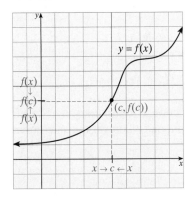

Figure 2.19 The geometric interpretation of continuity

WARNING   Do not rely on graphing calculators to test for continuity. For example, the graph of $y = \dfrac{x - 2}{x - 2}$ looks like a constant function when done with a calculator; it does not show the discontinuity at $x = 2$.

EXAMPLE 1   Testing continuity

Test the continuity of each of the following functions at $x = 1$. If the given function is not continuous at $x = 1$, explain.

a. $f(x) = \dfrac{x^2 + 2x - 3}{x - 1}$

b. $g(x) = \dfrac{x^2 + 2x - 3}{x - 1}$ if $x \neq 1$ and $g(x) = 6$ if $x = 1$

c. $h(x) = \dfrac{x^2 + 2x - 3}{x - 1}$ if $x \neq 1$ and $h(x) = 4$ if $x = 1$

d. $F(x) = \dfrac{x + 3}{x - 1}$ if $x \neq 1$ and $F(x) = 4$ if $x = 1$

e. $G(x) = 7x^3 + 3x^2 - 2$

f. $H(x) = 2 \sin x - \tan x$

Solution

a. The function f is not continuous at $x = 1$ [hole; $f(1)$ not defined] because it is not defined at this point.

b. 1. $g(1)$ is defined; $g(1) = 6$.

2. $\lim\limits_{x \to 1} g(x) = \lim\limits_{x \to 1} \dfrac{x^2 + 2x - 3}{x - 1}$

$= \lim\limits_{x \to 1} \dfrac{(x - 1)(x + 3)}{x - 1}$

$= \lim\limits_{x \to 1} (x + 3) = 4$

3. $\lim\limits_{x \to 1} g(x) \neq g(1)$, so g is not continuous at $x = 1$ [hole; $g(1)$ defined].

c. Compare h with g of part **b.** We see that all three conditions of continuity are satisfied, so h is continuous at $x = 1$.

d. 1. $F(1)$ is defined; $F(1) = 4$.

2. $\displaystyle\lim_{x \to 1} F(x) = \lim_{x \to 1} \frac{x + 3}{x - 1}$; the limit does not exist.

The function F is not continuous at $x = 1$ (pole).

e. 1. $G(1)$ is defined; $G(1) = 8$.

2. $\displaystyle\lim_{x \to 1} G(x) = 7 \left(\lim_{x \to 1} x \right)^3 + 3 \left(\lim_{x \to 1} x \right)^2 - \lim_{x \to 1} 2 = 8$.

3. $\displaystyle\lim_{x \to 1} G(x) = G(1)$.

Because the three conditions of continuity are satisfied, G is continuous at $x = 1$.

f. 1. $H(1)$ is defined; $H(1) = 2 \sin 1 - \tan 1$.

2. $\displaystyle\lim_{x \to 1} H(x) = 2 \lim_{x \to 1} \sin x - \lim_{x \to 1} \tan x = 2 \sin 1 - \tan 1$.

3. $\displaystyle\lim_{x \to 1} H(x) = H(1)$.

Because the three conditions of continuity are satisfied, H is continuous at $x = 1$.

CONTINUITY THEOREMS

It is often difficult to determine whether a given function is continuous at a specified number. However, many common functions are continuous wherever they are defined.

THEOREM 2.4   Continuity theorem

If f is a polynomial or a rational function, a power function, a trigonometric function, or an inverse trigonometric function, then f is continuous at any number $x = c$ for which $f(c)$ is defined.

Proof   The proof of the continuity theorem is based on the limit properties stated in the previous section. For instance, a polynomial is a function of the form

$$P(x) = a_n x^n + a_{n-1} x^{n-1} + \cdots + a_1 x + a_0$$

where $a_0, a_1, \ldots, a_n$ are constants. We know that $\displaystyle\lim_{x \to c} a_0 = a_0$ and that $\displaystyle\lim_{x \to c} x^m = c^m$ for $m = 1, 2, \ldots, n$. This is precisely the statement that the function $g(x) = ax^m$ is continuous at any number $x = c$, or simply continuous. Because P is a sum of functions of this form and a constant function, it follows from the limit properties that P is continuous.

The proofs of the other parts follow similarly.   ❑

The limit properties of the previous section can also be used to prove a second continuity theorem. This theorem tells us that continuous functions may be combined in various ways *without creating a discontinuity*.

THEOREM 2.5 **Properties of continuous functions**

If f and g are functions that are continuous at $x = c$, then the following functions are also continuous at $x = c$.

Scalar multiple	sf	for any constant s (called a *scalar*)
Sum and difference	$f + g$ and $f - g$	
Product	fg	
Quotient	$\dfrac{f}{g}$	provided $g(c) \neq 0$
Composition	$f \circ g$	provided g is continuous at c and f is continuous at $g(c)$

Proof The first four properties in this theorem follow directly from the basic limit rules given in Section 2.2. For instance, to prove the product property, note that since f and g are given to be continuous at $x = c$, we have

$$\lim_{x \to c} f(x) = f(c) \quad \text{and} \quad \lim_{x \to c} g(x) = g(c)$$

If $P(x) = (fg)(x) = f(x)g(x)$, then

$$\lim_{x \to c} P(x) = \lim_{x \to c} f(x)g(x)$$
$$= \left[\lim_{x \to c} f(x)\right]\left[\lim_{x \to c} g(x)\right]$$
$$= f(c)g(c) = P(c)$$

so $P(x)$ is continuous at $x = c$, as required. ❏

The continuous composition property is proved in a similar fashion, but requires the following limit rule, which we state without proof.

Composition Limit Rule

If $\lim_{x \to c} g(x) = L$ and f is a function continuous at L, then $\lim_{x \to c} f[g(x)] = f(L)$. That is,

$$\lim_{x \to c} f[g(x)] = f(L) = f\left(\lim_{x \to c} g(x)\right)$$

This property applies in the same way to other kinds of limits, in particular to one-sided limits.

Now we can prove the continuous composition property of Theorem 2.5. Let $h(x) = (f \circ g)(x)$. Then we have

$$\lim_{x \to c}(f \circ g)(x) = \lim_{x \to c} f[g(x)] \qquad \text{Definition of composition}$$
$$= f\left[\lim_{x \to c} g(x)\right] \qquad \text{Composition limit rule}$$
$$= f[g(c)] \qquad g \text{ is continuous at } x = c.$$
$$= (f \circ g)(c) \qquad \text{Definition of composition} \quad ❏$$

➡ **What This Says** The limit of a continuous function is the function of the limiting value. The continuous composition property says that a continuous function of a continuous function is continuous.

We need to talk about a function being continuous on an interval. To do so, we must first know how to handle continuity at the endpoints of the interval, which leads to the following definition.

One-Sided Continuity

> The function f is **continuous from the right at** a if and only if
>
> $$\lim_{x \to a^+} f(x) = f(a)$$
>
> and it is **continuous from the left at** b if and only if
>
> $$\lim_{x \to b^-} f(x) = f(b)$$

CONTINUITY ON AN INTERVAL

The function f is said to be **continuous on the open interval** (a, b) if it is continuous at each number in this interval. Also note that the endpoints are not part of open intervals. If f is also continuous from the right at a, we say it is **continuous on the half-open interval** $[a, b)$. Similarly, f is **continuous on the half-open interval** $(a, b]$ if it is continuous at each number between a and b and is continuous from the left at the endpoint b. Finally, f is **continuous on the closed interval** $[a, b]$ if it is continuous at each number between a and b and is both continuous from the right at a and continuous from the left at b.

EXAMPLE 2   Testing for continuity on an interval

Find the intervals on which each of the given functions is continuous.

a. $f_1(x) = \dfrac{x^2 - 1}{x^2 - 4}$ **b.** $f_2(x) = |x^2 - 4|$ **c.** $f_3(x) = \csc x$

d. $f_4(x) = \sin \dfrac{1}{x}$ **e.** $f_5(x) = \begin{cases} x \sin \frac{1}{x} & \text{if } x \neq 0 \\ 0 & \text{if } x = 0 \end{cases}$

Solution

a. Function f_1 is not defined when $x^2 - 4 = 0$; that is, when $x = 2$ or $x = -2$. The function is continuous on $(-\infty, -2) \cup (-2, 2) \cup (2, \infty)$.

b. Function f_2 is continuous on $(-\infty, \infty)$.

c. The cosecant function is not defined at $x = n\pi$, n an integer. At all other points it is continuous.

d. Because $1/x$ is continuous except at $x = 0$ and the sine function is continuous everywhere, we need only check continuity at $x = 0$

$$\lim_{x \to 0} \sin \frac{1}{x} \quad \text{does not exist}$$

Therefore, $f(x) = \sin \frac{1}{x}$ is continuous on $(-\infty, 0) \cup (0, \infty)$.

e. It can be shown that since

$$-1 \le \ \sin \frac{1}{x} \le 1$$

then

$$-|x| \le x \sin \frac{1}{x} \le |x|, \quad x \ne 0$$

We can now use the squeeze rule. Because $\lim_{x \to 0} |x| = 0$ and $\lim_{x \to 0} (-|x|) = 0$, it follows that $\lim_{x \to 0} x \sin \frac{1}{x} = 0$. Because $f_5(0) = 0$ we see that f is continuous at $x = 0$ and therefore is continuous on $(-\infty, \infty)$. ∎

Usually there are only a few points in the domain of a given function f where a discontinuity can occur. We use the term **suspicious point** for a number c where either

1. The defining rule for f changes.

2. Substitution of $x = c$ causes division by 0 in the function.

For Example 2, the suspicious points can be listed:

a. $\dfrac{x^2 - 1}{x^2 - 4}$ has suspicious points for division by zero when $x = 2$ and $x = -2$.

b. $|x^2 - 4| = x^2 - 4$ when $x^2 - 4 \ge 0$ and $|x^2 - 4| = 4 - x^2$ when $x^2 - 4 < 0$. This means the definition of the function changes when $x^2 - 4 = 0$, namely, when $x = 2$ and $x = -2$.

c. There are no suspicious points; we know the function cannot be continuous at places where the function is not defined.

d. $\sin \frac{1}{x}$ has a suspicious point when $x = 0$ (division by 0).

e. $x \sin \frac{1}{x}$ has a suspicious point when $x = 0$ (division by 0).

EXAMPLE 3 Checking continuity at suspicious points

Let $f(x) = \begin{cases} 3 - x & \text{if } -5 \le x < 2 \\ x - 2 & \text{if } 2 \le x < 5 \end{cases}$ and $g(x) = \begin{cases} 2 - x & \text{if } -5 \le x < 2 \\ x - 2 & \text{if } 2 \le x < 5 \end{cases}$.

Find the intervals on which f and g are continuous.

Solution

We find that the domain for both functions is $[-5, 5)$; the continuity theorem tells us both functions are continuous everywhere on that interval except possibly at the suspicious points. Consider the graphs of f and g as shown in Figure 2.20. Examining f, we see

$$\lim_{x \to 2^-} f(x) = \lim_{x \to 2^-} (3 - x) = 1 \quad \text{and} \quad \lim_{x \to 2^+} f(x) = \lim_{x \to 2^+} (x - 2) = 0$$

so $\lim_{x \to 2} f(x)$ does not exist and f is discontinuous at $x = 2$. Thus, f is continuous for $-5 \le x < 2$ and for $2 < x < 5$.

For g, the domain is also $-5 \le x < 5$, and again, the only suspicious point is $x = 2$. We have $g(2) = 0$ and

$$\lim_{x \to 2^-} g(x) = \lim_{x \to 2^-} (2 - x) = 0 \quad \text{and} \quad \lim_{x \to 2^+} g(x) = \lim_{x \to 2^+} (x - 2) = 0$$

Therefore, $\lim_{x \to 2} g(x) = 0 = g(2)$, and g is continuous at $x = 2$. Hence, g is continuous throughout the interval $[-5, 5)$. ∎

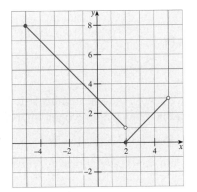

Although a graph is not part of the derivation, it can often be helpful in finding suspicious points. This is the graph of the function f.

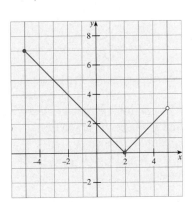

This is the graph of g. Does the graph reinforce our conclusions?

Figure 2.20 Graphs of f and g

THE INTERMEDIATE VALUE THEOREM

Intuitively, if f is continuous throughout an entire interval, its graph on that interval may be drawn "without the pencil leaving the paper." That is, if $f(x)$ varies continuously from $f(a)$ to $f(b)$ as x increases from a to b, then it must hit every number L between $f(a)$ and $f(b)$, as shown in Figure 2.21.

Figure 2.21   If L lies between $f(a)$ and $f(b)$, then $f(c) = L$ for some c between a and b.

To illustrate the property shown in Figure 2.21, suppose f is a function defined as the weight of a person at age x. If we assume that weight varies continuously with time, a person who weighs 50 pounds at age 6 and 120 pounds at age 15 must weigh 100 pounds at some time between ages 6 and 15.

This feature of continuous functions is known as the *intermediate value property.* A formal statement of this property is contained in the following theorem.

THEOREM 2.6   The intermediate value theorem

If f is a continuous function on the closed interval $[a, b]$ and L is some number strictly between $f(a)$ and $f(b)$, then there exists at least one number c on the open interval (a, b) such that $f(c) = L$.

Proof   This theorem is intuitively obvious, but it is not at all easy to prove. A proof may be found in most advanced calculus textbooks.   ❑

> → **What This Says**   If f is a continuous function (with emphasis on the word *continuous*) on some *closed* interval $[a, b]$, then $f(x)$ must take on all values between $f(a)$ and $f(b)$.

Figure 2.22 Because $f(a) > 0$ and $f(b) < 0$, then $f(c) = 0$ for some c between a and b.

The intermediate value theorem can be used to estimate roots of the equation $f(x) = 0$. Suppose $f(a) > 0$ and $f(b) < 0$, so the graph of f is above the x-axis at $x = a$ and below for $x = b$. Then, if f is continuous on the closed interval $[a, b]$, there must be a point $x = c$ between a and b where the graph crosses the x-axis, that is, where $f(c) = 0$. The same conclusion would be drawn if $f(a) < 0$ and $f(b) > 0$. The key is that $f(x)$ changes sign between $x = a$ and $x = b$. This is shown in Figure 2.22 and is summarized in the following theorem.

THEOREM 2.7   Root location theorem

If f is continuous on the closed interval $[a, b]$ and if $f(a)$ and $f(b)$ have opposite algebraic signs (one positive and the other negative), then $f(c) = 0$ for at least one number c on the open interval (a, b).

Proof This follows directly from the intermediate value theorem (see Figure 2.22). The details of this proof are left as a problem. ❑

EXAMPLE 4 Using the root location theorem

Show that $\cos x = x^3 - x$ has at least one solution on the interval $[\frac{\pi}{4}, \frac{\pi}{2}]$.

Solution

We can use a graphing utility to estimate the point of intersection, as shown in Figure 2.23.

We will now use the root location theorem to confirm the intersection point shown in Figure 2.23. By calculator, we estimate $c \approx 1.16$.

Notice that the function $f(x) = \cos x - x^3 + x$ is continuous on $[\frac{\pi}{4}, \frac{\pi}{2}]$. We find that

$$f(\tfrac{\pi}{4}) = \cos \tfrac{\pi}{4} - \left(\tfrac{\pi}{4}\right)^3 + \tfrac{\pi}{4} \approx 1.008$$

$$f(\tfrac{\pi}{2}) = \cos \tfrac{\pi}{2} - \left(\tfrac{\pi}{2}\right)^3 + \tfrac{\pi}{2} \approx -2.305$$

Therefore, by the root location theorem there is at least one number c on $(\frac{\pi}{4}, \frac{\pi}{2})$ for which $f(c) = 0$, and it follows that $\cos c = c^3 - c$. ■

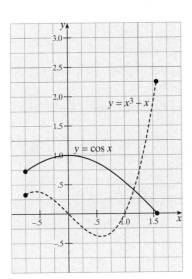

Figure 2.23 Estimate the point of intersection to be $x \approx 1.16$.

Since the definition of continuity involves the limit, formal proofs of continuity theorems can be given in terms of the ϵ-δ definition of limit. Here is an example.

EXAMPLE 5 Formal proof of the sum rule for continuity

Show that if f and g are continuous at $x = x_0$, then $f + g$ is also continuous at $x = x_0$.

Solution

The continuity of f and g at $x = x_0$ says that

$$\lim_{x \to x_0} f(x) = f(x_0) \quad \text{and} \quad \lim_{x \to x_0} g(x) = g(x_0)$$

This means that if $\epsilon > 0$ is given, there exist numbers δ_1 and δ_2 such that

$$|f(x) - f(x_0)| < \frac{\epsilon}{2} \quad \text{whenever} \quad 0 < |x - x_0| < \delta_1$$

and

$$|g(x) - g(x_0)| < \frac{\epsilon}{2} \quad \text{whenever} \quad 0 < |x - x_0| < \delta_2$$

Let δ be the smaller of δ_1 and δ_2; that is, $\delta = \min(\delta_1, \delta_2)$. Then

$$0 < |x - x_0| < \delta_1 \quad \text{and} \quad 0 < |x - x_0| < \delta_2$$

must both be true whenever $0 < |x - x_0| < \delta$ so that (by the triangle inequality)

$$
\begin{aligned}
|[f(x) + g(x)] - [f(x_0) + g(x_0)]| &= |[f(x) - f(x_0)] + [g(x) - g(x_0)]| \\
&\leq |f(x) - f(x_0)| + |g(x) - g(x_0)| \\
&< \frac{\epsilon}{2} + \frac{\epsilon}{2} = \epsilon
\end{aligned}
$$

Thus, $|(f + g)(x) - (f + g)(x_0)| < \epsilon$ whenever $0 < |x - x_0| < \delta$, and it follows from the definition of limit that

$$\lim_{x \to x_0} [f(x) + g(x)] = f(x_0) + g(x_0)$$

In other words, $f + g$ is continuous at $x = x_0$, as claimed. ■

2.3 PROBLEM SET

Ⓐ *Which of the functions described in Problems 1–4 represent continuous functions? State the domain, if possible, for each example.*

1. the temperature on a specific day at a given location considered as a function of time

2. the humidity on a specific day at a given location considered as a function of time

3. the charges for a taxi ride across town considered as a function of mileage

4. the charges to mail a package as a function of its weight

Identify all suspicious points and determine all points of discontinuity in Problems 5–14.

5. $f(x) = x^3 - 7x + 3$

6. $f(x) = \dfrac{3x + 5}{2x - 1}$

7. $f(x) = \dfrac{3x}{x^2 - x}$

8. $f(t) = 3 - (5 + 2t)^3$

9. $h(x) = \sqrt{x} + \dfrac{3}{x}$

10. $f(u) = \sqrt[3]{u^2 - 1}$

11. $f(x) = \begin{cases} x^2 - 2 & \text{if } x > 1 \\ 2x - 3 & \text{if } x \leq 1 \end{cases}$

12. $g(t) = \begin{cases} 3t + 2 & \text{if } t \leq 1 \\ 5 & \text{if } 1 < t \leq 3 \\ 3t^2 - 1 & \text{if } t > 3 \end{cases}$

13. $f(x) = 3 \tan x - 5 \sin x \cos x$

14. $g(x) = \dfrac{\cot x}{\sin x - \cos x}$

In Problems 15–20, the given function is defined for all $x > 0$, except at $x = 2$. In each case, find the value that should be assigned to $f(2)$, if any, to guarantee that f will be continuous at 2.

15. $f(x) = \dfrac{x^2 - x - 2}{x - 2}$

16. $f(x) = \sqrt{\dfrac{x^2 - 4}{x - 2}}$

17. $f(x) = \dfrac{\sin(\pi x)}{x - 2}$

18. $f(x) = \dfrac{\cos \frac{\pi}{x}}{x - 2}$

19. $f(x) = \begin{cases} 2x + 5 & \text{if } x > 2 \\ 15 - x^2 & \text{if } x < 2 \end{cases}$

20. $f(x) = \dfrac{\dfrac{1}{x} - 1}{x - 2}$

In Problems 21–25, determine whether or not the given function is continuous on the prescribed interval.

21. **a.** $f(x) = \dfrac{1}{x}$ on $[1, 2]$ **b.** $f(x) = \dfrac{1}{x}$ on $[0, 1]$

22. $f(x) = \begin{cases} x^2 & \text{if } 0 \leq x < 2 \\ 3x + 1 & \text{if } 2 \leq x < 5 \end{cases}$

23. $g(t) = \begin{cases} 15 - t^2 & \text{if } -3 < t \leq 0 \\ 2t & \text{if } 0 < t \leq 1 \end{cases}$

24. $f(x) = x \sin x$ on $(0, \pi)$

25. $f(x) = \dfrac{\cos x - 1}{x}$ on $[-\pi, \pi]$

Ⓑ 26. Let $f(x) = \begin{cases} x & \text{if } x \neq 0 \\ 2 & \text{if } x = 0 \end{cases}$ and $g(x) = \begin{cases} 3x & \text{if } x \neq 0 \\ -2 & \text{if } x = 0 \end{cases}$

Show that $f + g$ is continuous at $x = 0$ even though f and g are both discontinuous there.

In Problems 27–32, show that the given equation has at least one solution on the indicated interval.

27. $\sqrt[3]{x} = x^2 + 2x - 1$ on $(0, 1)$

28. $\dfrac{1}{x + 1} = x^2 - x - 1$ on $(1, 2)$

29. $\sqrt[3]{x - 8} + 9x^{2/3} = 29$ on $(0, 8)$

30. $\tan x = 2x^2 - 1$ on $(-\frac{\pi}{4}, 0)$

31. $\cos x - \sin x = x$ on $(0, \frac{\pi}{2})$

32. $\cos x = x^2 - 1$ on $(0, \pi)$

33. Let $f(x) = \begin{cases} x^2 & \text{if } x > 2 \\ x + 1 & \text{if } x \leq 2 \end{cases}$

Show that f is continuous from the left at 2, but not from the right.

34. Find constants a and b such that $f(2) + 3 = f(0)$ and f is continuous at $x = 1$.

$$f(x) = \begin{cases} ax + b & \text{if } x > 1 \\ 3 & \text{if } x = 1 \\ x^2 - 4x + b + 3 & \text{if } x < 1 \end{cases}$$

35. **Interpretation Problem** Use the intermediate value theorem to explain why the hands of a clock coincide at least once every hour.

36. **Interpretation Problem** The graph shown in Figure 2.24 models how the growth rate of a bacterial colony changes with temperature.*

Figure 2.24 Growth rate of a bacterial colony

---

*Michael D. La Grega, Philip L. Buckingham, and Jeffery C. Evans, *Hazardous Waste Management*, New York: McGraw-Hill, 1994, pp. 565–566.

What happens when the temperature reaches 45°C? Does it make sense to compute $\lim_{t\to 50} R(t)$? Write a paragraph describing how temperature affects the growth rate of a species.

37. A fish swims upstream at a constant speed v relative to the water, which in turn flows at a constant speed v_w $(v_w < v)$ relative to the ground. The energy expended by the fish in traveling to a point upstream is given by

$$E(v) = \frac{Cv^k}{v - v_w}$$

where $C > 0$ is a physical constant and $k > 0$ is a number that depends on the type of fish.*

a. Compute $\lim_{v\to v_w} E(v)$. Interpret your result in words.

b. What happens to $E(v)$ as $v \to +\infty$? Interpret your result in words.

38. The population (in thousands) of a colony of bacteria t minutes after the introduction of a toxin is given by the function

$$P(t) = \begin{cases} t^2 + 1 & \text{if } 0 \le t < 5 \\ -8t + 66 & \text{if } t \ge 5 \end{cases}$$

a. When does the colony die out?

b. Show that at some time between $t = 2$ and $t = 7$, the population is 9,000.

In Problems 39–44, find constants a and b so that the given function will be continuous for all x.

39. $f(x) = \begin{cases} ax + 3 & \text{if } x > 5 \\ 8 & \text{if } x = 5 \\ x^2 + bx + 1 & \text{if } x < 5 \end{cases}$

40. $f(x) = \begin{cases} \dfrac{ax - 4}{x - 2} & \text{if } x \ne 2 \\ b & \text{if } x = 2 \end{cases}$

41. $f(x) = \begin{cases} \dfrac{\sqrt{x} - a}{x - 1} & \text{if } x > 1 \\ b & \text{if } x \le 1 \end{cases}$

42. $f(x) = \begin{cases} 2\sin(a\cos^{-1}x) & \text{if } x > 0 \\ \sqrt{3} & \text{if } x = 0 \\ ax + b & \text{if } x < 0 \end{cases}$

43. $g(x) = \begin{cases} \dfrac{\sin ax}{x} & \text{if } x < 0 \\ 5 & \text{if } x = 0 \\ x + b & \text{if } x > 0 \end{cases}$

44. $f(x) = \begin{cases} \dfrac{\tan ax}{\tan bx} & \text{if } x < 0 \\ 4 & \text{if } x = 0 \\ ax + b & \text{if } x > 0 \end{cases}$

45. HISTORICAL QUEST The first modern formulation of the notion of continuity appeared in a pamphlet published by Bernard Bolzano, a Czechoslovakian priest whose mathematical work was, for the most part, overlooked by his contemporaries. In explaining the concept of continuity, Bolzano said that one must understand the phrase, "A function $f(x)$ (that) varies according to the law of continuity for all values of x which lie inside or outside certain limits, is nothing other than this: If x is any such value, the difference

BERNARD BOLZANO
1781–1848

$$f(x + \omega) - f(x)$$

can be made smaller than any given quantity, if one makes ω as small as one wishes."

In his book, *Cours d'Analyse*, Cauchy (see Historical Quest Problem 49, Section 2.1) introduces the concept of continuity for a function defined on an interval in essentially the same way as Bolzano. In this book Cauchy points out that the continuity of many functions is easily verified. As an example, he argues that $\sin x$ is continuous on every interval because "the numerical value of $\sin\frac{1}{2}\alpha$, and consequently that of the difference

$$\sin(x + \alpha) - \sin x = 2\sin\tfrac{1}{2}\alpha\cos(x + \tfrac{1}{2}\alpha)$$

decreases indefinitely with that of α."[†]

Show that $\sin x$ is continuous, and also show that the given expression decreases as claimed by Cauchy.

We conclude this Historical Quest by noting that the formal ϵ-δ definition of limits and continuity that we use today was first done by Karl Weierstrass (1815–1897), a German secondary school teacher who did his research at night. The historian David Burton describes Weierstrass as "the world's greatest analyst during the last third of the nineteenth century—the father of modern analysis."

46. Let f be a continuous function and suppose that $f(a)$ and $f(b)$ have opposite signs. The root location theorem tells us that at least one root of $f(x) = 0$ lies between $x = a$ and $x = b$.

a. Let $c = (a + b)/2$ be the midpoint of the interval $[a, b]$. Explain how the root location theorem can be used to determine whether the root lies in $[a, c]$ or in $[c, b]$. Does anything special have to be said about the case where the interval $[a, b]$ contains more than one root?

b. Based on your observation in part **a**, describe a procedure for approximating a root of $f(x) = 0$ more and more accurately. This is called the *bisection method* for root location.

c. Apply the bisection method to locate at least one root of $x^3 + x - 1 = 0$ on $[0, 1]$. Check your answer using your calculator.

47. Apply the bisection method (see Problem 46) to locate at least one root of each of the given equations, and then check your answer using your calculator.

a. $\sin x + \cos x = x$ on $[0, \frac{\pi}{2}]$

b. $\cos(x + 1) = \sin(x - 1)$ on $[0, 1]$

---

*E. Batschelet, *Introduction to Mathematics for Life Scientists*, 2nd ed., New York: Springer-Verlag, 1976, p. 280.

[†]A. L. Cauchy, *Cours d'Analyse de l'Ecole Royale Polytechnique*, *Oeuvres*, Ser. 2, Vol. 3. Paris: Gauthier-Villars, 1897, p. 44.

C 48. **Counterexample Problem**   Find functions f and g such that f is discontinuous at $x = 1$ but fg is continuous there.

49. **Counterexample Problem**   Give an example of a function defined for all real numbers that is continuous at only one point.

50. Prove the root location theorem, assuming the intermediate value theorem.

51. Show that f is continuous at c if and only if it is both contin-uous from the right and continuous from the left at c.

52. Show that if f and g are both continuous at $x = c$ and $g(c) \neq 0$, then f/g must also be continuous there.

53. Let $u(x) = x$ and $f(x) = \begin{cases} 0 & \text{if } x \neq 0 \\ 1 & \text{if } x = 0 \end{cases}$.

Show that $\lim\limits_{x \to 0} f[u(x)] \neq f\left[\lim\limits_{x \to 0} u(x)\right]$.

2.4   Exponential and Logarithmic Functions

IN THIS SECTION   exponential functions, logarithmic functions, natural exponential and logarithmic functions, continuous compounding of interest

In Chapter 1, we introduced polynomial and rational functions, power functions, trigonometric functions, and inverse trigonometric functions. In this section, we will complete our list of the elementary functions typically used in calculus by using limits to introduce exponential functions and their inverses, the logarithmic functions.

EXPONENTIAL FUNCTIONS

Recall that if n is a natural number, then

$$b^n = \underbrace{b \cdot b \cdot b \cdot \,\cdots\, \cdot b}_{n \text{ factors}}$$

If $b \neq 0$, then $b^0 = 1$, $b^{-n} = \dfrac{1}{b^n}$, and furthermore, if $b > 0$, then $b^{1/n} = \sqrt[n]{b}$. Also, if m and n are any integers, and m/n is a reduced fraction, then

$$b^{m/n} = (b^{1/n})^m$$

Figure 2.25 Graph of $y = 2^x$ for rational exponents

This definition tells us what b^x means for rational values of x. However, we now wish to enlarge the domain of x to include all real numbers. To get a feeling for what is involved in this problem, let us examine the special case where $b = 2$. In Figure 2.25, we have plotted several points with coordinates $(r, 2^r)$, where r is a rational number. We must now attach meaning to b^x if x is not a rational number by filling those holes or spaces to make the graph continuous for all x. We use the following fundamental fact about the real number system, called the **completeness property**.

For any real number x, there exist rational numbers r_n such that

$$x = \lim_{n \to \infty} r_n$$

which means that for any number $\epsilon > 0$, there exists a number N such that $|x - r_n| < \epsilon$ whenever $n > N$.

Another way of expressing the completeness property is to say that *each real number x can be approximated to any degree of accuracy by a rational number.* Using the completeness property, we can now define an exponential function.

Exponential Function

> Let x be a real number, and let r_n be a sequence of rational numbers such that $x = \lim\limits_{n \to \infty} r_n$. Then the **exponential function** with base $b > 0$ $(b \neq 1)$ is given by
>
> $$b^x = \lim_{n \to \infty} b^{r_n}$$

For example, the number $\sqrt{3}$ has an infinite decimal representation

$$\sqrt{3} \approx 1.732050808 \cdots$$

which means that $\sqrt{3}$ can be approximated to any desired degree of accuracy by members of the sequence of rational numbers

$$r_1 = 1, \ r_2 = 1.7, \ r_3 = 1.73, \ r_4 = 1.732, \ r_5 = 1.7320, \ r_6 = 1.73205, \ \ldots$$

Therefore, the exponential number $2^{\sqrt{3}}$ is given by the limit of the sequence of numbers

$$2^1, \ 2^{1.7}, \ 2^{1.73}, \ 2^{1.732}, \ 2^{1.7320}, \ 2^{1.73205}, \ \ldots$$

In this fashion, we can fill the holes and gaps shown in Figure 2.25 to obtain the continuous graph of $y = 2^x$, as shown in Figure 2.26.

The shape of the graph of $y = b^x$ for any $b > 1$ is essentially the same as that of $y = 2^x$. The graph of $y = b^x$ for a typical base $b > 1$ is shown in Figure 2.27a and is rising (when viewed from left to right). For a base b where $0 < b < 1$ the graph is falling, as shown in Figure 2.27b.

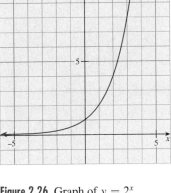

Figure 2.26 Graph of $y = 2^x$

a. $b > 1$

b. $0 < b < 1$

Figure 2.27   Graph of $y = b^x$

We summarize the basic properties of exponential functions with the following theorem. Many of these properties for bases with rational exponents should be familiar from previous courses.

THEOREM 2.8   Properties of exponential functions

Let x, y be real numbers, a and b positive real numbers.

Equality rule	If $b \neq 1$, then $b^x = b^y$ if and only if $x = y$.
Inequality rules	If $x > y$ and $b > 1$, then $b^x > b^y$.
	If $x > y$ and $0 < b < 1$, then $b^x < b^y$.
Product rule	$b^x b^y = b^{x+y}$
Quotient rule	$\dfrac{b^x}{b^y} = b^{x-y}$
Power rules	$(b^x)^y = b^{xy}$; $\ (ab)^x = a^x b^x$; $\ \left(\dfrac{a}{b}\right)^x = \dfrac{a^x}{b^x}$

> **Graphical Properties**
>
> The function $y = b^x$ is continuous for all x. Furthermore, its graph is always above the x-axis ($b^x > 0$), and is always rising for $b > 1$ and always falling for $0 < b < 1$.

Several parts of this theorem are used in the following example.

EXAMPLE 1 Exponential equations

Solve each of the following exponential equations.

a. $2^{x^2+3} = 16$ **b.** $2^x 3^{x+1} = 108$ **c.** $(\sqrt{2})^{x^2} = \dfrac{8^x}{4}$

Solution

a. $2^{x^2+3} = 16$

$\quad 2^{x^2+3} = 2^4$ *Write 16 as 2^4 so that the equality rule can be used.*

$\quad x^2 + 3 = 4, \quad$ or $\quad x = \pm 1$

b. $2^x 3^{x+1} = 108$

$\quad 2^x 3^x 3 = 3 \cdot 36$ *Product rule*

$\quad (2 \cdot 3)^x = 36$ *Divide both sides by 3 and then use the power rule.*

$\quad 6^x = 6^2$

$\quad x = 2$ *Equality rule*

c. $(\sqrt{2})^{x^2} = \dfrac{8^x}{4}$

$\quad (2^{1/2})^{x^2} = \dfrac{(2^3)^x}{2^2}$

$\quad 2^{x^2/2} = 2^{3x-2}$

$\quad \dfrac{x^2}{2} = 3x - 2$ *Equality rule*

$\quad x^2 - 6x + 4 = 0$

$\quad x = \dfrac{6 \pm \sqrt{36 - 4(1)(4)}}{2} = 3 \pm \sqrt{5}$ ■

LOGARITHMIC FUNCTIONS

Since the exponential function $y = b^x$ for $b > 0$, $b \neq 1$ is monotonic (its graph is either always rising for all x or always falling), it must have an inverse that is itself monotonic. This inverse function is called the **logarithm of x to the base b**. Here is a definition, with some notation.

Logarithmic Function

> If $b > 0$ and $b \neq 1$, the **logarithm of x to the base b** is the function $y = \log_b x$ that satisfies $b^y = x$; that is,
>
> $$y = \log_b x \quad \text{means} \quad b^y = x$$

> ➡ **What This Says** It is useful to think of a logarithm as an exponent. That is, consider the following sequence of interpretations:
>
> $\quad y = \log_b x$
>
> $\quad y$ is the logarithm to the base b of x.
>
> $\quad y$ is the **exponent to the base b** that gives x.
>
> $\quad b^y = x$

Notice that $y = \log_b x$ is defined only for $x > 0$ because $b^y > 0$ for all y. We have sketched the graph of $y = \log_b x$ for $b > 1$ in Figure 2.28 by reflecting the graph of $y = b^x$ in the line $y = x$.

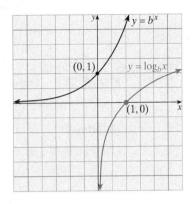

Figure 2.28   The graph of $y = \log_b x$ for $b > 1$

Because $y = b^x$ is a continuous, increasing function that satisfies $b^x > 0$ for all x, $\log_b x$ must also be continuous and increasing, and its graph must lie entirely to the right of the y-axis.

Also, because $b^0 = 1$ and $b^1 = b$, we have

$$\log_b 1 = 0 \quad \text{and} \quad \log_b b = 1$$

so $(1, 0)$ and $(b, 1)$ lie on the logarithmic curve.

Theorem 2.9 lists several general properties of logarithms.

WARNING   Remember that $\log_b x$ is defined only for $x > 0$. Any undefined values for $\log_b x$ are excluded.

THEOREM 2.9   Basic properties of logarithmic functions

Assume $b > 0$ and $b \neq 1$.

Equality rule	$\log_b x = \log_b y$ if and only if $x = y$.
Inequality rules	If $x > y$ and $b > 1$, then $\log_b x > \log_b y$. If $x > y$ and $0 < b < 1$, then $\log_b x < \log_b y$.
Product rule	$\log_b(xy) = \log_b x + \log_b y$
Quotient rule	$\log_b\left(\dfrac{x}{y}\right) = \log_b x - \log_b y$
Power rule	$\log_b x^p = p \log_b x$ for any real number p
Inversion rules	$b^{\log_b x} = x$ and $\log_b b^x = x$
Special values	$\log_b b = 1$ and $\log_b 1 = 0$

Proof   Each part of this theorem can be derived by using the definition of the logarithm in conjunction with a suitable property of exponentials. (See Problems 73–75.)   ❑

EXAMPLE 2   Evaluating a logarithmic expression

Evaluate $\log_2\left(\frac{1}{8}\right) + \log_2 128$.

Solution

Using Theorem 2.9, we find that

$$\log_2\left(\tfrac{1}{8}\right) + \log_2 128 = \log_2(2)^{-3} + \log_2 2^7$$
$$= \log_2[2^{-3}(2^7)] = \log_2 2^4 = 4\log_2 2 = 4(1) = 4$$

What we have shown here is longer than what you would do in your work, which would probably be shortened as follows:

$$\log_2\left(\tfrac{1}{8}\right) + \log_2 128 = \log_2 16 \quad \text{Because } \left(\tfrac{1}{8}\right)(128) = 16$$
$$= 4 \quad \text{Because 4 is the exponent on 2 that yields 16}$$

EXAMPLE 3   Logarithmic equation

Solve the equation $\log_3(2x + 1) - 2\log_3(x - 3) = 2$.

Solution

Use Theorem 2.9, and remember that both $2x + 1 > 0$ and $x - 3 > 0$.

$$\log_3(2x + 1) - 2\log_3(x - 3) = 2$$
$$\log_3(2x + 1) - \log_3(x - 3)^2 = 2$$
$$\log_3 \frac{2x + 1}{(x - 3)^2} = 2$$
$$3^2 = \frac{2x + 1}{(x - 3)^2}$$
$$9(x - 3)^2 = 2x + 1$$
$$9x^2 - 56x + 80 = 0$$
$$(x - 4)(9x - 20) = 0$$
$$x = 4, \tfrac{20}{9}$$

Notice that $x - 3 < 0$ if $x = \tfrac{20}{9}$. The given logarithmic equation has only $x = 4$ as a solution because we cannot take the logarithm of a negative number. ∎

NATURAL EXPONENTIAL AND LOGARITHMIC FUNCTIONS

In elementary algebra you probably studied logarithms to base 10 or possibly to base 2, but in calculus, it turns out to be convenient to use the number e defined by the limit*

$$e = \lim_{n \to \infty}\left(1 + \frac{1}{n}\right)^n$$

which can be thought of as the number approached by expressions of the form $(1 + \frac{1}{n})^n$ as n grows large without bound.

"But wait," you may be thinking, "as n grows large without bound, the expression $1 + 1/n$ tends toward 1 and since $1^n = 1$ for all n, no matter how large the number n, e must equal 1." However, the limit process does not work that way, and as you can see from Table 2.1, e is a number slightly less than 3. In fact, to twelve decimal places,

$$e \approx 2.71828182845 \cdots$$

The letter e was chosen to represent this number in honor of the great Swiss mathematician Leonhard Euler (1707–1783), who discovered many of its special properties and investigated applications in which e plays a vital role.

The number e is called the **natural exponential base** because many formulas in calculus assume an especially simple or "natural" form when expressed in terms of exponentials or logarithms to the base e. For this reason, e^x is called the **natural exponential function**, and $\log_e x$ is the **natural logarithm.** We will have more to say about what makes these functions "natural" in Chapter 3. Here is some special notation and terminology associated with these functions.

WARNING You must be careful when using software to reach conclusions in calculus. For example, on most software or calculators, if you input

$$\left(1 + \frac{1}{10^{20}}\right)^{10^{20}}$$

the output is 1 (which, of course, is incorrect). We know (from the definition of e) that $\lim_{n \to \infty}(1 + \frac{1}{n})^n = e > 1$.

Y₁⊟(1+1/X)^X

X	Y1
1	2
2	2.25
3	2.3704
4	2.4414
5	2.4883
10	2.5937
50	2.6916
100	2.7048
1000	2.7169
5000	2.718
10000	2.7181
1E6	2.7183
N=2	2.7183

TABLE 2.1 Calculator approximations of e

---

*Limits such as $n \to \infty$ are called "limits at infinity" and will be discussed in detail in Chapter 4.

EXP NOTATION

The natural exponential function e^x is sometimes denoted by $\exp(x)$.

COMMON LOGARITHM

The **common logarithm**, $\log_{10} x$, is denoted by $\log x$.

NATURAL LOGARITHM

The **natural logarithm**, $\log_e x$, is denoted by $\ln x$.
($\ln x$ is pronounced "ell n x" or "lawn x.")

The "exp" notation for the natural exponential function is especially useful for representing composite functions. For instance, the awkward expression

$$e^{3x^2 - 2\sin x + 8}$$

can be expressed more compactly as

$$\exp(3x^2 - 2\sin x + 8)$$

The following theorem summarizes the key properties of natural exponential and logarithmic functions.

THEOREM 2.10   Basic properties of the natural logarithm

a. $\ln 1 = 0$ **b.** $\ln e = 1$

c. $e^{\ln x} = x$ for all $x > 0$ **d.** $\ln e^y = y$ for all y

e. $b^x = e^{x \ln b}$ for any $b > 0$ $(b \neq 1)$

Proof   Parts **a** and **b** follow immediately from the definitions of $\ln x$ and e^x. Parts **c** and **d** are just the inversion rules for base e.

We show the proof of part **e.**

$$e^{x \ln b} = e^{\ln b^x} \qquad \text{Power rule (Theorem 2.9)}$$
$$= b^x \qquad \text{Property of natural log (Theorem 2.10c)} \qquad \square$$

WARNING   If you are using a software package such as *Mathematica*, *Derive*, or *Maple*, sometimes referred to as CAS (Computer Algebra System), be careful about the notation. Some versions do not distinguish between $\log x$ (common logarithm) and $\ln x$ (natural logarithm). All logarithms on these versions are assumed to be natural logarithms, so to evaluate a common logarithm requires using the change-of-base theorem.

On calculators, there are usually two logarithm keys, $\boxed{\text{LOG}}$ and $\boxed{\text{LN}}$. For other logarithms, you will need the useful conversion formula contained in the following theorem.

THEOREM 2.11   Change-of-base theorem

$$\log_b x = \frac{\ln x}{\ln b} \qquad \text{for any } b > 0 \ (b \neq 1)$$

Proof   Let $y = \log_b x$. Then

$$b^y = x \qquad \text{Definition of logarithm}$$
$$\ln b^y = \ln x \qquad \text{Equality rule of logarithms}$$
$$y \ln b = \ln x \qquad \text{Power rule of logarithms}$$
$$y = \frac{\ln x}{\ln b} \qquad \text{Divide both sides by } \ln b. \qquad \square$$

EXAMPLE 4   Solving an exponential equation using the change-of-base theorem

Solve $6^x = 200$.

Solution

$$6^x = 200$$

$$x = \log_6 200 \qquad \text{Definition of exponent}$$

$$= \frac{\ln 200}{\ln 6} \qquad \text{Change-of-base theorem}$$

$$\approx 2.957047225 \qquad \text{Approximate with a calculator.}$$

EXAMPLE 5   Finding a velocity

An object moves along a straight line in such a way that after t seconds, its velocity is given by

$$v(t) = 10 \log_5 t + 3 \log_2 t$$

ft/s. How long will it take for the velocity to reach 20 ft/s?

Solution

We want to solve the equation

$$10 \log_5 t + 3 \log_2 t = 20$$

$$10 \frac{\ln t}{\ln 5} + 3 \frac{\ln t}{\ln 2} = 20 \qquad \text{Change-of-base formula}$$

$$\left(\frac{10}{\ln 5} + \frac{3}{\ln 2} \right) \ln t = 20 \qquad \text{Factor.}$$

$$\ln t = 20 \left(\frac{10}{\ln 5} + \frac{3}{\ln 2} \right)^{-1} \approx 1.89727$$

$$t \approx e^{1.89727} \approx 6.6677$$

Thus, it takes about 6.67 seconds for the velocity to reach 20 ft/s.

EXAMPLE 6   Changing an exponential from base b to base e

Change 10^{2x} to an exponential to the natural base.

Solution

We wish to find N for

$$10^{2x} = e^N$$

$$\ln 10^{2x} = \ln e^N \qquad \text{Log of both sides}$$

$$2x \ln 10 = N \ln e \qquad \text{Power rule of logarithms}$$

$$2x \ln 10 = N \qquad \ln e = 1$$

Thus, $10^{2x} = e^{2x \ln 10}$.

EXAMPLE 7   Solving an exponential equation with base e

Solve $\frac{1}{2} = e^{-0.000425t}$.

Solution

We have

$$\frac{1}{2} = e^{-0.000425t}$$

$$-0.000425t = \ln 0.5 \qquad \text{Remember that } -0.000425t \text{ is the exponent on base } e \text{ that equals } \frac{1}{2} = 0.5$$

$$t = \frac{\ln 0.5}{-0.000425} \approx 1630.934542$$

EXAMPLE 8   Exponential growth

A biological colony grows in such a way that at time t (in minutes), the population is

$$P(t) = P_0 e^{kt}$$

where P_0 is the initial population and k is a positive constant. Suppose the colony begins with 5,000 individuals and contains a population of 7,000 after 20 min. Find k and determine the population (rounded to the nearest hundred individuals) after 30 min.

Solution

Because $P_0 = 5,000$, the population after t minutes will be

$$P(t) = 5,000 e^{kt}$$

In particular, because the population is 7,000 after 20 min,

$$P(20) = 5,000 e^{k(20)}$$
$$7,000 = 5,000 e^{20k}$$
$$\tfrac{7}{5} = e^{20k}$$
$$20k = \ln(\tfrac{7}{5})$$
$$k = \tfrac{1}{20} \ln(\tfrac{7}{5})$$
$$\approx 0.0168236$$

Finally, to determine the population after 30 min, substitute this value for k to find

$$P(30) = 5,000 e^{30k}$$
$$\approx 8,282.5117$$

The expected population is approximately $8,300$.   ∎

CONTINUOUS COMPOUNDING OF INTEREST

One reason e is called the "natural" exponential base is that many natural growth phenomena can be described in terms of e^x. As an illustration of this fact, we close this section by showing how e^x can be used to describe the accounting procedure called *continuous compounding of interest*.

Suppose a sum of money is invested and the interest is compounded once during a particular period. If P is the initial investment (the **present value** or **principal**) and i is the **interest rate** during the period, then the **future value** (in dollars), A, after the interest is added will be

$$A = \underset{\uparrow}{P} + \underset{\uparrow}{Pi} = P(1+i)$$
$$\text{principal}\quad\text{interest}$$

Thus, to compute the balance at the end of an interest period, we multiply the balance at the beginning of the period by the expression $1 + i$.

Interest is usually compounded more than once a year. The interest added to the account during one period will itself earn interest during subsequent periods. If the annual interest rate is r and the interest is compounded n times per year, then the year is divided into n equal interest periods and the interest rate during each such period is $i = \frac{r}{n}$. To compute the balance at the end of any period, you multiply the balance at the beginning of that period by the expression $1 + \frac{r}{n}$, so the balance at the end of the first period is

$$A_1 = P\left(1 + \frac{r}{n}\right)$$

At the end of the second period, the balance is

$$A_2 = A_1\left(1 + \frac{r}{n}\right) = \left[P\left(1 + \frac{r}{n}\right)\right]\left(1 + \frac{r}{n}\right) = P\left(1 + \frac{r}{n}\right)^2$$

At the end of the third period, the balance is

$$A_3 = A_2\left(1 + \frac{r}{n}\right) = \left[P\left(1 + \frac{r}{n}\right)^2\right]\left(1 + \frac{r}{n}\right) = P\left(1 + \frac{r}{n}\right)^3$$

and so on. At the end of t years, the interest has been compounded nt times, and the future value is

$$A(t) = P\left(1 + \frac{r}{n}\right)^{nt}$$

The first two graphs in Figure 2.29 show how an amount of money in an account over a one-year period of time grows, first with quarterly compounding and then with monthly compounding. Notice that these are "step" graphs, with jumps occurring at the end of each compounding period.

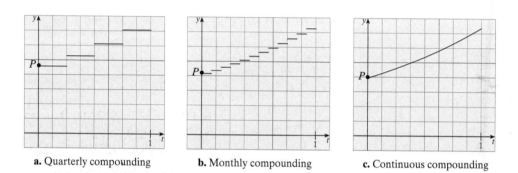

a. Quarterly compounding   **b.** Monthly compounding   **c.** Continuous compounding

Figure 2.29   The growth of an account over a one-year period with different compounding frequencies

With continuous compounding, we compound interest not quarterly, or monthly, or daily, or even every second, but *instantaneously*, so that the future amount of money A in the account grows continuously, as shown in Figure 2.29c. In other words, for a one-year period we compute A as the limiting value of

$$P\left(1 + \frac{r}{n}\right)^n$$

as the number of compounding periods n grows without bound. We denote by $A(t)$, the future value after t years, as with the definition of e, by the limit notation $n \to \infty$:

$$A(t) = \lim_{n \to \infty} P\left(1 + \frac{r}{n}\right)^{nt} \qquad \text{Let } k = \tfrac{n}{r}, \text{ so that } krt = nt \text{ and } \tfrac{r}{n} = \tfrac{1}{k};$$
$$\text{also } k \to \infty \text{ as } n \to \infty.$$

$$= \lim_{k \to \infty} P\left(1 + \frac{1}{k}\right)^{krt}$$

$$= \lim_{k \to \infty} P\left[\left(1 + \frac{1}{k}\right)^{k}\right]^{rt}$$

$$= P\left[\lim_{k \to \infty}\left(1 + \frac{1}{k}\right)^{k}\right]^{rt} \qquad \text{Scalar rule for limits}$$

$$= Pe^{rt} \qquad \text{Definition of } e$$

These observations are now summarized.

Future Value

If P dollars are compounded n times per year at an annual rate r, then the future value after t years is given by

$$A(t) = P\left(1 + \frac{r}{n}\right)^{nt}$$

and if the compounding is continuous, the future value is

$$A(t) = Pe^{rt}$$

EXAMPLE 9   Continuous compounding of interest

If \$12,000 is invested for 5 years at 18%, find the future value at the end of 5 years if interest is compounded

a. monthly.   b. continuously.
c. If the interest is compounded continuously, how long will it take for the money to double?

Solution

P = \$12,000; $t = 5$; and $r = 18\% = 0.18$ are given.

a. $n = 12$; $A = P\left(1 + \frac{r}{n}\right)^{nt} = \$12,000\left(1 + \frac{0.18}{12}\right)^{12 \cdot 5} \approx \$29,318.64$
b. $A = \$12,000e^{0.18(5)} \approx \$29,515.24$
c. The account doubles when $A(t) = 2P = \$24,000$; that is, when

$$12,000e^{0.18t} = 24,000$$

$$e^{0.18t} = 2$$

$$0.18t = \ln 2 \qquad \text{Definition of logarithm}$$

$$t = \frac{\ln 2}{0.18} \approx 3.851$$

This is about 3 years and 10 months.   ∎

2.4   PROBLEM SET

A *Sketch the graph of the functions in Problems 1–4.*

1. $y = 3^x$

2. $y = 4^{-x}$

3. $y = -e^{-x}$

4. $y = -e^x$

Evaluate the expressions (with calculator accuracy) in Problems 5–12.

5. $32^{2/5} + 9^{3/2}$

6. $(1 + 4^{3/2})^{-1/2}$

7. $e^3 e^{2.3}$

8. $\dfrac{(e^{1.3})^2}{e^{1.3} + \sqrt{e^{-1.4}}}$

9. $5,000\left(1 + \frac{0.135}{12}\right)^{12(5)}$

10. $145,000\left(1 + \frac{0.073}{365}\right)^{-365(5)}$

11. $2,589e^{0.45(6)}$

12. $850,000e^{-0.04(10)}$

Evaluate the expressions given in Problems 13–22.

13. $\log_2 4 + \log_3 \frac{1}{9}$

14. $2^{\log_2 3 - \log_2 5}$

15. $5\log_3 9 - 2\log_2 16$

16. $\left(\log_2 \frac{1}{8}\right)(\log_3 27)$

17. $(3^{\log_7 1})(\log_5 0.04)$

18. $e^{5\ln 2}$

19. $\log_3 3^4 - \ln e^{0.5}$

20. $\ln(\log 10^e)$

21. $\exp(\ln 3 - \ln 10)$

22. $\exp(\log_{e^2} 25)$

Solve the logarithmic and exponential equations to calculator accuracy in Problems 23–40.

23. $\log_x 16 = 2$

24. $\log x = 5.1$

25. $e^{-3x} = 0.5$

26. $\ln x^2 = 9$

27. $7^{-x} = 15$

28. $e^{2x} = \ln(4 + e)$

29. $\frac{1}{2}\log_3 x = \log_2 8$

30. $\log_2(x^{\log_2 x}) = 4$

31. $3^{x^2 - x} = 9$

32. $4^{x^2 + x} = 16$

33. $2^x 5^{x+2} = 25,000$

34. $3^x 4^{x+(1/2)} = 3,456$

35. $(\sqrt[3]{2})^{x+10} = 2^{x^2}$

36. $(\sqrt[3]{5})^{x+2} = 5^{x^2}$

37. $e^{2x+3} = 1$

38. $\dfrac{e^{x^2}}{e^{x+6}} = 1$

39. $\log_3 x + \log_3(2x + 1) = 1$

40. $\ln\left(\dfrac{x^2}{1-x}\right) = \ln x + \ln\left(\dfrac{2x}{1+x}\right)$

In Problems 41–46, evaluate the given limits.

41. a. $\lim\limits_{x \to 0^+} x^2 e^{-x}$ **b.** $\lim\limits_{x \to 1} x^2 e^{-x}$

42. a. $\lim\limits_{x \to 0} e^{-x^3}$ **b.** $\lim\limits_{x \to 1} e^{-x^3}$

43. a. $\lim\limits_{x \to 0^+} (1 + x)^{1/x}$ **b.** $\lim\limits_{x \to 1} (1 + x)^{1/x}$

44. a. $\lim\limits_{x \to 0^-} e^{1/x}$ **b.** $\lim\limits_{x \to 0^+} \ln x$

45. $\lim\limits_{x \to e} x \ln x^2$ **46.** $\lim\limits_{x \to e} \dfrac{\ln \sqrt{x}}{x}$

Ⓑ 47. If $\log_b 1{,}296 = 4$, what is $\left(\frac{3}{2}b\right)^{3/2}$?

48. If $\log_{\sqrt{b}} 106 = 2$, what is $\sqrt{b - 25}$?

49. Solve $\log_2 x + \log_5(2x + 1) = \ln x$ correct to the nearest tenth.

50. Solve $\log_x 2 = \log_3 x$ correct to the nearest tenth.

Data points with a curve fit to those points are shown in Problems 51–54. Decide whether the data are better modeled by an exponential or a logarithmic function.

51.

52.

53.

54.

55. According to the *Bouguer–Lambert law*, a beam of light that strikes the surface of a body of water with intensity I_0 will have intensity I at a depth of x meters, where

$$I = I_0 e^{kx}, \qquad k > 0$$

The constant k, called the *absorption coefficient*, depends on such things as the wavelength of the beam of light and the purity of the water. Suppose a given beam of light is only 5% as intense at a depth of 2 meters as it is at the surface. Find k and determine at what depth (to the nearest meter) the intensity is 1% of the intensity at the surface. (This explains why plant life exists only in the top 10 m of a lake or sea.)

56. A certain bank pays 6% interest compounded continuously. How long will it take for $835 to double?

57. If an amount of money is compounded continuously at an annual rate r, how long will it take for that amount of money to double?

58. First National Bank pays 7% interest compounded monthly, and World Savings pays 6.95% interest compounded continuously. Which bank offers a better deal?

59. A person invests $8,500 in a bank that compounds interest continuously. If the investment doubles in 10 years, what is the (annual) rate of interest (correct to the nearest hundredth percent)?

60. In 1626, Peter Minuit traded trinkets worth $24 for land on Manhattan Island. Assume that in 2000 the same land was worth $43.9 billion. Find the annual rate of interest compounded continuously at which the $24 would have had to be invested during this time to yield the same amount.

61. Biologists estimate that the population of a bacterial colony is

$$P(t) = P_0 2^{kt}$$

at time t (in minutes). Suppose the population is found to be 1,000 after 20 minutes and that it doubles every hour.

a. Find P_0 and k.

b. When (to the nearest minute) will the population be 5,000?

62. Spy Problem   Having escaped from the smugglers (Problem 49, Section 1.2), the Spy receives an e-mail message that his best friend, Sigmund ("Siggy") Leiter, has been found murdered, the body stuffed unceremoniously in a freezer. Fighting back tears of grief, he remembers that t hours after death, a body has temperature

$$T = A + (B - A)e^{-0.03t}$$

where A is the air temperature and B is the temperature of the body at the time of death. The police inform the Spy that at the time of discovery, 1:00 P.M. on Thursday, the corpse had temperature $40°F$, and the freezer $10°F$. The Spy knows the deed was done by either Coldfinger or André Scélérat. If Coldfinger was in jail from Monday until Wednesday noon and Scélérat was at a villains' convention in Las Vegas from noon on Wednesday until Friday, who "iced" Siggy and when? (By the way, one of the few normal things about Siggy was his body temperature, $98.6°F$.)

63. The *Richter scale* measures the intensity of earthquakes. Specifically, if E is the energy (watt/cm³) released by a quake, then it is said to have *magnitude M*, where

$$M = \frac{\log E - 11.4}{1.5}$$

a. Express E in terms of M.

b. How much more energy is released in an $M = 8.5$ earthquake (such as the devastating Alaska quake of 1964) than in an average quake of magnitude $M = 6.5$ (the Los Angeles quake of 1994)?

64. Exploration Problem   In the definition of the exponential function $f(x) = b^x$, we require that b be a positive constant. What happens if $b < 0$, for example, if $b = -2$? For what values of x is f defined? Describe the graph of f in this case.

65. Let $E(x) = 2^{x^2 - 2x}$.

a. Use a calculator to sketch $E(x)$.

b. Where does the graph cross the y-axis? What happens to the graph as $x \to +\infty$ and as $x \to -\infty$?

c. Use a calculator utility to find the value of x that minimizes $E(x)$. What is the minimum value?

66. A manufacturer of car batteries estimates that p percent of the batteries will work for at least t months, where

$$p(t) = 100e^{-0.03t}$$

a. What percent of the batteries can be expected to last at least 40 months?

b. What percent can be expected to *fail* before 50 months?

c. What percent can be expected to *fail* between the 40th and 50th months?

67. If \$3,600 is invested at 15% compounded daily, how much money will there be in 7 years?

a. Use a 365-day year; this is known as *exact interest*.

b. Use a 360-day year; this is known as *ordinary interest*.

68. If P dollars are borrowed for a total of N months compounded monthly at an annual interest rate of r, then the monthly payment is found by the formula

$$m = \frac{P\left(\frac{r}{12}\right)}{1 - \left(1 + \frac{r}{12}\right)^{-N}}$$

a. Use this formula to determine the monthly car payment for a new car costing \$17,487 with a down payment of \$7,487. The car is financed for 4 years at 12%.

b. A home loan is made for \$210,000 at 8% interest for 30 years. What is the monthly payment with a 20% down payment?

69. A cool drink is removed from a refrigerator and is placed in a room where the temperature is $70°$F. According to a result in physics known as *Newton's law of cooling*, the temperature of the drink in t minutes will be

$$F(t) = 70 - Ae^{-kt}$$

where A and k are positive constants. Suppose the temperature of the drink was $35°$F when it left the refrigerator, and 30 minutes later, it was $50°$F [that is, $F(0) = 35$ and $F(30) = 50$].

a. Find A and e^{-30k}.

b. What will the temperature of the drink be after one hour (to the nearest degree)?

c. What would you expect to happen to the temperature as $t \to +\infty$?

70. HISTORICAL QUEST   John Napier was a Scottish landowner who was the Isaac Asimov of his day, having envisioned the tank, the machine gun, and the submarine. He also predicted that the end of the world would occur between 1688 and 1700. He is best known today as the inventor of logarithms, which until the advent of the calculator were used extensively with complicated calculations.

JOHN NAPIER
1550–1617

Napier's logarithms are not identical to the logarithms we use today. Napier chose to use $1 - 10^{-7}$ as his given number, and then he multiplied by 10^7. That is, if

$$N = 10^7 \left(1 - \frac{1}{10^7}\right)^L$$

then L is Napier's logarithm of the number N; that is, $L = \text{nog } N$ means $N = 10^7 (1 - 10^{-7})^L$. ∎

One difference between Napier's logarithms (which for clarity we will call a nog) and modern logarithms is apparent when stating the product, quotient, and power rules for logarithms.

a. Show that if $L_1 = \text{nog } N_1$ and $L_2 = \text{nog } N_2$, then

$$\text{nog } N_1 + \text{nog } N_2 = \text{nog}\left(\frac{N_1 N_2}{10^7}\right).$$

b. State and prove similar results for a quotient rule and a power rule for Napier logarithms.

c. What is nog 10^7?

d. If nog $N = 10^7$, then $N \approx e^r$ for some rational number r. What is r?

e. Fortunately, Napier's 1614 paper on logarithms was read by a true mathematician, Henry Briggs (1561–1630), and together they decided that base 10 made a lot more sense. In the year Napier died, Briggs published a table of common logarithms (base 10) that at the time was a major accomplishment. In this paper he used the words "mantissa" and "characteristic." What is the meaning that Briggs gave to these words?

f. Who was the first person to publish a table of natural logarithms (base e)?

◯ 71. For $b > 0$ and all positive integers m and n, show that

a. $b^m b^n = b^{m+n}$ **b.** $\dfrac{b^m}{b^n} = b^{m-n}$

72. For $b > 0$ and all positive integers m and n, show that

a. $(b^m)^n = b^{mn}$ **b.** $(\sqrt[n]{b})^m = \sqrt[n]{b^m}$

73. Prove:

a. $\log_b x + \log_b y = \log_b (xy)$

b. $\log_b x - \log_b y = \log_b \dfrac{x}{y}$

74. Let b be any positive number other than 1. Show that

$$x^x = b^{x \log_b x}$$

75. Let a and b be any positive numbers other than 1. Show that

a. $\log_a x = \dfrac{\log_b x}{\log_b a}$ **b.** $(\log_a b)(\log_b a) = 1$

CHAPTER 2 REVIEW

Proficiency Examination

CONCEPT PROBLEMS

1. State the informal definition of a limit of a function. Discuss this informal definition.
2. State the formal definition of a limit.
3. State the following basic rules for limits:
   a. limit of a constant   **b.** multiple rule
   c. sum rule   **d.** difference rule
   e. product rule   **f.** quotient rule
   g. power rule
   h. limit of a polynomial function
   i. limit of a rational function
   j. limit of a transcendental function
4. State the squeeze rule.
5. What are the values of the given limits?
   a. $\lim\limits_{x \to 0} \dfrac{\sin x}{x}$ **b.** $\lim\limits_{x \to 0} \dfrac{\cos x - 1}{x}$
6. Define the continuity of a function at a point and discuss.
7. State the continuity theorem.
8. State the composition limit rule.
9. State the intermediate value theorem.
10. State the root location theorem.
11. State the completeness property.
12. **a.** What is an exponential function?
    b. How is such a function related to a logarithmic function?
13. Define e.
14. **a.** What is a logarithmic function?
    b. What is a common logarithm?
    c. What is a natural logarithm?
15. Draw a quick sketch of:
    a. An exponential function b^x for $b > 1$
    b. An exponential function b^x for $0 < b < 1$
16. Draw a quick sketch of:
    a. A logarithmic function $\log_b x$ for $b > 1$.
    b. A logarithmic function $\log_b x$ for $0 < b < 1$.
17. State each of the following exponential properties:
    a. Equality rule   **b.** Inequality rules
    c. Product rule   **d.** Quotient rule
    e. Power rules
18. Complete the statement of each of the following natural logarithm properties:
    a. $\ln 1 =$ _____   **b.** $\ln e =$ _____
    c. $e^{\ln x} =$ _____   **d.** $\ln e^y =$ _____
    e. $b^x =$ _____   (write as a power of e)

PRACTICE PROBLEMS

Evaluate the limits in Problems 19–24.

19. $\lim\limits_{x \to 3} \dfrac{x^2 - 4x + 9}{x^2 + x - 8}$ 20. $\lim\limits_{x \to 4} \dfrac{\sqrt{x} - 2}{x - 4}$

21. $\lim\limits_{x \to 2} \dfrac{x^2 - 5x + 6}{x^2 - 4}$ 22. $\lim\limits_{x \to 0} \dfrac{1 - \cos x}{2 \tan x}$

23. $\lim\limits_{x \to 0} \dfrac{\sin 9x}{\sin 5x}$ 24. $\lim\limits_{x \to \frac{1}{2}^-} \dfrac{|2x - 1|}{2x - 1}$

25. Decide whether each of the curves is exponential or logarithmic.

a.   **b.**

c.   **d.**

Determine whether the functions given in Problems 26–27 are continuous on the interval $[-5, 5]$.

26. $f(t) = \dfrac{1}{t} - \dfrac{3}{t + 1}$ 27. $g(x) = \dfrac{x^2 - 1}{x^2 + x - 2}$

28. How quickly will \$2,000 grow to \$5,000 when invested at an annual rate of 8% if interest is compounded:
    a. quarterly?   **b.** monthly?   **c.** continuously?

29. Find constants A and B such that f is continuous for all x:
$$f(x) = \begin{cases} Ax + 3 & \text{if } x < 1 \\ 2 & \text{if } x = 1 \\ x^2 + B & \text{if } x > 1 \end{cases}$$

30. Show that the equation
$$x + \sin x = \dfrac{1}{\sqrt{x} + 3}$$
has at least one solution on the interval $(0, \pi)$.

Supplementary Problems*

1. Evaluate each of the given numbers (calculator approximations for parts **a** and **b**).
   a. $\ln 4.5$ **b.** $e^{2.8}$ **c.** $e^{\ln \pi}$ **d.** $\ln e^{\sqrt{2}}$

Solve each equation in Problems 2–8 for x.

2. $4^{x-1} = 8$ 3. $2^{x^2 + 4x} = \frac{1}{16}$

4. $\log_2 2^{x^2} = 4$ 5. $\log_4 \sqrt{x(x - 15)} = 1$

6. $\log_2 x + \log_2(x - 15) = 4$ 7. $3^{2x-1} = 6^x 3^{1-x}$

8. $\ln(x - 1) + \ln(x + 1) = 2 \ln \sqrt{12}$

---

*The supplementary problems are presented in a somewhat random order, not necessarily in order of difficulty.

In Problems 9–28, evaluate the given limit.

9. $\lim\limits_{x\to 2} \dfrac{3x^2 - 7x + 2}{x - 2}$

10. $\lim\limits_{x\to -3} \dfrac{4x^2 + 11x - 3}{x^2 - x - 12}$

11. $\lim\limits_{x\to 1^+} \sqrt{\dfrac{x^2 - x}{x - 1}}$

12. $\lim\limits_{x\to 1} \dfrac{x^3 - 1}{x^2 - 1}$

13. $\lim\limits_{x\to 4} |4 - x|$

14. $\lim\limits_{x\to 0^-} \dfrac{|x|}{x}$

15. $\lim\limits_{x\to 1} \dfrac{x^2 - 3x + 2}{x^2 - 1}$

16. $\lim\limits_{x\to 1/\pi} \dfrac{1 + \cos\frac{1}{x}}{\pi x - 1}$

17. $\lim\limits_{x\to 0^+} (1 + x)^{4/x}$

18. $\lim\limits_{x\to 0^+} (1 + 2x)^{1/x}$

19. $\lim\limits_{x\to 1} \dfrac{x^5 - 1}{x - 1}$

20. $\lim\limits_{x\to e^2} \dfrac{(\ln x)^3 - 8}{\ln x - 2}$

21. $\lim\limits_{x\to 0} \dfrac{\sin 3x}{\sin 2x}$

22. $\lim\limits_{t\to 0} \dfrac{\tan^{-1} t}{\sin^{-1} t}$

23. $\lim\limits_{x\to 0} \dfrac{\sin(\cos x)}{\sec x}$

24. $\lim\limits_{x\to 0} \tan 2x \cot x$

25. $\lim\limits_{x\to 0} \dfrac{1 - \sin x}{\cos^2 x}$

26. $\lim\limits_{x\to 0} \dfrac{1 - 2\cos x}{\sqrt{3} - 2\sin x}$

27. $\lim\limits_{x\to 0} \dfrac{e^{3x} - 1}{e^x - 1}$

28. $\lim\limits_{x\to 3} \dfrac{\frac{1}{x} - \frac{1}{3}}{x - 3}$

Evaluate $\lim\limits_{\Delta x\to 0} \dfrac{f(x + \Delta x) - f(x)}{\Delta x}$ *for the function f in Problems 29–36.*

29. $f(x) = 7$

30. $f(x) = 3x + 5$

31. $f(x) = \sqrt{2x}$

32. $f(x) = x(x + 1)$

33. $f(x) = \dfrac{4}{x}$

34. $f(x) = \sin x$

35. $f(x) = e^x$

36. $f(x) = \ln x$

Decide whether the functions in Problems 37–40 are continuous on the given intervals. If not, define or redefine the function at one point to make it continuous everywhere on the given interval, or else explain why that cannot be done.

37. $f(x) = \dfrac{x + 4}{x - 8}$ on $[-5, 5]$

38. $f(x) = \dfrac{x + 4}{x - 8}$ on $[0, 10]$

39. $f(x) = \dfrac{\sqrt{x} - 8}{x - 64}$ on $\mathbb{R}$

40. $f(x) = \dfrac{\sqrt{x} - 6}{x - 36}$ on $[-5, 5]$

Decide whether the functions in Problems 41–44 are continuous on the given intervals. Check all of the suspicious points.

41. $r(x) = \begin{cases} 1 & \text{if } x \text{ is rational} \\ -1 & \text{if } x \text{ is irrational} \end{cases}$

42. $f(x) = |x - 2|$ on $[-5, 5]$

43. a. $f(x) = \dfrac{x^2 - x - 6}{x + 2}$ on $[0, 5]$

b. $f(x) = \dfrac{x^2 - x - 6}{x + 2}$ on $[-5, 5]$

c. $f(x) = \begin{cases} \dfrac{x^2 - x - 6}{x + 2} & \text{on } [-5, 5], x \neq -2 \\ -4 & \text{for } x = -2 \end{cases}$

d. $f(x) = \begin{cases} \dfrac{x^2 - x - 6}{x + 2} & \text{on } [-5, 5], x \neq -2 \\ -5 & \text{for } x = -2 \end{cases}$

44. a. $g(x) = \dfrac{x^2 - 3x - 10}{x + 2}$ on $[0, 5]$

b. $g(x) = \dfrac{x^2 - 3x - 10}{x + 2}$ on $[-5, 5]$

c. $g(x) = \begin{cases} \dfrac{x^2 - 3x - 10}{x + 2} & \text{for } -5 \leq x < -2 \\ -7 & \text{for } x = -2 \\ x - 5 & \text{for } -2 < x \leq 5 \end{cases}$

*The **greatest integer function** $f(x) = [\![x]\!]$ is the largest integer that is less than or equal to x. This definition is used in Problems 45–47.*

45. a. Graph $f(x) = [\![x]\!]$ on $[-3, 6]$.

b. Find $\lim\limits_{x\to 3} f(x)$.

c. For what values of a does $\lim\limits_{x\to a} f(x)$ exist?

46. Repeat Problem 45 for $f(x) = [\![\frac{x}{2}]\!]$.

47. Let $f(x) = [\![x^2 + 1]\!]^{[\![x + 1]\!]}$; find $\lim\limits_{x\to 1} f(x)$.

48. Find constants A and B such that f is continuous for all x:

$$f(x) = \begin{cases} Ax + 3 & \text{if } x < 1 \\ 5 & \text{if } x = 1 \\ x^2 + B & \text{if } x > 1 \end{cases}$$

49. Find numbers a and b so that

$$\lim\limits_{x\to 0} \dfrac{\sqrt{ax + b} - 1}{x} = 1$$

50. Find a number c so that

$$\lim\limits_{x\to 3} \dfrac{x^3 + cx^2 + 5x + 12}{x^2 - 7x + 12}$$

exists. Then find the corresponding limit.

51. A manufacturer of lightbulbs estimates that the fraction $F(t)$ of bulbs that remain burning after t weeks is given by

$$F(t) = e^{-kt}$$

where k is a positive constant. Suppose twice as many bulbs are burning after 5 weeks as after 9 weeks.

a. Find k and determine the fraction of bulbs still burning after 7 weeks.

b. What fraction of the bulbs burn out before 10 weeks?

c. What fraction of the bulbs can be expected to burn out between the 4th and 5th weeks?

52. How much should you invest now at an annual interest rate of 6.25% so that your balance 10 years from now will be $2,000 if interest is compounded

a. quarterly?   **b.** continuously?

53. A *decibel* (named for Alexander Graham Bell) is the smallest increase of the loudness of a sound that is detectable by the human ear. In physics, it is shown that when two sounds of intensity I_1 and I_2 (watts/cm^3) occur, the difference in loudness is D decibels, where

$$D = 10 \log\left(\dfrac{I_1}{I_2}\right)$$

When sound is rated in relation to the threshold of human hearing ($I_0 = 10^{-16}$), the level of normal conversation is 50 decibels, whereas that of a rock concert is 110 decibels. Show that a rock concert is 60 times as loud as normal conversation but a million times as intense.

54. If a function f is not continuous at $x = c$, but can be made continuous at c by being given a new value at the point, it is said to have a *removable discontinuity* at $x = c$. Which of the following functions has a removable discontinuity at $x = c$?

a. $f(x) = \dfrac{2x^2 + x - 15}{x + 3}$ at $c = -3$

b. $f(x) = \dfrac{x - 2}{|x - 2|}$ at $c = 2$

c. $f(x) = \dfrac{2 - \sqrt{x}}{4 - x}$ at $c = 4$

55. If $f(x) = x^3 - x^2 + x + 1$, show that there is a number c such that $f(c) = 0$.

56. Prove that $\sqrt{x + 3} = e^x$ has at least one real root, and then use a graphing utility to find the root correct to the nearest tenth.

57. Show that the tangent line to the circle $x^2 + y^2 = r^2$ at the point (x_0, y_0) has the equation $y_0 y + x_0 x = r^2$. *Hint*: Recall that the tangent line at a point P on a circle with center C is the line that passes through P and is perpendicular to the line segment $\overline{CP}$.

58. Counterexample Problem It is not necessarily true that $\lim_{x \to c} f(x)$ and $\lim_{x \to c} g(x)$ exist whenever $\lim_{x \to c} [f(x) \cdot g(x)]$ exists. Find functions f and g such that $\lim_{x \to c} [f(x) \cdot g(x)]$ exists and $\lim_{x \to c} g(x) = 0$, but $\lim_{x \to c} f(x)$ does not exist.

59. Counterexample Problem It is not necessarily true that $\lim_{x \to c} f(x)$ and $\lim_{x \to c} g(x)$ exist whenever $\lim_{x \to c} \dfrac{f(x)}{g(x)}$ exists. Find functions f and g such that $\lim_{x \to 0} \dfrac{f(x)}{g(x)}$ exists, but neither $\lim_{x \to 0} f(x)$ nor $\lim_{x \to 0} g(x)$ exists.

60. When analyzing experimental data involving two variables, a useful procedure is to pass a smooth curve through a number of plotted data points and then perform computations as if the curve were the graph of an equation relating the variables. Suppose the following data are gathered as a result of a physiological experiment in which skin tissue is subjected to external heat for t seconds, and a measurement is made of the change in temperature $\triangle T$ required to cause a change of $2.5°C$ at a depth of 0.5 mm in the skin:

t (sec)	1	2	3	4	10	20	25	30
$\triangle T$ (°C)	12.5	7.5	5.8	5.0	4.5	3.5	2.9	2.8

Plot the data points in a coordinate plane and then answer these questions.

a. What temperature difference $\triangle T$ would be expected if the exposure time is 2.2 sec?

b. Approximately what exposure time t corresponds to a temperature difference $\triangle T = 8.0°C$?

61. Journal Problem The theorem

$$\lim_{h \to 0} \frac{\sin h}{h} = 1$$

stated in this chapter has been the subject of much discussion in mathematical journals:

W. B. Gearhart and H. S. Schultz, "The Function $\sin x / x$," *College Mathematics Journal* (1990): 90–99.

L. Gillman, "π and the Limit of $(\sin \alpha)/\alpha$," *American Mathematical Monthly* (1991): 345–348.

F. Richman, "A Circular Argument," *College Mathematics Journal* (1993): 160–162.

D. A. Rose, "The Differentiability of $\sin x$," *College Mathematics Journal* (1991): 139–142.

P. Ungar, "Reviews," *American Mathematical Monthly* (1986): 221–230.

Some of these articles argue that the demonstration shown in the text is circular, since we use the fact that the area of a circle is πr^2. How do we know the area of a circle? The answer, of course, is that we learned it in elementary school, but that does not constitute a proof. On the other hand, some of these articles argue that the reasoning is not necessarily circular. Read one or more of these journal articles and write a report.

62. a. Show that $\sin x < x$ if $0 < x < \frac{\pi}{2}$. Refer to Figure 2.30. *Hint*: Compare the area of an appropriate triangle and sector.

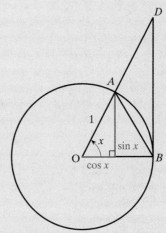

Figure 2.30 Figure for Problems 62 and 63

b. Show that $|\sin x| < |x|$ if $0 < |x| < \frac{\pi}{2}$.

c. Use the definition of continuity to show that $\sin x$ is continuous at $x = 0$.

d. Use the formula

$$\sin(\alpha + \beta) = \sin \alpha \cos \beta + \cos \alpha \sin \beta$$

to show that $\sin x$ is continuous for all real x.

63. Follow the procedure outlined in Problem 62 to show that $\cos x$ is continuous for all x. *Hint*: After showing that

$$\lim_{x \to 0} \cos x = 1$$

you may need the identity

$$\cos(\alpha + \beta) = \cos \alpha \cos \beta - \sin \alpha \sin \beta$$

to show that $\cos x$ is continuous for $x \neq 0$.

64. A regular polygon of n sides is inscribed in a circle of radius R, as shown in Figure 2.31.

Figure 2.31   A regular polygon of seven sides with central angle θ (Problem 64)

a. Show that the perimeter of the polygon is given by $P(\theta) = \dfrac{4\pi R}{\theta}\sin\dfrac{\theta}{2}$, where $\theta = \dfrac{2\pi}{n}$ is the central angle subtended by one side of the polygon.

b. Use the formula in part **a** together with the fact that $\displaystyle\lim_{x\to 0}\frac{\sin x}{x} = 1$ to show that a circle of radius R has circumference $2\pi R$.

c. Modify the approach suggested in parts **a** and **b** to show that a circle of radius R has area πR^2. *Hint*: First express the area of the shaded polygon in Figure 2.31 as a function of θ.

65. **Exploration Problem**   A cylindrical tank containing 50 L of water is drained into an empty rectangular trough that can hold 75 L. Explain why there must be a time when the height of water in the tank is the same as that in the trough.

66. The radius of the earth is roughly 4,000 mi, and an object located x miles from the center of the earth weighs w lb, where

$$w(x) = \begin{cases} Ax & \text{if } x \le 4{,}000 \\ \dfrac{B}{x^2} & \text{if } x > 4{,}000 \end{cases}$$

where A and B are positive constants. Assuming that w is continuous for all x, what must be true about A and B?

67. The windchill temperature in degrees is a function of the air temperature T (in degrees Fahrenheit) and the wind speed v (in mi/h).* If we hold T constant and consider the windchill as a function of v, we have

$$W(v) = \begin{cases} T & \text{if } 0 \le v \le 4 \\ 91.4 + (91.4 - T)(0.0203v - 0.304\sqrt{v} - 0.474) \\ & \text{if } 4 < v < 45 \\ 1.6T - 55 & \text{if } v \ge 45 \end{cases}$$

a. If $T = 30$, what is the windchill for $v = 20$? What is it for $v = 50$?

b. For $T = 30$, what wind speed corresponds to a windchill temperature of $0°$ Fahrenheit?

c. For what value of T is the windchill function continuous at $v = 4$? At $v = 45$?

68. Based on the estimate that there are 10 billion acres of arable land on the earth and that each acre can produce enough food to feed 4 people, some demographers believe that the earth can support a population of no more than 40 billion people.

*From William Bosch and L. G. Cobb, "Windchill," *UMAP Module No. 658*, (1984), pp. 244–247.

The population of the earth reached approximately 5 billion in 1986 and 6 billion in 1999. If the population of the earth were growing according to the formula

$$P(t) = P_0 e^{rt}$$

where t is the time after the population is P_0 and r is the growth rate, when would the population reach the theoretical limit of 40 billion?

69. A function f is said to satisfy a *Lipschitz condition* (named for the nineteenth century mathematician Rudolf Lipschitz, 1832–1903) on a given interval if there is a positive constant M such that

$$|f(x) - f(y)| < M\,|x - y|$$

for all x and y in the interval (with $x \neq y$). Suppose f satisfies a Lipschitz condition on an interval and let c be a fixed number chosen arbitrarily from the interval. Use the formal definition of limit to prove that f is continuous at c.

70. A population model employed at one time by the U.S. Census Bureau uses the formula

$$P(t) = \frac{202.31}{1 + e^{3.98 - 0.314t}}$$

to estimate the population of the United States (in millions) for every tenth year from the base year of 1790. For example, if $t = 0$, then the year is 1790, and if $t = 20$, then the year is 1990.

a. Draw the graph using this population model and predict the population in the year 2000.

b. Consult an almanac or some other source to find the actual population figures for the years from 1790 to present.

c. What happens to the population P "in the long run," that is, as t increases without bound?

*71. **Putnam Examination Problem**   Evaluate

$$\lim_{x\to +\infty}\left[\frac{1}{x}\cdot\frac{a^x - 1}{a - 1}\right]^{1/x} \quad \text{where } a > 0,\, a \neq 1$$

*72. **Putnam Examination Problem**   Figure 2.32 shows a rectangle of base b and height h inscribed in a circle of radius 1, surmounted by an isosceles triangle. Find $\displaystyle\lim_{h\to 2/5}\frac{A(h)}{bh}$.

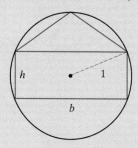

Figure 2.32   Putnam problem

*The Putnam Examination is a national annual examination given under the auspices of the Mathematical Association of America and is designed to recognize mathematically talented college and university students. We include Putnam examination problems, which should be considered optional, to encourage students to consider taking the examination. Putnam problems used by permission of the American Mathematical Association. Solutions to Putnam Examinations generally appear in *The American Mathematical Monthly*, approximately one year after the examination.

3

Differentiation

CONTENTS

PREVIEW

In this chapter, we develop the main ideas of differential calculus. We begin by defining the *derivative*, which is the central concept of differential calculus. Then we develop a list of rules and formulas for finding the derivative of a variety of expressions involving polynomials, rational and root functions, and trigonometric, logarithmic, and inverse trigonometric functions. Along the way, we will see how the derivative can be used to find slopes of tangent lines and rates of change.

PERSPECTIVE

A *calculus* is a body of calculation or reasoning associated with a certain concept. For *differential calculus*, that concept is the *derivative*, one of the fundamental ideas in all mathematics and, arguably, a cornerstone of modern scientific thought. The basic ideas of what we now call calculus had been fermenting in intellectual circles throughout much of the seventeenth century. The genius of Newton and Leibniz (see the guest essay at the end of Chapter 1) centered not so much on the discovery of those ideas as on their systematization.

In this chapter, we will consider various ways of efficiently computing derivatives. We will also see how the derivative can be used to find rates of change and to measure the direction (slope) at each point on a graph. Falling body problems in physics and marginal analysis from economics are examples of applied topics to be discussed, and we will also explore several topics involving basic concepts. It is fair to say that your success with differential calculus hinges on understanding this material.

3.1 An Introduction to the Derivative: Tangents

IN THIS SECTION tangent lines, the derivative, relationship between the graphs of f and f', existence of derivatives, continuity and differentiability, derivative notation

TANGENT LINES

Since classical times, mathematicians have known that the tangent line at a given point on a circle has the property that it is the only line that intersects the circle exactly once, but the same principle cannot be applied to finding tangent lines to more general curves. Indeed, as shown in Figure 3.1, the line we may intuitively think of as being tangent to the graph of $y = f(x)$ at point P can intersect the curve at other points, and for that matter, there may be many lines that intersect the curve only at P, but are clearly not tangent lines.

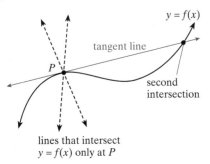

Figure 3.1 Tangent line at P

We begin this section by computing the slope of a tangent line as a limit, using a procedure originally developed by the French mathematician Pierre de Fermat (1601–1665). Fermat's use of the "dynamic" limit process rather than the "static" procedures of classical geometry and algebra was a breakthrough in thinking that eventually led to the development of differential calculus by Isaac Newton (1642–1727) and Gottfried Leibniz (1646–1716) in the latter half of the seventeenth century. Following in the footsteps of these giants, we, too, will use what we discover about tangent lines as a springboard for introducing the derivative and studying its basic properties.

Suppose we wish to find the slope of the tangent line to $y = f(x)$ at the point $P(x_0, f(x_0))$. The strategy is to approximate the tangent line by other lines whose slopes can be computed directly. In particular, consider the line joining the given point P to the neighboring point Q on the graph of f, as shown in Figure 3.2. This line is called a **secant line.** Compare the secant lines shown in Figure 3.2.

WARNING Do not confuse *secant line* (a line that intersects a curve in two or more points) with the *secant function* of trigonometry.

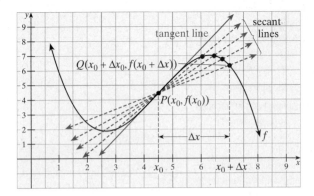

Figure 3.2 The secant line $\overline{PQ}$

Notice that a secant line is a good approximation to the tangent line at point P as long as Q is close to P.

To compute the slope of a secant line, first label the coordinates of the neighboring point Q, as indicated in Figure 3.2. In particular, let Δx denote the change in the x-coordinate between the given point $P(x_0, f(x_0))$ and the neighboring point $Q(x_0 + \Delta x, f(x_0 + \Delta x))$.

The slope of this secant line, m_{sec}, is easy to calculate:

$$m_{\text{sec}} = \frac{\Delta y}{\Delta x} = \frac{f(x_0 + \Delta x) - f(x_0)}{\Delta x}$$

To bring the secant line closer to the tangent line, let Q approach P *on the graph of f* by letting Δx approach 0. As this happens, the slope of the secant line should approach the slope of the tangent line at P. We denote the slope of the tangent line by m_{tan} to distinguish it from the slope of a secant line. These observations suggest the following definition.

Slope of a Line Tangent to a Graph at a Point

At the point $P(x_0, f(x_0))$, the tangent line to the graph of f has **slope** given by the formula

$$m_{\text{tan}} = \lim_{\Delta x \to 0} \frac{f(x_0 + \Delta x) - f(x_0)}{\Delta x}$$

provided this limit exists.

EXAMPLE 1   Slope of a tangent line at a particular point

Find the slope of the tangent line to the graph of $f(x) = x^2$ at the point $P(-1, 1)$.

Solution

Figure 3.3 shows the tangent line to f at $x = -1$. The slope of the tangent line is given by

Tangent line at $(-1, 1)$

Figure 3.3 Tangent line to the graph of $y = x^2$ at $(-1, 1)$

$$
\begin{aligned}
m_{\text{tan}} &= \lim_{\Delta x \to 0} \frac{f(-1 + \Delta x) - f(-1)}{\Delta x} \qquad \text{\textit{Because } } f(x) = x^2, \\
&\qquad\qquad\qquad\qquad\qquad\qquad\qquad\quad f(-1 + \Delta x) = (-1 + \Delta x)^2. \\
&= \lim_{\Delta x \to 0} \frac{(-1 + \Delta x)^2 - (-1)^2}{\Delta x} \\
&= \lim_{\Delta x \to 0} \frac{1 - 2\Delta x + (\Delta x)^2 - 1}{\Delta x} \\
&= \lim_{\Delta x \to 0} \frac{-2\Delta x + (\Delta x)^2}{\Delta x} \\
&= \lim_{\Delta x \to 0} \frac{(-2 + \Delta x)\Delta x}{\Delta x} \qquad \text{\textit{Factor out } } \Delta x \text{ \textit{and reduce.}} \\
&= \lim_{\Delta x \to 0} (-2 + \Delta x) = -2
\end{aligned}
$$

In Example 1, we found the slope of the tangent line to the graph of $y = x^2$ at the point $(-1, 1)$. In Example 2, we perform the same calculation again, this time representing the given point algebraically as (x, x^2). This is the situation shown in Figure 3.4 for the slope of the tangent line to $y = x^2$ at *any* point (x, x^2).

EXAMPLE 2   Slope of a tangent line at an arbitrary point

Derive a formula for the slope of the tangent line to the graph of $f(x) = x^2$, and then use the formula to compute the slope at $(4, 16)$.

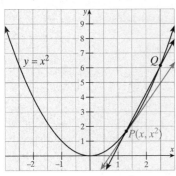

Tangent line at (x, x^2)

Figure 3.4 Tangent line to the graph of $y = x^2$ at (x, x^2)

Solution

Figure 3.4 shows a tangent line at an arbitrary point $P(x, x^2)$ on the curve. From the definition of slope of the tangent line,

$$m_{\tan} = \lim_{\Delta x \to 0} \frac{f(x + \Delta x) - f(x)}{\Delta x}$$

Because $f(x) = x^2$,
$f(x + \Delta x) = (x + \Delta x)^2$.

$$= \lim_{\Delta x \to 0} \frac{(x + \Delta x)^2 - x^2}{\Delta x}$$

$$= \lim_{\Delta x \to 0} \frac{x^2 + 2x\,\Delta x + (\Delta x)^2 - x^2}{\Delta x}$$

$$= \lim_{\Delta x \to 0} \frac{2x\,\Delta x + (\Delta x)^2}{\Delta x}$$

$$= \lim_{\Delta x \to 0} \frac{(2x + \Delta x)\Delta x}{\Delta x} \qquad \text{Factor and reduce.}$$

$$= \lim_{\Delta x \to 0} (2x + \Delta x) = 2x$$

At the point $(4, 16)$, $x = 4$, so $m_{\tan} = 2(4) = 8$. ∎

The result of Example 2 gives a general formula for the slope of a line tangent to the graph of $f(x) = x^2$, namely, $m_{\tan} = 2x$. The answer from Example 1 can now be verified using this formula; if $x = -1$, then $m_{\tan} = 2(-1) = -2$.

THE DERIVATIVE

The expression

$$\frac{f(x + \Delta x) - f(x)}{\Delta x}$$

which gives a formula for the slope of a secant line to the graph of a function f, is called the **difference quotient** of f. The limit of the difference quotient

$$\lim_{\Delta x \to 0} \frac{f(x + \Delta x) - f(x)}{\Delta x}$$

which gives a formula for the slope of the tangent line to the graph of f at the point $(x, f(x))$, is called the **derivative** of f and is frequently denoted by the symbol $f'(x)$ (read "eff prime of x"). To **differentiate** a function f at x means to find its derivative at the point $(x, f(x))$.

Derivative

> The **derivative** of f at x is given by
>
> $$f'(x) = \lim_{\Delta x \to 0} \frac{f(x + \Delta x) - f(x)}{\Delta x}$$
>
> provided this limit exists.

Alternatively, we can write

$$f'(x) = \lim_{t \to x} \frac{f(t) - f(x)}{t - x}$$

where $t = x + \Delta x$. Occasionally, this form is easier to use in computations.

The derivative is one of the fundamental concepts in calculus, and it is important to make some observations regarding this definition.

1. If the limit for the difference quotient exists, then we say that the function f is **differentiable at** x.

2. The value of a derivative depends only on the limit process and not on the symbols used in that process. In Example 2, we found that if $f(x) = x^2$, then $f'(x) = 2x$. This means that we also know

$$\text{If } g(t) = t^2, \text{ then } g'(t) = 2t.$$
$$\text{If } h(u) = u^2, \text{ then } h'(u) = 2u.$$
$$\vdots$$

3. Notice that the derivative of a function is itself a function.

Finding the slope of a tangent line is just one of several applications of the derivative that we will discuss in this chapter. In Section 3.4, we will examine rectilinear motion and other rates of change, and in Section 3.8, marginal analysis from economics. In Chapter 4 we will examine more complex applications such as curve sketching and optimization.

EXAMPLE 3 Derivative using the definition

Differentiate $f(t) = \sqrt{t}$.

WARNING Notice that $f(t) = \sqrt{t}$ is defined for all $t \geq 0$, whereas its derivative $f'(t) = 1/(2\sqrt{t})$ is defined for all $t > 0$. This shows that a function need not be differentiable throughout its entire domain.

Solution

$$
\begin{aligned}
f'(t) &= \lim_{\Delta t \to 0} \frac{f(t + \Delta t) - f(t)}{\Delta t} \\[2mm]
&= \lim_{\Delta t \to 0} \frac{\sqrt{t + \Delta t} - \sqrt{t}}{\Delta t} \\[2mm]
&= \lim_{\Delta t \to 0} \frac{\sqrt{t + \Delta t} - \sqrt{t}}{\Delta t} \left(\frac{\sqrt{t + \Delta t} + \sqrt{t}}{\sqrt{t + \Delta t} + \sqrt{t}} \right) \qquad \text{Rationalize numerator.} \\[2mm]
&= \lim_{\Delta t \to 0} \frac{(t + \Delta t) - t}{\Delta t (\sqrt{t + \Delta t} + \sqrt{t})} \\[2mm]
&= \lim_{\Delta t \to 0} \frac{\Delta t}{\Delta t (\sqrt{t + \Delta t} + \sqrt{t})} \\[2mm]
&= \lim_{\Delta t \to 0} \frac{1}{\sqrt{t + \Delta t} + \sqrt{t}} \qquad \text{Reduce fraction.} \\[2mm]
&= \frac{1}{2\sqrt{t}} \qquad \text{For } t > 0 \qquad \blacksquare
\end{aligned}
$$

EXAMPLE 4 Estimating a derivative using a table

Estimate the derivative of $f(x) = \cos x$ at $x = \frac{\pi}{6}$ by evaluating the difference quotient

$$\frac{\Delta y}{\Delta x} = \frac{f(x + \Delta x) - f(x)}{\Delta x}$$

near the point $x = \frac{\pi}{6}$.

Solution $\quad \dfrac{\Delta y}{\Delta x} = \dfrac{\cos(\frac{\pi}{6} + \Delta x) - \cos \frac{\pi}{6}}{\Delta x}$

Choose a sequence of values for $\Delta x \to 0$: say $1, \frac{1}{2}, \frac{1}{4}, \frac{1}{8}, \ldots$ and use a calculator (or a computer) to estimate the difference quotient by table. We show selected elements from this sequence of calculations:

A similar table for negative values of Δx should also be considered. From the table, we would guess that $f'(\frac{\pi}{6}) = -0.5$. $\blacksquare$

Δx	$\dfrac{\pi}{6} + \Delta x$	Difference quotient $\dfrac{\Delta y}{\Delta x}$
1	1.523598776	-0.81885
0.5	1.023598776	-0.69146
0.125	0.648598776	-0.55276
0.0625	0.586098776	-0.52673
0.015625	0.539223776	-0.50675
0.00195313	0.525551906	-0.50085
0.00012207	0.523720846	-0.50005
0.000007629	0.523606405	-0.50001
$\downarrow$	$\downarrow$	$\downarrow$
0	$\frac{\pi}{6} \approx$	$-\frac{1}{2}$
0	0.5235987756	$-\frac{1}{2}$

PREVIEW

An encouraging word is necessary after reading this first example of finding a derivative. By now you are aware of the fact that the derivative concept is one of the main ideas of calculus, and the previous example indicates that finding a derivative can be long and tedious. In the next section, we will begin to simplify the *process* of finding derivatives, so that you can quickly and efficiently find the derivatives of many functions (without using this definition directly). For now, however, we focus on the derivative *concept* and *definition*.

THEOREM 3.1   Equation of a line tangent to a curve at a point

If f is a differentiable function at x_0, the graph of $y = f(x)$ has a tangent line at the point $P(x_0, f(x_0))$ with slope $f'(x_0)$ and equation

$$y = f'(x_0)(x - x_0) + f(x_0)$$

Proof   To find the equation of the tangent line to the curve $y = f(x)$ at the point $P(x_0, y_0)$, we use the fact that the slope of the tangent line is the derivative $f'(x_0)$ and apply the point-slope formula for the equation of a line:

$$y - k = m(x - h) \qquad \text{Point-slope formula}$$
$$y - y_0 = m(x - x_0) \qquad \text{Given point } (x_0, y_0)$$
$$y - f(x_0) = f'(x_0)(x - x_0) \qquad y_0 = f(x_0) \text{ and } m_{\tan} = f'(x_0)$$
$$y = f'(x_0)(x - x_0) + f(x_0) \qquad \text{Add } f(x_0) \text{ to both sides.} \qquad \square$$

EXAMPLE 5   Finding the equation of a tangent line

Find an equation for the tangent line to the graph of $f(x) = \frac{1}{x}$ at the point where $x = 2$.

Solution

The graph of the function $y = \frac{1}{x}$, the point where $x = 2$, and the tangent line at the point are shown in Figure 3.5. First, find $f'(x)$:

$$f'(x) = \lim_{\Delta x \to 0} \frac{f(x + \Delta x) - f(x)}{\Delta x} \qquad \text{Definition of derivative}$$

$$= \lim_{\Delta x \to 0} \frac{\dfrac{1}{x + \Delta x} - \dfrac{1}{x}}{\Delta x} \qquad f(x) = \frac{1}{x}; f(x + \Delta x) = \frac{1}{x + \Delta x}$$

$$= \lim_{\Delta x \to 0} \frac{x - (x + \Delta x)}{x \, \Delta x (x + \Delta x)} \qquad \text{Simplify the fraction.}$$

$$= \lim_{\Delta x \to 0} \frac{-1}{x(x + \Delta x)}$$

$$= \frac{-1}{x^2}$$

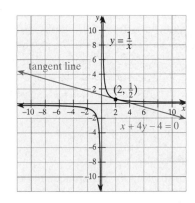

Figure 3.5 Tangent line to $y = \frac{1}{x}$ at $(2, \frac{1}{2})$

WARNING   The slope of f at x_0 is not the derivative f' but the value of the derivative at x_0. In Example 5 the function is defined by $f(x) = 1/x$, the derivative is $f'(x) = -1/x^2$, and the slope at $x = 2$ is the number $f'(2) = -\frac{1}{4}$.

Next, find the slope of the tangent line at $x = 2$: $m_{\tan} = f'(2) = -\frac{1}{4}$. Since $f(2) = \frac{1}{2}$, the equation of the tangent line can now be found by using Theorem 3.1:

$$y = -\tfrac{1}{4}(x - 2) + \tfrac{1}{2}, \quad \text{or in standard form,} \quad x + 4y - 4 = 0 \qquad \blacksquare$$

EXAMPLE 6   Finding a line that is perpendicular to a tangent line

Find the equation of the line that is perpendicular to the tangent line to $f(x) = \frac{1}{x}$ at $x = 2$ and intersects it at the point of tangency.

Solution

In Example 5, we found that the slope of the tangent line is $f'(2) = -\frac{1}{4}$ and that the point of tangency is $(2, \frac{1}{2})$. In Section 1.2, we saw that two lines are perpendicular if and only if their slopes are negative reciprocals of each other. Thus, the perpendicular line we seek has slope 4 (the negative reciprocal of $m = -\frac{1}{4}$). The desired equation is

$$y - \tfrac{1}{2} = 4(x - 2) \qquad \textit{Point-slope formula}$$

In standard form, $4x - y - \frac{15}{2} = 0$; compare the coefficients of the variables in the tangent and perpendicular lines. ∎

The perpendicular line we found in Example 6 has a name.

Normal Line to a Graph

> The **normal line** to the graph of f at the point P is the line that is perpendicular to the tangent line to the graph at P.

Figure 3.6 shows the graph of the function from Example 6, along with the graph of its normal line.

RELATIONSHIP BETWEEN THE GRAPHS OF f AND f'

It is important to take some time to study the relationship between the graph of a function and its derivative. Since slope is measured by f', it follows that at points x where $f'(x) > 0$, the tangent line must be tilted upward, and the graph is rising. Similarly, where $f'(x) < 0$, the tangent line is tilted downward, and the graph is falling. If $f'(x) = 0$, the tangent line is horizontal at x, so the graph "flattens." These observations are illustrated in Figure 3.7.

Figure 3.6 Graph of $y = \frac{1}{x}$ along with its normal $y = 4(x - 2) + \frac{1}{2}$

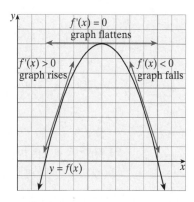

Figure 3.7   How the sign of f determines whether the tangent line slants up, down, or is horizontal

EXAMPLE 7   Sketching the graph of f', given the graph of f

The graph of a function f is shown in Figure 3.8. Sketch a possible graph for the derivative f'.

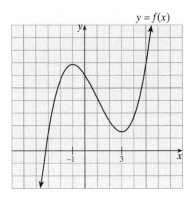

Figure 3.8   Graph of a function f

Solution

Notice that the graph of f is rising for $x < -1$ and for $x > 3$; it is falling for $-1 < x < 3$; and it has horizontal tangent lines at $x = -1$ and $x = 3$. Thus, the graph of f' is above the x-axis ($f'(x) > 0$) for $x < -1$ and $x > 3$, below the axis for $-1 < x < 3$ ($f'(x) < 0$), and crosses the axis at $x = -1$ and at $x = 3$. One possible graph with these features is shown in Figure 3.9.   ■

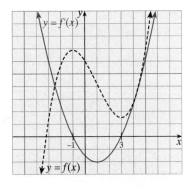

Figure 3.9 A possible graph for the derivative of the function whose graph is given in Figure 3.8

EXISTENCE OF DERIVATIVES

We observed that a function is differentiable only if *the limit* in the definition of derivative exists. At points where a function f is not differentiable, we say that *the derivative of f does not exist*. Three common ways for a derivative to fail to exist at a point $(c, f(c))$ in the domain of f are shown in Figure 3.10.

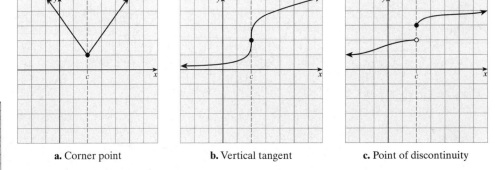

a. Corner point   **b.** Vertical tangent   **c.** Point of discontinuity

Figure 3.10   Common examples where a derivative does not exist

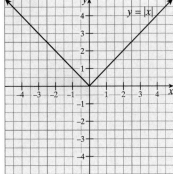

Figure 3.11 $f(x) = |x|$ is not differentiable at $x = 0$ because the slope from the left does not equal the slope from the right

EXAMPLE 8   A function that does not have a derivative because of a corner

Show that the absolute value function $f(x) = |x|$ is not differentiable at $x = 0$.

Solution

The graph of $f(x) = |x|$ is shown in Figure 3.11. Note that because the slope "from the left" at $x = 0$ is -1 while the slope "from the right" is $+1$, the graph has a corner at the origin, which prevents a unique tangent line from being drawn there.

We can show this algebraically by using the definition of derivative:

$$f'(0) = \lim_{\Delta x \to 0} \frac{f(0 + \Delta x) - f(0)}{\Delta x}$$

$$= \lim_{\Delta x \to 0} \frac{f(\Delta x) - f(0)}{\Delta x}$$

$$= \lim_{\Delta x \to 0} \frac{|\Delta x|}{\Delta x}$$

We must now consider one-sided limits, because

$$|\Delta x| = \Delta x \text{ when } \Delta x > 0, \text{ and } |\Delta x| = -\Delta x \text{ when } \Delta x < 0$$

$$\lim_{\Delta x \to 0^-} \frac{|\Delta x|}{\Delta x} = \lim_{\Delta x \to 0^-} \frac{-\Delta x}{\Delta x} = -1 \qquad \text{Derivative from the left}$$

$$\lim_{\Delta x \to 0^+} \frac{|\Delta x|}{\Delta x} = \lim_{\Delta x \to 0^+} \frac{\Delta x}{\Delta x} = 1 \qquad \text{Derivative from the right}$$

The left- and right-hand limits are not the same; therefore, the limit does not exist. This means that the derivative does not exist at $x = 0$. ∎

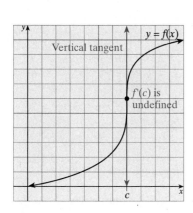

Figure 3.12 Vertical tangent line at $x = c$

The continuous function $f(x) = |x|$ in Example 8 failed to be differentiable at $x = 0$ because the one-sided limits of its difference quotient were unequal. A continuous function may also fail to be differentiable at $x = c$ if its difference quotient diverges to infinity. In this case, the function is said to have a *vertical tangent line* at $x = c$, as illustrated in Figure 3.12. We will have more to say about such functions when we discuss curve sketching with derivatives in Chapter 4.

CONTINUITY AND DIFFERENTIABILITY

If the graph of a function has a tangent line at a point, we would expect to be able to draw the graph continuously (without the pencil leaving the paper). In other words, we expect the following theorem to be true.

THEOREM 3.2   Differentiability implies continuity

If a function f is differentiable at c, then it is also continuous at c.

Proof   Recall that for f to be continuous at $x = c$: (1) $f(c)$ must be defined; (2) $\lim_{x \to c} f(x)$ exists; and (3) $\lim_{x \to c} f(x) = f(c)$. Thus, continuity can be established by showing that $\lim_{\Delta x \to 0} f(c + \Delta x) = f(c)$ or, equivalently,

$$\lim_{\Delta x \to 0} [f(c + \Delta x) - f(c)] = 0$$

Because f is a differentiable function at $x = c$, $f'(c)$ exists and

$$\lim_{\Delta x \to 0} \frac{f(c + \Delta x) - f(c)}{\Delta x} = f'(c)$$

Therefore, by applying the product rule for limits, we find that

$$\lim_{\Delta x \to 0} [f(c + \Delta x) - f(c)] = \lim_{\Delta x \to 0} \left[\frac{f(c + \Delta x) - f(c)}{\Delta x} \cdot \Delta x \right]$$

$$= \left[\lim_{\Delta x \to 0} \frac{f(c + \Delta x) - f(c)}{\Delta x} \right] \left[\lim_{\Delta x \to 0} \Delta x \right]$$

$$= f'(c) \cdot 0 = 0$$

Thus, $\lim_{x \to c} f(x) = f(c)$, and we see that the conditions for continuity are satisfied. ❑

WARNING   Be sure you understand what we have just shown with Example 8 and Theorem 3.2: If a function is differentiable at $x = c$, then it must be continuous at that point. The converse is not true: If a function is continuous at $x = c$, then it may or may not be differentiable at that point. Finally, if a function is discontinuous at $x = c$, then it cannot possibly have a derivative at that point. (See Figure 3.13c.)

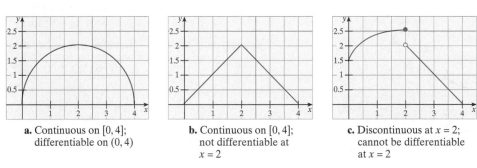

a. Continuous on $[0, 4]$; differentiable on $(0, 4)$

b. Continuous on $[0, 4]$; not differentiable at $x = 2$

c. Discontinuous at $x = 2$; cannot be differentiable at $x = 2$

Figure 3.13   A function continuous at $x = 2$ may or may not be differentiable at $x = 2$. A function discontinuous at $x = 2$ cannot be differentiable at $x = 2$.

DERIVATIVE NOTATION

In certain situations, it is convenient or suggestive to denote the derivative of $y = f(x)$ by $\dfrac{dy}{dx}$ instead of $f'(x)$. This notation is called the *Leibniz notation* because Leibniz was the first to use it.

For example, if $y = x^2$, the derivative is $y' = 2x$ or, using Leibniz notation, $\dfrac{dy}{dx} = 2x$. The symbol $\dfrac{dy}{dx}$ is read "the derivative of y with respect to x." When we wish to denote the value of the derivative at c in the Leibniz notation, we will write

$$\frac{dy}{dx}\bigg|_{x=c}$$

For instance, we would evaluate $\dfrac{dy}{dx} = 4x^2$ at $x = 3$ by writing

$$\frac{dy}{dx}\bigg|_{x=3} = 4x^2\big|_{x=3} = 4(3)^2 = 36$$

Another notation omits reference to y and f altogether, and we can write

$$\frac{d}{dx}(x^2) = 2x$$

which is read "the derivative of x^2 with respect to x is $2x$."

WARNING   Despite its appearance, $\frac{dy}{dx}$ is a single symbol and is *not a fraction*. In Section 3.8, we introduce a concept called a *differential* that will provide independent meaning to symbols like dy and dx, but for now, these symbols have meaning only in connection with the Leibniz derivative symbol $\frac{dy}{dx}$.

EXAMPLE 9   Derivative at a point with Leibniz notation

Find $\dfrac{dy}{dx}\bigg|_{x=-1}$ if $y = x^3$.

Solution

$$\frac{dy}{dx} = \frac{d}{dx}(x^3)$$

$$= \lim_{\Delta x \to 0} \frac{(x + \Delta x)^3 - x^3}{\Delta x}$$

$$= \lim_{\Delta x \to 0} \frac{[x^3 + 3x^2 \Delta x + 3x(\Delta x)^2 + (\Delta x)^3] - x^3}{\Delta x}$$

$$= \lim_{\Delta x \to 0} [3x^2 + 3x(\Delta x) + (\Delta x)^2]$$

$$= 3x^2$$

At $x = -1$, $\left.\dfrac{dy}{dx}\right|_{x=-1} = 3x^2\big|_{x=-1} = 3.$ ∎

```
nDeriv(X^3,X,-1)
        3.000001
```

Figure 3.14 Typical calculator output for $\left.\frac{dy}{dx}\right|_{x=-1}$, where $y = x^3$

If you are using technology to find derivatives, most formats require that you input not only the function, but the variable in the expression and the value at which you wish to find the derivative. The usual format is nDeriv(*expression, variable, value*). The symbol "nDeriv" depends on the calculator or software, and sometimes is "nDer" (TI-85/86), "d" (TI-92), "diff" (Maple V), or "Dif" (Derive). This notation is used in Figure 3.14. Some software and calculators will find the exact value of a derivative, whereas some will do a numerical evaluation (as shown in Figure 3.14).

3.1 PROBLEM SET

A **1. WHAT DOES THIS SAY?** Describe the process of finding a derivative using the definition.

2. WHAT DOES THIS SAY? What is the definition of a derivative?

3. WHAT DOES THIS SAY? Discuss the truth or falsity of the following statements:

If a function f is continuous on (a, b), then it is differentiable on (a, b).

If a function f is differentiable on (a, b), then it is continuous on (a, b).

4. WHAT DOES THIS SAY? Discuss the relationship between the derivative of a function f at a point $x = x_0$ and the tangent line at that same point.

In each of Problems 5–10, the graph of a function f is given. Sketch a graph of f'.

5.

6.

7.

8.

9.

10.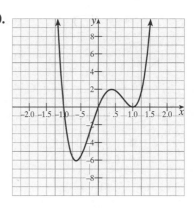

In each of Problems 11–16, a function f is given along with a number c in its domain.

  a. *Find the difference quotient of f.*

  b. *Find $f'(c)$ by computing the limit of the difference quotient.*

11. $f(x) = 3$ at $c = -5$ **12.** $f(x) = x$ at $c = 2$

13. $f(x) = 2x$ at $c = 1$ **14.** $f(x) = 2x^2$ at $c = 1$

15. $f(x) = 2 - x^2$ at $c = 0$ **16.** $f(x) = -x^2$ at $c = 2$

Use the definition to differentiate the functions given in Problems 17–28, and then describe the set of all numbers for which the function is differentiable.

17. $f(x) = 5$ **18.** $g(x) = 3x$

19. $f(x) = 3x - 7$ **20.** $g(x) = 4 - 5x$

21. $g(x) = 3x^2$ **22.** $h(x) = 2x^2 + 3$

23. $f(x) = x^2 - x$ **24.** $g(t) = 4 - t^2$

25. $f(s) = (s - 1)^2$ **26.** $f(x) = \dfrac{1}{2x}$

27. $f(x) = \sqrt{5x}$ **28.** $f(x) = \sqrt{x + 1}$

Find an equation for the tangent line to the graph of the function at the specified point in Problems 29–34.

29. $f(x) = 3x - 7$ at $(3, 2)$ **30.** $g(x) = 3x^2$ at $(-2, 12)$

31. $f(s) = s^3$ at $s = -\frac{1}{2}$ **32.** $g(t) = 4 - t^2$ at $t = 0$

33. $f(x) = \dfrac{1}{x + 3}$ at $x = 2$ **34.** $g(x) = \sqrt{x - 5}$ at $x = 9$

Find an equation of the normal line to the graph of the function at the specified point in Problems 35–38.

35. $f(x) = 3x - 7$ at $(3, 2)$ **36.** $g(x) = 4 - 5x$ at $(0, 4)$

37. $f(x) = \dfrac{1}{x + 3}$ at $x = 3$ **38.** $f(x) = \sqrt{5x}$ at $x = 5$

Find $\dfrac{dy}{dx}\Big|_{x=c}$ for the functions and values of c given in Problems 39–42.

39. $y = 2x, c = -1$ **40.** $y = 4 - x, c = 2$

41. $y = 1 - x^2, c = 0$ **42.** $y = \dfrac{4}{x}, c = 1$

B **43.** Suppose $f(x) = x^2$.

  a. Compute the slope of the secant line joining the points on the graph of f whose x-coordinates are -2 and -1.9.

  b. Use calculus to compute the slope of the line that is tangent to the graph when $x = -2$ and compare this slope with your answer in part **a.**

44. Suppose $f(x) = x^3$.

  a. Compute the slope of the secant line joining the points on the graph of f whose x-coordinates are 1 and 1.1.

  b. Use calculus to compute the slope of the line that is tangent to the graph when $x = 1$ and compare this slope to your answer from part **a.**

45. Sketch the graph of the function $y = x^2 - x$. Determine the value(s) of x for which the derivative is 0. What happens to the graph at the corresponding point(s)?

46. **a.** Find the derivative of $f(x) = x^2 - 3x$.

  b. Show that the parabola whose equation is $y = x^2 - 3x$ has one horizontal tangent line. Find the equation of this line.

  c. Find a point on the graph of f where the tangent line is parallel to the line $3x + y = 11$.

  d. Sketch the graph of the parabola whose equation is $y = x^2 - 3x$. Display the horizontal tangent line and the tangent line found in part **c.**

47. **a.** Find the derivative of $f(x) = 4 - 2x^2$.

  b. The graph of f has one horizontal tangent line. What is its equation?

  c. At what point on the graph of f is the tangent line parallel to the line $8x + 3y = 4$?

48. Show that the function $f(x) = |x - 2|$ is not differentiable at $x = 2$.

49. Is the function $f(x) = 2|x + 1|$ differentiable at $x = 1$?

50. Let $f(x) = \begin{cases} -x^2 & \text{if } x < 0 \\ x^2 & \text{if } x \geq 0 \end{cases}$.

  Does $f'(0)$ exist? *Hint*: Find the difference quotient and take the limit as $\Delta x \to 0$ from the left and from the right.

51. Let $f(x) = \begin{cases} -2x & \text{if } x < 1 \\ \sqrt{x} - 3 & \text{if } x \geq 1 \end{cases}$.

  a. Sketch the graph of f.

  b. Show that f is continuous but not differentiable at $x = 1$.

52. **Counterexample Problem** Give an example of a function that is continuous on $(-\infty, \infty)$ but is not differentiable at $x = 5$.

Estimate the derivative $f'(c)$ in Problems 53–58 by evaluating the difference quotient

$$\frac{\Delta y}{\Delta x} = \frac{f(c + \Delta x) - f(c)}{\Delta x}$$

at a succession of numbers near c.

53. $f(x) = (2x - 1)^2$ for $c = 1$ **54.** $f(x) = \dfrac{1}{x+1}$ for $c = 2$

55. $f(x) = \sin x$ for $c = \frac{\pi}{3}$ **56.** $f(x) = \cos x$ for $c = \frac{\pi}{3}$

57. $f(x) = \sqrt{x}$ for $c = 4$ **58.** $f(x) = \sqrt[3]{x}$ for $c = 8$

59. Show that the tangent line to the parabola $y = Ax^2$ (for $A \neq 0$) at the point where $x = c$ will intersect the x-axis at the point $(\frac{c}{2}, 0)$. Where does it intersect the y-axis?

60. Find the point(s) on the graph of $f(x) = -x^2$ such that the tangent line at that point passes through the point $(0, -9)$.

61. Interpretation Problem

  a. Find the derivatives of the functions $y = x^2$ and $y = x^2 - 3$ and account geometrically for their similarity.

  b. Without further computation, find the derivative of $y = x^2 + 5$.

62. Interpretation Problem

  a. Find the derivative of $f(x) = x^2 + 3x$.

  b. Find the derivatives of the functions $g(x) = x^2$ and $h(x) = 3x$ separately. How are these derivatives related to the derivative in part **a**?

  c. In general, if $f(x) = g(x) + h(x)$, what would you guess is the relationship between the derivative of f and the derivatives of g and h?

63. Consider a graph of the function defined by $f(x) = x^{2/3}$.

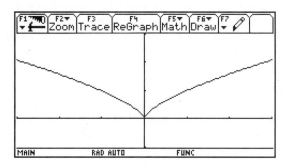

There is a tangent utility on many calculators that will draw tangent lines at a given point. Use this utility or use the preceding simulated graph to draw tangent lines as $x \to 0^-$. Next, draw tangent lines as $x \to 0^+$. Describe what happens, and use this description to support the conclusion that there is no tangent line at $x = 0$.

64. Consider the function

$$f(x) = (x - 2)^{2/3} + 2x^3$$

which gives trouble in seeking the tangent line at $(2, f(2))$. Attempt to compute $f'(2)$, either "by hand" or using a calculator. Describe what happens; in particular, do you see why the tangent line there is meaningless?

65. Compute the difference quotient for the function defined by

$$f(x) = \begin{cases} \dfrac{\sin x}{x} & \text{if } x \neq 0 \\ 1 & \text{if } x = 0 \end{cases}$$

Do you think $f(x)$ is differentiable at $x = 0$? If so, what is the equation of the tangent line at $x = 0$?

66. HISTORICAL QUEST In the guest essay at the end of Chapter 1, it was noted that Isaac Newton, who invented calculus at the same time as Leibniz, considered Fermat as "one of the giants" on whose shoulders he stood. Fermat was a lawyer by profession, but he liked to do mathematics in his spare time. He wrote well over 3,000 mathematical papers and notes. Fermat developed a general procedure for finding tangent lines that is a precursor to the methods of Newton and Leibniz. ■

PIERRE DE FERMAT
1601–1665

We will explore this procedure by finding a tangent line to the curve

$$x^3 + y^3 - 2xy = 0$$

A graph is shown in Figure 3.15.

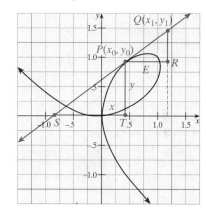

Figure 3.15 Fermat's method of subtangents

  a. Let $P(x_0, y_0)$ be a given point on the curve, and let S be the x-axis intercept of the tangent line at P. Let $Q(x_1, y_1)$ be another point on the tangent line and let T and R be located so $\triangle STP$ and $\triangle PRQ$ are right triangles. Finally, let $A = |\overline{ST}|$ and $E = |\overline{PR}|$. Express x_1 and y_1 in terms of x_0, y_0, A, and E.

  b. Fermat reasoned that if E were very small, then Q would "almost" be on the curve. Substitute the values for x_1 and y_1 you found in part **a** into the equation

$$x^3 + y^3 - 2xy = 0$$

That is, substitute the (x, y) values you found in part **a** into this equation.

  c. With the answer from part **b**, you can follow the steps of Fermat by dividing both sides of the equation by E. Fermat reasoned that since E was close to 0, this step should be permitted. Now, after doing this, Fermat further reasoned that since E was close to 0, he could now set $E = 0$ and solve for A. Carry out these steps to write A in terms of x_0 and y_0.

  d. Fermat then constructed the tangent line at P by joining P to S. Draw the tangent line for the given curve at the point $(0.5, 0.93)$ by plotting the point corresponding to the calculated value of A. Find an equation for the tangent line to the curve

$$x^3 + y^3 - 2xy = 0$$

at the point $P(0.5, 0.93)$.

67. HISTORICAL QUEST   The groundwork for much of the mathematics we do today, and certainly a necessity for calculus, is the development of analytic geometry by Descartes and Pierre de Fermat (see Problem 66). Descartes' ideas for analytic geometry were published in 1637 as one of three appendices to his *Discourse on the Method* (of Reasoning Well and Seeking Truth in the Sciences). In that same year, Fermat sent an essay entitled *Introduction to Plane and Solid Loci* to Paris, and in this essay he laid the foundation for analytic geometry. Fermat's paper was more complete and systematic, but Descartes' was published first. Descartes is generally credited with the discovery of analytic geometry, and we speak today of the Cartesian coordinate system and Cartesian geometry to honor Descartes' discovery. Today we describe analytic geometry from two viewpoints: (1) Given a curve, describe it by an equation (Descartes' viewpoint); and (2) given an equation, describe it by a curve (Fermat's viewpoint).

RENÉ DESCARTES
1596–1650

In this HISTORICAL QUEST we will describe Descartes' circle method for finding a tangent line to a given curve. This method uses algebra and geometry rather than limits. Descartes' method for finding the tangent line to the curve $y = f(x)$ at the point $P(x_0, f(x_0))$ involved first finding the point $Q(x_1, 0)$, which is the point of intersection of the normal line to the curve at P with the x-axis, as shown in Figure 3.16.

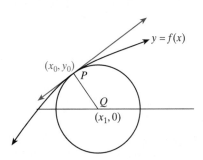

Figure 3.16   Descartes' circle method for finding tangent lines

Descartes then wrote the equation of the circle with center Q passing through P:

$$(x - x_1)^2 + y^2 = (x_0 - x_1)^2 + y_0^2$$

Descartes' next step was to use the equation of the given curve $y = f(x)$ (actually any equation involving two variables) to eliminate one of the variables (usually y) from the equation of the circle. Descartes reasoned that the circle will cut the given curve in two places except when $\overline{PQ}$ is normal, in which case the two intersection points will coincide and the circle will be tangent line to the curve at P. That is, Descartes imposed the condition that the resulting equation (after substitution) has only one root to solve for x_1. The point Q is thus found and the normal line was then known to Descartes. The tangent line can be taken as the perpendicular through P to the normal line. ∎

Carry out Descartes circle method for the parabola $y^2 = 4x$ at the point $(4, 4)$.

68. HISTORICAL QUEST   Pierre de Fermat (see Problem 66) also obtained the first method for differentiating polynomials, but his real love was number theory. His most famous problem has come to be known as Fermat's last theorem. He wrote in the margin of a text: *"To divide a cube into two cubes, a fourth power, or in general, any power whatever above the second, into powers of the same denominations, is impossible, and I have assuredly found an admirable proof of this, but the margin is too narrow to contain it."* ∎

In 1993 an imaginative, but lengthy, proof was constructed by Andrew Wiles, and in 1997 Wiles was awarded the Wolfskehl Prize for proving this theorem. An engaging book, *Fermat's Enigma*, shows how a great mathematical puzzle of our age was solved: Lock yourself in a room and emerge seven years later. Write a paper on the history of Fermat's last theorem, including some discussion of the current status of its proof. [You might begin with "Fermat's Last Theorem, The Four Color Conjecture, and Bill Clinton for April Fools' Day," by Edward B. Burger and Frank Morgan, *American Mathematical Monthly*, March 1997, pp. 246–250. Next, see the NOVA film, *The Proof*, which was shown in 1997 on many PBS stations (check with your library or PBS Online). Finally, see "Fermat's Last Stand" by Simon Singh and Kenneth A. Ribet, *Scientific American*, November 1997, pp. 68–73.]

C 69. If $f'(c) \neq 0$, what is the equation of the normal line to $y = f(x)$ at the point $P(c, f(c))$? What is the equation if $f'(c) = 0$?

70. Suppose a parabola is given in the plane along with its axis of symmetry. Explain how you could construct the tangent line at a given point P on the parabola using only compass and straightedge methods. *Hint*: You may assume that the parabola has an equation of the form $y = Ax^2$ in which the y-axis is the axis of symmetry and the vertex of the parabola is at the origin. Then use the result of Problem 59.

3.2   Techniques of Differentiation

IN THIS SECTION   derivative of a constant function, derivative of a power function, procedural rules for finding derivatives, higher-order derivatives

If we had to compute the limit of a difference quotient every time we wished to find a derivative, differentiation would not be the valuable tool that it is. Fortunately, there is a better way, and in this section, we will derive several formulas that will enable us to compute derivatives indirectly, without evaluating any limits.

DERIVATIVE OF A CONSTANT FUNCTION

We begin by proving that the derivative of any constant function is zero. Notice that this is plausible because the graph of the constant function $f(x) = k$ is a horizontal line, and its slope is zero. Thus, for example, if $f(x) = 5$, then $f'(x) = 0$.

THEOREM 3.3 Constant rule

A constant function $f(x) = k$ has derivative $f'(x) = 0$; in Leibniz notation,

$$\frac{d}{dx}(k) = 0$$

Proof   Note that if $f(x) = k$, then $f(x + \Delta x) = k$ for all Δx. Therefore, the difference quotient is

$$\frac{f(x + \Delta x) - f(x)}{\Delta x} = \frac{k - k}{\Delta x} = 0$$

and

$$f'(x) = \lim_{\Delta x \to 0} \frac{f(x + \Delta x) - f(x)}{\Delta x} = \lim_{\Delta x \to 0} 0 = 0$$

WARNING   Remember $\Delta x \neq 0$ even though $\Delta x \to 0$.

as claimed.   ❏

DERIVATIVE OF A POWER FUNCTION

Recall that a **power function** is a function of the form $f(x) = x^n$, where n is a real number. For example, $f(x) = x^2$, $g(x) = x^{-3}$, $h(x) = x^{1/2}$ are all power functions. So are

$$F(x) = \frac{1}{x^2} = x^{-2} \quad \text{and} \quad G(x) = \sqrt[3]{x^2} = x^{2/3}$$

Here is a simple rule for finding the derivative of any power function.

THEOREM 3.4 Power rule

For any real number n, the power function $f(x) = x^n$ has the derivative $f'(x) = nx^{n-1}$; in Leibniz notation,

$$\frac{d}{dx}(x^n) = nx^{n-1}$$

Proof   If the exponent n is a positive integer, we can prove the power rule by using the binomial theorem with the definition of derivative. Begin with the difference quotient:

$$\frac{f(x + \Delta x) - f(x)}{\Delta x} = \frac{(x + \Delta x)^n - x^n}{\Delta x}$$

$$= \frac{\left[x^n + nx^{n-1}\Delta x + \frac{n(n-1)}{2}x^{n-2}(\Delta x)^2 + \cdots + (\Delta x)^n\right] - x^n}{\Delta x}$$

$$= \frac{nx^{n-1}\Delta x + \frac{n(n-1)}{2}x^{n-2}(\Delta x)^2 + \cdots + (\Delta x)^n}{\Delta x}$$

$$= nx^{n-1} + \frac{n(n-1)}{2}x^{n-2}\Delta x + \cdots + (\Delta x)^{n-1}$$

Note that Δx is a factor of every term in this expression except the first. Hence, as $\Delta x \to 0$, we have

$$f'(x) = \lim_{\Delta x \to 0} \frac{f(x + \Delta x) - f(x)}{\Delta x}$$

$$= \lim_{\Delta x \to 0} \left[nx^{n-1} + \frac{n(n-1)}{2}x^{n-2}\Delta x + \cdots + (\Delta x)^{n-1}\right]$$

$$= nx^{n-1}$$

If $n = 0$, then $f(x) = x^0 = 1$, so $f'(x) = 0$. We will prove the power rule for negative integer exponents later in this section, and we will deal with the case in which the exponent is any real number in Section 3.6. Note, however, that we have already verified the power rule for $n = 3$ in Example 9 of Section 3.1 and for the rational exponent $\frac{1}{2}$ in Example 3 of Section 3.1, when we showed that the derivative of $f(t) = \sqrt{t} = t^{1/2}$ is

$$f'(t) = \frac{1}{2}t^{-1/2} = \frac{1}{2\sqrt{t}} \quad \text{for } t > 0$$

For the following examples, and the problems at the end of this section, you may assume that the power rule is valid when the exponent n is any real number. ❏

EXAMPLE 1 Using the power rule to find a derivative

Differentiate each of the following functions.

a. $f(x) = x^8$ **b.** $g(x) = x^{3/2}$ **c.** $h(x) = \dfrac{\sqrt[3]{x}}{x^2}$

Solution

a. Applying the power rule with $n = 8$, we find that

$$\frac{d}{dx}(x^8) = 8x^{8-1} = 8x^7$$

b. Applying the power rule with $n = \frac{3}{2}$, we get

$$\frac{d}{dx}(x^{3/2}) = \tfrac{3}{2}x^{3/2-1} = \tfrac{3}{2}x^{1/2} = \tfrac{3}{2}\sqrt{x}$$

c. For this part you need to recognize that $h(x) = \dfrac{x^{1/3}}{x^2} = x^{-5/3}$. Applying the power rule with $n = -\frac{5}{3}$, we find that

$$\frac{d}{dx}(x^{-5/3}) = -\tfrac{5}{3}x^{-5/3-1} = -\tfrac{5}{3}x^{-8/3} \qquad ∎$$

PROCEDURAL RULES FOR FINDING DERIVATIVES

The next theorem expands the class of functions that we can differentiate easily by giving rules for differentiating certain combinations of functions, such as sums, differences, products, and quotients. We will see that the derivative of a sum (difference) is the sum (difference) of derivatives, but the derivative of a product (or quotient) does not have such a simple form. For example, to convince yourself that the derivative of a product is not the product of the separate derivatives, consider the power functions

$$f(x) = x \quad \text{and} \quad g(x) = x^2$$

and their product

$$p(x) = f(x)g(x) = x^3$$

WARNING Note the derivative of a product is not the product of derivatives. A similar warning applies to the derivative of a quotient. (See Theorem 3.5.)

Because $f'(x) = 1$ and $g'(x) = 2x$, the product of the derivatives is

$$f'(x)g'(x) = (1)(2x) = 2x$$

whereas the actual derivative of $p(x) = x^3$ is $p'(x) = 3x^2$. The product rule tells us how to find the derivative of a product.

THEOREM 3.5 Basic rules for combining derivatives—Procedural forms

If f and g are differentiable functions at all x, and a, b, and c are any real numbers, then the functions cf, $f + g$, fg, and f/g (for $g(x) \neq 0$) are also differentiable, and their derivatives satisfy the following formulas:

Name of Rule	Function Notation	Leibniz Notation
Constant multiple	$[cf(x)]' = cf'(x)$	$\dfrac{d}{dx}(cf) = c\dfrac{df}{dx}$
Sum rule	$[f(x) + g(x)]' = f'(x) + g'(x)$	$\dfrac{d}{dx}(f + g) = \dfrac{df}{dx} + \dfrac{dg}{dx}$
Difference rule	$[f(x) - g(x)]' = f'(x) - g'(x)$	$\dfrac{d}{dx}(f - g) = \dfrac{df}{dx} - \dfrac{dg}{dx}$

The constant multiple, sum, and difference rules can be combined into a single rule, which is called the *linearity rule*.

Linearity rule	$[af(x) + bg(x)]' = af'(x) + bg'(x)$	$\dfrac{d}{dx}(af + bg) = a\dfrac{df}{dx} + b\dfrac{dg}{dx}$
Product rule	$[f(x)g(x)]' = f(x)g'(x) + f'(x)g(x)$	$\dfrac{d}{dx}(fg) = f\dfrac{dg}{dx} + g\dfrac{df}{dx}$
Quotient rule	$\left[\dfrac{f(x)}{g(x)}\right]' = \dfrac{g(x)f'(x) - f(x)g'(x)}{[g(x)]^2}$	$\dfrac{d}{dx}\left(\dfrac{f}{g}\right) = \dfrac{g\dfrac{df}{dx} - f\dfrac{dg}{dx}}{g^2}$

WARNING Note that the order of terms in the quotient rule matters because

$$gf' - fg' \neq fg' - gf'$$

Proof We will prove the product rule in detail, leaving the other rules as problems.
Let $f(x)$ and $g(x)$ be differentiable functions of x and let $p(x) = f(x)g(x)$. We will add and subtract the term $f(x + \Delta x)g(x)$ to the numerator of the difference quotient for $p(x)$ to create difference quotients for $f(x)$ and $g(x)$. Thus,

$$p'(x) = \frac{dp}{dx} = \lim_{\Delta x \to 0} \frac{p(x + \Delta x) - p(x)}{\Delta x}$$

$$= \lim_{\Delta x \to 0} \frac{f(x + \Delta x)g(x + \Delta x) - f(x)g(x)}{\Delta x}$$

$$= \lim_{\Delta x \to 0} \frac{f(x + \Delta x)g(x + \Delta x) - f(x + \Delta x)g(x) + f(x + \Delta x)g(x) - f(x)g(x)}{\Delta x}$$

$$= \lim_{\Delta x \to 0} \left[f(x + \Delta x) \left[\frac{g(x + \Delta x) - g(x)}{\Delta x} \right] + g(x) \left[\frac{f(x + \Delta x) - f(x)}{\Delta x} \right] \right]$$

$$= \lim_{\Delta x \to 0} f(x + \Delta x) \underbrace{\lim_{\Delta x \to 0} \left[\frac{g(x + \Delta x) - g(x)}{\Delta x} \right]}_{\text{This is the derivative of } g.} + \underbrace{\lim_{\Delta x \to 0} g(x) \lim_{\Delta x \to 0} \left[\frac{f(x + \Delta x) - f(x)}{\Delta x} \right]}_{\text{This is the derivative of } f.}$$

$$= f(x)g'(x) + g(x)f'(x) \qquad \lim_{\Delta x \to 0} f(x + \Delta x) = f(x) \text{ because } f \text{ is continuous.} \qquad \square$$

Note that in each part of Theorem 3.5, we prove the differentiability of the appropriate functional combination at the same time we are establishing the differentiation formula.

EXAMPLE 2   Using the basic rules to find derivatives

Differentiate each of the following functions.

a. $f(x) = 2x^2 - 5\sqrt{x}$ **b.** $p(x) = (3x^2 - 1)(7 + 2x^3)$

c. $q(x) = \dfrac{4x - 7}{3 - x^2}$ **d.** $g(x) = (4x + 3)^2$

e. $F(x) = \dfrac{2}{3x^2} - \dfrac{x}{3} + \dfrac{4}{5} + \dfrac{x + 1}{x}$

Solution

a. Apply the linearity rule (constant multiple, sum, and difference) and power rules:

$$f'(x) = 2(x^2)' - 5(x^{1/2})' = 2(2x) - 5\left(\tfrac{1}{2}\right)(x^{-1/2}) = 4x - \tfrac{5}{2}x^{-1/2}$$

b. Apply the product rule; then apply the linearity and power rules:

$$\begin{aligned}
p'(x) &= (3x^2 - 1)(7 + 2x^3)' + (3x^2 - 1)'(7 + 2x^3) \\
&= (3x^2 - 1)[0 + 2(3x^2)] + [3(2x) - 0](7 + 2x^3) \\
&= (3x^2 - 1)(6x^2) + (6x)(7 + 2x^3) \\
&= 6x(5x^3 - x + 7)
\end{aligned}$$

c. Apply the quotient rule, then the linearity and power rules:

$$\begin{aligned}
q'(x) &= \frac{(3 - x^2)(4x - 7)' - (4x - 7)(3 - x^2)'}{(3 - x^2)^2} \\
&= \frac{(3 - x^2)(4 - 0) - (4x - 7)(0 - 2x)}{(3 - x^2)^2} \\
&= \frac{12 - 4x^2 + 8x^2 - 14x}{(3 - x^2)^2} = \frac{4x^2 - 14x + 12}{(3 - x^2)^2}
\end{aligned}$$

d. Apply the product rule:

$$\begin{aligned}
g'(x) &= (4x + 3)(4x + 3)' + (4x + 3)'(4x + 3) \\
&= (4x + 3)(4) + (4)(4x + 3) = 8(4x + 3)
\end{aligned}$$

Sometimes when the exponent is 2, it is easier to expand before differentiating:

$$g(x) = (4x + 3)^2 = 16x^2 + 24x + 9$$
$$g'(x) = 32x + 24$$

e. Write the function using negative exponents for reciprocal powers:

$$F(x) = \tfrac{2}{3}x^{-2} - \tfrac{1}{3}x + \tfrac{4}{5} + 1 + x^{-1}$$

Then apply the power rule term by term to obtain

$$\begin{aligned}
F'(x) &= \tfrac{2}{3}(-2x^{-3}) - \tfrac{1}{3} + 0 + 0 + (-1)x^{-2} \\
&= -\tfrac{4}{3}x^{-3} - \tfrac{1}{3} - x^{-2}
\end{aligned}$$ ∎

In applying the power rule term by term in Example 2e, we really used the following generalization of the linearity rule.

COROLLARY TO THEOREM 3.5   The extended linearity rule

If $f_1, f_2, \ldots, f_n$ are differentiable functions and $a_1, a_2, \ldots, a_n$ are constants, then

$$\frac{d}{dx}[a_1 f_1 + a_2 f_2 + \cdots + a_n f_n] = a_1 \frac{df_1}{dx} + a_2 \frac{df_2}{dx} + \cdots + a_n \frac{df_n}{dx}$$

Proof   The proof is a straightforward extension (using mathematical induction) of the proof of the linearity rule of Theorem 3.5.   ❑

Example 3 illustrates how the extended linearity rule can be used to differentiate a polynomial.

EXAMPLE 3   Derivative of a polynomial function

Differentiate the polynomial function $p(x) = 2x^5 - 3x^2 + 8x - 5$.

Solution

$$
\begin{aligned}
p'(x) &= \frac{d}{dx}[2x^5 - 3x^2 + 8x - 5] \\
&= 2\frac{d}{dx}(x^5) - 3\frac{d}{dx}(x^2) + 8\frac{d}{dx}(x) - \frac{d}{dx}(5) \qquad \textit{Extended linearity rule} \\
&= 2(5x^4) - 3(2x) + 8(1) - 0 \qquad\qquad\qquad \textit{Power rule; constant rule} \\
&= 10x^4 - 6x + 8 \qquad\qquad\qquad\qquad\qquad\qquad\quad \blacksquare
\end{aligned}
$$

EXAMPLE 4   Derivative of a product of polynomials

Differentiate $p(x) = (x^3 - 4x + 7)(3x^5 - x^2 + 6x)$.

Solution

We could expand the product function $p(x)$ as a polynomial and proceed as in Example 3, but it is easier to use the product rule.

$$
\begin{aligned}
p'(x) &= (x^3 - 4x + 7)(3x^5 - x^2 + 6x)' + (x^3 - 4x + 7)'(3x^5 - x^2 + 6x) \\
&= (x^3 - 4x + 7)(15x^4 - 2x + 6) + (3x^2 - 4)(3x^5 - x^2 + 6x)
\end{aligned}
$$

This form is an acceptable answer, but if you are using software or an algebraic calculator, more than likely you will obtain the expanded formula

$$
p'(x) = 24x^7 - 72x^5 + 100x^4 + 24x^3 + 12x^2 - 62x + 42 \qquad \blacksquare
$$

EXAMPLE 5   Equation of a tangent line

Find the standard form equation for the line tangent to the graph of

$$
f(x) = \frac{3x^2 + 5}{2x^2 + x - 3}
$$

at the point where $x = -1$.

Solution

Evaluating $f(x)$ at $x = -1$, we find that $f(-1) = -4$ (verify); therefore, the point of tangency is $(-1, -4)$. The slope of the tangent line at $(-1, -4)$ is $f'(-1)$. Find $f'(x)$ by applying the quotient rule:

$$
\begin{aligned}
f'(x) &= \frac{(2x^2 + x - 3)(3x^2 + 5)' - (3x^2 + 5)(2x^2 + x - 3)'}{(2x^2 + x - 3)^2} \\
&= \frac{(2x^2 + x - 3)(6x) - (3x^2 + 5)(4x + 1)}{(2x^2 + x - 3)^2}
\end{aligned}
$$

The slope of the tangent line is

$$
f'(-1) = \frac{(2 - 1 - 3)(-6) - (3 + 5)(-4 + 1)}{(2 - 1 - 3)^2} = \frac{(-2)(-6) - (8)(-3)}{(-2)^2} = 9
$$

From the formula (Theorem 3.1) $y = f'(x_0)(x - x_0) + f(x_0)$, we find that an equation for the tangent line at $(-1, -4)$ is

$$
y = 9(x + 1) + (-4)
$$

or in standard form $9x - y + 5 = 0$. The graphs of both f and its tangent line at $(-1, -4)$ are shown in Figure 3.17. $\blacksquare$

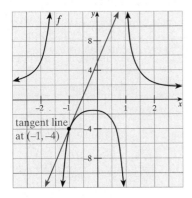

Figure 3.17 Graph of f and the tangent line at the point $(-1, -4)$

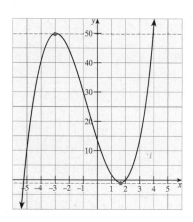

Figure 3.18 Graph of curve and the horizontal tangent lines

EXAMPLE 6   Finding horizontal tangent lines

Let $y = (x - 2)(x^2 + 4x - 7)$. Find all points on this curve where the tangent line is horizontal.

Solution

The tangent line will be horizontal when $dy/dx = 0$, because the derivative dy/dx measures the slope and a horizontal line has slope 0 (see Figure 3.18). Applying the product rule, we find

$$\frac{dy}{dx} = (x-2)(x^2+4x-7)' + (x-2)'(x^2+4x-7)$$
$$= (x-2)(2x+4) + (1)(x^2+4x-7)$$
$$= 2x^2 - 8 + x^2 + 4x - 7$$
$$= 3x^2 + 4x - 15 = (3x-5)(x+3)$$

Thus, $\dfrac{dy}{dx} = 0$ when $x = \frac{5}{3}$ or $x = -3$. The corresponding points $(\frac{5}{3}, -\frac{22}{27})$ and $(-3, 50)$ are the points on the curve at which the tangent line is horizontal.   ■

In the following example, we use the quotient rule to extend the proof of the power rule to the case in which the exponent n is a negative integer.

EXAMPLE 7   Proof of the power rule for negative exponents

Show that $\dfrac{d}{dx}(x^n) = nx^{n-1}$ if $n = -m$, where m is a positive integer.

Solution

We have $f(x) = x^n = x^{-m} = 1/x^m$, so apply the quotient rule:

$$\frac{d}{dx}(x^n) = \frac{d}{dx}\left(\frac{1}{x^m}\right) = \frac{x^m(1)' - (1)(x^m)'}{(x^m)^2} = \frac{x^m(0) - mx^{m-1}}{x^{2m}}$$
$$= -mx^{(m-1)-2m} = -mx^{-m-1} = nx^{n-1} \qquad \text{Substitute } -m = n. \quad ■$$

HIGHER-ORDER DERIVATIVES

Occasionally, it is useful to differentiate the derivative of a function. In this context, we will refer to f' as the **first derivative** of f and to the derivative of f' as the **second derivative** of f. We could denote the second derivative by $(f')'$, but for simplicity we write f''. Other higher-order derivatives are defined and denoted similarly. Thus, the **third derivative** of f is the derivative of f'' and is denoted by f'''. In general, for $n > 3$, the ***n*th derivative** of f is denoted by $f^{(n)}$; $f^{(4)}$ or $f^{(5)}$, for example. In Leibniz notation, higher-order derivatives for $y = f(x)$ are denoted as follows:

			Leibniz notation		
First derivative:	y'	$f'(x)$	$\dfrac{dy}{dx}$	**or**	$\dfrac{d}{dx}f(x)$
Second derivative:	y''	$f''(x)$	$\dfrac{d}{dx}\left(\dfrac{dy}{dx}\right) = \dfrac{d^2y}{dx^2}$	**or**	$\dfrac{d^2}{dx^2}f(x)$
Third derivative:	y'''	$f'''(x)$	$\dfrac{d}{dx}\left(\dfrac{d^2y}{dx^2}\right) = \dfrac{d^3y}{dx^3}$	**or**	$\dfrac{d^3}{dx^3}f(x)$

Leibniz notation

Fourth derivative: $y^{(4)}$ $f^{(4)}(x)$ $\dfrac{d^4 y}{dx^4}$ or $\dfrac{d^4}{dx^4} f(x)$

$\vdots$ $\qquad$ $\vdots$ $\quad$ $\vdots$ $\quad$ $\vdots$ $\qquad$ $\vdots$

nth derivative $y^{(n)}$ $f^{(n)}(x)$ $\dfrac{d^n y}{dx^n}$ or $\dfrac{d^n}{dx^n} f(x)$

> ➡ **What This Says** Because the derivative of a function is a function, differentiation can be applied over and over, as long as the derivative itself is a differentiable function. That is, we can take derivatives of derivatives.
> Notice also that for derivatives of higher order than the third, the parentheses distinguish a derivative from a power. For example, $f^4 \neq f^{(4)}$.

You should note that all higher-order derivatives of a polynomial $p(x)$ will also be polynomials, and if p has degree n, then $p^{(k)}(x) = 0$ for $k \geq n + 1$, as illustrated in the following example.

EXAMPLE 8 Higher-order derivatives for a polynomial function

Find the derivatives of all orders of

$$p(x) = -2x^4 + 9x^3 - 5x^2 + 7$$

Solution

$$p'(x) = -8x^3 + 27x^2 - 10x; \quad p''(x) = -24x^2 + 54x - 10;$$
$$p'''(x) = -48x + 54; \quad p^{(4)}(x) = -48; \quad p^{(5)}(x) = 0; \dots \quad p^{(k)}(x) = 0 \ (k \geq 5)$$ ■

3.2 PROBLEM SET

Ⓐ *To demonstrate the power of the theorems of this section, Problems 1–4 ask you to go back and rework some problems in Section 3.1, using the material of this section instead of the definition of derivative.*

1. Find the derivatives of the functions given in Problems 11–16 of Problem Set 3.1.
2. Find the derivatives of the functions given in Problems 17–22 of Problem Set 3.1.
3. Find the derivatives of the functions given in Problems 23–27 of Problem Set 3.1.
4. Find the derivatives of the functions given in Problems 39–42 of Problem Set 3.1.

Differentiate the functions given in Problems 5–20. Assume that C is a constant.

5. **a.** $f(x) = 3x^4 - 9$ **b.** $g(x) = 3(9)^4 - x$
6. **a.** $f(x) = 5x^2 + x$ **b.** $g(x) = \pi^3$
7. **a.** $f(x) = x^3 + C$ **b.** $g(x) = C^2 + x$
8. **a.** $f(t) = 10t^{-1}$ **b.** $g(t) = \dfrac{7}{t}$
9. $r(t) = t^2 - \dfrac{1}{t^2} + \dfrac{5}{t^4}$ 10. $f(x) = \pi^3 - 3\pi^2$

11. $f(x) = \dfrac{7}{x^2} + x^{2/3} + C$ 12. $g(x) = \dfrac{1}{2\sqrt{x}} + \dfrac{x^2}{4} + C$

13. $f(x) = \dfrac{x^3 + x^2 + x - 7}{x^2}$ 14. $g(x) = \dfrac{2x^5 - 3x^2 + 11}{x^3}$

15. $f(x) = (2x + 1)(1 - 4x^3)$ 16. $g(x) = (x + 2)(2\sqrt{x} + x^2)$

17. $f(x) = \dfrac{3x + 5}{x + 9}$ 18. $f(x) = \dfrac{x^2 + 3}{x^2 + 5}$

19. $g(x) = x^2(x + 2)^2$ 20. $f(x) = x^2(2x + 1)^2$

In Problems 21–24, find f', f'', f''', and $f^{(4)}$.

21. $f(x) = x^5 - 5x^3 + x + 12$
22. $f(x) = \frac{1}{4}x^8 - \frac{1}{2}x^6 - x^2 + 2$

23. $f(x) = \dfrac{-2}{x^2}$ 24. $f(x) = \dfrac{4}{\sqrt{x}}$

25. Find $\dfrac{d^2 y}{dx^2}$, where $y = 3x^3 - 7x^2 + 2x - 3$.

26. Find $\dfrac{d^2 y}{dx^2}$, where $y = (x^2 + 4)(1 - 3x^3)$.

In Problems 27–32, find the standard form equation for the tangent line to $y = f(x)$ at the specified point.

27. $f(x) = x^2 - 3x - 5$, where $x = -2$

28. $f(x) = x^5 - 3x^3 - 5x + 2$, where $x = 1$

29. $f(x) = (x^2 + 1)(1 - x^3)$, where $x = 1$

30. $f(x) = \dfrac{x + 1}{x - 1}$, where $x = 0$

31. $f(x) = \dfrac{x^2 + 5}{x + 5}$, where $x = 1$

32. $f(x) = 1 - \dfrac{1}{x} + \dfrac{2}{\sqrt{x}}$, where $x = 4$

Find the coordinates of each point on the graph of the given function where the tangent line is horizontal in Problems 33–39.

33. $f(x) = 2x^3 - 7x^2 + 8x - 3$

34. $f(t) = t^4 + 4t^3 - 8t^2 + 3$ **35.** $g(x) = (3x - 5)(x - 8)$

36. $f(t) = \dfrac{1}{t^2} - \dfrac{1}{t^3}$ **37.** $f(x) = \sqrt{x}(x - 3)$

38. $h(u) = \dfrac{1}{\sqrt{u}}(u + 9)$ **39.** $h(x) = \dfrac{4x^2 + 12x + 9}{2x + 3}$

B 40. a. Differentiate the function $f(x) = 2x^2 - 5x - 3$.

  b. Factor the function in part **a** and differentiate by using the product rule. Show that the two answers are the same.

41. a. Use the quotient rule to differentiate $f(x) = \dfrac{2x - 3}{x^3}$.

  b. Rewrite the function in part **a** as $f(x) = x^{-3}(2x - 3)$ and differentiate by using the product rule.

  c. Rewrite the function in part **a** as $f(x) = 2x^{-2} - 3x^{-3}$ and differentiate.

  d. Show that the answers to parts **a**, **b**, and **c** are all the same.

42. Find numbers a, b, and c that guarantee that the graph of the function $f(x) = ax^2 + bx + c$ will have x-intercepts at $(0, 0)$ and $(5, 0)$ and a tangent line with slope 1 where $x = 2$.

43. Find the equation for the tangent line to the curve with equation $y = x^4 - 2x + 1$ that is parallel to the line $2x - y - 3 = 0$.

44. Find equations for two tangent lines to the graph of $f(x) = \dfrac{3x + 5}{1 + x}$ that are perpendicular to the line $2x - y = 1$.

45. Let $f(x) = (x^3 - 2x^2)(x + 2)$.

  a. Find an equation for the tangent line to the graph of f at the point where $x = 1$.

  b. Find an equation for the normal line to the graph of f at the point where $x = 0$.

46. Find an equation for a normal line to the graph of $f(x) = (x^3 - 2x^2)(x + 2)$ that is parallel to the line $x - 16y + 17 = 0$.

47. Find all points (x, y) on the graph of $y = 4x^2$ with the property that the tangent line at (x, y) passes through the point $(2, 0)$.

48. Find the equations of all the tangent lines to the graph of the function $f(x) = x^2 - 4x + 25$ that pass through the origin.

Determine which (if any) of the functions $y = f(x)$ given in Problems 49–52 satisfy the equation

$$y''' + y'' + y' = x + 1$$

49. $f(x) = x^2 + 2x - 3$ **50.** $f(x) = x^3 + x^2 + x$

51. $f(x) = \frac{1}{2}x^2 + 3$ **52.** $f(x) = 2x^2 + x$

53. HISTORICAL QUEST   When working with rational expressions, we need to be careful about division by zero. One of the earliest recorded treatments of division by zero is attributed to the Hindu mathematician Āryabhata (476–550), in whose honor the first Indian satellite was named. He also gave rules for approximations of square roots and sums of arithmetic progressions as well as rules for basic algebraic manipulations. One example of his work is the following calculation for π: "Add four to one hundred, multiply by eight and add again sixty-two thousand; the result is the approximate value of the circumference of a circle whose diameter is twenty-thousand."

Follow the steps of Āryabhata's approximation for π. After you have completed this demonstration, discuss the procedure and technology you used and contrast it with the tools that Āryabhata must have had available.

C 54. What is the relationship between the degree of a polynomial function P and the value of k for which $P^{(k)}(x)$ is first equal to 0?

55. Prove the constant multiple rule $(cf)' = cf'$.

56. Prove the sum rule $(f + g)' = f' + g'$.

57. Use the definition of the derivative to find the derivative of f^2, given that f is a differentiable function.

58. Prove the product rule by using the result of Problem 57 and the identity

$$fg = \tfrac{1}{2}\left[(f + g)^2 - f^2 - g^2\right]$$

59. Prove the quotient rule

$$\left(\frac{f}{g}\right)' = \frac{gf' - fg'}{g^2}$$

where $g(x) \neq 0$. *Hint:* First show that the difference quotient for f/g can be expressed as

$$\frac{\frac{f}{g}(x + \Delta x) - \frac{f}{g}(x)}{\Delta x} = \frac{f(x + \Delta x)g(x) - f(x)g(x + \Delta x)}{(\Delta x)g(x + \Delta x)g(x)}$$

and then subtract and add the term $g(x)f(x)$ in the numerator.

60. Show that the reciprocal function $r(x) = 1/f(x)$ has the derivative $r'(x) = -f'(x)/[f(x)]^2$ at each point x where f is differentiable and $f(x) \neq 0$.

61. If f, g, and h are differentiable functions, show that the product fgh is also differentiable and

$$(fgh)' = fgh' + fg'h + f'gh$$

62. Let f be a function that is differentiable at x.

  a. If $g(x) = [f(x)]^3$, show that $g'(x) = 3[f(x)]^2 f'(x)$. *Hint:* Write $g(x) = [f(x)]^2 f(x)$ and use the product rule.

  b. Show that $p(x) = [f(x)]^4$ has the derivative

$$p'(x) = 4[f(x)]^3 f'(x).$$

63. Find constants A, B, and C so that $y = Ax^3 + Bx + C$ satisfies the equation

$$y''' + 2y'' - 3y' + y = x$$

3.3 Derivatives of Trigonometric, Exponential, and Logarithmic Functions

IN THIS SECTION derivatives of the sine and cosine functions, differentiation of the other trigonometric functions, derivatives of exponential and logarithmic functions

DERIVATIVES OF THE SINE AND COSINE FUNCTIONS

In calculus we assume that the trigonometric functions are functions of real numbers or of angles measured in radians.

We make this assumption because the trigonometric differentiation formulas rely on limit formulas that become complicated if degree measurement is used instead of radian measure.

Before stating the theorem for the derivatives of the sine and cosine functions, suppose we look at the graph of the difference quotient. Consider $f(x) = \sin x$. Then

$$\frac{\sin(x + \Delta x) - \sin x}{\Delta x} = \frac{\sin(x + 0.01) - \sin x}{0.01}$$

is the difference quotient for $\Delta x = 0.01$. To graph this difference quotient, shown in Figure 3.19, we input to a calculator as

$$Y1 = (\text{SIN}(X+.01) - \text{SIN } X)/.01$$

From the graph of this difference quotient, it appears that the derivative of $f(x) = \sin x$ is $f'(x) = \cos x$. We now verify this with the following theorem, which uses the following limits established earlier in Theorem 2.3:

$$\lim_{h \to 0} \frac{\sin h}{h} = 1 \qquad \lim_{h \to 0} \frac{\cos h - 1}{h} = 0$$

Y1**B**(sin (X+.01)−
sin X)/.01
Xmin=-9.424777…
Xmax=9.4247779…
Xscl=1.5707963…
Ymin=-1
Ymax=1
Yscl=.1

Figure 3.19 Calculator graph of the difference quotient for $y = \sin x$

THEOREM 3.6 Derivatives of the sine and cosine functions

The functions $\sin x$ and $\cos x$ are differentiable for all x and

$$\frac{d}{dx} \sin x = \cos x \qquad \frac{d}{dx} \cos x = -\sin x$$

Proof The proofs of these two formulas are similar. We will prove the first using the trigonometric identity

$$\sin(\alpha + \beta) = \sin \alpha \cos \beta + \cos \alpha \sin \beta$$

and leave the proof of the second formula as a problem. From the definition of the derivative

$$\frac{d}{dx} \sin x = \lim_{\Delta x \to 0} \frac{\sin(x + \Delta x) - \sin x}{\Delta x}$$

$$= \lim_{\Delta x \to 0} \frac{\sin x \cos \Delta x + \cos x \sin \Delta x - \sin x}{\Delta x}$$

$$= \lim_{\Delta x \to 0} \left[\sin x \left(\frac{\cos \Delta x}{\Delta x} \right) + \cos x \left(\frac{\sin \Delta x}{\Delta x} \right) - \frac{\sin x}{\Delta x} \right]$$

$$= (\sin x) \lim_{\Delta x \to 0} \left(\frac{\cos \Delta x - 1}{\Delta x} \right) + (\cos x) \lim_{\Delta x \to 0} \frac{\sin \Delta x}{\Delta x}$$

$$= (\sin x)(0) + (\cos x)(1)$$

$$= \cos x \qquad \qquad \square$$

EXAMPLE 1 Derivative involving a trigonometric function

Differentiate $f(x) = 2x^4 + 3 \cos x + \sin a$, for constant a.

Solution

$$f'(x) = \frac{d}{dx}(2x^4 + 3\cos x + \sin a)$$

$$= 2\frac{d}{dx}(x^4) + 3\frac{d}{dx}(\cos x) + \frac{d}{dx}(\sin a) \quad \text{Extended linearity rule}$$

$$= 2(4x^3) + 3(-\sin x) + 0 \quad \text{Power rule, derivative of cosine, and derivative of a constant}$$

$$= 8x^3 - 3\sin x$$

EXAMPLE 2   Derivative of a trigonometric function with product rule

Differentiate $f(x) = x^2 \sin x$.

Solution

$$f'(x) = \frac{d}{dx}(x^2 \sin x)$$

$$= x^2\frac{d}{dx}(\sin x) + \sin x\frac{d}{dx}(x^2) \quad \text{Product rule}$$

$$= x^2 \cos x + 2x \sin x \quad \text{Power rule and derivative of sine}$$

EXAMPLE 3   Derivative of a trigonometric function with quotient rule

Differentiate $h(t) = \dfrac{\sqrt{t}}{\cos t}$.

Solution

Write $\sqrt{t}$ as $t^{1/2}$. Then

$$h'(t) = \frac{d}{dt}\left[\frac{t^{1/2}}{\cos t}\right]$$

$$= \frac{\cos t \dfrac{d}{dt}(t^{1/2}) - t^{1/2}\dfrac{d}{dt}\cos t}{\cos^2 t} \quad \text{Quotient rule}$$

$$= \frac{\frac{1}{2}t^{-1/2}\cos t - t^{1/2}(-\sin t)}{\cos^2 t} \quad \text{Power rule and derivative of sine}$$

$$= \frac{\frac{1}{2}t^{-1/2}(\cos t + 2t \sin t)}{\cos^2 t} \quad \text{Common factor } \frac{1}{2}t^{-1/2}$$

$$= \frac{\cos t + 2t \sin t}{2\sqrt{t}\cos^2 t}$$

DIFFERENTIATION OF THE OTHER TRIGONOMETRIC FUNCTIONS

You will need to be able to differentiate not only the sine and cosine functions, but also the other trigonometric functions. To find the derivatives of these functions you will need the following identities, which are given in the *Student Mathematics Handbook* and Appendix E.

$$\tan x = \frac{\sin x}{\cos x} \qquad \cot x = \frac{\cos x}{\sin x}$$

$$\sec x = \frac{1}{\cos x} \qquad \csc x = \frac{1}{\sin x}$$

You will also need the following identities:

$$\cos^2 x + \sin^2 x = 1, \quad 1 + \tan^2 x = \sec^2 x, \quad \cot^2 x + 1 = \csc^2 x$$

THEOREM 3.7   Derivatives of the trigonometric functions

The six basic trigonometric functions $\sin x$, $\cos x$, $\tan x$, $\csc x$, $\sec x$, and $\cot x$ are all differentiable wherever they are defined, and

$$\frac{d}{dx} \sin x = \cos x \qquad\qquad \frac{d}{dx} \cos x = -\sin x$$

$$\frac{d}{dx} \tan x = \sec^2 x \qquad\qquad \frac{d}{dx} \cot x = -\csc^2 x$$

$$\frac{d}{dx} \sec x = \sec x \tan x \qquad\qquad \frac{d}{dx} \csc x = -\csc x \cot x$$

Proof   The derivatives for the sine and cosine functions were given in Theorem 3.6. All the other derivatives in this theorem are proved by using the quotient rule along with formulas for the derivatives of the sine and cosine functions. We will obtain the derivative of the tangent function and leave the rest as problems (see Problems 56–59).

$$\frac{d}{dx} \tan x = \frac{d}{dx} \frac{\sin x}{\cos x} \qquad\qquad \text{Trigonometric identity}$$

$$= \frac{\cos x \dfrac{d}{dx} \sin x - \sin x \dfrac{d}{dx} \cos x}{\cos^2 x} \qquad \text{Quotient rule}$$

$$= \frac{\cos x (\cos x) - \sin x (-\sin x)}{\cos^2 x} \qquad \text{Derivative of } \sin x \text{ and } \cos x$$

$$= \frac{\cos^2 x + \sin^2 x}{\cos^2 x}$$

$$= \frac{1}{\cos^2 x} \qquad\qquad \cos^2 x + \sin^2 x = 1$$

$$= \sec^2 x \qquad\qquad 1/\cos^2 x = \sec^2 x \qquad \Box$$

EXAMPLE 4   Derivative of a trigonometric function with the product rule

Differentiate $f(\theta) = 3\theta \sec \theta$.

Solution
$$f'(\theta) = \frac{d}{d\theta}(3\theta \sec \theta)$$
$$= 3\theta \frac{d}{d\theta} \sec \theta + \sec \theta \frac{d}{d\theta}(3\theta) \qquad \text{Product rule}$$
$$= 3\theta \sec \theta \tan \theta + 3 \sec \theta \qquad\qquad \blacksquare$$

EXAMPLE 5   Derivative of a product of trigonometric functions

Differentiate $f(x) = \sec x \tan x$.

Solution
$$f'(x) = \frac{d}{dx}(\sec x \tan x)$$
$$= \sec x \frac{d}{dx} \tan x + \tan x \frac{d}{dx} \sec x \qquad \text{Product rule}$$
$$= \sec x (\sec^2 x) + \tan x (\sec x \tan x)$$
$$= \sec^3 x + \sec x \tan^2 x \qquad\qquad \blacksquare$$

EXAMPLE 6   Equation of a tangent line involving a trigonometric function

Find the equation of the tangent line to the curve $y = \cot x - 2 \csc x$ at the point where $x = \frac{2\pi}{3}$.

Solution

When $x = \frac{2\pi}{3}$, we have

$$\cot \frac{2\pi}{3} - 2 \csc \frac{2\pi}{3} = \frac{-\sqrt{3}}{3} - 2\left(\frac{2\sqrt{3}}{3}\right) = \frac{-5\sqrt{3}}{3}$$

so the point of tangency is $P\left(\frac{2\pi}{3}, \frac{-5\sqrt{3}}{3}\right)$. To find the slope of the tangent line at P, we first compute the derivative $\frac{dy}{dx}$:

$$\frac{dy}{dx} = \frac{d}{dx}(\cot x - 2\csc x)$$

$$= \frac{d}{dx}\cot x - 2\frac{d}{dx}\csc x \qquad \text{Linearity rule}$$

$$= -\csc^2 x - 2(-\csc x \cot x)$$

$$= 2\csc x \cot x - \csc^2 x$$

Then the slope of the tangent line is given by

$$\left.\frac{dy}{dx}\right|_{x=2\pi/3} = 2\csc \frac{2\pi}{3} \cot \frac{2\pi}{3} - \csc^2 \frac{2\pi}{3}$$

$$= 2\left(\frac{2}{\sqrt{3}}\right)\left(\frac{-1}{\sqrt{3}}\right) - \left(\frac{2}{\sqrt{3}}\right)^2$$

$$= \frac{-8}{3}$$

so the tangent line at $\left(\frac{2\pi}{3}, \frac{-5\sqrt{3}}{3}\right)$ with slope $-\frac{8}{3}$ is

$$y + \frac{5\sqrt{3}}{3} = -\frac{8}{3}\left(x - \frac{2\pi}{3}\right)$$

$$24x + 9y + 15\sqrt{3} - 16\pi = 0 \qquad \blacksquare$$

DERIVATIVES OF EXPONENTIAL AND LOGARITHMIC FUNCTIONS

The next theorem, which is easy to prove and remember, is one of the most important results in all of differential calculus.

THEOREM 3.8   Derivative rule for the natural exponential

The natural exponential function e^x is differentiable for all x, with derivative

$$\frac{d}{dx}(e^x) = e^x$$

Proof   We will proceed informally. Recall the definition of e:

$$\lim_{n \to \infty}\left(1 + \frac{1}{n}\right)^n = e$$

Let $n = \frac{1}{\Delta x}$, so that $\lim_{\Delta x \to 0}(1 + \Delta x)^{1/\Delta x} = e$. This means that for Δx very small,

$e \approx (1 + \Delta x)^{1/\Delta x}$ or $e^{\Delta x} \approx 1 + \Delta x$, so that $e^{\Delta x} - 1 \approx \Delta x$. Thus, $\lim_{\Delta x \to 0} \dfrac{e^{\Delta x} - 1}{\Delta x} = 1$.

Finally, using the limit in the definition of derivative for e^x, we obtain

$$\frac{d}{dx}(e^x) = \lim_{\Delta x \to 0} \frac{e^{(x+\Delta x)} - e^x}{\Delta x}$$

$$= \lim_{\Delta x \to 0} \frac{e^x(e^{\Delta x} - 1)}{\Delta x}$$

$$= e^x \lim_{\Delta x \to 0} \frac{e^{\Delta x} - 1}{\Delta x}$$

$$= e^x(1) = e^x \qquad \square$$

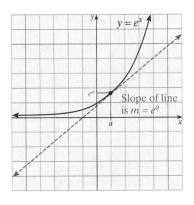

Figure 3.20 The slope of $y = e^x$ at each point (a, e^a) is $m = e^a$.

Note: An easier proof of the derivative formula in Theorem 3.8 is given in Example 6 of Section 3.6 using methods developed in that section.

The fact that $\dfrac{d}{dx}(e^x) = e^x$ means that the slope of the graph of $y = e^x$ at any point (a, e^a) is $m = e^a$, the y-coordinate of the point, as shown in Figure 3.20. This is one of the features of the exponential function $y = e^x$ that makes it "natural."

EXAMPLE 7 A derivative involving e^x

Differentiate $f(x) = x^2 e^x$. For what values of x does $f'(x) = 0$?

Solution

Using the product rule, we find

$$f'(x) = \frac{d}{dx}(x^2 e^x)$$

$$= x^2 \left[\frac{d}{dx} e^x \right] + \left[\frac{d}{dx} x^2 \right] e^x$$

$$= x^2 e^x + 2x e^x \qquad \text{Exponential and power rules}$$

To find where $f'(x) = 0$, we solve

$$x^2 e^x + 2x e^x = 0$$

$$x(x + 2) e^x = 0$$

$$x(x + 2) = 0 \qquad \text{Since } e^x \neq 0 \text{ for all } x$$

$$x = 0, -2$$
∎

EXAMPLE 8 A second derivative involving e^x

For $f(x) = e^x \sin x$, find $f'(x)$ and $f''(x)$.

Solution

Apply the product rule twice:

$$f'(x) = e^x (\sin x)' + (e^x)' \sin x$$

$$= e^x (\cos x) + e^x (\sin x)$$

$$= e^x (\cos x + \sin x)$$

$$f''(x) = e^x (\cos x + \sin x)' + (e^x)'(\cos x + \sin x)$$

$$= e^x (-\sin x + \cos x) + e^x (\cos x + \sin x)$$

$$= 2e^x \cos x$$
∎

THEOREM 3.9 Derivative rule for the natural logarithmic function

The natural logarithmic function $\ln x$ is differentiable for all $x > 0$, with derivative

$$\frac{d}{dx}(\ln x) = \frac{1}{x}$$

Proof According to the definition of the derivative, we have

$$\frac{d}{dx}(\ln x) = \lim_{\Delta x \to 0} \frac{\ln(x + \Delta x) - \ln x}{\Delta x}$$

$$= \lim_{\Delta x \to 0} \frac{1}{\Delta x} \ln \left(\frac{x + \Delta x}{x} \right)$$

$$= \lim_{\Delta x \to 0} \frac{1}{\Delta x} \ln \left(1 + \frac{\Delta x}{x} \right) \qquad \text{Let } h = \frac{x}{\Delta x}, \text{ so } \Delta x = \frac{x}{h}.$$

$$= \lim_{h \to +\infty} \frac{h}{x} \ln\left(1 + \frac{1}{h}\right) \qquad h \to +\infty \text{ as } \Delta x \to 0$$

$$= \frac{1}{x}\left[\lim_{h \to +\infty} h \ln\left(1 + \frac{1}{h}\right)\right]$$

$$= \frac{1}{x}\left[\lim_{h \to +\infty} \ln\left(1 + \frac{1}{h}\right)^h\right] \qquad \text{Power rule for logarithms}$$

$$= \frac{1}{x} \ln\left[\lim_{h \to +\infty} \left(1 + \frac{1}{h}\right)^h\right] \qquad \begin{array}{l}\text{Since } \ln x \text{ is continuous,}\\ \text{use the composition limit rule.}\end{array}$$

$$= \frac{1}{x} \ln e \qquad \text{Definition of } e$$

$$= \frac{1}{x} \qquad \ln e = 1 \qquad \square$$

EXAMPLE 9   Derivative of a quotient involving a natural logarithm

Differentiate $f(x) = \dfrac{\ln x}{\sin x}$.

Solution   We use the quotient rule.

$$f'(x) = \frac{(\sin x)\frac{d}{dx}(\ln x) - (\ln x)\frac{d}{dx}(\sin x)}{\sin^2 x}$$

$$= \frac{(\sin x)\left(\frac{1}{x}\right) - (\ln x)(\cos x)}{\sin^2 x}$$

$$= \frac{\sin x - x \ln x \cos x}{x \sin^2 x} \qquad \blacksquare$$

TECHNOLOGY NOTE

When you are using calculators or computer programs such as *Mathematica*, *Derive*, or *Maple*, the form of the derivative may vary. For example, you might obtain

$$\frac{d}{dx}(\tan x) = \tan^2 x + 1$$

instead of $\sec^2 x$ as found in this section;

$$\frac{d}{dx}(\cot x) = -\cot^2 x - 1$$

instead of $-\csc^2 x$;

$$\frac{d}{dx}(\sec x) = \frac{\sin x}{\cos^2 x}$$

instead of $\sec x \tan x$; and

$$\frac{d}{dx}(\csc x) = -\frac{\cos x}{\sin^2 x}$$

instead of $-\csc x \cot x$. A solution to Example 9, using a TI-92, is shown in Figure 3.21.

Figure 3.21   Sample output for Example 9. Notice that the form of the answer differs from that shown in the text.

Although the form of this calculator solution is different from the form shown in the example, you should notice that they are equivalent by using algebra (and recalling some of the fundamental identities from trigonometry).

A good test for your calculator is to try to find the derivative of $|x|$ at $x = 0$, which we know does not exist. Many calculators will give the incorrect answer of 0.

3.3   PROBLEM SET

A *Differentiate the functions given in Problems 1–32.*

1. $f(x) = \sin x + \cos x$
2. $f(x) = 2 \sin x + \tan x$
3. $g(t) = t^2 + \cos t + \cos \frac{\pi}{4}$
4. $g(t) = 2 \sec t + 3 \tan t - \tan \frac{\pi}{3}$
5. $f(t) = \sin^2 t$ *Hint*: Use the product rule.
6. $g(x) = \cos^2 x$ *Hint*: Use the product rule.
7. $f(x) = \sqrt{x} \cos x + x \cot x$ 8. $f(x) = 2x^3 \sin x - 3x \cos x$
9. $p(x) = x^2 \cos x$ 10. $p(t) = (t^2 + 2) \sin t$
11. $q(x) = \dfrac{\sin x}{x}$ 12. $r(x) = \dfrac{e^x}{\sin x}$
13. $h(t) = e^t \csc t$ 14. $f(\theta) = \dfrac{\sec \theta}{2 - \cos \theta}$
15. $f(x) = x^2 \ln x$ 16. $g(x) = \dfrac{\ln x}{x^2}$
17. $h(x) = e^{2x}(\sin x - \cos x)$ 18. $f(x) = \dfrac{\ln x}{x}$
19. $f(x) = \dfrac{\sin x}{e^x}$ 20. $g(x) = \dfrac{x \cos x}{e^x}$
21. $f(x) = \dfrac{\tan x}{1 - 2x}$ 22. $g(t) = \dfrac{1 + \sin t}{\sqrt{t}}$
23. $f(t) = \dfrac{2 + \sin t}{t + 2}$ 24. $f(\theta) = \dfrac{\theta - 1}{2 + \cos \theta}$
25. $f(x) = \dfrac{\sin x}{1 - \cos x}$ 26. $f(x) = \dfrac{x}{1 - \sin x}$
27. $f(x) = \dfrac{1 + \sin x}{2 - \cos x}$ 28. $g(x) = \dfrac{\cos x}{1 + \cos x}$
29. $f(x) = \dfrac{\sin x + \cos x}{\sin x - \cos x}$ 30. $f(x) = \dfrac{x^2 + \tan x}{3x + 2 \tan x}$
31. $g(x) = \sec^2 x - \tan^2 x + \cos x$
32. $g(x) = \cos^2 x + \sin^2 x + \sin x$

Find the second derivative of each function given in Problems 33–44.

33. $f(\theta) = \sin \theta$ 34. $f(\theta) = \cos \theta$
35. $f(\theta) = \tan \theta$ 36. $f(\theta) = \cot \theta$
37. $f(\theta) = \sec \theta$ 38. $f(\theta) = \csc \theta$
39. $f(x) = \sin x + \cos x$ 40. $f(x) = x \sin x$
41. $f(x) = e^x \cos x$ 42. $g(t) = t^3 e^t$
43. $h(t) = \sqrt{t} \ln t$ 44. $f(t) = \dfrac{\ln t}{t}$

B *Find an equation for the tangent line at the prescribed point for each function in Problems 45–52.*

45. $f(\theta) = \tan \theta$ at $(\frac{\pi}{4}, 1)$ 46. $f(\theta) = \sec \theta$ at $(\frac{\pi}{3}, 2)$
47. $f(x) = \sin x$, where $x = \frac{\pi}{6}$
48. $f(x) = \cos x$, where $x = \frac{\pi}{3}$
49. $y = x + \sin x$, where $x = 0$
50. $y = x \sec x$, where $x = 0$
51. $y = e^x \cos x$, where $x = 0$
52. $y = x \ln x$, where $x = 1$
53. Which of the following functions satisfy $y'' + y = 0$?
   a. $y_1 = 2 \sin x + 3 \cos x$ b. $y_2 = 4 \sin x - \pi \cos x$
   c. $y_3 = x \sin x$ d. $y_4 = e^x \cos x$
54. For what values of A and B does $y = A \cos x + B \sin x$ satisfy $y'' + 2y' + 3y = 2 \sin x$?
55. For what values of A and B does $y = Ax \cos x + Bx \sin x$ satisfy $y'' + y = -3 \cos x$?

C 56. Complete the proof of Theorem 3.6 by showing that $\dfrac{d}{dx} \cos x = -\sin x$. *Hint*: You will need to use the identity $\cos(\alpha + \beta) = \cos \alpha \cos \beta - \sin \alpha \sin \beta$.

Prove the requested parts of Theorem 3.7 in Problems 57–59.

57. $\dfrac{d}{dx} \cot x = -\csc^2 x$
58. $\dfrac{d}{dx} \sec x = \sec x \tan x$
59. $\dfrac{d}{dx} \csc x = -\csc x \cot x$
60. Use the limit of a difference quotient to prove that

$$\frac{d}{dx} \tan x = \sec^2 x$$

61. **Exploration Problem**   Write a short paper on the difficulties of differentiating trigonometric functions measured in degrees.*

---

*See, for example, "Fallacies, Flaws, and Flimflam," *The College Mathematics Journal*, Vol. 23, No. 3, May 1992, and Vol. 24, No. 4, September 1993. Another very understandable article, "Why Use Radians in Calculus?", by Carl E. Crockett, can be found in *The AMATYC Review*, Vol.19, No. 2, Spring 1998, pp. 44–47.

3.4   Rates of Change: Modeling Rectilinear Motion

IN THIS SECTION   average and instantaneous rate of change, introduction to mathematical modeling, rectilinear motion (modeling in physics), falling body problems

The speed of a car or airplane; interest rates; the growth rate of a population; the drip rate of an intravenous injection—these are examples of the many situations in which rates of change are an important consideration in practical problems. In this section, we begin by showing how rates of change can be computed using differentiation. Then we discuss the general notion of mathematical modeling and conclude by demonstrating modeling principles in a variety of practical problems involving motion along a line.

AVERAGE AND INSTANTANEOUS RATE OF CHANGE

The graph of a linear function $f(x) = ax + b$ is the line $y = ax + b$, whose slope $m = a$ can be thought of as the rate at which y is changing with respect to x (see Figure 3.22a). However, for another function g that is *not* linear, the average (and instantaneous) rate of change of $y = g(x)$ with respect to x varies from point to point, as shown in Figure 3.22b.

a. Linear function: rate of change $\Delta y / \Delta x$ is constant.

b. Nonlinear function: rate of change $\Delta y / \Delta x$ depends on chosen points.

Figure 3.22   Rate of change is measured by the slope of a tangent line

Because the slope of the tangent line is given by the derivative of the function, the preceding geometric observations suggest that the rate of change of a function is measured by its derivative. This connection will be made more precise in the following discussion.

Suppose y is a function of x, say $y = f(x)$. Corresponding to a change from x to $x + \Delta x$ the variable y changes from $f(x)$ to $f(x + \Delta x)$. The change in y is $\Delta y = f(x + \Delta x) - f(x)$, and the **average rate of change of y with respect to x** is

$$\text{AVERAGE RATE OF CHANGE} = \frac{\text{CHANGE IN } y}{\text{CHANGE IN } x} = \frac{\Delta y}{\Delta x} = \frac{f(x + \Delta x) - f(x)}{\Delta x}$$

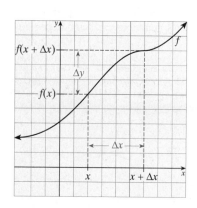

Figure 3.23 A change in Δy corresponding to a change Δx

WARNING   This formula for the change in y is important. Find Δy in Figure 3.23.

As the interval over which we are averaging becomes shorter (that is, as $\Delta x \to 0$), the average rate of change approaches what we would intuitively call the **instantaneous rate of change of y with respect to x**, and the difference quotient approaches the derivative $\dfrac{dy}{dx}$. Thus, we have

$$\begin{array}{l} \text{INSTANTANEOUS} \\ \text{RATE OF CHANGE} \end{array} = \lim_{\Delta x \to 0} \frac{\Delta y}{\Delta x} = \lim_{\Delta x \to 0} \frac{f(x + \Delta x) - f(x)}{\Delta x} = f'(x)$$

To summarize,

Instantaneous Rate of Change

Suppose $f(x)$ is differentiable at $x = x_0$. Then the **instantaneous rate of change** of $y = f(x)$ with respect to x at x_0 is the value of the derivative of f at x_0. That is,

$$\text{INSTANTANEOUS RATE OF CHANGE} = f'(x_0) = \left.\frac{dy}{dx}\right|_{x=x_0}$$

EXAMPLE 1 Instantaneous rate of change

Find the rate at which the function $y = x^2 \sin x$ is changing with respect to x when $x = \pi$.

Solution

For any x, the instantaneous rate of change is the derivative,

$$\frac{dy}{dx} = 2x \sin x + x^2 \cos x$$

Thus, the rate when $x = \pi$ is

$$\left.\frac{dy}{dx}\right|_{x=\pi} = 2\pi \sin \pi + \pi^2 \cos \pi = 2\pi(0) + \pi^2(-1) = -\pi^2$$

The negative sign indicates that when $x = \pi$, the function is *decreasing* at the rate of $\pi^2 \approx 9.9$ units of y for each one-unit increase in x. ∎

Let us consider an example comparing the average rate of change with the instantaneous rate of change.

EXAMPLE 2 Comparison between average rate and instantaneous rate of change

Let $f(x) = x^2 - 4x + 7$.

a. Find the average rate of change of f with respect to x between $x = 3$ and 5.
b. Find the instantaneous rate of change of f at $x = 3$.

Solution

a. The (average) rate of change from $x = 3$ to $x = 5$ is found by dividing the change in f by the change in x. The change in f from $x = 3$ to $x = 5$ is

$$f(5) - f(3) = [5^2 - 4(5) + 7] - [3^2 - 4(3) + 7] = 8$$

Thus, the average rate of change is

$$\frac{f(5) - f(3)}{5 - 3} = \frac{8}{2} = 4$$

The slope of the secant line is 4, as shown in Figure 3.24.

b. The derivative of the function is

$$f'(x) = 2x - 4$$

Thus, the instantaneous rate of change of f at $x = 3$ is

$$f'(3) = 2(3) - 4 = 2$$

The tangent line at $x = 3$ has slope 2, as shown in Figure 3.24. ∎

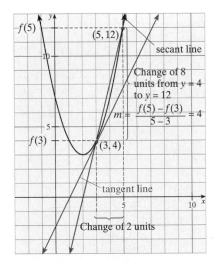

Figure 3.24 Comparison of the instantaneous rate of change and the average rate of change from 3 to 5

Often we are not as interested in the instantaneous rate of change of a quantity as its relative rate of change, defined as follows:

$$\text{RELATIVE RATE} = \frac{\text{INSTANTANEOUS RATE OF CHANGE}}{\text{SIZE OF QUANTITY}}$$

For instance, if you are earning \$15,000/yr and receive a \$3,000 raise, you would probably be very pleased. However, that \$3,000 raise might be much less meaningful if you were earning \$100,000/yr. In both cases, the change is the quantity \$3,000, but in the first case the relative rate of change is $3,000/15,000 = 20\%$, whereas in the second, it is a much smaller $3,000/100,000 = 3\%$.

In terms of the derivative, the relative rate of change may be defined as follows:

Relative Rate of Change

The **relative rate of change** of $y = f(x)$ at $x = x_0$ is given by the ratio

$$\text{RELATIVE RATE OF CHANGE} = \frac{f'(x_0)}{f(x_0)}$$

EXAMPLE 3   Relative rate of change

The aerobic rate is the rate of a person's oxygen consumption and is sometimes modeled by the function A defined by

$$A(x) = 110\left[\frac{\ln x - 2}{x}\right]$$

where x is the person's age. What is the relative rate of change of the aerobic rating of a 20-year-old person? A 50-year-old person?

Solution

Applying the quotient rule, we obtain the derivative of $A(x)$:

$$A'(x) = 110\left[\frac{x(1/x) - (\ln x - 2)(1)}{x^2}\right] = 110\left[\frac{3 - \ln x}{x^2}\right]$$

Thus, the relative rate of change of A at age x is

$$\text{RELATIVE RATE} = \frac{A'(x)}{A(x)} = \frac{110\left[\dfrac{3 - \ln x}{x^2}\right]}{110\left[\dfrac{\ln x - 2}{x}\right]} = \frac{3 - \ln x}{(\ln x - 2)x}$$

In particular, when $x = 20$, the relative rate is

$$\frac{A'(20)}{A(20)} = \frac{3 - \ln 20}{(\ln 20 - 2)20} \approx 0.0002143$$

In other words, at age 20, the person's aerobic rating is increasing at the rate of 0.02%/yr. However, at age 50, the relative rate is

$$\frac{A'(50)}{A(50)} = \frac{3 - \ln 50}{(\ln 50 - 2)50} \approx -0.0095399$$

and the person's aerobic rating is *decreasing* (because of the negative sign) at the relative rate of 0.95%/yr. ∎

INTRODUCTION TO MATHEMATICAL MODELING

A real-life situation is usually far too complicated to be precisely and mathematically defined. When confronted with a problem in the real world, therefore, it is usually necessary to develop a mathematical framework based on certain assumptions about the real world. This framework can then be used to find a solution to the real-world problem. The process of developing this body of mathematics is referred to as **mathematical modeling**.

Some mathematical models are quite accurate, particularly many used in the physical sciences. For example, one of the models we will consider in calculus is a model for the path of a projectile. Other rather precise models predict such things as the time of sunrise and sunset, or the distance which an object falls in a vacuum. Some mathematical models, however, are less accurate, especially those that involve examples from the life sciences and social sciences. Only recently has modeling in these disciplines become precise enough to be expressed in terms of calculus.

What, exactly, is a mathematical model? Sometimes, mathematical modeling can mean nothing more than a textbook word problem. But mathematical modeling can also mean choosing appropriate mathematics to solve a problem that has previously been unsolved. In this book, we use the term "mathematical modeling" to mean something between these two extremes. That is, it is a process we will apply to some real-life problem that does not have an obvious solution. It usually cannot be solved by applying a single formula.

Learning to use modeling techniques is an important part of your mathematical education. The first step in the modeling process involves *abstraction*, in which certain assumptions about a given practical problem are made in order to formulate a version of the problem in abstract terms. Next, the abstract version of the problem is *analyzed* using appropriate mathematical methods such as calculus, and finally, results derived from the abstract analysis are *interpreted* in terms of the original problem. The results obtained from the model often lead to predictions about the "real-world" situations related to the given problem. Data can be gathered to check the accuracy of these predictions and to modify the initial assumptions of the model to make it more precise. The process continues until a model is obtained that fits "reality" with acceptable accuracy. The steps in the modeling process are demonstrated in Figure 3.25. You may also find it interesting to read the article on modeling weather from *Scientific American* quoted in the margin.

HOW GLOBAL CLIMATE IS MODELED

We find a good example of mathematical modeling by looking at the work being done with weather prediction. In theory, if the correct assumptions could be programmed into a computer, along with appropriate mathematical statements of the ways global climate conditions operate, we would have a model to predict the weather throughout the world. In the global climate model, a system of equations calculates time-dependent changes in wind as well as temperature and moisture changes in the atmosphere and on the land. The model may also predict alterations in the temperature of the ocean's surface. At the National Center for Atmospheric Research, a CRAY supercomputer is used to do this modeling.

From *Scientific American*, March 1991

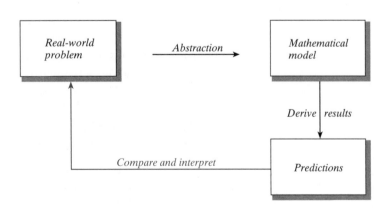

Figure 3.25 Mathematical modeling

EXAMPLE 4   Identifying models

It is common to gather data when studying most phenomena. Classifying the data by comparing with known models is an important mathematical skill. Identify each of the data sets as an example of a linear model, a quadratic model, a cubic model, a logarithmic model, an exponential model, or a sinusoidal model.

a.

x	y
−4	2.00
−3	1.28
−2	0.72
−1	0.32
0	0.08
1	0.00
2	0.08
3	0.32
4	0.72

b.

x	y
−4	−20.7
−3	−16.5
−2	−12.3
−1	−8.2
0	−3.8
1	0.4
2	4.7
3	8.9
4	13.12

c.

x	y
0.0	0.0
1.0	1.7
1.5	2.0
2.5	1.2
3.5	−0.7
4.5	−1.9
5.5	−1.4
6.0	−0.6
6.5	0.4

d.

x	y
−4	0.27
−3	0.30
−2	0.33
−1	0.36
0	0.40
1	0.44
2	0.49
3	0.52
4	0.60

e.

x	y
3	0.00
4	0.69
5	1.10
6	1.39
7	1.61
8	1.79
9	1.95
10	2.08
11	2.20
12	2.30

Solution

Sketch each graph and observe the pattern (if any) of the graphs.

a. Quadratic model

b. Linear model

c. The data points seem to form a pattern, and we hypothesize that the data points are periodic, namely a sinusoidal model.

d. The data points look almost linear (but not quite), so we hypothesize that the data points may indicate an exponential model.

e. The data points seem to indicate a logarithmic model.

Mathematical models will appear in both examples and exercises throughout this text. In the next subsection, we begin our work with modeling by studying rectilinear motion, an important topic from physics.

RECTILINEAR MOTION (MODELING IN PHYSICS)

Rectilinear motion is motion along a straight line. For example, the up and down motion of a yo-yo may be regarded as rectilinear, as may the motion of a rocket early in its flight.

When studying rectilinear motion, we may assume that the object is moving along a coordinate line. The *position* of the object on the line is a function of time t and is often expressed as $s(t)$. The rate of change of $s(t)$ with respect to time is the object's **velocity** $v(t)$, and the rate of change of the velocity with respect to t is its **acceleration** $a(t)$. The absolute value of the velocity is called **speed**. Thus,

$$\text{Velocity is } v(t) = \frac{ds}{dt}$$

$$\text{Speed is } |v(t)| = \left|\frac{ds}{dt}\right|$$

$$\text{Acceleration is } a(t) = \frac{dv}{dt} = \frac{d^2s}{dt^2}$$

If $v(t) > 0$, we say that the object is *advancing*, and if $v(t) < 0$, it is *retreating*. If $v(t) = 0$, the object is neither advancing nor retreating, and we say it is *stationary*. The object is *accelerating* when $a(t) > 0$ and is *decelerating* when $a(t) < 0$. The significance of the acceleration is that it gives the rate at which the velocity is changing. When you are riding in a car, you do not feel the velocity, but you do feel the acceleration. That is, you feel *changes* in the velocity.

These ideas are summarized in the following box.

Rectilinear Motion

An object that moves along a straight line with *position* $s(t)$ has *velocity* $v(t) = \dfrac{ds}{dt}$ and *acceleration* $a(t) = \dfrac{dv}{dt} = \dfrac{d^2s}{dt^2}$ when these derivatives exist. The *speed* of the object is $|v(t)|$.

WARNING   Rectilinear motion involves *position, velocity,* and *acceleration.* Sometimes there is confusion between the words "speed" and "velocity." Because speed is the absolute value of the velocity, it indicates how fast an object is moving, whereas velocity indicates both speed and direction (relative to a given coordinate system).

Notational comment: If distance is measured in meters and time in seconds, then velocity is measured in meters per second (m/s). Acceleration is recorded in meters per second per second (m/s/s). The notation m/s/s is awkward, so m/s^2 is more commonly used. Similarly, if distance is measured in feet, then velocity is measured in feet per second (ft/s) and acceleration in feet per second per second (ft/s^2).

EXAMPLE 5   The position, velocity, and acceleration of a moving object

Assume that the position at time t of an object moving along a line is given by

$$s(t) = 3t^3 - 40.5t^2 + 162t$$

for t on $[0, 8]$. Find the initial position, velocity, and acceleration for the object and discuss the motion. Compute the total distance traveled.

Solution

The position at time t is given by the function s. The initial position occurs at time $t = 0$, so

$$s(0) = 0 \qquad \text{The object starts at the origin.}$$

The velocity is determined by finding the derivative of the position function.

$$v(t) = s'(t) = 9t^2 - 81t + 162$$
$$= 9(t^2 - 9t + 18)$$
$$= 9(t - 3)(t - 6) \qquad \text{The initial velocity is } v(0) = 162.$$

When $t = 3$ and when $t = 6$, the velocity v is 0, which means the *object is stationary* at those times. Furthermore,

$$v(t) > 0 \quad \text{on} \quad [0, 3) \qquad \text{Object is advancing.}$$
$$v(t) < 0 \quad \text{on} \quad (3, 6) \qquad \text{Object is retreating.}$$
$$v(t) > 0 \quad \text{on} \quad (6, 8] \qquad \text{Object is advancing.}$$

For the acceleration,

$$a(t) = s''(t) = v'(t) = 18t - 81$$
$$= 18(t - 4.5) \qquad \text{The initial acceleration is } a(0) = -81.$$

We see that

$$a(t) < 0 \quad \text{on} \quad [0, 4.5) \qquad \text{Velocity is decreasing;}$$
$$\text{that is, the object is decelerating.}$$

$$a(t) = 0 \quad \text{at} \quad t = 4.5 \qquad \text{Velocity is not changing.}$$

$$a(t) > 0 \quad \text{on} \quad (4.5, 8] \qquad \text{Velocity is increasing;}$$
$$\text{so the object is accelerating.}$$

The table in the margin gives certain values for s, v, and a. We use these values to plot a few points, as shown in Figure 3.26. The actual path of the object is back and forth on the axis and the figure is for clarification only.

t	s	v	a
0	0	162	−81
1	124.5	90	−63
2	186	36	−45
3	202.5	0	−27
4	192	−18	−9
4.5	182.25	−20.25	0
5	172.5	−18	9
6	162	0	27
7	178.5	36	45
8	240	90	63

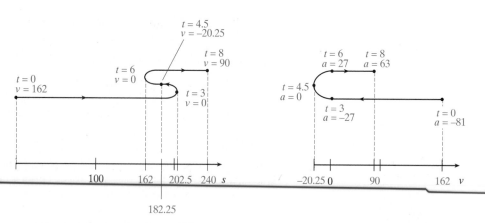

a. Position of the object: when $v > 0$, the object advances, and when $v < 0$, the object retreats.

b. Velocity of the object: when $a > 0$, the velocity increases, and when $a < 0$, the velocity decreases.

Figure 3.26 Analysis of rectilinear motion

Recall that the *speed* of the object is the absolute value of its velocity. The speed decreases from 162 to 0 between $t = 0$ and $t = 3$ and increases from 0 to 20.25 as the velocity becomes negative between $t = 3$ and $t = 4.5$. Then for $4.5 < t < 6$, the object slows down again, from 20.25 to 0, after which it speeds up.

The total distance traveled is

$$|s(3) - s(0)| + |s(6) - s(3)| + |s(8) - s(6)| = 202.5 + 40.5 + 78 = 321 \qquad \blacksquare$$

A common mistake is to think that a particle moving on a straight line speeds up when its acceleration is positive and slows down when the acceleration is negative, but this is not quite correct. Instead, the following is generally true:

The speed *increases* (particle speeds up) when the velocity and acceleration have the same signs.

The speed *decreases* (particle slows down) when the velocity and acceleration have opposite signs.

FALLING BODY PROBLEMS

As a second example of rectilinear motion, we will consider a *falling body problem*. In such a problem, it is assumed that an object is projected (that is, thrown, fired, dropped, etc.) vertically in such a way that the only acceleration acting on the object is the constant downward acceleration g due to gravity, which on the earth near sea level is approximately 32 ft/s^2 or 9.8 m/s^2.* Thus, we disregard air resistance. At time t, the height of the projectile is given by the following formula:

Formula for the height of a projectile
---

v_0 is the initial velocity.
$\downarrow$

$$h(t) = -\tfrac{1}{2}gt^2 + v_0 t + s_0 \leftarrow s_0 \text{ is the initial height.}$$

$\uparrow$
g is the acceleration due to gravity.

where s_0 and v_0 are the projectile's initial height and velocity, respectively. You are asked to derive this formula for earth's gravity and a particular initial velocity and a particular initial position in a problem in Section 5.1.

EXAMPLE 6 Position, velocity, and acceleration of a falling object

Suppose a person standing at the top of the Tower of Pisa (176 ft high) throws a ball directly upward with an initial speed of 96 ft/s.

a. Find the ball's height, its velocity, and acceleration at time t.
b. When does the ball hit the ground, and what is its impact velocity?
c. How far does the ball travel during its flight?

Solution

First, draw a picture such as the one shown in Figure 3.27 to help you understand the problem.

a. Substitute the known values into the formula for the height of an object:

$v_0 = 96$ ft/s: Initial velocity
$\downarrow$

$$h(t) = -\tfrac{1}{2}(32)t^2 + 96t + 176 \leftarrow h_0 = 176 \text{ ft is the height of tower}$$

$\uparrow$
$g = 32$ ft/s^2: Constant downward gravitational acceleration

$$h(t) = -16t^2 + 96t + 176 \qquad \text{This is the position function.}$$
It gives the height of the ball.

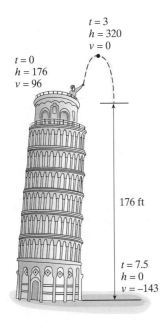

$t = 3$
$h = 320$
$v = 0$

$t = 0$
$h = 176$
$v = 96$

176 ft

$t = 7.5$
$h = 0$
$v = -143$

Figure 3.27 The motion of a ball thrown upward from the Tower of Pisa

---
*Although 32 ft/s^2 is the particular number we will use in this text, engineers in earthbound vehicular activity use 32.174 ft/s^2 as a better approximation for gravity.

The velocity at time t is the derivative:

$$v(t) = \frac{dh}{dt} = -32t + 96$$

The acceleration is the derivative of the velocity function:

$$a(t) = \frac{dv}{dt} = \frac{d^2h}{dt^2} = -32$$

This means that the acceleration of the ball is always decreasing at the rate of 32 ft/s^2.

b. The ball hits the ground when $h(t) = 0$. Solve the equation

$$-16t^2 + 96t + 176 = 0$$

to find that this occurs when $t \approx -1.47$ and $t \approx 7.47$. Disregarding the negative value, we see that impact occurs at $t \approx 7.47$ sec. The impact velocity is

$$v(7.47) \approx -143 \text{ ft/s}$$

The negative sign here means the ball is coming down at the moment of impact.

c. The ball travels upward for some time and then falls downward to the ground, as shown in Figure 3.27. We need to find the distance it travels upward plus the distance it falls to the ground. The turning point at the top (the highest point) occurs when the velocity is zero. Solve the equation

$$-32t + 96 = 0$$

to find that this occurs when $t = 3$. The ball starts at $h(0) = 176$ ft and rises to a maximum height when $t = 3$. Thus, the maximum height is

$$h(3) = -16(3)^2 + 96(3) + 176 = 320$$

and the total distance traveled is

$$\underbrace{(320 - 176)}_{\substack{\uparrow \\ \text{Initial height}}} + \underbrace{320}_{} = 464$$

Upward distance   Downward distance

The total distance traveled is 464 ft. ∎

3.4   PROBLEM SET

A **1. WHAT DOES THIS SAY?**   What is a mathematical model?

2. Interpretation Problem   Why are mathematical models necessary or useful?

For each function f given in Problems 3–16, find the rate of change with respect to x at $x = x_0$.

3. $f(x) = x^2 - 3x + 5$ when $x_0 = 2$

4. $f(x) = 14 + x - x^2$ when $x_0 = 1$

5. $f(x) = -2x^2 + x + 4$ for $x_0 = 1$

6. $f(x) = \dfrac{-2}{x+1}$ for $x_0 = 1$

7. $f(x) = \dfrac{2x-1}{3x+5}$ when $x_0 = -1$

8. $f(x) = (x^2 + 2)(x + \sqrt{x})$ when $x_0 = 4$

9. $f(x) = x \cos x$ when $x_0 = \pi$

10. $f(x) = (x+1)\sin x$ when $x_0 = \frac{\pi}{2}$

11. $f(x) = x \ln \sqrt{x}$ when $x_0 = 1$

12. $f(x) = \dfrac{x^2}{e^x}$ when $x_0 = 0$

13. $f(x) = \sin x \cos x$ when $x_0 = \frac{\pi}{2}$

14. $f(x) = \dfrac{x^2}{x^2 + 1}$ when $x_0 = 1$

15. $f(x) = \left(x - \dfrac{2}{x}\right)^2$ when $x_0 = 1$

16. $f(x) = \sin^2 x$ when $x_0 = \frac{\pi}{4}$

The function $s(t)$ in Problems 17–24 gives the position of an object moving along a line. In each case,

a. *Find the velocity at time t.*

b. *Find the acceleration at time t.*

c. *Describe the motion of the object; that is, tell when it is advancing and when it is retreating. Compute the total distance traveled by the object during the indicated time interval.*

d. *Tell when the object is accelerating and when it is decelerating.*

 For a review of solving equations, see Sections 2.5 and 4.7 of the Student Mathematics Handbook.

17. $s(t) = t^2 - 2t + 6$ on $[0, 2]$

18. $s(t) = 3t^2 + 2t - 5$ on $[0, 1]$

19. $s(t) = t^3 - 9t^2 + 15t + 25$ on $[0, 6]$

20. $s(t) = t^4 - 4t^3 + 8t$ on $[0, 4]$

21. $s(t) = \dfrac{2t + 1}{t^2}$ for $1 \le t \le 3$

22. $s(t) = t^2 + t \ln t$ for $1 \le t \le e$

23. $s(t) = 3 \cos t$ for $0 \le t \le 2\pi$

24. $s(t) = 1 + \sec t$ for $0 \le t \le \frac{\pi}{4}$

Modeling Problems *Identify each of the data sets in Problems 25–34 as an example of a linear model, a quadratic model, a cubic model, a logarithmic model, an exponential model, or a sinusoidal model.*

25.

x	y
−4	−140
−3	−96
−2	−60
−1	−32
0	−12
1	0
2	4
3	0
4	−12

26.

x	y
−4	−10.2
−3	−8.0
−2	−5.8
−1	−3.6
0	−1.4
1	0.8
2	3.0
3	5.2
4	7.4

27.

x	y
−4	−0.019531
−3	−0.078125
−2	−0.3125
−1	−1.25
0	−5
1	−20
2	−80
3	−320
4	−1280

28.

x	y
−4	−0.0003
−3	−0.0009
−2	−0.0027
−1	−0.0082
0	−0.0247
1	−0.0741
2	−0.2222
3	−0.6667
4	−2.0000

29.

x	y
−4	238
−3	173
−2	118
−1	73
0	38
1	13
2	−2
3	−7
4	−2

30.

x	y
−4	4.84
−3	2.46
−2	−1.41
−1	−4.42
0	−4.76
1	−2.20
2	1.68
3	4.55
4	4.66

31.

x	y
−2.0	−28.0
−1.5	−8.5
−1.0	1.0
−0.5	3.5
0.0	2.0
0.5	−0.5
1.0	−1.0
1.5	3.5
2.0	16.0

32.

x	y
−4	−3.5
−3	−2.4
−2	−1.3
−1	−0.2
0	0.9
1	2.0
2	3.1
3	4.2
4	5.3

33.

x	y
2	0.0
3	−1.2
4	−1.9
5	−2.4
6	−2.8
7	−3.1
8	−3.4
9	−3.6
10	−3.8

34.

x	y
1	4.2
2	5.6
3	6.4
4	7.0
5	7.4
6	7.8
7	8.1
8	8.4
9	8.6

B 35. It is estimated that x years from now, $0 \le x \le 10$, the average SAT score of the incoming students at a certain eastern liberal arts college will be

$$f(x) = -6x + 1{,}082$$

a. Derive an expression for the rate at which the average SAT score will be changing with respect to time x years from now.

b. What is the significance of the fact that the expression in part **a** is a negative constant?

36. A particle moving on the x-axis has position

$$x(t) = 2t^3 + 3t^2 - 36t + 40$$

after an elapsed time of t seconds.

a. Find the velocity of the particle at time t.

b. Find the acceleration at time t.

c. What is the total distance traveled by the particle during the first 3 seconds?

37. An object moving on the x-axis has position

$$x(t) = t^3 - 9t^2 + 24t + 20$$

after t seconds. What is the total distance traveled by the object during the first 8 seconds?

38. A car has position

$$s(t) = 50t \ln t + 80$$

ft after t seconds of motion. What is its acceleration (to the nearest hundredth ft/s^2) after 10 seconds?

39. An object moves along a straight line so that after t minutes, its position relative to its starting point (in meters) is

$$s(t) = 10t + \frac{t}{e^t}$$

a. At what speed (to the nearest thousandth m/min) is the object moving at the end of 4 min?

b. How far (to the nearest thousandth m) does the object actually travel during the fifth minute?

40. A bucket containing 5 gal of water has a leak. After t seconds, there are

$$Q(t) = 5\left(1 - \frac{t}{25}\right)^2$$

gallons of water in the bucket.

a. At what rate (to the nearest hundredth gal) is water leaking from the bucket after 2 seconds?

b. How long does it take for all the water to leak out of the bucket?

c. At what rate is the water leaking when the last drop leaks out?

Modeling Problems *In Problems 41–48, set up and solve an appropriate model to answer the given question. Be sure to state your assumptions.*

41. A person standing at the edge of a cliff throws a rock directly upward. It is observed that 2 seconds later the rock is at its maximum height (in ft) and that 5 seconds after that, it hits the ground at the base of the cliff.

a. What is the initial velocity of the rock?

b. How high is the cliff?

c. What is the velocity of the rock at time t?

d. With what velocity does the rock hit the ground?

42. A projectile is shot upward from the earth with an initial velocity of 320 ft/s.

a. What is its velocity after 5 seconds?

b. What is its acceleration after 3 seconds?

43. A rock is dropped from a height of 90 ft. One second later another rock is dropped from height H. What is H (to the nearest foot) if the two rocks hit the ground at the same time?

44. A ball is thrown vertically upward from the ground with an initial velocity of 160 ft/s.

a. When will the ball hit the ground?

b. With what velocity will the ball hit the ground?

c. When will the ball reach its maximum height?

45. An object is dropped (initial velocity $v_0 = 0$) from the top of a building and falls 3 seconds before hitting the pavement below. Determine the height of the building in ft.

46. An astronaut standing at the edge of a cliff on the moon throws a rock directly upward and observes that it passes her on the way down exactly 4 seconds later. Three seconds after that, the rock hits the ground at the base of the cliff. Use this information to determine the initial velocity v_0 and the height of the cliff. *Note*: $g = 5.5$ ft/s^2 on the moon.

47. Answer the question in Problem 46 assuming the astronaut is on Mars, where $g = 12$ ft/s^2.

48. A car is traveling at 88 ft/s (60 mi/h) when the driver applies the brakes to avoid hitting a child. After t seconds, the car is $s(t) = 88t - 8t^2$ feet from the point where the brakes were first applied. How long does it take for the car to come to a stop, and how far does it travel before stopping?

49. It is estimated that t years from now, the circulation of a local newspaper can be modeled by the formula

$$C(t) = 100t^2 + 400t + 50t \ln t$$

a. Find an expression for the rate at which the circulation will be changing with respect to time t years from now.

b. At what rate will the circulation be changing with respect to time 5 years from now?

c. By how much will the circulation actually change during the sixth year?

50. An efficiency study of the morning shift at a certain factory indicates that an average worker who arrives on the job at 8:00 A.M. will have assembled

$$f(x) = -\tfrac{1}{3}x^3 + \tfrac{1}{2}x^2 + 50x$$

units x hours later.

a. Find a formula for the rate at which the worker will be assembling the units after x hours.

b. At what rate will the worker be assembling units at 9:00 A.M.?

c. How many units will the worker actually assemble between 9:00 A.M. and 10:00 A.M.?

51. An environmental study of a certain suburban community suggests that t years from now, the average level of carbon monoxide in the air can be modeled by the formula

$$q(t) = 0.05t^2 + 0.1t + 3.4$$

parts per million.

a. At what rate will the carbon monoxide level be changing with respect to time one year from now?

b. By how much will the carbon monoxide level change in the first year?

c. By how much will the carbon monoxide level change over the next (second) year?

52. According to *Newton's law of universal gravitation*, if an object of mass M is separated by a distance r from a second object of mass m, then the two objects are attracted to one another by a force that acts along the line joining them and has magnitude

$$F = \frac{GmM}{r^2}$$

where G is a positive constant. Show that the rate of change of F with respect to r is inversely proportional to r^3.

53. The population of a bacterial colony is approximately

$$P(t) = P_0 + 61t + 3t^2$$

thousand t hours after observation begins, where P_0 is the initial population. Find the rate at which the colony is growing after 5 hours.

54. The gross domestic product (GDP) of a certain country is

$$g(t) = t^2 + 5t + 106$$

billion dollars t years after 2001.

a. At what rate does the GDP change in 2003?

b. At what percentage rate does the GDP change in 2003?

55. It is projected that x months from now, the population of a certain town will be

$$P(x) = 2x + 4x^{3/2} + 5,000$$

a. At what rate will the population be changing with respect to time 9 months from now?

b. At what percentage rate will the population be changing with respect to time 9 months from now?

56. Assume that your starting salary is $30,000 and you get a raise of $3,000 each year.

a. Express the percentage rate of change of your salary as a function of time.

b. At what percentage rate will your salary be increasing after one year?

c. What will happen to the percentage rate of change of your salary in the long run?

57. The GDP of a certain country is growing at a constant rate. In 1999 the GDP was 125 billion dollars, and in 2001 it was 155 billion dollars. At what percentage rate did the GDP grow in 2000?

58. If y is a linear function of x, what will happen to the percentage rate of change of y with respect to x as x increases without bound?

59. According to Debye's formula in physical chemistry, the orientation polarization P of a gas satisfies

$$P = \frac{4}{3}\pi N \left(\frac{\mu^2}{3kT} \right)$$

where μ, k, and N are constants and T is the temperature of the gas. Find the rate of change of P with respect to T.

60. A disease is spreading in such a way that after t weeks, for $0 \le t \le 6$, it has affected

$$N(t) = 5 - t^2(t - 6)$$

hundred people. Health officials declare that this disease will reach epidemic proportions when the percentage rate of increase of $N(t)$ at the start of a particular week is at least 30% per week. The epidemic designation level is dropped when the percentage rate falls below this level.

a. Find the percentage rate of change of $N(t)$ at time t.

b. Between what weeks is the disease at the epidemic level?

61. An object attached to a helical spring is pulled down from its equilibrium position and then released, as shown in Figure 3.28.

Figure 3.28 Helical spring

Suppose that t seconds later, its position (in centimeters, measured relative to the equilibrium position) is modeled by

$$s(t) = 7 \cos t$$

a. Find the velocity and acceleration of the object at time t.

b. Find the length of time required for one complete oscillation. This is called the *period* of the motion.

c. What is the distance between the highest point reached by the object and the lowest point? Half of this distance is called the *amplitude* of the motion.

62. Two cars leave a town at the same time and travel at constant speeds along straight roads that meet at an angle of 60° in the town. If one car travels twice as fast as the other and the distance between them increases at the rate of 45 mi/h, how fast is the slower car traveling?

63. Spy Problem In an attempt to divert the Spy from seeking revenge for the death of his friend Siggy (Problem 62, Section 2.4), his superior, Lord Newton Fleming (affectionately known as the "Flamer"), assigns him to a mission in space. However, an encounter with an enemy agent leaves him with a mild concussion that causes him to forget where he is. Fortunately, he remembers the values of g for various heavenly bodies (32 ft/s^2 on earth, 5.5 ft/s^2 on the moon, 12 ft/s^2 on Mars, and 28 ft/s^2 on Venus). To deduce his whereabouts, he throws a rock directly upward (from ground level) and notes that it reaches a maximum height of 37.5 ft and hits the ground 5 s after leaving his hand. Where is he?

64. Find the rate of change of the volume of a cube with respect to the length of one of its edges. How is this rate related to the surface area of the cube?

65. Show that the rate of change of the volume of a sphere with respect to its radius is equal to its surface area.

66. Van der Waal's equation states that a gas that occupies a volume V at temperature T (Kelvin) exerts pressure P, where

$$\left(P + \frac{A}{V^2} \right)(V - B) = kT$$

and A, B, and k are physical constants. Find the rate of change of pressure with respect to volume, assuming fixed temperature.

3.5   The Chain Rule

IN THIS SECTION   introduction to the chain rule, extended derivative formulas, justification of the chain rule

INTRODUCTION TO THE CHAIN RULE

Suppose it is known that the carbon monoxide pollution in the air is changing at the rate of 0.02 ppm (parts per million) for each person in a town whose population is growing at the rate of 1,000 people per year. To find the rate at which the level of pollution is increasing with respect to time, we form the product

$$(0.02 \text{ ppm/person}) \, (1,000 \text{ people/year}) = 20 \text{ ppm/year}$$

In this example, the level of pollution L is a function of the population P, which is itself a function of time t. Thus, L is a composite function of t, and

$$\begin{bmatrix} \text{RATE OF CHANGE OF } L \\ \text{WITH RESPECT TO } t \end{bmatrix} = \begin{bmatrix} \text{RATE OF CHANGE OF } L \\ \text{WITH RESPECT TO } P \end{bmatrix} \begin{bmatrix} \text{RATE OF CHANGE OF } P \\ \text{WITH RESPECT TO } t \end{bmatrix}$$

Expressing each of these rates in terms of an appropriate derivative in Leibniz form, we obtain the following equation:

$$\frac{dL}{dt} = \frac{dL}{dP}\frac{dP}{dt}$$

These observations anticipate the following important theorem.

THEOREM 3.10   Chain rule

If $y = f(u)$ is a differentiable function of u and u in turn is a differentiable function of x, then $y = f(u(x))$ is a differentiable function of x and its derivative is given by the product

$$\frac{dy}{dx} = \frac{dy}{du}\frac{du}{dx}$$

Proof   A rigorous proof involves a few details that make it inappropriate to include at this point in the text. A justification (partial proof) is given at the end of this section, and a full proof of the chain rule is included in Appendix B.

WARNING   Recall our earlier warning in Section 3.1 against thinking of $\frac{dy}{dx}$ as a fraction. That said, there are certain times when this incorrect reasoning can be used as a mnemonic device. For instance, "canceling du," as indicated below, makes it easy to remember the chain rule:

$$\frac{dy}{dx} = \frac{dy \, \cancel{du}}{\cancel{du} \, dx}$$ ❏

EXAMPLE 1   The chain rule

Find $\dfrac{dy}{dx}$ if $y = u^3 - 3u^2 + 1$ and $u = x^2 + 2$.

Solution

Because $\dfrac{dy}{du} = 3u^2 - 6u$ and $\dfrac{du}{dx} = 2x$, it follows from the chain rule that

$$\frac{dy}{dx} = \frac{dy}{du}\frac{du}{dx} = (3u^2 - 6u)(2x)$$

Notice that this derivative is expressed in terms of the variables x and u. To express dy/dx in terms of x alone, we substitute $u = x^2 + 2$ as follows:

$$\frac{dy}{dx} = [3(x^2 + 2)^2 - 6(x^2 + 2)](2x) = 6x^3(x^2 + 2)$$ ∎

The chain rule is actually a rule for differentiating composite functions. In particular, if $y = f(u)$ and $u = u(x)$, then y is the composite function $y = (f \circ u)(x) = f[u(x)]$, and the chain rule can be rewritten as follows:

THEOREM 3.10a   The chain rule (alternate form)

If u is differentiable at x and f is differentiable at $u(x)$, then the composite function $f \circ u$ is differentiable at x and

$$\frac{d}{dx} f[u(x)] = \frac{d}{du} f(u) \frac{du}{dx}$$

or

$$(f \circ u)'(x) = f'[u(x)]u'(x)$$

> ◆ **What This Says**   In Section 1.3 when we introduced composite functions, we talked about the "inner" and "outer" functions. With this terminology, the chain rule says that the derivative of the composite function $f[u(x)]$ is equal to the derivative of the inner function u times the derivative of the outer function f evaluated at the inner function.

❑

EXAMPLE 2   The chain rule applied to a power

Differentiate $y = (3x^4 - 7x + 5)^3$.

Solution

Here, the "inner" function is $u(x) = 3x^4 - 7x + 5$ and the "outer" function is u^3, so we have

$$y' = (u^3)'[u(x)]'$$
$$= (3u^2)(12x^3 - 7)$$
$$= 3(3x^4 - 7x + 5)^2(12x^3 - 7)$$ ∎

With a lot of work, you could have found the derivative in Example 2 without using the chain rule either by expanding the polynomial or by using the product rule. The answer would be the same but would involve much more algebra. The chain rule allows us to find derivatives that would otherwise be very difficult to handle.

EXAMPLE 3   Differentiation with quotient rule inside the chain rule

Differentiate $g(x) = \sqrt[4]{\dfrac{x}{1 - 3x}}$.

Solution

Write $g(x) = \left(\dfrac{x}{1 - 3x}\right)^{1/4} = u^{1/4}$, where $u = \dfrac{x}{1 - 3x}$ is the inner function and $u^{1/4}$ is the outer function. Then

$$g'(x) = (u^{1/4})'u'(x) = \tfrac{1}{4} u^{-3/4} u'(x)$$

and we have

$$g'(x) = \frac{1}{4}\left(\frac{x}{1 - 3x}\right)^{-3/4}\left(\frac{x}{1 - 3x}\right)'$$
$$= \frac{1}{4}\left(\frac{x}{1 - 3x}\right)^{-3/4}\left[\frac{(1 - 3x)(1) - x(-3)}{(1 - 3x)^2}\right] \quad \text{Quotient rule}$$

$$= \frac{1}{4}\left(\frac{x}{1-3x}\right)^{-3/4}\left[\frac{1}{(1-3x)^2}\right]$$

$$= \frac{1}{4x^{3/4}(1-3x)^{5/4}} \qquad \blacksquare$$

EXTENDED DERIVATIVE FORMULAS

The chain rule can be used to obtain generalized differentiation formulas for the standard functions, as displayed in the following box.

If u is a differentiable function of x, then

Extended Power Rule

$$\frac{d}{dx}u^n = nu^{n-1}\frac{du}{dx}$$

Extended Trigonometric Rules

$$\frac{d}{dx}\sin u = \cos u\frac{du}{dx} \qquad\qquad \frac{d}{dx}\cos u = -\sin u\frac{du}{dx}$$

$$\frac{d}{dx}\tan u = \sec^2 u\frac{du}{dx} \qquad\qquad \frac{d}{dx}\cot u = -\csc^2 u\frac{du}{dx}$$

$$\frac{d}{dx}\sec u = \sec u\tan u\frac{du}{dx} \qquad\qquad \frac{d}{dx}\csc u = -\csc u\cot u\frac{du}{dx}$$

Extended Exponential and Logarithmic Rules

$$\frac{d}{dx}e^u = e^u\frac{du}{dx} \qquad\qquad \frac{d}{dx}\ln u = \frac{1}{u}\frac{du}{dx}$$

EXAMPLE 4   Chain rule with a trigonometric function

Differentiate $f(x) = \sin(3x^2 + 5x - 7)$.

Solution

Think of this as $f(u) = \sin u$, where $u = 3x^2 + 5x - 7$, and apply the chain rule:

$$f'(x) = \cos(3x^2 + 5x - 7)\cdot(3x^2 + 5x - 7)'$$

$$= (6x + 5)\cos(3x^2 + 5x - 7) \qquad \blacksquare$$

EXAMPLE 5   Chain rule with other rules

Differentiate $g(x) = \cos x^2 + 5\left(\frac{3}{x} + 4\right)^6$.

Solution
$$\frac{dg}{dx} = \frac{d}{dx}\cos x^2 + 5\frac{d}{dx}(3x^{-1} + 4)^6$$

$$= -\sin x^2\frac{d}{dx}(x^2) + 5\left[6(3x^{-1} + 4)^5\frac{d}{dx}(3x^{-1} + 4)\right]$$

$$= (-\sin x^2)(2x) + 30(3x^{-1} + 4)^5(-3x^{-2})$$

$$= -2x\sin x^2 - 90x^{-2}(3x^{-1} + 4)^5 \qquad \blacksquare$$

EXAMPLE 6   Extended power and cosine rules

Differentiate $y = \cos^4(3x + 1)^2$.

Solution

$$\frac{dy}{dx} = 4\cos^3(3x+1)^2 \frac{d}{dx}\cos(3x+1)^2 \qquad \text{Extended power rule}$$

$$= 4\cos^3(3x+1)^2 \cdot [-\sin(3x+1)^2] \cdot \frac{d}{dx}(3x+1)^2 \qquad \text{Extended cosine rule}$$

$$= -4\cos^3(3x+1)^2 \sin(3x+1)^2 \cdot 2(3x+1)(3) \qquad \text{Extended power rule}$$

$$= -24(3x+1)\cos^3(3x+1)^2 \sin(3x+1)^2 \qquad\blacksquare$$

EXAMPLE 7 **Extended power rule inside quotient rule**

Differentiate $p(x) = \dfrac{\tan 7x}{(1-4x)^5}$.

Solution

$$\frac{dp}{dx} = \frac{(1-4x)^5 \left[\dfrac{d}{dx}\tan 7x\right] - \tan 7x \left[\dfrac{d}{dx}(1-4x)^5\right]}{[(1-4x)^5]^2}$$

$$= \frac{(1-4x)^5(\sec^2 7x)\dfrac{d}{dx}(7x) - (\tan 7x)[5(1-4x)^4 \dfrac{d}{dx}(1-4x)]}{(1-4x)^{10}}$$

$$= \frac{(1-4x)^5(\sec^2 7x)(7) - (\tan 7x)(5)(1-4x)^4(-4)}{(1-4x)^{10}}$$

$$= \frac{(1-4x)^4[7(1-4x)\sec^2 7x + 20\tan 7x]}{(1-4x)^{10}}$$

$$= \frac{7(1-4x)\sec^2 7x + 20\tan 7x}{(1-4x)^6} \qquad\blacksquare$$

EXAMPLE 8 **Extended exponential rule within product rule**

Differentiate $e^{-3x}\sin x$.

Solution

Use the product rule:

$$\frac{d}{dx}[e^{-3x}\sin x] = e^{-3x}\left[\frac{d}{dx}(\sin x)\right] + \left[\frac{d}{dx}(e^{-3x})\right]\sin x \qquad \text{Product rule}$$

$$= e^{-3x}(\cos x) + [e^{-3x}(-3)]\sin x$$

$$= e^{-3x}(\cos x - 3\sin x) \qquad\blacksquare$$

EXAMPLE 9 **Finding a horizontal tangent line**

Find the x-coordinate of each point on the graph of

$$f(x) = x^2(4x+5)^3$$

where the tangent line is horizontal.

Solution

Because horizontal tangent lines have zero slope, we need to solve the equation $f'(x) = 0$. We find that

$$f'(x) = [x^2(4x+5)^3]'$$

$$= [x^2]'(4x+5)^3 + x^2[(4x+5)^3]' \qquad \text{Product rule}$$

$$= [2x](4x+5)^3 + x^2[3(4x+5)^2(4)] \qquad \textit{Extended power rule}$$

$$= 2x(4x+5)^2[4x+5+6x] \qquad \textit{Common factor}$$

$$= 2x(4x+5)^2(10x+5)$$

$$= 10x(4x+5)^2(2x+1)$$

From the factored form of the derivative, we see that $f'(x) = 0$ when $x = 0$, $x = -\frac{5}{4}$, and $x = -\frac{1}{2}$, so these are the x-coordinates of the points on the graph of f where horizontal tangent lines occur.   ∎

EXAMPLE 10   A modeling problem using the chain rule

An environmental study of a certain suburban community suggests that the average daily level of carbon monoxide in the air may be modeled by the formula

$$C(p) = \sqrt{0.5p^2 + 17}$$

parts per million when the population is p thousand. It is estimated that t years from now, the population of the community will be

$$p(t) = 3.1 + 0.1t^2$$

thousand. At what rate will the carbon monoxide level be changing with respect to time 3 years from now?

Solution

$$\frac{dC}{dt} = \frac{dC}{dp}\frac{dp}{dt} = \left[\frac{1}{2}(0.5p^2+17)^{-1/2}(0.5)(2p)\right][0.2t]$$

When $t = 3$, $p(3) = 3.1 + 0.1(3)^2 = 4$, so

$$\left.\frac{dC}{dt}\right|_{t=3} = \left[\frac{1}{2}(0.5 \cdot 4^2 + 17)^{-1/2}(4)\right][0.2(3)] = 0.24$$

The carbon monoxide level will be changing at the rate of 0.24 parts per million. It will be increasing because the sign of dC/dt is positive.   ∎

JUSTIFICATION OF THE CHAIN RULE

To get a better feel for why the chain rule is true, suppose x is changed by a small amount Δx. This will cause u to change by an amount Δu, which, in turn, will cause y to change by an amount Δy. *If Δu is not zero*, we can write

$$\frac{\Delta y}{\Delta x} = \frac{\Delta y}{\Delta u}\frac{\Delta u}{\Delta x}$$

By letting $\Delta x \to 0$, we force Δu to approach zero as well, since

$$\Delta u = \left(\frac{\Delta u}{\Delta x}\right)\Delta x \quad \text{so} \quad \lim_{\Delta x \to 0}\Delta u = \left(\frac{du}{dx}\right)(0) = 0$$

It follows that

$$\lim_{\Delta x \to 0}\frac{\Delta y}{\Delta x} = \left(\lim_{\Delta u \to 0}\frac{\Delta y}{\Delta u}\right)\left(\lim_{\Delta x \to 0}\frac{\Delta u}{\Delta x}\right)$$

or, equivalently,

$$\frac{dy}{dx} = \frac{dy}{du}\frac{du}{dx}$$

Unfortunately, there is a flaw in this "proof" of the chain rule. At the beginning we assumed that $\Delta u \neq 0$. However, it is theoretically possible for a small change in x to produce no change in u so that $\Delta u = 0$. This is the case that we consider in the proof in Appendix B.

Historical Note

A calculus book written by the famous mathematician G. H. Hardy (1877–1947) contained essentially this "incorrect" proof rather than the one given in Appendix B. It is even more remarkable that the error was not noticed until the fourth edition.

3.5   PROBLEM SET

A 1. **WHAT DOES THIS SAY?**   What is the chain rule?

2. **WHAT DOES THIS SAY?**   When do you need to use the chain rule?

In Problems 3–8, use the chain rule to compute the derivative dy/dx and write your answer in terms of x only.

3. $y = u^2 + 1; u = 3x - 2$

4. $y = 2u^2 - u + 5; u = 1 - x^2$

5. $y = \dfrac{2}{u^2}; u = x^2 - 9$

6. $y = \cos u; u = x^2 + 7$

7. $y = u \tan u; u = 3x + \dfrac{6}{x}$

8. $y = u^2; u = \ln x$

Differentiate each function in Problems 9–12 with respect to the given variable of the function.

9. **a.** $g(u) = u^3$ **b.** $u(x) = x^2 + 1$
   c. $f(x) = (x^2 + 1)^3$

10. **a.** $g(u) = u^5$ **b.** $u(x) = 3x - 1$
    c. $f(x) = (3x - 1)^5$

11. **a.** $g(u) = u^7$ **b.** $u(x) = 5 - 8x - 12x^2$
    c. $f(x) = (5 - 8x - 12x^2)^7$

12. **a.** $g(u) = u^{15}$ **b.** $u(x) = 3x^2 + 5x - 7$
    c. $f(x) = (3x^2 + 5x - 7)^{15}$

In Problems 13–40, find the derivative of the given function.

13. $f(x) = (5x - 2)^5$

14. $f(x) = (x^4 - 7x)^{15}$

15. $f(x) = (3x^2 - 2x + 1)^4$

16. $f(x) = (3 - x^2 - x^4)^{11}$

17. $s(\theta) = \sin(4\theta + 2)$

18. $c(\theta) = \cos(5 - 3\theta)$

19. $f(x) = e^{-x^2 + 3x}$

20. $y = e^{x^3 - \pi}$

21. $y = e^{\sec x}$

22. $g(x) = e^{\sin x}$

23. $f(t) = \exp(t^2 + t + 5)$

24. $g(t) = t^2 e^{-t} + (\ln t)^2$

25. $g(x) = x \sin 5x$

26. $h(x) = \dfrac{\tan 3x}{x^2}$

27. $f(x) = \left(\dfrac{1}{1 - 2x}\right)^3$

28. $f(x) = \sqrt[3]{\dfrac{1}{2 - 3x}}$

29. $f(x) = xe^{1 - 2x}$

30. $g(x) = \ln(3x^4 + 5x)$

31. $p(x) = \sin x^2 \cos x^2$

32. $f(x) = \csc^2(\sqrt{x})$

33. $f(x) = x^3(2 - 3x)^2$

34. $f(x) = x^4(2 - x - x^2)^3$

35. $f(x) = \sqrt{\dfrac{x^2 + 3}{x^2 - 5}}$

36. $f(x) = \sqrt{\dfrac{2x^2 - 1}{3x^2 + 2}}$

37. $f(x) = \sqrt[3]{x + \sqrt{2x}}$

38. $g(x) = \ln(\ln x)$

39. $f(x) = \ln(\sin x + \cos x)$

40. $T(x) = \ln(\sec x + \tan x)$

Find the x-coordinate of each point in Problems 41–46 where the graph of the given function has a horizontal tangent line.

41. $f(x) = x\sqrt{1 - 3x}$

42. $g(x) = x^2(2x + 3)^2$

43. $q(x) = \dfrac{(x - 1)^2}{(x + 2)^3}$

44. $f(x) = (2x^2 - 7)^3$

45. $T(x) = x^2 e^{1 - 3x}$

46. $V(x) = \dfrac{\ln \sqrt{x}}{x^2}$

B 47. The graphs of $u = g(x)$ and $y = f(u)$ are shown in Figure 3.29.

a. $u = g(x)$ **b.** $y = f(u)$

Figure 3.29   Find the slope of $y = f[g(x)]$.

a. Find the approximate value of u at $x = 2$. What is the slope of the tangent line at that point?

b. Find the approximate value of y at $x = 5$. What is the slope of the tangent line at that point?

c. Find the slope of $y = f[g(x)]$ at $x = 2$.

48. The graphs of $y = f(x)$ and $y = g(x)$ are shown in Figure 3.30.

a. $y = f(x)$ **b.** $y = g(x)$

Figure 3.30   Problem 48

Let $h(x) = f[g(x)]$.

a. Estimate $h'(-1)$ **b.** $h'(1)$ **c.** $h'(3)$

49. Repeat Problem 48 for $h(x) = g[f(x)]$.

50. If $g(x) = f[f(x)]$, use the table to find the value of $g'(2)$.

x	0.0	1.0	2.0	3.0	4.0	5.0
$f(x)$	18.5	9.4	4.0	2.6	8.3	14.0

51. If $h(x) = f[g(x)]$, use the table to find the value of $h'(1)$.

x	0.0	1.0	2.0	3.0	4.0	5.0
$f(x)$	6.9	4.3	3.1	2.8	2.2	2.0
$g(x)$	0.8	2.0	2.5	1.8	0.9	0.4

52. Assume that a spherical snowball melts in such a way that its radius decreases at a constant rate (that is, the radius is a linear function of time). Suppose it begins as a sphere with radius 10 cm and takes 2 hours to disappear.

a. What is the rate of change of its volume after 1 hour? (Recall $V = \frac{4}{3}\pi r^3$.)

b. At what rate is its surface area changing after 1 hour? (Recall $S = 4\pi r^2$.)

53. It is estimated that t years from now, the population of a certain suburban community is modeled by the formula

$$p(t) = 20 - \frac{6}{t+1}$$

where $p(t)$ is in thousands of people. A separate environmental study indicates that the average daily level of carbon monoxide in the air will be

$$L(p) = 0.5\sqrt{p^2 + p + 58}$$

ppm when the population is p thousand. Find the rate at which the level of carbon monoxide will be changing with respect to time two years from now.

54. At a certain factory, the total cost of manufacturing q units during the daily production run is

$$C(q) = 0.2q^2 + q + 900$$

dollars. From experience, it has been determined that approximately

$$q(t) = t^2 + 100t$$

units are manufactured during the first t hours of a production run. Compute the rate at which the total manufacturing cost is changing with respect to time one hour after production begins.

55. An importer of Brazilian coffee estimates that local consumers will buy approximately

$$D(p) = \frac{4,374}{p^2}$$

pounds of the coffee per week when the price is p dollars per pound. It is estimated that t weeks from now, the price of Brazilian coffee will be

$$p(t) = 0.02t^2 + 0.1t + 6$$

dollars per pound. At what rate will the weekly demand for the coffee be changing with respect to time 10 weeks from now? Will the demand be increasing or decreasing?

56. When electric blenders are sold for p dollars apiece, local consumers will buy

$$D(p) = \frac{8,000}{p}$$

blenders per month. It is estimated that t months from now, the price of the blenders will be

$$p(t) = 0.04t^{3/2} + 15$$

dollars. Compute the rate at which the monthly demand for the blenders will be changing with respect to time 25 months from now. Will the demand be increasing or decreasing?

When a point source of light of luminous intensity K (candles) shines directly on a point on a surface s meters away, the illuminance on the surface is given by the formula $I = Ks^{-2}$. Use this formula to answer the questions in Problems 57–58. (Note that 1 lux = 1 candle/m^2.)

57. Suppose a person carrying a 20-candlepower light walks toward a wall in such a way that at time t (seconds) the distance to the wall is $s(t) = 28 - t^2$ meters.

a. How fast is the illuminance on the wall increasing when the person is 19 m from the wall?

b. How far is the person from the wall when the illuminance is changing at the rate of 1 lux/s?

58. A lamp of luminous intensity 40 candles is 20 m above the floor of a room and is being lowered at the constant rate of 2 m/s. At what rate will the illuminance at the point on the floor directly under the lamp be increasing when the lamp is 15 m above the floor? Your answer should be in lux/s rounded to the nearest hundredth.

59. To form a *simple pendulum*, a weight is attached to a rod that is then suspended by one end in such a way that it can swing freely in a vertical plane. Let θ be the angular displacement of the rod from the vertical, as shown in Figure 3.31.

Figure 3.31 Problem 59

It can be shown that as long as the maximum displacement θ_M is small, it is reasonable to assume that $\theta = \theta_M \sin kt$ at time t, where k is a constant that depends on the length of the rod and θ_M is a constant. Show that

$$\frac{d^2\theta}{dt^2} + k^2\theta = 0$$

60. A lighthouse is located 2 km directly across the sea from a point O on the shoreline, as shown in Figure 3.32.

Figure 3.32 Problem 60

A beacon in the lighthouse makes 3 complete revolutions (that is, 6π radians) each minute, and during part of each revolution, the light sweeps across the face of a row of cliffs lining the shore.

a. Show that t minutes after it passes point O, the beam of light is at a point P located $s(t) = 2\tan(6\pi t)$ km from O.

b. How fast is the beam of light moving at the time it passes a point on the cliff that is 4 km from the lighthouse?

61. The average Fahrenheit temperature t hours after midnight in a certain city is modeled by the formula

$$T(t) = 58 + 17 \sin\left(\frac{\pi t}{10} - \frac{5}{6}\right)$$

At what rate is the temperature changing with respect to time at 2 A.M.? Is it getting hotter or colder at that time?

62. In physics, it is shown that when an object is viewed at depth d under water from an angle of incidence θ, then refraction causes the apparent depth of the object to be

$$s(\theta) = \frac{3d \cos \theta}{\sqrt{7 + 9\cos^2 \theta}}$$

(see Figure 3.33). At what rate is the apparent depth changing with respect to θ when $d = 2$ m and $\theta = \frac{\pi}{6}$?

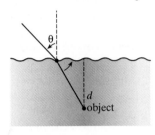

Figure 3.33   Light refraction

63. In the study of Frauenhofer diffraction in optics, a light beam of intensity I_0 from a source L passes through a narrow slit and is diffracted onto a screen (see Figure 3.34.)

Incoming   Screen
light wave

Figure 3.34   Optical diffraction

Experiments indicate that the intensity $I(\theta)$ of light on the screen depends on the diffraction angle θ shown in Figure 3.34 in such a way that

$$I(\beta) = I_0 \left(\frac{\sin \beta}{\beta}\right)^2$$

where $\beta = \pi a \sin(\theta/\lambda)$, λ is the wavelength of the light, and a is a constant related to the width of the screen. Use the chain rule to find $dI/d\theta$, the rate of change of intensity with respect to θ.

64. A worker pushes a heavy crate across a warehouse floor at a constant velocity. If the crate weighs W pounds and the worker applies a force F at an angle θ, as shown in Figure 3.35, then

$$F(\theta) = \frac{\mu W \sec \theta}{1 + \mu \tan \theta}$$

where μ is a constant (the coefficient of friction). For the case where $W = 100$ lb and $\mu = 0.25$, find $\frac{dF}{d\theta}$ when $\theta = 27°$.

Figure 3.35   Pushing a crate with variable force

C 65. Using only the formula

$$\frac{d}{dx} \sin u = \cos u \frac{du}{dx}$$

and the identities $\cos x = \sin(\frac{\pi}{2} - x)$ and $\sin x = \cos(\frac{\pi}{2} - x)$, show that

$$\frac{d}{dx} \cos u = -\sin u \frac{du}{dx}$$

66. Let $g(x) = f[u(x)]$, where $u(-3) = 5$, $u'(-3) = 2$, $f(5) = 3$, and $f'(5) = -3$. Find an equation for the tangent line to the graph of g at the point where $x = -3$.

67. Let f be a function for which

$$f'(x) = \frac{1}{x^2 + 1}$$

a. If $g(x) = f(3x - 1)$, what is $g'(x)$?

b. If $h(x) = f\left(\frac{1}{x}\right)$, what is $h'(x)$?

68. Let f be a function for which $f(2) = -3$ and $f'(x) = \sqrt{x^2 + 5}$. If $g(x) = x^2 f\left(\frac{x}{x - 1}\right)$, what is $g'(2)$?

69. a. If $F(x) = \ln|\cos x|$, show that $F'(x) = -\tan x$.

b. If $F(x) = \ln|\sec x + \tan x|$, show that $F'(x) = \sec x$.

70. If $\frac{df}{dx} = \frac{\sin x}{x}$ and $u(x) = \cot x$, what is $\frac{df}{du}$?

71. Use the chain rule to find

$$\frac{d}{dx} f'[f(x)] \quad \text{and} \quad \frac{d}{dx} f[f'(x)]$$

assuming these derivatives exist.

72. Show that if a particle moves along a straight line with position $s(t)$ and velocity $v(t)$, then its acceleration satisfies

$$a(t) = v(t)\frac{dv}{ds}$$

Use this formula to find $\frac{dv}{ds}$ in the case where

$$s(t) = -2t^3 + 4t^2 + t - 3$$

3.6 Implicit Differentiation

IN THIS SECTION general procedure for implicit differentiation, derivative formulas for the inverse trigonometric functions, logarithmic differentiation

GENERAL PROCEDURE FOR IMPLICIT DIFFERENTIATION

The equation $y = \sqrt{1-x^2}$ **explicitly** defines $y = f(x) = \sqrt{1-x^2}$ as a function of x for $-1 \leq x \leq 1$, but the same function can also be defined **implicitly** by the equation $x^2 + y^2 = 1$, as long as we restrict y by $0 \leq y \leq 1$ so that the vertical line test is satisfied. To find the derivative of the explicit form, we use the chain rule:

$$\frac{d}{dx}\sqrt{1-x^2} = \frac{d}{dx}(1-x^2)^{1/2} = \frac{1}{2}(1-x^2)^{-1/2}(-2x) = \frac{-x}{\sqrt{1-x^2}}$$

To obtain the derivative of the same function in its implicit form, we simply differentiate across the equation $x^2 + y^2 = 1$, remembering that y is a function of x:

$$\frac{d}{dx}(x^2+y^2) = \frac{d}{dx}(1) \qquad \text{Derivative of both sides}$$

$$2x + 2y\frac{dy}{dx} = 0 \qquad \text{Chain rule for the derivative of } y$$

$$\frac{dy}{dx} = -\frac{x}{y} \qquad \text{Solve for } \frac{dy}{dx}.$$

$$= -\frac{x}{\sqrt{1-x^2}} \qquad \text{Write as a function of } x, \text{ if desired.}$$

The procedure we have just illustrated is called **implicit differentiation**. Our illustrative example was simple, but consider a differentiable function $y = f(x)$ defined by the equation

$$x^2y^3 - 6 = 5y^3 + x$$

Implicit differentiation tells us that

$$x^2\frac{d}{dx}(y^3) + y^3\frac{d}{dx}(x^2) - \frac{d}{dx}(6) = 5\frac{d}{dx}(y^3) + \frac{d}{dx}(x)$$

$$x^2\left(3y^2\frac{dy}{dx}\right) + y^3(2x) - 0 = 5\left(3y^2\frac{dy}{dx}\right) + 1$$

$$(3x^2y^3 - 15y^2)\frac{dy}{dx} = 1 - 2xy^3$$

$$\frac{dy}{dx} = \frac{1-2xy^3}{3x^2y^2 - 15y^2}$$

You may think that we are not finished since the derivative involves both x and y, but for many applications, that is enough as we may know both x and y simultaneously at a point. In this example, we could have found y as an explicit function of x, namely

$$y = \left(\frac{x+6}{x^2-5}\right)^{1/3}$$

and then found dy/dx by the chain rule.

Now consider a differentiable function $y = f(x)$ defined by the equation

$$x^2y + 2y^3 = 3x + 2y$$

Finding y as an explicit function of x is very difficult in this case, and it is not at all hard to imagine similar situations where solving for y in terms of x would be impossible or at least not worth the effort. We begin our work with implicit differentiation by using it to find the slope of a tangent line at a point on a circle.

EXAMPLE 1   Slope of a tangent line using implicit differentiation

Find the slope of the tangent line to the circle $x^2 + y^2 = 10$ at the point $P(-1, 3)$.

Solution

y is a function of *x* in this neighborhood around *P*.

Figure 3.36 Graph of $x^2 + y^2 = 10$ showing a neighborhood about $P(-1, 3)$

We recognize the graph of $x^2 + y^2 = 10$ as not being the graph of a function. However, if we look at a small neighborhood around the point $(-1, 3)$, as shown in Figure 3.36, we see that this part of the graph *does pass the vertical line test* for functions. Thus, the required slope can be found by evaluating the derivative dy/dx at $(-1, 3)$. Instead of solving for y and finding the derivative, we *take the derivative of both sides of the equation*:

$$x^2 + y^2 = 10 \qquad \text{Given equation}$$

$$\frac{d}{dx}(x^2 + y^2) = \frac{d}{dx}(10) \qquad \text{Derivative of both sides}$$

$$2x + 2y\frac{dy}{dx} = 0 \qquad \text{Do not forget that } y \text{ is a function of } x.$$

$$\frac{dy}{dx} = -\frac{x}{y} \qquad \text{Solve the equation for } \frac{dy}{dx}.$$

The slope of the tangent line at $P(-1, 3)$ is

$$\left.\frac{dy}{dx}\right|_P = -\left.\frac{x}{y}\right|_{(-1,3)} = -\frac{-1}{3} = \frac{1}{3} \qquad \blacksquare$$

Here is a general description of the procedure for implicit differentiation.

Procedure for Implicit Differentiation

Suppose an equation defines y implicitly as a differentiable function of x. To find $\dfrac{dy}{dx}$:

Step 1.   Differentiate both sides of the equation with respect to x. Remember that y is really a function of x for part of the curve and use the chain rule when differentiating terms containing y.

Step 2.   Solve the differentiated equation algebraically for $\dfrac{dy}{dx}$. $\qquad \blacksquare$

EXAMPLE 2   Implicit differentiation

If $y = f(x)$ is a differentiable function of x such that

$$x^2 y + 2y^3 = 3x + 2y$$

find $\dfrac{dy}{dx}$.

Solution

The process is to differentiate both sides of the given equation with respect to x. To help remember that y is a function of x, replace y by the symbol $y(x)$:

$$x^2 y(x) + 2[y(x)]^3 = 3x + 2y(x)$$

Differentiate both sides of this equation term by term with respect to x:

$$\frac{d}{dx}\left\{x^2 y(x) + 2[y(x)]^3\right\} = \frac{d}{dx}\{3x + 2y(x)\}$$

$$\frac{d}{dx}[x^2 y(x)] + 2\frac{d}{dx}[y(x)]^3 = 3\frac{d}{dx}x + 2\frac{d}{dx}y(x)$$

$$\underbrace{x^2 \frac{d}{dx}y(x) + y(x)\frac{d}{dx}x^2}_{\text{Product rule}} + 2\underbrace{\left\{3[y(x)]^2 \frac{d}{dx}y(x)\right\}}_{\text{Extended power rule}} = 3 + 2\frac{d}{dx}y(x)$$

$$x^2 \frac{d}{dx}y(x) + 2xy(x) + 6[y(x)]^2 \frac{d}{dx}y(x) = 3 + 2\frac{d}{dx}y(x)$$

Now replace $y(x)$ by y and $\frac{d}{dx}y(x)$ by $\frac{dy}{dx}$ and rewrite the equation:

$$x^2 \frac{dy}{dx} + 2xy + 6y^2 \frac{dy}{dx} = 3 + 2\frac{dy}{dx}$$

Finally, solve this equation for $\frac{dy}{dx}$:

$$x^2 \frac{dy}{dx} + 6y^2 \frac{dy}{dx} - 2\frac{dy}{dx} = 3 - 2xy$$

$$(x^2 + 6y^2 - 2)\frac{dy}{dx} = 3 - 2xy$$

$$\frac{dy}{dx} = \frac{3 - 2xy}{x^2 + 6y^2 - 2}$$

Notice that the formula for dy/dx contains both the independent variable x and the dependent variable y. This is usual when derivatives are computed implicitly. ∎

WARNING It is important to realize that implicit differentiation is a technique for finding dy/dx that is valid only if y is a differentiable function of x, and careless application of the technique can lead to errors. For example, there is clearly no real-valued function $y = f(x)$ that satisfies the equation $x^2 + y^2 = -1$, yet formal application of implicit differentiation yields the "derivative" $dy/dx = -x/y$. To be able to evaluate this "derivative," we must find some values of x and y for which $x^2 + y^2 = -1$. Because no such values exist, the derivative does not exist.

In Example 2 it was suggested that you temporarily replace y by $y(x)$, so you would not forget to use the chain rule when first learning implicit differentiation. In the following example, we eliminate this unnecessary step and differentiate the given equation directly. Just keep in mind that y is really a function of x and remember to use the chain rule (or extended power rule) when it is appropriate.

EXAMPLE 3   Implicit differentiation; simplified notation

Find $\dfrac{dy}{dx}$ if y is a differentiable function of x that satisfies

$$\sin(x^2 + y) = y^2(3x + 1)$$

Solution

There is no obvious way to solve the given equation explicitly for y. Differentiate implicitly to obtain

$$\frac{d}{dx}[\sin(x^2 + y)] = \frac{d}{dx}[y^2(3x + 1)]$$

$$\underbrace{\cos(x^2 + y)\frac{d}{dx}(x^2 + y)}_{\text{Chain rule}} = \underbrace{y^2\frac{d}{dx}(3x + 1) + (3x + 1)\frac{d}{dx}y^2}_{\text{Product rule}}$$

$$\cos(x^2 + y)\left(2x + \frac{dy}{dx}\right) = y^2(3) + (3x + 1)\left(2y\frac{dy}{dx}\right)$$

Finally, solve for $\dfrac{dy}{dx}$:

$$2x\cos(x^2 + y) + \cos(x^2 + y)\frac{dy}{dx} = 3y^2 + 2y(3x + 1)\frac{dy}{dx}$$

$$[\cos(x^2 + y) - 2y(3x + 1)]\frac{dy}{dx} = 3y^2 - 2x\cos(x^2 + y)$$

$$\frac{dy}{dx} = \frac{3y^2 - 2x\cos(x^2 + y)}{\cos(x^2 + y) - 2y(3x + 1)} \qquad \blacksquare$$

EXAMPLE 4 Slope of a tangent line using implicit differentiation

Find the slope of a line tangent to the circle $x^2 + y^2 = 5x + 4y$ at the point $P(5, 4)$.

Solution

The slope of a curve $y = f(x)$ is $\dfrac{dy}{dx}$, which we find implicitly.

$$x^2 + y^2 = 5x + 4y$$

$$\frac{d}{dx}(x^2 + y^2) = \frac{d}{dx}(5x + 4y)$$

$$2x + 2y\frac{dy}{dx} = 5 + 4\frac{dy}{dx}$$

$$2y\frac{dy}{dx} - 4\frac{dy}{dx} = 5 - 2x$$

$$(2y - 4)\frac{dy}{dx} = 5 - 2x$$

$$\frac{dy}{dx} = \frac{5 - 2x}{2y - 4}$$

At $(5, 4)$, the slope of the tangent line is

$$\frac{dy}{dx}\bigg|_{(5,4)} = \frac{5 - 2x}{2y - 4}\bigg|_{(5,4)} = \frac{5 - 2(5)}{2(4) - 4} = -\frac{5}{4}$$

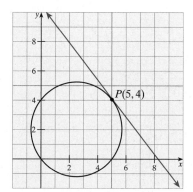

Figure 3.37 Graph of circle and tangent line

Note that the expression is undefined at $y = 2$; this makes sense when you see that the tangent line is vertical there. Look at the graph in Figure 3.37 and see whether you should exclude any other values. $\blacksquare$

EXAMPLE 5 Second derivative by implicit differentiation

Find $\dfrac{d^2y}{dx^2}$ if $x^2 + y^2 = 10$.

Solution

In Example 1 we found (implicitly) that $\dfrac{dy}{dx} = -\dfrac{x}{y}$. Thus,

$$\frac{d^2y}{dx^2} = \underbrace{\frac{d}{dx}\left(\frac{-x}{y}\right) = \frac{y\frac{d}{dx}(-x) - (-x)\frac{d}{dx}y}{y^2}}_{\text{Quotient rule}} = \frac{-y + x\frac{dy}{dx}}{y^2}$$

Note that the expression for the second derivative contains the first derivative dy/dx. To simplify the answer, substitute the algebraic expression previously found for dy/dx:

$$\frac{-y + x\dfrac{dy}{dx}}{y^2} = \frac{-y + x\left(\dfrac{-x}{y}\right)}{y^2} \qquad \text{Substitute } \frac{dy}{dx} = -\frac{x}{y}.$$

$$= \frac{-y^2 - x^2}{y^3}$$

$$= \frac{-(x^2 + y^2)}{y^3}$$

$$= \frac{-10}{y^3} \qquad \text{Substitute } x^2 + y^2 = 10.$$

Thus, $\dfrac{d^2y}{dx^2} = \dfrac{-10}{y^3}$. ∎

In Section 3.3, we found the derivative of $f(x) = e^x$ (Theorem 3.8). This derivative is more easily found using implicit differentiation, as shown in the following example.

EXAMPLE 6   Derivative rule for the natural exponential

Show that

$$\frac{d}{dx}(e^x) = e^x$$

Solution

Let $v = e^x$, so that $x = \ln v$. Then

$$\frac{d}{dx}(x) = \frac{d}{dx}(\ln v) \qquad \text{Derivative of both sides of } x = \ln v$$

$$1 = \frac{1}{v}\frac{dv}{dx} \qquad \text{Implicit differentiation}$$

$$v = \frac{dv}{dx}$$

$$e^x = \frac{dv}{dx} = \frac{d}{dx}(e^x) \qquad \text{Because } v = e^x \qquad ∎$$

Implicit differentiation is a valuable theoretical tool. For example, in Section 3.2 we proved the power rule for the case where the exponent is an integer. Implicit differentiation now allows us to extend the proof for all real exponents.

EXAMPLE 7   Proof of power rule for real (rational and irrational) exponents

Prove that $\dfrac{d}{dx}(x^r) = rx^{r-1}$ holds for all real numbers r if $x > 0$.

Solution

If $y = x^r$, then $y = e^{r \ln x}$, so that

$$\ln y = r \ln x \qquad \text{Definition of exponent}$$

$$\frac{1}{y}\frac{dy}{dx} = r\left(\frac{1}{x}\right) \qquad \text{Implicit differentiation}$$

$$\frac{dy}{dx} = y\left(\frac{r}{x}\right) \qquad \text{Solve for } \frac{dy}{dx}.$$

$$= x^r\left(\frac{r}{x}\right) \qquad \text{Substitute } y = x^r.$$

$$= rx^{r-1} \qquad \text{Property of exponents} \qquad \blacksquare$$

Notice in Example 7 that $y = x^r = e^{r \ln x}$ is differentiable because it is defined as the composition of the differentiable functions $y = e^u$ and $u = r \ln x$.

DERIVATIVE FORMULAS FOR THE INVERSE TRIGONOMETRIC FUNCTIONS

Next, we will use implicit differentiation to obtain differentiation formulas for the six inverse trigonometric functions. Note that these derivatives are not inverse trigonometric functions or even trigonometric functions, but are instead rational functions or roots of rational functions. In fact, the usefulness of inverse trigonometric functions in many areas is related to the simplicity of their derivatives.

THEOREM 3.11   Differentiation formulas for six inverse trigonometric functions

If u is a differentiable function of x, then

$$\frac{d}{dx}(\sin^{-1} u) = \frac{1}{\sqrt{1-u^2}}\frac{du}{dx} \qquad\qquad \frac{d}{dx}(\cos^{-1} u) = \frac{-1}{\sqrt{1-u^2}}\frac{du}{dx}$$

$$\frac{d}{dx}(\tan^{-1} u) = \frac{1}{1+u^2}\frac{du}{dx} \qquad\qquad \frac{d}{dx}(\cot^{-1} u) = \frac{-1}{1+u^2}\frac{du}{dx}$$

$$\frac{d}{dx}(\sec^{-1} u) = \frac{1}{|u|\sqrt{u^2-1}}\frac{du}{dx} \qquad\qquad \frac{d}{dx}(\csc^{-1} u) = \frac{-1}{|u|\sqrt{u^2-1}}\frac{du}{dx}$$

WARNING Note that the derivative of each inverse trigonometric function $y = \cos^{-1} x$, $y = \cot^{-1} x$, and $y = \csc^{-1} x$ is the opposite of the derivative of the corresponding inverse cofunction $y = \sin^{-1} x$, $y = \tan^{-1} x$, and $y = \sec^{-1} x$.

Proof   We will prove the first formula and leave the others as problems. Let $\alpha = \sin^{-1} x$, so $x = \sin \alpha$. Because the sine function is one-to-one and differentiable on $[-\pi/2, \pi/2]$, the inverse sine function is also differentiable. To find its derivative, we proceed implicitly:

$$\sin \alpha = x$$

$$\frac{d}{dx}(\sin \alpha) = \frac{d}{dx}(x)$$

$$\cos \alpha \frac{d\alpha}{dx} = 1$$

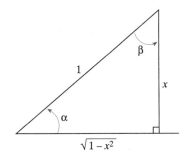

Figure 3.38 $\sin\alpha = x$ and $\cos\alpha = \sqrt{1-x^2}$

Since $-\frac{\pi}{2} \le \alpha \le \frac{\pi}{2}$, $\cos\alpha \ge 0$, so

$$\frac{d\alpha}{dx} = \frac{1}{\cos\alpha} = \frac{1}{\sqrt{1-\sin^2\alpha}} = \frac{1}{\sqrt{1-x^2}}$$ *Note the reference triangle in Figure 3.38.*

If u is a differentiable function of x, then the chain rule gives

$$\frac{d}{dx}(\sin^{-1}u) = \frac{d}{dx}(\sin^{-1}u)\frac{du}{dx} = \frac{1}{\sqrt{1-u^2}}\frac{du}{dx}$$ ❑

EXAMPLE 8   **Derivatives involving inverse trigonometric functions**

Differentiate each of the following functions.

a. $f(x) = \tan^{-1}\sqrt{x}$ **b.** $g(t) = \sin^{-1}(1-t)$ **c.** $h(x) = \sec^{-1}e^{2x}$

Solution

a. Let $u = \sqrt{x}$ in the formula for $\dfrac{d}{dx}(\tan^{-1}u)$:

$$f'(x) = \frac{d}{dx}(\tan^{-1}\sqrt{x})$$

$$= \frac{1}{1+(\sqrt{x})^2}\frac{d}{dx}(\sqrt{x}) = \frac{1}{1+x}\left(\frac{1}{2}\frac{1}{\sqrt{x}}\right) = \frac{1}{2\sqrt{x}(1+x)}$$

b. Let $u = (1-t)$:

$$g'(t) = \frac{d}{dt}\left[\sin^{-1}(1-t)\right]$$

$$= \frac{1}{\sqrt{1-(1-t)^2}}\frac{d}{dt}(1-t) = \frac{-1}{\sqrt{1-(1-t)^2}} = \frac{-1}{\sqrt{2t-t^2}}$$

c. Let $u = e^{2x}$:

$$h'(x) = \frac{d}{dx}\left[\sec^{-1}e^{2x}\right] = \frac{1}{|e^{2x}|\sqrt{(e^{2x})^2-1}}\frac{d}{dx}(e^{2x})$$

$$= \frac{1}{e^{2x}\sqrt{e^{4x}-1}}(2e^{2x}) = \frac{2}{\sqrt{e^{4x}-1}}$$ ■

The derivatives of b^u and $\log_b u$ for a differentiable function $u = u(x)$ and a base b other than e can be obtained using the chain rule and the change-of-base formulas from Chapter 2. We summarize the results in the following theorem.

THEOREM 3.12   **Derivatives of exponential and logarithmic functions with base b**

Let u be a differentiable function of x and b be a positive number (other than 1). Then

$$\frac{d}{dx}b^u = (\ln b)b^u\frac{du}{dx}$$

$$\frac{d}{dx}(\log_b u) = \frac{1}{\ln b}\cdot\frac{1}{u}\frac{du}{dx}$$

WARNING $\dfrac{d}{dx}b^u \ne ub^{u-1}\dfrac{du}{dx}$ since b^u is a constant to a variable power, not a variable to a constant power.

Proof   Because $b^u = e^{u\ln b}$, we can apply the chain rule as follows:

$$\frac{d}{dx}b^u = \frac{d}{dx}(e^{u\ln b}) = e^{u\ln b}\frac{d}{dx}(u\ln b) = e^{u\ln b}\left(\ln b\frac{du}{dx}\right) = (\ln b)b^u\frac{du}{dx}$$

To differentiate the logarithm, recall the change-of-base formula $\log_b u = \dfrac{\ln u}{\ln b}$ so that

$$\frac{d}{dx}\log_b u = \frac{d}{dx}\left(\frac{\ln u}{\ln b}\right) = \frac{1}{\ln b}\cdot\frac{1}{u}\frac{du}{dx}$$ ❑

> **→ What This Says** The derivatives of b^x and $\log_b x$ are the same as the derivatives of e^x and $\ln x$, respectively, except for a factor of $\ln b$ that appears as a multiplier in the formula
>
> $$\frac{d}{dx}(b^x) = (\ln b)b^x$$
>
> and as a divisor in the formula
>
> $$\frac{d}{dx}(\log_b x) = \frac{1}{(\ln b)x}$$

EXAMPLE 9 Derivative of an exponential function with base $b \neq e$

Differentiate $f(x) = x(2^{1-x})$.

Solution

Apply the product rule:

$$f'(x) = \frac{d}{dx}(x2^{1-x}) = x\frac{d}{dx}(2^{1-x}) + 2^{1-x}\frac{d}{dx}(x)$$
$$= x(\ln 2)(2^{1-x})(-1) + 2^{1-x}(1) = 2^{1-x}(1 - x\ln 2) \qquad\blacksquare$$

The following theorem will prove useful in Chapter 5.

THEOREM 3.13 Derivative of $\ln|u|$

If $f(x) = \ln|x|$, $x \neq 0$, then $f'(x) = \frac{1}{x}$. Also, if u is a differentiable function of x, then

$$\frac{d}{dx}\ln|u| = \frac{1}{u}\frac{du}{dx}$$

Proof Using the definition of absolute value,

$$f(x) = \begin{cases} \ln x & \text{if } x > 0 \\ \ln(-x) & \text{if } x < 0 \end{cases}$$

so

$$f'(x) = \begin{cases} \dfrac{1}{x} & \text{if } x > 0 \\ \dfrac{1}{-x}(-1) = \dfrac{1}{x} & \text{if } x < 0 \end{cases}$$

Thus, $f'(x) = \frac{1}{x}$ for all $x \neq 0$.

The second part of the theorem (for u, a differentiable function of x) follows from the chain rule. ❑

LOGARITHMIC DIFFERENTIATION

Logarithmic differentiation is a procedure in which logarithms are used to trade the task of differentiating products and quotients for that of differentiating sums and differences. It is especially valuable as a means for handling complicated product or quotient functions and exponential functions where variables appear in both the base and the exponent.

EXAMPLE 10   Logarithmic differentiation

Find the derivative of $y = \dfrac{e^{2x}(2x-1)^6}{(x^3+5)^2(4-7x)}$ if $y > 0$.

Solution

The procedure called logarithmic differentiation requires that we first take the logarithm of both sides and then apply properties of logarithms before attempting to take the derivative.

$$y = \frac{e^{2x}(2x-1)^6}{(x^3+5)^2(4-7x)}$$

$$\ln y = \ln\left[\frac{e^{2x}(2x-1)^6}{(x^3+5)^2(4-7x)}\right]$$

$$= \ln e^{2x} + \ln(2x-1)^6 - \ln(x^3+5)^2 - \ln(4-7x)$$

$$= 2x + 6\ln(2x-1) - 2\ln(x^3+5) - \ln(4-7x)$$

Next, differentiate both sides with respect to x and then solve for $\dfrac{dy}{dx}$:

$$\frac{1}{y}\frac{dy}{dx} = 2 + 6\left[\frac{1}{2x-1}(2)\right] - 2\left[\frac{1}{x^3+5}(3x^2)\right] - \left[\frac{1}{4-7x}(-7)\right]$$

$$\frac{dy}{dx} = y\left[2 + \frac{12}{2x-1} - \frac{6x^2}{x^3+5} + \frac{7}{4-7x}\right]$$

This is the derivative in terms of x and y. If we want the derivative in terms of x alone, we can substitute the expression for y:

$$\frac{dy}{dx} = \frac{e^{2x}(2x-1)^6}{(x^3+5)^2(4-7x)}\left[2 + \frac{12}{2x-1} - \frac{6x^2}{x^3+5} + \frac{7}{4-7x}\right] \qquad \blacksquare$$

EXAMPLE 11   Derivative with variables in both the base and the exponent

Find $\dfrac{dy}{dx}$, where $y = (x+1)^{2x}$.

Solution $\qquad\qquad\qquad\qquad y = (x+1)^{2x}$

$$\ln y = \ln\left[(x+1)^{2x}\right] = 2x\ln(x+1)$$

Differentiate both sides of this equation:

$$\frac{1}{y}\frac{dy}{dx} = 2x\left\{\frac{d}{dx}[\ln(x+1)]\right\} + \left[\frac{d}{dx}(2x)\right]\ln(x+1) \qquad \text{Product rule}$$

$$= 2x\left[\frac{1}{x+1}(1)\right] + 2\ln(x+1)$$

$$= \frac{2x}{x+1} + 2\ln(x+1)$$

Finally, multiply both sides by $y = (x+1)^{2x}$:

$$\frac{dy}{dx} = \left[\frac{2x}{x+1} + 2\ln(x+1)\right](x+1)^{2x} \qquad \blacksquare$$

3.6 PROBLEM SET

A *Find* $\dfrac{dy}{dx}$ *by implicit differentiation in Problems 1–14.*

1. $x^2 + y^2 = 25$ **2.** $x^2 + y = x^3 + y^3$

3. $xy = 25$ **4.** $xy(2x + 3y) = 2$

5. $x^2 + 3xy + y^2 = 15$ **6.** $x^3 + y^3 = x + y$

7. $\dfrac{1}{y} + \dfrac{1}{x} = 1$ **8.** $(2x + 3y)^2 = 10$

9. $\sin(x + y) = x - y$ **10.** $\tan \dfrac{x}{y} = y$

11. $\cos xy = 1 - x^2$ **12.** $e^{xy} + 1 = x^2$

13. $\ln(xy) = e^{2x}$ **14.** $e^{xy} + \ln y^2 = x$

In Problems 15–18, find $\dfrac{dy}{dx}$ *in two ways:*

a. *By implicit differentiation of the equation*

b. *By differentiating an explicit formula for y*

15. $x^2 + y^3 = 12$ **16.** $xy + 2y = x^2$

17. $x + \dfrac{1}{y} = 5$ **18.** $xy - x = y + 2$

Find the derivative $\dfrac{dy}{dx}$ *in Problems 19–32.*

19. $y = \sin^{-1}(2x + 1)$ **20.** $y = \cos^{-1}(4x + 3)$

21. $y = \tan^{-1}\sqrt{x^2 + 1}$ **22.** $y = \cot^{-1}x^2$

23. $y = (\sin^{-1}2x)^3$ **24.** $y = (\tan^{-1}x^2)^4$

25. $y = \sec^{-1}(e^{-x})$ **26.** $y = \ln\left|\sin^{-1}x\right|$

27. $y = \tan^{-1}\left(\dfrac{1}{x}\right)$ **28.** $y = \cos^{-1}(\sin x), \sin x \geq 0$

29. $y = \sin^{-1}(\cos x), \cos x \geq 0$ **30.** $y = \ln[\sin^{-1}(e^x)]$

31. $x \sin^{-1}y + y \tan^{-1}x = x$ **32.** $\sin^{-1}y + y = 2xy$

In Problems 33–38, find an equation of the tangent line to the graph of each equation at the prescribed point.

33. $x^2 + y^2 = 13$ at $(-2, 3)$

34. $x^3 + y^3 = y + 21$ at $(3, -2)$

35. $\sin(x - y) = xy$ at $(0, \pi)$

36. $3^x + \log_2(xy) = 10$ at $(2, 1)$

37. $x \tan^{-1}y = x^2 + y$ at $(0, 0)$

38. $\sin^{-1}(xy) + \dfrac{\pi}{2} = \cos^{-1}y$ at $(1, 0)$

Find the slope of the tangent line to the graph at the points indicated in Problems 39–42.

39. bifolium:
$(x^2 + y^2)^2 = 4x^2y$
at $(1, 1)$

40. lemniscate of Bernoulli:
$(x^2 + y^2)^2 = \frac{25}{3}(x^2 - y^2)$
at $(2, 1)$

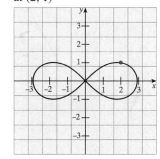

41. folium of Descartes:
$x^3 + y^3 - \frac{9}{2}xy = 0$
at $(2, 1)$

42. cissoid of Diocles:
$y^2(6 - x) = x^3$ at $(3, 3)$

43. Find an equation of the normal line to the curve $x^2 + 2xy = y^3$ at $(1, -1)$.

44. Find an equation of the normal line to the curve $x^2\sqrt{y - 2} = y^2 - 3x - 5$ at $(1, 3)$.

Use implicit differentiation to find the second derivative y'' of the functions given in Problems 45–46.

45. $7x + 5y^2 = 1$ **46.** $x^2 + 2y^3 = 4$

B **47. Interpretation Problem** Compare and contrast the derivatives of the following functions:

a. $y = x^2$ **b.** $y = 2^x$ **c.** $y = e^x$ **d.** $y = x^e$

48. Interpretation Problem Compare and contrast the derivatives of the following functions:

a. $y = \log x$ **b.** $y = \ln x$

49. Interpretation Problem Discuss logarithmic differentiation.

Use logarithmic differentiation in Problems 50–55 to find dy/dx. You may express your answer in terms of both x and y, and you do not need to simplify the resulting rational expressions.

50. $y = \sqrt[18]{(x^{10} + 1)^3(x^7 - 3)^8}$

51. $y = \dfrac{(2x - 1)^5}{\sqrt{x - 9}(x + 3)^2}$

52. $y = \dfrac{e^{2x}}{(x^2 - 3)^2 \ln \sqrt{x}}$

53. $y = \dfrac{e^{3x^2}}{(x^3 + 1)^2(4x - 7)^{-2}}$

54. $y = x^x$ **55.** $y = x^{\ln \sqrt{x}}$

56. Let $\dfrac{u^2}{a^2} + \dfrac{v^2}{b^2} = 1$, where a and b are nonzero constants. Find

a. $\dfrac{du}{dv}$ **b.** $\dfrac{dv}{du}$

57. Show that the tangent line at the point (a, b) on the curve whose equation is $2x^2 + 3xy + y^2 = -2$ is horizontal if $4a + 3b = 0$. Find two such points on the curve.

58. Find two points on the curve whose equation is $x^2 - 3xy + 2y^2 = -2$, where the tangent line is vertical.

59. Let g be a differentiable function of x that satisfies $g(x) < 0$ and $x^2 + g^2(x) = 10$ for all x.

a. Use implicit differentiation to show that $\dfrac{dg}{dx} = \dfrac{-x}{g(x)}$.

b. Show that $g(x) = -\sqrt{10 - x^2}$ satisfies the given requirements. Then use the chain rule to verify that

$$\frac{dg}{dx} = \frac{-x}{g(x)}$$

60. Find the equation of the tangent line and the normal line to the curve

$$x^3 + y^3 = 2Axy$$

at the point (A, A), where A is a nonzero constant.

61. Counterexample Problem

a. If $x^2 + y^2 = 6y - 10$ and $\dfrac{dy}{dx}$ exists, show that $\dfrac{dy}{dx} = \dfrac{x}{3 - y}$.

b. Show that there are no real numbers x, y that satisfy the equation $x^2 + y^2 = 6y - 10$.

c. What can you conclude from the result found in part **a** in light of the observation in part **b**?

62. Find all points on the lemniscate

$$(x^2 + y^2)^2 = 4(x^2 - y^2)$$

where the tangent line is horizontal. (See Figure 3.39.)

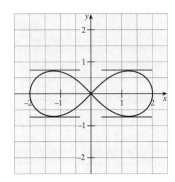

Figure 3.39 Lemniscate $(x^2 + y^2)^2 = 4(x^2 - y^2)$

63. Find all points on the cardioid

$$x^2 + y^2 = \sqrt{x^2 + y^2} + x$$

where the tangent line is vertical. (See Figure 3.40)

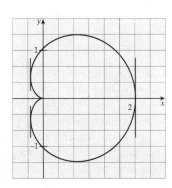

Figure 3.40 Cardioid $x^2 + y^2 = \sqrt{x^2 + y^2} + x$

64. The tangent line to the curve $x^{2/3} + y^{2/3} = 8$ at the point $(8, 8)$ and the coordinate axes form a triangle, as shown in Figure 3.41. What is the area of this triangle?

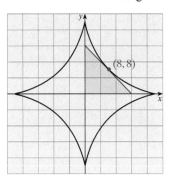

Figure 3.41 Graph of $x^{2/3} + y^{2/3} = 8$

65. Modeling Problem A worker stands 4 m from a hoist being raised at the rate of 2 m/s, as shown in Figure 3.42. Model the worker's angle of sight θ using an inverse trigonometric function, and then determine how fast θ is changing at the instant when the hoist is 1.5 m above eye level.

Figure 3.42 Modeling an angle of elevation

C 66. Find two differentiable functions f that satisfy the equation

$$x - [f(x)]^2 = 9$$

Give the explicit form of each function, and sketch its graph.

67. Show that the tangent line to the ellipse

$$\frac{x^2}{a^2} + \frac{y^2}{b^2} = 1$$

at the point (x_0, y_0) is

$$\frac{x_0 x}{a^2} + \frac{y_0 y}{b^2} = 1$$

(See Figure 3.43.)

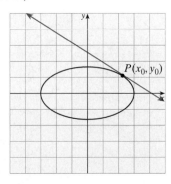

Figure 3.43 Ellipse $\dfrac{x^2}{a^2} + \dfrac{y^2}{b^2} = 1$

68. Find an equation of the tangent line to the hyperbola $\dfrac{x^2}{a^2} - \dfrac{y^2}{b^2} = 1$ at the point (x_0, y_0). (See Figure 3.44.)

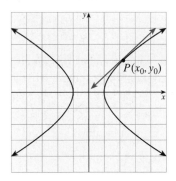

Figure 3.44 Hyperbola $\dfrac{x^2}{a^2} - \dfrac{y^2}{b^2} = 1$

69. Use implicit differentiation to find the second derivative y'', where y is a differentiable function of x that satisfies $ax^2 + by^2 = c$ (a, b, and c are constants).

70. Show that the sum of the x-intercept and the y-intercept of any tangent line to the curve $\sqrt{x} + \sqrt{y} = C$ is equal to C^2.

*The **angle between curves C_1 and C_2** at the point of intersection P is defined as the angle $0 \le \theta \le \frac{\pi}{2}$ between the tangent lines at P. Specifically, the angle between C_1 and C_2 is the angle between the tangent line to C_1 at P and the tangent line to C_2 at P as shown in Figure 3.45.*

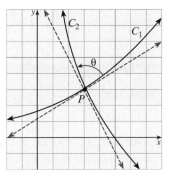

Figure 3.45 The angle θ between C_1 and C_2 at P

Use this information for Problems 71–72.

71. If θ is the angle between curve C_1 and curve C_2 at P and the tangent lines to C_1 and C_2 at P have slopes m_1 and m_2, respectively, show that

$$\tan \theta = \frac{|m_2 - m_1|}{1 + m_1 m_2}$$

72. Find the angle between the circle $x^2 + y^2 = 1$ and the circle $x^2 + (y - 1)^2 = 1$ at each of the two points of intersection.

73. **Journal Problem** *The Pi Mu Epsilon Journal* by Bruce W. King* : ■ When a professor asked the calculus class to find the derivative of y^2 with respect to x^2 for the function $y = x^2 - x$, one student found

$$\frac{dy}{dx} \cdot \frac{y}{x}$$

Was this answer correct? Suppose

$$y = x^3 - 3x^2 + \frac{7}{x}$$

What is the derivative of y^2 with respect to x^2 for this function?

74. Use implicit differentiation to obtain the differentiation formulas for $y = \tan^{-1} u$ and $y = \sec^{-1} u$.

75. Use the identity $\sin^{-1} u + \cos^{-1} u = \frac{\pi}{2}$ to obtain the differentiation formula for $y = \cos^{-1} u$. Similar identities can be used to obtain differentiation formulas for $y = \cot^{-1} u$ and $y = \csc^{-1} u$.

3.7 Related Rates and Applications

Certain practical problems involve a functional relationship $y = f(x)$ in which both x and y are themselves functions of another variable, such as time t. Implicit differentiation is then used to relate the rate of change dy/dt to the rate dx/dt. In this section, we will examine a variety of such **related rate problems**.

When working a related rate problem, you must distinguish between the general situation and the specific situation. The *general situation* comprises properties that are true at *every* instant of time, whereas the *specific situation* refers to those properties that are guaranteed to be true only at the *particular* instant of time that the problem investigates. Here is an example.

EXAMPLE 1 An application involving related rates

A spherical balloon is being filled with a gas in such a way that when the radius is 2 ft, the radius is increasing at the rate of 1/6 ft/min. How fast is the volume changing at this time?

Solution

The general situation: Let V denote the volume and r the radius, both of which are functions of time t (minutes). Since the container is a sphere, its volume is given by

$$V = \tfrac{4}{3}\pi r^3$$

Differentiating both sides implicitly with respect to time t yields

$$\frac{dV}{dt} = \frac{d}{dt}\left(\frac{4}{3}\pi r^3\right)$$

$$= 4\pi r^2 \frac{dr}{dt} \qquad \text{Do not forget to use the chain rule}$$
$$\text{because } r \text{ is also a function of time.}$$

The specific situation: Our goal is to find $\dfrac{dV}{dt}$ at the time when $r = 2$ ft and $\dfrac{dr}{dt} = \dfrac{1}{6}$.

$$\left.\frac{dV}{dt}\right|_{r=2} = 4\pi(2)^2\left(\frac{1}{6}\right) = \frac{8\pi}{3}$$

This means that the volume of the container is increasing at about 8.38 ft³/min when the radius is 2 ft. ∎

Although each related rate problem has its own "personality," many can be handled by the following summary:

Procedure for Solving Related Rate Problems

The General Situation

Step 1. *Draw a figure, if appropriate, and assign variables to the quantities that vary.* Be careful not to label a quantity with a number unless it *never* changes in the problem.

Step 2. *Find a formula or equation that relates the variables.* Eliminate unnecessary variables; some of these "extra" variables may be constants, but others may be eliminated because of given relationships among the variables.

Step 3. *Differentiate the equations.* You will usually differentiate implicitly with respect to time.

The Specific Situation

Step 4. *Substitute specific numerical values and solve algebraically for any required rate.* List the known quantities; list as unknown the quantity you wish to find. Substitute all values into the formula. The only remaining variable should be the unknown, which may be a variable or a rate. Solve for the unknown.

EXAMPLE 2 Moving shadow problem

A person 6 ft tall is walking away from a streetlight 20 ft high at the rate of 7 ft/s. At what rate is the length of the person's shadow increasing?

Figure 3.46 A person walking away from a streetlamp

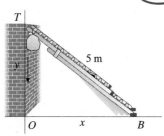

Figure 3.47 A ladder sliding down a wall

SMH

Solution

The general situation: (See *Student Mathematics Handbook*, Problem 37, Problem Set 1.)

Step 1. Let x denote the length (in feet) of the person's shadow, and y, the distance between the person and the street light, as shown in Figure 3.46. Let t denote the time (in seconds).

Step 2. Since $\triangle ABC$ and $\triangle DEC$ are similar, we have

$$\frac{x+y}{20} = \frac{x}{6}$$

Step 3. Write this equation as $x + y = \frac{20}{6}x$, or $y = \frac{7}{3}x$, and differentiate both sides with respect to t.

$$\frac{dy}{dt} = \frac{7}{3}\frac{dx}{dt}$$

The specific situation:

Step 4. List the known quantities. We know that $dy/dt = 7$. Our goal is to find dx/dt. Substitute and then solve for the unknown value:

$$\frac{dy}{dt} = \frac{7}{3}\frac{dx}{dt}$$

$$7 = \frac{7}{3}\frac{dx}{dt} \qquad \text{Substitute.}$$

$$3 = \frac{dx}{dt} \qquad \text{Multiply both sides by } \tfrac{3}{7}.$$

The length of the person's shadow is increasing at the rate of 3 ft/s. ∎

EXAMPLE 3 Leaning ladder problem

A bag is tied to the top of a 5-m ladder resting against a vertical wall. Suppose the ladder begins sliding down the wall in such a way that the foot of the ladder is moving away from the wall. How fast is the bag descending at the instant the foot of the ladder is 4 m from the wall and the foot is moving away at the rate of 2 m/s?

SMH

Solution

The general situation: Let x and y be the distances from the base of the wall to the foot and top of the ladder, respectively, as shown in Figure 3.47. (See *Student Mathematics Handbook*, Problem 38, Problem Set 1.)

Notice that $\triangle TOB$ is a right triangle, so a relevant formula is the Pythagorean theorem:

$$x^2 + y^2 = 25$$

Differentiate both sides of this equation with respect to t:

$$2x\frac{dx}{dt} + 2y\frac{dy}{dt} = 0$$

The specific situation: At the particular instant in question, $x = 4$ and $y = \sqrt{25 - 4^2} = 3$. We also know that $\dfrac{dx}{dt} = 2$, and the goal is to find $\dfrac{dy}{dt}$ at this instant. We have

$$2(4)(2) + 2(3)\frac{dy}{dt} = 0$$

$$\frac{dy}{dt} = -\frac{8}{3}$$

This tells us that, at the instant in question, the bag is descending (since dy/dt is negative) at the rate of $8/3 \approx 2.7$ m/sec. ∎

EXAMPLE 4   **Modeling a physical application involving related rates**

When air expands *adiabatically* (that is, with no change in heat), the pressure P and the volume V satisfy the relationship

$$PV^{1.4} = C$$

where C is a constant. At a certain instant, the pressure is 20 lb/in.^{2} and the volume is 280 in.^{3}. If the volume is decreasing at the rate of 5 in.^{3}/s at this instant, what is the rate of change of the pressure?

Solution

The general situation: The required equation was given, so we begin by differentiating both sides with respect to t. Remember, because C is a constant, its derivative with respect to t is zero.

$$1.4PV^{0.4}\frac{dV}{dt} + V^{1.4}\frac{dP}{dt} = 0 \qquad \text{Product rule}$$

The specific situation: At the instant in question, $P = 20$, $V = 280$, and $dV/dt = -5$ (negative because the volume is decreasing). The goal is to find dP/dt. First substitute to obtain

$$(20)(1.4)(280)^{0.4}(-5) + (280)^{1.4}\frac{dP}{dt} = 0$$

Now, solve for $\dfrac{dP}{dt}$:

$$\frac{dP}{dt} = \frac{5(20)(1.4)(280)^{0.4}}{(280)^{1.4}} = 0.5$$

Thus, at the instant in question, the pressure is increasing (because its derivative is positive) at the rate of 0.5 lb/in.^{2} per second. ∎

EXAMPLE 5   **The water level in a cone-shaped tank**

A tank filled with water is in the shape of an inverted cone 20 ft high with a circular base (on top) whose radius is 5 ft. Water is running out of the bottom of the tank at the constant rate of 2 ft^3/min. How fast is the water level falling when the water is 8 ft deep?

Solution

The general situation: Consider a conical tank with height 20 ft and circular base of radius 5 ft, as shown in Figure 3.48. Suppose that the water level is h ft and that the radius of the surface of the water is r.

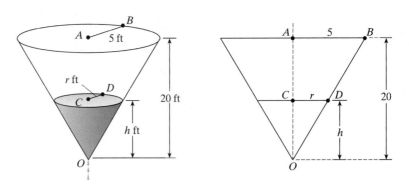

Figure 3.48   A conical water tank

Let V denote the volume of water in the tank after t minutes. We know that

$$V = \tfrac{1}{3}\pi r^2 h$$

(See *Student Mathematics Handbook*, Problem 39, Problem Set 1.) Once again, we use similar triangles (see Figure 3.48) to write $\frac{5}{20} = \frac{r}{h}$, or $r = \frac{h}{4}$. We substitute this into the formula to obtain

$$V = \frac{1}{3}\pi \left(\frac{h}{4}\right)^2 h = \frac{1}{48}\pi h^3$$

and then differentiate both sides of this equation with respect to t.

$$\frac{dV}{dt} = \frac{\pi}{16}h^2\frac{dh}{dt}$$

WARNING A common error in solving related rate problems is to substitute numerical values too soon or, equivalently, to use relationships that apply only at a particular moment in time. This is the reason we have separated related rate problems into two distinct parts. Be careful to work with general relationships among the variables and substitute specific numerical values only after you have found general rate relationships by differentiation.

The specific situation: Begin with the known quantities: We know that $dV/dt = -2$ (negative, because the volume is decreasing). The goal is to find dh/dt. At the particular instant in question, $h = 8$; we substitute to find

$$-2 = \frac{\pi}{16}(8)^2\frac{dh}{dt}$$

$$\frac{-1}{2\pi} = \frac{dh}{dt}$$

At the instant when the water is 8 ft deep, the water level is falling (since dh/dt is negative) at a rate of $\frac{1}{2\pi} \approx 0.16$ ft/min ≈ 2 in./min. ∎

EXAMPLE 6 Modeling with an angle of elevation

Every day, a flight to Los Angeles flies directly over my home at a constant altitude of 4 mi. If I assume that the plane is flying at a constant speed of 400 mi/h, at what rate is the angle of elevation of my line of sight changing with respect to time when the horizontal distance between the approaching plane and my location is exactly 3 mi?

Solution

The general situation: Let x denote the horizontal distance between the plane and the observer, as shown in Figure 3.49. The height of the observer is insignificant when compared to the height of the plane.

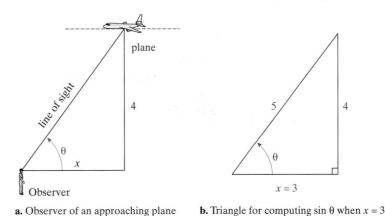

a. Observer of an approaching plane   **b.** Triangle for computing sin θ when $x = 3$

Figure 3.49 Angle of elevation problem

Then the angle of observation θ can be modeled by

$$\cot \theta = \frac{x}{4} \qquad \text{or} \qquad \theta = \cot^{-1}\left(\frac{x}{4}\right)$$

Differentiate both sides of this last equation with respect to t to obtain

$$\frac{d\theta}{dt} = \frac{-1}{1+(\frac{x}{4})^2}\left(\frac{1}{4}\right)\frac{dx}{dt} = \frac{-4}{16+x^2}\frac{dx}{dt}$$

The specific situation: At the instant when $x = 3$, we are given that $\dfrac{dx}{dt} = -400$ (negative because the distance is decreasing). Thus, the angle of elevation is changing at the rate of

$$\frac{d\theta}{dt} = \frac{-4}{16 + 3^2}(-400) = 64 \text{ rad/hr}$$

The angle of elevation is changing at the rate of 64 radians per hour or, equivalently,

$$(64 \text{ rad/hr})\left(\frac{360 \text{ deg}}{2\pi \text{ rad}}\right)\left(\frac{1 \text{ hr}}{3,600 \text{ s}}\right) \approx 1.02 \text{ deg/sec} \qquad \blacksquare$$

3.7   PROBLEM SET

Ⓐ *Find the indicated rate in Problems 1–9, given the other information. Assume $x > 0$ and $y > 0$.*

1. If $x^2 + y^2 = 25$ and $\dfrac{dx}{dt} = 4$, find $\dfrac{dy}{dt}$ when $x = 3$.

2. If $x^2 + y^2 = 25$ and $\dfrac{dy}{dt} = 2$, find $\dfrac{dx}{dt}$ when $x = 4$.

3. If $5x^2 - y = 100$ and $\dfrac{dx}{dt} = 10$, find $\dfrac{dy}{dt}$ when $x = 10$.

4. If $4x^2 - y = 100$ and $\dfrac{dy}{dt} = -6$, find $\dfrac{dx}{dt}$ when $x = 1$.

5. If $y = 2\sqrt{x} - 9$ and $\dfrac{dy}{dt} = 5$, find $\dfrac{dx}{dt}$ when $x = 9$.

6. If $y = 5\sqrt{x + 9}$ and $\dfrac{dx}{dt} = 2$, find $\dfrac{dy}{dt}$ when $x = 7$.

7. If $xy = 10$, and $\dfrac{dx}{dt} = -2$, find $\dfrac{dy}{dt}$ when $x = 5$.

8. If $5xy = 10$, and $\dfrac{dx}{dt} = -2$, find $\dfrac{dy}{dt}$ when $x = 1$.

9. If $x^2 + xy - y^2 = 11$ and $\dfrac{dy}{dt} = 5$, find $\dfrac{dx}{dt}$ when $x = 4$ and $y > 0$.

In physics, Hooke's law says that when a spring is stretched x units beyond its natural length, the elastic force $F(x)$ exerted by the spring is $F(x) = -kx$, where k is a constant that depends on the spring. Assume $k = 12$ in Problems 10 and 11.

10. If a spring is stretched at the constant rate of $\frac{1}{4}$ in./s, how fast is the force $F(x)$ changing when $x = 2$ in.?

11. If a spring is stretched at the constant rate of $\frac{1}{4}$ in./s, how fast is the force $F(x)$ changing when $x = 3$ in.?

12. A particle moves along the parabolic path given by $y^2 = 4x$ in such a way that when it is at the point $(1, -2)$, its horizontal velocity (in the direction of the x-axis) is 3 ft/s. What is its vertical velocity (in the direction of the y-axis) at this instant?

13. A particle moves along the elliptical path given by $4x^2 + y^2 = 4$ in such a way that when it is at the point $(\sqrt{3}/2, 1)$, its x-coordinate is increasing at the rate of 5 units per second. How fast is the y-coordinate changing at that instant?

14. A rock is dropped into a lake and an expanding circular ripple results. When the radius of the ripple is 8 in., the radius is increasing at a rate of 3 in./s. At what rate is the area enclosed by the ripple changing at this time?

15. A pebble dropped into a pond causes a circular ripple. Find the rate at which the radius of the ripple is changing at a time when the radius is one foot and the area enclosed by the ripple is increasing at the rate of 4 ft²/s.

16. An environmental study of a certain community indicates that there will be $Q(p) = p^2 + 3p + 1,200$ units of a harmful pollutant in the air when the population is p thousand. The population is currently 30,000 and is increasing at a rate of 2,000 per year. At what rate is the level of air pollution increasing currently?

17. It is estimated that the annual advertising revenue received by a certain newspaper will be $R(x) = 0.5x^2 + 3x + 160$ thousand dollars when its circulation is x thousand. The circulation of the paper is currently 10,000 and is increasing at a rate of 2,000 per year. At what rate will the annual advertising revenue be increasing with respect to time 2 years from now?

18. Hospital officials estimate that approximately $N(p) = p^2 + 5p + 900$ people will seek treatment in the emergency room each year if the population of the community is p thousand. The population is currently 20,000 and is growing at the rate of 1,200 per year. At what rate is the number of people seeking emergency room treatment increasing?

19. Boyle's law states that when gas is compressed at constant temperature, the pressure P of a given sample satisfies the equation $PV = C$, where V is the volume of the sample and C is a constant. Suppose that at a certain time the volume is 30 in.³, the pressure is 90 lb/in.², and the volume is increasing at the rate of 10 in.³/s. How fast is the pressure changing at this instant? Is it increasing or decreasing?

Ⓑ 20. **Interpretation Problem**   What do we mean by a related rate problem?

21. **Interpretation Problem**   Outline a procedure for solving related-rate problems.

22. The volume of a spherical balloon is increasing at a constant rate of 3 in.³/s. At what rate is the radius of the balloon increasing when the radius is 2 in.?

23. The surface area of a sphere is decreasing at the constant rate of 3π cm²/s. At what rate is the volume of the sphere decreasing at the instant its radius is 2 cm?

24. A person 6 ft tall walks away from a streetlight at the rate of 5 ft/s. If the light is 18 ft above ground level, how fast is the person's shadow lengthening?

25. A ladder 13 ft long rests against a vertical wall and is sliding down the wall at the rate of 3 ft/s at the instant the foot of the ladder is 5 ft from the base of the wall. At this instant, how fast is the foot of the ladder moving away from the wall?

26. A car traveling north at 40 mi/h and a truck traveling east at 30 mi/h leave an intersection at the same time. At what rate will the distance between them be changing 3 hours later?

27. A person is standing at the end of a pier 12 ft above the water and is pulling in a rope attached to a rowboat at the waterline at the rate of 6 ft of rope per minute, as shown in Figure 3.50. How fast is the boat moving in the water when it is 16 ft from the pier?

Figure 3.50   Problem 27

28. One end of a rope is fastened to a boat and the other end is wound around a windlass located on a dock at a point 4 m above the level of the boat. If the boat is drifting away from the dock at the rate of 2 m/min, how fast is the rope unwinding at the instant when the length of the rope is 5 m?

29. A ball is dropped from a height of 160 ft. A light is located at the same level, 10 ft away from the initial position of the ball. How fast is the ball's shadow moving along the ground one second after the ball is dropped?

30. A person 6 ft tall stands 10 ft from point P directly beneath a lantern hanging 30 ft above the ground, as shown in Figure 3.51.

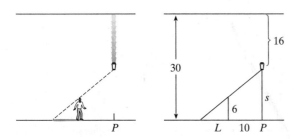

Figure 3.51   Problem 30

The lantern starts to fall, thus causing the person's shadow to lengthen. Given that the lantern falls $16t^2$ ft in t seconds, how fast will the shadow be lengthening when $t = 1$?

31. A race official is watching a race car approach the finish line at the rate of 200 km/h. Suppose the official is sitting at the finish line, 20 m from the point where the car will cross, and let θ be the angle between the finish line and the official's line of sight to the car, as shown as Figure 3.52. At what rate is θ changing when the car crosses the finish line? Give your answer in terms of rad/s.

Figure 3.52   Problem 31

Modeling Problems: *In Problems 32–35, set up and solve an appropriate model to answer the given question. Be sure to state your assumptions.*

32. Consider a piece of ice in the shape of a sphere that is melting at the rate of 5 in.^{3}/min. Model the volume of ice by a function of the radius r. How fast is the radius changing at the instant when the radius is 4 in.? How fast is the surface area of the sphere changing at the same instant?

33. A certain medical procedure requires that a balloon be inserted into the stomach and then inflated. Model the shape of the balloon by a sphere of radius r. If r is increasing at the rate of 0.3 cm/min, how fast is the volume changing when the radius is 4 cm?

34. Model a water tank by a cone 40 ft high with a circular base of radius 20 ft at the top. Water is flowing into the tank at a constant rate of 80 ft^3/min. How fast is the water level rising when the water is 12 ft deep? Give your answer to the nearest hundredth of a foot per minute.

35. In Problem 34, suppose that water is also flowing out the bottom of the tank. At what rate should the water be allowed to flow out so that the water level will be rising at a rate of only 0.05 ft/min when the water is 12 ft deep? Give your answer to the nearest tenth of a cubic foot per minute.

36. The air pressure $p(s)$ at a height of s meters above sea level is modeled by the formula $p(s) = e^{-0.000125s}$ atmospheres. An instrument box carrying a device for measuring pressure is dropped into the ocean from a plane and falls in such a way that after t seconds it is

$$s(t) = 3{,}000 - 49t - 245(e^{-t/5} - 1)$$

meters above the ocean's surface.

   a. Find ds/dt and then use the chain rule to find dp/dt. How fast is the air pressure changing 2 seconds after the box begins to fall?

   b. When (to the nearest second) does the box hit the water? How fast is the air pressure changing at the time of impact?

   What assumptions are you making in this model?

37. At noon on a certain day, a truck is 250 mi due east of a car. The truck is traveling west at a constant speed of 25 mi/h, while the car is traveling north at 50 mi/h.

   a. At what rate is the distance between them changing at time t?

   b. At what time is the distance between the car and the truck neither increasing nor decreasing?

   c. What is the minimal distance between the car and the truck? *Hint*: This distance must occur at the time found in part **b**. Do you see why?

38. A weather balloon is rising vertically at the rate of 10 ft/s. An observer is standing on the ground 500 ft horizontally from the point where the balloon was released. At what rate is the distance between the observer and the balloon changing when the balloon is 400 ft high?

39. An observer watches a plane approach at a speed of 500 mi/h and an altitude of 3 mi. At what rate is the angle of elevation of the observer's line of sight changing with respect to time when the horizontal distance between the plane and the observer is 4 mi? Give your answer in radians per minute.

40. A person 6 ft tall is watching a streetlight 18 ft high while walking toward it at a speed of 5 ft/s, as shown in Figure 3.53. At what rate is the angle of elevation of the person's line of sight changing with respect to time when the person is 9 ft from the base of the light?

Figure 3.53 Streetlight problem

41. A revolving searchlight in a lighthouse 2 mi offshore is following a beachcomber along the shore, as shown in Figure 3.54.

Figure 3.54 Problem 41

When the beachcomber is 1 mi from the point on the shore that is closest to the lighthouse, the searchlight is turning at the rate of 0.25 rev/h. How fast is the beachcomber walking at that moment? *Hint*: Note that 0.25 rev/h is the same as $\frac{\pi}{2}$ rad/h.

42. A water trough is 2 ft deep and 10 ft long. It has a trapezoidal cross section with base lengths 2 ft and 5 ft, as shown in Figure 3.55.

Figure 3.55 Water trough

a. Find a relationship between the volume of water in the trough at any given time and the depth of the water at that time.

b. If water enters the trough at the rate of 10 ft³/min, how fast is the water level rising (to the nearest $\frac{1}{2}$ in./min) when the water is 1 ft deep?

43. At noon, a ship sails due north from a point P at 8 knots (nautical miles per hour). Another ship, sailing at 12 knots, leaves the same point 1 h later on a course 60° east of north. How fast is the distance between the ships increasing at 2 P.M.? At 5 P.M.? *Hint*: Use the law of cosines.

44. A swimming pool is 60 ft long and 25 ft wide. Its depth varies uniformly from 3 ft at the shallow end to 15 ft at the deep end, as shown in Figure 3.56.

Figure 3.56 Swimming pool

Suppose the pool is being filled with water at the rate of 800 ft³/min. At what rate is the depth of water increasing at the deep end when it is 5 ft deep at that end?

45. Suppose a water bucket is modeled by the frustum of a cone with height 1 ft and upper and lower radii of 1 ft and 9 in., respectively, as shown in Figure 3.57.

Figure 3.57 Water bucket

If water is leaking from the bottom of the bucket at the rate of 8 in.³/min, at what rate is the water level falling when the depth of water in the bucket is 6 in.? *Hint*: The volume of the frustum of a cone with height h and base radii r and R is

$$V = \frac{\pi h}{3}(R^2 + rR + r^2).$$

46. A lighthouse is located 2 km directly across the sea from a point S on the shoreline. A beacon in the lighthouse makes 3 complete revolutions (6π radians) each minute, and during part of each revolution, the light sweeps across the face of a row of cliffs lining the shore.

a. Show that t minutes after it passes point S, the beam of light is at a point P located $s(t) = 2\tan 6\pi t$ km from S.

b. How fast is the beam of light moving at the time it passes a point on the cliff located 4 km from the lighthouse?

47. Modeling Problem A car is traveling at the rate of 40 ft/s along a straight, level road that parallels the seashore. A rock with a family of seals is located 50 yd offshore.

a. Model the angle θ between the road and the driver's line of sight as a function of the distance x to the point P directly opposite the rock.

b. As the distance x in Figure 3.58 approaches 0, what happens to $d\theta/dt$?

c. Suppose the car is traveling at v ft/s. Now what happens to $d\theta/dt$ as $x \to 0$? What effect does this have on a passenger looking at the seals if the car is traveling at a high rate of speed?

Figure 3.58 Problem 47

3.8 Linear Approximation and Differentials

IN THIS SECTION tangent line approximation, the differential, error propagation, marginal analysis in economics, the Newton–Raphson method for approximating roots

TANGENT LINE APPROXIMATION

If $f(x)$ is differentiable at $x = a$, the tangent line at a point $P(a, f(a))$ on the graph of $y = f(x)$ has slope $m = f'(a)$ and equation

$$\frac{y - f(a)}{x - a} = f'(a) \quad \text{or} \quad y = f(a) + f'(a)(x - a)$$

In the immediate vicinity of P, the tangent line closely approximates the shape of the curve $y = f(x)$. For instance, if $f(x) = x^3 - 2x + 5$, the tangent line at $P(1, 4)$ has slope $f'(1) = 3(1)^2 - 2 = 1$ and equation

$$y = 4 + (1)(x - 1) = x + 3$$

The graph of $y = f(x)$, the tangent line at $P(1, 4)$, and two enlargements showing how the tangent line approximates the graph of f near P are shown in Figure 3.59.

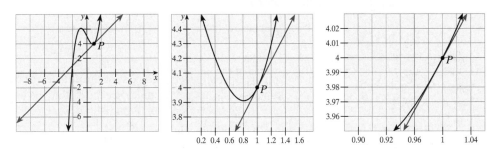

Figure 3.59 Tangent line approximation of $f(x) = x^3 - 2x + 5$ at $P(1, 4)$

Our observation about tangent lines suggests that if x_1 is near a, then $f(x_1)$ must be close to the point on the tangent line to $y = f(x)$ at $x = x_1$. That is,

$$f(x_1) \approx f(a) + f'(a)(x_1 - a)$$

We refer to this as a *linear approximation* of $f(x)$ at $x = a$, and the function

$$L(x) = f(a) + f'(a)(x - a)$$

is called a **linearization** of the function at a point $x = a$. We can use this line as an approximation of f as long as the line remains close to the graph of f, as shown in Figure 3.60.

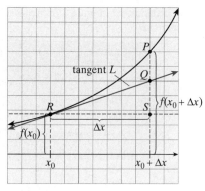

a. Tangent line to f at R **b.** Tangent line approximation

Figure 3.60   Tangent line approximation

Recall from Section 1.1, that for a line, we used the notation Δx for the horizontal change and Δy for the vertical change. Then, the linear approximation formula measures the vertical change from point R to point Q even though we really want the vertical change from the point R to the point P. If the distance Δx is small, then these two vertical distances should be approximately equal. This version of linear approximation is sometimes called the *incremental approximation formula*.

$$f(x_1) - f(a) \approx f'(a)(x_1 - a)$$
$$\Delta y \approx f'(a)\Delta x$$

EXAMPLE 1   Incremental approximation

Show that if $f(x) = \sin x$, the function $\dfrac{\Delta f}{\Delta x}$ approximates the function $f'(x) = \cos x$ for small values of Δx.

Solution

The approximation formula $\Delta f = f(x_0 + \Delta x) - f(x_0) \approx f'(x_0)\Delta x$ implies that
$$\frac{\Delta f}{\Delta x} = \frac{f(x_0 + \Delta x) - f(x_0)}{\Delta x} \approx f'(x_0)$$
Because $f(x) = \sin x$, $f'(x) = \cos x$, and
$$\frac{\sin(x + \Delta x) - \sin x}{\Delta x} \approx \cos x$$
Figure 3.61 shows the graphs for three different choices of Δx. Notice as Δx becomes smaller, it is more difficult to see the difference between f and g; in fact, for very small Δx, the graphs are virtually indistinguishable.

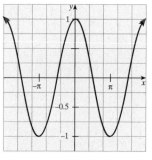

a. $\Delta x = 0.5$ **b.** $\Delta x = 0.1$ **c.** $\Delta x = 0.01$

Figure 3.61   Graphs of $f'(x) = \cos x$ and $g(x) = \dfrac{f(x + \Delta x) - f(x)}{\Delta x}$

THE DIFFERENTIAL

We have already observed that writing the derivative of $f(x)$ in the Leibniz notation df/dx suggests that the derivative may be incorrectly regarded as a quotient of "df" by "dx." It is a tribute to the genius of Leibniz that this erroneous interpretation of his notation often turns out to make good sense.

To give dx and dy meaning as separate quantities, let x be fixed and define dx to be an independent variable equal to Δx, the change in x. That is, define dx, called the **differential of x**, to be an independent variable equal to Δx, the change in x. Then, if f is differentiable at x, we define dy, called the **differential of y**, by the formula

$$dy = f'(x)\, dx \quad \text{or, equivalently,} \quad df = f'(x)\, dx$$

If we relate differentials to Figure 3.62, we see that $dx = \Delta x$ and that Δy is the rise of f that occurs for a change of Δx, whereas dy is the rise of a tangent line relative to the same change in x (Δy and dy are not the same thing). This is shown in Figure 3.62.

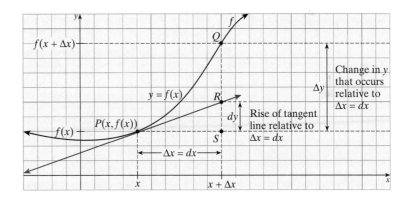

Figure 3.62   Geometrical definition of dx and dy

We now restate the standard rules and formulas for differentiation in terms of differentials. Remember that a and b are constants, while f and g are functions.

Differential rules

Linearity rule $d(af + bg) = a\, df + b\, dg$

Product rule $d(fg) = f\, dg + g\, df$

Quotient rule $d\left(\dfrac{f}{g}\right) = \dfrac{g\, df - f\, dg}{g^2} \qquad (g \neq 0)$

Power rule $d(x^n) = nx^{n-1}\, dx$

Trigonometric rules

$$d(\sin x) = \cos x\, dx \qquad\qquad d(\cos x) = -\sin x\, dx$$
$$d(\tan x) = \sec^2 x\, dx \qquad\qquad d(\cot x) = -\csc^2 x\, dx$$
$$d(\sec x) = \sec x \tan x\, dx \qquad\quad d(\csc x) = -\csc x \cot x\, dx$$

Exponential and logarithmic rules

$$d(e^x) = e^x\, dx \qquad\qquad d(\ln x) = \frac{1}{x}\, dx$$

Differential rules (Continued)

Inverse trigonometric rules

$$d(\sin^{-1} x) = \frac{dx}{\sqrt{1 - x^2}} \qquad d(\cos^{-1} x) = \frac{-dx}{\sqrt{1 - x^2}}$$

$$d(\tan^{-1} x) = \frac{dx}{1 + x^2} \qquad d(\cot^{-1} x) = \frac{-dx}{1 + x^2}$$

$$d(\sec^{-1} x) = \frac{dx}{|x| \sqrt{x^2 - 1}} \qquad d(\csc^{-1} x) = \frac{-dx}{|x| \sqrt{x^2 - 1}}$$

Note that each of the above differential formulas resembles the earlier derivative formula, and they can be remembered by "multiplying through" the corresponding derivative formula by "dx."

EXAMPLE 2   Differential involving a product and a trigonometric function

Find $d(x^2 \sin x)$.

Solution

You can work with the differential, as follows:

$$\begin{aligned}
d(x^2 \sin x) &= x^2 \, d(\sin x) + \sin x \, d(x^2) \\
&= x^2 (\cos x \, dx) + \sin x (2x \, dx) \\
&= (x^2 \cos x + 2x \sin x) \, dx
\end{aligned}$$

Or, you can work with the derivative, as shown:

$$\frac{d}{dx}(x^2 \sin x) = x^2 \cos x + 2x \sin x$$

so

$$d(x^2 \sin x) = (x^2 \cos x + 2x \sin x) \, dx \qquad \blacksquare$$

We can use differentials to approximate functional values, as shown in the following example.

EXAMPLE 3   Comparing differential and calculator approximations

Approximate $\frac{1}{3.98}$ using differentials, and compare with a calculator approximation.

Solution

Let $f(x) = \frac{1}{x} = x^{-1}$, so $f'(x) = -x^{-2}$. We know

$$f(x_0 + \Delta x) \approx f(x_0) + f'(x_0) \, dx$$

so we let $x_0 = 4$ and $\Delta x = dx = -0.02$ to find

$$\begin{aligned}
\frac{1}{3.98} &= \frac{1}{4 + (-0.02)} \\
&\approx f(4) + f'(4)(-0.02) \\
&= \frac{1}{4} + \frac{-1}{16}(-0.02) \\
&= 0.25 + 0.00125 \\
&= 0.25125
\end{aligned}$$

By calculator, we find $\frac{1}{3.98} \approx 0.2512562814$. $\qquad \blacksquare$

ERROR PROPAGATION

In the next example, the approximation formula is used to study **propagation of error**, which is the term used to describe error that accumulates from other errors in an approximation. In particular, in the next example, the derivative is used to estimate the maximum error in a calculation that is based on figures obtained by imperfect measurement.

EXAMPLE 4   Propagation of error in a volume measurement

You measure the side of a cube and find it to be 10 cm long. From this you conclude that the volume of the cube is $10^3 = 1,000$ cm^3. If your original measurement of the side is accurate to within 2%, approximately how accurate is your calculation of the volume?

Solution

The volume of the cube is $V(x) = x^3$, where x is the length of a side. If you take the length of a side to be 10 when it is really $10 + \Delta x$, your error is Δx; and your corresponding error when computing the volume will be ΔV, given by

$$\Delta V = V(10 + \Delta x) - V(10) \approx V'(10)\Delta x$$

Now, $V'(x) = 3x^2$, so $V'(10) = 300$. Also, your measurement of the side can be off by as much as 2%—that is, by as much as $0.02(10) = 0.2$ cm in either direction. Substituting $\Delta x = \pm 0.2$ in the incremental approximation formula for ΔV, we get

$$\Delta V = 3(10)^2(\pm 0.2) \approx \pm 60$$

Thus, the propagated error in computing the volume is approximately ± 60 cm^3. Hence the maximum error in your measurement of the side is $|\Delta x| = 0.2$ and the corresponding maximum error in your calculation for the volume is

$$|\Delta V| \approx V'(10)\,|\Delta x| = 300(0.2) = 60$$

This says that, at worst, your calculation of the volume as $1,000$ cm^3 is off by 60 cm^3, or 6% of the calculated volume, when your maximum error in measuring the side is 2%.   ■

Error Propagation

> If x_0 represents the measured value of a variable and $x_0 + \Delta x$ represents the exact value, then Δx is the **error in measurement**. The difference between $f(x+\Delta x)$ and $f(x)$ is called the **propagated error** at x and is defined by
>
> $$\Delta f = f(x + \Delta x) - f(x)$$
>
> **Relative Error**
>
> The **relative error** is $\dfrac{\Delta f}{f} \approx \dfrac{df}{f}$.
>
> **Percentage Error**
>
> The **percentage error** is $100\left(\dfrac{\Delta f}{f}\right)$ %.

In Example 4, the approximate propagated error in measuring volume is ± 60, and the approximate relative error is $\Delta V/V = \pm 60/10^3 = \pm 0.06$.

EXAMPLE 5   Estimating relative error and percentage error

A certain container is modeled by a right circular cylinder whose height is twice the radius of the base. The radius is measured to be 17.3 cm, with a maximum measurement error of 0.02 cm. Estimate the corresponding propagated error, the relative error, and the percentage error when calculating the surface area S.

$S = 6\pi r^2$

Figure 3.63 Surface area of a right circular cylinder

Solution

Figure 3.63 shows the container. We have

$$S = \underbrace{\underbrace{2\pi r}_{\text{Circumference}} \overbrace{2r}^{\text{Height}}}_{\text{Lateral side}} + \underbrace{\pi r^2}_{\text{Top}} + \underbrace{\pi r^2}_{\text{Bottom}} = 6\pi r^2$$

where r is the radius of the cylinder's base. Then the approximate *propagated error* is

$$\Delta S \approx S'(r)\Delta r = 12\pi r \Delta r = 12\pi(17.3)(\pm 0.02) \approx \pm 13.0438927$$

Thus, the maximum error in the measurement of the surface area is about 13.04 cm². Is this a large or a small error? The *relative error* is found by computing the ratio

$$\frac{\Delta S}{S} = \frac{12\pi r \Delta r}{6\pi r^2} = 2r^{-1}\Delta r = 2(17.3)^{-1}(\pm 0.02) \approx \pm 0.0023121$$

This tells us that the maximum error of approximately 13.04 is fairly small relative to the surface area S. The corresponding *percentage error* is found by

$$100\left(\frac{\Delta S}{S}\right)\% = 100(\pm 0.0023121387)\% = \pm 0.23121387\%$$

This means that the percentage error is about $\pm 0.23\%$. ∎

MARGINAL ANALYSIS IN ECONOMICS

Marginal analysis is an area of economics concerned with estimating the effect on quantities such as cost, revenue, and profit when the level of production is changed by a unit amount. For example, if $C(x)$ is the cost of producing x units of a certain commodity, then the **marginal cost**, $MC(x)$, is the additional cost of producing one more unit and is given by the difference $MC(x) = C(x + 1) - C(x)$. Using the linear approximation formula with $\Delta x = 1$, we see that

$$MC(x) = C(x + 1) - C(x) \approx C'(x)(1)$$

and for this reason, we will compute marginal cost by the derivative $C'(x)$.

Similarly, if $R(x)$ is the revenue obtained from producing x units of a commodity, then the **marginal revenue**, $MR(x)$, is the additional revenue obtained from producing one more unit, and we compute MR by the derivative $R'(x)$. To summarize:

Marginal Cost And Marginal Revenue

The marginal cost of producing x units of a commodity is computed by the derivative $MC(x) = C'(x)$ and the marginal revenue of producing x units is computed by $MR(x) = R'(x)$.

EXAMPLE 6   Modeling change in cost and revenue

A manufacturer models the total cost (in dollars) of a particular commodity by the function

$$C(x) = \tfrac{1}{8}x^2 + 3x + 98$$

and the price per item (in dollars) by

$$p(x) = \tfrac{1}{3}(75 - x)$$

where x is the number of items produced ($0 \le x \le 50$). The function p is called the **demand function** which is the market price per unit when x units are produced.

a. Find the marginal cost and the marginal revenue.

b. Use marginal cost to estimate the cost of producing the 9th unit. What is the actual cost of producing the 9th unit?

c. Use marginal revenue to estimate the revenue derived from producing the 9th unit. What is the actual revenue derived from producing the 9th unit?

Solution

You can compare the values on [0, 50] using a calculator in order to get a sense of the meaning of the problem.

No.	Total cost; Avg. cost	Price per item
0	$98.00	$25.00
5	$116.13; $23.23	$23.33
10	$140.50; $14.05	$21.67
15	$171.13; $11.41	$20.00
20	$208.00; $10.40	$18.33
25	$251.13; $10.05	$16.67
30	$300.50; $10.02	$15.00
35	$356.13; $10.18	$13.33
40	$418.00; $10.45	$11.67
45	$486.13; $10.80	$10.00
50	$560.50; $11.21	$8.33

a. The marginal cost is

$$C'(x) = \tfrac{1}{4}x + 3$$

To find the marginal revenue, we must first find the revenue function:

$$R(x) = xp(x) = x\left(\tfrac{1}{3}\right)(75 - x) = -\tfrac{1}{3}x^2 + 25x$$

Thus, the marginal revenue is

$$R'(x) = -\tfrac{2}{3}x + 25$$

b. The cost of producing the 9th unit is the change in cost as x increases from 8 to 9 and is estimated by

$$C'(8) = \tfrac{1}{4}(8) + 3 = 5$$

We estimate the cost of producing the 9th unit to be $5. The actual cost is

$$\Delta C = C(9) - C(8) = \left[\tfrac{1}{8}(9)^2 + 3(9) + 98\right] - \left[\tfrac{1}{8}(8)^2 + 3(8) + 98\right]$$
$$= 5\tfrac{1}{8} = 5.125 \quad \text{(that is, $5.13)}$$

> **WARNING** Remember that the marginal revenue from the sale of the 9th item is not the revenue derived from selling 9 items. Rather, it is the additional revenue the company has earned by selling the 9th item—that is, the total revenue of 9 items minus the total revenue of 8 items.

c. The revenue (to the nearest cent) obtained from the sale of the 9th unit is approximated by the marginal revenue:

$$R'(8) = -\tfrac{2}{3}(8) + 25 = \tfrac{59}{3} \approx 19.67 \quad \text{(that is, $19.67)}$$

The actual revenue (to the nearest cent) obtained from the sale of the 9th unit is

$$\Delta R = R(9) - R(8) = \tfrac{58}{3} \approx 19.33 \quad \text{(that is, $19.33)} \quad \blacksquare$$

THE NEWTON-RAPHSON METHOD FOR APPROXIMATING ROOTS

The Newton–Raphson method is a different kind of tangent line approximation, one that uses tangent lines as a means for estimating roots of equations. The basic idea behind the procedure is illustrated in Figure 3.64. In this figure, r is a root of the equation $f(x) = 0$, x_0 is an approximation to r, and x_1 is a better approximation obtained by taking the x-intercept of the line that is tangent to the graph of f at $(x_0, f(x_0))$.

a. Estimating a root, r, of $y = f(x)$

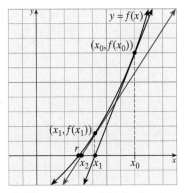

b. First, second, and third estimates ($x_0, x_1,$ and x_2, respectively)

Figure 3.64 The Newton–Raphson method

THEOREM 3.14   The Newton–Raphson method

To approximate a root of the equation $f(x) = 0$, start with a preliminary estimate x_0 and generate a sequence $x_1, x_2, x_3, \ldots$ using the formula

$$x_{n+1} = x_n - \frac{f(x_n)}{f'(x_n)} \qquad f'(x_n) \neq 0$$

Either this sequence of approximations will approach a limit that is a root of the equation or else the sequence does not have a limit.

Proof   Rather than present a formal proof, we will present a geometric description of the procedure to help you understand what is happening. Let x_0 be an initial approximation such that $f'(x_0) \neq 0$. To find a formula for the improved approximation x_1, recall that the slope of the tangent line through $(x_0, f(x_0))$ is the derivative $f'(x_0)$. Therefore (see Figure 3.64b),

$$\underbrace{f'(x_0)}_{\text{Slope of the tangent line through } (x_0, \, f(x_0))} = \frac{\Delta y}{\Delta x} = \frac{f(x_0) - 0}{x_0 - x_1}$$

or, equivalently (by solving the equation for x_1),

$$x_1 = x_0 - \frac{f(x_0)}{f'(x_0)} \qquad f'(x_0) \neq 0$$

If this procedure is repeated using x_1 as the initial approximation, an even better approximation may often be obtained (see Figure 3.64b). This approximation, x_2, is related to x_1 as x_1 was related to x_0. That is,

$$x_2 = x_1 - \frac{f(x_1)}{f'(x_1)} \qquad f'(x_1) \neq 0$$

If this process produces a limit, it can be continued until the desired degree of accuracy is obtained. In general, the nth approximation x_n is related to the $(n-1)$st by the formula

$$x_n = x_{n-1} - \frac{f(x_{n-1})}{f'(x_{n-1})} \qquad f'(x_{n-1}) \neq 0 \qquad \qquad \square$$

Here is a step-by-step procedure for applying the Newton–Raphson method. A flowchart for the method appears in Figure 3.65.

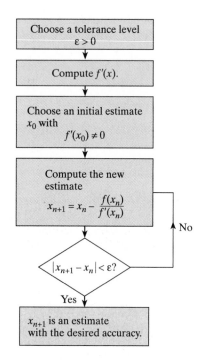

Figure 3.65 Flowchart for the Newton–Raphson method

Procedure for Applying the Newton–Raphson Method to Solve the Equation $f(x) = 0$

1. Choose a number $\epsilon > 0$ that determines the allowable tolerance for estimated solutions.
2. Compute $f'(x)$ and choose a number x_0 (with $f'(x_0) \neq 0$) "close" to a solution of $f(x) = 0$ as an initial estimate.
3. Compute a new approximation with the formula

$$x_{n+1} = x_n - \frac{f(x_n)}{f'(x_n)} \qquad f'(x_n) \neq 0$$

4. Repeat step 3 until $|x_{n+1} - x_n| < \epsilon$. The estimate $\overline{x} = x_{n+1}$ then has the required accuracy.

EXAMPLE 7   Estimating a root with the Newton–Raphson method

Approximate a real root of the equation $x^3 + x + 1 = 0$ on $[-2, 2]$.

Solution

Let $f(x) = x^3 + x + 1$. Our goal is to find a root of the equation $f(x) = 0$. The derivative of f is $f'(x) = 3x^2 + 1$, and so

$$x - \frac{f(x)}{f'(x)} = x - \frac{x^3 + x + 1}{3x^2 + 1} = \frac{2x^3 - 1}{3x^2 + 1}$$

Thus, for $n = 0, 1, 2, 3, \ldots,$

$$x_{n+1} = x_n - \frac{f(x_n)}{f'(x_n)} = \frac{2x_n^3 - 1}{3x_n^2 + 1}$$

A convenient choice for the preliminary estimate is $x_0 = -1$. Then

$$x_1 = \frac{2x_0^3 - 1}{3x_0^2 + 1} = -0.75 \qquad \text{You will need a calculator or a spreadsheet to help you with these calculations.}$$

$$x_2 = \frac{2x_1^3 - 1}{3x_1^2 + 1} \approx -0.6860465$$

$$x_3 = \frac{2x_2^3 - 1}{3x_2^2 + 1} \approx -0.6823396$$

So, to two decimal places, the root seems to be approximately $x_n \approx -0.68$. ∎

In general, we will stop finding new estimates when successive approximations x_n and x_{n+1} are within a desired tolerance of each other. Specifically, if we wish to have the solutions to be within ϵ ($\epsilon > 0$) of each other, we compute approximations until $|x_{n+1} - x_n| < \epsilon$ is satisfied. Using the Newton–Raphson method (Theorem 3.14), we see that this condition is equivalent to

$$|x_{n+1} - x_n| = \left| \frac{-f(x_n)}{f'(x_n)} \right| < \epsilon$$

3.8 PROBLEM SET

A *Find the differentials indicated in Problems 1–16.*

1. $d(2x^3)$

2. $d(3 - 5x^2)$

3. $d(2\sqrt{x})$

4. $d(x^5 + \sqrt{x^2 + 5})$

5. $d(x \cos x)$

6. $d(x \sin 2x)$

7. $d\left(\dfrac{\tan 3x}{2x}\right)$

8. $d(xe^{-2x})$

9. $d(\ln|\sin x|)$

10. $d(x \tan^{-1} x)$

11. $d(e^x \ln x)$

12. $d\left(\dfrac{\sin^{-1} x}{e}\right)$

13. $d\left(\dfrac{x^2 \sec x}{x - 3}\right)$

14. $d(x\sqrt{x^2 - 1})$

15. $d\left(\dfrac{x - 5}{\sqrt{x + 4}}\right)$

16. $d\left(\dfrac{\ln\sqrt{x}}{x}\right)$

17. WHAT DOES THIS SAY? What is a differential?

18. Interpretation Problem Discuss error propagation, including relative error, and percentage error.

B *Use differentials to approximate the requested values in Problems 19–22 and then determine the error as compared to the calculator value.*

19. $\sqrt{0.99}$

20. $\cos(\frac{\pi}{2} + 0.01)$

21. $(3.01)^5 - 2(3.01)^3 + 3(3.01)^2 - 2$

22. $\sqrt[4]{4,100} + \sqrt[3]{4,100} + 3\sqrt{4,100}$

23. You measure the radius of a circle to be 12 cm and use the formula $A = \pi r^2$ to calculate the area. If your measurement of the radius is accurate to within 3%, approximately how accurate (to the nearest percent) is your calculation of the area?

24. Exploration Problem Suppose a 12-oz can of Coke® has a height of 4.5 in. If your measurement of the radius has an accuracy to within 1%, how accurate is your measurement for volume? Check your answer by examining a Coke can.

25. You measure the radius of a sphere to be 6 in. Use the formula $V = \frac{4}{3}\pi r^3$ to calculate the volume. If your measurement of the radius is accurate to within 1%, approximately how accurate (to the nearest percent) is your calculation of the volume?

26. It is projected that t years from now the circulation of a local newspaper will be

$$C(t) = 100t^2 + 400t + 5,000$$

Estimate the amount by which the circulation will increase during the next 6 months.

27. An environmental study suggests that t years from now, the average level of carbon monoxide in the air will be

$$Q(t) = 0.05t^2 + 0.1t + 3.4$$

parts per million (ppm). By approximately how much will the carbon monoxide level change during the next 6 months?

28. A manufacturer's total cost (in dollars) is

$$C(q) = 0.1q^3 - 0.5q^2 + 500q + 200$$

when the level of production is q units. The current level of production is 4 units, and the manufacturer is planning to decrease this to 3.9 units. Estimate how the total cost will change as a result.

29. At a certain factory, the daily output is

$$Q(L) = 60,000L^{1/3}$$

units, where L denotes the size of the labor force measured in worker-hours. Currently 1,000 worker-hours of labor are used each day. Estimate the effect on output that will be produced if the labor force is cut to 940 worker-hours.

30. **Modeling Problem**   In a model developed by John Helms, the water evaporation $E(T)$ for a ponderosa pine is modeled by

$$E(T) = 4.6e^{17.3T/(T+237)}$$

where T (degrees Celsius) is the surrounding air temperature.[*] If the temperature is increased by 5% from 30°C, use differentials to estimate the corresponding percentage change in $E(T)$.

31. A soccer ball made of leather $\frac{1}{8}$ in. thick has an inner diameter of $8\frac{1}{2}$ in. Model the ball as a hollow sphere and estimate the volume of its leather shell.

32. A cubical box is to be constructed from three kinds of building materials. The material used in the four sides of the box costs 2¢/in.², the material in the bottom costs 3¢/in.², and the material used for the lid costs 4¢/in.². Estimate the additional total cost of all the building materials if the length of a side is increased from 20 in. to 21 in.

33. In a healthy person of height x in., the average pulse rate in beats per minute is modeled by the formula

$$P(x) = \frac{596}{\sqrt{x}} \qquad 30 \leq x \leq 100$$

Estimate the change in pulse rate that corresponds to a height change from 59 to 60 in.

34. A drug is injected into a patient's bloodstream. The concentration of the drug in the bloodstream t hours after the drug is injected is modeled by the formula

$$C(t) = \frac{0.12t}{t^2 + t + 1}$$

milligrams per cubic centimeter. Estimate the change in concentration over the time period from 30 to 35 minutes after injection.

35. **Modeling Problem**   According to Poiseuille's law, the speed of blood flowing along the central axis of an artery of radius R is modeled by the formula $S(R) = cR^2$, where c is a constant.[†] What percentage error (rounded to the nearest percent) will you make in the calculation of $S(R)$ from this formula if you make a 1% error in the measurement of R?

36. **Modeling Problem**   One of the laws attributed to Poiseuille models the volume of a fluid flowing through a small tube in unit time under fixed pressure by the formula $V = kR^4$, where k is a positive constant and R is the radius of the tube. This formula is used in medicine to determine how wide a clogged artery must be opened to restore a healthy flow of blood. Suppose the radius of a certain artery is increased by 5%. Approximately what effect does this have on the volume of the blood flowing through the artery?[‡]

37. **Modeling Problem**   A certain cell is modeled as a sphere. If the formulas $S = 4\pi r^2$ and $V = \frac{4}{3}\pi r^3$ are used to compute the surface area and volume of the sphere, respectively, estimate the effect on S and V produced by a 1% increase in the radius r.

38. The period of a pendulum is given by the formula

$$T = 2\pi \sqrt{\frac{L}{g}}$$

where L is the length of the pendulum in feet, $g = 32$ ft/s² is the acceleration due to gravity, and T is time in seconds. If the pendulum has been heated enough to increase its length by 0.4%, what is the approximate percentage change in its period?

39. The *thermal expansion coefficient* of an object is defined to be

$$\sigma = \frac{L'(T)}{L(T)}$$

where $L(T)$ is the length of the object when the temperature is T. Suppose a 75-ft span of a bridge is built with steel with $\sigma = 1.4 \times 10^{-5}$ per degree Celsius. Approximately how much will the length change during a year when the temperature varies from -10°C in winter to 40°C in summer?

40. The radius R of a spherical ball is measured as 14 in.

   a. Use differentials to estimate the maximum propagated error in computing volume V if R is measured with a maximum error of $\frac{1}{8}$ inch.

   b. With what accuracy must the radius R be measured to guarantee an error of at most 2 in.³ in the calculated volume?

41. A thin horizontal beam of alpha particles strikes a thin vertical foil, and the scattered alpha particles will travel along a cone of vertex angle θ, as shown in Figure 3.66.

---

[*]John A. Helms, "Environmental Control of Net Photosynthesis in Naturally Grown Pinus Ponderosa Nets," *Ecology* (Winter 1972), p. 92.

[†]See *Introduction to Mathematics for Life Scientists*, 2nd edition. New York: Springer-Verlag (1976), pp. 102–103.

[‡]Ibid.

Time 1: Stream is focused.

Time 2: Stream hits foil.

Time 3: Stream is scattered after colliding with foil and disperses in the shape of a cone.

Figure 3.66   Paths of alpha particles

A vertical screen is placed at a fixed distance from the point of scattering. Physical theory predicts that the number N of alpha particles falling on a unit area of the screen is inversely proportional to $\sin^4\left(\frac{\theta}{2}\right)$. Suppose N is modeled by the formula

$$N = \frac{1}{\sin^4\left(\frac{\theta}{2}\right)}$$

Estimate the change in the number of alpha particles per unit area of the screen if θ changes from 1 to 1.1.

42. Suppose the total cost of manufacturing q units is

$$C(q) = 3q^2 + q + 500$$

dollars.

a. Use marginal analysis to estimate the cost of manufacturing the 41st unit.

b. Compute the actual cost of manufacturing the 41st unit.

43. A manufacturer's total cost is

$$C(q) = 0.1q^3 - 5q^2 + 500q + 200$$

dollars, where q is the number of units produced.

a. Use marginal analysis to estimate the cost of manufacturing the 4th unit.

b. Compute the actual cost of manufacturing the 4th unit.

44. Suppose the total cost of producing x units of a particular commodity is modeled by

$$C(x) = \tfrac{1}{7}x^2 + 4x + 100$$

and that each unit of the commodity can be sold for

$$p(x) = \tfrac{1}{4}(80 - x)$$

dollars.

a. What is the marginal cost?

b. What is the price when the marginal cost is 10?

c. Estimate the cost of producing the 11th unit.

d. Find the actual cost of producing the 11th unit.

45. At a certain factory, the daily output is modeled by the formula

$$Q(L) = 360L^{1/3}$$

units, where L is the size of the labor force measured in worker-hours. Currently, 1,000 worker-hours of labor are used each day. Use differentials to estimate the effect that one additional worker-hour will have on the daily output.

46. Approximate $\sqrt{2}$ to four decimal places by using the Newton–Raphson method.

47. Approximate $-\sqrt{2}$ to four decimal places by using the Newton–Raphson method.

48. Use the Newton–Raphson method to estimate a root of the equation $\cos x = x$. You may start with $x_0 = 1$.

49. Use the Newton–Raphson method to estimate a root of the equation $e^{-x} = x$. You may start with $x_0 = 1$.

50. Use the Newton–Raphson method to estimate a root of the equation

$$x^6 - x^5 + x^3 = 3$$

51. Let $f(x) = -2x^4 + 3x^2 + \frac{11}{8}$.

a. Show that the equation $f(x) = 0$ has at least two solutions. *Hint*: Use the intermediate value theorem.

b. Use $x_0 = 2$ in the Newton–Raphson method to find a root of the equation $f(x) = 0$.

c. Show that the Newton–Raphson method fails if you choose $x_0 = \frac{1}{2}$ as the initial estimate. *Hint*: You should obtain $x_1 = -x_0, x_2 = x_0, \ldots$.

52. It can be shown that the volume of a spherical segment is given by

$$V = \frac{\pi}{3}H^2(3R - H)$$

where R is the radius of the sphere and H is the height of the segment, as shown in Figure 3.67.

Figure 3.67   Problem 52

If $V = 8$ and $R = 2$, use the Newton–Raphson method to estimate the corresponding H.

53. HISTORICAL QUEST   The Greek geometer Archimedes (see Essay Question 4 in Chapter 1) is acknowledged to be one of the greatest mathematicians of all time. Ten treatises of Archimedes have survived the rigors of time (as well as traces of some lost works) and are masterpieces of mathematical exposition. In one of these works, *On the Sphere and Cylinder*, Archimedes asks where a sphere should be cut in order to divide it into two pieces whose volumes have a given ratio. ∎

ARCHIMEDES
287–212 B.C.

Show that if a plane at distance x_c from the center of a sphere with $R = 1$ divides the sphere into two parts, one with volume twice that of the other, then

$$3x_c^3 - 9x_c^2 + 2 = 0$$

Use the Newton–Raphson method to estimate x_c.

54. Suppose the plane described in Problem 53 is located so that it divides the sphere in the ratio 1:3. Find an equation for x_c, and estimate the value of x_c by the Newton–Raphson method. *Hint*: You may need the result of Problem 52.

C **55.** Show that if h is sufficiently small, then

    a. $\sqrt{1+h}$ is approximately equal to $1 + \frac{h}{2}$.

    b. $\dfrac{1}{1+h}$ is approximately equal to $1 - h$.

56. If h is sufficiently small, find an approximate value for $\sqrt[n]{A^n + h}$ for constant A.

57. Tangent line approximations are useful only if Δx is small. Illustrate this fact by trying to approximate $\sqrt{97}$ by regarding 97 as being near 81 (instead of 100).

58. Let $f(x) = -x^4 + x^2 + A$ for constant A. What value of A should be chosen that guarantees that if $x_0 = \frac{1}{3}$ is chosen as the initial estimate, the Newton–Raphson method produces $x_1 = -x_0, x_2 = x_0, x_3 = -x_0, \dots$?

59. Suppose that when x units of a certain commodity are produced, the total cost is $C(x)$ and the total revenue is $R(x)$. Let $P(x) = R(x) - C(x)$ denote the total profit, and let

$$A(x) = \frac{C(x)}{x}$$

be the average cost.

    a. Show that $P'(x) = 0$ when marginal revenue equals marginal cost.

    b. Show that $A'(x) = 0$ when average cost equals marginal cost.

60. Can you solve the equation $\sqrt{x} = 0$ using the Newton–Raphson method with the initial estimate $x_0 = 0.05$? Does it make any difference if we choose another initial estimate (other than $x_0 = 0$)?

61. Interpretation Problem Suppose that when we try to use the Newton–Raphson method to approximate a solution of $f(x) = 0$, we find that $f(x_n) = 0$ but $f'(x_n) \neq 0$ for some x_n. What does this imply about $x_{n+1}, x_{n+2}, \dots$? Explain.

62. HISTORICAL QUEST Among the peoples of the region between the Tigris and Euphrates rivers (at different times) during the period 2000–600 B.C. were Sumerians, Akkadians, Chaldeans, and Assyrians. Since the middle half of the nineteenth century, archeologists have found well over 50,000 clay tablets describing these great civilizations. Records show that they had highly developed religion, history, science (including alchemy, astronomy, botany, chemistry, mathematics, and zoology). ▪

    Mesopotamian culture had iterative formulas for computing algebraic quantities such as roots. In particular, they approximated $\sqrt{N}$ by repeatedly applying the formula

$$x_{n+1} = \frac{1}{2}\left(x_n + \frac{N}{x_n}\right) \quad \text{for } n = 0, 1, 2, 3, \dots$$

    a. Apply the Newton–Raphson method to $f(x) = x^2 - N$ to justify this formula.

    b. Apply the formula to estimate $\sqrt{1,265}$ correct to five decimal places.

CHAPTER 3 REVIEW

Proficiency Examination

CONCEPT PROBLEMS

1. What is the slope of a tangent line? How does this compare to the slope of a secant line?
2. Define the derivative of a function.
3. What is a normal line to a graph?
4. What is the relationship between continuity and differentiability?
5. List and explain some of the notations for derivative.
6. State the following procedural rules for finding derivatives:
   - **a.** constant multiple
   - **b.** sum rule
   - **c.** difference rule
   - **d.** linearity rule
   - **e.** product rule
   - **f.** quotient rule
   - **g.** State each of these rules again, this time in differential form.
7. State the following derivative rules:
   - **a.** constant rule
   - **b.** power rule
   - **c.** trigonometric rules
   - **d.** exponential rule
   - **e.** logarithmic rule
   - **f.** inverse trigonometric rules
8. What is a higher-order derivative? List some of the different notations for higher-order derivatives.
9. What is meant by rate of change? Distinguish between average and instantaneous rate of change.
10. What is relative rate of change?
11. How do you find the velocity and the acceleration for an object with position $s(t)$? What is speed?
12. State the chain rule.
13. Outline a procedure for logarithmic differentiation.
14. Outline a procedure for implicit differentiation.
15. Outline a procedure for solving related rate problems.
16. What is meant by tangent line approximation?
17. Define the differential of x and the differential of y for a function $y = f(x)$. Draw a sketch showing Δx, Δy, dx, and dy.
18. Define the terms propagated error, relative error, and percentage error.
19. What is meant by marginal analysis?
20. What is the Newton–Raphson method?

PRACTICE PROBLEMS

Find $\dfrac{dy}{dx}$ in Problems 21–30.

21. $x^3 + x\sqrt{x} + \cos 2x$
22. $y = \sqrt{3x} + \dfrac{3}{x^2}$
23. $y = \sqrt{\sin(3 - x^2)}$
24. $xy + y^3 = 10$
25. $y = x^2 e^{-\sqrt{x}}$
26. $y = \dfrac{\ln 2x}{\ln 3x}$
27. $y = \sin^{-1}(3x + 2)$
28. $y = \tan^{-1} 2x$
29. $y = \sin^2(x^{10} + \sqrt{x}) + \cos^2(x^{10} + \sqrt{x})$
30. $y = \dfrac{\ln(x^2 - 1)}{\sqrt[3]{x}(1 - 3x)^3}$
31. Find $\dfrac{d^2y}{dx^2}$, the second derivative of $y = x^2(2x - 3)^3$.

32. Use the definition of the derivative to find $\dfrac{d}{dx}(x - 3x^2)$.
33. Find the equation of the tangent line to the graph of

$$y = (x^2 + 3x - 2)(7 - 3x)$$

at the point where $x = 1$.

34. Let $f(x) = \sin^2\left(\dfrac{\pi x}{4}\right)$. Find equations of the tangent line and the normal line to the graph of f at $x = 1$.
35. A rock tossed into a stream causes a circular ripple of water whose radius increases at a constant rate of 0.5 ft/s. How fast is the area contained inside of the ripple changing when the radius is 2 ft?

Supplementary Problems

Find dy/dx in Problems 1–36.

1. $y = x^4 + 3x^2 - 7x + 5$
2. $y = x^5 + 3x^3 - 11$
3. $y = \sqrt{\dfrac{x^2 - 1}{x^2 - 5}}$
4. $y = \dfrac{\cos x}{x + \sin x}$
5. $2x^2 - xy + 2y = 5$
6. $y = (x^2 + 3x - 5)^7$
7. $y = (x^3 + x)^{10}$
8. $y = \sqrt{x}(x^2 + 5)^{10}$
9. $y = \sqrt[3]{x}(x^3 + 1)^5$
10. $y = (x^2 + 3)^5(x^3 - 5)^8$
11. $y = (x^4 - 1)^{10}(2x^4 + 3)^7$
12. $y = \sqrt{\sin 5x}$
13. $y = \sqrt{\cos \sqrt{x}}$
14. $y = (\sin x + \cos x)^3$
15. $y = (\sqrt{x} + \sqrt[3]{x})^5$
16. $y = \sqrt{\dfrac{x^3 - x}{4 - x^2}}$
17. $y = \exp(2x^2 + 5x - 3)$
18. $y = \ln(x^2 - 1)$
19. $y = x3^{2-x}$
20. $y = \log_3(x^2 - 1)$
21. $e^{xy} + 2 = \ln\dfrac{y}{x}$
22. $y = \sqrt{x}\sin^{-1}(3x + 2)$
23. $y = e^{\sin x}$
24. $y = 2^x \log_2 x$
25. $y = e^{-x} \log_5 3x$
26. $x2^y + y2^x = 3$
27. $\ln(x + y^2) = x^2 + 2y$
28. $y = e^{-x}\sqrt{\ln 2x}$
29. $y = \sin(\sin x)$
30. $y = \cos(\sin x)$
31. $x^{1/2} + y^{1/2} = x$
32. $4x^2 - 16y^2 = 64$
33. $\sin xy = y + x$
34. $\sin(x + y) + \cos(x - y) = xy$
35. $y = \dfrac{x}{\sin^{-1} x} + \dfrac{\tan^{-1} x}{x}$
36. $y = (\sin x)(\sin^{-1} x) + x\cot^{-1} x$

Find d^2y/dx^2 in Problems 37–41.

37. $y = x^5 - 5x^4 + 7x^3 - 3x^2 + 17$
38. $y = \dfrac{x - 5}{2x + 3} + (3x - 1)^2$
39. $x^2 + y^3 = 10$
40. $x^2 + \sin y = 2$
41. $x^2 + \tan^{-1} y = 2$

In Problems 42–54, find an equation of the tangent line to the curve at the indicated point.

42. $y = x^4 - 7x^3 + x^2 - 3$ at $(0, -3)$
43. $y = (3x^2 + 5x - 7)^3$, where $x = 1$
44. $y = (x^3 - 3x^2 + 3)^3$, where $x = -1$
45. $y = x\cos x$, where $x = \dfrac{\pi}{2}$

46. $y = (\cos x)^2$, where $x = \pi$
47. $xy^2 + x^2 y = 2$ at $(1, 1)$
48. $y = x \ln(ex)$, where $x = 1$
49. $y = xe^{2x-1}$, where $x = \frac{1}{2}$
50. $e^{xy} = x - y$ at $(1, 0)$
51. $y = (1 - x)^x$, where $x = 0$
52. $y = 2^x - \log_2 x$, where $x = 1$
53. $y = \dfrac{3x - 4}{3x^2 + x - 5}$, where $x = 1$
54. $x^{2/3} + y^{2/3} = 2$ at $(1, 1)$
55. Let $f(x) = (x^3 - x^2 + 2x - 1)^4$. Find equations of the tangent and normal lines to the graph of f at $x = 1$.
56. Find equations for the tangent and normal lines to the graph of $y = \left(2x + \dfrac{1}{x}\right)^3$ at the point $(1, 27)$.
57. Use the chain rule to find $\dfrac{dy}{dt}$ when $y = x^3 - 7x$ and $x = t \sin t$.
58. Find $f''(x)$ if $f(x) = x^2 \sin x^2$.
59. Find $f^{(4)}(x)$ if $f(x) = x^4 - \dfrac{1}{x^4}$.
60. Find $f'''(x)$ if $f(x) = x(x^2 + 1)^{7/2}$.
61. Let $f(x) = \sqrt[3]{\dfrac{x^4 + 1}{x^4 - 2}}$. Find $f'(x)$ by implicitly differentiating $[f(x)]^3$.
62. Find y' if $x^3 y^3 + x - y = 1$. Leave your answer in terms of x and y.
63. Find y' and y'' if $x^2 + 4xy - y^2 = 8$. Your answer may involve x and y, but not y'.
64. Find the derivative of $f(x) = \begin{cases} x^2 + 5x + 4 & \text{for } x \le 0 \\ 5x + 4 & \text{for } 0 < x < 6 \\ x^2 - 2 & \text{for } x \ge 6 \end{cases}$.

65. **Advanced Placement Question** *The following problem is found on the 1982 BC AP Exam.* Let f be the function defined by

$$f(x) = \begin{cases} x^2 \sin \dfrac{1}{x} & \text{for } x \ne 0 \\ 0 & \text{for } x = 0 \end{cases}$$

   a. Using the definition of the derivative, prove that f is differentiable at $x = 0$.
   b. Find $f'(x)$ for $x \ne 0$.
   c. Show that f' is not continuous at $x = 0$.

66. Use differentials to approximate $(16.01)^{3/2} + 2\sqrt{16.01}$.
67. Use differentials to approximate $\cos \frac{101\pi}{600}$.
68. Use differentials to estimate the change in the volume of a cone if the height of the cone is increased from 10 cm to 10.01 cm while the radius of the base stays fixed at 2 cm.
69. On New Year's Eve, a network TV camera is focusing on the descent of a lighted ball from the top of a building that is 600 ft away. The ball is falling at the rate of 20 ft/min. At what rate is the angle of elevation of the camera's line of sight changing with respect to time when the ball is 800 ft from the ground?
70. Suppose f is a differentiable function whose derivative satisfies $f'(x) = 2x^2 + 3$. Find $\dfrac{d}{dx} f(x^3 - 1)$.

71. Suppose f is a differentiable function such that $f'(x) = x^2 + x$. Find $\dfrac{d}{dx} f(x^2 + x)$.
72. Let $f(x) = 3x^2 + 1$ for all x. Use the chain rule to find $\dfrac{d}{dx}(f \circ f)(x)$.
73. Let $f(x) = \sin 2x + \cos 3x$ and $g(x) = x^2$. Use the chain rule to find $\dfrac{d}{dx}(f \circ g)(x)$.
74. Let $f(x) = \begin{cases} x \sin \dfrac{1}{x} & \text{if } x \ne 0 \\ 0 & \text{if } x = 0 \end{cases}$. Use the definition of the derivative to find $f'(0)$, if it exists.
75. A car and a truck leave an intersection at the same time. The car travels north at 60 mi/h and the truck travels east at 45 mi/h. How fast is the distance between them changing after 45 minutes?
76. A spherical balloon is being filled with air in such a way that its radius is increasing at a constant rate of 2 cm/s. At what rate is the volume of the balloon increasing at the instant when its surface has area 4π cm^2?
77. Suppose the total cost of producing x units of a particular commodity is

$$C(x) = \tfrac{2}{5}x^2 + 3x + 10$$

and that each unit of the commodity can be sold for

$$p(x) = \tfrac{1}{5}(45 - x)$$

dollars.
   a. What is the marginal cost?
   b. What is the price when the marginal cost is 23?
   c. Estimate the cost of producing the 11th unit.
   d. Find the actual cost of producing the 11th unit.

78. A block of ice in the shape of a cube originally having volume 1,000 cm^3 is melting in such a way that the length of each of its edges is decreasing at the rate of 1 cm/hr. At what rate is its surface area decreasing at the time its volume is 27 cm^3? Assume that the block of ice maintains its cubical shape.
79. Show that the rate of change of the area of a circle with respect to its radius is equal to the circumference.
80. A charged particle is projected into a linear accelerator. The particle undergoes a constant acceleration that changes its velocity from 1,200 m/s to 6,000 m/s in 2×10^{-3} seconds. Find the acceleration of the particle.
81. A rocket is launched vertically from a point on level ground that is 3,000 feet from an observer with binoculars. If the rocket is rising vertically at the rate of 750 ft/s at the instant it is 4,000 ft above the ground, how fast must the observer change the angle of elevation of her line of sight in order to keep the rocket in sight at that instant?
82. **Modeling Problem** Assume that a certain artery in the body is modeled by a circular tube whose cross section has radius 1.2 mm. Fat deposits are observed to build up uniformly on the inside wall of the artery. Find the rate at which the cross-sectional area of the artery is decreasing relative to the thickness of the fat deposit at the instant when the deposit is 0.3 mm thick.

83. A processor who sells a certain raw material has analyzed the market and determined that the unit price should be modeled by the formula

$$p(x) = 60 - x^2$$

(thousand dollars) for x tons ($0 \le x \le 7$) produced. Estimate the change in the unit price that accompanies each change in sales:

a. from 2 tons to 2.05 tons

b. from 1 ton to 1.1 tons

c. from 3 tons to 2.95 tons

84. A company sends out a truck to deliver its products. To estimate costs, the manager models gas consumption by the formula

$$G(x) = \frac{1}{300}\left(\frac{1,500}{x} + x\right)$$

gal/mi, under the assumption that the truck travels at a constant rate of x mi/h ($x \ge 5$). The driver is paid $16 per hour to drive the truck 300 mi. Gasoline costs $2 per gallon.

a. Find an expression for the total cost $C(x)$ of the trip.

b. Use differentials to estimate the additional cost if the truck is driven at 57 mi/h instead of 55 mi/h.

85. A viewer standing at ground level and 30 ft from a platform watches a balloon rise from that platform (at the same height as the viewer's eyes) at the constant rate of 3 ft/s. (See Figure 3.68) How fast is the angle of sight between the viewer and the object changing at the instant when $\theta = \frac{\pi}{4}$?

Figure 3.68   Problem 85

86. A lighthouse is 4,000 ft from a straight shore. Watching the beam on the shore from the point P on the shore that is closest to the lighthouse, an observer notes that the light is moving at the rate of 3 ft/s when it is 1,000 ft from P. How fast is the light revolving at this instant (in radians per second)?

87. A light is 4 miles from a straight shoreline. The light revolves at the rate of 2 rev/min. Find the speed of the spot of light along the shore when the light spot is 2 miles past the point on the shore closest to the source of light.

88. A particle of mass m moves along the x-axis. The velocity $v = \dfrac{dx}{dt}$ and position $x = x(t)$ satisfy the equation

$$m(v^2 - v_0^2) = k(x_0^2 - x^2)$$

where k, x_0, and v_0 are positive constants. The force F acting on the object is defined by $F = ma$, where a is the object's acceleration. Show that $F = -kx$.

89. The equation $\dfrac{d^2s}{dt^2} + ks = 0$ is called a **differential equation of simple harmonic motion.** Let A be any number. Show that the function $s(t) = A \sin 2t$ satisfies the equation

$$\frac{d^2s}{dt^2} + 4s = 0$$

90. Suppose $L(x)$ is a function with the property that $L'(x) = x^{-1}$. Use the chain rule to find the derivatives of the following functions.

a. $f(x) = L(x^2)$ **b.** $f(x) = L\left(\dfrac{1}{x}\right)$

c. $f(x) = L\left(\dfrac{2}{3\sqrt{x}}\right)$ **d.** $f(x) = L\left(\dfrac{2x+1}{1-x}\right)$

91. Let f and g be differentiable functions such that $f[g(x)] = x$ and $g[f(x)] = x$ for all x (that is, f and g are inverses). Show that

$$\frac{dg}{dx} = \frac{1}{\dfrac{df}{dx}}$$

92. A baseball player is stealing second base, as shown in Figure 3.69.

Figure 3.69   Baseball diamond

He runs at 30 ft/s and when he is 25 ft from second base, the catcher, while standing at home plate, throws the ball toward the base at a speed of 120 ft/s. At what rate is the distance between the ball and the player changing at the time the ball is thrown?

93. A connection rod $\overline{OA}$ 2 m long is rotating counterclockwise in a plane about O at the rate of 3 rev/s, as shown in Figure 3.70.

Figure 3.70   Connection rod

The rod $\overline{AB}$ is attached to $\overline{OA}$ at A, and the end B slides along the x-axis. Suppose $\overline{AB}$ is 5 meters long. What are the velocity and the acceleration of the motion of the point B along the x-axis?

94. The ideal gas law states that for an ideal gas, pressure P is given by the formula $P = kT/V$, where V is the volume, T is the temperature, and k is a constant. Suppose the temperature is kept fixed at $100°C$, and the pressure decreases at the rate of 7 lb/in.^{2} per min. At what rate is the volume changing at the instant the pressure is 25 lb/in.^{2} and the volume is 30 in.^{3}?

95. **Putnam Examination Problem** Suppose $f(x) = ax^2 + bx + c$, where a, b, and c are real numbers and $|f(x)| \le 1$ for $x \le 1$.

Show that $|f'(x)| \le 4$ for $|x| \le 1$.

96. **Putnam Examination Problem** A point P is taken on the curve $y = x^3$. The tangent line at P meets the curve again at Q. Prove that the slope of the curve at Q is four times the slope at P.

97. **Putnam Examination Problem** A particle of unit mass moves on a straight line under the action of a force that is a function $f(v)$ of the velocity v of the particle, but the form of this function is not known. A motion is observed, and the distance x covered in time t is found to be related to t by the formula $x = at + bt^2 + ct^3$, where a, b, and c have numerical values determined by observation of the motion. Find the function $f(v)$ for the range of v covered by this experiment.

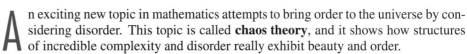

Group Research Project

Chaos

An exciting new topic in mathematics attempts to bring order to the universe by considering disorder. This topic is called **chaos theory**, and it shows how structures of incredible complexity and disorder really exhibit beauty and order.

Water flowing through a pipe offers one of the simplest physical models of chaos. Pressure is applied to the end of the pipe and the water flows in straight lines. More pressure increases the speed of the laminar flow until the pressure reaches a critical value, and a radically new situation evolves—turbulence. A simple laminar flow suddenly changes to a flow of beautiful complexity consisting of swirls within swirls. Before turbulence, the path of any particle was quite predictable. After a minute change in the pressure, turbulence occurs and predictability is lost. **Chaos** is concerned with systems in which minute changes suddenly transform predictability into unpredictability.*

For this research project begin with

$$f(x) = x^3 - x = x(x-1)(x+1)$$

This fractal image is an example of what is called mathematical chaos.

Use the Newton–Raphson method to investigate what happens as you change the starting value, x_0, to solve $f(x) = 0$. What would you select as a starting value? Two important x-values in our study are

$$s_3 = \frac{1}{\sqrt{3}} \qquad \text{and} \qquad s_5 = \frac{1}{\sqrt{5}}$$

One would not want to pick x_0 as either $\pm s_3$. Why not? Also, what happens if one picks $x_0 = s_5$? That is, what are $x_1, x_2, \dots$?

Generate a good plot of $f(x)$ on $[-2, 2]$. Explain from the plots why you would *expect* that an initial value $x_0 > s_3$ would lead to $\lim_{n \to +\infty} x_n = 1$. (Also, by symmetry, $x_0 < -s_3$ leads to $\lim_{n \to +\infty} x_n = -1$.)

Explain why if $|x_0| < s_5$, you would expect $x_n \to 0$. Numerically verify your assertions. Now to see the "chaos," use the following x_0 values and take 6 to 10 iterations, until convergence occurs, and report what happens: 0.448955; 0.447503; 0.447262; 0.447222; 0.4472215; 0.4472213. Write a paper telling what you see and why.

MATHEMATICS is one of the oldest of the sciences; it is also one of the most active, for its strength is the vigor of perpetual youth.

—— A. R. Forsythe
Nature 84 (1910): 285.

THE WHOLE OF MATHEMATICS consists in the organization of a series of aids to the imagination in the process of reasoning.

——A. N. Whitehead
Universal Algebra
(Cambridge, 1898), p. 12.

Extended paper for further study   Attempt to define chaos. Your paper should include, but not be limited to, answers to the questions on this page. Some references (to get you started) are listed:

Chaos: Making a New Science, James Gleick, Penguin Books, 1988.

Chaos and Fractals: New Frontiers of Science, H. O. Peitgen, H. Jürgens, and D. Saupe, Springer-Verlag, 1992.

Newton's Method and Fractal Patterns, Phillip D. Straffin, Jr., UMAP Module 716, COMAP, 1991.

---

*Thanks to Jack Wadhams of Golden West College for this paragraph.

4

Additional Applications of the Derivative

PREVIEW

In Chapter 3, we used the derivative to find tangent lines and to compute rates of change. The primary goal of this chapter is to examine the use of calculus in curve sketching, optimization, and other applications.

PERSPECTIVE

Homing pigeons and certain other birds are known to avoid flying over large bodies of water whenever possible. The reason for this behavior is not entirely known. However, it is reasonable to speculate that it may have something to do with minimizing the energy expended in flight, because the air over a lake is often "heavier" than that over land. Suppose a pigeon is released from a boat at point B on the lake shown in the accompanying figure.* It will fly to its loft at point L on the lakeshore by heading across water to a point P on the shore and then flying directly from P to L along the shore. If the pigeon expends e_ω units of energy per mile over water and e_L units over land, where should P be located to minimize the total energy expended in flight?[†]

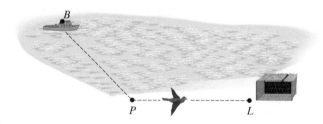

Optimization (finding the maximum or minimum values) is one of the most important applications we will study in calculus, from maximizing profit and minimizing cost to maximizing the strength of a structure and minimizing the distance traveled. Optimization problems are considered in depth in Sections 4.6 and 4.7.

---

*Edward Batschelet, *Introduction to Mathematics for Life Scientists*, 2nd ed. (New York: Springer-Verlag, 1979), pp. 276–277.

[†]See Problem 32, Section 4.7.

4.1   Extreme Values of a Continuous Function

IN THIS SECTION   extreme value theorem, relative extrema, absolute extrema, optimization

EXTREME VALUE THEOREM

One of the principal goals of calculus is to investigate the behavior of various functions. As part of this investigation, we will be laying the groundwork for solving a large class of problems that involve finding the maximum or minimum value of a function, if one exists. Such problems are called **optimization problems**. We begin by introducing some useful terminology.

Absolute Maximum and Minimum

> Let f be a function defined on an interval I that contains the number c. Then
>
> $f(c)$ is an **absolute maximum** of f on I if $f(c) \geq f(x)$ for all x in I
>
> $f(c)$ is an **absolute minimum** of f on I if $f(c) \leq f(x)$ for all x in I

Sometimes we just use the terms *maximum* and/or *minimum* if the context is clear. Together, the absolute maximum and minimum of f on the interval I are called the **extreme values**, or the **absolute extrema,** of f on I. A function does not necessarily have extreme values on a given interval. For instance, the continuous function $g(x) = x$ has neither a maximum nor a minimum on the open interval $(0, 1)$, as shown in Figure 4.1a.

The discontinuous function defined by

$$h(x) = \begin{cases} x^2 & \text{for } x \neq 0 \\ 1 & \text{for } x = 0 \end{cases}$$

has a maximum which is on the closed interval $[-1, 1]$, but no minimum, as shown in Figure 4.1b. Incidentally, this graph also illustrates the fact that a function may assume an absolute extremum at more than one point. In this case, the maximum occurs at the points $(-1, 1)$, $(0, 1)$, and $(1, 1)$. If a function f is continuous and the interval I is closed and bounded, it can be shown that both an absolute maximum and an absolute minimum *must* occur. This result, called the **extreme value theorem**, plays an important role in our work.

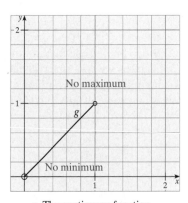

a. The continuous function
$g(x) = x$ has no extrema
on the open interval $(0, 1)$.

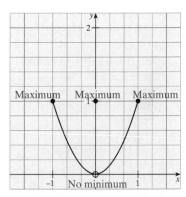

b. The discontinuous function h
has a maximum, but not a
minimum on the closed
interval $[-1, 1]$.

Figure 4.1   Functions that lack one or both extreme values

THEOREM 4.1   The extreme value theorem

A function f has both an absolute maximum and an absolute minimum on any closed, bounded interval $[a, b]$ where it is continuous.

Proof   Even though this result may seem quite reasonable (see Figure 4.2), its proof requires concepts beyond the scope of this text and will be omitted.

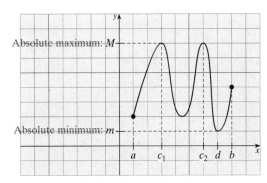

Absolute maximum at $x = c_1$ and $x = c_2$
Absolute minimum at $x = d$

Figure 4.2   Extreme value theorem

If f is discontinuous or the interval is not both closed and bounded, you cannot conclude that f has both a largest and smallest value. You will be asked for appropriate counterexamples in the problem set. Sometimes there are extreme values even when the conditions of the theorem are not satisfied, but if the conditions hold, the extreme values are guaranteed.

Note that the maximum of a function occurs at the highest point on its graph and the minimum occurs at the lowest point. These properties are illustrated in Example 1.

EXAMPLE 1   Extreme values of a continuous function

The graph of a function f is shown in Figure 4.3. Locate the extreme values of f defined on the closed interval $[a, b]$.

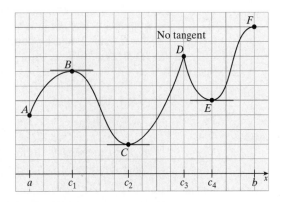

Figure 4.3   A continuous function on a closed interval $[a, b]$

Solution

The highest point on the graph occurs at the right endpoint F, and the lowest point occurs at C. Thus, the absolute maximum is $f(b)$, and the absolute minimum is $f(c_2)$. ∎

In Example 1, the existence of maxima and minima as required by the extreme value theorem may seem obvious, but there are times when it seems that the extreme value theorem fails. If this occurs, you need to see which of the conditions of the extreme value theorem are not satisfied. Consider Example 2.

EXAMPLE 2   Conditions of the extreme value theorem

In each case, explain why the given function does not contradict the extreme value theorem:

a. $f(x) = \begin{cases} 2x & \text{if } 0 \leq x < 1 \\ 1 & \text{if } 1 \leq x \leq 2 \end{cases}$

b. $g(x) = x^2$ on $0 < x \leq 2$

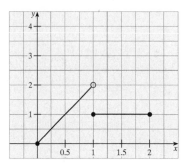

Does not have a maximum value.

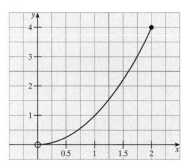

g does not have a minimum value
(but it does have a maximum value).

Solution

a. The function f has no maximum. It takes on all values less than, but arbitrarily close to, 2. However, it never reaches the value 2. This function does not contradict the extreme value theorem because f is not continuous on $[0, 2]$.

b. Although the functional values of $g(x)$ become arbitrarily small as x approaches 0, $g(x)$ never reaches the value 0 so g has no minimum. The function g is continuous on the interval $(0, 2]$, but the extreme value theorem is not contradicted because the interval is not closed. ■

RELATIVE EXTREMA

Typically, the extrema of a continuous function occur either at endpoints of the interval or at points where the graph has a "peak" or a "valley" (points where the graph is higher or lower than all nearby points). For example, the function f in Figure 4.3 has "peaks" at B, D, and "valleys" at C, E. Peaks and valleys are what we call *relative extrema*.

**Relative Maximum
and Relative Minimum**

> The function f is said to have a **relative maximum** at the point c if $f(c) \geq f(x)$ for all x in an open interval containing c. Likewise, f is said to have a **relative minimum** at d if $f(d) \leq f(x)$ for all x in an open interval containing d. Collectively, relative maxima and relative minima are called **relative extrema**.

WARNING   Note that an extremum at an endpoint is, by definition, not a relative extremum, since there is not an open interval around the endpoint entirely contained in the domain of the function.

Next we will formulate a procedure for finding relative extrema. By looking at Figure 4.3, we see that there are horizontal tangents at B, C, and E, while no tangent can be defined at the point D. This suggests that the relative extrema of f occur either where the derivative is zero (horizontal tangent) or where the derivative does not exist (no tangent). This notion leads us to the following definition.

Critical Numbers and Critical Points

Suppose f is defined at c and either $f'(c) = 0$ or $f'(c)$ does not exist. Then the number c is called a **critical number** of f, and the point $P(c, f(c))$ on the graph of f is called a **critical point**.

<u>**WARNING**</u>   Note that if $f(c)$ is not defined, then c **cannot** be a critical number.

EXAMPLE 3   Finding critical numbers

Find the critical numbers for the given functions.

a. $f(x) = 4x^3 - 5x^2 - 8x + 20$

b. $f(x) = \dfrac{e^x}{x - 2}$ Note that $x \neq 2$.

c. $f(x) = 2\sqrt{x}(6 - x)$ Note that $x \geq 0$.

Solution

a. $f'(x) = 12x^2 - 10x - 8$ is defined for all values of x. Solve

$$12x^2 - 10x - 8 = 0$$
$$2(3x - 4)(2x + 1) = 0$$
$$x = \tfrac{4}{3}, \ -\tfrac{1}{2} \quad \text{These are the critical numbers.}$$

b. $f'(x) = \dfrac{(x - 2)e^x - e^x(1)}{(x - 2)^2} = \dfrac{e^x(x - 3)}{(x - 2)^2}$ Note that $x \neq 2$.

The derivative is not defined at $x = 2$, but f is not defined at 2 either, so $x = 2$ is not a critical number. The actual critical numbers are found by solving $f'(x) = 0$:

$$\frac{e^x(x - 3)}{(x - 2)^2} = 0$$

$$x = 3 \quad \text{This is the only critical number since } e^x > 0.$$

c. Write $f(x) = 12x^{1/2} - 2x^{3/2}$ so $f'(x) = 6x^{-1/2} - 3x^{1/2}$. The derivative is not defined at $x = 0$. We have $f(0) = 12(0)^{1/2} - 2(0)^{3/2} = 0$, so we see that f is defined at $x = 0$, which means that $x = 0$ is a critical number. For other critical numbers, solve $f'(x) = 0$:

$$6x^{-1/2} - 3x^{1/2} = 0$$
$$3x^{-1/2}(2 - x) = 0$$
$$x = 2$$

The critical numbers are $x = 0, 2$. ∎

EXAMPLE 4   Critical numbers and critical points

Find the critical numbers and the critical points for the function

$$f(x) = (x - 1)^2(x + 2)$$

Solution

Because the function f is a polynomial, we know that it is continuous and that its derivative exists for all x. Thus, we find the critical numbers by using the product rule and extended power rule to solve the equation $f'(x) = 0$:

$$f'(x) = (x - 1)^2(1) + 2(x - 1)(1)(x + 2)$$
$$= (x - 1)[(x - 1) + 2(x + 2)]$$
$$= 3(x - 1)(x + 1)$$

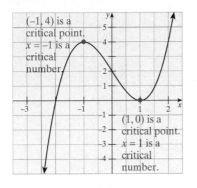

Figure 4.4 The graph of
$f(x) = (x-1)^2(x+2)$

The critical numbers are $x = \pm 1$. To find the critical points, we need to find the y-coordinate for each critical number.

$$f(1) = (1-1)^2(1+2) = 0$$
$$f(-1) = (-1-1)^2(-1+2) = 4$$

Thus, the critical points are $(1, 0)$ and $(-1, 4)$. The graph of $f(x) = (x-1)^2(x+2)$ is shown in Figure 4.4. ∎

Note how the relative extrema occur at the critical points. Our observation that the relative extrema occur only at points on a graph where there is either a horizontal tangent line or no tangent at all is equivalent to the following result.

THEOREM 4.2   Critical number theorem

If a continuous function f has a relative extremum at c, then c must be a critical number of f.

> ➤ **What This Says**   If a point is a relative maximum or a relative minimum value for a function, then either the derivative is 0 or the derivative does not exist at that point. We do not claim the converse.

Proof   Since f has a relative extremum at c, $f(c)$ is defined. If $f'(c)$ does not exist, then c is a critical number by definition. We will show that if $f'(c)$ exists and a relative maximum occurs at c, then $f'(c) = 0$. Our approach will be to examine the difference quotient. (The case where $f'(c)$ exists and a relative minimum occurs at c is handled similarly in Problem 61.)

Because a relative maximum occurs at c, we have $f(c) \geq f(x)$ for every number x in an open interval (a, b) containing c. Therefore, if Δx not equal to zero is small enough so $c + \Delta x$ is in (a, b), then

$$f(c) \geq f(c + \Delta x) \qquad \text{Because a relative maximum occurs at } c$$

$$f(c) - f(c + \Delta x) \geq 0$$
$$f(c + \Delta x) - f(c) \leq 0 \qquad \text{Multiply both sides by } -1, \text{ reversing the inequality.}$$

For the next step we want to divide both sides by Δx (to write the left side as a difference quotient). However, as this is an inequality, we need to consider two possibilities:

1. Suppose $\Delta x > 0$ (the inequality does not reverse):

$$\frac{f(c + \Delta x) - f(c)}{\Delta x} \leq 0 \qquad \text{Divide both sides by } \Delta x.$$

Now we take the limit of both sides as Δx approaches 0 from the right (because Δx is positive).

$$\underbrace{\lim_{\Delta x \to 0^+} \frac{f(c + \Delta x) - f(c)}{\Delta x}}_{f'(c)} \leq \underbrace{\lim_{\Delta x \to 0^+} 0}_{0}$$

$$f'(c) \leq 0$$

2. Next, suppose $\Delta x < 0$ (the inequality reverses). Then

$$\frac{f(c + \Delta x) - f(c)}{\Delta x} \geq 0$$

This time we take the limit of both sides as Δx approaches 0 from the left (because Δx is negative).

$$\lim_{\Delta x \to 0^-} \frac{f(c + \Delta x) - f(c)}{\Delta x} \geq \lim_{\Delta x \to 0^-} 0$$

$$f'(c) \geq 0$$

Because we have shown that $f'(c) \leq 0$ and $f'(c) \geq 0$, it follows that $f'(c) = 0$. ❏

WARNING Theorem 4.2 tells us that a relative extremum of a continuous function f can occur *only* at a critical number, but it does not say that a relative extremum must occur at each critical number.

For example, if $f(x) = x^3$, then $f'(x) = 3x^2$ and $f'(0) = 0$, so 0 is a critical number. But there is no relative extremum at $c = 0$ on the graph of f because the graph is rising for $x < 0$ and also for $x > 0$, as shown in Figure 4.5a. It is also quite possible for a continuous function g to have no relative extremum at a point c where $g'(x)$ does not exist (see Figure 4.5b).

a. The graph of $f(x) = x^3$
No relative extremum occurs
at $c = 0$ even though $f'(0) = 0$.

b. Although $g'(1)$ does not exist,
no relative extremum occurs
at $c = 1$.

Figure 4.5 A relative extremum may not occur at each critical number.

EXAMPLE 5 Critical numbers where the derivative does not exist

Find the critical numbers for $f(x) = |x + 1|$ on $[-5, 5]$. The graph is shown in Figure 4.6.

Solution

If $x > -1$, then $f(x) = x + 1$ and $f'(x) = 1$. However, if $x < -1$, then $f(x) = -x - 1$ and $f'(x) = -1$. We need to check what happens at $x = -1$.

$$f'(-1) = \lim_{\Delta x \to 0} \frac{f(-1 + \Delta x + 1) - f(-1 + 1)}{\Delta x}$$

$$= \lim_{\Delta x \to 0} \frac{|\Delta x| - |0|}{\Delta x}$$

$$= \lim_{\Delta x \to 0} \frac{|\Delta x|}{\Delta x}$$

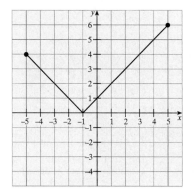

Figure 4.6 Graph of
$f(x) = |x + 1|$ on $[-5, 5]$

We consider the left and right limits:

$$\lim_{\Delta x \to 0^+} \frac{|\Delta x|}{\Delta x} = \lim_{\Delta x \to 0^+} \frac{\Delta x}{\Delta x} = 1$$

and

$$\lim_{\Delta x \to 0^-} \frac{|\Delta x|}{\Delta x} = \lim_{\Delta x \to 0^-} \frac{-\Delta x}{\Delta x} = -1$$

Because these are not the same, this limit does not exist and thus the function $f(x) = |x + 1|$ is not differentiable at $x = -1$. Because $f(-1)$ is defined, it follows that -1 is the only critical number. ∎

ABSOLUTE EXTREMA

Suppose we are looking for the absolute extrema of a continuous function f on the closed, bounded interval $[a, b]$. Theorem 4.1 tells us that these extrema exist and Theorem 4.2 enables us to narrow the list of "candidates" for points where extrema can occur from the entire interval $[a, b]$ to just the endpoints $x = a$, $x = b$, and the critical numbers between a and b. This suggests the following procedure.

Procedure for Finding Absolute Extrema

To find the absolute extrema of a continuous function f on $[a, b]$:

Step 1. Compute $f'(x)$ and find all critical numbers of f on $[a, b]$.
Step 2. Evaluate f at the endpoints a and b and at each critical number c.
Step 3. Compare the values in step 2.
    The largest value of f is the absolute maximum of f on $[a, b]$.
    The smallest value of f is the absolute minimum of f on $[a, b]$.

Figure 4.7 shows some of the possibilities in the application of this procedure.

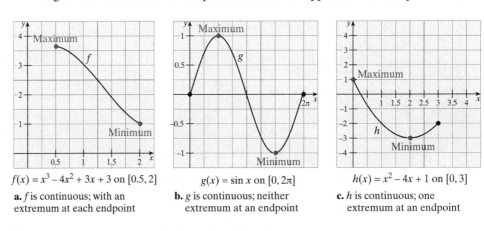

$f(x) = x^3 - 4x^2 + 3x + 3$ on $[0.5, 2]$

a. f is continuous; with an extremum at each endpoint

$g(x) = \sin x$ on $[0, 2\pi]$

b. g is continuous; neither extremum at an endpoint

$h(x) = x^2 - 4x + 1$ on $[0, 3]$

c. h is continuous; one extremum at an endpoint

Figure 4.7   Absolute extrema

EXAMPLE 6   Absolute extrema of a polynomial function

Find the absolute extrema of the function defined by the equation $f(x) = x^4 - 2x^2 + 3$ on the closed interval $[-1, 2]$.

Solution

Because f is a polynomial function, it is continuous on the closed interval $[-1, 2]$. Theorem 4.1 tells us that there must be an absolute maximum and an absolute minimum on the interval.

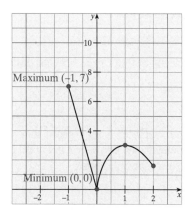

Figure 4.8 The graph of
$f(x) = x^4 - 2x^2 + 3$ on $[-1, 2]$

Step 1. $f'(x) = 4x^3 - 2(2x)$
$$= 4x(x^2 - 1)$$
$$= 4x(x - 1)(x + 1)$$
The critical numbers are $x = 0, 1$, and -1.

Step 2.   Values at endpoints: $f(-1) = 2$
$$f(2) = 11$$
Critical numbers: $f(0) = 3$
$$f(1) = 2$$

Step 3.   The absolute maximum of f occurs at $x = 2$ and is $f(2) = 11$; the absolute minimum of f occurs at $x = 1$ and $x = -1$ and are $f(1) = f(-1) = 2$. The graph of f is shown in Figure 4.8.   ∎

EXAMPLE 7   Absolute extrema when the derivative does not exist

Find the absolute extrema of $f(x) = x^{2/3}(5 - 2x)$ on the interval $[-1, 2]$.

Solution

Step 1.   To find the derivative, rewrite the given function as $f(x) = 5x^{2/3} - 2x^{5/3}$. Then
$$f'(x) = \tfrac{10}{3}x^{-1/3} - \tfrac{10}{3}x^{2/3} = \tfrac{10}{3}x^{-1/3}(1 - x)$$
Critical numbers are found by solving $f'(x) = 0$ and by locating the places where the derivative does not exist. First,
$$f'(x) = 0 \quad \text{when} \quad x = 1$$
Even though $f(0)$ exists, we note that $f'(x)$ does not exist at $x = 0$ (notice the division by zero when $x = 0$). Thus, the critical numbers are $x = 0$ and $x = 1$.

Step 2.   Values at endpoints: $f(-1) = 7$
$$f(2) = 2^{2/3} \approx 1.587401052$$
Critical numbers: $f(0) = 0$
$$f(1) = 3$$

Step 3.   The absolute maximum of f occurs at $x = -1$ and is $f(-1) = 7$; the absolute minimum of f occurs at $x = 0$ and is $f(0) = 0$. The graph of f is shown in Figure 4.9.   ∎

Figure 4.9 Graph of
$f(x) = 5x^{2/3} - 2x^{5/3}$ on $[-1, 2]$

EXAMPLE 8   Absolute extrema for a trigonometric function

Find the absolute extrema of the continuous function
$$T(x) = \tfrac{1}{2}(\sin^2 x + \cos x) + 2\sin x - x$$
on the interval $[0, \tfrac{\pi}{2}]$.

Solution

Step 1.   To find where $T'(x) = 0$, we begin by finding $T'(x)$:
$$T'(x) = \tfrac{1}{2}(2\sin x \cos x - \sin x) + 2\cos x - 1$$
$$= \tfrac{1}{2}(2\sin x \cos x - \sin x + 4\cos x - 2)$$
$$= \tfrac{1}{2}[2(\cos x)(\sin x + 2) - (\sin x + 2)]$$
$$= \tfrac{1}{2}[(\sin x + 2)(2\cos x - 1)]$$
Since the factor $(\sin x + 2)$ is never zero, it follows that $T'(x) = 0$ only when $2\cos x - 1 = 0$; that is, when $x = \tfrac{\pi}{3}$. This is the only critical number in $[0, \tfrac{\pi}{2}]$.

Step 2.   Evaluate the function at the endpoints:

$$T(0) = \tfrac{1}{2}(\sin^2 0 + \cos 0) + 2\sin 0 - 0 = \tfrac{1}{2}(0 + 1) + 2(0) - 0 = 0.5$$

$$T\left(\tfrac{\pi}{2}\right) = \tfrac{1}{2}\left(\sin^2 \tfrac{\pi}{2} + \cos \tfrac{\pi}{2}\right) + 2\sin\tfrac{\pi}{2} - \tfrac{\pi}{2}$$
$$= \tfrac{1}{2}(1 + 0) + 2(1) - \tfrac{\pi}{2} = \tfrac{5}{2} - \tfrac{\pi}{2} \approx 0.9292036732$$

Evaluate the function at the critical number:

$$T\left(\tfrac{\pi}{3}\right) = \tfrac{1}{2}\left(\sin^2\tfrac{\pi}{3} + \cos\tfrac{\pi}{3}\right) + 2\sin\tfrac{\pi}{3} - \tfrac{\pi}{3}$$
$$= \tfrac{1}{2}\left(\tfrac{3}{4} + \tfrac{1}{2}\right) + 2\left(\tfrac{\sqrt{3}}{2}\right) - \tfrac{\pi}{3} = \tfrac{5}{8} + \sqrt{3} - \tfrac{\pi}{3} \approx 1.309853256$$

Step 3.   The absolute maximum of T is approximately 1.31 at $x = \tfrac{\pi}{3}$ and the absolute minimum of T is 0.5 at $x = 0$.   ■

OPTIMIZATION

In our next two examples, we examine applications involving optimization. Such problems are investigated in more depth in Sections 4.6 and 4.7.

EXAMPLE 9   An applied maximum value problem

A box with a square base is constructed so that the length of one side of the base plus the height is 10 in. What is the largest possible volume of such a box?

Solution

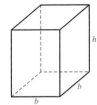

We let b be the length of one side of the base and h be the height of the box, as shown in Figure 4.10. The volume, V, is

$$V = b^2 h$$

Because our methods apply only to functions of one variable, it may seem that we cannot deal with V as a function of two variables. However, we know that $b + h = 10$; therefore, $h = 10 - b$, and we can now write V as a function of b alone:

$$V(b) = b^2(10 - b)$$

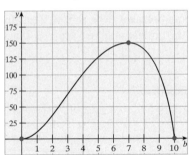

Figure 4.10 Volume of a box

The domain is not stated, but for physical reasons we must have $b \geq 0$ and $10 - b = h \geq 0$, so that $0 \leq b \leq 10$. Figure 4.10 shows the volume for various choices of b.

Now, find the value of b that produces the maximum volume. First, we find the critical numbers. Note that V is a polynomial function, so the derivative exists everywhere in the domain. Write $V(b) = 10b^2 - b^3$ and find $V'(b) = 20b - 3b^2$. Then

$$V'(b) = 0$$
$$20b - 3b^2 = 0$$
$$b(20 - 3b) = 0$$
$$b = 0, \ \tfrac{20}{3}$$

Checking the endpoints and the critical numbers, we have

$$V(0) = 0$$
$$V(10) = 10^2(10 - 10) = 0$$
$$V\left(\tfrac{20}{3}\right) = \left(\tfrac{20}{3}\right)^2 \left(10 - \tfrac{20}{3}\right) = \tfrac{4,000}{27}$$

Alternatively, we could have solved the problem with V as a function of h: $V(h) = (10 - h)^2 h, 0 \leq h \leq 10$, with the same result. We must write V as a function of one variable so that we can find its derivative.

Thus, the largest value for the volume V is $\tfrac{4,000}{27} \approx 148.1$ in.^{3}. It occurs when the square base has a side of length $\tfrac{20}{3}$ in. and the height is $h = 10 - \tfrac{20}{3} = \tfrac{10}{3}$ in.   ■

EXAMPLE 10 Maximum and minimum velocity of a moving particle

A particle moves along the s-axis with position

$$s(t) = t^4 - 8t^3 + 18t^2 + 60t - 8$$

Find the largest and smallest values of its velocity for $1 \leq t \leq 5$.

Solution

The velocity is

$$v(t) = s'(t) = 4t^3 - 24t^2 + 36t + 60$$

(**Note**: v is a polynomial function, so it has a derivative for all t.) To find the largest value of $v(t)$, we compute the derivative of v:

$$v'(t) = 12t^2 - 48t + 36$$
$$= 12(t - 1)(t - 3)$$

Setting $v'(t) = 0$, we find that the critical numbers of $v(t)$ are $t = 1, 3$. Now we evaluate v at the critical numbers and endpoints.

$t = 1$ is both a critical number and an endpoint:

$$v(1) = 4(1)^3 - 24(1)^2 + 36(1) + 60 = 76$$

$t = 5$ is an endpoint:

$$v(5) = 4(5)^3 - 24(5)^2 + 36(5) + 60 = 140$$

$t = 3$ is a critical number:

$$v(3) = 4(3)^3 - 24(3)^2 + 36(3) + 60 = 60$$

The largest value of the velocity is 140 at the endpoint where $t = 5$, and the smallest value is 60 when $t = 3$. ∎

4.1 PROBLEM SET

A *In Problems 1–14, find the critical numbers for each continuous function on the given closed, bounded interval, and then tell whether each yields a minimum, maximum, or neither.*

1. $f(x) = 5 + 10x - x^2$ on $[-3, 3]$
2. $f(x) = 10 + 6x - x^2$ on $[-4, 4]$
3. $f(x) = x^3 - 3x^2$ on $[-1, 3]$
4. $f(t) = t^4 - 8t^2$ on $[-3, 3]$
5. $f(x) = x^3$ on $\left[-\frac{1}{2}, 1\right]$
6. $g(x) = x^3 - 3x$ on $[-2, 2]$
7. $f(x) = x^5 - x^4$ on $[-1, 1]$
8. $g(t) = 3t^5 - 20t^3$ on $[-1, 2]$
9. $h(t) = te^{-t}$ on $[0, 2]$
10. $s(x) = \dfrac{\ln \sqrt{x}}{x}$ on $[1, 3]$
11. $f(x) = |x|$ on $[-1, 1]$
12. $f(x) = |x - 3|$ on $[-4, 4]$
13. $f(u) = \sin^2 u + \cos u$ on $[0, 2]$
14. $g(u) = \sin u - \cos u$ on $[0, \pi]$
15. **WHAT DOES THIS SAY?** Outline a procedure for finding the absolute extrema of a continuous function on a closed, bounded interval. Include in your outline a discussion of what is meant by critical numbers.

16. **Exploration Problem** Why is it important to check endpoints when finding an optimum value?

17. In Example 7, we found that the maximum value of $f(x) = x^{2/3}(5 - 2x)$ on $[-1, 2]$ is 7 and occurs at $x = -1$. The graph of this function on a leading brand of graphing calculator is shown:

```
Y₁■X^(2/3)(5-2X)
Xmin=-10 Ymin=-10
Xmax=10  Ymax=10
Xscl=1   Yscl=1
```

This graph is not correct. Can you explain the discrepancy? It is not our intent to "make up" problems to use with a calculator, but *whenever* you use a calculator or computer to assist you with calculus, you must understand the nature of the functions with which you are working, and not rely only on the calculator or computer output.

In Problems 18–29, find the absolute maximum and absolute minimum (largest and smallest values, respectively) of each continuous function on the closed, bounded interval. If the function is not continuous on the interval, then the extreme value theorem does not apply; if this is the case, so state.

18. $f(u) = 1 - u^{2/3}$ on $[-1, 1]$

19. $g(t) = (50 + t)^{2/3}$ on $[-50, 14]$

20. $g(x) = 2x^3 - 3x^2 - 36x + 4$ on $[-4, 4]$

21. $g(x) = x^3 + 3x^2 - 24x - 4$ on $[-4, 4]$

22. $f(x) = \frac{8}{3}x^3 - 5x^2 + 8x - 5$ on $[-4, 4]$

23. $f(x) = \frac{1}{6}(x^3 - 6x^2 + 9x + 1)$ on $[0, 2]$

24. $h(x) = \tan x + \sec x$ on $[0, 2\pi]$

25. $s(t) = t \cos t - \sin t$ on $[0, 2\pi]$

26. $f(x) = e^{-x} \sin x$ on $[0, 2\pi]$

27. $g(x) = \cot^{-1}\left(\frac{x}{9}\right) - \cot^{-1}\left(\frac{x}{5}\right)$ on $[0, 10]$

28. $f(x) = \begin{cases} 9 - 4x & \text{if } x < 1 \\ -x^2 + 6x & \text{if } x \geq 1 \end{cases}$ on $[0, 4]$

29. $f(x) = \begin{cases} 8 - 3x & \text{if } x < 2 \\ -x^2 + 3x & \text{if } x \geq 2 \end{cases}$ on $[-1, 4]$

Find the required extremum in Problems 30–35, or explain why it does not exist.

30. the smallest value of $f(x) = \dfrac{1}{x(x+1)}$ on $[-0.5, 0)$

31. the smallest value of $g(x) = \dfrac{9}{x} + x - 3$ on $[1, 9]$

32. the smallest value of $g(x) = \dfrac{x^2 - 1}{x^2 + 1}$ on $[-1, 1]$

33. the largest value of $f(t) = \begin{cases} -t^2 - t + 2 & \text{if } t < 1 \\ 3 - t & \text{if } t \geq 1 \end{cases}$

on $[-2, 3]$

34. the smallest value of $f(x) = e^x + e^{-x} - x$ on $[0, 2]$

35. the largest value of $g(x) = \dfrac{\ln x}{\cos x}$ on $[2, 3]$ correct to the nearest tenth.

B *In Problems 36–43, find the extrema (that is, the absolute maxima and minima).*

36. $f(\theta) = \cos^3 \theta - 4 \cos^2 \theta$ on $[-0.1, \pi + 0.1]$

37. $g(\theta) = \theta \sin \theta$ on $[-2, 2]$

38. $f(x) = 20 \sin(378\pi x)$ on $[-1, 1]$

39. $g(u) = 98u^3 - 4u^2 + 72u$ on $[0, 4]$

40. $f(w) = \sqrt{w}(w - 5)^{1/3}$ on $[0, 4]$

41. $h(x) = \sqrt[3]{x}\sqrt[3]{(x - 3)^2}$ on $[-1, 4]$

42. $h(x) = \cos^{-1} x \tan^{-1} x$ on $[0, 1]$

43. $f(x) = e^{-x}(\cos x + \sin x)$ on $[0, 2\pi]$

In Problems 44–47, find functions that satisfy the stated conditions and each of the following side conditions, if possible:

  a. *a minimum but no maximum*

  b. *a maximum but no minimum*

  c. *both a maximum and a minimum*

  d. *neither a maximum nor a minimum*

Note that in each problem, there may be a different solution function for each side condition.

44. Counterexample Problem   For each of the four given conditions, find a function that is discontinuous and defined on an open interval.

45. Counterexample Problem   For each of the four given conditions, find a function that is discontinuous and defined on a closed interval.

46. Counterexample Problem   For each of the four given conditions, find a function that is continuous and defined on an open interval.

47. Counterexample Problem   For each of the four given conditions, find a function that is continuous and defined on a closed interval.

48. Counterexample Problem   Give a counterexample to show that the extreme value theorem does not necessarily apply if one disregards the condition that f be continuous.

49. Counterexample Problem   Give a counterexample to show that the extreme value theorem does not necessarily apply if one disregards the condition that f be defined on a closed interval.

50. An object moves along the s-axis with position

$$s(t) = t^3 - 6t^2 - 15t + 11$$

Find the largest value of its velocity on $[0, 4]$.

51. An object moves along the s-axis with position

$$s(t) = t^4 - 2t^3 - 12t^2 + 60t - 10$$

Find the largest value of its velocity on $[0, 3]$.

52. Find two nonnegative numbers whose sum is 8 and the product of whose squares is as large as possible.

53. Find two nonnegative numbers such that the sum of one and twice the other is 12 if it is required that their product be as large as possible.

54. Under the condition that $3x + y = 80$, maximize xy^3 when $x \geq 0, y \geq 0$.

55. Under the condition that $3x + y = 126$, maximize xy when $x \geq 0$ and $y \geq 0$.

56. Under the condition that $2x - 5y = 18$, minimize x^2y when $x \geq 0$ and $y \leq 0$.

57. Show that if a rectangle with fixed perimeter P is to enclose the greatest area, it must be a square.

58. Find all points on the circle $x^2 + y^2 = a^2$ ($a \geq 0$) such that the product of the x-coordinate and the y-coordinate is as large as possible.

C 59. a. Show that $\frac{1}{2}$ is the number that is greater than or equal to its own square by the greatest amount.

  b. Which nonnegative number is greater than or equal to its own cube by the greatest amount?

  c. Which nonnegative number is greater than or equal to its nth power ($n > 0$) by the greatest amount?

60. Given the constants $a_1, a_2, \ldots, a_n$, find the value of x that guarantees that the sum

$$S(x) = (a_1 - x)^2 + (a_2 - x)^2 + \cdots + (a_n - x)^2$$

will be as small as possible.

61. Without using Theorem 4.2, show that if $f'(c)$ exists and a relative minimum occurs at c, then $f'(x) = 0$.

62. Explain why the function

$$f(\theta) = \frac{8}{\sin \theta} + \frac{27}{\cos \theta}$$

must attain a minimum in the open interval $(0, \frac{\pi}{2})$.

4.2   The Mean Value Theorem

IN THIS SECTION   Rolle's theorem, statement and proof of the mean value theorem, the zero-derivative theorem

If a car travels smoothly down a straight, level road with average velocity 60 mi/h, we would expect the speedometer reading to be exactly 60 mi/h at least once during the trip. After all, if the car's velocity were always above 60 mi/h, the average velocity would also be above that level, and the same reasoning applies if the velocity were always below 60 mi/h. More generally, if $s(t)$ is the car's position at time t during a trip over the time interval $a \le t \le b$, then there should be a time $t = c$ when the velocity $s'(c)$ equals the average velocity between times $t = a$ and $t = b$. That is, for some c with $a < c < b$,

$$\underbrace{\frac{s(b) - s(a)}{b - a}}_{\text{Average velocity}} = \underbrace{s'(c)}_{\text{Instantaneous velocity}}$$

This example illustrates a result called the mean value theorem for derivatives (abbreviated MVT) that is fundamental in the study of calculus. Our proof of the MVT is based on the following special case, which is named for the French mathematician, Michel Rolle (1652–1719), who gave a version of the result in an algebra text published in 1690.*

ROLLE'S THEOREM

The key to the proof of the mean value theorem is the following result, which is really just the MVT in the special case where $f(a) = f(b)$. In terms of our car example, Rolle's theorem says that if a moving car begins and ends at the same place, then somewhere during its journey, it must reverse direction, since $\dfrac{f(b) - f(a)}{b - a} = 0$ for $f(a) = f(b)$.

THEOREM 4.3   Rolle's theorem

Suppose f is continuous on the closed interval $[a, b]$ and differentiable on the open interval (a, b). If $f(a) = f(b)$, then there exists at least one number c between a and b such that $f'(c) = 0$.

Proof   To construct a formal proof, note that since f is continuous on the closed interval $[a, b]$, it follows from the extreme value theorem (Theorem 4.1) that f must have both a largest value and a smallest value on $[a, b]$. The case where f is constant on $[a, b]$ is easy since then $f'(x) = 0$ for all x in the interval (see Figure 4.11a).

---

*Incidentally, Rolle was a number theorist at heart and thoroughly distrusted the methods of calculus, which he regarded as a "collection of ingenious fallacies."

In the case where f is not constant throughout $[a, b]$, the largest and smallest values cannot be the same, and since $f(a) = f(b)$, we can conclude that at least one extreme value occurs at a number c that is not an endpoint; that is, $a < c < b$. Finally, since c is in the open interval (a, b) throughout which $f'(x)$ exists, it follows from the critical number theorem (Theorem 4.2) that $f'(c) = 0$. A graph illustrating this case is shown in Figure 4.11b.

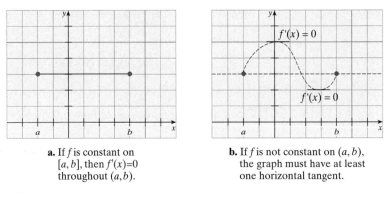

a. If f is constant on $[a, b]$, then $f'(x)=0$ throughout (a,b).

b. If f is not constant on (a, b), the graph must have at least one horizontal tangent.

Figure 4.11   A geometrical interpretation of Rolle's theorem   ❏

Note: Rolle's theorem asserts only that at least one number c exists between a and b such that $f'(c) = 0$. As illustrated in Figure 4.11b, there may be more than one such number.

STATEMENT AND PROOF OF THE MEAN VALUE THEOREM

Rolle's theorem can be interpreted as saying that there is at least one number c between a and b such that the tangent line to the graph of f at the point $(c, f(c))$ is parallel to the horizontal line through the points $(a, f(a))$ and $(b, f(b))$. It is reasonable to expect a similar result to hold if the endpoints of the graph are not necessarily at the same level of the graph; that is, if the graph in Figure 4.12a is tilted as shown in Figure 4.12b. This "tilted" version can be interpreted as showing that the line segment joining points $(a, f(a))$ and $(b, f(b))$ on the graph of a differentiable function $y = f(x)$ has the same slope as the tangent line to the graph at some point c between a and b. This observation leads to the following statement of the mean value theorem.

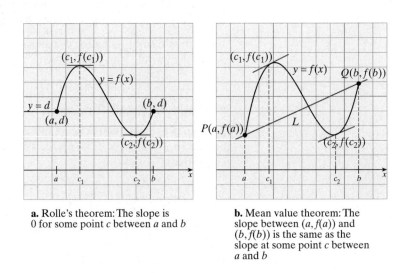

a. Rolle's theorem: The slope is 0 for some point c between a and b

b. Mean value theorem: The slope between $(a, f(a))$ and $(b, f(b))$ is the same as the slope at some point c between a and b

Figure 4.12   A geometrical comparison of Rolle's theorem with the mean value theorem

THEOREM 4.4   The mean value theorem for derivatives (MVT)

If f is continuous on the closed interval $[a, b]$ and differentiable on the open interval (a, b), then there exists in (a, b) at least one number c such that

$$\frac{f(b) - f(a)}{b - a} = f'(c)$$

Proof   Our strategy will be to apply Rolle's theorem to a function g related to f; namely the function

$$g(x) = \frac{f(b) - f(a)}{b - a}(x - a) + f(a) - f(x)$$

for $a \leq x \leq b$. Because f satisfies the hypotheses of the MVT, the function g is also continuous on the closed interval $[a, b]$ and differentiable on the open interval (a, b). In addition, we find that

$$g(a) = \left[\frac{f(b) - f(a)}{b - a}\right](a - a) + f(a) - f(a) = 0$$

$$g(b) = \left[\frac{f(b) - f(a)}{b - a}\right](b - a) + f(a) - f(b)$$

$$= [f(b) - f(a)] + f(a) - f(b) = 0$$

Thus, g satisfies the hypotheses of Rolle's theorem, so there exists at least one number c between a and b for which $g'(c) = 0$. Differentiating the function g, we find that

$$g'(x) = \frac{f(b) - f(a)}{b - a} - f'(x)$$

Because $g'(c) = 0$, we have

$$0 = g'(c) = \frac{f(b) - f(a)}{b - a} - f'(c)$$

This means that

$$f'(c) = \frac{f(b) - f(a)}{b - a}$$

as required.   ❏

The MVT asserts the existence of a number, c, but does not tell us how to find it. The following example illustrates one method for finding such a number c.

EXAMPLE 1   Finding the number c specified by the MVT

Show that the function $f(x) = x^3 + x^2$ satisfies the hypotheses of the MVT on the closed interval $[1, 2]$, and find a number c between 1 and 2 so that

$$f'(c) = \frac{f(2) - f(1)}{2 - 1}$$

Solution

Because f is a polynomial function, it is differentiable and also continuous on the entire interval $[1, 2]$. Thus, the hypotheses of the MVT are satisfied.

By differentiating f, we find that

$$f'(x) = 3x^2 + 2x$$

for all x. Therefore, we have $f'(c) = 3c^2 + 2c$, and the MVT equation

$$f'(c) = \frac{f(2) - f(1)}{2 - 1}$$

is satisfied when

$$3c^2 + 2c = \frac{f(2) - f(1)}{2 - 1} = \frac{12 - 2}{1} = 10$$

Solving the resulting equation $3c^2 + 2c - 10 = 0$ by the quadratic formula, we obtain

$$c = \frac{-1 \pm \sqrt{31}}{3}$$

The negative value is not in the open interval $(1, 2)$, but the positive value

$$c = \frac{-1 + \sqrt{31}}{3} \approx 1.522588121$$

satisfies the requirements of the MVT. ∎

EXAMPLE 2   Using the MVT to prove a trigonometric inequality

Show that $|\sin x - \sin y| \le |x - y|$ for all numbers x and y by applying the mean value theorem.

Solution

The inequality is true if $x = y$. Suppose $x \ne y$; then $f(\theta) = \sin \theta$ is differentiable and hence continuous for all θ, with $f'(\theta) = \cos \theta$. By applying the MVT to f on the closed interval with endpoints x and y, we see that

$$\frac{f(x) - f(y)}{x - y} = f'(c)$$

for some c between x and y. Because $f'(c) = \cos c$ and

$$f(x) - f(y) = \sin x - \sin y$$

it follows that

$$\frac{\sin x - \sin y}{x - y} = \cos c$$

Finally, we take the absolute value of the expression on each side, remembering that $|\cos c| \le 1$ for any number c:

$$\left| \frac{\sin x - \sin y}{x - y} \right| = |\cos c| \le 1$$

Thus, $|\sin x - \sin y| \le |x - y|$, as claimed. ∎

THE ZERO-DERIVATIVE THEOREM

The primary use of the MVT is as a tool for proving certain key theoretical results of calculus. For example, we know that the derivative of a constant is 0, and in Theorem 4.5, we prove a partial converse of this result by showing that a function whose derivative is 0 throughout the interval must be constant on that interval. This apparently simple result and its corollary (Theorem 4.6) turn out to be crucial to our development of integration in Chapter 5.

THEOREM 4.5   Zero-derivative theorem

Suppose f is a continuous function on the closed interval $[a, b]$ and is differentiable on the open interval (a, b), with $f'(x) = 0$ for all x on (a, b). Then the function f is constant on $[a, b]$.

Proof Let x_1 and x_2 be two distinct numbers ($x_1 \neq x_2$) chosen arbitrarily from the closed interval $[a, b]$. The function f satisfies the requirements of the MVT on the interval with endpoints x_1 and x_2, which means that there exists a number c between x_1 and x_2 such that

$$\frac{f(x_2) - f(x_1)}{x_2 - x_1} = f'(c)$$

By hypothesis, $f'(x) = 0$ throughout the open interval (a, b), and because c lies within this interval, we have $f'(c) = 0$. Thus, by substitution, we have

$$\frac{f(x_2) - f(x_1)}{x_2 - x_1} = 0$$
$$f(x_2) = f(x_1)$$

Because x_1 and x_2 were chosen arbitrarily from $[a, b]$, we conclude that $f(x) = k$, a constant, for all x, as required. ❏

THEOREM 4.6 Constant difference theorem

Suppose the functions f and g are continuous on the closed interval $[a, b]$ and differentiable on the open interval (a, b). Then if $f'(x) = g'(x)$ for all x in (a, b), there exists a constant C such that

$$f(x) = g(x) + C$$

for all x on $[a, b]$.

Proof Let $h(x) = f(x) - g(x)$; then

$$h'(x) = f'(x) - g'(x)$$
$$= 0 \qquad \text{Because } f'(x) = g'(x)$$

Thus, by Theorem 4.5, $h(x) = C$ for some constant C and all x on $[a, b]$, and because $h(x) = f(x) - g(x)$, it follows that

$$f(x) - g(x) = C$$
$$f(x) = g(x) + C \qquad \text{Add } g(x) \text{ to both sides.} \qquad ❏$$

➡ **What This Says** Two functions with equal derivatives on an open interval differ by a constant on that interval.

4.2 PROBLEM SET

Ⓐ 1. Exploration Problem What does Rolle's theorem say, and why is it important?

2. Exploration Problem State the hypotheses used in the proof of the MVT. How are the hypotheses used in the proof? Can the conclusion of the MVT be true if any or all of the hypotheses are not satisfied?

In Problems 3–20, verify that the given function f satisfies the hypotheses of the MVT on the given interval $[a, b]$. Then find all numbers c between a and b for which

$$\frac{f(b) - f(a)}{b - a} = f'(c)$$

3. $f(x) = 2x^2 + 1$ on $[0, 2]$ **4.** $f(x) = -x^2 + 4$ on $[-1, 0]$

5. $f(x) = x^3 + x$ on $[1, 2]$ **6.** $f(x) = 2x^3 - x^2$ on $[0, 2]$

7. $f(x) = x^4 + 2$ on $[-1, 2]$ **8.** $f(x) = x^5 + 3$ on $[2, 4]$

9. $f(x) = \sqrt{x}$ on $[1, 4]$ **10.** $f(x) = \dfrac{1}{\sqrt{x}}$ on $[1, 4]$

11. $f(x) = \dfrac{1}{x + 1}$ on $[0, 2]$ **12.** $f(x) = 1 + \dfrac{1}{x}$ on $[1, 4]$

13. $f(x) = \cos x$ on $[0, \frac{\pi}{2}]$

14. $f(x) = \sin x + \cos x$ on $[0, 2\pi]$

15. $f(x) = e^x$ on $[0, 1]$

16. $f(x) = \frac{1}{2}(e^x + e^{-x})$ on $[0, 1]$

17. $f(x) = \ln x$ on $[\frac{1}{2}, 2]$ **18.** $f(x) = \dfrac{\ln \sqrt{x}}{x}$ on $[1, 3]$

19. $f(x) = \tan^{-1} x$ on $[0, 1]$ **20.** $f(x) = x \sin^{-1} x$ on $[0, 1]$

Decide whether Rolle's theorem can be applied to f on the interval indicated in Problems 21–30.

21. $f(x) = |x - 2|$ on $[0, 4]$ **22.** $f(x) = \tan x$ on $[0, 2\pi]$

23. $f(x) = \sin x$ on $[0, 2\pi]$ **24.** $f(x) = |x| - 2$ on $[0, 4]$

25. $f(x) = \sqrt[3]{x} - 1$ on $[-8, 8]$ **26.** $f(x) = \dfrac{1}{x - 2}$ on $[-1, 1]$

27. $f(x) = \dfrac{1}{x - 2}$ on $[1, 2]$

28. $f(x) = 3x + \sec x$ on $[-\pi, \pi]$

29. $f(x) = \sin^2 x$ on $[-\frac{\pi}{2}, \frac{\pi}{2}]$

30. $f(x) = \sqrt{\ln x}$ on $[1, 2]$

31. Let $g(x) = 8x^3 - 6x + 8$. Find a function f with $f'(x) = g'(x)$ and $f(1) = 12$.

32. Let $g(x) = \sqrt{x^2 + 5}$. Find a function f with $f'(x) = g'(x)$ and $f(2) = 1$.

33. Show that $f(x) = \dfrac{x + 4}{5 - x}$ and $g(x) = \dfrac{-9}{x - 5}$ differ by a constant. Are the conditions of the constant difference theorem satisfied? Does $f'(x) = g'(x)$?

34. Let $f(x) = (x - 2)^3$ and $g(x) = (x^2 + 12)(x - 6)$. Use f and g to demonstrate the constant difference theorem.

35. Let $f(x) = (x - 1)^3$ and $g(x) = (x^2 + 3)(x - 3)$. Use f and g to demonstrate the constant difference theorem.

36. Let f be defined as shown in Figure 4.13.

Figure 4.13 Function f on $[0, 9]$

Use the graph of f to estimate the values of c that satisfy the conclusion of Rolle's theorem on $[0, 9]$. What theorem would apply for the interval $[0, 5]$?

37. Let g be defined as shown in Figure 4.14.

Figure 4.14 Function g on $[0, 9]$

Use the graph of g to estimate the values of c that satisfy the conclusion of the mean value theorem on $[0, 9]$. What theorem would apply for the interval $[4, 8]$?

B 38. Alternative form of the mean value theorem: If f is continuous on $[a, b]$ and differentiable on (a, b), then there exists a number c in (a, b) such that

$$f(b) = f(a) + (b - a)f'(c)$$

Prove this alternative form of the MVT.

39. Let u and v be any two numbers between $-\frac{\pi}{2}$ and $\frac{\pi}{2}$. Use the MVT to show that

$$|\tan u - \tan v| \geq |u - v|$$

40. If $f(x) = \dfrac{1}{x}$ on $[-1, 1]$, does the mean value theorem apply? Why or why not?

41. If $g(x) = |x|$ on $[-2, 2]$, does the mean value theorem apply? Why or why not?

42. Counterexample Problem Is it true that

$$|\cos x - \cos y| \leq |x - y|$$

for all x and y? Either prove that the inequality is always valid or find a counterexample.

43. Counterexample Problem Consider

$$f(x) = \begin{cases} 1 & \text{if } x \geq 0 \\ -1 & \text{if } x < 0 \end{cases}$$

$f'(x) = 0$ for all x in the domain, but f is not a constant. Does this example contradict the zero-derivative theorem? Why or why not?

44. a. Let n be a positive integer. Show that there is a number c between 0 and x for which

$$\frac{(1 + x)^n - 1}{x} = n(1 + c)^{n-1}$$

b. Use part **a** to evaluate

$$\lim_{x \to 0} \frac{(1 + x)^n - 1}{x}$$

45. a. Show that there is a number w between 0 and x for which

$$\frac{\cos x - 1}{x} = -\sin w$$

b. Use part **a** to evaluate

$$\lim_{x \to 0} \frac{\cos x - 1}{x}$$

46. Use the MVT to evaluate

$$\lim_{x \to \pi^+} \frac{\cos x + 1}{x - \pi}$$

47. Let $f(x) = 1 + \dfrac{1}{x}$. If a and b are constants such that $a < 0$ and $b > 0$, show that there is no number w between a and b for which

$$f(b) - f(a) = f'(w)(b - a)$$

48. Show that for any $x > 4$, there is a number w between 4 and x such that

$$\frac{\sqrt{x} - 2}{x - 4} = \frac{1}{2\sqrt{w}}$$

Use this fact to show that if $x > 4$, then

$$\sqrt{x} < 1 + \frac{x}{4}$$

49. Show that if an object moves along a straight line in such a way that its velocity is the same at two different times (that is, for a differentiable function v, we are given $v(t_1) = v(t_2)$ for $t_1 \neq t_2$), then there is some intermediate time when the acceleration is zero.

50. Modeling Problem Two radar patrol cars are located at fixed positions 6 mi apart on a long, straight road where the speed limit is 55 mi/h. A sports car passes the first patrol car traveling at 53 mi/h, and then 5 min later, it passes the second patrol car going 48 mi/h. Analyze a model of this situation to show that at some time between the two clockings, the sports car exceeded the speed limit. *Hint*: Use the MVT.

51. Modeling Problem Suppose two race cars begin at the same time and finish at the same time. Analyze a model to show that at some point in the race they had the same speed.

52. Use Rolle's theorem with

$$f(x) = (x - 1) \sin x$$

to show that the equation $\tan x = 1 - x$ has at least one solution for $0 < x < 1$.

53. Use the MVT to show that

$$\sqrt{1 + x} - 4 < \tfrac{1}{8}(x - 15)$$

if $x > 15$. *Hint*: Let $f(x) = \sqrt{1 + x}$.

54. Use the MVT to show that

$$\frac{1}{2x + 1} > \frac{1}{5} + \frac{2}{25}(2 - x)$$

if $0 \le x \le 2$.

Ⓒ 55. Let $f(x) = \tan x$. Note that

$$f(\pi) = f(0) = 0$$

Show that there is no number w between 0 and π for which $f'(w) = 0$. Why does this fact not contradict the MVT?

56. Use Rolle's theorem or the MVT to show that there is no number a for which the equation

$$x^3 - 3x + a = 0$$

has *two* distinct solutions in the interval $[-1, 1]$.

57. If $a > 0$ is a constant, show that the equation

$$x^3 + ax - 1 = 0$$

has exactly one real solution. *Hint*: Let $f(x) = x^3 + ax - 1$ and use the intermediate value theorem to show that there is at least one root. Then assume there are two roots, and use Rolle's theorem to obtain a contradiction.

58. For constants a and b, $a > 0$, and n a positive integer, use Rolle's theorem or the MVT to show that the polynomial

$$p(x) = x^{2n+1} + ax + b$$

can have at most one real root.

59. Show that if $f''(x) = 0$ for all x, then f is a linear function. (That is, $f(x) = Ax + B$ for constants $A \neq 0$ and B.)

60. Show that if $f'(x) = Ax + B$ for constants $A \neq 0$ and B, then $f(x)$ is a quadratic function. (That is, $f(x) = ax^2 + bx + c$ for constants a, b, and c, where $a \neq 0$.)

4.3 Using Derivatives to Sketch the Graph of a Function

IN THIS SECTION increasing and decreasing functions, the first-derivative test, concavity and inflection points, the second-derivative test, curve sketching using the first and second derivatives

Our next goal is to see how information about the derivative f' and the second derivative f'' can be used to determine the shape of the graph of f. We begin by showing how the sign of f' is related to whether the graph of f is rising or falling.

INCREASING AND DECREASING FUNCTIONS

Suppose an ecologist has modeled the size of a population of a certain species as a function f of time t (months). If it turns out that the population is increasing until the end of the first year and is decreasing thereafter, it is reasonable to expect the population to be maximized at time $t = 12$ months and for the population curve to have a high point at $t = 12$, as shown in Figure 4.15. If the graph of a function f, such as this population curve, is rising throughout the interval $0 < x < 12$, we say that f is *strictly increasing* on that interval. Similarly, the graph of the function in Figure 4.15 is *strictly decreasing* on the interval $12 < t < 20$. These terms may be defined more formally as follows:

Strictly Increasing and Strictly Decreasing Functions

The function f is **strictly increasing** on an interval I if

$$f(x_1) < f(x_2) \quad \text{whenever} \quad x_1 < x_2$$

for x_1 and x_2 on I. Likewise, f is **strictly decreasing** on I if

$$f(x_1) > f(x_2) \quad \text{whenever} \quad x_1 < x_2$$

for x_1 and x_2 on I. (See Figure 4.16.)

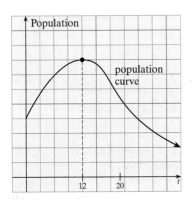

Figure 4.15 A population curve

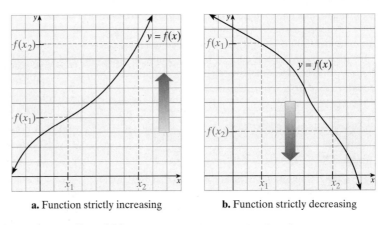

a. Function strictly increasing   **b.** Function strictly decreasing

Figure 4.16 Increasing and decreasing functions

Note that we do not use the word "strictly" every time we talk about a strictly monotonic function, or a strictly increasing function, or a strictly decreasing function.

A function f is said to be (strictly) **monotonic** on an interval I if it is either strictly increasing on all of I or strictly decreasing on all of I. The monotonic behavior is closely related to the sign of the derivative $f'(x)$. In particular, if the graph of a function has tangent lines with positive slope on I, the graph will be inclined upward and f will be increasing on I (see Figure 4.17). Since the slope of the tangent at each

WARNING   A note on usage: We say that functions are **increasing** and that graphs are **rising**. For example, if $f(x) = x^2$, we say that the function f is increasing for $x > 0$ and that the graph is rising for $x > 0$.

Figure 4.17 The graph is rising where $f' > 0$ and falling where $f' < 0$. Notice that the small flags indicate the slope at various points on the graph.

point on the graph is measured by the derivative f', it is reasonable to expect f to be increasing on intervals where $f' > 0$. Similarly, it is reasonable to expect f to be decreasing on an interval when $f' < 0$. These observations are established formally in Theorem 4.7.

THEOREM 4.7   Monotone function theorem

Let f be differentiable on the open interval (a, b).

   If $f'(x) > 0$ on (a, b), then f is strictly increasing on (a, b).

   If $f'(x) < 0$ on (a, b), then f is strictly decreasing on (a, b).

Proof   We will prove that f is strictly increasing on (a, b) if $f'(x) > 0$ throughout the interval. The strictly decreasing case is similar and is left as an exercise for the reader.

Suppose $f'(x) > 0$ throughout the interval (a, b), and let x_1 and x_2 be two numbers chosen arbitrarily from this interval, with $x_1 < x_2$. The MVT tells us that

$$\frac{f(x_2) - f(x_1)}{x_2 - x_1} = f'(c) \qquad \text{or} \qquad f(x_2) - f(x_1) = f'(c)(x_2 - x_1)$$

for some number c between x_1 and x_2. Because both $f'(c) > 0$ and $x_2 - x_1 > 0$, it follows that $f'(c)(x_2 - x_1) > 0$, and therefore

$$f(x_2) - f(x_1) > 0 \qquad \text{or} \qquad f(x_2) > f(x_1)$$

That is, if x_1 and x_2 are any two numbers in (a, b) such that $x_1 < x_2$, then $f(x_1) < f(x_2)$, which means that f is strictly increasing on (a, b).   ❏

To determine where a function f is increasing or decreasing, we begin by finding the critical numbers (where the derivative is zero or does not exist). These numbers divide the x-axis into intervals, and we test the sign of $f'(x)$ in each of these intervals. If $f'(x) > 0$ in an interval, then f is increasing in that same interval, and if $f'(x) < 0$ in an interval, then f is decreasing in that same interval.

To indicate where a given function f is increasing and where it is decreasing, we will mark the critical values on a number line and use an up arrow ($\uparrow$) to indicate an interval where f is increasing and a down arrow ($\downarrow$) to indicate an interval where f is decreasing. Sometimes, when the full graph of f is displayed, we indicate the sign of f' on an interval bounded by critical numbers with a string of $+$ signs if f is increasing on the interval or a string of $-$ signs if f is decreasing there. This notation is illustrated in the following example.

EXAMPLE 1   Finding intervals of increase and decrease

Determine where the function defined by $f(x) = x^3 - 3x^2 - 9x + 1$ is strictly increasing and where it is strictly decreasing.

Solution

First, we find the derivative:

$$f'(x) = 3x^2 - 6x - 9 = 3(x + 1)(x - 3)$$

Next, we determine the critical numbers: $f'(x)$ exists for all x and $f'(x) = 0$ at $x = -1$ and $x = 3$. These critical numbers divide the x-axis into three parts, as shown in Figure 4.18a, and we select any arbitrary number from each of these intervals. For example, we select -2, 0, and 4, evaluate the derivative at these numbers, and mark each interval as increasing ($\uparrow$) or decreasing ($\downarrow$), according to whether the derivative is positive or negative, respectively. This is shown in Figure 4.18b. The function f is increasing for $x < -1$ and for $x > 3$; f is decreasing for $-1 < x < 3$.   ∎

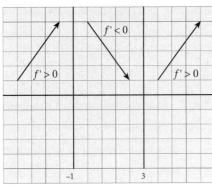

a. Intervals where f is increasing and where it is decreasing

b. The graph of f along with the sign graphs for f and f'

Figure 4.18 $f(x) = x^3 - 3x^2 - 9x + 1$

Figure 4.19 Graphs of f and f'

EXAMPLE 2   Comparing the graphs of a function and its derivative

Graph $f(x) = x^3 - 3x^2 - 9x + 1$ and $f'(x) = 3x^2 - 6x - 9$ and compare.

a. When $f' > 0$, what can be said about the graph of f?
b. When the graph of f is falling, what can be said about the graph of f'?
c. Where do the critical numbers of f appear on the graph of f?

Solution

The graphs of f and f' are shown in Figure 4.19.

a. When $f'(x) > 0$, the graph of f is rising. This occurs when $x < -1$ and when $x > 3$.
b. When the graph of f is falling (for $-1 < x < 3$), we have $f'(x) < 0$, so the graph of f' is below the x-axis.
c. The critical numbers of f are where $f'(x) = 0$; that is, at $x = -1$ and $x = 3$, so they are the x-intercepts of the graph of f'. ∎

THE FIRST-DERIVATIVE TEST

Every relative extremum is a critical point. However, as we saw in Section 4.1, not every critical point of a continuous function is necessarily a relative extremum. If the derivative is positive to the immediate left of a critical number and negative to its immediate right, the graph changes from increasing to decreasing and the critical point must be a relative maximum, as shown in Figure 4.20a. If the derivative is negative to the immediate left of a critical number and positive to its immediate right, the graph changes from decreasing to increasing and the critical point is a relative minimum (Figure 4.20b). However, if the sign of the derivative is the same on both immediate sides of the critical number, then it is neither a relative maximum nor a relative minimum (Figure 4.20c). These observations are summarized in a procedure called the *first-derivative test for relative extrema*.

a. A relative maximum

b. A relative minimum

c. No extremum

Figure 4.20   Three patterns of behavior near a critical number

The First-Derivative Test for Relative Extrema

Step 1.   Find all critical numbers of a continuous function f. That is, find all numbers c such that $f(c)$ is defined and either $f'(c) = 0$ or $f'(c)$ does not exist.

Step 2.   Classify each critical point $(c, f(c))$ as follows:

a. The point $(c, f(c))$ is a **relative maximum** if

$f'(x) > 0$ (graph rising) for all x in an open interval (a, c) to the left of c, and

$f'(x) < 0$ (graph falling) for all x in an open interval (c, b) to the right of c.

b. The point $(c, f(c))$ is a **relative minimum** if

$f'(x) < 0$ (graph falling) for all x in an open interval (a, c) to the left of c, and

$f'(x) > 0$ (graph rising) for all x in an open interval (c, b) to the right of c.

or

c. The point $(c, f(c))$ is **not an extremum** if the derivative $f'(x)$ has the same sign for all x in open intervals (a, c) and (c, b) on each side of c.

Suppose we apply this first-derivative test to the polynomial $f(x) = x^3 - 3x^2 - 9x + 1$ In Example 2 we found that this function has the critical numbers -1 and 3 and that f is increasing when $x < -1$ and $x > 3$ and decreasing when $-1 < x < 3$ (see the arrow pattern above). The first-derivative test tells us there is a relative maximum at -1 ($\uparrow \downarrow$ pattern) and a relative minimum at 3 ($\downarrow \uparrow$ pattern).

EXAMPLE 3   Relative extrema using the first-derivative test

Find all critical numbers of $g(t) = t - 2 \sin t$ for $0 \le t \le 2\pi$, and determine whether each corresponds to a relative maximum, a relative minimum, or neither. Sketch the graph of g.

Solution   Because $g'(t) = 1 - 2 \cos t$ exists for all t, the only critical numbers occur when $g'(t) = 0$; that is, when $\cos t = \frac{1}{2}$. Solving, we find that the critical numbers for $g(t)$ on the interval $[0, 2\pi]$ are $\frac{\pi}{3}$ and $\frac{5\pi}{3}$.

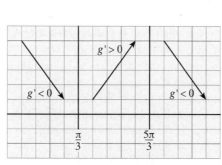

a. Intervals where $g(t) = t - 2 \sin t$ is increasing or decreasing

b. The graph of $g(t) = t - 2 \sin t$ for $0 \le t \le 2\pi$

Figure 4.21   The first-derivative test for $g(t) = t - 2 \sin t$

Next, we examine the sign of $g'(t)$. Thanks to the intermediate value theorem, because $g'(t)$ is continuous, it is enough to check the sign of $g'(t)$ at convenient numbers on each side of the critical numbers, as shown in Figure 4.21a. Notice that the arrows show the increasing and decreasing pattern for g. According to the first-derivative test, there is a relative minimum at $\frac{\pi}{3}$ and a relative maximum at $\frac{5\pi}{3}$. The graph of g is shown in Figure 4.21b.   ■

CONCAVITY AND INFLECTION POINTS

Knowing where a given graph is rising and falling gives only a partial picture of the graph. For example, suppose we wish to sketch the graph of $f(x) = x^3 + 3x + 1$. The derivative $f'(x) = 3x^2 + 3$ is positive for all x so the graph is always rising. But in what *way* is it rising? Each of the graphs in Figure 4.22 is a possible graph of f, but they are quite different from one another.

        a   b   c

Figure 4.22   Which curve is the graph of $f(x) = x^3 + 3x + 1$?

Figure 4.22 shows three possibilities for the graph of f. All three are always rising, but they differ in the way they "bend" as they rise. The "bending" of a curve is measured by its *concavity*. If a curve lies above its tangent lines on some interval (a, b), then we say the curve is *concave up*, and if the curve lies below its tangent lines, we say it is *concave down* on (a, b). We will examine concavity, and then return in Example 4 to determine which of the three candidates in Figure 4.22 is indeed the graph of $f(x) = x^3 + 3x + 1$.

Concavity

> If the graph of a function f lies above all of its tangents on an interval I, then it is said to be **concave up** on I. If the graph of f lies below all of its tangents on I, it is said to be **concave down**.

Figure 4.23 The slope of a graph increases or decreases, depending on its concavity.

To be specific, a portion of a graph that is cupped upward is called *concave up*, and a portion that is cupped downward is *concave down*. Figure 4.23 shows a graph that is concave up between A and C and concave down between C and E. At various points on the graph, the slope is indicated by "flags," and we observe that the slope increases from A to C and decreases from C to E. This is no accident! *The slope of a graph increases on an interval where the graph is concave up and decreases where the graph is concave down.*

Conversely, a graph will be concave up on any interval where the slope is increasing and concave down where the slope is decreasing. Because the slope is found by computing the derivative, it is reasonable to expect the graph of a given function f to be concave up where the derivative f' is strictly increasing. According to the monotone function theorem (Theorem 4.7), this occurs when $(f')' > 0$, which means that the graph of f is concave up where the *second derivative* f'' satisfies $f'' > 0$. Similarly, the graph is concave down where $f'' < 0$. We use this observation to characterize concavity.

Derivative Characterization of Concavity

The graph of a function f is **concave up** on any open interval I where $f''(x) > 0$, and it is **concave down** where $f''(x) < 0$.

When discussing the concavity of a function f, we will display a diagram in which a "cup" symbol ($\cup$) above an interval indicates that f is concave up on the interval and a "cap" symbol ($\cap$) indicates that f is concave down there. This convention is illustrated in the following example.

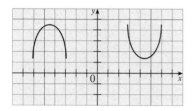

Figure 4.24 Intervals of concavity

EXAMPLE 4   Concavity for a polynomial function

Find where the graph of $f(x) = x^3 + 3x + 1$ is concave up and where it is concave down.

Solution

We find that $f'(x) = 3x^2 + 3$ and $f''(x) = 6x$. Therefore, $f''(x) < 0$ if $x < 0$ and $f''(x) > 0$ if $x > 0$, so the graph of f is concave down for $x < 0$ and concave up for $x > 0$, as indicated in Figure 4.24. Returning to Figure 4.22c, the graph is concave down to the left of $x = 0$ and concave up to the right. Hence, this is the graph of $f(x)$. ∎

In Figure 4.23, notice that the graph changes from concave up to concave down at the point C. It will be convenient to have a special name for such transition points.

Inflection Point

WARNING   An inflection point must be on the graph, meaning $f(c)$ must be defined if there is an inflection point at $x = c$.

A point $P(c, f(c))$ on a curve is called an **inflection point** if the graph is concave up on one side of P and concave down on the other side.

Returning to Example 4, notice that the graph of $f(x) = x^3 + 3x + 1$ has exactly one inflection point, at $(0, 1)$, where the concavity changes from down to up.

Various kinds of graphical behavior are illustrated in Figure 4.25. Note that the graph is rising on the interval $[a, c_1]$, falling on $[c_1, c_2]$, rising on $[c_2, c_3]$, falling on $[c_3, c_4]$, rising on $[c_4, c_5]$, and falling on $[c_5, b]$. The concavity is up on $[p_1, p_2]$, and down otherwise. In this figure, and elsewhere, when the graph of f is displayed, we will indicate the sign of f'' on an interval by a string of $+$ signs if $f'' > 0$ on the interval, and a string of $-$ signs if $f'' < 0$ there.

The graph has relative maxima at c_1, c_3, and c_5, and relative minima at c_2 and c_4. There are horizontal tangents ($f'(x) = 0$) at all of these points except at c_1 and c_3, where there are sharp points, called *corners* ($f'(c_1)$ and $f'(c_3)$ do not exist). There is a horizontal tangent at p_1; that is, $f'(p_1) = 0$, but no relative extremum appears there. Instead, we have points of inflection at p_1 and p_2, because the concavity changes direction at each of these points.

In general, the concavity of the graph of f will change only at points where $f''(x) = 0$ or $f''(x)$ does not exist—that is, at critical numbers of the derivative $f'(x)$. We will call the number c a **second-order critical number** if $f''(c) = 0$ or $f''(c)$ does not exist, and in this context an "ordinary" critical number (where $f'(c) = 0$ or $f'(c)$ does not exist) will be referred to as a **first-order critical number.** If we do not specify otherwise, a critical number is always a first-order critical number. Inflection points correspond to second-order critical numbers and must actually be on the graph

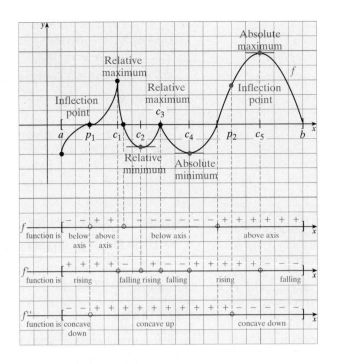

Figure 4.25 A graph of a function showing critical points and inflection points

of f. Specifically, a number c such that $f''(c)$ is not defined and the concavity of f changes at c will correspond to an inflection point if and only if $f(c)$ is defined.

Here is an example of one way inflection points may appear in applications.

EXAMPLE 5   Peak worker efficiency

An efficiency study of the morning shift at a factory indicates that the number of units produced by an average worker t hours after 8:00 A.M. may be modeled by the formula $Q(t) = -t^3 + 9t^2 + 12t$. At what time in the morning is the worker performing most efficiently?

Solution

We assume that the morning shift runs from 8:00 A.M. until noon and that worker efficiency is maximized when the rate of production

$$R(t) = Q'(t) = -3t^2 + 18t + 12$$

is as large as possible for $0 \leq t \leq 4$. The derivative of R is

$$R'(t) = Q''(t) = -6t + 18$$

which is zero when $t = 3$; this is the critical number. Using the optimization criterion of Section 4.1, we know that the extrema of $R(t)$ on the closed interval $[0, 4]$ must occur at either the interior critical number 3 or at one (or both) of the endpoints (which are 0 and 4). We find that

$$R(0) = 12 \quad R(3) = 39 \quad R(4) = 36$$

so the rate of production $R(t)$ is greatest and the worker is performing most efficiently when $t = 3$; that is, at 11:00 A.M. The graphs of the production function Q and its derivative, the rate-of-production function R, are shown in Figure 4.27. Notice that the production curve is steepest and the rate of production is greatest when $t = 3$.

Figure 4.26 The graph of $f(x) = x^4$ has no inflection point at $(0, 0)$ even though $f''(0) = 0$.

WARNING   A continuous function f need not have an inflection point at every number c where $f''(c) = 0$. For instance, if $f(x) = x^4$, we have $f''(0) = 0$, but the graph of f is always concave up (see Figure 4.26).

a. A production curve   **b.** Rate of production

Figure 4.27   Graph of a production curve showing the point of diminishing returns   ■

In Example 5, note how the rate of production, as measured by the slope of the graph of the average worker's output, increases from 0 to the inflection point I and then decreases from I to E, as shown in Figure 4.27a. Because the point I marks the point where the rate of production "peaks out," it is natural to refer to I as a point of **diminishing returns.** It is also an inflection point on the graph of Q. Knowing that this point occurs at 11:00 A.M., the manager of the factory might be able to increase the overall output of the labor force by scheduling a break near this time, since the production rate changes from increasing to decreasing at this time.

THE SECOND-DERIVATIVE TEST

It is often possible to classify a critical point $P(c, f(c))$ on the graph of f by examining the sign of $f''(c)$. Specifically, suppose $f'(c) = 0$ and $f''(c) > 0$. Then there is a horizontal tangent line at P and the graph of f is concave up in the neighborhood of P. This means that the graph of f is cupped upward from the horizontal tangent at P, and it is reasonable to expect P to be a relative minimum, as shown in Figure 4.28a. Similarly, we expect P to be a relative maximum if $f'(c) = 0$ and $f''(c) < 0$, because the graph is cupped down beneath the critical point P, as shown in Figure 4.28b.

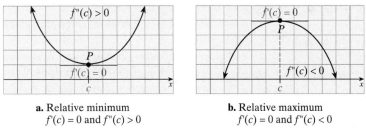

a. Relative minimum   **b.** Relative maximum
$f'(c) = 0$ and $f''(c) > 0$ $f'(c) = 0$ and $f''(c) < 0$

Figure 4.28   Second-derivative test for relative extrema

These observations lead to the *second-derivative test* for relative extrema.

The Second-Derivative Test for Relative Extrema

Let f be a function such that $f'(c) = 0$ and the second derivative exists on an open interval containing c.

If $f''(c) > 0$, there is a **relative minimum** at $x = c$.

If $f''(c) < 0$, there is a **relative maximum** at $x = c$.

If $f''(c) = 0$, then the second-derivative test **fails** (either a maximum, or a minimum, or neither may occur).

EXAMPLE 6   Using the second-derivative test

Use the second-derivative test to determine whether each critical number of the function $f(x) = 3x^5 - 5x^3 + 2$ corresponds to a relative maximum, a relative minimum, or neither.

Solution

Once again, we begin by finding the first and second derivatives:

$$f'(x) = 15x^4 - 15x^2 = 15x^2(x - 1)(x + 1)$$
$$f''(x) = 60x^3 - 30x = 30x(2x^2 - 1)$$

Solving $f'(x) = 0$, we find that the critical numbers are $x = 0$, $x = 1$, and $x = -1$. We plot these points on a number line, as shown in Figure 4.29. To apply the second-derivative test, we evaluate $f''(x)$ at each critical number.

$f''(0) = 0$; test fails at $x = 0$.

$f''(1) = 30 > 0$; positive, so the test tells us that there is a relative minimum at $x = 1$.

$f''(-1) = -30 < 0$; negative, so the test tells us that there is a relative maximum at $x = -1$.   ∎

When the second-derivative test fails, as at $x = 0$ in Example 6, the critical point can often be classified using the first-derivative test. For instance, the first derivative in Example 6 is

$$f'(x) = 15x^2(x - 1)(x + 1)$$

We see $f'(x) = 0$ at $x = -1$, 0, and 1. We can plot these points on a number line, shown in Figure 4.29, and then evaluate $f'(x)$ at test numbers just to the left and just to the right of each critical number.

Figure 4.29   Sign graph for the derivative of $f(x) = 3x^5 - 5x^3 + 2$

Direction:

$15x^2(x–1)(x+1)$;   Negative 0 Negative

Figure 4.30 The graph of $f(x) = 3x^5 - 5x^3 + 2$ is falling on both sides of $x = 0$.

We show a number line like this as part of the graph of many of our examples (as we did in Figure 4.25), but most often we show the derivative to the right and left of a critical number. For Example 6, the derivative is negative both to the immediate left and right of 0, which we illustrate as shown in Figure 4.30. Neither kind of extremum occurs at $x = 0$ ($\downarrow \downarrow$ pattern).

EXAMPLE 7   Finding inflection points

Find the inflection points for the function $f(x) = 3x^5 - 5x^3 + 2$, given in Example 6.

Solution

We begin with the second derivative (from Example 6):

$$f''(x) = 30x(2x^2 - 1)$$

To find the inflection points, we look at the second-order critical numbers, namely, where $f''(x) = 0$; that is, at $x = 0$ and $x = \pm\sqrt{\frac{1}{2}} = \pm\frac{\sqrt{2}}{2}$. We can show these on a sign graph of the second derivative, as in Figure 4.31. Notice that the inflection points occur where the sign of the second derivative changes.

The graph of f is shown in Figure 4.32, along with the sign graphs for both the first and second derivatives. Spend some time studying this figure to see how the signs of the first derivative indicate where the graph is rising and where it is falling, as well

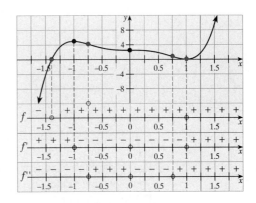

Figure 4.31   Sign graph for the second derivative of $f(x) = 3x^5 - 5x^3 + 2$

as how the signs of the second derivative show the concavity. For completeness, the figure also shows the sign graph for the function itself: The graph is above the x-axis where f is positive, below the x-axis where f is negative, and crosses the x-axis where $f(x) = 0$.

Figure 4.32   Graph of $f(x) = 3x^5 - 5x^3 + 2$ ■

Example 7 demonstrates both the strength and the weakness of the second-derivative test. In particular, when it is relatively easy to find the second derivative (as with a polynomial) and if the zeros of this function are easy to find, then the second-derivative test provides a quick means for classifying the critical points. However, if it is difficult to compute $f''(c)$ or if $f''(c) = 0$, it may be easier, or even necessary, to apply the first-derivative test.

CURVE SKETCHING USING THE FIRST AND SECOND DERIVATIVES

Now, we are ready to develop a procedure for curve sketching. The key ideas appear in Example 8.

EXAMPLE 8   Sketching the graph of a polynomial function

Determine where the function $f(x) = x^4 - 4x^3 + 10$ is increasing, where it is decreasing, where its graph is concave up, and where its graph is concave down. Find the relative extrema and inflection points, and sketch the graph of f.

Solution

The first derivative,

$$f'(x) = 4x^3 - 12x^2 = 4x^2(x - 3)$$

is zero when $x = 0$ and when $x = 3$. Because $4x^2 > 0$ for $x \neq 0$, we have $f'(x) < 0$ for $x < 3$ (except for $x = 0$) and $f'(x) > 0$ for $x > 3$. The pattern showing where f is increasing and where it is decreasing is displayed in Figure 4.33a.

Next, to determine the concavity of the graph we compute

$$f''(x) = 12x^2 - 24x = 12x(x - 2)$$

If $x < 0$ or $x > 2$, then $f''(x) > 0$ and the graph is concave up. It is concave down when $0 < x < 2$, because $f''(x) < 0$ on this interval. The concavity of the graph of f is shown in Figure 4.33b.

Direction:

$4x^2(x-3)$: Negative 0 Negative 3 Positive

a. First derivative signs.

Shape:

$12x(x-2)$: Positive 0 Negative 2 Positive

b. Second derivative signs.

Figure 4.33 Intervals of increase and decrease and concavity for $f(x) = x^4 - 4x^3 + 10$

The two diagrams in Figure 4.33 tell us that there is a relative minimum at $x = 3$ and inflection points at $x = 0$ and $x = 2$ (because the second derivative changes sign at these points).

To find the y-values of the critical points and the inflection points, evaluate f at $x = 0, 2,$ and 3:

$$f(0) = (0)^4 - 4(0)^3 + 10 = 10$$
$$f(2) = (2)^4 - 4(2)^3 + 10 = -6$$
$$f(3) = (3)^4 - 4(3)^3 + 10 = -17$$

Finally, to sketch the graph of f, we first place a "cup" ($\cup$) at the minimum point $(3, -17)$ and note that $(0, 10)$ and $(2, -6)$ are inflection points. Remember there is also a horizontal tangent at $(0, 10)$. The preliminary graph is shown in Figure 4.34a. Complete the sketch by passing a smooth curve through these points, using the two diagrams in Figure 4.33 as a guide for determining where the graph is rising and falling and where it is concave up and down. The completed graph is shown in Figure 4.34b.

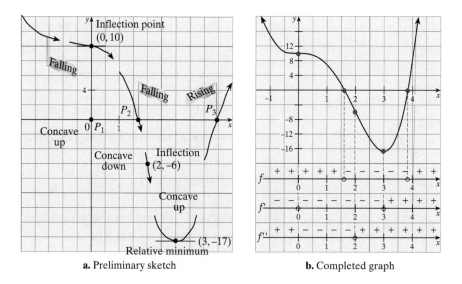

a. Preliminary sketch

b. Completed graph

Figure 4.34 Graphing $f(x) = x^4 - 4x^3 + 10$ ∎

EXAMPLE 9   Sketching the graph of an exponential function

Determine where the function

$$f(x) = \frac{1}{\sqrt{2\pi}} e^{-x^2/2}$$

is increasing, decreasing, concave up, and concave down. Find the relative extrema and inflection points and sketch the graph. This function plays an important role in statistics, where it is called the *standard normal density function*.

a. Intervals of increase and decrease for $f(x)$

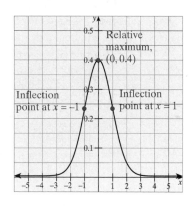

b. Concavity for $f(x)$

Figure 4.35 Intervals of increase, decrease, and concavity for $f(x) = \dfrac{1}{\sqrt{2\pi}} e^{-x^2/2}$

Solution

The first derivative is

$$f'(x) = \frac{-x}{\sqrt{2\pi}} e^{-x^2/2}$$

Because $e^{-x^2/2}$ is always positive, $f'(x) = 0$ if and only if $x = 0$. Hence, the corresponding point

$$\left(0, \frac{1}{\sqrt{2\pi}}\right) \approx (0, 0.4)$$

is the only critical point. Checking the sign of f' on each side of 0, we find that f is increasing for $x < 0$ and decreasing for $x > 0$, so there is a relative maximum at $x = 0$, as indicated in Figure 4.35a.

We find that the second derivative of f is

$$f''(x) = \frac{x^2}{\sqrt{2\pi}} e^{-x^2/2} - \frac{1}{\sqrt{2\pi}} e^{-x^2/2} = \frac{1}{\sqrt{2\pi}}(x^2 - 1)e^{-x^2/2}$$

which is zero when $x = \pm 1$. We find that $f(1) = f(-1) \approx 0.24$, and that the concavity of the graph of f is as indicated in Figure 4.35b.

Finally, we draw a smooth curve through the known points, as shown in Figure 4.36. The graph of f rises to the high point at approximately $(0, 0.4)$ and then falls indefinitely, approaching the x-axis arbitrarily closely because $e^{-x^2/2}$ approaches 0 as $|x|$ increases without bound. Note that the graph has no x-intercepts, because $e^{-x^2/2}$ is always positive.

Figure 4.36 Graph of $f(x) = \dfrac{1}{\sqrt{2\pi}} e^{-x^2/2}$ ∎

EXAMPLE 10 Sketching the graph of a trigonometric function

Sketch the graph of $T(x) = \sin x + \cos x$ on $[0, 2\pi]$.

Solution

You probably graphed this function in trigonometry by adding ordinates. However, with this example we wish to illustrate the power of calculus to draw the graph. Thus, we begin by finding the first and second derivatives.

$$T'(x) = \cos x - \sin x \qquad\qquad T''(x) = -\sin x - \cos x$$

We find the critical numbers (both T' and T'' are defined for all values of x):

$T'(x) = 0$ when $\cos x = \sin x$; thus, $x = \frac{\pi}{4}$ and $x = \frac{5\pi}{4}$.

$T''(x) = 0$ when $\cos x = -\sin x$; thus, $x = \frac{3\pi}{4}$ and $x = \frac{7\pi}{4}$.

The intervals of increase and decrease as well as the concavity pattern are shown in Figure 4.37.

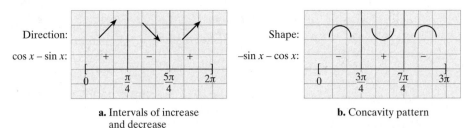

Direction:

$\cos x - \sin x$:   +   −   +

0 $\frac{\pi}{4}$ $\frac{5\pi}{4}$ 2π

a. Intervals of increase and decrease

Shape:

$-\sin x - \cos x$:   −   +   −

0 $\frac{3\pi}{4}$ $\frac{7\pi}{4}$ 3π

b. Concavity pattern

Figure 4.37   Preliminary work for sketching $T(x) = \sin x + \cos x$

Find the critical points, the points of inflection, and the endpoints:

Relative maximum: $T\left(\frac{\pi}{4}\right) = \sqrt{2}$; the critical point is $\left(\frac{\pi}{4}, \sqrt{2}\right)$.

Relative minimum: $T\left(\frac{5\pi}{4}\right) = -\sqrt{2}$; the critical point is $\left(\frac{5\pi}{4}, -\sqrt{2}\right)$.

Inflection: $T\left(\frac{3\pi}{4}\right) = 0$; the inflection point is $\left(\frac{3\pi}{4}, 0\right)$.

$T\left(\frac{7\pi}{4}\right) = 0$; the inflection point is $\left(\frac{7\pi}{4}, 0\right)$.

Endpoints: $T(0) = 1$ and $T(2\pi) = 1$.

Finally, pass a smooth curve through these key points as indicated in the preliminary sketch in Figure 4.38a to obtain the completed graph shown in Figure 4.38b.

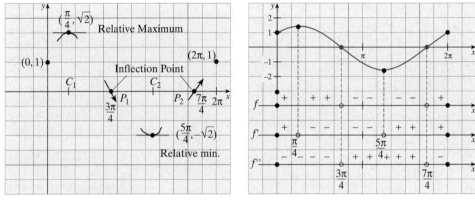

a. Preliminary sketch

b. Completed sketch

Figure 4.38   Graph of $T(x) = \sin x + \cos x$ on $[0, 2\pi]$ ■

4.3 PROBLEM SET

Ⓐ 1. **WHAT DOES THIS SAY?**   What is the first-derivative test?

2. **Exploration Problem**   What is the relationship between the graph of a function and the graph of its derivative?

3. **WHAT DOES THIS SAY?**   What is the second-derivative test?

4. **WHAT DOES THIS SAY?**   What is the relationship between concavity, points of inflection, and the second derivative?

5. **Exploration Problem**   The cartoon on page 209 exclaims, "Our prices are rising slower than any place in town." Restate using calculus.

In Problems 6–7, identify which curve represents a function f and which curve represents its derivative f'.

6.

7.

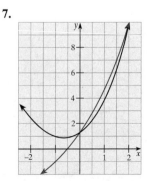

Draw a curve that represents the derivative of the function defined by the curves shown in Problems 8–11.

8.

9.

10.

11.

For the functions in Problems 12–19,

a. Find all critical numbers.

b. Find where the function is increasing and decreasing.

c. Find the critical points and identify each as a relative maximum, relative minimum, or neither.

d. Find the second-order critical numbers and tell where the graph is concave up and where it is concave down.

e. Sketch the graph.

12. $f(x) = \frac{1}{3}x^3 - 9x + 2$ **13.** $f(x) = x^3 + 3x^2 + 1$

14. $f(x) = x^5 + 5x^4 - 550x^3 - 2{,}000x^2 + 60{,}000x$

15. $f(x) = x^3 + 35x^2 - 125x - 9{,}375$

16. $f(x) = x + \dfrac{1}{x}$ **17.** $f(x) = \dfrac{x-1}{x^2+3}$

18. $f(x) = x \ln x$ **19.** $f(t) = (t+1)^2(t-5)$

In Problems 20–35, determine the intervals of increase and decrease and concavity for the given function, and then use those intervals to help you sketch its graph.

20. $f(x) = (x-12)^4 - 2(x-12)^3$

21. $f(x) = 1 + 2x + 18/x$

22. $f(u) = 3u^4 - 2u^3 - 12u^2 + 18u - 5$

23. $g(u) = u^4 + 6u^3 - 24u^2 + 26$

24. $f(x) = \sqrt{x^2+1}$ **25.** $g(t) = (t^3+t)^2$

26. $f(t) = (t^3 + 3t^2)^3$ **27.** $f(x) = \dfrac{x}{x^2+1}$

28. $f(t) = t - \ln t$ **29.** $f(t) = t^2 e^{-3t}$

30. $f(x) = e^x + e^{-x}$ **31.** $f(x) = (\ln x)^2$

32. $t(\theta) = \sin\theta - 2\cos\theta$ for $0 \le \theta \le 2\pi$

33. $t(\theta) = \theta + \cos 2\theta$ for $0 \le \theta \le \pi$

34. $f(x) = 2x - \sin^{-1} x$ for $-1 \le x \le 1$

35. $f(x) = x^3 + \sin x$ on $\left[-\frac{\pi}{2}, \frac{\pi}{2}\right]$

In Problems 36–39, use the first-derivative test to classify each of the given critical numbers as a relative minimum, a relative maximum, or neither.

36. $f(x) = (x^3 - 3x + 1)^7$ at $x = 1, x = -1$

37. $f(x) = \dfrac{e^{-x^2}}{3 - 2x}$ at $x = 1, x = \frac{1}{2}$

38. $f(x) = (x^2 - 4)^4(x^2 - 1)^3$ at $x = 1, x = 2$

39. $f(x) = \sqrt[3]{x^3 - 48x}$ at $x = 4$

In Problems 40–43, use the second-derivative test to classify each of the given critical numbers as a relative minimum, a relative maximum, or neither.

40. $f(x) = 2x^3 + 3x^2 - 12x + 11$ at $x = 1, x = -2$

41. $f(x) = \dfrac{x^2 - x + 5}{x + 4}$ at $x = -9, x = 1$

42. $f(x) = (x^2 - 3x + 1)e^{-x}$ at $x = 1, x = 4$

43. $f(x) = \sin x + \frac{1}{2}\cos 2x$ at $x = \frac{\pi}{6}, x = \frac{\pi}{2}$

B 44. Exploration Problem Sketch the graph of a function with the following properties:

$$f'(x) > 0 \quad \text{when} \quad x < -1$$
$$f'(x) > 0 \quad \text{when} \quad x > 3$$
$$f'(x) < 0 \quad \text{when} \quad -1 < x < 3$$
$$f''(x) < 0 \quad \text{when} \quad x < 2$$
$$f''(x) > 0 \quad \text{when} \quad x > 2$$

45. Exploration Problem Sketch the graph of a function with the following properties:

$$f'(x) > 0 \quad \text{when} \quad x < 2 \quad \text{and when} \quad 2 < x < 5$$
$$f'(x) < 0 \quad \text{when} \quad x > 5$$
$$f'(2) = 0$$
$$f''(x) < 0 \quad \text{when} \quad x < 2 \quad \text{and when} \quad 4 < x < 7$$
$$f''(x) > 0 \quad \text{when} \quad 2 < x < 4 \quad \text{and when} \quad x > 7$$

46. Exploration Problem Sketch the graph of a function with the following properties:

$$f'(x) > 0 \quad \text{when} \quad x < 1$$
$$f'(x) < 0 \quad \text{when} \quad x > 1$$
$$f''(x) > 0 \quad \text{when} \quad x < 1$$
$$f''(x) > 0 \quad \text{when} \quad x > 1$$

What can you say about the derivative of f when $x = 1$?

47. Sketch the graph of a function with the following properties: There are relative extrema at $(-1, 7)$ and $(3, 2)$. There is an inflection point at $(1, 4)$. The graph is concave down only when $x < 1$. The x-intercept is $(-4, 0)$ and the y-intercept is $(0, 5)$.

48. Sketch a graph of a function f that satisfies the following conditions:

(i) $f'(x) > 0$ when $x < -5$ and when $x > 1$.

(ii) $f'(x) < 0$ when $-5 < x < 1$.

(iii) $f(-5) = 4$ and $f(1) = -1$.

49. Sketch a graph of a function f that satisfies the following conditions:

(i) $f'(x) < 0$ when $x < -1$.

(ii) $f'(x) > 0$ when $-1 < x < 3$ and when $x > 3$.

(iii) $f'(-1) = 0$ and $f'(3) = 0$.

50. In physics, the formula

$$I(\theta) = I_0 \left(\frac{\sin \theta}{\theta} \right)^2 \quad \text{where} \quad I(0) = \lim_{\theta \to 0} I(\theta)$$

and I_0 is a constant, is used to model light intensity in the study of Fraunhofer diffraction.

a. Show that $I(0) = I_0$.

b. Sketch the graph for $[-3\pi, 3\pi]$. What are the critical points on this interval?

51. At a temperature of T (in degrees Celsius), the speed of sound in air is modeled by the formula

$$v = v_0 \sqrt{1 + \frac{1}{273} T}$$

where v_0 is the speed at $0°C$. Sketch the graph of v for $T > 0$, and use calculus to check for critical points.

Modeling Problems: *In Problems 52–53, set up an appropriate model to answer the given question. Be sure to state your assumptions.*

52. At noon on a certain day, Frank sets out to assemble five stereo sets. His rate of assembly increases steadily throughout the afternoon until 4:00 P.M., at which time he has completed three sets. After that, he assembles sets at a slower and slower rate until he finally completes the fifth set at 8:00 P.M. Sketch a rough graph of a function that represents the number of sets Frank has completed after t hours of work.

53. An industrial psychologist conducts two efficiency studies at the Chilco appliance factory. The first study indicates that the average worker who arrives on the job at 8:00 A.M. will have assembled

$$-t^3 + 6t^2 + 13t$$

blenders in t hours (without a break), for $0 \le t \le 4$. The second study suggests that after a 15-minute coffee break, the average worker can assemble

$$-\tfrac{1}{3}t^3 + \tfrac{1}{2}t^2 + 25t$$

blenders in t hours after the break for $0 < t \le 4$. *Note:* The 15-minute break is not part of the work time.

a. Verify that if the coffee break occurs at 10:00 A.M., the average worker will assemble 42 blenders before the break and $49\frac{1}{3}$ blenders for the two hours after the break.

b. Suppose the coffee break is scheduled to begin x hours after 8:00 A.M. Find an expression for the total number of blenders $N(x)$ assembled by the average worker during the morning shift (8 A.M. to 12:15 P.M.).

c. At what time should the coffee break be scheduled so that the average worker will produce the maximum number of blenders during the morning shift? How is this optimum time related to the point of diminishing returns?

54. Research indicates that the power P required by a bird to maintain flight is given by the formula

$$P = \frac{w^2}{2\rho S v} + \frac{1}{2} \rho A v^3$$

where v is the relative speed of the bird, w is its weight, ρ is the density of air, and S and A are constants associated

with the bird's size and shape.* What speed will minimize the power? You may assume that v, w, ρ, S, and A are all positive.

55. The deflection of a hardwood beam of length is given by

$$D(x) = \tfrac{9}{4}x^4 - 7\ell x^3 + 5\ell^2 x^2$$

where x is the distance from one end of the beam. What value of x yields the maximum deflection?

⊙ 56. HISTORICAL QUEST   One of the most famous women in the history of mathematics is Maria Gaëtana Agnesi (pronounced än yā´zē). She was born in Milan, the first of 21 children. Her first publication was at age 9, when she wrote a Latin discourse defending higher education for women. Her most important work was a now-classic calculus textbook published in 1748.

MARIA AGNESI
1718–1799

Maria Agnesi is primarily remembered for a curve defined by the equation

$$y = \frac{a^3}{x^2 + a^2}, \quad a \text{ a positive constant}$$

The curve was named *versiera* (from the Italian verb *to turn*) by Agnesi, but John Colson, an Englishman who translated her work, confused the word *versiera* with the word *avversiera*, which means "wife of the devil" in Italian; the curve has ever since been called the "witch of Agnesi." This was particularly unfortunate because Colson wanted Agnesi's work to serve as a model for budding young mathematicians, especially young women. ■

Graph this curve and find the critical numbers, extrema, and points of inflection.

57. Journal Problem: *Mathematics Magazine*[†] ■ Give an elementary proof that

$$f(x) = \frac{1}{\sin x} - \frac{1}{x}, \qquad 0 < x \le \frac{\pi}{2},$$

is positive and increasing.

58. An important formula in physical chemistry is *van der Waals' equation*, which says that

$$\left(P + \frac{a}{V^2} \right) (V - b) = nRT$$

where P, V, and T are the pressure, volume, and temperature, respectively, of a gas, and a, b, n, and R are positive constants. The *critical temperature* T_C of the gas is the highest temperature at which the gaseous and liquid phases can exist as separate states.

a. When $T = T_C$, the pressure P can be expressed as a function $P(V)$ of V alone. Show how this can be done, and then find $P'(V)$ and $P''(V)$.

---

*C. J. Pennycuick, "The Mechanics of Bird Migration," *Ibis* III (1969), pp. 525–556.

[†]Volume 55 (1982), p. 300. "Elementary proof" in the question means that you should use only techniques from beginning calculus. For our purposes, you simply need to give a reasonable argument to justify the conclusion.

b. The *critical volume* V_C is the volume that satisfies $P'(V_C) = 0$ and $P''(V_C) = 0$. Find V_C.

c. Find T_C, the point where $P''(V) = 0$, using the V_C from part **b** to write it in terms of a, b, n, and R. Finally, find the *critical pressure* $P_C = P(V_C)$ in terms of a, b, n, and R.

d. Sketch P as a function of V.

59. Prove or disprove that if the graphs of the functions f and g are both concave up on an interval, then the graph of their sum $f + g$ is also concave up on that interval.

60. Use calculus to prove that for constants a, b, and c, the vertex (relative extremum) of the quadratic function

$$y = ax^2 + bx + c \qquad (a \neq 0)$$

occurs at $x = -b/(2a)$.

61. Find constants A, B, and C that guarantee that the function

$$f(x) = Ax^3 + Bx^2 + C$$

will have a relative extremum at $(2, 11)$ and an inflection point at $(1, 5)$. Sketch the graph of f.

62. Find constants a, b, and c that guarantee that the graph of

$$f(x) = x^3 + ax^2 + bx + c$$

will have a relative maximum at $(-3, 18)$ and a relative minimum at $(1, -14)$.

63. Exploration Problem   Consider the graph of $y = x^3 + bx^2 + cx + d$ for constants b, c, and d. What happens to the graph as b changes?

64. Find constants A, B, C, and D that guarantee that the graph of

$$f(x) = 3x^4 + Ax^3 + Bx^2 + Cx + D$$

will have horizontal tangents at $(2, -3)$ and $(0, 7)$. There is a third point that has a horizontal tangent. Find this point. Then, for all three points, determine whether each corresponds to a relative maximum, a relative minimum, or neither.

4.4   Curve Sketching with Asymptotes: Limits Involving Infinity

IN THIS SECTION   limits to infinity, infinite limits, graphs with asymptotes, vertical tangents and cusps, a general graphing strategy

LIMITS TO INFINITY

In applications, we are often concerned with "long run" behavior of a function. To indicate such behavior, we write

$$\lim_{x \to +\infty} f(x) = L$$

to indicate that $f(x)$ approaches the number L as x increases without bound. Similarly, we write

$$\lim_{x \to -\infty} f(x) = M$$

to indicate that $f(x)$ approaches the number M as x decreases without bound. Here are the formal definitions of these *limits to infinity*.

Limits to Infinity

WARNING   Even though the symbols ∞ and $+\infty$ mean the same thing, for the time being we use $+\infty$ to help distinguish between $+\infty$ and $-\infty$.

> The limit statement $\lim\limits_{x \to +\infty} f(x) = L$ means that for any number $\epsilon > 0$, there exists a number N_1 such that
>
> $$|f(x) - L| < \epsilon \quad \text{whenever} \quad x > N_1$$
>
> for x in the domain of f. Similarly, $\lim\limits_{x \to -\infty} f(x) = M$ means that for any $\epsilon > 0$, there exists a number N_2 such that
>
> $$|f(x) - M| < \epsilon \quad \text{whenever} \quad x < N_2$$

This definition can be illustrated graphically, as shown in Figure 4.39.

$$\lim_{x \to +\infty} f(x) = L \qquad\qquad \lim_{x \to -\infty} f(x) = M$$

Figure 4.39   Graphical representation of limits to infinity

With this formal definition, we can show that all the rules for limits established in Chapter 2 also apply to $\lim\limits_{x \to +\infty} f(x)$ and $\lim\limits_{x \to -\infty} f(x)$.

Limit Rules

If $\lim\limits_{x \to +\infty} f(x)$ and $\lim\limits_{x \to +\infty} g(x)$ exist, then for constants a and b:

Power rule: $\lim\limits_{x \to +\infty} [f(x)]^n = \left[\lim\limits_{x \to +\infty} f(x) \right]^n$

Linearity rule: $\lim\limits_{x \to +\infty} [af(x) + bg(x)] = a \lim\limits_{x \to +\infty} f(x) + b \lim\limits_{x \to +\infty} g(x)$

Product rule: $\lim\limits_{x \to +\infty} [f(x)g(x)] = \left[\lim\limits_{x \to +\infty} f(x) \right]\left[\lim\limits_{x \to +\infty} g(x) \right]$

Quotient rule: $\lim\limits_{x \to +\infty} \dfrac{f(x)}{g(x)} = \dfrac{\lim\limits_{x \to +\infty} f(x)}{\lim\limits_{x \to +\infty} g(x)}$ if $\lim\limits_{x \to +\infty} g(x) \neq 0$

Analogous results hold for $\lim\limits_{x \to -\infty} f(x)$, if it exists.

The following theorem will allow us to evaluate certain limits to infinity with ease.

THEOREM 4.8   **Special limits to infinity**

If A is any real number and r is a positive rational number, then

$$\lim_{x \to +\infty} \frac{A}{x^r} = 0$$

Furthermore, if r is such that x^r is defined for $x < 0$, then

$$\lim_{x \to -\infty} \frac{A}{x^r} = 0$$

Proof   We begin by proving that $\lim\limits_{x \to +\infty} \dfrac{1}{x} = 0$. For $\epsilon > 0$, let $N = \dfrac{1}{\epsilon}$. Then for $x > N$ we have

$$x > N = \frac{1}{\epsilon} \qquad \text{so that} \qquad \frac{1}{x} < \epsilon$$

This means that $\left| \dfrac{1}{x} - 0 \right| < \epsilon$, so that, from the definition of limit we have $\lim\limits_{x \to +\infty} \dfrac{1}{x} = 0$. Now let r be a rational number, say $r = p/q$. Then

$$\lim_{x \to +\infty} \frac{A}{x^r} = \lim_{x \to +\infty} \frac{A}{x^{p/q}} = A \lim_{x \to +\infty} \left[\frac{1}{\sqrt[q]{x}} \right]^p$$

$$= A \left[\sqrt[q]{\lim_{x \to +\infty} \frac{1}{x}} \right]^p = A \left[\sqrt[q]{0} \right]^p = A \cdot 0 = 0$$

The proof for the analogous limit as $x \to -\infty$ follows similarly. ❑

When evaluating a limit of the form

$$\lim_{x \to +\infty} \frac{p(x)}{d(x)} \qquad \text{or} \qquad \lim_{x \to -\infty} \frac{p(x)}{d(x)}$$

where $p(x)$ and $d(x)$ are polynomials, it is often useful to divide both $p(x)$ and $d(x)$ by the highest power of x that occurs in either. The limit can then be found by applying Theorem 4.8. This process is illustrated by the following examples.

EXAMPLE 1   Evaluating a limit to infinity

Evaluate $\lim\limits_{x \to +\infty} \dfrac{3x^3 - 5x + 9}{5x^3 + 2x^2 - 7}$.

Solution

We may assume that $x \neq 0$, because we are interested only in very large values of x. Dividing both the numerator and denominator of the given expressions by x^3, the highest power of x appearing in the fraction, we find

$$\frac{3x^3 - 5x + 9}{5x^3 + 2x^2 - 7} = \frac{3x^3 - 5x + 9}{5x^3 + 2x^2 - 7} \cdot \frac{\dfrac{1}{x^3}}{\dfrac{1}{x^3}} = \frac{3 - \dfrac{5}{x^2} + \dfrac{9}{x^3}}{5 + \dfrac{2}{x} - \dfrac{7}{x^3}}$$

Thus,

$$\lim_{x \to +\infty} \frac{3x^3 - 5x + 9}{5x^3 + 2x^2 - 7} = \lim_{x \to +\infty} \frac{3 - \dfrac{5}{x^2} + \dfrac{9}{x^3}}{5 + \dfrac{2}{x} - \dfrac{7}{x^3}}$$

$$= \frac{\lim\limits_{x \to +\infty} \left(3 - \dfrac{5}{x^2} + \dfrac{9}{x^3} \right)}{\lim\limits_{x \to +\infty} \left(5 + \dfrac{2}{x} - \dfrac{7}{x^3} \right)}$$

$$= \frac{3 - 0 + 0}{5 + 0 - 0} = \frac{3}{5}$$

Figure 4.40 Graph of
$$y = \frac{3x^3 - 5x + 9}{5x^3 + 2x^2 - 7}$$

The graph of the given rational function is shown in Figure 4.40. Notice how the curve seems to approach $y = \frac{3}{5}$ as $x \to +\infty$ and as $x \to -\infty$. ∎

Example 2 illustrates how Theorem 4.8 can be used along with the other limit properties to evaluate limits to infinity.

EXAMPLE 2   Evaluating limits to infinity

Evaluate

$$\lim_{x \to +\infty} \sqrt{\frac{3x - 5}{x - 2}} \qquad \text{and} \qquad \lim_{x \to -\infty} \left(\frac{3x - 5}{x - 2} \right)^3$$

Solution

Notice that for $x \neq 0$,

$$\frac{3x - 5}{x - 2} = \frac{x\left(3 - \dfrac{5}{x}\right)}{x\left(1 - \dfrac{2}{x}\right)} = \frac{3 - \dfrac{5}{x}}{1 - \dfrac{2}{x}}$$

Also, according to Theorem 4.8, we know that

$$\lim_{x \to +\infty} \frac{5}{x} = 0 \quad \text{and} \quad \lim_{x \to +\infty} \frac{2}{x} = 0$$

We now find the limits using the quotient rule, the power rule, and Theorem 4.8:

$$\lim_{x \to +\infty} \sqrt{\frac{3x - 5}{x - 2}} = \lim_{x \to +\infty} \left(\frac{3x - 5}{x - 2}\right)^{1/2} = \left(\lim_{x \to +\infty} \frac{3x - 5}{x - 2}\right)^{1/2}$$

$$= \left(\frac{\lim\limits_{x \to +\infty}\left(3 - \frac{5}{x}\right)}{\lim\limits_{x \to +\infty}\left(1 - \frac{2}{x}\right)}\right)^{1/2} = \left(\frac{3 - 0}{1 - 0}\right)^{1/2} = \sqrt{3}$$

Similarly,

$$\lim_{x \to -\infty} \left(\frac{3x - 5}{x - 2}\right)^3 = 3^3 = 27 \qquad \blacksquare$$

EXAMPLE 3 Evaluating a limit to negative infinity

Evaluate $\displaystyle\lim_{x \to -\infty} \frac{95x^3 + 57x + 30}{x^5 - 1,000}$.

Solution

Dividing the numerator and the denominator by the highest power, x^5, we find that

$$\lim_{x \to -\infty} \frac{95x^3 + 57x + 30}{x^5 - 1,000} = \lim_{x \to -\infty} \frac{95x^3 + 57x + 30}{x^5 - 1,000} \cdot \frac{\dfrac{1}{x^5}}{\dfrac{1}{x^5}}$$

$$= \lim_{x \to -\infty} \frac{\dfrac{95}{x^2} + \dfrac{57}{x^4} + \dfrac{30}{x^5}}{1 - \dfrac{1,000}{x^5}}$$

$$= \frac{0 + 0 + 0}{1 - 0}$$

$$= 0$$

The graph of the rational function is shown in Figure 4.41. $\qquad \blacksquare$

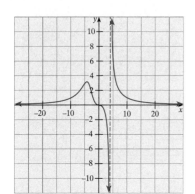

Figure 4.41 Graph of
$$y = \frac{95x^3 + 57x + 30}{x^5 - 1,000}$$

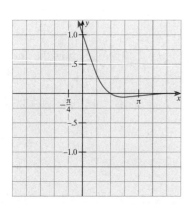

Figure 4.42 Graph of $y = e^{-x} \cos x$

EXAMPLE 4 Evaluating a limit to infinity involving e^x

Find $\displaystyle\lim_{x \to +\infty} e^{-x} \cos x$.

Solution

We cannot use the product rule for limits since $\displaystyle\lim_{x \to +\infty} \cos x$ does not exist (it diverges by oscillation—do you see why?). Note, however, that the magnitude of

$$e^{-x} \cos x = \frac{\cos x}{e^x}$$

must become smaller and smaller as $x \to +\infty$ since the numerator $\cos x$ is bounded between -1 and 1, while the denominator e^x grows relentlessly larger with x. Thus by the squeeze rule,

$$\lim_{x \to +\infty} e^{-x} \cos x = 0$$

The graph of $y = e^{-x} \cos x$ is shown in Figure 4.42. $\qquad \blacksquare$

Try drawing the graph of $y = e^{-x} \cos x$ on your calculator, or other graphing software, and you will see that what you get does not properly illustrate the oscillatory behavior of this graph. This demonstrates the importance of *not* relying on technology for graphing. In fact, the accompanying graph (without scale) shows more clearly the behavior of the graph of the function $y = e^{-x} \cos x$ for large, positive values of x.

INFINITE LIMITS

In mathematics, the symbol ∞ is not a number, but is used to describe either the process of unrestricted growth or the result of such growth. Thus, a limit statement such as

$$\lim_{x \to c} f(x) = +\infty$$

means that the function f increases without bound as x approaches c from either side, while

$$\lim_{x \to c} g(x) = -\infty$$

means that g decreases without bound as x approaches c. Such limits may be defined formally as follows.

Infinite Limits

We write $\lim\limits_{x \to c} f(x) = +\infty$ if for any number $N > 0$ (no matter how large), it is possible to find a number $\delta > 0$ such that $f(x) > N$ whenever $0 < |x - c| < \delta$. Similarly, $\lim\limits_{x \to c} f(x) = -\infty$ if for any $N > 0$, it is possible to find a number $\delta > 0$ so that $f(x) < -N$ when $0 < |x - c| < \delta$.

➡ **What This Says**   Remember, ∞ *is **not** a number*, so an infinite limit does not exist in the sense that limits were defined in Chapter 2. However, there are several ways for a limit not to exist (for example, $\lim\limits_{x \to \infty} \cos x$ fails to exist by oscillation), so saying that $\lim\limits_{x \to c} f(x) = +\infty$ or $\lim\limits_{x \to c} f(x) = -\infty$ conveys more information than simply observing that the limit does not exist.

EXAMPLE 5   Infinite limits

Find $\lim\limits_{x \to 2^-} \dfrac{3x - 5}{x - 2}$ and $\lim\limits_{x \to 2^+} \dfrac{3x - 5}{x - 2}$.

Solution

Notice that $\dfrac{1}{x - 2}$ increases without bound as x approaches 2 from the right and $\dfrac{1}{x - 2}$ decreases without bound as x approaches 2 from the left. That is,

$$\lim_{x \to 2^+} \frac{1}{x - 2} = +\infty \qquad \text{and} \qquad \lim_{x \to 2^-} \frac{1}{x - 2} = -\infty$$

We also have $\lim\limits_{x \to 2}(3x - 5) = 1$, and it follows that

$$\lim_{x \to 2^+} \frac{3x - 5}{x - 2} = +\infty \qquad \text{and} \qquad \lim_{x \to 2^-} \frac{3x - 5}{x - 2} = -\infty \qquad \blacksquare$$

GRAPHS WITH ASYMPTOTES

Figure 4.43 shows a graph that approaches the horizontal line $y = 2$ as $x \to +\infty$ and $y = -1$ as $x \to -\infty$, and the vertical line $x = 3$ as x approaches 3 from either side.

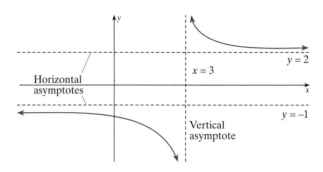

Figure 4.43   A typical graph with asymptotes

Vertical and Horizontal Asymptotes

> The line $x = c$ is a **vertical asymptote** of the graph of f if either of the one-sided limits
>
> $$\lim_{x \to c^-} f(x) \qquad \text{or} \qquad \lim_{x \to c^+} f(x)$$
>
> is infinite. The line $y = L$ is a **horizontal asymptote** of the graph of f if
>
> $$\lim_{x \to +\infty} f(x) = L \qquad \text{or} \qquad \lim_{x \to -\infty} f(x) = L$$

EXAMPLE 6   Graphing a rational function with asymptotes

Sketch the graph of $f(x) = \dfrac{3x - 5}{x - 2}$.

Solution

Vertical Asymptotes   First, make sure the rational function is written in simplified (reduced) form. Because vertical asymptotes for $f(x) = \dfrac{3x - 5}{x - 2}$ occur at values of c for which $\lim_{x \to c^-} f(x)$ or $\lim_{x \to c^+} f(x)$ is infinite, we look for values that cause the denominator to be zero (and the numerator not to be zero); that is, we solve $d(c) = 0$, where $d(x)$ is the denominator of $f(x)$, and then evaluate $\lim_{x \to c^-} f(x)$ and $\lim_{x \to c^+} f(x)$ to ascertain the behavior of the function at $x = c$. For this example, $x = 2$ is a value that causes division by zero, so we find

$$\lim_{x \to 2^+} \frac{3x - 5}{x - 2} = +\infty \qquad \text{and} \qquad \lim_{x \to 2^-} \frac{3x - 5}{x - 2} = -\infty$$

(We found these limits in Example 5.) This means that $x = 2$ is a vertical asymptote and that the graph is moving downward as $x \to 2$ from the left and upward as $x \to 2$ from the right. This information is recorded on the preliminary graph shown in Figure 4.44a by a dashed vertical line with upward ($\uparrow$) and downward ($\downarrow$) arrows.

Horizontal Asymptotes   To find the horizontal asymptotes we compute

$$\lim_{x \to +\infty} \frac{3x - 5}{x - 2} = \lim_{x \to +\infty} \frac{3x - 5}{x - 2} \cdot \frac{\frac{1}{x}}{\frac{1}{x}} = \lim_{x \to +\infty} \frac{3 - \frac{5}{x}}{1 - \frac{2}{x}} = \frac{3 - 0}{1 - 0} = 3$$

and

$$\lim_{x \to -\infty} \frac{3x - 5}{x - 2} = 3 \qquad \text{(The steps here are the same as for } x \to +\infty.)$$

This means that $y = 3$ is a horizontal asymptote. This information is recorded on the preliminary graph shown in Figure 4.44a by a dashed horizontal line with outbound arrows ($\leftarrow$, $\rightarrow$).

a. Preliminary sketch

b. Completed sketch

Figure 4.44 Graph of $f(x) = \dfrac{3x - 5}{x - 2}$

The preliminary sketch gives us some valuable information about the graph, but it does not present the entire picture. Next, we use calculus to find where the function is increasing and decreasing (first derivative) and where it is concave up and concave down (second derivative):

$$f'(x) = \frac{-1}{(x - 2)^2} \qquad \text{and} \qquad f''(x) = \frac{2}{(x - 2)^3}$$

Neither derivative is ever zero, and both are undefined at $x = 2$. Checking the signs of the first and second derivatives, we find that the graph is always falling and that it is concave up for $x > 2$ and concave down for $x < 2$. However, it does not have a point of inflection (the function is not defined at $x = 2$). This information is added to the preliminary sketch shown in Figure 4.44a. The completed graph is shown in Figure 4.44b, which also shows the x- and y-intercepts at $(\frac{5}{3}, 0)$ and $(0, \frac{5}{2})$. ∎

EXAMPLE 7   Sketching a curve with asymptotes

Discuss and sketch the graph of $f(x) = \dfrac{x^2 - x - 2}{x - 3}$.

Solution   We find that

$$f'(x) = \frac{x^2 - 6x + 5}{(x - 3)^2} \qquad \text{and} \qquad f''(x) = \frac{8}{(x - 3)^3}$$

Solving $f'(x) = 0$, we see that the critical numbers are $x = 1$ and $x = 5$. Testing on each side of the critical numbers and $x = 3$, where $f(x)$ is not defined, we obtain the intervals of increase and decrease and concavity shown in Figure 4.45.

Note that there is a relative maximum at $x = 1$ and a relative minimum at $x = 5$. The concavity changes (from $\downarrow$ to $\uparrow$) at $x = 3$, but this does not correspond to an inflection point since $f(3)$ is not defined. We look for vertical asymptotes where $f(x)$ (in reduced form) is not defined; in this case, at $x = 3$. Testing with limits, we find that

$$\lim_{x \to 3^-} \frac{x^2 - x - 2}{x - 3} = -\infty \qquad \text{and} \qquad \lim_{x \to 3^+} \frac{x^2 - x - 2}{x - 3} = +\infty$$

Positive 1 Negative 3 Negative 5 Positive x Negative 3 Positive x

a. Intervals of increase and decrease   **b.** Intervals of concavity

Figure 4.45   Intervals of increase and decrease and concavity for $f(x) = \dfrac{x^2 - x - 2}{x - 3}$

To check for horizontal asymptotes, we compute

$$\lim_{x \to +\infty} \frac{x^2 - x - 2}{x - 3} = +\infty \qquad \text{and} \qquad \lim_{x \to -\infty} \frac{x^2 - x - 2}{x - 3} = -\infty$$

Since neither limit of $f(x)$ at infinity is finite, there are no horizontal asymptotes.

We plot some points: the relative maximum $(1, 1)$; the relative minimum $(5, 9)$; the x-intercepts $(2, 0)$, $(-1, 0)$; and the y-intercept $(0, \frac{2}{3})$. We also show the vertical asymptote $x = 3$, and, using the intervals of increase and decrease and concavity indicated in Figure 4.45, we obtain the graph shown in Figure 4.46. ■

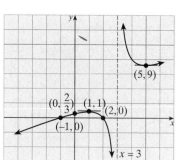

Figure 4.46 Graph of
$f(x) = \dfrac{x^2 - x - 2}{x - 3}$

VERTICAL TANGENTS AND CUSPS

Suppose the function f is continuous at the point P where $x = c$ and that $f'(x)$ becomes infinite as x approaches c. Then the graph of f has a *vertical tangent* at P if the graph turns smoothly through P and a *cusp* at P if the graph changes direction abruptly there. These graphical features can be defined in terms of limits, as follows:

Vertical Tangents and Cusps

Suppose the function f is continuous at the point $P(c, f(c))$. Then the graph of f has

a **vertical tangent** at P if $\lim\limits_{x \to c^-} f'(x)$ and $\lim\limits_{x \to c^+} f'(x)$ are either both $+\infty$ or both $-\infty$;

a **cusp** at P if $\lim\limits_{x \to c^-} f'(x)$ and $\lim\limits_{x \to c^+} f'(x)$ are both infinite with opposite signs (one $+\infty$ and the other $-\infty$).

These possibilities are shown in Figure 4.47.

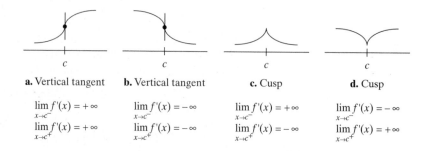

a. Vertical tangent   **b.** Vertical tangent   **c.** Cusp   **d.** Cusp

$\lim\limits_{x \to c^-} f'(x) = +\infty$ $\lim\limits_{x \to c^-} f'(x) = -\infty$ $\lim\limits_{x \to c^-} f'(x) = +\infty$ $\lim\limits_{x \to c^-} f'(x) = -\infty$

$\lim\limits_{x \to c^+} f'(x) = +\infty$ $\lim\limits_{x \to c^+} f'(x) = -\infty$ $\lim\limits_{x \to c^+} f'(x) = -\infty$ $\lim\limits_{x \to c^+} f'(x) = +\infty$

Figure 4.47   Vertical tangents and cusps

TECHNOLOGY NOTE

You can use a graphing calculator to help find many indeterminate-form limits. For instance, the limit in Example 3 can easily be checked by looking at the graph. If you have a graphing calculator, you can see that as $x \to 0$ from either the left or the right, the limit looks the same, namely, 0.

```
Y₁☰(1-cos X)cos
X
Xmin=-6.283185...
Xmax=6.2831853...
Xscl=1.5707963...
Ymin=-3
Ymax=1
Yscl=1
```

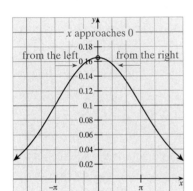

By looking at the graph, you can reinforce the result obtained using l'Hôpital's rule.

EXAMPLE 4   l'Hôpital's rule applied more than once

Evaluate $\displaystyle\lim_{x \to 0} \frac{x - \sin x}{x^3}$.

Solution

This is a 0/0 indeterminate form, and we find that

$$\lim_{x \to 0} \frac{x - \sin x}{x^3} = \lim_{x \to 0} \frac{1 - \cos x}{3x^2}$$

This is still the indeterminate form 0/0, so l'Hôpital's rule can be applied once again:

$$\lim_{x \to 0} \frac{1 - \cos x}{3x^2} = \lim_{x \to 0} \frac{-(-\sin x)}{6x} = \frac{1}{6} \lim_{x \to 0} \frac{\sin x}{x} = \frac{1}{6}(1) = \frac{1}{6} \qquad \blacksquare$$

EXAMPLE 5   l'Hôpital's rule with an ∞/∞ form

Evaluate $\displaystyle\lim_{x \to +\infty} \frac{2x^2 - 3x + 1}{3x^2 + 5x - 2}$.

Solution

Using the methods of Section 2.2, we could compute this limit by multiplying by $(1/x^2)/(1/x^2)$. Instead, we note that this is of the form ∞/∞ and apply l'Hôpital's rule:

$$\lim_{x \to +\infty} \frac{2x^2 - 3x + 1}{3x^2 + 5x - 2} = \lim_{x \to +\infty} \frac{4x - 3}{6x + 5} \qquad \text{Apply l'Hôpital's rule again.}$$

$$= \lim_{x \to +\infty} \frac{4}{6} = \frac{2}{3} \qquad \blacksquare$$

It may happen that even when l'Hôpital's rule applies to a limit, it is not the best way to proceed, as illustrated by the following example.

EXAMPLE 6   Using l'Hôpital's rule with other limit properties

Evaluate $\displaystyle\lim_{x \to 0} \frac{(1 - \cos x)\sin 4x}{x^3 \cos x}$.

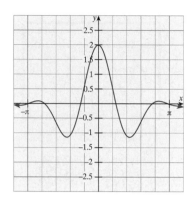

Solution

This limit has the form $0/0$, but direct application of l'Hôpital's rule leads to a real mess (try it!). Instead, we compute the given limit by using the product rule for limits first, followed by two simple applications of l'Hôpital's rule. Specifically, using the product rule for limits (assuming the limits exist), we have

$$\lim_{x \to 0} \frac{(1 - \cos x) \sin 4x}{x^3 \cos x} = \left[\lim_{x \to 0} \frac{1 - \cos x}{x^2} \right] \left[\lim_{x \to 0} \frac{\sin 4x}{x} \right] \left[\lim_{x \to 0} \frac{1}{\cos x} \right]$$

$$= \left[\lim_{x \to 0} \frac{\sin x}{2x} \right] \left[\lim_{x \to 0} \frac{4 \cos 4x}{1} \right] \left[\lim_{x \to 0} \frac{1}{\cos x} \right]$$

$$= \left(\frac{1}{2} \right) (4)(1) = 2 \qquad \blacksquare$$

EXAMPLE 7 Hypotheses of l'Hôpital's rule are not satisfied

Evaluate $\lim\limits_{x \to +\infty} \dfrac{x + \sin x}{x - \cos x}$.

Solution

This limit has the indeterminate form ∞/∞. If you try to apply l'Hôpital's rule, you find

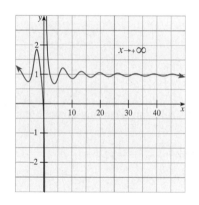

$$\lim_{x \to +\infty} \frac{x + \sin x}{x - \cos x} = \lim_{x \to +\infty} \frac{1 + \cos x}{1 + \sin x}$$

The limit on the right does not exist, because both $\sin x$ and $\cos x$ oscillate between -1 and 1 as $x \to +\infty$. Recall that l'Hôpital's rule applies only if $\lim\limits_{x \to c} \dfrac{f'(x)}{g'(x)} = L$ or is $\pm\infty$. This does not mean that the limit of the original expression does not exist or that we cannot find it; it simply means that we cannot apply l'Hôpital's rule. To find this limit, factor out an x from the numerator and denominator and proceed as follows:

$$\lim_{x \to +\infty} \frac{x + \sin x}{x - \cos x} = \lim_{x \to +\infty} \frac{x(1 + \frac{\sin x}{x})}{x(1 - \frac{\cos x}{x})} = \lim_{x \to +\infty} \frac{1 + \frac{\sin x}{x}}{1 - \frac{\cos x}{x}} = \frac{1 + 0}{1 - 0} = 1 \qquad \blacksquare$$

OTHER INDETERMINATE FORMS

Other indeterminate forms, such as 1^∞, 0^0, ∞^0, $\infty - \infty$, and $0 \cdot \infty$, can often be manipulated algebraically into one of the standard forms $0/0$ or ∞/∞ and then evaluated using l'Hôpital's rule. The following examples illustrate such procedures.

We begin by deriving a formula given without proof in Section 2.4.

WARNING l'Hôpital's rule itself applies only to the indeterminate form $0/0$ and ∞/∞.

EXAMPLE 8 Limit of the form 1^∞

Show that $\lim\limits_{x \to +\infty} \left(1 + \dfrac{1}{x} \right)^x = e$.

Solution

Note that this limit is indeed of the indeterminate form 1^∞. Let

$$L = \lim_{x \to +\infty} \left(1 + \frac{1}{x} \right)^x$$

Take the logarithm of both sides:

$$\ln L = \ln \left[\lim_{x \to +\infty} \left(1 + \frac{1}{x} \right)^x \right]$$

$$= \lim_{x \to +\infty} \ln \left(1 + \frac{1}{x} \right)^x \qquad \textit{ln x is continuous}$$

$$= \lim_{x \to +\infty} x \ln \left(1 + \frac{1}{x} \right) \qquad \textit{Property of logarithms}$$

$$= \lim_{x \to +\infty} \frac{\ln \left(1 + \frac{1}{x} \right)}{\frac{1}{x}} \qquad \textit{Form } \frac{0}{0}$$

$$= \lim_{x \to +\infty} \frac{\frac{1}{1 + 1/x} \left(-\frac{1}{x^2} \right)}{-\frac{1}{x^2}} \qquad \textit{l'Hôpital's rule}$$

$$= \lim_{x \to +\infty} \frac{1}{1 + \frac{1}{x}} \qquad \textit{Simplify}$$

$$= \frac{1}{1 + 0}$$

$$= 1$$

Thus, $\ln L = 1$ and $L = e^1 = e$.   ■

EXAMPLE 9   l'Hôpital's rule with the form $0 \cdot \infty$

Evaluate $\lim\limits_{x \to \pi/2^-} \left(x - \dfrac{\pi}{2} \right) \tan x$.

Solution

This limit has the form $0 \cdot \infty$, because

$$\lim_{x \to \pi/2^-} \left(x - \frac{\pi}{2} \right) = 0 \qquad \text{and} \qquad \lim_{x \to \pi/2^-} \tan x = +\infty$$

Write $\tan x = \dfrac{1}{\cot x}$ to obtain

$$\lim_{x \to \pi/2^-} \left(x - \frac{\pi}{2} \right) \tan x = \lim_{x \to \pi/2^-} \frac{x - \frac{\pi}{2}}{\cot x} \qquad \textit{Form } \frac{0}{0}$$

$$= \lim_{x \to \pi/2^-} \frac{1}{-\csc^2 x} \qquad \textit{l'Hôpital's rule}$$

$$= \lim_{x \to \pi/2^-} (-\sin^2 x) = -1 \qquad ■$$

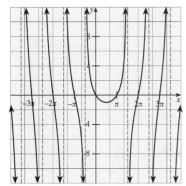

Look at the graph as $x \to \frac{\pi}{2}$ from both the left and the right.

EXAMPLE 10   Limit of the form 0^0

Find $\lim\limits_{x \to 0^+} x^{\sin x}$.

Solution

This is a 0^0 indeterminate form. From the graph shown in Figure 4.49, it looks as though the desired limit is 1. We can verify this conjecture analytically.

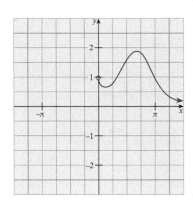

Figure 4.49 Graph of $x^{\sin x}$

As in Example 8, we begin by using properties of logarithms.

$$L = \lim_{x \to 0^+} x^{\sin x}$$

$$\ln L = \ln \lim_{x \to 0^+} x^{\sin x}$$

$$= \lim_{x \to 0^+} \ln x^{\sin x} \qquad \text{ln is continuous}$$

$$= \lim_{x \to 0^+} (\sin x) \ln x \qquad \text{This is } 0 \cdot \infty \text{ form.}$$

$$= \lim_{x \to 0^+} \frac{\ln x}{\csc x} \qquad \text{This is } \frac{\infty}{\infty} \text{ form.}$$

$$= \lim_{x \to 0^+} \frac{1/x}{-\csc x \cot x} \qquad \text{l'Hôpital's rule}$$

$$= \lim_{x \to 0^+} \frac{-\sin^2 x}{x \cos x}$$

$$= \lim_{x \to 0^+} \left(\frac{\sin x}{x} \right) \left(\frac{-\sin x}{\cos x} \right)$$

$$= (1)(0) = 0$$

Thus, $L = e^0 = 1$. ∎

EXAMPLE 11   Limit of the form ∞^0

Find $\displaystyle\lim_{x \to +\infty} x^{1/x}$.

Solution

This is a limit of the indeterminate form ∞^0.

If $L = \displaystyle\lim_{x \to +\infty} x^{1/x}$, then

$$\ln L = \ln \lim_{x \to +\infty} x^{1/x}$$

$$= \lim_{x \to +\infty} \ln x^{1/x} \qquad \text{The limit of a log is the log of the limit.}$$

$$= \lim_{x \to +\infty} \frac{1}{x} \ln x$$

$$= \lim_{x \to +\infty} \frac{\ln x}{x} \qquad \text{This is } \frac{\infty}{\infty} \text{ form.}$$

$$= \lim_{x \to +\infty} \frac{\frac{1}{x}}{1} \qquad \text{l'Hôpital's rule}$$

$$= 0$$

Thus, we have $\ln L = 0$; therefore, $L = e^0 = 1$. ∎

EXAMPLE 12   Horizontal asymptotes with l'Hôpital's rule

Find all horizontal asymptotes of the graph of $y = xe^{-2x}$.

Solution

To test for horizontal asymptotes, we compute

$$\lim_{x \to -\infty} xe^{-2x} \qquad \text{and} \qquad \lim_{x \to +\infty} xe^{-2x}$$

For the first limit, we find

$$\lim_{x \to -\infty} xe^{-2x} = -\infty$$

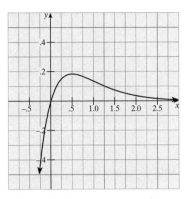

Figure 4.50 Graph of $y = xe^{-2x}$ showing the x-axis as a horizontal asymptote

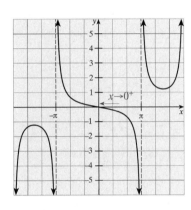

and for the second,

$$\lim_{x \to +\infty} xe^{-2x} = \lim_{x \to +\infty} \frac{x}{e^{2x}} \qquad \frac{\infty}{\infty} \text{ form}$$

$$= \lim_{x \to +\infty} \frac{1}{2e^{2x}} \qquad \text{l'Hôpital's rule}$$

$$= 0$$

Thus, $y = 0$ (the x-axis) is a horizontal asymptote. The graph of $y = xe^{-2x}$ is shown in Figure 4.50. ∎

EXAMPLE 13 l'Hôpital's rule with the form $\infty - \infty$

Evaluate $\lim\limits_{x \to 0^+} \left(\dfrac{1}{x} - \dfrac{1}{\sin x} \right)$.

Solution

As it stands, this has the form $\infty - \infty$, because $\dfrac{1}{x} \to +\infty$ and $\dfrac{1}{\sin x} \to +\infty$ as $x \to 0$ from the right. However, using a little algebra, we find

$$\lim_{x \to 0^+} \left(\frac{1}{x} - \frac{1}{\sin x} \right) = \lim_{x \to 0^+} \frac{\sin x - x}{x \sin x}$$

This limit is now of the form $0/0$, so the hypotheses of l'Hôpital's rule are satisfied. Thus,

$$\lim_{x \to 0^+} \frac{\sin x - x}{x \sin x} = \lim_{x \to 0^+} \frac{\cos x - 1}{\sin x + x \cos x} \qquad \text{Again, the form } \frac{0}{0}$$

$$= \lim_{x \to 0^+} \frac{-\sin x}{\cos x + x(-\sin x) + \cos x} = \frac{0}{2} = 0 \qquad ∎$$

WARNING Not all limits that appear indeterminate actually are indeterminate. For example,

$$\lim_{x \to 0^+} (\sin x)^{1/x} = 0 \qquad 0^{\infty} \text{ form}$$

$$\lim_{x \to 0^+} (\csc x - \ln x) = +\infty \qquad +\infty - (-\infty) \text{ form}$$

$$\lim_{x \to 0^+} \frac{\tan x}{\ln x} = 0 \qquad \frac{0}{\infty} \text{ form}$$

Other such "false indeterminate forms" include $+\infty + (+\infty)$, $\infty/0$, and $\infty \cdot \infty$, which are all actually infinite.

SPECIAL LIMITS INVOLVING e^x AND $\ln x$

We close this section with a theorem that summarizes the behavior of certain important special functions involving e^x and $\ln x$ near 0 and at $\pm\infty$.

THEOREM 4.10 Limits involving natural logarithms and exponentials

If k and n are positive numbers, then

$$\lim_{x \to 0^+} \frac{\ln x}{x^n} = -\infty \qquad \lim_{x \to +\infty} \frac{\ln x}{x^n} = 0$$

$$\lim_{x \to +\infty} \frac{e^{kx}}{x^n} = +\infty \qquad \lim_{x \to +\infty} x^n e^{-kx} = 0$$

Proof   These can all be verified directly or by applying l'Hôpital's rule. For example,

$$\lim_{x \to +\infty} \frac{\ln x}{x^n} = \lim_{x \to +\infty} \frac{\frac{1}{x}}{nx^{n-1}} = \lim_{x \to +\infty} \frac{1}{nx^n} = 0$$

The other parts are left for you to verify.   ❏

➤ **What This Says**   The limit statements

$$\lim_{x \to +\infty} \frac{e^{kx}}{x^n} = +\infty \qquad \text{and} \qquad \lim_{x \to +\infty} \frac{\ln x}{x^n} = 0$$

are especially important. They tell us that, in the long run, any exponential e^{kx} dominates any power x^n, for k and n positive, which, in turn, dominates any logarithm.

4.5 PROBLEM SET

Ⓐ 1. An incorrect use of l'Hôpital's rule is illustrated in the following limit computations. In each case, explain what is wrong and find the correct value of the limit.

a. $\displaystyle \lim_{x \to \pi} \frac{1 - \cos x}{x} = \lim_{x \to \pi} \frac{\sin x}{1} = 0$

b. $\displaystyle \lim_{x \to \pi/2} \frac{\sin x}{2x} = \lim_{x \to \pi/2} \frac{\cos x}{1} = 0$

2. **Exploration Problem**   Sometimes l'Hôpital's rule leads to faulty computations. For example, observe what happens when the rule is applied to

$$\lim_{x \to +\infty} \frac{x}{\sqrt{x^2 - 1}}$$

Use any method you wish to evaluate this limit.

Find each of the limits in Problems 3–36.

3. $\displaystyle \lim_{x \to 1} \frac{x^3 - 1}{x^2 - 1}$

4. $\displaystyle \lim_{x \to 2} \frac{x^3 - 27}{x^2 - 9}$

5. $\displaystyle \lim_{x \to 1} \frac{x^{10} - 1}{x - 1}$

6. $\displaystyle \lim_{x \to -1} \frac{x^{10} - 1}{x + 1}$

7. $\displaystyle \lim_{x \to 0} \frac{1 - \cos^2 x}{\sin^3 x}$

8. $\displaystyle \lim_{x \to 0} \frac{1 - \cos^2 x}{3 \sin x}$

9. $\displaystyle \lim_{x \to \pi} \frac{\cos \frac{x}{2}}{\pi - x}$

10. $\displaystyle \lim_{x \to 0} \frac{1 - \cos x}{x^2}$

11. $\displaystyle \lim_{x \to 0} \frac{\sin ax}{\cos bx}, ab \ne 0$

12. $\displaystyle \lim_{x \to 0} \frac{\tan 3x}{\sin 5x}$

13. $\displaystyle \lim_{x \to 0} \frac{x - \sin x}{\tan x - x}$

14. $\displaystyle \lim_{x \to 0} \frac{1 - \cos^2 x}{x \tan x}$

15. $\displaystyle \lim_{x \to \pi/2} \frac{3 \sec x}{2 + \tan x}$

16. $\displaystyle \lim_{x \to 0} \frac{x + \sin^3 x}{x^2 + 2x}$

17. $\displaystyle \lim_{x \to 0} \frac{\sin 3x \sin 2x}{x \sin 4x}$

18. $\displaystyle \lim_{x \to \pi/2} \frac{\sin 2x \cos x}{x \sin 4x}$

19. $\displaystyle \lim_{x \to 0} \frac{x^2 + \sin x^2}{x^2 + x^3}$

20. $\displaystyle \lim_{x \to 0} x^2 \sin \frac{1}{x}$

21. $\displaystyle \lim_{x \to +\infty} x^{3/2} \sin \frac{1}{x}$

22. $\displaystyle \lim_{x \to 0} \left(\cot x - \frac{1}{x} \right)$

23. $\displaystyle \lim_{x \to 1} \frac{(x - 1) \sin(x - 1)}{1 - \cos(x - 1)}$

24. $\displaystyle \lim_{\theta \to 0} \frac{\theta - 1 + \cos^2 \theta}{\theta^2 + 5\theta}$

25. $\displaystyle \lim_{x \to 0} \frac{x + \sin(x^2 + x)}{3x + \sin x}$

26. $\displaystyle \lim_{x \to \pi/2^-} \sec 3x \cos 9x$

27. $\displaystyle \lim_{x \to 0} \left(\frac{1}{\sin 2x} - \frac{1}{2x} \right)$

28. $\displaystyle \lim_{x \to +\infty} x^{-5} \ln x$

29. $\displaystyle \lim_{x \to 0^+} x^{-5} \ln x$

30. $\displaystyle \lim_{x \to 0^+} (\sin x) \ln x$

31. $\displaystyle \lim_{x \to +\infty} \frac{\ln(\ln x)}{x}$

32. $\displaystyle \lim_{x \to -\infty} \left(1 - \frac{3}{x} \right)^{2x}$

33. $\displaystyle \lim_{x \to +\infty} \left(1 + \frac{1}{2x} \right)^{3x}$

34. $\displaystyle \lim_{x \to +\infty} (\ln x)^{1/x}$

35. $\displaystyle \lim_{x \to 0^+} (e^x + x)^{1/x}$

36. $\displaystyle \lim_{x \to 0^+} (\sin x)^{1/\ln \sqrt{x}}$

Ⓑ *Find each of the limits in Problems 37–51.*

37. $\displaystyle \lim_{x \to (\pi/2)^-} \left(\frac{1}{\pi - 2x} + \tan x \right)$

38. $\displaystyle \lim_{x \to +\infty} \left(\sqrt{x^2 - x} - x \right)$

39. $\displaystyle \lim_{x \to +\infty} [x - \ln(x^3 - 1)]$ *Hint*: $\ln e^x = x$

40. $\displaystyle \lim_{x \to +\infty} [x - \ln(e^x + e^{-x})]$ *Hint*: $\ln e^x = x$

41. $\displaystyle \lim_{x \to 0^+} \left(\frac{1}{x^2} - \ln \sqrt{x} \right)$

42. $\displaystyle \lim_{x \to 0} (e^x - 1 - x)^x$

43. $\displaystyle \lim_{x \to 0^+} (\ln x)(\cot x)$

44. $\displaystyle \lim_{x \to 0^+} (e^x - 1)^{1/\ln x}$

45. $\displaystyle \lim_{x \to +\infty} \frac{x + \sin 3x}{x}$

46. $\displaystyle \lim_{x \to +\infty} \frac{x(\pi + \sin x)}{x^2 + 1}$

47. $\displaystyle \lim_{x \to 0^+} \left(\frac{2 \cos x}{\sin 2x} - \frac{1}{x} \right)$

48. $\displaystyle \lim_{x \to +\infty} \left(\frac{x^3}{x^2 - x + 1} - \frac{x^3}{x^2 + x - 1} \right)$

49. $\displaystyle \lim_{x \to 0} \frac{(2 - x)(e^x - x - 2)}{x^3}$

50. $\lim\limits_{x\to 0} \dfrac{\tan^{-1}(3x) - 3\tan^{-1} x}{x^3}$

51. $\lim\limits_{x\to +\infty} x^5 \left[\sin\left(\dfrac{1}{x}\right) - \dfrac{1}{x} + \dfrac{1}{6x^3} \right]$

52. Find A so that $\lim\limits_{x\to +\infty} \left(\dfrac{x+A}{x-2A} \right)^x = 5$.

In Problems 53–56, use l'Hôpital's rule to determine all horizontal asymptotes to the graph of the given function. You are NOT required to sketch the graph.

53. $f(x) = x^3 e^{-0.01x}$

54. $f(x) = \dfrac{\ln x^5}{x^{0.02}}$

55. $f(x) = (\ln \sqrt{x})^{2/x}$

56. $f(x) = \left(\dfrac{x+3}{x+2} \right)^{2x}$

57. Exploration Problem Write a paper on using technology to evaluate limits.

58. HISTORICAL QUEST The French mathematician Guillaume de l'Hôpital is best known today for the rule that bears his name, but the rule was discovered by l'Hôpital's teacher, Johann Bernoulli. Not only did l'Hôpital neglect to cite his sources in his book, but there is also evidence that he paid Bernoulli for his results and for keeping their arrangements for payment confidential. In a letter dated March 17, 1694, he asked Bernoulli "to communicate to me your discoveries ..."—with the request not to mention them to others—"... it would not please me if they were made public." L'Hôpital's argument, which was originally given without using functional notation, can easily be reproduced*:

$$\frac{f(a+dx)}{g(a+dx)} = \frac{f(a) + f'(a)\,dx}{g(a) + g'(a)\,dx}$$
$$= \frac{f'(a)\,dx}{g'(a)\,dx}$$
$$= \frac{f'(a)}{g'(a)} \quad \blacksquare$$

Supply reasons for this argument, and give necessary conditions for the functions f and g.

59. HISTORICAL QUEST The remarkable Bernoulli family of Switzerland produced at least eight noted mathematicians over three generations. Two brothers, Jacob (1654–1705) and Johann (1667–1748), were bitter rivals. These brothers were extremely influential advocates of the newly born calculus. Johann was the most prolific of the clan and was responsible for the discovery of l'Hôpital's rule (see Problem 58), Bernoulli numbers, Bernoulli polynomials, the lemniscate of Bernoulli, the Bernoulli equation, the Bernoulli theorem, and the Bernoulli

*D. J. Struik, A Source Book in Mathematics, 1200–1800. Cambridge, MA: Harvard University Press, 1969, pp. 313–316.

distribution. He did a great deal of work with differential equations. Johann was jealous and cantankerous; he tossed a son (Daniel) out of the house for winning an award he had expected to win himself. ■

Write a report on the Bernoulli family.

Find the limits in Problems 60–62 using the following methods.
  a. graphically **b.** analytically **c.** numerically
Compare, contrast, and reconcile the three methods.

60. $\lim\limits_{x\to 0^+} x^x$

61. $\lim\limits_{x\to 0^+} (x^x)^x$

62. $\lim\limits_{x\to +\infty} \left[x \sin^{-1}\left(\dfrac{1}{x}\right) \right]^{x^2}$

63. Find constants a and b so that
$$\lim_{x\to 0} \left(\frac{\sin 2x}{x^3} + \frac{a}{x^2} + b \right) = 1$$

64. Find all values of A and B so that
$$\lim_{x\to 0} \frac{\sin Ax + Bx}{x^3} = 36$$

65. For a certain value of C, the limit
$$\lim_{x\to +\infty} (x^4 + 5x^3 + 3)^C - x$$
is finite and nonzero. Find C and then compute the limit.

66. For which values of constants D and E is it true that
$$\lim_{x\to 0}(x^{-3}\sin 7x + Dx^{-2} + E) = -2?$$

67. A weight hanging by a spring is made to vibrate by applying a sinusoidal force, and the displacement at time t is given by
$$f(t) = \frac{C}{\beta^2 - \alpha^2}(\sin \alpha t - \sin \beta t)$$
where C, α, and β are constants such that $\alpha \neq \beta$. What happens to the displacement as $\beta \to \alpha$? You may assume that α is fixed.

68. For F and G positive constants, define
$$f(x) = (e^x + Fx)^{G/x}$$
  a. Compute $L_1 = \lim\limits_{x\to 0} f(x)$ and $L_2 = \lim\limits_{x\to +\infty} f(x)$.
  b. What is the largest value of F for which the equation $L_1 = GL_2$ has a solution? What are L_1 and L_2 in this case?

4.6   Optimization in the Physical Sciences and Engineering

IN THIS SECTION   optimization procedure, Fermat's principle of optics and Snell's law

We introduced the process of mathematical modeling in Chapter 3, and now that we have developed the necessary calculus skills, we will use mathematical modeling and calculus to solve optimization problems.

OPTIMIZATION PROCEDURE

NOTHING TAKES PLACE in the world whose meaning is not that of some maximum or minimum.

——Leonhard Euler

When light travels from one medium to another—say, from air into water—which path does it follow? In a mechanical system involving pulleys, which arrangement results in minimal potential energy? At what temperature does a given mass of water occupy the least volume (the answer is not 0°C)? These are typical *optimization problems* from the physical sciences and engineering and will be examined in the examples and exercises of this section. The process of finding the maximum or minimum of a function is called **optimization**. Then, in the next section, we will explore optimization involving applications to business, economics, and the social and life sciences. We begin by outlining a general optimization procedure based on ideas originally suggested by George Pólya (1887–1985), one of the great problem solvers of twentieth-century mathematics (see the Historical Quest in Problem 61 of Section 7.4).

Optimization Procedure

Step 1.   Draw a figure (if appropriate) and label all quantities relevant to the problem.

Step 2.   Focus on the quantity to be optimized. Name it. Find a formula for the quantity to be maximized or minimized.

Step 3.   Use conditions in the problem to eliminate variables in order to express the quantity to be maximized or minimized in terms of a single variable.

Step 4.   Find the practical domain for the variables in Step 3; that is, the interval of possible values determined from the physical restrictions in the problem.

Step 5.   If possible, use the methods of calculus to obtain the required optimum value.

EXAMPLE 1   Maximizing a constrained area

You need to fence a rectangular play zone for children. What is the maximum area for this play zone if it is to fit into a right-triangular plot with sides measuring 4 m and 12 m?

Solution

A picture of the play zone is shown in Figure 4.51. Let x and y denote the length and width of the inscribed rectangle. The appropriate formula for the area is

$$A = (\text{LENGTH})(\text{WIDTH}) = xy$$

We wish to maximize A, where $A = xy$, but first we must express A as a function of a single variable. To do this, note that, because $\triangle ABC$ is similar to $\triangle ADF$, the corresponding sides of these triangles are proportional,

$$\frac{4 - y}{4} = \frac{x}{12}$$
$$y = 4 - \tfrac{1}{3}x$$

Figure 4.51 Children's play zone

 SMH   *See Section 1.5 of the Student Mathematics Handbook* for similar triangles.

We can now write A as a function of x alone:

$$A(x) = x(4 - \tfrac{1}{3}x) = 4x - \tfrac{1}{3}x^2$$

The domain for A is $0 \leq x \leq 12$. The critical numbers for A are values such that $A'(x) = 0$ (since $A'(x)$ exists for all x). Since

$$A'(x) = 4 - \tfrac{2}{3}x$$

the only critical number is $x = 6$. Evaluate $A(x)$ at the endpoints and the critical number:

$$A(6) = 4(6) - \tfrac{1}{3}(6)^2 = 12; \quad A(0) = 0; \quad A(12) = 0$$

The maximum area occurs when $x = 6$. This means that

$$y = 4 - \tfrac{1}{3}(6) = 2$$

Thus, the largest rectangular play zone that can be built in the triangular plot is a rectangle 6 m long and 2 m wide. ■

EXAMPLE 2 Maximizing a volume

A carpenter wants to make an open-topped box out of a rectangular sheet of tin 24 in. wide and 45 in. long. The carpenter plans to cut congruent squares out of each corner of the sheet and then bend and solder the edges of the sheet upward to form the sides of the box, as shown in Figure 4.52.

Figure 4.52 A box cut from a 24 in. by 45 in. piece of tin

For what dimensions does the box have the greatest possible volume?

Solution

If each square corner cut out has side x, the box will be x inches deep, $45 - 2x$ inches long, and $24 - 2x$ inches wide. The volume of the box shown in Figure 4.52 is

$$V(x) = x(45 - 2x)(24 - 2x) = 4x^3 - 138x^2 + 1{,}080x$$

and this is the quantity to be maximized. To find the domain, we note that the dimensions must all be nonnegative; therefore, $x \geq 0$, $45 - 2x \geq 0$ (or $x \leq 22.5$), and $24 - 2x \geq 0$ (or $x \leq 12$). This implies that the domain is $[0, 12]$.

To find the critical numbers (the derivative is defined everywhere in the domain), we find values for which the derivative is 0:

$$V'(x) = 12x^2 - 276x + 1{,}080 = 12(x - 18)(x - 5)$$

The critical numbers are $x = 5$ and $x = 18$, but $x = 18$ is not in the domain, so the only relevant critical number is $x = 5$. Evaluating $V(x)$ at the critical number $x = 5$ and the endpoints $x = 0$, $x = 12$, we find

$$V(5) = 5(45 - 10)(24 - 10) = 2{,}450; \quad V(0) = 0; \quad V(12) = 0$$

Thus, the box with the largest volume is found when $x = 5$. Such a box has dimensions 5 in. × 14 in. × 35 in. ■

You will note in Example 2 we did not have to test whether the critical point was a maximum or a minimum, but knew that it was a maximum because the continuous function V is nonnegative on the interval $[0, 12]$ and is indeed zero at the endpoints. Since there is only one critical number $x = 5$, that must yield the maximum. Similar reasoning can be used in many examples, allowing us to forgo having to test for a maximum or minimum with the first- or second-derivative test. We summarize this procedure with the following definition.

EVT Convention

If the hypotheses of the extreme value theorem apply to a particular problem, we can often avoid testing a candidate to see whether it is a maximum or a minimum. Suppose we have three candidates and two are obviously minima—the third *must be a maximum*. We will refer to this principle of not testing a candidate as the **EVT convention** because it is shorthand for using the extreme value theorem.

EXAMPLE 3   Maximizing illumination

A lamp with adjustable height hangs directly above the center of a circular kitchen table that is 8 ft in diameter. Model the illumination I at the edge of the table to be directly proportional to the cosine of the angle θ and inversely proportional to the square of the distance d, where θ and d are as shown in Figure 4.53. How close to the table should the lamp be lowered to maximize the illumination at the edge of the table?

Solution

We note that

$$I(\theta) = \frac{k \cos \theta}{d^2}$$

where k is a (positive) constant of proportionality. Moreover, from the right triangle shown in Figure 4.53,

$$\sin \theta = \frac{4}{d} \quad \text{or} \quad d = \frac{4}{\sin \theta}$$

Hence,

$$I(\theta) = k \cos \theta \left[\frac{1}{(4/\sin \theta)^2} \right]$$

$$= \frac{k}{16} \cos \theta \sin^2 \theta$$

Only values of θ between 0 and $\frac{\pi}{2}$ are meaningful in the context of this problem. Hence, the goal is to find the absolute maximum of the function $I(\theta)$ on $[0, \frac{\pi}{2}]$.

Using the product rule and the chain rule to differentiate $I(\theta)$, we obtain

$$I'(\theta) = \frac{k}{16} [\cos \theta (2 \sin \theta \cos \theta) + \sin^2 \theta (-\sin \theta)]$$

$$= \frac{k}{16} (2 \cos^2 \theta \sin \theta - \sin^3 \theta)$$

$$= \frac{k}{16} \sin \theta (2 \cos^2 \theta - \sin^2 \theta)$$

This is zero when $\theta = 0$ (since $\sin 0 = 0$) and when

$$2 \cos^2 \theta - \sin^2 \theta = 0$$

$$\frac{\sin^2 \theta}{\cos^2 \theta} = 2$$

$$\tan^2 \theta = 2$$

$$\theta = \tan^{-1} \sqrt{2}$$

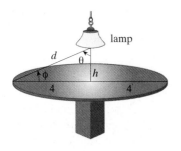

Figure 4.53 Illumination on a kitchen table

Evaluating $I(\theta)$ at the endpoints, $\theta = 0$ and $\theta = \frac{\pi}{2}$, and at the interior critical number $\tan^{-1}\sqrt{2} \approx 0.9553$, we find that

$$I(0) = \frac{k}{16}(\cos 0)(\sin^2 0) = 0$$

$$I\left(\frac{\pi}{2}\right) = \frac{k}{16}\left(\cos\frac{\pi}{2}\right)\left(\sin^2\frac{\pi}{2}\right) = 0$$

$$I(\tan^{-1}\sqrt{2}) \approx I(0.9553) \approx \frac{k}{16}(\cos 0.9553)(\sin^2 0.9553) \approx 0.0241k$$

Finally, to find the height h that maximizes the illumination I, observe from Figure 4.53 that

$$\tan\theta = \frac{4}{h}$$

Since $\tan\theta = \sqrt{2}$ when I is maximized, it follows that

$$h = \frac{4}{\tan\theta} = \frac{4}{\sqrt{2}} \approx 2.83$$

To maximize the illumination, the lamp should be placed about 2.8 ft (2 ft 10 in.) above the table. ∎

EXAMPLE 4 Minimizing time of travel

A dune buggy is on the desert at a point A located 40 km from a point B, which lies on a long, straight road, as shown in Figure 4.54. The driver can travel at 45 km/h on the desert and 75 km/h on the road. The driver will win a prize if she arrives at the finish line at point D, 50 km from B, in 84 min or less. What route should she travel to minimize the time of travel? Does she win the prize?

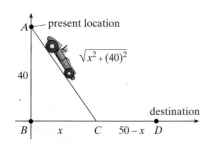

Figure 4.54 A path traveled by a dune buggy

Solution

Suppose the driver heads for a point C located x km down the road from B toward her destination, as shown in Figure 4.54. We want to minimize the time. We will need to remember the formula $d = rt$, or in terms of time, $t = \dfrac{d}{r}$.

$$\text{TIME} = \text{TIME FROM } A \text{ TO } C + \text{TIME FROM } C \text{ TO } D$$

$$= \frac{\text{DISTANCE FROM } A \text{ TO } C}{\text{RATE FROM } A \text{ TO } C} + \frac{\text{DISTANCE FROM } C \text{ TO } D}{\text{RATE FROM } C \text{ TO } D}$$

$$T(x) = \frac{\sqrt{x^2 + 1{,}600}}{45} + \frac{50 - x}{75}$$

The domain of T is $[0, 50]$. Next, find the derivative of time T with respect to x:

$$T'(x) = \frac{1}{45}\left[\frac{1}{2}(x^2 + 1{,}600)^{-1/2}(2x)\right] + \frac{1}{75}(-1)$$

$$= \frac{x}{45\sqrt{x^2 + 1{,}600}} - \frac{1}{75}$$

$$= \frac{5x - 3\sqrt{x^2 + 1{,}600}}{225\sqrt{x^2 + 1{,}600}}$$

The derivative exists for all x and is zero when

$$5x - 3\sqrt{x^2 + 1{,}600} = 0$$

Solving this equation, we find $x = 30$ (-30 is extraneous). Evaluating $T(x)$ here and at the endpoints, we find that

$$T(30) = \frac{\sqrt{30^2 + 1{,}600}}{45} + \frac{50 - 30}{75} \approx 1.3778 \text{ hr } \approx 83 \text{ min}$$

$$T(0) = \frac{\sqrt{0^2 + 1{,}600}}{45} + \frac{50 - 0}{75} \approx 1.5556 \text{ hr } \approx 93 \text{ min}$$

$$T(50) = \frac{\sqrt{50^2 + 1{,}600}}{45} + \frac{50 - 50}{75} \approx 1.4229 \text{ hr } \approx 85 \text{ min}$$

The driver can minimize the total driving time by heading for a point that is 30 miles from the point B and then traveling on the road to point D. She wins the prize because this minimal route requires only 83 minutes.   ■

EXAMPLE 5   Optimizing a constrained area

A wire of length L is to be cut into two pieces, one of which will be bent to form a circle and the other to form a square. Determine how the wire should be cut to

a. maximize the sum of the areas enclosed by the two pieces.
b. minimize the sum of the areas enclosed by the two pieces.

Solution

To understand the problem we draw a sketch, as shown in Figure 4.55, and label the radius of the circle r and the side of the square s.

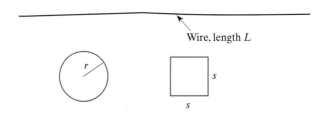

Figure 4.55   Forming a circle and a square from a wire with length L

We find that the combined area is

$$\text{AREA} = \text{AREA OF CIRCLE} + \text{AREA OF SQUARE} = \pi r^2 + s^2$$

We need to write the radius r and the side s in terms of the length of wire, L.

$$L = \text{CIRCUMFERENCE OF CIRCLE} + \text{PERIMETER OF SQUARE}$$
$$= 2\pi r + 4s$$

so $s = \frac{1}{4}(L - 2\pi r)$. (We could just as easily have solved for r.) By substitution,

$$A(r) = \pi r^2 + \left[\tfrac{1}{4}(L - 2\pi r)\right]^2 = \pi r^2 + \tfrac{1}{16}(L - 2\pi r)^2$$

To find the domain, we note that $r \geq 0$ and that $L - 2\pi r \geq 0$, so the domain is $0 \leq r \leq \dfrac{L}{2\pi}$. Note when $r = 0$ there is no circle, and when $r = \dfrac{L}{2\pi}$ there is no square. The derivative of $A(r)$ is

$$A'(r) = 2\pi r + \tfrac{1}{8}(L - 2\pi r)(-2\pi)$$
$$= 2\pi r - \tfrac{\pi}{4}(L - 2\pi r)$$
$$= \tfrac{\pi}{4}(8r - L + 2\pi r)$$

Solve $A'(r) = 0$ to find $r = \dfrac{L}{2(\pi + 4)}$.

Thus, the extreme values of the area function on $\left[0, \dfrac{L}{2\pi}\right]$ must occur either at the endpoints or at the critical number $\dfrac{L}{2\pi + 8}$. Evaluating $A(r)$ at each of these numbers, we find

$$A(0) = \pi (0)^2 + \frac{1}{16}[L - 2\pi(0)]^2 = \frac{L^2}{16}$$

$$A\left(\frac{L}{2\pi}\right) = \pi \left(\frac{L}{2\pi}\right)^2 + \frac{1}{16}\left[L - 2\pi\left(\frac{L}{2\pi}\right)\right]^2 = \frac{L^2}{4\pi}$$

$$A\left(\frac{L}{2\pi + 8}\right) = \pi \left(\frac{L}{2\pi + 8}\right)^2 + \frac{1}{16}\left[L - 2\pi\left(\frac{L}{2\pi + 8}\right)\right]^2 = \frac{L^2}{4(\pi + 4)}$$

Comparing these values, we see that the smallest area occurs at $r = \dfrac{L}{2\pi + 8}$ and the largest area occurs at $x = L/(2\pi)$. To summarize,

1. To maximize the sum of the areas, do not cut the wire at all. Bend the wire to form a circle of radius $r = \dfrac{L}{2\pi}$.

2. To minimize the sum of the areas, cut the wire at the point located $2\pi r = \dfrac{2\pi L}{2\pi + 8}$
$= \dfrac{\pi L}{\pi + 4}$ units from one end, and form the circular part with resulting radius $r = \dfrac{L}{2\pi + 8}$.

If you use a graphing calculator or a computer, you can verify this result (for $L = 1$) by looking at the graph of the area function, as shown in Figure 4.56. ∎

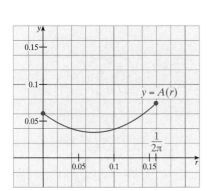

Figure 4.56 Graph of $A(r) = \pi r^2 + \frac{1}{16}(1 - 2\pi r)^2$

EXAMPLE 6   Optimizing an angle of observation

A painting is hung on a wall in such a way that its upper and lower edges are 10 ft and 7 ft above the floor, respectively. An observer whose eyes are 5 ft above the floor stands x feet from the wall, as shown in Figure 4.57. How far away from the wall should the observer stand to maximize the angle subtended by the painting?

Solution

In Figure 4.57, θ is the angle whose vertex occurs at the observer's eyes at a point O located x feet from the wall. Note that α is the angle between the line $\overline{OA}$ drawn horizontally from the observer's eyes to the wall and the line $\overline{OB}$ from the eyes to the bottom edge of the painting. In $\triangle OAB$, the angle at O is α, with $\cot \alpha = \frac{x}{2}$. In $\triangle OAC$, the angle at O is $(\alpha + \theta)$, and $\cot(\alpha + \theta) = \frac{x}{5}$. By using the definition of inverse cotangent, we have

$$\theta = (\alpha + \theta) - \alpha = \cot^{-1}\left(\frac{x}{5}\right) - \cot^{-1}\left(\frac{x}{2}\right)$$

To maximize θ, we first compute the derivative:

$$\frac{d\theta}{dx} = \frac{-1}{\left[1 + \left(\frac{x}{5}\right)^2\right]}\left(\frac{1}{5}\right) - \frac{-1}{\left[1 + \left(\frac{x}{2}\right)^2\right]}\left(\frac{1}{2}\right) = \frac{-5}{25 + x^2} + \frac{2}{4 + x^2}$$

Solving the equation $\dfrac{d\theta}{dx} = 0$ yields

$$-5(4 + x^2) + 2(25 + x^2) = 0$$
$$-3x^2 + 30 = 0$$
$$x = \pm\sqrt{10}$$

Because distance must be nonnegative, we reject the negative value. We apply the first-derivative test to show that the positive critical number $\sqrt{10}$ corresponds to a relative maximum (verify). Thus, the angle θ is maximized when the observer stands $\sqrt{10}$ ft away from the wall. ∎

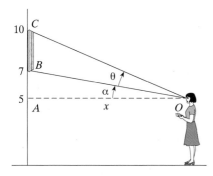

Figure 4.57 The angle θ is subtended by the observer's eye

FERMAT'S PRINCIPLE OF OPTICS AND SNELL'S LAW

Light travels at different rates in different media; the more optically dense the medium, the slower the speed of transit. Consider the situation shown in Figure 4.58, in which a beam of light originates at a point A in one medium, then strikes the upper surface of a second, denser medium at a point P, and is refracted to a point B in the second medium.

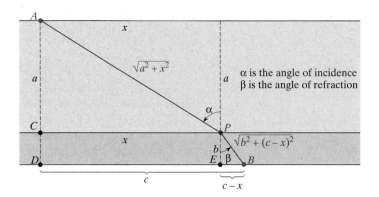

Figure 4.58 The path of a light beam through two media of different density

Suppose light travels with speed v_1 in the first medium and speed v_2 in the second. What can be said about the path followed by the beam of light?

Our method of investigating this question is based on the following optical property.

Fermat's Principle of Optics

Light travels between two points in such a way as to minimize the time of transit.

This problem is very similar to the dune buggy problem of Example 4: Light minimizes the time, and determining minimum time was the goal of that example. In Figure 4.58, let $a = |\overline{AC}|$, $b = |\overline{CD}|$, and $c = |\overline{DB}|$, and let x denote the distance from C to P. Because A and B are fixed points, the path APB is determined by the location of P, which, in turn, is determined by x. Because v_1 is a given constant, the time required for the light to travel from A to P is given by

$$T_1 = \frac{|\overline{AP}|}{v_1} = \frac{\sqrt{a^2 + x^2}}{v_1}$$

and the time required for the light to go from P to B is

$$T_2 = \frac{|\overline{PB}|}{v_2} = \frac{\sqrt{b^2 + (c - x)^2}}{v_2}$$

for another given constant v_2. Therefore, the total time of transit is

$$T = T_1 + T_2 = \frac{\sqrt{a^2 + x^2}}{v_1} + \frac{\sqrt{b^2 + (c - x)^2}}{v_2}$$

where it is clear that $0 \leq x \leq c$. According to Fermat's principle, the path followed by the beam of light is the one that corresponds to the smallest possible value of T; that is, we want to minimize T as a function of x.

Toward this end, we begin by finding $\dfrac{dT}{dx}$.

$$\frac{dT}{dx} = \frac{x}{v_1 \sqrt{a^2 + x^2}} - \frac{c - x}{v_2 \sqrt{b^2 + (c - x)^2}}$$

Note from Figure 4.58 that if α is the angle of incidence of the beam of light and β is the angle of refraction, then (from the definition of sine)

$$\sin \alpha = \frac{x}{\sqrt{a^2 + x^2}} \quad \text{and} \quad \sin \beta = \frac{c - x}{\sqrt{b^2 + (c - x)^2}}$$

Therefore (by substitution), we see the derivative of T can be expressed as

$$\frac{dT}{dx} = \frac{\sin \alpha}{v_1} - \frac{\sin \beta}{v_2}$$

and it follows that the only critical number occurs when

$$\frac{\sin \alpha}{\sin \beta} = \frac{v_1}{v_2}$$

By using the first-derivative test, it can be shown that this critical number corresponds to an absolute minimum. The corresponding value of x enables us to locate P and hence to determine the path followed by the beam of light. We have established the following law of optics.

Snell's Law of Refraction

If a beam of light strikes the boundary between two media with angle of incidence α and is refracted through an angle β, then

$$\frac{\sin \alpha}{\sin \beta} = \frac{v_1}{v_2}$$

where v_1 and v_2 are the rates at which light travels through the first and second medium, respectively. The constant ratio

$$n = \frac{\sin \alpha}{\sin \beta}$$

is called the **relative index of refraction** of the two media.

4.6 PROBLEM SET

A **1. WHAT DOES THIS SAY?**   Describe an optimization procedure.

2. WHAT DOES THIS SAY?   What is Fermat's principle of optics?

B **3.** A woman plans to fence off a rectangular garden whose area is 64 ft². What should be the dimensions of the garden if she wants to minimize the amount of fencing used?

4. The highway department is planning to build a rectangular picnic area for motorists along a major highway. It is to have an area of 5,000 yd² and is to be fenced off on the three sides not adjacent to the highway. What is the least amount of fencing that will be needed to complete the job?

5. Exploration Problem   Pull out a sheet of $8\frac{1}{2}$-in. by 11-in. engineering or binder paper. Cut squares from the corners and fold the sides up to form a container. Show that the maximum volume of such a container is about 1 liter.

6. Journal Problem: *Parabola.* ■ Farmer Jones has to build a fence to enclose a 1,200 m² rectangular area $ABCD$. Fencing

costs \$3 per meter, but Farmer Smith has agreed to pay half the cost of fencing $\overline{CD}$, which borders the property. Given x is the length of side $\overline{CD}$, what is the minimum amount (to the nearest cent) Jones has to pay?

7. Find the rectangle of largest area that can be inscribed in a semicircle of radius R, assuming that one side of the rectangle lies on the diameter of the semicircle.

8. A cylindrical container with no top is to be constructed to hold a fixed volume V_0 of liquid. The cost of the material used for the bottom is 50¢/in.², and the cost of the material used for the curved lateral side is 30¢/in.² Use calculus to find the radius (in terms of V_0) of the least expensive container.

9. Find the dimensions of the right circular cylinder of largest volume that can be inscribed in a sphere of radius R.

10. Given a sphere of radius R, find the radius r and altitude $2h$ of the right circular cylinder with largest lateral surface area that can be inscribed in the sphere.

11. Each edge of a square has length L, as shown in Figure 4.59.

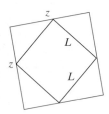

Figure 4.59 Square in a square

Determine the edge of the square of largest area that can be circumscribed about the given square.

12. Find the dimensions of the right circular cylinder of largest volume that can be inscribed in a right circular cone of radius R and altitude H. *Hint*: Begin with Figure 4.60.

Figure 4.60 Cylinder in a cone

13. A truck is 250 mi due east of a sports car and is traveling west at a constant speed of 60 mi/h. Meanwhile, the sports car is going north at 80 mi/h. When will the truck and the car be closest to each other? What is the minimum distance between them? *Hint*: Minimize the square of the distance.

14. Show that of all rectangles with a given perimeter, the square has the largest area.

15. Show that of all rectangles with a given area, the square has the smallest perimeter.

16. A closed box with square base is to be built to house an ant colony. The bottom of the box and all four sides are to be made of material costing \$1/ft², and the top is to be constructed of glass costing \$5/ft². What are the dimensions of the box of greatest volume that can be constructed for \$72?

17. According to postal regulations, the girth plus the length of a parcel sent by fourth-class mail may not exceed 108 in. What is the largest possible volume of a rectangular parcel with two square sides that can be sent by fourth-class mail?

18. The bottom of an 8-ft-high mural painted on a vertical wall is 13 ft above the ground. The lens of a camera fixed to a tripod is 4 ft above the ground. How far from the wall should the camera be placed to photograph the mural with the largest possible angle?

19. Missy Smith is at a point A on the north bank of a long, straight river 6 mi wide. Directly across from her on the south bank is a point B, and she wishes to reach a cabin C located s mi down the river from B, as shown in Figure 4.61.

Figure 4.61 Getting Missy home

Given that Missy can row at 6 mi/h (including the effect of the current) and run at 10 mi/h, what is the minimum time (to the nearest minute) required for her to travel from A to C in each case?

a. $s = 4$ **b.** $s = 6$

20. Modeling Problem Two towns A and B are 12.0 mi apart and are located 5.0 and 3.0 mi, respectively, from a long, straight highway, as shown in Figure 4.62.

Figure 4.62 Building the shortest road

A construction company has a contract to build a road from A to the highway and then to B. Analyze a model to determine the length (to the nearest tenth of a mile) of the *shortest* road that meets these requirements.

21. A poster is to contain 108 cm² of printed matter, with margins of 6 cm each at top and bottom and 2 cm on the sides. What is the minimum cost of the poster if it is to be made of material costing 20 ¢/cm²?

22. An isosceles trapezoid has a base of 14 cm and slant sides of 6 cm, as shown in Figure 4.63. What is the largest area of such a trapezoid?

Figure 4.63 Area of a trapezoid

23. Spy Problem It is noon. The Spy has returned from space (Problem 63 of Section 3.4) and is driving a jeep through the sandy desert in the tiny principality of Alta Loma. He is 32 km from the nearest point on a straight, paved road. Down the road 16 km is a power plant in which a band of international terrorists has placed a time bomb set to explode at 12:50 P.M. The jeep can travel at 48 km/h on the sand and at 80 km/h on the paved road. If he arrives at the power plant in the shortest possible time, how long will our hero have to defuse the bomb?

24. A storage bin is to be constructed by removing a sector with central angle θ from a circular piece of tin of radius 10 ft and folding the remainder of the tin to form a cone, as shown in Figure 4.64. What is the maximum volume of a storage bin formed in this fashion?

Figure 4.64 Fabricating a cone from a sheet of tin

25. Exploration Problem Use the fact that 12 oz $\approx$ 355 mL = 355 cm³ to find the dimensions of the 12-oz Coke® can that can be constructed by using the least amount of metal. Compare these dimensions with a Coke from your refrigerator. What do you think accounts for the difference? An interesting article that discusses a similar question regarding tuna fish cans and the resulting responses is "What Manufacturers Say about a Max/Min Application," by Robert F. Cunningham, Trenton State College, *The Mathematics Teacher*, March 1994, pp. 172–175.

26. A stained glass window in the form of an equilateral triangle is built on top of a rectangular window, as shown in Figure 4.65.

Figure 4.65 Maximizing the amount of light

The rectangular part of the window is of clear glass and transmits twice as much light per square foot as the triangular part, which is made of stained glass. If the entire window has a perimeter of 20 ft, find the dimensions (to the nearest ft) of the window that will admit the most light.

27. Figure 4.66 shows a thin lens located p cm from an object AB and q cm from the image RS.

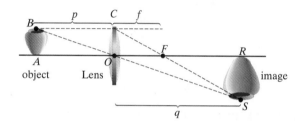

Figure 4.66 Image from a lens

The distance f from the center O of the lens to the point labeled F is called the focal length of the lens.

a. Using similar triangles, show that

$$\frac{1}{p} + \frac{1}{q} = \frac{1}{f}$$

b. Suppose a lens maker wishes to have $p + q = 24$. What is the largest value of f for which this condition can be satisfied?

28. One end of a cantilever beam of length L is built into a wall and the other end is supported by a single post. The deflection, or "sag," of the beam at a point located x units from the built-in end is modeled by the formula

$$D = k(2x^4 - 5Lx^3 + 3L^2x^2)$$

where k is a positive constant. Where does the maximum deflection occur on the beam?

29. When a mechanical system is at rest in an equilibrium position, its potential energy is minimized with respect to any small change in its position. Figure 4.67 shows a system involving a pulley, two small weights of mass m, and a larger weight of mass M.

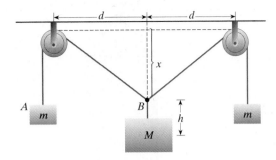

Figure 4.67   A pulley system

In physics, the total potential energy of the system is modeled by

$$E = -Mg(x + h) - 2mg(L - \sqrt{x^2 + d^2})$$

where x, h, and d are the distances shown in Figure 4.67, L is the length of the cord from A around one pulley to B, and g is the constant acceleration due to gravity. All the other symbols represent constants. Use this information to find the critical value(s) for E.

30. A resistor of R ohms is connected across a battery of E volts whose internal resistance is r ohms. According to the principles of electricity, the formula

$$I = \frac{E}{R + r}$$

gives the current (amperes) in the circuit, while

$$P = I^2 R$$

gives the power (watts) in the external resistor. Assuming that E and r are constant, what value of R will result in maximum power in the external resistor?

31. **Modeling Problem**   One model of a computer disk storage system uses the function

$$T(x) = N\left(k + \frac{c}{x}\right) p^{-x}$$

for the average time needed to send a file correctly by modem (including all retransmission of messages in which errors are detected), where x is the number of information bits, p is the (fixed) probability that any particular file will be received correctly, and N, k, and c are positive constants.[*]

a. Find $T'(x)$.

b. For what value of x is $T(x)$ minimized?

c. Sketch the graph of T.

32. **Modeling Problem**   In crystallography, a fundamental problem is the determination of the *packing fraction* of a crystal lattice, which is defined as the fraction of space occupied by the atoms in the lattice, assuming the atoms are hard spheres. When the lattice contains exactly two different kinds of atoms, the packing fraction is modeled by the formula

$$F(x) = \frac{K(1 + c^2 x^3)}{(1 + x)^3}$$

where $x = r/R$ is the ratio of the radii of the two kinds of atoms in the lattice and c and K are positive constants.[†]

a. The function F has exactly one critical number. Find it and use the second-derivative test to determine whether it corresponds to a relative maximum or a relative minimum.

b. The numbers c and K and the domain of F depend on the cell structure in the lattice. For ordinary rock salt, it turns out that $c = 1$, $K = 2\pi/3$, and the domain is $[\sqrt{2} - 1, 1]$. Find the largest and smallest values of F on this domain.

c. Repeat part **b** for β-cristobalite, for which $c = \sqrt{2}$, $K = \sqrt{3}\pi/6$, and the domain is $[0, 1]$.

33. If air resistance is neglected, it can be shown that the stream of water emitted by a fire hose will have height

$$y = -16(1 + m^2)(x/v)^2 + mx$$

above a point located x ft from the nozzle, where m is the slope of the nozzle, and v is the velocity of the stream of water as it leaves the nozzle. (See Figure 4.68.) Assume v is constant.

a. For fixed m, determine the distance x that results in maximum height.

b. If m is allowed to vary but the firefighter must stand $x = x_0$ ft from the base of a burning building, what is the highest point on the building that the firefighter can reach with the water from her hose?

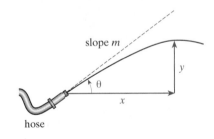

Figure 4.68   Stream of water from a fire hose

34. A telecommunications satellite is located

$$r(t) = \frac{4,831}{1 + 0.15 \cos 0.06t}$$

miles above the center of the earth t minutes after achieving orbit.

a. Sketch the graph of $r(t)$.

b. What are the lowest and highest points on the satellite's orbit? These are called the *perigee* and *apogee* positions, respectively.

---

[*]Paul J. Campbell, "Calculus Optimization in Information Technology," *UMAP module 1991: Tools for Teaching.* Lexington, MA: CUPM Inc., 1992; pp. 175–199.

[†]John C. Lewis and Peter P. Gillis, "Packing Factors in Diatomic Crystals," *American Journal of Physics,* Vol. 61, No. 5 (1993), pp. 434–438.

35. According to relativity theory, a particle's energy is related to its rest mass m_0, and its wavelength λ by the formula

$$E = \sqrt{\left(\frac{hc}{\lambda}\right)^2 + m_0^2 c^4}$$

where m_0 is the rest mass of the particle and h is a constant (Planck's constant). This equation is used mostly in atomic, nuclear, and particle physics where the masses are not very large. Sketch the graph of $E(\lambda)$. What happens to E as $\lambda \to +\infty$?

36. The theory of relativity models the Doppler shift s of an object traveling at a velocity v:

$$s = \sqrt{\frac{c + v}{c - v}} - 1$$

where c is the speed of light and $v < c$.

a. Express v as a function of s and sketch the graph of $v(s)$.

b. Certain stellar objects called quasars appear to be moving away at velocities approaching c. The fastest one has a redshift of $s = 3.78$. What is its velocity in relation to c?

c. How fast is the velocity changing with respect to s when $s = 3.78$?

○ 37. The lower right-hand corner of a piece of paper is folded over to reach the leftmost edge, as shown in Figure 4.69.

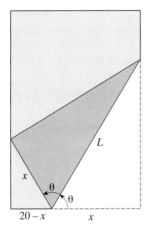

Figure 4.69 Paper folding problem

If the page is 20 cm wide and 30 cm long, what is the length L of the shortest possible crease? *Hint*: First express $\cos \theta$ and $\cos 2\theta$ in terms of x and then use the trigonometric identity

$$\cos 2\theta = 2\cos^2 \theta - 1$$

to eliminate θ and express L in terms of x alone.

38. Find the length of the longest pipe that can be carried horizontally around a corner joining two corridors that are $2\sqrt{2}$ ft wide. *Hint*: Show that the length L can be written as

$$L(\theta) = \frac{2\sqrt{2}}{\sin \theta} + \frac{2\sqrt{2}}{\cos \theta}$$

and find the absolute minimum of $L(\theta)$ on an appropriate interval, as shown in Figure 4.70.

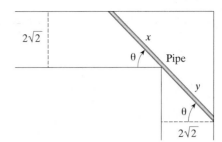

Figure 4.70 Corner problem

39. It is known that water expands and contracts according to its temperature. Physical experiments suggest that an amount of water that occupies 1 liter at $0°C$ will occupy

$$V(T) = 1 - 6.42 \times 10^{-5} T$$
$$+ 8.51 \times 10^{-6} T^2 - 6.79 \times 10^{-8} T^3$$

liters when the temperature is $T°C$. At what temperature is $V(T)$ minimized? How is this result related to the fact that ice forms only at the upper levels of a lake during winter?

40. Modeling Problem Light emanating from a source A is reflected by a mirror to a point B, as shown in Figure 4.71. Use Fermat's principle of optics to show that the angle of incidence α equals the angle of reflection β.

Figure 4.71 Angle of incidence is equal to the angle of reflection

41. Congruent triangles are cut out of a square piece of paper 20 cm on a side, leaving a starlike figure that can be folded to form a pyramid, as indicated in Figure 4.72.

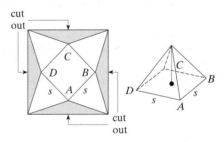

Figure 4.72 Constructing a pyramid

What is the largest volume of the pyramid that can be formed in this manner?

42. Modeling Problem A universal joint is a coupling used in cars and other mechanical systems to join rotating shafts together at an angle. In a model designed to explore the mechanics of the universal joint mechanism, the angular velocity $\beta(t)$ of the output (driven) shaft is modeled by the formula

$$\beta(t) = \frac{\alpha \cos \gamma}{1 - \sin^2 \gamma \sin^2(\alpha t)}$$

where α (in rad/s) is the angular velocity of the input (driving) shaft, and γ is the angle between the two shafts, as shown in Figure 4.73.*

Figure 4.73 Universal joint

a. Sketch the graph of $\beta(t)$ for the case where α and γ are both positive constants.

b. What are the largest and smallest values of $\beta(t)$?

43. **Modeling Problem** Rainbows are formed when sunlight traveling through the air is both reflected and refracted by raindrops.[†] Figure 4.74 shows a raindrop, which we assume to be a sphere for simplicity. An incoming beam of sunlight strikes the raindrop at point A with angle of incidence α, and some of the light is refracted through the angle β to point B. The process is then reversed, as indicated in the figure, and the light finally exits the drop at point C.

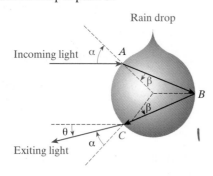

Figure 4.74 Light passing through a raindrop

---

*Thomas O'Neil, "A Mathematical Model of a Universal Joint," *UMAP Modules 1982: Tools for Teaching*. Lexington, MA: Consortium for Mathematics and Its Applications, Inc., 1983, pp. 393–405.

[†]This problem is based on an article by Steve Janke, "Somewhere Within the Rainbow," *UMAP Modules 1992: Tools for Teaching*. Lexington, MA: Consortium for Mathematics and Its Applications, Inc., 1993.

a. At the interface points, A, B, and C, the light beam is deflected (from a straight line path) and loses intensity. For instance, at A, the incoming beam is deflected through a clockwise rotation of $(\alpha - \beta)$ radians. Show that the total deflection (at all three interfaces) is

$$D = \pi + 2\alpha - 4\beta$$

b. For raindrops falling through the air, Snell's law of refraction has the form $\sin \alpha = 1.33 \sin \beta$. Use this fact to express the derivative $D'(\alpha)$ in terms of α and β.

c. The intensity of rainbow light reaching the eye of an observer will be greatest when the total deflection $D(\alpha)$ is minimized. If the minimum deflection occurs when $\alpha = \alpha_0$, the corresponding angle of observation $\theta = \pi - D(\alpha_0)$ is called the *rainbow angle* because the rainbow is brightest when viewed in that direction (see Figure 4.75).

Figure 4.75 A rainbow is brightest when viewed at the rainbow angle

Solve $D'(\alpha) = 0$ and show that the critical number you find minimizes $D(\alpha)$. Then find the rainbow angle θ. *Hint:* You may assume $0 \leq \alpha - \beta \leq \frac{\pi}{2}$.

4.7   Optimization in Business, Economics, and the Life Sciences

IN THIS SECTION **economics**: maximizing profit and marginal analysis, **business management**: an inventory model and optimal holding time, **physiology**: concentration of a drug in the bloodstream and optimal angle for vascular branching

In the previous section, we saw how differentiation can be used to solve optimization problems in the physical sciences and engineering. The methods of calculus are also used in business and economics and are becoming increasingly prominent in the social and life sciences. We will examine a selection of such applications, and we begin with two modeling problems involving maximization of profit.

ECONOMICS

Maximizing Profit

EXAMPLE 1 Maximizing profits

A manufacturer can produce a pair of earrings at a cost of $3. The earrings have been selling for $5 per pair and at this price, consumers have been buying 4,000 pairs per month. The manufacturer is planning to raise the price of the earrings and estimates that for each $1 increase in the price, 400 fewer pairs of earrings will be sold each month. At what price should the manufacturer sell the earrings to maximize profit?

Solution

Let x denote the number of $1 price increases, and let $P(x)$ represent the corresponding profit.

$$\begin{aligned}\text{PROFIT} &= \text{REVENUE} - \text{COST}\\ &= (\text{NUMBER SOLD})(\text{PRICE PER PAIR}) - (\text{NUMBER SOLD})(\text{COST PER PAIR})\\ &= (\text{NUMBER SOLD})(\text{PRICE PER PAIR} - \text{COST PER PAIR})\end{aligned}$$

Recall that 4,000 pairs of earrings are sold each month when the price is $5 per pair and 400 fewer pairs will be sold each month for each added dollar in the price. Thus,

$$\begin{aligned}\text{NUMBER OF PAIRS SOLD} &= 4{,}000 - 400(\text{NUMBER OF \$1 INCREASES})\\ &= 4{,}000 - 400x\end{aligned}$$

Knowing that the price per pair is $5 + x$, we can now write the profit as a function of x:

$$\begin{aligned}P(x) &= (\text{NUMBER SOLD})(\text{PRICE PER PAIR} - \text{COST PER PAIR})\\ &= (4{,}000 - 400x)[(5 + x) - 3]\\ &= 400(10 - x)(2 + x)\end{aligned}$$

To find the domain, we note that $x \geq 0$. And $400(10 - x)$, the number of pairs sold, should be nonnegative, so $x \leq 10$. Thus, the domain is $[0, 10]$.

The critical numbers are found when the derivative is 0 (P is a polynomial function, so there are no values for which the derivative is not defined):

$$\begin{aligned}P'(x) &= 400(10 - x)(1) + 400(-1)(2 + x)\\ &= 400(8 - 2x) = 800(4 - x)\end{aligned}$$

The critical number is $x = 4$ and the endpoints are $x = 0$ and $x = 10$. Checking for the maximum profit,

$$P(4) = 400(10 - 4)(2 + 4) = 14{,}400; \quad P(0) = 8{,}000; \quad P(10) = 0$$

The maximum possible profit is $14,400, which will be generated if the earrings are sold for $5.00 + $4.00 = $9.00 per pair. The graph of the profit function is shown in Figure 4.76. ∎

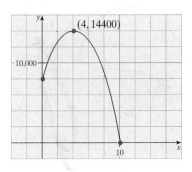

Figure 4.76 The profit function $P(x)$

Sometimes the function to be optimized has practical meaning only when its independent variable is a positive integer. Such functions are said to have a discrete domain and are called **discrete functions**. Technically, the methods of calculus cannot be applied to discrete functions because the theorems we have developed apply only to continuous functions. However, useful information about a discrete function $f(n)$ can often be obtained by analyzing the related continuous function $f(x)$ obtained by replacing the discrete variable n by a real variable x. Here is an example that illustrates the general procedure.

EXAMPLE 2   Maximizing a discrete revenue function

A travel company plans to sponsor a special tour to Africa. There will be accommodations for no more than 40 people, and the tour will be canceled if no more than 10 people book reservations. Based on past experience, the manager determines that if n people book the tour, the profit (in dollars) may be modeled by the function

$$P(n) = -n^3 + 27.6n^2 + 970.2n - 4,235$$

For what size tour group is profit maximized?

Solution

The domain of $P(n)$ is the set of integers n between 10 and 40. To optimize this function, we apply the methods developed in this chapter to the related continuous function

$$P(x) = -x^3 + 27.6x^2 + 970.2x - 4,235$$

where x is a real variable defined on the interval $10 \leq x \leq 40$. To determine the critical numbers of $P(x)$, we solve

$$P'(x) = -3x^2 + 55.2x + 970.2 = 0$$

and obtain $x = 29.4$ and $x = -11$. Only $x = 29.4$ is in the domain, and by testing the sign of $P'(x)$ on each side of this critical number, we find that the continuous function $P(x)$ is increasing for $0 < x < 29.4$ and decreasing for $29.4 < x < 40$. Therefore, the original discrete function $P(n)$ is maximized either at $n = 29$ or at $n = 30$. Since

$$P(29) = 22{,}723.4 \quad \text{and} \quad P(30) = 22{,}711$$

we conclude that the maximum profit is \$22,723.40, and occurs when 29 people book the tour. The graphs of $P(n)$ and the related continuous function $P(x)$ are shown in Figure 4.77. ∎

Figure 4.77 Discrete and continuous graphs of the profit function P

Marginal Analysis   In Section 3.8, we described *marginal analysis* as that branch of economics that is concerned with the way quantities such as price, cost, revenue, and profit vary with small changes in the level of production. Specifically, recall that if x is the number of units produced and brought to market, then the total cost of producing the units is denoted by $C(x)$. The **demand** function $p(x)$ is defined to be the price that consumers will pay for each unit of the commodity when x units are brought to market. Then $R(x) = xp(x)$ is the **total revenue** derived from the sale of the x units and $P(x) = R(x) - C(x)$ is the **total profit**. The relationship among revenue, cost, and profit is shown in Figure 4.78.

Figure 4.78   Cost, revenue, and profit functions

In Section 3.8, we worked with marginal quantities and their role as rates of change, namely, $C'(x)$, the *marginal cost*, and $R'(x)$, the *marginal revenue*.*

We will now consider these functions in optimization problems. For example, the manufacturer certainly would like to know what level of production results in maximum profit. To solve this problem, we want to maximize the profit function

$$P(x) = R(x) - C(x)$$

We differentiate with respect to x and find that

$$P'(x) = R'(x) - C'(x)$$

Thus, $P'(x) = 0$ when $R'(x) = C'(x)$, and by using economic arguments we can show that a maximum occurs at the corresponding critical point.

Maximum Profit

Profit is maximized when marginal revenue equals marginal cost.

EXAMPLE 3   Using marginal analysis to maximize profit

A manufacturer estimates that when x units of a particular commodity are produced each month, the total cost (in dollars) will be

$$C(x) = \tfrac{1}{8}x^2 + 4x + 200$$

and all units can be sold at a price of $p(x) = 49 - x$ dollars per unit. Determine the price that corresponds to the maximum profit.

Solution

The marginal cost is $C'(x) = \tfrac{1}{4}x + 4$. The revenue is

$$
\begin{aligned}
R(x) &= xp(x) \\
&= x(49 - x) \\
&= 49x - x^2
\end{aligned}
$$

The marginal revenue is $R'(x) = 49 - 2x$. The profit is maximized when $R'(x) = C'(x)$:

$$
\begin{aligned}
R'(x) &= C'(x) \\
49 - 2x &= \tfrac{1}{4}x + 4 \\
x &= 20
\end{aligned}
$$

Thus, the price that corresponds to the maximum profit is $p(20) = 49 - 20 = 29$ dollars/unit.   ∎

A second general principle of economics involves the following relationship between marginal cost and the *average cost* $A(x) = \dfrac{C(x)}{x}$.

Minimum Average Cost

Average cost is minimized at the level of production where the marginal cost equals the average cost.

---

*Recall (from Section 3.8) that marginal cost is the instantaneous rate of change of production cost C with respect to the number of units x_0 produced and the marginal revenue is the instantaneous rate of change of revenue with respect to the number of units x_0 produced.

To justify this second business principle, find the derivative of the average cost function:

$$A'(x) = \frac{xC'(x) - C(x)}{x^2} = \frac{C'(x) - \dfrac{C(x)}{x}}{x} = \frac{C'(x) - A(x)}{x}$$

Thus, $A'(x) = 0$ when $C'(x) = A(x)$. Once again, economic theory can be used to justify this result as a minimum.

EXAMPLE 4 Using marginal analysis to minimize average cost

A manufacturer estimates that when x units of a particular commodity are produced each month, the total cost (in dollars) will be

$$C(x) = \tfrac{1}{8}x^2 + 4x + 200$$

and they can all be sold at a price of $p(x) = 49 - x$ dollars per unit. Determine the level of production at which the average cost is minimized.

Solution

The average cost is $A(x) = \dfrac{C(x)}{x} = \dfrac{1}{8}x + 4 + 200x^{-1}$ and $A(x)$ is minimized when $C'(x) = A(x)$. Thus,

$$\tfrac{1}{4}x + 4 = \tfrac{1}{8}x + 4 + 200x^{-1}$$
$$x^2 = 1600 \qquad \text{Multiply both sides by } 8x \text{ and simplify.}$$

If we disregard the negative solution, it follows that the minimal average cost occurs when $x = 40$ units.

You might have noticed that we did not need the information $p(x) = 49 - x$ in arriving at the solution to this example. When doing real-world modeling, you must often make a choice about which parts of the available information are necessary to solve the problem. ■

BUSINESS MANAGEMENT

Sometimes mathematical methods can be used to assist managers in making certain business decisions. As an illustration, we will show how calculus can be applied to a problem involving inventory control.

An Inventory Model For each shipment of raw materials, a manufacturer must pay an ordering fee to cover handling and transportation. When the raw materials arrive, they must be stored until needed and storage costs result. If each shipment of raw materials is large, few shipments will be needed and ordering costs will be low, but storage costs will be high. If each shipment is small, ordering costs will be high because many shipments will be needed, but storage cost will be low. Managers want to determine the shipment size that will minimize the total cost. Here is an example of how such a problem may be solved using calculus.

EXAMPLE 5 Modeling Problem: Managing inventory to minimize cost

A retailer buys 6,000 calculator batteries a year from a distributor and is trying to decide how often to order the batteries. The ordering fee is $20 per shipment, the storage cost is $0.96 per battery per year, and each battery costs the retailer $0.25. Suppose that the batteries are sold at a constant rate throughout the year and that each shipment arrives uniformly throughout the year just as the preceding shipment has been used up. How many batteries should the retailer order each time to minimize the total cost?

Solution

We begin by writing the cost function:

$$\text{TOTAL COST} = \text{STORAGE COST} + \text{ORDERING COST} + \text{COST OF BATTERIES}$$

We need to find an expression for each of these unknowns. Assume that the same number of batteries must be ordered each time an order is placed; denote this number by x so that $C(x)$ is the corresponding total cost. The average number of batteries in storage during the year is half of a given order (that is, $x/2$), and we assume that the total yearly storage cost is the same as if the $x/2$ batteries were kept in storage for the entire year. This situation is shown in Figure 4.79.

Figure 4.79 Inventory graph

$$\text{STORAGE COST} = \begin{pmatrix} \text{AVERAGE NUMBER} \\ \text{IN STORAGE PER YR} \end{pmatrix} \begin{pmatrix} \text{COST OF STORING 1} \\ \text{BATTERY FOR 1 YR} \end{pmatrix}$$

$$= \left(\frac{x}{2}\right)(0.96)$$

$$= 0.48x$$

To find the total ordering cost, we can multiply the ordering cost per shipment by the number of shipments. We also note that because 6,000 batteries are ordered during the year and because each shipment contains x batteries, the number of shipments is $6,000/x$.

$$\text{ORDERING COST} = \begin{pmatrix} \text{ORDERING COST} \\ \text{PER SHIPMENT} \end{pmatrix} \begin{pmatrix} \text{NUMBER OF} \\ \text{SHIPMENTS} \end{pmatrix}$$

$$= (20)\left(\frac{6,000}{x}\right) = \frac{120,000}{x}$$

Finally, we must also include the cost of purchasing the batteries:

$$\text{COST OF BATTERIES} = \begin{pmatrix} \text{TOTAL NUMBER} \\ \text{OF BATTERIES} \end{pmatrix} \begin{pmatrix} \text{COST PER} \\ \text{BATTERY} \end{pmatrix}$$

$$= 6,000(0.25) = 1,500$$

Thus, we model the total cost by the function

$$C(x) = 0.48x + \frac{120,000}{x} + 1,500$$

The goal is to minimize $C(x)$ on $(0, 6000]$. To obtain the critical numbers, we find the derivative

$$C'(x) = 0.48 - 120,000x^{-2}$$

and then solve $C'(x) = 0$:

$$0.48 - 120,000x^{-2} = 0$$
$$0.48x^2 = 120,000$$
$$x = \pm 500$$

Figure 4.80 The total cost function $C(x) = 0.48x + 120,000x^{-1}$ $+1,500$

The root $x = -500$ does not lie in the interval $(0, 6000]$. It is easy to check that C is decreasing on $(0, 500)$ and increasing on $(500, 6000]$, as shown in Figure 4.80. Thus, the absolute minimum of C on the interval $(0, 6000]$ occurs when $x = 500$, and we conclude that to minimize cost, the manufacturer should order the batteries in lots of 500.   ■

Optimal Holding Time   Even with an asset that increases in value, there often comes a time when continuing to hold the asset is less advantageous than selling it and investing the proceeds of the sale. How should the investor decide when to sell? One way is to hold the asset until the time its present value at the current prevailing rate of interest is maximized. In other words, hold until today's dollar equivalent of the selling price is as large as possible, and then sell. Here is an example that illustrates this strategy.

EXAMPLE 6   Modeling Problem: Optimal holding time

Suppose you own an asset whose market price t years from now is modeled by $V(t) = 10,000e^{\sqrt{t}}$ dollars. If the prevailing rate of interest is 8% compounded continuously, when should the asset be sold?

Solution

The present value of the asset in t years is modeled by the function $P(t) = V(t)e^{-rt}$, where r is the annual interest rate and t is the time in years. Thus,

$$P(t) = 10,000e^{\sqrt{t}}e^{-0.08t}$$
$$= 10,000\exp(\sqrt{t} - 0.08t)$$

To maximize P, find $P'(t)$ and solve $P'(t) = 0$:

$$P'(t) = 10,000\exp(\sqrt{t} - 0.08t)\left(\frac{1}{2}\cdot\frac{1}{\sqrt{t}} - 0.08\right)$$

$P'(t) = 0$ when $\dfrac{1}{2\sqrt{t}} - 0.08 = 0$ or $t \approx 39.06$ years. Thus, the asset should be held for 39 years and then sold.   ■

PHYSIOLOGY

Calculus can be used to model a variety of situations in the biological and life sciences. We will consider two examples from physiology. In the first, we use a model for the concentration of a drug in the bloodstream of a patient to determine when the maximum concentration occurs.

Concentration of a Drug in the Bloodstream

EXAMPLE 7   Modeling Problem: Maximum concentration of a drug

Let $C(t)$ denote the concentration in the blood at time t of a drug injected into the body intramuscularly. In a classic paper by E. Heinz, it was observed that the concentration may be modeled by

$$C(t) = \frac{k}{b - a}(e^{-at} - e^{-bt})   \qquad t \geq 0$$

where a, b (with $b > a$), and k are positive constants that depend on the drug.* At what time does the largest concentration occur? What happens to the concentration as $t \to +\infty$?

Solution

To locate the extrema, we solve $C'(t) = 0$.

$$C'(t) = \frac{d}{dt}\left[\frac{k}{b - a}(e^{-at} - e^{-bt})\right]$$
$$= \frac{k}{b - a}[(-a)e^{-at} - (-b)e^{-bt}] = \frac{k}{b - a}(be^{-bt} - ae^{-at})$$

---

*E. Heinz, "Probleme bei der Diffusion kleiner Substanzmengen innerhalb des menschlichen Körpers," *Biochem.*, Vol. 319 (1949), pp. 482–492.

We see that $C'(t) = 0$ when

$$be^{-bt} = ae^{-at}$$

$$\frac{b}{a} = e^{bt}e^{-at} = e^{bt-at}$$

$$bt - at = \ln\frac{b}{a} \qquad \text{Definition of logarithm}$$

$$t = \frac{1}{b-a}\ln\frac{b}{a}$$

The second-derivative test can be used to show that the largest value of $C(t)$ occurs at this t (see Problem 49).

To see what happens to the concentration as $t \to +\infty$, we compute the limit

$$\lim_{t\to+\infty} C(t) = \lim_{t\to+\infty} \frac{k}{b-a}[e^{-at} - e^{-bt}]$$

$$= \frac{k}{b-a}\left[\lim_{t\to+\infty}\frac{1}{e^{at}} - \lim_{t\to+\infty}\frac{1}{e^{bt}}\right]$$

$$= \frac{k}{b-a}[0 - 0]$$

$$= 0$$

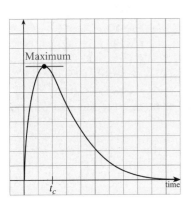

Figure 4.81 Graph of
$$C(t) = \frac{k}{b-a}(e^{-at} - e^{-bt})$$

This tells us that the longer the drug is in the blood, the closer the concentration is to 0. The graph of C is shown in Figure 4.81.

Intuitively, we would expect the Heinz concentration function to begin at 0, increase to a maximum, and then gradually drop off to 0 in a finite amount of time. Figure 4.81 indicates that $C(t)$ does not have these characteristics, because it does not quite get back to 0 in finite time. This suggests that the Heinz model may apply most reliably to the period of time right after the drug has been injected. ∎

Optimal Angle for Vascular Branching The blood vascular system operates in such a way that the circulation of blood—from the heart, through the organs of the body, and back to the heart—is accomplished with as little expenditure of energy as possible. Thus, it is reasonable to expect that when an artery branches, the angle between the "parent" artery and its "daughter" should minimize the total resistance to the flow of blood. Figure 4.82 shows a small artery of radius r branching from a larger artery of radius R. Blood flows in the direction of the arrows from point A to the branch at B and then to points C and D. For simplicity, we assume that C and D are located in such a way that $\overline{CD}$ is perpendicular to the main line through A, B, and D. We wish to find the value of the branching angle θ that minimizes the total resistance to the flow of blood as it moves from A to B and then to point C, which is located a fixed perpendicular distance h from the line through A and B.* (See Figure 4.82.)

Figure 4.82 Vascular branching

> **Poiseuille's Resistance To Flow Law**
>
> The resistance to the flow of blood in an artery is directly proportional to the artery's length and inversely proportional to the fourth power of its radius.

---

*The key to solving this problem is a result due to work of the nineteenth-century French physiologist Jean Louis Poiseuille (1789–1869). Our discussion of vascular branching is adapted from *Introduction to Mathematics for Life Scientists*, 2nd edition, by Edward Batschelet (New York: Springer-Verlag, 1976, pp. 278–280). In this excellent little book, Batschelet develops a number of interesting applications of calculus, several of which appear in the problem set.

According to Poiseuille's law, the resistance to flow, f_1, from A to B is

$$f_1 = \frac{ks_1}{R^4}$$

where R is the radius of the larger artery, and the resistance to flow, f_2, from B to C is

$$f_2 = \frac{ks_2}{r^4}$$

where r is the radius of the smaller artery, k is a viscosity constant, $s_1 = \left|\overline{AB}\right|$, and $s_2 = \left|\overline{BC}\right|$. Thus, the total resistance to flow may be modeled by the sum

$$f = f_1 + f_2 = \frac{ks_1}{R^4} + \frac{ks_2}{r^4} = k\left(\frac{s_1}{R^4} + \frac{s_2}{r^4}\right)$$

The next task is to write f as a function of θ. To do this, reconsider Figure 4.82 by labeling s_1, s_2, h, and ℓ as shown in Figure 4.83.

We want to find equations for s_1 and s_2 in terms of h and θ; to this end we notice that

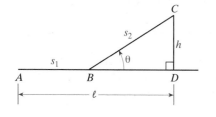

$$\sin\theta = \frac{h}{s_2} \qquad\qquad \tan\theta = \frac{h}{\ell - s_1}$$

$$s_2 = \frac{h}{\sin\theta} \qquad\qquad \ell - s_1 = \frac{h}{\tan\theta}$$

$$= h\csc\theta \qquad\qquad s_1 = \ell - \frac{h}{\tan\theta}$$

$$= \ell - h\cot\theta$$

Figure 4.83 Minimizing the resistance to blood flow

We can now write f as a function of θ:

$$f(\theta) = k\left(\frac{s_1}{R^4} + \frac{s_2}{r^4}\right) = k\left(\frac{\ell - h\cot\theta}{R^4} + \frac{h\csc\theta}{r^4}\right)$$

To minimize f, we need to find the critical numbers (remember that h and k are positive constants):

$$\frac{df}{d\theta} = k\left[\frac{-h(-\csc^2\theta)}{R^4} + \frac{h(-\csc\theta\cot\theta)}{r^4}\right]$$

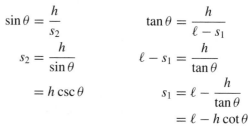

$$= kh\csc\theta\left(\frac{\csc\theta}{R^4} - \frac{\cot\theta}{r^4}\right)$$

$$\left(\frac{\csc\theta}{R^4} - \frac{\cot\theta}{r^4}\right) = 0$$

$$\frac{\csc\theta}{R^4} = \frac{\cot\theta}{r^4}$$

$$\frac{1}{R^4\sin\theta} = \frac{\cos\theta}{r^4\sin\theta}$$

$$\frac{r^4}{R^4} = \cos\theta$$

By finding the second derivative and noting that both r and R are positive, we can show that any value θ_m that satisfies this equation yields a value $f(\theta_m)$ that is a minimum. Thus, the required optimal angle is $\theta_m = \cos^{-1}(r^4/R^4)$.

4.7 PROBLEM SET

A *In Problems 1–3, we give the cost C of producing x units of a particular commodity and the selling price p when x units are produced. In each case, determine the level of production that maximizes profit.*

1. $C(x) = \frac{1}{8}x^2 + 5x + 98$ and $p(x) = \frac{1}{2}(75 - x)$

2. $C(x) = \frac{2}{5}x^2 + 3x + 10$ and $p(x) = \frac{1}{5}(45 - x)$

3. $C(x) = \frac{1}{5}(x + 30)$ and $p(x) = \dfrac{70 - x}{x + 30}$

4. Suppose the total cost of producing x units of a certain commodity is

$$C(x) = 2x^4 - 10x^3 - 18x^2 + x + 5$$

Determine the largest and smallest values of the marginal cost for $0 \le x \le 5$.

5. Suppose the total cost of manufacturing x units of a certain commodity is
$$C(x) = 3x^2 + x + 48$$
dollars. Determine the minimum average cost.

6. A toy manufacturer produces an inexpensive doll (Flopsy) and an expensive doll (Mopsy) in units of x hundreds and y hundreds, respectively. Suppose it is possible to produce the dolls in such a way that

$$y = \frac{82 - 10x}{10 - x} \qquad 0 \le x \le 8$$

and that the company receives *twice* as much for selling a Mopsy doll as for selling a Flopsy doll. Find the level of production for both x and y for which the total revenue derived from selling these dolls is maximized. What vital assumption must be made about sales in this model?

7. A business manager estimates that when p dollars are charged for every unit of a product, the sales will be $x = 380 - 20p$ units. At this level of production, the average cost is modeled by

$$A(x) = 5 + \frac{x}{50}$$

   a. Find the total revenue and total cost functions, and express the profit as a function of x.

   b. What price should the manufacturer charge to maximize profit? What is the maximum profit?

8. Suppose a manufacturer estimates that, when the market price of a certain product is p, the number of units sold will be

$$x = -6\ln\left(\frac{p}{40}\right)$$

It is also estimated that the cost of producing these x units will be

$$C(x) = 4xe^{-x/6} + 30$$

   a. Find the average cost, the marginal cost, and the marginal revenue for this production process.

   b. What level of production x corresponds to maximum profit?

9. A manufacturer can produce shoes at a cost of $50 a pair and estimates that if they are sold for x dollars a pair, consumers will buy approximately

$$s(x) = 1,000e^{-0.1x}$$

pairs of shoes per week. At what price should the manufacturer sell the shoes to maximize profit?

10. A certain industrial machine depreciates in such a way that its value (in dollars) after t years is

$$Q(t) = 20,000e^{-0.4t}$$

   a. At what rate is the value of the machine changing with respect to time after 5 years?

   b. At what percentage rate is the value of the machine changing with respect to time after t years?

11. It is projected that t years from now, the population of a certain country will be $P(t) = 50e^{0.02t}$ million.

   a. At what rate will the population be changing with respect to time 10 years from now?

   b. At what percentage rate will the population be changing with respect to time t years from now?

12. Some psychologists model a child's ability to memorize by a function of the form

$$g(t) = \begin{cases} t\ln t + 1 & \text{if } 0 < t \le 4 \\ 1 & \text{if } t = 0 \end{cases}$$

where t is time, measured in years. Determine when the largest and smallest values of g occur.

13. It is estimated that t years from now the population of a certain country will be

$$p(t) = \frac{160}{1 + 8e^{-0.01t}}$$

million. When will the population be growing most rapidly?

B 14. **Modeling Problem** The owner of the Pill Boxx drugstore expects to sell 600 bottles of hair spray each year. Each bottle costs $4, and the ordering fee is $30 per shipment. In addition, it costs 90¢ per year to store each bottle. Assuming that the hair spray sells at a uniform rate throughout the year and that each shipment arrives just as the last bottle from the previous shipment is sold, how frequently should shipments of hair spray be ordered to minimize the total cost?

15. **Modeling Problem** An electronics firm uses 18,000 cases of connectors each year. The cost of storing one case for a year is $4.50, and the ordering fee is $20 per shipment. Assume that the connectors are used at a constant rate throughout the year and that each shipment arrives just as the preceding shipment has been used up. How many cases should the firm order each time to keep total cost to a minimum?

16. Suppose the total cost (in dollars) of manufacturing x units of a certain commodity is

$$C(x) = 3x^2 + 5x + 75$$

   a. At what level of production is the average cost per unit the smallest?

   b. At what level of production is the average cost per unit equal to the marginal cost?

   c. Graph the average cost and the marginal cost functions on the same set of axes, for $x > 0$.

17. Suppose the total revenue (in dollars) from the sale of x units of a certain commodity is

$$R(x) = -2x^2 + 68x - 128$$

   a. At what level of sales is the average revenue per unit equal to the marginal revenue?

   b. Verify that the average revenue is increasing if the level of sales is less than the level in part **a** and decreasing if the level of sales is greater than the level in part **a**.

   c. On the same set of axes, graph the relevant portions of the average and marginal revenue functions.

18. A manufacturer finds that the demand function for a certain product is

$$x(p) = \frac{73}{\sqrt{p}}$$

Should the price p be raised or lowered to increase consumer expenditure? Explain your answer.

19. Suppose you own a rare book whose value t years from now is modeled as $300e^{\sqrt{3t}}$. If the prevailing rate of interest remains constant at 8% compounded continuously, when will be the most advantageous time to sell?

20. Suppose you own a parcel of land whose value t years from now is modeled as

$$P(t) = 200 \ln \sqrt{2t}$$

thousand dollars. If the prevailing rate of interest is 10% compounded continuously, when will be the most advantageous time to sell?

21. A store has been selling skateboards at the price of $40 per board, and at this price skaters have been buying 45 boards a month. The owner of the store wishes to raise the price and estimates that for each $1 increase in price, 3 fewer boards will be sold each month. If each board costs the store $29, at which price should the store sell the boards to maximize profit?

22. **Modeling Problem** As more and more industrial areas are constructed, there is a growing need for standards ensuring control of the pollutants released into the air. Suppose that the pollution at a particular location is based on the distance from the source of the pollution according to the principle that for distances greater than or equal to 1 mi, the concentration of particulate matter (in parts per million, ppm) decreases as the reciprocal of the distance from the source. This means that if you live 3 mi from a plant emitting 60 ppm, the pollution at your home is $\frac{60}{3} = 20$ ppm. On the other hand, if you live 10 mi from the plant, the pollution at your home is $\frac{60}{10} = 6$ ppm. Suppose that two plants 10 mi apart are releasing 60 and 240 ppm, respectively. At what point between the plants is the pollution a minimum? Where is it a maximum?

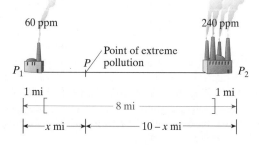

23. A tour agency is booking a tour and has 100 people signed up. The price of a ticket is $2,000 per person. The agency has booked a plane seating 150 people at a cost of $125,000. Additional costs to the agency are incidental fees of $500 per person. For each $10 that the price is lowered, a new person will sign up. How much should the price be lowered for all participants to maximize the profit to the tour agency?

24. A bookstore can obtain the best-seller *20,000 Leagues Under the Majors* from the publisher at a cost of $6 per book. The store has been offering the book at a price of $30 per copy and has been selling 200 copies per month at this price. The bookstore is planning to lower its price to stimulate sales and estimates that for each $2 reduction in the price, 20 more books will be sold per month. At what price should the bookstore sell the book to generate the greatest possible profit?

25. A Florida citrus grower estimates that if 60 orange trees are planted, the average yield per tree will be 400 oranges. The average yield will decrease by 4 oranges for each additional tree planted on the same acreage. How many trees should the grower plant to maximize the total yield?

26. Farmers can get $2 per bushel for their potatoes on July 1, and after that the price drops 2¢ per bushel per day. On July 1, a farmer has 80 bushels of potatoes in the field and estimates that the crop is increasing at the rate of 1 bushel per day. When should the farmer harvest the potatoes to maximize revenue?

27. A viticulturist estimates that if 50 grapevines are planted per acre, each grapevine will produce 150 lb of grapes. Each additional grapevine planted per acre (up to 20) reduces the average yield per vine by 2 lb. How many grapevines should be planted to maximize the yield per acre?

28. A commuter train carries 600 passengers each day from a suburb to a city. It now costs $5 per person to ride the train. A study shows that 50 additional people will ride the train for each 25¢ reduction in fare. What fare should be charged to maximize total revenue?

29. **Modeling Problem** To raise money, a service club has been collecting used bottles, which it plans to deliver to a local glass company for recycling. Since the project began 80 days ago, the club has collected 24,000 pounds of glass, for which the glass company offers 1¢ per pound. However, because bottles are accumulating faster than they can be recycled, the company plans to reduce the price it will pay by 1¢ per 100 pounds of used glass.

   a. What is the most advantageous time for the club to conclude its project and deliver all the bottles?

   b. What assumptions must be made in part **a** to solve the problem with the given information?

30. Suppose that the demand function for a certain commodity is expressed as

$$p(x) = \sqrt{\frac{120 - x}{0.1}} \quad \text{for} \quad 0 \le x \le 120$$

where x is the number of items sold.

   a. Find the total revenue function explicitly and use its first derivative to determine the price at which revenue is maximized.

   b. Graph the relevant portions of the demand and revenue functions.

31. Suppose the demand function for a certain commodity is linear, that is,

$$p(x) = \frac{b - x}{a} \quad \text{for} \quad 0 \leq x \leq b$$

where a and b are positive constants.

a. Find the total revenue function explicitly and use its first derivative to determine its intervals of increase and decrease.

b. Graph the relevant portions of the demand and revenue functions.

32. Modeling Problem Homing pigeons will rarely fly over large bodies of water unless forced to do so, presumably because it requires more energy to maintain altitude in flight over the cool water. Suppose a pigeon is released from a boat floating on a lake 3 mi from a point A on the shore and 10 mi away from the pigeon's loft, as shown in Figure 4.84.

Figure 4.84 Flight path for a pigeon

Assuming the pigeon requires twice as much energy to fly over water as over land and it follows a path that minimizes total energy, find the angle θ of its heading as it leaves the boat. (This is the situation in the Perspective section of the chapter introduction.)

33. Modeling Problem In an experiment, a fish swims s meters upstream at a constant velocity v m/sec relative to the water, which itself has velocity v_1 relative to the ground. The results of the experiment suggest that if the fish takes t seconds to reach its goal (that is, to swim s meters), the energy it expends may be modeled by

$$E = cv^k t$$

where $c > 0$ and $k > 2$ are physical constants. Assuming v_1 is known, what velocity v minimizes the energy?

34. In a learning model, two responses (A and B) are possible for each of a series of observations. If there is a probability p of getting response A in any observation, the probability of getting a response A exactly n times in a series of M observations is

$$F(p) = p^n (1 - p)^{M-n}$$

The *maximum likelihood estimate* is the value of p that maximizes $F(p)$ on $[0, 1]$. For what value of p does this occur?

35. Modeling Problem The production of blood cells plays an important role in medical research involving leukemia and other so-called *dynamical diseases*. In 1977, a mathematical model

was developed by A. Lasota that involved the cell production function

$$P(x) = Ax^s e^{-sx/r}$$

where A, s, and r are positive constants and x is the number of granulocytes (a type of white blood cell) present.[*]

a. Find the granulocyte level x that maximizes the production function P. How do you know it is a maximum?

b. If $s > 1$, show that the graph of P has two inflection points. Sketch the graph of P and give a brief interpretation of the inflection points.

c. Sketch the graph for the case where $0 < s < 1$. What is different about this case?

36. Modeling Problem During a cough, the diameter of the trachea decreases. The velocity v of air in the trachea during a cough may be modeled by the formula

$$v = Ar^2(r_0 - r)$$

where A is a constant, r is the radius of the trachea during the cough, and r_0 is the radius of the trachea in a relaxed state. Find the radius of the trachea when the velocity is greatest, and find the maximum velocity of air. (Notice that $0 \leq r \leq r_0$.)

37. Modeling Problem The work of V. A. Tucker and K. Schmidt-Koenig[†] models the energy expended in flight by a certain kind of bird by the function

$$E = \frac{1}{v}[a(v - b)^2 + c]$$

where a, b, and c are positive constants, v is the velocity of the bird, and the domain of v is [16, 60]. Which value of v will minimize the energy expenditure in the case where $a = 0.04$, $b = 36$, and $c = 9$?

38. Modeling Problem A plastics firm has received an order from the city recreation department to manufacture 8,000 special styrofoam kickboards for its summer swimming program. The firm owns 10 machines, each of which can produce 50 kickboards per hour. The cost of setting up the machines to produce the kickboards is $800 per machine. Once the machines have been set up, the operation is fully automated and can be overseen by a single production supervisor earning $35 per hour.

a. How many machines should be used to minimize the cost of production?

b. How much will the supervisor earn during the production run if the optimal number of machines is used?

39. Modeling Problem After hatching, the larva of the codling moth goes looking for food. The period between hatching and finding food is called the *searching period*. According to a model developed by P. L. Shaffer and H. J. Gold[‡] the length (in days) of the searching period is given by

$$S(T) = (-0.03T^2 + 1.67T - 13.67)^{-1}$$

[*]See "A Blood Cell Population Model, Dynamical Diseases, and Chaos," by W. B. Gearhart and M. Martelli, *UMAP Modules 1990: Tools for Teaching.* Arlington, MA: Consortium for Mathematics and Its Applications (CUPM) Inc., 1991.

[†]V. A. Tucker and K. Schmidt-Koenig, "Flight of Birds in Relation to Energetics and Wind Directions," *The Auk*, Vol. 88 (1971), pp. 97–107.

[‡]P. L. Shaffer and H. J. Gold, "A Simulation Model of Population Dynamics of the Codling Moth *Cydia Pomonella*," *Ecological Modeling*, Vol. 30 (1985) pp. 247–274.

where T (°C) is the air temperature ($20 \le T \le 30$), and the percentage of larvae that survive the searching period is

$$N(T) = -0.85T^2 + 45.45T - 547$$

a. Sketch the graph of N and find the largest and smallest survival percentages for the allowable range of temperatures, $20 \le T \le 30$.

b. Find $S'(T)$ and solve $S'(T) = 0$. Sketch the graph of S for $20 \le T \le 30$.

c. The percentage of codling moth eggs that hatch at a given temperature T is given by

$$H(T) = -0.53T^2 + 25T - 209$$

for $20 \le T \le 30$. Sketch the graph of H and determine the temperatures at which the largest and smallest hatching percentages occur.

40. According to a certain logistic model, the world's population (in billions) t years after 1960 is modeled by the function

$$P(t) = \frac{40}{1 + 12e^{-0.08t}}$$

a. If this model is correct, at what rate will the world's population be increasing with respect to time in the year 2010? At what percentage rate will it be increasing at this time?

b. Sketch the graph of P. What feature on the graph corresponds to the time when the population is growing most rapidly? What happens to $P(t)$ as $t \to +\infty$ (that is, "in the long run")?

C 41. **Modeling Problem**   Generalize Problem 38. Specifically, a manufacturing firm received an order for Q units of a certain commodity. Each of the firm's machines can produce n units per hour. The setup cost is S dollars per machine, and the operating cost is p dollars per hour.

a. Derive a formula for the number of machines that should be used to minimize the total cost of filling the order.

b. Show that when the total cost is minimal, the setup cost is equal to the cost of operating the machines.

42. **Modeling Problem**   (Continuation of Problem 39) Suppose you have 10,000 codling moth eggs. Find a function F for the number of moths that hatch and survive to have their first meal when the temperature is T. For what temperature T on [20, 30] is the number of dining moths the greatest? How large is the optimum dining party and how long did it take to find its meal?

43. An epidemic spreads through a community in such a way that t weeks after its outbreak, the number of residents who have been infected is modeled by a function of the form

$$f(t) = \frac{A}{1 + Ce^{-kt}}$$

where A is the total number of susceptible residents. Show that the epidemic is spreading most rapidly when half the susceptible residents have been infected.

44. In certain tissues, cells exist in the shape of circular cylinders. Suppose such a cylinder has radius r and height h. If the volume is fixed (say, at v_0), find the value of r that minimizes the total surface area ($S = 2\pi rh + 2\pi r^2$) of the cell.

45. **Modeling Problem**   A store owner expects to sell Q units of a certain commodity each year. It costs S dollars to order each new shipment of x units; and it costs t dollars to store each unit for a year. Assuming that the commodity is used at a constant rate throughout the year and that each shipment arrives just as the preceding shipment has been used up, show that the total cost of maintaining inventory is minimized when the ordering cost equals the storage cost.

46. **Modeling Problem**   Sometimes investment managers determine the optimal holding time of an asset worth $V(t)$ dollars by finding when the relative rate of change $V'(t)/V(t)$ equals the prevailing rate of interest r. Show that the optimal time determined by this criterion is the same as that found by maximizing the present value of $V(t)$.

47. An important quantity in economic analysis is *elasticity of demand*, defined by

$$E(x) = \frac{p}{x}\frac{dx}{dp}$$

where x is the number of units of a commodity demanded when the price is p dollars per unit.[*] Show that

$$\frac{dR}{dx} = \frac{R}{x}\left[1 + \frac{1}{E(x)}\right]$$

That is, marginal revenue is $[1 + 1/E(x)]$ times average revenue.

48. **Modeling Problem**   There are alternatives to the inventory model analyzed in Example 5. Suppose a company must supply N units/month at a uniform rate. Assume the storage cost/unit is S_1 dollars/month and that the setup cost is S_2 dollars. Further assume that production is at a uniform rate of m units/month (with no units left over in inventory at the end of the month). Let x be the number of items produced in each run.

a. Explain why the total average cost per month may be modeled by

$$C(x) = \frac{S_1 x}{2}\left(1 - \frac{N}{m}\right) + \frac{S_2 N}{x}$$

b. Find an expression for the number of items that should be produced in each run in order to minimize the total average cost C.

Economists refer to the optimum found in this inventory model as the *economic production quantity* (EPQ), whereas the optimum found in the model analyzed in Example 5 is called the *economic order quantity* (EOQ).

49. In this section we showed that the Heinz concentration function,

$$C(t) = \frac{k}{b - a}(e^{-at} - e^{-bt})$$

for $b > a$ has exactly one critical number, namely,

$$t_c = \frac{1}{b - a}\ln\left(\frac{b}{a}\right)$$

Find $C''(t)$ and use the second-derivative test to show that a relative maximum for the concentration $C(t)$ occurs at time $t = t_c$.

---

[*]See the module with the intriguing title, "Price Elasticity of Demand: Gambling, Heroin, Marijuana, Prostitution, and Fish," by Yves Nievergelt, *UMAP Modules 1987: Tools for Teaching*. Arlington, MA: CUPM Inc., 1988, pp. 153–181.

50. Let $f(\theta)$ be the resistance to blood flow obtained on page 257, namely

$$f(\theta) = k\left(\frac{\ell - h\cot\theta}{R^4} + \frac{h\csc\theta}{r^4}\right)$$

We found that $\theta_m = \cos^{-1}(r^4/R^4)$ is a critical number of f. Find $f''(\theta)$ and use the second-derivative test to verify that a relative minimum occurs at $\theta = \theta_m$. Explain why this must be an *absolute* minimum.

51. Beehives are formed by packing together cells that may be modeled as regular hexagonal prisms open at one end, as shown in Figure 4.85. It can be shown that a cell with hexagonal side of length s and prism height h has surface area

$$S(\theta) = 6sh + 1.5s^2(-\cot\theta + \sqrt{3}\csc\theta)$$

for $0 < \theta < \frac{\pi}{2}$. What is the angle θ (to the nearest degree) that

minimizes the surface area of the cell (assuming that s and h are fixed)?

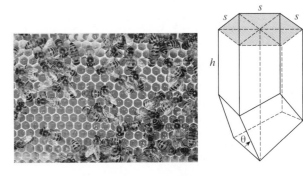

Figure 4.85   Beehive (left) and beehive cell (right)

CHAPTER 4 REVIEW

Proficiency Examination

CONCEPT PROBLEMS

1. What is the difference between absolute and relative extrema of a function?

2. State the extreme value theorem.

3. What are the critical numbers of a function? What is the difference between critical numbers and critical points?

4. Outline a procedure for finding the absolute extrema of a continuous function on a closed interval $[a, b]$.

5. State both Rolle's theorem and the mean value theorem, and discuss the relationship between them.

6. What is the zero-derivative theorem?

7. State the first-derivative and second-derivative tests, and discuss when and in which order you would use them.

8. What is a vertical asymptote? A horizontal asymptote?

9. How would you identify a cusp? A vertical tangent?

10. State l'Hôpital's rule.

11. Define $\lim\limits_{x\to+\infty} f(x) = L$ and $\lim\limits_{x\to c} f(x) = +\infty$.

12. Outline a graphing strategy for a function defined by $y = f(x)$.

13. What do we mean by optimization? Outline an optimization procedure.

14. What is Fermat's principle of optics?

15. What is Snell's law of refraction?

16. When is the profit maximized in terms of marginal revenue and marginal cost? When is the average cost minimized?

17. What is Poiseuille's law of resistance to flow?

PRACTICE PROBLEMS

Evaluate the limits in Problems 18–21.

18. $\lim\limits_{x\to\pi/2} \dfrac{\sin 2x}{\cos x}$

19. $\lim\limits_{x\to 1} \dfrac{1-\sqrt{x}}{x-1}$

20. $\lim\limits_{x\to+\infty} \left(\dfrac{1}{x} - \dfrac{1}{\sqrt{x}}\right)$

21. $\lim\limits_{x\to+\infty} \left(1 + \dfrac{2}{x}\right)^{3x}$

Sketch the graph of each function in Problems 22–27. Include all key features, such as relative extrema, inflection points, asymptotes, intercepts, cusps, and vertical tangents.

22. $f(x) = x^3 + 3x^2 - 9x + 2$

23. $f(x) = x^{1/3}(27 - x)$

24. $f(x) = \dfrac{x^2 - 1}{x^2 - 4}$

25. $f(x) = (x^2 - 3)e^{-x}$

26. $f(x) = x + \tan^{-1} x$

27. $f(x) = \sin^2 x - 2\cos x$ on $[0, 2\pi]$

28. Determine the largest and smallest values of

$$f(x) = x^4 - 2x^5 + 5$$

on the closed interval $[0, 1]$.

29. A box is to have a square base, an open top, and volume of 2 ft^3. Find the dimensions of the box (to the nearest inch) that uses the least amount of material.

30. The personnel manager of a department store estimates that if N temporary salespersons are hired for the holiday season, the total net revenue derived (in hundreds of dollars) from their efforts may be modeled by the function

$$R(N) = -3N^4 + 50N^3 - 261N^2 + 540N$$

for $0 \le N \le 9$. How many salespersons should be hired to maximize total net revenue?

Supplementary Problems

Sketch the graph of each function in Problems 1–20. Use as many as possible of the key features such as domain, relative extrema, inflection points, concavity, asymptotes, intercepts, cusps, and vertical tangents.

1. $f(x) = x^3 + 6x^2 + 9x - 1$
2. $f(x) = x^4 + 4x^3 + 4x^2 + 1$
3. $f(x) = 3x^4 - 4x^3 + 1$
4. $f(x) = 3x^4 - 4x^2 + 1$
5. $f(x) = 6x^5 - 15x^4 + 10x^3$
6. $f(x) = 3x^5 - 10x^3 + 15x + 1$

7. $f(x) = \dfrac{9 - x^2}{3 + x^2}$ 8. $f(x) = \dfrac{x}{1 - x}$

9. $f(x) = \sin 2x - \sin x$ on $[-\pi, \pi]$

10. $f(x) = \sin x \sin 2x$ on $[-\pi, \pi]$

11. $f(x) = \dfrac{x^2 - 4}{x^2}$ 12. $f(x) = \dfrac{x^2 + 2x - 3}{x^2 - 3x + 2}$

13. $f(x) = \dfrac{3x - 2}{(x + 1)^2(x - 2)}$ 14. $f(x) = \dfrac{x^3 + 3}{x(x + 1)(x + 2)}$

15. $f(x) = x^2 \ln \sqrt{x}$ 16. $f(x) = \sin^{-1} x + \cos^{-1} x$

17. $f(x) = x(e^{-2x} + e^{-x})$ 18. $f(x) = \ln\left(\dfrac{x - 1}{x + 1}\right)$

19. $f(x) = \dfrac{5}{1 + e^{-x}}$ 20. $f(x) = xe^{1/x}$

In Problems 21–24, the graph of the given function $f(x)$ for $x > 0$ is one of the six curves shown in Figure 4.86. In each case, match the function to a graph.

21. $f(x) = x2^{-x}$ 22. $f(x) = \dfrac{\ln \sqrt{x}}{x}$

23. $f(x) = \dfrac{e^x}{x}$ 24. $f(x) = e^{-x} \sin x$

a.

b.

c.

d.

e.

f.

Figure 4.86   Problems 21–24

Determine the absolute maximum and absolute minimum value of each function on the interval given in Problems 25–28.

25. $f(x) = x^4 - 8x^2 + 12$ on $[-1, 2]$

26. $f(x) = \sqrt{x}(x - 5)^{1/3}$ on $[0, 6]$

27. $f(x) = 2x - \sin^{-1} x$ on $[0, 1]$

28. $f(x) = e^{-x} \ln x$ on $\left[\frac{1}{2}, 2\right]$

Evaluate the limits in Problems 29–50.

29. $\displaystyle\lim_{x \to +\infty} \dfrac{x \sin^2 x}{x^2 + 1}$ 30. $\displaystyle\lim_{x \to +\infty} \dfrac{2x^4 - 7}{6x^4 + 7}$

31. $\displaystyle\lim_{x \to +\infty} (\sqrt{x^2 - x} - x)$ 32. $\displaystyle\lim_{x \to +\infty} [\sqrt{x(x + b)} - x]$

33. $\displaystyle\lim_{x \to 0} \dfrac{x \sin x}{x + \sin^3 x}$ 34. $\displaystyle\lim_{x \to 0} \dfrac{x \sin x}{x^2 - \sin^3 x}$

35. $\displaystyle\lim_{x \to 0} \dfrac{x \sin^2 x}{x^2 - \sin^2 x}$ 36. $\displaystyle\lim_{x \to 0} \dfrac{x - \sin x}{\tan^3 x}$

37. $\displaystyle\lim_{x \to 0} \dfrac{\sin^2 x}{\sin x^2}$ 38. $\displaystyle\lim_{x \to 0} \left(\dfrac{1}{x^2} - \dfrac{1}{x^2 \sec x}\right)$

39. $\displaystyle\lim_{x \to \pi/2^-} \dfrac{\sec^2 x}{\sec^2 3x}$ 40. $\displaystyle\lim_{x \to \pi/2^-} (1 - \sin x) \tan x$

41. $\displaystyle\lim_{x \to \pi/2^-} (\sec x - \tan x)$

42. $\displaystyle\lim_{x \to +\infty} \left(\sqrt{x^2 + 4} - \sqrt{x^2 - 4}\right)$

43. $\displaystyle\lim_{x \to 0^+} (1 + x)^{4/x}$ 44. $\displaystyle\lim_{x \to 1^-} \left(\dfrac{1}{1 - x}\right)^x$

45. $\displaystyle\lim_{x \to 0^+} x^{\tan x}$ 46. $\displaystyle\lim_{x \to 0} \dfrac{5^x - 1}{x}$

47. $\displaystyle\lim_{x \to 0} \dfrac{\ln(x^2 + 1)}{x}$ 48. $\displaystyle\lim_{x \to +\infty} \left(\dfrac{1}{x}\right)^x$

49. $\displaystyle\lim_{x \to +\infty} \left(4 - \dfrac{1}{x}\right)^x$ 50. $\displaystyle\lim_{x \to +\infty} \dfrac{e^x \cos x - 1}{x}$

51. **Journal Problem** *Mathematics Teacher.*[*] ■ Which of the graphs in Figure 4.87 is the derivative and which is the function?

Figure 4.87   A function and its derivative

52. Determine a, b, and c such that the graph of $f(x) = ax^3 + bx^2 + c$ has an inflection point and slope 1 at $(-1, 2)$.

53. **Exploration Problem**   Explain why the graph of a quadratic polynomial cannot have a point of inflection. How many inflection points does the graph of a cubic polynomial have?

54. Find the points on the hyperbola $x^2 - y^2 = 4$ that are closest to the point $(0, 1)$.

55. Find numbers A, B, C, and D that guarantee that the function
$$f(x) = Ax^3 + Bx^2 + Cx + D$$
will have a relative maximum at $(-1, 1)$ and a relative minimum at $(1, -1)$.

56. A Norman window consists of a rectangle with a semicircle surmounted on the top. What are the dimensions of the Norman window of largest area with a fixed perimeter of P_0 meters?

57. An apartment complex has 200 units. When the monthly rent for each unit is $600, all units are occupied. Experience indicates that for each $20-per-month increase in rent, 5 units will become vacant. Each rented apartment costs the owners of the complex $80 per month to maintain. What monthly rent should be charged to maximize the owner's profit? What is maximum profit? How many units are rented when profit is a maximum?

58. A farmer wishes to enclose a rectangular pasture with 320 ft of fence. Find the dimensions that give the maximum area in these situations:
a. The fence is on all four sides of the pasture.
b. The fence is on three sides of the pasture and the fourth side is bounded by a wall.

59. A peach grower has determined that if 30 trees are planted per acre, each tree will average 200 lb of peaches per season. However, for each tree grown in addition to the 30 trees, the average yield for each of the trees in the grove drops by 5 lb per tree. How many peach trees should be planted on each acre in order to maximize the yield of peaches per acre? What is the maximum yield?

60. Modeling Problem To obtain the maximum price, a shipment of fruit should reach the market as early as possible after the fruit has been picked. If a grower picks the fruit immediately for shipment, 100 cases can be shipped at a profit of $10 per case. By waiting, the grower estimates that the crop will yield an additional 25 cases per week, but because the competitor's yield will also increase, the grower's profit will decrease by $1 per case per week. Use calculus to determine when the grower should ship the fruit to maximize profit. What will be the maximum profit?

61. Modeling Problem Oil from an offshore rig located 3 mi from the shore is to be pumped to a location on the edge of the shore that is 8 mi east of the rig. The cost of constructing a pipe in the ocean from the rig to the shore is 1.5 times as expensive as the cost of construction on land. Set up and analyze a model to determine how the pipe should be laid to minimize cost.

62. The owner of a novelty store can obtain joy buzzers from the manufacturer for 40¢ apiece. It is estimated that 60 buzzers will be sold when the price is $1.20 per buzzer and that 10 more buzzers will be sold for every 10¢ decrease in price. What price should be charged to maximize profit?

63. Journal Problem *AMATYC Review** ■ You are to build a rectangular enclosure with front side made of material costing $10 per foot and with the other three sides (back, left, right) of material costing $5 per foot. If the enclosure is to contain exactly 600 square feet and to be built at minimum total cost, how long should the side be and what is the total cost?
Answer: $600.
Here is the real question: By coincidence, the answer matches the constrained area (both 600, aside from units, of course). *Question*: Characterize when this happens more generally, and thus describe the relationship that exists in such problems in which the resultant minimum cost matches the given area (all in appropriate units).

64. Journal Problem *Quantum*[†] ■ Two students are pondering the following problem: A tin can takes the form of a right circular cylinder with radius R and height H. An ant is sitting on the border circle of one of its bases (point A in Figure 4.88). It wants to crawl to the most distant point B at the border circle of the other base (symmetric to A with respect to the center of the tin). Find the shortest path for the ant.

Figure 4.88 Shortest path from A to B?

a. What is the length of the shortest path?
b. *After* finding a solution, compare your solution with that of the first student as follows: "It's a simple problem!" the first student says confidently. "We just have to consider the planar development of the tin. Let's say, for the sake of definiteness, that the ant first crawls along the side surface and then across the upper base (of course, it's possible that the ant takes the symmetric route: first crawling across the lower base, then along the side surface; but the length of this route is the same)."
c. Compare your solution with that of the second student, as follows: "Wait a second," the other student replies. "One can very comfortably develop the tin can in a different way! Just throw away the lids and spread the side surface on the plane so that we get a rectangle. Then the shortest path will be the segment connecting points A and B, where the image of the ant's route on the tin can will be part of a corkscrew line."
d. Either your solution matched one of the students' solutions or it did not. Which of the two (or possibly three) solutions provides the minimum? Before you answer, find out the conditions whereby the lengths of both of the students' routes are the equal. Critique the solutions given by the two students.

*Spring 1995, pp. 67–68.

[†] September/October, 1997, Vol. 7, No. 4, pp. 50–53.

65. Westel Corporation manufactures telephones and has developed a new cellular phone. Production analysis shows that its price must not be less than $50; if x units are sold, then the price per unit is given by the formula $p(x) = 150 - x$. The total cost of producing x units is given by the formula $C(x) = 2,500 + 30x$. Find the maximum profit, and determine the price that should be charged to maximize the profit.

66. Modeling Problem   A manufacturer receives an order for 5,000 items. There are 12 machines available, each of which can produce 25 items per hour. The cost of setting up a machine for a production run is $50. Once the machines are in operation, the procedure is fully automated and can be supervised by a single worker earning $20 per hour. Set up and analyze a model to determine the number of machines that should be used to minimize the total cost of filling the order. State any assumptions that must be made.

67. Show that the graph of a polynomial of degree n, with $n > 2$, has at most $n - 2$ inflection points.

68. Show that the graph of the function $f(x) = x^n$ with $n > 1$ has either one or no inflection points, depending on whether n is odd or even.

69. Each tangent line to the circle $x^2 + y^2 = 1$ at a point in the first quadrant will intersect the coordinate axes at points $(x_1, 0)$ and $(0, y_1)$. Determine the line for which $x_1 + y_1$ is a minimum.

70. An accelerated particle moving at speed close to the speed of light emits power P in the direction θ given by

$$P(\theta) = \frac{a \sin \theta}{(1 - b \cos \theta)^5} \quad \text{for} \quad 0 < b < 1 \quad \text{and} \quad a > 0$$

Find the value of θ ($0 \le \theta \le \pi$) for which P has the greatest value.

71. Suppose that f is a continuous function defined on the closed interval $[a, b]$ and that $f'(x) = c$ on the open interval (a, b) for some constant c. Use the MVT to show that

$$f(x) = c(x - a) + f(a)$$

for all x in the interval $[a, b]$.

72. Find the point of inflection of the curve

$$y = (x + 1) \tan^{-1} x$$

73. Find the critical numbers for the function

$$f(x) = \tan^{-1}\left(\frac{x}{a}\right) - \tan^{-1}\left(\frac{x}{b}\right), \quad a > b$$

Classify each as corresponding to a relative maximum, a relative minimum, or neither.

74. Suppose that $f''(x)$ exists for all x and that $f''(x) + c^2 f(x) = 0$ for some number c with $f'(0) = 1$. Show that for any $x = x_0$, there is a number w between 0 and x_0 for which $f'(x_0) + c^2 f(w)x_0 = 1$.

75. Modeling Problem   According to the *Mortality and Morbidity Report* of the U.S. Centers for Disease Control, the following table gives the number of annual deaths from acquired immune deficiency syndrome (AIDS) in the United States for the years 1982–1991.*

---

*For an interesting discussion of how mathematical modeling can be applied to a problem of great public and personal interest, see "Modeling the AIDS Epidemic," by Allyn Jackson, *Notices of the American Mathematical Society*, Vol. 36, No. 8 (October 1989), pp. 981–983.

Year	Deaths	Year	Deaths
1982	400	1987	13,900
1983	1,400	1988	17,300
1984	3,200	1989	32,000
1985	6,200	1990	39,000
1986	10,660	1991	45,000

a. To model the number of annual AIDS deaths, a data analysis program produced the cubic polynomial

$$N(x) = -8.58197x^3 + 732.727x^2 - 3,189.9x + 4,375.09$$

Sketch the graph of N and determine the time when its highest point occurs for $0 \le x \le 20$, where $x = 0$ represents 1982.

b. The same report gives the number of cases of reported AIDS for the period 1984–1991:

Year	Deaths	Year	Deaths
1984	4,445	1988	31,001
1985	8,249	1989	33,722
1986	12,932	1990	41,595
1987	21,070	1991	43,672

The same data analysis program used in part **a** yields

$$C(x) = -171.247x^3 + 3,770.90x^2 - 19,965.1x + 34,893.9$$

as a modeling polynomial for the number of cases as a function of time x. When does this model predict the number of reported cases will be the largest? When does it predict the number of reported cases will drop to 0?

c. When the data in part **b** are modeled exponentially, the data analysis program produces the function

$$y = 1,676e^{0.3256x}$$

Sketch the cubic modeling formula $y = C(x)$ from part **b** and this exponential formula on the same set of coordinate axes along with the data points from the table in part **b**. Which formula do you think does a better job of fitting the data?

d. Explore other modeling formulas for the data given in part **a**.

e. Call the Centers for Disease Control or check the World Wide Web (www.cdc.gov) for the most recent updates for the data in parts **a** and **b**. Do the formulas in this problem correctly model the current information? Explain the discrepancies.

f. Based on the information you obtained in part **e**, find new models for the information in parts **a** and **b** using all the data you have available.

Using the graphing and differentiation programs of your computer or calculator, graph $f(x)$, $f'(x)$, $f''(x)$ for each function in Problems 76–79. Print out a copy, if possible. On each graph, indicate where

a. $f'(x) > 0$ **b.** $f'(x) < 0$ **c.** $f''(x) > 0$
d. $f''(x) < 0$ **e.** $f'(x) = 0$ **f.** $f'(x)$ *does not exist*
g. $f''(x) = 0$

Describe how these inequalities qualitatively determine the shape of the graph of the functions over the given interval.

76. $f(x) = \sin 2x$ on $[-\pi, \pi]$

77. $f(x) = x^3 - x^2 - x + 1$ on $\left[-\frac{3}{2}, 2\right]$

78. $f(x) = x^4 - 2x^2$ on $[-2, 2]$

79. $f(x) = x^3 - x + \dfrac{1}{x} + 1$ on $[-2, 2]$

80. Consider a string 60 in. long that is formed into a rectangle. Using a graphing calculator or a graphing program, graph the area $A(x)$ enclosed by the string as a function of the length x of a given side $(0 < x < 30)$. From the graph, deduce that the maximum area is enclosed when the rectangle is a square. What is the minimum area enclosed? Compare this problem to the exact solution. What conclusions do you draw about the desirability of analytical solutions and the role of the computer?

81. Modeling Problem The beach of a lake follows contours that are approximated by the curve $4x^2 + y^2 = 1$, and a nearby road lies along the curve $y = 1/x$ for $x > 0$. Using your graphing calculator or computer software, determine the closest approach of the road to the lake in the north-south direction. Take the positive y-axis as pointing north.

82. Using your graphing and differentiation programs, locate and identify all the relative extrema for the function defined by

$$f(x) = \tfrac{1}{5}x^5 - \tfrac{5}{4}x^4 + 2x^2$$

This may require judicious choices of the window settings.

83. Modeling Problem Suppose you are a manager of a fleet of delivery trucks. Each truck is driven at a constant speed of x mi/h $(15 \le x \le 55)$, and gas consumption (gal/mi) is modeled by the function

$$\frac{1}{250}\left(\frac{750}{x} + x\right)$$

Using a graphing and differentiation program (and/or root-solving program on your computer or calculator), answer the following questions:

a. If gas costs \$1.70/gal, estimate the steady speed that will minimize the cost of fuel for a 500-mi trip.

b. Estimate the steady speed that minimizes the cost if the driver is paid \$28 per hour and the price of gasoline remains constant at \$1.70/gal.

84. HISTORICAL QUEST   Leonhard Euler is one of the giants in the history of mathematics. His name is attached to almost every branch of mathematics. He was the most prolific writer on the subject of mathematics, and his mathematical textbooks were masterfully written. His writing was not at all slowed down by his total blindness for the last 17 years of his life. He possessed a phenomenal memory, had

LEONHARD EULER
1707–1783

almost total recall, and could mentally calculate long and complicated problems. The basis for the historical development of calculus, as well as modern-day analysis, is the notion of a function. Euler's book *Introductio in analysin infinitorum* (1784) first used the function concept as the basic idea. It was the identification of functions, rather than curves, as the principal object of study, that permitted the advancement of mathematics in general, and calculus in particular. In Chapter 2, we noted that the number e is named in honor of Euler. Euler did not use the definition we use today, namely,

$$e = \lim_{n \to +\infty}\left(1 + \frac{1}{n}\right)^n$$

Instead, Euler used series in his work (which we will study in Chapter 8). ■

In this Historical Quest problem we will lay the groundwork for a Historical Quest in Chapter 8. Euler introduced $\log_a x$ (which he wrote as ℓx) as that exponent y such that $a^y = x$. This was done in 1748, which makes it the first appearance of a logarithm interpreted explicitly as an exponent. He does not define $a^0 = 1$, but instead writes

$$a^\epsilon = 1 + k\epsilon$$

for an infinitely small number ϵ. In other words,

$$k = \lim_{\epsilon \to 0}\frac{a^\epsilon - 1}{\epsilon}$$

Explain why $k = \ln a$.

When Euler was 13, he registered at the University of Basel and was introduced to another famous mathematician, Johann Bernoulli, who was an instructor there at the time (see Historical Quest, Section 4.5, Problems 58–59). If Bernoulli thought that a student was promising, he would provide, sometimes gratis, private instruction. Here is Euler's own account of this first encounter with Bernoulli

I soon found an opportunity to gain introduction to the famous professor Johann Bernoulli, whose good pleasure it was to advance me further in the mathematical sciences. True, because of his business he flatly refused me private lessons, but he gave me much wiser advice, namely, to get some more difficult mathematical books and work through them with all industry, and wherever I should find some check or difficulties, he gave me free access to him every Saturday afternoon and was so kind as to elucidate all difficulties, which happened with such greatly desired advantage that whenever he had obviated one check for me, because of that ten others disappeared right away, which is certainly the way to make a happy advance in the mathematical sciences.[*]

---

[*]From *Elements of Algebra* by Leonhard Euler, 1840. London: Longman, Orme, and Co. Reprinted by Springer-Verlag.

85. Putnam Examination Problem   Given the parabola $y^2 = 2mx$, what is the length of the shortest chord that is normal to the curve at one end? *Hint*: If $\overline{AB}$ is normal to the parabola, where A and B are the points $(2mt^2, 2mt)$ and $(2ms^2, 2ms)$, show that the slope of the tangent at A is $1/(2t)$.

86. Putnam Examination Problem   Prove that the polynomial

$$(a-x)^6 - 3a(a-x)^5 + \tfrac{5}{2}a^2(a-x)^4 - \tfrac{1}{2}a^4(a-x)^2$$

has only negative values for $0 < x < a$. *Hint*: Show that if $x = g(1-y)$, the polynomial becomes $a^6 y^2 g(y)$, where

$$g(y) = y^4 - 3y^3 + \tfrac{5}{2}y^2 - \tfrac{1}{2}$$

and then prove that $g(y) < 0$ for $0 < y < 1$.

87. Putnam Examination Problem   Find the maximum value of $f(x) = x^3 - 3x$ on the set of all real numbers x satisfying $x^4 + 36 \leq 13x^2$.

88. Putnam Examination Problem   Let T be an acute triangle. Inscribe a pair R and S of rectangles in T, as shown in Figure 4.89. Let $A(X)$ denote the area of polygon X. Find the maximum value, or show that no maximum exists, of the ratio

$$\frac{A(R) + A(S)}{A(T)}$$

where T ranges over all triangles, and R and S range over all rectangles as shown in Figure 4.89.

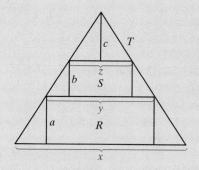

Figure 4.89   Problem 88

89. Putnam Examination Problem   Which is greater

$$(\sqrt{n})^{\sqrt{n+1}} \quad \text{or} \quad (\sqrt{n+1})^{\sqrt{n}}$$

where $n > 8$? *Hint*: Use the function $f(x) = \dfrac{\ln x}{x}$, and show that $x^y > y^x$ when $e \leq x < y$.

90. Putnam Examination Problem   The graph of the equation $x^y = y^x$ for $x > 0$, $y > 0$ consists of a straight line and a curve. Find the coordinates of the point where the line and the curve intersect.

Group Research Project

Wine Barrel Capacity*

This project is to be done in groups of three or four students. Each group will submit a single written report.

A wine barrel has a hole in the middle of its side called a **bung hole**. To determine the volume of wine in the barrel, a **bung rod** is inserted in the hole until it hits the lower seam. Determine how to calibrate such a rod so that it will measure the volume of the wine in the barrel. You should make the following assumptions:

1. The barrel is cylindrical.
2. The distance from the bung hole to the corner is λ.
3. The ratio of the height to the diameter of the barrel is t. This ratio should be chosen so that for a given λ value, the volume of the barrel is maximal.

Your paper is not limited to the following questions, but it should include these concerns: You should show that the volume of the cylindrical barrel is

$$V = 2\pi\lambda^3 t(4 + t^2)^{-3/2}$$

and you should find the approximate ideal value for t. Johannes Kepler was the first person to show mathematically why coopers were guided in their construction of wine barrels by one rule: *make the staves* (the boards that make up the sides of the barrel) *one and one-half times as long as the diameter.* (This is the approximate t-value.) You should provide dimensions for the barrel as well as for the bung rod.

JOHANNES KEPLER (1571–1630) is usually remembered for his work in astronomy, in particular for his three laws of planetary motion. Tycho Brahe (1546-1601) was working for Rodolf II, Holy Roman Emperor in Prague in 1599, and he asked Kepler to work with him. The historian Burton describes this as a fortunate alliance. "Tycho was a splendid observer but a poor mathematician, while Kepler was a splendid mathematician but a poor observer." Because Kepler was a Protestant during a time when most intellectuals were required to be Catholic, Kepler had trouble supporting himself, and consequently worked for many benefactors. While serving the Austrian emperor Matthew I, Kepler observed with admiration the ability of a young vintner to declare quickly and easily the capacities of a number of different wine casks. He describes how this can be done in his book *The New Stereometry of Wine Barrels, Mostly Austrian.*

bung rod

*The idea for this group research project comes from research done at Iowa State University as part of a National Science Foundation grant. Our thanks to Elgin Johnston of Iowa State University.

5

Integration

PREVIEW

A traveler, driving a car down a straight, level road at 40 mi/h, applies the brakes, which decelerate the car at the rate of 21 ft/s². How far does the car travel before coming to a full stop? An ecologist measures the width of an oil spill at intervals of 5 ft along its length. What is the approximate area of the spill? On Christmas day, it starts snowing at a steady rate. A snowplow starts out at noon, going 2 mi during the first hour and 1 mi during the second hour. When did it start to snow?

We will examine questions such as these in the examples and problems of this chapter, as we study **integral calculus**, the companion to the **differential calculus** developed in Chapters 3 and 4.

PERSPECTIVE

The key concept in integral calculus is *integration*, a procedure that involves computing a special kind of limit of sums called the *definite integral*. We will find that such limits can often be computed by reversing the process of differentiation; that is, given a function f, we find a function F such that $F' = f$. This is called *indefinite integration*, and the equation $F' = f$ is an example of a *differential equation*.

Finding integrals and solving differential equations are extremely important processes in calculus. We begin our study of these topics by defining definite and indefinite integration and showing how they are connected by a remarkable result called the *fundamental theorem of calculus*. Then we examine several techniques of integration and show how area, average value, and other quantities can be set up and analyzed by integration. Our study of differential equations begins in this chapter and will continue in appropriate sections throughout this text. We also establish a mean value theorem for integrals and develop numerical procedures for estimating the value of a definite integral.

5.1 Antidifferentiation

IN THIS SECTION reversing differentiation, antiderivative notation, antidifferentiation formulas, applications, area as an antiderivative

REVERSING DIFFERENTIATION

A physicist who knows the acceleration of a particle may want to determine its velocity or its position at a particular time. An ecologist who knows the rate at which a certain pollutant is being absorbed by a particular species of fish might want to know the actual amount of pollutant in the fish's system at a given time. In each of these cases, a derivative f' is given and the problem is that of finding the corresponding function f. Toward this end, we make the following definition.

Antiderivative

A function F is called an **antiderivative** of a given function f on an interval I if
$$F'(x) = f(x)$$
for all x in I.

Suppose we know $f(x) = 3x^2$. We wish to find a function $F(x)$ such that $F'(x) = 3x^2$. It is not difficult to use the power rule in reverse to discover that $F(x) = x^3$ is such a function. However, that is not the only possibility:

Given: $\quad F(x) = x^3 \qquad G(x) = x^3 - 5 \qquad H(x) = x^3 + \pi^2$

Find: $\quad F'(x) = 3x^2 \qquad G'(x) = 3x^2 \qquad H'(x) = 3x^2$

In fact, if F is an antiderivative of f, then so is $F + C$ for any constant C, because
$$[F(x) + C]' = F'(x) + 0 = f(x)$$

In the following theorem, we use the constant difference theorem of Section 4.2 to show that *any* antiderivative of f can be expressed in this form.

THEOREM 5.1 Antiderivatives of the same function differ by a constant

If F is an antiderivative of the continuous function f, then any other antiderivative, G, of f must have the form
$$G(x) = F(x) + C$$

> ➡ **What This Says** Two antiderivatives of the same function differ by a constant.

Proof If F and G are both antiderivatives of f, then $F' = f$ and $G' = f$ and Theorem 4.6 (the constant difference theorem) tells us that
$$G(x) - F(x) = C$$
so $G(x) = F(x) + C$. ❏

EXAMPLE 1 Finding antiderivatives

Find general antiderivatives for the given functions.

a. $f(x) = x^5$ **b.** $s(x) = \sin x$ **c.** $y'(x) = \dfrac{1}{x}$

Solution

a. If $F(x) = x^6$, then $F'(x) = 6x^5$, so we see that a particular antiderivative of f is $F(x) = \dfrac{x^6}{6}$ to obtain $F'(x) = \dfrac{6x^5}{6} = x^5$. By Theorem 5.1, the most general antiderivative is $G(x) = \dfrac{x^6}{6} + C$.

b. If $S(x) = -\cos x$, then $S'(x) = \sin x$, so $G(x) = -\cos x + C$.

c. If $y(x) = \ln|x|$, then $y'(x) = \dfrac{1}{x}$, so $G(x) = \ln|x| + C$. ∎

Recall that the slope of a function $y = f(x)$ at any point (x, y) on its graph is given by the derivative $f'(x)$. We can exploit this fact to obtain a "picture" of the graph of f. Reconsider Example 1c where $y' = 1/x$. There is an antiderivative $F(x)$ of $1/x$ such that the slope of F at each point $(x, F(x))$ is $1/x$ for each nonzero value of x. Let us draw a graph of these slopes:

If $x = 1$, then the slope is $\frac{1}{1} = 1$. Draw short line segments at $x = 1$, each with slope 1, for different y-values as shown in Figure 5.1a.

If $x = -3$, then the slope is $-1/3$, so draw short line segments at $x = -3$, each with slope $-1/3$, also shown in Figure 5.1a.

If we continue to plot these slope points for different values of x, we obtain many little slope lines. The resulting graph shown in Figure 5.1c is known as a **slope field** for the equation $y' = 1/x$. Beginning in Section 5.6 slope fields are also called *direction fields*.

Finally, notice the relationship between the slope field for $y' = 1/x$ and its antiderivative $y = \ln|x| + C$ (found in Example 1c). If we choose particular values for C, say $C = 0$, $C = -\ln 2$, or $C = 2$, and draw these particular antiderivatives in Figure 5.1c, we notice that these particular solutions are anticipated by the slope field drawn in part b. That is, the slope field shows the entire family of antiderivatives of the original equation.

a. Slopes for $x = 1$ and $x = -3$

b. Slope field

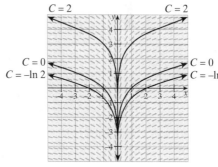

c. Sample antiderivatives

Figure 5.1 Slope field for the equation $y' = \dfrac{1}{x}$

Slope fields will be discussed in more detail in Section 5.6. In general, slope fields, and antiderivatives obtained by using slope fields, are usually generated using technology in computers and calculators.

Here is an example in which the antiderivatives cannot be obtained as elementary functions.

Figure 5.2 Slope field for $y' = e^{x^2}$

EXAMPLE 2 Finding an antiderivative using a slope field

Consider the slope field for $y' = e^{x^2}$, which is shown in Figure 5.2. Draw a possible graph of the antiderivative of e^{x^2} that passes through the point $(0, 0)$.

Figure 5.3 Antiderivative of $y' = e^{x^2}$ passing through $(0, 0)$

Indefinite Integral

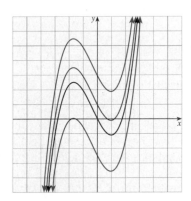

Figure 5.4 Several members of the family of curves $y = F(x) + C$

Solution

Each little segment represents the slope of e^{x^2} for a particular x value. For example, if $x = 0$, the slope is 1, if $x = 1$, the slope is e, and if $x = 2$, the slope is e^4 (quite steep). However, to draw the antiderivative, step back and take a large view of the graph, and "go with the flow." We sketch the apparent graph that passes through $(0, 0)$, as shown in Figure 5.3.   ∎

ANTIDERIVATIVE NOTATION

It is worthwhile to define a notation to indicate the operation of antidifferentiation.

The notation

$$\int f(x)\, dx = F(x) + C$$

where C is an arbitrary constant means that F is an antiderivative of f. It is called the **indefinite integral of** f and satisfies the condition that $F'(x) = f(x)$ for all x in the domain of f.

➡ **What This Says**   This is nothing more than the definition of the antiderivative, along with a convenient notation. We also agree that in the context of antidifferentiation, C is an arbitrary constant.

The graph of $F(x) + C$ for different values of C is called **a family of functions** (see Figure 5.4).

WARNING   It is important to remember that $\int f(x)\, dx$ represents a family of functions, not just a single function.

Because each member of the family $y = F(x) + C$ has the same derivative at x, the slope of the graph at x is the same. This means that the graph of all functions of the form $y = F(x) + C$ is a collection of parallel curves, as shown in Figure 5.4.

The process of finding indefinite integrals is called **indefinite integration**. Notice that this process amounts to finding an antiderivative of f and adding an arbitrary constant C, which is called the **constant of integration**.

EXAMPLE 3   Antidifferentiation

Find each of the following indefinite integrals.

a. $\displaystyle\int 5x^3\, dx$ **b.** $\displaystyle\int \sec^2 x\, dx$ **c.** $\displaystyle\int e^x\, dx$

Solution

a. Since $\dfrac{d}{dx}(x^4) = 4x^3$, it follows that $\dfrac{d}{dx}\left(\dfrac{5x^4}{4}\right) = 5x^3$. Thus,

$$\int 5x^3\, dx = \frac{5x^4}{4} + C$$

b. Because $\dfrac{d}{dx}(\tan x) = \sec^2 x$, we have $\displaystyle\int \sec^2 x\, dx = \tan x + C$.

c. Since $\dfrac{d}{dx}(e^x) = e^x$, we have $\displaystyle\int e^x\, dx = e^x + C$.   ∎

ANTIDIFFERENTIATION FORMULAS

Example 3 leads us to state formulas for antidifferentiation. Theorem 5.2 summarizes several fundamental properties of indefinite integrals, each of which can be derived by reversing an appropriate differentiation formula. Assume that f and g are functions; u is a variable; a, b, c are given constants; and C is an arbitrary constant.

THEOREM 5.2 **Basic integration rules**

	Differentiation Formulas	*Integration Formulas*

PROCEDURAL RULES

Constant multiple:
$$\frac{d}{du}(cf) = c\frac{df}{du}$$
$$\int cf(u)\, du = c\int f(u)\, du$$

Sum rule:
$$\frac{d}{du}(f+g) = \frac{df}{du} + \frac{dg}{du}$$
$$\int [f(u) + g(u)]\, du = \int f(u)\, du + \int g(u)\, du$$

Difference rule:
$$\frac{d}{du}(f-g) = \frac{df}{du} - \frac{dg}{du}$$
$$\int [f(u) - g(u)]\, du = \int f(u)\, du - \int g(u)\, du$$

Linearity rule:
$$\frac{d}{du}(af+bg) = a\frac{df}{du} + b\frac{dg}{du}$$
$$\int [af(u) + bg(u)]\, du = a\int f(u)\, du + b\int g(u)\, du$$

BASIC FORMULAS

Constant rule:
$$\frac{d}{du}(c) = 0$$
$$\int 0\, du = 0 + C$$

Exponential rule:
$$\frac{d}{du}(e^u) = e^u$$
$$\int e^u\, du = e^u + C$$

Power rule:
$$\frac{d}{du}(u^n) = nu^{n-1}$$
$$\int u^n\, du = \begin{cases} \dfrac{u^{n+1}}{n+1} + C; & n \neq -1 \\ \ln|u| + C; & n = -1 \end{cases}$$

Logarithmic rule:*
$$\frac{d}{du}(\ln|u|) = \frac{1}{u}$$

Trigonometric rules:
$$\frac{d}{du}(\cos u) = -\sin u$$
$$\int \sin u\, du = -\cos u + C$$

$$\frac{d}{du}(\sin u) = \cos u$$
$$\int \cos u\, du = \sin u + C$$

$$\frac{d}{du}(\tan u) = \sec^2 u$$
$$\int \sec^2 u\, du = \tan u + C$$

$$\frac{d}{du}(\sec u) = \sec u \tan u$$
$$\int \sec u \tan u\, du = \sec u + C$$

$$\frac{d}{du}(\csc u) = -\csc u \cot u$$
$$\int \csc u \cot u\, du = -\csc u + C$$

$$\frac{d}{du}(\cot u) = -\csc^2 u$$
$$\int \csc^2 u\, du = -\cot u + C$$

---

*Note the absolute value. This is the result of reversing Theorem 3.13, Section 3.6.

Inverse trigonometric rules:

$$\frac{d}{du}(\sin^{-1} u) = \frac{1}{\sqrt{1 - u^2}}$$
$$\int \frac{du}{\sqrt{1 - u^2}} = \sin^{-1} u + C$$

The other three inverse trigonometric rules are not needed since, for example, $\sin^{-1} u + \cos^{-1} u = \dfrac{\pi}{2}$, so that

$$\frac{d}{du}(\tan^{-1} u) = \frac{1}{1 + u^2}$$
$$\int \frac{du}{1 + u^2} = \tan^{-1} u + C$$

$$\frac{d}{du}(\sec^{-1} u) = \frac{1}{|u| \sqrt{u^2 - 1}}$$
$$\int \frac{du}{|u| \sqrt{u^2 - 1}} = \sec^{-1} u + C$$

$$\int \frac{du}{\sqrt{1 - u^2}} = \frac{\pi}{2} - \cos^{-1} u + C_1$$
$$= -\cos^{-1} u + C$$

Proof   Each of these parts can be derived by reversing the accompanying derivative formula. For example, to obtain the power rule, note that if n is any number other than -1, then

$$\frac{d}{du}\left[\frac{1}{n + 1} u^{n+1}\right] = \frac{1}{n + 1}\left[(n + 1)u^n\right] = u^n$$

so that $\dfrac{1}{n + 1} u^{n+1}$ is an antiderivative of u^n and

$$\int u^n \, du = \frac{1}{n + 1} u^{n+1} + C \quad \text{for} \quad n \neq -1 \qquad \square$$

Now we will use these rules to compute a number of indefinite integrals.

EXAMPLE 4   Indefinite integral of a polynomial function

Evaluate $\displaystyle\int (x^5 - 3x^2 - 7)\, dx$.

Solution

The first two steps are usually done mentally:

$$\int (x^5 - 3x^2 - 7)\, dx = \int x^5\, dx - \int 3x^2\, dx - \int 7\, dx \qquad \text{Sum and difference rules}$$

$$= \int x^5\, dx - 3\int x^2\, dx - 7\int dx \qquad \text{Constant multiple}$$

$$= \frac{x^{5+1}}{5 + 1} - 3\frac{x^{2+1}}{2 + 1} - 7x + C \qquad \text{Power rule}$$

$$= \tfrac{1}{6}x^6 - x^3 - 7x + C \qquad \blacksquare$$

EXAMPLE 5   Indefinite integral with a mixture of forms

Evaluate $\displaystyle\int (5\sqrt{x} + 4\sin x)\, dx$.

Solution

$$\int (5\sqrt{x} + 4\sin x)\, dx = 5\int x^{1/2}\, dx + 4\int \sin x\, dx \qquad \text{Sum and constant rules}$$

$$= 5\frac{x^{3/2}}{\frac{3}{2}} + 4(-\cos x) + C \qquad \text{Power and trig rules}$$

$$= \tfrac{10}{3}x^{3/2} - 4\cos x + C \qquad \blacksquare$$

Antiderivatives will be used extensively in integration in connection with a marvelous result called the fundamental theorem of calculus (Section 5.4).

APPLICATIONS

In Chapter 3, we used differentiation to compute the slope at each point on the graph of a function. Example 6 shows how this procedure can be reversed.

EXAMPLE 6   Finding the function with a given slope through a particular point

The graph of a certain function F has slope $4x^3 - 5$ at each point (x, y) and contains the point $(1, 2)$. Find the function F.

Solution

We will work this problem twice; first we find an analytic solution and the second time we approximate the solution using technology, which is often sufficient.

Analytic solution: Because the slope of the tangent at each point (x, y) is given by $F'(x)$, we have

$$F'(x) = 4x^3 - 5$$

and it follows that

$$\int F'(x)\,dx = \int (4x^3 - 5)\,dx$$
$$F(x) = 4\left(\frac{x^4}{4}\right) - 5x + C$$
$$= x^4 - 5x + C$$

The family of curves is $y = x^4 - 5x + C$. To find the one that passes through $(1, 2)$, substitute:

$$2 = 1^4 - 5(1) + C$$
$$6 = C$$

The curve is $y = x^4 - 5x + 6$.

Technology solution: Begin by drawing the slope field, as shown in Figure 5.5a. We are interested in drawing the particular solution passing through $(1, 2)$. Remember to "go with the flow," as shown in Figure 5.5b. If we compare the analytic solution and the graphical solution, we see that the graph of the equation in the analytic solution is the same as the one found by technology.

a. Slope field

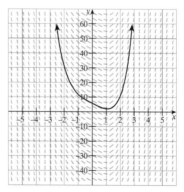
b. Particular solution

Figure 5.5   Slope field for $F'(x) = 4x^3 - 5$

In Section 3.4, we observed that an object moving along a straight line with position $s(t)$ has velocity $v(t) = \dfrac{ds}{dt}$ and acceleration $a(t) = \dfrac{dv}{dt}$. Thus, we have

$$v(t) = \int a(t)\, dt \qquad \text{and} \qquad s(t) = \int v(t)\, dt$$

These formulas are used in Examples 7 and 8.

EXAMPLE 7   Modeling Problem: The motion of a particle

A particle moves along a coordinate axis in such a way that its acceleration is modeled by $a(t) = 2t^{-2}$ for time $t > 0$. If the particle is at $s = 5$ when $t = 1$ and has velocity $v = -3$ at this time, where is it (to four decimal places) when $t = 4$?

Solution

Because $a(t) = v'(t)$, it follows that

$$v(t) = \int a(t)\, dt = \int 2t^{-2}\, dt = -2t^{-1} + C_1$$

and since $v(1) = -3$, we have

$$-3 = v(1) = \frac{-2}{1} + C_1 \qquad \text{so} \qquad C_1 = -3 + 2 = -1$$

We also know $v(t) = s'(t)$, so

$$s(t) = \int v(t)\, dt = \int (-2t^{-1} - 1)\, dt = -2\ln|t| - t + C_2$$

Since $s(1) = 5$, we have

$$5 = s(1) = -2\ln|1| - 1 + C_2 \qquad \text{or} \qquad C_2 = 6$$

Thus, $s(t) = -2\ln|t| - t + 6$ so that $s(4) \approx -0.7726$. The particle is at -0.7726 when $t = 4$.   ∎

EXAMPLE 8   Stopping distance for an automobile

The brakes of a certain automobile produce a constant deceleration of 22 ft/s². If the car is traveling at 60 mi/h (88 ft/s) when the brakes are applied, how far will it travel before coming to a complete stop?

Solution

Let $a(t)$, $v(t)$, and $s(t)$ denote the acceleration, velocity, and position of the car t seconds after the brakes are applied. We will assume that s is measured from the point where the brakes are applied, so that $s(0) = 0$.

$$v(t) = \int a(t)\, dt$$

$$= \int (-22)\, dt \qquad \text{\textit{Negative because the car is decelerating}}$$

$$= -22t + C_1 \qquad \text{\textit{$v(0) = -22(0) + C_1 = 88$, so that $C_1 = 88$}}$$
$$\text{\textit{Starting velocity is 88.}}$$

$$= -22t + 88$$

Similarly,

$$s(t) = \int v(t)\, dt$$

$$= \int (-22t + 88)\, dt$$

$$= -11t^2 + 88t + C_2 \qquad \text{\textit{$s(0) = -11(0)^2 + 88(0) + C_2 = 0$}}$$
$$\text{\textit{so that $C_2 = 0$}}$$

$$= -11t^2 + 88t \qquad \text{\textit{Starting distance is 0.}}$$

Finally, the car comes to rest when its velocity is 0, so we need to solve $v(t) = 0$ for t:

$$-22t + 88 = 0$$
$$t = 4$$

This means that the car decelerates for 4 sec before coming to rest, and in that time it travels

$$s(4) = -11(4)^2 + 88(4) = 176 \text{ ft}$$ ∎

Indefinite integration also has applications in business and economics. Recall from Section 4.7 that the *demand function* for a particular commodity is the function $p(x)$, which gives the price p that consumers will pay for each unit of the commodity when x units are brought to market. Then the total revenue is $R(x) = xp(x)$, and the marginal revenue is $R'(x)$. The next example shows how the demand function can be determined from the marginal revenue.

EXAMPLE 9   Finding the demand function given the marginal revenue

A manufacturer estimates that the marginal revenue of a certain commodity is $R'(x) = 240 + 0.1x$ when x units are produced. Find the demand function $p(x)$.

Solution

$$R(x) = \int R'(x)\,dx$$
$$= \int (240 + 0.1x)\,dx$$
$$= 240x + 0.1\left(\tfrac{1}{2}x^2\right) + C$$
$$= 240x + 0.05x^2 + C$$

Because $R(x) = xp(x)$, where $p(x)$ is the demand function, we must have $R(0) = 0$ so that

$$240(0) + 0.05(0)^2 + C = 0 \quad \text{or} \quad C = 0$$

Thus, $R(x) = 240x + 0.05x^2$. It follows that the demand function is

$$p(x) = \frac{R(x)}{x} = \frac{240x + 0.05x^2}{x} = 240 + 0.05x$$ ∎

AREA AS AN ANTIDERIVATIVE

In the next section, we will consider area as the limit of a sum, and we conclude this section by showing how area can be computed by antidifferentiation. The connection between area as a limit and area as an antiderivative is then made by a result called the fundamental theorem of calculus (see Section 5.4).

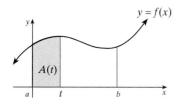

Figure 5.6 Area function $A(t)$

THEOREM 5.3   Area as an antiderivative

If f is a continuous function such that $f(x) \geq 0$ for all x on the closed interval $[a, b]$, then the area bounded by the curve $y = f(x)$, the x-axis, and the vertical lines $x = a$, $x = t$, viewed as a function of t, is an antiderivative of $f(t)$ on $[a, b]$.

Proof   Define an **area function**, $A(t)$, as the area of the region bounded by the curve $y = f(x)$, the x-axis, and the vertical lines $x = a$, $x = t$ for $a \leq t \leq b$, as shown in Figure 5.6. We need to show that $A(t)$ is an antiderivative of f on the interval $[a, b]$; that is, we need to show that $A'(t) = f(t)$.

Let $h > 0$ be small enough so that $t + h < b$ and consider the numerator of the difference quotient for $A(t)$, namely, the difference $A(t + h) - A(t)$. Geometrically this difference is the area under the curve $y = f(x)$ between $x = t$ and $x = t + h$, as shown in Figure 5.7a.

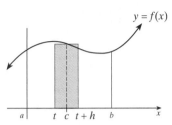

a. $A(t+h) - A(t)$ is the area over $[t, t+h]$.

b. $A(t+h) - A(t)$ is approximated by the area of a rectangle.

Figure 5.7   The area under the curve $y = f(x)$

If h is small enough, this area is approximately the same as the area of a rectangle with base h and height $f(c)$, where c is the midpoint of the interval $[t, t+h]$, as shown in Figure 5.7b. Thus, we have

$$\underbrace{A(t+h) - A(t)}_{\text{Area under the curve on } [t, t+h]} \approx \underbrace{hf(c)}_{\text{Area of rectangle}}$$

The difference quotient for $A(t)$ satisfies

$$\frac{A(t+h) - A(t)}{h} \approx f(c)$$

Finally, by taking the limit as $h \to 0^+$, we find the derivative of the area function $A(t)$ satisfies

$$\lim_{h \to 0^+} \frac{A(t+h) - A(t)}{h} = \lim_{h \to 0^+} f(c)$$
$$A'(t) = f(t)$$

The limit on the left is the definition of derivative, and on the right we see that since f is continuous and c is the midpoint of the interval $[t, t+h]$, c must approach t as $h \to 0^+$. A similar argument works as $h \to 0^-$. Thus, $A(t)$ is an antiderivative of $f(t)$. ❑

EXAMPLE 10   Area as an antiderivative

Find the area under the parabola $y = x^2$ over the interval $[0, 1]$. This area is shown in Figure 5.8.

Solution

Let $A(t)$ be the area function for this example—namely, the area under $y = x^2$ on $[0, t]$. Since f is continuous and $f(x) \geq 0$ on $[0, 1]$, Theorem 5.3 tells us that $A(t)$ is an antiderivative of $f(t) = t^2$ on $[0, 1]$. That is,

$$A(t) = \int t^2 \, dt = \tfrac{1}{3}t^3 + C$$

for all t in the interval $[0, 1]$. Clearly, $A(0) = 0$, so

$$A(0) = \tfrac{1}{3}(0)^3 + C \quad \text{or} \quad C = 0$$

and the area under the curve is

$$A(1) = \tfrac{1}{3}(1)^3 + 0 = \tfrac{1}{3}$$

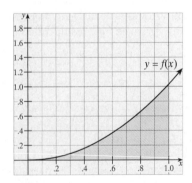

Figure 5.8 Area of $y = f(x)$ over $[0, 1]$

5.1   PROBLEM SET

A *Find the indefinite integral in Problems 1–30.*

1. $\displaystyle\int 2\,dx$

2. $\displaystyle\int -4\,dx$

3. $\displaystyle\int (2x+3)\,dx$

4. $\displaystyle\int (4-5x)\,dx$

5. $\displaystyle\int (4t^3+3t^2)\,dt$

6. $\displaystyle\int (-8t^3+15t^5)\,dt$

7. $\displaystyle\int \frac{dx}{2x}$

8. $\displaystyle\int 14e^x\,dx$

9. $\displaystyle\int (6u^2-3\cos u)\,du$

10. $\displaystyle\int (5t^3-\sqrt{t})\,dt$

11. $\displaystyle\int \sec^2\theta\,d\theta$

12. $\displaystyle\int \sec\theta\tan\theta\,d\theta$

13. $\displaystyle\int 2\sin\theta\,d\theta$

14. $\displaystyle\int \frac{\cos\theta}{3}\,d\theta$

15. $\displaystyle\int \frac{5}{\sqrt{1-y^2}}\,dy$

16. $\displaystyle\int \frac{dx}{10(1+x^2)}$

17. $\displaystyle\int (u^{3/2}-u^{1/2}+u^{-10})\,du$

18. $\displaystyle\int (x^3-3x+\sqrt[4]{x}-5)\,dx$

19. $\displaystyle\int x(x+\sqrt{x})\,dx$

20. $\displaystyle\int y(y^2-3y)\,dy$

21. $\displaystyle\int \left(\frac{1}{t^2}-\frac{1}{t^3}+\frac{1}{t^4}\right)dt$

22. $\displaystyle\int \frac{1}{t}\left(\frac{2}{t^2}-\frac{3}{t^3}\right)dt$

23. $\displaystyle\int (2x^2+5)^2\,dx$

24. $\displaystyle\int (3-4x^3)^2\,dx$

25. $\displaystyle\int \left(\frac{x^2+3x-1}{x^4}\right)dx$

26. $\displaystyle\int \frac{x^2+\sqrt{x}+1}{x^2}\,dx$

27. $\displaystyle\int \frac{x^2+x-2}{x^2}\,dx$

28. $\displaystyle\int \left(1+\frac{1}{x}\right)\left(1-\frac{4}{x^2}\right)dx$

29. $\displaystyle\int \frac{\sqrt{1-x^2}-1}{\sqrt{1-x^2}}\,dx$

30. $\displaystyle\int \frac{x^2}{x^2+1}\,dx$

The slope $F'(x)$ at each point on a graph is given in Problems 31–38 along with one point (x_0, y_0) on the graph. Use this information to find F both graphically and analytically.

31. $F'(x) = x^2+3x$
with point $(0,0)$

32. $F'(x) = (2x-1)^2$
with point $(1,3)$

33. $F'(x) = (\sqrt{x}+3)^2$
with point $(4,36)$

34. $F'(x) = 3-2\sin x$
with point $(0,0)$

35. slope $\dfrac{x+1}{x^2}$
with point $(1,-2)$

36. slope $\dfrac{2}{x\sqrt{x^2-1}}$
with point $(4,1)$

37. slope $x+e^x$
with point $(0,2)$

38. slope $\dfrac{x^2-1}{x^2+1}$
with point $(0,0)$

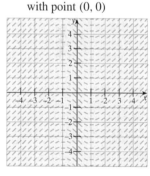

B **39. a.** If $F(x) = \displaystyle\int \left(\frac{1}{\sqrt{x}}-4\right)dx$, find F so that $F(1) = 0$.

   b. Sketch the graphs of $y = F(x)$, $y = F(x)+3$, and $y = F(x)-1$.

   c. Find a constant C_0 so that the largest value of $G(x) = F(x)+C_0$ is 0.

40. A ball is thrown directly upward from ground level with an initial velocity of 96 ft/s. Assuming that the ball's only acceleration is that due to gravity (that is, $a(t) = -32$ ft/s^2), determine the maximum height reached by the ball and the time it takes to return to ground level.

41. The marginal cost of a certain commodity is $C'(x) = 6x^2-2x+5$, where x is the level of production. If it costs \$5 to produce 1 unit, what is the total cost of producing 5 units?

42. The marginal revenue of a certain commodity is $R'(x) = -3x^2 + 4x + 32$, where x is the level of production (in thousands). Assume $R(0) = 0$.
    a. Find the demand function $p(x)$.
    b. Find the level of production that results in maximum revenue. What is the market price per unit at this level of production?

43. It is estimated that t months from now, the population of a certain town will be changing at the rate of $4 + 5t^{2/3}$ people per month. If the current population is 10,000, what will the population be 8 months from now?

44. A particle travels along the x-axis in such a way that its acceleration at time t is $a(t) = \sqrt{t} + t^2$. If it starts at the origin with an initial velocity of 2 (that is, $s(0) = 0$ and $v(0) = 2$), determine its position and velocity when $t = 4$.

45. An automobile starts from rest (that is, $v(0) = 0$) and travels with constant acceleration $a(t) = k$ in such a way that 6 sec after it begins to move, it has traveled 360 ft from its starting point. What is k?

46. The price of bacon is currently $1.80/lb in Styxville. A consumer service has conducted a study predicting that t months from now, the price will be changing at the rate of $0.084 + 0.012\sqrt{t}$ cents per month. How much will a pound of bacon cost 4 months from now?

47. An airplane has a constant acceleration while moving down the runway from rest. What is the acceleration of the plane at liftoff if the plane requires 900 ft of runway before lifting off at 88 ft/s (60 mi/h)?

48. After its brakes are applied, a certain sports car decelerates at a constant rate of 28 ft/s². Compute the stopping distance if the car is going 60 mi/h (88 ft/s) when the brakes are applied.

49. The brakes of a certain automobile produce a constant deceleration of k ft/s². The car is traveling at 60 mi/h (88 ft/s) when the driver is forced to hit the brakes, and it comes to rest at a point 121 ft from the point where the brakes were applied. What is k?

50. **Spy Problem**   The Spy, having defused the bomb in Problem 23, Section 4.6, is driving the sports car in Problem 48 at a speed of 60 mi/h on Highway 1 in the remote republic of San Dimas. Suddenly he sees a camel in the road 199 ft in front of the car. After a reaction time of 00.7 seconds, he steps on the brakes. Will he stop before hitting the camel?

51. A particle moves along the x-axis in such a way that at time $t > 0$, its velocity (in ft/s) is

    $$v(t) = t^{-1} + t$$

    How far does it move between times $t = 1$ and $t = e^2$?

52. A manufacturer estimates that the marginal cost in a certain production process is

    $$C'(x) = 0.1e^x + 21\sqrt{x}$$

    when x units are produced. If the cost of producing 1 unit is $100, what does it cost (to the nearest cent) to produce 4 units?

In Problems 53–58, find the area under the curve defined by the given equation, above the x-axis, and over the given interval.

53. $y = x^2$ over $[1, 4]$

54. $y = \sqrt{x}$ over $[1, 4]$

55. $y = e^x - x$ over $[0, 2]$

56. $y = \dfrac{x+1}{x}$ over $[1, 2]$

57. $y = \cos x$ over $\left[0, \frac{\pi}{2}\right]$

58. $y = (1 - x^2)^{-1/2}$ over $\left[0, \frac{1}{2}\right]$

59. Evaluate $\displaystyle\int \frac{dy}{2y\sqrt{y^2 - 1}}$ using the indicated methods.
    a. Use an inverse trigonometric differentiation rule.
    b. Use technology (TI-92, *Maple*, *Mathematica*, or *Derive*) to evaluate this integral.
    c. Reconcile your answers for parts **a** and **b**.

C 60. If a, b, and c are constants, use the linearity rule twice to show that

$$\int [af(x) + bg(x) + ch(x)]\, dx$$
$$= a \int f(x)\, dx + b \int g(x)\, dx + c \int h(x)\, dx$$

61. Use the area as an antiderivative theorem (Theorem 5.3) to find the area under the line $y = mx + b$ over the interval $[c, d]$, where $m > 0$ and $mc + b > 0$. Check your result by using geometry to find the area of a trapezoid.

5.2   Area as the Limit of a Sum

IN THIS SECTION   area as the limit of a sum, the general approximation scheme, summation notation, area using summation formulas

AREA AS THE LIMIT OF A SUM

Computing area has been a problem of both theoretical and practical interest since ancient times, but except for a few special cases, the problem is not easy. For example, you may know the formulas for computing the area of a rectangle, square, triangle, circle, and even a trapezoid. You have probably found the areas of regions that were more complicated but could be broken up into parts using these formulas. In Example 10 of the previous section we found the area between the parabola $y = x^2$ and the

x-axis on the interval $[0, 1]$ (see Figure 5.9a). We revisit this example to demonstrate a general procedure for computing area.

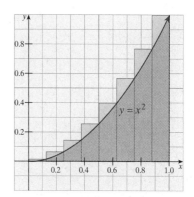

a. Problem: Compute the area under the curve $y = x^2$, above the x-axis, and between the lines $x = 0$ and $x = 1$.

b. The required area is approximately the same as the total area bounded by the shaded rectangles.

Figure 5.9 Example of the area problem

EXAMPLE 1   Estimating an area using rectangles and right endpoints

Estimate the area under the parabola $y = x^2$ on the interval $[0, 1]$.

Solution

In the previous section we found this area using the area function. In this example, we will *estimate* the area by adding the areas of approximating rectangles constructed on subintervals of $[0, 1]$, as shown in Figure 5.9b. To simplify computations, we will require all approximating rectangles to have the same width and will take the height of each rectangle to be the y-coordinate of the parabola above the *right endpoint* of the subinterval on which it is based.*

For the first estimate, we divide the interval $[0, 1]$ into 5 subintervals, as shown in Figure 5.10a. Because the approximating rectangles all have the same width, the right endpoints are $x_1 = 0.2$, $x_2 = 0.4$, $x_3 = 0.6$, $x_4 = 0.8$, and $x_5 = 1$. This subdivision is called a *partition* of the interval. The width of each subdivision is denoted by Δx and is found by dividing the length of the interval by the number of subintervals:

$$\Delta x = \frac{1 - 0}{5} = \frac{1}{5} = 0.2$$

Let S_n be the total area of n rectangles. For the case where $n = 5$,

$$
\begin{aligned}
S_5 &= f(x_1)\Delta x + f(x_2)\Delta x + f(x_3)\Delta x + f(x_4)\Delta x + f(x_5)\Delta x \\
&= [f(x_1) + f(x_2) + f(x_3) + f(x_4) + f(x_5)]\Delta x \\
&= [f(0.2) + f(0.4) + f(0.6) + f(0.8) + f(1)](0.2) \\
&= [0.2^2 + 0.4^2 + 0.6^2 + 0.8^2 + 1^2](0.2) \\
&= 0.44
\end{aligned}
$$

∎

Even though $S_5 = 0.44$ serves as a reasonable approximation of the area, we see from Figure 5.10b that this approximation seems too large. Let us rework Example

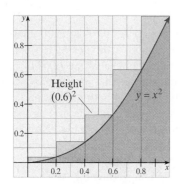

a. Partition into 5 subdivisions

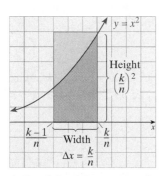

b. Detail showing one rectangle

Figure 5.10 Partitioning of Figure 5.8a into 5 subdivisions

---

*Actually, there is nothing special about right endpoints, and we could just as easily have used any other point in the base subinterval—say, the left endpoint or the midpoint.

1 using a general scheme rather than a specified number of rectangles. Partition the interval $[0, 1]$ into n equal parts, each with width

$$\Delta x = \frac{1 - 0}{n} = \frac{1}{n}$$

For $k = 1, 2, 3, \ldots, n$, the kth subinterval is $\left[\dfrac{k-1}{n}, \dfrac{k}{n}\right]$, and on this subinterval we then construct an approximating rectangle with width $\Delta x = \dfrac{1}{n}$ and height $\left(\dfrac{k}{n}\right)^2$, since $y = x^2$. The total area bounded by all n rectangles is

$$S_n = \left[\left(\frac{1}{n}\right)^2 + \left(\frac{2}{n}\right)^2 + \cdots + \left(\frac{n}{n}\right)^2\right]\left(\frac{1}{n}\right)$$

Consider different choices for n, as shown in Figure 5.11.

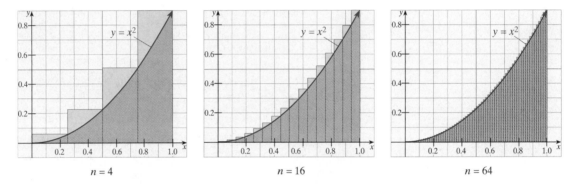

$n = 4$ $n = 16$ $n = 64$

Figure 5.11   The area estimate is improved by taking more rectangles.

n	S_n	
5	0.440	(Example 1 result)
4	0.469	(Figure 5.11a result)
16	0.365	(Figure 5.11b result)
64	0.341	(Figure 5.11c result)
100	0.338	
1,000	0.334	
5,000	0.333	

If we increase the number of subdivisions n, the width $\Delta x = \frac{1}{n}$ of each approximating rectangle will decrease, and we would expect the area estimates S_n to improve. Thus, it is reasonable to *define* the area A under the parabola to be the *limit* of S_n as $\Delta x \to 0$ or, equivalently, as $n \to +\infty$. We can attempt to predict its value by seeing what happens to the sum as n grows large without bound. It is both tedious and difficult to evaluate such sums by hand, but fortunately we can use a calculator or computer to obtain some (rounded) values for S_n, as shown in the following table. Notice that for $n = 5$, the value 0.44 corresponds to the calculation in Example 1 and that as n increases, S_n appears to approach $1/3$.

Figure 5.12 displays some computer-generated drawings showing the area under the curve $y = x^2$ on $[1, 5]$ as approximated by rectangles using left endpoints (as contrasted with right endpoints in Example 1). The sums in these outputs are called *Riemann sums*, which we will discuss in the next section.

THE GENERAL APPROXIMATION SCHEME

We now compute the area under any curve $y = f(x)$ on an interval $[a, b]$, where f is a nonnegative continuous function. We first partition the interval $[a, b]$ into n equal subintervals, each of width

$$\Delta x = \frac{b - a}{n}$$

For $k = 1, 2, 3, \ldots, n$ the kth subinterval is $[a + (k-1)\Delta x, a + k\Delta x]$, and the kth approximating rectangle is constructed with width Δx and height $f(a + k\Delta x)$ equal to the height of the curve $y = f(x)$ above the right endpoint of the subinterval. Adding the areas of these n rectangles, we obtain

$$S_n = \overbrace{f(a + \Delta x)\Delta x}^{\text{Area of first rectangle}} + \overbrace{f(a + 2\Delta x)\Delta x}^{\text{Area of second rectangle}} + \cdots + \overbrace{f(a + n\Delta x)\Delta x}^{\text{Area of }n\text{th rectangle}}$$

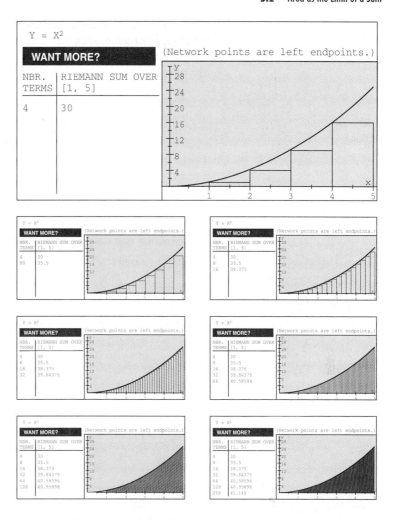

Figure 5.12 Sample of finding successive rectangular approximations

as an estimate of the area under the curve. In advanced calculus, it is shown that the continuity of f guarantees the existence of $\lim_{\Delta x \to 0} S_n$, and we use this limit to define the required area, as shown in the following box.

Area as the Limit of a Sum

Suppose f is continuous and $f(x) \geq 0$ throughout the interval $[a, b]$. Then the **area** of the region under the curve $y = f(x)$ over this interval is

$$A = \lim_{\Delta x \to 0} [f(a + \Delta x) + f(a + 2\Delta x) + \cdots + f(a + n\Delta x)]\Delta x$$

where $\Delta x = \dfrac{b - a}{n}$.

The definition of area as a limit of a sum is consistent with the area concept introduced in plane geometry, and with the area function defined in Section 5.1. For example, it would not be difficult to use this formula to show that a rectangle has area $A = \ell w$ or that a triangle has area $A = \frac{1}{2}bh$. You will also note that we maintained everyday usage in saying that the *formula* for the area of a rectangle is $A = \ell w$ or the *area* of the region under the curve $y = f(x)$ is a limit, but what we are really doing is *defining* area as a limit.

The problem we now face is how to implement this definition of area as the limit of a sum. The immediate answer (discussed in this section) is to use summation formulas and technology. The long-range goal is to develop integral calculus, which is discussed in the next section.

SUMMATION NOTATION

The expanded form of the sum for the definition of area makes it awkward to use. Therefore, we will digress to introduce a more compact notation for sums, and that notation will motivate integral notation. Using this **summation notation**, we express the sum $a_1 + a_2 + \cdots + a_n$ as follows:

$$a_1 + a_2 + \cdots + a_n = \sum_{k=1}^{n} a_k$$

The summation notation is sometimes called the **sigma notation** because the uppercase Greek letter sigma (Σ) is used to denote the summation process. The **index** k is called the **index of summation** (or *running index*). The terminology used in connection with the summation notation is shown:

$$\overset{\text{Upper limit of summation}}{\underset{\underset{\text{Lower limit of summation}}{\uparrow}}{\underset{k=1}{\overset{n}{\sum}}}} \overset{\overset{\text{General term}}{\downarrow}}{\overset{\frown}{a_k}} \underset{\text{Index of summation}}{\uparrow}$$

Note that in the summation process, the choice of summation index is immaterial. For example, the following sums are all exactly the same:

$$\sum_{k=3}^{7} k^2 = \sum_{j=3}^{7} j^2 = \sum_{i=3}^{7} i^2 = \sum_{\lambda=3}^{7} \lambda^2$$

In general, an index (k, j, i, or λ) that represents a process in which it has no direct effect on the result is called a **dummy variable**.

Several useful properties of sums and sum formulas are listed in Theorem 5.4. We will use the summation notation throughout the rest of this text, especially in this chapter, Chapter 6, and Chapter 8.

THEOREM 5.4   Basic rules for sums

For any numbers c and d and positive integers m and n,

1. **Constant term rule** $\displaystyle\sum_{k=1}^{n} c = \underbrace{c + c + \cdots + c}_{n \text{ terms}} = nc$

2. **Sum rule** $\displaystyle\sum_{k=1}^{n} (a_k + b_k) = \sum_{k=1}^{n} a_k + \sum_{k=1}^{n} b_k$

3. **Scalar multiple rule** $\displaystyle\sum_{k=1}^{n} c a_k = c \sum_{k=1}^{n} a_k = \left(\sum_{k=1}^{n} a_k \right) c$

4. **Linearity rule** $\displaystyle\sum_{k=1}^{n} (c a_k + d b_k) = c \sum_{k=1}^{n} a_k + d \sum_{k=1}^{n} b_k$

5. **Subtotal rule**   If $1 < m < n$, then $\displaystyle\sum_{k=1}^{n} a_k = \sum_{k=1}^{m} a_k + \sum_{k=m+1}^{n} a_k$

6. **Dominance rule**   If $a_k \leq b_k$ for $k = 1, 2, \ldots, n$, then $\displaystyle\sum_{k=1}^{n} a_k \leq \sum_{k=1}^{n} b_k$

Proof   These properties can all be established by applying well-known algebraic rules (see Problem 50).   ❏

AREA USING SUMMATION FORMULAS

Using summation notation, we can streamline the symbolism in the formula for the area under the curve $y = f(x)$, $f(x) \geq 0$ on the interval $[a, b]$. In particular, note that the approximating sum S_n is

$$S_n = [f(a + \Delta x) + f(a + 2\Delta x) + \cdots + f(a + n\Delta x)]\Delta x$$

$$= \sum_{k=1}^{n} f(a + k\Delta x)\Delta x$$

where $\Delta x = \dfrac{b - a}{n}$. Thus, the formula for the definition of area is shown:

n is the number of approximating rectangles

Area under $y = f(x)$ above the
x-axis between $x = a$ and $x = b$

Width of each rectangle

$$A = \lim_{n \to +\infty} S_n = \lim_{\Delta x \to 0} \sum_{k=1}^{n} f(a + k\Delta x) \; \Delta x$$

Height of the kth rectangle

From algebra we recall certain summation formulas (which can be proved using mathematical induction) that we will need in order to find areas using the limit definition.

Summation Formulas

$$\sum_{k=1}^{n} 1 = n$$

$$\sum_{k=1}^{n} k = 1 + 2 + 3 + \cdots + n = \frac{n(n + 1)}{2}$$

$$\sum_{k=1}^{n} k^2 = 1^2 + 2^2 + 3^2 + \cdots + n^2 = \frac{n(n + 1)(2n + 1)}{6}$$

$$\sum_{k=1}^{n} k^3 = 1^3 + 2^3 + 3^3 + \cdots + n^3 = \frac{n^2(n + 1)^2}{4}$$

EXAMPLE 2   Area using the definition and summation formulas

Use the summation definition of area to find the area under the parabola $y = x^2$ on the interval $[0, 1]$. You estimated this area in Example 1.

Solution

Partition the interval $[0, 1]$ into n subintervals with width $\Delta x = \dfrac{1 - 0}{n}$. The right endpoint of the kth subinterval is $a + k\Delta x = \dfrac{k}{n}$ and $f\left(\dfrac{k}{n}\right) = \dfrac{k^2}{n^2}$. Thus, from the definition of area we have

$$A = \lim_{\Delta x \to 0} \sum_{k=1}^{n} f(a + k\Delta x)\Delta x$$

$$= \lim_{n \to +\infty} \sum_{k=1}^{n} \left(\frac{k^2}{n^2}\right)\left(\frac{1}{n}\right)$$

$$= \lim_{n \to +\infty} \sum_{k=1}^{n} \frac{k^2}{n^3}$$ Note that $1/n^3$ is independent of the index of summation, k.

$$= \lim_{n \to +\infty} \frac{1}{n^3} \sum_{k=1}^{n} k^2$$ Scalar multiple rule

$$= \lim_{n \to +\infty} \frac{1}{n^3}\left[\frac{n(n+1)(2n+1)}{6}\right]$$ Summation formula for squares

$$= \lim_{n \to +\infty} \frac{1}{6}\left[2 + \frac{3}{n} + \frac{1}{n^2}\right] = \frac{1}{3}$$

This is the same as the answer found by antidifferentiation in Example 10 of Section 5.1 and is consistent with the table on page 284.   ∎

EXAMPLE 3   Tabular approach for finding area

Use a computer to estimate the area under the curve $y = \sin x$ on the interval $\left[0, \frac{\pi}{2}\right]$.

Solution

We see $a = 0$ and $b = \frac{\pi}{2}$, so $\Delta x = \frac{\frac{\pi}{2} - 0}{n} = \frac{\pi}{2n}$. The right endpoints are

$$a + \Delta x = 0 + \left(\frac{\pi}{2n}\right) = \frac{\pi}{2n}; a + 2\Delta x = \frac{2\pi}{2n} = \frac{\pi}{n}; a + 3\Delta x = \frac{3\pi}{2n}; \ldots$$

$$a + n\Delta x = \frac{n\pi}{2n} = \frac{\pi}{2} = b$$

Thus, $S_n = \sum_{k=1}^{n} f\left(0 + \frac{k\pi}{2n}\right)\left(\frac{\pi}{2n}\right) = \sum_{k=1}^{n} \left[\sin\left(\frac{k\pi}{2n}\right)\right]\left(\frac{\pi}{2n}\right)$. Now we know from the definition that the actual area is

$$S = \lim_{n \to +\infty} \sum_{k=1}^{n} \left[\sin\left(\frac{k\pi}{2n}\right)\right]\left(\frac{\pi}{2n}\right) = \frac{\pi}{2} \lim_{n \to +\infty} \sum_{k=1}^{n} \frac{1}{n}\sin\left(\frac{k\pi}{2n}\right)$$

n	S_n
10	1.07648
20	1.03876
50	1.01563
100	1.00783
500	1.00157

which we can estimate by computing S_n for successively large values of n, as summarized in the table in this margin. Note that the table suggests that

$$\lim_{\Delta x \to 0} S_n = \lim_{n \to +\infty} S_n = 1$$

Thus, we expect the actual area under the curve to be 1 square unit.   ∎

5.2   PROBLEM SET

Ⓐ *Evaluate the sums in Problems 1–8 by using the summation formulas.*

1. $\displaystyle\sum_{k=1}^{6} 1$

2. $\displaystyle\sum_{k=1}^{250} 2$

3. $\displaystyle\sum_{k=1}^{15} k$

4. $\displaystyle\sum_{k=1}^{10} (k+1)$

5. $\displaystyle\sum_{k=1}^{5} k^3$

6. $\displaystyle\sum_{k=1}^{7} k^2$

7. $\displaystyle\sum_{k=1}^{100} (2k-3)$

8. $\displaystyle\sum_{k=1}^{100} (k-1)^2$

Use the properties of summation notation in Problems 9–12 to evaluate the given limits.

9. $\displaystyle\lim_{n\to+\infty}\sum_{k=1}^{n}\frac{k}{n^2}$

10. $\displaystyle\lim_{n\to+\infty}\sum_{k=1}^{n}\frac{k^3}{n^4}$

11. $\displaystyle\lim_{n\to+\infty}\sum_{k=1}^{n}\left(1+\frac{k}{n}\right)\left(\frac{2}{n}\right)$

12. $\displaystyle\lim_{n\to+\infty}\sum_{k=1}^{n}\left(1+\frac{2k}{n}\right)^2\left(\frac{2}{n}\right)$

First sketch the region under the graph of $y = f(x)$ on the interval $[a, b]$ in Problems 13–21. Then approximate the area of each region by using right endpoints and the formula

$$S_n = \sum_{k=1}^{n} f(a + k\Delta x)\Delta x$$

for $\Delta x = \dfrac{b-a}{n}$ and the indicated values of n.

13. $f(x) = 4x + 1$ on $[0, 1]$ for
    a. $n = 4$ **b.** $n = 8$

14. $f(x) = 3 - 2x$ on $[0, 1]$ for
    a. $n = 3$ **b.** $n = 6$

15. $f(x) = x^2$ on $[1, 2]$ for
    a. $n = 4$ **b.** $n = 6$

16. $f(x) = \cos x$ on $\left[-\frac{\pi}{2}, 0\right]$ for $n = 4$

17. $f(x) = x + \sin x$ on $\left[0, \frac{\pi}{4}\right]$ for $n = 3$

18. $f(x) = \dfrac{1}{x^2}$ on $[1, 2]$ for $n = 4$

19. $f(x) = \dfrac{2}{x}$ on $[1, 2]$ for $n = 4$

20. $f(x) = \sqrt{x}$ on $[1, 4]$ for $n = 4$

21. $f(x) = \sqrt{1 + x^2}$ on $[0, 1]$ for $n = 4$

B *Find the exact area under the given curve on the interval prescribed in Problems 22–27 by using area as the limit of a sum and the summation formulas.*

22. $y = 4x^3 + 2x$ on $[0, 2]$

23. $y = 4x^3 + 2x$ on $[1, 2]$

24. $y = 6x^2 + 2x + 4$ on $[0, 3]$

25. $y = 6x^2 + 2x + 4$ on $[1, 3]$

26. $y = 3x^2 + 2x + 1$ on $[0, 1]$

27. $y = 4x^3 + 3x^2$ on $[0, 1]$

Counterexample Problems *Show that each statement about area in Problems 28–33 is generally true or provide a counterexample. It will probably help to sketch the indicated region for each problem.*

28. If $C > 0$ is a constant, the region under the line $y = C$ on the interval $[a, b]$ has area $A = C(b - a)$.

29. If $C > 0$ is a constant and $b > a \geq 0$, the region under the line $y = Cx$ on the interval $[a, b]$ has area $A = \frac{1}{2}C(b - a)$.

30. The region under the parabola $y = x^2$ on the interval $[a, b]$ has area less than $\frac{1}{2}(b^2 + a^2)(b - a)$.

31. The region under the curve $y = \sqrt{1 - x^2}$ on the interval $[-1, 1]$ has area $A = \frac{\pi}{2}$.

32. Let f be a function that satisfies $f(x) \geq 0$ for x in the interval $[a, b]$. Then the area under the curve $y = f^2(x)$ on the interval $[a, b]$ must always be greater than the area under $y = f(x)$ on the same interval.

33. Recall that a function f is said to be **even** if $f(-x) = f(x)$ for all x. If f is even and $f(x) \geq 0$ throughout the interval $[-a, a]$, then the area under the curve $y = f(x)$ on this interval is *twice* the area under $y = f(x)$ on $[0, a]$.

34. Show that the region under the curve $y = x^3$ on the interval $[0, 1]$ has area $\frac{1}{4}$ square units.

35. Use the definition of area to show that the area of a rectangle equals the product of its length ℓ and its width w.

36. Show that the triangle with vertices $(0, 0)$, $(0, h)$, and $(b, 0)$ has area $A = \frac{1}{2}bh$ using the area as the limit of a sum.

37. a. Compute the area under the parabola $y = 2x^2$ on the interval $[1, 2]$ as the limit of a sum.

    b. Let $f(x) = 2x^2$ and note that $g(x) = \frac{2}{3}x^3$ defines a function that satisfies $g'(x) = f(x)$ on the interval $[1, 2]$. Verify that the area computed in part **a** satisfies $A = g(2) - g(1)$.

    c. The function defined by

$$h(x) = \tfrac{2}{3}x^3 + C$$

    for any constant C also satisfies $h'(x) = f(x)$. Is it true that the area in part **a** satisfies $A = h(2) - h(1)$?

Use a tabular approach to compute the area under the curve $y = f(x)$ on each interval given in Problems 38–44 as the limit of a sum of terms.

38. $f(x) = 4x$ on $[0, 1]$

39. $f(x) = x^2$ on $[0, 4]$

40. $f(x) = \cos x$ on $\left[-\frac{\pi}{2}, 0\right]$ (Compare with Problem 16.)

41. $f(x) = x + \sin x$ on $\left[0, \frac{\pi}{4}\right]$ (Compare with Problem 17.)

42. $f(x) = \ln(x^2 + 1)$ on $[0, 3]$

43. $f(x) = e^{-3x^2}$ on $[0, 1]$

44. $f(x) = \cos^{-1}(x + 1)$ on $[-1, 0]$

45. a. Use the tabular approach to compute the area under the curve $y = \sin x + \cos x$ on the interval $\left[0, \frac{\pi}{2}\right]$ as the limit of a sum.

    b. Let $f(x) = \sin x + \cos x$ and note that $g(x) = -\cos x + \sin x$ satisfies $g'(x) = f(x)$ on the interval $\left[0, \frac{\pi}{2}\right]$. Verify that the area computed in part **a** satisfies $A = g(\frac{\pi}{2}) - g(0)$.

    c. The function $h(x) = -\cos x + \sin x + C$ for constant C also satisfies $h'(x) = f(x)$. Is it true that the area in part **a** satisfies $A = h(\frac{\pi}{2}) - h(0)$?

C 46. Derive the formula

$$\sum_{k=1}^{n} k = 1 + 2 + 3 + \cdots + n = \frac{n(n + 1)}{2}$$

by completing these steps:

a. Use the basic rules for sums to show that

$$\sum_{k=1}^{n} k = \frac{1}{2}\sum_{k=1}^{n}[k^2 - (k-1)^2] + \frac{1}{2}\sum_{k=1}^{n} 1$$

$$= \frac{1}{2}\sum_{k=1}^{n}[k^2 - (k-1)^2] + \frac{1}{2}n$$

b. Show that

$$\sum_{k=1}^{n}[k^2 - (k-1)^2] = n^2$$

Hint: Expand the sum by writing out a few terms. Note the internal cancellation.

c. Combine parts **a** and **b** to show that

$$\sum_{k=1}^{n} k = \frac{n(n+1)}{2}$$

47. a. First find constants a, b, c, and d such that

$$k^3 = a[k^4 - (k-1)^4] + bk^2 + ck + d$$

b. Modify the approach outlined in Problem 46 to establish the formula

$$\sum_{k=1}^{n} k^2 = \frac{n(n+1)(2n+1)}{6}$$

48. The purpose of this problem is to verify the results shown in Figure 5.11. Specifically, we will find the area A under the parabola $y = x^2$ on the interval $[0, 1]$ using approximating rectangles with heights taken at the *left* endpoints.

Verify that

$$\lim_{n \to +\infty} \sum_{k=1}^{n} \left(\frac{k-1}{n}\right)^2 \left(\frac{1}{n}\right) = \frac{1}{3}$$

Compare this with the procedure outlined in Example 1. Note that when the interval $[0, 1]$ is subdivided into n equal parts, the kth subinterval is

$$\left[\frac{k-1}{n}, \frac{k}{n}\right]$$

49. Develop a formula for area based on approximating rectangles with heights taken at the *midpoints* of subintervals.

50. Use the properties of real numbers to establish the summation formulas in Theorem 5.4. For example, to prove the linearity rule, use the associative, commutative, and distributive properties of real numbers to note that

$$\sum_{k=1}^{n}(ca_k + db_k)$$
$$= (ca_1 + db_1) + (ca_2 + db_2) + \cdots + (ca_n + db_n)$$
$$= (ca_1 + ca_2 + \cdots + ca_n) + (db_1 + db_2 + \cdots + db_n)$$
$$= c(a_1 + a_2 + \cdots + a_n) + d(b_1 + b_2 + \cdots + b_n)$$
$$= c\sum_{k=1}^{n} a_k + d\sum_{k=1}^{n} b_k$$

5.3   Riemann Sums and the Definite Integral

IN THIS SECTION   Riemann sums, the definite integral, area as an integral, properties of the definite integral, distance as an integral

We will soon discover that not just area, but other useful quantities such as distance, volume, mass, and work, can be first approximated by sums and then obtained exactly by taking a limit involving the approximating sums. The special kind of limit of a sum that appears in this context is called the *definite integral*, and the process of finding integrals is called *definite integration* or *Riemann integration* in honor of the German mathematician Georg Bernhard Riemann (1826–1866), who pioneered the modern approach to integration theory. We begin by introducing some special notation and terminology.

RIEMANN SUMS

Recall from Section 5.2 that to find the area under the graph of the function $y = f(x)$ on the closed interval $[a, b]$ using sums where f is continuous and $f(x) \geq 0$, we proceed as follows:

1. Partition the interval into n subintervals of equal width $\Delta x = \dfrac{b-a}{n}$.

2. Evaluate f at the right endpoint $a + k\Delta x$ of the kth subinterval for $k = 1, 2, \ldots, n$.

3. Form the approximating sum of the areas of the n rectangles, which we denote by

$$S_n = \sum_{k=1}^{n} f(a + k\Delta x)\Delta x.$$

4. Because we expect the estimates S_n to improve as Δx decreases, we *define* the area A under the curve, above the x-axis, and bounded by the lines $x = a$ and $x = b$, to be the limit of S_n as $\Delta x \to 0$. Thus, we write

$$A = \lim_{n \to +\infty} \sum_{k=1}^{n} f(a + k\Delta x)\Delta x$$

if this limit exists. This means that A can be estimated to any desired degree of accuracy by approximating the sum S_n with Δx sufficiently small (or, equivalently, n sufficiently large).

This approach to the area problem contains the essentials of integration, but there is no compelling reason for the partition points to be evenly spaced or to insist on evaluating f at right endpoints. These conventions are for convenience of computation, but to accommodate easily as many applications as possible, it is useful to consider a more general type of approximating sum and to specify what is meant by the limit of such sums. The approximating sums that occur in integration problems are called **Riemann sums**, and the following definition contains a step-by-step description of how such sums are formed.

Riemann Sum

Suppose a bounded function f is given along with a closed interval $[a, b]$ on which f is defined. Then

Step 1. Partition the interval $[a, b]$ into n subintervals by choosing points $\{x_0, x_1, \ldots, x_n\}$ arranged in such a way that

$$a = x_0 < x_1 < x_2 < \cdots < x_{n-1} < x_n = b$$

Call this partition P. For $k = 1, 2, \ldots, n$, the kth subinterval width is $\Delta x_k = x_k - x_{k-1}$. The largest of these widths is called the **norm** of the partition P and is denoted by $\|P\|$; that is,

$$\|P\| = \max_{k=1,2,\ldots,n} \{\Delta x_k\}$$

Step 2. Choose a number arbitrarily from each subinterval. For $k = 1, 2, \ldots, n$, the number x_k^* chosen from the kth subinterval $[x_{k-1}, x_k]$ is called the *kth subinterval representative* of the partition P.

Step 3. Form the sum

$$R_n = f(x_1^*)\Delta x_1 + f(x_2^*)\Delta x_2 + \cdots + f(x_n^*)\Delta x_n = \sum_{k=1}^{n} f(x_k^*)\Delta x_k$$

This is the **Riemann sum** associated with f, the given partition P, and the chosen subinterval representatives $x_1^*, x_2^*, \ldots, x_n^*$.

➡ **What This Says**   We will express quantities from geometry, physics, economics, and other applications in terms of a Riemann sum

$$\sum_{k=1}^{n} f(x_k^*)\Delta x_k$$

Riemann sums are generally used to model a quantity for a particular application. Note that the Riemann sum *does not* require that the function f be nonnegative, nor does it require that all the intervals must be the same length. In addition, x_k^* is *any* point in the kth subinterval and does not need to be something "nice" like the left or right endpoint, or the midpoint.

EXAMPLE 1   Formation of the Riemann sum for a given function

Suppose the interval $[-2, 1]$ is partitioned into 6 subintervals with subdivision points $a = x_0 = -2$, $x_1 = -1.6$, $x_2 = -0.93$, $x_3 = -0.21$, $x_4 = 0.35$, $x_5 = 0.82$, and

$x_6 = 1 = b$. Find the norm of this partition P and the Riemann sum associated with the function $f(x) = 2x$, the given partition, and the subinterval representatives $x_1^* = -1.81$, $x_2^* = -1.12$, $x_3^* = -0.55$, $x_4^* = -0.17$, $x_5^* = 0.43$, and $x_6^* = 0.94$.

Solution

Before we can find the norm of the partition or the required Riemann sum, we must compute the subinterval width Δx_k and evaluate f at each subinterval representative x_k^*. These values and computations are shown in Figure 5.13.

k	$x_k - x_{k-1} = \Delta x_k$ (Given)	x_k^* (Given)	$f(x_k^*) = 2(x_k^*)$
1	$-1.6 - (-2) = 0.40$	-1.81	$f(-1.81) = -3.62$
2	$-0.93 - (-1.6) = 0.67$	-1.12	$f(-1.12) = -2.24$
3	$-0.21 - (-0.93) = 0.72$	-0.55	$f(-0.55) = -1.10$
4	$0.35 - (-0.21) = 0.56$	-0.17	$f(-0.17) = -0.34$
5	$0.82 - 0.35 = 0.47$	0.43	$f(0.43) = 0.86$
6	$1.00 - 0.82 = 0.18$	0.94	$f(0.94) = 1.88$

Figure 5.13   Riemann sum

From the table, we see that the largest subinterval width is $\Delta x_3 = 0.72$, so the partition has norm $\|P\| = 0.72$. Finally, by using the definition, we compute the Riemann sum:

$$R_6 = (-3.62)(0.40) + (-2.24)(0.67) + (-1.10)(0.72) + (-0.34)(0.56)$$
$$+ (0.86)(0.47) + (1.88)(0.18)$$
$$\approx -3.1886 \approx -3.19 \qquad \blacksquare$$

WARNING   Notice from Example 1 that the Riemann sum does not necessarily represent an area. The sum found is negative (and area must be nonnegative).

THE DEFINITE INTEGRAL

By comparing the formula for the Riemann sum with that of area in the previous section, we recognize that the sum S_n used to approximate area is actually a special kind of Riemann sum that has

$$\Delta x_k = \Delta x = \frac{b-a}{n} \quad \text{and} \quad x_k^* = a + k\Delta x$$

for $k = 1, 2, \ldots, n$. Because the subintervals in the partition P associated with S_n are equally spaced, it is called a **regular partition**. When we express the area under the curve $y = f(x)$ as $A = \lim_{\Delta x \to 0} S_n$, we are actually saying that A can be estimated to any desired accuracy by finding a Riemann sum of the form S_n with norm

$$\|P\| = \frac{b-a}{n}$$

sufficiently small. We use this interpretation as a model for the following definition.

Definite Integral

If f is defined on the closed interval $[a, b]$, we say f is **integrable on $[a, b]$** if

$$I = \lim_{\|P\| \to 0} \sum_{k=1}^{n} f(x_k^*) \Delta x_k$$

exists. This limit is called the **definite integral** of f from a to b. The definite integral is denoted by

$$I = \int_a^b f(x)\, dx$$

➡ **What This Says** To say that f is *integrable* with definite integral I means the number I can be approximated to any prescribed degree of accuracy by *any* Riemann sum of f with norm sufficiently small. As long as the conditions of this definition are satisfied (that is, f is defined on $[a, b]$ and the Riemann sum exists), we can write

$$\int_a^b f(x)\, dx = \lim_{\|P\| \to 0} \sum_{k=1}^{n} f(x_k^*) \Delta x_k$$

Formally, I is the definite integral of f on $[a, b]$ if for each number $\epsilon > 0$, there exists a number $\delta > 0$ such that if

$$\sum_{k=1}^{n} f(x_k^*) \Delta x_k$$

is any Riemann sum of f whose norm satisfies $\|P\| < \delta$, then

$$\left| I - \sum_{k=1}^{n} f(x_k^*) \Delta x_k \right| < \epsilon$$

In advanced calculus, it is shown that when this limit exists, it is unique. Moreover, its value is independent of the particular way in which the partitions of $[a, b]$ and the subinterval representatives x_k^* are chosen.

WARNING Take a few minutes to make sure you understand the integral notation and terminology.

$$\int_a^b \overbrace{f(x)}^{\text{integrand}}{}^{\downarrow \text{upper limit of integration}} dx \qquad \text{The "}dx\text{" indicates the variable of integration.}$$

↑ lower limit of integration

The function f that is being integrated is called the **integrand**; the interval $[a, b]$ is the **interval of integration**; and the endpoints a and b are called, respectively, the **lower and upper limits of integration**.

In the special case where $a = b$, the interval of integration $[a, b]$ is really just a point, and the integral of any function on this "interval" is defined to be 0; that is,

$$\int_a^a f(x)\, dx = 0 \qquad \text{Why does this makes sense?}$$

Also, at times, we will consider integrals in which the lower limit of integration is a larger number than the upper limit. To handle this case, we specify that the integral from b to a is the negative of the integral from a to b:

$$\int_b^a f(x)\, dx = -\int_a^b f(x)\, dx$$

To summarize,

Definite Integral at a Point

$$\int_a^a f(x)\,dx = 0$$

Interchanging the Limits of a Definite Integral

$$\int_b^a f(x)\,dx = -\int_a^b f(x)\,dx$$

At first, the definition of the definite integral may seem rather imposing. How are we to tell whether a given function f is integrable on an interval $[a, b]$? If f is integrable, how are we supposed to actually compute the definite integral? Answering these questions is not easy, but in advanced calculus, it is shown that f is integrable on a closed interval $[a, b]$ if it is continuous on the interval except at a finite number of points and if it is bounded on the interval (that is, there is a number $A > 0$ such that $|f(x)| < A$ for all x in the interval). We will state a special case of this result as a theorem.

THEOREM 5.5 Integrability of a continuous function

If f is continuous on an interval $[a, b]$, then f is integrable on $[a, b]$.

Proof The proof requires the methods of advanced calculus and is omitted here. ❑

Our next example illustrates how to use the definition to find a definite integral.

EXAMPLE 2 Evaluating a definite integral using the definition

Evaluate $\displaystyle\int_{-2}^1 4x\,dx$.

Solution

The integral exists because $f(x) = 4x$ is continuous on $[-2, 1]$. Because the integral can be computed by any partition whose norm approaches 0 (that is, the integral is independent of the sequence of partitions *and* the subinterval representatives), we will simplify matters by choosing a partition in which the points are evenly spaced. Specifically, we divide the interval $[-2, 1]$ into n subintervals, each of width

$$\Delta x = \frac{1 - (-2)}{n} = \frac{3}{n}$$

For each k, we choose the kth subinterval representative to be the right endpoint of the kth subinterval; that is,

$$x_k^* = -2 + k\Delta x = -2 + k\left(\frac{3}{n}\right)$$

Finally, we form the Riemann sum

$$\int_{-2}^1 4x\,dx = \lim_{\|P\|\to 0} \sum_{k=1}^n f(x_k^*)\Delta x$$

$$= \lim_{n\to+\infty} \sum_{k=1}^n 4\left(-2 + \frac{3k}{n}\right)\left(\frac{3}{n}\right) \qquad n\to+\infty \text{ as } \|P\|\to 0$$

$$= \lim_{n\to+\infty} \frac{12}{n^2} \sum_{k=1}^n (-2n + 3k)$$

$$= \lim_{n \to +\infty} \frac{12}{n^2} \left(\sum_{k=1}^{n} (-2n) + \sum_{k=1}^{n} 3k \right)$$

$$= \lim_{n \to +\infty} \frac{12}{n^2} \left((-2n)n + 3 \left[\frac{n(n+1)}{2} \right] \right) \quad \text{Summation formulas}$$

$$= \lim_{n \to +\infty} \frac{12}{n^2} \left(\frac{-4n^2 + 3n^2 + 3n}{2} \right)$$

$$= \lim_{n \to +\infty} \frac{-6n^2 + 18n}{n^2} = -6 \qquad \blacksquare$$

AREA AS AN INTEGRAL

Because we have used the development of area in Section 5.2 as the model for our definition of the definite integral, it is no surprise to discover that the area under a curve can be expressed as a definite integral. However, integrals can be positive, zero, or negative (as in Example 2), and we certainly would not expect the area under a curve to be a negative number! The actual relationship between integrals and area under a curve is contained in the following observation, which follows from the definition of area as the limit of a sum, along with Theorem 5.5.

WARNING   We will find areas using a definite integral, but not every definite integral can be interpreted as an area.

Area as an Integral

Suppose f is continuous and $f(x) \geq 0$ on the closed interval $[a, b]$. Then the area under the curve $y = f(x)$ on $[a, b]$ is given by the definite integral of f on $[a, b]$. That is,

$$\text{AREA} = \int_a^b f(x) \, dx$$

Usually we find area by evaluating a definite integral, but sometimes area can be used to help us evaluate the integral. At this stage of our study, it is not easy to evaluate Riemann sums, so if you happen to recognize that the integral represents the area of some common geometric figure, you can use the known formula instead of the definite integral, as shown in Example 3.

EXAMPLE 3   Evaluating an integral using an area formula

Evaluate $\displaystyle\int_{-3}^{3} \sqrt{9 - x^2} \, dx$.

Solution

Let $f(x) = \sqrt{9 - x^2}$. The curve $y = \sqrt{9 - x^2}$ is a semicircle centered at the origin of radius 3, as shown in Figure 5.14. The given integral can be interpreted as the area under the semicircle on the interval $[-3, 3]$. From geometry, we know the area of the circle is $A = \pi r^2 = \pi(3)^2 = 9\pi$. Thus, the area of the semicircle is $\frac{9\pi}{2}$, and we conclude that

$$\int_{-3}^{3} \sqrt{9 - x^2} \, dx = \frac{9\pi}{2} \qquad \blacksquare$$

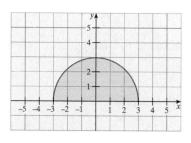

Figure 5.14 The curve $y = \sqrt{9 - x^2}$ is a semicircle of radius 3.

If f is continuous and $f(x) \leq 0$ on the closed interval $[a, b]$, then the area between the curve $y = f(x)$ and the x-axis on $[a, b]$ is the opposite of the definite integral of f on $[a, b]$, since $-f(x) \geq 0$. That is,

$$\text{AREA} = -\int_a^b f(x)\,dx$$

More generally, as illustrated in Figure 5.15, if a continuous function f is sometimes positive and sometimes negative, then

$$\int_a^b f(x)\,dx = A_1 - A_2$$

where

A_1 is the sum of all areas of the region above the x-axis and below the graph of f (that is, where $f(x) \geq 0$).

A_2 is the sum of all areas of the region below the x-axis and above the graph of f (that is, where $f(x) \leq 0$).

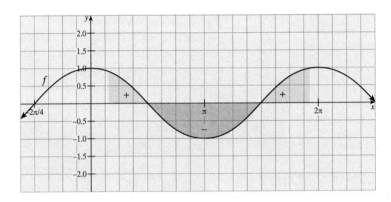

Figure 5.15 A_1 is the sum of the areas marked "+" and A_2 is the sum of the areas marked "−".

We just worked Example 2 (see Figure 5.16) using a Riemann sum, and it is easy to see from our knowledge of areas of triangles that the area of Triangle I (above the x-axis) is

$$A_1 = \tfrac{1}{2}bh = \tfrac{1}{2}(4)(1) = 2$$

and the area of Triangle II (below the x-axis) is

$$A_2 = \tfrac{1}{2}bh = \tfrac{1}{2}(8)(2) = 8$$

and we see that

$$A_1 - A_2 = 2 - 8 = -6$$

is the same as the value of the integral.

Figure 5.16 Area interpretation of Example 2

PROPERTIES OF THE DEFINITE INTEGRAL

In computations involving integrals, it is often helpful to use the three general properties listed in the following theorem.

THEOREM 5.6 General Properties of the Definite Integral

Linearity rule   If f and g are integrable on $[a, b]$, then so is $rf + sg$ for constants r, s, and

$$\int_a^b [rf(x) + sg(x)]\,dx = r\int_a^b f(x)\,dx + s\int_a^b g(x)\,dx$$

Dominance rule   If f and g are integrable on $[a, b]$ and $f(x) \leq g(x)$ throughout this interval, then

$$\int_a^b f(x)\,dx \leq \int_a^b g(x)\,dx$$

Subdivision rule   For any number c such that $a < c < b$,

$$\int_a^b f(x)\,dx = \int_a^c f(x)\,dx + \int_c^b f(x)\,dx$$

assuming all three integrals exist.

Proof   Each of these rules can be established by using a familiar property of sums or limits with the definition of the definite integral. For example, to derive the linearity rule, we note that any Riemann sum of the function $rf + sg$ can be expressed as

$$\sum_{k=1}^n [rf(x_k^*) + sg(x_k^*)]\Delta x_k = r\left[\sum_{k=1}^n f(x_k^*)\Delta x_k\right] + s\left[\sum_{k=1}^n g(x_k^*)\Delta x_k\right]$$

and the linearity rule then follows by taking the limit on each side of this equation as the norm of the partition tends to 0.   ❏

The dominance rule and the subdivision rule are interpreted geometrically for non-negative functions in Figure 5.17.

a. The dominance rule
If $g(x) \geq f(x)$, the area under
$y = g(x)$ can be no less than
the area under $y = f(x)$.

b. The subdivision rule
The area under $y = f(x)$ on
$[a, b]$ equals the sum of the
areas on $[a, c]$ and $[c, b]$.

Figure 5.17   The dominance and subdivision rules

Notice that if $g(x) \geq f(x) \geq 0$, the curve $y = g(x)$ is always above (or touching) the curve $y = f(x)$, as shown in Figure 5.17a. The dominance rule expresses the fact that the area under the upper curve $y = g(x)$ cannot be less than the area under $y = f(x)$. The subdivision rule says that the area under $y = f(x)$ above $[a, b]$ is the sum of the area on $[a, c]$ and the area on $[c, b]$, as shown in Figure 5.17b. The following example illustrates one way of using the subdivision rule.

EXAMPLE 4   Subdivision rule

If $\int_{-2}^1 f(x)\,dx = 3$ and $\int_{-2}^7 f(x)\,dx = -5$, what is $\int_1^7 f(x)\,dx$?

Solution

According to the subdivision rule, we have

$$\int_{-2}^7 f(x)\,dx = \int_{-2}^1 f(x)\,dx + \int_1^7 f(x)\,dx$$

Therefore,

$$\int_1^7 f(x)\,dx = \int_{-2}^7 f(x)\,dx - \int_{-2}^1 f(x)\,dx = (-5) - 3 = -8 \qquad \blacksquare$$

DISTANCE AS AN INTEGRAL

Many quantities other than area can be computed as the limit of a sum. For example, suppose an object moving along a line is known to have continuous velocity $v(t)$ for each time t between $t = a$ and $t = b$, and we wish to compute the total distance traveled by the object during this time period.

Let the interval $[a, b]$ be partitioned into n equal subintervals, each of length $\Delta t = \dfrac{b - a}{n}$ as shown in Figure 5.18.

Figure 5.18   The distance traveled during the kth time subinterval

The kth subinterval is $[a + (k-1)\Delta t, a + k\Delta t]$, and if Δt is small enough, the velocity $v(t)$ will not change much over the subinterval so it is reasonable to approximate $v(t)$ by the constant velocity of $v[a + (k-1)\Delta t]$ throughout the entire subinterval.

The corresponding change in the object's position will be approximated by the product

$$v[a + (k-1)\Delta t]\Delta t$$

and will be positive if $v[a + (k-1)\Delta t]$ is positive and negative if $v[a + (k-1)\Delta t]$ is negative. Both cases may be summarized by the formula

$$|v[a + (k-1)\Delta t]|\, \Delta t$$

and the total distance traveled by the object as t varies from $t = a$ to $t = b$ is given by the sum

$$S_n = \sum_{k=1}^{n} |v[a + (k-1)\Delta t]|\, \Delta t$$

which we recognize as a Riemann sum. We can make the approximation more precise by taking more refined partitions (that is, shorter and shorter time intervals Δt). Therefore, it is reasonable to *define* the exact distance S traveled as the *limit* of the sum S_n as $\Delta t \to 0$ or, equivalently, as $n \to +\infty$, so that

$$S = \lim_{n \to +\infty} \sum_{k=1}^{n} |v[a + (k-1)\Delta t]|\, \Delta t = \int_{a}^{b} |v(t)|\, dt$$

The reason that we use $|v|$ is that if we travel 300 miles east and then 300 miles west, for example, the *net* distance traveled is $300 - 300 = 0$ miles, but we are usually interested in the total distance traveled, $300 + 300 = 600$ miles.

Distance

> The **total distance traveled** by an object with continuous velocity $v(t)$ along a straight line from time $t = a$ to $t = b$ is
>
> $$S = \int_{a}^{b} |v(t)|\, dt$$

Note: There is a difference between the *total distance* traveled by an object over a given time interval $[a, b]$, and the **net distance** or **displacement** of the object over the same interval, which is defined as the difference between the object's final and initial positions. It is easy to see that displacement is given by

$$D = \int_a^b v(t)\, dt$$

EXAMPLE 5   Distance traveled by an object whose velocity is known

An object moves along a straight line with velocity $v(t) = t^2$ for $t > 0$. How far does the object travel between times $t = 1$ and $t = 2$?

Solution

We have $a = 1$, $b = 2$, and $\Delta t = \dfrac{2-1}{n} = \dfrac{1}{n}$; therefore, the required distance is

$$S = \int_1^2 |v(t)|\, dt$$

$$= \lim_{n \to +\infty} \sum_{k=1}^n \left| v\left[1 + (k-1)\left(\frac{1}{n}\right)\right] \right| \left(\frac{1}{n}\right)$$

$$= \lim_{n \to +\infty} \sum_{k=1}^n \left| v\left(\frac{n+k-1}{n}\right) \right| \left(\frac{1}{n}\right)$$

$$= \lim_{n \to +\infty} \sum_{k=1}^n \frac{(n+k-1)^2}{n^2} \left(\frac{1}{n}\right)$$

$$= \lim_{n \to +\infty} \frac{1}{n^3} \sum_{k=1}^n [(n^2 - 2n + 1) + k^2 + 2(n-1)k]$$

$$= \lim_{n \to +\infty} \frac{1}{n^3} \left[(n^2 - 2n + 1) \sum_{k=1}^n 1 + \sum_{k=1}^n k^2 + 2(n-1) \sum_{k=1}^n k \right]$$

$$= \lim_{n \to +\infty} \frac{1}{n^3} \left[(n^2 - 2n + 1)n + \frac{n(n+1)(2n+1)}{6} + 2(n-1)\frac{n(n+1)}{2} \right]$$

$$= \lim_{n \to +\infty} \frac{14n^3 - 9n^2 + n}{6n^3} = \frac{14}{6} = \frac{7}{3}$$

Thus, we expect the object to travel $\frac{7}{3}$ units during the time interval $[1, 2]$.   ∎

By considering the distance as an integral, we see again that zero or negative values of a definite integral can also be interpreted geometrically. When $v(t) > 0$ on a time interval $[a, b]$, then the total distance S traveled by the object between times $t = a$ and $t = b$ is the same as the area under the graph of $v(t)$ on $[a, b]$.

When $v(t) > 0$, the object moves forward (to the right), but when $v(t) < 0$, it reverses direction and moves backward (to the left). In the general case, where $v(t)$ changes sign on the time interval $[a, b]$, the integral

$$\int_a^b v(t)\, dt$$

measures the net distance or displacement of the object, taking into account both forward ($v > 0$) and backward ($v < 0$) motion. For instance, an object that moves forward 2 units and back 3 on a given time interval has moved a total distance of 5 units, but its displacement is -1 because it ends up 1 unit to the left of its initial position.

For example, for the velocity function $v(t)$ graphed in Figure 5.19a, the net displacement is 0 because the area above the t-axis is the same as the area below, but in Figure 5.19b, there is more area below the t-axis, which means the net displacement is negative, and the object ends up "behind" (to the left of) its starting position.

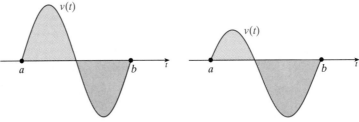

a. Net displacement of 0
There is as much area under
the positive part of the velocity
curve as there is above the
negative part.

b. Negative net displacement
There is more area above
the negative part of the velocity
curve than there is under the
positive part.

Figure 5.19   Definite integral in terms of displacement

PREVIEW

Students often become discouraged at this point, thinking they will have to compute all definite integrals using the limit of a Riemann sum definition, so we offer an encouraging word. Recall the definition of derivative in Chapter 3. It was difficult to find derivatives using the definition, but we soon proved some theorems to make it easier to find and evaluate derivatives. The same is true of integration. It is difficult to apply the definition of a definite integral, but you will soon discover that computing most definite integrals is no harder than finding an antiderivative.

As a preview of this result, consider an object moving along a straight line. We assume that its position is given by $s(t)$ and that its velocity $v(t) = s'(t)$ is positive ($v(t) > 0$) so that it is always moving forward. The total distance traveled by such an object between times $t = a$ and $t = b$ is clearly $S = s(b) - s(a)$, but earlier in this section, we showed that this distance is also given by the definite integral of $v(t)$ over the interval $[a, b]$. Thus we have

$$\text{TOTAL DISTANCE} = \int_a^b v(t)\, dt = s(b) - s(a)$$

where $s(t)$ is an antiderivative of $v(t)$.

This observation anticipates the *fundamental theorem of calculus*, which provides a vital link between differential and integral calculus. We will formally introduce the fundamental theorem in the next section. To illustrate, notice how our observation applies to Example 5, where $v(t) = t^2$ and the interval is $[1, 2]$. Since an antiderivative of $v(t)$ is $s(t) = \frac{1}{3}t^3$, we have

$$S = \int_1^2 t^2\, dt = s(2) - s(1) = \left[\frac{1}{3}(2)^3 - \frac{1}{3}(1)^3\right] = \frac{7}{3}$$

which coincides with the result found numerically in Example 5.

5.3 PROBLEM SET

A *In Problems 1–10, estimate (using right endpoints) the given integral $\int_a^b f(x)\,dx$ by using a Riemann sum*

$$S_n = \sum_{k=1}^{n} f(a + k\Delta x)\Delta x \text{ for } n = 4.$$

1. $\int_0^1 (2x + 1)\,dx$ **2.** $\int_0^1 (4x^2 + 2)\,dx$

3. $\int_1^3 x^2\,dx$ **4.** $\int_0^2 x^3\,dx$

5. $\int_0^1 (1 - 3x)\,dx$ **6.** $\int_1^3 (x^2 - x^3)\,dx$

7. $\int_{-\pi/2}^0 \cos x\,dx$ **8.** $\int_0^{\pi/4} (x + \sin x)\,dx$

9. $\int_0^1 e^x\,dx$ **10.** $\int_1^2 \frac{dx}{x}$

In Problems 11–16, $v(t)$ is the velocity of an object moving along a straight line. Use the formula

$$S_n = \sum_{k=1}^{n} |v(a + k\Delta t)|\,\Delta t$$

where $\Delta t = \dfrac{b - a}{n}$ to estimate (using right endpoints) the total distance traveled by the object during the time interval $[a, b]$. Let $n = 4$.

11. $v(t) = 3t + 1$ on $[1, 4]$ **12.** $v(t) = 1 + 2t$ on $[1, 2]$
13. $v(t) = \sin t$ on $[0, \pi]$ **14.** $v(t) = \cos t$ on $[0, \frac{\pi}{2}]$

15. $v(t) = e^{-t}$ on $[0, 1]$ **16.** $v(t) = \dfrac{1}{t + 1}$ on $[0, 1]$

Evaluate each of the integrals in Problems 17–22 by using the following information together with the linearity and subdivision properties:

$$\int_{-1}^2 x^2\,dx = 3; \quad \int_{-1}^0 x^2\,dx = \frac{1}{3}; \quad \int_{-1}^2 x\,dx = \frac{3}{2}; \quad \int_0^2 x\,dx = 2$$

17. $\int_0^{-1} x^2\,dx$ **18.** $\int_{-1}^2 (x^2 + x)\,dx$

19. $\int_{-1}^2 (2x^2 - 3x)\,dx$ **20.** $\int_0^2 x^2\,dx$

21. $\int_{-1}^0 x\,dx$ **22.** $\int_{-1}^0 (3x^2 - 5x)\,dx$

Use the dominance property of integrals to establish the given inequality in Problems 23–24.

23. $\int_0^1 x^3\,dx \leq \dfrac{1}{2}$ *Hint: Note that $x^3 \leq x$ on $[0, 1]$.*

24. $\int_0^\pi \sin x\,dx \leq \pi$ *Hint: $\sin x \leq 1$ for all x.*

B **25.** Given $\int_{-2}^4 [5f(x) + 2g(x)]\,dx = 7$ and $\int_{-2}^4 [3f(x) + g(x)]\,dx = 4$, find

$$\int_{-2}^4 f(x)\,dx \quad \text{and} \quad \int_{-2}^4 g(x)\,dx$$

26. Suppose $\int_0^2 f(x)\,dx = 3$, $\int_0^2 g(x)\,dx = -1$, and $\int_0^2 h(x)\,dx = 3$.

   a. Evaluate $\int_0^2 [2f(x) + 5g(x) - 7h(x)]\,dx$.

   b. Find a constant s such that

$$\int_0^2 [5f(x) + sg(x) - 6h(x)]\,dx = 0$$

27. Evaluate $\int_{-1}^2 f(x)\,dx$ given that

$$\int_{-1}^1 f(x)\,dx = 3; \quad \int_2^3 f(x)\,dx = -2; \quad \int_1^3 f(x)\,dx = 5$$

28. Let $f(x) = \begin{cases} 2 & \text{for } -1 \leq x \leq 1 \\ 3 - x & \text{for } 1 < x < 4 \\ 2x - 9 & \text{for } 4 \leq x \leq 5 \end{cases}$

Sketch the graph of f on the interval $[-1, 5]$ and show that f is continuous on this interval. Then use Theorem 5.6 to evaluate

$$\int_{-1}^5 f(x)\,dx$$

29. Let $f(x) = \begin{cases} 5 & \text{for } -3 \leq x \leq -1 \\ 4 - x & \text{for } -1 < x < 2 \\ 2x - 2 & \text{for } 2 \leq x \leq 5 \end{cases}$

Sketch the graph of f on the interval $[-3, 5]$ and show that f is continuous on this interval. Then use Theorem 5.6 to evaluate

$$\int_{-3}^5 f(x)\,dx$$

30. HISTORICAL QUEST
During the eighteenth century, integrals were considered simply as antiderivatives. That is, there were no underpinnings for the concept of an integral until Cauchy formulated the definition of the integral in 1823 (see Historical Quest 49, Problem Set 2.1). This formulation was later com-

GEORG FRIEDRICH RIEMANN
1826–1866

pleted by Georg Friedrich Riemann. In this section, we see that history honored Riemann by naming the process after him. In his personal life he was frail, bashful, and timid, but in his professional life he was one of the giants in mathematical history. Riemann used what are called topological methods in his theory of functions and in his work with surfaces and spaces. Riemann is remembered for his work in geometry (Riemann surfaces) and analysis. In his book *Space Through the Ages*, Cornelius Lanczos wrote, "Although Riemann's collected papers fill only one single volume of 538 pages, this volume weighs tons if measured intellectually. Every one of his many discoveries was destined to change the course of mathematical science." ∎

For this Historical Quest, investigate the Königsberg bridge problem, and its solution.

This famous problem was formulated by Leonhard Euler (1707–1783). The branch of mathematics known today as *topology* began with Euler's work on the bridge problem and other related questions and was extended in the nineteenth century by Riemann and others.

31. Generalize the subdivision property by showing that for $a \leq c \leq d \leq b$

$$\int_a^b f(x)\,dx = \int_a^c f(x)\,dx + \int_c^d f(x)\,dx + \int_d^b f(x)\,dx$$

whenever all these integrals exist.

32. If $Cx + D \geq 0$ for $a \leq x \leq b$, show that

$$\int_a^b (Cx + D)\,dx = (b - a)\left[\frac{C}{2}(b + a) + D\right]$$

Hint: Sketch the region under the line $y = Cx + D$, and express the integral as an area.

33. For $b > a > 0$, show that

$$\int_a^b x^2\,dx = \frac{1}{3}(b^3 - a^3)$$

In Problems 34–36, use the partition $P = \{-1, -0.2, 0.9, 1.3, 1.7, 2\}$ *on the interval* $[-1, 2]$.

34. Find the subinterval widths

$$\Delta x_k = x_k - x_{k-1}$$

for $k = 1, 2, \ldots, 5$. What is the norm of P?

35. Compute the Riemann sum on $[-1, 2]$ associated with $f(x) = 4 - 5x$; the partition P with $\Delta x_1 = 0.8$, $\Delta x_2 = 1.1$, $\Delta x_3 = 0.4$, $\Delta x_4 = 0.4$, and $\Delta x_5 = 0.3$; and the subinterval representatives $x_1^* = -0.5$, $x_2^* = 0.8$, $x_3^* = 1$, $x_4^* = 1.3$, $x_5^* = 1.8$. What is the norm of P?

36. Compute the Riemann sum on $[-1, 2]$ associated with $f(x) = x^3$; the partition P with $\Delta x_1 = 0.8$, $\Delta x_2 = 1.1$, $\Delta x_3 = 0.4$, $\Delta x_4 = 0.4$, and $\Delta x_5 = 0.3$; and subinterval representatives $x_1^* = -1$, $x_2^* = 0$, $x_3^* = 1$, $x_4^* = \frac{128}{81}$, and $x_5^* = \frac{125}{64}$. What is the norm of P?

37. If the numbers a_k and b_k satisfy $a_k \leq b_k$ for $k = 1, 2, \ldots, n$, then $\sum_{k=1}^n a_k \leq \sum_{k=1}^n b_k$. Use this dominance property of sums to establish the dominance property of integrals.

38. Counterexample Problem Either prove that the following result is generally true or find a counterexample: If f and g are continuous on $[a, b]$, then

$$\int_a^b f(x)g(x)\,dx = \left[\int_a^b f(x)\,dx\right]\left[\int_a^b g(x)\,dx\right]$$

39. Counterexample Problem Either prove that the following result is generally true or find a counterexample: There exists a nonzero continuous function f defined on $[a, b]$ with a nonzero area between the x-axis and the function f on the interval such that

$$\int_a^b f(x)\,dx = 0$$

40. Recall that a function f is **odd** if $f(-x) = -f(x)$ for all x. If f is odd, show that

$$\int_{-a}^a f(x)\,dx = 0$$

for all a. This problem is analogous to Problem 33 of the previous section.

41. Prove the *bounding rule* for definite integrals: If f is integrable on the closed interval $[a, b]$ and $m \leq f(x) \leq M$ for constants m, M, and all x in the closed interval, then

$$m(b - a) \leq \int_a^b f(x)\,dx \leq M(b - a)$$

42. Use the definition of the definite integral to prove that

$$\int_a^b C\,dx = C(b - a)$$

5.4 The Fundamental Theorems of Calculus

IN THIS SECTION the first fundamental theorem of calculus, the second fundamental theorem of calculus

THE FIRST FUNDAMENTAL THEOREM OF CALCULUS

In the previous section we observed that if $v(t)$ is the velocity of an object at time t as it moves along a straight line, then

$$\int_a^b v(t)\,dt = s(b) - s(a)$$

where $s(t)$ is the displacement of the object and satisfies $s'(t) = v(t)$. This result is an application of the following general theorem which was discovered by the English mathematician Isaac Barrow (1630–1677), Newton's mentor at Cambridge.

THEOREM 5.7 The first fundamental theorem of calculus

If f is continuous on the interval $[a, b]$ and F is any function that satisfies $F'(x) = f(x)$ throughout this interval, then

$$\int_a^b f(x)\, dx = F(b) - F(a)$$

Proof Let $P = \{x_0, x_1, x_2, \ldots, x_n\}$ be a regular partition of the interval, with subinterval widths $\Delta x = \dfrac{b - a}{n}$. Note that F satisfies the hypotheses of the mean value theorem (Theorem 4.4, Section 4.2) on each of the closed subintervals $[x_{k-1}, x_k]$. Thus, the MVT tells us that there is a point x_k^* in each open subinterval (x_{k-1}, x_k) for which

$$\frac{F(x_k) - F(x_{k-1})}{x_k - x_{k-1}} = F'(x_k^*)$$

$$F(x_k) - F(x_{k-1}) = F'(x_k^*)(x_k - x_{k-1})$$

Because $F'(x_k^*) = f(x_k^*)$ and $x_k - x_{k-1} = \Delta x = \dfrac{b - a}{n}$, we can write $F(x_k) - F(x_{k-1}) = f(x_k^*)\Delta x$, so that

$$F(x_1) - F(x_0) = f(x_1^*)\Delta x$$
$$F(x_2) - F(x_1) = f(x_2^*)\Delta x$$
$$\vdots$$
$$F(x_n) - F(x_{n-1}) = f(x_n^*)\Delta x$$

Thus, by adding both sides of all the equations, we obtain

$$\sum_{k=1}^n f(x_k^*)\Delta x = f(x_1^*)\Delta x + f(x_2^*)\Delta x + \cdots + f(x_n^*)\Delta x$$
$$= [F(x_1) - F(x_0)] + [F(x_2) - F(x_1)] + \cdots + [F(x_n) - F(x_{n-1})]$$
$$= F(x_n) - F(x_0)$$

Because $x_0 = a$ and $x_n = b$, we have

$$\sum_{k=1}^n f(x_k^*)\Delta x = F(b) - F(a)$$

Finally, we take the limit of the left side as $\|P\| \to 0$ (i.e., $n \to +\infty$), and because $F(b) - F(a)$ is a constant, we have

$$\int_a^b f(x)\, dx = F(b) - F(a) \qquad \square$$

➤ What This Says The definite integral $\displaystyle\int_a^b f(x)\, dx$ can be computed by finding an antiderivative on the interval $[a, b]$ and evaluating it at the limits of integration a and b. It is a consequence of the first fundamental theorem of calculus that our two definitions of area between the graph of $y = f(x)$ for $y \geq 0$ and the x-axis from the first two sections of this chapter agree. Also notice that this theorem does not say *how* to find the antiderivative, nor does it say that an antiderivative F *exists*. The existence of an antiderivative is asserted by the second fundamental theorem of calculus, which is examined later in this section (page 306).

To give you some insight into why this theorem is important enough to be named *the fundamental theorem of calculus*, we repeat Example 2 from Section 5.3 on the left and then work the same example using the fundamental theorem, on the right.

EXAMPLE 1   Evaluating a definite integral

Evaluate $\int_{-2}^{1} 4x \, dx$ using the definition of the definite integral and also using the fundamental theorem.

Solution

Solution from Section 5.3 using Riemann sums:

$$\int_{-2}^{1} 4x \, dx = \lim_{\|P\| \to 0} \sum_{k=1}^{n} f(x_k^*) \Delta x$$

$$= \lim_{n \to +\infty} \sum_{k=1}^{n} 4\left(-2 + \frac{3k}{n}\right)\left(\frac{3}{n}\right)$$

$$= \lim_{n \to +\infty} \frac{12}{n^2} \sum_{k=1}^{n} (-2n + 3k)$$

$$= \lim_{n \to +\infty} \frac{12}{n^2} \left(\sum_{k=1}^{n}(-2n) + \sum_{k=1}^{n} 3k\right)$$

$$= \lim_{n \to +\infty} \frac{12}{n^2} \left((-2n)n + 3\left[\frac{n(n+1)}{2}\right]\right)$$

$$= \lim_{n \to +\infty} \frac{12}{n^2} \left(\frac{-4n^2 + 3n^2 + 3n}{2}\right)$$

$$= \lim_{n \to +\infty} \frac{-6n^2 + 18n}{n^2}$$

$$= -6$$

Solution using the fundamental theorem of calculus:

If $F(x) = 2x^2$, then $F'(x) = 4x$, so F is an antiderivative of f. Thus,

$$\int_{-2}^{1} 4x \, dx = F(1) - F(-2)$$

$$= 2(1)^2 - 2(-2)^2 = -6$$

Note that if we choose a *different* antiderivative, say $G(x) = 2x^2 + 3$, then $G(1) - G(-2) = -6$, as well.

■

The variable used in a definite integral is a **dummy variable** in the sense that it can be replaced by any other variable with no effect on the value of the integral. For instance, we have just found that

$$\int_{-2}^{1} 4x \, dx = -6$$

and without further computation, it follows that

$$\int_{-2}^{1} 4t \, dt = -6 \qquad \int_{-2}^{1} 4u \, du = -6 \qquad \int_{-2}^{1} 4N \, dN = -6$$

Henceforth, when evaluating an integral by the fundamental theorem, we will denote the difference

$$F(b) - F(a) \quad \text{by} \quad F(x)\big|_a^b$$

Sometimes we also write $[F(x)]_a^b$, where $F'(x) = f(x)$ on $[a, b]$. This notation is illustrated in Example 2.

EXAMPLE 2   Finding the area under a curve using the fundamental theorem of calculus

Find the area under the curve $y = \cos x$ on $\left[-\frac{\pi}{2}, \frac{\pi}{2}\right]$.

Solution

Note that $f(x) = \cos x$ is continuous and $f(x) \geq 0$ on $\left[-\frac{\pi}{2}, \frac{\pi}{2}\right]$, as shown in Figure 5.20. Because the derivative of $\sin x$ is $\cos x$, it follows (from the definition of the antiderivative) that $\sin x$ is an antiderivative of $\cos x$. Thus, the required area is given by the integral

$$A = \int_{-\pi/2}^{\pi/2} \cos x \, dx = \sin x \big|_{-\pi/2}^{\pi/2} = \sin \tfrac{\pi}{2} - \sin\left(-\tfrac{\pi}{2}\right) = 1 - (-1) = 2$$

The region has area $A = 2$ square units.

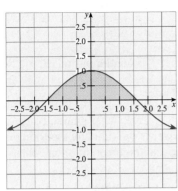

If you look at the graph of $f(x) = \cos x$ on $\left[-\frac{\pi}{2}, \frac{\pi}{2}\right]$, you can see that an area of 2 seems reasonable.

Figure 5.20 Graph to estimate area

WARNING   It is important to remember that, for fixed numerical values of a and b, the definite integral $\int_a^b f(x)\,dx$ is a number, whereas the indefinite integral $\int f(x)\,dx$ is a family of functions. The relationship between the indefinite and definite integral is given by

$$\int_a^b f(x)\,dx = \left[\int f(x)\,dx\right]\Big|_a^b \qquad \blacksquare$$

EXAMPLE 3   Evaluating an integral using the fundamental theorem

Evaluate

a. $\displaystyle\int_{-3}^{5} (-10)\,dx$ **b.** $\displaystyle\int_{4}^{9} \left[\frac{1}{\sqrt{x}} + x\right] dx$ **c.** $\displaystyle\int_{-2}^{2} |x|\,dx$

Solution

a. $\displaystyle\int_{-3}^{5} (-10)\,dx = -10x\big|_{-3}^{5} = -10(5+3) = -80$

b. $\displaystyle\int_{4}^{9} \left[\frac{1}{\sqrt{x}} + x\right] dx = \int_{4}^{9} \left[x^{-1/2} + x\right] dx = \left[\left(\frac{x^{1/2}}{\frac{1}{2}}\right) + \frac{x^2}{2}\right]\Big|_{4}^{9}$

$$= \left(6 + \tfrac{81}{2}\right) - \left(4 + \tfrac{16}{2}\right) = 34\tfrac{1}{2}$$

c. $\displaystyle\int_{-2}^{2} |x|\,dx = \int_{-2}^{0} |x|\,dx + \int_{0}^{2} |x|\,dx$ \qquad Subdivision rule

$$= \int_{-2}^{0} (-x)\,dx + \int_{0}^{2} x\,dx \qquad \text{Recall,} \quad \begin{aligned} |x| &= x & \text{if } x \geq 0 \\ |x| &= -x & \text{if } x < 0 \end{aligned}$$

$$= -\frac{x^2}{2}\Big|_{-2}^{0} + \frac{x^2}{2}\Big|_{0}^{2} = -\left(0 - \frac{4}{2}\right) + \left(\frac{4}{2} + 0\right) = 4$$

Note that we must use the subdivision rule because we do not know an antiderivative of the function $|x|$, where x is defined on an interval containing zero. $\blacksquare$

THE SECOND FUNDAMENTAL THEOREM OF CALCULUS

In certain circumstances, it is useful to consider an integral of the form

$$\int_a^x f(t)\,dt$$

where the upper limit of integration is a variable instead of a constant. As x varies, so does the value of the integral, and

$$F(x) = \int_a^x f(t)\,dt$$

is a function of the variable x. (Note that the integrand does not contain x.) For instance,

$$F(x) = \int_2^x t^2 \, dt = \tfrac{1}{3}x^3 - \tfrac{1}{3}(2)^3 = \tfrac{1}{3}x^3 - \tfrac{8}{3}$$

The first fundamental theorem of calculus tells us that if f is continuous on $[a, b]$, then

$$\int_a^b f(x) \, dx = F(b) - F(a)$$

where F is an antiderivative of f on $[a, b]$. But in general, what guarantee do we have that such an antiderivative even exists? The answer is provided by the following theorem, which is often referred to as the second fundamental theorem of calculus.

THEOREM 5.8   The second fundamental theorem of calculus

Let $f(t)$ be continuous on the interval $[a, b]$ and define the function G by the integral equation

$$G(x) = \int_a^x f(t) \, dt$$

for $a \le x \le b$. Then G is an antiderivative of f on $[a, b]$; that is,

$$G'(x) = \frac{d}{dx}\left[\int_a^x f(t) \, dt\right] = f(x)$$

on $[a, b]$.

Proof   Apply the definition of the derivative to G:

$$G'(x) = \lim_{\Delta x \to 0} \frac{G(x + \Delta x) - G(x)}{\Delta x} \qquad \textit{x is fixed, only } \Delta x \textit{ changes}$$

$$= \lim_{\Delta x \to 0} \frac{1}{\Delta x}[G(x + \Delta x) - G(x)]$$

$$= \lim_{\Delta x \to 0} \frac{1}{\Delta x}\left[\int_a^{x+\Delta x} f(t) \, dt - \int_a^x f(t) \, dt\right]$$

$$= \lim_{\Delta x \to 0} \frac{1}{\Delta x}\left[\int_a^{x+\Delta x} f(t) \, dt + \int_x^a f(t) \, dt\right]$$

$$= \lim_{\Delta x \to 0} \frac{1}{\Delta x}\left[\int_x^{x+\Delta x} f(t) \, dt\right] \qquad \textit{Subdivision rule}$$

Since f is continuous on $[a, b]$, it is continuous on $[x, x + \Delta x]$ for any x in $[a, b]$. Let $m(x)$ and $M(x)$ be the smallest and largest values, respectively, for $f(t)$ on $[x, x + \Delta x]$. (*Note:* In general, m and M depend on x and Δx but are constants as far as t-integration is concerned.) Since $m(x) \le f(t) \le M(x)$, we have (by the dominance rule)

$$\int_x^{x+\Delta x} m(x) \, dt \le \int_x^{x+\Delta x} f(t) \, dt \le \int_x^{x+\Delta x} M(x) \, dt$$

so

$$m(x)[(x + \Delta x) - x] \le \int_x^{x+\Delta x} f(t) \, dt \le M(x)[(x + \Delta x) - x]$$

or

$$m(x) \le \frac{1}{\Delta x}\int_x^{x+\Delta x} f(t) \, dt \le M(x)$$

Since

$$m(x) \le f(x) \le M(x)$$

on $[x, x + \Delta x]$, it follows that $m(x)$ and $M(x)$ are both "squeezed" toward $f(x)$ as $\Delta x \to 0$ and we have

$$G'(x) = \lim_{\Delta x \to 0} \frac{1}{\Delta x}\int_x^{x+\Delta x} f(t) \, dt = f(x) \qquad \square$$

We also note that the function G is continuous, since it is differentiable. If F is *any* antiderivative of f on the interval $[a, b]$, then the antiderivative

$$G(x) = \int_a^x f(t)\, dt$$

found in Theorem 5.8 satisfies $G(x) = F(x) + C$ for some constant C and all x on the interval $[a, b]$. In particular, when $x = a$, we have

$$0 = \int_a^a f(t)\, dt = G(a) = F(a) + C$$

so that $C = -F(a)$. Finally, by letting $x = b$, we find that

$$\int_a^b f(t)\, dt = G(b) = F(b) + C = F(b) + [-F(a)] = F(b) - F(a)$$

as claimed by the first fundamental theorem.

EXAMPLE 4   Using the second fundamental theorem

Differentiate $F(x) = \int_7^x (2t - 3)\, dt$.

Solution

From the second fundamental theorem, we can obtain $F'(x)$ by simply replacing t with x in the integrand $f(t) = 2t - 3$. Thus,

$$F'(x) = \frac{d}{dx}\left[\int_7^x (2t - 3)\, dt \right] = 2x - 3 \qquad \blacksquare$$

The second fundamental theorem of calculus can also be applied to an integral function with a variable *lower* limit of integration. For example, to differentiate

$$G(z) = \int_z^5 \frac{\sin u}{u}\, du$$

reverse the order of integration and apply the second fundamental theorem of calculus as before:

$$G'(z) = \frac{d}{dz}\left[\int_z^5 \frac{\sin u}{u}\, du \right] = \frac{d}{dz}\left[-\int_5^z \frac{\sin u}{u}\, du \right] = -\frac{d}{dz}\left[\int_5^z \frac{\sin u}{u}\, du \right] = -\frac{\sin z}{z}$$

EXAMPLE 5   Second fundamental theorem with the chain rule

Find the derivative of $F(x) = \int_0^{x^2} t^2\, dt$.

Solution

$$F'(x) = \frac{d}{dx}\left[\int_0^{x^2} t^2\, dt \right] = \frac{d}{du}\left[\int_0^u t^2\, dt \right] \frac{du}{dx} \qquad \text{Chain rule; } u = x^2$$

$$= u^2 \frac{du}{dx} \qquad \text{Second fundamental theorem of calculus}$$

$$= \left(x^2 \right)^2 (2x)$$

$$= 2x^5 \qquad \blacksquare$$

5.4 PROBLEM SET

In Problems 1–30, evaluate the definite integral.

1. $\int_{-10}^{10} 7\,dx$

2. $\int_{-5}^{7} (-3)\,dx$

3. $\int_{-3}^{5} (2x+a)\,dx$

4. $\int_{-2}^{2} (b-x)\,dx$

5. $\int_{-1}^{2} ax^3\,dx$

6. $\int_{-1}^{1} (x^3+bx^2)\,dx$

7. $\int_{1}^{2} \dfrac{c}{x^3}\,dx$

8. $\int_{-2}^{-1} \dfrac{p}{x^2}\,dx$

9. $\int_{0}^{9} \sqrt{x}\,dx$

10. $\int_{0}^{27} \sqrt[3]{x}\,dx$

11. $\int_{0}^{1} (5u^7+\pi^2)\,du$

12. $\int_{0}^{1} (7x^8+\sqrt{\pi})\,dx$

13. $\int_{1}^{2} x^{2a}\,dx,\ a\neq-\tfrac{1}{2}$

14. $\int_{1}^{2} (2x)^{\pi}\,dx$

15. $\int_{\ln 2}^{\ln 5} 5e^x\,dx$

16. $\int_{e^{-2}}^{e} \dfrac{dx}{x}$

17. $\int_{0}^{4} \sqrt{x}(x+1)\,dx$

18. $\int_{0}^{1} \sqrt{t}(t-\sqrt{t})\,dt$

19. $\int_{1}^{2} \dfrac{x^3+1}{x^2}\,dx$

20. $\int_{1}^{4} \dfrac{x^2+x-1}{\sqrt{x}}\,dx$

21. $\int_{1}^{\sqrt{3}} \dfrac{6a}{1+x^2}\,dx$

22. $\int_{0}^{0.5} \dfrac{b\,dx}{\sqrt{1-x^2}}$

23. $\int_{-2}^{3} (\sin^2 x+\cos^2 x)\,dx$

24. $\int_{0}^{\pi/4} (\sec^2 x-\tan^2 x)\,dx$

25. $\int_{0}^{1} (1-e^t)\,dt$

26. $\int_{1}^{2} \dfrac{x^3+1}{x}\,dx$

27. $\int_{0}^{1} \dfrac{x^2-4}{x-2}\,dx$

28. $\int_{0}^{1} \dfrac{x^2-1}{x^2+1}\,dx$

29. $\int_{-1}^{2} (x+|x|)\,dx$

30. $\int_{0}^{2} (x-|x-1|)\,dx$

In Problems 31–38, find the area of the region under the given curve over the prescribed interval.

31. $y=x^2+1$ on $[-1,1]$

32. $y=\sqrt{t}$ on $[0,1]$

33. $y=\sec^2 x$ on $\left[0,\tfrac{\pi}{4}\right]$

34. $y=\sin x+\cos x$ on $\left[0,\tfrac{\pi}{2}\right]$

35. $y=e^t-t$ on $[0,1]$

36. $y=(x^2+x+1)\sqrt{x}$ on $[1,4]$

37. $y=\dfrac{x^2-2x+3}{x}$ on $[1,2]$

38. $y=\dfrac{2}{1+t^2}$ on $[0,1]$

In Problems 39–44, find the derivative of the given function.

39. $F(x)=\int_{0}^{x} \dfrac{t^2-1}{\sqrt{t+1}}\,dt$

40. $F(x)=\int_{-2}^{x} (t+1)\sqrt[3]{t}\,dt$

41. $F(t)=\int_{1}^{t} \dfrac{\sin x}{x}\,dx$

42. $F(t)=\int_{t}^{2} \dfrac{e^x}{x}\,dx$

43. $F(x)=\int_{x}^{1} \dfrac{dt}{\sqrt{1+3t^2}}$

44. $F(x)=\int_{\pi/3}^{x} \sec^2 t\tan t\,dt$

The formulas in Problems 45–50 are taken from a table of integrals. In each case, use differentiation to verify that the formula is correct.

45. $\int \cos^2 au\,du = \dfrac{u}{2}+\dfrac{\sin 2au}{4a}+C$

46. $\int u\cos^2 au\,du = \dfrac{u^2}{4}+\dfrac{u\sin 2au}{4a}+\dfrac{\cos 2au}{8a^2}+C$

47. $\int \dfrac{u\,du}{(a^2-u^2)^{3/2}} = \dfrac{1}{\sqrt{a^2-u^2}}+C$

48. $\int \dfrac{du}{u^2-a^2} = \dfrac{1}{2a}\ln\left|\dfrac{u-a}{u+a}\right|+C$

49. $\int \dfrac{u\,du}{\sqrt{a^2-u^2}} = -\sqrt{a^2-u^2}+C$

50. $\int (\ln|u|)^2\,du = u(\ln|u|)^2 - 2u\ln|u| + 2u + C$

51. Exploration Problem What is the relationship between finding an area and evaluating an integral?

52. Journal Problem *FOCUS** ▪ The area of the shaded region in Figure 5.21a is 8 times the area of the shaded region in Figure 5.21b. What is c in terms of a?

a.

b.

Figure 5.21 Comparing areas

53. Exploration Problem If you use the first fundamental theorem for the following integral, you find

$$\int_{-1}^{1} \dfrac{dx}{x^2} = \left[-\dfrac{1}{x}\right]\Bigg|_{-1}^{1} = -1-1 = -2$$

But the function $y=\dfrac{1}{x^2}$ is never negative. What's wrong with this "evaluation"?

54. Evaluate

$$\int_{0}^{2} f(x)\,dx \quad \text{where} \quad f(x)=\begin{cases} x^3 & \text{if } 0\le x<1 \\ x^4 & \text{if } 1\le x\le 2 \end{cases}$$

55. Evaluate

$$\int_{0}^{\pi} f(x)\,dx \quad \text{where} \quad f(x)=\begin{cases} \cos x & \text{if } 0\le x<\pi/2 \\ x & \text{if } \pi/2\le x\le\pi \end{cases}$$

*February 1995, p. 15.

56. a. If $F(x) = \int \left(\dfrac{1}{\sqrt{x}} - 4 \right) dx$, find F so that $F(1) = 0$.

  b. Sketch the graphs of $y = F(x)$, $y = F(x) + 3$, and $y = F(x) - 1$.

  c. Find a constant K such that the largest value of $G(x) = F(x) + K$ is 0.

57. Let $g(x) = \displaystyle\int_0^x f(t)\,dt$, where f is the function defined by the following graph. Note that f crosses the x-axis at 3 points on $[0, 2]$; label these (from left to right), $x = a$, $x = b$, and $x = c$.

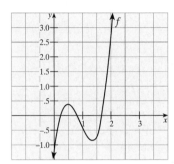

  a. What can you say about $g(a)$?
  b. Estimate $g(1)$.
  c. Where does g have a maximum value on $[0, 2]$?
  d. Sketch a rough graph of g.

58. Let $g(x) = \displaystyle\int_0^x f(t)\,dt$, where f is the function defined by the following graph.

  a. Where does g have a relative minimum on $[0, 5]$?

  b. Where does g have a relative maximum value on $[0, 5]$?
  c. If $g(1) = 1$, $g(2.5) = -2.5$, and $g(4) = 4$, sketch a rough graph of g.

C 59. Counterexample Problem   The purpose of this problem is to provide a counterexample showing that the integral of a product (or quotient) is not necessarily equal to the product (quotient) of the respective integrals.

  a. Show that $\int x\sqrt{x}\,dx \neq \left(\int x\,dx \right)\left(\int \sqrt{x}\,dx \right)$.

  b. Show that $\displaystyle\int \frac{\sqrt{x}}{x}\,dx \neq \frac{\int \sqrt{x}\,dx}{\int x\,dx}$.

60. Suppose f is a function with the property that $f'(x) = f(x)$ for all x.

  a. Show that $\displaystyle\int_a^b f(x)\,dx = f(b) - f(a)$.

  b. Show that $\displaystyle\int_a^b [f(x)]^2\,dx = \frac{1}{2}\{[f(b)]^2 - [f(a)]^2\}$.

61. Find $\dfrac{d}{dx} \displaystyle\int_0^{x^4} (2t - 3)\,dt$ using the second fundamental theorem of calculus. Check your work by evaluating the integral directly and then taking the derivative of your answer.

62. Find $\dfrac{d}{dx} \displaystyle\int_0^{x^2} \sin t\,dt$ using the second fundamental theorem of calculus. Check your work by evaluating the integral directly and then taking the derivative of your answer.

63. Find an equation for the tangent line to the curve $y = F(x)$ at the point P where $x = 1$ and

$$F(x) = \int_1^{\sqrt{x}} \frac{2t + 1}{t + 2}\,dt$$

64. Find an equation for the tangent line to the curve $y = F(x)$ at the point P where $x = 1$ and

$$F(x) = \int_1^{x^3} \frac{t^2 + 1}{t - 2}\,dt$$

65. Suppose $F(x)$ is the integral function $F(x) = \displaystyle\int_{u(x)}^{v(x)} f(t)\,dt$. What is $F'(x)$? This result is called *Leibniz' rule*.

66. Suppose $F(x) = \displaystyle\int_{-x}^x e^t\,dt$. Find $F'(x)$.

5.5   Integration by Substitution

IN THIS SECTION   substitution with indefinite integration, substitution with definite integration

SUBSTITUTION WITH INDEFINITE INTEGRATION

Recall that according to the chain rule, the derivative of $(x^2 + 3x + 5)^9$ is

$$\frac{d}{dx}(x^2 + 3x + 5)^9 = 9(x^2 + 3x + 5)^8(2x + 3)$$

Thus,

$$\int 9(x^2 + 3x + 5)^8(2x + 3)\,dx = (x^2 + 3x + 5)^9 + C$$

Note that the product is of the form $g(u)\dfrac{du}{dx}$ where, in this case, $g(u) = 9u^8$ and $u = x^2 + 3x + 5$. The following theorem generalizes this procedure by showing how products of the form $g(u)\dfrac{du}{dx}$ can be integrated by reversing the chain rule.

THEOREM 5.9   Integration by substitution

Let f, g, and u be differentiable functions of x such that

$$f(x) = g(u)\frac{du}{dx}$$

Then

$$\int f(x)\,dx = \int g(u)\frac{du}{dx}\,dx = \int g(u)\,du = G(u) + C$$

where G is an antiderivative of g.

Proof   If G is an antiderivative of g, then $G'(u) = g(u)$ and, by the chain rule,

$$f(x) = \frac{d}{dx}[G(u)] = G'(u)\frac{du}{dx} = g(u)\frac{du}{dx}$$

Integrating both sides of this equation, we obtain

$$\int f(x)\,dx = \int \left[g(u)\frac{du}{dx}\right]\,dx = \int \left[\frac{d}{dx}G(u)\right]\,dx = G(u) + C$$

as required.   ❑

EXAMPLE 1   Indefinite integral by substitution

Find $\displaystyle\int 9(x^2 + 3x + 5)^8(2x + 3)\,dx$.

Solution

Look at the integral and make the observations as shown in the boxes:

If $u = x^2 + 3x + 5$,

Let $u = x^2 + 3x + 5$.   ← then $du = (2x + 3)\,dx$.

$$\int \underbrace{9(x^2 + 3x + 5)^8}_{\text{This is } g(u).}\ \underbrace{(2x + 3)\,dx}_{\text{This is } du.} = \int 9u^8\,du$$

Now complete the integration of $g(u)$ and, when you are finished, back-substitute to express u in terms of x:

$$\int 9u^8\,du = 9\left(\frac{u^9}{9}\right) + C = (x^2 + 3x + 5)^9 + C \qquad ■$$

EXAMPLE 2   Finding an integral by substitution

Find $\displaystyle\int (2x + 7)^5\,dx$.

Solution

$$\int (2x + 7)^5\,dx = \int u^5\left(\tfrac{1}{2}\,du\right) \qquad \text{Note: If } du = 2\,dx, \text{ then } dx = \tfrac{1}{2}\,du.$$

Let $u = 2x + 7$, so $du = 2\,dx$.

$$= \tfrac{1}{2}\left[\tfrac{1}{6}u^6\right] + C \qquad \text{Power rule for integrals}$$

$$= \tfrac{1}{12}(2x + 7)^6 + C \qquad \text{Replace } u = 2x + 7. \qquad ■$$

When choosing a substitution to evaluate $\int f(x)\,dx$, where $f(x)$ is a composite function $f(x) = g(u(x))$, it is usually a good idea to pick the "inner" function $u = u(x)$. In Example 2, we chose $u = 2x + 7$, **NOT** the entire power $(2x + 7)^5$.

When making a substitution $u = u(x)$, remember that $du = u'(x)\,dx$, so you cannot simply replace dx by du in the transformed integral. For example, when you use the substitution $u = \sin x$, you have $du = \cos x\,dx$ and then replace the entire expression $\cos x\,dx$ by du in the transformed integral, as shown in the following example.

EXAMPLE 3 Substitution with a trigonometric function

Find $\int (4 - 2\cos\theta)^3 \sin\theta\,d\theta$.

Solution

Let $u = 4 - 2\cos\theta$, so $\dfrac{du}{d\theta} = -2(-\sin\theta) = 2\sin\theta$ and $d\theta = \dfrac{du}{2\sin\theta}$; substitute:

$$\int (4 - 2\cos\theta)^3 \sin\theta\,d\theta = \int u^3 \sin\theta\,\frac{du}{2\sin\theta} = \frac{1}{2}\int u^3\,du$$

$$= \frac{1}{2}\left(\frac{u^4}{4}\right) + C = \frac{1}{8}(4 - 2\cos\theta)^4 + C \qquad \blacksquare$$

Note: We will use a shorthand notation for substitution by placing a "box" under the integrand as follows:

$$\int (4 - 2\cos\theta)^3 \sin\theta\,d\theta = \int u^3 \sin\theta\,\frac{d\theta}{2\sin\theta}$$
$$\boxed{u = 4 - 2\cos\theta;\ du = -2(-\sin\theta)\,d\theta}$$

This notation is used in the following example.

EXAMPLE 4 Substitution in an exponential integral

Evaluate $\int xe^{4-x^2}\,dx$.

Solution

$$\int xe^{4-x^2}\,dx = \int e^u\left(-\tfrac{1}{2}\,du\right)$$
$$\boxed{u = 4 - x^2;\ du = -2x\,dx}$$
$$= -\tfrac{1}{2}e^u + C$$
$$= -\tfrac{1}{2}e^{4-x^2} + C \qquad \blacksquare$$

WARNING When using the method of substitution, all terms involving x in the original integral $\int f(x)\,dx$ must be converted into terms involving u in the transformed integral $\int g(u)\,du$. That is, after you have made your substitution and simplified, there should be no "leftover" x-values in the integrand.

EXAMPLE 5 Substitution with leftover x-values

Find $\int x(4x - 5)^3\,dx$.

Solution

$$\int x(4x-5)^3 \, dx = \int xu^3 \left(\frac{du}{4} \right) = \frac{1}{4} \int \underset{\uparrow}{xu^3} \, du$$

$$\boxed{u = 4x - 5; \, du = 4 \, dx}$$ There is a leftover x-value.

We are not ready to integrate until the leftover x-term has been eliminated. Because $u = 4x - 5$, we can solve for x:

$$x = \frac{u+5}{4}$$

so

$$\frac{1}{4} \int xu^3 \, du = \frac{1}{4} \int \left(\frac{u+5}{4} \right) u^3 \, du = \frac{1}{16} \int (u^4 + 5u^3) \, du$$

$$= \frac{1}{16} \left(\frac{u^5}{5} + 5\frac{u^4}{4} \right) + C = \frac{1}{80}(4x-5)^5 + \frac{5}{64}(4x-5)^4 + C \quad \blacksquare$$

Note: Always introduce the constant of integration immediately after the last integration.

The next example derives formulas that will be used many times throughout the text, often enough, in fact, that you should remember them or mark them for future reference.

EXAMPLE 6   Integral formulas for the tangent and cotangent functions

Show that $\displaystyle\int \tan x \, dx = -\ln|\cos x| + C$ and $\displaystyle\int \cot x \, dx = \ln|\sin x| + C$.

Solution

$$\int \tan x \, dx = \int \frac{\sin x}{\cos x} \, dx \qquad \text{Trigonometric identity } \tan x = \frac{\sin x}{\cos x}$$

$$= \int \left(\frac{1}{\cos x} \right) (\sin x \, dx) \qquad \text{Let } u = \cos x, \, du = -\sin x \, dx.$$

$$= -\int \frac{du}{u}$$

$$= -\ln|u| + C$$

$$= -\ln|\cos x| + C$$

This can also be written as $\ln\left|(\cos x)^{-1}\right| + C = \ln|\sec x| + C$. Similarly,

$$\int \cot x \, dx = \int \frac{\cos x}{\sin x} \, dx = \ln|\sin x| + C. \qquad \blacksquare$$

Here is a problem in which the rate of change of a quantity is known and we use the method of substitution to find an expression for the quantity itself.

EXAMPLE 7   Find the volume when the rate of flow is known

Water is flowing into a tank at the rate of $\sqrt{3t+1}$ ft^3/min. If the tank is empty when $t = 0$, how much water does it contain 5 min later?

Solution

Because the rate at which the volume V is changing is dV/dt,

$$\frac{dV}{dt} = \sqrt{3t+1}$$

$$V = \int \sqrt{3t+1}\,dt = \int \sqrt{u}\,\frac{du}{3}$$

$$\boxed{u = 3t+1;\ du = 3\,dt}$$

$$V = \frac{1}{3}\int u^{1/2}\,du$$

$$= \frac{1}{3}\cdot\frac{2}{3}u^{3/2} + C$$

$$V = \frac{2}{9}(3t+1)^{3/2} + C$$

$V(0) = \frac{2}{9}(3\cdot 0 + 1)^{3/2} + C = 0$ so that $C = -\frac{2}{9}$ The initial volume is 0.

$V(t) = \frac{2}{9}(3t+1)^{3/2} - \frac{2}{9}$ and $V(5) = \frac{2}{9}(16)^{3/2} - \frac{2}{9} = 14$

The tank contains 14 ft³ of water after 5 minutes.   ∎

SUBSTITUTION WITH DEFINITE INTEGRATION

Example 7 could be considered as a definite integral:

$$\int_0^5 \sqrt{3t+1}\,dt = \frac{2}{9}(3t+1)^{3/2}\Big|_0^5 = \frac{2}{9}(16)^{3/2} - \frac{2}{9}(1)^{3/2} = 14$$

Notice that the definite integral eliminates the need for finding C. Furthermore, the following theorem eliminates the need for returning to the original variable (even though sometimes it is more convenient to do so).

THEOREM 5.10   Substitution with the definite integral

If $f(u)$ is a continuous function of u and $u(x)$ is a differentiable function of x, then

$$\int_a^b f[u(x)]u'(x)\,dx = \int_{u(a)}^{u(b)} f(u)\,du$$

Proof   Let $F(u) = \int f(u)\,du$ be an antiderivative of $f(u)$. Then the first fundamental theorem of calculus tells us that

$$\int_a^b f[u(x)]u'(x)\,dx = F[u(x)]\big|_{x=a}^{x=b} = F[u(b)] - F[u(a)]$$

But since $F(u)$ is an antiderivative of $f(u)$, it also follows that

$$\int_{u(a)}^{u(b)} f(u)\,du = F(u)\big|_{u=u(a)}^{u=u(b)} = F[u(b)] - F[u(a)]$$

Thus,

$$\int_a^b f[u(x)]u'(x)\,dx = \int_{u(a)}^{u(b)} f(u)\,du$$ ❑

We now have two valid methods for evaluating a definite integral of the form $\int_a^b f[u(x)]u'(x)\,dx$:

1. Use substitution to evaluate the indefinite integral and then evaluate it between the limits of integration $x = a$ and $x = b$.
2. Change the limits of integration to conform with the change of variable $u = u(x)$ and evaluate the transformed definite integral between the limits of integration $u = u(a)$ and $u = u(b)$.

These two methods are illustrated in Example 8.

EXAMPLE 8   Substitution with the definite integral

Evaluate $\int_1^2 (4x - 5)^3\,dx$.

Solution

Method I: We first use a substitution with an indefinite integral.

$$\int (4x - 5)^3\,dx = \int u^3 \left(\frac{1}{4}\,du\right) = \frac{1}{4}\left(\frac{u^4}{4}\right) + C = \frac{1}{16}(4x - 5)^4 + C$$

$$\boxed{\text{Let } u = 4x - 5;\ du = 4\,dx}$$

Now evaluate the definite integral of $f(x)$ between $x = 1$ and $x = 2$ by using the fundamental theorem of calculus:

$$\int_1^2 (4x - 5)^3\,dx = \tfrac{1}{16}(4x - 5)^4\Big|_{x=1}^{x=2}$$

$$= \tfrac{1}{16}[4(2) - 5]^4 - \tfrac{1}{16}[4(1) - 5]^4$$

$$= \tfrac{1}{16}(81) - \tfrac{1}{16}(1)$$

$$= 5$$

Method II: This is similar to method I, except that as part of the integration process, we change the limits of integration when we are doing the substitution.

$$\boxed{\text{Let } u = 4x - 5;\ du = 4\,dx.}$$

If $x = 2$, then $u = 4(2) - 5 = 3$.

$$\int_1^2 (4x - 5)^3\,dx = \int_{-1}^3 u^3\,\frac{du}{4}$$

If $x = 1$, then $u = 4(1) - 5 = -1$.

WARNING   You cannot change variables and keep the orginal limits of integration because the limits of integration are expressed in terms of the variable of integration.

$$= \frac{1}{4}\cdot\frac{u^4}{4}\Big|_{-1}^3$$

$$= \tfrac{1}{16}(81 - 1) = 5$$ ∎

5.5 PROBLEM SET

A *Problems 1–8 present pairs of integration problems, one of which will require substitution and one of which will not. As you are working these problems, think about when substitution may be appropriate.*

1. a. $\int_0^4 (2t + 4)\, dt$ **b.** $\int_0^4 (2t + 4)^{-1/2}\, dt$

2. a. $\int_0^{\pi/2} \sin\theta\, d\theta$ **b.** $\int_0^{\pi/2} \sin 2\theta\, d\theta$

3. a. $\int_0^{\pi} \cos t\, dt$ **b.** $\int_0^{\sqrt{\pi}} t\cos t^2\, dt$

4. a. $\int_0^4 \sqrt{x}\, dx$ **b.** $\int_{-4}^0 \sqrt{-x}\, dx$

5. a. $\int_0^{16} \sqrt[4]{x}\, dx$ **b.** $\int_{-16}^0 \sqrt[4]{-x}\, dx$

6. a. $\int x(3x^2 - 5)\, dx$ **b.** $\int x(3x^2 - 5)^{50}\, dx$

7. a. $\int x^2\sqrt{2x^3}\, dx$ **b.** $\int x^2\sqrt{2x^3 - 5}\, dx$

8. a. $\int \dfrac{dx}{\sqrt{1 - x^2}}$ **b.** $\int \dfrac{x\, dx}{\sqrt{1 - x^2}}$

Use substitution to evaluate the indefinite integrals in Problems 9–34.

9. $\int (2x + 3)^4\, dx$ **10.** $\int \sqrt{3t - 5}\, dt$

11. $\int [\tan(x^2 + 5x + 3)(2x + 5)]\, dx$ **12.** $\int (11 - 2x)^{-4/5}\, dx$

13. $\int (x^2 - \cos 3x)\, dx$ **14.** $\int \csc^2 5t\, dt$

15. $\int \sin(4 - x)\, dx$ **16.** $\int \cot[\ln(x^2 + 1)]\dfrac{2x\, dx}{x^2 + 1}$

17. $\int \sqrt{t}(t^{3/2} + 5)^3\, dt$ **18.** $\int \dfrac{(6x - 9)\, dx}{(x^2 - 3x + 5)^3}$

19. $\int x\sin(3 + x^2)\, dx$ **20.** $\int \sin^3 t\cos t\, dt$

21. $\int \dfrac{x\, dx}{2x^2 + 3}$ **22.** $\int \dfrac{x^2\, dx}{x^3 + 1}$

23. $\int x\sqrt{2x^2 + 1}\, dx$ **24.** $\int \dfrac{4x\, dx}{2x + 1}$

25. $\int \sqrt{x}\, e^{x\sqrt{x}}\, dx$ **26.** $\int \dfrac{e^{\sqrt[3]{x}}\, dx}{x^{2/3}}$

27. $\int x(x^2 + 4)^{1/2}\, dx$ **28.** $\int x^3(x^2 + 4)^{1/2}\, dx$

29. $\int \dfrac{\ln x}{x}\, dx$ **30.** $\int \dfrac{\ln(x + 1)}{x + 1}\, dx$

31. $\int \dfrac{dx}{\sqrt{x}(\sqrt{x} + 7)}$ **32.** $\int \dfrac{dx}{x^{2/3}(\sqrt[3]{x} + 1)}$

33. $\int \dfrac{e^t\, dt}{e^t + 1}$ **34.** $\int \dfrac{e^{\sqrt{t}}\, dt}{\sqrt{t}(e^{\sqrt{t}} + 1)}$

Evaluate the definite integrals given in Problems 35–44. Approximate the answers to Problems 43 and 44 to two significant digits.

35. $\int_0^1 \dfrac{5x^2\, dx}{2x^3 + 1}$ **36.** $\int_1^4 \dfrac{e^{-\sqrt{x}}\, dx}{\sqrt{x}}$

37. $\int_{-\ln 2}^{\ln 2} \dfrac{1}{2}(e^x - e^{-x})\, dx$ **38.** $\int_0^2 (e^x - e^{-x})^2\, dx$

39. $\int_1^2 \dfrac{e^{1/x}}{x^2}\, dx$ **40.** $\int_0^2 x\sqrt{2x + 1}\, dx$

41. $\int_0^{\pi/6} \tan 2x\, dx$ **42.** $\int_0^1 x^2(x^3 + 9)^{1/2}\, dx$

43. $\int_0^5 \dfrac{0.58}{1 + e^{-0.2x}}\, dx$ **44.** $\int_0^{12} \dfrac{5,000}{1 + 10e^{-t/5}}\, dt$

45. HISTORICAL QUEST Johann Peter Gustav Lejeune Dirichlet was a professor of mathematics at the University of Berlin and is known for his role in formulating a rigorous foundation for calculus. He was not known as a good teacher. His nephew wrote that the mathematics instruction he received from Dirichlet was the most dreadful experience of his life. Howard Eves tells of the time Dirichlet was to deliver a lecture on definite integrals, but because of illness he posted the following note:

LEJEUNE DIRICHLET
1805–1859

Because of illness I cannot lecture today

Dirichlet

The students then doctored the note to read as follows:

Michaelmas

$\int$ *Because of illness I cannot lecture today* $d\,(1\ Frdor)$

Easter

Dirichlet

Michaelmas and Easter were school holidays, and 1 Frdor (Friedrichsd'or) was the customary honorarium for a semester's worth of lectures. ∎

a. What is the answer when you integrate the student-doctored note?

b. The so-called *Dirichlet function* is often used for counterexamples in calculus. Look up the definition of this function. What special property does it have?

B *Find the area of the region under the curves given in Problems 46–49.*

46. $y = t\sqrt{t^2 + 9}$ on $[0, 4]$ **47.** $y = \dfrac{1}{t^2}\sqrt{5 - \dfrac{1}{t}}$ on $[\tfrac{1}{5}, 1]$

48. $y = x(x-1)^{1/3}$ on $[2, 9]$ **49.** $y = \dfrac{x}{\sqrt{x^2+1}}$ on $[1, 3]$

50. a. In Section 5.3 (Problem 40), you were asked to prove that if f is continuous and **odd** [that is, $f(-x) = -f(x)$] on the interval $[-a, a]$, then

$$\int_{-a}^{a} f(x)\, dx = 0$$

b. In Section 5.2 (Problem 33), you were asked to find a counterexample or informally show that if f is continuous and **even** [$f(-x) = f(x)$] on the interval $[-a, a]$, then

$$\int_{-a}^{a} f(x)\, dx = 2 \int_{0}^{a} f(x)\, dx = 2 \int_{-a}^{0} f(x)\, dx$$

For this problem, use the properties of integrals to prove each of these statements.

Use the results of Problem 50 to evaluate the integrals given in Problems 51–54.

51. $\displaystyle\int_{-\pi}^{\pi} \sin x \, dx$ **52.** $\displaystyle\int_{-\pi/2}^{\pi/2} \cos x \, dx$

53. $\displaystyle\int_{-3}^{3} x\sqrt{x^4+1}\, dx$ **54.** $\displaystyle\int_{-1}^{1} \frac{\sin x \, dx}{x^2+1}$

55. In each of the following cases, determine whether the given relationship is true or false.

a. $\displaystyle\int_{-175}^{175} (7x^{1001} + 14x^{99})\, dx = 0$

b. $\displaystyle\int_{0}^{\pi} \sin^2 x \, dx = \int_{0}^{\pi} \cos^2 x \, dx$

c. $\displaystyle\int_{-\pi/2}^{\pi/2} \cos x \, dx = \int_{-\pi}^{0} \sin x \, dx$

56. The slope at each point (x, y) on the graph of $y = F(x)$ is given by $x(x^2 - 1)^{1/3}$, and the graph passes through the point $(3, 1)$. Use this information to find F. Sketch the graph of F.

57. The slope at each point (x, y) on the graph of $y = F(x)$ is given by $\dfrac{2x}{1 - 3x^2}$. What is $F(x)$ if the graph passes through $(0, 5)$?

58. A particle moves along the t-axis in such a way that at time t, its velocity is $v(t) = t^2(t^3 - 8)^{1/3}$.

a. At what time does the particle turn around?

b. If the particle starts at a position which we denote as 1, where does it turn around?

59. A rectangular storage tank has a square base 10 ft on a side. Water is flowing into the tank at the rate modeled by the function

$$R(t) = t(3t^2 + 1)^{-1/2} \quad \text{ft}^3/\text{s}$$

at time t seconds. If the tank is empty at time $t = 0$, how much water does it contain 4 sec later? What is the depth of the water (to the nearest quarter inch) at that time?

60. Journal Problem *College Mathematics Journal.** ■ Evaluate

$$\int [(x^2 - 1)(x + 1)]^{-2/3}\, dx$$

61. Modeling Problem A group of environmentalists model the rate at which the ozone level is changing in a suburb of Los Angeles by the function

$$L'(t) = \frac{0.24 - 0.03t}{\sqrt{36 + 16t - t^2}}$$

parts per million per hour (ppm/h) t hours after 7:00 A.M.

a. Express the ozone level $L(t)$ as a function of t if L is 4 ppm at 7:00 A.M.

b. Use the graphing utility of your calculator to find the time between 7:00 A.M. and 7:00 P.M. when the highest level of ozone occurs. What is the highest level?

c. Use your graphing utility or another utility of your calculator to determine a second time during the day when the ozone level is the same as it is at 11:00 A.M.

62. A *logistic* function is one of the form $Q(t) = \dfrac{B}{1 + Ae^{-rt}}$. Evaluate $\displaystyle\int Q(t)\, dt$.

*Vol. 20, No. 4, Sept. 1989, p. 343. Problem by Murray Klamkin.

5.6   Introduction to Differential Equations

IN THIS SECTION   introduction and terminology, direction fields, separable differential equations, modeling exponential growth and decay, orthogonal trajectories, modeling fluid flow through an orifice, modeling the motion of a projectile: escape velocity

The study of differential equations is as old as calculus itself. Today, it would be virtually impossible to make a serious study of physics, astronomy, chemistry, or engineering without encountering physical models based on differential equations. In addition, differential equations are beginning to appear more frequently in the biological and social sciences, especially in economics. We begin by introducing some basic terminology and examining a few modeling procedures.

INTRODUCTION AND TERMINOLOGY

Any equation that contains a derivative or differential is called a **differential equation**. For example, the equations

$$\frac{dy}{dx} = 3x^3 + 5, \quad \frac{dP}{dt} = kP^2, \quad \left(\frac{dy}{dx}\right)^2 + 3\frac{dy}{dx} + 2y = xy, \quad \frac{d^2x}{dt^2} + 2\frac{dx}{dt} + 5t = \sin t$$

are all differential equations.

Many practical situations, especially those involving rates, can be described mathematically by differential equations. For example, the assumption that population P grows at a rate proportional to its size can be expressed by the differential equation

$$\frac{dP}{dt} = kP$$

where t is time and k is the constant of proportionality.

A **solution** of a given differential equation is a function that satisfies the equation. A **general solution** is a characterization of all possible solutions of the equation. We say that the differential equation is **solved** when we find a general solution.

For example, $y = x^2$ is a solution of the differential equation

$$\frac{dy}{dx} = 2x$$

because

$$\frac{dy}{dx} = \frac{d}{dx}(y) = \frac{d}{dx}(x^2) = 2x$$

Moreover, because any solution of this equation must be an indefinite integral of $2x$, it follows that

$$y = \int 2x \, dx = x^2 + C$$

is the general solution of the differential equation.

EXAMPLE 1   Verifying that a given function is a solution to a differential equation

If $4x - 3y^2 = 10$, verify that $\dfrac{dy}{dx} = \dfrac{2}{3y}$.

Solution

$$4x - 3y^2 = 10$$

$$\frac{d}{dx}(4x - 3y^2) = \frac{d}{dx}(10) \qquad \text{Take the derivative of both sides.}$$

$$4 - 6y\frac{dy}{dx} = 0 \qquad \text{Don't forget the chain rule, since } y \text{ is a function of } x.$$

$$\frac{dy}{dx} = \frac{4}{6y} = \frac{2}{3y} \qquad \text{Solve for } \frac{dy}{dx}. \qquad \blacksquare$$

EXAMPLE 2   Finding future revenue

An oil well that yields 300 barrels of crude oil a day will run dry in 3 years. It is estimated that t days from now the price of the crude oil will be $p(t) = 30 + 0.3\sqrt{t}$ dollars per barrel. If the oil is sold as soon as it is extracted from the ground, what will be the total future revenue from the well?

Solution

Let $R(t)$ denote the total revenue up to time t. Then the rate of change of revenue is $\dfrac{dR}{dt}$, the number of dollars received per barrel is $p(t) = 30 + 0.3\sqrt{t}$, and the number of barrels sold per day is 300. Thus, we have

$$\begin{bmatrix} \text{RATE OF CHANGE} \\ \text{OF TOTAL REVENUE} \end{bmatrix} = \begin{bmatrix} \text{NUMBER OF DOLLARS} \\ \text{PER BARREL} \end{bmatrix} \begin{bmatrix} \text{NUMBER OF BARRELS} \\ \text{SOLD PER DAY} \end{bmatrix}$$

$$\frac{dR}{dt} = (30 + 0.3\sqrt{t})(300)$$

$$= 9{,}000 + 90\sqrt{t}$$

This is actually a statement of the chain rule: $\dfrac{dR}{dt} = \dfrac{dR}{dB} \cdot \dfrac{dB}{dt}$, where B denotes the number of barrels extracted and R denotes the revenue. It is often helpful to use the chain rule in this way when setting up differential equations.

We solve this differential equation by integration:

$$R = \int \frac{dR}{dt}\, dt = \int (9{,}000 + 90\sqrt{t})\, dt$$

$$R(t) = 9{,}000t + 60t^{3/2} + C$$

Since $R(0) = 0$, it follows that $C = 0$. We also are given that the well will run dry in 3 years or 1,095 days, so that the total revenue obtained during the life of the well is

$$R(1{,}095) = 9{,}000(1{,}095) + 60(1{,}095)^{3/2}$$

$$\approx 12{,}029{,}064.52$$

The total future revenue is approximately \$12 million.   ∎

DIRECTION FIELDS

We can use slope fields to help with a visualization of a differential equation. In Section 5.1 we looked at small segments with slope defined by the derivative of some function. The collection of all such line segments is called the *slope field* or **direction field** of the differential equation. Today, with the assistance of computers, we can sometimes draw the direction field in order to obtain a particular solution (or a family of solutions) as shown in the following example.

EXAMPLE 3   Finding a solution, given the direction field

The direction field for the differential equation

$$\frac{dy}{dx} = y - x^2$$

is shown in Figure 5.22.

a. Sketch a solution to the initial value problem passing through $(2, 1)$.
b. Sketch a solution to the initial value problem passing through $(0, 1)$.

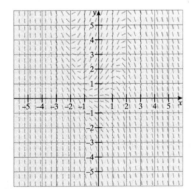

Figure 5.22 Direction field for $y' = y - x^2$

Solution

a. The initial value $y(2) = 1$ means the solution passes through $(2, 1)$ and is shown in Figure 5.23a.
b. Since $y(0) = 1$, the solution passes through $(0, 1)$, as shown in Figure 5.23b.

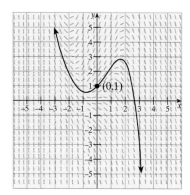

a. Solution passes through $(2, 1)$ **b.** Solution passes through $(0, 1)$

Figure 5.23   Particular solutions from a given direction field   ■

SEPARABLE DIFFERENTIAL EQUATIONS

Solving a differential equation is often a complicated process. However, many important equations can be expressed in the form

$$\frac{dy}{dx} = \frac{g(x)}{f(y)}$$

To solve such an equation, first separate the variables into the differential form

$$f(y)\, dy = g(x)\, dx$$

and then integrate both sides separately to obtain

$$\int f(y)\, dy = \int g(x)\, dx$$

A differential equation expressible in the form $dy/dx = g(x)/f(y)$ is said to be **separable**. This procedure is illustrated in Example 4.

EXAMPLE 4   Solving a separable differential equation

Solve $\dfrac{dy}{dx} = \dfrac{x}{y}$.

Solution

$$\frac{dy}{dx} = \frac{x}{y}$$
$$y\, dy = x\, dx$$
$$\int y\, dy = \int x\, dx$$
$$\tfrac{1}{2}y^2 + C_1 = \tfrac{1}{2}x^2 + C_2$$
$$x^2 - y^2 = C \qquad \text{where } C = 2(C_1 - C_2). \qquad ■$$

Notice the treatment of constants in Example 4. Because all constants can be combined into a single constant, it is customary not to write $2(C_1 - C_2)$ but rather to simply replace all the arbitrary constants in the problem by a single arbitrary constant, C, immediately after the last integral is found.

The remainder of this section is devoted to selected applications involving separable differential equations.

MODELING EXPONENTIAL GROWTH AND DECAY

A process is said to undergo **exponential change** if the relative rate of change of the process is modeled by a constant; in other words,

$$\frac{Q'(t)}{Q(t)} = k \quad \text{or} \quad \frac{dQ}{dt} = kQ(t)$$

If the constant k is positive, the exponential change is called **growth**, and if k is negative, it is called **decay**. Exponential growth occurs in certain populations, and exponential decay occurs in the disintegration of radioactive substances.

   To solve the growth/decay equation, separate the variables and integrate both sides.

$$\frac{dQ}{dt} = kQ$$

$$\int \frac{dQ}{Q} = \int k \, dt$$

$$\ln |Q| = kt + C_1$$

$$e^{kt+C_1} = Q \qquad \text{Definition of natural logarithm}$$

$$e^{kt} e^{C_1} = Q$$

Thus, $Q = Ce^{kt}$, where $C = e^{C_1}$. Finally, if we let Q_0 be the initial amount (that is, when $t = 0$), we see that

$$Q_0 = Ce^0 \quad \text{or} \quad C = Q_0, \quad \text{so} \quad Q(t) = Q_0 e^{kt}$$

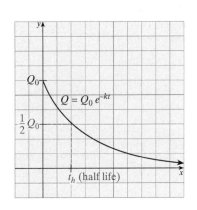

Figure 5.24 The decay curve for a radioactive substance

Growth/Decay Equation

The **growth/decay equation** of a substance is

$$Q(t) = Q_0 e^{kt}$$

where $Q(t)$ is the amount of the substance present at time t, Q_0 is the initial amount of the substance, and k is a constant that depends on the substance. If $k > 0$, it is a *growth* equation; if $k < 0$, it is a *decay* equation.

   The graph of $Q = Q_0 e^{kt}$ for $k < 0$ is shown in Figure 5.24. In Figure 5.24 we have also indicated the time t_h required for half of a given substance to disintegrate. The time t_h is called the **half-life** of the substance, and it provides a measure of the substance's rate of disintegration.

EXAMPLE 5   Amount of a radioactive substance present

A particular radioactive substance has a half-life of 600 yr. Find k for this substance and determine how much of a 50-g sample will remain after 125 yr.

Solution

From the decay equation,

$$Q(t) = Q_0 e^{kt}$$

$$\frac{Q(t)}{Q_0} = e^{kt}$$

$$\frac{1}{2} = e^{k(600)} \qquad \text{Half-life is 600 years.}$$

$$k(600) = \ln \tfrac{1}{2} \qquad \text{Definition of logarithm}$$

$$k = \frac{\ln \frac{1}{2}}{600} \approx -0.0011552453$$

Next, to see how much of a 50-g sample will remain after 125 yr, substitute $Q_0 = 50$, $k = -0.0011552453$, and $t = 125$ into the decay equation:

$$Q(125) \approx 50e^{-0.0011552453(125)} \approx 43.27682805$$

There will be about 43 g present. ∎

One of the more interesting applications of radioactive decay is a technique known as **carbon dating**, which is used by geologists, anthropologists, and archaeologists to estimate the age of various specimens.* The technique is based on the fact that all animal and vegetable systems (whether living or dead) contain both stable carbon ^{12}C and a radioactive isotope ^{14}C. Scientists assume that the ratio of ^{14}C to ^{12}C in the air has remained approximately constant throughout history. Living systems absorb carbon dioxide from the air, so the ratio of ^{14}C to ^{12}C in a living system is the same as that in the air itself. When a living system dies, the absorption of carbon dioxide ceases. The ^{12}C already in the system remains while the ^{14}C decays, and the ratio of ^{14}C to ^{12}C decreases exponentially. The half-life of ^{14}C is approximately 5,730 years. The ratio of ^{14}C to ^{12}C in a specimen t years after it was alive is approximately

$$R = R_0 e^{kt}$$

where $k = \dfrac{\ln(1/2)}{5,730}$ and R_0 is the ratio of ^{14}C to ^{12}C in the atmosphere. By comparing $R(t)$ with R_0, scientists can estimate the age of the object. Here is an example.

EXAMPLE 6 Modeling Problem: Carbon dating

An archaeologist has found a specimen in which the ratio of ^{14}C to ^{12}C is 20% the ratio found in the atmosphere. Approximately how old is the specimen?

Solution

The age of the specimen is the value of t for which $R(t) = 0.20R_0$:

$$0.20R_0 = R_0 e^{kt}$$
$$0.20 = e^{kt}$$
$$kt = \ln 0.20$$
$$t = \frac{1}{k}\ln 0.2 \qquad k = \frac{\ln(1/2)}{5,730}$$
$$\approx 13,304.64798$$

The specimen is approximately 13,000 yr old. ∎

ORTHOGONAL TRAJECTORIES

A curve that intersects each member of a given family of curves at right angles is called an **orthogonal trajectory** of that family and is shown in Figure 5.25. Orthogonal families arise in many applications. For example, in thermodynamics, the heat flow across a planar surface is orthogonal to the curves of constant temperature, called *isotherms*. In the theory of fluid flow, the flow lines are orthogonal trajectories of *velocity potential curves*. The basic procedure for finding orthogonal trajectories involves differential equations and is demonstrated in Example 7.

---

*Carbon dating is used primarily for estimating the age of relatively "recent" specimens. For example, it was used (along with other methods) to determine that the Dead Sea Scrolls were written and deposited in the Caves of Qumran approximately 2,000 years ago. For dating older specimens, it is better to use techniques based on radioactive substances with longer half-lives. In particular, potassium 40 is often used as a "clock" for events that occurred between 5 and 15 million years ago. Paleoanthropologists find this substance especially valuable because it often occurs in volcanic deposits and can be used to date specimens trapped in such deposits. See Mathematical Essay 5 on page 352.

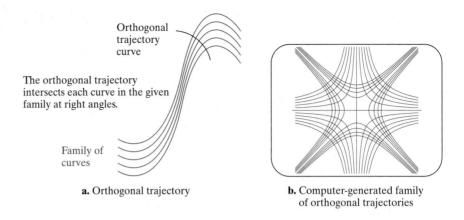

a. Orthogonal trajectory

b. Computer-generated family of orthogonal trajectories

Figure 5.25   Orthogonal trajectories

EXAMPLE 7   Finding orthogonal trajectories

Find the orthogonal trajectories of the family of curves of the form $xy = C$.

Solution

We are seeking a family of curves. Each curve in that family intersects each curve in the family $xy = C$ at right angles, as shown in Figure 5.26. Assume that a typical point on a given curve in the family $xy = C$ has coordinates (x, y) and that a typical point on the orthogonal trajectory curve has coordinates (X, Y).*

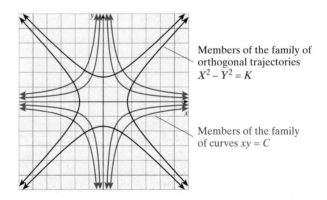

Members of the family of orthogonal trajectories $X^2 - Y^2 = K$

Members of the family of curves $xy = C$

Figure 5.26   The family of curves $xy = C$ and their orthogonal trajectories

Let P be a point where a particular curve of the form $xy = C$ intersects the orthogonal trajectory curve. At P, we have $x = X$ and $y = Y$, and the slope dY/dX of the orthogonal trajectory is the same as the negative reciprocal of the slope dy/dx of the curve $xy = C$. Using implicit differentiation, we find

$$xy = C$$

$$x\frac{dy}{dx} + y = 0 \qquad \text{Product rule}$$

$$\frac{dy}{dx} = \frac{-y}{x}$$

---

*We use uppercase letters for one kind of curve and lowercase for the other to make it easier to tell which curve is being mentioned at each stage of the following discussion.

Thus, at the point of intersection P, the slope $\dfrac{dY}{dX}$ of the orthogonal trajectory is

$$\frac{dY}{dX} = -\frac{1}{\dfrac{dy}{dx}} = -\frac{1}{\dfrac{-y}{x}} = \frac{x}{y} = \frac{X}{Y}$$

According to this equation, the coordinates (X, Y) of the orthogonal trajectory curve satisfy the separable differential equation

$$\frac{dY}{dX} = \frac{X}{Y}$$

discussed in Example 4. Using the result of that example, we see that the orthogonal trajectories of the family $xy = C$ are the curves in the family

$$X^2 - Y^2 = K$$

where K is a constant. The given family of curves $xy = C$ and the family of orthogonal trajectory curves $X^2 - Y^2 = K$ are shown in Figure 5.26. ∎

MODELING FLUID FLOW THROUGH AN ORIFICE

Consider a tank that is filled with a fluid being slowly drained through a small, sharp-edged hole in its base, as shown in Figure 5.27. By using a principle of physics known as Torricelli's law,[†] we can show that the rate of discharge dV/dt at time t is proportional to the square root of the depth h at that time. Specifically, if all dimensions are given in terms of feet, the drain hole has area A_0, and the height above the hole is h at time t (seconds), then

$$\frac{dV}{dt} = -4.8 A_0 \sqrt{h}$$

is the rate of flow of water in cubic feet per second. This formula is used in Example 8.

Fluid has height h above a hole of area A_0.

Figure 5.27 The flow of a fluid through an orifice

EXAMPLE 8   Fluid flow through an orifice

A cylindrical tank (with a circular base) is filled with a liquid that is draining through a small circular hole in its base. If the tank is 9 ft high with radius 4 ft and the drain hole has radius 1 in., how long does it take for the tank to empty?

Solution

Because the drain hole is a circle of radius $\frac{1}{12}$ ft (= 1 in.), its area is $\pi r^2 = \pi \left(\frac{1}{12}\right)^2$, and the rate of flow is

$$\frac{dV}{dt} = -4.8 \left(\frac{\pi}{144}\right) \sqrt{h}$$

Because the tank is cylindrical, the amount of fluid in the tank at any particular time will form a cylinder of radius 4 ft and height h. Also, because we are using $g = 32$ ft/s^2, it is implied that the time, t, is measured in seconds. The volume of such a liquid cylinder is

$$V = \pi r^2 h = \pi(4)^2 h = 16\pi h$$

[†]Torricelli's law says that the stream of liquid through the orifice has velocity $\sqrt{2gh}$, where $g = 32$ ft/s^2 is the acceleration due to gravity and h is the height of the liquid above the orifice. The factor 4.8 that appears in the rate of flow equation is required to compensate for the effect of friction.

and by differentiating both sides of this equation with respect to t, we obtain

$$\frac{dV}{dt} = 16\pi \frac{dh}{dt}$$

$$-4.8 \left(\frac{\pi}{144}\right) \sqrt{h} = 16\pi \frac{dh}{dt} \qquad \text{Torricelli's law}$$

$$\frac{dh}{dt} = \frac{-4.8\sqrt{h}}{144(16)} \approx -0.0021\sqrt{h}$$

$$\frac{1}{\sqrt{h}} dh = -0.0021\, dt \qquad \text{Separate the variables.}$$

$$\int h^{-1/2} dh = \int (-0.0021)\, dt \qquad \text{Integrate both sides.}$$

$$2h^{1/2} + C_1 = -0.0021t + C_2$$

$$2\sqrt{h} = -0.0021t + C$$

To evaluate C, recall that the tank is full at time $t = 0$. Thus, $h = 9$ when $t = 0$, so

$$2\sqrt{9} = -0.0021(0) + C, \quad \text{or} \quad C = 6$$

and the general formula is

$$2\sqrt{h} = -0.0021t + 6$$

where time t is in seconds (because g is 32 ft/s^2).

Now we can find the depth of the fluid at any given time or the time at which a prescribed depth occurs. In particular, the tank is empty at the time t_e when $h = 0$. By substituting $h = 0$ into this formula we find

$$2\sqrt{0} = -0.0021t_e + 6 \quad \text{so that} \quad t_e \approx 2,880 \text{ sec}$$

Thus, roughly 48 min are required to drain the tank. ■

MODELING THE MOTION OF A PROJECTILE: ESCAPE VELOCITY

Figure 5.28 A projectile launched from the surface of a planet

Consider a projectile that is launched with initial velocity v_0 from a planet's surface along a direct line through the center of the planet, as shown in Figure 5.28. We will find a general formula for the velocity of the projectile and the minimal value of v_0 required to guarantee that the projectile will escape the planet's gravitational attraction.

We assume that the only force acting on the projectile is that due to gravity, although in practice, factors such as air resistance must also be considered. With this assumption, Newton's law of gravitation can be used to show that when the projectile is at a distance s from the center of the planet, its acceleration is given by the formula

$$a = \frac{-gR^2}{s^2}$$

where R is the radius of the planet and g is the acceleration due to gravity at the planet's surface (see Problem 69).*

Our first goal is to express the velocity v of the projectile in terms of the height s. Because the projectile travels along a straight line, we know that

$$a = \frac{dv}{dt} \quad \text{and} \quad v = \frac{ds}{dt}$$

and by applying the chain rule, we see that

$$a = \frac{dv}{dt} = \frac{dv}{ds} \cdot \frac{ds}{dt} = \frac{dv}{ds}v$$

---

*According to Newton's law of gravitation, the force of gravity acting on a projectile of mass m has magnitude $F = mk/s^2$, where k is constant. If this is the only force acting on the projectile, then $F = ma$, and we have $ma = F = mk/s^2$. By canceling the m's on each side of the equation, we obtain $a = k/s^2$.

Therefore, by substitution for a we have

$$\frac{dv}{ds}v = \frac{-gR^2}{s^2}$$

$$v\,dv = -gR^2s^{-2}\,ds$$

$$\int v\,dv = \int -gR^2s^{-2}\,ds$$

$$\tfrac{1}{2}v^2 + C_1 = gR^2s^{-1} + C_2$$

$$v^2 = 2gR^2s^{-1} + C$$

To evaluate the constant C, recall that the projectile was fired from the planet's surface with initial velocity v_0. Thus, $v = v_0$ when $s = R$, and by substitution

$$v_0^2 = 2gR^2R^{-1} + C, \quad \text{which implies} \quad C = v_0^2 - 2gR$$

so

$$v^2 = 2gR^2s^{-1} + v_0^2 - 2gR$$

Because the projectile is launched in a direction away from the center of the planet, we would expect it to keep moving in that direction until it stops. In other words, *the projectile will keep moving away from the planet until it reaches a point where $v = 0$.* Because $2gR^2s^{-1} > 0$ for all $s > 0$, v^2 will always be positive if $v_0^2 - 2gR \geq 0$. On the other hand, if $v_0^2 - 2gR < 0$, then sooner or later v will become 0 and the projectile will eventually fall back to the surface of the planet.

Therefore, we conclude that the projectile will escape from the planet's gravitational attraction if $v_0^2 \geq 2gR$; that is, $v_0 \geq \sqrt{2gR}$. For this reason, the minimum speed for which this can occur, namely,

$$v_0 = \sqrt{2gR}$$

is called the **escape velocity** of the planet. In particular, for the earth, $R = 3,956$ mi and $g = 32$ ft/s^2 $= 0.00606$ mi/s^2, and the escape velocity is

$$v_0 = \sqrt{2gR} \approx \sqrt{2(0.00606)(3,956)} \approx 6.924357$$

The escape velocity for the earth is 6.92 mi/s.

5.6 PROBLEM SET

A *Verify in Problems 1–8 that if y satisfies the prescribed relationship with x, then it will be a solution of the given differential equation.*

1. If $x^2 + y^2 = 7$, then $\dfrac{dy}{dx} = -\dfrac{x}{y}$.

2. If $5x^2 - 2y^2 = 3$, then $\dfrac{dy}{dx} = \dfrac{5x}{2y}$.

3. If $xy = C$, then $\dfrac{dy}{dx} = \dfrac{-y}{x}$.

4. If $x^2 - 3xy + y^2 = 5$, then $(2x - 3y)\,dx + (2y - 3x)\,dy = 0$.

5. If $y = \sin(Ax + B)$, then $\dfrac{d^2y}{dx^2} + A^2y = 0$.

6. If $y = \dfrac{x^4}{20} - \dfrac{A}{x} + B$, then $x\dfrac{d^2y}{dx^2} + 2\dfrac{dy}{dx} = x^3$.

7. If $y = 2e^{-x} + 3e^{2x}$, then $y'' - y' - 2y = 0$.

8. If $y = Ae^x + Be^x \ln x$, then $xy'' + (1 - 2x)y' - (1 - x)y = 0$.

Find the particular solution of the first-order linear differential equations in Problems 9–14. A graphical solution within its direction field is shown.

9. $\dfrac{dy}{dx} = -\dfrac{x}{y}$
passing through $(2, 2)$

10. $\dfrac{dy}{dx} - y = 10$
passing through $(-2, -9)$

11. $\dfrac{dy}{dx} - y^2 = 1$
passing through $(\pi, 1)$

12. $\dfrac{dy}{dx} = e^{x+y}$
passing through $(0, 0)$

13. $\dfrac{dy}{dx} = \sqrt{\dfrac{x}{y}}$
passing through $(4, 1)$

14. $\dfrac{dy}{dx} = y^2 \sqrt{x}$
passing through $\left(9, -\dfrac{1}{18}\right)$

In each of Problems 15–20, sketch the particular solution passing through the given point for the differential equation whose direction field is given.

15. $(0, 1)$

16. $(0, 1)$

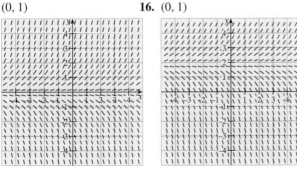

17. $(0, 0)$

18. $(3, 3)$

19. $(1, 0)$

20. $(0, 0)$

Find the general solution of the separable differential equations given in Problems 21–30. Note that for some problems, the general solution will be a relationship between x and y, rather than $y = f(x)$.

21. $\dfrac{dy}{dx} = 3xy$

22. $\dfrac{dy}{dx} = \sqrt{\dfrac{y}{x}}$

23. $\dfrac{dy}{dx} = \dfrac{x}{y}\sqrt{1 - x^2}$

24. $\dfrac{dy}{dx} = (y - 4)^2$

25. $xy\,dx + \sqrt{xy}\,dy = 0$

26. $\dfrac{dy}{dx} = \dfrac{y}{\sqrt{1 - 2x^2}}$

27. $\dfrac{dy}{dx} = \dfrac{\sin x}{\cos y}$

28. $x^2\,dy + \sec y\,dx = 0$

29. $xy\dfrac{dy}{dx} = \dfrac{\ln x}{\sqrt{1 - y^2}}$

30. $\dfrac{dy}{dx} = e^{y-x}$

In Problems 31–34, find the general solution of the given differential equation by using either the product or the quotient rule.

31. $x\,dy + y\,dx = 0$

32. $\dfrac{x\,dy - y\,dx}{x^2} = 0$

33. $y\,dx = x\,dy,\ x > 0,\ y > 0$ **34.** $x^2y\,dy + xy^2\,dx = 0$

B *Find the orthogonal trajectories of the family of curves given in Problems 35–42. In each case, sketch several members of the given family of curves and several members of the family of orthogonal trajectories on the same coordinate axes.*

35. the lines $2x - 3y = C$ **36.** the lines $y = x + C$

37. the cubic curves $y = x^3 + C$

38. the curves $y = x^4 + C$

39. the curves $xy^2 = C$ **40.** the parabolas $y^2 = 4kx$

41. the circles $x^2 + y^2 = r^2$

42. the exponential curves $y = Ce^{-x}$

Modeling Problems   *Write a differential equation to model the situation given in each of Problems 43–48. Be sure to define your unknown function first. Do not solve.*

43. The number of bacteria in a culture grows at a rate that is proportional to the number present.

44. A sample of radium decays at a rate that is proportional to the amount of radium present in the sample.

45. The rate at which the temperature of an object changes is proportional to the difference between its own temperature and the temperature of the surrounding medium.

46. When a person is asked to recall a set of facts, the rate at which the facts are recalled is proportional to the number of relevant facts in the person's memory that have not yet been recalled. *Hint:* Let Q denote the number of facts recalled and N the total number of relevant facts in the person's memory.

47. The rate at which an epidemic spreads through a community of P susceptible people is proportional to the product of the number of people who have caught the disease and the number who have not.

48. The rate at which people are implicated in a government scandal is proportional to the product of the number of people already implicated and the number of people involved who have not yet been implicated.

49. What do you think the orthogonal trajectories of the family of curves $x^2 - y^2 = C$ will be? Verify your conjecture. *Hint*: Take another look at Example 7.

50. The following is a list of six families of curves. Sketch several members of each family and then determine which pairs are orthogonal trajectories of one another.
   a. the circles $x^2 + y^2 = A$ **b.** the ellipses $2x^2 + y^2 = B^2$
   c. the ellipses $x^2 + 2y^2 = C$ **d.** the lines $y = Cx$
   e. the parabolas $y^2 = Cx$ **f.** the parabolas $y = Cx^2$

51. The Dead Sea Scrolls were written on parchment in about 100 B.C. What percentage of ^{14}C originally contained in the parchment remained when the scrolls were discovered in 1947?

52. Tests of an artifact discovered at the Debert site in Nova Scotia show that 28% of the original ^{14}C is still present. What is the probable age of the artifact?

53. HISTORICAL QUEST   The Shroud of Turin is a rectangular linen cloth kept in the Chapel of the Holy Shroud in the cathedral of St. John the Baptist in Turin, Italy. It shows the image of a man whose wounds correspond with the biblical accounts of the crucifixion.

In 1389, Pierre d'Arcis, the Bishop of Troyes, wrote a memo to the Pope, accusing a colleague of passing off "a certain cloth, cunningly painted" as the burial shroud of Jesus Christ. Despite this early testimony of forgery, this so-called Shroud of Turin has survived as a famous relic. In 1988, a small sample of the Shroud of Turin was taken and scientists from Oxford University, the University of Arizona, and the Swiss Federal Institute of Technology were permitted to test it. It was found that the cloth contained 92.3% of the original ^{14}C.

According to this information, how old is the Shroud?

54. A cylindrical tank of radius 3 ft is filled with water to a depth of 5 ft. Determine how long (to the nearest minute) it takes to drain the tank through a sharp-edged circular hole in the bottom with radius 2 in.

55. Rework Problem 54 for a tank with a sharp-edged drain hole that is square with side of length 1.5 in.

56. **Experiment**   A rectangular tank has a square base 2 ft on a side that is filled with water to a depth of 4 ft. It is being drained from the bottom of the tank through a sharp-edged square hole that is 2 in. on a side.
   a. Show that at time t, the depth h satisfies the differential equation
$$\frac{dh}{dt} = -\frac{1}{30}\sqrt{h}$$
   b. How long will it take to empty the tank?
   c. Construct this tank and then drain it out of the 4 in.^{2} hole. Is the time that it takes consistent with your answer to part **b**?

57. A toy rocket is launched from the surface of the earth with initial velocity $v_0 = 150$ ft/s. (The radius of the earth is roughly 3,956 mi, and $g = 32$ ft/s^2.)
   a. Determine the velocity (to the nearest ft/s) of the rocket when it is first 200 feet above the ground. (Remember, this is not the same as 200 ft from the center of the earth.)
   b. What is s when $v = 0$? Determine the maximum height above the ground that is attained by the rocket.

58. Determine the escape velocity of each of the following heavenly bodies:
   a. moon ($R = 1,080$ mi; $g = 5.5$ ft/s^2)
   b. Mars ($R = 2,050$ mi; $g = 12$ ft/s^2)
   c. Venus ($R = 3,800$ mi; $g = 28$ ft/s^2)

59. Population statistics indicate that t years after 1990, a certain city was growing at a rate of approximately $1,500t^{-1/2}$ people per year. In 1994, the population of the city was 39,000.
   a. What was the population in 1990?
   b. If this pattern continued, how many people were living in the city in the year 1999?

60. The radius of planet X is one-fourth that of planet Y, and the acceleration due to gravity at the surface of X is eight-ninths that at the surface of Y. If the escape velocity of planet X is 6 ft/s, what is the escape velocity of planet Y?

61. A survey indicates that the population of a certain town is growing in such a way that the rate of growth at time t is proportional to the square root of the population P at the time. If the population was 4,000 ten years ago and is observed to be 9,000 now, how long will it take before 16,000 people live in the town?

62. A scientist has discovered a radioactive substance that disintegrates in such a way that at time t, the rate of disintegration is proportional to the *square* of the amount present.
   a. If a 100-g sample of the substance dwindles to only 80 g in 1 day, how much will be left after 6 days?
   b. When will only 10 g be left?

63. The shape of a tank is such that when it is filled to a depth of h feet, it contains $V = 9\pi h^3$ ft^3 of water. The tank is being drained through a sharp-edged circular hole of radius 1 in. If the tank is originally filled to a depth of 4 ft, how long does it take for the tank to empty?

64. A rectangular tank has a square base 4 ft on a side and is 10 ft high. Originally, it was filled with water to a depth of 6 feet, but now is being drained from the bottom of the tank through a sharp-edged square hole 1 in. on a side.

  a. Find an equation involving the rate dh/dt.

  b. How long will it take to drain the tank?

65. The radioactive substance neptunium-139 decays to 73.36% of its original amount after 24 hours. How long would it take for 43% of the original neptunium to be present? What is the half-life of neptunium-139?

66. A certain artifact is tested by carbon dating and found to contain 73% of its original carbon-14. As a cross-check, it is also dated using radium and is found to contain 32% of the original amount. Assuming the dating procedures are accurate, what is the half-life of radium?

67. The radioactive isotope gallium-67 (symbol ^{67}Ga) used in the diagnosis of malignant tumors has a half-life of 46.5 hours. If we start with 100 mg of ^{67}Ga, what percent is lost between the 30th and 35th hours? Is this the same as the percent lost over any other 5-hour period?

C **68. Modeling Problem**   A projectile is launched from the surface of a planet whose radius is R and where the acceleration due to gravity at the surface is g.

  a. If the initial velocity v_0 of the projectile satisfies $v_0 < \sqrt{2gR}$, show that the maximum height above the surface of the planet reached by the projectile is

$$h = \frac{v_0^2 R}{2gR - v_0^2}$$

  b. On a certain planet, it is known that $g = 25$ ft/s^2. A projectile is fired with an initial velocity of $v_0 = 2$ mi/s and attains a maximum height of 450 mi. What is the radius of the planet?

69. A projectile is launched from the surface of a planet whose radius is R and where the constant acceleration due to gravity is g. According to Newton's law of gravitation, the force of gravity acting on a projectile of mass m has magnitude

$$F = \frac{mk}{s^2}$$

where k is a constant. If this is the only force acting on the projectile, then $F = ma$, where a is the acceleration of the projectile. Show that

$$a = \frac{-gR^2}{s^2}$$

70. Modeling Problem   In physics, it is shown that the amount of heat Q (calories) that flows through an object by conduction will satisfy the differential equation

$$\frac{dQ}{dt} = -kA\frac{dT}{ds}$$

where t (seconds) is the time of flow, k is a physical constant (the *thermal conductivity* of the object), A is the surface area of the object measured at right angles to the direction of flow, and T is the temperature at a point s centimeters within the object, measured in the direction of the flow, as shown in Figure 5.29.

Figure 5.29   Heat conduction through an object

Under certain conditions (equilibrium), the rate of heat flow dQ/dt will be constant. Assuming these conditions exist, find the number of calories that will flow each second across the face of a square pane of glass 2 cm thick and 50 cm on a side if the temperature on one side of the pane is 5°C and on the other side is 60°C. The thermal conductivity of glass is approximately $k = 0.0025$.

5.7   The Mean Value Theorem for Integrals; Average Value

IN THIS SECTION   mean value theorem for integrals, modeling average value of a function

MEAN VALUE THEOREM FOR INTEGRALS

In Section 4.2 we established a very useful theoretical tool called the mean value theorem, which said that under reasonable conditions, there is at least one number c in the interval (a, b) such that

$$\frac{f(b) - f(a)}{b - a} = f'(c)$$

The mean value theorem for integrals is similar, and in the special case where $f(x) \geq 0$ it has a geometric interpretation that makes the theorem easy to understand.

In particular, the theorem says that it is possible to find at least one number c on the interval (a, b) such that the area of the rectangle with height $f(c)$ and width $(b - a)$ has exactly the same area as the region under the curve $y = f(x)$ on $[a, b]$. This is illustrated in Figure 5.30.

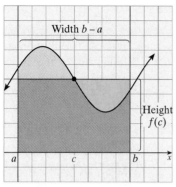

Width $b - a$

Height $f(c)$

$a \quad c \quad b \quad x$

The shaded rectangle has the same area as the region under the curve $y = f(x)$ on $[a, b]$.

Figure 5.30 Geometric interpretation of the mean value theorem for integrals

THEOREM 5.11   Mean value theorem for integrals

If f is continuous on the interval $[a, b]$, then there is at least one number c between a and b such that

$$\int_a^b f(x)\,dx = f(c)(b - a)$$

Proof   Suppose M and m are the absolute maximum and the absolute minimum of f, respectively, on $[a, b]$. This means that

$$m \le \qquad f(x) \qquad \le M \quad \text{when } a \le x \le b$$

$$\int_a^b m\,dx \le \quad \int_a^b f(x)\,dx \quad \le \int_a^b M\,dx \qquad \text{Dominance rule}$$

$$m(b - a) \le \quad \int_a^b f(x)\,dx \quad \le M(b - a)$$

$$m \le \frac{1}{b - a}\int_a^b f(x)\,dx \le M$$

Because f is continuous on the closed interval $[a, b]$ and because the number

$$I = \frac{1}{b - a}\int_a^b f(x)\,dx$$

lies between m and M, the intermediate value theorem (Theorem 2.6 of Section 2.3) says that there exists a number c between a and b for which $f(c) = I$; that is,

$$\frac{1}{b - a}\int_a^b f(x)\,dx = f(c)$$

$$\int_a^b f(x)\,dx = f(c)(b - a) \qquad \square$$

The mean value theorem for integrals does not specify how to determine c. It simply guarantees the existence of at least one number c in the interval. However, Example 1 shows how to find a value of c guaranteed by this theorem for a particular function and interval.

EXAMPLE 1   Finding c using the mean value theorem for integrals

Find a value of c guaranteed by the mean value theorem for integrals for $f(x) = \sin x$ on $[0, \pi]$.

Solution

$$\int_0^\pi \sin x\,dx = -\cos x \big|_0^\pi = -\cos \pi + \cos 0 = -(-1) + 1 = 2$$

The region bounded by f and the x-axis on $[0, \pi]$ is shaded in Figure 5.31.

The mean value theorem for integrals asserts the existence of a number c on $[0, \pi]$ such that $f(c)(b - a) = 2$. We can solve this equation to find this value:

$$f(c)(b - a) = 2$$

$$(\sin c)(\pi - 0) = 2$$

$$\sin c = \frac{2}{\pi}$$

$$c \approx 0.690107 \quad \text{or} \quad 2.451486$$

Because each choice of c is between 0 and π, we have found two values of c guaranteed by the mean value theorem for integrals.   ∎

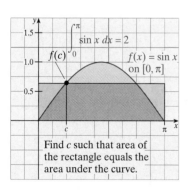

$\int_0^\pi \sin x\,dx = 2$

$f(c)$

$f(x) = \sin x$ on $[0, \pi]$

$c \qquad \pi \quad x$

Find c such that area of the rectangle equals the area under the curve.

Figure 5.31 Graph of $f(x) = \sin x$ illustrating the mean value theorem on $[0, \pi]$

MODELING AVERAGE VALUE OF A FUNCTION

There are many practical situations in which one is interested in the *average value* of a continuous function on an interval, such as the average level of air pollution over a 24-hour period, the average speed of a truck during a 3-hour trip, or the average productivity of a worker during a production run.

You probably know that the average value of n numbers $x_1, x_2, \ldots, x_n$ is

$$\frac{x_1 + x_2 + \cdots + x_n}{n}$$

but what if there are infinitely many numbers? Specifically, what is the average value of $f(x)$ on the interval $a \leq x \leq b$? To see how the definition of finite average value can be used, imagine that the interval $[a, b]$ is divided into n equal subintervals, each of width

$$\Delta x = \frac{b - a}{n}$$

For $k = 1, 2, \ldots, n$, let x_k^* be a number chosen arbitrarily from the kth subinterval. Then the average value AV of f on $[a, b]$ is estimated by the sum

$$S_n = \frac{f(x_1^*) + f(x_2^*) + \cdots + f(x_n^*)}{n} = \frac{1}{n} \sum_{k=1}^{n} f(x_k^*)$$

Because $\Delta x = \dfrac{b - a}{n}$, we know that $\dfrac{1}{n} = \dfrac{1}{b - a} \Delta x$ and

$$S_n = \frac{1}{n} \sum_{k=1}^{n} f(x_k^*) = \left[\frac{1}{b - a} \Delta x \right] \sum_{k=1}^{n} f(x_k^*) = \frac{1}{b - a} \sum_{k=1}^{n} f(x_k^*) \Delta x$$

The sum on the right is a Riemann sum with norm $\|P\| = \dfrac{b - a}{n}$. It is reasonable to expect the estimating average S_n to approach the "true" average value AV of $f(x)$ on $[a, b]$ as $n \to +\infty$. Thus, we model average value by

$$AV = \lim_{n \to +\infty} \frac{1}{b - a} \sum_{k=1}^{n} f(x_k^*) \Delta x = \frac{1}{b - a} \int_a^b f(x) \, dx$$

We use this integral as the definition of average value.

Average Value

> If f is continuous on the interval $[a, b]$, the **average value** (AV) of f on this interval is given by the integral
>
> $$AV = \frac{1}{b - a} \int_a^b f(x) \, dx$$

EXAMPLE 2   Modeling average speed of traffic

Suppose a study suggests that between the hours of 1:00 P.M. and 4:00 P.M. on a normal weekday the speed of the traffic at a certain expressway exit is modeled by the formula

$$S(t) = 2t^3 - 21t^2 + 60t + 20$$

kilometers per hour, where t is the number of hours past noon. Compute the average speed of the traffic between the hours of 1:00 P.M. and 4:00 P.M.

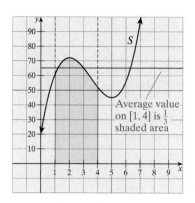

Figure 5.32 Graph of S and the average value

Solution

Our goal is to find the average value of $S(t)$ on the interval $[1, 4]$. The average speed is

$$\frac{1}{4-1} \int_1^4 (2t^3 - 21t^2 + 60t + 20)\, dt = \frac{1}{3} \left[\frac{1}{2}t^4 - 7t^3 + 30t^2 + 20t \right]_1^4$$

$$= \frac{1}{3}(240 - 43.5) = 65.5$$

The function, as well as the average value, is shown in Figure 5.32. ■

According to the mean value theorem for integrals, we have

$$\frac{1}{b-a} \int_a^b f(x)\, dx = f(c)$$

which says that the *average value of a continuous function f on $[a, b]$ equals the value of f for at least one number c between a and b.* This is quite reasonable since the intermediate value theorem for continuous functions assures us that a continuous function f assumes every value between its maximum M and minimum m, and we would expect the average value to be between these two extremes. Example 3 illustrates one way of using these ideas.

EXAMPLE 3 Modeling average temperature

Suppose that during a typical winter day in Minneapolis, the temperature (in degrees Celsius) x hours after midnight is modeled by the formula

$$T(x) = 2 - \tfrac{1}{7}(x - 13)^2$$

A graph of this formula is shown in Figure 5.33. Find the average temperature over the time period from 2:00 A.M. to 2:00 P.M., and find a time when the average temperature actually occurs.

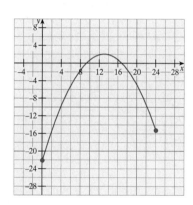

Figure 5.33 Graph of temperatures in Minneapolis over a 24-hour period

Solution

We wish to find the average temperature T on the interval $[2, 14]$ (because 2 P.M. is 14 hours after midnight). The average value is

$$T = \frac{1}{14-2} \int_2^{14} \left[2 - \frac{1}{7}(x-13)^2 \right] dx = \frac{1}{12} \left[2x - \frac{1}{7} \cdot \frac{1}{3}(x-13)^3 \right]\Big|_2^{14}$$

$$= \frac{1}{12}\left[\frac{587}{21} - \frac{1,415}{21} \right] \approx -3.2857143$$

Thus, the average temperature on the given time period is approximately $3.3°$C below zero. To determine when this temperature actually occurs, solve the equation

AVERAGE TEMPERATURE = TEMPERATURE AT TIME x

$$-3.2857143 = 2 - \tfrac{1}{7}(x - 13)^2$$

$$37 = (x - 13)^2$$

$$x = 13 \pm \sqrt{37} \approx 19.082763 \quad \text{or} \quad 6.9172375$$

The first value is to be rejected because it is not in the interval $[2, 14]$, so we find that the average temperature occurs 6.917 hr after midnight, or at approximately 6:55 A.M. ■

EXAMPLE 4   Modeling average population

The logistic formula

$$P(t) = \frac{202.31}{1 + e^{3.938 - 0.314t}}$$

was developed by the United States Bureau of the Census to represent the population of the United States (in millions) during the period 1790–1990. Time t in the formula is the number of decades after 1790. Thus, $t = 0$ for 1790, $t = 20$ for 1990. Use this formula to compute the average population of the United States between 1790 and 1990. When did the average population actually occur?

Solution

The average population is given by the integral

$$A = \frac{1}{20 - 0} \int_0^{20} \frac{202.31\, dt}{1 + e^{3.938 - 0.314t}} \qquad \text{Multiply by } \frac{e^{0.314t}}{e^{0.314t}}.$$

$$= \frac{1}{20} \int_0^{20} \frac{202.31 e^{0.314t}}{e^{0.314t} + e^{3.938}}\, dt \qquad \begin{array}{l} \text{Let } u = e^{0.314t} + e^{3.938}; \\ du = 0.314 e^{0.314t}\, dt. \end{array}$$

$$= \frac{1}{20} \left(\frac{202.31}{0.314} \right) \int_{t=0}^{t=20} \frac{du}{u}$$

$$= \frac{1}{20} \left(\frac{202.31}{0.314} \right) \ln \left| e^{0.314t} + e^{3.938} \right| \Big|_0^{20} \approx 77.7827445296$$

To find when the average population of 77.7827 million actually occurred, solve $P(t) = 77.7827$:

$$\frac{202.31}{1 + e^{3.938 - 0.314t}} = 77.7827$$

$$e^{3.938 - 0.314t} = \frac{202.31}{77.7827} - 1$$

$$3.938 - 0.314t = \ln \left(\frac{202.31}{77.7827} - 1 \right)$$

$$t \approx 11.043$$

The average population occurred approximately 11 decades after 1790, around the year 1900. ∎

5.7   PROBLEM SET

A *In Problems 1–10, find c such that*

$$\int_a^b f(x)\, dx = f(c)(b - a)$$

as guaranteed by the mean value theorem for integrals. If you cannot find such a value, explain why the theorem does not apply.

1. $f(x) = 4x^3$ on $[1, 2]$ **2.** $f(x) = x^2 + 4x + 1$ on $[0, 2]$

3. $f(x) = 15x^{-2}$ on $[1, 5]$ **4.** $f(x) = 12x^{-3}$ on $[-3, 3]$

5. $f(x) = 2\csc^2 x$ on $\left[-\frac{\pi}{3}, \frac{\pi}{3} \right]$ **6.** $f(x) = \cos x$ on $\left[-\frac{\pi}{2}, \frac{\pi}{2} \right]$

7. $f(x) = e^{2x}$ on $\left[-\frac{1}{2}, \frac{1}{2} \right]$ **8.** $f(x) = \dfrac{x}{1 + x}$ on $[0, 1]$

9. $f(x) = \dfrac{x + 1}{1 + x^2}$ on $[-1, 1]$ **10.** $f(x) = \tan x$ on $[0, 2]$

Determine the area of the indicated region in Problems 11–16 and then draw a rectangle with base $(b - a)$ and height $f(c)$ for some c on $[a, b]$ so that the area of the rectangle is equal to the area of the given region.

11. $y = \frac{1}{2} x$ on $[0, 10]$ **12.** $y = x^2$ on $[0, 3]$

13. $y = x^2 + 2x + 3$ on $[0, 2]$ **14.** $y = \dfrac{1}{x^2}$ on $[0.5, 2]$

15. $y = \cos x$ on $[-1, 1.5]$ **16.** $y = x + \sin x$ on $[0, 10]$

Find the average value of the function given in Problems 17–30 on the prescribed interval.

17. $f(x) = x^2 - x + 1$ on $[-1, 2]$

18. $f(x) = x^3 - 3x^2$ on $[-2, 1]$

19. $f(x) = e^x - e^{-x}$ on $[-1, 1]$

20. $f(x) = \dfrac{x}{2x + 3}$ on $[0, 1]$

21. $f(x) = \sin x$ on $\left[0, \frac{\pi}{4}\right]$

22. $f(x) = 2 \sin x - \cos x$ on $\left[0, \frac{\pi}{2}\right]$

23. $f(x) = \sqrt{4 - x}$ on $[0, 4]$

24. $f(x) = \sqrt[3]{1 - x}$ on $[-7, 0]$

25. $f(x) = (2x - 3)^3$ on $[0, 1]$

26. $f(x) = x\sqrt{2x^2 + 7}$ on $[0, 1]$

27. $f(x) = x(x^2 + 1)^3$ on $[-2, 1]$

28. $f(x) = \dfrac{x}{\sqrt{x^2 + 1}}$ on $[0, 3]$

29. $f(x) = \sqrt{9 - x^2}$ on $[-3, 3]$ *Hint: The integral can be evaluated as the area of part of a circle.*

30. $f(x) = \sqrt{2x - x^2}$ on $[0, 2]$ *Hint: The integral can be evaluated as the area of part of a circle.*

B **31.** If an object is propelled upward from ground level with an initial velocity v_0, then its height at time t is given by

$$s = -\tfrac{1}{2}gt^2 + v_0 t$$

where g is the constant acceleration due to gravity. Show that between times t_0 and t_1, the average height of the object is

$$s = -\tfrac{1}{6}g[t_1^2 + t_1 t_0 + t_0^2] + \tfrac{1}{2}v_0(t_1 + t_0)$$

32. What is the average velocity for the object described in Problem 31 during the same time period?

33. Records indicate that t hours past midnight, the temperature at the local airport was

$$f(t) = -0.1t^2 + t + 50$$

degrees Fahrenheit. What was the average temperature at the airport between 9:00 A.M. and noon?

34. Suppose a study indicates that t years from now, the level of carbon dioxide in the air of a certain city will be

$$L(t) = te^{-0.01t^2}$$

parts per million (ppm) for $0 \le t \le 20$.

  a. What is the average level of carbon dioxide in the first 3 years?

  b. At what time (or times) does the average level of carbon dioxide actually occur? Answer to the nearest month.

35. The number of bacteria (in thousands) present in a certain culture after t minutes is modeled by

$$Q(t) = \frac{2,000}{1 + 0.3e^{-0.276t}}$$

  a. What was the average population during the *second* ten minutes ($10 \le t \le 20$)?

  b. At what time during the period $10 \le t \le 20$ is the average population actually attained?

C **36.** Let $f(t)$ be a function that is continuous and satisfies $f(t) \ge 0$ on the interval $\left[0, \frac{\pi}{2}\right]$. Suppose it is known that for any number x between 0 and $\frac{\pi}{2}$, the region under the graph of f on $[0, x]$ has area $A(x) = \tan x$.

  a. Explain why $\displaystyle\int_0^x f(t)\, dt = \tan x$ for $0 \le x \le \frac{\pi}{2}$.

  b. Differentiate both sides of the equation in part **a** and deduce the identity of f.

37. Exploration Problem Suppose that $f(t)$ is continuous for all t and that for any number x it is known that the average value of f on $[-1, x]$ is

$$A(x) = \sin x$$

Use this information to deduce the identity of f.

38. Modeling Problem In Example 7 of Section 4.7, we gave the Heinz function

$$f(t) = \frac{k}{b - a}(e^{-at} - e^{-bt}) \qquad t \ge 0$$

where $f(t)$ is the concentration of a drug in a person's bloodstream t hours after an intramuscular injection. The coefficients a and b ($b > a$) are characteristics of the drug and the patient's metabolism and are called the absorption and diffusion rates, respectively.

  a. Show that for each fixed t, $f(t)$ can be thought of as the average value of a function of the form

$$g(\lambda) = (At^2 + Bt + C)e^{-\lambda t}$$

over the interval $a \le \lambda \le b$. Find A, B, and C.

  b. Exploration Problem So what? In particular, can you see any value in the interpretation of the Heinz concentration function as an average value? Explain.

5.8 Numerical Integration: The Trapezoidal Rule and Simpson's Rule

IN THIS SECTION approximation by rectangles, trapezoidal rule, Simpson's rule, error estimation

The fundamental theorem of calculus can be used to evaluate an integral whenever an appropriate antiderivative is known. However, certain functions, such as $f(x) = e^{x^2}$, have no simple antiderivatives. To find a definite integral of such a function, it is often necessary to use numerical approximation.

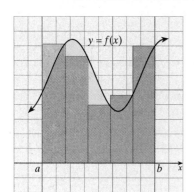

Figure 5.34 Approximation by rectangles

APPROXIMATION BY RECTANGLES

If $f(x) \geq 0$ on the interval $[a, b]$, the definite integral $\int_a^b f(x)\,dx$ is equal to the area under the graph of f on $[a, b]$. As we saw in Section 5.2, one way to approximate this area is to use n rectangles, as shown in Figure 5.34. In particular, divide the interval $[a, b]$ into n subintervals, each of width $\Delta x = \dfrac{b - a}{n}$, and let x_k^* denote the right end-point of the kth subinterval. The base of the kth rectangle is the kth subinterval, and its height is $f(x_k^*)$. Hence, the area of the kth rectangle is $f(x_k^*)\Delta x$. The sum of the areas of all n rectangles is an approximation for the area under the curve and hence an approximation for the corresponding definite integral. Thus,

$$\int_a^b f(x)\,dx \approx f(x_1^*)\Delta x + f(x_2^*)\Delta x + \cdots + f(x_n^*)\Delta x$$

This approximation improves as the number of rectangles increases, and we can estimate the integral to any desired degree of accuracy by taking n large enough. However, because fairly large values of n are usually required to achieve reasonable accuracy, approximation by rectangles is rarely used in practice.

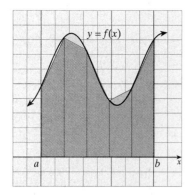

Figure 5.35 Approximation by trapezoids

TRAPEZOIDAL RULE

The accuracy of the approximation generally improves significantly if trapezoids are used instead of rectangles. Figure 5.35 shows the area from Figure 5.34 approximated by n trapezoids instead of rectangles. Even from these rough illustrations you can see how much better the approximation is in this case.

Suppose the interval $[a, b]$ is partitioned into n equal parts by the subdivision points $x_0, x_1, \ldots, x_n$, where $x_0 = a$ and $x_n = b$. The kth trapezoid is shown in greater detail in Figure 5.36. Recall the formula for the area of a trapezoid:

$$A = \tfrac{1}{2}(b_1 + b_2)h$$

If we let T_n denote the sum of the areas of n trapezoids, we see that

$$T_n = \tfrac{1}{2}[f(x_0) + f(x_1)]\Delta x + \tfrac{1}{2}[f(x_1) + f(x_2)]\Delta x + \cdots + \tfrac{1}{2}[f(x_{n-1}) + f(x_n)]\Delta x$$
$$= \tfrac{1}{2}[f(x_0) + 2f(x_1) + 2f(x_2) + \cdots + 2f(x_{n-1}) + f(x_n)]\Delta x$$

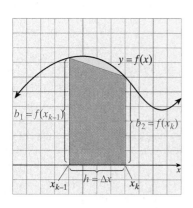

Figure 5.36 The kth trapezoid has area $\tfrac{1}{2}[f(x_{k-1}) + f(x_k)]\Delta x$.

The sum T_n estimates the total area under the curve $y = f(x)$ on the interval $[a, b]$ and hence also estimates the integral

$$\int_a^b f(x)\,dx$$

This approximation formula is known as the *trapezoidal rule* and applies as a means of approximating the integral, even if the function f is not positive.

> **Trapezoidal Rule**
>
> Let f be continuous on $[a, b]$. The **trapezoidal rule** is
>
> $$\int_a^b f(x)\,dx \approx \tfrac{1}{2}[f(x_0) + 2f(x_1) + 2f(x_2) + \cdots + 2f(x_{n-1}) + f(x_n)]\Delta x$$
>
> where $\Delta x = \dfrac{b-a}{n}$ and, for the kth subinterval, $x_k = a + k\Delta x$. Moreover, the larger the value for n, the better the approximation.

Our first example uses the trapezoidal rule to estimate the value of an integral that we can compute exactly by using the fundamental theorem.

EXAMPLE 1   Trapezoidal rule approximation

Use the trapezoidal rule with $n = 4$ to estimate $\displaystyle\int_{-1}^{2} x^2\,dx$.

Solution

The interval is $[a, b] = [-1, 2]$, so $a = -1$ and $b = 2$. Then $\Delta x = \dfrac{2-(-1)}{4} = \dfrac{3}{4} = 0.75$. Thus,

$$x_0 = a = -1 \qquad\qquad\qquad f(x_0) = f(-1) = (-1)^2 = 1$$

$$x_1 = a + 1 \cdot \Delta x = -1 + \tfrac{3}{4} = -\tfrac{1}{4} \qquad 2f(x_1) = 2\left(-\tfrac{1}{4}\right)^2 = \tfrac{1}{8} = 0.125$$

$$x_2 = a + 2 \cdot \Delta x = -1 + \tfrac{6}{4} = \tfrac{1}{2} \qquad 2f(x_2) = 2\left(\tfrac{1}{2}\right)^2 = \tfrac{1}{2} = 0.5$$

$$x_3 = a + 3 \cdot \Delta x = -1 + \tfrac{9}{4} = \tfrac{5}{4} \qquad 2f(x_3) = 2\left(\tfrac{5}{4}\right)^2 = \tfrac{25}{8} = 3.125$$

$$x_4 = a + 4 \cdot \Delta x = b = 2 \qquad\qquad f(x_4) = 2^2 = 4$$

$$T_4 = \tfrac{1}{2}[1 + 0.125 + 0.5 + 3.125 + 4](0.75) = 3.28125 \qquad\blacksquare$$

The exact value of the integral in Example 1 is

$$\int_{-1}^{2} x^2\,dx = \frac{x^3}{3}\Big|_{-1}^{2} = \frac{8}{3} - \frac{-1}{3} = 3$$

Therefore, the trapezoidal estimate T_4 involves an error, which we denote by E_4. We find that

$$E_4 = \int_{-1}^{2} x^2\,dx - T_4 = 3 - 3.28125 = -0.28125$$

The negative sign indicates that the trapezoidal formula *overestimated* the true value of the integral in Example 1.

Software for numerical evaluation of integrals is common (see the *Technology Manual*). The output for Example 1 is shown in the margin.

Trapezoidal Rule to calculate estimate		
Type of estimate	# of sub-intervals	Estimate over [-1, 2]
Trapezoid	4	3.28125
Trapezoid	10	3.045
Trapezoid	100	3.00045
Trapezoid	1000	3.000005

SIMPSON'S RULE

Roughly speaking, the accuracy of a procedure for estimating the area under a curve depends on how well the upper boundary of each approximating area fits the shape of the given curve. Trapezoidal strips often result in a better approximation than rectangular strips, and it is reasonable to expect even greater accuracy to occur if the approximating regions have curved upper boundaries, as shown in Figure 5.37.

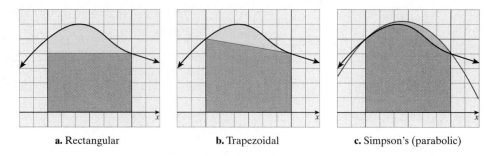

a. Rectangular   b. Trapezoidal   c. Simpson's (parabolic)

Figure 5.37 A comparison of approximation methods

The name given to the procedure in which the approximating strip has a parabolic arc for its upper boundary is called **Simpson's rule**.*

As with the trapezoidal rule, we will derive Simpson's rule as a means for approximating the area under the curve $y = f(x)$ on the interval $[a, b]$, where f is continuous and satisfies $f(x) \geq 0$. First, we partition the given interval into a number of equal subintervals, but this time, we require the number of subdivisions to be an *even* number (because this requirement will simplify the formula associated with the final result).

If $x_0, x_1, \ldots, x_n$ are the subdivision points in our partition (with $x_0 = a$ and $x_n = b$), we pass a parabolic arc through the points, three at a time (the points with x-coordinates x_0, x_1, x_2, then those with x_2, x_3, x_4, and so on). It can be shown (see Problem 42) that the region under the parabolic curve $y = f(x)$ on the interval $[x_{2k-2}, x_{2k}]$ has area

$$\tfrac{1}{3}[f(x_{2k-2}) + 4f(x_{2k-1}) + f(x_{2k})]\Delta x$$

where $\Delta x = \dfrac{b-a}{n}$. This procedure is illustrated in Figure 5.38.

Figure 5.38 Approximation using parabolas

By adding the area of the approximating parabolic strips and combining terms, we obtain the sum S_n of n parabolic regions:

$$S_n = \tfrac{1}{3}[f(x_0) + 4f(x_1) + f(x_2)]\Delta x + \tfrac{1}{3}[f(x_2) + 4f(x_3) + f(x_4)]\Delta x$$
$$+ \cdots + \tfrac{1}{3}[f(x_{n-2}) + 4f(x_{n-1}) + f(x_n)]\Delta x$$
$$= \tfrac{1}{3}[f(x_0) + 4f(x_1) + 2f(x_2) + 4f(x_3) + \cdots + 4f(x_{n-1}) + f(x_n)]\Delta x$$

These observations are summarized in the following box:

---

*This rule is named for Thomas Simpson (1710–1761), an English mathematician who, curiously, neither discovered nor made any special use of the formula that bears his name.

Simpson's Rule

Let f be continuous on $[a, b]$. **Simpson's rule** is

$$\int_a^b f(x)\,dx \approx \tfrac{1}{3}[f(x_0) + 4f(x_1) + 2f(x_2) + 4f(x_3) + 2f(x_4) + \cdots$$

$$+ 4f(x_{n-1}) + f(x_n)]\Delta x$$

where $\Delta x = \dfrac{b - a}{n}$, $x_k = a + k\Delta x$, k an integer and n an even integer. Moreover, the larger the value for n, the better the approximation.

EXAMPLE 2   Approximation by Simpson's rule

Use Simpson's rule with $n = 10$ to approximate $\displaystyle\int_1^2 \dfrac{dx}{x}$.

Solution

We have $\Delta x = \dfrac{2 - 1}{10} = 0.1$, and $x_0 = a = 1$, $x_1 = 1.1$, $x_2 = 1.2, \ldots, x_9 = 1.9$, $x_{10} = b = 2$.

$$\int_1^2 \frac{1}{x}\,dx \approx S_{10}$$

$$= \frac{1}{3}\left(\frac{1}{1} + \frac{4}{1.1} + \frac{2}{1.2} + \frac{4}{1.3} + \frac{2}{1.4} + \frac{4}{1.5} + \frac{2}{1.6} + \frac{4}{1.7} + \frac{2}{1.8} + \frac{4}{1.9} + \frac{1}{2}\right)(0.1)$$

$$\approx 0.6931502$$

This estimate compares well with the value found directly by applying the fundamental theorem of calculus:

$$\int_1^2 \frac{dx}{x} = \ln 2 - \ln 1 \approx 0.6931472$$

An example of the output of a computer program for this type of problem is shown in the margin.   ∎

Comparison of Simpson's and trapezoidal rules		
Type of estimate	# of sub-intervals	Estimate over [1, 2]
Trapezoid	10	.693771403175
Simpson	10	.693150230689
Trapezoid	100	.693153430482
Simpson	100	.693147180872
Trapezoid	1000	.693147243060
Simpson	1000	.693147100560

Comparison of Simpson's and trapezoidal rules

ERROR ESTIMATION

The difference between the value of an integral and its estimated value is called its error. Since this error is a function of n we denote it by E_n.

THEOREM 5.12   Error in the trapezoidal rule and Simpson's rule

If f has a continuous second derivative on $[a, b]$, then the error E_n in approximating $\int_a^b f(x)\,dx$ by the trapezoidal rule satisfies the following:

Trapezoidal error: $|E_n| \leq \dfrac{(b - a)^3}{12n^2} M$, where M is the maximum value of $\left|f''(x)\right|$ on $[a, b]$.

Moreover, if f has a continuous fourth derivative on $[a, b]$, then the error E_n (n even) in approximating $\int_a^b f(x)\,dx$ by Simpson's rule satisfies the following:

Simpson's error: $|E_n| \leq \dfrac{(b - a)^5}{180n^4} K$, where K is the maximum value of $\left|f^{(4)}(x)\right|$ on $[a, b]$.

Proof   The proofs of these error estimates are beyond the scope of this book and can be found in many advanced calculus textbooks and in most numerical analysis books.   ❑

EXAMPLE 3 **Estimate of error when using Simpson's rule**

Estimate the accuracy of the approximation of $\int_1^2 \dfrac{dx}{x}$ by Simpson's rule with $n = 10$ in Example 2.

Solution

If $f(x) = \dfrac{1}{x}$, we find that $f^{(4)}(x) = 24x^{-5}$. The maximum value of this function will occur at a critical number (there is none on $[1, 2]$) or at an endpoint. Thus, the largest value of $\left|f^{(4)}(x)\right|$ on $[1, 2]$ is $\left|f^{(4)}(1)\right| = 24$. Now, apply the error formula with $K = 24$, $a = 1$, $b = 2$, and $n = 10$ to obtain

$$|E_{10}| \le \frac{k(b-a)^5}{180n^4} = \frac{24(2-1)^5}{180(10)^4} \approx 0.0000133$$

That is, the error in the approximation in Example 2 is guaranteed to be no greater than 0.0000133. ∎

With the aid of the error estimates, we can decide in advance how many subintervals to use to achieve a desired degree of accuracy.

EXAMPLE 4 **Choosing the number of subintervals to guarantee given accuracy**

How many subintervals are required to guarantee that the error will be correct to four decimal places in the approximation of

$$\int_1^2 \frac{dx}{x}$$

on $[1, 2]$ using the trapezoidal rule?

Solution

To be correct to four decimal places means that $|E_n| < 0.00005$. Because $f(x) = x^{-1}$, we have $f''(x) = 2x^{-3}$. On $[1, 2]$ the largest value of $\left|f''(x)\right|$ is $\left|f''(1)\right| = 2$, so $M = 2$, $a = 1$, $b = 2$, and

$$|E_n| \le \frac{2(2-1)^3}{12n^2} = \frac{1}{6n^2}$$

The goal is to find the smallest positive integer n for which

$$\frac{1}{6n^2} < 0.00005$$

$$10{,}000 < 3n^2 \qquad \text{Multiply by the positive number } 60{,}000n^2.$$

$$10{,}000 - 3n^2 < 0$$

$$(100 - \sqrt{3}n)(100 + \sqrt{3}n) < 0$$

$$n < -\frac{100}{\sqrt{3}} \quad \text{or} \quad n > \frac{100}{\sqrt{3}} \approx 57.735$$

The smallest positive integer that satisfies this condition is $n = 58$; therefore, 58 subintervals are required to ensure the desired accuracy. ∎

If f is the linear function $f(x) = Ax + B$, then $f''(x) = 0$ and we can take $M = 0$ as the error estimate. In this case, the error in applying the trapezoidal rule satisfies $|E_n| \le 0$. That is, the trapezoidal rule is exact for a linear function, which is what we would expect, because the region under a line on an interval is a trapezoid.

In discussing the accuracy of the trapezoidal rule as a means of estimating the value of the definite integral

$$I = \int_a^b f(x)\,dx$$

we have focused attention on the "error term," but this only measures the error that comes from estimating I by the trapezoidal or Simpson approximation sum. There are other kinds of errors that must be considered in this or any other method of approximation. In particular, each time we cut off digits from a decimal, we incur what is known as a round-off error. For example, a round-off error occurs when we use 0.66666667 in place of $\frac{2}{3}$ or 3.1415927 for the number π. Round-off errors occur even in large computers and can accumulate to cause real problems. Specialized methods for dealing with these and other errors are studied in numerical analysis.

The examples of numerical integration examined in this section are intended as illustrations and thus involve relatively simple computations, whereas in practice, such computations often involve hundreds of terms and can be quite tedious. Fortunately, these computations are extremely well suited to automatic computing, and the reader who is interested in pursuing computer methods in numerical integration will find an introduction to this topic in the *Technology Manual*.

5.8   PROBLEM SET

A *Approximate the integrals in Problems 1–2 using the trapezoidal rule and Simpson's rule with the specified number of subintervals and then compare your answers with the exact value of the definite integral.*

1. $\int_1^2 x^2\, dx$ with $n = 4$ **2.** $\int_0^4 \sqrt{x}\, dx$ with $n = 6$

Approximate the integrals given in Problems 3–8 with the specified number of subintervals using

   a. *the trapezoidal rule*   **b.** *Simpson's rule*

3. $\int_0^1 \dfrac{dx}{1 + x^2}$ with $n = 4$

4. $\int_{-1}^0 \sqrt{1 + x^2}\, dx$ with $n = 4$

5. $\int_2^4 \sqrt{1 + \sin x}\, dx$ with $n = 4$

6. $\int_0^2 x \cos x\, dx$ with $n = 6$

7. $\int_0^2 x e^{-x}\, dx$ with $n = 6$

8. $\int_1^2 \ln x\, dx$ with $n = 6$

B **9. WHAT DOES THIS SAY?**   Describe the trapezoidal rule.

10. WHAT DOES THIS SAY?   Describe Simpson's rule.

Estimate the value of the integrals in Problems 11–18 to within the prescribed accuracy.

11. $\int_0^1 \dfrac{dx}{x^2 + 1}$ with error less than 0.05. Use the trapezoidal rule.

12. $\int_{-1}^2 \sqrt{1 + x^2}\, dx$ with error less than 0.05. Use Simpson's rule.

13. $\int_0^1 \cos 2x\, dx$ accurate to three decimal places. Use Simpson's rule.

14. $\int_1^2 x^{-1}\, dx$ accurate to three decimal places. Use Simpson's rule.

15. $\int_0^2 x\sqrt{4 - x}\, dx$ with error less than 0.01. Use the trapezoidal rule.

16. $\int_0^\pi \theta \cos^2 \theta\, d\theta$ with error less than 0.01. Use Simpson's rule.

17. $\int_0^1 \tan^{-1} x\, dx$ with error less than 0.01. Use Simpson's rule.

18. $\int_0^\pi e^{-x} \sin x\, dx$ to three decimal places. Use Simpson's rule.

In Problems 19–23, determine how many subintervals are required to guarantee accuracy to within 0.00005 using

   a. *the trapezoidal rule*   **b.** *Simpson's rule*

19. $\int_1^3 x^{-1}\, dx$ **20.** $\int_{-1}^4 (x^3 + 2x^2 + 1)\, dx$

21. $\int_1^4 \dfrac{dx}{\sqrt{x}}$ **22.** $\int_0^2 \cos x\, dx$

23. $\int_1^2 \ln \sqrt{x}\, dx$ **24.** $\int_0^1 e^{-2x}\, dx$

25. A quarter-circle of radius 1 has the equation $y = \sqrt{1 - x^2}$ for $0 \le x \le 1$, which implies that

$$\int_0^1 \sqrt{1 - x^2}\, dx = \frac{\pi}{4}$$

   a. Estimate π correct to one decimal place by applying the trapezoidal rule to this integral.

   b. Estimate π correct to one decimal place by applying Simpson's rule to this integral.

26. Find the smallest value of n for which the trapezoidal rule estimates the value of the integral

$$\int_1^2 x^{-1}\, dx$$

with six-decimal-place accuracy.

27. The width of an irregularly shaped dam is measured at 5-m intervals, with the results indicated in Figure 5.39. Use the trapezoidal rule to estimate the area of the face of the dam.

Figure 5.39 Area of the face of a dam

28. Jack and Jill are traveling in a car with a broken odometer. In order to determine the distance they traveled between noon and 1:00 P.M., Jack (the passenger) takes a speedometer reading every 5 minutes.

Minutes (after noon)	0	5	10	15	20
Speedometer reading	54	57	50	51	55

Minutes	25	30	35	40	45	50	55	60
Speedometer	60	49	53	47	39	42	48	53

Use the trapezoidal rule to estimate the total distance traveled by the couple from noon to 1:00 P.M.

29. An industrial plant spills pollutant into a lake. The pollutant spread out to form the pattern shown in Figure 5.40. All distances are in feet.

Figure 5.40 Pollutant spill

Use Simpson's rule to estimate the area of the spill.

30. Apply the trapezoidal rule to estimate $\int_0^3 f(x)\,dx$, where the values for f are found on the following spreadsheet (table):

	A	B
1	x	$f(x)$
2	0	3.7
3	0.3	3.9
4	0.6	4.1
5	0.9	4.1
6	1.2	4.2
7	1.5	4.4
8	1.8	4.6
9	2.1	4.9
10	2.4	5.2
11	2.7	5.5
12	3	6

31. Apply Simpson's rule to estimate $\int_0^5 f(x)\,dx$, where the values for f are found on the following spreadsheet (table):

	A	B
1	x	$f(x)$
2	0	10
3	0.5	9.75
4	1	10
5	1.5	10.75
6	2	12
7	2.5	13.75
8	3	16
9	3.5	18.75
10	4	22
11	4.5	25.75
12	5	30

32. In this problem, we explore the "order of convergence" of three numerical integration methods. A method is said to have "order of convergence n^k" if $E_n \cdot n^k = C$, where n is the number of intervals in the approximation and k is a constant power. In other words, the error $E_n \to 0$ as $C/n^k \to 0$. In each of the following cases, use the fact that

$$I = \int_0^\pi \sin x\,dx = -\cos x\big|_0^\pi = 2$$

a. Use the trapezoidal rule to estimate I for $n = 10, 20, 40$, and 80. Compute the error E_n in each case and compute $E_n \cdot n^k$ for $k = 1, 2, 3, 4$. Based on your results, you should be able to conclude that the order of convergence of the trapezoidal approximation is n^2.

b. Repeat part **a** using Simpson's rule. Based on your results, what is the order of convergence for Simpson's rule?

c. Repeat part **a** using a rectangular approximation with right endpoints. What is the order of convergence for this method?

33. Let $I = \int_0^\pi (9x - x^3)\,dx$.

a. Estimate I using rectangles, the trapezoidal rule, and Simpson's rule, for $n = 10, 20, 40, 80$. Something interesting happens with Simpson's rule. Explain.

b. **Simpson struggles!** In contrast to part **a**, here is an example where Simpson's rule does not live up to expectations. For $n = 10, 20, 40$, and 80, use Simpson's rule and the rectangular rule (select midpoints) for this integral and make a table of errors. Then try to explain Simpson's poor performance. *Hint:* Look at the formula for the error.

$$\int_0^2 \sqrt{4 - x^2}\,dx = \pi$$

34. HISTORICAL QUEST The mathematician Seki Kōwa was born in Fujioka, Japan, the son of a samurai, but was adopted by a patriarch of the Seki family. Seki invented and used an early form of determinants for solving systems of equations, and he also invented a method for approximating areas that is very similar to the rectangular

TAKAKAZU SEKI KŌWA
1642–1708

method introduced in this section. This method, known as the yenri (circle principle), found the area of a circle by dividing the circle into small rectangles, as shown in Figure 5.41.

Figure 5.41   Early Asian calculus

The sample shown in Figure 5.41 was drawn by a student of Seki Kōwa. ∎

For this quest, draw a circle with radius 10 cm. Draw vertical chords through each centimeter on a diameter (you should have 18 rectangles). Measure the heights of the rectangles and approximate the area of the circle by adding the areas of the rectangles. Compare this with the formula for the area of this circle.

35. HISTORICAL QUEST   In 1670, a predecessor of Seki Kōwa (see Problem 34), Kazuyuki Sawaguchi wrote seven volumes that concluded with fifteen problems that he believed were unsolvable. In 1674, Seki Kōwa published solutions to all fifteen of Kazuyuki's unsolvable problems. One of the "unsolvable" problems was the following: Three circles are inscribed in a circle, each tangent to the other two and to the original circle. All three cover all but 120 square units of the circumscribing circle. The diameters of the two smaller inscribed circles are equal, and each is five units less than the diameter of the larger inscribed circle. Find the diameters of the three inscribed circles.

The solution to this "impossible" problem is beyond the scope of this course (it involves solving a 6th degree equation with a horrendous amount of computation), but we can replace it with a simpler problem: Consider two circles inscribed in a circle, each tangent to the other and to the original circle. The diameter of the smaller circle is five units less than the diameter of the larger inscribed circle. The sum of the areas of the larger circle and twice the smaller circle cover all but 120 square units of the circumscribing circles. Find the diameters of the two inscribed circles.

36. HISTORICAL QUEST   Isaac Newton (see Historical Quest essay at the end of Chapter 1 on page 48) invented a preliminary version of Simpson's rule. In 1779, Newton wrote an article to an addendum to *Methodus Differentials* (1711) in which he gave the following example: If there are four ordinates at equal intervals, let A be the sum of the first and fourth, B the sum of the second and third, and R the interval

ROGER COTES
1682–1716

between the first and fourth; then ... the area between the first and fourth ordinates is approximated by $\frac{1}{8}(A + 3B)R$. This is known today as the "Newton-Cotes three-eighths rule," which can be expressed in the form

$$\int_{x_0}^{x_3} f(x)\, dx \approx \tfrac{3}{8}(y_0 + 3y_1 + 3y_2 + y_3)\Delta x$$

Roger Cotes and James Stirling (1692–1770) both knew this formula, as well as what we called in this section Simpson's rule. In 1743, this rule was rediscovered by Thomas Simpson (1710–1761). ∎

Estimate the integral

$$\int_0^3 \tan^{-1} x\, dx$$

using the Newton-Cotes three-eighths rule, and then compare with approximation using left endpoints (rectangles) and trapezoids with $n = 4$. Which of the three rules gives the most accurate estimate?

37. Show that if $p(x)$ is any polynomial of degree less than or equal to 3, then

$$\int_a^b p(x)\, dx = \frac{b-a}{6}\left[p(a) + 4p\left(\frac{a+b}{2}\right) + p(b)\right]$$

This result is often called the *prismoidal rule*.

38. Use the prismoidal rule (Problem 37) to evaluate

$$\int_{-1}^2 (x^3 - 3x + 4)\, dx$$

39. Use the prismoidal rule (Problem 37) to evaluate

$$\int_{-1}^3 (x^3 + 2x^2 - 7)\, dx$$

40. Let $p(x)$ be a polynomial of degree at most 3.
    a. Find a number c between 0 and 1 such that

$$\int_{-1}^1 p(x)\, dx = p(c) + p(-c)$$

    b. Find a number c between $-\frac{1}{2}$ and $\frac{1}{2}$ such that

$$\int_{-1/2}^{1/2} p(x)\, dx = \tfrac{1}{3}[p(-c) + p(0) + p(c)]$$

41. Let $p(x) = Ax^3 + Bx^2 + Cx + D$ be a cubic polynomial. Show that Simpson's rule gives the exact value for

$$\int_a^b p(x)\, dx$$

42. The object of this exercise is to prove Simpson's rule for the special case involving three points.
    a. Let $P_1(-h, f(-h))$, $P_2(0, f(0))$, $P_3(h, f(h))$. Find the equation of the form $y = Ax^2 + Bx + C$ for the parabola through the points P_1, P_2, and P_3.

b. If $y = p(x)$ is the quadratic function found in part **a**, show that

$$\int_{-h}^{h} p(x)\, dx = \frac{h}{3}[p(-h) + 4p(0) + p(h)]$$

c. Let $Q_1(x_1, f(x_1))$, $Q_2(x_2, f(x_2))$, $Q_3(x_3, f(x_3))$ be points with $x_2 = x_1 + h$, and $x_3 = x_1 + 2h$. Explain why

$$\int_{x_1}^{x_3} p(x)\, dx = \frac{h}{3}[p(x_1) + 4p(x_2) + p(x_3)]$$

5.9 An Alternative Approach: The Logarithm as an Integral

IN THIS SECTION   natural logarithm as an integral, geometric interpretation, the natural exponential function

NATURAL LOGARITHM AS AN INTEGRAL

You may have noticed in Section 2.4 that we did not prove the properties of exponential functions for all real number exponents. To treat exponentials and logarithms *rigorously*, we use the alternative approach provided in this section. Specifically, we use a definite integral to introduce the *natural logarithmic function* and then use this function to *define* the *natural exponential function*.

> "When I use a word," Humpty Dumpty said, "it means just what I choose it to mean—nothing more or less."
> ——Lewis Carroll (1832–1898)

Natural Logarithm

> The **natural logarithm** is the function defined by
> $$\ln x = \int_1^x \frac{dt}{t} \qquad x > 0$$

At first glance, it appears there is nothing "natural" about this definition, but if this integral function has the properties of a logarithm, why should we not call it a logarithm? We begin with a theorem that shows that $\ln x$ does indeed have the properties we would expect of a logarithm.

THEOREM 5.13 Properties of a logarithm as defined by $\ln x = \displaystyle\int_1^x \frac{dt}{t}$

Let $x > 0$ and $y > 0$ be positive numbers. Then

a. $\ln 1 = 0$

b. $\ln xy = \ln x + \ln y$

c. $\ln \dfrac{x}{y} = \ln x - \ln y$

d. $\ln x^p = p \ln x$ for all rational numbers p

Proof

a. Let $x = 1$. Then $\displaystyle\int_1^1 \frac{dt}{t} = 0$.

b. For fixed positive numbers x and y, we use the additive property of integrals as follows:

$$\ln(xy) = \int_1^{xy} \frac{dt}{t} = \int_1^x \frac{dt}{t} + \int_x^{xy} \frac{dt}{t}$$

$$= \ln x + \int_1^y \frac{x\, du}{ux}$$

$$= \ln x + \int_1^y \frac{du}{u}$$

$$= \ln x + \ln y$$

> Let $u = \dfrac{t}{x}$, so $t = ux$;
> $dt = x\, du$.
> If $t = x$, then $u = 1$;
> if $t = xy$, then $u = y$.

c. and **d.** The proofs are outlined in the problem set.   ❏

GEOMETRIC INTERPRETATION

An advantage of defining the logarithm by the integral formula is that calculus can be used to study the properties of $\ln x$ from the beginning. For example, note that if $x > 1$, the integral

$$\ln x = \int_1^x \frac{dt}{t}$$

Figure 5.42 If $x > 1$, $\ln x = \int_1^x \frac{dt}{t}$ is the area under $y = \frac{1}{t}$ on $[1, x]$.

may be interpreted geometrically as the area under the graph of $y = \frac{1}{t}$ from $t = 1$ to $t = x$, as shown in Figure 5.42. If $x > 1$, then $\ln x > 0$. On the other hand, if $0 < x < 1$, then

$$\ln x = \int_1^x \frac{dt}{t} = -\int_x^1 \frac{dt}{t} < 0$$

so that

$$\ln x > 0 \quad \text{if } x > 1$$
$$\ln 1 = 0$$
$$\ln x < 0 \quad \text{if } 0 < x < 1$$

The definition $\ln x = \int_1^x \frac{dt}{t}$ makes it easy to differentiate $\ln x$. Recall from Section 5.4 that according to the second fundamental theorem of calculus, if f is continuous on $[a, b]$, then

$$F(x) = \int_a^x f(t)\, dt$$

is a differentiable function of x with derivative $\dfrac{dF}{dx} = f(x)$ on any interval $[a, x]$. Therefore, because $\frac{1}{t}$ is continuous for all $t > 0$, it follows that $\ln x = \int_1^x \frac{dt}{t}$ is differentiable for all $x > 0$ with derivative $\dfrac{d}{dx}(\ln x) = \dfrac{1}{x}$. By applying the chain rule, we also find that

$$\frac{d}{dx}(\ln u) = \frac{1}{u}\frac{du}{dx}$$

for any differentiable function u of x with $u > 0$.

To analyze the graph of $f(x) = \ln x$, we use the curve-sketching methods of Chapter 4:

a. $\ln x$ is continuous for all $x > 0$ (because it is differentiable), so its graph is "unbroken."

b. The graph of $\ln x$ is always *rising*, because the derivative

$$\frac{d}{dx}(\ln x) = \frac{1}{x}$$

is positive for $x > 0$. (Recall that the natural logarithm is defined only for $x > 0$.)

c. The graph of $\ln x$ is *concave down*, because the second derivative

$$\frac{d^2}{dx^2}(\ln x) = \frac{d}{dx}\left(\frac{1}{x}\right) = \frac{-1}{x^2}$$

is negative for all $x > 0$.

d. Note that $\ln 2 > 0$ (because $\int_1^2 \frac{dt}{t} > 0$) and because $\ln 2^p = p \ln 2$, it follows that $\lim\limits_{p \to +\infty} \ln 2^p = +\infty$. But the graph of $f(x) = \ln x$ is always rising, and thus $\lim\limits_{x \to +\infty} \ln x = +\infty$. Similarly, it can be shown that $\lim\limits_{x \to 0^+} \ln x = -\infty$.

e. If b is any positive number, there is exactly one number a such that $\ln a = b$ (because the graph of $\ln x$ is always rising for $x > 0$). In particular, we define $x = e$ as the unique number that satisfies $\ln x = 1$.

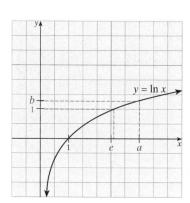

Figure 5.43 The graph of the natural logarithm function, $\ln x$

These features are shown in Figure 5.43.

THE NATURAL EXPONENTIAL FUNCTION

Originally, we introduced the natural exponential function e^x and then defined the natural logarithm $\ln x$ as the inverse of e^x. In this alternative approach, we note that because the natural logarithm is an increasing function, it must be one-to-one. Therefore, it has an inverse function, which we denote by $E(x)$.

Because $\ln x$ and $E(x)$ are inverses, we have

$$E(x) = y \quad \text{if and only if} \quad \ln y = x$$

From the definition of inverse formulas we have

$$E(\ln x) = x \quad \text{and} \quad \ln[E(x)] = x$$

We call these formulas the **inversion formulas**. Therefore,

$$\begin{aligned} E(0) &= E(\ln 1) && \textit{Because } \ln 1 = 0 \\ &= 1 && \textit{Because } E(\ln x) = x \end{aligned}$$

and

$$\begin{aligned} E(1) &= E(\ln e) && \textit{Because } \ln e = 1 \\ &= e && \textit{Because } E(\ln e) = e \end{aligned}$$

To obtain the graph of $E(x)$, we reflect the graph of $y = \ln x$ in the line $y = x$. The graph is shown in Figure 5.44. Notice that $\ln x = 1$ has the unique solution $x = e$.

The following algebraic properties of the natural exponential function can be obtained by using the properties of the natural logarithm given in Theorem 5.13 along with the inversion formulas.

Figure 5.44 The graph of $E(x)$ is the reflection of the graph of $\ln x$ in the line $y = x$.

THEOREM 5.14 Properties of the exponential as defined by $E(x)$

For any numbers x and y,

a. $E(0) = 1$ **b.** $E(x + y) = E(x)E(y)$

c. $E(x - y) = \dfrac{E(x)}{E(y)}$ **d.** $[E(x)]^p = E(px)$

Proof

a. Proved above (inversion formula).

b. We use the fact that $\ln AB = \ln A + \ln B$ to show that

$$\begin{aligned} \ln[E(x)E(y)] &= \ln E(x) + \ln E(y) && \textit{Property of logarithms} \\ &= x + y && \textit{Inversion formula} \\ &= \ln E(x + y) && \textit{Inversion formula} \end{aligned}$$

Because $\ln x$ is a one-to-one function, we conclude that

$$E(x)E(y) = E(x + y)$$

c. and **d.** The proofs are similar and are left as exercises. ❏

Next, we use implicit differentiation along with the differentiation formula

$$\frac{d}{dx}(\ln x) = \frac{1}{x}$$

to obtain a differentiation formula for $E(x)$.

THEOREM 5.15 Derivative of $E(x)$

The function defined by $E(x)$ is differentiable and

$$\frac{dE}{dx} = E(x)$$

Proof   It can be shown that the inverse of any differentiable function is also differentiable so long as the derivative is never zero (see Appendix B). Since $\ln x$ is differentiable, it follows that its inverse $E(x)$ is also differentiable, and using implicit differentiation, we find that

$$\ln y = \ln[E(x)] \qquad y = E(x)$$

$$\ln y = x \qquad \text{Inversion formula}$$

$$\frac{1}{y}\frac{dy}{dx} = 1 \qquad \text{Differentiate implicitly.}$$

$$\frac{dy}{dx} = y$$

$$\frac{dE}{dx} = E(x) \qquad \text{Because } \frac{dy}{dx} = \frac{dE}{dx} \text{ and } E(x) = y \qquad \square$$

Finally, we observe that if r is a rational number, then

$$\ln(e^r) = r \ln e = r$$

so that

$$E(r) = e^r$$

This means that $E(x) = e^x$ when x is a rational number, and we **define** $E(x)$ to be e^x for irrational x as well. In particular, we now have an alternative definition for e: $E(1) = e^1 = e$.

5.9   PROBLEM SET

C **1.** Use the integral definition

$$\ln x = \int_1^x \frac{dt}{t}$$

to show that $\ln x \to -\infty$ as $x \to 0^+$. *Hint*: What happens to $\ln(2^{-N})$ as N grows large without bound?

2. Use Simpson's rule with $n = 8$ subintervals to estimate

$$\ln 3 = \int_1^3 \frac{dt}{t}$$

Compare your estimate with the value of $\ln 3$ obtained from your calculator.

3. Use the error estimate for Simpson's rule to determine the accuracy of the estimate in Problem 2. How many subintervals should be used in Simpson's rule to estimate $\ln 3$ with an error not greater than 0.00005?

4. Prove the quotient rule for logarithms,

$$\ln\left(\frac{x}{y}\right) = \ln x - \ln y$$

for all $x > 0$, $y > 0$. *Hint*: Use the product rule for logarithms.

5. Show that $\ln(x^p) = p \ln x$ for $x > 0$ and all rational exponents p by completing the following steps.

a. Let $F(x) = \ln(x^p)$ and $G(x) = p \ln x$. Show that $F'(x) = G'(x)$ for $x > 0$. Conclude that $F(x) = G(x) + C$.

b. Let $x = 1$ and conclude that $F(x) = G(x)$. That is, $\ln(x^p) = p \ln x$.

6. Use Rolle's theorem to show that

$$\ln M = \ln N$$

if and only if $M = N$. *Hint*: Show that if $M \neq N$, Rolle's theorem implies that

$$\frac{d}{dx}(\ln x) = 0$$

for some number c between M and N. Why is this impossible?

7. The product rule for logarithms states

$$\log_b(MN) = \log_b M + \log_b N$$

For this reason, we want the natural logarithm function $\ln x$ to satisfy the functional equation

$$f(xy) = f(x) + f(y)$$

Suppose $f(x)$ is a function that satisfies this equation throughout its domain D.

a. Show that if $f(1)$ is defined, then $f(1) = 0$.

b. Show that if $f(-1)$ is defined, then $f(-1) = 0$.

c. Show that $f(-x) = f(x)$ for all x in D.

d. If $f'(x)$ is defined for each $x \neq 0$, show that

$$f'(x) = \frac{f'(1)}{x}$$

Then show that

$$f(x) = f'(1) \int_1^x \frac{dt}{t} \qquad \text{for} \quad x > 0$$

e. Conclude that any solution of

$$f(xy) = f(x) + f(y)$$

that is not identically 0 and has a derivative for all $x \neq 0$ must be a multiple of

$$L(x) = \int_1^{|x|} \frac{dt}{t} \qquad \text{for } x \neq 0$$

8. Show that for each number A, there is only one number x for which $\ln x = A$. *Hint*: If not, then $\ln x = \ln y = A$ for $x \neq y$. Why is this impossible?

9. a. Use an area argument to show that $\ln 2 < 1$.

   b. Show that $\ln 3 > 1$, and then explain why $2 < e < 3$.

10. Prove $E(x - y) = \dfrac{E(x)}{E(y)}$.

11. Prove $E(x)^p = E(px)$.

CHAPTER 5 REVIEW

Proficiency Examination

CONCEPT PROBLEMS

1. What is an antiderivative?
2. State the integration rule for powers.
3. State the exponential rule for integration.
4. State the integration rules we have learned for the trigonometric functions.
5. State the integration rules for the inverse trigonometric functions.
6. What is an area function? What do we mean by area as an antiderivative? What conditions are necessary for an integral to represent an area?
7. What is the formula for area as the limit of a sum?
8. What is a Riemann sum?
9. Define a definite integral.
10. Complete these statements summarizing the general properties of the definite integral.

    a. Definite integral at a point: $\displaystyle\int_a^a f(x)\,dx = $ _____

    b. Switching limits of a definite integral: $\displaystyle\int_b^a f(x)\,dx = $ _____

11. How can distance traveled by an object be expressed as an integral?
12. State the first fundamental theorem of calculus.
13. State the second fundamental theorem of calculus.
14. Describe in your own words the process of integration by substitution.
15. What is a differential equation?
16. What is a separable differential equation? How do you solve such an equation?
17. What is the growth/decay equation? Describe carbon dating.
18. What is an orthogonal trajectory?
19. State the mean value theorem for integrals.
20. What is the formula for the average value of a continuous function?
21. State the following approximation rules:

    a. rectangular approximation

    b. trapezoidal rule

    c. Simpson's rule

PRACTICE PROBLEMS

22. Given that $\displaystyle\int_0^1 [f(x)]^4\,dx = \frac{1}{5}$ and $\displaystyle\int_0^1 [f(x)]^2\,dx = \frac{1}{3}$, find

$$\int_0^1 [f(x)]^2 \{2[f(x)]^2 - 3\}\,dx$$

23. Find $F'(x)$ if $F(x) = \displaystyle\int_3^x t^5 \sqrt{\cos(2t + 1)}\,dt$.

Evaluate the integrals in Problems 24–29.

24. $\displaystyle\int \frac{dx}{1 + 4x^2}$ 25. $\displaystyle\int x e^{-x^2}\,dx$

26. $\displaystyle\int_1^4 (\sqrt{x} + x^{-3/2})\,dx$

27. $\displaystyle\int_0^1 (2x - 6)(x^2 - 6x + 2)^2\,dx$

28. $\displaystyle\int_0^{\pi/2} \frac{\sin x\,dx}{(1 + \cos x)^2}$

29. $\displaystyle\int_{-2}^1 (2x + 1)\sqrt{2x^2 + 2x + 5}\,dx$

30. Find the area under the curve $f(x) = 3x^2 + 2$ over $[-1, 3]$.

31. Approximate the area under the curve $g(x) = e^{x^2}$ over $[1, 2]$.[*]

32. Find the average value of $y = \cos 2x$ on the interval $\left[0, \frac{\pi}{2}\right]$.

33. **Modeling Problem** In 1995, a team of archaeologists led by Michel Brunet of the University of Poitiers announced the identification in Chad of *Australopithecus* specimens believed to be about 3.5 Myr (million years) old.[†] Using the exponential decay equation, explain why the archaeologists were reluctant to use carbon-14 to date their find.

---

[*]There is an interesting discussion of this integral in the May 1999 issue of *The American Mathematical Monthly* in an article entitled, "What Is a Closed-Form Number?" The function $\int e^{x^2}$ is an example of a function that is **not** an elementary function (that we mentioned in Section 1.3).

[†]"Early Hominid Fossils from Africa," by Meave Leakey and Alan Walker, *Scientific American*, June 1997, p. 79.

34. A slope field for the differential equation

$$y' = x + y$$

is given in Figure 5.45.

Figure 5.45 Direction field

Draw the particular solutions passing through the points requested in parts **a–d**.

a. $(0, -4)$ **b.** $(0, 2)$ **c.** $(4, 0)$ **d.** $(-4, 0)$

e. All of the curves you have drawn in parts **a–d** seem to have a common asymptote. From the slope field, write the equation for this asymptote.

35. a. Find the necessary n to estimate the value of $\displaystyle\int_0^{\pi/2} \cos x \, dx$ to within 0.0005 of its correct value using the trapezoidal rule.

b. What is n if Simpson's rule is used?

Supplementary Problems

Find the definite integrals of the functions in Problems 1–3.

1. Given that $\displaystyle\int_{-1}^{0} f(x) \, dx = 3$, $\displaystyle\int_{0}^{1} f(x) \, dx = -1$, and $\displaystyle\int_{-1}^{1} g(x) \, dx = 7$, find $\displaystyle\int_{-1}^{1} [3g(x) + 2f(x)] \, dx$.

2. Use the definition of the definite integral to find $\displaystyle\int_{0}^{1} (3x^2 + 2x - 1) \, dx$.

3. Use the definition of the definite integral to find $\displaystyle\int_{0}^{1} (4x^3 + 6x^2 + 3) \, dx$.

Find the definite and indefinite integrals in Problems 4–31. If you do not have a technique for finding a closed (exact) answer, approximate the integral using numerical integration.

4. $\displaystyle\int_{0}^{1} (5x^4 - 8x^3 + 1) \, dx$

5. $\displaystyle\int_{-1}^{2} 30(5x - 2)^2 \, dx$

6. $\displaystyle\int_{0}^{1} (x\sqrt{x} + 2)^2 \, dx$

7. $\displaystyle\int_{1}^{2} \frac{x^2 \, dx}{\sqrt{x^3 + 1}}$

8. $\displaystyle\int_{2}^{2} (x + \sin x)^3 \, dx$

9. $\displaystyle\int_{-1}^{0} \frac{dx}{\sqrt{1 - 2x}}$

10. $\displaystyle\int_{1}^{2} \frac{dx}{\sqrt{3x - 1}}$

11. $\displaystyle\int_{-1}^{0} \frac{dx}{\sqrt[3]{1 - 2x}}$

12. $\displaystyle\int \frac{1}{x^3} \, dx$

13. $\displaystyle\int \frac{5x^2 - 2x + 1}{\sqrt{x}} \, dx$

14. $\displaystyle\int \frac{x + 1}{2x} \, dx$

15. $\displaystyle\int \frac{3 - x}{\sqrt{1 - x^2}} \, dx$

16. $\displaystyle\int (e^{-x} + 1)e^x \, dx$

17. $\displaystyle\int \frac{\sin x - \cos x}{\sin x + \cos x} \, dx$

18. $\displaystyle\int \sqrt{x}(x^2 + \sqrt{x} + 1) \, dx$

19. $\displaystyle\int (x - 1)^2 \, dx$

20. $\displaystyle\int \frac{x^2 + 1}{x^2} \, dx$

21. $\displaystyle\int (\sin^2 x + \cos^2 x) \, dx$

22. $\displaystyle\int x(x + 4)\sqrt{x^3 + 6x^2 + 2} \, dx$

23. $\displaystyle\int x(2x^2 + 1)\sqrt{x^4 + x^2} \, dx$

24. $\displaystyle\int \frac{dx}{\sqrt{x}(\sqrt{x} + 1)^2}$

25. $\displaystyle\int x\sqrt{1 - 5x^2} \, dx$

26. $\displaystyle\int \sqrt{\sin x - \cos x}(\sin x + \cos x) \, dx$

27. $\displaystyle\int x^3 \cos(\tan^{-1} x) \, dx$

28. $\displaystyle\int (\cos^{11} x \sin^9 x - \cos^9 x \sin^{11} x) \, dx$

29. $\displaystyle\int \frac{\ln x}{x} \sqrt{1 + \ln(2x)} \, dx$

30. $\displaystyle\int_{-10}^{10} [3 + 7x^{73} - 100x^{101}] \, dx$

31. $\displaystyle\int_{-\pi/4}^{\pi/4} [\sin(4x) + 2\cos(4x)] \, dx$

In Problems 32–35, draw the indicated particular solution for the differential equations whose direction fields are given.

32. $y(4) = 0$

33. $y(10) = 1$

34. $y(0) = 2$

35. $y(0) = 0$

36. Find $F'(x)$, where $F(x) = \displaystyle\int_{5}^{x} t^2 \cos^4 t \, dt$

37. Find the area under $f(x) = x^{-1}$ on $[1, 4]$.

38. Find the area under $f(x) = 2 + x - x^2$ on $[-1, 1]$.

39. Find the area under $f(x) = e^{4x}$ on $[0, 2]$.

40. Find the area bounded by the curve $y = x\sqrt{x^2 + 5}$, the x-axis, and the vertical lines $x = -1$ and $x = 2$.

41. Find $f(t)$ if $f''(t) = \sin 4t - \cos 2t$ and $f\left(\frac{\pi}{2}\right) = f'\left(\frac{\pi}{2}\right) = 1$.

42. Find $f(x)$ if $f'''(x) = 2x^3 + x^2$, given that $f''(1) = 2$, $f'(1) = 1$, and $f(1) = 0$.

Solve the differential equations in Problems 43–50.

43. $\dfrac{dy}{dx} = (1 - y)^2$

44. $\dfrac{dy}{dx} = \dfrac{\cos 4x}{y}$

45. $\dfrac{dy}{dx} = \left(\dfrac{\cos y}{\sin x}\right)^2$

46. $\dfrac{dy}{dx} = \dfrac{x}{y}$

47. $\dfrac{dy}{dx} = y(x^2 + 1)$

48. $\dfrac{dy}{dx} = \dfrac{x}{y}\sqrt{\dfrac{y^2 + 2}{x^2 + 1}}$

49. $\dfrac{dy}{dx} = \dfrac{\cos^2 y}{\cot x}$

50. $\dfrac{dy}{dx} = \sqrt{\dfrac{x}{y}}$

51. Find the average value of $f(x) = \dfrac{\sin x}{\cos^2 x}$ on the interval $\left[0, \frac{\pi}{4}\right]$.

52. Find the average value of $f(x) = \sin x$
  a. on $[0, \pi]$ **b.** on $[0, 2\pi]$

53. Use the trapezoidal rule with $n = 6$ to approximate $\displaystyle\int_0^{\pi} \sin x \, dx$. Compare your result with the exact value of this integral.

54. Estimate $\displaystyle\int_0^1 \sqrt{1 + x^3}\, dx$ using the trapezoidal rule with $n = 6$.

55. Estimate $\displaystyle\int_0^1 \dfrac{dx}{\sqrt{1 + x^3}}$ using the trapezoidal rule with $n = 8$.

56. Estimate $\displaystyle\int_0^1 \sqrt{1 + x^3}\, dx$ using Simpson's rule with $n = 6$.

57. Use the trapezoidal rule to approximate $\displaystyle\int_0^1 \dfrac{x^2\, dx}{1 + x^2}$ with an error no greater than 0.005.

58. Use the trapezoidal rule to estimate to within 0.00005 the value of the integral

$$\int_1^2 \sqrt{x + \frac{1}{x}}\, dx$$

59. HISTORICAL QUEST Karl Gauss is considered to be one of the four greatest mathematicians of all time, along with Archimedes (Historical Quest, Problem 53, Section 3.8), Newton (Historical Quest, Problem 1, Mathematical Essays of Chapter 1, p. 48), and Euler (Historical Quest, Chapter 4 supplementary Problem 84). Gauss graduated from college at the age of 15 and KARL GAUSS 1777–1855 proved what was to become the fundamental theorem of algebra for his doctoral thesis at the age of 22. He published only a small portion of the ideas that seemed to storm his mind, because he believed that each published result had to be complete, concise, polished, and convincing. His motto was "Few, but ripe." Carl B. Boyer, in *A History of Mathematics*, describes Gauss as the last mathematician to know everything in his subject, and we could relate Gauss to nearly every topic of this book. Such a generalization is bound to be inexact, but it does emphasize the breadth of interest Gauss displayed.

A **prime number**, p, is a counting number that has exactly two divisors. **The prime number theorem**, conjectured by Gauss in 1793, says that the number of primes $\pi(x)$ less than a real number x is a function that behaves like $x / \ln x$ as $x \to \infty$; that is,

$$\lim_{x \to +\infty} \frac{\pi(x)}{x / \ln x} = 1$$

a. $\pi(n)$ for $0 \le n \le 100$

b. $\pi(n)$ for $0 \le n \le 1{,}000$

Figure 5.46 Graphs of $\pi(n)$

Gauss estimated the prime distribution function $\pi(x)$ by the integral

$$G(x) = \int_2^x \frac{dt}{\ln t}$$

known as the *integral logarithm* function. History records many who constructed tables for $\pi(n)$, and Figure 5.46 shows values for $\pi(n)$ for various choices of n. ∎

Your Quest is to examine this estimate by using the trapezoidal rule to approximate $G(x)$ for $x = 100$, $1{,}000$, and $10{,}000$. The actual value of $\pi(x)$ for selected values of x are as follows:

n	$\pi(n)$
10^3	168
10^4	1,229
10^5	9,592
10^6	78,498
10^{10}	455,052,511

60. HISTORICAL QUEST There are many ways to estimate the prime distribution function $\pi(x)$ defined in Problem 59 other than by the integral function $G(x)$ used by Gauss. Adrien-Marie Legendre is best known for his work with elliptic integrals and mathematical physics. In 1794, he proved that π was an irrational number and formulated a conjecture that is equivalent to the prime number theorem. He made his estimate using the function

ADRIEN-MARIE
LEGENDRE
1752–1833

$$\pi(x) \approx L(x) = \frac{x}{\ln x - 1.08366} \quad \blacksquare$$

For this Quest, use $L(x)$ to estimate $\pi(x)$ for $x = 100$, $1,000$, and $10,000$, and then compare with the results found using $G(x)$ in Problem 59.

61. Approximate the average value of the function defined by

$$f(x) = \frac{\cos x}{1 - \frac{x^2}{2}}$$

on $[0, 1]$. *Hint*: Use the trapezoidal rule with $n = 6$.

62. The brakes of a certain automobile produce a constant deceleration of k m/s^2. The car is traveling at 25 m/s when the driver is forced to hit the brakes. If it comes to rest at a point 50 m from the point where the brakes are applied, what is k?

63. A particle moves along the t-axis in such a way that $a(t) = -4s(t)$, where $s(t)$ and $a(t)$ are its position and acceleration, respectively, at time t. The particle starts from rest at $s = 5$.

  a. Show that $v^2 + 4s^2 = 100$, where $v(t)$ is the velocity of the particle. *Hint*: First use the chain rule to show that $a(t) = v(dv/ds)$.

  b. What is the velocity when the particle first reaches $s = 3$? *Note*: At the time in question, the sign of v is determined by the direction the particle is moving.

64. An object experiences linear acceleration given by

$$a(t) = 2t + 1 \quad \text{ft/s}^2$$

Find the velocity and position of the object, given that it starts its motion (at $t = 0$) at $s = 4$ with initial velocity 2 ft/s.

65. When it is x years old, a certain industrial machine generates revenue at the rate of $R'(x) = 1,575 - 5x^2$ thousand dollars per year. Find a function that measures the amount of revenue, and find the revenue for the first five years.

66. A manufacturer estimates marginal revenue to be $100x^{-1/3}$ dollars per unit when the level of production is x units. The corresponding marginal cost is found to be $0.4x$ dollars per unit. Suppose the manufacturer's profit is \$520 when the level of production is 16 units. What is the manufacturer's profit when the level of production is 25 units?

67. A tree has been transplanted and after t years is growing at a rate of

$$1 + \frac{1}{(t+1)^2}$$

feet per year. After 2 years it has reached a height of 5 ft. How tall was the tree when it was transplanted?

68. A manufacturer estimates that the marginal revenue of a certain commodity is

$$R'(x) = \sqrt{x}(x^{3/2} + 1)^{-1/2}$$

dollars per unit when x units are produced. Assuming no revenue is obtained when $x = 0$, how much revenue is obtained from producing $x = 4$ units?

69. Find a function whose tangent has slope $x\sqrt{x^2 + 5}$ for each value of x and whose graph passes through the point $(2, 10)$.

70. A particle moves along the x-axis in such a way that after t seconds its acceleration is $a(t) = 12(2t + 1)^{-3/2}$. If it starts at rest at $x = 3$, where will it be 4 seconds later?

71. An environmental study of a certain community suggests that t years from now, the level of carbon monoxide in the air will be changing at the rate of $0.1t + 0.2$ parts per million per year. If the current level of carbon monoxide in the air is 3.4 parts per million, what will the level be 3 years from now?

72. A woman, driving on a straight, level road at the constant speed v_0, is forced to apply her brakes to avoid hitting a cow. The car comes to a stop 3 seconds later, s_0 ft from the point where the brakes were applied. Continuing on her way, she increases her speed by 20 ft/s to make up time but is again forced to hit the brakes, and this time it takes her 5 seconds and s_1 feet to come to a full stop. Assuming that her brakes supplied a constant deceleration k ft/s^2 each time they were used, find k and determine v_0, s_0, and s_1.

73. A study indicates that x months from now the population of a certain town will be increasing at the rate of $10 + 2\sqrt{x}$ people per month. By how much will the population of the town increase over the next 9 months?

74. It is estimated that t days from now a farmer's crop will be increasing at the rate of $0.3t^2 + 0.6t + 1$ bushels per day. By how much will the value of the crop increase during the next 6 days if the market price remains fixed at \$2 per bushel?

75. Records indicate that t months after the beginning of the year, the price of turkey in local supermarkets was

$$P(t) = 0.06t^2 - 0.2t + 1.2$$

dollars per pound. What was the average price of turkey during the first 6 months of the year?

76. Modeling Problem V. A. Tucker and K. Schmidt-Koenig have investigated the relationship between the velocity v (km/h) of a bird in flight and the energy $E(v)$ expended by the bird.* Their study showed that for a certain kind of parakeet, the rate of change of the energy expended with respect to velocity is modeled (for $v > 0$) by

$$\frac{dE}{dv} = \frac{0.074v^2 - 112.65}{v^2}$$

  a. What is the most economical velocity for the parakeet? That is, find the velocity v_0 that minimizes the energy.

  b. Suppose it is known that $E = E_0$ when $v = v_0$. Express E in terms of v_0 and E_0.

  c. Express the average energy expended as the parakeet's velocity ranges from $v = \frac{1}{2}v_0$ to $v = v_0$ in terms of E_0.

*Adapted from "Flight Speeds of Birds in Relation to Energies and Wind Directions," by V. A. Tucker and K. Schmidt-Koenig. *The Auk*, Vol. 88 (1971), pp. 97–107.

77. **Modeling Problem** A toxin is introduced into a bacterial culture, and t hours later, the population $P(t)$ of the culture is found to be changing at the rate

$$\frac{dP}{dt} = -(\ln 2)2^{5-t}$$

If there were 1 million bacteria in the culture when the toxin was introduced, when will the culture die out?

78. **Modeling Problem** *The Snowplow Problem of R. P. Agnew*[*] ■ (This problem was mentioned in the preview for this chapter.) One day it starts snowing at a steady rate sometime before noon. At noon, a snowplow starts to clear a straight, level section of road. If the plow clears 1 mile of road during the first hour but requires 2 hr to clear the second mile, at what time did it start snowing? Answer this question by completing the following steps:

   a. Let t be the time (in hours) from noon. Let h be the depth of the snow at time t, and let s be the distance moved by the plow. If the plow has width w and clears snow at the constant rate p, explain why $wh\dfrac{ds}{dt} = p$.

   b. Suppose it started snowing t_0 hours before noon. Let r denote the (constant) rate of snowfall. Explain why

   $$h(t) = r(t + t_0)$$

   By combining this equation with the differential equation in part **a**, note that

   $$wr(t + t_0)\frac{ds}{dt} = p$$

   Solve this differential equation (with appropriate conditions) to obtain t_0 and answer the question posed in the problem.

79. a. If f is continuous on $[a, b]$, show that

   $$\left| \int_a^b f(x)\,dx \right| \leq \int_a^b |f(x)|\,dx$$

   b. Show that $\left| \displaystyle\int_1^4 \dfrac{\sin x}{x}\,dx \right| \leq \dfrac{3}{2}$.

80. A company plans to hire additional advertising personnel. Suppose it is estimated that if x new people are hired, they will bring in additional revenue of $R(x) = \sqrt{2x}$ thousand dollars and that the cost of adding these x people will be $C(x) = \frac{1}{3}x$ thousand dollars. How many new people should be hired? How much total net revenue (that is, revenue minus cost) is gained by hiring these people?

81. The half-life of the radioactive isotope cobalt-60 is 5.25 years.

   a. What percentage of a given sample of cobalt-60 remains after 5 years?

   b. How long will it take for 90% of a given sample to disintegrate?

82. The rate at which salt dissolves in water is directly proportional to the amount that remains undissolved. If 8 lb of salt is placed in a container of water and 2 lb dissolves in 30 min, how long will it take for 1 lb to dissolve?

83. Scientists are observing a species of insect in a certain swamp region. The insect population is estimated to be 10 million and is expected to grow at the rate of 2% per year. Assuming that the growth is exponential and stays that way for a period of years, what will the insect population be in 10 yr? How long will it take to double?

84. Solve the system of differential equations

$$\begin{cases} 2\dfrac{dx}{dt} + 5\dfrac{dy}{dt} = t \\[2mm] \dfrac{dx}{dt} + 3\dfrac{dy}{dt} = 7\cos t \end{cases}$$

Hint: Solve for $\dfrac{dx}{dt}$ and $\dfrac{dy}{dt}$ algebraically, and then integrate.

85. **Spy Problem** While lunching on cassoulet de chameau at his favorite restaurant in San Dimas (recall Problem 50, Section 5.1), the Spy finds a message from his esteemed superior, Lord Newton Fleming, spelled out in his alphabet soup, "The average of the temperature $F(t) = at^3 + bt^2 + ct + d$ over the time period from 9 A.M. to 3 P.M. is the average temperature between two fixed times t_1 and t_2. Mother waits at the well."

   The Spy knows that "Mother" is the codename of Lord Newton himself, and "the well" is a sleazy bar at the edge of town. He decides that time t is measured from noon because the soup was served at that time. The cryptic nature of the message suggests that t_1 and t_2 are independent of a, b, c, and d and that one of them is the rendezvous time. When should he arrive for the meeting? (Remember, it's not wise to go to the well too often.)

86. **Putnam Examination Problem** Where on the parabola $4ay = x^2$ ($a > 0$) should a chord be drawn so that it will be normal to the parabola and cut off a parabolic sector of minimum area? That is, find P so that the shaded area in Figure 5.47 is as small as possible.

Figure 5.47 Minimum area problem

87. **Putnam Examination Problem** If $a_0, a_1, \ldots, a_n$ are real numbers that satisfy

$$\frac{a_0}{1} + \frac{a_1}{2} + \cdots + \frac{a_n}{n+1} = 0$$

show that the equation $a_0 + a_1 x + a_2 x^2 + \cdots + a_n x^n = 0$ has at least one real root. *Hint*: Use the mean value theorem for integrals.

---

[*]There is an interesting discussion of this problem, along with a BASIC computer simulation, in the November 1995 (Vol. 26, No. 5) issue of *The College Mathematics Journal* in the article "The Meeting of the Plows: A Simulation" by Jerome Lewis (pp. 395–400).

Kinematics of Jogging

Ralph Boas (1912–1992), who was a professor of mathematics at Northwestern University, wrote this guest essay. Professor Boas was well known for his many papers and professional activities. In addition to his work in real and complex analysis, he wrote many expository articles, such as this guest essay, about teaching or using mathematics.

NATURE HERSELF EXHIBITS to us measurable and observable quantities in definite mathematical dependence; the conception of a function is suggested by all the processes of nature where we observe natural phenomena varying according to distance or to time. Nearly all the "known" functions have presented themselves in the attempt to solve geometrical, mechanical, or physical problems.

——J. T. Mertz
*A History of European Thought
in the Nineteenth Century*
(Edinburgh and London, 1903), p. 696.

Some people think that calculus is dull, but it did not seem so three centuries ago, when it was invented. Then, it produced unexpected results; and, now and then, it still does. This essay is about such a result.

You have learned about the intermediate value theorem (see Section 2.3), which tells you, for instance, that if you jog at 8 min per mile, there must be some instant when your speed is exactly $\frac{1}{8}$ mi per minute—assuming, as is only natural in a course in calculus, that your elapsed time is a continuous function of the distance covered. This principle is very intuitive and was recognized before calculus was invented: Galileo was aware of it in 1638, and thought that it had been known to Plato. On the other hand, there is a question with a much less intuitive answer that was noticed only recently (and, as happens more often than mathematicians like to admit, by a physicist). Suppose that you average 8 min/mi; must you cover some one continuous mile (such as a "measured mile" on a highway) in exactly 8 min? The answer is not intuitive at all: It depends on whether or not your total distance was an integral number of miles. More precisely, if you cover an integral number of miles, then you cover exactly one mile in some 8 min. However, if you cover a nonintegral number of miles, there is not necessarily any one continuous mile that you cover in 8 min.

To prove this, let x be the distance (in miles) covered at any point during your trip, and suppose that when you stop you have covered an integral number of n miles. Let $f(x)$ be the time (in minutes) that it took to cover the first x miles; we will suppose that f is a continuous function. If you averaged 8 min/mi, then $f(x) - 8x = 0$ when $x = 0$ and when $x = n$. Now suppose that you never did cover any consecutive mile in 8 min; in mathematical terms,

$$f(x + 1) - f(x) \neq 8$$

Because

$$f(x + 1) - f(x) - 8$$

is continuous and never 0, it must either always be positive, or else always negative; let us suppose the former. Write the corresponding facts for $x = 0, 1, \ldots, n$:

$$f(1) - f(0) > 8$$
$$f(2) - f(1) > 8$$
$$\vdots$$
$$f(n) - f(n - 1) > 8$$

If we add these inequalities, we obtain

$$f(n) - f(0) > 8n$$

But we started with the assumption that $f(n) = 8n$ and $f(0) = 0$, so assuming that $f(x + 1) - f(x)$ is never 8 leads to a contradiction.

351

It is somewhat harder to show that only integral values of n will work. Suppose you jog so that your time to cover x miles is

$$J(x) = k \sin^2 \frac{\pi x}{n} + 8x$$

where n is not an integer and k is a small number. This is a legitimate assumption, because J is an increasing function (as a time has to be), if k is small enough. To be sure of this, we calculate $J'(x)$—and here we actually have to use some calculus (or have a calculator that will do it for us). We find

$$J'(x) = \frac{k\pi}{n} \sin \frac{2\pi x}{n} + 8$$

If k is small enough ($k < \frac{8n}{\pi}$), then $J'(x) > 0$. This shows not only that J increases but also that

$$J(x+1) - J(x)$$

cannot be eight. Because $J(x+1) - J(x)$ is never negative, if you jog so that your time is $J(x)$, you will never cover a whole mile in less than 8 min.

Mathematical Essays

Use the Internet, a library, or references other than this textbook to research the information necessary to answer questions in Problems 1–11

1. HISTORICAL QUEST   The derivative is one of the great ideas of calculus. Write an essay of at least 500 words about some application of the derivative that is not discussed in this text.

2. HISTORICAL QUEST   The concept of the integral is one of the great ideas of calculus. Write an essay of at least 500 words about the relationship of integration and differentiation as it relates to the history of calculus.

3. HISTORICAL QUEST   Write a report on Georg Riemann. (See Historical Quest in Section 5.3.) Include the 1984 development in the solution of the Riemann hypothesis.

4. HISTORICAL QUEST   As we saw in a Historical Quest in Section 5.8, the mathematician Seki Kōwa was doing a form of integration at about the same time that Newton and Leibniz were inventing the calculus. Write a paper on the history of calculus from the Eastern viewpoint.

5. "Lucy," the famous prehuman whose skeleton was discovered in East Africa, has been found to be approximately 3.8 million years old. About what percentage of ^{14}C would you expect to find if you tried to "date" Lucy by the usual carbon dating procedure? The answer you get to this question illustrates why it is reasonable to use carbon dating only on more recent artifacts, usually less than 50,000 years old (roughly the time since the last major ice age). Read an article on alternative dating procedures such as potassium-argon and rubidium-strontium dating. Write a paper comparing and contrasting such methods.

6. In the guest essay it was assumed that the time to cover x miles is

$$J(x) = k \sin^2 \frac{\pi x}{n} + 8x$$

Suppose that $n = 5$. What choices for k seem reasonable?

7. Suppose that the time to cover x miles is given by

$$J(x) = \sin^2 \frac{\pi x}{5} + 8x$$

Graph this function on $[0, 8]$.

8. Use calculus to find how small k (from the guest essay) needs to be in the expression

$$J(x) = k \sin^2 \frac{\pi x}{n} + 8x$$

so that $J'(x) > 0$.

9. **a.** Find a number x, $x \neq 0$, that, when divided by 2, gives a display of 0 on your calculator. Write a paper describing your work as well as the processes on your calculator.

   b. Calculate $\sqrt{2}$ using a calculator. Next, repeatedly subtract the integer part of the displayed number and multiply the result by 10. Describe the outcome, and then devise a method for finding $\sqrt{2}$ using calculus. Write a paper comparing these answers.

10. **Book Report**   Eli Maor, a native of Israel, has a long-standing interest in the relations between mathematics and the arts. Read the fascinating book *To Infinity and Beyond, A Cultural History of the Infinite* (Boston: Birkhäuser, 1987), and prepare a book report.

11. **Book Report**   In Section 3.1, we told the story of Fermat's last theorem. The book *Fermat's Enigma* by Simon Singh (New York: Walker & Company, 1997) tells the story of how Andrew Wiles solved the greatest mathematical puzzle of our age: He locked himself in a room and emerged seven years later. Read this book and prepare a book report.

12. Make up a word problem involving an application of the integral. Send your problem and solution to

   Strauss, Bradley, and Smith
   Prentice Hall Publishing Company
   1 Lake Street
   Upper Saddle River, NJ 07458

The best one submitted will appear in the next edition (along with credit to the problem poser).

CHAPTERS 1–5   Cumulative Review

1. **WHAT DOES THIS SAY?**   Define limit.   Explain what this definition is saying using your own words.

2. **WHAT DOES THIS SAY?**   Define derivative.   Explain what this definition is saying using your own words.

3. **WHAT DOES THIS SAY?**   Define a definite integral.   Explain what this definition is saying using your own words.

4. **WHAT DOES THIS SAY?**   Define a differential equation, and in your own words describe the procedure for solving a separable differential equation.

Evaluate the limits in Problems 5–13.

5. $\lim\limits_{x \to 2} \dfrac{3x^2 - 5x - 2}{3x^2 - 7x + 2}$

6. $\lim\limits_{x \to +\infty} \dfrac{3x^2 + 7x + 2}{5x^2 - 3x + 3}$

7. $\lim\limits_{x \to +\infty} (\sqrt{x^2 + x} - x)$

8. $\lim\limits_{x \to \pi/2} \dfrac{\cos^2 x}{\cos x^2}$

9. $\lim\limits_{x \to 0} \dfrac{x \sin x}{x + \sin^2 x}$

10. $\lim\limits_{x \to 0} \dfrac{\sin 3x}{x}$

11. $\lim\limits_{x \to +\infty} (1 + x)^{2/x}$

12. $\lim\limits_{x \to 0} \dfrac{\tan^{-1} x - x}{x^3}$

13. $\lim\limits_{x \to 0^+} x^{\sin x}$

Find the derivatives in Problems 14–22.

14. $y = 6x^3 - 4x + 2$

15. $y = (x^2 + 1)^3 (3x - 4)^2$

16. $y = \dfrac{x^2 - 4}{3x + 1}$

17. $y = \dfrac{x}{x + \cos x}$

18. $x^2 + 3xy + y^2 = 0$

19. $y = \sec^2 3x$

20. $y = e^{5x - 4}$

21. $y = \ln(5x^2 + 3x - 2)$

22. $y = \tan^{-1}(x^2 - 3)$

Find the integrals in Problems 23–28.

23. $\displaystyle\int_4^9 d\theta$

24. $\displaystyle\int_{-1}^1 50(2x - 5)^3 \, dx$

25. $\displaystyle\int_0^1 \dfrac{x \, dx}{\sqrt{9 + x^2}}$

26. $\displaystyle\int \csc 3\theta \cot 3\theta \, d\theta$

27. $\displaystyle\int \dfrac{e^x \, dx}{e^x + 2}$

28. $\displaystyle\int \dfrac{x^3 + 2x - 5}{x} \, dx$

29. Approximate $\displaystyle\int_0^4 \dfrac{dx}{\sqrt{1 + x^3}}$ using Simpson's rule with $n = 6$.

30. Sketch the graph of $y = x^3 - 5x^2 + 2x + 8$.

31. Sketch the graph of $y = \dfrac{4 + x^2}{4 - x^2}$.

32. Find the largest value of $f(x) = \frac{1}{3}x^3 - 2x^2 + 3x - 10$ on $[0, 6]$.

Solve the differential equations in Problems 33–34.

33. $\dfrac{dy}{dx} = x^2 y^2 \sqrt{4 - x^3}$

34. $(1 + x^2) \, dy = (x + 1) y \, dx$

Find the particular solution of the differential equations in Problems 35–36.

35. $\dfrac{dy}{dx} = 2(5 - y)$; $y = 3$ when $x = 0$

36. $\dfrac{dy}{dx} = e^y \sin x$; $y = 5$ when $x = 0$

37. The graph of a function f consists of a semicircle of radius 3 and two line segments as shown in Figure 5.48. Let F be the function defined by

$$F(x) = \int_0^x f(t) \, dt$$

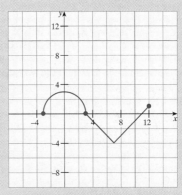

Figure 5.48   Graph of f

a. Find $F(7)$.

b. Find all values on the interval $(-3, 12)$ at which F has a relative maximum.

c. Write an equation for the line tangent to the graph of F at $x = 7$.

d. Find the x-coordinate of each point of inflection of the graph of F on the interval $(-3, 12)$.

38. An electric charge Q_0 is distributed uniformly over a ring of radius R. The electric field intensity at any point x along the axis of the ring is given by

$$E(x) = \dfrac{Q_0 x}{(x^2 + R^2)^{3/2}}$$

At what point on the axis is $E(x)$ maximized?

39. A rocket is launched vertically from a point on the ground 5,000 ft from a TV camera. If the rocket is rising vertically at the rate of 850 ft/s at the instant the rocket is 4,000 ft above the ground, how fast must the TV camera change the angle of elevation to keep the rocket in the picture?

40. A particle moves along the x-axis so that its velocity at any time $t \geq 0$ is given by $v(t) = 6t^2 - 2t - 4$. It is known that the particle is at position $x = 6$ for $t = 2$.

a. Find the position of the particle at any time $t \geq 0$.

b. For what value(s) of t, $0 \leq t \leq 3$, is the particle's instantaneous velocity the same as its average velocity over the interval $[0, 3]$?

c. Find the total distance traveled by the particle from time $t = 0$ to $t = 3$.

41. Let f be the function defined by $f(x) = 2 \cos x$. Let $P(0, 2)$ and $Q\left(\frac{\pi}{2}, 0\right)$ be the points where f crosses the y-axis and x-axis, respectively.

a. Write an equation for the secant line passing through points P and Q.

b. Write an equation for the tangent line of f at point Q.

c. Find the x-coordinate of the point on the graph of f, between point P and Q, at which the line tangent to the graph of f is parallel to $\overline{PQ}$. Cite a theorem or result that assures the existence of such a tangent.

42. Let f be the function defined by $f(x) = \sqrt{x-2}$.

    a. Sketch f and shade the region R enclosed by the graph of f, the x-axis, and the vertical line $x = 5$.

    b. Find the area of R.

43. Let f be the function defined by $f(x) = x^3 - 6x^2 + k$ for k an arbitrary constant.

    a. Find the relative maximum and minimum of f in terms of k.

    b. For which values of k does f have three distinct roots?

    c. Find the values of k for which the average value of f over $[-1, 2]$ is 2.

44. An object is thrown directly downward with velocity 5 ft/s from an airplane, and t seconds later is falling with acceleration

$$a(t) = 32 - 0.08v$$

where $v(t)$ is the object's velocity at time t.

    a. Find an expression for v in terms of t.

    b. If the terminal velocity is defined as $\lim_{t \to +\infty} v(t)$, find the terminal velocity of the object (to the nearest ft/s).

    c. The object can land safely if the impact velocity is no greater than 60 ft/s. What is the maximum altitude the airplane can fly to guarantee a safe landing for the object?

45. Solve the differential equation $\dfrac{dy}{dx} = y^2 \sin 3x$.

6

Additional Applications of the Integral

PREVIEW

In Chapter 5, we found that area and average value can be expressed in terms of the definite integral. The goal of this chapter is to consider various other applications of integration, such as computing volume, arc length, surface area, work, hydrostatic force, centroids of planar regions, and applications to business, economics, and life sciences.

PERSPECTIVE

What is the volume of the material that remains when a hole of radius of $0.5R$ is drilled into a sphere of radius R? The "center" of a rectangle is the point where its diagonals intersect, but what is the center of a quarter circle? If a reservoir is filled to the top of a dam shaped like a parabola, what is the total force of the water on the face of the dam? If a worker is carrying a bag of sand up a ladder and the bag is leaking sand through a hole in such a way that all the sand is gone when the worker reaches the top, how much work is done by the worker? An investor asks a question seeking the future value of an income flow, or a physician asks how to measure the flow of blood through an artery. These questions are typical of the kind we will answer in this chapter.

6.1   Area Between Two Curves

IN THIS SECTION   area between curves, area using vertical strips, area using horizontal strips

AREA BETWEEN CURVES

In some practical problems, you may have to compute the area between two curves. Suppose f and g are functions such that $f(x) \geq g(x)$ on the interval $[a, b]$, as shown in Figure 6.1. We do not insist that both f and g be nonnegative functions, but we begin by showing that case in Figure 6.1.

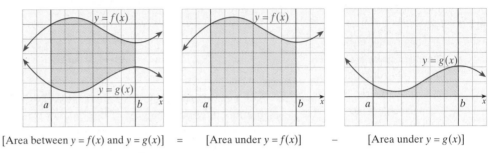

[Area between $y = f(x)$ and $y = g(x)$]   =   [Area under $y = f(x)$]   −   [Area under $y = g(x)$]

Figure 6.1   Area between two curves

To find the area of the region R between the curves defined by $y = f(x)$ and $y = g(x)$ from $x = a$ to $x = b$, we subtract the area between the lower curve $y = g(x)$ and the x-axis from the area between the upper curve $y = f(x)$ and the x-axis; that is,

$$\text{AREA OF } R = \int_a^b f(x)\, dx - \int_a^b g(x)\, dx = \int_a^b [f(x) - g(x)]\, dx$$

This formula seems obvious in the situation where both $f(x) \geq 0$ and $g(x) \geq 0$, as shown in Figure 6.1. However, the following derivation requires only that f and g be continuous and satisfy $f(x) \geq g(x)$ on the interval $[a, b]$. We wish to find the area between the curves $y = f(x)$ and $y = g(x)$ on this interval. Choose a partition

$$\{x_0 = a, x_1, x_2, \ldots, x_n = b\}$$

of the interval $[a, b]$ and a representative number x_k^* from each subinterval $[x_{k-1}, x_k]$. Next, for each index k, with $k = 1, 2, \ldots, n$, construct a rectangle of width

$$\Delta x_k = x_k - x_{k-1}$$

and height

$$f(x_k^*) - g(x_k^*)$$

equal to the vertical distance between the two curves at $x = x_k^*$. A typical approximating rectangle is shown in Figure 6.2b. We refer to this approximating rectangle as a **vertical strip**.

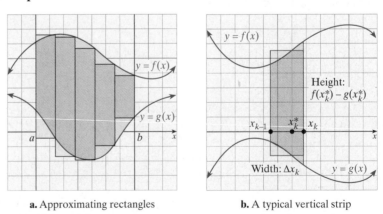

a. Approximating rectangles   **b.** A typical vertical strip

Figure 6.2   Using Riemann sums to find the area between two curves

The representative rectangle has area

$$\Delta A_k = [f(x_k^*) - g(x_k^*)]\Delta x_k$$

and the total area between the curves $y = f(x)$ and $y = g(x)$ can be estimated by the Riemann sum

$$A_n = \sum_{k=1}^{n}[f(x_k^*) - g(x_k^*)]\Delta x_k$$

It is reasonable to expect this estimate to improve if we increase the number of subdivision points in the partition P in such a way that the norm $\|P\|$ approaches 0. Thus, the region between the two curves has area

$$A = \lim_{\|P\|\to 0}\sum_{k=1}^{n}[f(x_k^*) - g(x_k^*)]\Delta x_k$$

which we recognize as the integral of the function $f(x) - g(x)$ over the interval $[a, b]$. These observations may be used to define the area between the curves.

Area Between Two Curves

If f and g are continuous and satisfy $f(x) \geq g(x)$ on the closed interval $[a, b]$, then the **area between the two curves** $y = f(x)$ and $y = g(x)$ is given by

$$A = \int_a^b [f(x) - g(x)]\,dx$$

➡ **What This Says**   To find the area between two curves on a given closed interval $[a, b]$ use the formula

$$A = \int_a^b [\text{TOP CURVE} - \text{BOTTOM CURVE}]\,dx$$

It is no longer necessary to require either curve to be above the x-axis. In fact, we will see later in this section that the curves might cross somewhere in the domain so that one curve is on top for part of the interval and the other curve is on top for the rest.

EXAMPLE 1   Area between two curves

Find the area of the region between the curves $y = x^3$ and $y = x^2 - x$ on the interval $[0, 1]$.

Solution

The region is shown in Figure 6.3. We need to know which curve is the *top curve* on $[0, 1]$. Solve

$$x^3 = x^2 - x \quad \text{or} \quad x(x^2 - x + 1) = 0$$

to find the points of intersection. The only real root is $x = 0$ ($x^2 - x + 1 = 0$ has no real roots). Thus, by the intermediate value theorem, the same curve is on top throughout the interval $[0, 1]$. To see which curve is on top, take some representative value in $[0, 1]$, such as $x = 0.5$, and note that because $(0.5)^3 > 0.5^2 - 0.5$, the curve $y = x^3$ must be above $y = x^2 - x$. Thus, the required area is given by

$$A = \int_0^1 [\underbrace{x^3}_{\text{Top curve}} - \underbrace{(x^2 - x)}_{\text{Bottom curve}}]\,dx = \left(\tfrac{1}{4}x^4 - \tfrac{1}{3}x^3 + \tfrac{1}{2}x^2\right)\Big|_0^1 = \tfrac{5}{12} \quad ∎$$

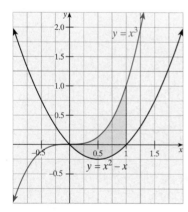

Figure 6.3 The area between the curves $y = x^3$ and $y = x^2 - x$ on $[0, 1]$

In Section 5.1, we defined area for a continuous function f with the restriction that $f(x) \geq 0$. Example 1 shows us that when considering the area between two curves, we need to be concerned no longer with the nonnegative restriction but only with whether $f(x) \geq g(x)$. We now use this idea to find the area for a function that is negative.

EXAMPLE 2   Area with a negative function

Find the area of the region bounded by the curve $y = e^{2x} - 3e^x + 2$ and the x-axis.

Solution

We find the points of intersection of the curve and the x-axis:

$$e^{2x} - 3e^x + 2 = 0 \qquad \text{The curve intersects the } x\text{-axis where } y = 0.$$
$$(e^x - 1)(e^x - 2) = 0 \qquad \text{so that} \quad x = 0, \ln 2$$

The graph of $f(x) = e^{2x} - 3e^x + 2$ is shown in Figure 6.4. We see that on the interval $[0, \ln 2]$, $f(x) \leq 0$, but we can find the area of the given region by considering the area between the curves defined by equations $y = 0$ and $y = e^{2x} - 3e^x + 2$.

$$A = \int_0^{\ln 2} [\underbrace{0}_{\text{Top curve}} - \underbrace{(e^{2x} - 3e^x + 2)}_{\text{Bottom curve}}] \, dx = \left(-\frac{e^{2x}}{2} + 3e^x - 2x \right) \Big|_0^{\ln 2}$$

$$= \frac{3}{2} - 2 \ln 2 \approx 0.114 \qquad \blacksquare$$

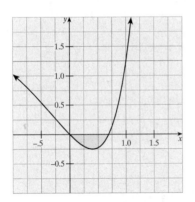

Figure 6.4 Area of region

AREA USING VERTICAL STRIPS

Although the only mathematically correct way to establish a formula involving integrals is to form Riemann sums and take the limit according to the definition of the definite integral, we can simulate this procedure with approximating strips. This simplification is especially useful for finding the area of a complicated region formed when two curves intersect one or more times, as shown in Figure 6.5. Note that the vertical strip has height $f(x) - g(x)$ if $y = f(x)$ is above $y = g(x)$, and height $g(x) - f(x)$ if $y = g(x)$ is above $y = f(x)$. In either case, the height can be represented by $|f(x) - g(x)|$, and the area of the vertical strip is

$$\Delta A = |f(x) - g(x)| \, \Delta x = |f(x) - g(x)| \, dx$$

Thus, we have a new integration formula for area; namely,

$$A = \int_a^b |f(x) - g(x)| \, dx$$

WARNING You cannot use the formula $A = \int [f - g] \, dx$ directly here because the hypothesis $f \geq g$ is not satisfied. In order to use $A = \int |f - g| \, dx$ over the entire interval, you must remember that $|f - g|$ might be $f - g$ over part of the interval and $g - f$ over another part (see Figure 6.5b). Make sure you check to see which curve is on top. Because the curves cross in Example 3, the interval must be subdivided accordingly.

a. A typical vertical strip

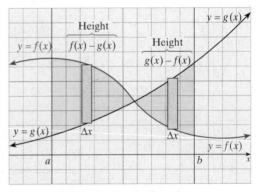

b. Approximation by strips

Figure 6.5   Area by vertical strips

EXAMPLE 3   **Area using vertical strips**

Find the area of the region bounded by the line $y = 3x$ and the curve $y = x^3 + 2x^2$.

Solution

The region between the curve and the line is the shaded portion of Figure 6.6. Part of the process of graphing these curves is to find which is the top curve and which is the bottom. To do this we need to find where the curves intersect:

$$x^3 + 2x^2 = 3x \quad \text{or} \quad x(x + 3)(x - 1) = 0$$

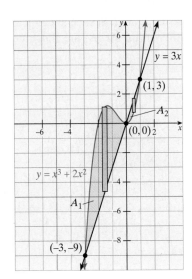

The points of intersection occur at $x = -3, 0$, and 1. In the subinterval $[-3, 0]$, labeled A_1 in Figure 6.6, the curve $y = x^3 + 2x^2$ is on top (test a typical point in the subinterval, such as $x = -1$), and on $[0, 1]$, the region labeled A_2, curve $y = 3x$ is on top. The representative vertical strips are shown in Figure 6.6, and the area between the curve and the line is given by the sum

$$A = \int_{-3}^{0} [(x^3 + 2x^2) - (3x)]\, dx + \int_{0}^{1} [(3x) - (x^3 + 2x^2)]\, dx$$

$$= \left(\tfrac{1}{4}x^4 + \tfrac{2}{3}x^3 - \tfrac{3}{2}x^2\right)\Big|_{-3}^{0} + \left(\tfrac{3}{2}x^2 - \tfrac{1}{4}x^4 - \tfrac{2}{3}x^3\right)\Big|_{0}^{1}$$

$$= 0 - \left(\tfrac{81}{4} - \tfrac{54}{3} - \tfrac{27}{2}\right) + \left(\tfrac{3}{2} - \tfrac{1}{4} - \tfrac{2}{3}\right) - 0$$

$$= \frac{71}{6} \qquad \blacksquare$$

Figure 6.6 The area between the curve $y = x^3 + 2x^2$ and the line $y = 3x$

EXAMPLE 4   **Area using symmetry**

Find the area bounded by the curve $y = \sin x$ and the x-axis between $x = -\pi/2$ and $x = \pi/2$.

Solution

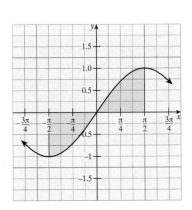

We sketch the region as shown in Figure 6.7. The area we seek is found by

$$\int_{-\pi/2}^{0} (0 - \sin x)\, dx + \int_{0}^{\pi/2} (\sin x - 0)\, dx = \cos x\Big|_{-\pi/2}^{0} + (-\cos x)\Big|_{0}^{\pi/2}$$

$$= (1 - 0) + (0 + 1) = 2$$

There is an easier procedure using symmetry of the curve $y = \sin x$ with respect to the origin. Symmetry uses the fact that the area bounded by the lines $x = -\pi/2$ and $x = 0$ is the same as the area bounded by the lines $x = 0$ and $x = \pi/2$. As a result, we often work this type of problem as shown here:

Figure 6.7 Area bounded by $y = \sin x$, $y = 0$, $x = -\pi/2$ and $x = \pi/2$

$$A = 2 \int_{0}^{\pi/2} (\sin x - 0)\, dx \qquad \text{By symmetry}$$

$$= 2(-\cos x)\Big|_{0}^{\pi/2}$$

$$= 2(0 + 1)$$

$$= 2 \qquad \blacksquare$$

WARNING A common mistake is to work this problem by evaluating the integral

$$\int_{-\pi/2}^{\pi/2} \sin x\, dx = (-\cos x)\Big|_{-\pi/2}^{\pi/2} = -0 + 0 = 0$$

forgetting that the integral does not give the area unless the function is nonnegative.

AREA USING HORIZONTAL STRIPS

For many regions, it is easier to form horizontal strips rather than vertical strips. The procedure for horizontal strips duplicates the procedure for vertical strips. If we want to find the area between two curves of the form $x = F(y)$ and $x = G(y)$ on the interval $[c, d]$ on the y-axis, we form horizontal strips. In Figure 6.8, such a region and its typical horizontal approximating rectangles, or **horizontal strips**, are illustrated. Note that the width of the horizontal strip is denoted by Δy.

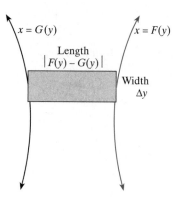

a. Approximation by horizontal strips of width Δy **b.** A typical horizontal strip

Figure 6.8   Area by horizontal strips

Regardless of which curve is "ahead" or "behind," the horizontal strip has length $|F(y) - G(y)|$ and area

$$\Delta A = |F(y) - G(y)|\, \Delta y$$

However, in practice, you must make sure to find the points of intersection of the curves and divide the integrals so that in each region one curve is the *leading curve* ("right curve") and the other is the *trailing curve* ("left curve"). Suppose the curves intersect where $y = b$ for b on the interval $[c, d]$, as shown in Figure 6.8. Then

$$A = \int_c^b \underbrace{[G(y) - F(y)]}_{G \text{ leading } F}\, dy + \int_b^d \underbrace{[F(y) - G(y)]}_{F \text{ leading } G}\, dy$$

EXAMPLE 5   Area using horizontal strips

Find the area of the region between the parabola $x = 4y - y^2$ and the line $x = 2y - 3$.

Solution

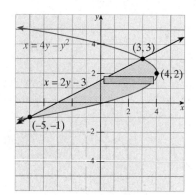

Figure 6.9 The area between the curves $x = 4y - y^2$ and $x = 2y - 3$ using horizontal strips

Figure 6.9 shows the region between the parabola and the line, together with a typical horizontal strip. To find where the line and the parabola intersect, solve

$$4y - y^2 = 2y - 3 \quad \text{to obtain} \quad y = -1 \text{ and } y = 3$$

Throughout the interval $[-1, 3]$ the parabola is to the right of the line (test a typical point between -1 and 3, such as $y = 0$). Thus, the horizontal strip has area

$$\Delta A = [\underbrace{(4y - y^2)}_{\text{Right curve}} - \underbrace{(2y - 3)}_{\text{Left curve}}]\Delta y$$

and the area between the parabola and the line is given by

$$A = \int_{-1}^3 [(4y - y^2) - (2y - 3)]\, dy = \int_{-1}^3 (3 + 2y - y^2)\, dy$$

$$= \left(3y + y^2 - \tfrac{1}{3}y^3\right)\Big|_{-1}^3 = (9 + 9 - 9) - \left(-3 + 1 + \tfrac{1}{3}\right) = \tfrac{32}{3} \qquad \blacksquare$$

Top half of
$y^2 - 4y + x = 0$
(3, 3)
(4, 2)
$y = \frac{1}{2}(x + 3)$
(-5, -1)
Bottom half of
$y^2 - 4y + x = 0$

Figure 6.10 The area between the
curves $x = 4y - y^2$ and
$x = 2y - 3$ using vertical strips

In Example 5, the area can also be found by using vertical strips, but the procedure is more complicated. Note in Figure 6.10 that on the interval $[-5, 3]$, a representative vertical strip would extend from the bottom half of the parabola $y^2 - 4y + x = 0$ to the line $y = \frac{1}{2}(x + 3)$, whereas on the interval $[3, 4]$, a typical vertical strip would extend from the bottom half of the parabola $y^2 - 4y + x = 0$ to the top half. Thus, the area is given by the sum of two integrals, the second of which requires solving the equation $y^2 - 4y + x = 0$ for y using the quadratic formula. With some effort, it can be shown that the computation of area by vertical strips gives the same result as that found by horizontal strips in Example 5.

6.1 PROBLEM SET

A *Sketch a representative vertical or horizontal strip and find the area of the given regions bounded by the specified curves in Problems 1–6.*

1. $y = -x^2 + 6x - 5,$
   $y = \frac{3}{2}x - \frac{3}{2}$

2. $y = x^2 - 8x,$
   $y = 0$

3. $y = \sin 2x$ on $[0, \pi],$
   $y = 0$

4. $y = (x - 1)^3,$
   $y = x - 1$

5. $x = y^2 - 5y,$
   $x = 0$

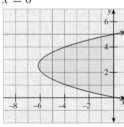

6. $x = y^2 - 6y,$
   $x = -y$

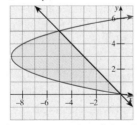

Sketch the region bounded between the given curves and then find the area of each region in Problems 7–25.

7. $y = x^2, y = x, x = -1, x = 1$

8. $y = x^3, y = x, x = -1, x = 1$

9. $y = x^2, y = x^3$

10. $y = x^2, y = \sqrt[3]{x}$

11. $y = x^2 - 1, x = -1, x = 2, y = 0$

12. $y = 4x^2 - 9, x = 3, x = 0, y = 0$

13. $y = x^4 - 3x^2, y = 6x^2$

14. $x = 8 - y^2, x = y^2$

15. $x = 2 - y^2, x = y$

16. $y = x^2 + 3x - 5, y = -x^2 + x + 7$

17. $y = 2x^3 + x^2 - x - 1, y = x^3 + 2x^2 + 5x - 1$

18. $y = \sin x, y = \cos x, x = 0, x = \frac{\pi}{4}$

19. $y = \sin x, y = \sin 2x, x = 0, x = \pi$

20. $y = |x|, y = x^2 - 6$

21. $y = |4x - 1|, y = x^2 - 5, x = 0, x = 4$

22. x-axis, $y = x^3 - 2x^2 - x + 2$

23. y-axis, $x = y^3 - 3y^2 - 4y + 12$

24. $y = e^x, y = \frac{1}{2}e^x + \frac{1}{2}, x = -2, x = 2$

25. $y = \dfrac{1}{\sqrt{1 - x^2}}, y = \dfrac{2}{x + 1}, y$-axis

B 26. WHAT DOES THIS SAY? When finding the area between two curves, discuss criteria for deciding between vertical and horizontal strips.

27. Find the number k (correct to two decimal places) so that the line $y = k$ bisects the area under the curve $y = \sin^{-1} x$ for $0 \le x \le 1$.

28. Find the area of the region that contains the origin and is bounded by the lines $2y = 11 - x$ and $y = 7x + 13$ and the curve $y = x^2 - 5$.

29. Show that the region defined by the inequalities $x^2 + y^2 \le 8$, $x \ge y$, and $y \ge 0$ has area π.

30. Find the area of the region bounded by the curve $\sqrt{x} + \sqrt{y} = 1$ and the coordinate axes.

C **31. Modeling Problem**   Imagine a cylindrical fuel tank of length L lying on its side; the ends are circular with radius b. Determine the amount of fuel in the tank for a given level by completing these steps:

a. Explain why the volume of the tank may be modeled by

$$V = 2L \int_{-b}^{b} \sqrt{b^2 - y^2} \, dy$$

b. Explain why the volume of fuel at level h ($-b \leq h \leq b$)

may be modeled by

$$V(h) = 2L \int_{-b}^{h} \sqrt{b^2 - y^2} \, dy$$

c. Finally, for $b = 4$ and $L = 20$, numerically compute $V(h)$ for $h = -3, -2, \ldots, 4$. *Note*: $V(0)$ and $V(4)$ will serve as a check on your work.

32. The curve with the equation $(x^2 + y^2)^3 = x^2$ has a loop for $0 \leq x \leq 1$. Find the area of the loop. *Hint*: Use the substitution $u = x^{1/3}$ in the area integral.

6.2   Volume

IN THIS SECTION   method of cross sections, method of disks and washers, method of cylindrical shells

The methods used in Section 6.1 to compute area by integration can be modified to compute the volume of a solid region. We will develop methods for computing volume of several kinds of solids in this section, beginning with the case where the solid has a known cross section.

METHOD OF CROSS SECTIONS

Let S be a solid and suppose that for $a \leq x \leq b$, the cross section of S that is perpendicular to the x-axis at x has area $A(x)$. Think of cutting the solid with a knife at $x = x_k^*$ and removing a very thin slab whose face has area $A(x_k^*)$ and whose thickness is Δx_k, as shown in Figure 6.11.

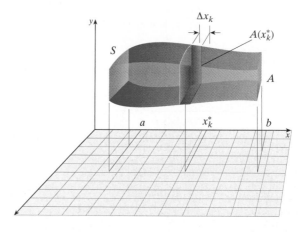

Figure 6.11   Volumes of solids with known cross sections

To find the volume of S, we first take a partition $x_0 = a, x_1, x_2, \ldots, x_n = b$ of the interval $[a, b]$ and choose a representative number x_k^* in each subinterval $[x_{k-1}, x_k]$. Next, we consider a slab with width

$$\Delta x_k = x_k - x_{k-1}$$

and cross-sectional area $A(x_k^*)$, as indicated in Figure 6.11. This slab has volume

$$\Delta V_k = A(x_k^*) \Delta x_k$$

and by adding up the volumes of all such slabs, we obtain an approximation to the volume of the solid S:

$$V_n = \sum_{k=1}^{n} A(x_k^*) \Delta x_k$$

The approximation improves as the number of partition points increases, and it is reasonable to *define* the volume V of the solid S as the limit of V_n as the norm of the partition $\|P\|$ tends to 0. That is,

$$V = \lim_{\|P\| \to 0} \sum_{k=1}^{n} A(x_k^*) \Delta x_k$$

which we recognize as the definite integral $\int_a^b A(x)\, dx$. To summarize:

Volume of a Solid with Known Cross-Sectional Area

A solid S with cross-sectional area $A(x)$ perpendicular to the x-axis at each point on the interval $[a, b]$ has **volume**

$$V = \int_a^b A(x)\, dx$$

EXAMPLE 1   Volume of a solid using square cross sections

The base of a solid is the region in the xy-plane bounded by the y-axis and the lines $y = 1 - x$, $y = 2x + 5$, and $x = 3$. Each cross section perpendicular to the x-axis is a square. Find the volume of the solid.

Solution

The solid resembles a tapered brick, and it may be constructed by "gluing" together a number of thin slabs with square cross sections, like the one shown in Figure 6.12c. We begin by drawing the base in two dimensions and then find the volume of the kth slice.

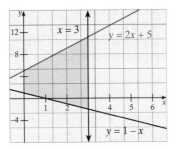

a. Two-dimensional graph of the given base

b. Three-dimensional solid

c. Cross-sectional representative element: $\Delta V = L^2 \Delta x$

Figure 6.12   A solid with a square cross section

To model this construction mathematically, we subdivide the interval $[0, 3]$, form a vertical approximating rectangle on each resulting subinterval, and then construct a slab with square cross section on each approximating rectangle. If we choose the thickness of a typical slab to be Δx and the height and width to be L, the slab will have volume ΔV, where

$$
\begin{aligned}
\Delta V &= L^2 \Delta x \\
&= L^2(x) \Delta x \\
&= [(2x + 5) - (1 - x)]^2 \Delta x \\
&= (3x + 4)^2 \Delta x
\end{aligned}
$$

The volume of the entire solid is obtained by integrating to "add up" all the volumes ΔV, and we find that the volume of the solid is

$$
\begin{aligned}
V &= \int_0^3 (3x + 4)^2 \, dx \\
&= \int_0^3 (9x^2 + 24x + 16) \, dx \\
&= (3x^3 + 12x^2 + 16x)\big|_0^3 = 237
\end{aligned}
$$

$\blacksquare$

EXAMPLE 2   Volume of a regular pyramid with a square base

SMH See Example 1.7, Section 1.5, of the *Student Mathematics Handbook*.

A regular pyramid has a square base of side L and its apex located H units above the center of its base. Derive the formula $V = \frac{1}{3}HL^2$.

Solution

The pyramid is shown in Figure 6.13.

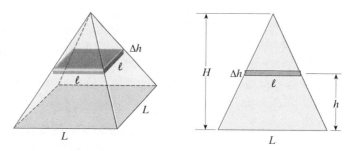

Figure 6.13   The volume of a pyramid

This pyramid can be constructed by stacking a number of thin square slabs. Suppose that a representative slab has side ℓ and thickness Δh, and that it is located h units above the base of the pyramid as shown in Figure 6.13. By creating a proportion from corresponding parts of similar triangles, we see that

$$
\frac{\ell}{L} = \frac{H - h}{H} \quad \text{so that} \quad \ell = \left(1 - \frac{h}{H}\right)L
$$

Therefore, the volume of the representative slab is

$$
\begin{aligned}
\Delta V &= \ell^2 \Delta h \\
&= \left(1 - \frac{h}{H}\right)^2 L^2 \Delta h
\end{aligned}
$$

To compute the volume V of the entire pyramid, we integrate with respect to h from the base of the pyramid ($h = 0$) to the apex ($h = H$). Thus,

$$
\begin{aligned}
V &= \int_0^H \left(1 - \frac{h}{H}\right)^2 L^2 \, dh = L^2 \int_0^H \left[1 - \frac{2}{H}h + \frac{1}{H^2}h^2\right] dh \\
&= L^2 \left(h - \frac{h^2}{H} + \frac{h^3}{3H^2}\right)\bigg|_0^H = L^2 \left(H - \frac{H^2}{H} + \frac{H^3}{3H^2}\right) = \frac{1}{3}HL^2
\end{aligned}
$$

$\blacksquare$

Other volume formulas can be found in a similar fashion (see Problems 61 and 62).

METHOD OF DISKS AND WASHERS

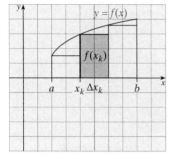

a. A representative vertical strip has height $f(x_k)$ and width Δx_k.

b. The representative disk is formed by revolving the representative strip about the x-axis. This disk has radius $f(x_k)$ and thickness Δx_k.

Figure 6.14 The disk method

A **solid of revolution** is a solid figure S obtained by revolving a region R in the xy-plane about a line L (called the **axis of revolution**) that lies outside or on the boundary of R. Note that such a solid S may be thought of as having circular cross sections in the direction perpendicular to L.

Suppose the function f is continuous and satisfies $f(x) \geq 0$ on the interval $[a, b]$, and suppose we wish to find the volume of the solid S generated when the region R under the curve $y = f(x)$ on $[a, b]$ is revolved about the x-axis. That is, *the axis of revolution is horizontal and it is a boundary of the region R.* Our strategy will be to form vertical strips and revolve them about the x-axis, generating what are called **disks** (that is, thin right circular cylinders) that approximate a portion of the solid of revolution S, as shown in Figure 6.14.

Now we can compute the total volume of S by using integration to sum the volumes of all the approximating disks. Recall that the formula for the volume of a cylinder of height h and cross-sectional area A is Ah. Figure 6.14a shows a typical vertical strip with height $f(x_k)$ and width Δx_k. You might notice that the width of the strip is the same as the thickness of the disk. The solid of revolution can be thought of as having cross sections perpendicular to the x-axis that are circular disks of volume

$$\Delta V_k(x) = \underbrace{\pi [f(x_k)]^2}_{\text{Area of circular cross section}} \Delta x_k$$

The total volume may be found by integration:

$$V = \int_a^b \overbrace{A(x)}^{\text{Area of cross section}} \; \overbrace{dx}^{\text{Thickness}} = \int_a^b \pi [f(x)]^2 \, dx$$

This procedure may be summarized as follows:

The Disk Method

The **disk method** is used to find a volume generated when a region R is revolved about an axis L that is *perpendicular* to a typical approximating strip in R. Suppose R is the region bounded by the curve $y = f(x)$, the x-axis, and the vertical lines $x = a$ and $x = b$. Then if R is revolved about the x-axis, it generates a solid with volume

$$V = \int_a^b \pi y^2 \, dx = \int_a^b \pi [f(x)]^2 \, dx$$

→ **What This Says** The following diagram may help you remember the key ideas behind the disk method.

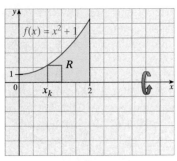

a. The region under
$y = x^2 + 1$ on $[0, 2]$

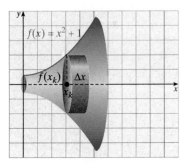

b. The solid of revolution

Figure 6.15 Volume of a solid of
revolution: Disk method

EXAMPLE 3   Volume by disks

Find the volume of the solid S formed by revolving the region under the curve
$y = x^2 + 1$ on the interval $[0, 2]$ about the x-axis.

Solution

The region is shown in Figure 6.15.

$$V = \pi \int_0^2 (x^2 + 1)^2 \, dx = \pi \int_0^2 (x^4 + 2x^2 + 1) \, dx$$

$$= \pi \left(\frac{1}{5}x^5 + \frac{2}{3}x^3 + x \right)\Big|_0^2 = \frac{206}{15}\pi \approx 43.14453911 \qquad \blacksquare$$

With a small modification of the disk method, we can find the volume of a solid
figure generated by revolving about the x-axis the region between two curves $y = f(x)$
and $y = g(x)$, where $f(x) \geq g(x)$ for $a \leq x \leq b$. When a typical vertical strip is
revolved about the x-axis, a "washer" with cross-sectional area $\pi([f(x)]^2 - [g(x)]^2)$
is formed, as shown in Figure 6.16. This can be remembered by $\pi R^2 - \pi r^2$, where R
is the outer radius and r is the inner radius.

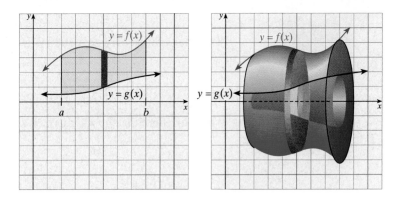

Figure 6.16   The washer method

The volume of the solid of revolution is found by the formula given in the following box.

The Washer Method

The **washer method** is used to find volume when a region between two
curves is revolved about an external axis perpendicular to the approximating
strip. In particular, suppose f and g are continuous functions on $[a, b]$ with
$f(x) \geq g(x) \geq 0$. If R is the radius of the (outer) curve $y = f(x)$ and r is
the radius of the (inner) curve $y = g(x)$, and the region bounded by $y = f(x)$,
$y = g(x)$, $x = a$, and $x = b$ is revolved about the x-axis, then the volume thus
formed is

$$V = \int_a^b \pi \left(\underbrace{[f(x)]^2}_{\text{Outer radius}} - \underbrace{[g(x)]^2}_{\text{Inner radius}} \right) dx$$

WARNING $f^2 - g^2 \neq (f - g)^2$

The disk method and the washer method also apply when the axis of revolution is
a line other than the x-axis. In Example 4, we consider what happens when a particular
region R is revolved not only about the x-axis, but also about other axes.

EXAMPLE 4   Volume by washers

Let D be the solid region bounded by the parabola $y = x^2$ and the line $y = x$. Find the volume of the solid generated when D is revolved about the

a. x-axis   **b.** y-axis   **c.** line $y = 2$

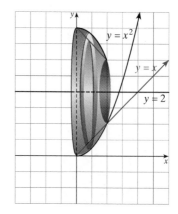

Solution

The region D to be rotated is shown in Figure 6.17. Note that since $x = x^2$ at $x = 0$ and at $x = 1$, the line and parabola intersect at the origin and at $(1, 1)$.

a. The line is always above the parabola on the interval $[0, 1]$, so when we form a washer to approximate the volume of revolution, the outer radius is $R = x$ and the inner radius is $r = x^2$, as shown in Figure 6.17a. Thus, the required volume is

$$V = \pi \int_0^1 [x^2 - (x^2)^2]\, dx$$

$$= \pi \int_0^1 (x^2 - x^4)\, dx$$

$$= \pi \left(\tfrac{1}{3}x^3 - \tfrac{1}{5}x^5\right)\Big|_0^1 = \frac{2\pi}{15}$$

a. The vertical strip is rotated about the x-axis

b. The horizontal strip is rotated about the y-axis

c. The vertical strip is rotated about the line $y = 2$

Figure 6.17   Volume obtained by revolving a region about different lines

b. Because we are revolving R about the y-axis, we use *horizontal* strips to approximate the solid of revolution, as shown in Figure 6.17b. Note that the parabola $x = \sqrt{y}$ is to the right of the line $x = y$ on the interval $[0, 1]$, so the approximating washer has outer radius $R = \sqrt{y}$ and inner radius $r = y$; thus,

$$V = \pi \int_0^1 [(\sqrt{y})^2 - (y)^2] \, dy$$

$$= \pi \int_0^1 (y - y^2) \, dy$$

$$= \pi \left[\frac{y^2}{2} - \frac{y^3}{3} \right]\Big|_0^1 = \frac{\pi}{6}$$

c. Since the axis of rotation is the line $y = 2$, the outer radius of a typical approximating washer is $R = 2 - x^2$, and the inner radius is $r = 2 - x$, as shown in Figure 6.17c. Thus, the volume of revolution is

$$V = \pi \int_0^1 [(2 - x^2)^2 - (2 - x)^2] \, dx$$

$$= \pi \int_0^1 (x^4 - 5x^2 + 4x) \, dx$$

$$= \pi \left[\frac{x^5}{5} - \frac{5x^3}{3} + \frac{4x^2}{2} \right]\Big|_0^1 = \frac{8\pi}{15} \qquad \blacksquare$$

METHOD OF CYLINDRICAL SHELLS

Sometimes it is easier (or even necessary) to compute a volume by taking the approximating strip parallel to the axis of rotation instead of perpendicular to the axis as in the disk and washer methods. Figure 6.18a shows a region R under the curve $y = f(x)$ on the interval $[a, b]$, together with a representative vertical strip. When this strip is revolved about the y-axis, it forms a solid called a **cylindrical shell** of height $f(x)$ and thickness Δx, as shown in Figure 6.18b.

a. A vertical strip in the region R under the curve $y = f(x)$ on the interval $[a, b]$

b. When the strip is revolved about the y-axis, a shell is generated.

c. The unwrapped "flattened" shell has volume $\Delta V = 2\pi x f(x) \Delta x$.

Figure 6.18   Method of cylindrical shells

Because the strip is x units from the axis of rotation and is assumed to be very thin, the cross section of the shell (perpendicular to the y-axis) will be a circle of radius x and circumference $2\pi x$. If we imagine the shell to be cut and flattened out, it is seen to be a rectangular slab of volume

$$\Delta V = \underbrace{2\pi x f(x)}_{\text{Area of the rectangular slab}} \overbrace{\Delta x}^{\text{Thickness}}$$

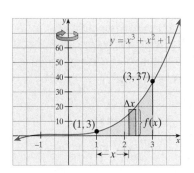

Figure 6.19 Approximating a solid of revolution by cylindrical shells

(See Figure 6.18c.) Thus, the total volume of the solid is given by the integral

$$V = \int_a^b 2\pi x f(x) \, dx$$

The approximating shells are shown in Figure 6.19 and the formula is repeated in the following box.

Method Of Cylindrical Shells

The **shell method** is used to find a volume generated when a region R is revolved about an axis L *parallel* to a typical approximating strip in R. In particular, if R is the region, as shown in Figure 6.20, bounded by the curve $y = f(x)$, the x-axis, and the vertical lines $x = a$ and $x = b$, where $0 \le a \le b$, then the solid generated by revolving R about the y-axis has volume

$$V = \int_a^b 2\pi x f(x) \, dx$$

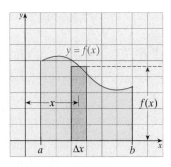

Figure 6.20   Shell Method

EXAMPLE 5   Volume using cylindrical shells

Find the volume of the solid formed by revolving the region bounded by the x-axis and the graphs of $y = x^3 + x^2 + 1$, $x = 1$, and $x = 3$ about the y-axis.

Solution

The region to be rotated is shown in Figure 6.21, along with a typical vertical strip of height $f(x) = x^3 + x^2 + 1$ and width Δx. The volume, by the method of shells, is

$$V = 2\pi \int_1^3 x(x^3 + x^2 + 1) \, dx = 2\pi \int_1^3 (x^4 + x^3 + x) \, dx$$
$$= 2\pi \left(\tfrac{1}{5}x^5 + \tfrac{1}{4}x^4 + \tfrac{1}{2}x^2 \right)\Big|_1^3 = 144.8\pi \approx 454.9 \qquad \blacksquare$$

Figure 6.21 A volume of revolution by the shell method

When the axis of rotation is a line $x = L$ parallel to the y-axis, the distance of a typical vertical strip from the axis will no longer be x. Here is an example that illustrates how to compute volume of revolution in such a case.

EXAMPLE 6   Volume of revolution about a horizontal or vertical line

Find the volume of the solid formed by revolving the region R bounded by the curve $y = x^{-2}$ and the x-axis for $1 \le x \le 2$ about the line $x = -1$.

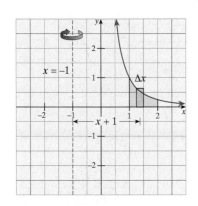

Figure 6.22 Volume by shells (rotating about $x = -1$)

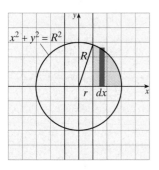

Figure 6.23 Drill a hole in a sphere

Figure 6.24 Volume of solid by cylindrical shells; strip parallel to axis of rotation

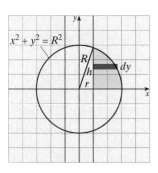

Figure 6.25 Volume of solid by washer method; strip perpendicular to axis of rotation

Solution

The region R is shown in Figure 6.22, along with a typical vertical strip of height $f(x) = x^{-2}$. If we were revolving this strip about the y-axis, the distance from the axis to the strip would be $L = x$, but since the axis of rotation is $x = -1$, the distance is $L = x + 1$.

$$V = 2\pi \int_1^2 [x - (-1)] \left(\frac{1}{x^2}\right) dx$$

$$= 2\pi \int_1^2 (x^{-1} + x^{-2}) \, dx$$

$$= 2\pi \left[\ln x - x^{-1}\right]\Big|_1^2$$

$$= 2\pi \left[\ln 2 + \tfrac{1}{2}\right]$$

$$\approx 7.4968 \qquad \blacksquare$$

Sometimes you have a choice between using shells or washers to compute a volume of revolution. The decision as to which method to use often comes down to which leads to the easier integration. Which method would you choose in the following example?

EXAMPLE 7 Comparing the methods of shells and washers

A cylindrical hole of radius r is bored through the center of a solid sphere of radius R $(r < R)$, as shown in Figure 6.23. Find the volume of the solid that remains by using

a. cylindrical shells   **b.** washers

Solution

a. The required volume can be thought of as twice the volume generated when the shaded region of Figure 6.24 is revolved about the y-axis.

If we use vertical strips, a typical strip has height $y = \sqrt{R^2 - x^2}$ and is x units from the axis of rotation. The volume is given by

$$V = 2\int_r^R 2\pi x y \, dx = 4\pi \int_r^R x\sqrt{R^2 - x^2} \, dx$$

$$= 4\pi \left[-\frac{1}{3}(R^2 - x^2)^{3/2}\right]\Big|_r^R = \frac{4\pi}{3}(R^2 - r^2)^{3/2}$$

where we have used the substitution $u = R^2 - x^2$ to evaluate the integral.

b. For the washer method, we use horizontal strips and note that a typical washer has outer radius $x = \sqrt{R^2 - y^2}$ and inner radius r (see Figure 6.25). We integrate from 0 to h, where $h = \sqrt{R^2 - r^2}$, so the volume (again by symmetry) is given by

$$V = 2\int_0^h \pi \left[\left(\sqrt{R^2 - y^2}\right)^2 - r^2\right] dy$$

$$= 2\pi \int_0^{\sqrt{R^2 - r^2}} \left[(R^2 - y^2) - r^2\right] dy$$

$$= 2\pi \left[(R^2 - r^2)y - \frac{y^3}{3}\right]\Big|_0^{\sqrt{R^2 - r^2}}$$

$$= 2\pi \left[(R^2 - r^2)^{3/2} - \frac{1}{3}(R^2 - r^2)^{3/2}\right]$$

$$= \frac{4\pi}{3}(R^2 - r^2)^{3/2} \qquad \blacksquare$$

Table 6.1 compares and contrasts the disk, washer, and shell methods for computing volume.

TABLE 6.1   Volumes of revolution when the axis of revolution is either the *x*-axis or the *y*-axis.

Summary

a. Disk method: A representative rectangle is **perpendicular** to the axis of revolution. The axis of revolution is a boundary of the region.

b. Washer method: A representative rectangle is **perpendicular** to the axis of revolution. The axis of revolution is *not* part of the boundary.

Horizontal axis of revolution

Width of rectangle is Δx.

$$V = \pi \int_a^b \underbrace{[f(x)]^2 \, dx}_{\text{Length of rectangle}}$$

Vertical axis of revolution

Width of rectangle is Δy.

$$V = \pi \int_c^d \underbrace{[g(y)]^2 \, dy}_{\text{Length of rectangle}}$$

Horizontal axis of revolution

$$V = \pi \int_a^b \left(\overset{\text{Top curve}}{\overset{\downarrow}{[f(x)]^2}} - \underset{\underset{\text{Bottom curve}}{\uparrow}}{[g(x)]^2}\right) \overset{\text{Width}}{\overbrace{dx}}$$

Vertical axis of revolution

$$V = \pi \int_c^d \left(\overset{\text{Right curve}}{\overset{\downarrow}{[f(y)]^2}} - \underset{\underset{\text{Left curve}}{\uparrow}}{[g(y)]^2}\right) \overset{\text{Width}}{\overbrace{dy}}$$

c. Shell method: A representative rectangle is **parallel** to the axis of revolution.

Horizontal axis of revolution

$$V = 2\pi \int_c^d y \underbrace{[g(y)]}_{\uparrow\text{Length}} \overset{\text{Width}}{\overbrace{dy}}$$
Distance to axis

Vertical axis of revolution

$$V = 2\pi \int_a^b x \underbrace{[f(x)]}_{\uparrow\text{Length}} \overset{\text{Width}}{\overbrace{dx}}$$
Distance to axis

6.2   PROBLEM SET

Ⓐ *In Problems 1–4, sketch the given region and then find the volume of the solid whose base is the given region and which has the property that each cross section perpendicular to the x-axis is a* **square.**

1. the triangular region bounded by the coordinate axes and the line $y = 3 - x$

2. the region bounded by the x-axis and the semicircle $y = \sqrt{16 - x^2}$

3. the region bounded by the line $y = x + 1$ and the curve $y = x^2 - 2x + 3$

4. the region bounded above by $y = \sqrt{\sin x}$ and below by the x-axis on the interval $[0, \pi]$

In Problems 5–8, sketch the region and then find the volume of the solid whose base is the given region and which has the property that each cross section perpendicular to the x-axis is an **equilateral triangle.**

5. the region bounded by the circle $x^2 + y^2 = 9$

6. the region bounded by the curves $y = x^3$ and $y = x^2$

7. the region bounded above by $y = \sqrt{\cos x}$ and below by the x-axis on the interval $\left[-\frac{\pi}{2}, \frac{\pi}{2}\right]$

8. the triangular region with vertices $A(1, 1)$, $B(3, 5)$, and $C(3, -2)$

In Problems 9–12, sketch the region and then find the volume of the solid where the base is the given region and which has the property that each cross section perpendicular to the x-axis is a semicircle.

9. the region bounded by the y-axis, the parabola $y = x^2$, and the line $2x + y - 3 = 0$

10. the region bounded above by $y = \cos x$, below by $y = \sin x$, and on the left by the y-axis

11. the region bounded above by the curve $y = \tan x$ and below by the x-axis, on the interval $\left[0, \frac{\pi}{4}\right]$

12. the region bounded by the x-axis and the curve $y = e^x$ between $x = 1$ and $x = 3$

In Problems 13–20, find the volume of the solid formed when the region described is revolved about the x-axis using washers or disks.

13. the region under the curve $y = \sqrt{x}$ on the interval $0 \le x \le 1$

14. the region under the curve $y = \sqrt[3]{x}$ on the interval $0 \le x \le 8$

15. the region bounded by the lines $y = x$, $y = 2x$, and $x = 1$

16. the region bounded by the lines $x = 0$, $x = 1$, $y = x + 1$, and $y = x + 2$

17. the region under the curve $y = x^2 + x^3$ on the interval $0 \le x \le \pi$; approximate your answer to the nearest unit

18. the region under the curve $y = \sqrt{2 \sin x}$ on the interval $0 \le x \le \pi$

19. the region between $y = \sin x$ and $y = \cos x$ on $0 \le x \le \pi/4$

20. the region bounded by the curves $y = e^x$ and $y = e^{-x}$ on $[0, 2]$

In Problems 21–26, use shells to find the volume of the solid formed by revolving the given region about the y-axis.

21. the region bounded by the lines $y = 2x$, the y-axis, and $y = 1$

22. the region bounded by the curve $y = \sqrt{x}$, the y-axis, and the line $y = 1$

23. the region bounded by the parabola $y = 1 - x^2$, the y-axis, and the positive x-axis

24. the region bounded by the parabolas $y = x^2$, $y = 1 - x^2$, and the y-axis for $x \ge 0$

25. the region bounded by the curve $y = e^{-x^2}$, the y-axis, and the line $y = \frac{1}{2}$

26. the region inside the ellipse $2(x - 3)^2 + 3(y - 2)^2 = 6$ about the y-axis

In Problems 27–34, set up but do not evaluate an integral using the shaded strips for the volume generated when the given region is revolved about:

   a. the x-axis

   b. the y-axis

   c. the line $y = -1$

   d. the line $x = -2$

27. $y = 4 - x$, $0 \le x \le 4$

28. $y = 4 - x^2$, $0 \le x \le 2$

29. $y = \sqrt{4 - x^2}$, $0 \le x \le 2$

30. $y = \sqrt{4 - x^2}$, $0 \le x \le 2$

31. $y = e^{-x}$, $0 \le x \le 1$

32. $y = \dfrac{x}{x + 1}$, $0 \le x \le 1$

33. $y = \sin^{-1} x$, $0 \le x \le 1$

34. $y = \ln x$, $1 \le x \le 2$

In Problems 35–40, draw a representative strip and set up an integral for the volume of the solid formed by revolving the given region:

a. about the x-axis   b. about the y-axis

Set up the integral only; DO NOT EVALUATE.

35. the region bounded by $x = y^2$ and $y = x^2$

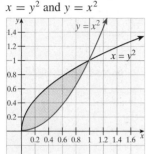

36. the region bounded by $y = \frac{1}{3}x$ and $x = y^2$

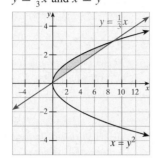

37. the region bounded by the lines $y = 1$, $x = 2$, and the curve $y = x^2 + 1$

38. the region bounded by the curves $y = x^2$ and $y = -x^2 - 4x$

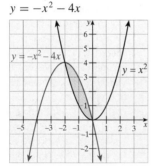

39. the region bounded by $y = 0.1x^2$ and $y = \ln x$

40. the region bounded by $y = e^x - 1$, $y = 2e^{-x}$, and $x = 0$

B In each of Problems 41–44, find the volume by the

a. disk/washer method   b. shell method

41. Revolve the region bounded by $y = x^2$ and $y = x^3$ about the x-axis.

42. Revolve the region bounded by $y = x^2$ and $y = x^3$ about the y-axis.

43. Revolve the region bounded by $y = x$, $y = 2x$, and $y = 1$ about the x-axis.

44. Revolve the region bounded by $y = x$, $y = 2x$, and $y = 1$ about the y-axis.

45. Derive the formula for the volume of a sphere of radius r by revolving the region bounded by the semicircle $y = \sqrt{r^2 - x^2}$ about the x-axis.

In Problems 46–49, find the volume of the solid whose base is bounded by the circle $x^2 + y^2 = 9$ with the indicated cross sections taken perpendicular to the x-axis.

46. squares

47. equilateral triangles

48. isosceles right triangles

49. semicircles

In Problems 50–51, find the volume V of the solid with the given information regarding its cross section.

50. The base of the solid is the hyperbola

$$\frac{x^2}{4} - \frac{y^2}{9} = 1$$

for $2 \le x \le 5$, and the cross sections perpendicular to the x-axis are squares.

51. The base of the solid is an equilateral triangle, each side of which has length 4. The cross sections perpendicular to a given altitude of the triangle are squares.

52. The base of the solid is an isosceles right triangle whose legs are each 4 units long. Each cross section perpendicular to a side is a semicircle.

53. The great pyramid of Cheops is approximately 480 ft tall and 750 ft square at the base. Find the volume of this pyramid by using the cross section method.

54. When viewed from above, a swimming pool has the shape of the ellipse

$$\frac{x^2}{900} + \frac{y^2}{400} = 1$$

The cross sections of the pool perpendicular to the ground and parallel to the y-axis are squares. If the units are in feet, what is the volume of the pool?

55. Find the volume of the solid generated when the region $y = x^{-1/2}$ on the interval $[1, 4]$ is revolved about
  a. the x-axis   **b.** the y-axis   **c.** the line $y = -2$

56. Let R be the region in the first quadrant bounded by the curve $y = kx$ and the line $x = k$, $k > 0$. Find the volume when R is revolved about
  a. the y-axis   **b.** the line $x = 2k$

57. The portion on the ellipse $\dfrac{x^2}{9} + \dfrac{y^2}{4} = 1$ with $x \geq 0$ is rotated about the y-axis to form a solid S. A hole of radius 1 is drilled through the center of S, along the y-axis. Find the volume of the part of S that remains.

58. Cross-sectional areas are measured at 1-foot intervals along the length of an irregularly shaped object, with the results listed in the following table (x in ft and A in ft^2):

x	0	1	2	3	4	5
A	1.12	1.09	1.05	1.03	0.99	1.01

x	6	7	8	9	10
A	0.98	0.99	0.96	0.93	0.91

Estimate the volume (correct to the nearest hundredth) by using the trapezoidal rule.

59. Cross-sectional areas are measured at 2-meter intervals along the length of an irregularly shaped object, with the results listed in the following table (x in meters and A in m^2):

x	0	2	4	6	8	10
A	1.12	1.09	1.05	1.03	0.99	1.01

x	12	14	16	18	20
A	0.98	0.99	0.96	0.93	0.91

Estimate the volume (correct to the nearest hundredth) by using Simpson's rule.

60. HISTORICAL QUEST   Johannes Kepler is usually remembered for his work in astronomy (see Section 10.3, and Historical Quests at the end of Chapter 10). Kepler also made interesting mathematical discoveries. In fact, he has been described as "number-intoxicated." He was looking for mathematical harmonies in the physical universe. He is quoted in *The World of Mathematics*[*]: "Nothing holds me; I will indulge my sacred fury; I will triumph over mankind by the honest confession that I have stolen the golden vases of the Egyptians to build up a tabernacle for my God far away from the confines of Egypt." ∎

JOHANNES KEPLER
1571–1630

In this Historical Quest we look at Kepler's derivation for the volume of a torus, generated by revolving a circle of radius a around a vertical axis at a distance b from its center ($b \geq a$). He found $V = (\pi a^2)(2\pi b) = 2\pi^2 a^2 b$. He derived this formula by dissecting a torus into infinitely many thin vertical circular slices by considering planes perpendicular to the axis of revolution, as shown in Figure 6.26.

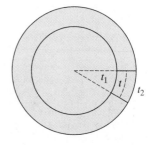

Figure 6.26 Torus cross section

Note that each slice is thinner (t_1) on the inside (nearer the axis) and thicker (t_2) on the outside. Kepler assumed that the volume of each slice is $\pi a^2 t$, where $t = \frac{1}{2}(t_1 + t_2)$. Because t is the average of its minimum and maximum thickness, it must be the thickness of the slice at its center. Use washers or shells to verify Kepler's formula for the volume of a torus.

C 61. A hemisphere of radius r may be regarded as a solid whose base is the region bounded by the circle $x^2 + y^2 = r^2$ and with the property that each cross section perpendicular to the x-axis is a semicircle with a diameter in the base. Use this characterization and the method of cross sections to show that a sphere of radius r has volume $V = \frac{4}{3}\pi r^3$.

62. Use the method of cross sections to show that the volume of a regular tetrahedron of side a is $\frac{1}{12}\sqrt{2}a^3$.

63. A student, Frank Kornercutter, conjectures that a tetrahedron is nothing more than a solid figure with an equilateral triangular base and cross sections perpendicular to that base that are also equilateral triangles. Is Frank correct this time? Either prove that he is or show that his conjecture must be false.

64. Find the volume of the football-shaped solid (called an *ellipsoid*) formed by revolving the ellipse

$$\frac{x^2}{a^2} + \frac{y^2}{b^2} = 1$$

about the x-axis. What is the volume if the ellipse is revolved about the y-axis?

65. A "cap" is formed by truncating a sphere of radius R at a point h units from the center, as shown in Figure 6.27. Find the volume of the cap.

Figure 6.27 Volume of the cap of a sphere

66. The *frustum of a cone* is the solid region bounded by a cone and two parallel planes, as shown in Figure 6.28. Suppose the planes are h units apart and intersect the cone in plane regions of area A_1 and A_2, respectively. Use integration to show that the frustum has volume

$$V = \frac{h}{3}(A_1 + \sqrt{A_1 A_2} + A_2)$$

Figure 6.28 Volume of the frustum of a cone

6.3 Polar Forms and Area

IN THIS SECTION the polar coordinate system, polar graphs, intersection of polar-form curves, polar area

THE POLAR COORDINATE SYSTEM

In the **polar coordinate system**, points are plotted in relation to a fixed point O, called the origin or **pole** and a fixed ray emanating from the origin called the **polar axis**. We then associate with each point P in the plane an ordered pair of numbers $P(r, \theta)$, where r is the distance from O to P and θ is the angle measured from the polar axis to the ray OP, as shown in Figure 6.29. The number r is called the **radial coordinate** of P, and θ is the **polar angle**. The polar angle is regarded as positive if measured counterclockwise up from the polar axis, and negative if measured clockwise. The origin O has radial coordinate 0, and it is convenient to say that O has polar coordinates $(0, \theta)$, for all angles θ.

If the point P has polar coordinates (r, θ), we say that the point Q obtained by reflecting P in the origin O has coordinates $(-r, \theta)$. Thus, if you think of a pencil lying along the directed line segment $\overline{OP}$, with its midpoint at O and tip at $P(r, \theta)$, then the eraser will be at $Q(-r, \theta)$, as illustrated in Figure 6.29.

While each point in the plane is associated with exactly one pair of rectangular coordinates, a given point in polar coordinates has an infinite number of representations. For example, $\left(5, \frac{3\pi}{2}\right)$, $\left(-5, \frac{\pi}{2}\right)$, and $\left(5, -\frac{\pi}{2}\right)$ all represent the same point in polar coordinates. The nonuniqueness of representation in the polar coordinate system causes some difficulties, but they can be handled by exercising a little care. We will point out situations in our examples where the multiplicity of polar representation must be taken into account.

Figure 6.29 Polar-form points

(SMH) For a more complete discussion of the polar coordinate system, see Chapter 5 of the *Student Mathematics Handbook*.

There is a simple trigonometric relationship linking polar and rectangular coordinates. Specifically, if we assume that the origin of the rectangular coordinate system is the pole and the positive x-axis is the polar axis, then the rectangular coordinates (x, y) of each point are related to the polar coordinates (r, θ) of the same point by the formulas summarized in the following box.

Relationship between Rectangular and Polar Coordinates

1. To change *from polar to rectangular*:

$$x = r \cos \theta \qquad y = r \sin \theta$$

2. To change *from rectangular to polar*:

$$r = \sqrt{x^2 + y^2}$$

$$\tan \theta = \frac{y}{x} \quad \text{if } x \neq 0$$

Note that the preceding formula gives r as the distance from the pole to the point. This does not mean that ordered pairs in polar form need to have positive first components. For example $\left(1, \frac{3\pi}{2}\right)$ is the same point as $\left(-1, \frac{\pi}{2}\right)$.

EXAMPLE 1   Converting from polar to rectangular coordinates

Change the polar coordinates $\left(-3, \frac{5\pi}{4}\right)$ to rectangular coordinates.

Solution

$$x = -3 \cos \frac{5\pi}{4} = -3 \left(-\frac{\sqrt{2}}{2}\right) = \frac{3\sqrt{2}}{2}$$

$$y = -3 \sin \frac{5\pi}{4} = -3 \left(-\frac{\sqrt{2}}{2}\right) = \frac{3\sqrt{2}}{2}$$

The rectangular coordinates are $\left(\dfrac{3\sqrt{2}}{2}, \dfrac{3\sqrt{2}}{2}\right)$. ∎

EXAMPLE 2   Converting from rectangular coordinates to polar coordinates

Write polar-form coordinates for the point with rectangular coordinates $\left(\dfrac{5\sqrt{3}}{2}, -\dfrac{5}{2}\right)$.

Solution
$$r = \sqrt{\left(\frac{5\sqrt{3}}{2}\right)^2 + \left(-\frac{5}{2}\right)^2} = \sqrt{\frac{75}{4} + \frac{25}{4}} = 5$$

Note that θ is in Quadrant IV because x is positive and y is negative.

$$\bar{\theta} = \tan^{-1}\left|\frac{-\frac{5}{2}}{\frac{5\sqrt{3}}{2}}\right| = \tan^{-1}\left(\frac{1}{\sqrt{3}}\right) = \frac{\pi}{6}; \quad \text{thus,} \quad \theta = \frac{11\pi}{6} \qquad \text{(Quadrant IV)}$$

Polar-form coordinates are $\left(5, \frac{11\pi}{6}\right)$. A representation with a negative r is $\left(-5, \frac{5\pi}{6}\right)$. ∎

POLAR GRAPHS

The **graph** of an equation in polar coordinates is the set of all points P whose polar coordinates (r, θ) satisfy the given equation. Circles, lines through the origin, and rays emanating from the origin have particularly simple equations in polar coordinates.*

EXAMPLE 3   Graphing circles, lines, and rays

Graph

a. $r = 6$ **b.** $\theta = \frac{\pi}{6}$

Solution

a. The graph is the set of all points (r, θ) such that the first component is 6 for any angle θ. This is a circle with radius 6 centered at the origin, as shown in Figure 6.30a.

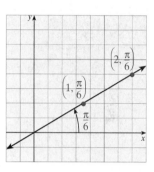

a. The circle $r = 6$

b. The line $\theta = \frac{\pi}{6}$

Figure 6.30   Two polar graphs

b. The graph is the set of all polar points (r, θ) with polar angle $\theta = \frac{\pi}{6}$. This is a line through the origin (pole) that makes an angle $\pi/6$ with the positive x-axis, as shown in Figure 6.30b. If we further require that $r \geq 0$, then the graph is a *ray* (sometimes called a *half-line*).   ■

EXAMPLE 4   Graphing a polar curve by converting to rectangular coordinates

Graph $r = 4 \cos \theta$ by first converting the polar form to rectangular form.

Solution

$$r = 4 \cos \theta \qquad \text{Given}$$
$$r^2 = 4r \cos \theta \qquad \text{Multiply both sides by } r.$$
$$x^2 + y^2 = 4x \qquad r^2 = x^2 + y^2 \text{ and } x = r \cos \theta$$
$$x^2 - 4x + y^2 = 0$$
$$(x - 2)^2 + y^2 = 2^2 \qquad \text{Complete the square; add } 2^2 \text{ to both sides.}$$

This is a circle with center at $(2, 0)$ and radius 2, as shown in Figure 6.31.   ■

In our next example, we sketch a more complicated polar graph by plotting points in a polar coordinate plane. The graph is called a **cardioid** because of its heart-like shape.

Figure 6.31 Graph of $r = 4 \cos \theta$

---

*Many books use what is called **polar graph paper**, but such paper is not really necessary. It also obscures the fact that polar curves and rectangular curves are both plotted as ordered pairs, only with a different meaning attached to the ordered pairs. You can estimate the angles as necessary without polar graph paper.

EXAMPLE 5   Graphing a polar curve by plotting points

Graph $r = 2(1 - \cos\theta)$.

Solution
Construct a table of values by choosing values for θ and approximating the corresponding values for r.

θ	r
0	0
1	0.9193954
2	2.832294
3	3.979985
4	3.307287
5	1.432676
6	0.079659

The points are connected as shown in Figure 6.32.

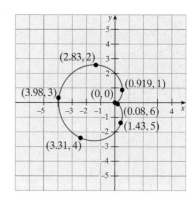

Figure 6.32   Graph of $r = 2(1 - \cos\theta)$

The graph of any polar equation of the general form

$$r = b \pm a\cos\theta \quad \text{or} \quad r = b \pm a\sin\theta$$

is called a **limaçon** (derived from the Latin word *limax*, which means "slug"; in French, it is the word for *snail*). The special case where $a = b$ is the *cardioid*. Figure 6.33 shows four different kinds of limaçons that can occur. Note how the appearance of the graph depends on the ratio b/a.

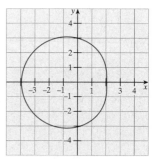

a. $\frac{b}{a} < 1$; inner loop   Case I $(r = 1 - 2\cos\theta)$

b. $\frac{b}{a} = 1$; cardioid   Case II $(r = 2 - 2\cos\theta)$

c. $1 < \frac{b}{a} < 2$; dimple   Case III $(r = 3 - 2\cos\theta)$

d. $\frac{b}{a} \geq 2$; convex   Case IV $(r = 3 - \cos\theta)$

Figure 6.33   Limaçons: $r = b \pm a\cos\theta$

Summary of Polar-Form Curves   There are several polar-form curves which you can graph by plotting points or by using a graphing calculator (be sure to convert to polar form). Table 6.2 gives you the names of some of the special types of polar-form curves you will encounter. There are many others, some of which are represented in the problems. Others may be found in Chapter 5 of the *Student Mathematics Handbook*.

TABLE 6.2 Directory of Polar-Form Curves

Limaçons $\quad r = b \pm a \cos\theta \quad$ **or** $\quad r = b \pm a \sin\theta$

$r = b - a \cos\theta, \dfrac{b}{a} < 1$
standard form
with inner loop

$r = b - a \cos\theta, 1 < \dfrac{b}{a} < 2$
standard form
with a dimple

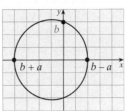

$r = b - a \cos\theta, \dfrac{b}{a} \geq 2$
standard form,
convex

$r = b - a \sin\theta, \dfrac{b}{a} < 1$
$\dfrac{\pi}{2}$ rotation
with inner loop

Cardioids $\quad r = a(1 \pm \cos\theta) \quad$ **or** $\quad r = a(1 \pm \sin\theta)$

Limaçons in which $a = b$

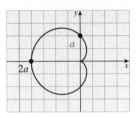

$r = a - a \cos\theta$
standard form

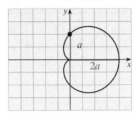

$r = a + a \cos\theta$
π rotation

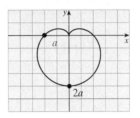

$r = a - a \sin\theta$
$\dfrac{\pi}{2}$ rotation

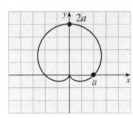

$r = a + a \sin\theta$
$\dfrac{3\pi}{2}$ rotation

Rose Curves

$r = a \cos n\theta \quad$ **or** $\quad r = a \sin n\theta$

If n is odd, the rose has n petals (leaves); if n is even, it has $2n$ petals.

Lemniscates

$r^2 = a^2 \cos 2\theta \quad$ **or** $\quad r^2 = a^2 \sin 2\theta$

Two loops

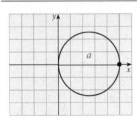

$r = a \cos\theta$
standard form
one (circular) petal

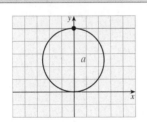

$r = a \sin\theta$
$\dfrac{\pi}{2}$ rotation,
one (circular) petal

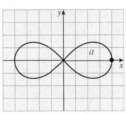

$r^2 = a^2 \cos 2\theta$
standard form

$r^2 = a^2 \sin 2\theta$
$\dfrac{\pi}{4}$ rotation

$r = a \cos 3\theta$
standard form,
three petals

$r = a \sin 3\theta$
$\dfrac{\pi}{6}$ rotation,
three petals

$r = a \cos 2\theta$
standard form,
four petals

$r = a \sin 2\theta$
$\dfrac{\pi}{4}$ rotation,
four petals

INTERSECTION OF POLAR-FORM CURVES

To find the points of intersection of graphs in rectangular form, you need only find the simultaneous solution of the equations that define those graphs. It is not even necessary to draw the graphs, because there is a one-to-one correspondence between ordered pairs satisfying an equation and points on its graph. However, in polar form, this one-to-one property is lost, so that without drawing the graphs you may fail to find all points of intersection. For example, $(-1, \pi)$ and $(1, 0)$ both represent the same point, in polar coordinates. Therefore, our method for finding the intersection of polar-form curves will include sketching the graphs.

Graphical Solution of the Intersection of Polar Curves

Step 1. Find all simultaneous solutions of the given system of equations.
Step 2. Determine whether the pole lies on the two graphs.
Step 3. Graph the curves to look for other points of intersection.

EXAMPLE 6 Intersection of polar-form curves

Find the points of intersection of the curves $r = \frac{3}{2} - \cos\theta$ and $\theta = \frac{2\pi}{3}$.

Solution

Step 1. Solve the system by substitution:
$$r = \frac{3}{2} - \cos\frac{2\pi}{3} = \frac{3}{2} - \left(-\frac{1}{2}\right) = 2$$
The solution is $\left(2, \frac{2\pi}{3}\right)$.

Step 2. If $r = 0$, the first equation has no solution because
$$0 = \frac{3}{2} - \cos\theta \quad \text{or} \quad \cos\theta = \frac{3}{2}$$
and a cosine cannot be larger than 1.

Step 3. Now look at the graphs, as shown in Figure 6.34. We see that $\left(-1, \frac{2\pi}{3}\right)$ may also be a point of intersection. It satisfies the equation $\theta = \frac{2\pi}{3}$, but what about $r = \frac{3}{2} - \cos\theta$? Check $\left(-1, \frac{2\pi}{3}\right)$:
$$-1 \overset{?}{=} \frac{3}{2} - \cos\left(\frac{2\pi}{3}\right)$$
$$= \frac{3}{2} - \left(-\frac{1}{2}\right)$$
$$= 2 \qquad \text{Not satisfied}$$

However, if you check an alternative representation of $\left(-1, \frac{2\pi}{3}\right)$, namely $\left(1, \frac{5\pi}{3}\right)$:
$$1 \overset{?}{=} \frac{3}{2} - \cos\left(\frac{5\pi}{3}\right)$$
$$= \frac{3}{2} - \frac{1}{2}$$
$$= 1 \qquad \text{Satisfied}$$

Be sure to check for points of intersection that you may have missed. ■

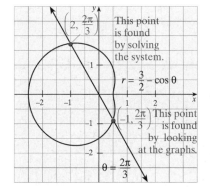

Figure 6.34 Graphs of $r = \frac{3}{2} - \cos\theta$ and $\theta = \frac{2\pi}{3}$

POLAR AREA

To find the area of a region bounded by a polar graph, we use Riemann sums in much the same way as when we developed the integral formula for the area of a region described in rectangular form. However, instead of using rectangular areas as the basic units being summed, in polar form, we sum areas of *circular sectors*. A typical circular sector is shown in Figure 6.35. Recall the formula for the area of such a sector from trigonometry.

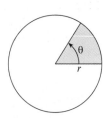

Figure 6.35 Sector of a circle

> ### Area of a Sector
> The area of a circular sector of radius r is given by
> $$A = \tfrac{1}{2}r^2\theta$$
> where θ is the central angle of the sector measured in radians.

Using this formula for the area of a sector, we now state a theorem that gives us a formula for finding the area enclosed by a polar curve. If $r = f(\theta)$, then we call r a **polar function** of θ.

THEOREM 6.1 Area in polar coordinates

Let $r = f(\theta)$ define a polar curve, where f is continuous and $f(\theta) \geq 0$ on the closed interval $\alpha \leq \theta \leq \beta$ for $0 \leq \beta - \alpha \leq 2\pi$. Then the region bounded by the curve $r = f(\theta)$ and the rays $\theta = \alpha$ and $\theta = \beta$ has area

$$A = \frac{1}{2}\int_{\alpha}^{\beta} r^2\, d\theta = \frac{1}{2}\int_{\alpha}^{\beta} [f(\theta)]^2\, d\theta$$

Proof   The region is shown in Figure 6.36. To find the area bounded by the graphs of the polar functions, partition the region between $\theta = \alpha$ and $\theta = \beta$ by a collection of rays, say $\theta_0, \theta_1, \ldots, \theta_n$. Let $\alpha = \theta_0$ and $\beta = \theta_n$.

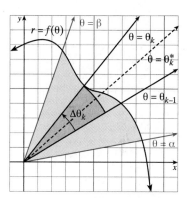

a. The region bounded by the polar curve $r = f(\theta)$ and the rays $\theta = a$ and $\theta = b$

b. The area A can be estimated by adding the area of "small" (gray) circular sectors.

Figure 6.36 Area in polar form

Pick any ray $\theta = \theta_k^*$, with $\theta_{k-1} \leq \theta_k^* \leq \theta_k$. Then the area ΔA_k of the circular sector is approximately the same as the area of the region bounded by the graph of f and the lines $\theta = \theta_{k-1}$ and $\theta = \theta_k$. Because this circular sector has radius $f(\theta_k^*)$ and central angle $\Delta \theta_k$, its area is

$$\Delta A_k \approx \tfrac{1}{2}(\text{radius})^2(\text{central angle}) = \tfrac{1}{2}\left[f\left(\theta_k^*\right)\right]^2 \Delta \theta_k$$

The sum $\displaystyle\sum_{k=1}^{n} \Delta A_k$ is an approximation to the total area A bounded by the polar curve, and by taking the limit as $n \to \infty$, we obtain

$$A = \lim_{n \to \infty} \sum_{k=1}^{n} \Delta A_k = \lim_{n \to \infty} \frac{1}{2} \sum_{k=1}^{n} \left[f\left(\theta_k^*\right)\right]^2 \Delta \theta_k \qquad \text{Riemann sum}$$

$$= \frac{1}{2}\int_{\alpha}^{\beta} [f(\theta)]^2\, d\theta \qquad\qquad\qquad \square$$

You will probably find that the most difficult part of the problem is deciding on the limits of integration. A sketch of the region will help with this.

EXAMPLE 7   Finding area of part of a cardioid

Find the area of the top half ($0 \leq \theta \leq \pi$) of the cardioid $r = 1 + \cos \theta$.

Solution

The cardioid is shown in Figure 6.37. Note that the top half of the graph lies between the rays $\theta = 0$ and $\theta = \pi$. Hence the required area is given by

$$A = \frac{1}{2} \int_0^\pi (1 + \cos \theta)^2 \, d\theta = \frac{1}{2} \int_0^\pi (1 + 2 \cos \theta + \cos^2 \theta) \, d\theta$$

$$= \frac{1}{2} \left[\theta + 2 \sin \theta + \frac{\theta}{2} + \frac{\sin 2\theta}{4} \right] \Big|_0^\pi \qquad \text{Integration table, formula 317}$$

$$= \frac{1}{2} \left[\pi + 2(0) + \frac{\pi}{2} + \frac{0}{4} - 0 \right] = \frac{3\pi}{4} \qquad \blacksquare$$

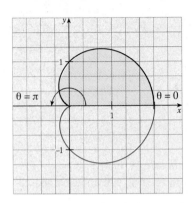

Figure 6.37 Area of the top half of the cardioid $r = 1 + \cos \theta$

EXAMPLE 8   Finding area enclosed by a four-leaved rose

Find the area enclosed by the four-leaved rose $r = \cos 2\theta$.

Solution

The rose curve is shown in Figure 6.38. We will find the area of the top half of the right loop (shaded portion) to make sure $f(\theta) \geq 0$. We see that this corresponds to angles with measures from $\theta = 0$ to $\theta = \pi/4$. If you have a calculator, the easiest way to see this is to plot the graph and then use the *trace* feature to find the θ-values for the top half of the first leaf. By symmetry, the entire area enclosed by the four-leaved rose is 8 times the shaded region. Thus, the required area is given by

$$A = 8 \left[\frac{1}{2} \int_0^{\pi/4} \cos^2 2\theta \, d\theta \right]$$

$$= 4 \left[\frac{\theta}{2} + \frac{\sin 4\theta}{8} \right] \Big|_0^{\pi/4} \qquad \text{Integration table, formula 317}$$

$$= 4 \left[\frac{\pi}{8} + 0 - 0 \right]$$

$$= \frac{\pi}{2} \qquad \blacksquare$$

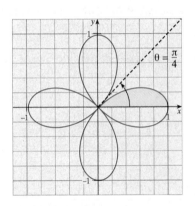

Figure 6.38 Area enclosed by the four-leaved rose $r = \cos 2\theta$

EXAMPLE 9   Finding the area of a region between two polar curves

Find the area of the region common to the circles $r = a \cos \theta$ and $r = a \sin \theta$.

Solution

The circles are shown in Figure 6.39. Solve $a \cos \theta = a \sin \theta$ to find that the circles intersect at $\theta = \pi/4$ and at the pole. To set up the integrals properly, remember to think in terms of polar coordinates and not in terms of rectangular coordinates. Specifically, we must **scan radially**. This means that we need to find the area of intersection by adding the areas of the regions marked R_1 and R_2. Region R_1 is bounded by the circle $r = a \sin \theta$ and the rays $\theta = 0$, $\theta = \frac{\pi}{4}$. Region R_2 is bounded by the circle $r = a \cos \theta$ and $\theta = \frac{\pi}{4}$, $\theta = \frac{\pi}{2}$. Note that the rays $\theta = 0$ and $\theta = \frac{\pi}{2}$ are not necessary as geometric boundaries for R_1 and R_2, but they are necessary to describe which part of each circle

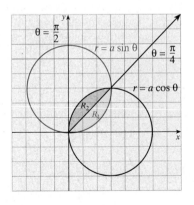

Figure 6.39 Area bounded by $r = a \cos \theta$ and $r = a \sin \theta$

is being calculated.

$$A = \text{AREA OF } R_1 + \text{AREA OF } R_2$$

$$= \frac{1}{2}\int_0^{\pi/4} a^2 \sin^2\theta \, d\theta + \frac{1}{2}\int_{\pi/4}^{\pi/2} a^2 \cos^2\theta \, d\theta$$

$$= \frac{a^2}{2}\left[\frac{\theta}{2} - \frac{\sin 2\theta}{4}\right]\Big|_0^{\pi/4} + \frac{a^2}{2}\left[\frac{\theta}{2} + \frac{\sin 2\theta}{4}\right]\Big|_{\pi/4}^{\pi/2} \qquad \text{Integration table,}$$
$$\text{formulas 348 and 317}$$

$$= \frac{a^2}{2}\left[\frac{\pi}{8} - \frac{1}{4} - 0\right] + \frac{a^2}{2}\left[\frac{\pi}{4} + 0 - \frac{\pi}{8} - \frac{1}{4}\right] = \frac{a^2}{2}\left[\frac{\pi}{4} - \frac{1}{2}\right]$$

$$= \frac{1}{8}a^2(\pi - 2) \qquad \blacksquare$$

EXAMPLE 10 Finding the area between a circle and a limaçon

Find the area between the circle $r = 5\cos\theta$ and the limaçon $r = 2 + \cos\theta$. Round your answer to the nearest hundredth of a square unit.

Solution

As usual, begin by drawing the graphs, as shown in Figure 6.40. Note that both the limaçon and circle are symmetric with respect to the x-axis, so we can find the area in the first quadrant and multiply by 2.

Next, we need to find the points of intersection. We see that the curves do not intersect at the pole. Now solve

$$5\cos\theta = 2 + \cos\theta$$

$$\cos\theta = \frac{1}{2}$$

$$\theta = \frac{\pi}{3} \qquad \text{Solution in the first quadrant}$$

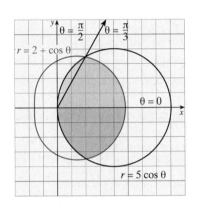

Figure 6.40 Area between $r = 5\cos\theta$ and $r = 2 + \cos\theta$

To find the area, divide the region into two parts, along the ray $\theta = \pi/3$. The right part (gray region above the x-axis) is bounded by the limaçon $r = 2 + \cos\theta$ and the rays $\theta = 0$ and $\theta = \pi/3$. The left part (shaded region shown in color above the x-axis) is bounded by the circle $r = 5\cos\theta$ and the rays $\theta = \pi/3$ and $\theta = \pi/2$. Using this preliminary information, we can now set up and evaluate a sum of integrals for the required area

$$A = 2[\text{AREA OF RIGHT PART} + \text{AREA OF LEFT PART}]$$

$$= 2\left[\frac{1}{2}\int_0^{\pi/3} (2 + \cos\theta)^2 \, d\theta + \frac{1}{2}\int_{\pi/3}^{\pi/2} (5\cos\theta)^2 \, d\theta\right]$$

$$= \int_0^{\pi/3} (4 + 4\cos\theta + \cos^2\theta) \, d\theta + \int_{\pi/3}^{\pi/2} 25\cos^2\theta \, d\theta$$

$$= \left[4\theta + 4\sin\theta + \frac{\theta}{2} + \frac{\sin 2\theta}{4}\right]\Big|_0^{\pi/3} + 25\left[\frac{\theta}{2} + \frac{\sin 2\theta}{4}\right]\Big|_{\pi/3}^{\pi/2}$$

$$= \left[\frac{4\pi}{3} + \frac{4\sqrt{3}}{2} + \frac{\pi}{6} + \frac{\sqrt{3}}{8} - 0\right] + 25\left[\frac{\pi}{4} + 0 - \frac{\pi}{6} - \frac{\sqrt{3}}{8}\right]$$

$$= \frac{43\pi}{12} - \sqrt{3} \approx 9.53 \qquad \blacksquare$$

6.3 PROBLEM SET

Ⓐ 1. **WHAT DOES THIS SAY?** Discuss a procedure for finding the intersection of polar-form curves.

2. **WHAT DOES THIS SAY?** Discuss a procedure for finding the area enclosed by a polar curve.

3. Identify each of the curves as a cardioid, rose curve (state number of petals), lemniscate, limaçon, circle, line, or none of the above.
   a. $r^2 = 9\cos 2\theta$
   b. $r = 2\sin\frac{\pi}{6}$
   c. $r = 3\sin 3\theta$
   d. $r = 3\theta$
   e. $r = 2 - 2\cos\theta$
   f. $\theta = \frac{\pi}{6}$
   g. $r^2 = \sin 2\theta$
   h. $r - 2 = 4\cos\theta$

4. Identify each of the curves as a cardioid, rose curve (state number of petals), lemniscate, limaçon, circle, line, or none of the above.
   a. $r = 2\sin 2\theta$
   b. $r^2 = 2\cos 2\theta$
   c. $r = 5\cos 60°$
   d. $r = 5\sin 8\theta$
   e. $r\theta = 3$
   f. $r^2 = 9\cos(2\theta - \frac{\pi}{4})$
   g. $r = \sin 3(\theta + \frac{\pi}{6})$
   h. $\cos\theta = 1 - r$

5. Identify each of the curves as a cardioid, rose curve (state number of petals), lemniscate, limaçon, circle, line, or none of the above.
   a. $r = 2\cos 2\theta$
   b. $r = 4\sin 30°$
   c. $r + 2 = 3\sin\theta$
   d. $r + 3 = 3\sin\theta$
   e. $\theta = 4$
   f. $\theta = \tan\frac{\pi}{4}$
   g. $r = 3\cos 5\theta$
   h. $r\cos\theta = 2$

Graph the polar-form curves given in Problems 6–21.

6. $r = 3, 0 \le \theta \le \frac{\pi}{2}$
7. $\theta = -\frac{\pi}{2}, 0 \le r \le 3$
8. $r = \theta + 1, 0 \le \theta \le \pi$
9. $r = 2\theta, \theta \ge 0$
10. $r = 2\cos 2\theta$
11. $r = 5\sin 3\theta$
12. $r^2 = 16\cos 2\theta$
13. $r = 3\cos 3(\theta - \frac{\pi}{3})$
14. $r = 5\cos 3(\theta - \frac{\pi}{4})$
15. $r = \sin(2\theta + \frac{\pi}{3})$
16. $r^2 = 16\cos 2(\theta - \frac{\pi}{6})$
17. $r = 2 + \cos\theta$
18. $r = 1 + \sin\theta$
19. $r\cos\theta = 2$
20. $r = 1 + 3\cos\theta$
21. $r = -2\sin\theta$

Find the points of intersection of the curves given in Problems 22–29.

22. $\begin{cases} r = 4\cos\theta \\ r = 4\sin\theta \end{cases}$
23. $\begin{cases} r = 4\sin\theta \\ r = 2 \end{cases}$
24. $\begin{cases} r^2 = 9\cos 2\theta \\ r = 3 \end{cases}$
25. $\begin{cases} r = 2(1 + \sin\theta) \\ r = 2(1 - \sin\theta) \end{cases}$
26. $\begin{cases} r = 3\theta \\ \theta = \frac{\pi}{3} \end{cases}$
27. $\begin{cases} r^2 = \sin 2\theta \\ r = \sqrt{2}\sin\theta \end{cases}$
28. $\begin{cases} r = 2(1 - \cos\theta) \\ r = 4\sin\theta \end{cases}$
29. $\begin{cases} r = \dfrac{4}{1 - \cos\theta} \\ r = 2\cos\theta \end{cases}$

Ⓑ *Find the area of each polar region enclosed by $f(\theta)$ for $a \le \theta \le b$ in Problems 30–37.*

30. $f(\theta) = \sin\theta, 0 \le \theta \le \frac{\pi}{6}$
31. $f(\theta) = \cos\theta, 0 \le \theta \le \frac{\pi}{6}$
32. $f(\theta) = \sec\theta, -\frac{\pi}{4} \le \theta \le \frac{\pi}{4}$
33. $f(\theta) = \sqrt{\sin\theta}, \frac{\pi}{6} \le \theta \le \frac{\pi}{2}$
34. $f(\theta) = e^{\theta/2}, 0 \le \theta \le 2\pi$

35. $f(\theta) = \sin\theta + \cos\theta, 0 \le \theta \le \frac{\pi}{4}$
36. $f(\theta) = \dfrac{\theta}{\pi}, 0 \le \theta \le 2\pi$
37. $f(\theta) = \dfrac{\theta^2}{\pi}, 0 \le \theta \le 2\pi$

38. **Spirals** are interesting mathematical curves. There are three special types of spirals:
   a. A **spiral of Archimedes** has the form $r = a\theta$. Graph $r = 2\theta$.
   b. A **hyperbolic spiral** has the form $r\theta = a$. Graph $r\theta = 2$.
   c. A **logarithmic spiral** has the form $r = a^{k\theta}$. Graph $r = 2^{\theta}$.

39. The **strophoid** is a curve of the form $r = a\cos 2\theta \sec\theta$. Graph this curve for $a = 2$.

40. The **bifolium** has the form $r = a\sin\theta\cos^2\theta$. Graph this curve for $a = 1$.

41. The **folium of Descartes** has the form $r = \dfrac{3a\sin\theta\cos\theta}{\sin^3\theta + \cos^3\theta}$. Graph this curve for $a = 2$.

42. Find the area of one loop of the four-leaved rose $r = 2\sin 2\theta$.

43. Find the area enclosed by the three-leaved rose $r = a\sin 3\theta$.

44. Find the area of the region that is inside the circle $r = 4\cos\theta$ and outside the circle $r = 2$.

45. Find the area of the region that is inside the circle $r = a$ and outside the cardioid $r = a(1 - \cos\theta)$.

46. Find the area of the region that is inside the circle $r = \sin\theta$ and outside the cardioid $r = 1 - \cos\theta$.

47. Find the area of the region that is inside the circle $r = 6\cos\theta$ and outside the cardioid $r = 2(1 + \cos\theta)$.

48. Find the area of the portion of the lemniscate $r^2 = 8\cos 2\theta$ that lies in the region $r \ge 2$.

49. Find the area between the inner and outer loops of the limaçon $r = 2 - 4\sin\theta$.

50. Find the area to the right of the line $r\cos\theta = 1$ and inside the lemniscate $r^2 = 2\cos 2\theta$.

51. Find the area to the right of the line $r\cos\theta = 1$ and inside the lemniscate $r^2 = 2\sin 2\theta$.

52. Find the maximum value of the y-coordinate of points on the limaçon $r = 2 + 3\cos\theta$.

53. Find the maximum value of the x-coordinate of points on the cardioid $r = 3 + 3\sin\theta$.

Ⓒ 54. a. Show that if the polar curve $r = f(\theta)$ is rotated about the pole through an angle α, the equation for the new curve is $r = f(\theta - \alpha)$.
   b. Use a rotation to sketch $r = 2\sec(\theta - \frac{\pi}{3})$.

55. The **ovals of Cassini** have the polar form $r^4 + b^4 - 2b^2 r^2 \cos 2\theta = k^4$. Graph the curve where $b = 2$, $k = 3$.

56. Sketch the graph of
$$r = \frac{\theta}{\cos\theta} \quad \text{for} \quad 0 \le \theta \le \frac{\pi}{2}$$
In particular, show that the graph has a vertical asymptote at $x = \frac{\pi}{2}$.

57. If f is a differentiable function of θ, then show that the tangent line to the polar curve $r = f(\theta)$ at the point $P(r, \theta)$ has slope

$$m = \frac{f(\theta)\cos\theta + f'(\theta)\sin\theta}{-f(\theta)\sin\theta + f'(\theta)\cos\theta}$$

whenever the denominator is not zero. *Hint*: $\dfrac{dy}{dx} = \dfrac{dy/d\theta}{dx/d\theta}$.

58. Because the formula for the slope in polar form is complicated (see Problem 57), it is more convenient to measure the inclination of the tangent line to a polar curve in terms of the angle α extended from the radial line to the tangent line, as shown in Figure 6.41.

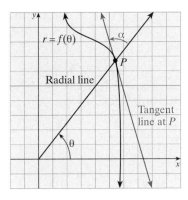

Figure 6.41   Problem 58

Let P be a point on the polar curve $r = f(\theta)$, and let α be the angle extending from the radial line to the tangent line, as shown in Figure 6.41. Assuming that $f'(\theta) \neq 0$, show that

$$\tan\alpha = \frac{f(\theta)}{f'(\theta)}$$

Hint: Use the formula $\tan\alpha = \tan(\phi - \theta) = \dfrac{\tan\phi - \tan\theta}{1 + \tan\phi\tan\theta}$.

59. Use the formula in Problem 58 to find $\tan\alpha$ for each of the following:
  a. the circle $r = a\cos\theta$
  b. the cardioid $r = 2(1 - \cos\theta)$
  c. the logarithmic spiral $r = 2e^{3\theta}$

60. Journal Problem*   *School Science and Mathematics* ■ Graph the polar curve $r = 4 + 2\sin\dfrac{5\theta}{2}$ and find the total area interior to this curve. Then find the area of the "star."

*Problem 3949 by V. C. Bailey, Vol. 83, 1983, p. 356.

6.4   Arc Length and Surface Area

IN THIS SECTION   the arc length of a curve, the area of a surface of revolution, polar arc length and surface area

THE ARC LENGTH OF A CURVE

If a function f has a derivative that is continuous on some interval, then f is said to be **continuously differentiable** on the interval. The portion of the graph of a continuously differentiable function f that lies between $x = a$ and $x = b$ is called the **arc** of the graph on the interval $[a, b]$. To find, and define, the length of this arc, let P be a partition of the interval $[a, b]$, with subdivision points $x_0, x_1, \ldots, x_n$, where $x_0 = a$ and $x_n = b$. Let P_k denote the point (x_k, y_k) on the graph, where $y_k = f(x_k)$. By joining the points $P_0, P_1, \ldots, P_n$, we obtain a polygonal path whose length approximates that of the arc. Figure 6.42 demonstrates the labeling for the case where $n = 6$.

The length of the polygonal path connecting the points $P_0, P_1, \ldots, P_n$ on the graph of f is the sum

$$\sum_{k=1}^{n} \Delta s_k$$

where Δs_k is the length of the segment joining P_{k-1} to P_k. By applying the distance formula and rearranging terms with $\Delta x_k = x_k - x_{k-1}$ and $\Delta y_k = y_k - y_{k-1}$, we find

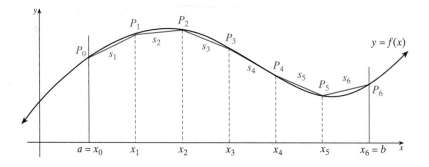

Figure 6.42   A polygonal path approximating an arc of a curve

that

$$\Delta s_k = \sqrt{(x_k - x_{k-1})^2 + (y_k - y_{k-1})^2}$$

$$= \sqrt{(\Delta x_k)^2 + (\Delta y_k)^2}$$

$$= \sqrt{\frac{(\Delta x_k)^2 + (\Delta y_k)^2}{(\Delta x_k)^2}}\, \Delta x_k$$

$$= \sqrt{1 + \left(\frac{\Delta y_k}{\Delta x_k}\right)^2}\, \Delta x_k$$

It is reasonable to expect a connection between the ratio $\Delta y_k/\Delta x_k$ and the derivative $dy/dx = f'(x)$. Indeed, it can be shown (see Problem 42) that

$$\Delta s_k = \sqrt{1 + [f'(x_k^*)]^2}\, \Delta x_k$$

for some number x_k^* between x_{k-1} and x_k. Therefore, the arc length of the graph of f on the interval $[a, b]$ may be estimated by the Riemann sum

$$\sum_{k=1}^{n} \sqrt{1 + [f'(x_k^*)]^2}\, \Delta x_k$$

This estimate may be improved by increasing the number of subdivision points in the partition P of the interval $[a, b]$ in such a way that the subinterval lengths tend to zero. Thus, it is reasonable to define the actual arc length to be the limit

$$\lim_{\|P\| \to 0} \sum_{k=1}^{n} \sqrt{1 + [f'(x_k^*)]^2}\, \Delta x_k$$

Notice that if f' is continuous on the interval $[a, b]$, then so is $\sqrt{1 + [f'(x)]^2}$. Thus, this limit exists and is the integral of $\sqrt{1 + [f'(x)]^2}$ with respect to x on the interval $[a, b]$. These observations lead us to the following definition.*

---

*We have used integration to obtain a meaningful definition of arc length for a function $f(x)$ with a continuous derivative $f'(x)$. It is possible to define arc length for certain curves $y = f(x)$ when $f(x)$ does not have a continuous derivative, but we will not pursue this more general topic.

Arc Length

Let f be a function whose derivative f' is continuous on the interval $[a, b]$ and differentiable on (a, b). Then the **arc length**, s, of the graph of $y = f(x)$ between $x = a$ and $x = b$ is given by the integral

$$s = \int_a^b \sqrt{1 + [f'(x)]^2}\, dx$$

Similarly, for the graph of $x = g(y)$, where g' is continuous on the interval $[c, d]$, the arc length from $y = c$ to $y = d$ is

$$s = \int_c^d \sqrt{1 + [g'(y)]^2}\, dy$$

EXAMPLE 1   Arc length of a curve

Find the arc length (rounded to two decimal places) of the curve $y = x^{3/2}$ on the interval $[0, 4]$.

Solution

The graph is shown in Figure 6.43. Let $f(x) = x^{3/2}$; therefore, $f'(x) = \frac{3}{2}x^{1/2}$.

$$s = \int_0^4 \sqrt{1 + \left[\frac{3}{2}x^{1/2}\right]^2}\, dx$$

$$= \int_0^4 \sqrt{1 + \frac{9}{4}x}\, dx = \left[\frac{4}{9} \cdot \frac{2}{3}\left(1 + \frac{9}{4}x\right)^{3/2}\right]\Big|_0^4$$

$$= \frac{8}{27}\left[(10)^{3/2} - (1)^{3/2}\right] \approx 9.07 \quad\blacksquare$$

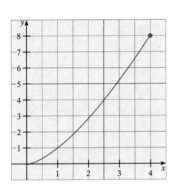

Figure 6.43 What is the length of this curve?

EXAMPLE 2   Arc length of a curve $x = g(y)$

Find the arc length of the curve $x = \frac{1}{3}y^3 + \frac{1}{4}y^{-1}$ from $y = 1$ to $y = 3$.

Solution

Because $g(y) = \frac{1}{3}y^3 + \frac{1}{4}y^{-1}$, we have $g'(y) = y^2 - \frac{1}{4}y^{-2} = \dfrac{4y^4 - 1}{4y^2}$, which is continuous throughout the interval from $y = 1$ to $y = 3$. Therefore, the arc length is

$$\int_1^3 \sqrt{1 + [g'(y)]^2}\, dy = \int_1^3 \sqrt{1 + \left(\frac{4y^4 - 1}{4y^2}\right)^2}\, dy$$

$$= \int_1^3 \sqrt{1 + \frac{16y^8 - 8y^4 + 1}{16y^4}}\, dy$$

$$= \int_1^3 \sqrt{\frac{16y^4 + 16y^8 - 8y^4 + 1}{16y^4}}\, dy$$

$$= \int_1^3 \sqrt{\frac{(4y^4 + 1)^2}{(4y^2)^2}}\, dy$$

$$= \int_1^3 \frac{4y^4 + 1}{4y^2}\, dy$$

$$= \int_1^3 \left(y^2 + \frac{1}{4}y^{-2}\right)\, dy$$

$$= \left(\frac{1}{3}y^3 + \frac{1}{4} \cdot \frac{y^{-1}}{-1}\right)\Big|_1^3 = \frac{53}{6} \quad\blacksquare$$

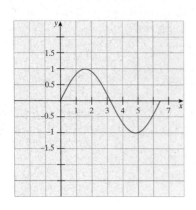

Figure 6.44 Finding the length of an arc

EXAMPLE 3   Estimating arc length using numerical integration

Find the length of the curve defined by $y = \sin x$ on $[0, 2\pi]$.

Solution

The region is shown in Figure 6.44. Because $y' = \cos x$, we have, from the arc length formula,

$$s = \int_0^{2\pi} \sqrt{1 + \cos^2 x}\, dx$$

We do not have techniques to allow us to evaluate this integral, so we turn to numerical integration. If we consider four rectangles, we find, by various methods:

Method ($n = 4$)	Approximation
Rectangles	
Left endpoints	7.584476
Right endpoints	7.584476
Midpoints	7.695299
Trapezoidal rule	7.584476
Simpson's rule	7.150712
Calculator (n not specified)	7.640396

In practice, you would choose just *one* of the methods whose approximate values are given. You might also note that, for this example, Simpson's rule performs worse than the trapezoidal, midpoint, or even the left- and right-endpoint methods. For more accurate results, you might wish to use a computer program. A simple output is shown in Figure 6.45.

Figure 6.45   Using technology to find an arc length

The actual arc length (rounded to six places) computed as an area is 7.640396.   ■

THE AREA OF A SURFACE OF REVOLUTION

When the arc of a curve is revolved about a line L, it generates a surface called a **surface of revolution**, as shown in Figure 6.46a. In particular, if the generating arc is a line segment, the frustum of a cone is generated (Figure 6.46b).

We will obtain a formula for the lateral area of the frustum of a cone, which will then be used as part of an approximation scheme for the area of a general surface of revolution. Accordingly, first note that a cone with slant height s and base radius r has lateral surface area $A = \pi r s$. To see this, cut the cone along a dashed line as shown in Figure 6.47a and flatten it out as shown in Figure 6.47b. The angle θ is determined by noting that the circumference of the circular base of the cone is $2\pi r$, which is the

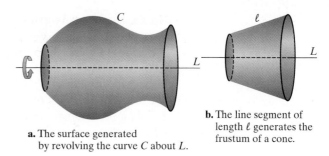

a. The surface generated by revolving the curve C about L.

b. The line segment of length ℓ generates the frustum of a cone.

Figure 6.46   Surfaces of revolution

same as the arclength $s\theta$ after the cone is flattened. Thus, $2\pi r = s\theta$ and $\theta = \dfrac{2\pi r}{s}$. The lateral surface area of the cone is the area of sector COD; namely,

$$A = \frac{1}{2}s^2\theta = \frac{1}{2}s^2\left(\frac{2\pi r}{s}\right) = \pi r s$$

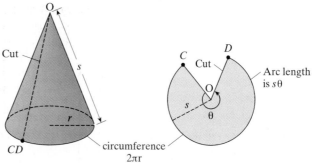

a. A cone with base radius r and slant height s is cut along CD.

b. The flattened cone is a sector with central angle θ in a circle of radius s.

Figure 6.47   A cone with slant height s and base radius r has area $A = \pi r s$

Finally, the frustum of the cone shown in Figure 6.48 has slant height ℓ, top radius r_1 and bottom radius r_2 and can be formed by removing a small cone of base radius r_1 and slant height ℓ_1 from the larger cone of base radius r_2 and slant height $\ell_2 = \ell_1 + \ell$. Thus, the lateral surface area F of the frustum can be found by subtracting the area of the smaller cone from that of the larger:

$$F = \pi r_2 \ell_2 - \pi r_1 \ell_1$$
$$= \pi r_2(\ell_1 + \ell) - \pi r_1 \ell_1$$
$$= \pi[(r_2 - r_1)\ell_1 + r_2\ell]$$

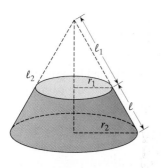

Figure 6.48 Frustum of cone

Using similar triangles, we see that

$$\frac{\ell_1}{r_1} = \frac{\ell_1 + \ell}{r_2} \qquad \text{or} \qquad r_1\ell = (r_2 - r_1)\ell_1$$

so that

$$F = \pi[(r_2 - r_1)\ell_1 + r_2\ell]$$
$$= \pi[r_1\ell + r_2\ell]$$
$$= 2\pi\left(\frac{r_1 + r_2}{2}\right)\ell$$

where $\frac{1}{2}(r_1 + r_2)$ is the average of the radii of the frustum.

Next, assuming that the function $f(x)$ is continuous with $f(x) \geq 0$ on the interval $[a, b]$, we will derive a formula for the area of the surface generated by revolving the curve $y = f(x)$ about the x-axis. Rather than argue formally with Riemann sums, we will proceed heuristically. Figure 6.49a shows an approximating line segment with length Δs, whose endpoints are, respectively, distances r_1 and r_2 above the x-axis. Let $f(x)$ be the average height of the segment above the x-axis; that is, $f(x) = \frac{1}{2}(r_1 + r_2)$. By revolving this segment about the x-axis, we generate the frustum of a cone, with surface area

$$\Delta S = 2\pi \left(\frac{r_1 + r_2}{2} \right) \Delta s = 2\pi f(x) \sqrt{1 + [f'(x)]^2}\, \Delta x$$

where Δx is the projection of the line segment on the x-axis.

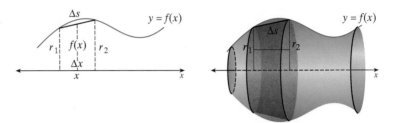

a. The approximating line segment of length Δs is at average height $f(x) = \frac{1}{2}(r_1 + r_2)$ above the x-axis.

b. Revolving the segment about the x-axis generates the frustum of a cone.

Figure 6.49   Approximating the area of a surface of revolution by the area of the frustum of a cone.

The area of the entire surface of revolution S can be obtained by integrating this expression over the interval $[a, b]$. These considerations lead us to the following definition.

Surface Area

> Suppose f' is continuous on the interval $[a, b]$ and differentiable on (a, b). Then the surface generated by revolving about the x-axis the arc of the curve $y = f(x)$ on $[a, b]$ has **surface area**
>
> $$S = 2\pi \int_a^b f(x) \sqrt{1 + [f'(x)]^2}\, dx$$

WARNING   An easy-to-remember form for surface area is $S = 2\pi \int y\, ds$ since $ds = \sqrt{1 + [f'(x)]^2}\, dx$.

You may find it instructive to derive this formula by the more rigorous approach outlined in Problem 43. We close with two examples that illustrate the use of the surface area formula.

EXAMPLE 4   Area of a surface of revolution

Find the area of the surface generated by revolving about the x-axis the arc of the curve $y = x^3$ on $[0, 1]$.

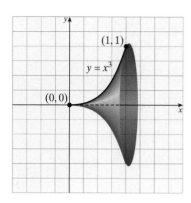

Figure 6.50 Graph of surface of revolution

Solution

The graph is shown in Figure 6.50. Because $f(x) = x^3$, we have $f'(x) = 3x^2$, which is certainly continuous on the interval $[0, 1]$.

$$S = 2\pi \int_0^1 x^3 \sqrt{1 + (3x^2)^2}\, dx = 2\pi \int_0^1 x^3 \sqrt{1 + 9x^4}\, dx = 2\pi \int_1^{10} u^{1/2} \left(\frac{du}{36} \right)$$

> Let $u = 1 + 9x^4$; $du = 36x^3\, dx$.
> If $x = 1$, then $u = 10$.
> If $x = 0$, then $u = 1$.

$$= \frac{\pi}{18} \left(\frac{2}{3} u^{3/2} \right) \bigg|_1^{10} = \frac{\pi}{27}[10\sqrt{10} - 1] \approx 3.5631 \qquad \blacksquare$$

EXAMPLE 5   Derive the formula for the surface area of a sphere

Find a formula for the surface area of a sphere of radius r.

Solution

We can generate the surface of the sphere by revolving the semicircle $y = \sqrt{r^2 - x^2}$ about the x-axis (see Figure 6.51). We find that

$$y' = -x(r^2 - x^2)^{-1/2} = \frac{-x}{\sqrt{r^2 - x^2}}$$

Because the semicircle intersects the x-axis at $x = r$ and $x = -r$, the interval of integration will be $[-r, r]$. Finally, by applying the formula for the surface area, we find

Figure 6.51 Graph of semicircle

$$S = 2\pi \int_{-r}^{r} \sqrt{r^2 - x^2} \sqrt{1 + \left(\frac{-x}{\sqrt{r^2 - x^2}} \right)^2}\, dx$$

$$= 2\pi \int_{-r}^{r} \sqrt{(r^2 - x^2)\left(\frac{r^2 - x^2 + x^2}{r^2 - x^2} \right)}\, dx$$

$$= 2\pi \int_{-r}^{r} r\, dx = 2\pi r(2r) = 4\pi r^2 \qquad \blacksquare$$

To generalize the formula for surface area to apply to any vertical or horizontal axis of revolution, suppose that an axis of revolution is $R(x)$ units from a typical element of arc on the graph of $y = f(x)$, as shown in Figure 6.52.

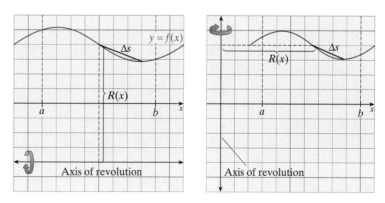

Figure 6.52   Surface of revolution about a general vertical or horizontal axis

Then $2\pi R(x)$ is the circumference of a circle of radius $R(x)$, and it can be shown that an element of surface area is

$$\Delta S = 2\pi R(x) \Delta s = 2\pi R(x) \sqrt{1 + [f'(x)]^2}\, \Delta x$$

Thus, the surface of revolution on the interval $[a, b]$ has area

$$S = 2\pi \int_a^b R(x) \sqrt{1 + [f'(x)]^2}\, dx$$

In particular, if the graph of $y = f(x)$ is revolved about the y-axis, an element of the arc is $R(x) = x$ units from the y-axis, and the resulting surface has area

$$S = 2\pi \int_a^b x \sqrt{1 + [f'(x)]^2}\, dx$$

POLAR ARC LENGTH AND SURFACE AREA

Sometimes we wish to find arc length and surface area in polar form. Figure 6.53 shows the portion of a polar curve subtended by a central angle $\Delta\theta$. If $\Delta\theta$ is small, the corresponding arc length Δs may be thought of as the hypotenuse of a right triangle with sides Δr and $r\Delta\theta$, the length of the circular arc AB with radius r. We have,

$$\Delta s = \sqrt{(r\Delta\theta)^2 + (\Delta r)^2} \qquad \text{Pythagorean theorem}$$

$$= \sqrt{r^2 + \left(\frac{\Delta r}{\Delta\theta}\right)^2}\, \Delta\theta$$

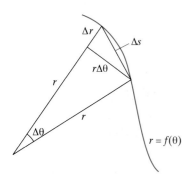

Figure 6.53 Polar arc length

By formalizing these observations, we are led to compute polar arc length by the following formula.

> ### Arc Length in Polar Coordinates
>
> The length of a polar curve $r = f(\theta)$ for $\alpha \le \theta \le \beta$ is given by the integral
>
> $$s = \int_\alpha^\beta \sqrt{r^2 + \left(\frac{dr}{d\theta}\right)^2}\, d\theta$$

EXAMPLE 6 Computing polar arc length

Find the length of the circle $r = 2\sin\theta$.

Solution

This circle is shown in Figure 6.54, and since this is a unit circle, we know the circumference to be 2π. We will verify this conclusion using the arc length formula. Note that $0 \le \theta \le \pi$.

$$
\begin{aligned}
s &= \int_0^\pi \sqrt{r^2 + \left(\frac{dr}{d\theta}\right)^2}\, d\theta && \text{Arc length formula} \\
&&& \text{where } \alpha = 0 \text{ and } \beta = \pi \\
&= \int_0^\pi \sqrt{(2\sin\theta)^2 + (2\cos\theta)^2}\, d\theta && r = 2\sin\theta; \frac{dr}{d\theta} = 2\cos\theta \\
&= \int_0^\pi 2\, d\theta && \sin^2\theta + \cos^2\theta = 1 \\
&= 2\pi
\end{aligned}
$$

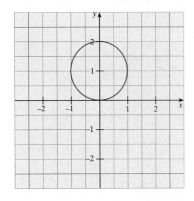

Figure 6.54 Circle $r = 2\sin\theta$

If a polar curve $r = f(\theta)$ between $\theta = \alpha$ and $\theta = \beta$ is revolved about the x-axis, it generates a surface of area

$$A = \int_\alpha^\beta 2\pi y \, ds$$

$$= 2\pi \int_\alpha^\beta (r \sin\theta) \sqrt{r^2 + \left(\frac{dr}{d\theta}\right)^2} \, d\theta$$

We conclude this section with an example showing the use of this formula.

EXAMPLE 7   Computing surface area in polar coordinates

Find the area of the surface generated by revolving about the x-axis the top half of the cardioid $r = 1 + \cos\theta$ (that is, for $0 \le \theta \le \pi$).

Solution

The cardioid is shown in Figure 6.55. Since $r = 1 + \cos\theta$ and $r' = -\sin\theta$, the surface area is given by

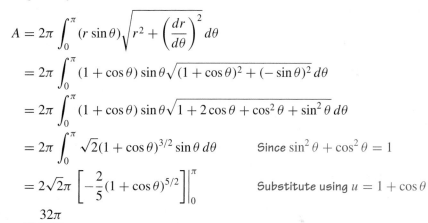

$$A = 2\pi \int_0^\pi (r \sin\theta) \sqrt{r^2 + \left(\frac{dr}{d\theta}\right)^2} \, d\theta$$

$$= 2\pi \int_0^\pi (1 + \cos\theta) \sin\theta \sqrt{(1 + \cos\theta)^2 + (-\sin\theta)^2} \, d\theta$$

$$= 2\pi \int_0^\pi (1 + \cos\theta) \sin\theta \sqrt{1 + 2\cos\theta + \cos^2\theta + \sin^2\theta} \, d\theta$$

$$= 2\pi \int_0^\pi \sqrt{2}(1 + \cos\theta)^{3/2} \sin\theta \, d\theta \qquad \text{Since } \sin^2\theta + \cos^2\theta = 1$$

$$= 2\sqrt{2}\pi \left[-\frac{2}{5}(1 + \cos\theta)^{5/2} \right]_0^\pi \qquad \text{Substitute using } u = 1 + \cos\theta$$

$$= \frac{32\pi}{5} \qquad\qquad\qquad\qquad \blacksquare$$

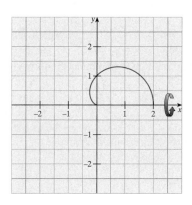

Figure 6.55 Cardioid
$r = 1 + \cos\theta$ for $0 \le \theta \le \pi$

6.4   PROBLEM SET

A *Find the length of the arc of the curve $y = f(x)$ on the intervals given in Problems 1–10.*

1. $f(x) = 3x + 2$ on $[-1, 2]$

2. $f(x) = 5 - 4x$ on $[-2, 0]$

3. $f(x) = 1 - 2x$ on $[1, 3]$

4. $f(x) = x^{3/2}$ on $[0, 4]$

5. $f(x) = \frac{2}{3}x^{3/2} + 1$ on $[0, 4]$

6. $f(x) = \frac{1}{3}(2 + x^2)^{3/2}$ on $[0, 3]$

7. $f(x) = \frac{1}{12}x^5 + \frac{1}{5}x^{-3}$ on $[1, 2]$

8. $f(x) = \frac{1}{3}x^3 + \frac{1}{4}x^{-1}$ on $[1, 4]$

9. $f(x) = \frac{1}{4}x^4 + \frac{1}{8}x^{-2}$ on $[1, 2]$

10. $f(x) = \sqrt{e^{2x} - 1} - \sec^{-1}(e^x)$ on $[0, \ln 2]$

11. Find the length of the curve defined by $9x^2 = 4y^3$ between the points $(0, 0)$ and $(2\sqrt{3}, 3)$.

12. Find the length of the curve defined by $(y + 1)^2 = 4x^3$ between the points $(0, -1)$ and $(1, 1)$.

Find the surface area generated when the graph of each function given in Problems 13–16 on the prescribed interval is revolved about the x-axis. Give your answer to the nearest hundredth.

13. $f(x) = 2x + 1$ on $[0, 2]$

14. $f(x) = \sqrt{x}$ on $[2, 6]$

15. $f(x) = \frac{1}{3}x^3 + \frac{1}{4}x^{-1}$ on $[1, 2]$

16. $f(x) = \frac{1}{4}x^4 + \frac{1}{8}x^{-2}$ on $[1, 2]$

Find the length of the polar curves given in Problems 17–22.

17. $r = \cos\theta$ **18.** $r = \sin\theta + \cos\theta$

19. $r = e^{3\theta}, 0 \le \theta \le \frac{\pi}{2}$ **20.** $r = e^{1-\theta}, 0 \le \theta \le 1$

21. $r = \theta^2, 0 \le \theta \le 1$ **22.** $r = \sin^2\frac{\theta}{2}$

In Problems 23–26, find the surface area generated when the given polar curve is revolved about the x-axis.

23. $r = 5, 0 \le \theta \le \frac{\pi}{3}$ **24.** $r = 1 - \cos\theta, 0 \le \theta \le \pi$

25. $r = \csc\theta, \frac{\pi}{4} \le \theta \le \frac{\pi}{3}$ **26.** $r = \cos^2\frac{\theta}{2}, 0 \le \theta \le \pi$

B *In Problems 27 and 28, use numerical integration to find the arc length.*

27. $f(x) = \sin x$ on $[0, \pi]$ **28.** $f(x) = \tan x$ on $[0, 1]$

29. Find the area of the surface generated when the arc of the curve

$$y = \tfrac{1}{3}x^3 + (4x)^{-1}$$

between $x = 1$ and $x = 3$ is revolved about:

 a. the x-axis   **b.** the y-axis

30. Find the area of the surface generated when the arc of the curve

$$x = \tfrac{3}{5}y^{5/3} - \tfrac{3}{4}y^{1/3}$$

between $y = 0$ and $y = 1$ is revolved about:

 a. the y-axis   **b.** the x-axis   **c.** the line $y = -1$

Find the surface area generated when the graph of each function given in Problems 31–34 on the prescribed interval is revolved about the y-axis. Give your answers to the nearest hundredth.

31. $f(x) = \tfrac{1}{3}(12 - x)$ on $[0, 3]$ **32.** $f(x) = \tfrac{2}{3}x^{3/2}$ on $[0, 3]$

33. $f(x) = \tfrac{1}{3}\sqrt{x}(3 - x)$ on $[1, 3]$ **34.** $f(x) = 2\sqrt{4 - x}$ on $[1, 3]$

35. The graph of the equation $x^{2/3} + y^{2/3} = 1$ is an **astroid**.

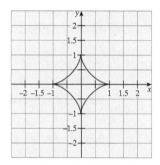

Find the length of this particular astroid by finding the length of the first quadrant portion, $y = (1 - x^{2/3})^{3/2}$ on $[0, 1]$. By symmetry, the length of the entire curve is four times the length of the arc in the first quadrant.

36. Find the area of the surface generated by revolving the portion of the astroid with $y \geq 0$ (see Problem 35) $x^{2/3} + y^{2/3} = 1$ on $[-1, 1]$ about the x-axis.

37. Estimate the area of the surface obtained by revolving $y = \tan x$ about the x-axis on the interval $[0, 1]$.

38. The following table gives the slope $f'(x_k)$ for points x_k located at 0.3-unit intervals on a certain curve defined by $y = f(x)$. Use the trapezoidal rule to estimate the arc length of $y = f(x)$ over the interval $[0, 3]$.

x_k	$f'(x_k)$	x_k	$f'(x_k)$
0.0	3.7	1.8	4.6
0.3	3.9	2.1	4.9
0.6	4.1	2.4	5.2
0.9	4.1	2.7	5.5
1.2	4.2	3.0	6.0
1.5	4.4		

39. Use the table in Problem 38 to estimate the surface area generated when $y = f(x)$ is revolved about the y-axis over the interval $[0, 3]$.

40. Show that a cone of radius r and height h has lateral surface area

$$S = \pi r \sqrt{r^2 + h^2}$$

Hint: Revolve part of the line $hy = rx$ about the x-axis.

C **41.** Show that when the arc of the graph of $f(x)$ between $x = a$ and $x = b$ is revolved about the y-axis, the surface generated has area

$$S = 2\pi \int_a^b x\sqrt{1 + [f'(x)]^2}\, dx$$

42. If $f'(x)$ exists throughout the interval $[x_{k-1}, x_k]$, show that there exists a number x_k^* in this interval for which

$$\sqrt{(\Delta x_k)^2 + (\Delta y_k)^2} = \sqrt{1 + [f'(x_k^*)]^2}\,\Delta x_k$$

where $\Delta x_k = x_k - x_{k-1}$ and $\Delta y_k = f(x_k) - f(x_{k-1})$. *Hint:* Use the mean value theorem.

43. Verify the surface area formula by completing the following steps (see Figure 6.56).

 a. Let $P = \{x_0, x_1, \ldots, x_n\}$ be a partition of the interval $[a, b]$, and for $k = 0, 1, \ldots, n$, let P_k denote the point (x_k, y_k), where $y_k = f(x_k)$. Show that when the line segment between P_{k-1} and P_k is revolved about the x-axis, it generates a frustum of a cone with surface area

$$\pi(y_{k-1} + y_k)\ell_k$$

where ℓ_k is the distance between P_{k-1} and P_k.

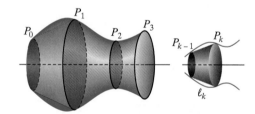

Figure 6.56 Surface area formula

 b. Notice that $\tfrac{1}{2}(y_{k-1} + y_k)$ is a number between y_{k-1} and y_k. Use the intermediate value theorem (Section 2.3) to show that

$$y_{k-1} + y_k = 2f(c_k)$$

for some number c_k between x_{k-1} and x_k.

 c. Explain why the surface generated when the arc of the graph of f between $x = a$ and $x = b$ is revolved about the x-axis has area

$$\lim_{\|P\| \to 0} \sum_{k=1}^n 2\pi f(c_k)\sqrt{1 + [f'(x_k^*)]^2}\, \Delta x_k$$

 d. Notice that as $\|P\| \to 0$, the numbers c_k and x_k^* are "squeezed" together. Use this observation to show that the surface area is given by the integral

$$S = 2\pi \int_a^b f(x)\sqrt{1 + [f'(x)]^2}\, dx$$

44. The center of a circle of radius r is located at $(R, 0)$, where $R > r$. When the circle is revolved about the y-axis, it generates a *torus* (a doughnut-shaped solid). What is the surface area of the torus?

45. Let n be a positive integer and C, D be constants such that

$$n(n-1) = \frac{1}{16CD}$$

Find the length of the curve

$$y = Cx^{2n} + Dx^{2(1-n)}$$

from $x = a$ to $x = b$.

46. Find the length of

$$y = \frac{x^3}{24} + \frac{2}{x}$$

from $x = 2$ to $x = 3$. Find a general formula for the length of the curve

$$y = Cx^n + Dx^{2-n}$$

from $x = a$ to $x = b$ for

$$n(n-2) = \frac{1}{4CD}$$

47. Suppose it is known that the arc length of the curve $y = f(t)$ on the interval $0 \le t \le x$ is

$$L(x) = \ln(\sec x + \tan x)$$

for every x on $0 \le x \le 1$. If the curve $y = f(x)$ passes through the origin, what is $f(x)$?

6.5   Physical Applications: Work, Liquid Force, and Centroids

IN THIS SECTION   work, modeling fluid pressure and force, modeling the centroid of a plane region, volume theorem of Pappus

WORK

In physics, "force" is an influence that tends to cause motion in a body. When a constant force of magnitude F is applied to an object through a distance d, it performs **work** measured by the product $W = Fd$.

Work Done by a Constant Force

> If a body moves a distance d in the direction of an applied constant force F, the **work** W done is
>
> $$W = Fd$$

TABLE 6.3   Common Units of Work and Force

Mass	Dist.	Force	Work
kg	m	newton (N)	joule
g	cm	dyne (dyn)	erg
slug	ft	pound	ft-lb

If there is no movement, there is no work!

For example, the work done in lifting a 90-lb bag of concrete 3 ft is $W = Fd = (90 \text{ lb})(3 \text{ ft}) = 270 \text{ ft·lb}$. Notice that this definition does not conform to everyday use of the word *work*. If you work at lifting the concrete all day, but you are not able to move the sack of concrete, then no work has been done. Table 6.3 shows some common units of work and force.

To find the work done in moving an object against a variable force, calculus is required. Suppose F is a continuous function and $F(x)$ is a variable force that acts on an object moving along the x-axis from $x = a$ to $x = b$. We will heuristically define the work done by this force. Partition the interval $[a, b]$ into n subintervals, and let Δx_k be the length of the kth subinterval I_k. If Δx_k is sufficiently small, we can expect the force to be essentially constant on I_k, equal, for instance, to $F(x_k^*)$, where x_k^* is a point chosen arbitrarily from the interval I_k. It is reasonable to expect that the work ΔW_k required to move the object on the interval I_k is approximately $\Delta W_k = F(x_k^*)\Delta x_k$, and the total work is estimated by the sum

$$\sum_{k=1}^{n} F(x_k^*)\Delta x_k$$

as indicated in Figure 6.57.

Figure 6.57   Work performed by a variable force

By taking the limit as the norm of the partition approaches 0, we find that the total work may be modeled by

$$W = \lim_{\|P\| \to 0} \sum_{k=1}^{n} F(x_k^*) \Delta x_k = \int_a^b F(x)\,dx$$

Work Done by a Variable Force

The work done by the variable force $F(x)$ in moving an object along the x-axis from $x = a$ to $x = b$ is given by

$$W = \int_a^b F(x)\,dx$$

EXAMPLE 1   Work done by a variable force

An object located x ft from a fixed starting position is moved along a straight road by a force of $F(x) = (3x^2 + 5)$ lb. What work is done by the force to move the object

a. through the first 4 ft?   **b.** from 1 ft to 4 ft?

Solution

a. $W = \displaystyle\int_0^4 (3x^2 + 5)\,dx = (x^3 + 5x)\big|_0^4 = 84$ ft-lb

b. $W = \displaystyle\int_1^4 (3x^2 + 5)\,dx = (x^3 + 5x)\big|_1^4 = 78$ ft-lb   ∎

Hooke's law, named for the English physicist Robert Hooke (1635–1703), states that *when a spring is pulled x units past its equilibrium (rest) position, there is a restoring force $F(x) = kx$ that pulls the spring back toward equilibrium.** (See Figure 6.58.) The constant k in this formula is called the **spring constant**.

A spring at rest
(equilibrium position)

A spring stretched x units
past equilibrium

Hooke's law: A force $F(x) = kx$ acts to restore the spring to its equilibrium position.

Figure 6.58   Hooke's law

---

*Actually, Hooke's law applies only in ideal circumstances, and a spring for which the law applies is sometimes called an ideal (or linear) spring.

EXAMPLE 2   Modeling work using Hooke's formula

Figure 6.59 Stretching a spring

The natural length of a certain spring is 10 cm. If it requires 2 ergs of work to stretch the spring to a total length of 18 cm, how much work will be performed in stretching the spring to a total length of 20 cm? (See Figure 6.59.)

Solution

Assume that the point of equilibrium is at 0 on a number line, and let x be the length the free end of the spring is extended past equilibrium. Because the stretching force of the spring is $F(x) = kx$, the work done in stretching the spring b cm beyond equilibrium is

$$W = \int_0^b F(x)\,dx = \int_0^b kx\,dx = \tfrac{1}{2}kb^2$$

We are given that $W = 2$ when $b = 8$ (since 18 cm is 8 cm beyond equilibrium), so

$$2 = \tfrac{1}{2}k(8)^2 \quad \text{implying} \quad k = \tfrac{1}{16}$$

Thus, the work done in stretching the spring b cm beyond equilibrium is

$$W = \tfrac{1}{2}\left(\tfrac{1}{16}\right)b^2 = \tfrac{1}{32}b^2 \text{ ergs}$$

In particular, when the total length of the spring is 20 cm, it is extended $b = 10$ cm, and the required work is

$$W = \tfrac{1}{32}(10)^2 = \tfrac{25}{8} = 3.125 \text{ ergs}$$ ■

EXAMPLE 3   Modeling the work performed in pumping water out of a tank

A tank in the shape of a right circular cone of height 12 ft and radius 3 ft is inserted into the ground with its vertex pointing down and its top at ground level, as shown in Figure 6.60a. If the tank is filled with water (density $\rho = 62.4$ lb/ft^3) to a depth of 6 ft, how much work is performed in pumping all the water in the tank to ground level? What changes if the water is pumped to a height of 3 ft above ground level?

a. A conical water tank   **b.** A disk of water is $12 - y_k^*$ units from the top.

Figure 6.60   Work in pumping water

Solution

Set up a coordinate system with the origin at the vertex of the cone and the y-axis as the axis of symmetry (Figure 6.60b). Partition the interval $0 \le y \le 6$ (remember, the water only comes up 6 ft). Choose a representative point y_k^* in the kth subinterval, and construct a thin disklike slab of water y_k^* units above the vertex of the cone. Our plan is to think of the water in the tank as a collection of these slabs of water piled on top of each other. We will find the work done to raise a typical water disk and then compute the total work by using integration to add the contributions of all such slabs.

Note that the force required to lift the slab is equal to the weight of the slab, which equals its volume multiplied by the weight per cubic foot of water. Let x_k^* be the radius of the kth slab. By similar triangles (Figure 6.60b), we have

$$\frac{x_k^*}{y_k^*} = \frac{3}{12}, \qquad \text{which implies} \qquad x_k^* = \frac{y_k^*}{4}$$

Hence, the volume of the cylindrical slab is

$$\Delta V = \pi x_k^{*2} \Delta y_k = \pi \underbrace{\left(\frac{y_k^*}{4}\right)^2}_{\text{Radius}} \overbrace{\Delta y_k}^{\text{Thickness}}$$

so that the weight of the kth slab is modeled by

$$62.4\pi \left(\frac{y_k^*}{4}\right)^2 \Delta y_k \text{ lb}$$

From Figure 6.60b, we see that the slab of water must be raised $12 - y_k^*$ feet, which means that the work ΔW required to raise this kth slab is

$$\Delta W = \underbrace{62.4\pi \left(\frac{y_k^*}{4}\right)^2 \Delta y_k}_{\text{Weight (force)}} \underbrace{(12 - y_k^*)}_{\text{Distance}} = 62.4\pi \left(\frac{y_k^*}{4}\right)^2 (12 - y_k^*) \Delta y_k$$

Finally, to compute the total work, we add the work required to lift each thin slab and take the limit of the sum as the norm of the partition approaches 0.

$$W = \lim_{n \to \infty} 62.4\pi \sum_{k=1}^{n} \left(\frac{y_k^*}{4}\right)^2 (12 - y_k^*) \Delta y_k$$

$$= 62.4\pi \int_0^6 \left(\frac{y}{4}\right)^2 (12 - y) \, dy = \frac{62.4}{16}\pi \int_0^6 (12y^2 - y^3) \, dy$$

$$= 3.9\pi \left(4y^3 - \frac{1}{4}y^4\right)\Big|_0^6 = 2{,}106\pi \approx 6{,}616 \text{ ft-lb}$$

If the water is pumped to a height of 3 ft above ground level, all that changes is the distance moved by the kth slab of water. It becomes $12 + 3 - y_k^* = 15 - y_k^*$, and the work is given by

$$W = 62.4\pi \int_0^6 \left(\frac{y}{4}\right)^2 (15 - y) \, dy \approx 9{,}263 \text{ ft-lb} \qquad \blacksquare$$

MODELING FLUID PRESSURE AND FORCE

Anyone who has dived into water has probably noticed that the pressure (that is, the force per unit area) due to the water's weight increases with depth. Careful observations show that the water pressure at any given point is directly proportional to the depth at that point. The same principle applies to other fluids.

In physics, **Pascal's principle** (named for Blaise Pascal, 1623–1662, a French mathematician and scientist) states that fluid pressure is the same in all directions (see Figure 6.61). This means that the pressure must be the same at all points on a surface submerged horizontally, and in this case, the fluid force on the surface is given by the formula in the following box.

Figure 6.61 Pascal's principle: In an equilibrium state, the fluid pressure is the same in all directions.

Fluid Force

> If a surface of area A is submerged horizontally at a depth h in a fluid, the weight of the fluid exerts a force of
>
> $$F = (\text{PRESSURE})(\text{AREA}) = \rho h A$$
>
> on the surface, where ρ is the fluid density (weight per unit volume). This is called **fluid force** or **hydrostatic force**.

TABLE 6.4 Weight-density, ρ (lb/ft³)

Water	62.4
Seawater	64.0
Gasoline	42.0
Kerosene	51.2
Milk	64.5
Mercury	849.0

Table 6.4 shows the fluid density for some common fluids. It turns out that this fluid force does not depend on the shape of the container or its size. For instance, each of the containers in Figure 6.62 has the same pressure on its base because the fluid depth h and the base area A are the same in each case.

Figure 6.62 The fluid force $F = \rho h A$ does not depend on the shape or size of the container

When the surface is submerged vertically or at an angle, however, this simple formula does not apply because different parts of the surface are at different depths, as shown in Figure 6.63. We will use integration to compute the fluid force in such cases. Consider a plate that is submerged vertically in a fluid of density ρ as shown in Figure 6.64a. We set up a coordinate system with the horizontal axis on the surface of the fluid and the positive h-axis (depth) pointing down. Thus, greater depths correspond to larger values of h (see Figure 6.64b). For simplicity, we assume the plate is oriented so its top and bottom are located a units and b units below the surface, respectively.

Figure 6.63 Pressure on a submerged object

a. A plate submerged vertically in a fluid of density ρ

b. A typical slab has area $A = L(h)\Delta h$.

Figure 6.64 Hydrostatic force

Our strategy will be to think of the plate as a pile of subplates or slabs, each so thin that we may regard it as being at a constant depth below the surface. A typical slab (a horizontal strip) is shown in Figure 6.64b. Note that it has thickness Δh and length $L = L(h)$, so that its area is $\Delta A = L\Delta h$. Because the slab is at a constant depth h below the surface, the fluid force on its surface is

$$\Delta F = (\text{PRESSURE})(\text{AREA}) = \rho h \Delta A = \rho h L(h)\Delta h$$

and by integrating as h varies from a to b, we can model the total force on the plate.

Fluid (Hydrostatic) Force

Suppose a flat surface (a plate) is submerged vertically in a fluid of weight density ρ (lb/ft^3) and that the submerged portion of the plate extends from $h = a$ to $h = b$ on a vertical axis. Then the total force F exerted by the fluid is given by

$$F = \int_a^b \rho h L(h) \, dh$$

where h is the depth and $L(h)$ is the corresponding length of a typical horizontal approximating strip.

EXAMPLE 4   Fluid force on a vertical surface

The cross sections of a certain trough are inverted isosceles triangles with height 6 ft and base 4 ft, as shown in Figure 6.65. Suppose the trough contains water to a depth of 3 ft. Find the total fluid force on one end.

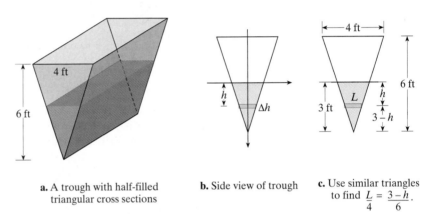

a. A trough with half-filled   **b.** Side view of trough   **c.** Use similar triangles
   triangular cross sections   to find $\dfrac{L}{4} = \dfrac{3-h}{6}$.

Figure 6.65   Fluid force

Solution

First, we set up a coordinate system in which the horizontal axis lies at the fluid surface and the positive vertical axis (the h-axis) points down (see Figure 6.65b). *Next*, we find expressions for the length and depth of a typical thin slab (horizontal strip) in terms of the variables. We assume the slab has thickness Δh and length L. Then by similar triangles (see Figure 6.65c) we have

$$\frac{L}{4} = \frac{3-h}{6} \quad \text{so that} \quad L = \frac{2}{3}(3-h)$$

We see that the approximating slab has area $\Delta A = L \Delta h = \frac{2}{3}(3-h)\Delta h$. *Finally*, we multiply the product of the fluid's weight-density at a given depth (ρh) and the area of the approximating slab (ΔA) and integrate on the interval of depths occupied by the vertical plate.

$$F = \underbrace{\int_0^3 \ \tfrac{2}{3}\rho h(3-h)\,dh}_{\Delta F\,=\,\text{force on a typical slab}} \qquad \textit{Note}: \underset{\downarrow}{\text{(density)}}\underset{\downarrow}{\text{(depth)}}\underset{\downarrow}{\text{(area of strip)}}$$

$$\qquad\qquad\qquad\qquad\qquad\qquad\qquad\qquad \rho \qquad h \qquad \tfrac{2}{3}(3-h)\Delta h$$

$$= \tfrac{2}{3}(62.4)\int_0^3 (3h - h^2)\,dh \qquad\qquad \rho = 62.4 \text{ lb/ft}^3 \text{ for water}$$

$$= 41.6 \left(\tfrac{3}{2}h^2 - \tfrac{1}{3}h^3\right)\Big|_0^3 = 187.2 \text{ lb} \qquad\qquad\qquad \blacksquare$$

EXAMPLE 5   Modeling the force on one face of a dam

A reservoir is filled with water to the top of a dam. If the dam is in the shape of a parabola 40 ft high and 20 ft wide at the top, as shown in Figure 6.66, what is the total fluid force on the face of the dam?

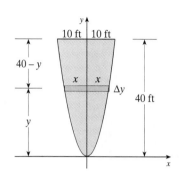

a. Cross section of a dam

b. A typical horizontal strip is 40 − y ft below the water surface.

Figure 6.66   Force on one face of a dam

Solution

Instead of putting the x-axis on the surface of the water (at the top of the dam), we place the origin at the vertex of the parabola and the positive y-axis along the parabola's axis of symmetry (Figure 6.66b). The advantage of this choice of axes is that the parabola can be represented by an equation of the form $y = cx^2$. Because we know that $y = 40$ when $x = 10$, it follows that

$$40 = c(10)^2 \quad \text{so that} \quad c = \tfrac{2}{5} \quad \text{and thus} \quad y = \tfrac{2}{5}x^2$$

The typical horizontal strip on the face of the dam is located y feet above the x-axis, which means it is $h = 40 - y$ feet below the surface of the water. The strip is Δy feet wide and $L = 2x$ feet long. Write L as a function of y:

$$y = \tfrac{2}{5}x^2 \quad \text{so that} \quad x = \sqrt{\tfrac{5}{2}y}$$

and

$$L(y) = 2x = 2\left(\sqrt{\tfrac{5}{2}y}\right) = \sqrt{10y}$$

Therefore,

$$\Delta A = L\,\Delta y = \sqrt{10y}\,\Delta y$$

and the total force may be modeled by the integral

$$F = \int_0^{40} \rho h \underbrace{L(y)\,dy}_{\substack{\uparrow \\ \text{Area of the strip}}} = \int_0^{40} (62.4)(40 - y)\sqrt{10y}\,dy$$

Depth below water: $h = 40 - y$

Density of water = 62.4

$$= 62.4\sqrt{10} \int_0^{40} (40y^{1/2} - y^{3/2})\,dy$$

$$= 62.4\sqrt{10} \left(\frac{80}{3}y^{3/2} - \frac{2}{5}y^{5/2}\right)\Big|_0^{40}$$

$$= 62.4\sqrt{10}\,(y^{3/2})\left(\frac{80}{3} - \frac{2}{5}y\right)\Big|_0^{40}$$

$$= 532{,}480 \text{ lb} \qquad \blacksquare$$

MODELING THE CENTROID OF A PLANE REGION

In mechanics, its often important to determine the point where an irregularly shaped plate will balance. The **moment of a force** measures its tendency to produce rotation in an object and depends on the magnitude of the force and the point on the object where it is applied. Since the time of Archimedes (287–212 B.C.), it has been known that the balance point of an object occurs where all its moments cancel out (so there is no rotation).

The **mass** of an object is a measure of its **inertia**; that is, its propensity to maintain a state of rest or uniform motion. A thin plate whose material is distributed uniformly, so that its density ρ (mass per unit area) is constant, is called a **homogeneous lamina**. The balance point of such a lamina is called its **centroid** and may be thought of as its geometrical center. We will see how centroids can be computed by integration in this section, and will examine the topic even further in Section 12.6.

Consider a homogeneous lamina that covers a region R bounded by the curves $y = f(x)$ and $y = g(x)$ on the interval $[a, b]$, and consider a thin, vertical approximating strip within R, as shown in Figure 6.67. The mass of the strip is given by

$$\Delta m = \underbrace{\rho}_{\substack{\uparrow \\ \text{Density}}} \underbrace{[f(x) - g(x)]\Delta x}_{\text{Area of strip}}$$

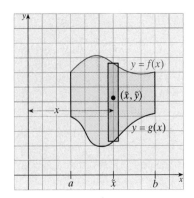

Theoretically, the lamina should balance on a point placed at its centroid.

and the total mass of the lamina may be found by integration:

$$m = \rho \int_a^b [f(x) - g(x)]\, dx$$

Note that $m = \text{AREA}$ if $\rho = 1$.

The centroid (geometrical center) of the approximating strip is $(\tilde{x}, \tilde{y})$, where $\tilde{x} = x$ and $\tilde{y} = \frac{1}{2}[f(x) + g(x)]$. The **moment of the approximating strip about the y-axis** is defined to be the product

$$\Delta M_y = \tilde{x} \underbrace{\Delta m}_{\substack{\uparrow \\ \text{Mass of strip}}} = \tilde{x}\{\rho[f(x) - g(x)]\Delta x\}$$

$$\underset{\substack{\text{Distance of the} \\ \text{strip from } y\text{-axis}}}{}$$

Figure 6.67 Homogeneous lamina with a vertical approximating strip

This product provides a measure of the tendency of the strip to rotate about the y-axis. Similarly, the **moment of the strip about the x-axis** is defined to be the product

$$\Delta M_x = \tilde{y}\Delta m = \underbrace{\tfrac{1}{2}[f(x) + g(x)]}_{\text{Average distance of the strip from } x\text{-axis}}\{\rho[f(x) - g(x)]\Delta x\}$$

$$= \tfrac{1}{2}\rho\{[f(x)]^2 - [g(x)]^2\}\Delta x$$

This product provides a measure of the tendency of the strip to rotate about the x-axis.

Integrating on $[a, b]$, we find the moments of the entire lamina R about the y-axis and x-axis may be modeled by

$$M_y = \rho \int_a^b x[f(x) - g(x)]\, dx \quad \text{and} \quad M_x = \tfrac{1}{2}\rho \int_a^b \{[f(x)]^2 - [g(x)]^2\}\, dx$$

If the entire mass m of the lamina R were located at the point $(\bar{x}, \bar{y})$, then its moments about the x-axis and y-axis would be $m\bar{y}$ and $m\bar{x}$, respectively. Thus, we have $m\bar{x} = M_y$ and $m\bar{y} = M_x$, so that

$$\bar{x} = \frac{M_y}{m} = \frac{\rho \displaystyle\int_a^b x[f(x) - g(x)]\, dx}{\rho \displaystyle\int_a^b [f(x) - g(x)]\, dx} \quad \text{and} \quad \bar{y} = \frac{M_x}{m} = \frac{\tfrac{1}{2}\rho \displaystyle\int_a^b \{[f(x)]^2 - [g(x)]^2\}\, dx}{\rho \displaystyle\int_a^b [f(x) - g(x)]\, dx}$$

We can now summarize these results.

Let f and g be continuous and satisfy $f(x) \geq g(x)$ on the interval $[a, b]$, and consider a thin plate (lamina) of uniform density ρ that covers the region R between the graphs of $y = f(x)$ and $y = g(x)$ on the interval $[a, b]$. Then

MASS

The **mass** of R is $m = \rho \int_a^b [f(x) - g(x)] \, dx$

CENTROID

The **centroid** of R is the point $(\bar{x}, \bar{y})$ such that

$$\bar{x} = \frac{M_y}{m} = \frac{\displaystyle\int_a^b x[f(x) - g(x)] \, dx}{\displaystyle\int_a^b [f(x) - g(x)] \, dx}$$

and

$$\bar{y} = \frac{M_x}{m} = \frac{\frac{1}{2} \displaystyle\int_a^b \{[f(x)]^2 - [g(x)]^2\} \, dx}{\displaystyle\int_a^b [f(x) - g(x)] \, dx}$$

EXAMPLE 6   Centroid of a thin plate

A homogeneous lamina R has constant density $\rho = 1$ and is bounded by the parabola $y = x^2$ and the line $y = x$. Find the mass and the centroid of R.

Solution

We see that the line and the parabola intersect at the origin and at the point $(1, 1)$, as shown in Figure 6.68. Because $\rho = 1$, the mass is the same as the area of the region R. That is,

$$m = A = \int_0^1 (x - x^2) \, dx = \left(\frac{x^2}{2} - \frac{x^3}{3} \right) \Big|_0^1 = \frac{1}{6}$$

and we find that the region has moments M_y and M_x about the y-axis and x-axis, respectively, where

$$M_y = \int_0^1 x(x - x^2) \, dx = \left(\frac{x^3}{3} - \frac{x^4}{4} \right) \Big|_0^1 = \frac{1}{12}$$

and

$$M_x = \frac{1}{2} \int_0^1 (x + x^2)(x - x^2) \, dx = \frac{1}{2} \int_0^1 (x^2 - x^4) \, dx$$

$$= \frac{1}{2} \left(\frac{x^3}{3} - \frac{x^5}{5} \right) \Big|_0^1 = \frac{1}{2} \left(\frac{1}{3} - \frac{1}{5} \right) = \frac{1}{15}$$

Thus, the centroid of the region R has coordinates

$$\bar{x} = \frac{M_y}{m} = \frac{\frac{1}{12}}{\frac{1}{6}} = \frac{1}{2} \quad \text{and} \quad \bar{y} = \frac{M_x}{m} = \frac{\frac{1}{15}}{\frac{1}{6}} = \frac{2}{5}$$

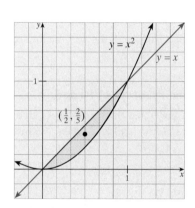

Figure 6.68 The centroid of a planar region

VOLUME THEOREM OF PAPPUS

There is a remarkable result attributed to Pappus of Alexandria (ca. 300 A.D.), whom many regard as the last great Greek geometer, which gives a connection between centroids and volumes of revolution.

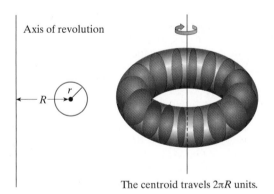

The approximating solid of revolution has volume $\Delta V = 2\pi x(\Delta A)$.

Area ΔA

Figure 6.69 The volume theorem of Pappus

THEOREM 6.2   Volume theorem of Pappus

The solid generated by revolving a region R about a line outside its boundary (but in the same plane) has volume $V = As$, where A is the area of R and s is the distance traveled by the centroid of R.

Proof   First, choose a coordinate system in which the y-axis coincides with the axis of revolution. Figure 6.69 shows a typical region R together with a vertical approximating rectangle. We will assume that this rectangle has area ΔA and that it is located x units from the y-axis.

Notice that when the rectangle is revolved about the y-axis, it generates a shell of volume $\Delta V = 2\pi x \Delta A$. Thus, by partitioning the region R into a number of rectangles and taking the limit of the sum of the volumes of all the related approximating shells as the norm of the partition approaches 0, we find that

$$V = \lim_{\|P\|\to 0} \sum_{k=1}^{n} 2\pi x_k \Delta A_k = \int 2\pi x \, dA = 2\pi \int x \, dA$$

$$= 2\pi \bar{x} \int dA \quad \text{Because } \bar{x} = \frac{\int x \, dA}{\int dA}$$

$$= 2\pi \bar{x} A \quad \text{Because } \int dA = A$$

$$= As \quad \begin{array}{l}\text{Where } s = 2\pi \bar{x} \text{ is the distance traveled by the}\\ \text{centroid (the circumference of a circle of radius } \bar{x})\end{array} \quad \square$$

We close this section with an example that illustrates the use of the volume theorem of Pappus.

EXAMPLE 7   Volume of a torus using the volume theorem of Pappus

When a circle of radius r is revolved about a line in the plane of the circle located R units from its center ($R > r$), the solid figure so generated is called a **torus**, as shown in Figure 6.70. Show that the torus has volume $V = 2\pi^2 r^2 R$.

Axis of revolution

The centroid travels $2\pi R$ units.

Figure 6.70   The volume of a torus

Solution

The circle has area $A = \pi r^2$ and its center (which is its centroid) travels $2\pi R$ units. Comparing this example to the volume theorem of Pappus, we see $s = 2\pi R$. Therefore, the torus has volume

$$V = (2\pi R)A = 2\pi R(\pi r^2) = 2\pi^2 r^2 R \qquad \blacksquare$$

Incidentally, the volume found in Example 7 using the theorem of Pappus was worked in Problem 60 of Section 6.2 by using washers.

6.5 PROBLEM SET

A 1. **WHAT DOES THIS SAY?** Discuss work and how to find it. (*Hint*: The answer is *not* to apply at the unemployment office!)

2. **WHAT DOES THIS SAY?** Discuss fluid force and how to find it.

3. **WHAT DOES THIS SAY?** What is meant by a centroid? Outline a procedure for finding both the mass and the centroid.

4. What is the work done in lifting a 50-lb bag of salt 5 ft?

5. What is the work done in lifting an 850-lb billiard table 15 ft?

6. Suppose it takes 4 erg of work to stretch a spring 10 cm beyond its natural length. How much more work is needed to stretch it 4 cm further?

7. A bucket weighing 75 lb when filled and 10 lb when empty is pulled up the side of a 100-ft building. How much more work is done in pulling up the full bucket than the empty bucket?

8. A spring whose natural length is 10 cm exerts a force of 30 m/s when stretched to a length of 15 cm. (*Note*: 15 cm − 10 cm = 5 cm = 0.05 m.) Find the spring constant, and then determine the work done in stretching the spring 7 cm (= 0.07 m) beyond its natural length.

9. A 5-lb force will stretch a spring 9 in. (= 0.75 ft) beyond its natural length. How much work is required to stretch it 1 ft beyond its natural length?

10. A 30-ft rope weighing 0.4 lb/ft hangs over the edge of a building 100 ft high. How much work is done in pulling the rope to the top of the building? Assume that the top of the rope is flush with the top of the building and the lower end of the rope is swinging freely.

11. A 30-lb ball hangs at the bottom of a cable that is 50 ft long and weighs 20 lb. The entire length of cable hangs over a cliff. Find the work done to raise the cable and get the ball to the top of the cliff.

12. An object moving along the x-axis is acted upon by a force $F(x) = x^4 + 2x^2$. Find the total work done by the force in moving the object from $x = 1$ cm to $x = 2$ cm. *Note*: Distance is in cm, so force is in dynes.

13. An object moving along the x-axis is acted upon by a force $F(x) = |\sin x|$. Find the total work done by the force in moving the object from $x = 0$ cm to $x = 2\pi$ cm. *Note*: Distance is in cm, so force is in dynes.

In Problems 14–17, the given figure is the vertical cross section of a tank containing the indicated fluid. Find the fluid force against the end of the tank. The weight densities are given in Table 6.4 on page 399.

14. water

15. seawater

16. gasoline

17. kerosene

In Problems 18–21, set up, but do not evaluate, an integral for the fluid force against the indicated vertical plate.

18. water

19. milk (half cookie)

20. mercury

21. kerosene

Find the centroid ($\rho = 1$) of each planar region in Problems 22–27.

22. the region in the first quadrant bounded by the curves $y = x^3$ and $y = \sqrt[3]{x}$

23. the region bounded by the parabola $y = x^2 - 9$ and the x-axis

24. the region bounded by the parabola $y = 4 - x^2$ and the line $y = x + 2$

25. the region bounded by the curve $y = x^{-1}$, the x-axis, and the lines $x = 1$ and $x = 2$

26. the region bounded by the curve $y = x^{-1}$ and the line $2x + 2y = 5$

27. the region bounded by $y = \dfrac{1}{\sqrt{x}}$ and the x-axis on $[1, 4]$

Use the theorem of Pappus to compute the volume of the solids generated by revolving the regions given in Problems 28–31 about the prescribed axes.

28. the region bounded by the parabola $y = \sqrt{x}$, the x-axis, and the line $x = 4$ about the line $x = -1$

29. the triangular region with vertices $(-3, 0)$, $(0, 5)$, and $(2, 0)$, about the line $y = -1$

30. the semicircular region $y = \sqrt{1 - x^2}$, about the line $y = -1$

31. the semicircular region $x = \sqrt{4 - y^2}$, about the line $x = -2$

B 32. The centroid of any triangle lies at the intersection of the medians, two-thirds the distance from each vertex to the midpoint of the opposite side. Locate the centroid of the triangle with vertices $(0, 0)$, $(7, 3)$, and $(7, -2)$ using this method, and then check your result by calculus.

33. A tank in the shape of an inverted right circular cone of height 6 ft and top radius 3 ft is filled to a depth of 3 ft.

   a. How much work is performed in pumping all the water over the top edge of the tank?

   b. How much work is performed in draining all the water from the tank through a hole at the tip?

34. A tank is formed by rotating the curve $y = x^2$ about the y-axis for $0 \le x \le 4$, where x is in feet. If the tank is filled with water to a depth of 12 feet, how much work is performed in pumping all the water to a height of 3 ft above the top of the tank?

35. A hemispherical bowl with radius 10 ft is filled with water to a level of 6 ft. Find the work done to the nearest ft-lb to pump all the water to the top of the bowl.

36. A holding tank has the shape of a rectangular parallelepiped 20 ft by 30 ft by 10 ft.
   a. How much work is done in pumping all the water to the top of the tank?
   b. How much work is done in pumping all the water out of the tank to a height of 2 ft above the top of the tank?

37. A cylindrical tank of radius 3 ft and height 10 ft is filled to a depth of 2 ft with a liquid of density $\rho = 40$ lb/ft^3. Find the work done in pumping all the liquid to a height of 2 ft above the top of the tank.

38. An oil can in the shape of a rectangular parallelepiped is filled with oil of density 0.87 g/cm^3. What is the total force on a side of the can if it is 20 cm high and has a square base 15 cm on a side?

39. A swimming pool 20 ft by 15 ft by 10 ft deep is filled with water. A cube 1 ft on a side lies on the bottom of the pool. Find the total fluid force on the five exposed faces of the cube.

40. Suppose a log of radius 1 ft lies at the bottom of the pool in Problem 39. What is the total fluid force on one end of the log?

41. An object weighing 800 lb on the surface of the earth is propelled to a height of 200 mi above the earth. How much work is done against gravity? *Hint:* Assume the radius of the earth is 4,000 mi, and use Newton's law, $F = -k/x^2$, for the force on an object x miles from the center of the earth.

42. A dam is in the shape of an isosceles trapezoid 200 ft at the top, 100 ft at the bottom, and 75 ft high. When water is up to the top of the dam, what is the force on its face?

43. Modeling Problem   According to Coulomb's law in physics, two similarly charged particles repel each other with a force inversely proportional to the square of the distance between them. Suppose the force is 12 dynes when they are 5 cm apart.
   a. How much work is done in moving one particle from a distance of 10 cm to a distance of 8 cm from the other?

b. Set up and analyze a model to determine the amount of work performed in moving one particle from an "infinite" distance to a distance of 8 cm from the other. State any assumptions that you must make.

44. Modeling Problem   Figure 6.71 shows a dam whose face against the water is an inclined rectangle.
   a. Set up and analyze a model for determining the fluid force against the face of the dam when the water is level with the top.
   b. What would the fluid force be if the face of the dam were a vertical rectangle (not inclined)?

Figure 6.71   Dam with face at an inclined rectangle

45. Modeling Problem   Figure 6.72 shows a swimming pool, part of whose bottom is an inclined plane. Set up and analyze a model for determining the fluid force on the bottom when the pool is filled to the top with water.

Figure 6.72   Swimming pool dimensions

46. An irregular plate is measured at 0.2-ft intervals from top to bottom, and the corresponding widths noted. The plate is then submerged vertically in water with its top 1 ft below the surface. Use Simpson's rule to estimate the total force on the face of the plate.

Depth, x_k	Width, L_k
1.0	0.0
1.2	3.3
1.4	3.6
1.6	3.7
1.8	3.7
2.0	3.6
2.2	3.4
2.4	2.9
2.6	0.0

*A region R with uniform density is said to have **moments of inertia** I_x about the x-axis and I_y about the y-axis, where*

$$I_x = \int y^2 \, dA \quad and \quad I_y = \int x^2 \, dA$$

The dA represents the area of an approximating strip. In Problems 47–48, find I_x and I_y for the given region.

47. the region bounded by the parabola $y = \sqrt{x}$, the x-axis, and the line $x = 4$

48. the region bounded by the curve $y = \sqrt[3]{1 - x^3}$ and the coordinate axes

49. Modeling Problem If a region has area A and moment of inertia I_y about the y-axis, then it is said to have a **radius of gyration**

$$\rho_y = \sqrt{\frac{I_y}{A}}$$

with respect to the y-axis. Set up and analyze a model to determine the radius of gyration for the triangular region bounded by the line $y = 4 - x$ and the positive coordinate axes.

C 50. Show that the centroid of a rectangle is the point of intersection of its diagonals.

51. A right circular cone is generated when the region bounded by the line $y = x$ and the vertical lines $x = 0$ and $x = r$ is revolved about the x-axis. Use Pappus' theorem to show that the volume of this cone is $V = \frac{1}{3}\pi r^3$.

52. Let R be the triangular region bounded by the line $y = x$, the x-axis, and the vertical line $x = r$. When R is rotated about the x-axis, it generates a cone of volume $V = \frac{1}{3}\pi r^3$. Use Pappus' theorem to determine $\bar{y}$, the y-coordinate of the centroid of R. Then use similar reasoning to find $\bar{x}$.

53. Find the volume of the solid figure generated when a square of side L is revolved about a line that is outside the square, parallel to two of its sides, and located s units from the closer side.

54. Find the volume of the solid figure generated by revolving an equilateral triangle of side L about one of its sides.

55. Let R be a region in the plane, and suppose that R can be partitioned into two subregions R_1 and R_2. Let $(\bar{x}_1, \bar{y}_1)$ and $(\bar{x}_2, \bar{y}_2)$ be the centroids of R_1 and R_2, respectively, and let A_1 and A_2 be the areas of the subregions. Show that $(\bar{x}, \bar{y})$ is the centroid of R, where

$$\bar{x} = \frac{A_1 x_1 + A_2 x_2}{A_1 + A_2} \quad \text{and} \quad \bar{y} = \frac{A_1 y_1 + A_2 y_2}{A_1 + A_2}$$

6.6 Applications to Business, Economics, and Life Sciences

IN THIS SECTION future and present value of a flow of income; cumulative change: net earnings; consumer's and producer's surplus; survival and renewal; flow of blood through an artery

In Section 6.5, we saw how integration can be applied to the physical sciences. The goal of this section is to explore applications to a variety of other areas, especially business, economics, and the life sciences.

FUTURE AND PRESENT VALUE OF A FLOW OF INCOME

Recall from Chapter 2 that when P_0 dollars are invested at the continuous compounding annual rate r, then after t years, the account is worth $A = P_0 e^{rt}$ dollars. The amount A is called the *future value* of the initial value P_0 dollars, but what is the future value if the money were not deposited just once, in a lump sum P_0, but continuously over the entire time period? The difficulty is that at any given time, the "old" money deposited earlier in the time period has been accumulating interest longer than "new" money deposited more recently. To compute future value, in this case, our strategy will be to approximate the continuous income flow by a sequence of discrete deposits called an **annuity**. Then, the amount of the approximating annuity is a sum whose limit is a definite integral that gives the future value of the income flow. The general procedure is illustrated in Example 1.

EXAMPLE 1 Computing future value of a continuous income flow

Money is transferred into an account continuously at the rate of 1,500 dollars/yr. If the account earns interest at the rate of 7% compounded continuously, what is the future value of the account 5 years from now?

Solution

To approximate the future value of the income flow, divide the 5-yr time period $0 \le t \le 5$ into n intervals, each of length $\Delta_n t = (5 - 0)/n$, and for $k = 1, 2, \ldots, n$, let t_{k-1} denote the beginning of the kth subinterval $[t_{k-1}, t_k]$. The amount of money deposited during the kth subinterval is $1,500\Delta_n t$. Therefore, if as an approximation, we assume that all this money is deposited at time t_{k-1}, then since $5 - t_{k-1}$ years remain in the 5-yr investment period, it follows that this money would grow to $(1,500\Delta_n t)e^{0.07(5-t_{k-1})}$ dollars; that is,

$$\begin{bmatrix} \text{FUTURE VALUE OF MONEY DEPOSITED} \\ \text{DURING THE } k\text{TH SUBINTERVAL} \end{bmatrix} = 1,500e^{0.07(5-t_{k-1})}\Delta_n t$$

This approximation is illustrated in Figure 6.73.

Figure 6.73 Approximate future value of money deposited during the kth subinterval

The future value of the entire income stream can be approximated by the sum of the future values of the money deposited during all n time subintervals. Hence, we have

$$[\text{FUTURE VALUE OF INCOME FLOW}] \approx \sum_{k=1}^{n} 1,500e^{0.07(5-t_{k-1})}\Delta_n t$$

As n increases without bound, the length of each subinterval tends toward zero, and there is less and less error in our assumption that all the money in the kth subinterval is deposited at time $t = t_{k-1}$. Therefore, it is reasonable to compute the total future value of the income flow by taking the limit of the approximating sum as $n \to +\infty$. We recognize this sum as a Riemann sum, and the limit as the definite integral. That is

$$
\begin{aligned}
[\text{FUTURE VALUE OF INCOME FLOW}] &= \lim_{n \to +\infty} \sum_{k=1}^{n} 1,500e^{0.07(5-t_{k-1})}\Delta_n t \\
&= \int_0^5 1,500e^{0.07(5-t)}\, dt \\
&= 1,500e^{0.35} \int_0^5 e^{-0.07t}\, dt \\
&= 1,500e^{0.35} \left[\frac{e^{-0.07t}}{-0.07} \right]\Big|_0^5 \\
&= 8,980.02 \qquad\blacksquare
\end{aligned}
$$

Recall from Chapter 2, that the *present value*, PV, of A dollars invested at the annual interest rate r compounded continuously for a term T years is the amount of money $PV = Ae^{-rt}$ that must be deposited today at the prevailing interest rate to generate A dollars at the end of the term. Likewise, the present value of an income stream $f(t)$ generated continuously at a certain rate over a specified term is the amount of money that would have to be deposited at the beginning of the term at the prevailing interest rate to generate the income stream over the entire term. As with future value, the strategy is to approximate the continuous income stream by a sequence of discrete payments; that is, by an annuity. The present value of the approximating annuity is a sum, and by taking the limit of that sum, we obtain an integral

$$\int_0^T f(t)e^{-rt}\, dt$$

that represents the present value of the entire income flow. The details of this approximation scheme are outlined in the problem set. Here is a summary of the formulas to be used for computing future and present value of an income flow.

Future Value and Present Value of an Income Flow

Let $f(t)$ be the amount of money deposited at time t over the time period $[0, T]$ in an account that earns interest at the annual rate r compounded continuously. Then the **future value** of the income flow over the time period is given by

$$FV = \int_0^T f(t)e^{rt}\, dt$$

and the **present value** of the same income flow over the time period is

$$PV = \int_0^T f(t)e^{-rt}\, dt$$

CUMULATIVE CHANGE: NET EARNINGS

In certain applications, we are given the rate of change $Q'(x)$ of a quantity $Q(x)$ and are required to find the net change $Q(b) - Q(a)$ in $Q(x)$ as x varies from $x = a$ to $x = b$. But because $Q(x)$ is an antiderivative of $Q'(x)$, the fundamental theorem of calculus tells us that the net change is given by the definite integral

$$Q(b) - Q(a) = \int_a^b Q'(x)\, dx$$

Since the definite integral on the right "adds" all the instantaneous changes in $Q(x)$ over the time period $[a, b]$, the integral can also be thought of as **cumulative change**. For example, if $N(t)$ is the population of a certain species that is growing at the rate $N'(t)$ at time t, then the cumulative change in population over the time period $[t_1, t_2]$ is given by the integral

$$N(t_2) - N(t_1) = \int_{t_1}^{t_2} N'(t)\, dt$$

A business application involves the **net earnings** of an industrial process. Suppose $C(x)$ represents the total cost of producing x units and $R(x)$ represents the revenue generated by the sale of those units. Then $E(x) = R(x) - C(x)$ represents earnings, and if the marginal cost $C'(x)$ and marginal revenue $R'(x)$ are known, the net earnings for the production range $a \le x \le b$ are given by the definite integral

$$E(b) - E(a) = \int_a^b E'(x)\, dx$$
$$= \int_a^b [R'(x) - C'(x)]\, dx$$

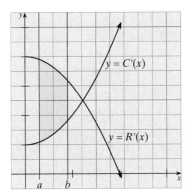

Figure 6.74 Net earnings as an area

Net earnings can be interpreted geometrically as an area between the marginal revenue and marginal cost curves as shown in Figure 6.74. These ideas are illustrated in Example 2.

EXAMPLE 2   Computing net earnings

For a certain industrial process, the marginal cost and marginal revenue (in thousands of dollars) associated with producing x units are

$$C'(x) = 0.1x^2 + 4x + 10 \quad \text{and} \quad R'(x) = 70 - x$$

respectively. What are the net earnings of the process as x ranges from $x = 0$ to $x = x_m$, where x_m is the level of production at which marginal cost equals marginal revenue?

Solution

First, note that marginal cost equals marginal revenue when

$$0.1x^2 + 4x + 10 = 70 - x$$
$$0.1x^2 + 5x - 60 = 0$$
$$x^2 + 50x - 600 = 0$$
$$(x + 60)(x - 10) = 0$$
$$x = 10, -60$$

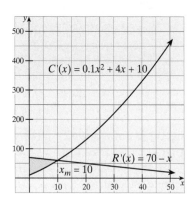

Reject the negative value, giving the solution $x_m = 10$, so that the net earnings as x ranges from $x = 0$ to $x = 10$ are given by

$$E(10) - E(0) = \int_0^{10} [R'(x) - C'(x)]\,dx$$
$$= \int_0^{10} [70 - x - (0.1x^2 + 4x + 10)]\,dx$$
$$= \int_0^{10} [-0.1x^2 - 5x + 60]\,dx = \frac{950}{3} = 316.\bar{6}$$

Figure 6.75 Net earnings for the marginal cost and marginal revenue

That is, the net earnings are $316,667, and are shown as the area between the marginal cost and marginal revenue curves in Figure 6.75. ∎

CONSUMER'S AND PRODUCER'S SURPLUS

In a competitive economy, the total amount that consumers actually spend on a commodity is usually less than the total amount they would have been willing to spend. The difference between the two amounts can be thought of as a savings realized by consumers and is known in economics as the **consumer's surplus**.

To get a better feel for the concept of consumer's surplus, consider an example of a couple who are willing to spend $500 for their first TV set, $300 for a second set, and only $50 for a third set. If the market price is $300, then the couple would buy only two sets and would spend a total of $2 \times \$300 = \$600$. This is less than the $\$500 + \$300 = \$800$ that the couple would have been willing to spend to get the two sets. The savings of $\$800 - \$600 = \$200$ is the couple's consumer's surplus. Consumer's surplus has a simple geometric interpretation, which is illustrated in Figure 6.76.

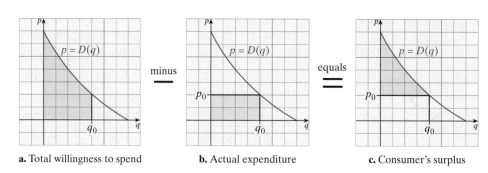

a. Total willingness to spend   **b.** Actual expenditure   **c.** Consumer's surplus

Figure 6.76 Geometric interpretation of consumer's surplus

Note that p and q denote the market price and the corresponding demand, respectively. Figure 6.76a shows the region under the demand curve $p = D(q)$ from $q = 0$ to $q = q_0$, and the area of this region represents the total amount that consumers are willing to spend to get q_0 units of the commodity. The rectangle in Figure 6.76b has an area of $p_0 q_0$ and represents the actual consumer expenditure for q_0 units at p_0 dollars per unit. The difference between these two areas (Figure 6.76c) represents the consumer's surplus. That is, consumer's surplus is the area of the region between the demand curve $p = D(q)$ and the horizontal line $p = p_0$ so that

$$\int_0^{q_0} [D(q) - p_0]\, dq = \int_0^{q_0} D(q)\, dq - \int_0^{q_0} p_0\, dq = \int_0^{q_0} D(q)\, dq - p_0 q_0$$

Consumer's Surplus

If q_0 units of a commodity are sold at a price of p_0 dollars per unit, and if $p = D(q)$ is the consumer's demand function for the commodity, then

$$\begin{bmatrix} \text{CONSUMER'S} \\ \text{SURPLUS} \end{bmatrix} = \begin{bmatrix} \text{TOTAL AMOUNT} \\ \text{CONSUMERS ARE} \\ \text{WILLING TO SPEND} \\ \text{FOR } q_0 \text{ UNITS} \end{bmatrix} - \begin{bmatrix} \text{ACTUAL CONSUMER} \\ \text{EXPENDITURE FOR} \\ q_0 \text{ UNITS} \end{bmatrix}$$

$$\text{CS} = \int_0^{q_0} D(q)\, dq - p_0 q_0$$

Producer's surplus is the other side of the coin from consumer's surplus. In particular, a supply function $p = S(q)$ gives the price per unit that producers are willing to accept in order to supply q units to the marketplace. However, any producer who is willing to accept less than $p_0 = S(q_0)$ dollars for q_0 units gains from the fact that the price is actually p_0. Then **producer's surplus** is the difference between what producers would be willing to accept for supplying q units and the price they actually receive. Reasoning as we did with consumer's surplus, we obtain the following formula for producer's surplus.

Producer's Surplus

If q_0 units of a commodity are sold at a price of p_0 dollars per unit, and if $p = S(q)$ is the producer's supply function for the commodity, then

$$\begin{bmatrix} \text{PRODUCER'S} \\ \text{SURPLUS} \end{bmatrix} = \begin{bmatrix} \text{ACTUAL CONSUMER} \\ \text{EXPENDITURE FOR} \\ q_0 \text{ UNITS} \end{bmatrix} - \begin{bmatrix} \text{TOTAL AMOUNT} \\ \text{PRODUCERS RECEIVE} \\ \text{WHEN } q_0 \text{ UNITS} \\ \text{ARE SUPPLIED} \end{bmatrix}$$

$$\text{PS} = p_0 q_0 - \int_0^{q_0} S(q)\, dq$$

EXAMPLE 3 Computing consumer's and producer's surplus

A tire manufacturer estimates that q (thousand) radial tires will be purchased (that is, demanded) by wholesalers when the price is

$$p = D(q) = -0.1q^2 + 90$$

dollars/tire, and the same number of tires will be supplied when the price is

$$p = S(q) = 0.2q^2 + q + 50$$

dollars/tire.

a. Find the equilibrium price (that is, where supply equals demand) and the quantity supplied and demanded at that price.

b. Determine the consumer's and producer's surplus at the equilibrium price.

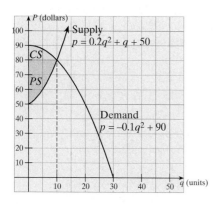

Figure 6.77 Consumer's and producer's surplus at equilibrium for the given supply and demand functions

Solution

a. The supply and demand curves are shown in Figure 6.77. Supply equals demand when

$$0.2q^2 + q + 50 = -0.1q^2 + 90$$

$$0.3q^2 + q - 40 = 0$$

$$q = 10 \qquad (\text{reject } q = -\tfrac{40}{3})$$

and $p = S(10) = D(10) = 80$. Thus, equilibrium occurs at a price of $80/tire, when 10,000 tires are supplied and demanded.

b. Using $p_0 = 80$ and $q_0 = 10$, we find that the consumer's surplus is

$$\int_0^{10} (-0.1q^2 + 90)\, dq - (80)(10) = \left[-0.1\left(\frac{q^3}{3}\right) + 90q \right]\Big|_0^{10} - (80)(10)$$

$$= \frac{2{,}600}{3} - (80)(10)$$

$$= \frac{200}{3} = 66.\bar{6}$$

The producer's surplus is

$$(80)(10) - \int_0^{10} (0.2q^2 + q + 50)\, dq = (80)(10) - \left[0.2\left(\frac{q^3}{3}\right) + \left(\frac{q^2}{2}\right) + 50q \right]\Big|_0^{10}$$

$$= (80)(10) - \frac{1{,}850}{3}$$

$$= 183.\bar{3}$$

Since q is in thousands, the consumer's surplus is about $66,667, and the producer's surplus is $183,333. The consumer's surplus and producer's surplus are the regions labeled CS and PS, respectively, in Figure 6.77. ∎

SURVIVAL AND RENEWAL

There is an important category of problems in which a survival function gives the fraction of individuals in a group of population that can be expected to remain in the group and a renewal function gives the rate at which new members can be expected to join the group. Such **survival and renewal** problems arise in many fields, including sociology, ecology, demography, and even finance, where the "population" is the number of dollars in an investment account and "survival and renewal" refers to results of an investment strategy. The following example illustrates the general procedure for modeling with survival and renewal functions.

EXAMPLE 4   Modeling enrollment using survival and renewal

A new county mental health clinic has just opened. Statistics from similar facilities suggest that the fraction of patients who will still be receiving treatment at the clinic t months after their initial visit is given by the function $f(t) = e^{-t/20}$. The clinic initially accepts 300 people for treatment and plans to accept new patients at the rate of 10 per month. Approximately how many people will be receiving treatment at the clinic 15 months from now?

Solution

Since $f(15)$ is the fraction of patients whose treatment continues at least 15 months, it follows that of the current 300 patients, only $300 f(15)$ will still be receiving treatment 15 months from now.

To approximate the number of *new* patients who will be receiving treatment 15 months from now, divide the 15-month time interval $0 \le t \le 15$ into n equal subintervals of length Δt months and let t_{j-1} denote the beginning of the jth subinterval. Since new patients are accepted at the rate of 10/mo, the number of new patients accepted during the jth subinterval is $10\Delta t$. Fifteen months from now, approximately $15 - t_{j-1}$ months will have elapsed since these $10\Delta t$ new patients had their initial visits, and so approximately $(10\Delta t)f(15 - t_{j-1})$ of them will still be receiving treatment at that time (see Figure 6.78).

Figure 6.78 New members arriving during the jth subinterval

It follows that the total number of new patients still receiving treatment 15 months from now can be approximated by the sum

$$\sum_{j=1}^{n} 10f(15 - t_{j-1})\Delta t$$

Adding this to the number of current patients who will still be receiving treatment in 15 months, we obtain

$$P \approx 300f(15) + \sum_{j=1}^{n} 10f(15 - t_{j-1})\Delta t$$

where P is the total number of patients who will be receiving treatment 15 months from now. As n increases without bound, the approximation improves and approaches the true value of P. That is,

$$P = 300f(15) + \lim_{n \to +\infty} \sum_{j=1}^{n} 10f(15 - t_{j-1})\Delta t$$

$$= 300f(15) + \int_{0}^{15} 10f(15 - t)\,dt$$

Since $f(t) = e^{-t/20}$, we have $f(15) = e^{-3/4}$ and $f(15 - t) = e^{-(15-t)/20} = e^{-3/4}e^{t/20}$. Thus,

$$P = 300e^{-3/4} + 10e^{-3/4} \int_{0}^{15} e^{t/20}\,dt$$

$$= 300e^{-3/4} + 200e^{-3/4}e^{t/20}\Big|_{0}^{15}$$

$$= 300e^{-3/4} + 200(1 - e^{-3/4})$$

$$\approx 247.24$$

That is, 15 months from now, the clinic will be treating approximately 247 patients. ∎

THE FLOW OF BLOOD THROUGH AN ARTERY

In biology, Poiseuille's law of flow* asserts that the speed of blood in an artery is a function of the distance of the blood from the artery's central axis. That is, the speed (in cm/s) of blood that is r centimeters from the central axis of the artery may be modeled by

$$S(r) = k(R^2 - r^2)$$

where R is the radius of the artery and k is a constant. In the following example, you will see how to use this information to compute the rate (in cm^3/s) at which blood flows through the artery.

---

*The law of flow is named for the same French physician, Jean Louis Poiseuille (1799–1869), responsible for the resistance to flow law cited in Section 4.7.

EXAMPLE 5   Modeling blood flow

Find an expression for the volume rate (in cm^3/s) at which blood flows through an artery of radius R if the speed of the blood r cm from the central axis is $S(r) = k(R^2 - r^2)$, where k is a constant.

Solution

To find the approximate volume of blood that flows through a cross section of the artery each second, divide the interval $0 \leq r \leq R$ into n equal subintervals of width Δr cm and let r_j denote the beginning of the jth subinterval. These subintervals determine n concentric rings, as illustrated in Figure 6.79.

If Δr is small, the area of the jth ring is approximately the area of a rectangle whose length is the circumference of the (inner) boundary of the ring and whose width is Δr. That is,

$$\text{AREA OF THE } j\text{TH RING} \approx 2\pi r_j \Delta r$$

Figure 6.79 Subdividing a cross section of an artery into concentric rings

If we multiply the area of the jth ring (cm^2) by the speed (cm/s) of the blood flowing through this ring, we obtain the volume rate (cm^3/s) at which blood flows through the jth ring. Since the speed of blood flowing through the jth ring is approximately $S(r_j)$ cm/s, it follows that

$$\begin{aligned}\text{VOLUME RATE OF FLOW} \atop \text{THROUGH THE } j\text{TH RING} &\approx (\text{AREA OF } j\text{TH RING})\left(\text{SPEED OF BLOOD} \atop \text{THROUGH THE } j\text{TH RING}\right) \\ &= (2\pi r_j \Delta r) S(r_j) \\ &= (2\pi r_j \Delta r)\left[k(R^2 - r_j^2)\right] \\ &= 2\pi k(R^2 r_j - r_j^3)\Delta r\end{aligned}$$

The volume rate of flow of blood through the entire cross section is the sum of n such terms, one for each of the n concentric rings. That is,

$$\text{VOLUME RATE OF FLOW} \approx \sum_{j=1}^{n} 2\pi k(R^2 r_j - r_j^3)\Delta r$$

As n increases without bound, this approximation approaches the true value of the volume rate of flow. That is,

$$\begin{aligned}\text{VOLUME RATE OF FLOW} &= \lim_{n \to +\infty} \sum_{j=1}^{n} 2\pi k(R^2 r_j - r_j^3)\Delta r \\ &= \int_0^R 2\pi k(R^2 r - r^3)\, dr \\ &= 2\pi k \left(\frac{R^2}{2}r^2 - \frac{1}{4}r^4\right)\Big|_0^R \\ &= \frac{\pi k R^4}{2} \quad \text{in } cm^3/s\end{aligned}$$ ■

6.6   PROBLEM SET

Ⓐ **1. WHAT DOES THIS SAY?** What is cumulative change for a quantity $Q(x)$?

2. WHAT DOES THIS SAY? What are the future and present values of an income flow?

3. WHAT DOES THIS SAY? What are consumer's and producer's surplus?

Find the consumer's surplus for the given demand function defined by $D(q)$ at the point that corresponds to the sales level q as given in Problems 4–7.

4. $D(q) = 3.5 - 0.5q$
   a. $q_0 = 1$ **b.** $q_0 = 1.5$

5. $D(q) = 2.5 - 1.5q$
   a. $q_0 = 1$ **b.** $q_0 = 0$

6. $D(q) = 100 - 8q$
   a. $q_0 = 4$ **b.** $q_0 = 10$

7. $D(q) = 150 - 6q$
   a. $q_0 = 5$ **b.** $q_0 = 12$

In Problems 8–11, $p = D(q)$ is the price (in dollars/unit) at which q units of a particular commodity will be demanded by the market (that is, all q units will be sold at this price), and q_0 is a specified level of production. In each case, find the price $p_0 = D(q_0)$ at which q_0 units will be demanded and compute the corresponding consumer's surplus. Sketch the demand curve $y = D(q)$ and shade the region whose area represents the consumer's surplus.

8. $D(q) = 2(64 - q^2)$; $q_0 = 3$ units

9. $D(q) = 150 - 2q - 3q^2$; $q_0 = 6$ units

10. $D(q) = 40e^{-0.05q}$; $q_0 = 5$ units

11. $D(q) = 75e^{-0.04q}$; $q_0 = 3$ units

In Problems 12–15, $p = S(q)$ is the price (in dollars/unit) at which q units of a particular commodity will be supplied to the market by producers, and q_0 is a specified level of production. In each case, find the price $p_0 = S(q_0)$ at which q_0 units will be supplied and compute the corresponding producer's surplus. Sketch the supply curve $y = S(q)$ and shade the region whose area represents the producer's surplus.

12. $S(q) = 0.3q^2 + 30$; $q_0 = 4$ units

13. $S(q) = 0.5q + 15$; $q_0 = 5$ units

14. $S(q) = 10 + 15e^{0.03q}$; $q_0 = 3$ units

15. $S(q) = 17 + 11e^{0.01q}$; $q_0 = 7$ units

B *In Problems 16–19, find the consumer's surplus at the point of market equilibrium (the level of production where supply $S(q)$ equals demand $D(q)$).*

Demand function	**Supply function**
16. $D(q) = 14 - q^2$	$S(q) = 2q^2 + 2$
17. $D(q) = 25 - q^2$	$S(q) = 5q^2 + 1$
18. $D(q) = 32 - 2q^2$	$S(q) = \frac{1}{3}q^2 + 2q + 5$
19. $D(q) = 27 - q^2$	$S(q) = \frac{1}{4}q^2 + \frac{1}{2}q + 5$

20. A manufacturer has determined that when q units of a particular commodity are produced, the price at which all the units can be sold is $p = D(q)$ dollars per unit, where D is the demand function given by

$$D(q) = \frac{300}{(0.1q + 1)^2}$$

a. How many units can the manufacturer expect to sell if the commodity is priced at $p_0 = \$12$/unit?

b. Find the consumer's surplus that corresponds to the level of production q_0 found in part **a**.

21. Money is transferred continuously into an account at the constant rate of $3,000/yr. The account earns interest at the annual rate of 8% compounded continuously.

a. What is the future value of the income flow in three years?

b. What is the present value of the income flow over a three-year term?

22. The management of a small firm expects to generate income at the rate of $f(t) = 2,000e^{0.03t}$ dollars/yr. The prevailing rate of interest is 7% compounded continuously.

a. What is the future value for the firm over a five-year term?

b. What is the present value for the firm over the same five-year term?

23. Suppose an industrial machine that is x years old generates revenue at the rate of

$$R'(x) = 6,025 - 10x^2$$

dollars per year and costs accumulate at the rate of

$$C'(x) = 4,000 + 15x^2$$

dollars per year for operation and maintenance.

a. The machine will be sold when earnings $E(x) = R(x) - C(x)$ are maximized. When does this occur?

b. What are the net earnings generated by the machine between now and when it is sold? Interpret this amount as the area between two curves.

24. After t hours on the job, Joan is producing units at the rate of

$$Q_1'(t) = 60 - 2(t - 1)^2$$

units per hour, and Jill is producing at the rate of

$$Q_2'(t) = 50 - 5t$$

units per hour. If both arrive on the job at 8:00 A.M., how many more units (to the nearest unit) will Joan have produced by noon than Jill? Interpret your answer as the area between two curves.

25. Suppose that x years from now, one investment plan will be generating profit at the rate of

$$P_1'(x) = 100 + x^2$$

dollars per year, whereas a second plan will be generating profit at the rate of

$$P_2'(x) = 190 + 2x$$

dollars per year. Neither plan generates a profit in the beginning; that is, $P_1(0) = P_2(0) = 0$.

a. For how many years will the second plan be more profitable?

b. How much more total profit will you earn if you invest in the second plan instead of the first for the period of time in part **a**? Interpret the extra profit as the area between two curves.

26. Parts for a piece of heavy machinery are sold in units of 1,000. The demand for the parts (in dollars) is given by $p(x) = 110 - x$. The total cost is given by $C(x) = x^3 - 25x^2 + 2x + 30$ (dollars).

a. For what value of x is the profit

$$P(x) = xp(x) - C(x)$$

maximized?

b. Find the consumer's surplus with respect to the price that corresponds to maximum profit.

27. Repeat Problem 26 with $p(x) = 124 - 2x$ and $C(x) = 2x^3 - 59x^2 + 4x + 76$.

28. Suppose when q units of a commodity are produced, the demand is $p(q) = 45 - q^2$ dollars per unit, and the marginal cost is

$$\frac{dC}{dq} = 6 + \frac{1}{4}q^2$$

Assume there is no overhead (so $C(0) = 0$).

a. Find the total revenue and the marginal revenue.

b. Find the value of q (to the nearest unit) that maximizes profit.

c. Find the consumer's surplus at the value of q where profit is maximized. (Use the exact value of q.)

29. Repeat Problem 28 with

$$p(q) = \tfrac{1}{4}(10-q)^2 \quad \text{and} \quad \frac{dC}{dq} = \tfrac{3}{4}q^2 + 5$$

30. A company plans to hire additional personnel. Suppose it is estimated that as x new people are hired, it will cost $C(x) = 0.2x$ thousand dollars and that these x people will bring in $R(x) = \sqrt{3x}$ thousand dollars in additional revenue. How many new people would be hired to maximize the profit? How much net revenue would the company gain by hiring these people?

31. Suppose that the demand function for a certain commodity is $D(q) = 20 - 4q^2$, and that the marginal cost is $C'(q) = 2q + 6$, where q is the number of units produced. Find the consumer's surplus at the sales level q_0 where profit is maximized.

32. The resale value of a certain industrial machine decreases over a ten-year period at a rate that changes with time. When the machine is x years old, its value is decreasing at the rate of $220(x-10)$ dollars/yr. By how much does the machine depreciate during the second year?

33. The promoters of a country fair estimate that t hours after the gates open at 9:00 A.M. visitors will be entering the fair at the rate of $-4(t+2)^3 + 54(t+2)^2$ people/hr. How many people will enter the fair between 10:00 A.M. and noon?

34. At a certain factory, the marginal cost is $6(q-5)^2$ dollars/unit when the level of production is q units. By how much will the total manufacturing cost increase if the level of production is raised from 10 to 13 units?

35. A certain oil well that yields 400 barrels of crude oil a month will run dry in two years. The price of crude oil is currently $18/barrel and is expected to rise at a constant rate of $0.03 per barrel per month. If the oil is sold as soon as it is extracted from the ground, what will be the total future revenue from the well?

36. It is estimated that t days from now a farmer's crop (in bushels) will be increasing at the rate of $0.3t^2 + 0.6t + 1$ bu/day. By how much will the value of the crop increase during the next five days if the market price remains fixed at $3/bu?

37. It is estimated that the demand for a manufacturer's product is increased exponentially at the rate of 2%/yr. If the current demand is 5,000 units/yr and if the price remains fixed at $400/unit, how much revenue will the manufacturer receive from the sale of the product over the next two years?

38. After t hours on the job, a factory worker can produce $100(2 + \sin t)$ units/hr. How many units does a worker who arrives on the job at 8:00 A.M. produce between 10:00 A.M. and noon?

39. Suppose that t years from now, one investment plan will be generating profit at the rate of $P_1'(t) = 100 + t^2$ hundred dollars per year, while a second investment will be generating profit at the rate of $P_2'(t) = 220 + 2t$ hundred dollars per year.

a. For how many years does the rate of profitability of the second investment exceed that of the first?

b. Compute the net excess profit assuming that you invest in the second plan for the time period determined in part **a**.

c. Sketch the rate of profitability curves $y = P_1'(t)$ and $y = P_2'(t)$ and shade the region whose area represents the net excess profit computed in part **b**.

40. Answer the questions in Problem 39 for two investments with respective rates of profitability $P_1'(t) = 130 + t^2$ hundred dollars per year and $P_2'(t) = 306 + 5t$ hundred dollars per year.

41. Answer the questions in Problem 39 for two investments with respective rates of profitability $P_1'(t) = 60e^{0.12t}$ thousand dollars per year and $P_2'(t) = 160e^{0.08t}$ thousand dollars per year.

42. Answer the questions in Problem 39 for two investments with respective rates of profitability $P_1'(t) = 90e^{0.1t}$ thousand dollars per year and $P_2'(t) = 140e^{0.07t}$ thousand dollars per year.

43. Suppose that when it is t years old, a particular industrial machine generates revenue at the rate $R'(t) = 6,025 - 8t^2$ dollars/yr and that operating and servicing costs accumulate at the rate of $C'(t) = 4,681 + 13t^2$ dollars/yr.

a. How many years pass before the profitability of the machine begins to decline?

b. Compute the net earnings generated by the machine over the time period determined in part **a**.

c. Sketch the revenue rate curve $y = R'(t)$ and the cost rate curve $y = C'(t)$ and shade the region whose area represents the net earnings computed in part **b**.

44. Answer the questions in Problem 43 for a machine that generates revenue at the rate $R'(t) = 7,250 - 18t^2$ dollars/yr and for which costs accumulate at the rate $C'(t) = 3,620 + 12t^2$ dollars/yr.

45. After t hours on the job, one factory worker is producing $Q_1'(t) = 60 - 2(t-1)^2$ units/hr, while a second worker is producing $Q_2'(t) = 50 - 5t$ units/hr.

a. If both arrive on the job at 8:00 A.M., how many more units will the first worker have produced by noon than the second worker?

b. Interpret the answer in part **a** as the area between two curves.

46. A national consumers' association has compiled statistics suggesting that the fraction of its members who are still active t months after joining is given by the function $f(t) = e^{-0.2t}$. A new local chapter has 200 charter members and expects to attract new members at the rate of 10/mo. How many members can the chapter expect to have at the end of 8 months?

47. The population density r miles from the center of a certain city is $D(r) = 25,000e^{-0.05r^2}$ people/mi^2. How many people live between 1 and 2 miles from the city center?

48. The population density r miles from the center of a certain city is $D(r) = 5,000e^{-0.1r^2}$ people/mi^2. How many people live within 3 miles from the city center?

49. Calculate the volume rate (in cm^3/s) at which blood flows through an artery of radius 0.1 cm if the speed of the blood r cm from the central axis is $8(1 - 100r^2)$ cm/s.

50. Modeling Problem Blood flows through an artery of radius R. At a distance r cm from the central axis of the artery, the speed of the blood is given by $S(r) = k(R^2 - r^2)$. Show that the average speed of the blood is one-half the maximum speed.

51. A study indicates that x months from now the population of a certain town will be increasing at the rate of $5 + 3x^{2/3}$

people/mo. By how much will the population of the town increase over the next 8 months?

52. An object is moving so that its speed after t minutes is $5 + 2t + 3t^2$ m/min. How far does the object travel during the second minute?

53. It is estimated that the demand for oil is increasing exponentially at the rate of 10%/yr. If the demand for oil is currently 30 billion barrels per year, how much oil will be consumed during the next 10 years?

54. It is estimated that t weeks from now, contributions in response to a fund-raising campaign will be coming in at the rate of $R'(t) = 5,000e^{-0.2t}$ dollars/wk, while campaign expenses are expected to accumulate at the constant rate of $676/wk.
   a. For how many weeks does the rate of revenue exceed the rate of cost?
   b. What net earnings will be generated by the campaign during the period of time determined in part **a**.
   c. Interpret the net earnings in part **b** as an area between two curves.

55. Spy Problem The Spy arrives at the sleazy bar at the appointed time for his meeting with his superior (recall Supplementary Problem 85, Chapter 5), but, as usual, his peerless leader is fashionably late. While waiting, the Spy reflects on the dangerous life he leads. Ten years ago he took his spying oath, followed on the same day by 6 friends from college. On average, 30 new people join each year, and he estimates that the fraction of spies still alive after t years of service is $F(t) = e^{-t/20}$. Approximately how many people who joined the service after the Spy are still alive?

C 56. The administrators of a town estimate that the fraction of people who will still be residing in the town t years after they arrive is given by the function $f(t) = e^{-0.04t}$. If the current population is 20,000 people and new townspeople arrive at the rate of 500/yr, what will the population be ten years from now? *Hint*: Divide the interval $0 \le t \le 10$ into n equal subintervals and form a Riemann sum.

57. The operators of a new computer dating service estimate that the fraction of people who will retain their membership in the service for at least t months is given by the function $f(t) = e^{-t/10}$. There are 8,000 charter members, and the operators expect to attract 200 new members/mo. How many

members will the service have ten months from now? *Hint*: Divide the interval $0 \le t \le 10$ into n equal subintervals and form a Riemann sum.

58. Modeling Problem A certain nuclear power plant produces radioactive waste in the form of strontium-90 at the constant rate of 500 pounds/yr. The waste decays exponentially with a half-life of 28 years. How much of the radioactive waste from the nuclear plant will be present after 140 years. *Hint*: Think of this as a survival and renewal problem and form a Riemann sum.

59. Modeling Problem A certain investment generates income continuously over a period of N years. After t years, the investment will be generating income at the rate of $f(t)$ dollars/yr. The prevailing annual rate of interest, compounded continuously, is r. Derive the formula

$$\int_0^N f(t)e^{-rt}\, dt$$

for the present value of the income flow by completing these steps.
   a. Subdivide the time interval $[0, N]$ into n equal subintervals, the kth of which is $[t_{k-1}, t_k]$ for $k = 1, 2, \ldots, n$. Assuming that the income generated during the entire kth subinterval is $f(t_k)$, what is the present value of this income?
   b. Approximate the present value of the total income flow by an appropriate sum.
   c. Note that the sum in part **b** is a Riemann sum. Make the approximation precise by taking a limit to obtain the required integral formula for present value.

60. Modeling Problem A manufacturer receives N units of a certain raw material that are initially placed in storage and then withdrawn and used at a constant rate until the supply is exhausted one year later. Suppose storage costs remain fixed at p dollars per unit per year. Use definite integration to find an expression for the total storage costs the manufacturer will pay during the year. *Hint*: Let $Q(t)$ denote the number of units in storage after t years and find an expression for $Q(t)$. Then subdivide the interval $0 \le t \le 1$ into n equal subintervals and express the total storage cost as the limit of a sum.

CHAPTER 6 REVIEW

Proficiency Examination

CONCEPT PROBLEMS

1. Describe the process for finding the area between two curves.

2. Describe the process for finding volumes of solids with a known cross-sectional area.

3. Compare and contrast the methods of finding volumes by disks, washers, and shells.

4. What are the formulas for changing from polar to rectangular form? From rectangular to polar form?

5. Draw one example of each of the given polar-form curves.
   a. cardioid   **b.** limaçon   **c.** lemniscate

   b. rose curve with three petals

6. Identify and then sketch each of the given curves.
   a. $r = 2\cos\theta$ **b.** $r = \cos\theta - 1$
   c. $r = \theta, r \ge 0$ **d.** $r^2 = \cos 2\theta$

7. What is the procedure for finding the intersection of polar-form curves?

8. Outline a procedure for finding the area bounded by a polar graph.

9. What is the formula for arc length of a graph?

10. What is the formula for area of a surface of revolution?

11. How do you find the work done by a variable force?

12. How do you find the force exerted by a liquid on an object submerged vertically?

13. What are the formulas for the mass and centroid of a homogeneous lamina?

14. State the volume theorem of Pappus.

15. What formulas are used to compute future value and present value of an income flow?

16. How do you find the net change of a function f whose derivative f' is given?

17. Describe the process for finding consumer's and producer's surplus.

18. Describe a typical survival and renewal problem.

19. What formula is used for computing the flow of blood through an artery?

PRACTICE PROBLEMS

20. **Journal Problem** *Mathematics Teacher** ■ *Use the integrals A–G to answer the given questions.*

    A. $\pi \displaystyle\int_a^b [f(x)]^2 \, dx$ **B.** $\pi \displaystyle\int_c^d [f(y)]^2 \, dy$

    C. $\displaystyle\int_a^b A(x) \, dx$ **D.** $\displaystyle\int_c^d A(y) \, dy$

    E. $\pi \displaystyle\int_a^b [f(x) - g(x)] \, dx$

    F. $\pi \displaystyle\int_a^b \{[f(x)]^2 - [g(x)]^2\} \, dx$

    G. $\pi \displaystyle\int_c^d \{[f(y)]^2 - [g(y)]^2\} \, dy$

    a. Which formula does not seem to belong to this list?

    b. Which expressions represent volumes of solids of revolution?

    c. Which expressions represent volumes of solids containing holes?

    d. Which expressions represent volumes of a solid with cross sections that are perpendicular to an axis?

    e. Which expressions represent volumes of solids of revolution revolved about the x-axis?

    f. Which expressions represent volumes of solids of revolution revolved about the y-axis?

21. Find the area above the curve $f(x) = 3x^2 - 6$ and below the x-axis.

22. Find the area between the curves $y = x^2$ and $y^3 = x$.

23. The base of a solid in the xy-plane is the circular region given by $x^2 + y^2 = 4$. Every cross section perpendicular to the x-axis is a rectangle whose height is twice the length of the side that lies in the xy-plane. Find the volume of this solid.

24. Let R be a region bounded by the graphs of the equations $y = (x - 2)^2$ and $y = 4$.

    a. Sketch the graph of R and find its area.

    b. Find the volume of a solid of revolution of R about the x-axis.

    c. Find the volume of a solid of revolution of R about the y-axis.

25. Find the area of the intersection of the circles $r = 2a \cos \theta$ and $r = 2a \sin \theta$, where $a > 0$ is constant.

26. Find the arc length of the curve $y = 2 - x^{3/2}$ from $(0, 2)$ to $(1, 1)$. Give the exact value.

27. Find the surface area obtained by revolving the parabolic arc $y^2 = x$ from $(0, 0)$ to $(1, 1)$ about the x-axis. Give the answer to the nearest hundredth unit.

28. Find the centroid of the homogeneous region between $y = x^2 - x$ and $y = x - x^3$ on $[0, 1]$.

29. An observation porthole on a ship is circular with diameter of 2 feet and is located so its center is 3 feet below the waterline. Find the fluid force on the porthole, assuming that the boat is in seawater (density is 64 lb/ft^3). Set up the integral only.

30. Money is transferred continuously into an account at the constant rate of $\$1,200$/yr. The account earns interest at the annual rate of 8% compounded continuously. How much money will be in the account at the end of 5 years?

Supplementary Problems

Name and sketch the curves whose equations are given in Problems 1–8.

1. $r = 2 \sin \theta - 3 \cos \theta$
2. $r = -2\theta, \theta \ge 0$
3. $r = -\csc \theta$
4. $r = \sin 3\theta$
5. $r = \cos(\theta - \frac{\pi}{3})$
6. $r^2 = 3 \cos 2\theta$
7. $r = \dfrac{4}{1 - \cos \theta}$
8. $r = \dfrac{5}{2 + 3 \sin \theta}$

Find the points of intersection of the polar-form curves in Problems 9–12.

9. $\begin{cases} r = 2 \cos \theta \\ r = 1 + \cos \theta \end{cases}$
10. $\begin{cases} r = 1 + \sin \theta \\ r = 1 + \cos \theta \end{cases}$
11. $\begin{cases} r \cos \theta + 2r \sin \theta = 4 \\ r = 2 \sec \theta \end{cases}$
12. $\begin{cases} r = 2 \sin 2\theta \\ r = 1 \text{ for } 0 \le \theta \le \frac{\pi}{2} \end{cases}$

Find the area of the regions bounded by the curves and lines in Problems 13–21.

13. $4y^2 = x$ and $2y = x - 2$ 14. $y = x^3, y = x^4$
15. $x = y^{2/3}, x = y^2$
16. $y = \sqrt{3} \sin x$, $y = \cos x$, and the vertical lines $x = 0$ and $x = \pi/2$
17. the region bounded by the polar curve $r = 4 \cos 2\theta$
18. the region inside both the circles $r = 1$ and $r = 2 \sin \theta$
19. the region inside one loop of the lemniscate $r^2 = \cos 2\theta$
20. the region inside one petal of the rose curve $r = \sin 2\theta$
21. the region inside the circle $r = 2a \sin \theta$ and outside the circle $r = a$

Find the volume of a solid of revolution obtained by revolving the region R about the indicated axis in Problems 22–25.

22. R is bounded by $y = \sqrt{x}$, $x = 9$, $y = 0$; about the x-axis
23. R is bounded by $y = x^2$, $y = 9 - x^2$, and $x = 0$; about the y-axis
24. R is bounded by $y = x^2$ and $y = x^4$; about the x-axis
25. R is bounded by $y = \sqrt{\cos x}$, $x = \frac{\pi}{4}$, $x = \frac{\pi}{3}$, and $y = 0$; about the x-axis.
26. The base of a solid is the region in the xy-plane that is bounded by the curve $y = 4 - x^2$ and the x-axis. Every cross section of the solid perpendicular to the y-axis is a rectangle whose height is twice the length of the side that lies in the xy-plane. Find the volume of the solid.

27. The base of a solid is the region R in the xy-plane bounded by the curve $y^2 = x$ and the line $y = x$. Find the volume of the solid if each cross section perpendicular to the x-axis is a semicircle with a diameter in the xy-plane.

28. Find the centroid of a homogeneous lamina covering the region R that is bounded by the parabola $y = 2 - x^2$ and the line $y = x$.

29. A force of 100 lb is required to compress a spring from its natural length of 10 in. to a length of 8 in. Find the work required to compress it to a length of 7 in. from its natural length.

30. A bowl is formed by revolving the region bounded by the curve $y = 2x^2$ and the lines $x = 0$ and $y = 1$ about the y-axis (units are in ft). The bowl is filled with water to a height of 9 in.

    a. How much work is done in pumping the water to the top of the bowl?

    b. How much work is done in pumping the water to a point 0.5 ft above the top of the bowl?

31. Two solids S_1 and S_2 have as their base the region in the xy-plane bounded by the parabola $y = x^2$ and the line $y = 1$. For solid S_1, each cross section perpendicular to the x-axis is a square, and for S_2, each cross section perpendicular to the y-axis is a square. Which solid has the greater volume?

32. Use calculus to find the lateral surface area of a right circular cylinder with radius r and height h.

33. Use calculus to find the surface area of a right circular cone with (top) radius r and height h.

34. A 20-lb bucket in a well is attached to the end of a 60-ft chain that weighs 16 lb. If the bucket is filled with 50 lb of water, how much work is done in lifting the bucket and chain to the top of the well?

35. A swimming pool is 3 ft deep at one end and 8 ft deep at the other, as shown in Figure 6.80. The pool is 30 ft long and 25 ft wide with vertical sides. Set up and analyze a model to find the fluid force against one of the 30-ft sides.

Figure 6.80   Cross section of a swimming pool

36. The ends of a container filled with water have the shape of the region R bounded by the curve $y = x^4$ and the line $y = 16$. Find the fluid force (to two decimal places) exerted by the water on one end of the container.

37. A conical tank whose vertex points down is 12 ft high with (top) diameter 6 ft. The tank is filled with a fluid of density $\rho = 22$ lb/ft^3. If the tank is filled to a depth of 6 ft, find the work required to pump all the fluid to a height of 2 ft above the top of the tank.

38. Each end of a container filled with water has the shape of an isosceles triangle, as shown in Figure 6.81. Find the force exerted by the water on an end of the container.

Figure 6.81   Water pressure in a tank

39. A container has the shape of the solid formed by rotating the region bounded by the curve $y = x^3$ and the lines $x = 0$ and $y = 1$ about the y-axis, where x and y are measured in feet. If the container is filled with water, how much work is required to pump all the water out of the container?

40. A gate in a water main has the shape of a circle with diameter 10 ft. If water stands 2 ft high in the main, what is the force exerted by the water on the gate?

*Suppose a particle of mass m moves along a number line under the influence of a variable force F which always acts in a direction parallel to the line of motion. At a given time the particle is said to have **kinetic energy***

$$K = \tfrac{1}{2}mv^2$$

*where v is the particle's velocity, and it has **potential energy***

$$P = -\int_{s_0}^{s} F\, dx$$

relative to the point s on the line of motion. Use these definitions in Problems 41–42.

41. A spring at rest has potential energy $P = 0$. If the spring constant is 30 lb/ft, how far must the spring be stretched so that its potential energy is 20 ft-lb?

42. A particle with mass m falls freely near the earth's surface. At time $t = 0$, the particle is s_0 units above the ground and is falling with velocity v_0.

    a. Find the potential energy P of the particle in terms of position, s, if the potential energy is 0 when $s = s_0$.

    b. Find the potential energy P and the kinetic energy K of the particle in terms of time t.

43. An object weighs $w = 1,000r^{-2}$ grams, where r is the object's distance from the center of the earth. What work is required to lift the object from the surface of the earth ($r = 6,400$ km) to a point 3,600 km above the surface of the earth?

44. **Modeling Problem**   A certain species has population $P(t)$ at time t whose growth rate is affected by seasonal variations in the food supply. Suppose the population may be modeled by the differential equation

$$\frac{dP}{dt} = k(P - A)\cos t$$

where k and A are physical constants. Solve the differential equation and express $P(t)$ in terms of k, A, and P_0, the initial population.

45. Find the center of mass of a thin plate of constant density if the plate occupies the region bounded by the lines $2y = x$, $x + 3y = 5$, and the x-axis.

46. The portion of the ellipse (Figure 6.82)

$$\frac{x^2}{a^2} + \frac{y^2}{b^2} = 1$$

that lies in the first quadrant between $x = r$ and $x = a$, for $0 < r < a$, is revolved about the x-axis to form a solid S. Find the volume of S.

Figure 6.82   Ellipse

47. What is the present value of an investment that will generate income continuously at a constant rate of $1,000/yr for 10 years if the prevailing annual interest rate remains fixed at 7% compounded continuously?

48. In a certain community the fraction of the homes placed on the market that remain unsold for at least t weeks is approximately $f(t) = e^{-0.2t}$. If 200 homes are currently on the market and if additional homes are placed on the market at the rate of 8 per week, approximately how many homes will be on the market 10 weeks from now?

49. The population density r miles from the center of a certain city is $D(r) = 5,000e^{-0.02r^2}$ people/mi^2. How many people live within 5 miles of the center of the city?

50. A certain oil well that yields 900 barrels of crude oil per month will run dry in 3 years. The price of crude oil is currently $16/barrel and is expected to rise at the constant rate of 8¢ per barrel per month. If the oil is sold as soon as it is extracted from the ground, what will be the total future revenue from the well?

51. Suppose that the consumer's demand function for a certain commodity is $D(q) = 50 - 3q - q^2$ dollars per unit.
   a. Find the number of units that will be bought if the market price is $32/unit.
   b. Compute the consumer's surplus when the market price is $32/unit.
   c. Graph the demand curve and interpret the consumer's surplus as an area in relation to this curve.

52. Let R be the part of the ellipse

$$\frac{x^2}{a^2} + \frac{y^2}{b^2} = 1$$

that lies in the first quadrant. Use the theorem of Pappus to find the coordinates of the centroid of R. You may use the facts that R has area $A = \frac{1}{4}\pi ab$, and the semiellipsoid it generates when it is revolved about the x-axis has volume $V = \frac{2}{3}\pi ab^2$.

53. Modeling Problem   A container of a certain fluid is raised from the ground on a pulley. While resting on the ground, the container and fluid together weigh 200 lb. However, as the container is raised, fluid leaks out in such a way that at height x feet above the ground, the total weight of the container and fluid is $200 - 0.5x$ lb. Set up and analyze a model to find the work done in raising the container 100 ft.

54. A bag of sand is being lifted vertically upward at the rate of 31 ft/min. If the bag originally weighed 40 lb and leaks at the rate of 0.2 lb/s, how much work is done in raising it until it is empty?

55. The vertical end of a trough is an equilateral triangle, 3 ft on a side. Assuming the trough is filled with a fluid of density $\rho = 40$ lb/ft^3, find the force on the end of the trough.

56. The radius of a circular water main is 2 ft. Assuming the main is half full of water, find the total force exerted by the water on a gate that crosses the main at one end (see Figure 6.83).

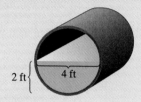

2 ft   4 ft

Figure 6.83   Problem 56

57. The force acting on an object moving along a straight line is known to be a linear function of its position. Find the force function F if it is known that 13 ft-lb of work is required to move the object 6 ft and 44 ft-lb of work is required to move it 12 ft.

58. A tank has the shape of a rectangular parallelepiped with length 10 ft, width 6 ft, and height 8 ft. The tank contains a liquid of density $\rho = 20$ lb/ft^3. Find the level of the liquid in the tank if it will require 2,400 ft-lb of work to move all the liquid in the tank to the top.

59. A tank is constructed by placing a right circular cylinder of height 10 ft and radius 2 ft on top of a hemisphere with the same radius (see Figure 6.84).

2 ft

10 ft

Figure 6.84   Problem 59

If the tank is filled with a fluid of density $\rho = 24$ lb/ft^3, how much work is done in bringing all the fluid to the top of the tank?

60. Modeling Problem   A piston in a cylinder causes a gas either to expand or to contract. Assume the pressure P and the volume of gas V in the cylinder satisfy the adiabatic law (no exchange of heat)

$$P(V^{1.4}) = C$$

where C is a constant.

a. If the gas goes from volume V_1 to volume V_2, show that the work done may be modeled by

$$W = \int_{V_1}^{V_2} CV^{-1.4} \, dV$$

b. How much work is done if 0.6 m³ of steam at a pressure of 2,500 newtons/m² expands 50%?

61. A thin sheet of tin is in the shape of a square, 16 cm on a side. A rectangular corner of area 156 cm² is cut from the square, and the centroid of the portion that remains is found to be 4.88 cm from one side of the original square, as shown in Figure 6.85. How far is it from the other three sides?

16 cm

156 cm²

?

4.88 cm

?

?

16 cm

Figure 6.85 Centroid of a region

62. Let R be the region bounded by the semicircle $y = \sqrt{4 - x^2}$ and the x-axis, and let S be the solid with base R and trapezoidal cross sections perpendicular to the x-axis. Assume that the two slant sides and the shorter parallel side are all half the length that lies in the base, as shown in Figure 6.86. Set up and evaluate an integral for the volume of S.

$y = \sqrt{4 - x^2}$

h

ℓ

ℓ ℓ

h

Cross section

Figure 6.86 Volume of a region S

63. A child is building a sand structure at the beach.

a. Find the work done if the density of the sand is $\rho = 140$ lb/ft³ and the structure is a cone of height 1 ft and diameter 2 ft.

b. What if the structure is a tower (cylinder) of height 1 ft and radius 4 in.?

64. Let S be the solid generated by revolving about the x-axis the region bounded by the parabola $y = x^2$ and the line $y = x$. Find the number A so that the plane perpendicular to the x-axis at $x = A$ divides S in half. That is, the volume on one side of A equals the volume on the other side.

65. A plate in the shape of an isosceles triangle is submerged vertically in water, as shown in Figure 6.87.

D

A A

B

Figure 6.87 Hydrostatic force

a. Find the force on one side of the plate if the two equal sides have length 3, the third side has length 5, and the top vertex of the triangle is 4 units below the surface.

b. Find the fluid force for the general case where the triangle has equal sides of length A, the third side has length B, and the top vertex is D units below the surface.

66. Find the arc length of the curve

$$y = Ax^3 + \frac{B}{x}$$

on the interval $[a, b]$, where A and B are positive constants with $AB = 1/12$.

67. Counterexample Problem When the line segment $y = x - 2$ between $x = 1$ and $x = 3$ is revolved about the x-axis, it generates part of a cone. The formula obtained in Section 6.4 says that the surface area of the cone is

$$\int_1^3 2\pi(x - 2)\sqrt{2} \, dx$$

but this integral is equal to 0. What is wrong? What is the actual surface area?

68. HISTORICAL QUEST

AMALIE ("EMMY") NOETHER
1882–1935

Emmy Noether has been called the greatest woman mathematician in the history of mathematics. Once again, we see a woman overcoming great obstacles in achieving her education. At the time she received her Ph.D. at the University of Erlangen (in 1907), the Academic Senate declared that the admission of women students would "overthrow all academic order." Emmy Noether is famous for her work in physics and mathematics. In mathematics, she opened up entire areas for study in the theory of ideals, and in 1921 she published a paper on rings, which have since been called Noetherian rings. Rings and ideals are objects of study in modern algebra. She was not only a renowned mathematician but also an excellent teacher. It has been reported that she often lectured by "working it out as she went." Nevertheless, her students flocked around her "as if following the Pied Piper." In the fall of 1934, pressure from the Nazi regime forced her to immigrate to Bryn Mawr College in Pennsylvania. In the years that followed, there were mass dismissals of "racially undesirable" professors, and many noted European Jewish mathematicians immigrated to American universities. ∎

For this Historical Quest compile a list of notable mathematicians who immigrated to the United States from 1933 to 1940.

69. Modeling Problem   An object below the surface of the earth is attracted by a gravitational force that is directly proportional to the square of the distance from the object to the center of the earth. Set up and analyze a model to find the work done in lifting an object weighing P lb from a depth of s feet (below the surface) to the surface. (Assume the earth's radius is 4,000 mi.)

In Problems 70–71, consider a continuous curve $y = f(x)$ that passes through $(-1, 4)$, $(0, 3)$, $(1, 4)$, $(2, 6)$, $(3, 5)$, $(4, 6)$, $(5, 7)$, $(6, 6)$, $(7, 4)$, $(8, 3)$ *and* $(9, 4)$.

70. Use Simpson's rule to estimate the volume of the solid formed by revolving this curve about the x-axis for $[-1, 9]$. Assume the curve passes through $(9, 4)$.

71. Use the trapezoidal rule to estimate the volume of the solid formed by revolving this curve about the line $y = -1$ on $[-1, 9]$.

72. Modeling Problem   In this computational problem we revisit a problem similar to the group research project at the end of Chapter 4, but add a couple of complications. Imagine a cylindrical fuel tank of length L lying on its side; its ends are elliptically shaped and are defined by the equation

$$\left(\frac{x}{a}\right)^2 + \left(\frac{y}{b}\right)^2 = 1$$

We are concerned with the amount of fuel in the tank for a given level.

a. Explain why the area of an end of the tank can be modeled by

$$A = 2a \int_{-b}^{b} \sqrt{1 - \left(\frac{y}{b}\right)^2}\, dy$$

b. By making the right substitution in the integral in part **a**, we can obtain an integral representing half the area inside the circle (with unit radius). Do this and show that the area inside the above ellipse is πab.

c. Explain why the volume of the fuel at level h $(-b \le h \le b)$ may be modeled by

$$V(h) = 2La \int_{-b}^{h} \sqrt{1 - \left(\frac{y}{b}\right)^2}\, dy$$

d. For $a = 5$, $b = 4$, and $L = 20$, numerically compute $V(h)$ for $h = -3, -2, \ldots, 3, 4$. *Note:* $V(0)$ and $V(4)$ will serve as a check on your work.

73. We think of the last part of Problem 72 as the "direct" problem: Given a value of h, compute $V(h)$ by a formula. Now we turn to the "inverse" problem: For a set value of V, say V_0, find h such that $V(h) = V_0$. Typically, the inverse problem is more difficult than the direct problem and requires some sort of iterative method to approximate the solution (for example, the Newton-Raphson method). For example, if $V_0 = 500$ ft^3, we could define $f(h) = V(h) - 500$ and seek the relevant zero of f. Set up the necessary computer program to do this and find the h values, to two-decimal-place accuracy, corresponding to these values of V:

a. 500   **b.** 800

Hint: A key to this problem is thinking about $V'(h)$, where $V(h)$ is expressed in Problem 72**c**. Also, for starting values h_0, refer to your work in Problem 72**d**.

74. Find the centroid of a right triangle whose legs have lengths a and b. *Hint*: Use the volume theorem of Pappus.

75. The largest of the pyramids at Giza in Egypt is roughly 480 ft high and has a square base 750 ft on a side. How much work was required to lift the stones to build this pyramid if we assume that the rock in the pyramid had an average density of 160 lb/ft^3?

76. Find the length of the curve $y = e^x$ on the interval $[0, 1]$.

77. Find the volume of the solid generated when the region under the curve

$$y = \frac{1}{\sqrt{9 - x^2}}$$

on the interval $[0, 2]$ is revolved about the y-axis.

78. Find the volume of the solid formed by revolving about the x-axis the region bounded by the curve

$$y = \frac{2}{\sqrt{3x - 2}}$$

the x-axis, and the lines $x = 1$ and $x = 2$.

79. Find the area between the curves $y = \tan^2 x$ and $y = \sec^2 x$ on the interval $\left[0, \frac{\pi}{4}\right]$.

80. Find the surface area of the solid generated by revolving the region bounded by the x-axis and the curve

$$y = e^x + \tfrac{1}{4}e^{-x}$$

on $[0, 1]$ about the x-axis.

81. Find the volume of the solid whose base is the region R bounded by the curve $y = e^x$ and the lines $y = 0$, $x = 0$, $x = 1$, if cross sections perpendicular to the x-axis are equilateral triangles.

82. Find the length of the curve

$$x = \frac{ay^2}{b^2} - \frac{b^2}{8a} \ln \frac{y}{b}$$

between $y = b$ and $y = 2b$.

83. Putnam Examination Problem   Find the length of the curve $y^2 = x^3$ from the origin to the point where the tangent makes an angle of $45°$ with the x-axis.

84. Putnam Examination Problem   A solid is bounded by two bases in the horizontal planes $z = h/2$ and $z = -h/2$, and by a surface with the property that every horizontal cross section has area given by an expression of the form

$$A = a_0 z^3 + a_1 z^2 + a_2 z + a_3$$

a. Show that the solid has volume $V = \tfrac{1}{6}h(B_1 + B_2 + 4M)$, where B_1 and B_2 are the area of the bases and M is the area of the middle horizontal cross section.

b. Show that the formulas for the volume of a cone and a sphere can be obtained from the formula in part **a** for certain special choices of a_0, a_1, a_2, and a_3.

85. **Putnam Examination Problem**   The horizontal line $y = c$ intersects the curve $y = 2x - 3x^3$ in the first quadrant as shown in Figure 6.88. Find c so that the areas of the two shaded regions are equal.

Figure 6.88   Putnam Examination Problem

Group Research Project[*]

"Houdini's Escape"

This project is to be done in groups of three or four students. Each group will submit a single written report.

Harry Houdini (born Ehrich Weiss)
(1874–1926)

H arry Houdini was a famous escape artist. In this project we relive a trick of his that challenged his mathematical prowess, as well as his skill and bravery. It may challenge these qualities in you as well.

Houdini had his feet shackled to the top of a concrete block which was placed on the bottom of a giant laboratory flask. The cross-sectional radius of the flask, measured in feet, was given as a function of height z from the ground by the formula $r(z) = 10z^{-1/2}$, with the bottom of the flask at $z = 1$ ft. The flask was then filled with water at a steady rate of 22π ft^3/min. Houdini's job was to escape the shackles before he was drowned by the rising water in the flask.

Now Houdini knew it would take him exactly 10 minutes to escape the shackles. For dramatic impact, he wanted to time his escape so it was completed precisely at the moment the water level reached the top of his head. Houdini was exactly 6 ft tall. In the design of the apparatus, he was allowed to specify only one thing: the height of the concrete block he stood on.

Your paper is not limited to the following questions but should include these concerns: How high should the block be? (You can neglect Houdini's volume and the volume of the block.) How fast is the water level changing when the flask first starts to fill? How fast is the water level changing at the instant when the water reaches the top of his head? You might also help Houdini with any size flask by generalizing the derivation: Consider a flask with cross-sectional radius $r(z)$, an arbitrary function of z with a constant inflow rate of $dV/dt = A$. Can you find dh/dt as a function of $h(t)$?

[*]*MAA Notes* 17 (1991), "Priming the Calculus Pump: Innovations and Resources," by Marcus S. Cohen, Edward D. Gaughan, R. Arthur Knoebel, Douglas S. Kurtz, and David J. Pengelley.

7

Methods of Integration

PREVIEW

In this chapter we increase the number of techniques and procedures for integrating a function. One of the most important techniques, substitution, is reviewed in the first section and is then expanded in several different contexts. Other important integration procedures include using tables, integrating by parts, and using partial fractions. In addition, improper integrals, hyperbolic functions, and inverse hyperbolic functions are discussed, along with first-order differential equations.

PERSPECTIVE

It is possible to differentiate most functions that arise in practice by applying a fairly short list of rules and formulas, but integration is a more complicated process. The purpose of this chapter is to increase your ability to integrate a variety of different functions. Learning to integrate is like learning to play a musical instrument: At first, it may seem impossibly complicated, but if you persevere, after a while music starts to happen. It should be noted that as more powerful technology becomes available, *techniques* become less important and *ideas* become more important. If you have such technology available, you might wish to consult the *Technology Manual*.

7.1   Review of Substitution and Integration by Table

IN THIS SECTION   review of substitution, using tables of integrals

In Chapter 5, we derived a number of integration formulas and examined various algebraic procedures for reducing a given integral to a form that can be handled by these formulas. In this section, we review those formulas as well as the important method of integration by substitution. Today's technology has greatly enhanced the ability of professional mathematicians and scientists in techniques of integration, thereby minimizing the need for spending hours learning rarely used integration techniques. Along with this new technology, the use of integral tables has increased in importance. Most mathematical handbooks contain extensive integral tables, and we have included such an integration table in the accompanying *Student Mathematics Handbook*. For your convenience, we have also included a short table of integrals in Appendix D and will refer to these integrals by number throughout the rest of the book.

For easy reference, the most important integration rules and formulas are listed inside the back cover. Notice that the first four integration formulas are the procedural rules, which allow us to break up integrals into simpler forms, whereas those on the remainder of the page form the building blocks of integration.

REVIEW OF SUBSTITUTION

Remember that when doing substitution you must choose u, calculate du, and then substitute to make the form you are integrating look *exactly like* one of the known integration formulas. We will review substitution by looking at different situations and special substitutions that prove useful.

EXAMPLE 1   Integration by substitution

Find $\displaystyle\int \frac{x^2\, dx}{(x^3 - 2)^5}$.

Solution

Let u be the value in parentheses; that is, let $u = x^3 - 2$. Then $du = 3x^2\, dx$, so by substitution,

$$\int \frac{x^2\, dx}{(x^3 - 2)^5} = \int \frac{\frac{du}{3}}{u^5} = \frac{1}{3}\int u^{-5}\, du \qquad \text{All } x\text{'s must be eliminated.}$$

$$= \frac{1}{3}\frac{u^{-4}}{-4} + C = -\frac{1}{12}(x^3 - 2)^{-4} + C$$

You can check your work by finding the derivative of your answer to see if you obtain the integrand. For this example,

$$\frac{d}{dx}\left[-\frac{1}{12}(x^3 - 2)^{-4} + C \right] = -\frac{1}{12}\left[-4(x^3 - 2)^{-5}(3x^2) + 0 \right]$$

$$= \frac{x^2}{(x^3 - 2)^5} \qquad\blacksquare$$

EXAMPLE 2   Fitting the form of a known integration formula by substitution

Find $\displaystyle\int \frac{t\, dt}{\sqrt{1 - t^4}}$.

Solution

We notice the similarity between this and the formula for inverse sine if we let $a = 1$ and $u = t^2$. Then $du = 2t\,dt$ and

Formula 22, Appendix D (inverse sine):

$$\int \frac{du}{\sqrt{a^2 - u^2}} = \sin^{-1} \frac{u}{a}$$

$$\int \frac{t\,dt}{\sqrt{1 - t^4}} = \int \frac{\frac{du}{2}}{\sqrt{1 - u^2}} \qquad \boxed{\text{If } u = t^2, \text{ then } du = 2t\,dt, t\,dt = \frac{du}{2}, \text{ and } \sqrt{1 - t^4} = \sqrt{1 - (t^2)^2} = \sqrt{1 - u^2}.}$$

$$= \frac{1}{2} \int \frac{du}{\sqrt{1 - u^2}} \qquad \text{Don't forget the factor } \tfrac{1}{2}.$$

$$= \tfrac{1}{2} \sin^{-1} u + C = \tfrac{1}{2} \sin^{-1} t^2 + C \qquad \blacksquare$$

The method of substitution (Section 5.5) is very important, because many of the techniques developed in this chapter will be used in conjunction with substitution. Examples 3 and 4 illustrate additional ways substitution can be used in integration problems.

EXAMPLE 3   Multiplication by 1 to derive an integration formula

Find $\displaystyle\int \sec x\,dx$.

Solution

Multiply the integrand $\sec x$ by $\sec x + \tan x$ and divide by the same quantity:

$$\int \sec x\,dx = \int \frac{\sec x(\sec x + \tan x)}{\sec x + \tan x}\,dx = \int \frac{\sec^2 x + \sec x \tan x}{\sec x + \tan x}\,dx$$

TECHNOLOGY NOTE

Certain computer algebra systems (CAS) evaluate this integral as $\ln \left| \tan \left(\dfrac{\pi + 2x}{4} \right) \right|$. You could show these are the same by evaluating the derivatives of each and then using some trigonometric identities. Note while an indefinite integral represents *all possible* antiderivatives, popular software provide only a single antiderivative with no C.

The advantage of this rearrangement is that the numerator is now the derivative of the denominator. That is, using the substitution

$$\boxed{\begin{aligned} u &= \tan x + \sec x \\ du &= (\sec^2 x + \sec x \tan x)\,dx \end{aligned}}$$

we find

$$\int \frac{\sec^2 x + \sec x \tan x}{\sec x + \tan x}\,dx = \int \frac{du}{u} = \ln |u| + C = \ln |\sec x + \tan x| + C \qquad \blacksquare$$

You may wonder why anyone would think to multiply and divide the integrand $\sec x$ in Example 3 by $\sec x + \tan x$. To say that we do it "because it works" is probably not a very satisfying answer. However, techniques like this are passed on from generation to generation, and it should be noted that multiplication by 1 is a common method in mathematics for changing the form of an expression to one that can be manipulated more easily.

EXAMPLE 4   Substitution after an algebraic manipulation and polynomial division

Find $\displaystyle\int \frac{dx}{1 + e^x}$.

Solution

The straightforward substitution $u = 1 + e^x$ does not work:

$$\int \frac{dx}{1 + e^x} = \int \frac{\frac{du}{e^x}}{u} = \int \frac{du}{e^x u} \qquad \text{This is not an appropriate form because } x \text{ has not been eliminated.}$$

$$\boxed{u = 1 + e^x; \quad du = e^x\,dx}$$

Instead rewrite the integrand as follows:

$$\frac{1}{1+e^x} = \frac{e^{-x}}{e^{-x}}\left(\frac{1}{1+e^x}\right) = \frac{e^{-x}}{e^{-x}+1}; \quad \text{then make the substitution}$$

$$\boxed{u = e^{-x} + 1; \quad du = -e^{-x}\,dx}$$

$$\int \frac{dx}{1+e^x} = \int \frac{e^{-x}\,dx}{e^{-x}+1} = \int \frac{-du}{u} = -\ln|u| + C = -\ln(e^{-x}+1) + C$$

Remember, $e^{-x} + 1 > 0$ for all x, so $\ln\left|e^{-x}+1\right| = \ln(e^{-x}+1)$.   ■

When the integrand involves terms with fractional exponents, it is usually a good idea to choose the substitution $x = u^n$, where n is the smallest integer that is divisible by all the denominators of the exponents (that is, the least common multiple of the denominators). For example, if the integrand involves terms such as $x^{1/4}$, $x^{2/3}$, and $x^{1/6}$, then the substitution $x = u^{12}$ is suggested, because 12 is the smallest integer divisible by the exponential denominators 4, 3, and 6. The advantage of this policy is that it guarantees that each fractional power of x becomes an integral power of u. Thus,

$$x^{1/6} = (u^{12})^{1/6} = u^2, \quad x^{1/4} = (u^{12})^{1/4} = u^3, \quad x^{2/3} = (u^{12})^{2/3} = u^8$$

EXAMPLE 5   Substitution with fractional exponents

Find $\displaystyle\int \frac{dx}{x^{1/3} + x^{1/2}}$.

Solution

Because 6 is the smallest integer divisible by the denominators 2 and 3, we set $u = x^{1/6}$, so that $u^6 = x$ and $6u^5\,du = dx$. We now use substitution

$$\int \frac{dx}{x^{1/3} + x^{1/2}} = \int \frac{6u^5\,du}{(u^6)^{1/3} + (u^6)^{1/2}} \qquad \text{Let } x = u^6.$$

$$= \int \frac{6u^5\,du}{u^2 + u^3} = \int \frac{6u^5\,du}{u^2(1+u)} = \int \frac{6u^3\,du}{1+u}$$

Substitution does not guarantee an obvious integrable form. When the degree of the numerator is greater than or equal to the degree of the denominator, division is often helpful. By long division

$$\frac{6u^3}{1+u} = 6u^2 - 6u + 6 + \frac{-6}{1+u}$$

$$\int \frac{6u^3\,du}{1+u} = \int \left(6u^2 - 6u + 6 + \frac{-6}{1+u}\right) du$$

$$= 2u^3 - 3u^2 + 6u - 6\ln|1+u| + C$$

$$= 2(x^{1/6})^3 - 3(x^{1/6})^2 + 6x^{1/6} - 6\ln\left|1+x^{1/6}\right| + C \quad \text{Because } u = x^{1/6}$$

$$= 2x^{1/2} - 3x^{1/3} + 6x^{1/6} - 6\ln(1+x^{1/6}) + C \qquad \text{Note: } 1+x^{1/6} > 0$$

■

USING TABLES OF INTEGRALS

Example 5 (especially the part involving long division) seems particularly lengthy. When faced with the necessity of integrating a function such as

$$\int \frac{6u^3\,du}{1+u}$$

most people would turn to a computer, calculator, or a table of integrals. If you look, for example, at the short integration table (Appendix D), you will find (Formula 37)

$$\int \frac{u^3 \, du}{au+b} = \frac{(au+b)^3}{3a^4} - \frac{3b(au+b)^2}{2a^4} + \frac{3b^2(au+b)}{a^4} - \frac{b^3}{a^4} \ln|au+b|$$

If we let $a = 1$ and $b = 1$, we find

$$\int \frac{6u^3 \, du}{1+u} = 6 \left[\frac{(u+1)^3}{3} - \frac{3(u+1)^2}{2} + \frac{3(u+1)}{1} - \ln|u+1| \right] + C$$

$$= 2(u+1)^3 - 9(u+1)^2 + 18(u+1) - 6\ln|u+1| + C$$

When using integration tables, we note that the algebraic form does not always match the form we obtain by direct integration. You might wish to show algebraically that the form we have just found and the form

$$2u^3 - 3u^2 + 6u - 6\ln|1+u| + C$$

from Example 5 differ by a constant and hence are equivalent.

To use an integral table, first classify the integral by form. To facilitate substitution of forms, we use u as the variable of integration, and let a, b, c, m, and n represent constants. The forms listed in the table of integrals included in the *Student Mathematics Handbook* are as follows:

SMH

Elementary forms (Formulas 1–29; these were developed in the text)

Linear and quadratic forms (Formulas 30–134)
Forms involving $au+b$; u^2+a^2; u^2-a^2; a^2-u^2; $au+b$ and $pu+q$; au^2+bu+c

Radical forms (Formulas 135–270)
Forms involving $\sqrt{au+b}$; both $\sqrt{au+b}$ and $pu+q$; both $\sqrt{au+b}$ and $\sqrt{pu+q}$; $\sqrt{u^2+a^2}$; $\sqrt{u^2-a^2}$; $\sqrt{a^2-u^2}$; $\sqrt{au^2+bu+c}$

Higher degree binomials (Formulas 271–310)
Forms involving u^3+a^3; $u^4 \pm a^4$, $u^n \pm a^n$

Trigonometric forms (Formulas 311–444)
Forms involving $\cos au$; $\sin au$; both $\sin au$ and $\cos au$; $\tan au$; $\cot au$; $\sec au$; $\csc au$

Inverse trigonometric forms (Formulas 445–482)

Exponential and logarithmic forms (Formulas 483–513)
Forms involving e^{au}; $\ln|u|$

Hyperbolic forms (Formulas 514–619; we study these in Section 7.8)
Forms involving $\cosh au$; $\sinh au$; $\sinh au$ and $\cosh au$; $\tanh au$; $\coth au$; $\mathrm{sech}\, au$; $\mathrm{csch}\, au$

Inverse hyperbolic forms (Formulas 620–650; Section 7.8)

A condensed version of this table is provided in Appendix D.

There is a common misconception that integration will be easy if a table is provided, but even with a table available there can be a considerable amount of work. After deciding which form applies, match the individual type with the problem at hand by making appropriate choices for the arbitrary constants. More than one form may apply, but the results derived by using different formulas will be the same (except for the constants) even though they may look quite different. We will not include the constants in the table listing, but you should remember to include them with your answers when using the table for integration.

Take a few moments to look at the integration table in the *Student Mathematics Handbook* (or Appendix D). Notice that the table has two basic types of integration formulas. The first gives a formula that is the antiderivative, whereas the second, called a *reduction formula*, simply rewrites the integral in another form. The following illustration demonstrates each of these forms.

SMH

Table of Integrals

Integrals Involving $au + b$

30. $\displaystyle\int (au + b)^n \, du = \frac{(au + b)^{n+1}}{(n + 1)a}$

31. $\displaystyle\int u(au + b)^n \, du = \frac{(au + b)^{n+2}}{(n + 2)a^2} - \frac{b(au + b)^{n+1}}{(n + 1)a^2}$

32. $\displaystyle\int u^2(au + b)^n \, du = \frac{(au + b)^{n+3}}{(n + 3)a^3} - \frac{2b(au + b)^{n+2}}{(n + 2)a^3} + \frac{b^2(au + b)^{n+1}}{(n + 1)a^3}$

33. $\displaystyle\int u^m(au + b)^n \, du = \begin{cases} \dfrac{u^{m+1}(au + b)^n}{m + n + 1} + \dfrac{nb}{m + n + 1}\displaystyle\int u^m(au + b)^{n-1}\, du \\[2ex] \dfrac{u^m(au + b)^{n+1}}{(m + n + 1)a} - \dfrac{mb}{(m + n + 1)a}\displaystyle\int u^{m-1}(au + b)^n\, du \\[2ex] \dfrac{-u^{m+1}(au + b)^{n+1}}{(n + 1)b} - \dfrac{m + n + 2}{(n + 1)b}\displaystyle\int u^m(au + b)^{n+1}\, du \end{cases}$

34. $\displaystyle\int \frac{du}{au + b} = \frac{1}{a}\ln|au + b|$

EXAMPLE 6   Integration using a table of integrals

Find $\displaystyle\int x^2(3 - x)^5 \, dx$.

Solution

We can evaluate this integral using substitution:

$$\int x^2(3 - x)^5 \, dx = \int (3 - u)^2 u^5 \, (-du)$$

$$\boxed{u = 3 - x;\ du = -dx}$$

$$= \int (-u^7 + 6u^6 - 9u^5) \, du$$

$$= -\frac{u^8}{8} + \frac{6u^7}{7} - \frac{9u^6}{6} + C$$

$$= -\tfrac{1}{8}(3 - x)^8 + \tfrac{6}{7}(3 - x)^7 - \tfrac{3}{2}(3 - x)^6 + C$$

Even though this was not too difficult, it is a bit tedious, so we might think to evaluate this integral by using a table of integrals. This is an integral involving an expression of the form $au + b$; we find that this is Formula 32, where $u = x$, $a = -1$, $b = 3$, and $n = 5$.

$$\int x^2(3 - x)^5 \, dx = \frac{(3 - x)^{5+3}}{(5 + 3)(-1)^3} - \frac{2(3)(3 - x)^{5+2}}{(5 + 2)(-1)^3} + \frac{3^2(3 - x)^{5+1}}{(5 + 1)(-1)^3} + C$$

$$= -\tfrac{1}{8}(3 - x)^8 + \tfrac{6}{7}(3 - x)^7 - \tfrac{3}{2}(3 - x)^6 + C$$

One of the difficult considerations when using tables or computer software to carry out integration is recognizing the variety of different forms for acceptable answers. Note that it is not easy to show that these forms are algebraically equivalent. If we use computer software for this evaluation, we find yet another algebraically equivalent form:

$$\int x^2(3-x)^5\,dx = -\frac{x^8}{8} + \frac{15x^7}{7} - 15x^6 + 54x^5 - \frac{405x^4}{4} + 81x^3$$

Once again, remember that you must add the $+C$ to the computer software form. ■

EXAMPLE 7 Integration using a reduction formula from a table of integrals

Find $\int (\ln x)^4\,dx$ for $x > 0$.

Solution

The integrand is in logarithmic form; from the table of integrals we see that Formula 501 applies, where $u = x$ and $n = 4$. Take a close look at Formulas 499–502, and note that Formula 501 gives another integral as part of the result.

Integrals Involving $\ln|u|$
499. $\displaystyle\int \ln|u|\,du = u\ln|u| - u$
500. $\displaystyle\int (\ln|u|)^2\,du = u\,(\ln|u|)^2 - 2u\ln|u| + 2u$
501. $\displaystyle\int (\ln|u|)^n\,du = u\,(\ln|u|)^n - n\int (\ln|u|)^{n-1}\,du$
502. $\displaystyle\int u\ln|u|\,du = \frac{u^2}{2}\left(\ln|u| - \frac{1}{2}\right)$

This is called a **reduction formula** because it enables us to compute the given integral in terms of an integral of a similar type, only with a lower power in the integral.

$$\int (\ln x)^4\,dx = x(\ln x)^4 - 4\int (\ln x)^{4-1}\,dx \qquad \text{Formula 501}$$

$$= x(\ln x)^4 - 4\left[x(\ln x)^3 - 3\int (\ln x)^{3-1}\,dx\right] \qquad \text{Formula 501 again}$$

$$= x(\ln x)^4 - 4x(\ln x)^3 + 12\int (\ln x)^2\,dx$$

$$= x(\ln x)^4 - 4x(\ln x)^3 + 12\left[x(\ln x)^2 - 2x\ln x + 2x\right] + C$$

$$\text{This is Formula 500.}$$

$$= x(\ln x)^4 - 4x(\ln x)^3 + 12x(\ln x)^2 - 24x\ln x + 24x + C \qquad ■$$

Reduction formulas in a table of integrals, such as the one illustrated in Example 7, are usually obtained using substitution and integration by parts (which we will do in Section 7.2).

Note from the previous example that we follow the convention of adding the constant C only after eliminating the last integral sign (rather than being technically correct and writing $C_1, C_2, \ldots$ for *each* integral). The reason we can do this is that $C_1 + C_2 + \cdots = C$, for arbitrary constants.

It is often necessary to make substitutions before using one of the integration formulas, as shown in the following example.

EXAMPLE 8 **Using an integral table after substitution**

Find $\displaystyle\int \frac{x\,dx}{\sqrt{8-5x^2}}$.

Solution

(SMH)

This is an integral of the form $\sqrt{a^2-u^2}$, but it does not exactly match any of the formulas.

Integrals Involving $\sqrt{a^2-u^2}$
224. $\displaystyle\int \frac{du}{\sqrt{a^2-u^2}} = \sin^{-1}\frac{u}{a}$
225. $\displaystyle\int \frac{u\,du}{\sqrt{a^2-u^2}} = -\sqrt{a^2-u^2}$
226. $\displaystyle\int \frac{u^2\,du}{\sqrt{a^2-u^2}} = -\frac{u\sqrt{a^2-u^2}}{2} + \frac{a^2}{2}\sin^{-1}\frac{u}{a}$

Note, however, that except for the coefficient of 5, it is like Formula 225. Let $u = \sqrt{5}\,x$ (so $u^2 = 5x^2$); then $du = \sqrt{5}\,dx$:

$$\int \frac{x\,dx}{\sqrt{8-5x^2}} = \int \frac{\frac{u}{\sqrt{5}} \cdot \frac{du}{\sqrt{5}}}{\sqrt{8-u^2}}$$

Let $u = \sqrt{5}\,x$ so that $x = \dfrac{u}{\sqrt{5}}$;

$du = \sqrt{5}\,dx$ so that $dx = \dfrac{du}{\sqrt{5}}$

$$= \frac{1}{5}\int \frac{u\,du}{\sqrt{8-u^2}}$$

Now apply Formula 225, where $a^2 = 8$:

$$\frac{1}{5}\int \frac{u\,du}{\sqrt{8-u^2}} = \frac{1}{5}(-\sqrt{8-u^2}) + C = -\frac{1}{5}\sqrt{8-5x^2} + C \qquad \blacksquare$$

As you can see from Example 8, using an integral table is not a trivial task. In fact, other methods of integration may be preferable. For Example 8, you can let $u = 8-5x^2$ and integrate by substitution:

Let $u = 8 - 5x^2$; $du = -10x\,dx$

$$\int \frac{x\,dx}{\sqrt{8-5x^2}} = \int \frac{x\left(\dfrac{du}{-10x}\right)}{\sqrt{u}} = -\frac{1}{10}\int u^{-1/2}\,du$$

$$= -\frac{1}{10}(2u^{1/2}) + C = -\frac{1}{5}\sqrt{8-5x^2} + C$$

Of course, this answer is the same as the one we obtained in Example 8. The point of this calculation is to emphasize that you should try simple methods of integration before turning to the table of integrals.

EXAMPLE 9 **Integration by table (multiple forms with substitution)**

Find $\displaystyle\int 5x^2\sqrt{3x^2+1}\,dx$.

Solution

This is similar to Formula 170, but you must take care of the 5 (constant multiple) and the 3 (by making a substitution).

Integrals Involving $\sqrt{u^2 + a^2}$
168. $\displaystyle \int \sqrt{u^2 + a^2}\, du = \frac{u\sqrt{u^2 + a^2}}{2} + \frac{a^2}{2} \ln\left\| u + \sqrt{u^2 + a^2} \right\|$
169. $\displaystyle \int u\sqrt{u^2 + a^2}\, du = \frac{(u^2 + a^2)^{3/2}}{3}$
170. $\displaystyle \int u^2\sqrt{u^2 + a^2}\, du = \frac{u(u^2 + a^2)^{3/2}}{4} - \frac{a^2 u\sqrt{u^2 + a^2}}{8} - \frac{a^4}{8} \ln\left\| u + \sqrt{u^2 + a^2} \right\|$

$\boxed{\text{Let } u = \sqrt{3}\,x; \; du = \sqrt{3}\,dx}$

$$\int 5x^2\sqrt{3x^2 + 1}\, dx = 5\int \left(\frac{u^2}{3}\right)\sqrt{u^2 + 1}\,\frac{du}{\sqrt{3}}$$

$\qquad\qquad u^2 = 3x^2$ so that $x^2 = \dfrac{u^2}{3}$

$$= \frac{5}{3\sqrt{3}}\int u^2\sqrt{u^2 + 1}\, du \qquad \text{Use Formula 170, where } a = 1.$$

$$= \frac{5}{3\sqrt{3}}\left[\frac{u(u^2 + 1)^{3/2}}{4} - \frac{u\sqrt{u^2 + 1}}{8} - \frac{1}{8}\ln\left| u + \sqrt{u^2 + 1} \right|\right] + C$$

$$= \frac{5}{24\sqrt{3}}\left[2\sqrt{3}\,x(3x^2 + 1)^{3/2} - \sqrt{3}\,x\sqrt{3x^2 + 1} - \ln\left|\sqrt{3}\,x + \sqrt{3x^2 + 1}\right|\right] + C$$

$$= \frac{5}{24}\left[2x(3x^2 + 1)^{3/2} - x\sqrt{3x^2 + 1} - \frac{1}{\sqrt{3}}\ln\left(\sqrt{3}\,x + \sqrt{3x^2 + 1}\right)\right] + C$$

$\qquad\qquad$ You might want to show that $u + \sqrt{u^2 + 1} > 0$.

If you use a calculator or computer, you will probably obtain an alternate, but equivalent, form:

$$-\frac{5\sqrt{3}\ln\left|\sqrt{3x^2 + 1} + \sqrt{3}\,x\right|}{72} - \frac{5x\sqrt{3x^2 + 1}\,(6x^2 + 1)}{24} \qquad ■$$

7.1   PROBLEM SET

Ⓐ *Find each integral in Problems 1–12.*

1. $\displaystyle \int \frac{2x + 5}{\sqrt{x^2 + 5x}}\, dx$

2. $\displaystyle \int \frac{\ln x}{x}\, dx$

3. $\displaystyle \int \frac{dx}{x\ln x}$

4. $\displaystyle \int \cos x\, e^{\sin x}\, dx$

5. $\displaystyle \int \frac{x\, dx}{4 + x^4}$

6. $\displaystyle \int \frac{t^2\, dt}{9 + t^6}$

7. $\displaystyle \int (1 + \cot x)^4 \csc^2 x\, dx$

8. $\displaystyle \int \frac{4x^3 - 4x}{x^4 - 2x^2 + 3}\, dx$

9. $\displaystyle \int \frac{x^3 - x}{(x^4 - 2x^2 + 3)^2}\, dx$

10. $\displaystyle \int \frac{2x + 4}{x^2 + 4x + 3}\, dx$

11. $\displaystyle \int \frac{2x + 1}{x^2 + x + 1}\, dx$

12. $\displaystyle \int \frac{2x - 1}{(4x^2 - 4x)^2}\, dx$

Integrate the expressions in Problems 13–24 using the short table of integrals given in Appendix D.

13. $\displaystyle \int \frac{dx}{x^2\sqrt{x^2 - a^2}}$

14. $\displaystyle \int \frac{dx}{x^2\sqrt{a^2 - x^2}}$

15. $\displaystyle \int x\ln x\, dx$

16. $\displaystyle \int \ln x\, dx$

17. $\displaystyle \int xe^{ax}\, dx$

18. $\displaystyle \int \frac{dx}{a + be^{2x}}$

19. $\int \dfrac{x^2\,dx}{\sqrt{x^2+1}}$

20. $\int \dfrac{dx}{x^2\sqrt{x^2+16}}$

21. $\int \dfrac{x\,dx}{\sqrt{4x^2+1}}$

22. $\int \dfrac{dx}{x\sqrt{1-9x^2}}$

23. $\int e^{-4x}\sin 5x\,dx$

24. $\int x\sin^{-1}x\,dx$

Evaluate the integrals in Problems 25–33. If you use the integral table, state the number of the formula used, and if you use substitution, show each step. If you use an alternative table of integrals, then cite the source as well as the formula.

25. $\int (1+bx)^{-1}\,dx$

26. $\int \dfrac{x\,dx}{\sqrt{a^2-x^2}}$

27. $\int x(1+x)^3\,dx$

28. $\int x\sqrt{1+x}\,dx$

29. $\int xe^{4x}\,dx$

30. $\int x\ln 2x\,dx$

31. $\int \dfrac{dx}{1+e^{2x}}$

32. $\int \ln^3 x\,dx$

33. $\int \dfrac{x^3\,dx}{\sqrt{4x^4+1}}$

B *Use the Student Mathematics Handbook or other available integration tables to find the integrals given in Problems 34–39. Cite the formula number or source you are using.*

34. $\int \dfrac{\sqrt{4x^2+1}}{x}\,dx$

35. $\int \sec^3\left(\dfrac{x}{2}\right)\,dx$

36. $\int \sin^6 x\,dx$

37. $\int \dfrac{dx}{9x^2+6x+1}$

38. $\int (9-x^2)^{3/2}\,dx$

39. $\int \dfrac{\sin^2 x}{\cos x}\,dx$

40. Derive the **sine squared formula** shown on the inside back cover (Formula 348):

$$\int \sin^2 x\,dx = \tfrac{1}{2}x - \tfrac{1}{4}\sin 2x + C$$

Hint: Use the identity $\sin^2 x = \dfrac{1-\cos 2x}{2}$.

41. Derive the **cosine squared formula** shown on the inside back cover (Formula 317):

$$\int \cos^2 x\,dx = \tfrac{1}{2}x + \tfrac{1}{4}\sin 2x + C$$

Hint: Use the identity $\cos^2 x = \dfrac{1+\cos 2x}{2}$.

Problems 42–44 use substitution to integrate certain powers of sine and cosine.

42. $\int \sin^4 x\cos x\,dx$ *Hint:* Let $u = \sin x$.

43. $\int \sin^3 x\cos^4 x\,dx$ *Hint:* Let $u = \cos x$.

44. $\int \sin^2 x\cos^2 x\,dx$ *Hint:* Use the identities shown in Problems 40 and 41.

45. Exploration Problem Using Problems 42–44 formulate a procedure for integrals of the form

$$\int \sin^m x\,\cos^n x\,dx$$

Find each integral in Problems 46–51.

46. $\int \dfrac{e^x\,dx}{1+e^{x/2}}$

47. $\int \dfrac{dx}{x^{1/2}+x^{1/4}}$

48. $\int \dfrac{4\,dx}{x^{1/3}+2x^{1/2}}$

49. $\int \dfrac{18\tan^2 t\sec^2 t}{(2+\tan^3 t)^2}\,dt$

50. $\int \dfrac{dx}{\left(x+\tfrac{1}{2}\right)\sqrt{4x^2+4x}}$

51. $\int \dfrac{e^{-x}-e^x}{e^{2x}+e^{-2x}+2}\,dx$

52. Find the area of the region bounded by the graphs of $y = \dfrac{2x}{\sqrt{x^2+9}}$ and $y = 0$ from $x = 0$ to $x = 4$.

53. Find the volume of the solid generated when the region under the curve

$$y = \dfrac{x^{3/2}}{\sqrt{x^2+9}}$$

between $x = 0$ and $x = 9$ is revolved about the x-axis.

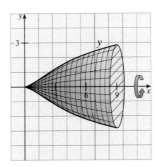

54. Find the volume of the solid generated when the curve $y = x(1-x^2)^{1/4}$ from $x = 0$ to $x = 1$ is revolved about the x-axis.

55. Find the volume of the solid generated when the curve

$$y = \dfrac{1}{\sqrt{x}}(1+\sqrt{x})^{1/3}$$

between $x = 1$ and $x = 4$ is revolved about the y-axis.

56. Find the volume of the solid generated when the curve $x = \sqrt[4]{4-y^2}$ between $y = 1$ and $y = 2$ is revolved about the y-axis.

57. Let $y = f(x)$ be a function that satisfies the differential equation

$$xy' = \sqrt{(\ln x)^2 - x^2}$$

Find the arc length of $y = f(x)$ between $x = \tfrac{1}{4}$ and $x = \tfrac{1}{2}$.

58. Find the arc length of the curve $y = \ln(\cos x)$ on the interval $[0, \pi/4]$.

59. Find the area of the surface generated when the curve $y = x^2$ on the interval $[0, 1]$ is revolved about the x-axis.

60. Find the area of the surface generated when the curve $y = x^2$ on the interval $[0, 1]$ is revolved about the y-axis.

61. Show that $\int \csc x\,dx = -\ln|\csc x + \cot x| + C$.

Hint: Multiply the integrand by $\dfrac{\csc x + \cot x}{\csc x + \cot x}$.

62. Find $\int 2\sin x\cos x\,dx$ by using the indicated substitution.

   a. Let $u = \cos x$.

b. Let $u = \sin x$.

c. Write $2 \sin x \cos x = \sin 2x$ and carry out the integration.

d. Show that the answers you obtained for parts **a–c** are the same.

63. Derive the formula

$$\int_0^\pi x f(\sin x)\, dx = \frac{\pi}{2} \int_0^\pi f(\sin x)\, dx$$

64. Find the surface area of the torus generated when the circle

$$x^2 + (y - b)^2 = 1, \qquad b > 1$$

is revolved about the x-axis.

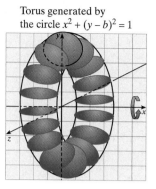

Torus generated by the circle $x^2 + (y - b)^2 = 1$

7.2   Integration by Parts

IN THIS SECTION   integration by parts formula, repeated use of integration by parts, definite integration by parts

Integration by parts is a procedure based on reversing the product rule for differentiation. We present this section not only as a technique of integration but also as a procedure that is necessary in a variety of useful applications.

INTEGRATION BY PARTS FORMULA

Recall the formula for differentiation of a product. If u and v are differentiable functions, then

$$d(uv) = u\, dv + v\, du$$

Integrate both sides of this equation to find the formula for integration by parts:

$$\int d(uv) = \int u\, dv + \int v\, du$$

$$uv = \int u\, dv + \int v\, du$$

If we rewrite this last equation, we obtain the formula summarized in the following box.

Formula for Integration by Parts

$$\int u\, dv = uv - \int v\, du$$

EXAMPLE 1   Integration by parts

Find $\int x e^x\, dx$.

Solution

To use integration by parts, we must choose u and dv so that the new integral is easier to integrate than the original.

$$\int \underbrace{x}_{u}\,\underbrace{e^x\,dx}_{dv}$$

$$
\boxed{
\begin{array}{ll}
\text{Let } \; u = x & \text{and} \quad dv = e^x\,dx \\
\quad\;\; du = dx & \qquad\quad v = \int e^x\,dx = e^x
\end{array}
}
$$

$$\int xe^x\,dx = \underbrace{x}_{u}\,\underbrace{e^x}_{v} - \int \underbrace{e^x}_{v}\,\underbrace{dx}_{du} = xe^x - e^x + C$$

You can check your work with integration by parts by differentiating, using software, or using the integration table found in Appendix D or the *Student Mathematics Handbook* (Formula 484, with $a = 1$). ∎

SMH

Note: You may wonder why an arbitrary constant (call it K) was not included when performing the integration associated with $\int dv$ in integration by parts. The reason is that when applying integration by parts, we need just *one* function v whose derivative is dv, so we take the simplest one—the one with $K = 0$. You may find it instructive to see that taking $v = e^x + K$ gives the same result.

Integration by parts is often difficult the first time you try to do it because there is no absolute choice for u and dv. In Example 1, you might have chosen

$$
\begin{array}{ll}
u = e^x & \text{and} \quad dv = x\,dx \\
du = e^x\,dx & \qquad\quad v = \int x\,dx = \dfrac{x^2}{2}
\end{array}
$$

Then

$$\int xe^x\,dx = \underbrace{e^x}_{u}\,\underbrace{\frac{x^2}{2}}_{v} - \int \underbrace{\frac{x^2}{2}}_{v}\,\underbrace{e^x\,dx}_{du}$$

$$= \frac{1}{2}x^2 e^x - \frac{1}{2}\int x^2 e^x\,dx$$

Note, however, that this choice of u and dv leads to a more complicated form than the original. In general, when you are integrating by parts, if you make a choice for u and dv that leads to a more complicated form than when you started, consider going back and making another choice for u and dv.

Generally, you want to choose dv to be as difficult as possible (and still be something you can integrate), with the remainder being left for the u-factor.

EXAMPLE 2   When the differentiable part is the entire integrand

Find $\displaystyle\int \ln x\,dx$ for $x > 0$.

Solution

$$
\boxed{
\begin{array}{ll}
\text{Let } \; u = \ln x & \text{and} \quad dv = dx \\
\quad\;\; du = \dfrac{1}{x}\,dx & \qquad\quad v = x
\end{array}
}
$$

$$\int \ln x\,dx = (\ln x)x - \int x\,\frac{1}{x}\,dx = x\ln x - \int dx = x\ln x - x + C$$

Check with Formula 499 (Appendix D), where $a = 1$, which we worked as Problem 16 of Problem Set 7.1. ∎

REPEATED USE OF INTEGRATION BY PARTS

Sometimes integration by parts must be applied several times to evaluate a given integral.

EXAMPLE 3   Repeated integration by parts

Find $\int x^2 e^{-x}\, dx$.

Solution

$$\begin{array}{ll} \text{Let} \quad u = x^2 & \text{and} \quad dv = e^{-x}\, dx \\ \quad\quad du = 2x\, dx & \quad\quad v = -e^{-x} \end{array}$$

$$\int x^2 e^{-x}\, dx = x^2(-e^{-x}) - \int (-e^{-x})(2x\, dx)$$

$$= -x^2 e^{-x} + 2\int x e^{-x}\, dx$$

$$\begin{array}{ll} \text{Let} \quad u = x & \text{and} \quad dv = e^{-x}\, dx \\ \quad\quad du = dx & \quad\quad v = -e^{-x} \end{array}$$

$$= -x^2 e^{-x} + 2\left[x(-e^{-x}) - \int (-e^{-x})\, dx \right]$$

Do not forget the grouping symbols here.

$$= -x^2 e^{-x} - 2x e^{-x} - 2e^{-x} + C$$

$$= -e^{-x}(x^2 + 2x + 2) + C$$

Check with Formula 485, where $a = -1$. ∎

In the following example, it is necessary to apply integration by parts more than once, but as you will see, when we do so a second time we return to the original integral. Note carefully how this situation can be handled algebraically.

EXAMPLE 4   Repeated integration by parts with algebraic manipulation

Find $\int e^{2x} \sin x\, dx$.

Solution   For this problem you will see that it will be useful to call the original integral I. That is, we let

$$I = \int e^{2x} \sin x\, dx$$

$$\begin{array}{ll} \text{Let} \quad u = e^{2x} & \text{and} \quad dv = \sin x\, dx \\ \quad\quad du = 2e^{2x}\, dx & \quad\quad v = -\cos x \end{array}$$

$$I = \int e^{2x} \sin x\, dx = e^{2x}(-\cos x) - \int (-\cos x)(2e^{2x}\, dx)$$

$$= -e^{2x} \cos x + 2\int e^{2x} \cos x\, dx$$

$$\begin{array}{ll} u = e^{2x} & \text{and} \quad dv = \cos x\, dx \\ \quad du = 2e^{2x}\, dx & \quad\quad v = \sin x \end{array}$$

$$= -e^{2x} \cos x + 2 \left[e^{2x}(\sin x) - \int \sin x (2e^{2x} \, dx) \right]$$

$$= -e^{2x} \cos x + 2e^{2x} \sin x - 4 \int e^{2x} \sin x \, dx$$

Notice that this last integral is I. Thus, we have

$$I = -e^{2x} \cos x + 2e^{2x} \sin x - 4I \quad \text{Solve this equation for } I.$$

$$5I = -e^{2x} \cos x + 2e^{2x} \sin x \quad \quad \text{Add } 4I \text{ to both sides.}$$

$$I = \tfrac{1}{5} e^{2x} (2 \sin x - \cos x)$$

Thus, $\displaystyle\int e^{2x} \sin x \, dx = \tfrac{1}{5} e^{2x} (2 \sin x - \cos x) + C$.

Check with the integration table (Formula 492, where $a = 2$ and $b = 1$), or by taking the derivative. ■

DEFINITE INTEGRATION BY PARTS

The integration by parts formula can be used for definite integrals, as summarized in the following box.

Integration by Parts for Definite Integrals

$$\int_a^b u \, dv = uv \Big|_a^b - \int_a^b v \, du$$

You should recognize that this formula for definite integrals is the same as the formula for indefinite integrals, where the first term after the equal sign has been evaluated at the appropriate limits of integration. This is illustrated by the following example.

EXAMPLE 5   Integration by parts with a definite integral

Evaluate $\displaystyle\int_0^1 x e^{2x} \, dx$.

Solution

$$\text{Let} \quad u = x \quad \text{and} \quad dv = e^{2x} \, dx$$
$$du = dx \quad \quad \quad v = \tfrac{1}{2} e^{2x}$$

$$\int_0^1 x e^{2x} \, dx = \frac{1}{2} x e^{2x} \Big|_0^1 - \frac{1}{2} \int_0^1 e^{2x} \, dx$$

It is sometimes easier to simplify the algebra and then do one evaluation here at the end of the integration part of the problem.
↓

$$= \left(\frac{1}{2} x e^{2x} - \frac{1}{4} e^{2x} \right) \Big|_0^1 = \frac{1}{4} e^2 + \frac{1}{4}$$

Check in Appendix D (Formula 484, with $a = 2$). ■

TECHNOLOGY NOTE

You can use computer software or many calculators to evaluate definite integrals. Some calculators give the exact form just shown, while others only give a decimal approximation such as 2.097264025.

EXAMPLE 6   Integration by parts followed by substitution

Evaluate $\displaystyle\int_0^{\pi/4} \tan^{-1} x \, dx$.

Solution

$$\boxed{\begin{array}{ll} \text{Let} \quad u = \tan^{-1} x & \text{and} \quad dv = dx \\ \qquad du = \dfrac{dx}{1+x^2} & \qquad\quad v = x \end{array}}$$

$$\int_0^{\pi/4} \underbrace{\tan^{-1} x}_{u} \underbrace{dx}_{dv} = \underbrace{(\tan^{-1} x)}_{u} \underbrace{x}_{v} \Big|_0^{\pi/4} - \int_0^{\pi/4} \underbrace{\dfrac{x\, dx}{1+x^2}}_{v\,du}$$

$$= \left[x \tan^{-1} x - \dfrac{1}{2} \ln(1+x^2) \right]\Big|_0^{\pi/4}$$

$$\boxed{\begin{array}{l} \text{Use substitution where } w = 1 + x^2 \text{ and} \\ dw = 2x\, dx. \\[4pt] \displaystyle\int \dfrac{x\, dx}{1+x^2} = \int \dfrac{dw}{2w} = \dfrac{1}{2} \ln w = \dfrac{1}{2} \ln(1+x^2) \end{array}}$$

$$= \left[\dfrac{\pi}{4}\left(\tan^{-1}\dfrac{\pi}{4}\right) - \dfrac{1}{2}\ln\left(1+\dfrac{\pi^2}{16}\right)\right] - \left[0 - \dfrac{1}{2}\ln 1\right]$$

$$= \dfrac{\pi}{4}\tan^{-1}\dfrac{\pi}{4} - \dfrac{1}{2}\ln\left(1+\dfrac{\pi^2}{16}\right) \approx 0.2827$$

Check with an integration table (Formula 457, for example, with $a = 1$). ∎

7.2 PROBLEM SET

A *Find each integral in Problems 1–16.*

1. $\displaystyle\int xe^{-2x}\, dx$

2. $\displaystyle\int x \sin x\, dx$

3. $\displaystyle\int x \ln x\, dx$

4. $\displaystyle\int x \tan^{-1} x\, dx$

5. $\displaystyle\int \sin^{-1} x\, dx$

6. $\displaystyle\int x^2 \sin x\, dx$

7. $\displaystyle\int e^{-3x} \cos 4x\, dx$

8. $\displaystyle\int e^{2x} \sin 3x\, dx$

9. $\displaystyle\int x^2 \ln x\, dx$

10. $\displaystyle\int (x + \sin x)^2\, dx$

11. $\displaystyle\int \sin(\ln x)\, dx$

12. $\displaystyle\int x \sin x \cos x\, dx$

13. $\displaystyle\int \ln(x^2 + 1)\, dx$

14. $\displaystyle\int \sin \sqrt{x}\, dx$

15. $\displaystyle\int \dfrac{xe^{-x}}{(x-1)^2}\, dx$

16. $\displaystyle\int \dfrac{\ln(\sin x)\, dx}{\tan x}$

Find the exact value of the definite integrals in Problems 17–22 using integration by parts, and then check by using a calculator to find an approximate answer correct to four decimal places.

17. $\displaystyle\int_1^4 \sqrt{x}\, \ln x\, dx$

18. $\displaystyle\int_1^e x^3 \ln x\, dx$

19. $\displaystyle\int_1^e (\ln x)^2\, dx$

20. $\displaystyle\int_{1/3}^e 3(\ln 3x)^2\, dx$

21. $\displaystyle\int_0^\pi e^{2x} \cos 2x\, dx$

22. $\displaystyle\int_0^\pi x(\sin x + \cos x)\, dx$

B **23.** **WHAT DOES THIS SAY?** Describe the process known as *integration by parts.*

In Problems 24–27, first use an appropriate substitution and then integrate by parts to evaluate the integral. Remember to give your answers in terms of x.

24. $\displaystyle\int \dfrac{\ln x \sin(\ln x)}{x}\, dx$

25. $\displaystyle\int [\sin 2x \ln(\cos x)]\, dx$

26. $\displaystyle\int e^{2x} \sin e^x\, dx$

27. $\displaystyle\int [\sin x \ln(2 + \cos x)]\, dx$

28. a. Evaluate $\displaystyle\int \dfrac{x^3}{x^2 - 1}\, dx$ using integration by parts.

  b. Evaluate the integral in part **a** by first dividing the integrand.

29. Evaluate $\displaystyle\int \cos^2 x\, dx$.

30. Evaluate $\displaystyle\int \dfrac{x\, dx}{\sqrt{x^2 + 1}}$.

31. Use Problem 29 to evaluate $\displaystyle\int x \cos^2 x\, dx$ using integration by parts.

32. Use Problem 30 to evaluate $\displaystyle\int \dfrac{x^3\, dx}{\sqrt{x^2 + 1}}$ using integration by parts.

33. Find $\displaystyle\int x^n \ln x\, dx$, where n is any positive real number.

34. After t seconds, an object is moving along a line with velocity of $te^{-t/2}$ meters per second. Express the position of the moving object as a function of time.

35. After t hours on the job, a factory worker can produce $100te^{-0.5t}$ units per hour. How many units does the worker produce during the first 3 hours?

36. After t weeks, contributions in response to a local fund-raising campaign were coming in at the rate of $2,000te^{-0.2t}$ dollars per week. How much money was raised during the first 5 weeks?

37. Find the volume of the solid generated when the region under the curve $y = e^{-x}$ on the interval $[0, 2]$ is revolved about the y-axis.

38. Find the volume of the solid generated when the region under the curve $y = \sin x + \cos x$ on the interval $[0, \frac{\pi}{4}]$ is revolved about the y-axis.

39. Find the volume of the solid generated when the region under the curve $y = \ln x$ on the interval $[1, e]$ is revolved about the indicated axis:

   a. x-axis   **b.** y-axis

40. Find the centroid (with coordinates rounded to the nearest hundredth) of the region in the first quadrant bounded by the curves $y = \sin x$ and $y = \cos x$ and the y-axis.

41. Find the centroid (with coordinates rounded to the nearest hundredth) of the region bounded by the curves $y = e^x$, $y = e^{-x}$, and the line $x = 1$.

In Problems 42–43, solve the given separable differential equations.

42. $\dfrac{dy}{dx} = xe^{y-x}$ **43.** $\dfrac{dy}{dx} = \sqrt{xy} \ln x$

44. Find a function $y = f(x)$ whose graph passes through $(1, 1)$ and has the property that at each point (x, y) on the graph, the slope of the tangent line is $y \tan^{-1} x$.

45. Find a function $y = f(x)$ whose graph passes through $(0, 1)$ and has the property that the normal line at each point (x, y) on the graph has slope $\dfrac{\sec x}{xy}$.

46. Suppose it is known that $f(0) = 3$ and

$$\int_0^\pi [f(x) + f''(x)] \sin x \, dx = 0$$

What is $f(\pi)$?

47. Because a rocket burns fuel in flight, its mass decreases with time, and this in turn affects its velocity. It can be shown that the velocity $v(t)$ of the rocket at time t in its flight is given by

$$v(t) = -r \ln \frac{w - kt}{w} - gt$$

where w is the initial weight of the rocket (including its fuel) and r and k are, respectively, the expulsion speed and the rate of consumption of the fuel, which are assumed to be constant. As usual, $g = 32$ ft/s^2 is the constant acceleration due to gravity. Suppose $w = 30,000$ lb, $r = 8,000$ ft/s, and $k = 200$ lb/sec. What is the height of the rocket after 2 minutes (120 seconds)?

48. In physics, it is known that loudness L of a sound is related to its intensity I by the equation

$$L = 10 \log \frac{I}{I_0}$$

decibels, where $I_0 = 10^{-12}$ watt/m^2 is the threshold of audibility (the lowest intensity that can be heard). What is the average value of L as the intensity of a TV show ranges between I_0 and $I_1 = 3 \cdot 10^{-5}$ watt/m^2?

49. The displacement from equilibrium of a mass oscillating at the end of a spring hanging from the ceiling is given by

$$y = 2.3e^{-0.25t} \cos 5t$$

feet. What is the average displacement (rounded to the nearest hundredth) of the mass between times $t = 0$ and $t = \pi/5$ seconds?

equilibrium point

50. A photographer is taking a picture of a clever sign on the back of a truck. The sign is 5 ft high and its lower edge is 1 ft above the lens of the camera. At first the truck is 4 ft away from the photographer, but then it begins to move away. What is the average value of the angle θ (correct to two decimal places) subtended by the camera lens as the truck moves from 4 ft to 20 ft away from the photographer?

51. If n moles of an ideal gas expand at constant temperature T, then its pressure p and volume V satisfy the equation $pV = nRT$, for constant R. It can be shown that the work done by the gas in expanding from volume V_1 to V is

$$W = nRT \ln \frac{V}{V_1}$$

What is the average work done as V increases from V_1 to $V_2 = 10V_1$?

52. Evaluate $\displaystyle\int \frac{(2x - 1)}{x^2} e^{2x} \, dx$.

53. Derive the reduction formula

$$\int x^n e^x \, dx = x^n e^x - n \int x^{n-1} e^x \, dx$$

(This is Formula 486, with $a = 1$.)

54. Derive the reduction formula

$$\int (\ln x)^n \, dx = x(\ln x)^n - n \int (\ln x)^{n-1} \, dx$$

(This is Formula 501.)

C 55. Wallis's formula If n is an even positive integer, use reduction formulas to show

$$\int_0^{\pi/2} \sin^n x \, dx = \int_0^{\pi/2} \cos^n x \, dx$$

$$= \left[\frac{1 \cdot 3 \cdot 5 \cdot \cdots \cdot (n-1)}{2 \cdot 4 \cdot 6 \cdot \cdots \cdot n} \right] \frac{\pi}{2}$$

56. Exploration Problem State and prove a result similar to the one in Problem 55 for the case where n is an odd positive integer.

7.3 Trigonometric Methods

IN THIS SECTION powers of sine and cosine, powers of secant and tangent, trigonometric substitutions, quadratic-form integrals

POWERS OF SINE AND COSINE

Problems 40–45 of Problem Set 7.1 anticipated the results of this section. We begin by considering a product of powers of sine and cosine, which we represent in the form

$$\int \sin^m x \cos^n x \, dx$$

There are essentially two cases that must be considered, depending on whether the powers m and n are both even or not. We will state the general strategy for handling each case and then illustrate the procedure with an example.

Case I: Either m or n is odd (or both)
General strategy: Suppose, for simplicity, that m is odd. Separate a factor of $\sin x$ from the rest of the integrand, so the remaining power of $\sin x$ is even; use the identity $\sin^2 x = 1 - \cos^2 x$ to express everything but the term $(\sin x \, dx)$ in terms of $\cos x$. Substitute $u = \cos x$, $du = -\sin x \, dx$ to convert the integral into a polynomial in u and integrate using the power rule. The case where n is odd is handled in analogous fashion, as shown in the following example by reversing the roles of $\sin x$ and $\cos x$.

EXAMPLE 1 Power of cosine is odd

Evaluate $\int \sin^4 x \cos^3 x \, dx$.

Solution
Since $n = 3$ is odd, peel off a factor of $\cos x$ and use $\cos^2 x = 1 - \sin^2 x$ to express the integral as a polynomial in $\sin x$.

$$\int \sin^4 x \cos^3 x \, dx = \int \sin^4 x \cos^2 x (\cos x \, dx)$$

$$= \int \sin^4 x (1 - \sin^2 x)(\cos x \, dx)$$

$$= \int u^4 (1 - u^2) \, du \quad \boxed{u = \sin x; \quad du = \cos x \, dx}$$

$$= \tfrac{1}{5} u^5 - \tfrac{1}{7} u^7 + C$$

$$= \tfrac{1}{5} \sin^5 x - \tfrac{1}{7} \sin^7 x + C \qquad \blacksquare$$

Case II: Neither m nor n is odd
General strategy: Convert this into Case I by using the identities $\sin^2 x = \frac{1}{2}(1 - \cos 2x)$ and $\cos^2 x = \frac{1}{2}(1 + \cos 2x)$. These are sometimes called *half-angle identities*.

EXAMPLE 2   All powers are even

Evaluate $\displaystyle\int \sin^2 x \cos^4 x\, dx$.

Solution

$$\int \sin^2 x \cos^4 x\, dx = \int \tfrac{1}{2}(1 - \cos 2x)\left(\tfrac{1}{4}\right)(1 + \cos 2x)^2\, dx$$

$$= \tfrac{1}{8}\int (1 + \cos 2x - \cos^2 2x - \cos^3 2x)\, dx$$

Half-angle identities

$$= \tfrac{1}{8}\int \left[1 + \cos 2x - \tfrac{1}{2}(1 + \cos 4x) - (1 - \sin^2 2x)(\cos 2x)\right] dx$$

Half-angle identities again

$$= \tfrac{1}{8}\int \left[1 + \cos 2x - \tfrac{1}{2} - \tfrac{1}{2}\cos 4x - \cos 2x + (\sin^2 2x)\cos 2x\right] dx$$

$$= \tfrac{1}{8}\int \left(\tfrac{1}{2} - \tfrac{1}{2}\cos 4x\right) dx + \tfrac{1}{8}\int \sin^2 2x(\cos 2x)\, dx$$

$$\boxed{u = \sin 2x; \quad du = 2\cos 2x\, dx}$$

$$= \tfrac{1}{16}x - \tfrac{1}{64}\sin 4x + \tfrac{1}{48}\sin^3 2x + C \qquad \blacksquare$$

TECHNOLOGY NOTE

One computer algebra system applied to this example yields

$$\int \sin^2 x \cos^4 x\, dx = -\tfrac{1}{6}\sin x \cos^5 x + \tfrac{1}{24}\sin x \cos^3 x + \tfrac{1}{16}\sin x \cos x + \tfrac{1}{16}x$$

(Recall, technology does not show the "$+C$" term.) Both answers are correct. They are related by trigonometric identities, and it can be shown that they differ by a constant.

POWERS OF SECANT AND TANGENT

The simplest integrals of this form are

$$\int \tan x\, dx = \ln |\sec x| + C$$

$$\int \sec x\, dx = \ln |\sec x + \tan x| + C$$

For the more general situation, which we write as

$$\int \tan^m x \sec^n x\, dx,$$

there are three essentially different cases to consider.

Case I: n is even

General strategy:　Peel off a factor of $\sec^2 x$ from the integrand and use the identity $\sec^2 x = \tan^2 x + 1$ to express the integrand in powers of $\tan x$ except for $(\sec^2 x \, dx)$; substitute $u = \tan x$, $du = \sec^2 x \, dx$, and integrate by using the power rule.

EXAMPLE 3　Power of the secant is even

Evaluate $\displaystyle\int \tan^2 x \sec^4 x \, dx$.

Solution

$$\int \tan^2 x \sec^4 x \, dx = \int \tan^2 x \sec^2 x (\sec^2 x \, dx)$$

$$= \int \tan^2 x (\tan^2 x + 1) \sec^2 x \, dx$$

$$= \int u^2 (u^2 + 1) \, du \qquad \boxed{u = \tan x; \quad du = \sec^2 x \, dx}$$

$$= \tfrac{1}{5} u^5 + \tfrac{1}{3} u^3 + C$$

$$= \tfrac{1}{5} \tan^5 x + \tfrac{1}{3} \tan^3 x + C \qquad\blacksquare$$

Case II: m is odd

General strategy:　Peel off a factor of $\sec x \tan x$ from the integrand and use the identity $\tan^2 x = \sec^2 x - 1$ to express the integrand in powers of $\sec x$, except for $(\sec x \tan x \, dx)$; substitute $u = \sec x$, $du = \sec x \tan x \, dx$, and integrate using the power rule.

EXAMPLE 4　Power of the tangent is odd

Evaluate $\displaystyle\int \tan x \sec^6 x \, dx$.

Solution

$$\int \tan x \sec^6 x \, dx = \int \sec^5 x (\sec x \tan x \, dx)$$

$$= \int u^5 \, du \qquad \boxed{u = \sec x; \quad du = \sec x \tan x \, dx}$$

$$= \tfrac{1}{6} \sec^6 x + C \qquad\blacksquare$$

Case III: m is even and n is odd

General strategy:　Use the identity $\tan^2 x = \sec^2 x - 1$ to express the integrand in terms of powers of $\sec x$; then use the reduction Formula 428:

$$\int \sec^n x \, dx = \frac{\sec^{n-2} x \tan x}{n-1} + \frac{n-2}{n-1} \int \sec^{n-2} x \, dx$$

EXAMPLE 5　Power of the tangent is even and power of the secant is odd

Evaluate $\displaystyle\int \tan^2 x \sec^3 x \, dx$.

Solution

$$\int \tan^2 x \sec^3 x \, dx = \int (\sec^2 x - 1) \sec^3 x \, dx$$

$$= \int \sec^5 x \, dx - \int \sec^3 x \, dx$$

$$= \left[\frac{\sec^3 x \tan x}{4} + \frac{3}{4} \int \sec^3 x \, dx \right] - \int \sec^3 x \, dx$$

$$= \frac{\sec^3 x \tan x}{4} - \frac{1}{4} \int \sec^3 x \, dx$$

$$= \frac{\sec^3 x \tan x}{4} - \frac{1}{4} \left[\frac{\sec x \tan x}{2} + \frac{1}{2} \int \sec x \, dx \right]$$

$$= \frac{\sec^3 x \tan x}{4} - \frac{\sec x \tan x}{8} - \frac{1}{8} \ln |\sec x + \tan x| + C \quad \blacksquare$$

TRIGONOMETRIC SUBSTITUTIONS

Trigonometric substitutions can also be useful. For instance, suppose an integrand contains the term $\sqrt{a^2 - u^2}$, where $a > 0$. Then by setting $u = a \sin \theta$ for an acute angle θ, and using the identity $\cos^2 \theta = 1 - \sin^2 \theta$, we obtain

$$\sqrt{a^2 - u^2} = \sqrt{a^2 - a^2 \sin^2 \theta} = a\sqrt{1 - \sin^2 \theta} = a \cos \theta$$

Thus, the substitution $u = a \sin \theta$, $du = a \cos \theta \, d\theta$ eliminates the square root and may convert the given integral into one involving only sine and cosine. This substitution can best be remembered by setting up a reference triangle. This process is illustrated in the following example.

EXAMPLE 6 **Trigonometric substitution with form $\sqrt{a^2 - u^2}$**

Find $\int \sqrt{4 - x^2} \, dx$.

Solution

First, using a table of integration, we find (Formula 231; $a = 2$)

$$\int \sqrt{4 - x^2} \, dx = \frac{x\sqrt{4 - x^2}}{2} + 2 \sin^{-1} \left(\frac{x}{2} \right) + C$$

Our goal with this example is to show how we might obtain this formula with a trigonometric substitution. Refer to the triangle shown in Figure 7.1.

Let $x = 2 \sin \theta$, so $dx = 2 \cos \theta \, d\theta$. Then,

$$\int \sqrt{4 - x^2} \, dx = \int \sqrt{4 - 4 \sin^2 \theta} \, (2 \cos \theta \, d\theta)$$

$$= 4 \int \cos^2 \theta \, d\theta \qquad \text{Since } \sqrt{1 - \sin^2 \theta} = \cos \theta$$

$$= 4 \int \frac{1 + \cos 2\theta}{2} \, d\theta \qquad \text{Half-angle identity}$$

$$= 2\theta + \sin 2\theta + C$$

$$= 2\theta + 2 \sin \theta \cos \theta + C$$

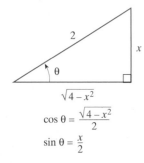

$$\cos \theta = \frac{\sqrt{4 - x^2}}{2}$$

$$\sin \theta = \frac{x}{2}$$

Figure 7.1 Reference triangle with form $\sqrt{a^2 - u^2}$

The final step is to convert this answer back to terms involving x. Using the reference triangle in Figure 7.1, we find that

$$\sin\theta = \frac{x}{2} \quad \text{and} \quad \cos\theta = \frac{1}{2}\sqrt{4-x^2}$$

Thus, we have $\theta = \sin^{-1}\frac{x}{2}$

$$\int \sqrt{4-x^2}\, dx = 2\sin^{-1}\left(\frac{x}{2}\right) + 2\left(\frac{x}{2}\right)\left(\frac{\sqrt{4-x^2}}{2}\right) + C$$

$$= 2\sin^{-1}\left(\frac{x}{2}\right) + \frac{x}{2}\sqrt{4-x^2} + C \qquad\blacksquare$$

Similar methods can be used to convert integrals that involve terms of the form $\sqrt{a^2+u^2}$ or $\sqrt{u^2-a^2}$ into trigonometric integrals, as shown in Table 7.1. For this table, we require $0 \le \theta < \frac{\pi}{2}$.

TABLE 7.1 Trigonometric substitution for integrands involving a radical form

If the integrand involves ...	substitute ...	to obtain ...
$\sqrt{a^2-u^2}$	$u = a\sin\theta$	$\sqrt{a^2-u^2} = a\cos\theta$
$\sqrt{a^2+u^2}$	$u = a\tan\theta$	$\sqrt{a^2+u^2} = a\sec\theta$
$\sqrt{u^2-a^2}$	$u = a\sec\theta$	$\sqrt{u^2-a^2} = a\tan\theta$

EXAMPLE 7 **Trigonometric substitution with form $\sqrt{a^2+u^2}$**

Evaluate $\displaystyle\int x^2\sqrt{9+x^2}\, dx$.

Solution

Let $x = 3\tan\theta$, $dx = 3\sec^2\theta\, d\theta$; then

$$\int x^2\sqrt{9+x^2}\, dx = \int (3\tan\theta)^2\sqrt{9+9\tan^2\theta}\,(3\sec^2\theta\, d\theta)$$

$$= \int (9\tan^2\theta)(3\sec\theta)(3\sec^2\theta\, d\theta) \quad 1+\tan^2\theta = \sec^2\theta$$

$$= 81\int \tan^2\theta\sec^3\theta\, d\theta$$

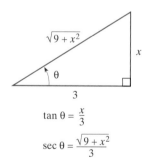

$$\tan\theta = \frac{x}{3}$$

$$\sec\theta = \frac{\sqrt{9+x^2}}{3}$$

Figure 7.2 Reference triangle with form $\sqrt{a^2+u^2}$

This integration can now be completed by looking at Example 5. In order to express the antiderivative in terms of the original variable x, you can use the reference triangle in Figure 7.2.

Since $\tan\theta = x/3$, we have $\sec\theta = (\sqrt{9+x^2})/3$, and by substituting into the solution shown in Example 5, we find

$$\int x^2\sqrt{9+x^2}\, dx = 81\left[\frac{\sec^3\theta\tan\theta}{4} - \frac{\sec\theta\tan\theta}{8} - \frac{1}{8}\ln|\sec\theta+\tan\theta|\right] + C$$

$$= \frac{81}{4}\left(\frac{\sqrt{9+x^2}}{3}\right)^3\left(\frac{x}{3}\right) - \frac{81}{8}\left(\frac{\sqrt{9+x^2}}{3}\right)\left(\frac{x}{3}\right) - \frac{81}{8}\ln\left|\frac{\sqrt{9+x^2}}{3} + \frac{x}{3}\right| + C$$

$$= \frac{x}{4}(9+x^2)^{3/2} - \frac{9x}{8}(9+x^2)^{1/2} - \frac{81}{8}\ln\left|\frac{(9+x^2)^{1/2}}{3} + \frac{x}{3}\right| + C \qquad\blacksquare$$

EXAMPLE 8 **Trigonometric substitution with form $\sqrt{u^2 - a^2}$**

Evaluate $\displaystyle\int x^3 \sqrt{x^2 - 1}\, dx$.

Solution

Let $x = \sec\theta$, $dx = \sec\theta \tan\theta\, d\theta$; we use the reference triangle shown in Figure 7.3.

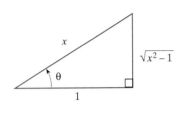

$$\int x^3 \sqrt{x^2 - 1}\, dx = \int \sec^3\theta \sqrt{\sec^2\theta - 1}\,(\sec\theta \tan\theta\, d\theta)$$

$$= \int \sec^4\theta \tan^2\theta\, d\theta \qquad \tan^2\theta = \sec^2\theta - 1$$

$$= \int \sec^2\theta \tan^2\theta\,(\sec^2\theta\, d\theta)$$

$$= \int (\tan^2\theta + 1)\tan^2\theta\,(\sec^2\theta\, d\theta)$$

$$= \int (\tan^4\theta + \tan^2\theta)(\sec^2\theta\, d\theta)$$

$$= \int (u^4 + u^2)\, du \qquad \text{Where } u = \tan\theta$$

$$= \tfrac{1}{5}u^5 + \tfrac{1}{3}u^3 + C$$

$$= \tfrac{1}{5}\tan^5\theta + \tfrac{1}{3}\tan^3\theta + C$$

$$= \tfrac{1}{5}(x^2 - 1)^{5/2} + \tfrac{1}{3}(x^2 - 1)^{3/2} + C \qquad\blacksquare$$

Figure 7.3 Reference triangle with form $\sqrt{x^2 - a^2}$

$\sec\theta = x,\ \tan\theta = \sqrt{x^2 - 1}$

(labels on triangle: x, $\sqrt{x^2-1}$, 1, θ)

QUADRATIC-FORM INTEGRALS

An integral involving an expression of the form $Ax^2 + Bx + C$, with $A \neq 0$, $B \neq 0$, can often be evaluated by completing the square and making an appropriate substitution to convert it to one of the forms we have previously analyzed.

EXAMPLE 9 **Integration by completing the square**

Evaluate $\displaystyle\int \sqrt{16x - 2x^2 - 23}\, dx$.

Solution

Complete the square within the radicand:

$$16x - 2x^2 - 23 = -2(x^2 - 8x \qquad) - 23$$
$$= -2(x^2 - 8x + 4^2) + 2\cdot 4^2 - 23$$
$$= -2(x - 4)^2 + 9$$

Thus,

$$\int \sqrt{16x - 2x^2 - 23}\, dx = \int \sqrt{9 - 2(x - 4)^2}\, dx$$

$$= \int \sqrt{9 - u^2}\left(\frac{du}{\sqrt{2}}\right) \qquad \text{Where } u = \sqrt{2}(x - 4)$$

$$= \int \sqrt{9 - (3\sin\theta)^2}\left(\frac{3}{\sqrt{2}}\cos\theta\, d\theta\right) \qquad \text{Where } u = 3\sin\theta$$

$$= \frac{3}{\sqrt{2}}\int 3\sqrt{1 - \sin^2\theta}\,\cos\theta\, d\theta$$

$$= \frac{9}{\sqrt{2}}\int \cos^2\theta\, d\theta$$

$$= \frac{9}{2\sqrt{2}}\left[\theta + \frac{\sin 2\theta}{2}\right] + C \qquad \text{Half-angle identity}$$

$$= \frac{9}{2\sqrt{2}} \sin^{-1}\left[\frac{\sqrt{2}}{3}(x-4)\right] + \frac{x-4}{2}\sqrt{16x - 2x^2 - 23} + C$$

This last step requires back-substituting from θ to u and then from u to x. Details are left as an exercise. ∎

7.3 PROBLEM SET

A **1. WHAT DOES THIS SAY?** Explain how to integrate $\int \sin^m x \cos^n x \, dx$ when m and n are both even.

2. WHAT DOES THIS SAY? Explain how to integrate $\int \tan^m x \sec^n x \, dx$ when n is even.

3. WHAT DOES THIS SAY? Explain the process of using a trigonometric substitution on integrals of the form $\sqrt{a^2 + u^2}$.

4. WHAT DOES THIS SAY? Explain the process of using a trigonometric substitution on integrals of the form $\sqrt{a^2 - u^2}$. How is this different from handling an integral involving $\sqrt{u^2 - a^2}$?

Evaluate the integrals in Problems 5–50.

5. $\int \cos^3 x \, dx$

6. $\int \sin^5 x \, dx$

7. $\int \sin^2 x \cos^3 x \, dx$

8. $\int \sin^3 x \cos^3 x \, dx$

9. $\int \sqrt{\cos t} \, \sin t \, dt$

10. $\int \frac{\cos x \, dx}{1 + 3\sin x}$

11. $\int e^{\cos x} \sin x \, dx$

12. $\int \cos^2(2t) \, dt$

13. $\int \sin^2 x \cos^2 x \, dx$

14. $\int \frac{\sin x \, dx}{\cos^5 x}$

15. $\int \tan 2\theta \, d\theta$

16. $\int \sec\left(\frac{x}{2}\right) dx$

17. $\int \tan^3 x \sec^4 x \, dx$

18. $\int \sec^5 x \tan x \, dx$

19. $\int (\tan^2 x + \sec^2 x) \, dx$

20. $\int (\sin x + \cos x)^2 \, dx$

21. $\int \tan^2 u \sec u \, du$

22. $\int \sec^4 x \, dx$

23. $\int \sqrt[3]{\tan x} \, \sec^2 x \, dx$

24. $\int e^x \sec(e^x) \, dx$

25. $\int x \sin x^2 \cos x^2 \, dx$

26. $\int x \sec^2 x \, dx$

27. $\int \tan^4 t \sec t \, dt$

28. $\int \csc(2\theta) \, d\theta$

29. $\int \csc^3 x \cot x \, dx$

30. $\int \csc^2 x \cot^2 x \, dx$

31. $\int \csc^2 x \cos x \, dx$

32. $\int \tan x \csc^3 x \, dx$

33. $\int \sqrt{4 - t^2} \, dt$

34. $\int \frac{dx}{\sqrt{9 - x^2}}$

35. $\int \frac{x+1}{\sqrt{4 + x^2}} \, dx$

36. $\int \sqrt{9 + x^2} \, dx$

37. $\int \frac{dx}{\sqrt{x^2 - 7}}$

38. $\int \frac{dx}{5 + 2x^2}$

39. $\int \frac{dx}{\sqrt{5 - x^2}}$

40. $\int \frac{dx}{x\sqrt{7x^2 - 4}}$

41. $\int \frac{dx}{x^2\sqrt{4 - x^2}}$

42. $\int \frac{dx}{x\sqrt{x^2 + 9}}$

43. $\int \frac{\sqrt{x^2 - 4}}{x} \, dx$

44. $\int \frac{dx}{(x-1)^2 + 4}$

45. $\int \frac{x \, dx}{9 - (x+1)^2}$

46. $\int \sqrt{2x - x^2} \, dx$

47. $\int \frac{dx}{\sqrt{x^2 - 2x + 6}}$

48. $\int \frac{dx}{\sqrt{x^2 + 8x + 3}}$

49. $\int \frac{\sin^3 u \, du}{\cos^5 u}$

50. $\int \frac{\sec^2 x \, dx}{\tan^2 x + \sec^2 x}$

B **51.** Find the average value of $f(x) = \sin^2 x$ over the interval $[0, \pi]$.

52. Find the centroid of the region (correct to two decimal places) bounded by the curve $y = \cos^2 x$, the x-axis, and the vertical lines $x = \frac{\pi}{4}$ and $x = \frac{\pi}{3}$.

53. Find the volume (correct to four decimal places) of the solid generated when the region bounded by the curve $y = \sin^2 x$ and the x-axis is revolved about the y-axis $0 \le x \le \pi$.

54. A particle moves along the x-axis in such a way that the acceleration at time t is $a(t) = \sin^2 t$. What is the total distance traveled by the particle over the time interval $[0, \pi]$ if its initial velocity is $v(0) = 2$ units per second?

55. Evaluate $\int \sqrt{1 + \cos x} \, dx$.

Hint: Use the identity $\cos x = 2\cos^2 \frac{x}{2} - 1$.

In Problems 56–59, use the following identities:

$$\sin A \cos B = \frac{1}{2}[\sin(A - B) + \sin(A + B)]$$
$$\sin A \sin B = \frac{1}{2}[\cos(A - B) - \cos(A + B)]$$
$$\cos A \cos B = \frac{1}{2}[\cos(A - B) + \cos(A + B)]$$

56. $\int \sin 3x \sin 5x \, dx$

57. $\int \cos \frac{x}{2} \sin 2x \, dx$

58. $\int \cos 7x \cos(-3x) \sin 4x \, dx$

59. $\int \sin^2 3x \cos 4x \, dx$

60. Let f be a twice differentiable function that satisfies the initial value problem

$$f''(x) = \frac{1}{2}(\tan x)f'(x) \qquad f'(0) = f(0) = 1$$

on the interval $\left[0, \frac{\pi}{2}\right]$. Find the arc length of the curve $y = f(x)$ over this interval.

C **61.** Evaluate $\int \frac{x \, dx}{9 - x^2 - \sqrt{9 - x^2}}$.

7.4 Method of Partial Fractions

IN THIS SECTION partial fraction decomposition, integrating rational functions, rational functions of sine and cosine

PARTIAL FRACTION DECOMPOSITION

You are familiar with the algebraic procedure of adding a string of rational expressions to form a combined rational function with a common denominator. For example,

$$\frac{2}{x+1} + \frac{-3}{x+2} = \frac{2(x+2)+(-3)(x+1)}{(x+1)(x+2)} = \frac{-x+1}{x^2+3x+2}$$

In **partial fraction decomposition**, we do just the opposite: We start with the reduced fraction

$$\frac{-x+1}{x^2+3x+2}$$

and write it as the sum of fractions

$$\frac{2}{x+1} + \frac{-3}{x+2}$$

This procedure has great value for integration because the terms $\dfrac{2}{x+1}$ and $\dfrac{-3}{x+2}$ are easy to integrate. In particular,

$$\int \frac{-x+1}{x^2+3x+2}\,dx = \int \frac{2}{x+1}\,dx + \int \frac{-3}{x+2}\,dx$$
$$= 2\ln|x+1| - 3\ln|x+2| + C$$

In the following discussion, we will consider rational functions

$$f(x) = \frac{P(x)}{D(x)}$$

that are reduced in the sense that $P(x)$ and $D(x)$ are polynomials in x which have no common factors and the degree of P is less than the degree of D. In algebra, it is shown that if $P(x)/D(x)$ is such an expression, then

$$\frac{P(x)}{D(x)} = F_1(x) + F_2(x) + \cdots + F_N(x)$$

where the $F_k(x)$ are expressions of the form

$$\frac{A}{(x-r)^n} \quad \text{or} \quad \frac{Ax+B}{(x^2+sx+t)^m}$$

If $\dfrac{P(x)}{D(x)}$ is not reduced, then we simply divide until a reduced form is obtained. For example ($x \neq 1$),

$$\frac{(2x^3+7x^2+6x+3)(x-1)}{(x-1)(x^2+3x+2)} = 2x+1+\frac{-x+1}{x^2+3x+2}$$
$$= 2x+1+\frac{2}{x+1}+\frac{-3}{x+2}$$

TECHNOLOGY NOTE

Software programs will decompose rational expressions quite handily using the direction "expand."

We will examine a nonreduced rational expression in Example 5.

We begin by focusing on the case where $D(x)$ can be expressed as a product of linear powers.

Partial Fraction Decomposition: A Single Linear Power

Let $f(x) = P(x)/(x-r)^n$, where $P(x)$ is a polynomial of degree less than n and $P(r) \neq 0$. Then $f(x)$ can be **decomposed** into partial fractions in the following "cascading form"

$$\frac{A_1}{x-r} + \frac{A_2}{(x-r)^2} + \cdots + \frac{A_n}{(x-r)^n}$$

EXAMPLE 1 Partial fraction decomposition with a single linear power

Decompose $\dfrac{x^2 - 6x + 3}{(x-2)^3}$ into a sum of partial fractions.

Solution $\dfrac{x^2 - 6x + 3}{(x-2)^3} = \dfrac{A_1}{x-2} + \dfrac{A_2}{(x-2)^2} + \dfrac{A_3}{(x-2)^3}$

Multiply both sides by $(x-2)^3$ to obtain

$$x^2 - 6x + 3 = A_1(x-2)^2 + A_2(x-2) + A_3$$
$$\text{Let } x = 2: \quad 2^2 - 6(2) + 3 = A_1(0) + A_2(0) + A_3$$
$$-5 = A_3$$

Substitute $A_3 = -5$ and expand the right side to obtain

$$x^2 - 6x + 3 = A_1 x^2 + (-4A_1 + A_2)x + (4A_1 - 2A_2 - 5)$$

This implies (by equating the coefficients of the similar terms) that

$$
\begin{aligned}
1 &= A_1 & & x^2 \textbf{ terms} \\
-6 &= -4A_1 + A_2 & & x \textbf{ terms} \\
3 &= 4A_1 - 2A_2 - 5 & & \textbf{Constants}
\end{aligned}
$$

Because $A_1 = 1$, we find $A_2 = -2$, so the decomposition is

$$\frac{x^2 - 6x + 3}{(x-2)^3} = \frac{1}{x-2} + \frac{-2}{(x-2)^2} + \frac{-5}{(x-2)^3} \qquad\blacksquare$$

If there are two or more linear factors in the factorization of $D(x)$, there must be a separate cascade for each power. In particular, if $D(x)$ can be expressed as the product of n distinct linear factors, then

$$\frac{P(x)}{(x-r_1)(x-r_2)\cdots(x-r_n)}$$

is decomposed into separate terms

$$\frac{A_1}{x-r_1} + \frac{A_2}{x-r_2} + \cdots + \frac{A_n}{x-r_n}$$

This process is illustrated with the following example.

EXAMPLE 2 Partial fraction decomposition with distinct linear factors

Decompose $\dfrac{8x - 1}{x^2 - x - 2}$.

Solution

$$\frac{8x-1}{x^2-x-2}=\frac{8x-1}{(x-2)(x+1)}$$ First factor the denominator.

$$=\frac{A_1}{x-2}+\frac{A_2}{x+1}$$ Break up the fraction into parts, each with a linear denominator. The task is to find A_1 and A_2.

$$=\frac{A_1(x+1)+A_2(x-2)}{(x-2)(x+1)}$$ Obtain the least common denominator on the right.

WARNING   Note that the degree of the denominator is the same number as the number of arbitrary constants, A_1 and A_2. This provides a quick intermediate check on the correct procedure.

Now, multiply both sides of this equation by the least common denominator, which is $(x-2)(x+1)$ for this example.

$$8x-1=A_1(x+1)+A_2(x-2)$$

Substitute, one at a time, the values that cause each of the factors in the least common denominator to be zero.

Let $x=-1$:
$$8x-1=A_1(x+1)+A_2(x-2)$$
$$8(-1)-1=A_1(-1+1)+A_2(-1-2)$$
$$-9=-3A_2$$
$$3=A_2$$

Let $x=2$:
$$8x-1=A_1(x+1)+A_2(x-2)$$
$$8(2)-1=A_1(2+1)+A_2(2-2)$$
$$15=3A_1$$
$$5=A_1$$

Thus, $\dfrac{8x-1}{x^2-x-2}=\dfrac{5}{x-2}+\dfrac{3}{x+1}$.

If there is a mixture of distinct and repeated linear factors, we combine the procedures illustrated in the preceding examples. For example,

$$\frac{5x^2+21x+4}{(x+1)^3(x-3)}\quad\text{is decomposed as}\quad\frac{A_1}{x+1}+\frac{A_2}{(x+1)^2}+\frac{A_3}{(x+1)^3}+\frac{A_4}{x-3}$$

Note that the degree of the denominator is 4, so we use four constants.

If the denominator $D(x)$ in the rational expression $P(x)/D(x)$ contains an irreducible quadratic power, the partial fraction decomposition contains a different kind of cascading sum.

Partial Fraction Decomposition: A Single Irreducible Quadratic Factor

Let $f(x)=P(x)/(x^2+sx+t)^m$, where $P(x)$ is a polynomial of degree less than $2m$. Then $f(x)$ can be decomposed into partial fractions in the following "cascading form":

$$\frac{A_1x+B_1}{x^2+sx+t}+\frac{A_2x+B_2}{(x^2+sx+t)^2}+\cdots+\frac{A_mx+B_m}{(x^2+sx+t)^m}$$

Because the degree of the denominator is $2m$, we have $2m$ arbitrary constants, namely, $A_1,A_2,\ldots,A_m,B_1,B_2,\ldots,B_m$.

WARNING   Note that the numerator in each term of a linear cascade is a constant A_k, while each numerator in a quadratic cascade is a linear term of the form A_kx+B_k.

EXAMPLE 3   **Partial fraction decomposition with a single quadratic power**

Decompose $\dfrac{-3x^3-x}{(x^2+1)^2}$.

Solution

The decomposition gives

$$\frac{-3x^3 - x}{(x^2 + 1)^2} = \frac{A_1 x + B_1}{x^2 + 1} + \frac{A_2 x + B_2}{(x^2 + 1)^2}$$

Multiply both sides of this equation by $(x^2 + 1)^2$, and simplify algebraically.

$$\begin{aligned}
-3x^3 - x &= (A_1 x + B_1)(x^2 + 1) + (A_2 x + B_2) \\
&= A_1 x^3 + B_1 x^2 + (A_1 + A_2)x + (B_1 + B_2)
\end{aligned}$$

Next, equate the corresponding coefficients on each side of this equation and solve the resulting system of equations to find $A_1 = -3$, $A_2 = 2$, $B_1 = 0$, and $B_2 = 0$. This means that

$$\frac{-3x^3 - x}{(x^2 + 1)^2} = \frac{-3x}{x^2 + 1} + \frac{2x}{(x^2 + 1)^2} \qquad \blacksquare$$

Many of the examples we encounter will offer a mixture of linear and quadratic factors. For example,

$$\frac{x^2 + 4x - 23}{(x^2 + 4)(x + 3)^2} \quad \text{is decomposed as} \quad \frac{A_1 x + B_1}{x^2 + 4} + \frac{A_2}{x + 3} + \frac{A_3}{(x + 3)^2}$$

The degree of the denominator is four, and there are four constants.

In algebra, the theory of equations tells us that any polynomial P with real coefficients can be expressed as a product of linear and irreducible quadratic powers, some of which may be repeated. This fact can be used to justify the following general procedure for obtaining the partial fraction decomposition of a rational function.

Partial Fraction Decomposition of the Rational Function $P(x)/D(x)$

Let $f(x) = \dfrac{P(x)}{D(x)}$, where $P(x)$ and $D(x)$ are polynomials with no common factors and $D(x) \neq 0$.

Step 1.   If the degree of P is greater than or equal to the degree of D, use long (or synthetic) division to express $\dfrac{P(x)}{D(x)}$ as the sum of a polynomial and a fraction $\dfrac{R(x)}{D(x)}$ in which the degree of the remainder polynomial $R(x)$ is less than the degree of the denominator polynomial $D(x)$.

Step 2.   Factor the denominator $D(x)$ into the product of linear and irreducible quadratic powers.

Step 3.   Express $\dfrac{P(x)}{D(x)}$ as a cascading sum of partial fractions of the form

$$\frac{A_i}{(x - r)^n} \quad \text{and} \quad \frac{A_k x + B_k}{(x^2 + sx + t)^m}.$$

INTEGRATING RATIONAL FUNCTIONS

We will now apply the procedure of partial fraction decomposition to integration.

EXAMPLE 4   Integrating a rational function with a repeated linear factor

Find $\displaystyle\int \frac{x^2 - 6x + 3}{(x - 2)^3}\, dx.$

Solution

From Example 1, we have

$$\int \frac{x^2 - 6x + 3}{(x-2)^3}\, dx = \int \left[\frac{1}{x-2} + \frac{-2}{(x-2)^2} + \frac{-5}{(x-2)^3} \right] dx$$

$$= \int (x-2)^{-1}\, dx - 2 \int (x-2)^{-2}\, dx - 5 \int (x-2)^{-3}\, dx$$

$$= \ln|x-2| + \frac{2}{x-2} + \frac{5}{2(x-2)^2} + C \qquad \blacksquare$$

EXAMPLE 5 **Integrating a rational expression with distinct linear factors**

Find $\displaystyle \int \frac{x^4 + 2x^3 - 4x^2 + x - 3}{x^2 - x - 2}\, dx.$

Solution

We have $P(x) = x^4 + 2x^3 - 4x^2 + x - 3$ and $D(x) = x^2 - x - 2$; because the degree of P is greater than the degree of D, we carry out the long division and write

$$\int \frac{x^4 + 2x^3 - 4x^2 + x - 3}{x^2 - x - 2}\, dx = \int \left(x^2 + 3x + 1 + \frac{8x - 1}{x^2 - x - 2} \right) dx$$

The polynomial part is easy to integrate. The rational expression was decomposed into partial fractions in Example 2.

$$\int \frac{x^4 + 2x^3 - 4x^2 + x - 3}{x^2 - x - 2}\, dx = \int \left(x^2 + 3x + 1 + \frac{8x - 1}{x^2 - x - 2} \right) dx$$

$$= \int \left(x^2 + 3x + 1 + \frac{5}{x-2} + \frac{3}{x+1} \right) dx$$

$$= \int x^2\, dx + 3 \int x\, dx + \int dx + 5 \int (x-2)^{-1}\, dx + 3 \int (x+1)^{-1}\, dx$$

$$= \frac{x^3}{3} + \frac{3x^2}{2} + x + 5 \ln|x-2| + 3 \ln|x+1| + C \qquad \blacksquare$$

EXAMPLE 6 **Integrating a rational function with repeated quadratic factors**

Find $\displaystyle \int \frac{-3x^3 - x}{(x^2 + 1)^2}\, dx.$

Solution

From Example 3,

$$\int \frac{-3x^3 - x}{(x^2 + 1)^2}\, dx = \int \frac{2x}{(x^2 + 1)^2}\, dx + \int \frac{-3x}{x^2 + 1}\, dx$$

$$\boxed{u = x^2 + 1;\; du = 2x\, dx}$$

$$= \int u^{-2}\, du - \frac{3}{2} \int u^{-1}\, du$$

$$= -u^{-1} - \frac{3}{2} \ln|u| + C$$

$$= \frac{-1}{x^2 + 1} - \frac{3}{2} \ln(x^2 + 1) + C \qquad \blacksquare$$

EXAMPLE 7 Repeated linear factors

Find $\displaystyle\int \frac{5x^2 + 21x + 4}{(x+1)^2(x-3)}\, dx$.

Solution

The partial fraction decomposition of the integrand is

$$\frac{5x^2 + 21x + 4}{(x+1)^2(x-3)} = \frac{A_1}{(x+1)^2} + \frac{A_2}{x+1} + \frac{A_3}{x-3}$$

Multiply both sides by $(x+1)^2(x-3)$:

$$5x^2 + 21x + 4 = A_1(x-3) + A_2(x+1)(x-3) + A_3(x+1)^2$$

As in Example 2, we substitute $x = -1$ and $x = 3$ on both sides of the equation to obtain $A_1 = 3$ and $A_3 = 7$, but A_2 cannot be obtained in this fashion. To find A_2, multiply out the polynomial on the right:

$$5x^2 + 21x + 4 = (A_2 + A_3)x^2 + (A_1 - 2A_2 + 2A_3)x + (-3A_1 - 3A_2 + A_3)$$

Equate the coefficients of x^2, x, and 1 (the constant term) on each side of the equation:

$$
\begin{aligned}
5 &= A_2 + A_3 && x^2 \text{ terms} \\
21 &= A_1 - 2A_2 + 2A_3 && x \text{ terms} \\
4 &= -3A_1 - 3A_2 + A_3 && \text{Constants}
\end{aligned}
$$

Because we already know that $A_1 = 3$ and $A_3 = 7$, we use the equation $5 = A_2 + A_3$ to obtain $A_2 = -2$. We now turn to the integration:

$$\int \frac{5x^2 + 21x + 4}{(x+1)^2(x-3)}\, dx = \int \frac{3\, dx}{(x+1)^2} + \int \frac{-2\, dx}{x+1} + \int \frac{7\, dx}{x-3}$$

$$= -3(x+1)^{-1} - 2\ln|x+1| + 7\ln|x-3| + C \quad \blacksquare$$

EXAMPLE 8 Distinct linear and quadratic factors

Find $\displaystyle\int \frac{x^2 + 4x - 23}{(x^2+4)(x+3)}\, dx$.

Solution

The partial fraction decomposition of the integrand has the form

$$\frac{x^2 + 4x - 23}{(x^2+4)(x+3)} = \frac{A_1 x + B_1}{x^2+4} + \frac{A_2}{x+3}$$

Multiply both sides by $(x^2+4)(x+3)$ and then combine the terms on the right:

$$
\begin{aligned}
x^2 + 4x - 23 &= (A_1 x + B_1)(x+3) + A_2(x^2+4) \\
&= (A_1 + A_2)x^2 + (3A_1 + B_1)x + (3B_1 + 4A_2)
\end{aligned}
$$

Equate the coefficients to set up the following system of equations:

$$
\begin{cases}
A_1 + A_2 = 1 & x^2 \text{ terms} \\
3A_1 + B_1 = 4 & x \text{ terms} \\
3B_1 + 4A_2 = -23 & \text{Constants}
\end{cases}
$$

Solve this system (the details are not shown) to find $A_1 = 3$, $B_1 = -5$, and $A_2 = -2$. We now turn to the integration:

$$\int \frac{x^2 + 4x - 23}{(x^2+4)(x+3)}\, dx = \int \frac{3x-5}{x^2+4}\, dx + \int \frac{-2\, dx}{x+3}$$

$$= 3 \int \frac{x \, dx}{x^2 + 4} - 5 \int \frac{dx}{x^2 + 4} - 2 \int \frac{dx}{x + 3}$$

$$\uparrow$$

$$\boxed{u = x^2 + 4; \; du = 2x \, dx}$$

$$= 3 \left[\frac{1}{2} \ln(x^2 + 4) \right] - 5 \left[\frac{1}{2} \tan^{-1} \left(\frac{x}{2} \right) \right] - 2 \ln |x + 3| + C \quad \blacksquare$$

RATIONAL FUNCTIONS OF SINE AND COSINE

The German mathematician Karl Weierstrass (1815–1897) noticed that the substitution
$$u = \tan \frac{x}{2} \qquad -\pi < x < \pi$$
will convert any rational function of $\sin x$ and $\cos x$ into a rational function of x. To see why this is true, we use the double-angle identities for sine and cosine (see Figure 7.4):

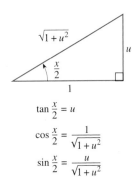

$$\tan \frac{x}{2} = u$$

$$\cos \frac{x}{2} = \frac{1}{\sqrt{1 + u^2}}$$

$$\sin \frac{x}{2} = \frac{u}{\sqrt{1 + u^2}}$$

Figure 7.4 Weierstrass substitution

$$\sin x = 2 \sin \frac{x}{2} \cos \frac{x}{2} \qquad \text{and} \qquad \cos x = \cos^2 \frac{x}{2} - \sin^2 \frac{x}{2}$$

$$= 2 \left(\frac{u}{\sqrt{1 + u^2}} \right) \left(\frac{1}{\sqrt{1 + u^2}} \right) \qquad\qquad = \left(\frac{1}{\sqrt{1 + u^2}} \right)^2 - \left(\frac{u}{\sqrt{1 + u^2}} \right)^2$$

$$= \frac{2u}{1 + u^2} \qquad\qquad\qquad\qquad\qquad = \frac{1 - u^2}{1 + u^2}$$

Finally, because $u = \tan \frac{x}{2}$, we have $x = 2 \tan^{-1} u$, so $dx = \frac{2}{1 + u^2} \, du$.

Weierstrass Substitution

For $-\pi < x < \pi$, let $u = \tan \frac{x}{2}$ so that

$$\sin x = \frac{2u}{1 + u^2}, \quad \cos x = \frac{1 - u^2}{1 + u^2}, \quad \text{and} \quad dx = \frac{2}{1 + u^2} \, du$$

This is called the **Weierstrass substitution**.

The following example illustrates how this substitution can be used along with partial fractions to integrate a rational trigonometric function.

EXAMPLE 9 Integrating a rational trigonometric function

Find $\displaystyle\int \frac{dx}{3 \cos x - 4 \sin x}$.

Solution

Use the Weierstrass substitution—that is, let $u = \tan \frac{x}{2}$. Remember, $dx = \frac{2 \, du}{1 + u^2}$;
$\cos x = \frac{1 - u^2}{1 + u^2}$; and $\sin x = \frac{2u}{1 + u^2}$.

$$\int \frac{dx}{3 \cos x - 4 \sin x} = \int \frac{\dfrac{2 \, du}{1 + u^2}}{3 \left(\dfrac{1 - u^2}{1 + u^2} \right) - 4 \left(\dfrac{2u}{1 + u^2} \right)} \qquad \text{Simplify.}$$

$$= \int \frac{-2 \, du}{3u^2 + 8u - 3} = \int \frac{-2 \, du}{(3u - 1)(u + 3)}$$

This integral can be handled by the method of partial fractions.

$$\frac{-2}{(3u-1)(u+3)} = \frac{A_1}{3u-1} + \frac{A_2}{u+3}$$

Solve this to find $A_1 = -\frac{3}{5}$ and $A_2 = \frac{1}{5}$. We continue with the integration.

$$\int \frac{dx}{3\cos x - 4\sin x} = \int \frac{-2\,du}{(3u-1)(u+3)}$$

$$= \int \frac{-\frac{3}{5}\,du}{3u-1} + \int \frac{\frac{1}{5}\,du}{u+3}$$

$$= -\frac{3}{5}\cdot\frac{1}{3}\ln|3u-1| + \frac{1}{5}\cdot\ln|u+3| + C$$

$$= -\frac{1}{5}\ln\left|3\tan\frac{x}{2}-1\right| + \frac{1}{5}\ln\left|\tan\frac{x}{2}+3\right| + C \qquad ■$$

Once again, observe that when carrying out integration, you may obtain very different forms for the result. For Example 9, you might use an integration table (Formula 393 in the *Student Mathematics Handbook*, for example) to find

$$\int \frac{du}{p\sin au + q\cos au} = \frac{1}{a\sqrt{p^2+q^2}}\ln\left|\tan\left(\frac{au + \tan^{-1}\left(\frac{q}{p}\right)}{2}\right)\right|$$

Let $p = -4$, $q = 3$, $a = 1$, so that

$$\int \frac{dx}{3\cos x - 4\sin x} = \frac{1}{5}\ln\left|\tan\left(\frac{x + \tan^{-1}\left(-\frac{3}{4}\right)}{2}\right)\right| + C$$

Problem 67 asks you to derive the formula

$$\int \sec x\,dx = \ln|\sec x + \tan x| + C$$

from scratch. You might recall that we derived this formula using an unusual algebraic step in Example 3 of Section 7.1. You can now derive it by using a Weierstrass substitution.

7.4 PROBLEM SET

A *Write each rational function given in Problems 1–14 as a sum of partial fractions.*

1. $\dfrac{1}{x(x-3)}$

2. $\dfrac{3x-1}{x^2-1}$

3. $\dfrac{3x^2+2x-1}{x(x+1)}$

4. $\dfrac{2x^2+5x-1}{x(x^2-1)}$

5. $\dfrac{4}{2x^2+x}$

6. $\dfrac{x^2-x+3}{x^2(x-1)}$

7. $\dfrac{4x^3+4x^2+x-1}{x^2(x+1)^2}$

8. $\dfrac{x^2-5x-4}{(x^2+1)(x-3)}$

9. $\dfrac{x^3+3x^2+3x-4}{x^2(x+3)^2}$

10. $\dfrac{1}{x^3-1}$

11. $\dfrac{1}{1-x^4}$

12. $\dfrac{x^4-x^2+2}{x^2(x-1)}$

13. $\dfrac{x^2+x-1}{x(x+1)(2x-1)}$

14. $\dfrac{x^3-2x^2+x-5}{x(x^2-1)(3x+5)}$

Compute the integrals given in Problems 15–20. Notice that in each case, the integrand is a rational function decomposed into partial fractions in Problems 1–6.

15. $\displaystyle\int \frac{dx}{x(x-3)}$

16. $\displaystyle\int \frac{3x-1}{x^2-1}\,dx$

17. $\displaystyle\int \frac{3x^2+2x-1}{x(x+1)}\,dx$

18. $\displaystyle\int \frac{2x^2+5x-1}{x(x^2-1)}\,dx$

19. $\displaystyle\int \frac{4\,dx}{2x^2+x}$

20. $\displaystyle\int \frac{x^2-x+3}{x^2(x-1)}\,dx$

Find the indicated integrals in Problems 21–36.

21. $\displaystyle\int \frac{2x^3 + 9x - 1}{x^2(x^2 - 1)}\,dx$ **22.** $\displaystyle\int \frac{x^4 - x^2 + 2}{x^2(x - 1)}\,dx$

23. $\displaystyle\int \frac{x^2 + 1}{x^2 + x - 2}\,dx$ **24.** $\displaystyle\int \frac{dx}{x^3 - 8}$

25. $\displaystyle\int \frac{x^4 + 1}{x^4 - 1}\,dx$ **26.** $\displaystyle\int \frac{x^3 + 1}{x^3 - 1}\,dx$

27. $\displaystyle\int \frac{x\,dx}{(x + 1)^2}$ **28.** $\displaystyle\int \frac{2x\,dx}{(x - 2)^2}$

29. $\displaystyle\int \frac{dx}{x(x + 1)(x - 2)}$ **30.** $\displaystyle\int \frac{x + 2}{x(x - 1)^2}\,dx$

31. $\displaystyle\int \frac{x\,dx}{(x + 1)(x + 2)^2}$ **32.** $\displaystyle\int \frac{x + 1}{x(x^2 + 2)}\,dx$

33. $\displaystyle\int \frac{5x + 7}{x^2 + 2x - 3}\,dx$ **34.** $\displaystyle\int \frac{5x\,dx}{x^2 - 6x + 9}$

35. $\displaystyle\int \frac{3x^2 - 2x + 4}{x^3 - x^2 + 4x - 4}\,dx$ **36.** $\displaystyle\int \frac{3x^2 + 4x + 1}{x^3 + 2x^2 + x - 2}\,dx$

B 37. WHAT DOES THIS SAY? Describe the process of partial fraction decomposition.

Find the indicated integrals in Problems 38–55.

38. $\displaystyle\int \frac{\cos x\,dx}{\sin^2 x - \sin x - 2}$ **39.** $\displaystyle\int \frac{e^x\,dx}{2e^{2x} - 5e^x - 3}$

40. $\displaystyle\int \frac{e^x\,dx}{e^{2x} - 1}$ **41.** $\displaystyle\int \frac{\sin x\,dx}{(1 + \cos x)^2}$

42. $\displaystyle\int \frac{\tan x\,dx}{\sec^2 x + 4}$ **43.** $\displaystyle\int \frac{\sec^2 x\,dx}{\tan x + 4}$

44. $\displaystyle\int \frac{dx}{x^{1/4} - x}$ **45.** $\displaystyle\int \frac{dx}{x^{2/3} - x^{1/2}}$

46. $\displaystyle\int \frac{dx}{\sin x - \cos x}$ **47.** $\displaystyle\int \frac{dx}{3\cos x + 4\sin x}$

48. $\displaystyle\int \frac{dx}{5\sin x + 4}$ **49.** $\displaystyle\int \frac{\sin x - \cos x}{\sin x + \cos x}\,dx$

50. $\displaystyle\int \frac{dx}{4\cos x + 5}$ **51.** $\displaystyle\int \frac{dx}{\sec x - \tan x}$

52. $\displaystyle\int \frac{dx}{3\sin x + 4\cos x + 5}$ **53.** $\displaystyle\int \frac{dx}{4\sin x - 3\cos x - 5}$

54. $\displaystyle\int \frac{dx}{2\csc x - \cot x + 2}$ **55.** $\displaystyle\int \frac{dx}{x(3 - \ln x)(1 - \ln x)}$

56. Find the area under the curve $y = \dfrac{1}{x^2 + 5x + 4}$ between $x = 0$ and $x = 3$.

57. Find the area of the region bounded by the curve $y = \dfrac{1}{6 - 5x + x^2}$ and the lines $x = \frac{4}{3}$, $x = \frac{7}{4}$, and $y = 0$.

58. Find the volume (to four decimal places) of the solid generated when the curve
$$y = \frac{1}{x^2 + 5x + 4}, \qquad 0 \le x \le 1$$
is revolved about
a. the y-axis **b.** the x-axis **c.** the line $x = -1$

59. Find the volume (to four decimal places) of the solid generated when the region under the curve
$$y = \frac{1}{\sqrt{x^2 + 4x + 3}}$$
on the interval $[0, 3]$ is revolved about
a. the x-axis **b.** the y-axis

60. Find both the exact and approximate volume of the solid generated when the region under the curve $y^2 = x^2\left(\dfrac{4 - x}{4 + x}\right)$ on the interval $[0, 4]$ is revolved about the x-axis.

61. HISTORICAL QUEST George Pólya was born in Hungary and attended universities in Budapest, Vienna, Göttingen, and Paris. He was a professor of mathematics at Stanford University. Pólya's research and winning personality earned him a place of honor not only among mathematicians, but among students and teachers as well. His discoveries spanned an impressive range of mathematics, including real and complex analysis, probability, combinatorics, number theory, and geometry. Pólya's book, *How to Solve It*, has been translated into 20 languages. His books have a clarity and elegance seldom seen in mathematics, making them a joy to read. For example, here is his explanation of why he was a mathematician: "It is a little shortened but not quite wrong to say: I thought I am not good enough for physics and I am too good for philosophy. Mathematics is in between." ∎

GEORGE PÓLYA
1887–1985

A story told by Pólya provides our next Quest. "A number of years ago," Pólya related with his lovable accent, "I deliberately put the problem

$$\int \frac{x\,dx}{x^2 - 9}$$

as the first problem on a test of techniques of integration, to give my students a boost as they began the exam. With the substitution $u = x^2 - 9$, which I expected the students to use, you can knock the problem off in just a few seconds. Half of the students did this, and got off to a good start. But a fourth of them used the correct but time-consuming procedure of partial fractions—and because they spent so much time on the problem, they did poorly on the exam. Half of the rest used the trig substitution $x = 3\sin\theta$—also correct but so time-consuming that they wound up very far behind and bombed the exam. It is interesting that the students who used the harder techniques showed they knew 'more,' or at least more difficult mathematics than the ones who used the easy technique. But they showed that 'it's not just what you know; it's how and when you use it.' It's nice when what you do is right, but it's much better when it's also *appropriate*."* Carry out all three methods of solution of the given integral Pólya described in this quotation.

---

*"Pólya, Problem Solving, and Education," by Alan H. Schoenfeld, *Mathematics Magazine*, Vol. 60, No. 5, December 1987, p. 290.

62. Spy Problem After waiting hours at the sleazy bar for his superior, Lord Newton Fleming (recall Problem 55, Section 6.6), the Spy is about to leave when a courier arrives with shocking news—Lord Newton has been kidnapped by Ernst Stavro Blohardt! The Spy hurries to the village outside the thug's château disguised as an old duck plucker. On the day he arrives, his true identity is known only to the Redselig twins, Hans and Franz, but the next day, Hans' girlfriend, Blabba, finds out. Soon, word begins to circulate among the 60 citizens of the village as to the Spy's true identity. He determines that the rate at which the rumor concerning his identity will spread is jointly proportional to the product of the number N of those who already know and the number $60 - N$ of those who do not, so that

$$\frac{dN}{dt} = kN(60 - N)$$

for a constant k. He needs a week to get the information to break into the château to free Lord Newton, but figures that as soon as 20 or more villagers know his identity, Blohardt is sure to find out, too. Does he complete his mission, or is he about to be a dead duck plucker?

63. Two substances, A and B, are being converted into a single compound C. In the laboratory, it is shown that the time rate of change of the amount x of compound C is proportional to the product of the amount of unconverted substances A and B. Thus,

$$\frac{dx}{dt} = k(a - x)(b - x)$$

for initial concentrations a and b of A and B, respectively.

a. Solve this differential equation to express x in terms of time t. What happens to $x(t)$ as $t \to +\infty$ if $b > a$? What if $a > b$?

b. Suppose $a = b$. What is $x(t)$? What happens to $x(t)$ as $t \to +\infty$ in this case?

64. Use partial fractions to derive the integration formula

$$\int \frac{dx}{a^2 - x^2} = \frac{1}{2a} \ln \left| \frac{a + x}{a - x} \right| + C$$

65. Use partial fractions to derive the integration formula

$$\int \frac{dx}{x(ax + b)} = \frac{1}{b} \ln \left| \frac{x}{ax + b} \right| + C$$

66. Exploration Problem Consider the integral $\int \sqrt{1 - \cos x}\, dx$.

a. If you have access to a CAS, report on your attempts to find this integral.

b. Use a Weierstrass substitution to find this integral.

c. Can you think of an even more clever way to perform the integration?

C **67.** Derive the formula

$$\int \sec x\, dx = \ln |\sec x + \tan x| + C$$

using a Weierstrass substitution.

68. Derive the formula

$$\int \csc x\, dx = -\ln |\csc x + \cot x| + C$$

using a Weierstrass substitution.

7.5 Summary of Integration Techniques

IN THIS SECTION integration strategy

We conclude our study of integration techniques with a table summarizing integration techniques.

TABLE 7.2 Integration Strategy

Step 1. Simplify.	Simplify the integrand, if possible, and use one of the procedural rules (see the inside back cover or the table of integrals).
Step 2. Use basic formulas.	Use the basic integration formulas (1–29 in the integration table in Appendix D and the *Student Mathematics Handbook*). These are the fundamental building blocks for integration. Almost every integration will involve some basic formulas somewhere in the process, which means that you should memorize these formulas.
Step 3. Substitute.	Make any substitution that will transform the integral into one of the basic forms.
Step 4. Classify.	Classify the integrand according to form in order to use a table of integrals. You may need to use substitution to transform the integrand into a form contained in the integration table. Some special types of substitution are contained in the following list:

TABLE 7.2   Integration Strategy (Continued)

I. Integration by parts

    A. Forms $\int x^n e^{ax}\, dx$, $\int x^n \sin ax\, dx$, $\int x^n \cos ax\, dx$
       Let $u = x^n$.

    B. Forms $\int x^n \ln x\, dx$, $\int x^n \sin^{-1} ax\, dx$, $\int x^n \tan^{-1} ax\, dx$
       Let $dv = x^n\, dx$.

    C. Forms $\int e^{ax} \sin bx\, dx$, $\int e^{ax} \cos bx\, dx$
       Let $dv = e^{ax}\, dx$.

II. Trigonometric forms

    A. $\int \sin^m x \cos^n x\, dx$
       m odd: Peel off a factor of $(\sin x\, dx)$ and **let $u = \cos x$.**
       n odd: Peel off a factor of $(\cos x\, dx)$ and **let $u = \sin x$.**
       m and n both even: **Use half-angle identities:**

$$\cos^2 x = \tfrac{1}{2}(1 + \cos 2x) \quad \text{and} \quad \sin^2 x = \tfrac{1}{2}(1 - \cos 2x)$$

    B. $\int \tan^m x \sec^n x\, dx$
       n even: Peel off a factor of $(\sec^2 x\, dx)$ and **let $u = \tan x$.**
       m odd: Peel off a factor of $(\sec x \tan x\, dx)$ and **let $u = \sec x$.**
       m even, n odd: Write using powers of secant and **use a reduction formula** (Formula
       428 in the integration tables).

    C. For a rational trigonometric integral, try the Weierstrass substitution.

       Let $u = \tan \frac{x}{2}$, so that $\sin x = \dfrac{2u}{1+u^2}$, $\cos x = \dfrac{1-u^2}{1+u^2}$, and $dx = \dfrac{2}{1+u^2}\, du$.

III. Radical forms; try a trigonometric substitution.

    A. Form $\sqrt{a^2 - u^2}$: **Let $u = a \sin \theta$.**
    B. Form $\sqrt{a^2 + u^2}$: **Let $u = a \tan \theta$.**
    C. Form $\sqrt{u^2 - a^2}$: **Let $u = a \sec \theta$.**

IV. Rational forms; try partial fraction decomposition.

Step 5. Try again.

Still stuck?

1. Manipulate the integrand:
   Multiply by 1 (clever choice of numerator or denominator).
   Rationalize the numerator.
   Rationalize the denominator.
2. Relate the problem to a previously worked problem.
3. Look at another table of integrals or consult computer software that does integration.
4. Some integrals do not have simple antiderivatives, so all these methods may fail. We will
   look at some of these forms later in the text.
5. If dealing with a definite integral, an approximation may suffice. It may be appropriate to use
   a calculator, computer, or the techniques introduced in Section 5.8.

EXAMPLE 1   Deciding on an integration procedure

Indicate a procedure to set up each integral. It is not necessary to carry out the integration.

a. $\displaystyle\int \frac{\sin \sqrt{x}}{\sqrt{x}}\, dx$ **b.** $\displaystyle\int (1 + \tan^2 \theta)\, d\theta$ **c.** $\displaystyle\int \sin^3 x \cos^2 x\, dx$

d. $\displaystyle\int 4x^2 \cos 3x\, dx$ **e.** $\displaystyle\int \frac{x\, dx}{\sqrt{9 - x^2}}$ **f.** $\displaystyle\int \frac{x^2\, dx}{\sqrt{9 - x^2}}$

g. $\displaystyle\int e^{3x} \sin 2x\, dx$ **h.** $\displaystyle\int \frac{\cos^4 x\, dx}{1 - \sin^2 x}$

Solution

Keep in mind that there may be several correct approaches, and the way you proceed from problem to solution is often a matter of personal preference. However, to give you some practice, we will present several hints and suggestions in the context of this example.

a. $\displaystyle\int \frac{\sin \sqrt{x}}{\sqrt{x}}\, dx$

The integrand is simplified and is not a fundamental type. Substitution will work well for this problem if you let $u = \sqrt{x}$. After you make this substitution you can integrate using a basic formula.

b. $\displaystyle\int (1 + \tan^2 \theta)\, d\theta$

The integrand can be simplified using a trigonometric identity $(1+\tan^2 \theta = \sec^2 \theta)$, so that it can now be integrated using a basic formula.

c. $\displaystyle\int \sin^3 x \cos^2 x\, dx$

The integrand is simplified, it is not a basic formula, and there does not seem to be an easy substitution. Next, try to classify the integrand. We see that it involves powers of trigonometric functions of the type $\int \sin^m x \cos^n x\, dx$, where $m = 3$ (odd) and $n = 2$ (even), so we make the substitution $u = \cos x$.

d. $\displaystyle\int 4x^2 \cos 3x\, dx$

The integrand is simplified (you can bring the 4 out in front of the integral, if you wish), it is not a basic formula, and there is no easy substitution. You might try integration by parts (several steps) or you can check the table of integrals and find it to be Formula 313.

e. $\displaystyle\int \frac{x\, dx}{\sqrt{9 - x^2}}$

The integrand is simplified and is not a basic formula. However, we might try the substitution $u = 9 - x^2$. It looks like this will work because the degree of the numerator is one less than the degree of the denominator.

f. $\displaystyle\int \frac{x^2\, dx}{\sqrt{9 - x^2}}$

The integrand is simplified, it is not a basic formula, and it looks as though an algebraic substitution will not work because of the degree of the numerator. Since the integrand involves $\sqrt{9 - x^2}$, we use the trigonometric substitution $x = 3 \sin \theta$, $dx = 3 \cos \theta\, d\theta$. If the integral had been $\displaystyle\int \frac{x^2\, dx}{\sqrt{x^2 - 9}}$, you would have used the substitution $x = 3 \sec \theta$, $dx = 3 \tan \theta \sec \theta\, d\theta$.

g. $\displaystyle\int e^{3x} \sin 2x\, dx$

The integrand is simplified and it is not a basic formula. If we try to classify the integrand, we see that integration by parts will work with $u = \sin 2x$ and $dv = e^{3x}\, dx$. We also find that this form is Formula 492 in the integration table.

h. $\displaystyle\int \frac{\cos^4 x\, dx}{1 - \sin^2 x}$

The integration can be simplified by writing $1 - \sin^2 x = \cos^2 x$. After doing this substitution, you will obtain

$$\int \frac{\cos^4 x\, dx}{1 - \sin^2 x} = \int \frac{\cos^4 x\, dx}{\cos^2 x} = \int \cos^2 x\, dx$$

This form can be integrated by using a half-angle identity or Formula 317 in the table of integrals. It can also be integrated by parts. ∎

7.5 PROBLEM SET

A *Find each integral in Problems 1–54.*

1. $\displaystyle\int \frac{2x - 1}{(x - x^2)^3}\, dx$

2. $\displaystyle\int \frac{2x + 3}{\sqrt{x^2 + 3x}}\, dx$

3. $\displaystyle\int (x \sec 2x^2)\, dx$

4. $\displaystyle\int (x^2 \csc^2 2x^3)\, dx$

5. $\displaystyle\int (e^x \cot e^x)\, dx$

6. $\displaystyle\int \frac{\tan \sqrt{x}\, dx}{\sqrt{x}}$

7. $\displaystyle\int \frac{\tan(\ln x)\, dx}{x}$

8. $\displaystyle\int \sqrt{\cot x}\, \csc^2 x\, dx$

9. $\displaystyle\int \frac{(3 + 2 \sin t)}{\cos t}\, dt$

10. $\displaystyle\int \frac{2 + \cos x}{\sin x}\, dx$

11. $\displaystyle\int \frac{e^{2t}\, dt}{1 + e^{4t}}$

12. $\displaystyle\int \frac{\sin 2x\, dx}{1 + \sin^4 x}$

13. $\displaystyle\int \frac{x^2 + x + 1}{x^2 + 9}\, dx$

14. $\displaystyle\int \frac{3x + 2}{\sqrt{4 - x^2}}\, dx$

15. $\displaystyle\int \frac{1 + e^x}{1 - e^x}\, dx$

16. $\displaystyle\int \frac{e^{1 - \sqrt{x}}\, dx}{\sqrt{x}}$

17. $\displaystyle\int \frac{2t^2\, dt}{\sqrt{1 - t^6}}$

18. $\displaystyle\int \frac{t^3\, dt}{2^8 + t^8}$

19. $\displaystyle\int \frac{dx}{1 + e^{2x}}$

20. $\displaystyle\int \frac{dx}{4 - e^{-x}}$

21. $\displaystyle\int \frac{dx}{x^2 + 2x + 2}$

22. $\displaystyle\int \frac{dx}{x^2 + x + 4}$

23. $\displaystyle\int \frac{dx}{x^2 + x + 1}$

24. $\displaystyle\int \frac{dx}{x^2 - x + 1}$

25. $\displaystyle\int \tan^{-1} x\, dx$

26. $\displaystyle\int x^3 \sin x^2\, dx$

27. $\displaystyle\int e^{-x} \cos x\, dx$

28. $\displaystyle\int e^{2x} \sin 3x\, dx$

29. $\displaystyle\int \cos^{-1}(-x)\, dx$

30. $\displaystyle\int x \sec^{-1} x\, dx,\ x > 0$

31. $\displaystyle\int \sin^3 x\, dx$

32. $\displaystyle\int \cos^5 x\, dx$

33. $\displaystyle\int \sin^3 x \cos^2 x\, dx$

34. $\displaystyle\int \sin^3 x \cos^3 x\, dx$

35. $\displaystyle\int \sin^2 x \cos^4 x\, dx$

36. $\displaystyle\int \sin^2 x \cos^5 x\, dx$

37. $\displaystyle\int \sin^5 x \cos^4 x\, dx$

38. $\displaystyle\int \sin^4 x \cos^2 x\, dx$

39. $\displaystyle\int \tan^5 x \sec^4 x\, dx$

40. $\displaystyle\int \tan^4 x \sec^4 x\, dx$

41. $\displaystyle\int \frac{\sqrt{1 - x^2}}{x}\, dx$

42. $\displaystyle\int \frac{dx}{\sqrt{x^2 - 16}}$

43. $\displaystyle\int \frac{2x + 3}{\sqrt{2x^2 - 1}}\, dx$

44. $\displaystyle\int \frac{dx}{x\sqrt{x^2 - 1}}$

45. $\displaystyle\int \frac{dx}{x\sqrt{x^2 + 1}}$

46. $\displaystyle\int x\sqrt{x^2 + 1}\, dx$

47. $\displaystyle\int \frac{(2x + 1)\, dx}{\sqrt{4x - x^2 - 2}}$

48. $\displaystyle\int \sqrt{3 + 4x - 4x^2}\, dx$

49. $\displaystyle\int \frac{\cos x\, dx}{\sqrt{1 + \sin^2 x}}$

50. $\displaystyle\int \frac{\sec^2 x\, dx}{\sqrt{\sec^2 x - 2}}$

51. $\displaystyle\int \sin^5 x\, dx$

52. $\displaystyle\int \cos^6 x\, dx$

53. $\displaystyle\int \tan^4 x\, dx$

54. $\displaystyle\int \sec^4 x\, dx$

Find the exact value of the definite integrals in Problems 55–62.

55. $\displaystyle\int_0^2 \sqrt{4 - x^2}\, dx$

56. $\displaystyle\int_0^1 \frac{dx}{\sqrt{9 - x^2}}$

57. $\displaystyle\int_0^{\ln 2} e^t \sqrt{1 + e^{2t}}\, dt$

58. $\displaystyle\int_0^1 \frac{dt}{4t^2 + 4t + 5}$

59. $\displaystyle\int_1^2 \frac{dx}{x^4 \sqrt{x^2 + 3}}$

60. $\displaystyle\int_0^2 \frac{x^3}{(3 + x^2)^{3/2}}\, dx$

61. $\displaystyle\int_{-2}^{2\sqrt{3}} x^3 \sqrt{x^2 + 4}\, dx$

62. $\displaystyle\int_0^{\sqrt{5}} x^2 \sqrt{5 - x^2}\, dx$

B *Find each integral in Problems 63–72.*

63. $\displaystyle\int \frac{e^x\, dx}{\sqrt{1 + e^{2x}}}$

64. $\displaystyle\int \frac{(2x + 1)\, dx}{\sqrt{4x^2 + 4x + 2}}$

65. $\displaystyle\int \frac{x^2 + 4x + 3}{x^3 + x^2 + x}\, dx$

66. $\displaystyle\int \frac{5x^2 + 3x - 2}{x^3 + 2x^2}\, dx$

67. $\displaystyle\int \frac{5x^2 + 18x + 34}{(x - 7)(x + 2)^2}\, dx$

68. $\displaystyle\int \frac{-3x^2 + 9x + 21}{(x + 2)^2(2x + 1)}\, dx$

69. $\displaystyle\int \frac{3x + 5}{x^2 + 2x + 1}\, dx$

70. $\displaystyle\int \frac{3x^2 + 2x + 1}{x^3 + x^2 + x}\, dx$

71. $\displaystyle\int \frac{x\, dx}{(x + 1)(x + 2)(x + 3)}$

72. $\displaystyle\int \frac{5x^2 - 4x + 9}{x^3 - x^2 + 4x - 4}\, dx$

73. Find the average value (to the nearest hundredth) of the function $f(x) = x \sin^3 x^2$ between $x = 0$ and $x = 1$.

74. An object moves along the x-axis in such a way that its velocity at time t is $v(t) = \sin t + \sin^2 t \cos^3 t$. Find the distance moved by the object between times $t = 0$ and $t = \frac{\pi}{3}$.

75. **Exploration Problem**   Integrals of the general form

$$\int \cot^m x \csc^n x \, dx$$

are handled in much the same way as those of the form

$$\int \tan^m x \sec^n x \, dx$$

    a. What substitution would you use in the case where m is odd?

    b. What substitution would you use if n is even?

    c. What would you do if n is odd and m is even?

76. Find the arc length (correct to four decimal places) of the curve $y = \ln x$ from $x = 2$ to $x = 3$.

77. Find the arc length (correct to four decimal places) of the curve $y = x^2$ from $x = -1$ to $x = 1$.

78. Find the volume of the solid generated when the region under the curve $y = \cos x$ between $x = 0$ and $x = \frac{\pi}{2}$ is revolved about the x-axis.

79. Find the area of the region bounded by the graphs of $y = \dfrac{2x}{\sqrt{x^2 + 9}}$, $y = 0$, $x = 0$, and $x = 4$.

80. What is the volume of the solid obtained when the region bounded by $y = \sqrt{x}e^{-x^2}$, $y = 0$, $x = 0$, and $x = 2$ is revolved about the x-axis?

81. What is the volume of the solid obtained when the region bounded by $y = x\sqrt{9 - x^2}$ and $y = 0$ is revolved about the x-axis?

82. Find the length of the curve $y = \ln(\sec x)$ from $x = 0$ to $x = \frac{\pi}{4}$.

83. Find the centroid (to the nearest hundredth) of the region bounded by the curve $y = x^2 e^{-x}$ and the x-axis, between $x = 0$ and $x = 1$.

C 84. Generalize the result of Example 4, Section 7.2 by showing that

$$\int e^{ax} \sin bx \, dx = \frac{(a \sin bx - b \cos bx)e^{ax}}{a^2 + b^2} + C$$

for constants a and b. (This is Formula 492.)

85. Let f'' be continuous on the closed interval $[a, b]$. Use integration by parts to show that

$$\int_a^b x f''(x) \, dx = bf'(b) - f(b) + f(a) - af'(a)$$

86. Derive the reduction formula

$$\int x^m (\ln x)^n \, dx = \frac{x^{m+1}(\ln x)^n}{m+1} - \frac{n}{m+1} \int x^m (\ln x)^{n-1} \, dx$$

where m and n are positive integers. (This is Formula 510 in the *Student Mathematics Handbook*.) Use the formula to find

$$\int x^2 (\ln x)^3 \, dx$$

87. Derive the reduction formula

$$\int \sin^n Ax \, dx = \frac{-\sin^{n-1} Ax \cos Ax}{An} + \frac{n-1}{n} \int \sin^{n-2} Ax \, dx$$

(This is Formula 352 in the *Student Mathematics Handbook*.) Use this formula to evaluate

$$\int \sin^4 4x \, dx$$

7.6   First-Order Differential Equations

IN THIS SECTION   first-order linear differential equations, applications of first-order equations

FIRST-ORDER LINEAR DIFFERENTIAL EQUATIONS

We introduced separable differential equations in Section 5.6, but not all first-order differential equations have the separable form

$$\frac{dy}{dx} = \frac{f(x)}{g(y)}$$

The goal of this section is to examine a second class of differential equations, called **first-order linear**, that have the general form

$$\frac{dy}{dx} + P(x)y = Q(x)$$

For example, the differential equation

$$\frac{dy}{dx} - \frac{y}{x} = e^{-x}$$

is first-order linear, with $P(x) = -\dfrac{1}{x}$ and $Q(x) = e^{-x}$, and

$$x^2 \frac{dy}{dx} - (x^2 + 2)y = x^5$$

can be put into this form by dividing by x^2, the coefficient of $\dfrac{dy}{dx}$, to obtain

$$\frac{dy}{dx} - \left(\frac{x^2 + 2}{x^2}\right) y = \frac{x^5}{x^2}$$

(Note that $P(x) = -\left(\dfrac{x^2 + 2}{x^2}\right) = -1 - 2x^{-2}$ and $Q(x) = \dfrac{x^5}{x^2} = x^3$.)

First-order linear equations can be used to model a variety of important situations in the physical, social, and life sciences. We will examine a few such models as well as additional models involving separable equations later in this section, but first we need to know how first-order linear differential equations may be solved.

THEOREM 7.1 General solution of a first-order linear differential equation

The general solution of the first-order linear differential equation

$$\frac{dy}{dx} + P(x)y = Q(x)$$

is given by

$$y = \frac{1}{I(x)}\left[\int Q(x)I(x)\,dx + C\right]$$

where $I(x) = e^{\int P(x)\,dx}$, and C is an arbitrary constant.

Proof First, note that

$$\frac{dI(x)}{dx} = e^{\int P(x)\,dx}\left[\frac{d}{dx}\int P(x)\,dx\right] = e^{\int P(x)\,dx}P(x) = I(x)P(x)$$

because $\dfrac{d}{dx}e^u = e^u \dfrac{du}{dx}$. Thus,

$$\frac{d}{dx}[I(x)y] = I(x)\frac{dy}{dx} + y\frac{d}{dx}I(x) = I(x)\frac{dy}{dx} + yI(x)P(x) \quad \text{Product rule}$$

We use this computation as the third step in the following.

$$\frac{dy}{dx} + P(x)y = Q(x) \qquad\qquad\qquad \text{Given}$$

$$I(x)\left[\frac{dy}{dx} + P(x)y\right] = I(x)Q(x) \qquad\qquad \text{Multiply both sides by } I(x).$$

$$I(x)\frac{dy}{dx} + I(x)P(x)y = I(x)Q(x)$$

$$\frac{d}{dx}[I(x)y] = I(x)Q(x) \qquad\qquad\qquad \text{Substitute.}$$

$$\int \frac{d}{dx}[I(x)y]\,dx = \int I(x)Q(x)\,dx \qquad\quad \text{Integrate both sides.}$$

$$I(x)y = \int I(x)Q(x)\,dx + C$$

As usual, the constant of integration C is added after the last integration.

$$y = \frac{1}{I(x)}\left[\int I(x)Q(x)\,dx + C\right] \qquad\qquad\qquad\qquad \square$$

Because multiplying both sides of the differential equation

$$\frac{dy}{dx} + P(x)y = Q(x)$$

by $I(x)$ makes the left side an exact derivative, the function $I(x)$ is called an **integrating factor** of the differential equation.

A first-order **initial value problem** involves a first-order equation and the value of y at a particular value $x = x_0$. Here is an example of such a problem.

EXAMPLE 1   First-order linear initial value problem

Solve $\dfrac{dy}{dx} = e^{-x} - 2y$, $x \geq 0$, subject to the initial condition $y = 2$ when $x = 0$.

Solution

A graphical solution (using technology) is found by looking at the direction field as shown in Figure 7.5.

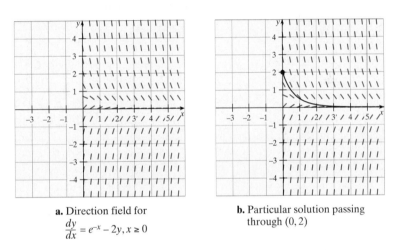

a. Direction field for
$\frac{dy}{dx} = e^{-x} - 2y, x \geq 0$

b. Particular solution passing through $(0, 2)$

Figure 7.5   Graphical solution using a direction field

To find an analytic solution, we use Theorem 7.1. The differential equation can be expressed in first-order linear form by adding $2y$ to both sides:

$$\frac{dy}{dx} + 2y = e^{-x} \qquad \text{for} \quad x \geq 0$$

We have $P(x) = 2$ and $Q(x) = e^{-x}$. Both $P(x)$ and $Q(x)$ are continuous in the domain $x \geq 0$. An integrating factor is given by

$$I(x) = e^{\int P(x)\,dx} = e^{\int 2\,dx} = e^{2x} \qquad \text{for} \quad x \geq 0$$

We now use the first-order linear differential equation theorem, where $I(x) = e^{2x}$ and $Q(x) = e^{-x}$, to find y.

$$y = \frac{1}{e^{2x}} \left[\int e^{2x} e^{-x}\,dx + C \right]$$

$$= \frac{1}{e^{2x}} \left[e^x + C \right] = e^{-x} + Ce^{-2x} \qquad \text{for} \quad x \geq 0$$

To find C, note that because $y = 2$ when $x = 0$, it follows that $1 = C$. Thus, $y = e^{-x} + e^{-2x}$, $x \geq 0$.

You might wish to compare the analytic and graphical solutions by noting that the graph of the equation matches the solution shown in Figure 7.5b.   ∎

Integration by parts is often used in solving first-order linear differential equations. Here is an example.

EXAMPLE 2   First-order linear differential equation

Find the general solution of the differential equation

$$x\frac{dy}{dx} + y = xe^{2x}, \qquad x > 0$$

Solution

As a first step, divide each term in the given equation by x so that we can use integrating factors, as in Theorem 7.1.

$$\frac{dy}{dx} + \frac{y}{x} = e^{2x}$$

For this equation, $P(x) = \dfrac{1}{x}$ and $Q(x) = e^{2x}$. The integrating factor is

$$I(x) = e^{\int (1/x)\,dx} = e^{\ln x} = x$$

Thus, the general solution is

$$
\begin{aligned}
y &= \frac{1}{x}\left[\int xe^{2x}\,dx + C\right] \qquad &&\text{Integrate } \int xe^{2x}\,dx \text{ by parts.}\\[2mm]
&= \frac{1}{x}\left[\frac{1}{2}xe^{2x} - \frac{1}{4}e^{2x} + C\right] \qquad &&\text{See Example 5, Section 7.2.}\\[2mm]
&= \frac{1}{2}e^{2x} - \frac{e^{2x}}{4x} + \frac{C}{x} \qquad && \blacksquare
\end{aligned}
$$

APPLICATIONS OF FIRST-ORDER LINEAR EQUATIONS

Modeling Logistic Growth   When the population $Q(t)$ of a colony of living organisms (humans, bacteria, etc.) is small, it is reasonable to expect the relative rate of change of the population to be constant. In other words,

$$\frac{\dfrac{dQ}{dt}}{Q} = k \quad \text{or} \quad \frac{dQ}{dt} = kQ$$

where k is a constant (the **unrestricted growth rate**). This is called **exponential growth**. As long as the colony has plenty of food and living space, its population will obey this unrestricted growth rate formula and $Q(t)$ will grow exponentially.

However, in practice, there often comes a time when environmental factors begin to restrict the further expansion of the colony, and at this point, the growth ceases to be purely exponential in nature. To construct a population model that takes into account the effect of diminishing resources and crowding, we assume that the population of the species has a "cap" B, called the **carrying capacity** of the species. We further assume that the relative rate of change in population is jointly proportional to the current population $Q(t)$ and the difference $B - Q$, which can be thought of as the potential unborn population of the species. That is,

$$\frac{\dfrac{dQ}{dt}}{Q} = k(B - Q) \quad \text{or} \quad \frac{dQ}{dt} = kQ(B - Q)$$

This is called a **logistic equation**, and it arises not only in connection with population models, but also in a variety of other situations. The following example illustrates one way such an equation can arise.

EXAMPLE 3   Logistic equation for the spread of an epidemic

The rate at which an epidemic spreads through a community is proportional to the product of the number of residents who have been infected and the number of susceptible uninfected residents. Express the number of residents who have been infected as a function of time.

Solution

Let $Q(t)$ denote the number of residents who have been infected by time t and B the total number of susceptible residents. Then the number of susceptible residents who have not been infected is $B - Q$, and the differential equation describing the spread of the epidemic is

$$\frac{dQ}{dt} = kQ(B - Q) \qquad \text{k is the constant of proportionality.}$$

$$\frac{dQ}{Q(B - Q)} = k\,dt \qquad \text{Use separation of variables.}$$

$$\int \frac{dQ}{Q(B - Q)} = \int k\,dt$$

The integral on the right causes no difficulty, but the integral on the left requires partial fractions:

$$\frac{1}{Q(B - Q)} = \frac{1}{B}\left[\frac{1}{Q} + \frac{1}{B - Q}\right]$$

We now substitute the form on the left into the equation and complete the integration:

$$\frac{1}{B}\int \frac{dQ}{Q} + \frac{1}{B}\int \frac{dQ}{B - Q} = \int k\,dt$$

$$\frac{1}{B}\ln|Q| - \frac{1}{B}\ln|B - Q| = kt + C \qquad \text{Integrate each term.}$$

$$\frac{1}{B}\ln\left|\frac{Q}{B - Q}\right| = kt + C \qquad \text{Division property of logs}$$

$$\ln\left(\frac{Q}{B - Q}\right) = Bkt + BC \qquad \begin{array}{l}\text{Multiply both sides by B}\\ \text{(note that $Q > 0$, $B > Q$).}\end{array}$$

$$\frac{Q}{B - Q} = e^{Bkt + BC} \qquad \text{Definition of ln}$$

$$Q = (B - Q)e^{Bkt}e^{BC} \qquad \begin{array}{l}\text{Because e^{BC} is a constant,}\\ \text{we can let $A_1 = e^{BC}$.}\end{array}$$

$$Q = A_1Be^{Bkt} - A_1Qe^{Bkt}$$

$$Q + A_1Qe^{Bkt} = A_1Be^{Bkt}$$

$$Q = \frac{A_1Be^{Bkt}}{1 + A_1e^{Bkt}}$$

To simplify the equation, we make another substitution; let $A = \dfrac{1}{A_1}$, so that (after several simplification steps) we have

$$Q = \frac{\dfrac{Be^{Bkt}}{A}}{1 + \dfrac{e^{Bkt}}{A}} = \frac{B}{1 + Ae^{-Bkt}}$$

The graph of $Q(t)$ is shown in Figure 7.6. Note that the curve has an inflection point where

$$Q(t) = \frac{B}{2}$$

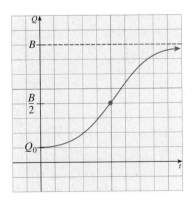

Figure 7.6 A logistic curve:
$$Q(t) = \frac{B}{1 + Ae^{-Bkt}}$$

since

$$\frac{d^2Q}{dt^2} = kQ(-1) + k(B - Q)$$

$$= k(B - 2Q)$$

$$= 0 \qquad \text{when } Q = \frac{B}{2}.$$

This corresponds to the fact that the epidemic is spreading most rapidly when half the susceptible residents have been infected.

Note also in Example 3 that $y = Q(t)$ approaches the line $y = B$ asymptotically as $t \to +\infty$. Thus, the number of infected people approaches the number of those susceptible in the long run. ∎

A summary of growth models is given in Table 7.3.

TABLE 7.3 Summary of Growth Models

Graph	Growth Model	Equation and Solution	Sample Applications

$Q(t)$ is the population at time t, Q_0 is the initial population, and k is a constant of proportionality. If $Q(t)$ has a limiting value, let it be denoted by B.

Graph	Growth Model	Equation and Solution	Sample Applications
J-curve	**Uninhibited growth** $(k > 0)$ Rate is proportional to the amount present.	$\dfrac{dQ}{dt} = kQ$ *Solution:* $Q = Q_0 e^{kt}$	Exponential growth; interest compounded continuously; inflation; price/supply curves
L-curve	**Uninhibited decay** $(k < 0)$ Rate is proportional to the amount present.	$\dfrac{dQ}{dt} = kQ$ *Solution:* $Q = Q_0 e^{kt}$	Radioactive decay; depletion of natural resources; price/demand curves; electric condenser discharge
S-curve	**Inhibited (logistic) growth** Rate is jointly proportional to the amount present and to the difference between the amount present and the maximum amount possible B (called the carrying capacity).	$\dfrac{dQ}{dt} = kQ(B - Q)$ *Solution:* $Q = \dfrac{B}{1 + Ae^{-Bkt}}$ where A is an arbitrary constant	Long-term restricted population growth; spread of a disease in a population; growth of a business; spread of a rumor
C-curve	**Limited growth** $(k > 0)$ Rate is proportional to the difference between the amount present and a fixed limiting value B.	$\dfrac{dQ}{dt} = k(B - Q)$ *Solution:* $Q = B - Ae^{-kt}$ where A is an arbitrary constant.	Learning curve; diffusion of information by mass media; amount of drug retained after intraveneous injection; Newton's law of cooling; sales of new products; growth of a business

Compartmental Analysis: Modeling Dilution

EXAMPLE 4 A dilution problem

A tank contains 20 lb of salt dissolved in 50 gal of water. Suppose 3 gal of brine containing 2 lb of dissolved salt per gallon runs into the tank every minute and that the mixture (kept uniform by stirring) runs out of the tank at the rate of 2 gal/min. Find the amount of salt in the tank at any time t. How much salt is in the tank at the end of one hour?

Solution

Let $S(t)$ denote the amount of salt in the tank at the end of t minutes. Because 3 gal of brine flows into the tank each minute and each gallon contains 2 lb of salt, it follows that $(3)(2) = 6$ lb of salt flows into the tank each minute (see Figure 7.7). This is the inflow rate.

For the outflow, note that at time t, there are $S(t)$ lb of salt and $50+(3-2)t$ gallons of solution (because solution flows in at 3 gal/min and out at 2 gal/min). Thus, the concentration of salt in the solution at time t is

$$\frac{S(t)}{50+t} \text{ lb/gal}$$

and the outflow rate of salt is

$$\left[\frac{S(t)}{50+t} \text{ lb/gal}\right]\left[2 \text{ gal/min}\right] = \frac{2S(t)}{50+t} \text{ lb/min}$$

Combining these observations, we see that the net rate of change of salt dS/dt is given by

$$\frac{dS}{dt} = \underbrace{6}_{\text{Inflow}} - \underbrace{\frac{2S}{50+t}}_{\text{Outflow}}$$

or

$$\frac{dS}{dt} + \frac{2S}{50+t} = 6$$

which we recognize as a first-order linear differential equation with

$$P(t) = \frac{2}{50+t}$$

and $Q(t) = 6$. An integrating factor is

$$I(t) = e^{\int P(t)\,dt} = e^{\int (2\,dt)/(50+t)} = e^{2\ln|50+t|} = (50+t)^2$$

and the general solution is

$$S(t) = \frac{1}{(50+t)^2}\left[\int 6(50+t)^2\,dt + C\right]$$

$$= \frac{1}{(50+t)^2}\left[2(50+t)^3 + C\right]$$

$$= 2(50+t) + \frac{C}{(50+t)^2}$$

To evaluate C, we recall that there are 20 lb of salt initially in the solution. This means $S(0) = 20$, so that

$$S(0) = 2(50+0) + \frac{C}{(50+0)^2}$$

$$20 = 100 + \frac{C}{50^2}$$

$$-80(50^2) = C$$

Inflow rate of salt: 6 lb/min

Outflow rate of salt: $\dfrac{S(t)}{50+t}$ lb / min

Figure 7.7 Rate of flow equals inflow rate minus outflow rate.

Thus, the solution to the given differential equation, subject to the initial condition $S(0) = 20$, is

$$S(t) = 2(50 + t) - \frac{80(50)^2}{(50 + t)^2}$$

Specifically, at the end of 1 hour (60 min), the tank contains

$$S(60) = 2(50 + 60) - \frac{80(50)^2}{(50 + 60)^2} \approx 203.4710744$$

The tank contains about 200 lb of salt.   ∎

Modeling RL Circuits   Another application of first-order linear differential equations involves the current in an RL electric circuit. An RL circuit is one with a constant resistance, R, and a constant inductance, L. Figure 7.8 shows an electric circuit with an electromotive force (EMF), a resistor, and an inductor connected in series.

Figure 7.8   An RL circuit diagram

The EMF source, which is usually a battery or generator, supplies voltage that causes a current to flow in the circuit. According to Kirchhoff's second law, if the circuit is closed at time $t = 0$, then the applied electromotive force is equal to the sum of the voltage drops in the rest of the circuit. It can be shown that this implies that the current $I(t)$ that flows in the circuit at time t must satisfy the first-order linear differential equation

$$L\frac{dI}{dt} + RI = E$$

where L (the inductance) and R (the resistance) are positive constants.* Write

$$\frac{dI}{dt} + \frac{R}{L}I = \frac{E}{L}$$

to see that $P(t) = \dfrac{R}{L}$ (a constant) and $Q(t) = \dfrac{E}{L}$. In the case where E is a constant, we have

$$I(t) = e^{-\int (R/L)\,dt} \left[\int \frac{E}{L} e^{\int (R/L)\,dt}\, dt + C \right]$$

$$= e^{-(R/L)t} \left[\frac{E}{L} \int e^{(R/L)t}\, dt + C \right]$$

$$= e^{-(R/L)t} \left[\frac{E}{L} \cdot \frac{L}{R} e^{(R/L)t} + C \right]$$

$$= \frac{E}{R} + Ce^{-(R/L)t}$$

---

*It is common practice in physics and applied mathematics to use I as the symbol for current. Of course, this has nothing to do with the concept of integrating factor introduced earlier in this section.

It is reasonable to assume that no current flows when $t = 0$. That is, $I = 0$ when $t = 0$, so we have $0 = E/R + C$ or $C = -E/R$. The solution of the initial value problem is

$$I = \frac{E}{R}\left(1 - e^{-Rt/L}\right)$$

Notice that because $e^{-(Rt/L)} \to 0$ as $t \to +\infty$, we have

$$\lim_{t \to +\infty} I(t) = \lim_{t \to +\infty} \frac{E}{R}\left(1 - e^{-Rt/L}\right)$$
$$= \frac{E}{R}(1 - 0)$$
$$= \frac{E}{R}$$

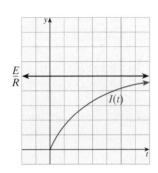

This means that, in the long run, the current I must approach $\frac{E}{R}$. The solution of the differential equation consists of two parts, which are given special names:

$\frac{E}{R}$ is the **steady-state current**.

$-\frac{E}{R}e^{-Rt/L}$ is the **transient current.**

Figure 7.9 The current in an RL circuit with constant EMF

Figure 7.9 shows how the current $I(t)$ varies with time t.

7.6 PROBLEM SET

Ⓐ *Solve the differential equations in Problems 1–10.*

1. $\dfrac{dy}{dx} + \dfrac{3y}{x} = x$

2. $\dfrac{dy}{dx} - \dfrac{2y}{x} = \sqrt{x} + 1$

3. $x^4 \dfrac{dy}{dx} + 2x^3 y = 5$

4. $x^2 \dfrac{dy}{dx} + xy = 2$

5. $x\dfrac{dy}{dx} + 2y = xe^{x^3}$

6. $\dfrac{dy}{dx} + \left(\dfrac{2x + 1}{x}\right)y = e^{-2x}$

7. $\dfrac{dy}{dx} + \dfrac{y}{x} = \tan^{-1} x$

8. $\dfrac{dy}{dx} + \dfrac{2y}{x} = \dfrac{\ln x}{x}$

9. $\dfrac{dy}{dx} + (\tan x)y = \sin x$

10. $\dfrac{dy}{dx} + (\sec x)y = \sin 2x$

In Problems 11–14, solve the initial value problems. A graphical solution is shown as a check for your work.

11. $\dfrac{dy}{dx} + \dfrac{xy}{1 + x} = x(1 + x)$
with $y = -1$ when $x = 0$

12. $\dfrac{dy}{dx} + \dfrac{2xy}{1 + x^2} = \sin x$
with $y = 1$ when $x = 0$

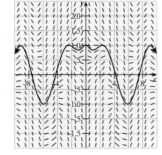

13. $x\dfrac{dy}{dx} - 2y = 2x^3$, $x > 0$
with $y = 0$ when $x = 3$

14. $y' - \dfrac{y}{x} = x^2 e^{x^2}$, $x > 0$
with $y = 2$ when $x = 2$

Ⓑ *In Problems 15–18, find the orthogonal trajectories of the given family of curves. Recall from Section 5.6 that a curve is an orthogonal trajectory of a given family if it intersects each member of that family at right angles.*

15. the family of parabolas $y^2 = 4kx$

16. the family of hyperbolas $xy = c$

17. the family of circles $x^2 + y^2 = r^2$

18. the family of exponential curves $y = Ce^{-x}$

Modeling Problems. *In Problems 19–30, set up and solve an appropriate model to answer the given question. Be sure to state your assumptions.*

19. In 2000 the gross domestic product (GDP) of the United States was \$9,382 billion (in constant 1996 dollars). Suppose the growth rate from 1999 to 2000 was 6.42%. Predict the GDP in 2010.

20. In 1992 the gross domestic product (GDP) of the United States in constant 1996 dollars was \$6,803 billion. Suppose the growth rate from 1992 to 2000 was 4.0% per year. Predict the GDP in 2002. Consult an almanac to see if this prediction is correct.

21. According to the Department of Health and Human Services, the divorce rate in 1990 in the United States was 4.7% and there were 1,175,000 divorces that year. How many divorces will there be in 2005 if the divorce rate is constant?

22. According to the Department of Health and Human Services, the marriage rate in 1990 in the United States was 9.8% and there were 2,448,000 marriages that year. How many marriages will there be in 2005 if the marriage rate is constant?

23. A tank contains 10 lb of salt dissolved in 30 gal of water. Suppose 2 gal of brine containing 1 lb of dissolved salt per gallon runs into the tank every minute and that the mixture (kept uniform by stirring) runs out at the same rate.
    a. Find the amount of salt in the tank at time t.
    b. How long does it take (to the nearest second) for the tank to contain 15 lb of salt?

24. In Problem 23, suppose the tank has a capacity of 100 gal and that the mixture flows out at the rate of 1 gal/min (instead of 2 gal/min).
    a. How long will it take for the tank to fill?
    b. How much salt will be in the tank when it is full?

25. The rate at which a drug is absorbed into the blood system is given by

    $$\frac{db}{dt} = \alpha - \beta b$$

    where $b(t)$ is the concentration of the drug in the bloodstream at time t. What does $b(t)$ approach in the long run (that is, as $t \to +\infty$)? At what time is $b(t)$ equal to half this limiting value? Assume that $b(0) = 0$.

26. An RL circuit has a resistance of R ohms, inductance of L henries, and EMF of E volts, where R, L, and E are constant. Suppose no current flows in the circuit at time $t = 0$. If L is doubled and E and R are held constant, what effect does this have on the "long run" current in the circuit (that is, the current as $t \to +\infty$)?

27. The 1990 census recorded a population of 16.1 million Hispanics, while in 2000 the figure was 32.8 million. Assuming that the rate of population growth is proportional to the population, predict the Hispanic population in the year 2005.

28. A population of animals on Catalina Island is limited by the amount of food available. Studies show there were 1,800 animals present in 1980 and 2,000 in 1986 and suggest that 5,000 animals can be supported by the conditions present on the island. Use a logistic model to predict the animal population in the year 2005.

29. In 1986 the Chernobyl nuclear disaster in the Soviet Union contaminated the atmosphere. The buildup of radioactive material in the atmosphere satisfies the differential equation

    $$\frac{dM}{dt} = r\left(\frac{k}{r} - M\right), \qquad M = 0 \text{ when } t = 0$$

    where M = mass of radioactive material in the atmosphere after time t (in years); k is the rate at which the radioactive material is introduced into the atmosphere; r is the annual decay rate of the radioactive material. Find the solution, $M(t)$, of this differential equation in terms of k and r.

30. **The Motion of a Body Falling through a Resisting Medium** A body of mass m is dropped from a great height and falls in a straight line. Assume that the only forces acting on the body are the earth's gravitational attraction mg and air resistance kv. (Recall $g = -32$ ft/s^2.)
    a. According to Newton's second law,

    $$m\frac{dv}{dt} = mg - kv$$

    Solve this equation, assuming the object has velocity $v_0 = 0$ at time $t = 0$.
    b. Find the distance $s(t)$ the body has fallen at time t. Assume $s = 0$ at time $t = 0$.
    c. If the body weighs $W = 100$ lb and $k = 0.35$, how long does it take for the body to reach the ground from a height of 10,000 ft?

Recall from Section 5.6 that a tank filled with water drains at the rate

$$\frac{dV}{dt} = -4.8A_0\sqrt{h}$$

where h is the height of water at time t (in seconds) and A_0 is the area (in ft^2) of the drain hole. This formula, called Torricelli's law, is used in Problems 31 and 32.

31. A full tank of water has a drain with area 0.07 ft^2. If the tank has a constant cross-sectional area of $A = 5$ ft^2 and height of $h_0 = 4$ ft, how long does it take to empty?

32. A full tank of water of height 5 ft and constant cross-sectional area $A = 3$ ft^2 has two drains, both of area 0.02 ft^2. One drain is at the bottom and the other at height 2 ft. How long does it take for the tank to drain?

33. **The Euler Beam Model** For a rigid beam with uniform loading, the deflection $y(x)$ is modeled by the differential equation $y^{(4)} = -k$, where $k > 0$ is a constant and x measures the distance along the beam from one of its ends. Assume that $y(0) = y(L) = 0$, where L is the length of the beam, and that $y''(0) = y''(L) = 0$.
    a. Solve the beam equation to find $y(x)$.
    b. Where does the maximum deflection occur? What is the maximum deflection?
    c. Suppose the beam is cantilevered, so that $y(0) = y(L) = 0$ and $y''(0) = y'(L) = 0$. Now where does the maximum deflection occur? Is the maximum deflection greater or less than the case considered in part **b**? Would you expect the graph of the deflection $y = y(x)$ to be concave up or concave down on $[0, L]$? Prove your conjecture.

34. **Modeling Problem** A tank initially contains 5 lb of salt in 50 ft^3 of solution. At time $t = 0$, brine begins to enter the tank at the rate of 2 ft^3/h, and the mixed solution drains at the same rate. The brine coming into the tank has concentration

    $$C(t) = 1 - e^{-0.2t} \text{ lb/ft}^3$$

    t hours after the dilution begins.
    a. Set up and solve a differential equation for the amount of salt $S(t)$ in the solution at time t.
    b. How much salt is eventually in the tank (as $t \to +\infty$)?
    c. At what time is $S(t)$ minimized? What is the minimum amount of salt in the tank?

35. Modeling Problem Two 100-gallon tanks initially contain pure water. Brine containing 2 lb of salt per gallon enters the first tank at the rate of 1 gal/min, and the mixed solution drains into the second tank at the same rate. There, it is again thoroughly mixed and drains at the same rate, 1 gal/min.

  a. Set up and solve a differential equation of the amount of salt $S_1(t)$ in the first tank at time t (minutes).

  b. Set up and solve a second differential equation for the amount of salt $S_2(t)$ in the second tank at time t.

  c. Let $S(t) = S_1 - S_2$. Intuitively, $S(t) \geq 0$ for all t. At what time is the excess $S(t)$ maximized? What is the maximum excess?

36. Modeling Problem A chemical in a solution diffuses from a compartment with known concentration $C_1(t)$ across a membrane to a second compartment whose concentration $C_2(t)$ changes at a rate proportional to the difference $C_1 - C_2$. Set up and solve the differential equation for $C_2(t)$ in the following cases:

  a. $C_1(t) = 5e^{-2t}; k = 1.7; C_2(0) = 0$

  b. $k = 4; C_2(0) = 3$, and

$$C_1(t) = \begin{cases} 4 & \text{if } 0 \leq t \leq 3 \\ 5 & \text{if } t > 3 \end{cases}$$

37. The Gompertz equation for a population $P(t)$ is

$$\frac{dP}{dt} = kP(B - \ln P)$$

where k and B are positive constants. If the initial population is $P(0) = P_0$, and the ultimate population is

$$\lim_{t \to +\infty} P(t) = P_\infty$$

find $P(t)$.

38. Modeling Problem Uranium-234 (half-life 2.48×10^5 yr) decays to Thorium-230 (half-life 80,000 yr).

  a. If $U(t)$ and $T(t)$ are the amounts of uranium and thorium at time t, then

$$\frac{dU}{dt} = -k_1 U, \quad \frac{dT}{dt} = -k_2 T + k_1 U$$

Solve this system of differential equations to obtain $U(t)$ and $T(t)$.

  b. If we start with 100 g of pure U-234, how much Th-230 will there be after $t = 5,000$ yr?

39. Solve the differential equation

$$\frac{dy}{dx} = \frac{1 + y}{xy + e^y(1 + y)}$$

by regarding y as the independent variable (i.e., reverse the roles of x and y).

40. Certain biological processes occur periodically over the 24 hours of a day. Suppose a patient's metabolic exertion rate is modeled by

$$R(t) = \frac{1}{4} - \frac{1}{8} \cos \frac{\pi}{12}(t - 6)$$

and the rate of intake is $I(t)$ for $0 \leq t \leq 24$. The patient's body contains 200 g of the substance when $t = 0$.

  a. If $I(t) = 0$ (the patient intakes only water), the amount of substance $Q(t)$ in the patient's body at time t satisfies

$$\frac{dQ}{dt} = 0 - R(t)$$

Solve this equation for $Q(t)$.

  b. Suppose the patient intakes the substance at a constant rate for part of the day. Specifically,

$$I(t) = \begin{cases} 0.4 \text{ g/h} & \text{for } 10 \leq t \leq 20 \\ 0 & \text{otherwise} \end{cases}$$

Set up and solve a differential equation for the amount of substance $Q(t)$ in the patient's body at time t. When is $Q(t)$ maximized?

41. An RL circuit has a resistance of $R = 10$ ohms and an inductance of $L = 5$ henries. Find the current $I(t)$ in the circuit at time t if $I(0) = 0$ and the electromotive force (EMF) is:

  a. $E = 15$ volts   **b.** $E = 5e^{-2t} \sin t$

42. An RL circuit has an inductance of $L = 3$ henries and a resistance $R = 6$ ohms in series with an EMF of $E = 50 \sin 30t$. Assume $I(0) = 0$.

  a. What is the current $I(t)$ at time t?

  b. What is the transient current? The steady-state current?

43. Modeling Problem A lake has a volume of 6 billion ft³, and its initial pollutant content is 0.22%. A river whose waters contain only 0.06% pollutants flows into the lake at the rate of 350 million ft³/day, and another river flows out of the lake also carrying 350 million ft³/day. Assume that the water in the two rivers and the lake is always well mixed. How long does it take for the pollutant content to be reduced to 0.15%?

⊙ 44. A Bernoulli equation is a differential equation of the form

$$y' + P(x)y = Q(x)y^n$$

where n is a real number, $n \neq 0, n \neq 1$.

  a. Show that the change of variable $u = y^{1-n}$ transforms such an equation into one of the form

$$u'(x) + (1 - n)P(x)u(x) = (1 - n)Q(x)$$

This transformed equation is a first-order linear differential equation in u and can be solved by the methods of this section, yielding a solution to the given Bernoulli equation.

  b. Use the change of variable suggested to solve the Bernoulli equation

$$y' + \frac{y}{x} = 2y^2$$

45. HISTORICAL QUEST Daniel Bernoulli was a member of the famous Bernoulli family (see Historical Quest Problem 59, Section 4.5). Between 1725 and 1749 he won ten prizes for his work in astronomy, gravity, tides, magnetism, ocean currents, and the behavior of ships at sea. While modeling the effects of a smallpox epidemic, he obtained the differential equation

DANIEL BERNOULLI
1700–1782

$$\frac{dS}{dt} = -pS + \left(\frac{S}{N}\right)\frac{dN}{dt} + \frac{pS^2}{mN}$$

In this equation $S(t)$ is the number of people at age t that are susceptible to smallpox; $N(t)$ is the number of people at age t who survive; p is the probability of a susceptible person getting the disease, and $1/m$ is the proportion of those who die from the disease. ∎

Let $y = S/N$, and solve the resulting equation for $y(t)$. Then write N as a function of S. (*Hint*: You will need the result of Problem 44.)

46. Consider a curve with the property that when horizontal and vertical lines are drawn from each point $P(x, y)$ on the curve to the coordinate axes, the area A_1 under the curve is twice the area A_2 above the curve, as shown in Figure 7.10.

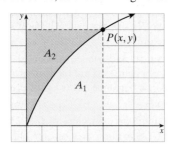

Figure 7.10   Problem 46

Show that x and y satisfy the differential equation

$$\frac{dy}{dx} = \frac{y}{2x}$$

Solve the equation and characterize the family of all curves that satisfy the given geometric condition.

47. An object of mass m is projected upward from ground level with initial velocity v_0 against air resistance proportional to its velocity $v(t)$. Thus,

$$m\frac{dv}{dt} = -mg - kv$$

a. Solve this equation for $v(t)$, and then find $s(t)$, the height of the object above the ground at time t.

b. When does the object reach its maximum height? What is the maximum height?

c. Suppose $k = \frac{3}{4}$. For an object weighing 20 lb with initial velocity $v_0 = 150$ ft/s, what is the maximum height? When does the object hit the ground?

d. Suppose the same object as in part **c** is launched with the same initial velocity but in a vacuum, where there is no air resistance ($k = 0$). Would you expect the object to hit the ground sooner or later than the object in part **c**? Prove your conjecture.

48. A body of mass m falls from a height of s_0 ft against air resistance modeled to be proportional to the *square* of its velocity v, so that

$$m\frac{dv}{dt} = mg - kv^2$$

a. Solve this differential equation to find $v(t)$. Then find the height $s(t)$ of the object at time t (in seconds).

b. If the object weighs 1,000 lb, $s_0 = 100$ ft, and $k = 0.01$, how long does it take for the object to hit the ground? Interpret this result.

7.7   Improper Integrals

IN THIS SECTION   improper integrals with infinite limits of integration, improper integrals with unbounded integrands

We have defined the definite integral $\displaystyle\int_a^b f(x)\,dx$ on a closed bounded interval $[a, b]$ where the integrand $f(x)$ is bounded. In this section, we extend the concept of integral to the case where the interval of integration is infinite and also to the case where f is unbounded at a finite number of points on the interval of integration. Collectively, these are called **improper integrals**.

IMPROPER INTEGRALS WITH INFINITE LIMITS OF INTEGRATION

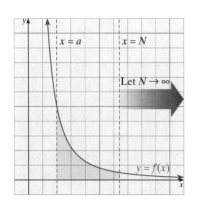

Figure 7.11 Finding the area under a curve on an unbounded region

In physics, economics, probability and statistics, and other applied areas, it is useful to have a concept of an integral that is defined on the entire real line or on half-lines of the form $x \geq a$ or $x \leq a$. If $f(x) \geq 0$, the integral of f on the interval $x \geq a$ can be thought of as the area under the curve $y = f(x)$ on this unbounded interval, as shown in Figure 7.11. A reasonable strategy for finding this area is first to use a definite integral to compute the area from $x = a$ to some finite number $x = N$ and then to let N approach infinity in the resulting expression. Here is a definition.

Improper Integrals (First Type)

Let a be a fixed number and assume $\int_a^N f(x)\,dx$ exists for all $N \ge a$. Then if

$$\lim_{N \to +\infty} \int_a^N f(x)\,dx \text{ exists, we define the \textbf{improper integral} } \int_a^{+\infty} f(x)\,dx \text{ by}$$

$$\int_a^{+\infty} f(x)\,dx = \lim_{N \to +\infty} \int_a^N f(x)\,dx$$

The improper integral is said to **converge** if this limit is a finite number and to **diverge** otherwise.

Figure 7.12 Graph of $y = \dfrac{1}{x^2}$

EXAMPLE 1 Convergent improper integral

Evaluate $\displaystyle\int_1^{+\infty} \frac{dx}{x^2}$.

Solution

The region under the curve $y = 1/x^2$ for $x \ge 1$ is shown in Figure 7.12. This region is unbounded, so it might seem reasonable to conclude that the area is also infinite. Begin by computing the integral from 1 to N:

$$\int_1^N \frac{dx}{x^2} = -\frac{1}{x}\Big|_1^N = -\frac{1}{N} + 1$$

Let us consider the situation more carefully, as shown in Figure 7.13.

Let $N = 2, A = 1/2$.

Let $N = 3, A = 2/3$.

Let $N = 10, A = 9/10$.

Let $N = 100, A = 99/100$.

Figure 7.13 Area enclosed by $y = \dfrac{1}{x^2}$, the x-axis, $x = 1$, and $y = N$

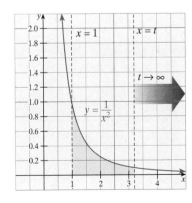

Figure 7.14 The region to the right of $x = 1$, bounded by the curve $y = 1/x^2$ and the x-axis, has area 1.

For $N = 2, 3, 10$, or 100, this integral has values $\frac{1}{2}, \frac{2}{3}, \frac{9}{10}$, and $\frac{99}{100}$, respectively, suggesting that the region under $y = 1/x^2$ for $x \ge 1$ actually has finite area approaching 1 as N gets larger and larger.

To find this area analytically, we take the limit as $N \to +\infty$, as shown in Figure 7.14.

$$\int_1^{+\infty} \frac{dx}{x^2} = \lim_{N \to +\infty} \int_1^N \frac{dx}{x^2}$$
$$= \lim_{N \to +\infty} \left[-x^{-1}\right]\Big|_1^N$$
$$= \lim_{N \to +\infty} \left[-N^{-1} + 1\right]$$
$$= 1$$

Thus, the improper integral converges and has the value 1. ∎

EXAMPLE 2 **A divergent improper integral**

Evaluate $\int_1^{+\infty} \dfrac{dx}{x}$.

Solution

The graph of $y = 1/x$, shown in Figure 7.15, looks very much like the graph of $y = 1/x^2$ in Figure 7.14. We compute the integral from 1 to N and then let N go to infinity.

$$\int_1^{+\infty} \frac{dx}{x} = \lim_{N \to +\infty} \int_1^N \frac{dx}{x}$$

$$= \lim_{N \to +\infty} \ln|x|\big|_1^N$$

$$= \lim_{N \to +\infty} \ln N = +\infty$$

The limit is not a finite number, which means the improper integral diverges. ∎

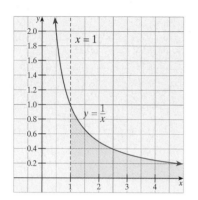

Figure 7.15 The area to the right of the line $x = 1$ bounded by the curve $y = 1/x$ and the x-axis is infinite.

We have shown that the improper integral $\int_1^{+\infty} \dfrac{dx}{x^2}$ converges and that the improper integral $\int_1^{+\infty} \dfrac{dx}{x}$ diverges. In geometric terms, this says that the area to the right of $x = 1$ under the curve $y = 1/x^2$ is finite, whereas the corresponding area under the curve $y = 1/x$ is infinite. The reason for the difference is that as x increases, $1/x^2$ approaches zero more quickly than does $1/x$, as shown in Figure 7.16.

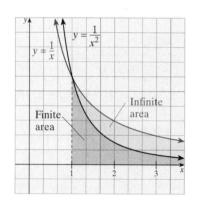

Figure 7.16 An unbounded region may have either a finite or an infinite area.

EXAMPLE 3 **Determine convergence of an improper integral**

Show that the improper integral $\int_1^{+\infty} \dfrac{dx}{x^p}$ converges only for constant $p > 1$.

Solution

For this example, remember that p is a constant and x is the variable. We have seen in Example 1 that the integral converges for $p = 2$. We also know from Example 2 that the integral diverges for $p = 1$, so we can assume $p \neq 1$ in the following computation:

$$\int_1^{+\infty} \frac{dx}{x^p} = \lim_{N \to +\infty} \int_1^N \frac{dx}{x^p} = \lim_{N \to +\infty} \left[\frac{x^{-p+1}}{-p+1}\right]\Bigg|_1^N = \lim_{N \to +\infty} \frac{1}{1-p}\left[N^{1-p} - 1\right]$$

If $p > 1$, then $1 - p < 0$, so $N^{1-p} \to 0$ as $N \to +\infty$, and

$$\int_1^{+\infty} \frac{dx}{x^p} = \lim_{N \to +\infty} \frac{1}{1-p}\left[N^{1-p} - 1\right] = \frac{1}{1-p}[-1] = \frac{1}{p-1}$$

If $p < 1$, then $1 - p > 0$, so $N^{1-p} \to +\infty$ as $N \to +\infty$, and

$$\int_1^{+\infty} \frac{dx}{x^p} = \lim_{N \to +\infty} \frac{1}{1-p}\left[N^{1-p} - 1\right] = +\infty$$

Thus, the improper integral diverges for $p \leq 1$ and converges for $p > 1$. ∎

➤ **What This Says** The improper integral

$$\int_1^{+\infty} \frac{dx}{x^p} = \begin{cases} \dfrac{1}{p-1} & \text{if } p > 1 \\[2mm] \infty & \text{if } p \leq 1 \end{cases}$$

EXAMPLE 4   Improper integral using l'Hôpital's rule and integration by parts

Evaluate $\int_0^{+\infty} x e^{-2x}\, dx$.

Solution

$$\int_0^{+\infty} x e^{-2x}\, dx = \lim_{N \to +\infty} \int_0^N \underbrace{x}_{u}\, \underbrace{e^{-2x}\, dx}_{dv} \qquad \boxed{\begin{array}{l} \text{By parts:} \\ \text{Let } u = x \quad \text{and} \quad dv = e^{-2x}\, dx \\ \quad\; du = dx \qquad\qquad v = \frac{1}{-2}e^{-2x} \end{array}}$$

$$= \lim_{N \to +\infty} \left[\left(\underbrace{\frac{x}{-2}}_{u}\, \underbrace{e^{-2x}}_{v} \right) \Big|_0^N - \int_0^N \underbrace{\frac{1}{-2}e^{-2x}}_{v}\, \underbrace{dx}_{du} \right]$$

$$= \lim_{N \to +\infty} \left[\frac{-x e^{-2x}}{2} - \frac{e^{-2x}}{4} \right] \Big|_0^N$$

$$= \lim_{N \to +\infty} \left[-\frac{1}{2} N e^{-2N} - \frac{1}{4} e^{-2N} + 0 + \frac{1}{4} \right]$$

$$= -\frac{1}{2} \lim_{N \to +\infty} \left(\frac{N}{e^{2N}} \right) + \frac{1}{4}$$

$$= -\frac{1}{2} \lim_{N \to +\infty} \left(\frac{1}{2 e^{2N}} \right) + \frac{1}{4} \qquad \text{l'Hôpital's rule}$$

$$= \frac{1}{4}$$ ∎

EXAMPLE 5   Gabriel's horn: a solid with a finite volume but infinite surface area

Gabriel's horn is the name given to the solid formed by revolving about the x-axis the unbounded region under the curve $y = \frac{1}{x}$ for $x \geq 1$. Show that this solid has finite volume but infinite surface area.

Solution

We will find the volume, V, by using disks, as shown in Figure 7.17.

$$V = \pi \int_1^{+\infty} (x^{-1})^2\, dx = \pi \lim_{N \to +\infty} \int_1^N x^{-2}\, dx = \pi \lim_{N \to +\infty} \left(-\frac{1}{N} + 1 \right) = \pi$$

We also find the surface area, S:

$$S = 2\pi \int_1^{+\infty} f(x)\sqrt{1 + [f'(x)]^2}\, dx = 2\pi \int_1^{+\infty} \frac{1}{x}\sqrt{1 + \frac{1}{x^4}}\, dx$$

$$= 2\pi \int_1^{+\infty} \frac{1}{x}\sqrt{\frac{x^4 + 1}{x^4}}\, dx = 2\pi \int_1^{+\infty} \frac{\sqrt{x^4 + 1}}{x^3}\, dx \qquad \boxed{\begin{array}{l} \text{Let } u = x^2 \\ \quad du = 2x\, dx \end{array}}$$

$$= 2\pi \int_1^{+\infty} \frac{\sqrt{u^2 + 1}}{2u^2}\, du = \pi \lim_{N \to +\infty} \int_1^N \frac{\sqrt{u^2 + 1}}{u^2}\, du$$

$$= \pi \lim_{N \to +\infty} \left[\frac{-\sqrt{u^2 + 1}}{u} + \ln \left| u + \sqrt{u^2 + 1} \right| \right] \Big|_1^N \qquad \text{Formula 180}$$

$$= +\infty \quad \text{(The details are left for the reader.)}$$ ∎

We can also define an improper integral on an interval that is unbounded to the left or on the entire real number line.

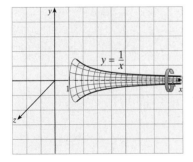

Figure 7.17 Gabriel's horn

You can fill Gabriel's horn with a finite amount of paint, but it takes an infinite amount of paint to color its inside surface!

Let b be a fixed number and assume $\int_t^b f(x)\,dx$ exists for all $t < b$. Then if

$$\lim_{t \to -\infty} \int_t^b f(x)\,dx \text{ exists, we define the } \textbf{improper integral}$$

$$\int_{-\infty}^b f(x)\,dx = \lim_{t \to -\infty} \int_t^b f(x)\,dx$$

The improper integral $\int_{-\infty}^b f(x)\,dx$ is said to **converge** if this limit is a finite number and to **diverge** otherwise. If both

$$\int_a^{+\infty} f(x)\,dx \quad \text{and} \quad \int_{-\infty}^a f(x)\,dx$$

converge for some number a, the improper integral of $f(x)$ on the entire x-axis is defined by

$$\int_{-\infty}^{+\infty} f(x)\,dx = \int_{-\infty}^a f(x)\,dx + \int_a^{+\infty} f(x)\,dx$$

EXAMPLE 6   Improper integral to negative infinity

a. Evaluate $\int_{-\infty}^0 \dfrac{dx}{1 + x^2}$.

b. How would you expect the value of $\int_{-\infty}^{\infty} \dfrac{dx}{1 + x^2}$ to be related to the improper integral in **a**? Verify your conjecture by evaluating this integral.

Solution

a.
$$\int_{-\infty}^0 \frac{dx}{1 + x^2} = \lim_{t \to -\infty} \int_t^0 \frac{dx}{1 + x^2}$$
$$= \lim_{t \to -\infty} \tan^{-1} x \Big|_t^0$$
$$= \lim_{t \to -\infty} \left[\tan^{-1} 0 - \tan^{-1} t \right]$$
$$= 0 - \left(-\frac{\pi}{2} \right)$$
$$= \frac{\pi}{2}$$

b. Since $y(x) = y(-x)$, we see that $y = \dfrac{1}{1 + x^2}$ is symmetric about the y-axis, and it is reasonable to conjecture that the integral in part **b** is twice the integral in part **a**. We have

$$\int_{-\infty}^{+\infty} \frac{dx}{1 + x^2} = \lim_{s \to -\infty} \int_s^0 \frac{dx}{1 + x^2} + \lim_{N \to +\infty} \int_0^N \frac{dx}{1 + x^2}$$
$$= \frac{\pi}{2} + \lim_{N \to +\infty} (\tan^{-1} N - \tan^{-1} 0)$$
$$= \frac{\pi}{2} + \left(\frac{\pi}{2} - 0 \right)$$
$$= \pi$$

We see this integral is indeed twice that in part **a**, as conjectured.   ■

EXAMPLE 7   An improper polar integral

Evaluate $\displaystyle\int_0^{+\infty} \frac{1}{2} e^{-2\theta}\, d\theta$.

Solution

$$\int_0^{+\infty} \frac{1}{2} e^{-2\theta}\, d\theta = \lim_{M\to+\infty} \int_0^M \frac{1}{2} e^{-2\theta}\, d\theta$$

$$= \lim_{M\to+\infty} \left. -\frac{1}{4} e^{-2\theta} \right|_0^M$$

$$= \lim_{M\to+\infty} \left[-\frac{1}{4} e^{-2M} - \left(-\frac{1}{4} \right) \right]$$

$$= \frac{1}{4} \qquad\blacksquare$$

IMPROPER INTEGRALS WITH UNBOUNDED INTEGRANDS

A function f is **unbounded** at c if it has arbitrarily large values near c. Geometrically, this occurs when the line $x = c$ is a vertical asymptote to the graph of f at c, as shown in Figure 7.18.

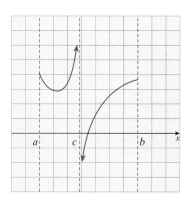

Figure 7.18   Two functions that are unbounded at c

If f is unbounded at c and $a \le c \le b$, the Riemann integral $\int_a^b f(x)\, dx$ is not even defined (only bounded functions are Riemann integrable). However, it may still be possible to define $\int_a^b f(x)\, dx$ as an improper integral in certain cases.

Let us examine a specific problem. Suppose

$$f(x) = \frac{1}{\sqrt{x}} \quad \text{for} \quad 0 < x \le 1$$

Then f is unbounded at $x = 0$ and $\int_0^1 f(x)\, dx$ is not defined. However,

$$f(x) = \frac{1}{\sqrt{x}}$$

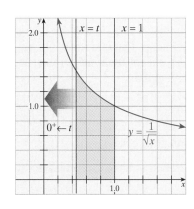

Figure 7.19   Graph of $y = \dfrac{1}{\sqrt{x}}$

is continuous on every interval $[t, 1]$ for $t > 0$, as shown in Figure 7.19. For any such interval $[t, 1]$, we have

$$\int_t^1 \frac{dx}{\sqrt{x}} = \int_t^1 x^{-1/2}\, dx = \left. 2\sqrt{x} \right|_t^1 = 2 - 2\sqrt{t}$$

If we let $t \to 0$ through positive values, we see that

$$\lim_{t\to 0^+} \int_t^1 \frac{dx}{\sqrt{x}} = \lim_{t\to 0^+} (2 - 2\sqrt{t}) = 2$$

This is called a **convergent improper integral** with value 2, and it seems reasonable to say

$$\int_0^1 \frac{dx}{\sqrt{x}} = \lim_{t \to 0^+} \int_t^1 \frac{dx}{\sqrt{x}} = 2$$

In this example, f is unbounded at the left endpoint of the interval of integration, but similar reasoning would apply if it were unbounded at the right endpoint or at an interior point. Here is a definition of this kind of improper integral.

Improper Integral (Second Type)

If f is unbounded at a and $\int_t^b f(x)\,dx$ exists for all t such that $a < t \le b$, then

$$\int_a^b f(x)\,dx = \lim_{t \to a^+} \int_t^b f(x)\,dx$$

If the limit exists (as a finite number), we say that the improper integral **converges**; otherwise, the improper integral **diverges**. Similarly, if f is unbounded at b and $\int_a^t f(x)\,dx$ exists for all t such that $a \le t < b$, then

$$\int_a^b f(x)\,dx = \lim_{t \to b^-} \int_a^t f(x)\,dx$$

If f is unbounded at c, where $a < c < b$, and the improper integrals $\int_a^c f(x)\,dx$ and $\int_c^b f(x)\,dx$ both converge, then

$$\int_a^b f(x)\,dx = \int_a^c f(x)\,dx + \int_c^b f(x)\,dx$$

We say that the integral on the left diverges if *either* (or both) of the integrals on the right diverges.

➤ **What This Says**   If a function is continuous on the open interval (a, b) but is unbounded at one of its endpoints of the interval, then replace that endpoint by t, evaluate the integral, and take the limit as t approaches that endpoint. On the other hand, if f is continuous on the interval $[a, b]$, except for some c in (a, b) where f has an infinite discontinuity, rewrite the integral as

$$\int_a^b f(x)\,dx = \int_a^c f(x)\,dx + \int_c^b f(x)\,dx$$

You will then need limits to evaluate each of the integrals on the right.

EXAMPLE 8   Improper integral at a right endpoint

Find $\int_0^1 \frac{dx}{(x-1)^{2/3}}$.

Solution

$f(x) = (x-1)^{-2/3}$ is unbounded at the right endpoint of the interval of integration and is continuous on $[0, t]$ for any t with $0 \le t < 1$. We find that

$$\int_0^1 \frac{dx}{(x-1)^{2/3}} = \lim_{t \to 1^-} \int_0^t \frac{dx}{(x-1)^{2/3}} = \lim_{t \to 1^-} [3(x-1)^{1/3}]\Big|_0^t$$

$$= 3 \lim_{t \to 1^-} [(t-1)^{1/3} - (-1)] = 3$$

That is, the improper integral converges and has the value 3. ∎

EXAMPLE 9 Improper integral at a left endpoint

Find $\displaystyle\int_{\pi/2}^{\pi} \sec x \, dx$.

Solution

Because $\sec x$ is unbounded at the left endpoint $\frac{\pi}{2}$ of the interval of integration and is continuous on $[t, \pi]$ for any t with $\frac{\pi}{2} < t \le \pi$, we find that

$$\int_{\pi/2}^{\pi} \sec x \, dx = \lim_{t \to (\pi/2)^+} \int_t^{\pi} \sec x \, dx = \lim_{t \to (\pi/2)^+} \ln|\sec x + \tan x|\big|_t^{\pi}$$

$$= \lim_{t \to (\pi/2)^+} [\ln|-1+0| - \ln|\sec t + \tan t|] = -\infty$$

Thus, the integral diverges. ∎

EXAMPLE 10 Improper integral at an interior point

Find $\displaystyle\int_0^3 (x-2)^{-1} \, dx$.

Solution

WARNING A common mistake is to fail to notice the discontinuity at $x = 2$. It is WRONG to write

$$\int_0^3 (x-2)^{-1} \, dx = \ln|x-2|\big|_0^3$$
$$= \ln 1 - \ln 2$$
$$= -\ln 2$$

Notice that this mistake leads to the conclusion that the integral converges, which, as you can see, is incorrect. You must also be cautious in using computer software with improper integrals, because it may not detect that the integral is improper.

The integral is improper because the integrand is unbounded at $x = 2$. If the improper integral converges, we have

$$\int_0^3 (x-2)^{-1} \, dx = \int_0^2 (x-2)^{-1} \, dx + \int_2^3 (x-2)^{-1} \, dx$$

$$= \lim_{t \to 2^-} \int_0^t (x-2)^{-1} \, dx + \lim_{t \to 2^+} \int_t^3 (x-2)^{-1} \, dx$$

If either of these limits fails to exist, then the original integral diverges. Because

$$\lim_{t \to 2^-} \int_0^t (x-2)^{-1} \, dx = \lim_{t \to 2^-} \ln|x-2|\big|_0^t$$
$$= \lim_{t \to 2^-} [\ln|t-2| - \ln 2]$$
$$= -\infty$$

we find that the original integral diverges. ∎

7.7 PROBLEM SET

Ⓐ **1. WHAT DOES THIS SAY?** What is an improper integral?

2. WHAT DOES THIS SAY? Discuss the different types of improper integrals.

In Problems 3–46, either show that the improper integral converges and find its value, or show that it diverges.

3. $\displaystyle\int_1^{+\infty} \frac{dx}{x^3}$

4. $\displaystyle\int_1^{+\infty} \frac{dx}{\sqrt[3]{x}}$

5. $\displaystyle\int_1^{+\infty} \frac{dx}{x^{0.99}}$

6. $\displaystyle\int_1^{+\infty} \frac{dx}{\sqrt{x}}$

7. $\displaystyle\int_1^{+\infty} \frac{dx}{x^{1.1}}$

8. $\displaystyle\int_1^{+\infty} x^{-2/3} \, dx$

9. $\displaystyle\int_3^{+\infty} \frac{dx}{2x-1}$

10. $\displaystyle\int_3^{+\infty} \frac{dx}{\sqrt[3]{2x-1}}$

11. $\displaystyle\int_3^{+\infty} \frac{dx}{(2x-1)^2}$

12. $\displaystyle\int_0^{+\infty} e^{-x} \, dx$

13. $\displaystyle\int_0^{+\infty} 5e^{-2x} \, dx$

14. $\displaystyle\int_1^{+\infty} e^{1-x} \, dx$

15. $\displaystyle\int_1^{+\infty} \frac{x^2\,dx}{(x^3+2)^2}$

16. $\displaystyle\int_1^{+\infty} \frac{x^2\,dx}{x^3+2}$

17. $\displaystyle\int_1^{+\infty} \frac{x^2\,dx}{\sqrt{x^3+2}}$

18. $\displaystyle\int_0^{+\infty} xe^{-x^2}\,dx$

19. $\displaystyle\int_1^{+\infty} \frac{e^{-\sqrt{x}}}{\sqrt{x}}\,dx$

20. $\displaystyle\int_0^{+\infty} xe^{-x}\,dx$

21. $\displaystyle\int_0^{+\infty} 5xe^{10-x}\,dx$

22. $\displaystyle\int_1^{+\infty} \frac{\ln x\,dx}{x}$

23. $\displaystyle\int_2^{+\infty} \frac{dx}{x\ln x}$

24. $\displaystyle\int_2^{+\infty} \frac{dx}{x\sqrt{\ln x}}$

25. $\displaystyle\int_0^{+\infty} x^2 e^{-x}\,dx$

26. $\displaystyle\int_0^{+\infty} x^3 e^{-x^2}\,dx$

27. $\displaystyle\int_{-\infty}^0 \frac{2x\,dx}{x^2+1}$

28. $\displaystyle\int_1^{+\infty} \frac{x\,dx}{(1+x^2)^2}$

29. $\displaystyle\int_{-\infty}^0 \frac{dx}{\sqrt{2-x}}$

30. $\displaystyle\int_{-\infty}^4 \frac{dx}{(5-x)^2}$

31. $\displaystyle\int_{-\infty}^{+\infty} xe^{-|x|}\,dx$

32. $\displaystyle\int_{-\infty}^{+\infty} \frac{dx}{x^2+1}$

33. $\displaystyle\int_0^1 \frac{dx}{x^{1/5}}$

34. $\displaystyle\int_0^4 \frac{dx}{x\sqrt{x}}$

35. $\displaystyle\int_0^1 \frac{dx}{(1-x)^{1/2}}$

36. $\displaystyle\int_0^2 \frac{dx}{(1-x)^2}$

37. $\displaystyle\int_{-1}^1 \frac{e^x}{\sqrt[3]{1-e^x}}\,dx$

38. $\displaystyle\int_{-\infty}^{+\infty} \frac{3x\,dx}{(3x^2+2)^3}$

39. $\displaystyle\int_0^1 \ln x\,dx$

40. $\displaystyle\int_1^{+\infty} \ln x\,dx$

41. $\displaystyle\int_e^{+\infty} \frac{dx}{x(\ln x)^2}$

42. $\displaystyle\int_0^1 \frac{x\,dx}{1-x^2}$

43. $\displaystyle\int_0^1 e^{-(1/2)\ln x}\,dx$

44. $\displaystyle\int_0^{+\infty} \frac{dx}{e^x+e^{-x}}$

45. $\displaystyle\int_0^{\pi/3} \frac{\sec^2 x\,dx}{1-\tan x}$

46. $\displaystyle\int_0^{\pi/2} \frac{\sin x\,dx}{\sqrt[3]{1-2\cos x}}$

Ⓑ 47. Find the area of the unbounded region between the x-axis and the curve

$$y = \frac{2}{(x-4)^3} \quad \text{for } x \geq 6$$

48. Find the area of the unbounded region between the x-axis and the curve

$$y = \frac{2}{(x-4)^3} \quad \text{for } x \leq 2$$

49. The total amount of radioactive material present in the atmosphere at time T is modeled by

$$A = \int_0^T Pe^{-rt}\,dt$$

where P is a constant and t is the number of years. Suppose a recent United Nations publication indicates that, at the present time, $r = 0.002$ and $P = 200$ millirads. Estimate the total future buildup of radioactive material in the atmosphere if these values remain constant.

50. Suppose that an oil well produces $P(t)$ thousand barrels of crude oil per month according to the formula

$$P(t) = 100e^{-0.02t} - 100e^{-0.1t}$$

where t is the number of months the well has been in production. What is the total amount of oil produced by the oil well?

51. Let $f(x) = \begin{cases} \dfrac{1}{x^2} & \text{for } x \geq 1 \\ 1 & \text{for } -1 < x < 1 \\ e^{x+1} & \text{for } x \leq -1 \end{cases}$

Sketch the graph of f and evaluate

$$\int_{-\infty}^{+\infty} f(x)\,dx$$

52. Find all values of p for which $\displaystyle\int_2^{+\infty} \frac{dx}{x(\ln x)^p}$ converges, and find the value of the integral when it exists.

53. Find all values of p for which $\displaystyle\int_0^1 \frac{dx}{x^p}$ converges, and find the value of the integral when it exists.

54. Find all values of p for which $\displaystyle\int_0^{1/2} \frac{dx}{x(\ln x)^p}$ converges, and find the value of the integral when it exists.

55. Counterexample Problem Discuss the calculation

$$\int_{-1}^1 \frac{dx}{x^2} = \left.\frac{-1}{x}\right|_{-1}^1 = -[1-(-1)] = -2$$

Is the calculation correct? Explain.

56. Journal Problem *College Mathematics Journal*[*] ∎ Peter Lindstrom of North Lake College in Irving, Texas, had a student who handled an ∞/∞ form as follows:

$$\int_1^{+\infty} (x-1)e^{-x}\,dx = \int_1^{+\infty} \frac{x-1}{e^x}\,dx$$
$$= \int_1^{+\infty} \frac{1}{e^x} \quad \text{l'Hôpital's rule}$$
$$= \frac{1}{e}$$

What is wrong, if anything, with this student's solution?

57. Find $\displaystyle\int_0^2 f(x)\,dx$, where

$$f(x) = \begin{cases} \dfrac{1}{\sqrt[4]{x^3}} & \text{for } 0 < x \leq 1 \\ \dfrac{1}{\sqrt[4]{(x-1)^3}} & \text{for } 1 < x < 2 \end{cases}$$

58. Evaluate the improper polar integral $\displaystyle\int_0^{+\infty} \theta e^{-\theta}\,d\theta$.

59. Find the total area between the spirals $r = e^{-2\theta}$ and $r = e^{-5\theta}$ for $\theta \geq 0$.

C *The **Laplace transform** of the function f is defined by the improper integral*

$$F(s) = \mathcal{L}\{f(t)\} = \int_0^{+\infty} e^{-st} f(t)\, dt$$

where s is a constant. This notation is used in Problems 60–67.

60. Show that for constant a (with $s - a > 0$)

  a. $\mathcal{L}\{e^{at}\} = \dfrac{1}{s-a}$ **b.** $\mathcal{L}\{a\} = \dfrac{a}{s}$

  c. $\mathcal{L}(t) = \dfrac{1}{s^2}$ **d.** $\mathcal{L}(t^n) = \dfrac{n!}{s^{n+1}}$

  e. $\mathcal{L}\{\cos at\} = \dfrac{s}{s^2 + a^2}$ **f.** $\mathcal{L}\{\sin at\} = \dfrac{a}{s^2 + a^2}$

61. Show that $\mathcal{L}\{af + bg\} = a\mathcal{L}\{f\} + b\mathcal{L}\{g\}$.

62. Let $\mathcal{L}^{-1}\{F(s)\}$ denote the inverse Laplace transform of $F(s)$; that is, the function $f(t)$ such that $\mathcal{L}\{f(t)\} = F(s)$. Find the following inverse Laplace transformations (Problems 60 and 61 may help).

  a. $\mathcal{L}^{-1}\left\{\dfrac{5}{s}\right\}$ **b.** $\mathcal{L}^{-1}\left\{\dfrac{s+2}{s^2+4}\right\}$

  c. $\mathcal{L}^{-1}\left\{\dfrac{2s^2 - 3s + 3}{s^2(s-1)}\right\}$

  Hint: Use partial fraction decomposition.

  d. $\mathcal{L}^{-1}\left\{\dfrac{3s^3 + 2s^2 - 3s - 17}{(s^2+4)(s^2-1)}\right\}$

63. If $F(s) = \mathcal{L}\{f(t)\}$, show that $\mathcal{L}\{e^{at} f(t)\} = F(s-a)$.

64. Use the result in Problem 63, along with those of Problem 60, to find the following Laplace transforms and inverse transforms.

  a. $\mathcal{L}\left\{t^3 e^{-2t}\right\}$ **b.** $\mathcal{L}\left\{e^{-3t} \cos 2t\right\}$

  c. $\mathcal{L}^{-1}\left\{\dfrac{5}{(s-1)^2}\right\}$

  d. $\mathcal{L}^{-1}\left\{\dfrac{4s}{s^2 + 4s + 5}\right\}$

  Hint: Complete the square.

65. a. If $F(s) = \mathcal{L}\{f(t)\}$, show that $\mathcal{L}\{tf(t)\} = -F'(s)$. You may assume that

$$\frac{d}{ds}\int_0^{+\infty} e^{-st} f(t)\, dt = \int_0^{+\infty} \frac{d}{ds}[e^{-st} f(t)\, dt]$$

  b. Use the result to find $\mathcal{L}\{t \cos 2t\}$.

66. What is $\mathcal{L}\{t^2 f(t)\}$? What about $\mathcal{L}\{t^n f(t)\}$?

67. Let $\mathcal{L}\{f\} = F(s)$. Suppose that f is continuous on $[0, \infty)$ and that f' is piecewise continuous on $[0, k]$ for every positive number k. Further suppose that $\lim\limits_{k\to\infty} e^{-sk} f(k) = 0$ if $s > 0$.

  a. Show that $\mathcal{L}\{f'\} = sF(s) - f(0)$. *Hint*: Use integration by parts.

  b. Apply the result in part **a** to find $\mathcal{L}\{y\}$, where y satisfies the initial value problem

$$y' + 3y = \sin 2t$$

  with $y(0) = 2$.

  c. Solve the initial value problem in part **b** by using inverse Laplace transforms to obtain y.

7.8   Hyperbolic and Inverse Hyperbolic Functions

IN THIS SECTION   hyperbolic functions, derivatives and integrals involving hyperbolic functions, inverse hyperbolic functions

HYPERBOLIC FUNCTIONS

In physics, it is shown that a heavy, flexible cable (for example, a power line) that is suspended between two points at the same height assumes the shape of a curve called a **catenary** (see Figure 7.20), with an equation of the form

$$y = \frac{a}{2}\left(e^{x/a} + e^{-x/a}\right)$$

This is one of several important applications that involve sums of exponential functions. The goal of this section is to study such sums and their inverses.

In certain ways, the functions we will study are analogous to the trigonometric functions, and they have essentially the same relationship to the hyperbola that the trigonometric functions have to the circle. For this reason, these functions are called **hyperbolic functions**. Three basic hyperbolic functions are the **hyperbolic sine** (denoted $\sinh x$ and pronounced "cinch"), the **hyperbolic cosine** ($\cosh x$; pronounced "kosh"), and the **hyperbolic tangent** ($\tanh x$; pronounced "tansh"). They are defined as follows.

Figure 7.20 The hanging cable problem

Hyperbolic Functions

$$\sinh x = \frac{e^x - e^{-x}}{2} \qquad \text{for all } x$$

$$\cosh x = \frac{e^x + e^{-x}}{2} \qquad \text{for all } x$$

$$\tanh x = \frac{\sinh x}{\cosh x} = \frac{e^x - e^{-x}}{e^x + e^{-x}} \qquad \text{for all } x$$

Graphs of the three basic hyperbolic functions are shown in Figure 7.21.

a. The hyperbolic sine
$y = \sinh x$

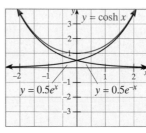

b. The hyperbolic cosine
$y = \cosh x$

c. The hyperbolic tangent
$y = \tanh x$

Figure 7.21   Graphs of the three basic hyperbolic functions

The list of properties in the following theorem suggests that the basic hyperbolic functions are analogous to the trigonometric functions.

THEOREM 7.2   Properties of the basic hyperbolic functions

$$\cosh^2 x - \sinh^2 x = 1$$
$$\sinh(-x) = -\sinh x \qquad (\sinh x \text{ is odd})$$
$$\cosh(-x) = \cosh x \qquad (\cosh x \text{ is even})$$
$$\tanh(-x) = -\tanh x \qquad (\tanh x \text{ is odd})$$
$$\sinh(x + y) = \sinh x \cosh y + \cosh x \sinh y$$
$$\cosh(x + y) = \cosh x \cosh y + \sinh x \sinh y$$

WARNING A major difference between the trigonometric and hyperbolic functions is that the trigonometric functions are periodic, but the hyperbolic functions are not.

Proof   We will verify the first identity and leave the others for the reader.

$$\cosh^2 x - \sinh^2 x = \left(\frac{e^x + e^{-x}}{2}\right)^2 - \left(\frac{e^x - e^{-x}}{2}\right)^2$$

$$= \frac{e^{2x} + 2 + e^{-2x}}{4} - \frac{e^{2x} - 2 + e^{-2x}}{4} = 1 \qquad \square$$

TECHNOLOGY NOTE

Some of the software packages show these in simplified form as:

$$\coth x = \frac{e^{2x} + 1}{e^{2x} - 1}$$

$$\text{sech } x = \frac{2e^x}{e^{2x} + 1}$$

$$\text{csch } x = \frac{2e^x}{e^{2x} - 1}$$

There are three additional hyperbolic functions: the **hyperbolic cotangent** (coth x), the **hyperbolic secant** (sech x), and the **hyperbolic cosecant** (csch x). These functions are defined as follows:

$$\coth x = \frac{1}{\tanh x} = \frac{e^x + e^{-x}}{e^x - e^{-x}}$$

$$\text{sech } x = \frac{1}{\cosh x} = \frac{2}{e^x + e^{-x}}$$

$$\text{csch } x = \frac{1}{\sinh x} = \frac{2}{e^x - e^{-x}}$$

Two identities involving these functions are

$$\operatorname{sech}^2 x = 1 - \tanh^2 x \quad \text{and} \quad \operatorname{csch}^2 x = \coth^2 x - 1$$

You will be asked to verify these identities in the problems.

DERIVATIVES AND INTEGRALS INVOLVING HYPERBOLIC FUNCTIONS

Rules for differentiating the hyperbolic functions are listed in Theorem 7.3.

THEOREM 7.3   Rules for differentiating hyperbolic functions

Let u be a differentiable function of x. Then

$$\frac{d}{dx}(\sinh u) = \cosh u \frac{du}{dx} \qquad\qquad \frac{d}{dx}(\cosh u) = \sinh u \frac{du}{dx}$$

$$\frac{d}{dx}(\tanh u) = \operatorname{sech}^2 u \frac{du}{dx} \qquad\qquad \frac{d}{dx}(\coth u) = -\operatorname{csch}^2 u \frac{du}{dx}$$

$$\frac{d}{dx}(\operatorname{sech} u) = -\operatorname{sech} u \tanh u \frac{du}{dx} \qquad \frac{d}{dx}(\operatorname{csch} u) = -\operatorname{csch} u \coth u \frac{du}{dx}$$

Proof   Each of these rules can be obtained by differentiating the exponential functions that make up the appropriate hyperbolic function. For example, to differentiate $\sinh x$, we use the definition of $\sinh x$:

$$\frac{d}{dx}(\sinh x) = \frac{d}{dx}\left(\frac{e^x - e^{-x}}{2}\right)$$

$$= \tfrac{1}{2}e^x - \tfrac{1}{2}e^{-x}(-1) = \tfrac{1}{2}(e^x + e^{-x}) = \cosh x$$

The proofs for the other derivatives can be handled similarly.   ❏

EXAMPLE 1   Derivatives involving hyperbolic functions

Find $\dfrac{dy}{dx}$ for each of the following functions:

a. $y = \cosh Ax$ (A is constant)   **b.** $y = \tanh(x^2 + 1)$ **c.** $y = \ln(\sinh x)$

Solution

a. We have

$$\frac{d}{dx}(\cosh Ax) = \sinh(Ax)\frac{d}{dx}(Ax) = A \sinh Ax$$

b. We find that

$$\frac{d}{dx}[\tanh(x^2 + 1)] = \operatorname{sech}^2(x^2 + 1)\frac{d}{dx}(x^2 + 1) = 2x \operatorname{sech}^2(x^2 + 1)$$

c. Using the chain rule, with $u = \sinh x$, we obtain

$$\frac{d}{dx}[\ln(\sinh x)] = \frac{1}{\sinh x}\frac{d}{dx}(\sinh x) = \frac{1}{\sinh x}(\cosh x) = \coth x \qquad ■$$

Each differentiation formula for hyperbolic functions corresponds to an integration formula. These formulas are listed in the following theorem.

THEOREM 7.4   Integration formulas involving the hyperbolic functions

$$\int \sinh x \, dx = \cosh x + C \qquad\qquad \int \cosh x \, dx = \sinh x + C$$

$$\int \operatorname{sech}^2 x \, dx = \tanh x + C \qquad\qquad \int \operatorname{csch}^2 x \, dx = -\coth x + C$$

$$\int \operatorname{sech} x \tanh x \, dx = -\operatorname{sech} x + C \qquad \int \operatorname{csch} x \coth x \, dx = -\operatorname{csch} x + C$$

Proof   The proof of each of these formulas follows directly from the corresponding derivative formula.   ❑

EXAMPLE 2   Integration involving hyperbolic forms

Find each of the following integrals:

a. $\displaystyle\int \cosh^3 x \sinh x\, dx$ **b.** $\displaystyle\int x \operatorname{sech}^2(x^2)\, dx$ **c.** $\displaystyle\int \tanh x\, dx$

Solution

a. $\displaystyle\int \cosh^3 x \sinh x\, dx = \int u^3\, du = \frac{u^4}{4} + C = \frac{1}{4}\cosh^4 x + C$

$\boxed{\text{Let } u = \cosh x;\ du = \sinh x\, dx}$

b. $\displaystyle\int x \operatorname{sech}^2(x^2)\, dx = \int \operatorname{sech}^2(x^2)(x\, dx)$

$\boxed{\text{Let } u = x^2;\ du = 2x\, dx}$

$$= \int \operatorname{sech}^2 u \left(\frac{1}{2}\, du\right)$$

$$= \tfrac{1}{2}\tanh u + C$$

$$= \tfrac{1}{2}\tanh x^2 + C$$

c. $\displaystyle\int \tanh x\, dx = \int \frac{\sinh x\, dx}{\cosh x}$

$\boxed{\text{Let } u = \cosh x;\ du = \sinh x\, dx}$

$$= \int \frac{du}{u} = \ln|u| + C = \ln(\cosh x) + C \qquad ■$$

INVERSE HYPERBOLIC FUNCTIONS

Inverse hyperbolic functions are also of interest, mainly because they enable us to express certain integrals in simple terms.

Because $\sinh x$ is continuous and strictly monotonic (increasing), it is one-to-one and has an inverse, which is defined by

$$y = \sinh^{-1} x \quad \text{if and only if} \quad x = \sinh y$$

for all x and y. This is called the **inverse hyperbolic sine** function, and its graph is obtained by reflecting the graph of $y = \sinh x$ in the line $y = x$, as shown in Figure 7.22. Other inverse hyperbolic functions are defined similarly (see Problem 60).

Because the hyperbolic functions are defined in terms of exponential functions, we may expect to be able to express the inverse hyperbolic functions in terms of logarithmic functions. We summarize these relationships in the following theorem.

The graph of $y = \sinh^{-1} x$ is obtained by reflecting $y = \sinh x$ in the line $y = x$.

Figure 7.22 The graph of $y = \sinh^{-1} x$

THEOREM 7.5   Logarithmic formulas for inverse hyperbolic functions

$$\sinh^{-1} x = \ln\!\left(x + \sqrt{x^2 + 1}\right),\ \text{all } x \qquad \operatorname{csch}^{-1} x = \ln\!\left(\frac{1}{x} + \frac{\sqrt{1 + x^2}}{|x|}\right),\ x \neq 0$$

$$\cosh^{-1} x = \ln\!\left(x + \sqrt{x^2 - 1}\right),\ x \geq 1 \qquad \operatorname{sech}^{-1} x = \ln\!\left(\frac{1 + \sqrt{1 - x^2}}{x}\right),\ 0 < x \leq 1$$

$$\tanh^{-1} x = \frac{1}{2}\ln\frac{1 + x}{1 - x},\ |x| < 1 \qquad \coth^{-1} x = \frac{1}{2}\ln\frac{x + 1}{x - 1},\ |x| > 1$$

Proof   We will prove the first formula and leave the next two formulas for you to verify (see Problem 61). The other formulas are proved similarly. Let $y = \sinh^{-1} x$; then its inverse is

$$x = \sinh y$$
$$= \tfrac{1}{2}(e^y - e^{-y})$$
$$2x = e^y - \frac{1}{e^y}$$
$$e^{2y} - 2xe^y - 1 = 0 \qquad \text{Quadratic formula with}$$
$$\qquad\qquad\qquad\qquad\qquad a = 1, b = -2x, c = -1$$
$$e^y = \frac{2x \pm \sqrt{4x^2 + 4}}{2}$$
$$= x \pm \sqrt{x^2 + 1}$$

Because $e^y > 0$ for all y, the only solution is $e^y = x + \sqrt{x^2 + 1}$, and from the definition of logarithms,

$$y = \ln(x + \sqrt{x^2 + 1})$$

Differentiation and integration formulas involving inverse hyperbolic functions are listed in Theorem 7.6.

THEOREM 7.6   Differentiation and integration formulas involving the inverse hyperbolic functions

$$\frac{d}{dx}(\sinh^{-1} u) = \frac{1}{\sqrt{1 + u^2}}\frac{du}{dx} \qquad \int \frac{du}{\sqrt{1 + u^2}} = \sinh^{-1} u + C$$

$$\frac{d}{dx}(\cosh^{-1} u) = \frac{1}{\sqrt{u^2 - 1}}\frac{du}{dx} \qquad \int \frac{du}{\sqrt{u^2 - 1}} = \cosh^{-1} u + C$$

$$\frac{d}{dx}(\tanh^{-1} u) = \frac{1}{1 - u^2}\frac{du}{dx} \qquad \int \frac{du}{1 - u^2} = \tanh^{-1} u + C$$

$$\frac{d}{dx}(\operatorname{csch}^{-1} u) = \frac{-1}{|u|\sqrt{1 + u^2}}\frac{du}{dx} \qquad \int \frac{du}{u\sqrt{1 + u^2}} = -\operatorname{csch}^{-1} |u| + C$$

$$\frac{d}{dx}(\operatorname{sech}^{-1} u) = \frac{-1}{u\sqrt{1 - u^2}}\frac{du}{dx} \qquad \int \frac{du}{u\sqrt{1 - u^2}} = -\operatorname{sech}^{-1} |u| + C$$

$$\frac{d}{dx}(\coth^{-1} u) = \frac{1}{1 - u^2}\frac{du}{dx} \qquad \int \frac{du}{1 - u^2} = \coth^{-1} u + C$$

Proof   The derivative of each inverse hyperbolic function can be found either by ferentiating the appropriate logarithmic function or by using the definition in ter hyperbolic functions. We will show how this is done for $y = \sinh^{-1} x$ and then it for you to apply the chain rule for u, a differentiable function of x. By definit inverse, we see

$$x = \sinh y \quad \text{and} \quad \frac{dx}{dy} = \cosh y$$

Thus,

$$\frac{d}{dx}(\sinh^{-1} x) = \frac{dy}{dx} = \frac{1}{\dfrac{dx}{dy}} = \frac{1}{\cosh y} = \frac{1}{\sqrt{1 + \sinh^2 y}} = \frac{1}{\sqrt{1 + x^2}}$$

since $\cosh^2 y - \sinh^2 y = 1$ and $\sinh y = x$. The integration formula that cor to this differentiation formula is

$$\int \frac{dx}{\sqrt{1 + x^2}} = \sinh^{-1} x + C$$

The other differentiation and integration formulas follow similarly.

EXAMPLE 3 Derivatives involving inverse hyperbolic functions

Find $\dfrac{dy}{dx}$ for   **a.** $y = \sinh^{-1}(ax + b)$ **b.** $y = \cosh^{-1}(\sec x), 0 \le x < \dfrac{\pi}{2}$

Solution

a. $\dfrac{d}{dx}\left[\sinh^{-1}(ax + b)\right] = \dfrac{1}{\sqrt{1 + (ax + b)^2}} \dfrac{d}{dx}(ax + b)$

$= \dfrac{a}{\sqrt{1 + (ax + b)^2}}$

b. $\dfrac{d}{dx}\left[\cosh^{-1}(\sec x)\right] = \dfrac{1}{\sqrt{\sec^2 x - 1}} \dfrac{d}{dx}(\sec x)$

$= \dfrac{\sec x \tan x}{\sqrt{\tan^2 x}} = \sec x$ ∎

Recall that the formula

$$\int \sec x\, dx = \ln|\sec x + \tan x| + C$$

was derived using algebraic procedures in Section 7.1.

The result of Example 3b implies

$$\int \sec x\, dx = \cosh^{-1}(\sec x) + C$$

$$= \ln(\sec x + \sqrt{\sec^2 x - 1}) + C$$

$$= \ln(\sec x + \tan x) + C$$

Notice that we do not need absolute values because $0 \le x < \frac{\pi}{2}$.

EXAMPLE 4 Integral involving an inverse hyperbolic function

Evaluate $\displaystyle\int_0^1 \dfrac{dx}{\sqrt{1 + x^2}}$.

Solution

$$\int_0^1 \dfrac{dx}{\sqrt{1 + x^2}} = [\sinh^{-1} x]\Big|_0^1$$

$$= \ln(x + \sqrt{x^2 + 1})\Big|_0^1$$

$$= \ln(1 + \sqrt{2}) - \ln(1)$$

$$= \ln(1 + \sqrt{2})$$

It may be a worthwhile exercise to evaluate this integral using a trigonometric substitution to obtain the same result. ∎

7.8 PROBLEM SET

Evaluate (correct to four decimal places) the indicated hyperbolic or inverse hyperbolic functions in Problems 1–12.

1. $\sinh 2$

2. $\cosh 3$

3. $\tanh(-1)$

4. $\sinh^{-1} 0$

5. $\coth 1.2$

6. $\tanh^{-1} 0$

7. $\cosh^{-1} 1.5$

8. $\sinh(\ln 2)$

9. $\cosh(\ln 3)$

10. $\operatorname{sech}^{-1} 0.2$

11. $\operatorname{sech} 1$

12. $\coth^{-1}(-3)$

Find $\dfrac{dy}{dx}$ in Problems 13–29.

13. $y = \sinh 3x$

14. $y = \cosh(1 - 2x^2)$

15. $y = \cosh(2x^2 + 3x)$

16. $y = \sinh \sqrt{x}$

17. $y = \sinh x^{-1}$

18. $y = \cosh^{-1}(x^2)$

19. $y = \sinh^{-1}(x^3)$

20. $y = x \tanh^{-1}(3x)$

21. $y = \sinh^{-1}(\tan x)$

22. $y = \cosh^{-1}(\sec x)$

23. $y = \tanh^{-1}(\sin x)$

24. $y = \operatorname{sech}\left(\dfrac{1 - x}{1 + x}\right)$

25. $y = \dfrac{\sinh^{-1} x}{x}$

26. $y = \sinh^{-1} x - \sqrt{1 + x^2}$

27. $y = x \cosh^{-1} x - \sqrt{x^2 - 1}$

28. $x \cosh y = y \sinh x + 5$

29. $e^x \sinh^{-1} x + e^{-x} \cosh^{-1} y = 1$

Compute the integrals in Problems 30–45.

30. $\displaystyle\int x \cosh(1 - x^2)\, dx$

31. $\displaystyle\int \dfrac{\sinh \frac{1}{x}\, dx}{x^2}$

32. $\displaystyle\int \frac{\text{sech}^2(\ln x)\,dx}{x}$

33. $\displaystyle\int \coth x\,dx$

34. $\displaystyle\int \frac{dx}{\sqrt{4x^2 + 16}}$

35. $\displaystyle\int \frac{dt}{\sqrt{9t^2 - 16}}$

36. $\displaystyle\int \frac{dt}{36 - 16t^2}$

37. $\displaystyle\int \frac{\cos x\,dx}{\sqrt{1 + \sin^2 x}}$

38. $\displaystyle\int \frac{x\,dx}{\sqrt{1 + x^4}}$

39. $\displaystyle\int \frac{x^2\,dx}{1 - x^6}$

40. $\displaystyle\int_0^{1/2} \frac{dx}{1 - x^2}$

41. $\displaystyle\int_2^3 \frac{dx}{1 - x^2}$

42. $\displaystyle\int_0^1 \frac{t^5\,dt}{\sqrt{1 + t^{12}}}$

43. $\displaystyle\int_1^2 \frac{e^x\,dx}{\sqrt{e^{2x} - 1}}$

44. $\displaystyle\int_0^{\ln 2} \sinh 3x\,dx$

45. $\displaystyle\int_0^1 x\,\text{sech}^2 x^2\,dx$

B 46. Show that

   a. $\tanh(x + y) = \dfrac{\tanh x + \tanh y}{1 + \tanh x \tanh y}$

   b. $\sinh 2x = 2 \sinh x \cosh x$

   c. $\cosh 2x = \cosh^2 x + \sinh^2 x$

47. Evaluate $\displaystyle\lim_{x \to +\infty} \frac{\sinh^{-1} x}{\cosh^{-1} x}$

48. Show that

   a. $-1 < \tanh x < 1$ for all x

   b. $\displaystyle\lim_{x \to +\infty} \tanh x = 1$ and $\displaystyle\lim_{x \to -\infty} \tanh x = -1$

49. Determine where the graph of $y = \tanh x$ is rising and falling and where it is concave up and concave down. Sketch the graph and compare with Figure 7.21c.

50. Sketch the graph of $y = \coth x$. Be sure to show key features such as intercepts, high and low points, and points of inflection.

51. First show that $\cosh x + \sinh x = e^x$, and then use this result to prove that $(\sinh x + \cosh x)^n = \cosh nx + \sinh nx$ for positive integers n.

52. If $x = a \cosh t$ and $y = b \sinh t$ where a, b are positive constants and t is any number, show that

$$\frac{x^2}{a^2} - \frac{y^2}{b^2} = 1$$

53. a. Verify that $y = a \cosh cx + b \sinh cx$ satisfies the differential equation

$$y'' - c^2 y = 0$$

   b. Use part **a** to find a solution of the differential equation

$$y'' - 4y = 0$$

   subject to the initial conditions $y(0) = 1$, $y'(0) = 2$.

54. Find the volume V generated when the region bounded by the curves $y = \sinh x$, $y = \cosh x$, the y-axis, and the line $x = c$ ($c > 0$) is revolved about the x-axis. For what value of c does $V = 1$?

55. Find the length of the catenary $y = a \cosh \dfrac{x}{a}$ between $x = -a$ and $x = a$.

56. Find the volume of the solid formed by revolving the region bounded by the curve $y = \tanh x$ on the interval $[0, 1]$ about the x-axis.

57. Find the surface area of the solid generated by revolving the curve $y = \cosh x$ on the interval $[-1, 1]$ about the x-axis.

C 58. Prove the following formulas

   a. $\text{sech}^2 x + \tanh^2 x = 1$ **b.** $\coth^2 x - \text{csch}^2 x = 1$

59. Derive the differentiation formulas for $\cosh u$, $\tanh u$, and $\text{sech}\,u$, where u is a differentiable function of x.

60. First give definitions for $\cosh^{-1} x$, $\tanh^{-1} x$, and $\text{sech}^{-1} x$, and then derive the differentiation formulas for these functions by using these definitions.

61. Prove

   a. $\cosh^{-1} x = \ln(x + \sqrt{x^2 - 1})$, $x \geq 1$

   b. $\tanh^{-1} x = \dfrac{1}{2} \ln \dfrac{1 + x}{1 - x}$, $|x| < 1$

62. Show that

$$\int \frac{dx}{\sqrt{x^2 + a^2}} = \sinh^{-1} \frac{x}{a} + C$$

for constant $a > 0$. State and prove a similar formula for

$$\int \frac{dx}{\sqrt{x^2 - a^2}}$$

with $x > a > 0$.

CHAPTER 7 REVIEW

Proficiency Examination

CONCEPT PROBLEMS

1. Discuss the method of integration by substitution for both indefinite and definite integrals.

2. **a.** What is the formula for integration by parts?

   b. What is a reduction formula for integration?

3. **a.** When should you consider a trigonometric substitution?

   b. What are the substitutions to use when integrating rational trigonometric integrals?

4. When should you consider the method of partial fractions?

5. Outline a strategy for integration.

6. What is a first-order linear differential equation? Outline a procedure for solving such an equation.

7. What is an improper integral?

8. Define the hyperbolic sine, hyperbolic cosine, and hyperbolic tangent.

9. State the rules for differentiating and integrating the hyperbolic functions.

10. What are the logarithmic formulas for the inverse hyperbolic formulas?

11. State the differentiation and integration formulas for the inverse hyperbolic functions.

PRACTICE PROBLEMS

12. Evaluate each of the given numbers.

    a. $\tanh^{-1}(0.5)$ **b.** $\sinh(\ln 3)$ **c.** $\coth^{-1}(2)$

Find the integrals in Problem 13–18.

13. $\displaystyle\int \frac{2x+3}{\sqrt{x^2+1}}\,dx$

14. $\displaystyle\int x\sin 2x\,dx$

15. $\displaystyle\int \sinh(1-2x)\,dx$

16. $\displaystyle\int \frac{dx}{\sqrt{4-x^2}}$

17. $\displaystyle\int \frac{x^2\,dx}{(x^2+1)(x-1)}$

18. $\displaystyle\int \frac{x^3\,dx}{x^2-1}$

Evaluate the definite integrals in Problems 19–22.

19. $\displaystyle\int_1^2 x\ln x^3\,dx$

20. $\displaystyle\int_2^3 \frac{dx}{(x-1)^2(x+2)}$

21. $\displaystyle\int_3^4 \frac{dx}{2x-x^2}$

22. $\displaystyle\int_0^{\pi/4} \sec^3 x\tan x\,dx$

Determine whether each improper integral in Problems 23–26 converges, and if it does, find its value.

23. $\displaystyle\int_0^{+\infty} xe^{-2x}\,dx$

24. $\displaystyle\int_0^{\pi/4} \frac{\sec^2 x\,dx}{\sqrt{\tan x}}$

25. $\displaystyle\int_0^1 \frac{2x+3}{x^2(x-2)}\,dx$

26. $\displaystyle\int_0^{+\infty} e^{-x}\sin x\,dx$

27. Find $\dfrac{dy}{dx}$ for $y=\sqrt{\tanh^{-1} 2x}$.

28. Find the volume of the solid generated when the region under the curve

$$y=\frac{1}{\sqrt{9-x^2}}$$

on the interval $[0,2]$ is revolved about the y-axis.

29. Solve the first-order linear differential equation

$$\frac{dy}{dx}+\frac{xy}{x+1}=e^{-x}$$

subject to the condition $y=1$ when $x=0$.

30. A tank contains 200 gal of saturated brine with 2 lb of salt per gallon. A salt solution containing 1.3 lb of salt per gallon flows in at the rate of 5 gal/min, and the uniform mixture flows out at 3 gal/min. Find the amount of salt in the tank after 1 hour.

Supplementary Problems

Find the derivative $\dfrac{dy}{dx}$ in Problems 1–5.

1. $y=\tanh^{-1}\dfrac{1}{x}$

2. $y=x\cosh^{-1}(3x+1)$

3. $y=\dfrac{\sinh x}{e^x}$

4. $y=\sinh x\cosh x$

5. $y=x\sinh x+\sinh(e^x+e^{-x})$

Find the integrals in Problems 6–43.

6. $\displaystyle\int \cos^{-1} x\,dx$

7. $\displaystyle\int \frac{x^2\,dx}{\sqrt{4-x^2}}$

8. $\displaystyle\int \frac{x\,dx}{x^2-2x+5}$

9. $\displaystyle\int \frac{3x-2}{x^3-2x^2}\,dx$

10. $\displaystyle\int \frac{dx}{(x^2+x+1)^{3/2}}$

11. $\displaystyle\int \frac{dx}{\sqrt{x}(1+\sqrt[4]{x})}$

12. $\displaystyle\int \frac{\sqrt{9x^2-1}\,dx}{x}$

13. $\displaystyle\int x^2\tan^{-1} x\,dx$

14. $\displaystyle\int \frac{dx}{\sin x+\tan x}$

15. $\displaystyle\int e^x\sqrt{4-e^{2x}}\,dx$

16. $\displaystyle\int \cos\frac{x}{2}\sin\frac{x}{3}\,dx$

17. $\displaystyle\int \frac{\sqrt{1+1/x^2}}{x^5}\,dx$

18. $\displaystyle\int \frac{\sin x\,dx}{\cos^5 x}$

19. $\displaystyle\int \sqrt{1+\sin x}\,dx$

20. $\displaystyle\int \cos x\ln(\sin x)\,dx$

21. $\displaystyle\int \sin(\ln x)\,dx$

22. $\displaystyle\int e^{2x}\operatorname{sech}(e^{2x})\,dx$

23. $\displaystyle\int \frac{\sinh x\,dx}{2+\cosh x}$

24. $\displaystyle\int \frac{\tanh^{-1} x\,dx}{1-x^2}$

25. $\displaystyle\int x^2\cot^{-1} x\,dx$

26. $\displaystyle\int x(1+x)^{1/3}\,dx$

27. $\displaystyle\int \frac{x^2+2}{x^3+6x+1}\,dx$

28. $\displaystyle\int \frac{\sin x-\cos x}{(\sin x+\cos x)^{1/4}}\,dx$

29. $\displaystyle\int \cos(\sqrt{x+2})\,dx$

30. $\displaystyle\int \sqrt{5+2\sin^2 x}\sin 2x\,dx$

31. $\displaystyle\int \frac{x^3+2x}{x^4+4x^2+3}\,dx$

32. $\displaystyle\int \frac{x\,dx}{\sqrt{5-x^2}}$

33. $\displaystyle\int \frac{\sqrt{5-x^2}\,dx}{x}$

34. $\displaystyle\int \frac{\sqrt{x^2+x}\,dx}{x}$

35. $\displaystyle\int x^3(x^2+4)^{-1/2}\,dx$

36. $\displaystyle\int \sin^2\left(\frac{x}{2}\right)\cos^2\left(\frac{x}{2}\right)\,dx$

37. $\displaystyle\int \sec^3 x\tan x\,dx$

38. $\displaystyle\int \sec^5 x\tan^2 x\,dx$

39. $\displaystyle\int \frac{x^{1/3}\,dx}{x^{1/2}+x^{2/3}}$

40. $\displaystyle\int \frac{\sin x\,dx}{\sin x+\cos x}$

41. $\displaystyle\int \frac{\cos x+\sin x}{1+\cos x-\sin x}\,dx$

42. $\displaystyle\int \frac{dx}{x^{1/4}+1}$

43. $\displaystyle\int \frac{e^x\,dx}{e^{x/3}-e^{x/2}}$

In Problems 44–48, solve the given differential equations.

44. $\dfrac{dy}{dx}=\sqrt{\dfrac{x^2-1}{y^2+1}}$

45. $\dfrac{dy}{dx}=\dfrac{1-y}{1+x}$

46. $\dfrac{dy}{dx}-\dfrac{y}{2x}=\dfrac{1}{x^{1/2}+1}$

47. $y'+(\tan x)y=\sec^3 x$

48. $(\cos^3 x)y'=\sin^3 x\cos y$

Determine whether the improper integral in Problems 49–52 converges, and if it does, find its value.

49. $\displaystyle\int_0^{\pi/2} \frac{\cos x\,dx}{\sqrt{\sin x}}$

50. $\displaystyle\int_{-\infty}^{+\infty} \frac{dx}{4+x^2}$

51. $\displaystyle\int_{-1}^1 \frac{dx}{(2x+1)^{1/3}}$

52. $\displaystyle\int_1^{+\infty} \frac{dx}{x^4+x^2}$

53. If n is a positive integer and a is a positive number, find

$$\int_0^{+\infty} x^n e^{-ax}\,dx$$

54. Find $\displaystyle\int \frac{\sin x\,dx}{\sqrt{\cos 2x}}$.

55. Show that $\int_0^1 x^m(1-x)^n\,dx = \int_0^1 x^n(1-x)^m\,dx$ for positive integers m, n.

56. Counterexample Problem Each of the following equations may be either true or false. In each case, either show that the equation is generally true or find a number x for which it fails.

a. $\tanh\left(\frac{1}{2}\ln x\right) = \frac{x-1}{x+1}, x > 0$

b. $\sinh^{-1}(\tan x) = \tanh^{-1}(\sin x)$ for $-\frac{\pi}{2} < x < \frac{\pi}{2}$

57. Compute $\int_0^1 xf''(3x)\,dx$, given that $f'(0)$ is defined, $f(0) = 1$, $f(3) = 4$, and $f'(3) = -2$.

58. What is $\int_0^{\pi/2} [f(x) + f''(x)]\cos x\,dx$ if $f(0)$ is defined, $f(\frac{\pi}{2}) = 5$, and $f'(0) = -1$?

59. Newton's law of cooling says that the temperature T of a body satisfies
$$\frac{dT}{dt} = k(T - T_0)$$
where T_0 is the temperature of the surrounding medium. Solve this equation for T.

60. Modeling Problem The psychologist L. L. Thurstone investigated the way people learn using the differential equation
$$\frac{dS}{dt} = \frac{2k}{\sqrt{m}}[S(1-S)]^{3/2}$$
where $S(t)$ is the state of the learner at time t, and k and m are positive constants depending on the learner and the nature of the task.* Solve this equation.

61. If $s > 0$, the integral function defined by
$$\Gamma(s) = \int_0^{+\infty} e^{-t}t^{s-1}\,dt$$
is called the **gamma function**. It was introduced by Leonhard Euler in 1729 and has some useful properties.

a. Show that $\Gamma(s)$ converges for all $s > 0$.

b. Show that $\Gamma(s+1) = s\Gamma(s)$.

c. Show that $\Gamma(n+1) = n!$ for any positive integer n.

62. Find the centroid of a thin plate of constant density that occupies the region bounded by the graph of $y = \sin x + \cos x$, the coordinate axes, and the line $x = \pi/4$.

63. Repeat Problem 62 for a plate that occupies the region bounded by $y = \sec^2 x$, the coordinate axes, and the line $x = \pi/3$.

64. Find the volume obtained by revolving about the y-axis the region bounded by the curve $y = \sinh x$, the x-axis, and the line $x = 1$.

65. Find the volume obtained by revolving about the y-axis the region bounded by the curve $y = \cosh x$, the y-axis, and the line $y = 2$.

66. The region bounded by the graph of $y = \sin x$ and the x-axis between $x = 0$ and $x = \pi$ is revolved about the x-axis to generate a solid. Find the volume of the solid and its surface area.

67. Find the length of the arc of the curve $y = \frac{4}{5}x^{5/4}$ that lies between $x = 0$ and $x = 1$.

68. Find the volume of the solid obtained by revolving about the x-axis the region bounded by the curve $y = (9 - x^2)^{1/4}$ and the x-axis.

69. Find the volume of the solid formed by revolving about the y-axis the region bounded by the curve
$$y = \frac{1}{1+x^4}$$
between $x = 0$ and $x = 4$.

70. Find the volume of the solid formed by revolving about the x-axis the region bounded by the curve
$$y = \frac{2}{\sqrt{3x-2}}$$
the x-axis, and the lines $x = 1$ and $x = 2$.

71. Find the surface area of the solid generated by revolving the region bounded by the curve
$$y = e^x + \frac{1}{4}e^{-x}$$
on $[0, 1]$ about the x-axis.

72. Find the volume of the solid whose base is the region R bounded by the curve $y = e^x$ and the lines $y = 0$, $x = 0$, $x = 1$, if cross sections perpendicular to the x-axis are equilateral triangles.

73. Show that the area under $y = \frac{1}{x}$ on the interval $[1, a]$ equals the area under the same curve on $[k, ka]$ for any number $k > 0$.

74. Find the length of the curve $y = \ln(\csc x)$ on the interval $[\frac{\pi}{3}, \frac{\pi}{2}]$.

75. Find the length of the curve
$$x = \frac{ay^2}{b^2} - \frac{b^2}{8a}\ln\frac{y}{b}$$
between $y = b$ and $y = 2b$.

76. Derive the reduction formula
$$\int (a^2 - x^2)^n\,dx = \frac{x(a^2-x^2)^n}{2n+1} + \frac{2a^2n}{2n+1}\int (a^2-x^2)^{n-1}\,dx$$
Use this formula to find $\int (9 - x^2)^{5/2}\,dx$.

77. Derive the reduction formula
$$\int \frac{\sin^n x\,dx}{\cos^m x} = \frac{1}{m-1}\frac{\sin^{n-1}x}{\cos^{m-1}x} - \frac{n-1}{m-1}\int \frac{\sin^{n-2}x\,dx}{\cos^{m-2}x}$$

78. Derive a reduction formula for $\int \frac{\cos^n x\,dx}{\sin^m x}$.

79. Derive a reduction formula for $\int x^n(x^2+a^2)^{-1/2}\,dx$.

80. Derive the reduction formula
$$\int \frac{dx}{x^n\sqrt{ax+b}} = \frac{-\sqrt{ax+b}}{(n-1)bx^{n-1}} - \frac{(2n-3)a}{(2n-2)b}\int \frac{dx}{x^{n-1}\sqrt{ax+b}}$$

81. The improper integral $\displaystyle\int_1^{+\infty}\left[\frac{2Ax^3}{x^4+1}-\frac{1}{x+1}\right]dx$ converges for exactly one value of the constant A. Find A and then compute the value of the integral.

82. Evaluate $\displaystyle\int_0^{+\infty}\frac{\sqrt{x}\,\ln x\,dx}{(x+1)(x^2+x+1)}$ if it exists. *Hint:* Let $u=\dfrac{1}{x}$.

83. Spectroscopic measurements based on the *Doppler effect* yield an observed mass m_0 for a certain type of binary star. The true mass m is then estimated by $m=m_0/I$, where

$$I=\int_0^{\pi/2}\sin^4 x\,dx$$

Determine the number I.

84. **Modeling Problem**   The residents of a certain community have voted to discontinue the fluoridation of their water supply. The local reservoir currently holds 200 million gallons of fluoridated water containing 1,600 lb of fluoride. The fluoridated water flows out of the reservoir at the rate of 4 million gallons per day and is replaced by unfluoridated water at the same rate. At all times, the remaining fluoride in the reservoir is evenly distributed. Set up and solve a differential equation for the amount $F(t)$ of fluoride in the reservoir at time t. When will 90% of the fluoride be replaced?

85. **Modeling Problem**   The Nuclear Regulatory Commission puts atomic waste into sealed containers and dumps them into the ocean. It is important to dump the containers in water shallow enough to ensure they do not break when they hit bottom. Suppose $s(t)$ is the depth of the container at time t and let W and B denote the container's weight and the constant buoyancy force, respectively. Assume there is a "drag force" proportional to velocity v.

a. Explain why the motion of the containers may be modeled by the differential equation

$$\frac{W}{g}\frac{dv}{dt}=W-B-kv$$

where g is the constant acceleration due to gravity. Solve this equation to express $v(t)$ in terms of W, k, B, and g.

b. Integrate the expression for $v(t)$ in part a to find $s(t)$.

c. Suppose $W=1{,}125$ newtons (about 250 lb), $B=1{,}100$ newtons, and $k=0.64$ kg/s. If the container breaks when the impact speed exceeds 10 m/s, what is the maximum depth for safe dumping?

86. **The Evans price-adjustment model** assumes that if there is an excess demand D over supply S in any time period, the price p changes at a rate proportional to the excess, $D-S$; that is,

$$\frac{dp}{dt}=k(D-S)$$

Suppose for a certain commodity, demand is linear,

$$D(p)=c-pd$$

and supply is cyclical

$$S(t)=a\sin(bt)$$

Solve the differential equation to express price $p(t)$ in terms of a, b, c, and d. What happens to $p(t)$ "in the long run" (as $t\to+\infty$)? Use $p_0=p(0)$.

87. **Modeling Problem**   A country has 5 billion dollars of paper currency. Each day about 10 million dollars comes into banks and 12 million is paid out. The government decides to issue new currency, and whenever an "old" bill comes into the bank, it is replaced by a "new" bill. Set up and solve a differential equation to model the currency replacement. How long will it take for 95% of the currency in circulation to be "new" style bills? How much total currency will be in circulation at this time?

88. **Putnam Examination Problem**   Show that

$$\frac{22}{7}-\pi=\int_0^1\frac{x^4(1-x)^4\,dx}{1+x^2}$$

89. **Putnam Examination Problem**   Evaluate

$$\int_0^{\pi/2}\frac{dx}{1+(\tan x)^{\sqrt{2}}}$$

90. **Putnam Examination Problem**   Evaluate

$$\int_0^{+\infty}t^{-1/2}\exp\left[-1985\left(t+\frac{1}{t}\right)\right]dt$$

You may assume that $\displaystyle\int_{-\infty}^{+\infty}e^{-x^2}\,dx=\sqrt{\pi}$.

Buoy Design

THE BOOK OF NATURE is written in the language of mathematics.

——Galileo

THE PSEUDOMATH is a person who handles mathematics as a monkey handles the razor. The creature tried to shave himself as he had seen his master do; but, not having any notion of the angle at which the razor was to be held, he cut his own throat. He never tried it a second time, poor animal! But the pseudomath keeps on in his work, proclaims himself clean shaved, and all the rest of the world hairy.

——A. De Morgan
Budget of Paradoxes
(LONDON, 1872), p. 473.

This project is to be done in groups of three or four students. Each group will submit a single written report.

You have been hired as a special consultant by the U.S. Coast Guard to evaluate some proposed new designs for navigational aids (buoys).

The buoys are floating cans that need to be visible from some distance away, without rising too far out of the water. Each buoy has a circular cross section and will be fitted with a superstructure that carries equipment such as lights and batteries.

To be acceptable, a fully equipped buoy must float with not less than 1.5 ft nor more than 3 ft of freeboard. (Freeboard is the distance from the water level to the top of the device.)

You should make the following assumptions:

a. A floating object will displace a volume of water whose weight equals the weight of the floating object. (This is Archimedes' principle.)

b. Your devices will be floating in salt water, which weighs 65.5500 lb/ft^3.

c. The buoys will be constructed of $\frac{1}{2}$ in.-thick sheet metal that weighs 490 lb/ft^3.

d. You can estimate an additional 20% in weight attributable to welds, bolts, and the like.

e. Each buoy will be fitted with a superstructure and equipment weighing a total of 2,000 lb.

f. Each design you are given to evaluate will be presented to you in the form of a curve $x = f(y)$, to be revolved about the y-axis.

---

*This project is adapted from a computer project used at the U.S. Coast Guard Academy.

8

Infinite Series

PREVIEW

Is it possible for the sum of infinitely many nonzero numbers to be finite? This concept, which may seem paradoxical at first, plays a central role in mathematics and has a variety of important applications. The goal of this chapter is to examine the theory and applications of infinite sums, which will be referred to as *infinite series*. Geometric series, introduced in Section 8.2, are among the simplest infinite series we will encounter and, in some ways, the most important. In Sections 8.3–8.6, we will develop *convergence tests*, which provide ways of determining quickly whether or not certain infinite series have a finite sum. Next, we will turn our attention to series in which the individual terms are functions instead of numbers. We will be especially interested in the properties of *power series*, which may be thought of as polynomials of infinite degree, although some of their properties are quite different from those of polynomials. We will find that many common functions, such as e^x, $\ln(x + 1)$, $\sin x$, $\cos x$, and $\tan^{-1} x$, can be represented as power series, and we will discuss some important theoretical and computational aspects of this kind of representation.

PERSPECTIVE

Series, or sums, arise in many different ways. For example, suppose it is known that a certain pollutant is released into the atmosphere at weekly intervals and is dissipated at the rate of 2% per week. If m grams of pollutant are released each week, then at the beginning of the first week, there will be $S_1 = m$ grams in the atmosphere, and at the beginning of the second, there will be $0.98m$ grams of "old" pollutant left plus m grams of "new" pollutant, to yield a total of $S_2 = m + 0.98m$ grams. Continuing, at the beginning of the nth week there will be

$$S_n = m + 0.98m + (0.98)^2 m + \cdots + (0.98)^{n-1} m$$

grams. It is natural to wonder how much pollutant will accumulate in the "long run" (as $n \to +\infty$), and the answer is given by the infinite sum

$$S_{+\infty} = m + 0.98m + (0.98)^2 m + \cdots$$

But just exactly what do we mean by such a sum, and if the total is a finite number, how can we compute its value? We seek answers to these questions in this chapter.

8.1 Sequences and Their Limits

IN THIS SECTION sequences; limits of sequences; bounded, monotonic sequences

These consecutive frames from a 16-mm movie film show a golfer from the moment he is at the top of his backswing until he hits the ball. If this film were projected at a rate of 24 frames per second, the viewer would have the illusion of seeing the golfer in action as he makes his downswing.

It is sometimes useful to start a sequence with a_0 instead of a_1.

The making of a motion picture is a complex process, and editing all the film into a movie requires that all the frames of the action be labeled in chronological order. For example, R21-435 might signify the 435th frame of the 21st reel. A mathematician might refer to the movie editor's labeling procedure by saying the frames are arranged in a *sequence*.

SEQUENCES

A sequence is a succession of numbers that are listed according to a given prescription or rule. Specifically, if n is a positive integer, the sequence whose nth term is the number a_n can be written as

$$a_1, a_2, \ldots, a_n, \ldots$$

or, more simply,

$$\{a_n\}$$

The number a_n is called the **general term** of the sequence. We will deal only with infinite sequences, so each term a_n has a **successor** a_{n+1} and for $n > 1$, a **predecessor** a_{n-1}. For example, by associating each positive integer n with its reciprocal $\frac{1}{n}$, we obtain the sequence denoted by

$$\left\{\frac{1}{n}\right\}, \text{ which represents the succession of numbers } 1, \frac{1}{2}, \frac{1}{3}, \cdots, \frac{1}{n}, \cdots$$

The general term is denoted by $a_n = \frac{1}{n}$.

The following examples illustrate the notation and terminology used in connection with sequences.

EXAMPLE 1 **Given the general term, find particular terms of a sequence**

Find the 1st, 2nd, and 15th terms of the sequence $\{a_n\}$, where the general term is

$$a_n = \left(\frac{1}{2}\right)^{n-1}$$

Solution

If $n = 1$, then $a_1 = \left(\frac{1}{2}\right)^{1-1} = 1$. Similarly,

$$a_2 = \left(\frac{1}{2}\right)^{2-1} = \frac{1}{2}$$

$$a_{15} = \left(\frac{1}{2}\right)^{15-1} = \left(\frac{1}{2}\right)^{14} = 2^{-14}$$

The reverse question, that of finding a general term given certain terms of a sequence, is a more difficult task, and even if we find a general term, we have no assurance that the general term is unique. For example, consider the sequence

$$2, 4, 6, 8, \ldots$$

This seems to have a general term $a_n = 2n$. However, the general term

$$a_n = (n-1)(n-2)(n-3)(n-4) + 2n$$

has the same first four terms, but $a_5 = 34$ (not 10, as we would expect from the sequence 2, 4, 6, 8).

So far, we have been discussing the concept of a sequence informally, without a definition. We have observed that a sequence $\{a_n\}$ associates the number a_n with the positive integer n. Hence, a sequence is really a special kind of function, one whose domain is the set of all positive (or possibly, nonnegative) integers.

Sequence

A **sequence** $\{a_n\}$ is a function whose domain is a set of nonnegative integers and whose range is a subset of the real numbers. The functional values $a_1, a_2, a_3, \ldots$ are called the **terms** of the sequence, and a_n is called the **nth term**, or **general term**, of the sequence.

LIMITS OF SEQUENCES

It is often desirable to examine the behavior of a given sequence $\{a_n\}$ as n gets arbitrarily large. For example, consider the sequence

$$a_n = \frac{n}{n+1}$$

WARNING Even though we write a_n, remember this is a function, $f(n) = \dfrac{n}{n+1}$, where the domain is the set of nonnegative integers.

a. Graphing a sequence in one dimension

Because $a_1 = \frac{1}{2}, a_2 = \frac{2}{3}, a_3 = \frac{3}{4}, \cdots$, we can plot the terms of this sequence on a number line, as shown in Figure 8.1a, or the sequence can be plotted in two dimensions, as shown in Figure 8.1b.

By looking at either graph in Figure 8.1, we see that it appears the terms of the sequence are approaching 1. In general, if the terms of the sequence approach the number L as n increases without bound, we say that the sequence *converges to the limit L* and write

$$L = \lim_{n \to \infty} a_n$$

(Note that for simplicity, the limiting behavior is denoted by $n \to \infty$ instead of the usual $n \to +\infty$.) For instance, in our example we would expect

$$\lim_{n \to \infty} \frac{n}{n+1} = 1$$

b. Graphing a sequence in two dimensions

Figure 8.1 Graphing the sequence $a_n = \dfrac{n}{n+1}$

This limiting behavior is analogous to the continuous case (discussed in Section 4.4), and may be defined formally as follows.

Convergent Sequence

The sequence $\{a_n\}$ **converges** to the number L, and we write

$$L = \lim_{n \to \infty} a_n$$

if for every $\epsilon > 0$, there is an integer N such that

$$|a_n - L| < \epsilon \quad \text{whenever} \quad n > N$$

Otherwise, the sequence **diverges**.

➡ **What This Says** The notation $L = \lim\limits_{n \to \infty} a_n$ means that eventually the terms of the sequence $\{a_n\}$ can be made as close to L as may be desired by taking n sufficiently large.

a. one dimension

b. two dimensions

Figure 8.2 Geometric interpretation of a converging sequence

A geometric interpretation of this definition is shown in Figure 8.2. Note that the numbers a_n may be practically anywhere at first (that is, for "small" n), but eventually, the a_n must "cluster" near the limiting value L.

The theorem on limits of functions carries over to sequences. We have the following useful result.

THEOREM 8.1 **Limit theorem for sequences**

If $\lim\limits_{n\to\infty} a_n = L$ and $\lim\limits_{n\to\infty} b_n = M$, then

Linearity rule for sequences	$\lim\limits_{n\to\infty} (ra_n + sb_n) = rL + sM$
Product rule for sequences	$\lim\limits_{n\to\infty} (a_n b_n) = LM$
Quotient rule for sequences	$\lim\limits_{n\to\infty} \dfrac{a_n}{b_n} = \dfrac{L}{M}$ provided $M \neq 0$
Root rule for sequences	$\lim\limits_{n\to\infty} \sqrt[m]{a_n} = \sqrt[m]{L}$ provided $\sqrt[m]{a_n}$ is defined for all n and $\sqrt[m]{L}$ exists

Proof The proof of these rules is similar to the limit rules that were stated in Section 4.4. ❏

EXAMPLE 2 **Convergent sequences**

Find the limit of each of these convergent sequences:

a. $\left\{ \dfrac{100}{n} \right\}$ **b.** $\left\{ \dfrac{2n^2 + 5n - 7}{n^3} \right\}$

c. $\left\{ \dfrac{3n^4 + n - 1}{5n^4 + 2n^2 + 1} \right\}$

Solution

a. As n grows arbitrarily large, $100/n$ gets smaller and smaller. Thus,

$$\lim_{n\to\infty} \frac{100}{n} = 0$$

Figure 8.3 Graphical representation of $a_n = 100/n$

A graphical representation is shown in Figure 8.3.

b. We cannot use the quotient rule of Theorem 8.1 because neither the limit in the numerator nor the one in the denominator exists. However,

$$\frac{2n^2 + 5n - 7}{n^3} = \frac{2}{n} + \frac{5}{n^2} - \frac{7}{n^3}$$

and by using the linearity rule, we find that

$$\lim_{n\to\infty} \frac{2n^2 + 5n - 7}{n^3} = \lim_{n\to\infty} \frac{2}{n} + 5\lim_{n\to\infty} \frac{1}{n^2} - 7\lim_{n\to\infty} \frac{1}{n^3}$$
$$= 0 + 0 + 0$$
$$= 0$$

Figure 8.4 Graph of
$$a_n = \frac{2n^2 + 5n - 7}{n^3}$$

A graph is shown in Figure 8.4.

Figure 8.5 Graph of
$$a_n = \frac{3n^4 + n - 1}{5n^4 + 2n^2 + 1}$$

c. Divide the numerator and denominator by n^4, the highest power of n that occurs in the expression, to obtain

$$\lim_{n \to \infty} \frac{3n^4 + n - 1}{5n^4 + 2n^2 + 1} = \lim_{n \to \infty} \frac{3 + \dfrac{1}{n^3} - \dfrac{1}{n^4}}{5 + \dfrac{2}{n^2} + \dfrac{1}{n^4}} = \frac{3}{5}$$

Note that this procedure is very similar to the one we used in Section 4.4 when we evaluated limits of functions to infinity. A graph of this sequence is shown in Figure 8.5. ∎

EXAMPLE 3 Divergent sequences

Show that the following sequences diverge:

a. $\{(-1)^n\}$ **b.** $\left\{\dfrac{n^5 + n^3 + 2}{7n^4 + n^2 + 3}\right\}$

Solution

a. The sequence defined by $\{(-1)^n\}$ is $-1, 1, -1, 1, \ldots$, and this sequence diverges by oscillation because the nth term is always either 1 or -1. Thus a_n cannot approach one specific number L as n grows large. The graph is shown in Figure 8.6a.

b. $\displaystyle\lim_{n \to \infty} \frac{n^5 + n^3 + 2}{7n^4 + n^2 + 3} = \lim_{n \to \infty} \frac{1 + \dfrac{1}{n^2} + \dfrac{2}{n^5}}{\dfrac{7}{n} + \dfrac{1}{n^3} + \dfrac{3}{n^5}}$

The numerator tends toward 1 as $n \to \infty$, and the denominator approaches 0. Hence the quotient increases without bound, and the sequence must diverge. The graph is shown in Figure 8.6b.

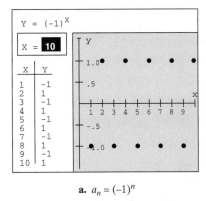

a. $a_n = (-1)^n$ **b.** $a_n = \dfrac{n^5 + n^3 + 2}{7n^4 + n^2 + 3}$

Figure 8.6 Graphs and values for two divergent sequences ∎

If $\displaystyle\lim_{n \to \infty} a_n$ does not exist because the numbers a_n become arbitrarily large as $n \to \infty$, we write $\displaystyle\lim_{n \to \infty} a_n = \infty$. We summarize this more precisely in the following box.

Limit Notation

$\displaystyle\lim_{n \to \infty} a_n = \infty$	means that for any real number A, we have $a_n > A$ for all sufficiently large n.
$\displaystyle\lim_{n \to \infty} b_n = -\infty$	means that for any real number B, we have $b_n < B$ for all sufficiently large n.

Rewriting the answer to Example 3b in this notation,

$$\lim_{n\to\infty} \frac{n^5 + n^3 + 2}{7n^4 + n^2 + 3} = \infty$$

EXAMPLE 4   Determining the convergence or divergence of a sequence

Determine the convergence or divergence of the sequence $\left\{ \sqrt{n^2 + 3n} - n \right\}$.

Solution

Consider

$$\lim_{n\to\infty} (\sqrt{n^2 + 3n} - n)$$

It would not be correct to apply the linearity property for sequences (because neither $\lim_{n\to\infty} \sqrt{n^2 + 3n}$ nor $\lim_{n\to\infty} n$ exists). It is also not correct to use this as a reason to say that the limit does not exist. You might even try some values of n (shown in Figure 8.7) to guess that there is some limit. To find the limit, however, we will rewrite the general term algebraically as follows:

$$\sqrt{n^2 + 3n} - n = (\sqrt{n^2 + 3n} - n) \frac{\sqrt{n^2 + 3n} + n}{\sqrt{n^2 + 3n} + n}$$

$$= \frac{n^2 + 3n - n^2}{\sqrt{n^2 + 3n} + n} \cdot \frac{\frac{1}{n}}{\frac{1}{n}}$$

$$= \frac{3}{\frac{\sqrt{n^2 + 3n}}{n} + 1} = \frac{3}{\sqrt{\frac{n^2 + 3n}{n^2}} + 1} = \frac{3}{\sqrt{1 + \frac{3}{n}} + 1}$$

Hence,

$$\lim_{n\to\infty} (\sqrt{n^2 + 3n} - n) = \lim_{n\to\infty} \frac{3}{\sqrt{1 + \frac{3}{n}} + 1} = \frac{3}{2} \qquad \blacksquare$$

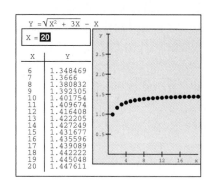

X	Y
6	1.348469
7	1.3666
8	1.380832
9	1.392305
10	1.401754
11	1.409674
12	1.416408
13	1.422205
14	1.427249
15	1.431677
16	1.435596
17	1.439089
18	1.442222
19	1.445048
20	1.447611

Figure 8.7 Graph of $a_n = \sqrt{n^2 + 3n} - n$; convergent or divergent?

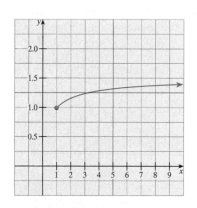

Figure 8.8 Graph of $f(x) = \sqrt{x^2 + 3x} - x, x \geq 1$

Note the graph of the sequence in Example 4 in Figure 8.7. The graph of a sequence consists of a succession of isolated points. This can be compared with the graph of $y = \sqrt{x^2 + 3x} - x$, $x \geq 1$, which is a continuous curve (shown in Figure 8.8). The only difference between $\lim_{n\to\infty} a_n = L$ and $\lim_{x\to\infty} f(x) = L$ is that n is required to be an integer. This is stated in the hypothesis of the following theorem.

THEOREM 8.2   Limit of a sequence from the limit of a continuous function

Given the sequence $\{a_n\}$, let f be a continuous function such that $a_n = f(n)$ for $n = 1, 2, \ldots$. If $\lim_{x\to\infty} f(x)$ exists and $\lim_{x\to\infty} f(x) = L$, the sequence $\{a_n\}$ converges and $\lim_{n\to\infty} a_n = L$.

Proof   Let $\epsilon > 0$ be given. Because $\lim_{x\to\infty} f(x) = L$, there exists a number $N > 0$ such that

$$|f(x) - L| < \epsilon \quad \text{whenever } x > N$$

In particular, if $n > N$, it follows that $|f(n) - L| = |a_n - L| < \epsilon$.   ❑

Be sure you read this theorem correctly. In particular, note that it *does not* say that if $\lim_{n\to\infty} a_n = L$, then $\lim_{x\to\infty} f(x) = L$ (see Figure 8.9b).

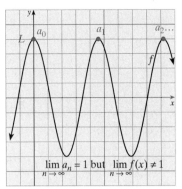

$$\lim_{n \to \infty} a_n = 1 \text{ but } \lim_{x \to \infty} f(x) \neq 1$$

a. If $\lim_{x \to \infty} f(x) = L$, then $\lim_{n \to \infty} a_n = L$.

b. ⚠ WARNING If $\lim_{n \to \infty} a_n = L$, then $\lim_{x \to \infty} f(x)$ does NOT necessarily equal L.

Figure 8.9   Graphical comparison of $\lim_{n \to \infty} a_n$ and $\lim_{x \to \infty} f(x)$, where $f(n) = a_n$ for $n = 1, 2, \dots$.

EXAMPLE 5   Evaluating a limit using l'Hôpital's rule

Given that the sequence $\left\{ \dfrac{n^2}{1 - e^n} \right\}$ converges, evaluate $\lim_{n \to \infty} \dfrac{n^2}{1 - e^n}$.

Solution

Let $L = \lim_{x \to \infty} f(x)$, where $f(x) = \dfrac{x^2}{1 - e^x}$. Because $f(n) = a_n$ for $n = 1, 2, \dots$, Theorem 8.2 tells us that $\lim_{n \to \infty} \dfrac{n^2}{1 - e^n}$ is the same as $\lim_{n \to \infty} f(x)$, provided this latter limit exists. Since the function f is continuous for all $x \neq 0$, we use l'Hôpital's rule.

$$\lim_{x \to \infty} \frac{x^2}{1 - e^x} = \lim_{x \to \infty} \frac{2x}{-e^x} \qquad \text{l'Hôpital's rule}$$

$$= \lim_{x \to \infty} \frac{2}{-e^x} \qquad \text{l'Hôpital's rule again}$$

$$= 0$$

Figure 8.10 From the graph, it looks like $\lim_{n \to \infty} a_n = 0$.

Thus, by Theorem 8.2, $\lim_{n \to \infty} \dfrac{n^2}{1 - e^n} = \lim_{x \to \infty} f(x) = L = 0$. This conclusion is confirmed geometrically, as shown in Figure 8.10. ■

The squeeze rule (Section 2.2) can be reformulated in terms of sequences.

THEOREM 8.3   Squeeze rule for sequences

If $a_n \leq b_n \leq c_n$ for all $n > N$, and $\lim_{n \to \infty} a_n = \lim_{n \to \infty} c_n = L$, then $\lim_{n \to \infty} b_n = L$.

Proof   The proof of this theorem follows the same steps as the proof of the squeeze rule given in Appendix A. ❏

EXAMPLE 6   Using l'Hôpital's rule and the squeeze rule

Show that the following sequences converge, and find their limits.

a. $\lim_{n \to \infty} n^{1/n}$ **b.** $\lim_{n \to \infty} \dfrac{n!}{n^n}$

Solution

a. Let $L = \lim\limits_{n\to\infty} n^{1/n}$; then we let $f(x) = x^{1/x}$. The function f is continuous for all $x > 0$, so we can apply l'Hôpital's rule.

$$\ln L = \lim_{n\to\infty} \frac{\ln n}{n} = \lim_{x\to\infty} \frac{\ln x}{x} = \lim_{x\to\infty} \frac{\dfrac{1}{x}}{1} \qquad \text{l'Hôpital's rule}$$
$$= 0$$

Thus, $L = e^0 = 1$.

b. We cannot use l'Hôpital's rule because $x!$ is not defined as an elementary function when x is not an integer. Instead, we use the squeeze rule for sequences. Accordingly, note that if we let $a_n = 0$ and $c_n = \dfrac{1}{n}$, for all n, we have

$$\overbrace{0}^{a_n} \le \overbrace{\frac{n!}{n^n}}^{b_n} \le \overbrace{\frac{1}{n}}^{c_n}$$

The right inequality is true because

$$\frac{n!}{n^n} = \frac{n(n-1)(n-2)\cdots 1}{n \cdot n \cdot n \cdot \cdots \cdot n}$$
$$= \left(\frac{n}{n}\right)\left(\frac{n-1}{n}\right)\left(\frac{n-2}{n}\right)\cdots\left(\frac{1}{n}\right)$$
$$< (1)\left(\frac{n}{n}\right)\left(\frac{n}{n}\right)\cdots\left(\frac{n}{n}\right)\left(\frac{1}{n}\right) = \frac{1}{n}$$

Because $\lim\limits_{n\to\infty} a_n = 0$ and $\lim\limits_{n\to\infty} c_n = 0$, it follows from the squeeze rule that $\lim\limits_{n\to\infty} \dfrac{n!}{n^n} = 0$. ∎

BOUNDED, MONOTONIC SEQUENCES

We introduce, along with a simple example, some terminology associated with sequences $\{a_n\}$.

Name	Condition	Example
Strictly increasing	$a_1 < a_2 < \cdots < a_{k-1} < a_k < \cdots$	$1, 2, 3, 4, 5, \ldots$
Increasing	$a_1 \le a_2 \le \cdots \le a_{k-1} \le a_k \le \cdots$	$1, 1, 2, 2, 3, 3, \ldots$
Strictly decreasing	$a_1 > a_2 > \cdots > a_{k-1} > a_k > \cdots$	$1, \frac{1}{2}, \frac{1}{3}, \frac{1}{4}, \ldots$
Decreasing	$a_1 \ge a_2 \ge \cdots \ge a_{k-1} \ge a_k \ge \cdots$	$1, 1, \frac{1}{2}, \frac{1}{2}, \frac{1}{3}, \frac{1}{3}, \ldots$
Bounded above by M	$a_n \le M$ for $n = 1, 2, 3, \ldots$	$M = 1$ for $1, \frac{1}{2}, \frac{1}{3}, \frac{1}{4}, \ldots$ $M = 2$, $M = 3$, are other choices for M.
Bounded below by m	$m \le a_n$ for $n = 1, 2, 3, \ldots$	$m = 1$ for $1, 2, 3, 4, \ldots$ $m = 0$, $m = -1$, are other choices for m.
Bounded	if it is bounded both above and below	$1, \frac{1}{2}, \frac{1}{3}, \frac{1}{4}, \ldots$ $m = 0$ and $M = 1$, so sequence is bounded.

We also say that a series is **monotonic** if it is increasing or decreasing or **strictly monotonic** if it is strictly increasing or strictly decreasing. In general, it is difficult to tell whether a given sequence converges or diverges, but because of the following theorem, it is easy to make this determination if we know the sequence is monotonic.

THEOREM 8.4   BMCT: The bounded, monotonic, convergence theorem

A monotonic sequence $\{a_n\}$ converges if it is bounded and diverges otherwise.

Proof   A formal proof of the BMCT is outlined in Problem 53. For the following informal argument, we will assume that $\{a_n\}$ is an increasing sequence. You might wish to see whether you can give a similar informal argument for the decreasing case. Because the terms of the sequence satisfy $a_1 \leq a_2 \leq a_3 \leq \cdots$, we know that the sequence is bounded from below by a_1 and that the graph of the corresponding points (n, a_n) will be rising in the plane. Two cases can occur, as shown in Figure 8.11.

a. If $a_n < M$ for $n = 1, 2, ...$, the graph of the points (n, a_n) will approach a horizontal "barrier" line $y = L$.

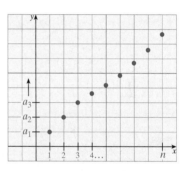

b. If $\{a_n\}$ is not bounded from above, the graph rises indefinitely.

Figure 8.11   Graphical possibilities for the BMCT theorem

Suppose the sequence $\{a_n\}$ is also bounded from above by a number M, so that $a_1 \leq a_n \leq M$ for $n = 1, 2, \ldots$. Then the graph of the points (n, a_n) must continually rise (because the sequence is monotonic) and yet it must stay below the line $y = M$. The only way this can happen is for the graph to approach a "barrier" line $y = L$ (where $L \leq M$), and we have $\lim\limits_{n \to \infty} a_n = L$, as shown in Figure 8.11a. However, if the sequence is not bounded from above, the graph will rise indefinitely (Figure 8.11b), and the terms in the sequence $\{a_n\}$ cannot approach any finite number L.   ❏

EXAMPLE 7   Convergence using the BMCT

Show that the sequence $\left\{ \dfrac{1 \cdot 3 \cdot 5 \cdot \cdots \cdot (2n-1)}{2 \cdot 4 \cdot 6 \cdots \cdots (2n)} \right\}$ converges.

Solution

The first few terms of this sequence are

$$a_1 = \frac{1}{2} \qquad a_2 = \frac{1 \cdot 3}{2 \cdot 4} = \frac{3}{8} \qquad a_3 = \frac{1 \cdot 3 \cdot 5}{2 \cdot 4 \cdot 6} = \frac{5}{16}$$

Because $\dfrac{1}{2} > \dfrac{3}{8} > \dfrac{5}{16}$, it appears that the sequence is decreasing (that is, it is monotonic). We can prove this by showing that $a_{n+1} < a_n$ for all $n > 0$, or equivalently,

$\dfrac{a_{n+1}}{a_n} < 1.$ (Note that $a_n \neq 0$ for all n.)

$$\frac{a_{n+1}}{a_n} = \frac{\dfrac{1 \cdot 3 \cdot 5 \cdots [2(n+1)-1]}{2 \cdot 4 \cdot 6 \cdots [2(n+1)]}}{\dfrac{1 \cdot 3 \cdot 5 \cdots (2n-1)}{2 \cdot 4 \cdot 6 \cdots (2n)}}$$

$$= \frac{1 \cdot 3 \cdot 5 \cdots (2n+1)}{2 \cdot 4 \cdot 6 \cdots (2n+2)} \cdot \frac{2 \cdot 4 \cdot 6 \cdots (2n)}{1 \cdot 3 \cdot 5 \cdots (2n-1)}$$

$$= \frac{2n+1}{2n+2} < 1$$

for any $n > 0$. Hence $a_{n+1} < a_n$ for all n, and $\{a_n\}$ is a decreasing sequence. Because $a_n > 0$ for all n, it follows that $\{a_n\}$ is bounded below by 0. Applying the BMCT, we see that $\{a_n\}$ converges, but the BMCT tells us nothing about the limit. ■

The BMCT also applies to sequences whose terms are *eventually* monotonic. That is, it can be shown that the sequence $\{a_n\}$ converges if it is bounded and there exists an integer N such that $\{a_n\}$ is monotonic for all $n > N$. This modified form of the BMCT is illustrated in the following example.

EXAMPLE 8   Convergence of a sequence that is eventually monotonic

Show that the sequence $\left\{ \dfrac{\ln n}{\sqrt{n}} \right\}$ converges.

Solution

We will apply the BMCT. A few values are shown in Figure 8.12. The succession of numbers suggests that the sequence increases at first and then gradually begins to decrease. To verify this behavior, we let

$$f(x) = \frac{\ln x}{\sqrt{x}}$$

and find that

$$f'(x) = \frac{\sqrt{x}\left(\frac{1}{x}\right) - (\ln x)\left(\frac{1}{2}x^{-1/2}\right)}{x} = \frac{2\sqrt{x} - \sqrt{x}\ln x}{2x^2}$$

Find the critical values:

$$\frac{2\sqrt{x} - \sqrt{x}\ln x}{2x^2} = 0$$

$$2\sqrt{x} = \sqrt{x}\ln x$$

$$\ln x = 2$$

$$e^2 = x$$

Thus, $x = e^2$ is the only critical value, and you can show that $f'(x) < 0$ for $x > e^2$. This means that f is a decreasing function for $x > e^2$. Thus, the sequence $\left\{ \dfrac{\ln n}{\sqrt{n}} \right\}$ must be decreasing for $n > 8$ (because e^2 is between 7 and 8). We see that the sequence $\left\{ \dfrac{\ln n}{\sqrt{n}} \right\}$ is bounded from below because

$$0 < \frac{\ln n}{\sqrt{n}} \quad \text{for all } n \geq 2$$

Therefore, the given sequence is bounded from below and is eventually decreasing, so it must converge. ■

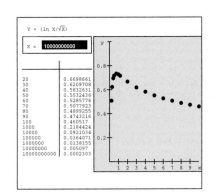

Y = (ln X/√X)

X = 10000000000

20	0.6698661
30	0.6209708
40	0.5832631
50	0.5532436
60	0.5285776
70	0.5077923
80	0.4899255
90	0.4743216
100	0.460517
1000	0.2184424
10000	0.0921034
100000	0.0364071
1000000	0.0138155
10000000	0.005097
10000000000	0.0002303

Figure 8.12 $a_n = \dfrac{\ln n}{\sqrt{n}}$

The BMCT is an extremely valuable theoretical tool. For example, in Section 2.4, we defined the number e by the limit

$$\lim_{n \to \infty} \left(1 + \frac{1}{n}\right)^n = e$$

but to do so we assumed that this limit exists. We can now show this assumption is warranted, because it turns out that the sequence $\left\{(1 + \frac{1}{n})^n\right\}$ is increasing and bounded from above by 3 (see Problem 80 in the Supplementary Problems). Thus, the BMCT assures us that the sequence converges, and this in turn guarantees the existence of the limit. We end this section with a result that will be useful in our subsequent work. The proof makes use of the formal definition of BMCT.

THEOREM 8.5 Convergence of a power sequence

If r is a fixed number such that $|r| < 1$, then $\lim_{n \to \infty} r^n = 0$.

Proof The case where $r = 0$ is trivial. We will prove the theorem for the case $0 < r < 1$ and leave the case $-1 < r < 0$ as an exercise.

$$0 \le r^{n+1} = r^n r < r^n$$

The sequence $\{r^n\}$ is monotonic and bounded, so the limit exists.
Either $\lim_{n \to \infty} r^n = 0$ or $\lim_{n \to \infty} r^n = L > 0$. Suppose

$$\lim_{n \to \infty} r^n = L$$

$$r \lim_{n \to \infty} r^n = rL \qquad \text{Multiply both sides by } r.$$

$$\lim_{n \to \infty} r r^n = rL$$

$$\lim_{n \to \infty} r^{n+1} = rL < L \qquad \text{Since } 0 < r < 1 \text{ and } L > 0$$

But $\lim_{n \to \infty} r^{n+1} = \lim_{n \to \infty} r^n = L$ by the definition of limit. Hence $L = rL < L$, which is impossible, so $\lim_{n \to \infty} r^n = 0$. ❑

8.1 PROBLEM SET

A **1. WHAT DOES THIS SAY?** What do we mean by the limit of a sequence?

2. WHAT DOES THIS SAY? What is meant by the bounded, monotonic, convergence theorem?

Write out the first five terms (beginning with $n = 1$) of the sequences given in Problems 3–11.

3. $\{1 + (-1)^n\}$

4. $\left\{\left(\frac{-1}{2}\right)^{n+2}\right\}$

5. $\left\{\dfrac{\cos 2n\pi}{n}\right\}$

6. $\left\{n \sin \dfrac{n\pi}{2}\right\}$

7. $\left\{\dfrac{3n + 1}{n + 2}\right\}$

8. $\left\{\dfrac{n^2 - n}{n^2 + n}\right\}$

9. $\{a_n\}$ where $a_1 = 256$ and $a_n = \sqrt{a_{n-1}}$ for $n \ge 2$

10. $\{a_n\}$ where $a_1 = -1$ and $a_n = n + a_{n-1}$ for $n \ge 2$

11. $\{a_n\}$ where $a_1 = 1$ and $a_n = (a_{n-1})^2 + a_{n-1} + 1$ for $n \ge 2$

Compute the limit of the convergent sequences in Problems 12–35.

12. $\left\{\dfrac{5n + 8}{n}\right\}$

13. $\left\{\dfrac{5n}{n + 7}\right\}$

14. $\left\{\dfrac{2n + 1}{3n - 4}\right\}$

15. $\left\{\dfrac{4 - 7n}{8 + n}\right\}$

16. $\left\{\dfrac{8n^2 + 800n + 5,000}{2n^2 - 1,000n + 2}\right\}$

17. $\left\{\dfrac{100n + 7,000}{n^2 - n - 1}\right\}$

18. $\left\{\dfrac{8n^2 + 6n + 4,000}{n^3 + 1}\right\}$

19. $\left\{\dfrac{n^3 - 6n^2 + 85}{2n^3 - 5n + 170}\right\}$

20. $\left\{\dfrac{2n}{n + 7\sqrt{n}}\right\}$

21. $\left\{\dfrac{8n - 500\sqrt{n}}{2n + 800\sqrt{n}}\right\}$

22. $\left\{\dfrac{3\sqrt{n}}{5\sqrt{n} + \sqrt[4]{n}}\right\}$

23. $\left\{\dfrac{\ln n}{n^2}\right\}$

24. $\{2^{5/n}\}$

25. $\{n^{3/n}\}$

26. $\left\{\left(1 + \dfrac{3}{n}\right)^n\right\}$

27. $\{(n + 4)^{1/n}\}$

28. $\left\{ n^{1/(n+2)} \right\}$

29. $\left\{ (\ln n)^{1/n} \right\}$

30. $\left\{ \int_0^\infty e^{-nx}\, dx \right\}$

31. $\left\{ \sqrt{n^2 + n} - n \right\}$

32. $\left\{ \sqrt{n + 5\sqrt{n}} - \sqrt{n} \right\}$

33. $\left\{ \sqrt[n]{n} \right\}$

34. $\{\ln n - \ln(n + 1)\}$

35. $\left\{ (an + b)^{1/n} \right\}$, a and b positive constants

B *Show that each sequence given in Problems 36–41 converges either by showing it is increasing with an upper bound or decreasing with a lower bound.*

36. $\left\{ \dfrac{n}{2^n} \right\}$

37. $\left\{ \ln\left(\dfrac{n+1}{n} \right) \right\}$

38. $\left\{ \dfrac{3n - 2}{n} \right\}$

39. $\left\{ \dfrac{4n + 5}{n} \right\}$

40. $\left\{ \dfrac{3n - 7}{2^n} \right\}$

41. $\left\{ \sqrt[n]{n} \right\}$

Explain why each sequence in Problems 42–45 diverges.

42. $\{1 + (-1)^n\}$

43. $\{\cos n\pi\}$

44. $\left\{ \dfrac{n^3 - 7n + 5}{100n^2 + 219} \right\}$

45. $\left\{ \sqrt{n} \right\}$

46. Suppose that a particle of mass m moves back and forth along a line segment of length $|a|$. In classical mechanics, the particle can move at any speed, and thus its energy can be any positive number. However, quantum mechanics replaces this continuous model of the particle's behavior with one in which the particle's energy level can have only certain discrete values, say, $E_1, E_2, \ldots$. Specifically, it can be shown that the nth term in this quantum sequence has the value

$$E_n = \frac{n^2 h^2}{8ma^2} \quad n = 1, 2, \ldots$$

where h is a physical constant known as *Planck's constant* ($h \approx 6.6254 \times 10^{-34}$ Js). List the first four values of E_n for a particle with mass $m = 0.008$ g moving along a segment of length $a = 1$ m.

47. **Modeling Problem** A drug is administered into the body. At the end of each hour, the amount of drug present is half what it was at the end of the previous hour. What percent of the drug is present at the end of 4 hours? At the end of n hours?

48. HISTORICAL QUEST   Leonardo de Pisa, also known as Fibonacci, was one of the best mathematicians of the Middle Ages. He played an important role in reviving ancient mathematics and introduced the Hindu-Arabic place-value decimal system to Europe. His book, Liber Abaci, published in 1202, introduced the Arabic numerals, as well as the

FIBONACCI
1170–1250

famous **rabbit problem**, for which Fibonacci is best remembered today. ∎

To describe Fibonacci's rabbit problem, we consider a sequence whose nth term is defined by a *recursion formula*—that is, a formula in which the nth term is given in terms of previous terms in the sequence. Suppose rabbits breed in such a way that each pair of adult rabbits produces a pair of baby rabbits each month.

Number of Months	Number of Pairs	Pairs of Rabbits (the blue rabbits are ready to reproduce)
Start		
1	1	
2	2	
3	3	
4	5	
5	8	
⋮	⋮	

Same pair
(rabbits never die)

The first month after birth, the rabbits are adolescents and produce no offspring. However, beginning with the second month the rabbits are adults, and each pair produces a pair of offspring every month. The sequence of numbers describing the number of rabbits is called the *Fibonacci sequence*, and it has applications in many areas, including biology and botany.

In this Quest you are to examine some properties of the Fibonacci sequence. Let a_n denote the number of pairs of rabbits in the "colony" at the end of n months.

a. Explain why $a_1 = 1$, $a_2 = 1$, $a_3 = 2$, $a_4 = 3$, and, in general, $a_{n+1} = a_n + a_{n-1}$ for $n = 2, 3, 4, \ldots$. The equation $a_{n+1} = a_n + a_{n-1}$ that specifies each term of the sequence $\{a_n\}$ for $n = 2, 3, \ldots$ is called a *recursion formula* because it defines later terms from those that precede them.

b. The *growth rate* of the colony during the $(n + 1)$th month is

$$r_n = \frac{a_{n+1}}{a_n}$$

Compute r_n for $n = 1, 2, 3, \ldots, 10$.

c. Assume that the growth rate sequence $\{r_n\}$ defined in part **b** converges. Find

$$L = \lim_{n \to \infty} r_n$$

Hint: Use the recursion formula in part **a**.

C 49. The claim is made that $\lim\limits_{n \to \infty} \dfrac{n}{n + 1} = 1$. If $\epsilon = 0.01$ is chosen, find N so that $\left| \dfrac{n}{n + 1} - 1 \right| < 0.01$ if $n > N$.

50. The claim is made that $\lim\limits_{n \to \infty} \dfrac{2n + 1}{n + 3} = 2$. If $\epsilon = 0.01$ is chosen, find N so that $\left| \dfrac{2n + 1}{n + 3} - 2 \right| < 0.01$ if $n > N$.

51. The claim is made that $\lim\limits_{n \to \infty} \dfrac{n^2 + 1}{n^3} = 0$. If $\epsilon = 0.001$ is chosen, find N so that $\left| \dfrac{n^2 + 1}{n^3} \right| < 0.001$ if $n > N$.

52. The claim is made that $\lim\limits_{n \to \infty} e^{-n} = 0$. If $\epsilon = 0.001$ is chosen, find N so that $\dfrac{1}{e^n} < 0.001$ if $n > N$.

53. Proof of the BMCT Suppose there exist a number M and a positive integer N so that $a_n \le a_{n+1} \le M$ for all $n > N$.

a. It is a fundamental property of the real numbers that a sequence with an upper bound must have a *least* upper bound; that is, there must be a number A with the property that $a_n \le A$ for all n but if c is a number such that $c < A$, then $a_m > c$ for at least one m. Use this property to show that if $\epsilon > 0$ is given, there exists a positive integer N so that

$$A - \epsilon < a_n < A$$

for all $n > N$.

b. Show that $|a_n - A| < \epsilon$ for all $n > N$ and conclude that $\lim_{n \to \infty} a_n = A$.

8.2 Introduction to Infinite Series; Geometric Series

IN THIS SECTION definition of infinite series, general properties of infinite series, geometric series, applications of geometric series

DEFINITION OF INFINITE SERIES

One way to add a list of numbers is to form subtotals until the end of the list is reached. Similarly, to give meaning to the infinite sum

$$S = a_1 + a_2 + a_3 + a_4 + \cdots$$

it is natural to examine the "partial sums"

$$a_1, \quad a_1 + a_2, \quad a_1 + a_2 + a_3, \quad a_1 + a_2 + a_3 + a_4, \ldots$$

If it is possible to attach a numerical value to the infinite sum, we would expect the partial sums $S_n = a_1 + a_2 + \cdots + a_n$ to approach that value as n increases without bound. These ideas lead us to the following definition.

Infinite Series

An **infinite series** is an expression of the form

$$a_1 + a_2 + a_3 + \cdots = \sum_{k=1}^{\infty} a_k$$

and the **nth partial sum** of the series is

$$S_n = a_1 + a_2 + \cdots + a_n = \sum_{k=1}^{n} a_k$$

The series is said to **converge with sum S** if the *sequence* of partial sums $\{S_n\}$ converges to S. In this case, we write

$$\sum_{k=1}^{\infty} a_k = \lim_{n \to \infty} S_n = S$$

If the sequence $\{S_n\}$ does not converge, the series

$$\sum_{k=1}^{\infty} a_k$$

diverges and has no sum.

➡ **What This Says** An infinite series converges if its sequence of partial sums converges and diverges otherwise. If it converges, its sum is defined to be the limit of the sequence of partial sums.

REMARK: We will use the symbol $\displaystyle\sum_{k=1}^{\infty} a_k$ to denote the series $a_1 + a_2 + a_3 + \cdots$ regardless of whether this series converges or diverges. If the sequence of partial sums $\{S_n\}$ converges, then

$$\sum_{k=1}^{\infty} a_k = \lim_{n\to\infty}\left(\sum_{k=1}^{n} a_k\right)$$

and the symbol $\displaystyle\sum_{k=1}^{\infty} a_k$ is used to represent *both* the series and its sum.

We will also consider certain series in which the starting point is not 1; for example, the series

$$\frac{1}{3} + \frac{1}{4} + \frac{1}{5} + \cdots \quad \text{can be denoted by} \quad \sum_{k=3}^{\infty} \frac{1}{k} \quad \text{or} \quad \sum_{k=2}^{\infty} \frac{1}{k+1}$$

EXAMPLE 1 Convergent series

Show that the series $\displaystyle\sum_{k=1}^{\infty} \frac{1}{2^k}$ converges.

Solution

This series has the following partial sums.

$$S_1 = \frac{1}{2} \qquad\qquad \text{Note:} \quad 1 - \frac{1}{2^1} = \frac{1}{2}$$
$$S_2 = \frac{1}{2} + \frac{1}{4} = \frac{3}{4} \qquad\qquad 1 - \frac{1}{2^2} = \frac{3}{4}$$
$$S_3 = \frac{1}{2} + \frac{1}{4} + \frac{1}{8} = \frac{7}{8} \qquad\qquad 1 - \frac{1}{2^3} = \frac{7}{8}$$
$$\vdots$$
$$S_n = \frac{1}{2} + \frac{1}{4} + \cdots + \frac{1}{2^n}$$

The sequence of partial sums is $\frac{1}{2}, \frac{3}{4}, \frac{7}{8}, \frac{15}{16}, \frac{31}{32}, \frac{63}{64}, \frac{127}{128}, \cdots$, and in general (by mathematical induction)

$$S_n = 1 - \frac{1}{2^n}$$

Because $\displaystyle\lim_{n\to\infty}\left(1 - \frac{1}{2^n}\right) = 1$, we conclude that the series converges and its sum is $S = 1$. ∎

EXAMPLE 2 Divergent series

Show that the series $\displaystyle\sum_{k=1}^{\infty}(-1)^k$ diverges.

Solution

The series can be **expanded** (written out) as

$$\sum_{k=1}^{\infty}(-1)^k = -1 + 1 - 1 + 1 - 1 + 1 - \cdots$$

and we see that the nth partial sum is

$$S_n = \begin{cases} -1 & \text{if } n \text{ is odd} \\ 0 & \text{if } n \text{ is even} \end{cases}$$

Because the sequence $\{S_n\}$ has no limit, the given series must diverge. ∎

A series is called a **telescoping series** (or collapsing series) if there is internal cancellation in the partial sums, as illustrated by the following example.

EXAMPLE 3 **A telescoping series**

Show that the series $\displaystyle\sum_{k=1}^{\infty} \frac{1}{k^2 + k}$ converges and find its sum.

Solution

Using partial fractions, we find that

$$\frac{1}{k^2 + k} = \frac{1}{k(k+1)} = \frac{1}{k} + \frac{-1}{k+1}$$

Thus, the nth partial sum of the given series can be represented as follows:

$$S_n = \sum_{k=1}^{n} \frac{1}{k^2 + k} = \sum_{k=1}^{n}\left[\frac{1}{k} - \frac{1}{k+1}\right]$$

$$= \left(1 - \frac{1}{2}\right) + \left(\frac{1}{2} - \frac{1}{3}\right) + \left(\frac{1}{3} - \frac{1}{4}\right) + \cdots + \left(\frac{1}{n} - \frac{1}{n+1}\right)$$

$$= 1 + \left(-\frac{1}{2} + \frac{1}{2}\right) + \left(-\frac{1}{3} + \frac{1}{3}\right) + \cdots + \left(-\frac{1}{n} + \frac{1}{n}\right) - \frac{1}{n+1}$$

$$= 1 - \frac{1}{n+1}$$

The limit of the sequence of partial sums is

$$\lim_{n\to\infty} S_n = \lim_{n\to\infty}\left[1 - \frac{1}{n+1}\right] = 1$$

so the series converges, with sum $S = 1$. ∎

GENERAL PROPERTIES OF INFINITE SERIES

Next, we will examine two general properties of infinite series. Here and elsewhere, *when the starting point of a series is not important, we may denote the series by writing*

$$\sum a_k \quad \text{instead of} \quad \sum_{k=1}^{\infty} a_k$$

THEOREM 8.6 **Linearity of infinite series**

If Σa_k and Σb_k are convergent series, then so is $\Sigma(ca_k + db_k)$ for constants c, d, and

$$\sum(ca_k + db_k) = c\sum a_k + d\sum b_k$$

Proof Compare this with the linearity property in Theorem 5.4. In Chapter 5, the limit properties are for finite sums, and this theorem is for infinite sums. The proof of this theorem is similar to the proof of Theorem 5.4, but in this case it follows from the linearity rule for sequences (Theorem 8.1). The details are left as a problem. ❏

EXAMPLE 4 **Linearity used to establish convergence**

Show that the series $\displaystyle\sum_{k=1}^{\infty}\left[\frac{4}{k^2 + k} - \frac{6}{2^k}\right]$ converges, and find its sum.

Solution

Because we know from Examples 1 and 3 that $\displaystyle\sum_{k=1}^{\infty} \frac{1}{k^2+k}$ and $\displaystyle\sum_{k=1}^{\infty} \frac{1}{2^k}$ both converge with sum 1, the linearity property allows us to write the given series as

$$4\sum_{k=1}^{\infty} \frac{1}{k^2+k} - 6\sum_{k=1}^{\infty} \frac{1}{2^k}$$

We can now conclude that the series converges and that the sum is

$$4\sum_{k=1}^{\infty} \frac{1}{k^2+k} - 6\sum_{k=1}^{\infty} \frac{1}{2^k} = 4(1) - 6(1) = -2 \qquad\blacksquare$$

The linearity property also provides useful information about a series of the form $\Sigma(ca_k + db_k)$ when exactly one of Σa_k or Σb_k diverges; that is, one converges and the other diverges.

THEOREM 8.7   Divergence of the sum of a convergent and a divergent series

If either Σa_k or Σb_k diverges and the other converges, then the series $\Sigma(a_k + b_k)$ must diverge.

Proof   Suppose Σa_k diverges and Σb_k converges. Then if the series $\Sigma(a_k + b_k)$ also converges, the linearity property tells us that the series

$$\sum[(a_k + b_k) - b_k] = \sum a_k$$

must converge, contrary to hypotheses. It follows that the series $\Sigma(a_k + b_k)$ diverges.   ❏

This theorem tells us, for example, that $\displaystyle\sum_{k=1}^{\infty} \left[\frac{1}{k^2+k} + (-1)^k\right]$ must diverge because even though $\displaystyle\sum_{k=1}^{\infty} \frac{1}{k^2+k}$ converges (Example 3), the series $\displaystyle\sum_{k=1}^{\infty} (-1)^k$ diverges (Example 2).

GEOMETRIC SERIES

We are still very limited in the available techniques for determining whether a given series is convergent or divergent. In fact, the purpose of a great part of this chapter is to develop efficient techniques for making this determination. We begin this quest by considering an important special kind of series.

Geometric Series

> A **geometric series** is an infinite series in which the ratio of successive terms in the series is constant. If this constant ratio is r, then the series has the form
>
> $$\sum_{k=0}^{\infty} ar^k = a + ar + ar^2 + ar^3 + \cdots + ar^n + \cdots \qquad a \neq 0$$

For example, $3 + \frac{3}{2} + \frac{3}{4} + \frac{3}{8} + \cdots$ is a geometric series because each term is one-half the preceding term. The ratio of a geometric series may be positive or negative. For example,

$$\sum_{k=0}^{\infty} \frac{2}{(-3)^k} = 2 - \frac{2}{3} + \frac{2}{9} - \frac{2}{27} + \cdots$$

is a geometric series with $r = -\frac{1}{3}$. The following theorem tells us how to determine whether a given geometric series converges or diverges and, if it does converge, what its sum must be.

THEOREM 8.8 Geometric series theorem

The geometric series $\displaystyle\sum_{k=0}^{\infty} ar^k$ with $a \neq 0$ diverges if $|r| \geq 1$ and converges if $|r| < 1$ with sum

$$\sum_{k=0}^{\infty} ar^k = \frac{a}{1-r}$$

Proof   Note that the nth partial sum of the geometric series is

$$S_n = a + ar + ar^2 + \cdots + ar^{n-1}$$

Thus,

$$rS_n = ar + ar^2 + ar^3 + \cdots + ar^n \qquad \text{Multiply both sides by } r.$$
$$rS_n - S_n = (ar + ar^2 + ar^3 + \cdots + ar^n) - (a + ar + ar^2 + \cdots + ar^{n-1})$$
$$(r-1)S_n = ar^n - a$$
$$S_n = \frac{a(r^n - 1)}{r - 1} \qquad \text{if } r \neq 1$$

If $|r| > 1$, the sequence of partial sums $\{S_n\}$ has no limit since r^n is unbounded in this case, so the geometric series must diverge. However, if $|r| < 1$, Theorem 8.5 tells us that $r^n \to 0$ as $n \to \infty$, so we have

$$\sum_{k=0}^{\infty} ar^k = \lim_{n\to\infty} S_n = \lim_{n\to\infty} a\left(\frac{r^n - 1}{r - 1}\right) = \frac{a(0 - 1)}{r - 1} = \frac{a}{1 - r}$$

To complete the proof, it must be shown that the geometric series diverges when $|r| = 1$, and we leave this final step as a problem.   ❑

EXAMPLE 5   Testing for convergence in a geometric series

Determine whether each of the following geometric series converges or diverges. If the series converges, find its sum.

a. $\displaystyle\sum_{k=0}^{\infty} \frac{1}{7}\left(\frac{3}{2}\right)^k$ **b.** $\displaystyle\sum_{k=2}^{\infty} 3\left(-\frac{1}{5}\right)^k$

Solution
a. Because $r = \frac{3}{2}$ satisfies $|r| \geq 1$, the series diverges.
b. We have $r = -\frac{1}{5}$ so $|r| < 1$, and the geometric series converges. The first value of k is 2 (not 0), so the value a (the first value) is $a = 3\left(-\frac{1}{5}\right)^2 = \frac{3}{25}$, and

$$\sum_{k=2}^{\infty} 3\left(-\frac{1}{5}\right)^k = \frac{a}{1-r} = \frac{\frac{3}{25}}{1 - \left(-\frac{1}{5}\right)} = \frac{1}{10} \qquad ∎$$

APPLICATIONS OF GEOMETRIC SERIES

Geometric series can be used in many different ways. Our next three examples illustrate several applications involving geometric series.

Recall that a rational number r is one that can be written as $r = p/q$ for integer p and nonzero integer q. It can be shown that any such number has a decimal representation in which a pattern of numbers repeats. For instance,

$$\frac{5}{10} = 0.5 = 0.5\overline{0} \qquad\qquad \frac{5}{11} = 0.454545\cdots = 0.\overline{45}$$

where the bar indicates that the pattern repeats indefinitely. Example 6 shows how geometric series can be used to reverse this process by writing a given repeating decimal as a rational number.

EXAMPLE 6 Expressing a repeating decimal as a rational number

Write $15.4\overline{23}$ as a rational number $\dfrac{p}{q}$.

Solution

The bar over the 23 indicates that this block of digits is to be repeated, that is, $15.4\overline{23} = 15.423232323\ldots$. The repeating part of the decimal can be written as a geometric series as follows.

$$
\begin{aligned}
15.4\overline{23} &= 15 + \frac{4}{10} + \frac{23}{10^3} + \frac{23}{10^5} + \frac{23}{10^7} + \cdots \\
&= 15 + \frac{4}{10} + \frac{23}{10^3}\left[1 + \frac{1}{10^2} + \frac{1}{10^4} + \cdots\right] \\
&= 15 + \frac{4}{10} + \frac{23}{10^3}\left[\frac{1}{1 - \frac{1}{100}}\right] \qquad \text{\textit{Sum of a geometric series with } } a = 1 \text{ \textit{and} } r = 1/100 \\
&= 15 + \frac{4}{10} + \frac{23}{10^3}\left[\frac{100}{99}\right] \\
&= 15 + \frac{4}{10} + \frac{23}{990} \\
&= \frac{15,269}{990}
\end{aligned}
$$

∎

A tax rebate that returns a certain amount of money to taxpayers can result in spending that is many times this amount. This phenomenon is known in economics as the *multiplier effect*. It occurs because the portion of the rebate that is spent by one individual becomes income for one or more others, who, in turn, spend some of it again, creating income for yet other individuals to spend. If the fraction of income that is saved remains constant as this process continues indefinitely, the total amount spent as a result of the rebate is the sum of a geometric series.*

EXAMPLE 7 Modeling Problem: Multiplier effect in economics

Suppose that nationwide approximately 90% of all income is spent and 10% is saved. How much total spending will be generated by a $40 billion tax rebate if savings habits do not change?

Solution

The amount (in billions) spent by original recipients of the rebate is 40. This becomes new income, of which 90%, or 0.9(40), is spent. This, in turn, generates additional spending of 0.9[0.9(40)], and so on. The total amount spent if this process continues indefinitely is

$$
40 + 0.9(40) + 0.9^2(40) + 0.9^3(40) + \cdots = 40[1 + 0.9 + 0.9^2 + \cdots]
$$

$$
= 40 \sum_{k=0}^{\infty} 0.9^k \qquad \text{\textit{Geometric series with} } a = 1, r = 0.9
$$

$$
= 40\left(\frac{1}{1 - 0.9}\right)
$$

$$
= 400 \text{ (billion)}
$$

∎

---

*The fraction s of income that is saved is called the *marginal propensity to save* and $c = 1 - s$ is the *marginal propensity to consume*. In Example 7, we have $s = 0.1$ and $c = 0.9$.

EXAMPLE 8 **Modeling Problem: Accumulation of medication in a body**

A patient is given an injection of 10 units of a certain drug every 24 hours. The drug is eliminated exponentially so that the proportion that remains in the patient's body after t days is $f(t) = e^{-t/5}$. If the treatment is continued indefinitely, approximately how many units of the drug will eventually be in the patient's body just prior to an injection?

Solution

Of the original dose of 10 units, only $10e^{-1/5}$ are left in the patient's body after the first day (just prior to the second injection). That is,

$$S_1 = 10e^{-1/5}$$

The medication in the patient's body after 2 days consists of what remains from the first two doses. Of the original dose, only $10e^{-2/5}$ units are left (because 2 days have elapsed), and of the second dose, $10e^{-1/5}$ units remain:

$$S_2 = 10e^{-1/5} + 10e^{-2/5}$$

Similarly, for n days,

$$S_n = 10e^{-1/5} + 10e^{-2/5} + \cdots + 10e^{-n/5} = \sum_{k=1}^{n} 10e^{-k/5}$$

The amount S of medication in the patient's body in the long run is the limit of S_n as $n \to \infty$. That is,

$$S = \lim_{n \to \infty} S_n$$

$$= \sum_{k=1}^{\infty} 10e^{-k/5} \qquad \text{\textit{Geometric series with}} \atop a = 10e^{-1/5} \text{ and } r = e^{-1/5}$$

$$= \frac{10e^{-1/5}}{1 - e^{-1/5}}$$

$$\approx 45.166556$$

We see that about 45 units remain in the patient's body. ■

WARNING As a final note, remember that a sequence is a succession of terms, whereas a series is a sum of such terms. Do not confuse the two concepts because they have very different properties. For example, a sequence of terms may converge, but the series of the same terms may diverge:

$$\left\{ 1 + \frac{1}{2^n} \right\} \text{ is the sequence } \frac{3}{2}, \frac{5}{4}, \frac{9}{8}, \frac{17}{16}, \dots, \text{ which converges to 1.}$$

$$\sum_{n=1}^{\infty} \left(1 + \frac{1}{2^n} \right) \text{ is the series } \frac{3}{2} + \frac{5}{4} + \frac{9}{8} + \cdots, \text{ which diverges.}$$

8.2 PROBLEM SET

A **1. WHAT DOES THIS SAY?** Explain the difference between sequences and series.

2. WHAT DOES THIS SAY? What is a geometric series?

Determine whether the geometric series given in Problems 3–22 converges or diverges, and find the sum of each convergent series.

3. $\displaystyle\sum_{k=0}^{\infty} \left(\frac{4}{5} \right)^k$

4. $\displaystyle\sum_{k=0}^{\infty} \left(-\frac{4}{5} \right)^k$

5. $\displaystyle\sum_{k=0}^{\infty} \frac{2}{3^k}$

6. $\displaystyle\sum_{k=0}^{\infty} \frac{2}{(-3)^k}$

7. $\displaystyle\sum_{k=1}^{\infty} \left(\frac{3}{2} \right)^k$

8. $\displaystyle\sum_{k=1}^{\infty} \frac{3}{2^k}$

9. $\displaystyle\sum_{k=0}^{\infty} \frac{(-2)^k}{5^{2k+1}}$

10. $\displaystyle\sum_{k=1}^{\infty} 5(0.9)^k$

11. $\displaystyle\sum_{k=1}^{\infty} e^{-0.2k}$

12. $\displaystyle\sum_{k=1}^{\infty} \frac{3^k}{4^{k+2}}$

13. $\displaystyle\sum_{k=2}^{\infty} \frac{(-2)^{k-1}}{3^{k+1}}$

14. $\displaystyle\sum_{k=2}^{\infty} (-1)^k \frac{2^{k+1}}{3^{k-3}}$

15. $\displaystyle\frac{1}{2} - \frac{1}{2^2} + \frac{1}{2^3} - \frac{1}{2^4} + \cdots$

16. $1 + \pi + \pi^2 + \pi^3 + \cdots$

17. $\displaystyle\frac{1}{4} + \left(\frac{1}{4}\right)^4 + \left(\frac{1}{4}\right)^7 + \left(\frac{1}{4}\right)^{10} + \cdots$

18. $\displaystyle\frac{2}{3} - \left(\frac{2}{3}\right)^3 + \left(\frac{2}{3}\right)^5 - \left(\frac{2}{3}\right)^7 + \cdots$

19. $2 + \sqrt{2} + 1 + \frac{1}{\sqrt{2}} + \cdots$

20. $3 - \sqrt{3} + 1 - \frac{1}{\sqrt{3}} + \cdots$

21. $(1 + \sqrt{2}) + 1 + (-1 + \sqrt{2}) + (3 - 2\sqrt{2}) + \cdots$

22. $(\sqrt{2} - 1) + 1 + (\sqrt{2} + 1) + (2\sqrt{2} + 3) + \cdots$

In Problems 23–30, each series telescopes. In each case, express the nth partial sum S_n in terms of n and determine whether the series converges or diverges by examining $\lim_{n\to\infty} S_n$. Notice that some of these problems may require partial fraction decomposition.

23. $\displaystyle\sum_{k=1}^{\infty} \left[\frac{1}{k^{0.1}} - \frac{1}{(k+1)^{0.1}}\right]$

24. $\displaystyle\sum_{k=1}^{\infty} \left[\frac{1}{2k+1} - \frac{1}{2k+3}\right]$

25. $\displaystyle\sum_{k=0}^{\infty} \frac{1}{(k+1)(k+2)}$

26. $\displaystyle\sum_{k=2}^{\infty} \frac{1}{k(k+1)}$

27. $\displaystyle\sum_{k=1}^{\infty} \ln\left(1 + \frac{1}{k}\right)$

28. $\displaystyle\sum_{k=1}^{\infty} \frac{1}{(2k-1)(2k+1)}$

29. $\displaystyle\sum_{k=1}^{\infty} \frac{2k+1}{k^2(k+1)^2}$

30. $\displaystyle\sum_{k=1}^{\infty} \frac{\sqrt{k+1} - \sqrt{k}}{\sqrt{k^2+k}}$. *Hint*: Note that $\sqrt{k^2+k} = \sqrt{k}\sqrt{k+1}$.

Express each repeating decimal given in Problems 31–34 as a common (reduced) fraction.

31. $0.\overline{01}$

32. $2.23\overline{1}$

33. $1.\overline{405}$

34. $41.\overline{2010}$

B **35. a.** Find numbers A and B such that
$$\frac{k-1}{2^{k+1}} = \frac{Ak}{2^k} - \frac{B(k+1)}{2^{k+1}}$$

b. Evaluate $\displaystyle\sum_{k=1}^{\infty} \frac{k-1}{2^{k+1}}$ as a telescoping series.

36. Evaluate $\displaystyle\sum_{k=1}^{\infty} \frac{2k-1}{3^{k+1}}$. *Hint*: Follow the procedure in Problem 35.

37. Evaluate $\displaystyle\sum_{n=1}^{\infty} \frac{\ln\left(\frac{n^{n+1}}{(n+1)^n}\right)}{n(n+1)}$. *Hint*: Use the properties of logarithms to show that the series telescopes.

38. Evaluate $\displaystyle\sum_{n=1}^{\infty} \frac{n}{(n+1)!}$. *Hint*: Express as a telescoping series.

39. Find $\displaystyle\sum_{k=0}^{\infty} (2a_k + 2^{-k})$ given that $\displaystyle\sum_{k=0}^{\infty} a_k = 3.57$.

40. Find $\displaystyle\sum_{k=0}^{\infty} \frac{b_k - 3^{-k}}{2}$ given that $\displaystyle\sum_{k=0}^{\infty} b_k = 0.54$.

41. Evaluate $\displaystyle\sum_{k=0}^{\infty} \left(\frac{1}{2^k} + \frac{1}{3^k}\right)^2$.

42. Evaluate $\displaystyle\sum_{k=0}^{\infty} \left[\left(\frac{2}{3}\right)^k + \left(\frac{3}{4}\right)^k\right]^2$.

43. Show that $\displaystyle\sum_{k=0}^{\infty} \frac{1}{(a+k)(a+k+1)} = \frac{1}{a}$ if $a > 0$.

44. Evaluate $\displaystyle\sum_{k=0}^{\infty} \frac{1}{(2+k)^2 + 2 + k}$.

45. If $\displaystyle\sum_{k=0}^{\infty} a_k^2 = \sum_{k=0}^{\infty} b_k^2 = 4$ and $\displaystyle\sum_{k=0}^{\infty} a_k b_k = 3$, what is
$$\sum_{k=0}^{\infty} (a_k - b_k)^2?$$

Modeling Problems *In Problems 46–56, set up an appropriate model to answer the given question.*

46. A pendulum is released through an arc of length 20 cm from a vertical position and is allowed to swing freely until it eventually comes to rest. Each subsequent swing of the bob of the pendulum is 90% as far as the preceding swing. How far will the bob travel before coming to rest?

47. A flywheel rotates at 500 rpm and slows in such a way that during each minute it revolves at two-thirds the rate of the preceding minute. How many total revolutions does the flywheel make before coming to rest?

48. A ball is dropped from a height of 10 ft. Each time the ball bounces, it rises 0.6 the distance it had previously fallen. What is the total distance traveled by the ball?

49. Suppose the ball in Problem 48 is dropped from a height of h feet and on each bounce rises 75% of the distance it had previously fallen. If it travels a total distance of 21 ft, what is h?

50. Suppose that nationwide approximately 92% of all income is spent and 8% is saved. How much total spending will be generated by a $50 billion tax cut if savings habits do not change?

51. Suppose that a piece of machinery costing $10,000 depreciates 20% of its current value each year. That is, the first year $10,000(0.20) = $2,000 is depreciated. The second year's depreciation is $8,000(0.20) = $1,600, because the value for the second year is $10,000 − $2,000 = $8,000. If the depreciation is calculated this way indefinitely, what is the total depreciation?

52. Winnie Winner wins $100 in a pie-baking contest run by the Hi-Do Pie Co. The company gives Winnie the $100. However, the tax collector wants 20% of the $100. Winnie pays the tax. But then she realizes that she didn't really win a $100 prize and tells her story to the Hi-Do Co. The friendly Hi-Do Co. gives Winnie the $20 she paid in taxes. Unfortunately, the tax collector now wants 20% of the $20. She pays the tax again and then goes back to the Hi-Do Co. with her story. Assume that this can go on indefinitely. How much money does the Hi-Do Co. have to give Winnie so that she will really win $100? How much does she pay in taxes?

53. A patient is given an injection of 20 units of a certain drug every 24 hr. The drug is eliminated exponentially, so the part that remains in the patient's body after t days is $f(t) = e^{-t/2}$. If the treatment is to continue indefinitely, approximately how many units of the drug will eventually be in the patient's body just before an injection?

54. A certain drug is injected into a body. It is known that the amount of drug in the body at the end of a given hour is $\frac{1}{4}$ the amount that was present at the end of the previous hour. One unit is injected initially, and to keep the concentration up, an additional unit is injected at the end of each subsequent hour. What is the highest possible amount of the drug that can ever be present in the body?

55. Each January 1, the administration of a certain private college adds six new members to its board of trustees. If the fraction of trustees who remain active for at least t years is $f(t) = e^{-0.2t}$, approximately how many active trustees will the college have in the long run? Assume that there were six members on the board on January 1 of the present year.

56. How much should you invest today at an annual interest rate of 15% compounded continuously so that, starting next year, you can make annual withdrawals of $2,000 forever?

57. HISTORICAL QUEST The Greek philosopher Zeno of Elea (495–435 B.C.) presented a number of paradoxes that caused a great deal of concern to the mathematicians of his period. These paradoxes were important in the development of the notion of infinitesimals. Perhaps the most famous is the *racecourse paradox*, which can be stated as follows:

A runner can never reach the end of a race: As he runs the track, he must first run $\frac{1}{2}$ the length of the track, then $\frac{1}{2}$ the remaining distance, so that at this point he has run $\frac{1}{2} + \frac{1}{4} = \frac{3}{4}$ of the length of the track, and $\frac{1}{4}$ remains to be run. After running half this distance, he finds he still has $\frac{1}{8}$ of the track to run, and so on, indefinitely.

Suppose the runner runs at a constant pace and that it takes him T minutes to run the first half of the track. Set up an infinite series that gives the total time required to run the track, and verify that the total time is $2T$, as intuition would lead us to expect.

58. HISTORICAL QUEST A professor of mathematics at the University of Warsaw, Waclaw Sierpinski was the first person to give a lecture course devoted entirely to set theory. He became interested in set theory in 1907 when he came across a theorem which stated that points in the plane could be specified with a single coordinate. He wrote to a friend and asked how such a result could be possible, and he received

WACLAW SIERPINSKI
1882–1969

a one word reply: "Cantor." Georg Cantor (1845–1918) wrote a series of eight papers that provided the basis for what we today call set theory. ∎

For this quest, we investigate an interesting geometrical figure named after Sierpinski, namely the *Sierpinski gasket*, shown in Figure 8.13.

Figure 8.13 Sierpinski gasket

To form the gasket, begin by drawing an equilateral triangle, T, with sides of length 1. Remove the middle triangle formed by joining the midpoints of the sides of T, as shown in Figure 8.14a. This leaves three equal triangular regions, and the next step is to remove the middle triangle from each of these, as shown in Figure 8.14b. If you continue this process indefinitely, what remains is the Sierpinski gasket. Show that the Sierpinski gasket has area 0. *Hint*: What is the sum of the areas of the regions removed?

a. First step   **b.** Second step   **c.** Third step   **d.** Fourth step

Figure 8.14 Construction of a Sierpinski gasket

59. HISTORICAL QUEST John von Neumann was one of the original six mathematics professors at the Institute for Advanced Study in Princeton, the other five also being world-class mathematicians or Nobel laureates: J. W. Alexander, Albert Einstein, Marston Morse, Oswald Veblen, and Hermann Weyl. Von Neumann, a pioneer in computer science, re-

JOHN VON NEUMANN
1903–1957

ceived two Presidential Awards, the Medal for Merit, and the Medal for Freedom. The fact that he had a brilliant mind is an understatement, as illustrated in the following anecdote, stated in the form of a quest problem. ∎

Two trains, each traveling at 10 mph approach each other on the same straight track. When the trains are 5 miles apart, a fly begins flying back and forth from train to train at the rate of 16 mi/h. Between the time the fly starts and the time the trains collide, how far does the fly fly?

a. Find the fly's total distance traveled by determining how long the fly flies.

b. When von Neumann gave the correct answer almost immediately, the poser of the problem was convinced he had used the "trick" suggested in part **a**. "What trick?" asked von Neumann. "I summed the series." Follow von Neumann's lead by summing an infinite series whose terms are the distances traveled by the fly as it flies between the trains.

C **60. a.** Let $\displaystyle\sum_{k=1}^{n} a_k = S_n$ be the nth partial sum of an infinite series. Show that $S_n - S_{n-1} = a_n$.

b. Use part **a** to find the nth term of the infinite series whose nth partial sum is $S_n = \dfrac{n}{2n+3}$.

61. Prove that if $\Sigma a_k = A$ and $\Sigma b_k = B$, then $\Sigma(a_k + b_k) = A + B$.

62. Show that if Σa_k converges and Σb_k diverges, then $\Sigma(a_k - b_k)$ diverges. *Hint*: Note that $b_k = a_k - (a_k - b_k)$.

63. Counterexample Problem Either prove or disprove (by finding a counterexample) each of the following statements:

a. If Σa_k and Σb_k are both divergent series, then $\Sigma(a_k + b_k)$ is also divergent.

b. If Σa_k and Σb_k are both convergent series, then $\Sigma(a_k + b_k)$ is also convergent.

64. Show that $\displaystyle\sum_{k=0}^{\infty} ar^k$ diverges if $|r| \geq 1$.

65. a. Show that $\displaystyle\sum_{k=1}^{n}(a_k - a_{k+2}) = (a_1 + a_2) - (a_{n+1} + a_{n+2})$.

b. Use part **a** to show that if

$$\lim_{k \to \infty} a_k = A, \quad \text{then} \quad \sum_{k=1}^{\infty}(a_k - a_{k+2}) = a_1 + a_2 - 2A$$

c. Use part **b** to evaluate $\displaystyle\sum_{k=1}^{\infty}\left[k^{1/k} - (k+2)^{1/(k+2)}\right]$.

d. Evaluate $\displaystyle\sum_{k=2}^{\infty}\dfrac{1}{k^2 - 1}$.

66. Square $ABCD$ has sides of length 1. Square $EFGH$ is formed by connecting the midpoints of the sides of the first square, as shown in Figure 8.15. Assume that the pattern of shaded regions in the square is continued indefinitely. What is the total area of the shaded regions?

Figure 8.15   Find the area of the shaded regions.

67. Evaluate $\displaystyle\sum_{k=1}^{\infty}\dfrac{12^k}{(4^k - 3^k)(4^{k+1} - 3^{k+1})}$

Hint: Find A and B so that

$$\frac{3^k A}{4^k - 3^k} + \frac{3^k B}{4^{k+1} - 3^{k+1}} = \frac{12^k}{(4^k - 3^k)(4^{k+1} - 3^{k+1})}$$

8.3   The Integral Test; p-Series

IN THIS SECTION   divergence test, series of nonnegative numbers, the integral test, p-series

The convergence or divergence of an infinite series is determined by the behavior of its nth partial sum, S_n, as $n \to \infty$. In Section 8.2, we saw how algebraic methods can sometimes be used to find formulas for the nth partial sum of a series. Unfortunately, it is often difficult or even impossible to find a usable formula for the nth partial sum S_n of a series, and other techniques must be used to determine convergence or divergence.

DIVERGENCE TEST

When investigating the infinite series Σa_k, it is easy to confuse the sequence of *general terms* $\{a_k\}$ with the sequence of *partial sums* $\{S_n\}$, where

$$S_n = \sum_{k=1}^{n} a_k$$

Because the sequence $\{a_k\}$ is usually more accessible than $\{S_n\}$, it would be convenient if the convergence of $\{S_n\}$ could be settled by examining

$$\lim_{k \to \infty} a_k$$

Even though we do not have one simple, definitive test for convergence, the following theorem tells us that if certain conditions are not met, the series must diverge.

THEOREM 8.9   The divergence test

WARNING
Theorem 8.9 gives a necessary but not a sufficient condition for convergence. For example, clouds are necessary for rain; clouds are not sufficient for rain.

If $\lim\limits_{k \to \infty} a_k \neq 0$, then the series Σa_k must diverge.

Proof   Suppose the sequence of partial sums $\{S_n\}$ converges with sum L, so that $\lim\limits_{n \to \infty} S_n = L$. Then we also have

$$\lim_{n \to \infty} S_{n-1} = L$$

Because $S_k - S_{k-1} = a_k$, it follows that

$$\lim_{k \to \infty} a_k = \lim_{k \to \infty} (S_k - S_{k-1}) = \lim_{k \to \infty} S_k - \lim_{k \to \infty} S_{k-1} = L - L = 0$$

If the series $\sum a_k$ converges, then $\lim\limits_{k \to \infty} a_k = 0$. Thus, if $\lim\limits_{k \to \infty} a_k \neq 0$, the series Σa_k must diverge.   ❏

> ➡ **What This Says**   The divergence test can only tell us that Σa_k diverges if $\lim\limits_{k \to \infty} a_k \neq 0$, **but it *cannot* be used to show convergence of a series.** In Example 2, we show that the series $\Sigma \frac{1}{k}$ *diverges* even though $\lim\limits_{k \to \infty} \frac{1}{k} = 0$.

EXAMPLE 1   Divergence test to show divergence

Show that the series $\displaystyle\sum_{k=0}^{\infty} \frac{k - 300}{4k + 750}$ diverges.

Solution

Taking the limit of the kth term as $k \to \infty$, we find

$$\lim_{k \to \infty} \frac{k - 300}{4k + 750} = \frac{1}{4}$$

Because this limit is not 0, the divergence test tells us that the series must diverge.   ∎

SERIES OF NONNEGATIVE NUMBERS, THE INTEGRAL TEST

Series whose terms are all nonnegative numbers play an important role in the general theory of infinite series and in applications. Our next goal is to develop methods for investigating the convergence of nonnegative-term series, and we begin by establishing the following general principle.

THEOREM 8.10   Convergence criterion for series with nonnegative terms

A series Σa_k with $a_k \geq 0$ for all k converges if its sequence of partial sums is bounded from above and diverges otherwise.

Proof   Suppose Σa_k is a series of nonnegative terms, and let S_n denote its nth partial sum; that is

$$S_n = \sum_{k=1}^{n} a_k = a_1 + a_2 + \cdots + a_n$$

Because $S_{n+1} = S_n + a_{n+1}$ and because $a_k \geq 0$ for all k, it follows that

$$S_{n+1} \geq S_n$$

for all n, so that $\{S_n\}$ is an increasing sequence. According to the BMCT, the increasing sequence $\{S_n\}$ converges if it is bounded from above and diverges otherwise. Hence, because the series Σa_k represents the limit of the sequence $\{S_n\}$, the series Σa_k converges if the sequence $\{S_n\}$ is bounded from above and diverges if it is unbounded.   ❑

This convergence criterion is often difficult to apply because it is not easy to determine whether the sequence of partial sums $\{S_n\}$ is bounded from above. Our next goal in this section and in Sections 8.4–8.6 is to examine various "tests for convergence." These are procedures that allow us to determine indirectly whether a given series converges or diverges without actually having to compute the limit of the partial sums. We begin with a convergence test that relates the convergence of a series to that of an improper integral.

THEOREM 8.11   The integral test

If $a_k = f(k)$ for $k = 1, 2, \ldots$, where f is a positive, continuous, and decreasing function of x for $x \geq 1$, then

$$\sum_{k=1}^{\infty} a_k \quad \text{and} \quad \int_{1}^{\infty} f(x)\, dx$$

either both converge or both diverge.

WARNING   When applying the integral test, the function f need not be decreasing for all $x \geq 1$, only for all $x \geq N$ for some number N. That is, f must be decreasing "in the long run."

Proof   We will use a geometric argument to show that the sequence of partial sums $\{S_n\}$ of the series $\sum_{k=1}^{\infty} a_k$ is bounded from above if the improper integral $\int_{1}^{\infty} f(x)\, dx$ converges and that it is unbounded if the integral diverges. Accordingly, Figures 8.16a and 8.16b show the graph of a continuous decreasing function f that satisfies $f(k) = a_k$ for $k = 1, 2, 3, \ldots$. Rectangles have been constructed at unit intervals in both figures, but in Figure 8.16a the kth rectangle has height $f(k+1) = a_{k+1}$, whereas in Figure 8.16b, the comparable rectangle has height $f(k) = a_k$. Notice that in Figure 8.16a, the rectangles are inscribed so that

AREA OF THE FIRST $n - 1$ RECTANGLES $<$ AREA UNDER $y = f(x)$ OVER $[1, n]$

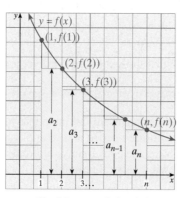

a. The kth rectangle has height $f(k+1) = a_{k+1}$.

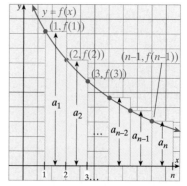

b. The kth rectangle has height $f(k) = a_k$.

Figure 8.16   Integral test

$$a_2 + a_3 + \cdots + a_n < \int_1^n f(x)\,dx$$

$$a_1 + a_2 + a_3 + \cdots + a_n < a_1 + \int_1^n f(x)\,dx \qquad \text{Add } a_1 \text{ to both sides.}$$

Similarly, in Figure 8.16b, the rectangles are circumscribed so that

AREA UNDER $y = f(x)$ OVER $[1, n+1]$ < AREA OF THE FIRST n RECTANGLES

$$\int_1^{n+1} f(x)\,dx < a_1 + a_2 + a_3 + \cdots + a_n$$

Let $S_n = a_1 + a_2 + a_3 + \cdots + a_n$ be the nth partial sum of the series so that

$$\int_1^{n+1} f(x)\,dx < S_n < a_1 + \int_1^n f(x)\,dx$$

Now, suppose the improper integral $\int_1^\infty f(x)\,dx$ converges and has the value I; that is, $\int_1^\infty f(x)\,dx = I$. Then

$$S_n < a_1 + \int_1^n f(x)\,dx < a_1 + \int_1^\infty f(x)\,dx = a_1 + I$$

It follows that the sequence of partial sums is bounded from above (by $a_1 + I$), and the convergence criterion for nonnegative term series tells us that the series $\sum_{k=1}^\infty a_k$ must converge.

On the other hand, if the improper integral $\int_1^\infty f(x)\,dx$ diverges, then

$$\lim_{n \to \infty} \int_1^{n+1} f(x)\,dx = \infty$$

It follows that $\lim_{n \to \infty} S_n = \infty$ also because $\int_1^{n+1} f(x)\,dx < S_n$. This means that the series must diverge. Thus, the series and the improper integral either both converge or both diverge, as claimed. ❑

A series that arises in connection with overtones produced by a vibrating string is

$$\sum_{k=1}^\infty \frac{1}{k} = 1 + \frac{1}{2} + \frac{1}{3} + \frac{1}{4} + \cdots$$

This is called the **harmonic series**, and in the next example, we use the integral test to show that this series diverges.

EXAMPLE 2   Harmonic series diverges

Test the series $\sum_{k=1}^\infty \frac{1}{k}$ for convergence.

Solution

Because $f(x) = \dfrac{1}{x}$ is positive, continuous, and decreasing for $x \geq 1$, the conditions of the integral test are satisfied.

$$\int_1^\infty \frac{1}{x}\,dx = \lim_{b \to \infty} \int_1^b \frac{1}{x}\,dx = \lim_{b \to \infty} [\ln b - \ln 1] = \infty$$

The integral diverges, so the harmonic series diverges. ■

EXAMPLE 3 **Integral test for convergence**

Test the series $\sum_{k=1}^{\infty} \dfrac{k}{e^{k/5}}$ for convergence.

Solution

The function $f(x) = \dfrac{x}{e^{x/5}} = xe^{-x/5}$ is positive and continuous for all $x > 0$. We find that

$$f'(x) = x\left(-\frac{1}{5}e^{-x/5}\right) + e^{-x/5} = \left(1 - \frac{x}{5}\right)e^{-x/5}$$

The critical number is found when $f'(x) = 0$, so we solve

$$\left(1 - \frac{x}{5}\right)e^{-x/5} = 0$$

$$1 - \frac{x}{5} = 0 \qquad \text{Since } e^{-x/5} \neq 0$$

$$x = 5$$

We see that $f'(x) < 0$ for $x > 5$, so it follows that f is decreasing for $x > 5$. Thus, the conditions for the integral test have been established, and the given series and the improper integral either both converge or both diverge. Computing the improper integral, we find

$$\int_5^{\infty} xe^{-x/5}\,dx = \lim_{b \to \infty} \int_5^b xe^{-x/5}\,dx$$

$$\boxed{\begin{array}{ll} \text{Let } u = x & dv = e^{-x/5}\,dx \\ du = dx & v = -5e^{-x/5} \end{array}}$$

$$= \lim_{b \to \infty}\left[-5xe^{-x/5}\Big|_5^b - \int_5^b (-5e^{-x/5})\,dx\right]$$

$$= \lim_{b \to \infty}\left[-5xe^{-x/5} - 25e^{-x/5}\right]\Big|_5^b$$

$$= \lim_{b \to \infty}\left[-5be^{-b/5} - 25e^{-b/5} + 25e^{-1} + 25e^{-1}\right]$$

$$= -5\lim_{b \to \infty}\frac{b+5}{e^{b/5}} + \lim_{b \to \infty} 50e^{-1}$$

$$= -5\lim_{b \to \infty}\frac{1}{\left(\frac{1}{5}\right)e^{b/5}} + 50e^{-1} \qquad \text{l'Hôpital's rule}$$

$$= 50e^{-1} \qquad\qquad \lim_{b \to \infty}\frac{1}{e^{b/5}} = 0$$

Thus, the improper integral converges, which in turn assures the convergence of the given series. ■

WARNING Note that the series converges because the integral converges to a finite number $50e^{-1}$, but this does not mean that the series

$$\sum_{k=1}^{\infty} \frac{k}{e^{k/5}}$$

converges to the same number $50e^{-1}$.

***p*-SERIES**

The harmonic series $\Sigma 1/k$ is a special case of a more general series form called a *p*-series.

p-Series

A series of the form

$$\sum_{k=1}^{\infty} \frac{1}{k^p} = \frac{1}{1^p} + \frac{1}{2^p} + \frac{1}{3^p} + \cdots$$

where p is a positive constant is called a ***p*-series.** The harmonic series (Example 2) is the case where $p = 1$.

The following theorem shows how the convergence of a *p*-series depends on the value of *p*.

THEOREM 8.12   The *p*-series test

The *p*-series $\displaystyle\sum_{k=1}^{\infty} \frac{1}{k^p}$ converges if $p > 1$ and diverges if $p \leq 1$.

Proof   We leave it for the reader to verify that $f(x) = \dfrac{1}{x^p}$ is continuous, positive, and decreasing for $x \geq 1$ and $p > 0$. We know the harmonic series ($p = 1$) diverges. For $p > 0$, $p \neq 1$, we have

$$\int_1^{\infty} \frac{dx}{x^p} = \lim_{b \to \infty} \int_1^b x^{-p}\, dx = \lim_{b \to \infty} \frac{b^{1-p} - 1}{1 - p} = \begin{cases} \dfrac{1}{p-1} & \text{if } p > 1 \\ \infty & \text{if } 0 < p < 1 \end{cases}$$

That is, this improper integral converges if $p > 1$ and diverges if $0 < p < 1$. For $p = 0$, the series becomes

$$\sum_{k=1}^{\infty} \frac{1}{k^0} = \frac{1}{1} + \frac{1}{1} + \frac{1}{1} + \cdots$$

and if $p < 0$, we have $\displaystyle\lim_{k \to \infty} \frac{1}{k^p} = \infty$, so the series diverges by the divergence test (Theorem 8.9). Thus, a *p*-series converges only when $p > 1$.   ❏

➡ **What This Says**   It should be clear that all functions of the form $y = 1/x^p$ with $p > 0$ are decreasing, but why should the series $\Sigma \frac{1}{k^p}$ converge only for $p > 1$? The answer lies in the rate of decrease for $y = 1/x^p$. If $p > 1$, the curve $y = 1/x^p$ falls fast enough to guarantee that the area under the curve for $p > 1$ is finite (see Figure 8.17a), while if $p \leq 1$, the curve $1/x^p$ falls more gradually, and the area under the curve is infinite (Figure 8.17b).

Areas are bounded.

a. $\displaystyle\int_1^{\infty} \frac{dx}{x^p} = \lim_{b \to \infty} \int_1^b \frac{dx}{x^p}$

converges if $p > 1$

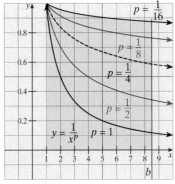

Areas are not bounded.

b. $\displaystyle\int_1^{\infty} \frac{dx}{x^p} = \lim_{b \to \infty} \int_1^b \frac{dx}{x^p}$

diverges if $p \leq 1$

Figure 8.17   Graphs for the *p*-series test

EXAMPLE 4 **p-series test**

Test each of the following series for convergence.

a. $\displaystyle\sum_{k=1}^{\infty}\frac{1}{\sqrt{k^3}}$ **b.** $\displaystyle\sum_{k=1}^{\infty}\left(\frac{1}{e^k}-\frac{1}{\sqrt{k}}\right)$

Solution

a. Here $\sqrt{k^3}=k^{3/2}$, so $p=\frac{3}{2}>1$, and the series converges.

b. $\displaystyle\sum_{k=1}^{\infty}\left(\frac{1}{e^k}-\frac{1}{\sqrt{k}}\right)$; We note that

$$\sum_{k=1}^{\infty}\frac{1}{e^k}\quad\text{converges, because it is geometric with }|r|=\frac{1}{e}<1$$

and

$$\sum_{k=1}^{\infty}\frac{1}{\sqrt{k}}\quad\text{diverges, because it is a }p\text{-series with }p=\frac{1}{2}<1$$

Because one series in the difference converges and the other diverges, the given series must diverge (Theorem 8.7). ∎

8.3 PROBLEM SET

Ⓐ **1. WHAT DOES THIS SAY?** What is a *p*-series?

2. WHAT DOES THIS SAY? What is the integral test?

For the p-series given in Problems 3–6, specify the value of p and tell whether the series converges or diverges.

3. $\displaystyle\sum_{k=1}^{\infty}\frac{1}{k^3}$ **4.** $\displaystyle\sum_{k=1}^{\infty}\frac{100}{\sqrt{k}}$

5. $\displaystyle\sum_{k=1}^{\infty}\frac{1}{\sqrt[3]{k}}$ **6.** $\displaystyle\sum_{k=1}^{\infty}\frac{1}{2k\sqrt{k}}$

Show that the function f determined by the nth term of each series given in Problems 7–12 satisfies the hypotheses of the integral test. Then use the integral test to determine whether the series converges or diverges.

7. $\displaystyle\sum_{k=1}^{\infty}\frac{1}{(2+3k)^2}$ **8.** $\displaystyle\sum_{k=1}^{\infty}(2+k)^{-3/2}$

9. $\displaystyle\sum_{k=2}^{\infty}\frac{\ln k}{k}$ **10.** $\displaystyle\sum_{k=2}^{\infty}\frac{1}{k(\ln k)^2}$

11. $\displaystyle\sum_{k=1}^{\infty}\frac{(\tan^{-1}k)^2}{1+k^2}$ **12.** $\displaystyle\sum_{k=1}^{\infty}ke^{-k^2}$

In Problems 13–18, either use the divergence test to show that the given series diverges or show that the divergence test does not apply.

13. $\displaystyle\sum_{k=1}^{\infty}\frac{k}{k+1}$ **14.** $\displaystyle\sum_{k=1}^{\infty}\left(1+\frac{1}{k}\right)^{-k}$

15. $\displaystyle\sum_{k=2}^{\infty}\frac{k}{\sqrt{k^2-1}}$ **16.** $\displaystyle\sum_{k=1}^{\infty}e^{-1/k}$

17. $\displaystyle\sum_{k=0}^{\infty}\cos(k\pi)$ **18.** $\displaystyle\sum_{k=1}^{\infty}ke^{-k}$

Test the series in Problems 19–42 for convergence.

19. $\displaystyle\sum_{k=1}^{\infty}\frac{\ln k}{k^2}$ **20.** $\displaystyle\sum_{k=1}^{\infty}e^{-k}$

21. $\displaystyle\sum_{k=1}^{\infty}\left(2+\frac{3}{k}\right)^k$ **22.** $\displaystyle\sum_{k=1}^{\infty}\left(1+\frac{2}{k}\right)^k$

23. $\displaystyle\sum_{k=1}^{\infty}\frac{1}{k^4}$ **24.** $\displaystyle\sum_{k=1}^{\infty}\frac{5}{\sqrt{k}}$

25. $\displaystyle\sum_{k=1}^{\infty}k^{-3/4}$ **26.** $\displaystyle\sum_{k=1}^{\infty}k^{-4/3}$

27. $\displaystyle\sum_{k=1}^{\infty}\frac{k}{k^2+1}$ **28.** $\displaystyle\sum_{k=1}^{\infty}\frac{k^2}{\sqrt{k^3+2}}$

29. $\displaystyle\sum_{k=1}^{\infty}ke^{-k^2}$ **30.** $\displaystyle\sum_{k=1}^{\infty}\frac{1}{(0.25)^k}$

31. $\displaystyle\sum_{k=2}^{\infty}\frac{\ln\sqrt{k}}{\sqrt{k}}$ **32.** $\displaystyle\sum_{k=2}^{\infty}\frac{1}{k\sqrt{\ln k}}$

33. $\displaystyle\sum_{k=1}^{\infty}\frac{k}{e^k}$ **34.** $\displaystyle\sum_{k=1}^{\infty}\frac{k^2}{e^k}$

35. $\displaystyle\sum_{k=1}^{\infty}\frac{\tan^{-1}k}{1+k^2}$ **36.** $\displaystyle\sum_{k=1}^{\infty}\cot^{-1}k$

37. $\displaystyle\sum_{k=1}^{\infty}\tan^{-1}k$ **38.** $\displaystyle\sum_{k=1}^{\infty}\frac{1}{e^k+e^{-k}}$

39. $\displaystyle\sum_{k=1}^{\infty}\frac{k-\sin k}{3k+2\sin k}$ **40.** $\displaystyle\sum_{k=1}^{\infty}\left(\sqrt{k^2+k}-k\right)$

41. $\displaystyle\sum_{k=1}^{\infty} k \sin \frac{1}{k}$

42. $\displaystyle\sum_{k=4}^{\infty} \ln\left(\frac{k}{k-3}\right)$

59. $\displaystyle\sum_{k=3}^{\infty} \frac{1}{k \ln k [\ln(\ln k)]^p}$

60. $\displaystyle\sum_{k=2}^{\infty} \frac{1}{k(\ln k)^p}$

Use the divergence test and the p-series test to determine which of the series in Problems 43–54 converge. Write "inconclusive" if there is not yet enough information to settle convergence.

43. $\displaystyle\sum_{k=1}^{\infty} \frac{k-1}{k+1}$

44. $\displaystyle\sum_{k=1}^{\infty} \frac{k^2+1}{k+5}$

45. $\displaystyle\sum_{k=1}^{\infty} \frac{k^2+1}{k^3}$

46. $\displaystyle\sum_{k=1}^{\infty} \frac{2k^4+3}{k^5}$

47. $\displaystyle\sum_{k=1}^{\infty} \frac{1}{k^2}$

48. $\displaystyle\sum_{k=1}^{\infty} \frac{-k^5+k^2+1}{k^5+2}$

49. $\displaystyle\sum_{k=1}^{\infty} \frac{k^{\sqrt{3}}+1}{k^{2.7321}}$

50. $\displaystyle\sum_{k=1}^{\infty} \frac{k^{\sqrt{5}}+1}{k^{2.236}}$

51. $\displaystyle\sum_{k=1}^{\infty} \left[\frac{1}{k}+\frac{k+1}{k+2}\right]$

52. $\displaystyle\sum_{k=1}^{\infty} \left[\frac{1}{2k}+\left(\frac{3}{2}\right)^k\right]$

53. $\displaystyle\sum_{k=1}^{\infty} \left[\frac{1}{2^k}+\frac{2k+3}{3k+4}\right]$

54. $\displaystyle\sum_{k=1}^{\infty} \left[\frac{1}{2^k}-\frac{1}{k}\right]$

B *Find all values of p for which each series given in Problems 55–60 converges.*

55. $\displaystyle\sum_{k=2}^{\infty} \frac{k}{(k^2-1)^p}$

56. $\displaystyle\sum_{k=1}^{\infty} \frac{k^2}{(k^2+4)^p}$

57. $\displaystyle\sum_{k=3}^{\infty} \frac{1}{k^p \ln k}$

58. $\displaystyle\sum_{k=2}^{\infty} \frac{\ln k}{k^p}$

61. **Counterexample Problem** If Σa_k converges, Σb_k diverges, and $\displaystyle\lim_{k\to\infty} b_k = 0$, then $\Sigma a_k b_k$ must diverge. Either prove this statement or provide a counterexample to show it is not generally true.

62. **Counterexample Problem** If $\displaystyle\lim_{k\to\infty} a_k = 0$, $\displaystyle\lim_{k\to\infty} b_k = 0$, and Σa_k and Σb_k both diverge, then $\Sigma a_k b_k$ must also diverge. Either prove this statement or provide a counterexample to show it is not generally true.

C 63. Let $f(x)$ be a positive, continuous, decreasing function for $x \geq 1$, and let $a_k = f(k)$ for $k = 1, 2, \dots$. If Σa_k converges, show that

$$\int_1^{\infty} f(x)\,dx \leq \sum_{k=1}^{\infty} a_k \leq a_1 + \int_1^{\infty} f(x)\,dx$$

64. Getting a feeling for a series by examining its graph is a good idea, but you should be careful because you cannot always obtain a graph. Decide whether

$$\sum \left(1-\frac{100}{n}\right)^n$$

converges or diverges, and then see if you can verify your result graphically.

8.4 Comparison Tests

IN THIS SECTION direct comparison test, limit comparison tests

DIRECT COMPARISON TEST

THEOREM 8.13 Direct comparison test

Suppose $0 \leq a_k \leq c_k$ for all $k \geq N$ for some N.

If $\displaystyle\sum_{k=1}^{\infty} c_k$ converges, then $\displaystyle\sum_{k=1}^{\infty} a_k$ also converges.

Let $0 \leq d_k \leq a_k$ for all $k \geq N$ for some N.

If $\displaystyle\sum_{k=1}^{\infty} d_k$ diverges, then $\displaystyle\sum_{k=1}^{\infty} a_k$ also diverges.

Proof Suppose Σa_k is a given nonnegative term series that is **dominated by** a convergent series Σc_k in the sense that for sufficiently large k, $0 \leq a_k \leq c_k$ for all k. Then, because the series Σc_k converges, its sequence of partial sums is bounded from above (say, by M), and we have

$$\sum_{k=1}^{n} a_k \leq \sum_{k=1}^{n} c_k < M \qquad \text{for all } n$$

Thus, the sequence of partial sums of the dominated series Σa_k is also bounded from above by M, and it, too, must converge.

On the other hand, suppose the given series Σa_k **dominates** a divergent series Σd_k so that $0 \leq d_k \leq a_k$. Then because the sequence of partial sums of Σd_k is unbounded, the same must be true for the partial sums of the dominating series Σa_k, and Σa_k must also diverge.   ❏

> ➡ **What This Says**   Let Σa_k, Σc_k and Σd_k be series with positive terms. The series Σa_k *converges* if it is "smaller" than (dominated by) a known convergent series Σc_k and *diverges* if it is "larger" than (dominates) a known divergent series Σd_k. That is, "smaller than convergent is convergent," and "bigger than divergent is divergent."

EXAMPLE 1   Convergence using the direct comparison test

Test the series $\displaystyle\sum_{k=1}^{\infty} \frac{1}{3^k + 1}$ for convergence.

Solution

For $k \geq 0$, we have $3^k + 1 > 3^k > 0$, and $0 < \dfrac{1}{3^k + 1} < \dfrac{1}{3^k}$. Thus, the given series is *dominated by* the convergent geometric series $\displaystyle\sum_{k=1}^{\infty} \frac{1}{3^k}$ (convergent because $r = \frac{1}{3}$). The direct comparison test tells us the given series must converge.   ∎

EXAMPLE 2   Divergence using the direct comparison test

Test the series $\displaystyle\sum_{k=2}^{\infty} \frac{1}{\sqrt{k} - 1}$ for convergence.

Solution

For $k \geq 2$, we have

$$\frac{1}{\sqrt{k} - 1} > \frac{1}{\sqrt{k}} > 0,$$

so the given series *dominates* the divergent p-series $\displaystyle\sum_{k=2}^{\infty} \frac{1}{k^{1/2}}$ (divergent because $p = \frac{1}{2} < 1$). The direct comparison test tells us the given series must diverge.   ∎

EXAMPLE 3   Applying the direct comparison test

Test the series $\displaystyle\sum_{k=1}^{\infty} \frac{1}{k!}$ for convergence.

Solution

We will compare the given series with a geometric series. Specifically, note that if $k \geq 2$, then $1/k! > 0$ and
$$k! = k(k-1)(k-2) \cdots 3 \cdot 2 \cdot 1 \geq \underbrace{2 \cdot 2 \cdot 2 \cdots 2 \cdot 2}_{k - 1 \text{ terms}} \cdot 1 = 2^{k-1}$$

Therefore, we have $0 < \dfrac{1}{k!} \leq \dfrac{1}{2^{k-1}}$, and because the given series is *dominated by* the convergent geometric series $\displaystyle\sum_{k=1}^{\infty} \frac{1}{2^{k-1}}$ (with $r = \frac{1}{2}$), it must also converge.   ∎

LIMIT COMPARISON TESTS

It is not always easy or even possible to make a suitable direct comparison between two similar series. For example, we would expect the series $\Sigma 1/(2^k - 5)$ to converge because it is so much like the convergent geometric series $\Sigma 1/2^k$. To compare the series, we must first note that $\dfrac{1}{2^k - 5}$ is negative for $k = 1$ and $k = 2$ and positive for $k \geq 3$. Thus, if $k \geq 3$,

$$0 \leq \frac{1}{2^k} < \frac{1}{2^k - 5}$$

That is, $\Sigma 1/(2^k - 5)$ *dominates* the convergent series $\Sigma 1/2^k$, which means the comparison test cannot be used to determine the convergence of $\Sigma 1/(2^k - 5)$ by comparing with $\Sigma 1/2^k$. In such situations, the following test is often useful.

THEOREM 8.14 Limit comparison test

Suppose $a_k > 0$ and $b_k > 0$ for all sufficiently large k and that

$$\lim_{k \to \infty} \frac{a_k}{b_k} = L$$

where L is finite and positive $(0 < L < \infty)$. Then

$$\Sigma a_k \text{ and } \Sigma b_k \text{ either both converge or both diverge.}$$

Proof   The limit comparison test is proved by using the direct comparison test. Details of the proof are given in Appendix B. ❏

➡ **What This Says**   We have the following procedure for testing the convergence of Σa_k.

Step 1.   Find a series Σb_k whose convergence properties are known and whose general term b_k is "essentially the same" as a_k.

Step 2.   Verify that $\lim_{k \to \infty} \dfrac{a_k}{b_k}$ exists and is positive.

Step 3.   Determine whether Σb_k converges or diverges. Then the limit comparison test tells us that Σa_k does the same.

EXAMPLE 4 Convergence using the limit comparison test

Test the series $\displaystyle\sum_{k=1}^{\infty} \frac{1}{2^k - 5}$ for convergence.

Solution

Because the given series has the same general appearance as the convergent geometric series $\Sigma 1/2^k$, compute the limit

$$\lim_{k \to \infty} \frac{\dfrac{1}{2^k - 5}}{\dfrac{1}{2^k}} = \lim_{k \to \infty} \frac{2^k}{2^k - 5} = 1$$

This limit is finite and positive, so the limit comparison test tells us that the given series will have the same convergence properties as $\Sigma 1/2^k$. Thus, the given series converges. ■

EXAMPLE 5 **Divergence using the limit comparison test**

Test the series $\displaystyle\sum_{k=1}^{\infty} \frac{3k+2}{\sqrt{k}(3k-5)}$ for convergence.

Solution

For large values of k, the general term of the given series

$$a_k = \frac{3k+2}{\sqrt{k}(3k-5)}$$

seems to be similar to

$$b_k = \frac{3k}{\sqrt{k}(3k)} = \frac{1}{\sqrt{k}}$$

To apply the limit comparison test, we first compute the limit

$$\lim_{k\to\infty} \frac{a_k}{b_k} = \lim_{k\to\infty} \frac{\dfrac{3k+2}{\sqrt{k}(3k-5)}}{\dfrac{1}{\sqrt{k}}} = \lim_{k\to\infty} \frac{3k+2}{3k-5} = 1$$

Because the limit is finite and positive, it follows that the given series behaves like the series $\Sigma 1/\sqrt{k}$, which is a divergent p-series ($p = \frac{1}{2}$). We conclude that the given series diverges. ∎

EXAMPLE 6 **Applying the limit comparison test**

Test the series $\displaystyle\sum_{k=1}^{\infty} \frac{7k+100}{e^{k/5}-70}$ for convergence.

Solution

For large k,

$$a_k = \frac{7k+100}{e^{k/5}-70} \quad \text{seems to behave like} \quad b_k = \frac{k}{e^{k/5}}$$

and indeed, we find that

$$\lim_{k\to\infty} \frac{\dfrac{7k+100}{e^{k/5}-70}}{\dfrac{k}{e^{k/5}}} = \lim_{k\to\infty} \frac{e^{k/5}(7k+100)}{k(e^{k/5}-70)} = 7$$

In Example 3, Section 8.3, we showed that the series $\Sigma k/e^{k/5}$ converges, and therefore the limit comparison test tells us that the given series also converges. ∎

Occasionally, two series Σa_k and Σb_k appear to have similar convergence properties, but it turns out that

$$\lim_{k\to\infty} \frac{a_k}{b_k}$$

is either 0 or ∞, and the limit comparison test therefore does not apply. In such cases, it is often useful to have the following generalization of the limit comparison test.

THEOREM 8.15 **The zero-infinity limit comparison test**

Suppose $a_k > 0$ and $b_k > 0$ for all sufficiently large k.

$$\text{If } \lim_{k \to \infty} \frac{a_k}{b_k} = 0 \text{ and } \Sigma b_k \text{ converges, then the series } \Sigma a_k \text{ converges.}$$

$$\text{If } \lim_{k \to \infty} \frac{a_k}{b_k} = \infty \text{ and } \Sigma b_k \text{ diverges, then the series } \Sigma a_k \text{ diverges.}$$

Proof The first part of the proof is outlined in Problem 61; the second part follows similarly. ❏

EXAMPLE 7 **Convergence of the q-log series**

Show that the series $\displaystyle\sum_{k=1}^{\infty} \frac{\ln k}{k^q}$ converges if $q > 1$ and diverges if $q \le 1$. We call this the **q-log** series.

Solution

We will carry out this proof in three parts:

Case I: $(q > 1)$
Let p be a number that satisfies $1 < p < q$, and let

$$a_k = \frac{\ln k}{k^q} \quad \text{and} \quad b_k = \frac{1}{k^p}$$

Then

$$\lim_{k \to \infty} \frac{a_k}{b_k} = \lim_{k \to \infty} \frac{\dfrac{\ln k}{k^q}}{\dfrac{1}{k^p}}$$

$$= \lim_{k \to \infty} \frac{\ln k}{k^{q-p}} \qquad \text{This is the indeterminate form } \infty/\infty \text{ because } q - p > 0.$$

$$= \lim_{k \to \infty} \frac{\dfrac{1}{k}}{(q-p)k^{q-p-1}} \qquad \text{l'Hôpital's rule}$$

$$= \lim_{k \to \infty} \frac{1}{(q-p)k^{q-p}} = 0 \qquad \text{Because } q - p > 0$$

Because $\Sigma 1/k^p$ converges (p-series with $p > 1$), the series converges by the zero-infinity limit comparison test.

Case II: $(q < 1)$
Now, let p satisfy $q < p < 1$. Then with a_k and b_k defined as in case I, we have

$$\lim_{k \to \infty} \frac{a_k}{b_k} = \lim_{k \to \infty} \frac{\ln k}{k^{q-p}} = \lim_{k \to \infty} \left[(\ln k)k^{p-q} \right] = \infty$$

Because $\displaystyle\sum_{k=1}^{\infty} b_k = \sum_{k=1}^{\infty} \frac{1}{k^p}$ diverges (we know $p < 1$), it follows from the zero-infinity comparison test that $\displaystyle\sum_{k=1}^{\infty} \frac{\ln k}{k^q}$ diverges when $q < 1$.

Case III: $(q = 1)$

Here, the series is $\displaystyle\sum_{k=1}^{\infty} \frac{\ln k}{k}$, which diverges by the integral test:

$$\int_1^{\infty} \frac{\ln x}{x}\, dx = \lim_{b\to\infty} \left[\frac{\ln^2 x}{2} \right]\Bigg|_1^b = \infty$$

$$\boxed{\text{Let } u = \ln x;\ du = \frac{1}{x}\, dx}$$

$\blacksquare$

8.4 PROBLEM SET

A *The most common series used for comparison are given in Problems 1 and 2. Tell when each converges and when it diverges.*

1. geometric series: $\displaystyle\sum_{k=0}^{\infty} r^k$ **2.** *p*-series: $\displaystyle\sum_{k=1}^{\infty} \frac{1}{k^p}$

Each series in Problems 3–12 can be compared to the geometric series or p-series given in Problems 1 and 2. State which, and then determine whether it converges or diverges.

3. $\displaystyle\sum_{k=1}^{\infty} \cos^k\left(\frac{\pi}{6}\right)$ **4.** $\displaystyle\sum_{k=0}^{\infty} 0.5^k$

5. $\displaystyle\sum_{k=0}^{\infty} 1.5^k$ **6.** $\displaystyle\sum_{k=0}^{\infty} 2^{k/2}$

7. $\displaystyle\sum_{k=1}^{\infty} \frac{1}{k}$ **8.** $\displaystyle\sum_{k=1}^{\infty} \frac{1}{k^{0.5}}$

9. $\displaystyle\sum_{k=1}^{\infty} \frac{1}{k^{3/2}}$ **10.** $\displaystyle\sum_{k=1}^{\infty} \sqrt{\frac{2}{k}}$

11. $\displaystyle\sum_{k=0}^{\infty} 1^k$ **12.** $\displaystyle\sum_{k=1}^{\infty} e^k$

Test the series in Problems 13–44 for convergence.

13. $\displaystyle\sum_{k=1}^{\infty} \frac{1}{k^2 + k}$ **14.** $\displaystyle\sum_{k=1}^{\infty} \frac{1}{k^2 + 3k + 2}$

15. $\displaystyle\sum_{k=1}^{\infty} \frac{1}{\sqrt{k}}$ **16.** $\displaystyle\sum_{k=1}^{\infty} \frac{1}{k\sqrt{k}}$

17. $\displaystyle\sum_{k=1}^{\infty} \frac{1}{\sqrt{2k+3}}$ **18.** $\displaystyle\sum_{k=1}^{\infty} \frac{1}{\sqrt{k(k+1)}}$

19. $\displaystyle\sum_{k=1}^{\infty} \frac{1}{\sqrt{k^3 + 2}}$ **20.** $\displaystyle\sum_{k=1}^{\infty} \frac{1}{\sqrt{k^2 + 1}}$

21. $\displaystyle\sum_{k=1}^{\infty} \frac{2k^2}{k^4 - 4}$ **22.** $\displaystyle\sum_{k=1}^{\infty} \frac{k+1}{k^2 + 1}$

23. $\displaystyle\sum_{k=1}^{\infty} \frac{(k+2)(k+3)}{k^{7/2}}$ **24.** $\displaystyle\sum_{k=1}^{\infty} \frac{(k+1)^3}{k^{9/2}}$

25. $\displaystyle\sum_{k=1}^{\infty} \frac{2k+3}{k^2 + 3k + 2}$ **26.** $\displaystyle\sum_{k=1}^{\infty} \frac{3k^2 + 2}{k^2 + 3k + 2}$

27. $\displaystyle\sum_{k=1}^{\infty} \frac{k}{(k+2)2^k}$ **28.** $\displaystyle\sum_{k=1}^{\infty} \frac{5}{4^k + 3}$

29. $\displaystyle\sum_{k=1}^{\infty} \frac{1}{k(k+2)}$ **30.** $\displaystyle\sum_{k=1}^{\infty} \frac{1}{(k+2)(k+3)}$

31. $\displaystyle\sum_{k=1}^{\infty} \frac{1}{\sqrt{k}\, 2^k}$ **32.** $\displaystyle\sum_{k=1}^{\infty} \frac{1,000}{\sqrt{k}\, 3^k}$

33. $\displaystyle\sum_{k=1}^{\infty} \frac{|\sin(k!)|}{k^2}$ **34.** $\displaystyle\sum_{k=2}^{\infty} \frac{1}{\sqrt{k}\, \ln k}$

35. $\displaystyle\sum_{k=1}^{\infty} \frac{2k^3 + k + 1}{k^3 + k^2 + 1}$ **36.** $\displaystyle\sum_{k=1}^{\infty} \frac{6k^3 - k - 4}{k^3 - k^2 - 3}$

37. $\displaystyle\sum_{k=1}^{\infty} \frac{k}{4k^3 - 5}$ **38.** $\displaystyle\sum_{k=1}^{\infty} \frac{\ln k}{\sqrt{2k+3}}$

39. $\displaystyle\sum_{k=1}^{\infty} \frac{k^2 + 1}{(k^2 + 2)k^2}$ **40.** $\displaystyle\sum_{k=1}^{\infty} \sin\frac{1}{k}$

41. $\displaystyle\sum_{k=1}^{\infty} \frac{6k^2 + 2k + 1}{k^{1.1}(4k^2 + k + 4)}$ **42.** $\displaystyle\sum_{k=1}^{\infty} \frac{6k^2 + 2k + 1}{k^{0.9}(4k^2 + k + 4)}$

43. $\displaystyle\sum_{k=1}^{\infty} \frac{\sqrt[6]{k}}{\sqrt[4]{k^3} + 2\sqrt[8]{k}}$ **44.** $\displaystyle\sum_{k=1}^{\infty} \frac{\sqrt{k}}{\sqrt[3]{k^3 + 1}\sqrt[6]{k^5}}$

B *Test the series given in Problems 45–52 for convergence.*

45. $\displaystyle\sum_{k=1}^{\infty} \frac{1}{k^3 + 4}$ **46.** $\displaystyle\sum_{k=2}^{\infty} \frac{\ln k}{k - 1}$

47. $\displaystyle\sum_{k=1}^{\infty} \frac{\ln(k+1)}{(k+1)^3}$ **48.** $\displaystyle\sum_{k=1}^{\infty} \frac{\ln k}{k^2}$

49. $\displaystyle\sum_{k=2}^{\infty} \frac{1}{(k+3)(\ln k)^{1.1}}$ **50.** $\displaystyle\sum_{k=2}^{\infty} \frac{1}{(k+3)(\ln k)^{0.9}}$

51. $\displaystyle\sum_{k=1}^{\infty} k^{(1-k)/k}$ **52.** $\displaystyle\sum_{k=1}^{\infty} k^{(1+k)/k}$

53. Show that the series

$$\sum_{k=1}^{\infty} \frac{k^2}{(k+3)!} = \frac{1}{4!} + \frac{4}{5!} + \frac{9}{6!} + \cdots$$

converges by using the limit comparison test.

54. Show that the series

$$1 + \frac{1}{1\cdot 3} + \frac{1}{1\cdot 3\cdot 5} + \frac{1}{1\cdot 3\cdot 5\cdot 7} + \cdots + \frac{2^k k!}{(2k+1)!} + \cdots$$

converges. *Hint*: Compare with the convergent series $\sum 1/k!$.

55. Use a comparison test to show that

$$\sum_{k=2}^{\infty} \frac{1}{(\ln k)^{\ln k}}$$

converges. *Hint*: Use the fact that $\ln k > e^2$ for sufficiently large k.

56. a. Let $\{a_n\}$ be a positive sequence such that

$$\lim_{k \to \infty} k^p a_k$$

exists. Use the limit comparison test to show that Σa_k converges if $p > 1$.

b. Show that Σe^{-k^2} converges. *Hint*: Show that $\lim_{k \to \infty} k^2 e^{-k^2}$ exists and use part **a**.

57. Let Σa_k be a series of positive terms and let $\{b_k\}$ be a sequence of positive numbers that converge to a positive number. Show that Σa_k converges if and only if $\Sigma a_k b_k$ converges. *Hint*: Use the limit comparison test.

58. Suppose $0 \le a_k \le A$ for some number A and $b_k \ge k^2$. Show that $\Sigma \dfrac{a_k}{b_k}$ converges.

59. Suppose $a_k > 0$ for all k and that Σa_k converges. Show that $\Sigma \dfrac{1}{a_k}$ diverges.

60. Counterexample Problem Suppose that $0 < a_k < 1$ for every k, and that Σa_k converges. Use a comparison test to show that Σa_k^2 converges. Find a counterexample where Σa_k^2 converges, but Σa_k diverges.

61. Show that if $a_k > 0$, $b_k > 0$, and

$$\lim_{k \to \infty} \frac{a_k}{b_k} = 0$$

then Σa_k converges whenever Σb_k converges. *Hint*: Show that there is an integer N such that $a_k < b_k$ if $k > N$. Then use a comparison test.

62. Show that $\displaystyle\sum_{k=2}^{\infty} \frac{1}{k(\ln k)^q}$ converges if and only if $q > 1$.

8.5 The Ratio Test and the Root Test

IN THIS SECTION ratio test, root test

RATIO TEST

Intuitively, a series of positive terms Σa_k converges if and only if the sequence $\{a_k\}$ decreases rapidly toward 0. One way to measure the rate at which the sequence $\{a_k\}$ is decreasing is to examine the ratio a_{k+1}/a_k as k grows large. This approach leads to the following result.

THEOREM 8.16 The ratio test

Given the series Σa_k with $a_k > 0$, suppose that

$$\lim_{k \to \infty} \frac{a_{k+1}}{a_k} = L$$

The **ratio test** states the following:

If $L < 1$, then Σa_k converges.

If $L > 1$ or if L is infinite, then Σa_k diverges.

If $L = 1$, the test is inconclusive.

Proof In a sense, the ratio test is a limit comparison test in which Σa_k is compared to itself. We will use the direct comparison test to show that Σa_k converges if $L < 1$. Choose R such that $0 < L < R < 1$. Then there exists some $N > 0$ such that

$$\frac{a_{k+1}}{a_k} < R \quad \text{for all } k > N; \qquad \text{Since the ratios } a_{k+1}/a_k$$
$$\text{are eventually very close to } L$$

Therefore,

$$a_{N+1} < a_N R$$
$$a_{N+2} < a_{N+1} R < a_N R^2$$
$$a_{N+3} < a_{N+2} R < a_{N+1} R^2 < a_N R^3$$

$$\vdots$$

The geometric series

$$\sum_{k=1}^{\infty} a_N R^k = a_N R + a_N R^2 + \cdots + a_N R^n + \cdots$$

converges because $0 < R < 1$, which means the "smaller" series

$$\sum_{k=1}^{\infty} a_{N+k} = a_{N+1} + a_{N+2} + \cdots + a_{N+n} + \cdots$$

also converges by the direct comparison test. Thus, Σa_k converges, because we can discard a finite number of terms ($k \leq N$).

The proof of the second part is similar, except now we choose R such that

$$\lim_{k \to \infty} \frac{a_{k+1}}{a_k} = L > R > 1$$

and show there exists some $M > 0$ so that $a_{M+k} > a_M R^k$.

To prove that the ratio test is inconclusive if $L = 1$, it is enough to note that the harmonic series

$$\sum_{k=1}^{\infty} \frac{1}{k} \quad \text{diverges} \quad \text{with} \quad \lim_{k \to \infty} \frac{a_{k+1}}{a_k} = \lim_{k \to \infty} \frac{\dfrac{1}{k+1}}{\dfrac{1}{k}} = 1$$

while the p-series

$$\sum_{k=1}^{\infty} \frac{1}{k^2} \quad \text{converges} \quad \text{with} \quad \lim_{k \to \infty} \frac{a_{k+1}}{a_k} = \lim_{k \to \infty} \frac{\dfrac{1}{(k+1)^2}}{\dfrac{1}{k^2}} = 1 \qquad \square$$

EXAMPLE 1 Convergence using the ratio test

WARNING You will find the ratio test most useful with series involving factorials or exponentials.

Test the series $\displaystyle\sum_{k=1}^{\infty} \frac{2^k}{k!}$ for convergence.

Solution

Let $a_k = \dfrac{2^k}{k!}$ and note that

$$L = \lim_{k \to \infty} \frac{a_{k+1}}{a_k} = \lim_{k \to \infty} \frac{\dfrac{2^{k+1}}{(k+1)!}}{\dfrac{2^k}{k!}} = \lim_{k \to \infty} \frac{k! 2^{k+1}}{(k+1)! 2^k} = \lim_{k \to \infty} \frac{2}{k+1} = 0$$

Thus $L < 1$, and the ratio test tells us that the given series converges. ∎

EXAMPLE 2 Divergence using the ratio test

Test the series $\displaystyle\sum_{k=1}^{\infty} \frac{k^k}{k!}$ for convergence.

Solution

Let $a_k = \dfrac{k^k}{k!}$ and note that

$$L = \lim_{k \to \infty} \frac{a_{k+1}}{a_k} = \lim_{k \to \infty} \frac{\dfrac{(k+1)^{k+1}}{(k+1)!}}{\dfrac{k^k}{k!}} = \lim_{k \to \infty} \frac{k!(k+1)^{k+1}}{k^k(k+1)!}$$

$$= \lim_{k \to \infty} \frac{(k+1)^k}{k^k} = \lim_{k \to \infty} \left(1 + \frac{1}{k}\right)^k = e > 1$$

Because $L > 1$, the given series diverges. ∎

It is instructive to compare and contrast two different ways you can use technology to help you with problems such as Example 2. First, as you have seen, you can use a calculator or a computer to calculate the partial sums directly. This table is shown below at the left.

A second procedure is to use the ratio test for Example 2. In this case, we are interested in

$$\lim_{k \to \infty} \frac{a_{k+1}}{a_k} = \lim_{k \to \infty} \frac{\dfrac{(k+1)^{k+1}}{(k+1)!}}{\dfrac{k^k}{k!}} = \lim_{k \to \infty} \left(1 + \frac{1}{k}\right)^k$$

We can look at terms of the sequence $a_k = \left(1 + \dfrac{1}{k}\right)^k$ to see whether this limit is greater than, less than, or equal to 1. This is shown below at the right.

By table:

$$\sum_{k=1}^{\infty} \frac{k^k}{k!}$$

By graph:

$$\lim_{k \to \infty} \left(1 + \frac{1}{k}\right)^k$$

N	F(1)+...+F(N)
1	1
2	3
3	7.5
4	18.16667
5	44.20833
6	109.0083
7	272.4097
8	688.5113
9	1756.138
10	4511.87
11	11659.53
12	30273.46
13	78912.3
14	206375.3
15	541239.9

The partial sums seem to diverge.

We see that the ratio test gives a limit greater than 1.

Finally, you might test this using a calculator to verify that it diverges.

EXAMPLE 3   Ratio test fails

Test the series $\displaystyle\sum_{k=2}^{\infty} \frac{1}{2k-3}$ for convergence.

Solution

Let $a_k = \dfrac{1}{2k-3}$ and find

$$L = \lim_{k \to \infty} \frac{\dfrac{1}{2(k+1)-3}}{\dfrac{1}{2k-3}} = \lim_{k \to \infty} \frac{2k-3}{2k-1} = 1$$

The ratio test is inconclusive. We can use the integral test or the limit comparison test to determine convergence. Note that

$$\sum_{k=2}^{\infty} \frac{1}{2k-3} \quad \text{is similar to the known divergent series} \quad \sum_{k=2}^{\infty} \frac{1}{2k}$$

so we conclude the given series is divergent. ∎

EXAMPLE 4   Convergence of a series of power functions

Find all numbers $x > 0$ for which the series

$$\sum_{k=1}^{\infty} k^3 x^k = x + 2^3 x^2 + 3^3 x^3 + \cdots$$

converges. This is called a *power series*, and it will be discussed in more detail in Sections 8.7 and 8.8.

Solution

Viewing x as fixed (that is, as a constant), we apply the ratio test:

$$L = \lim_{k \to \infty} \frac{(k+1)^3 x^{k+1}}{k^3 x^k} = \lim_{k \to \infty} \left(\frac{k+1}{k}\right)^3 x = x$$

Thus, the series converges if $L = x < 1$ and diverges if $x > 1$. When $x = 1$, the series becomes $\sum_{k=1}^{\infty} k^3$, which diverges by the divergence test. ∎

ROOT TEST

Of all the tests we have developed, the divergence test is perhaps the easiest to apply, because it involves simply computing $\lim_{k \to \infty} a_k$ and observing whether that limit is zero. Unfortunately, most "interesting" series have $\lim_{k \to \infty} a_k = 0$, so the divergence test cannot be used to determine whether they converge or diverge. However, the following result shows that it may be possible to say more about the convergence of Σa_k by examining what happens to $\sqrt[k]{a_k}$ as $k \to \infty$. This test will prove particularly useful with a series involving a kth power.

THEOREM 8.17   The root test

Given the series Σa_k with $a_k \geq 0$, suppose that $\lim_{k \to \infty} \sqrt[k]{a_k} = L$. The **root test** states the following:

    If $L < 1$, then $\sum a_k$ converges.

    If $L > 1$ or if L is infinite, then $\sum a_k$ diverges.

    If $L = 1$, the root test is inconclusive.

Proof   The proof of this theorem is similar to that of the ratio test and is outlined in Problem 57. ❑

EXAMPLE 5   Convergence with the root test

Test the series $\sum_{k=2}^{\infty} \frac{1}{(\ln k)^k}$ for convergence.

Solution

Let $a_k = \dfrac{1}{(\ln k)^k}$ and note that

$$L = \lim_{k \to \infty} \sqrt[k]{a_k} = \lim_{k \to \infty} \sqrt[k]{(\ln k)^{-k}} = \lim_{k \to \infty} \frac{1}{\ln k} = 0$$

Because $L < 1$, the root test tells us that the given series converges. ∎

EXAMPLE 6 Divergence with the root test

Test the series $\displaystyle\sum_{k=1}^{\infty}\left(1+\frac{1}{k}\right)^{k^2}$ for convergence.

Solution

$$L = \lim_{k\to\infty}\left[\left(1+\frac{1}{k}\right)^{k^2}\right]^{1/k} \qquad \text{Root test}$$

$$= \lim_{k\to\infty}\left(1+\frac{1}{k}\right)^{k}$$

$$= e > 1$$

Since $L > 1$, the series diverges. ∎

We have seen that the ratio test is especially useful for series Σa_k for which the general term a_k involves factorials or powers, and the root test applies naturally if a_k involves a power of k. However, both the ratio test and the root test fail with certain very simple series such as the p-series $\Sigma \frac{1}{k^p}$ since

$$\lim_{k\to\infty}\frac{\dfrac{1}{(k+1)^p}}{\dfrac{1}{k^p}} = \lim_{k\to\infty}\left(\frac{k}{k+1}\right)^p = 1^p = 1 \qquad \text{Inconclusive ratio test}$$

$$\lim_{k\to\infty}\sqrt[k]{\frac{1}{k^p}} = \left[\lim_{k\to\infty}k^{1/k}\right]^{-p} = 1^{-p} = 1 \qquad \begin{array}{l}\text{Inconclusive root test}\\ \text{where l'Hôpital's rule}\\ \text{is used to evaluate the limit.}\\ \text{See Example 11 of Section 4.5.}\end{array}$$

EXAMPLE 7 Testing for convergence

Test the following series for convergence:

$$\sum_{k=0}^{\infty}\frac{k!}{1\cdot4\cdot7\cdots(3k+1)} = 1 + \frac{1!}{1\cdot4} + \frac{2!}{1\cdot4\cdot7} + \frac{3!}{1\cdot4\cdot7\cdot10} + \cdots$$

Solution

Let $a_k = \dfrac{k!}{1\cdot4\cdot7\cdots(3k+1)}$. Because a_k involves $k!$, we try the ratio test.

$$\frac{a_{k+1}}{a_k} = \frac{\dfrac{(k+1)!}{1\cdot4\cdot7\cdots[3(k+1)+1]}}{\dfrac{k!}{1\cdot4\cdot7\cdots(3k+1)}} = \frac{(k+1)!\cdot[1\cdot4\cdot7\cdots(3k+1)]}{k!\cdot[1\cdot4\cdot7\cdots(3k+4)]}$$

$$= \frac{k+1}{3k+4}$$

Thus,

$$L = \lim_{k\to\infty}\frac{a_{k+1}}{a_k} = \lim_{k\to\infty}\frac{k+1}{3k+4} = \frac{1}{3}$$

Because $L < 1$, the given series converges. ∎

EXAMPLE 8 Testing for convergence

Test the series $\displaystyle\sum_{k=1}^{\infty}\frac{k!}{k^k}$ for convergence.

Solution

The general term $a_k = \dfrac{k!}{k^k}$ involves a kth power, which suggests using the root test, but the presence of the factorial $k!$ makes using the ratio test much more reasonable.

$$L = \lim_{k \to \infty} \frac{\dfrac{(k+1)!}{(k+1)^{k+1}}}{\dfrac{k!}{k^k}} = \lim_{k \to \infty} \frac{(k+1)! k^k}{k!(k+1)^{k+1}} = \lim_{k \to \infty} \frac{(k+1)k^k}{(k+1)^{k+1}}$$

$$= \lim_{k \to \infty} \left(\frac{k}{k+1} \right)^k = \lim_{k \to \infty} \frac{1}{\left(1 + \dfrac{1}{k} \right)^k} = \frac{1}{e} < 1$$

Since $L < 1$, the series converges. ∎

8.5 PROBLEM SET

Ⓐ 1. WHAT DOES THIS SAY?   Describe the ratio test.

2. WHAT DOES THIS SAY?   Describe the root test.

Use either the ratio test or the root test to determine the convergence of the series given in Problems 3–26.

3. $\displaystyle\sum_{k=1}^{\infty} \frac{1}{k!}$

4. $\displaystyle\sum_{k=1}^{\infty} \frac{k!}{2^k}$

5. $\displaystyle\sum_{k=1}^{\infty} \frac{k!}{2^{3k}}$

6. $\displaystyle\sum_{k=1}^{\infty} \frac{3^k}{k!}$

7. $\displaystyle\sum_{k=1}^{\infty} \frac{k}{2^k}$

8. $\displaystyle\sum_{k=1}^{\infty} \frac{2^k}{k^2}$

9. $\displaystyle\sum_{k=1}^{\infty} \frac{k^{100}}{e^k}$

10. $\displaystyle\sum_{k=1}^{\infty} k e^{-k}$

11. $\displaystyle\sum_{k=1}^{\infty} k \left(\frac{4}{3} \right)^k$

12. $\displaystyle\sum_{k=1}^{\infty} k \left(\frac{3}{4} \right)^k$

13. $\displaystyle\sum_{k=1}^{\infty} \left(\frac{2}{k} \right)^k$

14. $\displaystyle\sum_{k=1}^{\infty} \frac{k^{10} 2^k}{k!}$

15. $\displaystyle\sum_{k=1}^{\infty} \frac{k^5}{10^k}$

16. $\displaystyle\sum_{k=1}^{\infty} \frac{3^k}{k^2}$

17. $\displaystyle\sum_{k=1}^{\infty} \left(\frac{k}{3k+1} \right)^k$

18. $\displaystyle\sum_{k=1}^{\infty} \frac{3k+1}{2^k}$

19. $\displaystyle\sum_{k=1}^{\infty} \frac{k!}{(k+2)^4}$

20. $\displaystyle\sum_{k=1}^{\infty} \frac{k^5 + 100}{k!}$

21. $\displaystyle\sum_{k=1}^{\infty} \frac{(k!)^2}{(2k)!}$

22. $\displaystyle\sum_{k=1}^{\infty} \frac{(k!)^2}{[(2k)!]^2}$

23. $\displaystyle\sum_{k=1}^{\infty} k^2 2^{-k}$

24. $\displaystyle\sum_{k=1}^{\infty} k^4 3^{-k}$

25. $\displaystyle\sum_{k=1}^{\infty} \left(\frac{k-2}{k} \right)^{k^2}$

26. $\displaystyle\sum_{k=1}^{\infty} \left(\frac{k}{2k+1} \right)^k$

Test the series in Problems 27–44 for convergence. Justify your answers (that is, state explicitly which test you are using).

27. $\displaystyle\sum_{k=1}^{\infty} \frac{1,000}{k}$

28. $\displaystyle\sum_{k=1}^{\infty} \frac{5,000}{k\sqrt{k}}$

29. $\displaystyle\sum_{k=1}^{\infty} \frac{5k+2}{k2^k}$

30. $\displaystyle\sum_{k=1}^{\infty} \frac{(k!)^2}{k^k}$

31. $\displaystyle\sum_{k=1}^{\infty} \frac{\sqrt{k!}}{2^k}$

32. $\displaystyle\sum_{k=1}^{\infty} \frac{3k+5}{k3^k}$

33. $\displaystyle\sum_{k=1}^{\infty} \frac{2^k k!}{k^k}$

34. $\displaystyle\sum_{k=1}^{\infty} \frac{2^{2k} k!}{k^k}$

35. $\displaystyle\sum_{k=1}^{\infty} \frac{\sqrt{k+1}}{k^{k+0.5}}$

36. $\displaystyle\sum_{k=1}^{\infty} \frac{1}{k^k}$

37. $\displaystyle\sum_{k=1}^{\infty} \frac{k!}{(k+1)!}$

38. $\displaystyle\sum_{k=1}^{\infty} \frac{2^{1,000k}}{k^{k/2}}$

39. $\displaystyle\sum_{k=1}^{\infty} \left(1 + \frac{1}{k} \right)^{-k^2}$

40. $\displaystyle\sum_{k=1}^{\infty} \left(\frac{k+2}{k} \right)^{-k^2}$

41. $\displaystyle\sum_{k=1}^{\infty} \left| \frac{\cos k}{2^k} \right|$

42. $\displaystyle\sum_{k=1}^{\infty} \left| \frac{\sin k}{3^k} \right|$

43. $\displaystyle\sum_{k=2}^{\infty} \left(\frac{\ln k}{k} \right)^k$

44. $\displaystyle\sum_{k=2}^{\infty} \frac{1}{(\ln k)^k}$

Ⓑ *In Problems 45–52, assume $x > 0$ and find all x for which the given series converges.*

45. $\displaystyle\sum_{k=1}^{\infty} k^2 x^k$

46. $\displaystyle\sum_{k=1}^{\infty} k x^k$

47. $\displaystyle\sum_{k=1}^{\infty} \frac{(x+0.5)^k}{k\sqrt{k}}$

48. $\displaystyle\sum_{k=1}^{\infty} \frac{(3x-0.4)^k}{k^2}$

49. $\displaystyle\sum_{k=1}^{\infty} \frac{x^k}{k!}$

50. $\displaystyle\sum_{k=1}^{\infty} \frac{x^{2k}}{k}$

51. $\displaystyle\sum_{k=1}^{\infty}(ax)^k$ for $a > 0$ **52.** $\displaystyle\sum_{k=1}^{\infty}kx^{2k}$

53. Use the root test to show that $\Sigma k^p e^{-k}$ converges for any fixed positive real number p. What does this imply about the following improper integral?

$$\int_1^{\infty} x^p e^{-x}\,dx$$

54. Consider the series $\displaystyle\sum_{k=1}^{\infty} 2^{-k+(-1)^k}$. What can you conclude about the convergence of this series if you use each of the following?

  a. the ratio test   **b.** the root test

55. Let $\{a_k\}$ be a sequence of positive numbers and suppose that $\displaystyle\lim_{k\to\infty}\frac{a_{k+1}}{a_k} = L$, where $0 < L < 1$.

  a. Show that $\displaystyle\lim_{k\to\infty} a_k = 0$. *Hint*: Note that Σa_k converges.

  b. Show that $\displaystyle\lim_{k\to\infty}\frac{x^k}{k!} = 0$ for any $x > 0$.

56. Consider the series

$$1 + \frac{1}{2} + \frac{1}{2} + \frac{1}{4} + \frac{1}{4} + \frac{1}{8} + \frac{1}{8} + \frac{1}{16} + \cdots$$

  a. Show that the ratio test fails.

  b. What is the result of applying the root test?

57. Prove the root test by completing the steps in the following outline:

  a. Let $\displaystyle\lim_{n\to\infty}\sqrt[n]{a_n} = R$. If $R < 1$, choose x so $R < x < 1$. Explain why there is an N such that

$$0 \le \sqrt[n]{a_n} \le x$$

  for all $n \ge N$. Then use the direct comparison test to show that Σa_n converges.

  b. If $R > 1$, explain why $a_n > 1$ for all but a finite number of n. Use the divergence test to show Σa_n diverges.

  c. Find two series Σa_n and Σb_n with $R = 1$, so that Σa_n converges but Σb_n diverges.

58. Exploration Problem   Recall from Problem 48, Section 8.1, that the Fibonacci sequence $\{a_n\}$ is

$$1, 1, 2, 3, 5, 8, 13, 21, \ldots$$

where $a_1 = 1$, $a_2 = 1$, and the rest of the terms are defined recursively by the formula $a_{n+1} = a_n + a_{n-1}$ for $n = 2, 3, \ldots$. Determine whether the series of reciprocal Fibonacci numbers

$$\sum_{k=1}^{\infty}\frac{1}{a_k} = 1 + 1 + \frac{1}{2} + \frac{1}{3} + \frac{1}{5} + \cdots$$

converges or diverges.

8.6 Alternating Series; Absolute and Conditional Convergence

IN THIS SECTION   alternating series test, error estimates for alternating series, absolute and conditional convergence, summary of convergence tests for series, rearrangement of terms in an absolutely convergent series

There are two classes of series for which the successive terms alternate in sign, and each of these series is appropriately called an **alternating series**:

$$\text{odd-indexed terms are negative: } \sum_{k=1}^{\infty}(-1)^k a_k = -a_1 + a_2 - a_3 + \cdots$$

$$\text{even-indexed terms are negative: } \sum_{k=1}^{\infty}(-1)^{k+1} a_k = a_1 - a_2 + a_3 - \cdots$$

where in both cases $a_k > 0$.

ALTERNATING SERIES TEST

In general, just knowing that $\displaystyle\lim_{k\to\infty} a_k = 0$ tells us very little about the convergence properties of the series Σa_k, but it turns out that an alternating series must converge if the absolute value of its terms decreases monotonically toward zero. This result, first established by Leibniz in the 17th century, may be stated as shown in the following theorem.

THEOREM 8.18   Alternating series test

An alternating series

$$\sum_{k=1}^{\infty}(-1)^k a_k \quad \text{or} \quad \sum_{k=1}^{\infty}(-1)^{k+1} a_k \qquad \text{where } a_k > 0 \text{ for all } k$$

converges if both of the following two conditions are satisfied:

1. $\lim_{k \to \infty} a_k = 0$
2. $\{a_k\}$ is a decreasing sequence; that is, $a_{k+1} \le a_k$ for all k.

Proof   We will show that when an alternating series of the form

$$\sum_{k=1}^{\infty}(-1)^{k+1} a_k$$

satisfies the two required properties, it converges. The steps for the other type of alternating series are similar. For this proof we are given that

$$\lim_{k \to \infty} a_k = 0 \quad \text{and} \quad a_{k+1} \le a_k \quad \text{for all } k$$

We need to prove that the sequence of partial sums $\{S_n\}$ converges, where

$$S_n = \sum_{k=1}^{n}(-1)^{k+1} a_k = a_1 - a_2 + a_3 - a_4 + \cdots + (-1)^{n+1} a_n$$

Our strategy will be to show first that the sequence of *even*-indexed partial sums $\{S_{2n}\}$ is *increasing* and converges to a certain limit L. Then we will show that the sequence of odd-indexed partial sums $\{S_{2n-1}\}$ also converges to L.

To understand why we would think to break up S_n into even- and odd-indexed partial sums, consider Figure 8.18. Start at the origin. Because $a_1 > 0$, S_1 will be somewhere to the right of 0 as shown in Figure 8.18a. We know that $a_2 \le a_1$ (decreasing sequence) so S_2 is to the left of S_1 but to the right of 0, as shown in Figure 8.18b. As you continue this process you can see that the partial sums oscillate back and forth. Because $a_n \to 0$, the successive steps are getting smaller and smaller, as shown in Figure 8.18c.

Note that the even partial sums of the alternating series satisfy

$$S_2 = a_1 - a_2$$
$$S_4 = (a_1 - a_2) + (a_3 - a_4)$$
$$\vdots$$
$$S_{2n} = (a_1 - a_2) + (a_3 - a_4) + \cdots + (a_{2n-1} - a_{2n})$$

(a)

(b)

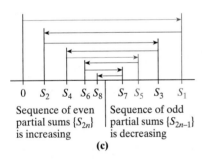

Sequence of even | Sequence of odd
partial sums $\{S_{2n}\}$ | partial sums $\{S_{2n-1}\}$
is increasing | is decreasing

(c)

Figure 8.18 Alternating series test

Because $\{a_k\}$ is a decreasing sequence, each of the quantities in parentheses $(a_{2j-1} - a_{2j})$ is nonnegative, and it follows that $\{S_{2n}\}$ is an increasing sequence:

$$S_2 \le S_4 \le S_6 \le \cdots$$

Moreover, because

$$S_2 = a_1 - a_2 \le a_1$$
$$S_4 = a_1 - (a_2 - a_3) - a_4 \le a_1$$
$$\vdots$$
$$S_{2n} = a_1 - (a_2 - a_3) - (a_4 - a_5) - \cdots - (a_{2n-2} - a_{2n-1}) - a_{2n} \le a_1$$

the sequence of even partial sums $\{S_{2n}\}$ is bounded above by a_1, and because it is also an increasing sequence, the BMCT tells us that this sequence must converge, say, to L; that is,

$$\lim_{n \to \infty} S_{2n} = L$$

Next, consider $\{S_{2n-1}\}$, the sequence of odd partial sums. Because $S_{2n} - S_{2n-1} = a_{2n}$ and $\lim_{n\to\infty} a_n = 0$, it follows that $S_{2n-1} = S_{2n} - a_{2n}$ and

$$\lim_{n\to\infty} S_{2n-1} = \lim_{n\to\infty} S_{2n} - \lim_{n\to\infty} a_{2n} = L - 0 = L$$

Since the sequence of even partial sums and the sequence of odd partial sums both converge to L, it follows that $\{S_n\}$ converges to L, and the alternating series must converge (in fact, to L). ❑

REMARK: When you are testing the alternating series $\Sigma(-1)^{k+1} a_k$ for convergence, it is wise to begin by computing $\lim_{k\to\infty} a_k$. If $\lim_{k\to\infty} a_k \neq 0$, you know immediately that the series diverges (by the divergence test), and no further computation is necessary. However, if this limit is 0, you can show that the alternating series converges by verifying that $\{a_k\}$ is a decreasing sequence.

EXAMPLE 1   Convergence using the alternating series test

Test the series $\displaystyle\sum_{k=1}^{\infty} \frac{(-1)^k}{k}$ for convergence. This series is called the **alternating harmonic series.**

Solution

The series can be expressed as $\Sigma(-1)^k a_k$, where $a_k = \dfrac{1}{k}$. We have $\lim_{k\to\infty} \dfrac{1}{k} = 0$ and because

$$\frac{1}{k+1} < \frac{1}{k}$$

for all $k > 0$, the sequence $\{a_k\}$ is decreasing. Thus, the alternating series test tells us that the given series must converge. ∎

EXAMPLE 2   Using l'Hôpital's rule with the alternating series test

Test the series $\displaystyle\sum_{k=1}^{\infty} \frac{(-1)^{k+1} \ln k}{k}$ for convergence.

Solution

Express the series in the form $\Sigma(-1)^{k+1} a_k$, where $a_k = \dfrac{\ln k}{k}$ (note that $a_k > 0$ for $k > 1$). The graph of $y = (\ln x)/x$ is shown in Figure 8.19. By applying l'Hôpital's rule, we find that

$$\lim_{x\to\infty} \frac{\ln x}{x} = \lim_{x\to\infty} \frac{\dfrac{1}{x}}{1} = 0$$

It remains to show that the sequence $\{a_k\}$ is decreasing. We can do this by setting $f(x) = \dfrac{\ln x}{x}$ and noting that the derivative

$$f'(x) = \frac{1 - \ln x}{x^2}$$

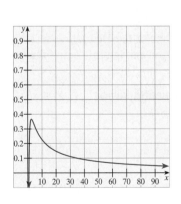

Figure 8.19 Graph of $y = \dfrac{\ln x}{x}$

satisfies $f'(x) < 0$ for all $x > e$. Thus, the sequence $\{a_k\}$ is (eventually) decreasing for all $k > 3 > e$. The conditions of the alternating series test are satisfied, and the given alternating series must converge. ∎

EXAMPLE 3   Divergence of an alternating series

Test the series $\displaystyle\sum_{k=1}^{\infty}(-1)^{k+1}e^{1/k}$ for convergence.

Solution

Since

$$\lim_{k\to\infty}e^{1/k} = \exp\left(\lim_{k\to\infty}\frac{1}{k}\right) = e^0 = 1 \neq 0$$

the series diverges by the divergence test.   ∎

Alternating p-Series

> The series $\displaystyle\sum_{k=1}^{\infty}\frac{(-1)^{k+1}}{k^p}$ is called the **alternating p-series**.

The following example shows that the alternating p-series converges for all $p > 0$.

EXAMPLE 4   Alternating p-series

Prove that the alternating p-series $\displaystyle\sum_{k=1}^{\infty}\frac{(-1)^{k+1}}{k^p}$ converges for $p > 0$.

Solution

If $a_k = \dfrac{1}{k^p}$, then $\lim_{k\to\infty} a_k = 0$ for $p > 0$. To show that the sequence $\{a_k\}$ is decreasing, we note that

$$\frac{a_{k+1}}{a_k} = \frac{\dfrac{1}{(k+1)^p}}{\dfrac{1}{k^p}} = \frac{k^p}{(k+1)^p} < 1$$

so $a_k > a_{k+1}$. Thus, the alternating p-series converges for all $p > 0$.   ∎

ERROR ESTIMATES FOR ALTERNATING SERIES

For any convergent series with sum L, the nth partial sum is an approximation to L that we expect to improve as n increases. In general, it is difficult to know how large an n to pick in order to ensure that the approximation will have a desired degree of accuracy. However, if the series in question satisfies the conditions of the alternating series test, the following theorem provides a convenient criterion for making this determination.

THEOREM 8.19   The error estimate for an alternating series

Suppose an alternating series

$$\sum_{k=1}^{\infty}(-1)^k a_k \quad\text{or}\quad \sum_{k=1}^{\infty}(-1)^{k+1}a_k$$

satisfies the conditions of the alternating series test; namely,

$$\lim_{k\to\infty} a_k = 0 \quad\text{and}\quad \{a_k\}\text{ is a decreasing sequence }(a_{k+1} \leq a_k)$$

If the series has sum S, then

$$|S - S_n| \leq a_{n+1}$$

where S_n is the nth partial sum of the series.

Proof   We will prove the result for the second form of the alternating series and leave the first form for the reader. That is, let

$$S = \sum_{k=1}^{\infty}(-1)^{k+1}a_k \quad \text{and} \quad S_n = \sum_{k=1}^{n}(-1)^{k+1}a_k$$

Begin with $S - S_n$:

$$S - S_n = \sum_{k=1}^{\infty}(-1)^{k+1}a_k - \sum_{k=1}^{n}(-1)^{k+1}a_k$$

$$= \sum_{k=n+1}^{\infty}(-1)^{k+1}a_k$$

$$= (-1)^{n+2}a_{n+1} + (-1)^{n+3}a_{n+2} + (-1)^{n+4}a_{n+3} + \cdots$$

$$= (-1)^n(-1)^2 a_{n+1} + (-1)^n(-1)^3 a_{n+2} + (-1)^n(-1)^4 a_{n+3} + \cdots$$

$$= (-1)^n[a_{n+1} - a_{n+2} + a_{n+3} - a_{n+4} + \cdots]$$

$$= (-1)^n[a_{n+1} - (a_{n+2} - a_{n+3}) - (a_{n+4} - a_{n+5}) - \cdots]$$

Because the sequence $\{a_n\}$ is decreasing, we have $a_k - a_{k+1} \geq 0$ for all k, and it follows that

$$|S - S_n| = |a_{n+1} - (a_{n+2} - a_{n+3}) - (a_{n+4} - a_{n+5}) - \cdots| \leq a_{n+1}$$

since every quantity in parentheses is positive.   ❏

➡ **What This Says**   If an alternating series satisfies the conditions of the alternating series test, you can approximate the sum of the series by using the nth partial sum (S_n), and your error will have an absolute value no greater than the first term left off (namely, a_{n+1}).

EXAMPLE 5   Error estimate for an alternating series

Consider the convergent alternating series $\displaystyle\sum_{k=1}^{\infty}\frac{(-1)^{k+1}}{k^4}$.

a. Estimate the sum of the series by taking the sum of the first four terms. How accurate is this estimate?
b. How many terms of the series are necessary to estimate its sum with three-decimal-place accuracy? What is this estimate?

Solution

a. Let $a_k = \dfrac{1}{k^4}$ and let S denote the actual sum of the series. The error estimate tells us that

$$|S - S_4| = a_5$$

where S_4 is the sum of the first four terms of the series. Using a calculator, we find

$$S_4 = \frac{1}{1^4} - \frac{1}{2^4} + \frac{1}{3^4} - \frac{1}{4^4} \approx 0.9459394$$

and $a_5 = \dfrac{1}{5^4} = 0.0016$. Thus, if we estimate S by $S_4 \approx 0.9459$, we incur an error of about 0.0016, which means that

$$|S - S_4| \leq 0.0016$$

$$0.9459394 - 0.0016 \leq S \leq 0.9459394 + 0.0016$$

$$0.9443394 \leq S \leq 0.9475394$$

b. Because we want to approximate S by a partial sum S_n with three-decimal-place accuracy, we can allow an error of no more than 0.0005. According to Theorem 8.19, we want to find an index n such that $a_{n+1} \leq 0.0005$; that is,

$$\frac{1}{(n+1)^4} \leq 0.0005$$

$$\frac{1}{0.0005} \leq (n+1)^4$$

$$\sqrt[4]{2,000} \leq n+1$$

$$6.687403 - 1 \leq n$$

Thus, $n \geq 5.687403$, so we need at least 6 terms to achieve the required accuracy. We find that

$$S_6 = \frac{1}{1^4} - \frac{1}{2^4} + \frac{1}{3^4} - \frac{1}{4^4} + \frac{1}{5^4} - \frac{1}{6^4} \approx 0.94677$$

The first term left off (namely $\frac{1}{7^4} \approx 0.00042$) is positive, so

$$0.94677 < S < 0.94677 + 0.00042$$

$$0.94677 < S < 0.94719$$

Both bounds round to 0.947, so $S \approx 0.947$, with three-decimal-place accuracy.   ■

ABSOLUTE AND CONDITIONAL CONVERGENCE

The convergence tests we have developed cannot be applied to a series that has mixed terms or does not strictly alternate. In such cases, it is often useful to apply the following result.

THEOREM 8.20   The absolute convergence test

A series of real numbers Σa_k must converge if the related absolute value series $\Sigma |a_k|$ converges.

Proof   Assume $\Sigma |a_k|$ converges and let $b_k = a_k + |a_k|$ for all k. Note that

$$b_k = \begin{cases} 2|a_k| & \text{if } a_k > 0 \\ 0 & \text{if } a_k \leq 0 \end{cases}$$

Thus, we have $0 \leq b_k \leq 2|a_k|$ for all k. Because the series $\Sigma |a_k|$ converges and both Σb_k and $\Sigma |a_k|$ are series of nonnegative terms, the direct comparison test tells us that the dominated series Σb_k also converges. Finally, because

$$a_k = b_k - |a_k|$$

and both Σb_k and $\Sigma |a_k|$ converge, it follows that Σa_k must also converge.   ❑

EXAMPLE 6   Convergence using the absolute convergence test

Test the series $1 + \frac{1}{4} + \frac{1}{9} - \frac{1}{16} - \frac{1}{25} + \frac{1}{36} + \frac{1}{49} + \frac{1}{64} - \frac{1}{81} - \frac{1}{100} \cdots$ for convergence. This is the series in which the absolute value of the general term is $\frac{1}{k^2}$ and the pattern of the signs is $+++--+++--\cdots$.

Solution

This is not a series of positive terms, nor is it strictly alternating. However, we find that the corresponding series of absolute values is the convergent p-series

$$1 + \frac{1}{4} + \frac{1}{9} + \frac{1}{16} + \cdots = \sum_{k=1}^{\infty} \frac{1}{k^2}$$

Because the absolute value series converges, the absolute convergence test assures us that the given series also converges.   ■

EXAMPLE 7   Convergence of a trigonometric series

Test the series $\displaystyle\sum_{k=1}^{\infty} \frac{\sin k}{2^k}$ for convergence.

Solution

Because $\sin k$ takes on both positive and negative values, the series cannot be analyzed by methods that apply only to series of positive terms. Moreover, note that the series is not strictly alternating:

$$\sum_{k=1}^{\infty} \frac{\sin k}{2^k} = 0.421 + 0.227 + 0.018 - 0.047 - 0.030 - 0.004 + 0.005 + \cdots$$

The corresponding series of absolute values is

$$\sum_{k=1}^{\infty} \left| \frac{\sin k}{2^k} \right|$$

which is dominated by the convergent geometric series $\Sigma \frac{1}{2^k}$ because $|\sin k| \leq 1$ for all k; that is,

$$0 \leq \left| \frac{\sin k}{2^k} \right| \leq \frac{1}{2^k} \quad \text{for all } k$$

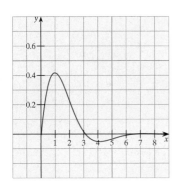

Figure 8.20 Graph of $y = \dfrac{\sin x}{2^x}$

Therefore, the absolute convergence test assures us that the given series converges. The graph of the related continuous function $y = (\sin x)/(2^x)$ is shown in Figure 8.20. ∎

If Σa_k converges, then $\Sigma |a_k|$ may either converge or diverge. The two cases that can occur are given the following special names.

Absolutely And Conditionally Convergent Series

WARNING Note that a convergent series is either absolutely convergent or conditionally convergent, but it cannot be both.

> The series Σa_k is **absolutely convergent** (or converges absolutely) if the related series $\Sigma |a_k|$ converges. The series Σa_k is **conditionally convergent** (or converges conditionally) if it converges but $\Sigma |a_k|$ diverges.

For example, the series in Examples 6 and 7 are both absolutely convergent, but the alternating harmonic series

$$\sum_{k=1}^{\infty} \frac{(-1)^{k+1}}{k}$$

is conditionally convergent because it converges (see Example 1), whereas the related series of absolute value terms $\Sigma 1/k$ (the harmonic series) diverges.

The ratio test and the root test for positive-term series can be generalized to apply to arbitrary series. The following is a statement and proof of the generalized ratio test.

THEOREM 8.21   The generalized ratio test

For the series $\sum a_k$, suppose $a_k \neq 0$ for $k \geq 1$ and that

$$\lim_{k \to \infty} \left| \frac{a_{k+1}}{a_k} \right| = L$$

where L is a real number or ∞. Then

If $L < 1$, the series Σa_k converges absolutely and hence converges.

If $L > 1$ or if L is infinite, the series Σa_k diverges.

If $L = 1$, the test is inconclusive.

Proof Assume $L < 1$; then the positive series $\Sigma |a_k|$ converges by the ratio test, and hence Σa_k converges absolutely. Assume $L > 1$; then the sequence $\{|a_k|\}$ is eventually increasing, which means that $\{a_k\}$ cannot converge to 0 as $k \to \infty$, so Σa_k diverges by the divergence test. Assume $L = 1$; see Problem 61 for this part of the proof. ❑

EXAMPLE 8 Convergence and divergence using the generalized ratio test

Find all values of x for which the series $\Sigma k x^k$ converges.

Solution

Let $a_k = k x^k$.

$$L = \lim_{k \to \infty} \left| \frac{a_{k+1}}{a_k} \right|$$

$$= \lim_{k \to \infty} \left| \frac{(k+1)x^{k+1}}{kx^k} \right|$$

$$= \lim_{k \to \infty} \left(\frac{k+1}{k} \right) \left| \frac{x^{k+1}}{x^k} \right|$$

$$= \lim_{k \to \infty} \left(1 + \frac{1}{k} \right) |x|$$

$$= |x|$$

Thus, by the generalized ratio test, since $L = |x|$ the series converges for $|x| < 1$ and diverges if $|x| > 1$. If $x = 1$, the series is $\Sigma k(1)^k$, which clearly diverges by the divergence test. Similarly, if $x = -1$, $\Sigma k(-1)^k$ diverges by the divergence test. Thus, the series converges for $|x| < 1$ and diverges for $|x| \geq 1$. ■

The statement and proof of the generalized root test is left to the problems (see Problem 60).

SUMMARY OF CONVERGENCE TESTS FOR SERIES

This concludes our basic study of convergence tests. There are no firm rules for deciding how to test the convergence of a given series Σa_k, but we can offer a few guidelines (Table 8.1).

TABLE 8.1 Guidelines for Determining Convergence of the Series Σa_k

Series with known convergence properties	
Geometric series, $\sum_{k=0}^{\infty} ar^k$	diverges if $|r| \geq 1$, and converges if $|r| < 1$ with sum $S = \dfrac{a}{1-r}$
p-series, $\sum_{k=1}^{\infty} \dfrac{1}{k^p}$	converges if $p > 1$ and diverges if $p \leq 1$
Harmonic series, $\sum_{k=1}^{\infty} \dfrac{1}{k}$	special case, $p = 1$, of a p-series (diverges)
q-log series $\sum_{k=1}^{\infty} \dfrac{\ln k}{k^q}$	converges if $q > 1$ and diverges if $q \leq 1$

TABLE 8.1 (continued)

Series with known convergence properties	Review of Convergence Tests
Divergence test	Compute $\lim\limits_{k\to\infty} a_k$. If this limit is anything but 0, the series diverges.
Tests for Series Σa_k with $a_k > 0$	
Limit comparison test	Check to see whether Σa_k is similar in appearance to a series Σb_k whose convergence properties are known, and apply the limit comparison test. If $\lim\limits_{k\to\infty} a_k/b_k = L$, where L is finite and positive, then Σa_k and Σb_k either both converge or both diverge.
Ratio test	If a_k involves $k!$, k^p, or c^k, try the ratio test, $\lim\limits_{k\to\infty} \dfrac{a_{k+1}}{a_k} = L$. Converges if $L < 1$, diverges if $L > 1$, test is inconclusive if $L = 1$.
Root test	If it is easy to find $\lim\limits_{k\to\infty} \sqrt[k]{a_k} = L$, try the root test. Converges if $L < 1$, diverges if $L > 1$, test is inconclusive if $L = 1$.
Integral test	If f is continuous, positive, and decreasing, and if $a_k = f(k)$ for all k, then $\displaystyle\sum_{k=1}^{\infty} a_k$ and the improper integral $\displaystyle\int_1^{\infty} f(x)\,dx$ either both converge or both diverge. Consider using this test when f is easy to integrate or if a_k involves a logarithm, a trigonometric function, or an inverse trigonometric function.
Direct comparison test	If $0 \le a_k \le c_k$ and $\displaystyle\sum_{k=1}^{\infty} c_k$ converges, then $\displaystyle\sum_{k=1}^{\infty} a_k$ converges. If $0 \le d_k \le a_k$ and $\displaystyle\sum_{k=1}^{\infty} d_k$ diverges, then $\displaystyle\sum_{k=1}^{\infty} a_k$ diverges.
Zero-infinity limit comparison test	If $\lim\limits_{k\to\infty} \dfrac{a_k}{b_k} = 0$ and Σb_k converges, then the series Σa_k converges. If $\lim\limits_{k\to\infty} \dfrac{a_k}{b_k} = \infty$ and Σb_k diverges, then the series Σa_k diverges.
Tests for Series Σa_k that do not require $a_k > 0$	
Absolute convergence test	Take the absolute value of each term and proceed to test for convergence of positive series, as summarized previously. 1. If $\Sigma\,\lvert a_k\rvert$ *converges*, then $\displaystyle\sum a_k$ converges. 2. If $\Sigma\,\lvert a_k\rvert$ diverges, then the series is *not absolutely convergent*; but may still converge *conditionally*.
Alternating series test (conditional convergence)	The alternating series $\Sigma(-1)^k a_k$ or $\Sigma(-1)^{k+1} a_k$ with $a_k > 0$ converges if $a_{k+1} \le a_k$ for $k = 1, 2, \ldots$ and $\lim\limits_{k\to\infty} a_k = 0$.

REARRANGEMENT OF TERMS IN AN ABSOLUTELY CONVERGENT SERIES

You may be surprised to learn that if the terms in a conditionally convergent series are rearranged (that is, the order of the summands is changed), the new series may not converge or it may converge to a different sum from that of the original series! For example, we know that the alternating harmonic series

$$\sum_{k=1}^{\infty} \frac{(-1)^{k+1}}{k}$$

converges conditionally, and it can be shown (see Problem 52) that

$$1 - \frac{1}{2} + \frac{1}{3} - \frac{1}{4} + \frac{1}{5} - \frac{1}{6} + \frac{1}{7} - \frac{1}{8} + \frac{1}{9} - \cdots = \ln 2$$

If we rearrange this series by placing two successive subtracted terms after each added term, we obtain

$$1 - \tfrac{1}{2} - \tfrac{1}{4} + \tfrac{1}{3} - \tfrac{1}{6} - \tfrac{1}{8} + \tfrac{1}{5} - \cdots = \left(1 - \tfrac{1}{2}\right) - \tfrac{1}{4} + \left(\tfrac{1}{3} - \tfrac{1}{6}\right) - \tfrac{1}{8} + \cdots$$
$$= \tfrac{1}{2} - \tfrac{1}{4} + \tfrac{1}{6} - \tfrac{1}{8} + \cdots$$
$$= \tfrac{1}{2}\left(1 - \tfrac{1}{2} + \tfrac{1}{3} - \tfrac{1}{4} + \cdots\right)$$
$$= \tfrac{1}{2}\ln 2$$

In general, it can be shown that if Σa_k is conditionally convergent, there is a rearrangement of the terms of Σa_k so that the sum is equal to any specified finite number. In Problem 53, for example, you are asked to rearrange the terms of the series

$$\sum_{k=1}^{\infty} \frac{(-1)^{k+1}}{k}$$

so that the sum is $\tfrac{3}{2}\ln 2$.

This information may be somewhat unsettling, because it is reasonable to expect a sum to be unaffected by the order in which the summands are taken. Absolutely convergent series behave more in the way we would expect. In fact, if the series Σa_k converges absolutely with sum S, then **any** rearrangement of the terms also converges absolutely to S. A detailed discussion of rearrangement of terms of a series requires the techniques of advanced calculus.

8.6   PROBLEM SET

A **1. WHAT DOES THIS SAY?** Discuss absolute versus conditional convergence.

2. WHAT DOES THIS SAY? Discuss the convergence of the alternating p-series.

3. WHAT DOES THIS SAY? Outline a procedure for determining the convergence of Σa_k for $a_k > 0$.

Determine whether each series in Problems 4–31 converges absolutely, converges conditionally, or diverges.

4. $\displaystyle\sum_{k=1}^{\infty} \frac{(-1)^{k+1}k}{k^2+1}$

5. $\displaystyle\sum_{k=1}^{\infty} \frac{(-1)^{k+1}k^2}{k^3+1}$

6. $\displaystyle\sum_{k=1}^{\infty} \frac{(-1)^{k+1}k}{2k+1}$

7. $\displaystyle\sum_{k=1}^{\infty} \frac{(-1)^{k+1}k^2}{k^2+1}$

8. $\displaystyle\sum_{k=1}^{\infty} \frac{(-1)^{k+1}}{k^{3/2}}$

9. $\displaystyle\sum_{k=1}^{\infty} \frac{(-1)^{k+1}k}{2^k}$

10. $\displaystyle\sum_{k=1}^{\infty} (-1)^{k+1}\frac{k^2}{e^k}$

11. $\displaystyle\sum_{k=1}^{\infty} \frac{(-1)^k}{\sqrt{k}}$

12. $\displaystyle\sum_{k=1}^{\infty} (-1)^k \frac{(1+k^2)}{k^3}$

13. $\displaystyle\sum_{k=1}^{\infty} \frac{(-1)^{k+1}k!}{k^k}$

14. $\displaystyle\sum_{k=2}^{\infty} (-1)^k \frac{k!}{\ln k}$

15. $\displaystyle\sum_{k=1}^{\infty} \frac{(-1)^k (2k)!}{k^k}$

16. $\displaystyle\sum_{k=1}^{\infty} (-1)^{k+1} \frac{2^k}{k!}$

17. $\displaystyle\sum_{k=1}^{\infty} \frac{(-3)^k (k+1)}{k!}$

18. $\displaystyle\sum_{k=1}^{\infty} (-1)^{k+1} \frac{2^{2k+1}}{k!}$

19. $\displaystyle\sum_{k=2}^{\infty} \frac{(-1)^{k+1}}{\ln k}$

20. $\displaystyle\sum_{k=1}^{\infty} \frac{(-1)^{k+1}k}{(k+1)(k+2)}$

21. $\displaystyle\sum_{k=2}^{\infty} \frac{(-1)^{k+1}}{(\ln k)^4}$

22. $\displaystyle\sum_{k=2}^{\infty} \frac{(-1)^{k+1}}{\ln(\ln k)}$

23. $\displaystyle\sum_{k=2}^{\infty} \frac{(-1)^{k+1}}{k \ln k}$

24. $\displaystyle\sum_{k=1}^{\infty} \frac{(-1)^{k+1} \ln k}{k}$

25. $\displaystyle\sum_{k=2}^{\infty} \frac{(-1)^{k+1}k}{\ln k}$

26. $\displaystyle\sum_{k=1}^{\infty} (-1)^{k+1} \frac{\ln k}{k^2}$

27. $\displaystyle\sum_{k=1}^{\infty} \frac{(-1)^{k+1}k}{(k+2)^2}$

28. $\displaystyle\sum_{k=1}^{\infty} (-1)^{k+1} \left(\frac{k}{k+1}\right)^k$

29. $\displaystyle\sum_{k=2}^{\infty} (-1)^{k+1} \frac{\ln(\ln k)}{k \ln k}$

30. $\displaystyle\sum_{k=1}^{\infty} (-1)^{k+1} \left(\frac{1}{k}\right)^{1/k}$

31. $\displaystyle\sum_{k=1}^{\infty} (-1)^{k+1} \frac{k^5 5^{k+2}}{2^{3k}}$

B *Given the series in Problems 32–37,*

    a. Estimate the sum of the series by taking the sum of the first four terms. How accurate is this estimate?

    b. How many terms of the series are necessary to estimate its sum with three-place accuracy? What is this estimate?

32. $\displaystyle\sum_{k=1}^{\infty} \frac{(-1)^{k+1}}{2^{2k-2}}$

33. $\displaystyle\sum_{k=1}^{\infty} \frac{(-1)^{k+1}}{k!}$

34. $\displaystyle\sum_{k=1}^{\infty} \frac{(-1)^k}{k^2}$

35. $\displaystyle\sum_{k=1}^{\infty} \left(\frac{-1}{3}\right)^{k+1}$

36. $\displaystyle\sum_{k=1}^{\infty} \frac{(-1)^{k+1}}{k^3}$

37. $\displaystyle\sum_{k=1}^{\infty} \left(\frac{-1}{5}\right)^k$

Use the generalized ratio test in Problems 38–43 to find all numbers x for which the given series converges.

38. $\displaystyle\sum_{k=1}^{\infty} \frac{x^k}{k}$

39. $\displaystyle\sum_{k=1}^{\infty} \frac{(2x)^k}{\sqrt{k}}$

40. $\displaystyle\sum_{k=1}^{\infty} \frac{2^k x^k}{k!}$

41. $\displaystyle\sum_{k=1}^{\infty} \frac{(k+2)x^k}{k^2(k+3)}$

42. $\displaystyle\sum_{k=1}^{\infty} (-1)^{k+1}\left(\frac{x}{k}\right)^k$

43. $\displaystyle\sum_{k=1}^{\infty} (-1)^k k^p x^k$, for $p > 0$

44. Find an upper bound for the error if the convergent alternating series $\displaystyle\sum_{k=1}^{\infty} \frac{(-1)^{k+1}}{k}$ is approximated by the partial sum

$$S_5 = 1 - \frac{1}{2} + \frac{1}{3} - \frac{1}{4} + \frac{1}{5}$$

45. Find an upper bound for the error if the convergent alternating series $\displaystyle\sum_{k=1}^{\infty} \frac{(-1)^{k+1}}{k^2}$ is approximated by the partial sum

$$S_5 = 1 - \frac{1}{2^2} + \frac{1}{3^2} - \frac{1}{4^2} + \frac{1}{5^2}$$

46. Find an upper bound for the error if the convergent alternating series $\displaystyle\sum_{k=2}^{\infty} \frac{(-1)^k}{\ln k}$ is approximated by the partial sum

$$S_7 = \frac{1}{\ln 2} - \frac{1}{\ln 3} + \frac{1}{\ln 4} - \frac{1}{\ln 5} + \frac{1}{\ln 6} - \frac{1}{\ln 7}$$

47. Find an upper bound for the error if the convergent alternating series $\displaystyle\sum_{k=1}^{\infty} \frac{(-1)^{k+1}k}{2^k}$ is approximated by the partial sum

$$S_6 = \frac{1}{2} - \frac{2}{2^2} + \frac{3}{2^3} - \frac{4}{2^4} + \frac{5}{2^5} - \frac{6}{2^6}$$

48. For what numbers p does the alternating series $\displaystyle\sum_{k=2}^{\infty} \frac{(-1)^{k+1}}{k(\ln k)^p}$ converge? For what numbers p does it converge absolutely?

49. Test the series $\displaystyle\sum_{k=1}^{\infty} \frac{\sin \sqrt[k]{2}}{k^2}$ for convergence.

50. Use series methods to evaluate $\displaystyle\lim_{k\to\infty} \frac{x^k}{k!}$ where x is a real number.

51. Show that the sequence $\{a_n\}$ converges, where

$$a_n = \frac{1}{1} + \frac{1}{2} + \frac{1}{3} + \cdots + \frac{1}{n} - \ln n$$

52. Show that $\displaystyle\sum_{k=1}^{\infty} \frac{(-1)^{k+1}}{k} = \ln 2$ by completing the following steps.

a. Let $\displaystyle S_m = \sum_{k=1}^{m} \frac{(-1)^{k+1}}{k}$ and let $\displaystyle H_m = \sum_{k=1}^{m} \frac{1}{k}$. Show that $S_{2m} = H_{2m} - H_m$.

b. It can be shown (see Problem 50)

$$\lim_{m\to\infty}\left(1 + \frac{1}{2} + \cdots + \frac{1}{m} - \ln m\right)$$

exists. The value of this limit is a number $\gamma \approx 0.57722$, called **Euler's constant.** Use this fact, along with the relationship in part **a**, to show that $\displaystyle\lim_{h\to\infty} S_n = \ln 2$. *Hint:* Note that

$$S_{2m} = H_{2m} - H_m$$
$$= [H_{2m} - \ln(2m)] - [H_m - \ln m] + \ln(2m) - \ln m$$

53. Consider the following rearrangement of the conditionally convergent alternating harmonic series $\displaystyle\sum_{k=1}^{\infty} \frac{(-1)^{k+1}}{k}$:

$$1 + \frac{1}{3} - \frac{1}{2} + \frac{1}{5} + \frac{1}{7} - \frac{1}{4} + \frac{1}{9} + \frac{1}{11} - \frac{1}{6} + \cdots$$

a. Let S_n and H_n denote the nth partial sum of the given rearranged series and the harmonic series $\displaystyle\sum_{k=1}^{\infty} \frac{1}{k}$, respectively. Show that if n is a multiple of 3, say $n = 3m$, then

$$S_{3m} = H_{4m} - \frac{1}{2}H_{2m} - \frac{1}{2}H_m$$

b. Show that $\displaystyle\lim_{n\to\infty} S_n = \frac{3}{2}\ln 2$. *Hint:* You many find it helpful to use

$$\lim_{n\to\infty}\left(1 + \frac{1}{2} + \cdots + \frac{1}{n} - \ln n\right) = \gamma$$

given in Problem 52**b**.

54. Journal Problem *School Science and Mathematics* by Michael Brozinsky* ■ Show that the series

$$1 + \frac{1}{2} + \frac{1}{3} - \frac{1}{4} - \frac{1}{5} - \frac{1}{6} + \frac{1}{7} + \frac{1}{8} + \frac{1}{9} - - - + + + \cdots$$

converges and find its sum.

55. Exploration Problem What (if anything) is wrong with the following computation?

$$1 - \frac{1}{2} + \frac{1}{3} - \frac{1}{4} + \frac{1}{5} - \frac{1}{6} + \cdots$$

$$= 1 + \left(\frac{1}{2} - 1\right) + \frac{1}{3} + \left(\frac{1}{4} - \frac{1}{2}\right) + \frac{1}{5} + \left(\frac{1}{6} - \frac{1}{3}\right) + \cdots$$

$$= \left(1 + \frac{1}{2} + \frac{1}{3} + \frac{1}{4} + \cdots\right) - 1 - \frac{1}{2} - \frac{1}{3} - \frac{1}{4} - \cdots$$

$$= \left(1 + \frac{1}{2} + \frac{1}{3} + \cdots\right) - \left(1 + \frac{1}{2} + \frac{1}{3} + \cdots\right)$$

$$= 0$$

56. Suppose $\{a_k\}$ is a sequence with the property that $|a_n| < A^n$ for some positive number A and all positive integers n. Show that the series $\Sigma a_k x^k$ converges absolutely for $|x| \le 1/A$.

57. Counterexample Problem Let Σa_k^2 be a series that converges. Either prove that Σa_k must diverge or find a counterexample.

58. Counterexample Problem Let Σa_k^2 be a series that converges. Either prove that Σa_k must also converge or find a counterexample.

59. Counterexample Problem Let Σa_k be a series that converges. Either prove that Σa_k^2 must also converge or find a counterexample.

60. Prove the generalized root test, which may be stated as follows: Suppose that $\lim_{k \to \infty} \sqrt[k]{|a_k|} = L$. Then

If $L < 1$, the series Σa_k converges absolutely.

If $L > 1$, the series Σa_k diverges.

If $L = 1$, the test is inconclusive.

61. Show that if

$$L = \lim_{k \to \infty} \left| \frac{a_{k+1}}{a_k} \right| = 1$$

the series Σa_k can either converge or diverge. *Hint:* Find a convergent series with $L = 1$ and a divergent series that satis-

fies the same condition.

62. Let $\Sigma (-1)^{k+1} a_k$ be an alternating series such that $\{a_k\}$ is decreasing and

$$\lim_{k \to \infty} a_k = 0$$

Show that the sequence of odd partial sums $\{S_{2n-1}\}$ is decreasing.

63. Exploration Problem   Construct a divergent alternating series $\sum_{k=1}^{\infty} (-1)^{k+1} a_k$ such that $\lim_{k \to \infty} a_k = 0$, but $\{a_k\}$ is not decreasing.

8.7   Power Series

IN THIS SECTION   convergence of a power series, term-by-term differentiation and integration of power series

An infinite series of the form

$$\sum_{k=0}^{\infty} a_k (x - c)^k = a_0 + a_1(x - c) + a_2(x - c)^2 + \cdots$$

is called a **power series** in $(x - c)$. The numbers $a_0, a_1, a_2, \ldots$ are the *coefficients* of the power series. If $c = 0$, the series has the form

$$\sum_{k=0}^{\infty} a_k x^k = a_0 + a_1 x + a_2 x^2 + a_3 x^3 + \cdots$$

A power series in x may be considered as an extension of a polynomial in x.

CONVERGENCE OF A POWER SERIES

For what numbers x does a given power series converge? The following theorem answers this question for the case where $c = 0$.

THEOREM 8.22   Convergence of a power series

For a power series $\sum_{k=0}^{\infty} a_k x^k$, exactly one of the following is true:

1. The series converges for all x.
2. The series converges only for $x = 0$.
3. The series **converges absolutely** for all x in an open interval $(-R, R)$ and **diverges** for $|x| > R$. It may either converge or diverge at the endpoints of the interval, $x = R$ and $x = -R$.

Proof   The proof is found in most advanced calculus textbooks.   ❑

The set of values for which a series converges is called the **convergence set** for the series. The following three examples show each of these possibilities.

EXAMPLE 1   When the convergence set is the entire x-axis

Show that the power series $\sum_{k=1}^{\infty} \frac{x^k}{k!}$ converges for all x.

Solution

If $x = 0$, then the series is trivial and converges. For $x \neq 0$, we use the generalized ratio test to find

$$L = \lim_{k \to \infty} \left| \frac{\dfrac{x^{k+1}}{(k+1)!}}{\dfrac{x^k}{k!}} \right| = \lim_{k \to \infty} \left| \frac{x^{k+1} k!}{(k+1)! x^k} \right| = \lim_{k \to \infty} \frac{|x|}{k+1} = 0$$

Because $L = 0$ and thus $L < 1$, the series converges for all x. ∎

EXAMPLE 2   Convergence only at the point $x = 0$

Show that the power series $\displaystyle\sum_{k=1}^{\infty} k! x^k$ converges only when $x = 0$.

Solution

We use the generalized ratio test to find

$$L = \lim_{k \to \infty} \left| \frac{(k+1)! x^{k+1}}{k! x^k} \right| = \lim_{k \to \infty} (k+1)|x|$$

For any x other than 0, we have $L = \infty$. Hence, the power series converges only when $x = 0$. ∎

EXAMPLE 3   Convergence set is a bounded interval

Find the convergence set for the power series

$$\sum_{k=1}^{\infty} \frac{x^k}{\sqrt{k}}$$

Solution

Using the generalized ratio test, we find

$$L = \lim_{k \to \infty} \left| \frac{\dfrac{x^{k+1}}{\sqrt{k+1}}}{\dfrac{x^k}{\sqrt{k}}} \right| = \lim_{k \to \infty} \left| \frac{\sqrt{k}}{\sqrt{k+1}} \right| |x| = |x|$$

The power series converges absolutely if $|x| < 1$, and diverges if $|x| > 1$. We must also check the endpoints of the interval $|x| < 1$, namely $x = -1$ and $x = 1$:

$$x = -1: \quad \sum_{k=1}^{\infty} \frac{(-1)^k}{\sqrt{k}} \quad \text{\textit{converges} by the alternating series test}$$

$$\left(\frac{1}{\sqrt{k+1}} < \frac{1}{\sqrt{k}} \quad \text{and} \quad \lim_{k \to \infty} \frac{1}{\sqrt{k}} = 0 \right)$$

$$x = 1: \quad \sum_{k=1}^{\infty} \frac{(1)^k}{\sqrt{k}} \quad \text{\textit{diverges} (p-series with $p = \tfrac{1}{2} < 1$)}$$

Thus, the given power series converges for $-1 \le x < 1$ and diverges otherwise. ∎

According to Theorem 8.22, the set of numbers for which the power series $\Sigma a_k x^k$ converges is an interval centered at $x = 0$. We call this the **interval of convergence** of the power series. If this interval has length $2R$, then R is called the **radius of convergence** of the series, as shown in Figure 8.21. If the series converges only for $x = 0$, the series has radius of convergence $R = 0$, and if it converges for all x, we say that $R = \infty$.

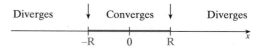

May converge or diverge at endpoints

Figure 8.21   The interval of convergence of a power series

In Example 1, we found that $\Sigma x^k/k!$ converges for all x, so $R = \infty$ and the interval of convergence is the entire real line. In Example 2, we found that the power series $\Sigma k! x^k$ has radius of convergence 0. In Example 3, we found that $\Sigma x^k/\sqrt{k}$ converges on $[-1, 1)$, so $R = 1$.

The procedure for determining the convergence set of a power series is further illustrated in the following examples.

EXAMPLE 4   Finding the interval of convergence for a power series

Find the interval of convergence for the power series $\displaystyle\sum_{k=1}^{\infty} \frac{2^k x^k}{k}$. What is the radius of convergence?

Solution
We apply the generalized ratio test to find

$$L = \lim_{k \to \infty} \left| \frac{\dfrac{2^{k+1} x^{k+1}}{k+1}}{\dfrac{2^k x^k}{k}} \right| = \lim_{k \to \infty} \left| \frac{2k}{k+1} \right| |x| = 2 |x|$$

Thus, the series converges absolutely for $2|x| < 1$; that is, for $-\frac{1}{2} < x < \frac{1}{2}$. The radius of convergence is $R = \frac{1}{2}$. Checking the endpoints, we find

$$x = \frac{1}{2}: \quad \sum_{k=1}^{\infty} \frac{2^k}{k} \left(\frac{1}{2} \right)^k = \sum_{k=1}^{\infty} \frac{1}{k} \text{ diverges} \qquad \textsf{Harmonic series}$$

$$x = -\frac{1}{2}: \quad \sum_{k=1}^{\infty} \frac{2^k}{k} \left(-\frac{1}{2} \right)^k = \sum_{k=1}^{\infty} \frac{(-1)^k}{k} \text{ converges} \qquad \textsf{Alternating harmonic series}$$

The interval of convergence is $-\frac{1}{2} \leq x < \frac{1}{2}$. ∎

EXAMPLE 5   Finding an interval of convergence

Find the radius of convergence and the interval of convergence for the power series

$$\sum_{k=1}^{\infty} \frac{2^k x^k}{k!}$$

Solution
Applying the generalized ratio test, we find

$$L = \lim_{k \to \infty} \left| \frac{\dfrac{2^{k+1} x^{k+1}}{(k+1)!}}{\dfrac{2^k x^k}{k!}} \right| = \lim_{k \to \infty} \left| \frac{2}{k+1} \right| |x| = 0$$

Thus, the power series converges absolutely for x, and the radius of convergence is infinite. ∎

EXAMPLE 6   **Radius of convergence using the root test**

Find the radius of convergence of the power series $\displaystyle\sum_{k=1}^{\infty} \left(\frac{k+1}{k}\right)^{k^2} x^k$.

Solution

Using the root test to examine absolute convergence, we find that

$$L = \lim_{k \to \infty} \sqrt[k]{\left|\left(\frac{k+1}{k}\right)^{k^2} x^k\right|} = \lim_{k \to \infty} \left|\left(\frac{k+1}{k}\right)^{k^2} x^k\right|^{1/k}$$

$$= \lim_{k \to \infty} \left(1 + \frac{1}{k}\right)^k |x| = e\,|x|$$

Thus, the power series converges absolutely for $e\,|x| < 1$; that is, for $-e^{-1} < x < e^{-1}$. It follows that the radius of convergence is $R = e^{-1}$. We do not need to test the endpoints because we are not finding the interval of convergence.   ■

In some applications, we will encounter power series of the form

$$\sum_{k=0}^{\infty} a_k (x - c)^k = a_0 + a_1(x - c) + a_2(x - c)^2 + \cdots$$

in which each term is a constant times a power of $x - c$, where c is a constant. The intervals of convergence are intervals of the form $-R < x - c < R$, including possibly one or both of the endpoints $x = c - R$ and $x = c + R$. The procedure for determining the interval of convergence of such power series is illustrated in the following example.

EXAMPLE 7   **Power series in** $(x - c)$

Find the interval of convergence of the power series

$$\sum_{k=0}^{\infty} \frac{(x+1)^k}{3^k}$$

Solution

Applying the generalized ratio rest, we obtain

$$L = \lim_{k \to \infty} \left|\frac{\dfrac{(x+1)^{k+1}}{3^{k+1}}}{\dfrac{(x+1)^k}{3^k}}\right| = \lim_{k \to \infty} \left|\frac{3^k}{3^{k+1}}\right| |x+1| = \lim_{k \to \infty} \frac{1}{3} |x+1|$$

Thus, the power series converges absolutely for

$$\tfrac{1}{3} |x + 1| < 1$$
$$|x + 1| < 3$$
$$-3 < x + 1 < 3$$
$$-4 < x < 2$$

Checking the endpoints, we find

$$x = -4: \quad \sum_{k=0}^{\infty} \frac{(-4+1)^k}{3^k} = \sum_{k=0}^{\infty} (-1)^k \text{ diverges (by oscillation)}$$

$$x = 2: \quad \sum_{k=0}^{\infty} \frac{(2+1)^k}{3^k} = \sum_{k=0}^{\infty} 1^k \text{ diverges (by divergence test)}$$

The interval of convergence is $(-4, 2)$.   ■

TERM-BY-TERM DIFFERENTIATION AND INTEGRATION OF POWER SERIES

The power series $\displaystyle\sum_{k=0}^{\infty} a_k x^k$ can be thought of as a function defined on its interval of convergence. It is reasonable to ask if this function is differentiable and if it is integrable and, if so, what are its derivative and integral. This is useful because we can form new series using differentiation or integration of known series. If we regard a power series as an "infinite polynomial," we would expect to be able to differentiate and integrate it term by term. The following theorem shows that this procedure is legitimate on the interval of absolute convergence.

THEOREM 8.23   Term-by-term differentiation and integration of a power series

A power series $\displaystyle\sum_{k=0}^{\infty} a_k x^k$ with radius of convergence $R > 0$ can be differentiated or integrated term by term on its interval of absolute convergence $-R < x < R$. More specifically, if $f(x) = \displaystyle\sum_{k=0}^{\infty} a_k x^k$ for $|x| < R$, then for $|x| < R$ we have

$$f'(x) = \sum_{k=1}^{\infty} k a_k x^{k-1} = a_1 + 2a_2 x + 3a_3 x^2 + 4a_4 x^3 + \cdots$$

and

$$\int f(x)\, dx = \int \left(\sum_{k=0}^{\infty} a_k x^k \right) dx = \sum_{k=0}^{\infty} \left(\int a_k x^k\, dx \right) = \sum_{k=0}^{\infty} \frac{a_k}{k+1} x^{k+1} + C$$

Proof   A proof can be found in most advanced calculus texts.   ❑

➡ **What This Says**   In many ways, a function defined by a power series behaves like a polynomial. It is differentiable, hence continuous, inside its interval of absolute convergence, and its derivative and integral can be determined by differentiating and integrating term by term, respectively.

EXAMPLE 8   Term-by-term differentiation of a power series

Let f be a function defined by the power series

$$f(x) = \sum_{k=0}^{\infty} \frac{x^k}{k!} \qquad \text{for all } x$$

Show that $f'(x) = f(x)$ for all x, and deduce that $f(x) = e^x$.

Solution

The given power series converges for all x (by the generalized ratio test—see Example 1), and Theorem 8.23 tells us that it is differentiable for all x. Differentiating term by term, we find

$$f'(x) = \frac{d}{dx}\left[1 + x + \frac{x^2}{2!} + \frac{x^3}{3!} + \frac{x^4}{4!} + \cdots\right]$$

$$= 0 + 1 + \frac{2x}{2!} + \frac{3x^2}{3!} + \frac{4x^3}{4!} + \cdots$$

$$= 1 + x + \frac{x^2}{2!} + \frac{x^3}{3!} + \cdots$$

$$= f(x)$$

In Chapter 5, we found that the differential equation $f'(x) = f(x)$ has the general solution $f(x) = Ce^x$. Substituting $x = 0$ into the power series for $f(x)$, we obtain

$$f(0) = 1 + 0 + \frac{0^2}{2!} + \frac{0^3}{3!} + \cdots = 1$$

and by solving the equation $1 = Ce^0$ for C, we find $C = 1$; therefore, $f(x) = e^x$. ∎

A power series can be differentiated term by term—not just once, but infinitely often in its interval of absolute convergence. The key to this fact lies in showing that if f satisfies

$$f(x) = \sum_{k=0}^{\infty} a_k x^k$$

and R is the radius of convergence of the power series on the right, then the derivative series

$$f'(x) = \sum_{k=1}^{\infty} k a_k x^{k-1}$$

also has the radius of convergence R. (You are asked to show this in Problem 44.) Therefore, Theorem 8.23 can be applied to the derivative series to obtain the second derivative

$$f''(x) = \sum_{k=2}^{\infty} k(k-1) a_k x^{k-2}$$

for $|x| < R$. Continuing in this fashion, we can apply Theorem 8.23 to $f'''(x)$, $f^{(4)}$, and all other higher derivatives of f.

For example, we know that the geometric series

$$\sum_{k=0}^{\infty} x^k \quad \text{converges absolutely to} \quad f(x) = \frac{1}{1-x} \quad \text{for } |x| < 1$$

Thus, the term-by-term derivative

$$\frac{d}{dx}\left[\sum_{k=0}^{\infty} x^k\right] = \sum_{k=1}^{\infty} k x^{k-1} = 1 + 2x + 3x^2 + \cdots$$

converges to $f'(x) = \dfrac{1}{(1-x)^2}$ for $|x| < 1$, and the term-by-term *second derivative*

$$\frac{d^2}{dx^2}\left[\sum_{k=0}^{\infty} x^k\right] = \frac{d}{dx}\left[\sum_{k=1}^{\infty} k x^{k-1}\right] = \sum_{k=2}^{\infty} k(k-1) x^{k-2}$$

converges to $f''(x) = \dfrac{2}{(1-x)^3}$, and so on. These ideas are illustrated in our next example.

EXAMPLE 9   Second derivative of a power series

Let f be the function defined by the power series

$$f(x) = \sum_{k=0}^{\infty} \frac{(-1)^k x^{2k}}{(2k)!}$$

for all x. Show that $f''(x) = -f(x)$ for all x.

Solution

First, we use the ratio test to verify that the given power series converges absolutely for all x.

$$L = \lim_{k \to \infty} \left| \frac{\dfrac{(-1)^{k+1} x^{2(k+1)}}{[2(k+1)]!}}{\dfrac{(-1)^k x^{2k}}{(2k)!}} \right| = \lim_{k \to \infty} \frac{x^{2k+2}}{[2(k+1)]!} \cdot \frac{(2k)!}{x^{2k}} = \lim_{k \to \infty} \frac{x^2}{(2k+2)(2k+1)} = 0$$

Because $L < 1$, the series converges for all x. Next, differentiate the series term by term:

$$f'(x) = \frac{d}{dx}\left[1 - \frac{x^2}{2!} + \frac{x^4}{4!} - \frac{x^6}{6!} + \cdots \right] = -\frac{2x}{2!} + \frac{4x^3}{4!} - \frac{6x^5}{6!} + \cdots$$

$$= -\frac{x}{1!} + \frac{x^3}{3!} - \frac{x^5}{5!} + \cdots$$

Finally, by differentiating term by term again, we obtain

$$f''(x) = \frac{d}{dx}\left[-\frac{x}{1!} + \frac{x^3}{3!} - \frac{x^5}{5!} + \cdots \right] = -1 + \frac{3x^2}{3!} - \frac{5x^4}{5!} + \cdots$$

$$= -\left[1 - \frac{x^2}{2!} + \frac{x^4}{4!} - \cdots \right] = -f(x) \qquad \blacksquare$$

EXAMPLE 10   Term-by-term integration of a power series

By integrating an appropriate geometric series term by term, show that

$$\sum_{k=0}^{\infty} \frac{x^{k+1}}{k+1} = -\ln(1-x) \quad \text{for } -1 < x < 1$$

Solution

Integrating the geometric series $\displaystyle\sum_{k=0}^{\infty} u^k = \frac{1}{1-u}$ term by term in the interval $-1 < u < 1$, we obtain

$$\int_0^x \frac{1}{1-u}\, du = \int_0^x \left[\sum_{k=0}^{\infty} u^k \right] du = \int_0^x [1 + u + u^2 + u^3 + \cdots]\, du$$

$$= x + \frac{x^2}{2} + \frac{x^3}{3} + \cdots = \sum_{k=0}^{\infty} \frac{x^{k+1}}{k+1} \quad \text{for } -1 < x < 1$$

We also know

$$\int_0^x \frac{du}{1-u} = -\ln(1-x) \qquad \text{for } -1 < x < 1$$

Thus,

$$-\ln(1-x) = \int_0^x \frac{du}{1-u} = \sum_{k=0}^{\infty} \frac{x^{k+1}}{k+1} \qquad \text{for } -1 < x < 1$$

If we check endpoints, we find the series actually converges for $-1 \le x < 1$. $\qquad \blacksquare$

EXAMPLE 11 Term-by-term integration of a power series

Find a power series for $\tan^{-1} x$.

Solution

We will use the fact that $\tan^{-1} x = \int_0^x \dfrac{dt}{1 + t^2}$. Consider the integrand as a geometric series:

$$\frac{1}{1 + t^2} = 1 - t^2 + t^4 - t^6 + \cdots = \sum_{k=0}^{\infty} (-1)^k t^{2k}$$

Thus,

$$
\begin{aligned}
\tan^{-1} x &= \int_0^x \frac{1}{1 + t^2}\, dt \\
&= \int_0^x \left[1 - t^2 + t^4 - t^6 + \cdots \right] dt \\
&= \sum_{k=0}^{\infty} \int_0^x (-1)^k t^{2k}\, dt \\
&= \sum_{k=0}^{\infty} (-1)^k \frac{t^{2k+1}}{2k+1} \Bigg|_0^x \\
&= x - \frac{x^3}{3} + \frac{x^5}{5} - \frac{x^7}{7} + \cdots \\
&= \sum_{k=0}^{\infty} \frac{(-1)^k x^{2k+1}}{2k+1} \quad \text{for } |x| < 1
\end{aligned}
$$

If we check the endpoints, we find the series converges for $|x| \le 1$. ∎

This series for $\tan^{-1} x$ is called **Gregory's series** after the British mathematician James Gregory (1638–1675), who developed it in 1671 (see Figure 8.22). Because it converges at $x = 1$ and $x = -1$, it can be used to represent $\tan^{-1} x$ on $[-1, 1]$. For example, at $x = 1$

$$\frac{\pi}{4} = \tan^{-1} 1 = 1 - \frac{1}{3} + \frac{1}{5} - \frac{1}{7} + \frac{1}{9} - \cdots$$

$$\pi = 4 \left[1 - \frac{1}{3} + \frac{1}{5} - \frac{1}{7} + \frac{1}{9} - \cdots \right]$$

This is the *Leibniz formula* for π. Unfortunately, as a means of estimating the actual value of π, it has a serious drawback—the convergence is very slow. Look at the partial sums for the 49th and 50th terms, and note that these are not very close to the limit (which is $\pi \approx 3.14159265359 \cdots$).

In 1706, the mathematician John Machin discovered another series that converges much faster. The identity

$$\tan(\alpha + \beta) = \frac{\tan \alpha + \tan \beta}{1 - \tan \alpha \tan \beta}$$

can be used to show that

$$\tan^{-1} 1 = 4 \tan^{-1} \frac{1}{5} - \tan^{-1} \frac{1}{239}$$

$$\frac{\pi}{4} = 4 \left[\frac{1}{5} - \frac{1}{3} \left(\frac{1}{5} \right)^3 + \frac{1}{5} \left(\frac{1}{5} \right)^5 - \cdots \right]$$

$$- \left[\frac{1}{239} - \frac{1}{3} \left(\frac{1}{239} \right)^3 + \frac{1}{5} \left(\frac{1}{239} \right)^5 - \cdots \right]$$

No. of terms	Gregory's Series term	partial sum
0	4	
1	−1.33333333333333	2.666666667
2	0.8	3.466666667
3	−0.57142857142857	2.895238095
4	0.44444444444444	3.33968254
5	−0.36363636363636	2.976046176
6	0.30769230769231	3.283738484
7	−0.26666666666667	3.017071817
8	0.23529411764706	3.252365935
9	−0.21052631578947	3.041839619
10	0.19047619047619	3.232315809
11	−0.17391304347826	3.058402766
12	0.16	3.218402766
13	−0.14814814814815	3.070254618
14	0.13793103448276	3.208185652
15	−0.12903225806452	3.079153394
16	0.12121212121212	3.200365515
17	−0.11428571428571	3.086079801
18	0.10810810810811	3.194187909
19	−0.1025641025641	3.091623807
20	0.09560975609756	3.189184782
21	−0.09302325581395	3.096161526
22	0.088888888888889	3.185050415
23	−0.08510638297872	3.099944032
24	0.081632653061224	3.181576685
25	−0.07843137254902	3.103145313
26	0.075471698113208	3.178617011
27	−0.07272727272727	3.105889738
28	0.070175438596491	3.176065177
29	−0.06779661016949	3.108268567
30	0.065573770491803	3.173842337
31	−0.06349206349206	3.110350274
32	0.061538461538462	3.171888735
33	−0.05970149253731	3.112187243
34	0.057971014492754	3.170158257
35	−0.05633802816901	3.113820229
36	0.054794520547945	3.16861475
37	−0.05333333333333	3.115281416
38	0.051948051948052	3.167229468
39	−0.05063291139241	3.116596557
40	0.049382716049383	3.165979273
41	−0.04819277108434	3.117786502
42	0.047058823529412	3.164845325
43	−0.04597701149425	3.118868314
44	0.044943820224719	3.163812134
45	−0.04395604395604	3.11985609
46	0.043010752688172	3.162866843
47	−0.04210526315789	3.12076158
48	0.041237113402062	3.161998693
49	−0.04040404040404	3.121594653
50	0.03960396039604	3.161198613

Figure 8.22 Approximation of π using Gregory's series

No. of terms	Machin's Series term	partial sum
0	3.1832635983264	
1	−0.0426665690003	3.140597029
2	0.001023999998974	3.141621029
3	−2.9257142857E−05	3.141591772
4	9.10222222222E−07	3.141592682
5	−2.9789090909E−08	3.141592653
6	1.00824615385E−09	3.141592654
7	−3.4952533333E−11	3.141592654

If we multiply both sides by 4, we find another series for π:

$$\pi = 16\left[\tfrac{1}{5} - \tfrac{1}{3}\left(\tfrac{1}{5}\right)^3 + \tfrac{1}{5}\left(\tfrac{1}{5}\right)^5 - \cdots\right]$$
$$- 4\left[\tfrac{1}{239} - \tfrac{1}{3}\left(\tfrac{1}{239}\right)^3 + \tfrac{1}{5}\left(\tfrac{1}{239}\right)^5 - \cdots\right]$$

By comparing the spreadsheet in the margin with the one in Figure 8.22, we see that Machin's formula for π converges much more rapidly than the Leibniz formula. Indeed, the sum of the first seven terms $M_7 = 3.141592654$ already coincides with π for the first eight decimal places.

8.7   PROBLEM SET

A *Find the convergence set for the power series given in Problems 1–28.*

1. $\displaystyle\sum_{k=1}^{\infty} \frac{kx^k}{k+1}$

2. $\displaystyle\sum_{k=1}^{\infty} \frac{k^2x^k}{k+1}$

3. $\displaystyle\sum_{k=1}^{\infty} \frac{k(k+1)x^k}{k+2}$

4. $\displaystyle\sum_{k=1}^{\infty} \sqrt{k-1}\,x^k$

5. $\displaystyle\sum_{k=1}^{\infty} k^2 3^k (x-3)^k$

6. $\displaystyle\sum_{k=1}^{\infty} \frac{k^2(x-2)^k}{3^k}$

7. $\displaystyle\sum_{k=0}^{\infty} \frac{3^k(x+3)^k}{4^k}$

8. $\displaystyle\sum_{k=0}^{\infty} \frac{4^k(x+1)^k}{3^k}$

9. $\displaystyle\sum_{k=0}^{\infty} \frac{k!(x-1)^k}{5^k}$

10. $\displaystyle\sum_{k=0}^{\infty} \frac{(x-15)^k}{\ln(k+1)}$

11. $\displaystyle\sum_{k=1}^{\infty} \frac{k^2}{2^k}(x-1)^k$

12. $\displaystyle\sum_{k=1}^{\infty} \frac{2^k(x-3)^k}{k(k+1)}$

13. $\displaystyle\sum_{k=1}^{\infty} \frac{k(3x-4)^k}{(k+1)^2}$

14. $\displaystyle\sum_{k=0}^{\infty} \frac{(2x+3)^k}{4^k}$

15. $\displaystyle\sum_{k=1}^{\infty} \frac{kx^k}{7^k}$

16. $\displaystyle\sum_{k=0}^{\infty} \frac{(2k)!x^k}{(3k)!}$

17. $\displaystyle\sum_{k=1}^{\infty} \frac{(k!)^2 x^k}{k^k}$

18. $\displaystyle\sum_{k=1}^{\infty} \frac{(-1)^k kx^k}{\ln(k+2)}$

19. $\displaystyle\sum_{k=2}^{\infty} \frac{(-1)^k x^k}{k(\ln k)^2}$

20. $\displaystyle\sum_{k=0}^{\infty} \frac{(3x)^k}{2^{k+1}}$

21. $\displaystyle\sum_{k=0}^{\infty} \frac{(2x)^{2k}}{k!}$

22. $\displaystyle\sum_{k=0}^{\infty} \frac{(x+2)^{2k}}{3^k}$

23. $\displaystyle\sum_{k=0}^{\infty} \frac{k!}{2^k}(3x)^{3k}$

24. $\displaystyle\sum_{k=1}^{\infty} \frac{(3x)^{3k}}{\sqrt{k}}$

25. $\displaystyle\sum_{k=0}^{\infty} \frac{2^k}{k!}(2x-1)^{2k}$

26. $\displaystyle\sum_{k=0}^{\infty} 2^k(3x)^{3k}$

27. $\displaystyle\sum_{k=1}^{\infty} \frac{x^k}{k\sqrt{k}}$

28. $\displaystyle\sum_{k=1}^{\infty} \frac{(\ln k)x^k}{k}$

B *Find the radius of convergence R in Problems 29–34.*

29. $\displaystyle\sum_{k=1}^{\infty} k^2(x+1)^{2k+1}$

30. $\displaystyle\sum_{k=1}^{\infty} 2^{\sqrt{k}}(x-1)^k$

31. $\displaystyle\sum_{k=1}^{\infty} \frac{k!x^k}{k^k}$

32. $\displaystyle\sum_{k=1}^{\infty} \frac{(k!)^2 x^k}{(2k)!}$

33. $\displaystyle\sum_{k=1}^{\infty} k(ax)^k$ for constant a

34. $\displaystyle\sum_{k=1}^{\infty} (a^2 x)^k$ for constant a

In Problems 35–38, find the derivative $f'(x)$ by differentiating term by term.

35. $f(x) = \displaystyle\sum_{k=0}^{\infty} \left(\frac{x}{2}\right)^k$

36. $f(x) = \displaystyle\sum_{k=1}^{\infty} \frac{x^k}{k}$

37. $f(x) = \displaystyle\sum_{k=0}^{\infty} (k+2)x^k$

38. $f(x) = \displaystyle\sum_{k=0}^{\infty} kx^k$

In Problems 39–42, find $\displaystyle\int_0^x f(u)\,du$ by integrating term by term.

39. $f(x) = \displaystyle\sum_{k=0}^{\infty} \left(\frac{x}{2}\right)^k$

40. $f(x) = \displaystyle\sum_{k=1}^{\infty} \frac{x^k}{k}$

41. $f(x) = \displaystyle\sum_{k=0}^{\infty} (k+2)x^k$

42. $f(x) = \displaystyle\sum_{k=1}^{\infty} kx^k$

43. Counterexample Problem Show that the series

$$S = \sum_{k=1}^{\infty} \frac{\sin(k!x)}{k^2}$$

converges for all x. Differentiate term by term to obtain the series

$$T = \sum_{k=1}^{\infty} \frac{k!\cos(k!x)}{k^2}$$

Show that this series diverges for all x. Why does this not violate Theorem 8.23?

C **44.** Suppose $\{a_k\}$ is a sequence for which

$$\lim_{k\to\infty} \sqrt[k]{|a_k|} = \frac{1}{R}$$

Show that the power series

$$\sum_{k=1}^{\infty} ka_k x^{k-1}$$

has radius of convergence R.

45. Show that if $f(x) = \sum_{k=1}^{\infty} a_k x^k$ has radius of convergence $R > 0$, then the series

$$\sum_{k=1}^{\infty} a_k x^{kp}$$

where p is a positive integer, has radius of convergence $R^{1/p}$.

46. For what values of x does the series $\sum_{k=1}^{\infty} \dfrac{x}{k+x}$ converge?

Note: This is not a power series.

47. For what values of x does the series $\sum_{k=1}^{\infty} \dfrac{k}{x^k}$ converge?

Note: This is not a power series.

48. For what values of x does the series $\sum_{k=1}^{\infty} \dfrac{1}{kx^k}$ converge?

Note: This is not a power series.

49. Use term-by-term differentiation of a geometric series to find a power series for

$$f(x) = \frac{1}{(1-x)^3}$$

For what values of x does this power series converge?

50. Find the radius of convergence for $\sum_{k=1}^{\infty} \dfrac{(k+3)!x^k}{k!(k+4)!}$.

51. Find the radius of convergence for $\sum_{k=1}^{\infty} \dfrac{1 \cdot 2 \cdot 3 \cdots k(-x)^{2k-1}}{1 \cdot 3 \cdot 5 \cdots (2k-1)}$.

52. Let f be the function defined by the power series $f(x) = \sum_{k=0}^{\infty} \dfrac{(-1)^k x^{2k+1}}{(2k+1)!}$ for all x. Show that $f''(x) = -f(x)$ for all x.

53. Let f be the function defined by the power series $f(x) = \sum_{k=0}^{\infty} \dfrac{x^{2k}}{(2k)!}$ for all x. Show that $f''(x) = f(x)$ for all x.

54. Find the radius of convergence for the power series $\sum_{k=0}^{\infty} \dfrac{(qk)!}{(k!)^q} x^k$, where q is a positive integer.

8.8 Taylor and Maclaurin Series

IN THIS SECTION   Taylor and Maclaurin polynomials, Taylor's theorem, Taylor and Maclaurin series, operations with Taylor and Maclaurin series

TAYLOR AND MACLAURIN POLYNOMIALS

Consider a function f that can be differentiated n times on some interval I. Our goal is to find a polynomial function that approximates f at a number c in its domain. For simplicity, we begin by considering an important special case where $c = 0$. For example, consider the function $f(x) = e^x$ at the point $x = 0$, as shown in Figure 8.23. To approximate f by a polynomial function $M(x)$, we begin by making sure that both the polynomial function and f pass through the same point. That is, $M(0) = f(0)$. We say that the polynomial is **expanded about $c = 0$** or that it is **centered at 0**.

There are many polynomial functions that we could choose to approximate f at $x = 0$, and we proceed by making sure that both f and M have the same slope at $x = 0$. That is,

$$M'(0) = f'(0)$$

The graph in Figure 8.23 shows that we have $f(x) = e^x$, so $f'(x) = e^x$ and $f'(0) = f'(0) = 1$. To find M, we let

$$M_1(x) = a_0 + a_1 x \qquad M_1'(x) = a_1$$

and require $M_1(0) = 1$ and $M_1'(0) = 1$. Because $M_1(0) = a_1 = 1$ and $M_1'(0) = a_0 = 1$, we find that

$$M_1(x) = 1 + x$$

We see from Figure 8.23 that close to $x = c$ the approximation of f by M_1 is good, but as we move away from $(0, 1)$, M_1 no longer serves as a good approximation. To improve the approximation, we impose the requirement that the values of the second derivatives of M and f agree at $x = 0$. Using this criterion, we find

$$M_2(x) = a_0 + a_1 x + a_2 x^2$$

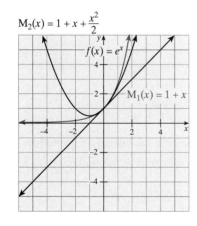

$M_2(x) = 1 + x + \dfrac{x^2}{2}$

Figure 8.23 Graph of $f(x) = e^x$ and approximating polynomials $M_1(x)$ and $M_2(x)$

so that

$$M_2'(x) = a_1 + 2a_2x \quad \text{and} \quad M_2''(x) = 2a_2$$

Because $f''(x) = e^x$ and $f''(0) = 1$, we want

$$2a_2 = 1 \quad \text{or} \quad a_2 = \tfrac{1}{2}$$

Thus,

$$M_2(x) = 1 + x + \tfrac{1}{2}x^2$$

To improve the approximation even further, we can require that the values of the approximating polynomials $M_3, M_4, \cdots, M_n$ at $x = 0$ have derivatives that match those of f at $x = 0$. We find that

$$M_3(x) = 1 + x + \tfrac{1}{2}x^2 + \tfrac{1}{6}x^3$$
$$M_4(x) = 1 + x + \tfrac{1}{2}x^2 + \tfrac{1}{6}x^3 + \tfrac{1}{24}x^4$$
$$M_5(x) = 1 + x + \tfrac{1}{2}x^2 + \tfrac{1}{6}x^3 + \tfrac{1}{24}x^4 + \tfrac{1}{120}x^5$$
$$M_n(x) = 1 + \frac{x}{1!} + \frac{x^2}{2!} + \frac{x^3}{3!} + \cdots + \frac{x^n}{n!}$$

This nth-degree polynomial approximation of f at $x = 0$ is called the **nth-degree Maclaurin polynomial for f.** If we repeat the steps for $x = c$ rather than for $x = 0$, we find

$$T_n(x) = e^c + \frac{(x-c)}{1!}e^c + \frac{(x-c)^2}{2!}e^c + \frac{(x-c)^3}{3!}e^c + \cdots + \frac{(x-c)^n}{n!}e^c$$

which is called the **nth-degree Taylor polynomial** of the function $f(x) = e^x$ at $x = c$.

TAYLOR'S THEOREM

Instead of stopping at the nth term, we will now approximate a function f by an infinite series. A function f is said to be represented by the power series

$$\sum_{k=0}^{\infty} a_k(x - c)^k$$

on an interval I if $f(x) = \displaystyle\sum_{k=0}^{\infty} a_k(x - c)^k$ for all x in I. Power series representation of functions are extremely useful, but before we can deal effectively with such representations, we must answer two questions.

1. **Existence:** Under what conditions does a given function have a power series representation?
2. **Uniqueness:** When f can be represented by a power series, is there only one such series, and, if so, what is it?

The uniqueness issue is addressed in the following theorem.

THEOREM 8.24 The uniqueness theorem for power series representation

Suppose an infinitely differentiable function f is known to have the power series representation

$$f(x) = \sum_{k=0}^{\infty} a_k(x - c)^k$$

for $-R < x - c < R$. Then there is exactly one such representation, and the coefficients a_k must satisfy

$$a_k = \frac{f^{(k)}(c)}{k!} \quad \text{for } k = 0, 1, 2, \ldots$$

Proof The uniqueness theorem may be established by differentiating the given power series term by term and evaluating successive derivatives at c. We start with

$$f(x) = \sum_{k=0}^{\infty} a_k(x - c)^k$$

and substitute $x = c$ to obtain

$$f(c) = a_0 + a_1(c - c) + a_2(c - c)^2 + \cdots = a_0$$

Next, differentiate the original series, term by term, to get

$$f'(x) = a_1 + 2a_2(x - c) + 3a_3(x - c)^2 + \cdots$$

and thus, $f'(c) = a_1 + 2a_2(0) + 3a_3(0)^2 + \cdots = a_1$. Differentiating once again and substituting $x = c$, we find

$$f''(c) = 2a_2 \quad \text{so that} \quad a_2 = \frac{f''(c)}{2}$$

In general, the kth derivative of f at $x = c$ is given by

$$f^{(k)}(c) = k!a_k \quad \text{so that} \quad a_k = \frac{f^{(k)}(c)}{k!} \qquad \square$$

The question of existence—that is, under what conditions a given function has a power series representation—is answered by considering what is called the **Taylor remainder function**:

$$R_n(x) = f(x) - T_n(x)$$

We see that f is represented by its Taylor series in I if and only if

$$\lim_{n \to \infty} R_n(x) = 0 \quad \text{for all } x \text{ in } I$$

This relationship among f, its Taylor polynomial $T_n(x)$, and the Taylor remainder function $R_n(x)$ is summarized in the following theorem.

THEOREM 8.25 Taylor's theorem

If f and all its derivatives exist in an open interval I containing c, then for each x in I,

$$f(x) = f(c) + \frac{f'(c)}{1!}(x - c) + \frac{f''(c)}{2!}(x - c)^2 + \cdots + \frac{f^{(n)}(c)}{n!}(x - c)^n + R_n(x)$$

where the remainder function $R_n(x)$ is given by

$$R_n(x) = \frac{f^{(n+1)}(z_n)}{(n + 1)!}(x - c)^{n+1}$$

for some z_n that depends on x and lies between c and x.

Proof The proof of this theorem is given in Appendix B. $\qquad \square$

The formula for $R_n(x)$ is called the *Lagrange form* of the Taylor remainder function, after the French/Italian mathematician Joseph Lagrange (1736–1813). There are other forms for the remainder, principally one developed by Cauchy (see the HISTORICAL QUEST in Problem 49 of Section 2.1), but we will consider only the Lagrange form in this text.

When applying Taylor's theorem, we do not expect to be able to find the exact value of z_n. However, we can often determine an upper bound for $|R_n(x)|$ in an open interval. We will illustrate this situation later in this section.

TAYLOR AND MACLAURIN SERIES

The uniqueness theorem tells us that if f has a power series representation at c, it must be the series

$$f(c) + \frac{f'(c)}{1!}(x - c) + \frac{f''(c)}{2!}(x - c)^2 + \frac{f'''(c)}{3!}(x - c)^3 + \cdots$$

and Taylor's theorem says that $f(x)$ is represented by such a series precisely where the remainder term $R_n(x) \to 0$ as $n \to \infty$. It is convenient to have the following terminology.

Taylor And Maclaurin Series

WARNING   The uniqueness theorem may be summarized by saying that the Taylor series of f at c is the only power series of the form $\Sigma a_k(x - c)^k$ that can possibly represent f on I. But it does not say that f is equal to the power series on I. All we really know is that if f has a power series representation at c, then that representation must be its Taylor series.

Suppose there is an open interval I containing c throughout which the function f and all its derivatives exist. Then the power series

$$f(c) + \frac{f'(c)}{1!}(x - c) + \frac{f''(c)}{2!}(x - c)^2 + \frac{f'''(c)}{3!}(x - c)^3 + \cdots$$

is called the **Taylor series of f at c.** The special case where $c = 0$ is called the **Maclaurin series of f:**

$$f(0) + \frac{f'(0)}{1!}x + \frac{f''(0)}{2!}x^2 + \frac{f'''(0)}{3!}x^3 + \cdots$$

EXAMPLE 1   Maclaurin series

Find the Maclaurin series for $f(x) = \cos x$.

Solution

First, note that f is infinitely differentiable at $x = 0$. We find

$$\begin{array}{ll} f(x) = \cos x & f(0) = 1 \\ f'(x) = -\sin x & f'(0) = 0 \\ f''(x) = -\cos x & f''(0) = -1 \\ f'''(x) = \sin x & f'''(0) = 0 \\ f^{(4)}(x) = \cos x & f^{(4)}(0) = 1 \\ \quad\vdots & \quad\vdots \end{array}$$

Thus, by using the definition of a Maclaurin series, we have

$$\cos x = 1 - \frac{x^2}{2!} + \frac{x^4}{4!} - \frac{x^6}{6!} + \cdots = \sum_{k=0}^{\infty} \frac{(-1)^k x^{2k}}{(2k)!} \qquad \blacksquare$$

You might ask about the relationship between a Maclaurin series and a Maclaurin polynomial. Figure 8.24 shows the function $f(x) = \cos x$ from Example 1 along with some successive Maclaurin polynomials. We use the notation $M_n(x)$ to represent the first $n + 1$ terms of the corresponding Maclaurin series. Note that the polynomials are quite close to the function near $x = 0$, but that they become very different as x moves away from the origin.

EXAMPLE 2   Maximum error when using a Maclaurin polynomial approximation

Find the Maclaurin polynomial $M_5(x)$ for the function $f(x) = e^x$ and use this polynomial to approximate e. Use Taylor's theorem to determine the accuracy of this approximation.

a. $M_0(x) = 1$, $M_1(x) = 1 - \dfrac{x^2}{2!}$,

$M_2(x) = 1 - \dfrac{x^2}{2!} + \dfrac{x^4}{4!}$,

$M_3(x) = 1 - \dfrac{x^2}{2!} + \dfrac{x^4}{4!} - \dfrac{x^6}{6!}$

b. $M_{10}(x) = 1 - \dfrac{x^2}{2!} + \dfrac{x^4}{4!} - \dfrac{x^6}{6!} + \dfrac{x^8}{8!}$

$- \dfrac{x^{10}}{10!} + \dfrac{x^{12}}{12!} - \dfrac{x^{14}}{14!} + \dfrac{x^{16}}{16!} - \dfrac{x^{18}}{18!}$

$+ \dfrac{x^{20}}{20!}$

Figure 8.24 Comparison of $\cos x$ and its Maclaurin polynomials

a. Comparison of f and M_5 shown on domain $[0, 7]$

b. Zoom showing the error at $x = 1$

Figure 8.25 Comparison of $f(x) = e^x$ and $M_5(x) = 1 + x + \dfrac{x^2}{2} + \dfrac{x^3}{6} + \dfrac{x^4}{24} + \dfrac{x^5}{120}$

Solution

First find the Maclaurin series for $f(x) = e^x$.

$$f(x) = e^x \qquad\qquad f(0) = 1$$
$$f'(x) = e^x \qquad\qquad f'(0) = 1$$
$$\vdots \qquad\qquad\qquad \vdots$$

The Maclaurin series for e^x is

$$e^x = 1 + \frac{1}{1!}x + \frac{1}{2!}x^2 + \frac{1}{3!}x^3 + \cdots = \sum_{k=0}^{\infty} \frac{x^k}{k!}$$

Then find $M_5(x)$:

$$M_5(x) = 1 + x + \frac{x^2}{2} + \frac{x^3}{6} + \frac{x^4}{24} + \frac{x^5}{120} \approx e^x$$

A comparison of f and $M_5(x)$ is shown in Figure 8.25. Notice from the graphs of these functions that the error seems to increase as x moves away from the origin.

To determine the accuracy, we use Taylor's theorem

$$e = 1 + 1 + \frac{1}{2} + \frac{1}{6} + \frac{1}{24} + \frac{1}{120} + R_5(1)$$
$$= 2.71\overline{6} + R_5(1)$$

The remainder term is given by

$$R_5(1) = \frac{e^{z_5}}{(5+1)!}(1)^{5+1}$$

for some z_5 between 0 and 1. Because $0 < z_5 < 1$, we have $e^{z_5} < e$, and since $e < 3$, it follows that

$$R_5(1) < \frac{e}{6!} < \frac{3}{6!} \approx 0.004167$$

Thus, the number e satisfies

$$2.716667 - 0.004167 < e < 2.716667 + 0.004167$$
$$2.712500 < e < 2.720834$$

EXAMPLE 3   Taylor series

Find the Taylor series for $f(x) = \ln x$ at $c = 1$.

Solution

Note that f is infinitely differentiable at $x = 1$. We find

$$f(x) = \ln x \qquad\qquad\qquad f(1) = 0$$

$$f'(x) = \frac{1}{x} \qquad\qquad\qquad f'(1) = 1$$

$$f''(x) = \frac{-1}{x^2} \qquad\qquad\qquad f''(1) = -1$$

$$f'''(x) = \frac{2}{x^3} \qquad\qquad\qquad f'''(1) = 2$$

$$f^{(4)}(x) = \frac{-6}{x^4} \qquad\qquad\qquad f^{(4)}(1) = -6$$

$$\vdots \qquad\qquad\qquad\qquad \vdots$$

$$f^{(k)}(x) = \frac{(-1)^{k+1}(k-1)!}{x^k} \qquad f^{(k)}(1) = (-1)^{k+1}(k-1)!$$

Then, use the definition of a Taylor series to write

$$\ln x = 0 + \frac{1}{1!}(x-1) - \frac{1}{2!}(x-1)^2 + \frac{2}{3!}(x-1)^3 - \frac{6}{4!}(x-1)^4 + \cdots$$

$$= (x-1) - \frac{1}{2}(x-1)^2 + \frac{1}{3}(x-1)^3 - \frac{1}{4}(x-1)^4 + \cdots$$

$$= \sum_{k=1}^{\infty} \frac{(-1)^{k+1}(x-1)^k}{k} \qquad\qquad\blacksquare$$

Suppose f is a function that is infinitely differentiable at c. Now we have two mathematical quantities, f and its Taylor series. There are several possibilities:

1. The Taylor series of f may converge to f on the interval of absolute convergence, $-R < x - c < R$ (that is, $|x - c| < R$).
2. The Taylor series may converge only at $x = c$, in which case it certainly does not represent f on any interval containing c.
3. The Taylor series of f may have a positive radius of convergence (even $R = \infty$), but it may converge to a function g that does not equal f on the interval $|x - c| < R$.

EXAMPLE 4   A function defined at points where its Taylor series does not converge

Show that the function $\ln x$ is defined at points for which its Taylor series at $c = 1$ does not converge.

Solution

We know that $\ln x$ is defined for all $x > 0$. From the previous example, we know the Taylor series for $\ln x$ at $c = 1$ is

$$\sum_{k=1}^{\infty} \frac{(-1)^{k-1}(x-1)^k}{k}$$

We find the interval of convergence for this series centered at $c = 1$ by computing

$$L = \lim_{k \to \infty} \left| \frac{\dfrac{(-1)^k(x-1)^{k+1}}{k+1}}{\dfrac{(-1)^{k-1}(x-1)^k}{k}} \right| = \lim_{k \to \infty} \left| \frac{\dfrac{(-1)^k}{k+1}}{\dfrac{(-1)^{k-1}}{k}} \right| |x-1|$$

$$= \lim_{k \to \infty} \left(\frac{k}{k+1} \right) |x-1| = |x-1|$$

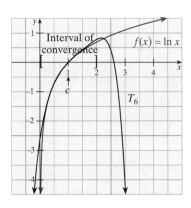

Figure 8.26 The graphs of $f(x) = \ln x$ and the Taylor polynomial

$$T_6(x) = \sum_{k=1}^{6} \frac{(-1)^{k-1}}{k}(x-1)$$

are very close near 1.

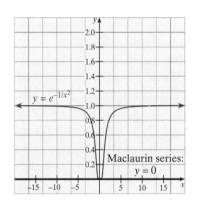

Figure 8.27 Graph of f and its Maclaurin series $y = 0$

This means that the power series converges absolutely for

$$|x - 1| < 1$$
$$0 < x < 2$$

Test the endpoints:

$$x = 0: \qquad \sum_{k=1}^{\infty} \frac{(-1)^{k-1}(0-1)^k}{k} = \sum_{k=1}^{\infty} \frac{-1}{k} \text{ diverges} \qquad \textit{Opposite of harmonic series}$$

$$x = 2: \sum_{k=1}^{\infty} \frac{(-1)^{k-1}(2-1)^k}{k} = \sum_{k=1}^{\infty} \frac{(-1)^{k-1}}{k} \text{ converges} \qquad \textit{Alternating harmonic series}$$

Thus, the series converges only on $(0, 2]$, but the function $\ln x$ is defined for all $x > 0$. The function is compared with the sixth degree Taylor approximation in Figure 8.26. ∎

The next example introduces a function whose Maclaurin series does not represent that function on *any* interval.

EXAMPLE 5   A function represented by its Maclaurin series at a single point

Let f be the function defined by

$$f(x) = \begin{cases} 0 & \text{if } x = 0 \\ e^{-1/x^2} & \text{if } x \neq 0 \end{cases}$$

Show that f is represented by its Maclaurin series only at $x = 0$.

Solution

It can be shown that f is infinitely differentiable at 0 and that $f^{(k)}(0) = 0$ for all k. Therefore, f has the Maclaurin series

$$0 + \frac{0}{1!}x + \frac{0}{2!}x^2 + \frac{0}{3!}x^3 + \cdots$$

which converges to the zero function for all x and thus represents $f(x)$ only at $x = 0$. The graph of f and the Maclaurin series are shown in Figure 8.27. ∎

OPERATIONS WITH TAYLOR AND MACLAURIN SERIES

According to the uniqueness theorem, the Taylor series for f at c is the only power series in $x - c$ that can satisfy $f(x) = \sum_{k=0}^{\infty} a_k(x - c)^k$ for all x in some interval containing c. The coefficients in this series can always be found by substituting into the formula $a_k = \dfrac{f^{(k)}(c)}{k!}$, but occasionally they can also be found by algebraic manipulation. In several of the examples that follow, we will use the geometric series equation

$$\frac{1}{1-u} = \sum_{k=0}^{\infty} u^k \qquad \text{for } |u| < 1$$

to obtain Maclaurin series for rational functions. The general procedure is illustrated in Example 6.

EXAMPLE 6   Maclaurin series using substitution in a geometric series

Find the Maclaurin series for $\dfrac{1}{1 + x^2}$.

Solution

We want a series of the form $\displaystyle\sum_{k=0}^{\infty} a_k u^k$ that represents $\dfrac{1}{1+x^2}$ on an interval containing 0. We could proceed directly using the definition of a Maclaurin series, but instead we will modify a known series. We know that if $|u| < 1$ we can write

$$\frac{1}{1-u} = 1 + u + u^2 + \cdots$$

so by substitution ($u = -x^2$) we have

$$\frac{1}{1+x^2} = \frac{1}{1-(-x^2)} = 1 - x^2 + x^4 - x^6 + \cdots$$

provided $\left|-x^2\right| < 1$; that is, $-1 < x < 1$. Hence, by the uniqueness theorem, the desired representation is

$$\frac{1}{1+x^2} = \sum_{k=0}^{\infty} (-1)^k x^{2k} \quad \text{for } -1 < x < 1 \qquad\blacksquare$$

EXAMPLE 7 Maclaurin series by modifying a geometric series

Find the Maclaurin series for $f(x) = \dfrac{5 - 2x}{3 + 2x}$ and determine the interval of convergence of this series.

Solution

Our strategy is to rewrite $f(x)$ in the form $\dfrac{1}{1-u}$ of a geometric series:

$$\frac{5 - 2x}{3 + 2x} = -1 + \frac{8}{3 + 2x} \qquad \text{Long division}$$

$$= -1 + \frac{\frac{8}{3}}{1 - \left(-\frac{2}{3}x\right)} \qquad \text{Writing in the form } \frac{1}{1-u}$$

$$= -1 + \sum_{k=0}^{\infty} \frac{8}{3}\left[-\frac{2}{3}x\right]^k$$

$$= -1 + \frac{8}{3} - \frac{16}{9}x + \frac{32}{27}x^2 - \frac{64}{81}x^3 + \cdots$$

The convergence set for $f(x)$ is the same as for the series $\displaystyle\sum_{k=1}^{\infty}\left(-\frac{2}{3}x\right)^k$; namely,

$$\left|-\tfrac{2}{3}x\right| < 1$$

$$-\tfrac{3}{2} < x < \tfrac{3}{2} \qquad\blacksquare$$

EXAMPLE 8 Using a trigonometric identity with a known series

Find the Maclaurin series for

a. $\cos x^2$ **b.** $\cos^2 x$

Solution

a. In Example 1 we found that

$$\cos u = 1 - \frac{u^2}{2!} + \frac{u^4}{4!} - \frac{u^6}{6!} + \cdots \qquad \text{for all } u$$

Therefore, by substituting $u = x^2$, we obtain

$$\cos x^2 = 1 - \frac{(x^2)^2}{2!} + \frac{(x^2)^4}{4!} - \frac{(x^2)^6}{6!} + \cdots$$

$$= 1 - \frac{x^4}{2!} + \frac{x^8}{4!} - \frac{x^{12}}{6!} + \cdots \qquad \text{for all } x$$

b. For $\cos^2 x$ we could use the definition of a Maclaurin series, but instead we will use a double-angle trigonometric identity.

$$\cos^2 x = \frac{1}{2} + \frac{1}{2}\cos 2x$$

$$= \frac{1}{2} + \frac{1}{2}\left[1 - \frac{(2x)^2}{2!} + \frac{(2x)^4}{4!} - \frac{(2x)^6}{6!} + \cdots\right] \qquad \text{Let } u = 2x.$$

$$= \frac{1}{2} + \frac{1}{2} - \frac{2x^2}{2!} + \frac{2^3 x^4}{4!} - \frac{2^5 x^6}{6!} + \cdots$$

$$= 1 - x^2 + \frac{1}{3}x^4 - \frac{2}{45}x^6 + \cdots \qquad \text{for all } x \qquad \blacksquare$$

EXAMPLE 9   Maclaurin series using substitution

Find the Maclaurin series for $f(x) = \ln\left(\dfrac{1+x}{1-x}\right)$ and use this series to compute $\ln 2$ correct to five decimal places.

Solution

For this function, we first use a property of logarithms:

$$f(x) = \ln\left(\frac{1+x}{1-x}\right) = \ln(1+x) - \ln(1-x)$$

From Example 3, we know

$$\ln x = (x-1) - \tfrac{1}{2}(x-1)^2 + \tfrac{1}{3}(x-1)^3 - \cdots$$

so that

$$\ln(1+x) = [1+x-1] - \tfrac{1}{2}[1+x-1]^2 + \tfrac{1}{3}[1+x-1]^3 - \cdots$$

$$= x - \frac{x^2}{2} + \frac{x^3}{3} - \frac{x^4}{4} + \cdots$$

$$\ln(1-x) = [1-x-1] - \tfrac{1}{2}[1-x-1]^2 + \tfrac{1}{3}[1-x-1]^3 - \cdots$$

$$= (-x) - \frac{(-x)^2}{2} + \frac{(-x)^3}{3} - \frac{(-x)^4}{4} + \cdots$$

$$= -x - \frac{x^2}{2} - \frac{x^3}{3} - \frac{x^4}{4} - \cdots$$

By subtracting series, we find

$$f(x) = \ln(1+x) - \ln(1-x)$$

$$= \left[x - \frac{x^2}{2} + \frac{x^3}{3} - \frac{x^4}{4} + \cdots\right] - \left[-x - \frac{x^2}{2} - \frac{x^3}{3} - \frac{x^4}{4} - \cdots\right]$$

$$= 2\left[x + \frac{x^3}{3} + \frac{x^5}{5} + \cdots\right]$$

Next, by solving the equation $\dfrac{1+x}{1-x} = 2$, we find $x = \frac{1}{3}$, so we are looking for $\ln\left(\dfrac{1+x}{1-x}\right)$ where $x = \frac{1}{3}$. It follows that

$$\ln 2 = 2\left[\frac{1}{3} + \frac{\left(\frac{1}{3}\right)^3}{3} + \frac{\left(\frac{1}{3}\right)^5}{5} + \cdots\right] = 2\left[\frac{1}{3} + \frac{1}{3}\left(\frac{1}{3}\right)^3 + \frac{1}{5}\left(\frac{1}{3}\right)^5 + \cdots\right]$$

According to Taylor's theorem, we find

$$\ln 2 = \frac{2}{3} + \frac{2}{3}\left(\frac{1}{3}\right)^3 + \frac{2}{5}\left(\frac{1}{3}\right)^5 + \cdots + \frac{2}{2n+1}\left(\frac{1}{3}\right)^{2n+1} + R_n\left(\frac{1}{3}\right)$$

where $R_n\left(\frac{1}{3}\right)$ is the remainder term. To estimate the remainder, we note that $R_n\left(\frac{1}{3}\right)$ is the tail of the infinite series for $\ln 2$. Thus,

$$\left|R_n\left(\frac{1}{3}\right)\right| = \frac{2}{2n+3}\left(\frac{1}{3}\right)^{2n+3} + \frac{2}{2n+5}\left(\frac{1}{3}\right)^{2n+5} + \frac{2}{2n+7}\left(\frac{1}{3}\right)^{2n+7} + \cdots$$

$$< \frac{2}{2n+3}\left(\frac{1}{3}\right)^{2n+3} + \frac{2}{2n+3}\left(\frac{1}{9}\right)\left(\frac{1}{3}\right)^{2n+3} + \cdots$$

$$= \frac{2}{2n+3}\left(\frac{1}{3}\right)^{2n+3}\left[1 + \frac{1}{9} + \frac{1}{81} + \cdots\right]$$

Because the geometric series in brackets converges to $\dfrac{1}{1 - \frac{1}{9}} = \dfrac{9}{8}$, we find that

$$\left|R_n\left(\frac{1}{3}\right)\right| < \frac{2}{2n+3}\left(\frac{9}{8}\right)\left(\frac{1}{3}\right)^{2n+3}$$

In particular, to achieve five-place accuracy, we must make sure the term on the right is less than 0.000005 (six places to account for round-off error). With a calculator we can see that if $n = 4$, we have

$$\frac{2}{2(4)+3}\left(\frac{9}{8}\right)\left(\frac{1}{3}\right)^{11} = 0.0000012$$

Thus, we approximate $\ln 2$ with $n = 4$:

$$\ln 2 \approx T_4\left(\frac{1}{3}\right) = \frac{2}{3} + \frac{2}{3}\left(\frac{1}{3}\right)^3 + \frac{2}{5}\left(\frac{1}{3}\right)^5 + \frac{2}{7}\left(\frac{1}{3}\right)^7 + \frac{2}{9}\left(\frac{1}{3}\right)^9 \approx 0.6931460$$

which is correct with an error of no more than 0.0000012; therefore,

$$0.6931460 - 0.0000012 < \ln 2 < 0.6931460 + 0.0000012$$

$$0.6931448 < \ln 2 < 0.6931472$$

Rounded to five-place accuracy, $\ln 2 = 0.69315$. We can check with calculator accuracy: $\ln 2 \approx 0.6931471806$. ∎

Some of the more common power series that we have derived are given in Table 8.2. For a more complete list, see the *Student Mathematics Handbook*. Before we present this table, there is one last result we should consider. It is a generalization of the binomial theorem that was discovered by Isaac Newton while he was still a student at Cambridge University.

THEOREM 8.26 Binomial series

If p is any real number and $-1 < x < 1$, then

$$(1+x)^p = 1 + px + \frac{p(p-1)}{2!}x^2 + \frac{p(p-1)(p-2)}{3!}x^3 + \cdots = \sum_{k=0}^{\infty}\binom{p}{k}x^k$$

where $\dbinom{p}{k} = \dfrac{p!}{k!(p-k)!}$ for $k = 1, 2, \ldots$.

Proof   We begin by showing how the Maclaurin series is found. Let $f(x) = (1+x)^p$.

$$f(x) = (1+x)^p \qquad\qquad\qquad f(0) = 1$$
$$f'(x) = p(1+x)^{p-1} \qquad\qquad f'(0) = p$$
$$f''(x) = p(p-1)(1+x)^{p-2} \qquad f''(0) = p(p-1)$$
$$f'''(x) = p(p-1)(p-2)(1+x)^{p-3} \qquad f'''(0) = p(p-1)(p-2)$$
$$\vdots \qquad\qquad\qquad\qquad\qquad \vdots$$
$$f^{(n)}(x) = p(p-1)\cdots(p-n+1)(1+x)^{p-n} \qquad f^{(n)}(0) = p(p-1)\cdots(p-n+1)$$

TABLE 8.2 Power Series for Elementary Functions

Name	Series	Interval of Convergence
	Taylor series at $x = c$ follows the Maclaurin series for each function.	
Exponential series	$e^u = 1 + u + \dfrac{u^2}{2!} + \dfrac{u^3}{3!} + \dfrac{u^4}{4!} + \cdots + \dfrac{u^k}{k!} + \cdots$	$(-\infty, \infty)$
	$e^x = e^c + e^c(x - c) + \dfrac{e^c(x - c)^2}{2!} + \dfrac{e^c(x - c)^3}{3!} + \cdots + \dfrac{e^c(x - c)^k}{k!} + \cdots$	$(-\infty, \infty)$
Cosine series	$\cos u = 1 - \dfrac{u^2}{2!} + \dfrac{u^4}{4!} - \dfrac{u^6}{6!} + \cdots + \dfrac{(-1)^k u^{2k}}{(2k)!} + \cdots$	$(-\infty, \infty)$
	$\cos x = \cos c - (x - c)\sin c - \dfrac{(x - c)^2}{2!}\cos c + \dfrac{(x - c)^3}{3!}\sin c + \cdots$	$(-\infty, \infty)$
Sine series	$\sin u = u - \dfrac{u^3}{3!} + \dfrac{u^5}{5!} - \dfrac{u^7}{7!} + \cdots + \dfrac{(-1)^k u^{2k+1}}{(2k + 1)!} + \cdots$	$(-\infty, \infty)$
	$\sin x = \sin c + (x - c)\cos c - \dfrac{(x - c)^2}{2!}\sin c - \dfrac{(x - c)^3}{3!}\cos c + \cdots$	$(-\infty, \infty)$
Geometric series	$\dfrac{1}{1 - u} = 1 + u + u^2 + u^3 + \cdots + u^k + \cdots$	$(-1, 1)$
Reciprocal series	$\dfrac{1}{u} = 1 - (u - 1) + (u - 1)^2 - (u - 1)^3 + (u - 1)^4 - \cdots$	$(0, 2)$
Logarithmic series	$\ln u = (u - 1) - \dfrac{(u - 1)^2}{2} + \dfrac{(u - 1)^3}{3} - \cdots + \dfrac{(-1)^{k-1}(u - 1)^k}{k} + \cdots$	$(0, 2]$
	$\ln(1 + x) = x - \dfrac{1}{2}x^2 + \dfrac{1}{3}x^3 - \dfrac{1}{4}x^4 + \cdots + \dfrac{(-1)^n x^{n+1}}{n + 1} + \cdots$	$(-1, 1]$
	$\ln x = \ln c + \dfrac{x - c}{c} - \dfrac{(x - c)^2}{2c^2} + \dfrac{(x - c)^3}{3c^3} - \cdots + \dfrac{(-1)^{k-1}(x - c)^k}{kc^k} + \cdots$	$(0, 2c]$
Inverse tangent series	$\tan^{-1} u = u - \dfrac{u^3}{3} + \dfrac{u^5}{5} - \dfrac{u^7}{7} + \cdots + \dfrac{(-1)^k u^{2k+1}}{2k + 1} + \cdots$	$[-1, 1]$
Inverse sine series	$\sin^{-1} u = u + \dfrac{u^3}{2 \cdot 3} + \dfrac{1 \cdot 3 u^5}{2 \cdot 4 \cdot 5} + \dfrac{1 \cdot 3 \cdot 5 u^7}{2 \cdot 4 \cdot 6 \cdot 7} + \cdots + \dfrac{1 \cdot 3 \cdot 5 \cdots (2k - 3)u^{2k-1}}{2 \cdot 4 \cdot 6 \cdots (2k - 2)(2k - 1)} + \cdots$	$[-1, 1]$
Binomial series	$(1 + u)^p = 1 + pu + \dfrac{p(p - 1)}{2!}u^2 + \dfrac{p(p - 1)(p - 2)}{3!}u^3 + \cdots + \binom{p}{q}u^q + \cdots$	

which produces the Maclaurin series. We could use the ratio test to show that this series converges on the interval $(-1, 1)$, but to show that it converges to $(1 + x)^p$ requires advanced calculus. ❑

The binomial series converges on the open interval $-1 < x < 1$, but convergence at the endpoints $x = -1$ and $x = 1$ of this interval depends on the exponent p. In particular, it can be shown that

if $-1 < p < 0$, the series converges at $x = 1$;

if $p \geq 0$, the series converges at both $x = 1$ and $x = -1$.

Moreover, if p is a nonnegative integer, the series terminates after only a finite number of terms (since $\binom{p}{k} = 0$ for $k > p$) and thus reduces to the ordinary binomial expansion, which, of course, converges for all x. The next example illustrates how the binomial series theorem can be used.

EXAMPLE 10 Using the binomial series theorem to obtain a Maclaurin series expansion

Determine the Maclaurin series for $f(x) = \sqrt{9+x}$ and find its interval of convergence.

Solution

Write $f(x) = \sqrt{9+x} = (9+x)^{1/2} = 3\left(1 + \dfrac{x}{9}\right)^{1/2}$. Thus,

$$
\sqrt{9+x} = 3\left(1 + \frac{x}{9}\right)^{1/2}
$$

$$
= 3\left[1 + \frac{1}{2}\left(\frac{x}{9}\right) + \frac{\frac{1}{2}\left(\frac{1}{2}-1\right)}{2!}\left(\frac{x}{9}\right)^2 + \frac{\frac{1}{2}\left(\frac{1}{2}-1\right)\left(\frac{1}{2}-2\right)}{3!}\left(\frac{x}{9}\right)^3 + \cdots\right]
$$

$$
= 3\left[1 + \frac{1}{18}x - \frac{1}{648}x^2 + \frac{1}{11{,}664}x^3 - \cdots\right]
$$

$$
= 3 + \frac{1}{6}x - \frac{1}{216}x^2 + \frac{1}{3{,}888}x^3 - \cdots
$$

Since the exponent $p = \frac{1}{2}$ satisfies $p > 0$, p not an integer, it follows from Theorem 8.26 that the series converges for $\left|\dfrac{x}{9}\right| \le 1$, that is, for $|x| \le 9$. ∎

8.8 PROBLEM SET

A 1. WHAT DOES THIS SAY? Compare and contrast Maclaurin and Taylor series.

2. WHAT DOES THIS SAY? Discuss the binomial series theorem.

Find the Maclaurin series for the functions given in Problems 3–30. Assume that a is any constant, and that all the derivatives of all orders exist at $x = 0$.

3. e^{2x}
4. e^{-x}
5. e^{x^2}
6. e^{ax}
7. $\sin x^2$
8. $\sin^2 x$
9. $\sin ax$
10. $\cos ax$
11. $\cos 2x^2$
12. $\cos x^3$
13. $x^2 \cos x$
14. $\sin \dfrac{x}{2}$
15. $x^2 + 2x + 1$
16. $x^3 - 2x^2 + x - 5$
17. xe^x
18. $e^{-x} + e^{2x}$
19. $e^x + \sin x$
20. $\sin x + \cos x$
21. $\dfrac{1}{1+4x}$
22. $\dfrac{1}{1-ax}, a \ne 0$
23. $\dfrac{1}{a+x}, a \ne 0$
24. $\dfrac{1}{a^2+x^2}, a \ne 0$
25. $\ln(3+x)$
26. $\log(1+x)$
27. $\tan^{-1}(2x)$
28. $\sqrt{1-x}$
29. e^{-x^2}

30. $f(x) = \begin{cases} \dfrac{\sin x}{x} & \text{if } x \ne 0 \\ 1 & \text{if } x = 0 \end{cases}$

Find the first four terms of the Taylor series of the functions in Problems 31–42 at the given value of c.

31. $f(x) = e^x$ at $c = 1$
32. $f(x) = \ln x$ at $c = 3$

33. $f(x) = \cos x$ at $c = \frac{\pi}{3}$
34. $f(x) = \sin x$ at $c = \frac{\pi}{4}$
35. $f(x) = \tan x$ at $c = 0$
36. $f(x) = x^2 + 2x + 1$ at $c = 200$
37. $f(x) = x^3 - 2x^2 + x - 5$ at $c = 2$
38. $f(x) = \sqrt{x}$ at $c = 9$
39. $f(x) = \dfrac{1}{2-x}$ at $c = 5$
40. $f(x) = \dfrac{1}{4-x}$ at $c = -2$
41. $f(x) = \dfrac{3}{2x-1}$ at $c = 2$
42. $f(x) = \dfrac{5}{3x+2}$ at $c = 2$

Expand each function in Problems 43–48 as a binomial series. Give the interval of convergence of the series.

43. $f(x) = \sqrt{1+x}$
44. $f(x) = \dfrac{1}{\sqrt{1+x^2}}$
45. $f(x) = (1+x)^{2/3}$
46. $f(x) = (4+x)^{-1/3}$
47. $f(x) = \dfrac{x}{\sqrt{1-x^2}}$
48. $f(x) = \sqrt[4]{2-x}$

B 49. Use term-by-term integration to show that

$$
\ln(1+x) = \sum_{k=1}^{\infty} \frac{(-1)^{k-1}x^k}{k} \quad \text{for } |x| < 1
$$

50. Use the Maclaurin series for e^x and e^{-x} to find the Maclaurin series for

$$
\cosh x = \frac{e^x + e^{-x}}{2}
$$

51. Use the Maclaurin series for e^x and e^{-x} to find the Maclaurin series for

$$
\sinh x = \frac{e^x - e^{-x}}{2}
$$

52. Use the Maclaurin series expansion of e^x to estimate $\sqrt{e}$ with three-decimal-place accuracy? How many terms are needed?

53. Use the Maclaurin series expansion of e^x to estimate $\sqrt[3]{e}$ with three-decimal-place accuracy. How many terms are needed?

Find the Maclaurin series for the functions given in Problems 54–56. Hint: Begin by expressing each function as a sum of partial fractions.

54. $f(x) = \dfrac{2x}{x^2 - 1}$

55. $f(x) = \dfrac{1}{x^2 - 3x + 2}$

56. $f(x) = \dfrac{x^2}{(x + 2)(x^2 - 1)}$

57. Use an appropriate identity to find the Maclaurin series for $f(x) = \sin x \cos x$.

58. a. It can be shown that $\cos 3x = 4\cos^3 x - 3\cos x$. Use this identity to find the Maclaurin series for $\cos^3 x$.

b. Find the Maclaurin series for $\sin^3 x$.

59. Use the identity $2\cos\alpha\cos\beta = \cos(\alpha + \beta) + \cos(\alpha - \beta)$ to find the Maclaurin series for

$$f(x) = \left(\cos\frac{3x}{2}\right)\left(\cos\frac{x}{2}\right)$$

60. Find the Maclaurin series for $\ln[(1 + 2x)(1 + 3x)]$.

61. Find the Maclaurin series for $\ln\left[\dfrac{1 + 2x}{1 - 3x + 2x^2}\right]$.

62. Find the Maclaurin series for $f(x) = x + \sin x$ and then use this series to find $\displaystyle\lim_{x \to 0} \frac{x + \sin x}{x}$.

63. Find the Maclaurin series for $g(x) = e^x - 1$ and then use this series to find $\displaystyle\lim_{x \to 0} \frac{e^x - 1}{x}$.

In Problems 64–66, the Maclaurin series of a function appears with its remainder. Verify that the remainder satisfies the given inequality.

64. $e^x = \displaystyle\sum_{k=0}^{\infty} \frac{x^k}{k!}$; $|R_n(x)| \le \dfrac{e^p |x|^{n+1}}{(n + 1)!}$, where p is some constant such that $0 < p < x$.

65. $\cos x = \displaystyle\sum_{k=0}^{\infty} \frac{(-1)^k x^{2k}}{(2k)!}$; $|R_n(x)| \le \dfrac{|x|^{n+1}}{(n + 1)!}$

66. $\ln(x + 1) = \displaystyle\sum_{k=0}^{\infty} \frac{(-1)^k x^{k+1}}{k + 1}$ for $0 \le x < 1$; $|R_n(x)| \le \dfrac{|x|^{n+1}}{n + 1}$

67. HISTORICAL QUEST The two mathematicians most closely associated with power series are Brook Taylor and Colin Maclaurin. The result stated in this section, called Taylor's theorem, was published by Taylor in 1712, but the result itself was discovered by the Scottish mathematician James Gregory (1638–1675) in 1670 and was obtained by Taylor in 1688, after Gregory's death.

BROOK TAYLOR
1685–1731

James Gregory was a Scottish mathematician who apparently anticipated some of the key calculus discoveries, but died prematurely and never received proper recognition for his work. In 1694, Johann Bernoulli (see Historical Quests 58 and 59, Section 4.5) published a series that was very similar to what we call Taylor's series. In fact, after Taylor published his work in 1714, Bernoulli accused him of plagiarism. ∎

In this Quest we seek to duplicate Bernoulli's work. He started by writing

$$n\,dx = n\,dx + \left(x\,dn - x\frac{dn}{dx}\,dx\right)$$

$$= n\,dx + \left(x\,dn - x\frac{dn}{dx}\,dx\right) - \left(\frac{x^2}{2!}\frac{d^2n}{dx^2}\,dx - \frac{x^2}{2!}\frac{d^2n}{dx^2}\,dx\right)$$

$$= n\,dx + \left(x\,dn - x\frac{dn}{dx}\,dx\right) - \left(\frac{x^2}{2!}\frac{d^2n}{dx^2}\,dx - \frac{x^2}{2!}\frac{d^2n}{dx^2}\,dx\right)$$

$$+ \left(\frac{x^3}{3!}\frac{d^3n}{dx^3}\,dx - \frac{x^3}{3!}\frac{d^3n}{dx^3}\,dx\right) - \cdots$$

$$= (n\,dx + x\,dn) - \left(x\frac{dn}{dx} + \frac{x^2}{2!}\frac{d^2n}{dx^2}\right)dx$$

$$+ \left(\frac{x^2}{2!}\frac{d^2n}{dx^2} + \frac{x^3}{3!}\frac{d^3n}{dx^3}\right)dx - \cdots$$

$$= d(nx) - d\left(\frac{x^2}{2!}\frac{dn}{dx}\right) + d\left(\frac{x^3}{3!}\frac{d^2n}{dx^2}\right) - \cdots$$

Integrate termwise to obtain what is known as Bernoulli's series. Next, see if you can show what angered Bernoulli by deriving Taylor's series from Bernoulli's series.

Incidentally, Bernoulli's assertion was not founded. The first known explicit statement of the general Taylor series was in *De Quadrature* (1691) by Isaac Newton.

68. HISTORICAL QUEST Colin Maclaurin was a mathematical prodigy. He entered the University of Glasgow at the age of 11. At 15 he took his master's degree and gave a remarkable public defense of his thesis on gravitational attraction. At 19 he was elected to the chair of mathematics at the Marischal College in Aberdeen and at 21 published his first important work,

COLIN MACLAURIN
1698–1746

Geometrica organica. At 27 he became deputy, or assistant, to the professor of mathematics at the University of Edinburgh. There was some difficulty in obtaining a salary to cover his assistantship, and Newton offered to bear the cost personally so that the university could secure the services of such an outstanding young man. In time Maclaurin succeeded Newton and at the age of 44, published the first systematic exposition of Newton's work. In his work, Maclaurin uses what were called at the time fluxions, rather than derivatives. Here is what he concluded:

"If the first fluxion of the ordinate, with its fluxions of several subsequent orders vanish, the ordinate is a minimum or maximum, when the number of all those fluxions that vanish is 1, 3, 5, or any odd number. The ordinate is a *minimum*, when the fluxion next to those that vanish is positive; but a *maximum* when the fluxion is negative .... But if the number of all the fluxions of the first and successive orders that vanish be an even number, the ordinate is then neither a *maximum* nor *minimum*."* ∎

What is Maclaurin saying in this quotation?

The Historical Development of the Calculus by C. H. Edwards, Jr., Springer-Verlag, 1979, p. 292.

C **69.** Use series methods to approximate $\int_0^{1/2} \dfrac{dx}{1+x^7}$ correct to six decimal places.

70. Express $\int_0^1 x^{0.2} e^x \, dx$ as an infinite series.

71. The functions $J_0(x)$ and $J_1(x)$ are defined as the following power series:

$$J_0(x) = \sum_{k=0}^{\infty} \frac{(-1)^k x^{2k}}{(k!)^2 2^{2k}} \quad \text{and} \quad J_1(x) = \sum_{k=0}^{\infty} \frac{(-1)^k x^{2k+1}}{k!(k+1)! 2^{2k+1}}$$

a. Show that $J_0(x)$ and $J_1(x)$ both converge for all x.

b. Show that $J_0'(x) = -J_1(x)$.

These functions are called **Bessel functions of the first kind.** Bessel functions were first used in analyzing Kepler's laws of planetary motion. These functions play an important role in physics and engineering.

72. Show that J_0, the Bessel function of the first kind defined in Problem 71, satisfies the differential equation $x^2 J_0''(x) + x J_0'(x) + x^2 J_0(x) = 0$.

73. For x in the open interval $(-1, 1)$, let

$$f(x) = x + \frac{x^2}{2} + \frac{x^3}{3} + \cdots + \frac{x^k}{k} + \cdots$$

Show that f is a logarithmic function. *Hint*: Find f' and then retrieve f by integrating f'.

74. Given that

$$\frac{1}{(1-x)(1-2x)} = \sum a_n x^n$$

and

$$\frac{1+2x}{(1-x)(1-2x)(1-4x)} = \sum b_n x^n$$

show that $b_n = a_n^2$ for all n. *Hint*: Use a partial fraction decomposition.

75. Journal Problem   *The Pi Mu Epsilon Journal* by Robert C. Gebhardt* ∎ For what x does $\displaystyle\sum_{n=0}^{\infty} \frac{x^n}{(2n)!}$ converge, and what is the sum?

———————

*Volume 8, 1984, Problem 586.

CHAPTER 8 REVIEW

Proficiency Examination

CONCEPT PROBLEMS

1. What is a sequence?
2. What is meant by the limit of a sequence?
3. Define the convergence and divergence of a sequence.
4. Explain each of the following terms:
   a. bounded sequence
   b. monotonic sequence
   c. strictly monotonic sequence
5. State the BMCT.
6. What is an infinite series?
7. Compare or contrast the convergence and divergence of sequences and series.
8. What is a telescoping series?
9. Describe the harmonic series. Does it converge or diverge?
10. Define a geometric series and give its sum.
11. State the divergence test.
12. State the integral test.
13. State the p-series test.
14. State the direct comparison test.
15. State the limit comparison test.
16. State the zero-infinity limit comparison test.
17. State the ratio test.
18. State the root test.
19. What is the alternating series test?

20. How do you make an error estimate for an alternating series?
21. State the absolute convergence test.
22. What is meant by absolute and conditional convergence?
23. What is the generalized ratio test?
24. What is a power series?
25. How do you find the convergence set of a power series?
26. What are the radius of convergence and interval of convergence for a power series?
27. What is a Taylor polynomial?
28. State Taylor's theorem.
29. What are a Taylor series and a Maclaurin series?
30. State the binomial series theorem.

PRACTICE PROBLEMS

31. Find the limit of the sequence $\left\{ \left(1 + \dfrac{1}{n}\right)^n \right\}$

32. **a.** Find the limit of the sequence $\left\{ \dfrac{e^n}{n!} \right\}$.

     b. Test the series $\displaystyle\sum_{k=1}^{\infty} \dfrac{e^k}{k!}$ for convergence.

     c. Discuss the similarities and/or differences of parts **a** and **b**.

In Problems 33–37, test the given series for convergence.

33. $\displaystyle\sum_{k=2}^{\infty} \dfrac{1}{k \ln k}$

34. $\displaystyle\sum_{k=1}^{\infty} \dfrac{\pi^k k!}{k^k}$

35. $\displaystyle\sum_{k=2}^{\infty} \frac{1}{(\ln k)^{1/k}}$ **36.** $\displaystyle\sum_{k=1}^{\infty} \frac{3k^2 - k + 1}{(1 - 2k)k}$

37. $1 - \dfrac{1}{4} + \dfrac{1}{9} - \dfrac{1}{16} + \dfrac{1}{25} - \dfrac{1}{36} + \dfrac{1}{49} - \dfrac{1}{64} + \cdots + \dfrac{(-1)^{n+1}}{n^2} + \cdots$

38. Find the convergence set for the series

$$1 - 2u + 3u^2 - 4u^3 + \cdots$$

39. Find the Maclaurin series for $f(x) = \sin 2x$.

40. Find the Taylor series for $f(x) = \dfrac{1}{x - 3}$ at $c = \frac{1}{2}$.

Supplementary Problems

Determine whether each sequence in Problems 1–16 converges or diverges. If it converges, find its limit.

1. $\left\{ \dfrac{(-2)^n}{n^2 + 1} \right\}$ **2.** $\left\{ \dfrac{(\ln n)^2}{\sqrt{n}} \right\}$

3. $\left\{ \left(1 - \dfrac{2}{n}\right)^n \right\}$ **4.** $\left\{ \dfrac{e^{0.1n}}{n^5 - 3n + 1} \right\}$

5. $\left\{ \dfrac{n + (-1)^n}{n} \right\}$ **6.** $\left\{ \dfrac{3^n}{3^n + 2^n} \right\}$

7. $\left\{ \dfrac{5n^4 - n^2 - 700}{3n^4 - 10n^2 + 1} \right\}$ **8.** $\left\{ \left(1 + \dfrac{e}{n}\right)^{2n} \right\}$

9. $\{\sqrt{n+1} - \sqrt{n}\}$ **10.** $\{\sqrt{n^4 + 2n^2} - n^2\}$

11. $\left\{ \dfrac{\ln n}{n} \right\}$ **12.** $\{1 + (-1)^n\}$

13. $\left\{ \left(1 + \dfrac{4}{n}\right)^n \right\}$ **14.** $\{5^{2/n}\}$

15. $\left\{ \dfrac{n^{3/4} \sin n^2}{n + 4} \right\}$

16. $\left\{ \displaystyle\sum_{k=1}^{n} \dfrac{n}{n^2 + k^2} \right\}$ *Hint:* This sequence is related to an integral.

17. Find the sum $\displaystyle\sum_{k=-123,456,788}^{123,456,789} \dfrac{k}{370,370,367}$

Find the sum of the given convergent geometric or telescoping series in Problems 18–27.

18. $\displaystyle\sum_{k=1}^{\infty} 4\left(\dfrac{2}{3}\right)^k$ **19.** $\displaystyle\sum_{k=1}^{\infty} \left(\dfrac{e}{3}\right)^k$

20. $\displaystyle\sum_{k=2}^{\infty} \dfrac{1}{k^2 - 1}$ **21.** $\displaystyle\sum_{k=1}^{\infty} \dfrac{1}{4k^2 - 1}$

22. $\displaystyle\sum_{k=0}^{\infty} \left[\left(\dfrac{-3}{8}\right)^k + \left(\dfrac{3}{4}\right)^{2k} \right]$ **23.** $\displaystyle\sum_{k=0}^{\infty} \dfrac{e^k + 3^{k-1}}{6^{k+1}}$

24. $\displaystyle\sum_{k=1}^{\infty} (-1)^{k+1}\left(\dfrac{1}{k} + \dfrac{1}{k+1}\right)$ **25.** $\displaystyle\sum_{k=2}^{\infty} \left[\dfrac{1}{\ln(k+1)} - \dfrac{1}{\ln k} \right]$

26. $\displaystyle\sum_{k=0}^{\infty} \left[3\left(\dfrac{2}{3}\right)^{2k} - \left(\dfrac{-1}{3}\right)^{4k} \right]$ **27.** $\displaystyle\sum_{k=1}^{\infty} \dfrac{k}{(k+1)(k+2)(k+3)}$

Test the series in Problems 28–45 for convergence.

28. $\displaystyle\sum_{k=1}^{\infty} \dfrac{5^k k!}{k^k}$ **29.** $\displaystyle\sum_{k=1}^{\infty} \dfrac{1}{\sqrt{k^2 + 4}}$

30. $\displaystyle\sum_{k=0}^{\infty} \dfrac{k!}{2^k}$ **31.** $\displaystyle\sum_{k=1}^{\infty} \dfrac{1}{2k - 1}$

32. $\displaystyle\sum_{k=0}^{\infty} ke^{-k}$ **33.** $\displaystyle\sum_{k=0}^{\infty} \dfrac{k^3}{k!}$

34. $\displaystyle\sum_{k=1}^{\infty} \dfrac{7^k}{k^2}$ **35.** $\displaystyle\sum_{k=1}^{\infty} \dfrac{k}{(2k - 1)!}$

36. $\displaystyle\sum_{k=0}^{\infty} \dfrac{k^2}{(k^3 + 1)^2}$ **37.** $\displaystyle\sum_{k=0}^{\infty} \dfrac{k^2}{3^k}$

38. $\displaystyle\sum_{k=2}^{\infty} \dfrac{1}{k(\ln k)^{1.1}}$ **39.** $\displaystyle\sum_{k=2}^{\infty} \dfrac{1}{k(\ln k)^2}$

40. $\displaystyle\sum_{k=0}^{\infty} (\sqrt{k^3 + 1} - \sqrt{k^3})$ **41.** $\displaystyle\sum_{k=1}^{\infty} \dfrac{1}{\sqrt{k(k + 1)}}$

42. $\displaystyle\sum_{k=1}^{\infty} \dfrac{k - 1}{k2^k}$ **43.** $\displaystyle\sum_{k=1}^{\infty} \dfrac{k!}{k^2(k + 1)^2}$

44. $\displaystyle\sum_{k=0}^{\infty} \dfrac{1}{1 + \sqrt{k}}$ **45.** $\displaystyle\sum_{k=1}^{\infty} \dfrac{k^2 3^k}{k!}$

Determine whether each series given in Problems 46–55 converges conditionally, converges absolutely, or diverges.

46. $\dfrac{1}{1 \cdot 2} - \dfrac{1}{3 \cdot 2} + \dfrac{1}{5 \cdot 2} - \dfrac{1}{7 \cdot 2} + \cdots$

47. $\dfrac{1}{1 \cdot 2} - \dfrac{1}{2 \cdot 3} + \dfrac{1}{3 \cdot 4} - \dfrac{1}{4 \cdot 5} + \cdots$

48. $\dfrac{3}{2} - \dfrac{4}{3} + \dfrac{5}{4} - \dfrac{6}{5} + \cdots$

49. $1 - \dfrac{1}{3} + \dfrac{1}{9} - \dfrac{1}{27} + \dfrac{1}{81} - \cdots$

50. $\dfrac{1}{5} - \dfrac{1}{7} + \dfrac{1}{9} - \dfrac{1}{11} + \cdots$

51. $-1 + \dfrac{1}{\sqrt{2}} - \dfrac{1}{\sqrt[3]{3}} + \dfrac{1}{\sqrt[4]{4}} - \cdots$

52. $\displaystyle\sum_{k=1}^{\infty} \dfrac{(-1)^k}{k[\ln(k + 1)]^2}$

53. $\displaystyle\sum_{k=1}^{\infty} (-1)^k \left(\dfrac{3k + 85}{4k + 1}\right)^k$

54. $\displaystyle\sum_{k=1}^{\infty} (-1)^k \tan^{-1}\left(\dfrac{1}{2k + 1}\right)$

55. $\displaystyle\sum_{k=1}^{\infty} \dfrac{(-1)^{k(k+1)/2}}{2^k}$

Find the interval of convergence for each power series in Problems 56–67.

56. $\displaystyle\sum_{k=1}^{\infty} \dfrac{kx^k}{3^k}$ **57.** $\displaystyle\sum_{k=1}^{\infty} k(x - 1)^k$

58. $\displaystyle\sum_{k=1}^{\infty} \dfrac{x^k}{k(k + 1)}$ **59.** $\displaystyle\sum_{k=1}^{\infty} \dfrac{\ln k(x^2)^k}{\sqrt{k}}$

60. $\displaystyle\sum_{k=1}^{\infty} \dfrac{k^2(x + 1)^{2k}}{2^k}$ **61.** $\displaystyle\sum_{k=1}^{\infty} \dfrac{(-1)^k x^k}{k^k}$

62. $\displaystyle\sum_{k=1}^{\infty} \dfrac{(-1)^k(2x - 1)^k}{k^2}$ **63.** $\displaystyle\sum_{k=1}^{\infty} \dfrac{(x + 2)^k}{k \ln(k + 1)}$

64. $\displaystyle\sum_{k=1}^{\infty} \frac{k(x-3)^k}{(k+3)!}$

65. $\displaystyle\sum_{n=1}^{\infty} \frac{(3n)!}{n!} x^n$

66. $\displaystyle\sum_{n=1}^{\infty} \frac{(-1)^n x^n}{9n^2 - 1}$

67. $x - \dfrac{x^3}{3!} + \dfrac{x^5}{5!} - \dfrac{x^7}{7!} + \cdots + \dfrac{(-1)^{k+1} x^{2k-1}}{(2k-1)!} + \cdots$

Find the Maclaurin series for each function given in Problems 68–73.

68. $f(x) = x^2 e^{-3x}$

69. $f(x) = x^3 \sin x$

70. $f(x) = 3^x$

71. $f(x) = x^2 + \tan^{-1} x$

72. $f(x) = \dfrac{5 + 7x}{1 + 2x - 3x^2}$

73. $f(x) = \dfrac{11x - 1}{2 + x - 3x^2}$

74. Compute the sum of the series

$$1 - \frac{1}{2} + \frac{1}{2!}\left(\frac{1}{2}\right)^2 - \frac{1}{3!}\left(\frac{1}{2}\right)^3 + \cdots$$

correct to three decimal places.

75. Find a bound on the error incurred by approximating the sum of the series

$$1 - \frac{1}{2!} + \frac{2}{3!} - \frac{3}{4!} + \cdots$$

by the sum of the first eight terms.

76. Use geometric series to write $12.342\overline{132}$ as a rational number.

77. A ball is dropped from a height of A feet. Each time it drops h feet, it rebounds $0.8h$ feet. If it travels a total distance of 20 feet before coming to a stop, what is A?

78. Find the first three nonzero terms of the Maclaurin series for $\tan x$ and then use term-by-term integration to find the first three nonzero terms of the Maclaurin series for $\ln|\cos x|$.

79. Spy Problem   The Spy, who is smarter than he looks, aborts his mission to the village after six days of duck plucking (recall Problem 62, Section 7.4). As he races away, he is confronted by four of Blohardt's thugs and soon finds himself in a gun battle using a revolver he managed to steal from Blabba. While dodging bullets, he recalls from his math class at Spy Academy that if the probability of a bullet being "good" is p, then on average, the number of bullets fired before a bad one jams the gun is given by the series $\displaystyle\sum_{n=0}^{\infty} np^{n-1}(1-p)$. If he takes five shots to dispose of each villain, what is the smallest value of p that will guarantee he gets off enough rounds to dispatch all four foes before a "bad" bullet gets him into another jam?

80. Convergence of the sequence $\left\{\left(1 + \dfrac{1}{n}\right)^n\right\}$

a. Suppose n is a positive integer. Use the binomial theorem to show that

$$\left(1 + \frac{1}{n}\right)^n = 1 + 1 + \frac{1}{2!}\left(1 - \frac{1}{n}\right)$$
$$+ \frac{1}{3!}\left(1 - \frac{1}{n}\right)\left(1 - \frac{2}{n}\right) + \cdots$$
$$+ \frac{1}{n!}\left(1 - \frac{1}{n}\right)\left(1 - \frac{2}{n}\right)\cdots\left(1 - \frac{n-1}{n}\right)$$

b. Use part **a** to show that $\left(1 + \dfrac{1}{n}\right)^n < 3$ for all n.

c. Use part **a** to show that $\left\{\left(1 + \dfrac{1}{n}\right)^n\right\}$ is an increasing sequence. Then use the BMCT to show that this sequence converges.

81. HISTORICAL QUEST
Ramanujan was that rarest of mathematicians, an instinctive genius with virtually no formal training. Like his Hindu predecessor, Bhaskara (see the Historical Quest in Problem 55 of Section 4.4), Ramanujan had an uncanny instinct for numerical "truth" and conceived of his results in much the same way that a sculptor "sees" a statue

SRINIVASA RAMANUJAN
1887–1920

in a raw block of stone. In his short life, he initiated new ways of thinking about number theory and made conjectures that are the subject of mathematical inquiry to this day.

A story related by his friend and mentor, the eminent British number theorist G. H. Hardy (1877–1947), serves to illustrate the resonance between Ramanujan's mind and the concept of number. Ramanujan was ill and Hardy came to visit him in the hospital. At a loss for how to begin the conversation, Hardy idly remarked that he had arrived in a taxi with the "dull" number 1729. Ramanujan immediately grew excited and exclaimed, "No, Hardy, it is a very interesting number, for it is the smallest integer that can be expressed as the sum of cubes in two different ways." ($1729 = 1^3 + 12^3 = 9^3 + 10^3$.)

In a letter from Ramanujan to G. H. Hardy at Cambridge University,* Ramanujan stated

$$1 - 5\left(\frac{1}{2}\right)^3 + 9\left(\frac{1 \cdot 3}{2 \cdot 4}\right)^3 - 13\left(\frac{1 \cdot 3 \cdot 5}{2 \cdot 4 \cdot 6}\right)^3 + \cdots = \frac{2}{\pi}$$

Hardy spent a great deal of time wondering how this sum could be equal to $2/\pi$. ∎

a. Find a_k so this sum can be expressed as $\displaystyle\sum_{k=1}^{\infty} a_k$.

b. There is no elementary way to prove this identity. Use technology to check that this formula is valid.

82. HISTORICAL QUEST   Leonhard Euler (1707–1783) was introduced in Chapter 4 Supplementary Problem 84, where we saw that Euler writes $a^\epsilon = 1 + k\epsilon$ for an "infinitely small" number ϵ. In this Historical Quest we will see why we define

$$\lim_{n \to +\infty} \left(1 + \frac{1}{n}\right)^n = e$$

by asking you to reconstruct the steps of Euler's work.

Let x be a (finite) number. Euler introduced the "infinitely large" number $N = x/\epsilon$. Explain each step.

*Quantum, March/April 1998.

$$a^x = a^{N\epsilon} = (a^\epsilon)^N$$

$$= (1 + k\epsilon)^N$$

$$= \left(1 + \frac{kx}{N}\right)^N$$

$$= 1 + N\left(\frac{kx}{N}\right) + \frac{N(N-1)}{2!}\left(\frac{kx}{N}\right)^2$$

$$+ \frac{N(N-1)(N-2)}{3!}\left(\frac{kx}{N}\right)^3 + \cdots$$

$$= 1 + kx + \frac{1}{2!}\frac{N(N-1)}{N^2}k^2x^2$$

$$+ \frac{1}{3!}\frac{N(N-1)(N-2)}{N^3}k^3x^3 + \cdots$$

Because N is infinitely large, Euler assumed that

$$1 = \frac{N-1}{N} = \frac{N-2}{N} = \cdots$$

Substitute these values into the given derivation and also substitute e for a to obtain

$$e = 1 + \frac{1}{1!} + \frac{1}{2!} + \frac{1}{3!} + \cdots$$

Euler calculated e to 26 places:

$$e \approx 2.71828182845904523536028$$

Show that Euler's derivation is equivalent to $\displaystyle\lim_{n\to+\infty}\left(1 + \frac{1}{n}\right)^n$.

83. Show that

$$\cos x = \cos c - (\sin c)(x - c) - (\cos c)\frac{(x-c)^2}{2!}$$

$$+ (\sin c)\frac{(x-c)^3}{3!} + \cdots$$

84. Show that

$$e^x = e^c + e^c(x - c) + e^c\frac{(x-c)^2}{2!} + \cdots + e^c\frac{(x-c)^k}{k!} + \cdots$$

85. Express the integral $\displaystyle\int \sqrt{x^3 + 1}\, dx$ as an infinite series by writing the first four nonzero terms. *Hint:* Use the binomial theorem and then integrate term by term.

86. Estimate the value of the integral

$$\int_0^{0.4} \frac{dx}{\sqrt{1 + x^3}}$$

with five-decimal-place accuracy.

87. Show that if $a > 0$, $a \neq 1$,

$$a^x = \sum_{k=0}^{\infty} \frac{(\ln a)^k}{k!}x^k \qquad \text{for all } x$$

88. Show that for $0 < x < 2c$,

$$\frac{1}{x} = \frac{1}{c} - \frac{1}{c^2}(x-c) + \frac{1}{c^3}(x-c)^2 - \frac{1}{c^4}(x-c)^3 + \cdots$$

89. Use series to find $\sin 0.2$ correct to five decimal places.

90. Use series to find $\ln 1.05$ correct to five decimal places.

91. Use a Maclaurin series to find a cubic polynomial that approximates e^{2x} for $|x| < 0.001$. Find an upper bound for the error in this approximation.

92. For what values of x can $\sin x$ be replaced by $x - \dfrac{x^3}{6}$ if the allowable error is 0.0005?

93. For what values of x can $\cos x$ be replaced by $1 - \dfrac{x^2}{2}$ if the allowable error is 0.00005?

94. Find the Maclaurin series for $f(x) = \dfrac{2\tan x}{1 + \tan^2 x}$.

95. Modeling Problem When we studied carbon dating in Section 5.6, it was pointed out that radiocarbon methods apply only to objects that are not too "old" (roughly 40,000 years). To date older objects, radioactive isotopes other than ^{14}C are often employed. Regardless of the radioactive substance used, it can be shown that the ratio of stable isotope to radioactive isotope at any given time may be modeled by

$$\frac{S(t) - S(0)}{R(t)} + 1 = e^{(\ln 2)t/\lambda}$$

where $R(t)$ is the number of atoms of radioactive material at time t, $S(t)$ is the number of atoms of the stable product of radioactive decay, $S(0)$ is the number of atoms of stable product initially present (at $t = 0$), and λ is the half-life of the radioactive isotope.*

a. Use the first two terms of the Maclaurin series for e^x to approximate $e^{(\ln 2)t/\lambda}$, and then solve the modeling equation for t.

b. Suppose a piece of mica is analyzed, and it is found that 5% of the atoms in the rock are radioactive rubidium-87 and 0.04% of the atoms are strontium-87. Assuming that the strontium was produced by decay of the rubidium-87 originally in the rock, what is the approximate age of the sample? Use the result in part **a**, with $\lambda = 4.86 \times 10^9$ years as the half-life of rubidium-87.

96. Putnam Examination Problem Express

$$\sum_{k=1}^{\infty} \frac{6^k}{(3^{k+1} - 2^{k+1})(3^k - 2^k)}$$

as a rational number.

97. Putnam Examination Problem Find the sum of the convergent alternating series $\displaystyle\sum_{n=0}^{\infty} \frac{(-1)^{n+1}}{3n - 2}$.

98. Putnam Examination Problem For positive real x, let

$$B_n(x) = 1^x + 2^x + 3^x + \cdots + n^x$$

Test for convergence of the series

$$\sum_{k=2}^{\infty} \frac{B_k(\log_k 2)}{(k\log_2 k)^2}$$

---

*This model is discussed in detail in the article, "How Old Is the Earth?" by Paul J. Campbell, *UMAP Modules 1992: Tools for Teaching*, Lexington, MA: Consortium for Mathematics and Its Applications, Inc., 1993, pp. 105–137.

Elastic Tightrope

This project is to be done in groups of three or four students. Each group will submit a single written report.

Suppose a cockroach starts at one end of a 100-m tightrope and runs toward the other end at a speed of 1 m/s. At the end of every minute, the tightrope stretches uniformly and instantaneously, increasing its length by 100 meters each time. Thus, after the first minute, the roach has traveled 60 meters and the length of the tightrope is 200 m. After the second minute, the roach has traveled another 60 meters and the tightrope has become 300 m long, and so on.

Does the roach ever reach the other end? If so, how long does it take?

Your paper is not limited to the following questions but should include these concerns. Suppose it is you standing on the elastic rope b meters from one end and c meters from the other end. Then suppose the entire rope stretches uniformly, increasing its length by d meters. Only the tightrope stretches; units of length and time remain constant. How far are you from each end? You might also try to generalize the roach problem.

THE INFINITE! No other question has ever moved so profoundly the spirit of man; no other idea has so fruitfully stimulated his intellect; yet no other concept stands in greater need of clarification than that of the infinite.

——David Hilbert
To Infinity and Beyond
by Eli Maor
(Birkhäuser, 1987), p. vii.

THERE IS NOTHING now which ever gives me any thought or care in algebra except divergent series, which I cannot follow the French in rejecting.

——Augustus De Morgan
Graves' Life of W.R. Hamilton
(New York: 1882–1889), p. 249

[*]This group project is courtesy of David Pengelley, New Mexico State University, Las Cruces.

CHAPTERS 6–8   Cumulative Review

1. **WHAT DOES THIS SAY?**   Describe a procedure for using integration to compute
   a. work done by a variable force
   b. fluid (or hydrostatic) force
2. **WHAT DOES THIS SAY?**   Describe integration by parts.
3. **WHAT DOES THIS SAY?**   Describe a geometric series. For what values does such a series converge? What is the sum of a convergent geometric series?
4. **WHAT DOES THIS SAY?**   Describe a procedure for determining the convergence set of the power series $\Sigma a_k x^k$.

Evaluate each of the integrals in Problems 5–10.

5. $\int \ln \sqrt{x}\,dx$

6. $\int \sin^2 x \cos^3 x\,dx$

7. $\int \dfrac{\tan^{-1} x}{1 + x^2}\,dx$

8. $\int \sec^{3/2} x \tan x\,dx$

9. $\int \sqrt{4 - x^2}\,dx$

10. $\int \dfrac{1 - x^2}{1 + x^2}\,dx$

11. Let $f(x) = \dfrac{x^2 + 17x - 8}{(2x + 3)(x - 1)^2}$.
   a. Express $f(x)$ as a partial fraction sum.
   b. Find $\int f(x)\,dx$.
   c. Either evaluate the improper integral $\int_2^\infty f(x)\,dx$ or show that it diverges.

12. In each of the following cases, either evaluate the given improper integral or show that it diverges.
   a. $\int_0^\infty x e^{-2x}\,dx$
   b. $\int_0^{\pi/2} \sec x\,dx$

In Problems 13–18, test the given series for convergence.

13. $\sum_{k=0}^\infty \dfrac{k^2 3^{k+1}}{4^k}$

14. $\sum_{k=1}^\infty \dfrac{k + 3k^{3/2}}{5k^2 - k + 3}$

15. $\sum_{k=1}^\infty \dfrac{(-1)^{k+1} k}{k + 1}$

16. $\sum_{k=1}^\infty \left(\dfrac{2k + 5}{3k - 2}\right)^k$

17. $\sum_{k=1}^\infty k e^{-k^2}$

18. $\sum_{k=1}^\infty \dfrac{k^2}{k!}$

In Problems 19 and 20, find the sum of the given convergent series.

19. $\sum_{k=1}^\infty 5 \left(\dfrac{-2}{3}\right)^{2k-1}$

20. $\sum_{k=1}^\infty \dfrac{2}{(k + 1)(k + 3)}$

In Problems 21 and 22, find the convergence set of the given power series.

21. $\sum_{k=1}^\infty \dfrac{x^k}{k^3}$

22. $\sum_{k=1}^\infty \dfrac{(-x)^k}{k + 1}$

23. A tank has the shape of a circular cone with height 10 ft and top radius 4 ft. If the tank is filled with water ($\rho = 62.4$ lb/ft³) to a depth of 6 ft, how much work is done (rounded to the nearest ft-lb) in pumping all the water in the tank to a height 2 ft above the top?

24. A triangular plate is submerged in oil ($\rho = 50$ lb/ft³) to a depth of 2 ft, as shown in Figure 8.28. Find the fluid force on the plate.

Figure 8.28   Submerged plate

25. Find the arc length of the curve $y = \frac{3}{2}x^{2/3}$ on the interval $\left[\frac{1}{8}, 8\right]$.

26. Let R be the region bounded by the x-axis, the curve $y = \dfrac{1}{\sqrt{x + 1}}$, and the lines $x = 3$ and $x = 8$. Find the volume of the solid generated when R is revolved about
   a. the x-axis   b. the y-axis   c. the line $y = -1$

27. Solve each of the given first-order linear differential equations.
   a. $y' - \dfrac{y}{x} = \dfrac{x}{1 + x^2}$
   b. $y' + (\tan x)y = \tan x$, where $y = 3$ when $x = 0$

28. A ball is dropped from a height of 20 ft. Each time it bounces, it rises to $\frac{3}{4}$ of its previous height. What is the total distance traveled by the ball?

29. Find the Maclaurin series for each of the following functions.
   a. $f(x) = \sin x^2$
   b. $f(x) = \sqrt{e^x}$
   c. $f(x) = \dfrac{3x - 2}{5 + 2x}$

30. A tank contains 15 lb of salt dissolved in 70 gallons of water. Suppose four gallons of brine containing 3 lb/gal of salt run into the tank every minute and the mixture (kept uniform by stirring) runs out of the tank at the rate of 2 gal/min.
   a. Set up and solve a differential equation for the amount of salt $S(t)$ in the tank at time t.
   b. Suppose the capacity of the tank is 100 gal. How much salt is in the tank at the time it just begins to overflow?

9

Vectors in the Plane and in Space

PREVIEW

In this chapter, we focus on various algebraic and geometric aspects of vector representations. Then in Chapter 10, we see how vectors can be combined with calculus to study motion in space and other applications.

PERSPECTIVE

Suppose a person is pulling a block of ice across the floor. The effect of this effort on the block of ice is dependent on the direction it is pulled, as well as the magnitude of the force that is applied when pulling. In this chapter, we consider the concept of a *vector*, which is defined as a quantity with both direction and magnitude. In Problem 43 in Section 9.3 we consider the amount of work done by dragging a large block of ice, which we compute using vectors and vector operations.

In general, a *scalar* quantity is one that can be described in terms of magnitude alone, whereas a *vector* quantity requires both magnitude and direction. Force, velocity, and acceleration are common vector quantities, and an important goal of this chapter is to develop methods for dealing with such quantities.

Finding the amount of work to pull a box across
the floor can be formulated as a vector calculation.

9.1   Vectors in $\mathbb{R}^2$

IN THIS SECTION   introduction to vectors, vectors in component form, standard representation of vectors in the plane

INTRODUCTION TO VECTORS

A **vector** is a quantity (such as velocity or force) that has both magnitude and direction. We sometimes represent a vector as a directed line segment, an "arrow" with **initial point** P and **terminal point** Q. The direction of the vector is that of the arrow, and its magnitude is represented by the arrow's length, as shown in Figure 9.1. We will indicate such a vector by writing **PQ** in boldface type, but in your work you may write an arrow over the designated points: $\overrightarrow{PQ}$. The order of letters you write down is important: **PQ** means that the vector is from P to Q, but **QP** means that the vector is from Q to P. The first letter is the initial point and the second letter is the terminal point.

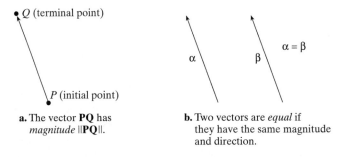

a. The vector **PQ** has *magnitude* ‖**PQ**‖.

b. Two vectors are *equal* if they have the same magnitude and direction.

Figure 9.1   Magnitude and equality of vectors

Figure 9.2 Some multiples of the vector **v**

We will denote the **magnitude** (length) of a vector **PQ** by ‖**PQ**‖. Two vectors are regarded as **equal** (or **equivalent**) if they have the same magnitude and the same direction, even if they are in different locations. Thus, the initial point of a given vector may be moved to any convenient location by translating the vector parallel to its original position.

A vector with magnitude 0 is called the **zero** (or **null**) **vector** and is denoted by **0**. The **0** vector has no specific direction, and we will adopt the convention of assigning it any direction that may be convenient in a particular problem.

Vectors can be multiplied by scalars and added to one another according to the rules originally developed in physics. A **scalar** quantity is one that has only magnitude and, in the context of vectors, is used to describe a real number. A scalar multiple $s\mathbf{v}$ of **v** is parallel to **v** with magnitude $|s|\,\|\mathbf{v}\|$ and points in the same direction as **v** if $s > 0$, and in the opposite direction if $s < 0$, as shown in Figure 9.2. If $s = 0$ or $\mathbf{v} = \mathbf{0}$, then $s\mathbf{v} = \mathbf{0}$.

Physical experiments indicate that force and velocity vectors can be added according to a **triangle rule** displayed in Figure 9.3a, and we use this rule as our definition

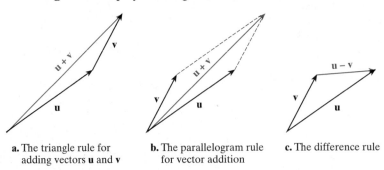

a. The triangle rule for adding vectors **u** and **v**

b. The parallelogram rule for vector addition

c. The difference rule

Figure 9.3   Vector addition and subtraction

of vector addition. In particular, to add the vector **v** to the vector **u**, we place the initial point of **v** at the tip (terminal point) of **u** and define the **sum**, also called the **resultant** **u** + **v**, to be the vector that extends from the initial point of **u** to the terminal point of **v**.

Equivalently, **u** + **v** is the diagonal of the parallelogram formed with sides **u** and **v**, as shown in Figure 9.3b. The **difference u** − **v** is just the vector **w** that satisfies **v** + **w** = **u**, and it may be found by placing the initial points of **u** and **v** together and extending a vector from the terminal point of **v** to the terminal point of **u** (see Figure 9.3c). Note that **u** + **v** = **v** + **u**. The vector **v** − **u** may be found similarly and has the same magnitude, but points in the opposite direction to **u** − **v**.

VECTORS IN COMPONENT FORM

Many issues involving vectors can be simplified by introducing a rectangular coordinate system. Suppose the vector **v** is positioned in a rectangular coordinate plane with its initial point at the origin $(0, 0)$ and its terminal point at (v_1, v_2), as shown in Figure 9.4. Then v_1 and v_2 are called the **standard components** of **v**, and we write **v** = $\langle v_1, v_2 \rangle$. The "dart" parentheses "$\langle \quad \rangle$" are used to distinguish the **component form** $\langle v_1, v_2 \rangle$ of **v** from the point (v_1, v_2) in $\mathbb{R}^2$.

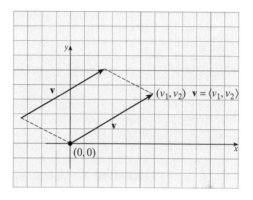

Figure 9.4 The standard component form of a vector **v**

Using the geometric relationships shown in Figure 9.5, the vector **PQ** with initial point $P(a, b)$ and terminal point $Q(c, d)$ equals the vector **OR**, where $O(0, 0)$ is the origin and R is the point with coordinates $(c - a, d - b)$.

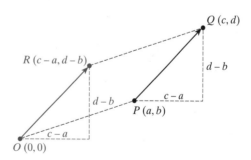

Figure 9.5 Given $P(a, b)$ and $Q(c, d)$, the vector **PQ** is $\langle c - a, d - b \rangle$.

Thus, we can make the following general statement.

Standard Components of a Vector in the Plane

If $P(a, b)$ and $Q(c, d)$ are points in a coordinate plane, then the vector **PQ** has the unique standard component form **PQ** = $\langle c - a, d - b \rangle$.

Vector properties are easily represented when vectors are given in component form. In particular, we have

$$\langle a_1, b_1 \rangle = \langle a_2, b_2 \rangle \qquad \text{if and only if } a_1 = a_2 \text{ and } b_1 = b_2$$
$$k \langle a, b \rangle = \langle ka, kb \rangle \qquad \text{for constant } k$$
$$\langle a, b \rangle + \langle c, d \rangle = \langle a + c, b + d \rangle$$
$$\langle a, b \rangle - \langle c, d \rangle = \langle a - c, b - d \rangle$$

These formulas may be verified geometrically. For instance, the rule for multiplication by a scalar can be obtained by using the relationships in Figure 9.6a, and Figure 9.6b illustrates the rule for vector addition.

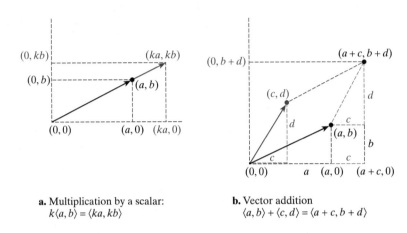

a. Multiplication by a scalar:
$k\langle a, b \rangle = \langle ka, kb \rangle$

b. Vector addition
$\langle a, b \rangle + \langle c, d \rangle = \langle a + c, b + d \rangle$

Figure 9.6 Vector operations

EXAMPLE 1 **Vector operations**

For the vectors $\mathbf{u} = \langle 2, -3 \rangle$ and $\mathbf{v} = \langle -1, 7 \rangle$, find

a. $\mathbf{u} + \mathbf{v}$ **b.** $\frac{3}{4}\mathbf{u}$ **c.** $3\mathbf{u} - \frac{1}{2}\mathbf{v}$

Solution

a. $\mathbf{u} + \mathbf{v} = \langle 2, -3 \rangle + \langle -1, 7 \rangle = \langle 2 + (-1), -3 + 7 \rangle = \langle 1, 4 \rangle$

b. $\frac{3}{4}\mathbf{u} = \frac{3}{4}\langle 2, -3 \rangle = \left\langle \frac{3}{4}(2), \frac{3}{4}(-3) \right\rangle = \left\langle \frac{3}{2}, \frac{-9}{4} \right\rangle$

c. $3\mathbf{u} - \frac{1}{2}\mathbf{v} = 3 \langle 2, -3 \rangle - \frac{1}{2} \langle -1, 7 \rangle$

$$= \langle 6, -9 \rangle + \left\langle \frac{1}{2}, -\frac{7}{2} \right\rangle \qquad \text{Scalar multiplication}$$
$$= \left\langle 6 + \frac{1}{2}, -9 - \frac{7}{2} \right\rangle \qquad \text{Vector addition}$$
$$= \left\langle \frac{13}{2}, -\frac{25}{2} \right\rangle \qquad \text{Simplify} \qquad \blacksquare$$

An expression of the form $a\mathbf{u} + b\mathbf{v}$ is called a **linear combination** of the vectors $\mathbf{u}$ and $\mathbf{v}$ and scalars a and b. Note that if $\mathbf{u} = \langle u_1, u_2 \rangle$ and $\mathbf{v} = \langle v_1, v_2 \rangle$, then

$$a\mathbf{u} + b\mathbf{v} = a\langle u_1, u_2 \rangle + b\langle v_1, v_2 \rangle = \langle au_1 + bv_1, au_2 + bv_2 \rangle$$

Vector addition and multiplication of a vector by a scalar inherit many properties of ordinary addition and multiplication. The following theorem lists several useful properties of these operations.

THEOREM 9.1 **Properties of vector operations**

For any vectors $\mathbf{u}$, $\mathbf{v}$, and $\mathbf{w}$ in the plane and scalars s and t:

Commutativity of vector addition $\mathbf{u} + \mathbf{v} = \mathbf{v} + \mathbf{u}$

Associativity of vector addition $(\mathbf{u} + \mathbf{v}) + \mathbf{w} = \mathbf{u} + (\mathbf{v} + \mathbf{w})$

Associativity of scalar multiplication	$(st)\mathbf{u} = s(t\mathbf{u})$
Identity for addition	$\mathbf{u} + \mathbf{0} = \mathbf{u}$
Inverse property for addition	$\mathbf{u} + (-\mathbf{u}) = \mathbf{0}$
Distributive laws	$(s + t)\mathbf{u} = s\mathbf{u} + t\mathbf{u}$
	$s(\mathbf{u} + \mathbf{v}) = s\mathbf{u} + s\mathbf{v}$

Proof   Each vector property can be established by using a corresponding property of real numbers. For example, to prove associativity of vector addition, let $\mathbf{u} = \langle u_1, u_2 \rangle$, $\mathbf{v} = \langle v_1, v_2 \rangle$ and $\mathbf{w} = \langle w_1, w_2 \rangle$. Then

$$
\begin{aligned}
(\mathbf{u} + \mathbf{v}) + \mathbf{w} &= (\langle u_1, u_2 \rangle + \langle v_1, v_2 \rangle) + \langle w_1, w_2 \rangle \\
&= \langle u_1 + v_1, u_2 + v_2 \rangle + \langle w_1, w_2 \rangle \\
&= \langle (u_1 + v_1) + w_1, (u_2 + v_2) + w_2 \rangle \\
&= \langle u_1 + (v_1 + w_1), u_2 + (v_2 + w_2) \rangle \\
&\qquad\qquad \textit{Associativity of addition for the real numbers} \\
&= \langle u_1, u_2 \rangle + (\langle v_1, v_2 \rangle + \langle w_1, w_2 \rangle) \\
&= \mathbf{u} + (\mathbf{v} + \mathbf{w})
\end{aligned}
$$

You are asked to prove other of these properties in the problem set.   ❏

Many results from geometry can be proved easily by using vector notation and vector properties. An illustration of this procedure is proved in Example 2.

EXAMPLE 2   Vector proof of a geometric property

Show that the line segment joining the midpoints of two sides of a triangle is parallel to the third side and has half its length.

Solution

Consider $\triangle ABC$, and let P and Q be the midpoints of sides $\overline{AC}$ and $\overline{BC}$, respectively, as shown in Figure 9.7.

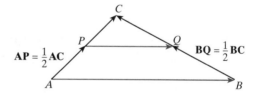

Figure 9.7   Vector proof of a geometric property

We have $\mathbf{AP} = \frac{1}{2}\mathbf{AC}$ and $\mathbf{BQ} = \frac{1}{2}\mathbf{BC}$, and we wish to show that $\mathbf{PQ}$ is parallel to $\mathbf{AB}$ with $\|\mathbf{PQ}\| = \frac{1}{2}\|\mathbf{AB}\|$. Equivalently, we will show that $\mathbf{PQ} = \frac{1}{2}\mathbf{AB}$. Toward this end, we begin by noting that $\mathbf{AB}$ can be expressed as the following vector sum:

$$
\begin{aligned}
\mathbf{AB} &= \mathbf{AP} + \mathbf{PQ} + \mathbf{QB} \\
&= \tfrac{1}{2}\mathbf{AC} + \mathbf{PQ} - \mathbf{BQ} && \textit{AP = $\frac{1}{2}$AC and QB = $-$BQ} \\
&= \tfrac{1}{2}(\mathbf{AB} + \mathbf{BC}) + \mathbf{PQ} - \tfrac{1}{2}\mathbf{BC} && \textit{AC = (AB + BC) and BQ = $\frac{1}{2}$BC} \\
&= \tfrac{1}{2}\mathbf{AB} + \tfrac{1}{2}\mathbf{BC} + \mathbf{PQ} - \tfrac{1}{2}\mathbf{BC} \\
&= \tfrac{1}{2}\mathbf{AB} + \mathbf{PQ}
\end{aligned}
$$

$$\tfrac{1}{2}\mathbf{AB} = \mathbf{PQ} \qquad\qquad \textit{Subtract $\frac{1}{2}$AB from both sides.} \qquad ■$$

When a vector $\mathbf{u}$ is represented in component form $\mathbf{u} = \langle u_1, u_2 \rangle$, its length is given by the formula

$$\|\mathbf{u}\| = \sqrt{u_1^2 + u_2^2}$$

This is a simple application of the Pythagorean theorem as shown in Figure 9.8a. Another important relationship involving the length of vectors is the *triangle inequality*

$$\|\mathbf{u} + \mathbf{v}\| \leq \|\mathbf{u}\| + \|\mathbf{v}\|$$

for any vectors **u** and **v**. Equality will occur precisely when **u** and **v** are nonnegative multiples of one another (that is, when **u** and **v** have the same direction). To establish the inequality, we observe that if **u** and **v** are positioned as two sides of a triangle in the plane, then the third side, **u** + **v**, has length $\|\mathbf{u} + \mathbf{v}\|$ and is "shorter" than the sum $\|\mathbf{u}\| + \|\mathbf{v}\|$ of the lengths of the other two sides, as shown in Figure 9.8b.

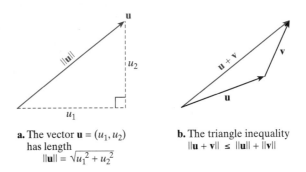

a. The vector $\mathbf{u} = (u_1, u_2)$ has length $\|\mathbf{u}\| = \sqrt{u_1^2 + u_2^2}$

b. The triangle inequality $\|\mathbf{u} + \mathbf{v}\| \leq \|\mathbf{u}\| + \|\mathbf{v}\|$

Figure 9.8   Geometric representation of two vector properties

EXAMPLE 3   Speed and heading of a motorboat

A river 4 mi wide flows south with a current of 5 mi/h. What speed and heading should a motorboat assume to travel directly across the river from east to west in 20 min?

Solution

Begin by drawing a diagram, like the one shown in Figure 9.9.

Figure 9.9   A velocity problem

Let **B** be the velocity vector of the boat in the direction of the angle θ. If the river's current has velocity **C**, the given information tells us that $\|\mathbf{C}\| = 5$ mi/h and that **C** points directly south. Moreover, because the boat is to cross the river from east to west in 20 min (that is, $\frac{1}{3}$ hr), its *effective velocity* after compensating for the current is a vector **V** that points west and has magnitude

$$\|\mathbf{v}\| = \frac{\text{WIDTH OF THE RIVER}}{\text{TIME OF CROSSING}} = \frac{4 \text{ mi}}{\frac{1}{3} \text{ h}} = 12 \text{ mi/h} \qquad \textit{Since } d = rt$$

The effective velocity **V** is the resultant of **B** and **C**; that is, **V** = **B** + **C**. Because **V** and **C** act in perpendicular directions, we can determine **B** by referring to the right triangle with sides $\|\mathbf{v}\| = 12$ and $\|\mathbf{C}\| = 5$ and hypotenuse $\|\mathbf{B}\|$. We find that

$$\|\mathbf{B}\| = \sqrt{\|\mathbf{V}\|^2 + \|\mathbf{C}\|^2} = \sqrt{12^2 + 5^2} = 13$$

The direction of the velocity vector **V** is given by the angle θ in Figure 9.9, and we find that

$$\tan \theta = \tfrac{5}{12} \quad \text{so that} \quad \theta = \tan^{-1}\left(\tfrac{5}{12}\right) \approx 0.3948$$

Thus, the boat should travel at 13 mi/h in a direction of approximately 0.3948 radian. In navigation, it is common to specify direction in degrees rather than radians (0.3948 ≈ 22.6°) and to measure the direction as an angle from either north or south. In this case, it is 90° − 22.6° = 57.4°, or N57.4°W (that is, 57.4° west of north). ∎

A **unit vector** is a vector with length 1, and a **direction vector** for a given nonzero vector **v** is a unit vector **u** that points in the same direction as **v**. Such a vector can be found by dividing **v** by its length $\|\mathbf{v}\|$; that is,

$$\mathbf{u} = \frac{\mathbf{v}}{\|\mathbf{v}\|}$$

EXAMPLE 4   Finding a direction vector

Find a direction vector for the vector $\mathbf{v} = \langle 2, -3 \rangle$.

Solution

The vector **v** has length (magnitude) $\|\mathbf{v}\| = \sqrt{2^2 + (-3)^2} = \sqrt{13}$. Thus, the required direction vector is the unit vector

$$\mathbf{u} = \frac{\mathbf{v}}{\|\mathbf{v}\|} = \frac{\langle 2, -3 \rangle}{\sqrt{13}} = \frac{1}{\sqrt{13}} \langle 2, -3 \rangle = \left\langle \frac{2}{\sqrt{13}}, \frac{-3}{\sqrt{13}} \right\rangle \quad \blacksquare$$

STANDARD REPRESENTATION OF VECTORS IN THE PLANE

The unit vectors $\mathbf{i} = \langle 1, 0 \rangle$ and $\mathbf{j} = \langle 0, 1 \rangle$ point in the directions of the positive x- and y-axes, respectively, and are called **standard basis vectors**. Any vector $\mathbf{v} = \langle v_1, v_2 \rangle$ in the plane can be expressed as a linear combination of the vectors **i** and **j**, because

$$\mathbf{v} = \langle v_1, v_2 \rangle = v_1 \langle 1, 0 \rangle + v_2 \langle 0, 1 \rangle = v_1 \mathbf{i} + v_2 \mathbf{j}$$

This is called the **standard representation** of the vector **v**, and it can be shown that the representation is unique in the sense that if $\mathbf{v} = a\mathbf{i} + b\mathbf{j}$, then $a = v_1$ and $b = v_2$. In this context, the scalars v_1 and v_2 are called the **horizontal and vertical components** of **v**, respectively. See Figure 9.10.

a. The standard basis vectors **i** and **j**

b. Any vector $\mathbf{v} = \langle v_1, v_2 \rangle$ can be expressed uniquely as $\mathbf{v} = v_1 \mathbf{i} + v_2 \mathbf{j}$

Figure 9.10   Standard representation of vectors in the plane

EXAMPLE 5   Finding the standard representation of a vector

If $\mathbf{u} = 3\mathbf{i} + 2\mathbf{j}$, $\mathbf{v} = -2\mathbf{i} + 5\mathbf{j}$, and $\mathbf{w} = \mathbf{i} - 4\mathbf{j}$, what is the standard representation of the vector $2\mathbf{u} + 5\mathbf{v} - \mathbf{w}$?

Solution

Using Theorem 9.1, we find that

$$\begin{aligned}
2\mathbf{u} + 5\mathbf{v} - \mathbf{w} &= 2(3\mathbf{i} + 2\mathbf{j}) + 5(-2\mathbf{i} + 5\mathbf{j}) - (\mathbf{i} - 4\mathbf{j}) \\
&= [2(3) + 5(-2) - 1]\mathbf{i} + [2(2) + 5(5) - (-4)]\mathbf{j} \\
&= -5\mathbf{i} + 33\mathbf{j}
\end{aligned}$$

∎

EXAMPLE 6 **Finding the standard representation of a vector connecting two points**

Find the standard representation of the vector **PQ** for the points $P(3, -4)$ and $Q(-2, 6)$.

Solution

The component form of **PQ** is

$$\mathbf{PQ} = \langle (-2) - 3, 6 - (-4) \rangle = \langle -5, 10 \rangle$$

This means that **PQ** has the standard representation $\mathbf{PQ} = -5\mathbf{i} + 10\mathbf{j}$. ■

EXAMPLE 7 **Computing a compensating force**

Two forces $\mathbf{F}_1$ and $\mathbf{F}_2$ act on the same body. It is known that $\mathbf{F}_1$ has magnitude 3 newtons and acts in the direction of $-\mathbf{i}$, whereas $\mathbf{F}_2$ has magnitude 2 newtons and acts in the direction of the unit vector

$$\mathbf{u} = \tfrac{3}{5}\mathbf{i} - \tfrac{4}{5}\mathbf{j}$$

Find the magnitude and direction of the additional force $\mathbf{F}_3$ that must be applied to keep the body at rest.

Solution

Figure 9.11 Find $\mathbf{F}_3$ so
$\mathbf{F}_1 + \mathbf{F}_2 + \mathbf{F}_3 = \mathbf{0}$

According to the given information, we have

$$\mathbf{F}_1 = 3(-\mathbf{i}) = -3\mathbf{i} \quad \text{and} \quad \mathbf{F}_2 = 2\left(\tfrac{3}{5}\mathbf{i} - \tfrac{4}{5}\mathbf{j}\right) = \tfrac{6}{5}\mathbf{i} - \tfrac{8}{5}\mathbf{j}$$

and we want to find $\mathbf{F}_3 = a\mathbf{i} + b\mathbf{j}$ so that $\mathbf{F}_1 + \mathbf{F}_2 + \mathbf{F}_3 = \mathbf{0}$ (see Figure 9.11). Substituting into this vector equation, we obtain

$$(-3\mathbf{i}) + \left(\tfrac{6}{5}\mathbf{i} - \tfrac{8}{5}\mathbf{j}\right) + (a\mathbf{i} + b\mathbf{j}) = 0\mathbf{i} + 0\mathbf{j}$$

By combining terms on the left, we find that

$$\left(-3 + \tfrac{6}{5} + a\right)\mathbf{i} + \left(-\tfrac{8}{5} + b\right)\mathbf{j} = 0\mathbf{i} + 0\mathbf{j}$$

Because the standard representation is unique, we must have

$$-3 + \tfrac{6}{5} + a = 0 \quad \text{and} \quad -\tfrac{8}{5} + b = 0$$
$$a = \tfrac{9}{5} \qquad\qquad b = \tfrac{8}{5}$$

The required force is $\mathbf{F}_3 = \tfrac{9}{5}\mathbf{i} + \tfrac{8}{5}\mathbf{j}$. This is a force of magnitude

$$\|\mathbf{F}_3\| = \sqrt{\left(\tfrac{9}{5}\right)^2 + \left(\tfrac{8}{5}\right)^2} = \tfrac{1}{5}\sqrt{145} \quad \text{newtons}$$

which acts in the direction of the unit vector

$$\mathbf{v} = \frac{\mathbf{F}_3}{\|\mathbf{F}_3\|} = \frac{5}{\sqrt{145}}\left(\tfrac{9}{5}\mathbf{i} + \tfrac{8}{5}\mathbf{j}\right) = \frac{9}{\sqrt{145}}\mathbf{i} + \frac{8}{\sqrt{145}}\mathbf{j} \quad ■$$

9.1 PROBLEM SET

Ⓐ *Sketch each vector given in Problems 1–4 assuming that its initial point is at the origin.*

1. $3\mathbf{i} - 4\mathbf{j}$

2. $-2\mathbf{i} - 3\mathbf{j}$

3. $-\tfrac{1}{2}\mathbf{i} + \tfrac{5}{2}\mathbf{j}$

4. $-2(-\mathbf{i} + 2\mathbf{j})$

*The initial point **P** and terminal point **Q** of a vector are given in Problems 5–8. Sketch each vector, write it in standard component form, and find $\|\mathbf{PQ}\|$.*

5. $P(3, -1), Q(7, 2)$

6. $P(5, -2), Q(5, 8)$

7. $P(3, 4), Q(-2, 4)$

8. $P\left(\tfrac{1}{2}, 6\right), Q(-3, -2)$

*Find the standard representation and the length of each vector **PQ** in Problems 9–12.*

9. $P(-1, -2), Q(1, -2)$

10. $P(5, 7), Q(6, 8)$

11. $P(-4, -3), Q(0, -1)$

12. $P(3, -5), Q(2, 8)$

Find a unit vector that points in the direction of each of the vectors given in Problems 13–16.

13. $\mathbf{i} + \mathbf{j}$

14. $\tfrac{1}{2}\mathbf{i} + \tfrac{1}{4}\mathbf{j}$

15. $3\mathbf{i} - 4\mathbf{j}$

16. $-4\mathbf{i} + 7\mathbf{j}$

Let $\mathbf{u} = \langle -3, 4 \rangle$ *and* $\mathbf{v} = \langle 1, -1 \rangle$. *Find scalars s and t so that the given equation in Problems 17–20 is satisfied.*

17. $s\mathbf{u} + t\mathbf{v} = \langle 6, 0 \rangle$

18. $s \langle 0, -3 \rangle + t\mathbf{u} = \mathbf{v}$

19. $s\mathbf{v} + t\langle -2, 1 \rangle = \mathbf{u}$

20. $s\mathbf{u} + \langle 8, 11 \rangle = t\mathbf{v}$

Suppose $\mathbf{u} = 3\mathbf{i} - 4\mathbf{j}$, $\mathbf{v} = 4\mathbf{i} - 3\mathbf{j}$, *and* $\mathbf{w} = \mathbf{i} + \mathbf{j}$. *Represent each of the expressions in Problems 21–24 in standard form.*

21. $2\mathbf{u} + 3\mathbf{v} - \mathbf{w}$

22. $\frac{1}{2}(\mathbf{u} + \mathbf{v}) - \frac{1}{4}\mathbf{w}$

23. $\|\mathbf{v}\|\mathbf{u} + \|\mathbf{u}\|\mathbf{v}$

24. $\|\mathbf{u}\| \|\mathbf{v}\|\mathbf{w}$

Find all real numbers x and y that satisfy the vector equations given in Problems 25–28.

25. $(x - y - 1)\mathbf{i} + (2x + 3y - 12)\mathbf{j} = \mathbf{0}$

26. $x\mathbf{i} - 4y^2\mathbf{j} = (5 - 3y)\mathbf{i} + (10 - 7x)\mathbf{j}$

27. $(x^2 + y^2)\mathbf{i} + y\mathbf{j} = 20\mathbf{i} + (x + 2)\mathbf{j}$

28. $(y - 1)\mathbf{i} + y\mathbf{j} = (\log x)\mathbf{i} + [\log 2 + \log(x + 4)]\mathbf{j}$

In Problems 29–32, find a unit vector $\mathbf{u}$ *with the given characteristics.*

29. $\mathbf{u}$ makes an angle of 30° measured from the positive x-axis.

30. $\mathbf{u}$ has the same direction as the vector $2\mathbf{i} - 3\mathbf{j}$.

31. $\mathbf{u}$ has the opposite direction of the vector $-4\mathbf{i} + \mathbf{j}$.

32. $\mathbf{u}$ has the direction of the vector from $P(-1, 5)$ to $Q(7, -3)$.

In Problems 33–34, let $\mathbf{u} = 4\mathbf{i} - \mathbf{j}$, $\mathbf{v} = \mathbf{i} + 2\mathbf{j}$, *and* $\mathbf{w} = -3\mathbf{i} + 4\mathbf{j}$.

33. Find a unit vector in the same direction as $\mathbf{u} + \mathbf{v}$.

34. Find a vector of length 3 with the same direction as $\mathbf{u} - 2\mathbf{v} + 2\mathbf{w}$.

35. Find the terminal point of the vector $5\mathbf{i} + 7\mathbf{j}$ if the initial point is $(-2, 3)$.

36. Find the initial point of the vector $-\mathbf{i} + 2\mathbf{j}$ if the terminal point is $(-1, -2)$.

37. a. Find the vector from the point P to the midpoint of the line segment joining the points $P(-3, -8)$ and $Q(9, -2)$.

  b. What is the vector from P to the point which is located $\frac{5}{6}$ of the distance from P to Q?

38. If $\|\mathbf{v}\| = 3$ and $-3 \leq r \leq 1$, what are the possible values of $r\|\mathbf{v}\|$?

39. Show that $\mathbf{v} = (\cos\theta)\mathbf{i} + (\sin\theta)\mathbf{j}$ is a unit vector for any angle θ.

B **40.** If $\mathbf{u}$ and $\mathbf{v}$ are nonzero vectors and $r = \dfrac{\|\mathbf{u}\|}{\|\mathbf{v}\|}$, what is $\|r\mathbf{v}\|$?

41. If $\mathbf{u}$ and $\mathbf{v}$ are nonzero vectors with $\|\mathbf{u}\| = \|\mathbf{v}\|$, does it follow that $\mathbf{u} = \mathbf{v}$? Explain.

42. If $\mathbf{u} = 2\mathbf{i} - 3\mathbf{j}$ and $\mathbf{v} = x\mathbf{i} + y\mathbf{j}$, describe the set of points in the plane whose coordinates (x, y) satisfy $\|\mathbf{v} - \mathbf{u}\| \leq 2$.

43. Let $\mathbf{u}_0 = x_0\mathbf{i} + y_0\mathbf{j}$ for constants x_0 and y_0, and let $\mathbf{u} = x\mathbf{i} + y\mathbf{j}$. Describe the set of all points in the plane whose coordinates satisfy

  a. $\|\mathbf{u} - \mathbf{u}_0\| = 1$ **b.** $\|\mathbf{u} - \mathbf{u}_0\| \leq r$

44. Let $\mathbf{u} = 3\mathbf{i} - \mathbf{j}$ and $\mathbf{v} = -6\mathbf{i} + 2\mathbf{j}$. Show that there are no numbers a, b for which $a\mathbf{u} + b\mathbf{v} = 2\mathbf{i} + 5\mathbf{j}$.

45. Suppose $\mathbf{u}$ and $\mathbf{v}$ are a pair of nonzero, nonparallel vectors. Find all numbers a, b, c such that $a\mathbf{u} + b(\mathbf{u} - \mathbf{v}) + c(\mathbf{u} + \mathbf{v}) = \mathbf{0}$.

46. Let $\mathbf{u} = \langle 2, 1 \rangle$ and $\mathbf{v} = \langle -3, 4 \rangle$.

  a. Sketch the vector $c\mathbf{u} + (1 - c)\mathbf{v}$ for the cases where $c = 0$, $c = \frac{1}{4}$, $c = \frac{1}{2}$, $c = \frac{3}{4}$, and $c = 1$.

  b. In general, if the initial point of $c\mathbf{u} + (1-c)\mathbf{v}$ for $0 \leq c \leq 1$ is at the origin, where is its terminal point (x, y)?

47. Two forces $\mathbf{F}_1 = 3\mathbf{i} + 4\mathbf{j}$ and $\mathbf{F}_2 = 3\mathbf{i} - 7\mathbf{j}$ act on an object. What additional force should be applied to keep the body at rest?

48. Three forces $\mathbf{F}_1 = \mathbf{i} - 2\mathbf{j}$, $\mathbf{F}_2 = 3\mathbf{i} - 7\mathbf{j}$, and $\mathbf{F}_3 = \mathbf{i} + \mathbf{j}$ act on an object. What additional force $\mathbf{F}_4$ should be applied to keep the body at rest?

49. A river 2.1 mi wide flows south with a current of 3.1 mi/h. What speed and heading should a motorboat assume to travel across the river from east to west in 30 min?

50. Four forces act on an object: $\mathbf{F}_1$ has magnitude 10 lb and acts at an angle of $\frac{\pi}{6}$ measured counterclockwise from the positive x-axis; $\mathbf{F}_2$ has magnitude 8 lb and acts in the direction of the vector $\mathbf{j}$; $\mathbf{F}_3$ has magnitude 5 lb and acts at an angle of $4\pi/3$ measured counterclockwise from the positive x-axis. What must the fourth force $\mathbf{F}_4$ be to keep the object at rest?

C **51.** Use vector methods to show that the diagonals of a parallelogram bisect each other.

52. In a triangle, let $\mathbf{u}$, $\mathbf{v}$, and $\mathbf{w}$ be the vectors from each vertex to the midpoint of the opposite side. Use vector methods to show that $\mathbf{u} + \mathbf{v} + \mathbf{w} = \mathbf{0}$.

53. Prove that the medians of a triangle (see Figure 9.12) intersect at a single point by completing the following argument.

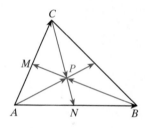

Figure 9.12 $\triangle ABC$

  a. Let M and N be the midpoints of sides $\overline{AC}$ and $\overline{AB}$, respectively. Show that

$$\mathbf{CN} = \tfrac{1}{2}\mathbf{AB} - \mathbf{AC} \quad \text{and} \quad \mathbf{BM} = \tfrac{1}{2}\mathbf{AC} - \mathbf{AB}$$

  b. Let P be the point where medians $\overline{BM}$ and $\overline{CN}$ intersect, and let r, s be constants such that

$$\mathbf{CP} = r(\mathbf{CN}) \quad \text{and} \quad \mathbf{BP} = s(\mathbf{BM})$$

  Note that $\mathbf{CP} + \mathbf{PB} = \mathbf{CB}$. Use this relationship to prove that $r = s = \frac{2}{3}$. Explain why this shows that any pair of medians meet at a point located $\frac{2}{3}$ the distance from each vertex to the midpoint of the opposite side. Why does this show that *all three* medians meet at a single point?

  c. The *centroid* of a triangle is the point where the medians meet. Show that a triangle with vertices $A(x_1, y_1)$, $B(x_2, y_2)$, $C(x_3, y_3)$ has centroid with coordinates

$$\left(\frac{x_1 + x_2 + x_3}{3}, \frac{y_1 + y_2 + y_3}{3} \right)$$

54. Prove the following parts of Theorem 9.1.

  a. Commutativity

  b. Identity

  c. Inverse property of addition

  d. The distributive law $s(\mathbf{u} + \mathbf{v}) = s\mathbf{u} + s\mathbf{v}$

55. Let A, B, C, and D be any four points in the plane. If M and N are midpoints of $\overline{AC}$ and $\overline{BD}$, show that

$$\mathbf{MN} = \tfrac{1}{4}(\mathbf{AB} + \mathbf{AD} + \mathbf{CB} + \mathbf{CD})$$

56. Let A, B, C, and D be vertices of a quadrilateral, and let M, N, P, and Q, be the midpoints of the sides $\overline{AB}$, $\overline{BC}$, $\overline{CD}$, and $\overline{AD}$, respectively, as shown in Figure 9.13. Use vector methods to show that $MNPQ$ is a parallelogram.

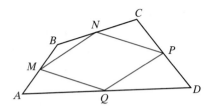

Figure 9.13   Quadrilateral $ABCD$

9.2   Coordinates and Vectors in $\mathbb{R}^3$

IN THIS SECTION   coordinates in $\mathbb{R}^3$; graphs in $\mathbb{R}^3$: spheres and cylinders; vectors in $\mathbb{R}^3$

COORDINATES IN $\mathbb{R}^3$

Figure 9.14   A "right-handed" rectangular coordinate system for $\mathbb{R}^3$

We have already considered ordered pairs in $\mathbb{R}^2$, and because we exist in a *three-dimensional* world, it is also important to consider a three-dimensional system. We call this *three-space* and denote it by $\mathbb{R}^3$. We introduce a coordinate system to three-space by choosing three mutually perpendicular axes to serve as a frame of reference. The orientation of our reference system will be *right-handed* in the sense that if you stand at the origin with your right arm along the positive x-axis and your left arm along the positive y-axis, as shown in Figure 9.14, your head will then point in the direction of the positive z-axis.

To orient yourself to a three-dimensional coordinate system, think of the x-axis and y-axis as lying in the plane of the floor and the z-axis as a line perpendicular to the floor. All the graphs that we have drawn in the first eight chapters of this book would now be drawn on the floor. If you orient yourself in a room as shown in Figure 9.15, you may notice some important planes. Assume that the room is 25 ft × 30 ft × 8 ft and fix the origin at a front corner (where the board hangs).

Figure 9.15   A typical classroom; assume the dimensions are 25 ft by 30 ft with an 8-foot ceiling

Floor:	**xy-plane**; equation is $z = 0$.
Ceiling:	plane parallel to the xy-plane; equation is $z = 8$.
Front wall:	**yz-plane**; equation is $x = 0$.
Back wall:	plane parallel to the yz-plane; equation is $x = 30$.
Left wall:	**xz-plane**; equation is $y = 0$.
Right wall:	plane parallel to the xz-plane; equation is $y = 25$.

The xy-, xz-, and yz-planes are called the **coordinate planes**. We will obtain equations for planes in space after we discuss vectors in $\mathbb{R}^3$. However, in beginning to visualize objects in $\mathbb{R}^3$, we do not want to ignore planes, because they are so common (for example, the walls, ceiling, and floor in Figure 9.15). We will show that the graph of $Ax + By + Cz = D$ is a plane if A, B, C, and D are real numbers (not all zero). Points in $\mathbb{R}^3$ are located by their position in relation to the three coordinate planes and are given appropriate coordinates. Specifically, the point P is assigned coordinates (a, b, c) to indicate that it is a, b, and c units, respectively, from the yz-, xz-, and xy-planes.

EXAMPLE 1 Points in three dimensions

Graph the following ordered triples:

a. $(10, 20, 10)$ **b.** $(-12, 6, 12)$ **c.** $(-12, -18, 6)$ **d.** $(20, -10, 18)$

Solution

The points are plotted as shown in Figure 9.16.

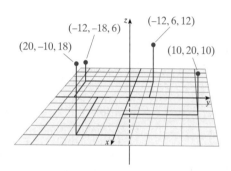

Figure 9.16 Plotting points in $\mathbb{R}^3$

In $\mathbb{R}^2$, the distance from the origin $(0, 0)$ to the point (a, b) is $d = \sqrt{a^2 + b^2}$, and in $\mathbb{R}^3$, the distance from $(0, 0, 0)$ to (a, b, c) is $d = \sqrt{a^2 + b^2 + c^2}$, as shown in Figure 9.17. We can now summarize.

> **Distance Formula**
> The distance $|P_1 P_2|$ between $P_1(x_1, y_1, z_1)$ and $P_2(x_2, y_2, z_2)$ is
> $$|P_1 P_2| = \sqrt{(x_2 - x_1)^2 + (y_2 - y_1)^2 + (z_2 - z_1)^2}$$

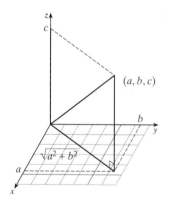

Figure 9.17 Distance from the origin $(0, 0, 0)$ to the point (a, b, c) is
$$d = \sqrt{\left(\sqrt{a^2 + b^2}\right)^2 + c^2}$$
$$= \sqrt{a^2 + b^2 + c^2}$$

EXAMPLE 2 Finding the distance between points in $\mathbb{R}^3$

Find the distance between $(10, 20, 10)$ and $(-12, 6, 12)$.

Solution
$$d = \sqrt{(-12 - 10)^2 + (6 - 20)^2 + (12 - 10)^2}$$
$$= \sqrt{684}$$
$$= 6\sqrt{19}$$

GRAPHS IN $\mathbb{R}^3$: SPHERES AND CYLINDERS

In $\mathbb{R}^3$, the collection of all points (x, y, z) whose coordinates satisfy a given equation is called the **graph of an equation**. Geometrically, the graph of an equation in x, y, z is a **surface** in $\mathbb{R}^3$. In this section, we will examine two kinds of surfaces, spheres and cylinders. Later in the chapter, we will examine a variety of other surfaces.

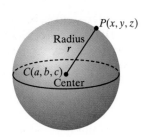

Figure 9.18 Graph of a sphere with center (a, b, c), radius r

Equation of a Sphere

Spheres A **sphere** (see Figure 9.18) is defined as the collection of all points located a fixed distance (the **radius**) from a fixed point (the **center**).

In particular, if $P(x, y, z)$ is a point on the sphere with radius r and center $C(a, b, c)$, then the distance from C to P is r. Thus,

$$r = \sqrt{(x - a)^2 + (y - b)^2 + (z - c)^2}$$

If you square both sides of this equation, you can see that it is equivalent to the equation of a sphere displayed in the following box. Conversely, if the point (x, y, z) satisfies an equation of this form, it must lie on a sphere with center (a, b, c) and radius r.

The graph of the equation

$$(x - a)^2 + (y - b)^2 + (z - c)^2 = r^2$$

is a sphere with center (a, b, c) and radius r, and any sphere has an equation of this form. This is called the **standard form of the equation of a sphere** (or simply *standard-form sphere*).

EXAMPLE 3 Center and radius of a sphere from a given equation

Show that the graph of the equation $x^2 + y^2 + z^2 + 4x - 6y - 3 = 0$ is a sphere, and find its center and radius.

Solution

By completing the square in both variables x and y, we have

$$(x^2 + 4x) + (y^2 - 6y) + z^2 = 3$$
$$(x^2 + 4x + 2^2) + [y^2 - 6y + (-3)^2] + z^2 = 3 + 4 + 9$$
$$(x + 2)^2 + (y - 3)^2 + z^2 = 16$$

Figure 9.19 Graph of a sphere

Comparing this equation with the standard form, we see that it is the equation of a sphere with center $(-2, 3, 0)$ and radius 4, as shown in Figure 9.19. ∎

Cylinders A **cross section** of a surface in $\mathbb{R}^3$ is a curve obtained by intersecting the surface with a plane. If parallel planes intersect a given surface in congruent cross-sectional curves, the surface is called a **cylinder**. We define a cylinder with *principal cross sections C and generating line L* to be the surface obtained by moving lines parallel to L along the boundary of the curve C, as shown in Figure 9.20. In this context, the curve C is called a **directrix** of the cylinder, and L is the **generatrix**.

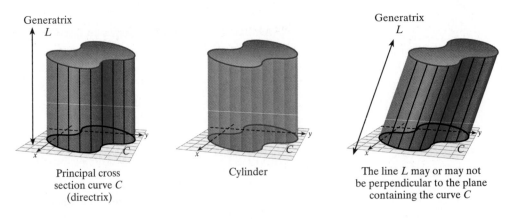

Principal cross section curve C (directrix)

Cylinder

The line L may or may not be perpendicular to the plane containing the curve C

Figure 9.20 A cylinder with directrix C and generatrix L

We will deal primarily with cylinders in which the directrix is a conic section and the generatrix L is one of the coordinate axes. Such a cylinder is often named for the type of conic section in its principal cross sections and is described by an equation involving only two of the variables x, y, z. In this case, the generating line L is parallel to the coordinate axis of the missing variable. Thus,

$x^2 + y^2 = 5$ is a **circular cylinder** with L parallel to the z-axis

$y^2 - z^2 = 9$ is a **hyperbolic cylinder** with L parallel to the x-axis

$x^2 + 2z^2 = 25$ is an **elliptic cylinder** with L parallel to the y-axis.

The graphs of these cylinders are shown in Figure 9.21.

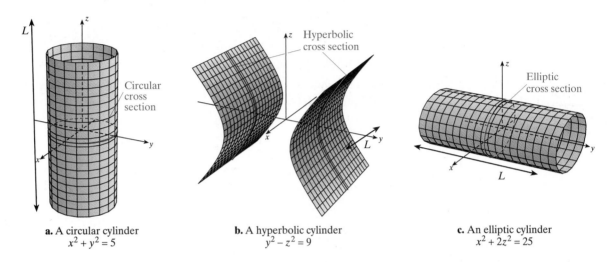

a. A circular cylinder
$x^2 + y^2 = 5$

b. A hyperbolic cylinder
$y^2 - z^2 = 9$

c. An elliptic cylinder
$x^2 + 2z^2 = 25$

Figure 9.21   Cylinders

VECTORS IN $\mathbb{R}^3$

A vector in $\mathbb{R}^3$ is a directed line segment (an "arrow") in space. The vector $\mathbf{P_1P_2}$ with initial point $P_1(x_1, y_1, z_1)$ and terminal point $P_2(x_2, y_2, z_2)$ has the component form analogous to the $\mathbb{R}^2$ case

$$\mathbf{P_1P_2} = \langle x_2 - x_1, y_2 - y_1, z_2 - z_1 \rangle$$

Vector addition and multiplication of a vector by a scalar are defined for vectors in $\mathbb{R}^3$ in the same way as these operations were defined for vectors in $\mathbb{R}^2$. In addition, the properties of vector algebra listed in Theorem 9.1 of Section 9.1 apply to vectors in $\mathbb{R}^3$ as well as to those in $\mathbb{R}^2$.

For example, we observed that each vector in $\mathbb{R}^2$ can be expressed as a unique linear combination of the standard basis vectors $\mathbf{i}$ and $\mathbf{j}$. This representation can be extended to vectors in $\mathbb{R}^3$ by adding a vector $\mathbf{k}$ defined to be the unit vector in the direction of the positive z-axis. In component form, we have in $\mathbb{R}^3$,

$$\mathbf{i} = \langle 1, 0, 0 \rangle \qquad \mathbf{j} = \langle 0, 1, 0 \rangle \qquad \mathbf{k} = \langle 0, 0, 1 \rangle$$

We call these the **standard basis vectors** in $\mathbb{R}^3$ (shown in Figure 9.22a). The **standard representation** of the vector with initial point at the origin O and terminal point $Q(a_1, a_2, a_3)$ is $\mathbf{OQ} = a_1\mathbf{i} + a_2\mathbf{j} + a_3\mathbf{k}$ as shown in Figure 9.22b. More generally, the vector $\mathbf{P_1P_2}$ with initial point $P_1(x_1, y_1, z_1)$ and terminal point $P_2(x_2, y_2, z_2)$ has the standard representation

$$\mathbf{P_1P_2} = (x_2 - x_1)\mathbf{i} + (y_2 - y_1)\mathbf{j} + (z_2 - z_1)\mathbf{k}$$

with magnitude

$$\|\mathbf{P_1P_2}\| = \sqrt{(x_2 - x_1)^2 + (y_2 - y_1)^2 + (z_2 - z_1)^2}$$

a. The standard basis vectors in $\mathbb{R}^3$

b. The vector from O to $Q(a_1, a_2, a_3)$ is $\mathbf{OQ} = a_1\mathbf{i} + a_2\mathbf{j} + a_3\mathbf{k}$.

Figure 9.22   Standard representation of vectors in $\mathbb{R}^3$

EXAMPLE 4   Standard representation of a vector in $\mathbb{R}^3$

Find the standard representation of the vector **PQ** with initial point $P(-1, 2, 2)$ and terminal point $Q(3, -2, 4)$.

Solution

We have

$$\mathbf{PQ} = [3 - (-1)]\mathbf{i} + [-2 - 2]\mathbf{j} + [4 - 2]k = 4\mathbf{i} - 4\mathbf{j} + 2\mathbf{k} \qquad \blacksquare$$

EXAMPLE 5   Magnitude of a vector in $\mathbb{R}^3$

Find the magnitude of the vector $\mathbf{v} = 2\mathbf{i} - 3\mathbf{j} + 5\mathbf{k}$ and the distance between the points $A(1, -1, -4)$ and $B(-2, 3, 8)$.

Solution $\|\mathbf{v}\| = \sqrt{2^2 + (-3)^2 + 5^2} = \sqrt{38}$ and

$$\|\overline{AB}\| = \sqrt{(-2 - 1)^2 + [3 - (-1)]^2 + [8 - (-4)]^2} = 13 \qquad \blacksquare$$

As in $\mathbb{R}^2$, if **v** is a given nonzero vector in $\mathbb{R}^3$, then the unit vector **u** that points in the same direction as **v** is

$$\mathbf{u} = \frac{\mathbf{v}}{\|\mathbf{v}\|}$$

EXAMPLE 6   Finding a direction vector

Find the unit vector that points in the direction of the vector **PQ** from $P(-1, 2, 5)$ to $Q(0, -3, 7)$.

Solution $\mathbf{PQ} = [0 - (-1)]\mathbf{i} + [-3 - 2]\mathbf{j} + [7 - 5]\mathbf{k} = \mathbf{i} - 5\mathbf{j} + 2\mathbf{k}$

$\|\mathbf{PQ}\| = \sqrt{1^2 + (-5)^2 + 2^2} = \sqrt{30}$

Thus,

$$\mathbf{u} = \frac{\mathbf{PQ}}{\|\mathbf{PQ}\|} = \frac{\mathbf{i} - 5\mathbf{j} + 2\mathbf{k}}{\sqrt{30}} = \frac{1}{\sqrt{30}}\mathbf{i} - \frac{5}{\sqrt{30}}\mathbf{j} + \frac{2}{\sqrt{30}}\mathbf{k} \qquad \blacksquare$$

As in $\mathbb{R}^2$, two vectors in $\mathbb{R}^3$ are **parallel** if they are scalar multiples of one another. Similarly, nonzero vectors **u** and **v** are parallel if and only if $\mathbf{u} = s\mathbf{v}$ for some nonzero scalar s.

EXAMPLE 7   Parallel vectors

A vector **PQ** has initial point $P(1, 0, -3)$ and length 3. Find **Q** so that **PQ** is parallel to $\mathbf{v} = 2\mathbf{i} - 3\mathbf{j} + 6\mathbf{k}$.

Solution

Let **Q** have coordinates (a_1, a_2, a_3). Then

$$\mathbf{PQ} = [a_1 - 1]\mathbf{i} + [a_2 - 0]\mathbf{j} + [a_3 - (-3)]\mathbf{k}$$
$$= (a_1 - 1)\mathbf{i} + a_2\mathbf{j} + (a_3 + 3)\mathbf{k}$$

Because **PQ** is parallel to **v**, we have $\mathbf{PQ} = s\mathbf{v}$ for some scalar s, that is,

$$(a_1 - 1)\mathbf{i} + a_2\mathbf{j} + (a_3 + 3)\mathbf{k} = s(2\mathbf{i} - 3\mathbf{j} + 6\mathbf{k})$$

Since the standard representation is unique, this implies that

$$\begin{array}{ccc} a_1 - 1 = 2s & a_2 = -3s & a_3 + 3 = 6s \\ a_1 = 2s + 1 & & a_3 = 6s - 3 \end{array}$$

Because **PQ** has length 3, we have

$$3 = \sqrt{(a_1 - 1)^2 + a_2^2 + (a_3 + 3)^2}$$
$$= \sqrt{[(2s + 1) - 1]^2 + (-3s)^2 + [(6s - 3) + 3]^2}$$
$$= \sqrt{4s^2 + 9s^2 + 36s^2} = \sqrt{49s^2} = 7\,|s|$$

Thus, $s = \pm\frac{3}{7}$ and for the two cases, we have

$$\begin{array}{llll} s = \frac{3}{7}: & a_1 = 2(\frac{3}{7}) + 1 = \frac{13}{7} & a_2 = -3(\frac{3}{7}) = -\frac{9}{7} & a_3 = 6(\frac{3}{7}) - 3 = -\frac{3}{7} \\ s = -\frac{3}{7}: & a_1 = 2(-\frac{3}{7}) + 1 & a_2 = -3(-\frac{3}{7}) & a_3 = 6(-\frac{3}{7}) - 3 \\ & a_1 = \frac{1}{7} & a_2 = \frac{9}{7} & a_3 = -\frac{39}{7} \end{array}$$

There are two points that satisfy the conditions for the required terminal point Q: $\left(\frac{13}{7}, -\frac{9}{7}, -\frac{3}{7}\right)$ and $\left(\frac{1}{7}, \frac{9}{7}, -\frac{39}{7}\right)$. ∎

9.2   PROBLEM SET

Ⓐ *For the given vectors in Problems 1–4, find*

  a. $\mathbf{u} + \mathbf{v}$ *b.* $\mathbf{u} - \mathbf{v}$ *c.* $\frac{5}{2}\mathbf{u}$ *d.* $2\mathbf{u} + 3\mathbf{v}$

  1. $\mathbf{u} = \langle 4, -3, 1\rangle$, $\mathbf{v} = \langle -2, 5, 3\rangle$
  2. $\mathbf{u} = \langle 2, -1, 0\rangle$, $\mathbf{v} = \langle 5, -3, 4\rangle$
  3. $\mathbf{u} = \langle 1, -2, 5\rangle$, $\mathbf{v} = \langle 0, -1, 3\rangle$
  4. $\mathbf{u} = \langle 0, 3, -1\rangle$, $\mathbf{v} = \langle 4, -2, 0\rangle$

In Problems 5–8, plot the points P and Q in $\mathbb{R}^3$, and find $\|\mathbf{PQ}\|$.

  5. $P(3, -4, 5)$, $Q(1, 5, -3)$ **6.** $P(3, 0, 0)$, $Q(-2, 5, 7)$
  7. $P(-3, -5, 8)$, $Q(3, 6, -7)$ **8.** $P(0, 5, -3)$, $Q(2, -1, 0)$

In Problems 9–12, find the standard form equation of the sphere with the given center C and radius r.

  9. $C(0, 0, 0)$, $r = 1$ **10.** $C(-3, 5, 7)$, $r = 2$
  11. $C(0, 4, -5)$, $r = 3$ **12.** $C(-2, 3, -1)$, $r = \sqrt{5}$

Find the center and radius of each sphere whose equations are given in Problems 13–16.

  13. $x^2 + y^2 + z^2 - 2y + 2z - 2 = 0$
  14. $x^2 + y^2 + z^2 + 4x - 2z - 8 = 0$
  15. $x^2 + y^2 + z^2 - 6x + 2y - 2z + 10 = 0$

  16. $x^2 + y^2 + z^2 - 2x - 4y + 8z + 17 = 0$

*Find the standard representation of the vector **PQ**, and then find $\|\mathbf{PQ}\|$ in Problems 17–20.*

  17. $P(1, -1, 3)$, $Q(-1, 1, 4)$ **18.** $P(0, 2, 3)$, $Q(2, 3, 0)$
  19. $P(1, 1, 1)$, $Q(-3, -3, -3)$ **20.** $P(3, 0, -4)$, $Q(0, -4, 3)$

In Problems 21–24, perform the indicated operations.

$$\mathbf{u} = 2\mathbf{i} - \mathbf{j} + 3\mathbf{k} \qquad \mathbf{v} = \mathbf{i} + \mathbf{j} - 5\mathbf{k} \qquad \mathbf{w} = 5\mathbf{i} + 7\mathbf{k}$$

  21. $\mathbf{u} + \mathbf{v} - 2\mathbf{w}$ **22.** $2\mathbf{u} - \mathbf{v} + 3\mathbf{w}$
  23. $4\mathbf{u} + \mathbf{w}$ **24.** $3\mathbf{u} - \mathbf{v} - 5\mathbf{w}$

*In Problems 25–28, find a unit vector that points in the same direction as the given vector **v**.*

  25. $\mathbf{v} = \langle 3, -2, 1\rangle$ **26.** $\mathbf{v} = \langle -1, 1, \sqrt{2}\rangle$
  27. $\mathbf{v} = \langle -5, 3, 4\rangle$ **28.** $\mathbf{v} = \langle 1, \sqrt{2}, 7\rangle$

Sketch the cylindrical surfaces given in Problems 29–32.

  29. $x^2 + y^2 = 4$ **30.** $z = 9 - x^2$
  31. $y = 3z^2$ **32.** $z = \sin x$

33. Find an equation for a sphere, given that the endpoints of a diameter of the sphere are $(1, 2, -3)$ and $(-2, 3, 3)$.

Evaluate the expressions given in Problems 34–37.

34. $\|\mathbf{i} + \mathbf{j} + \mathbf{k}\|$

35. $\|\mathbf{i} - \mathbf{j} + \mathbf{k}\|$

36. $\|2\mathbf{i} + \mathbf{j} - 3\mathbf{k}\|^2$

37. $\|2(\mathbf{i} - \mathbf{j} + \mathbf{k}) - 3(2\mathbf{i} + \mathbf{j} - \mathbf{k})\|^2$

Let $\mathbf{v} = \mathbf{i} - 2\mathbf{j} + 2\mathbf{k}$ and $\mathbf{w} = 2\mathbf{i} + 4\mathbf{j} - \mathbf{k}$; find the vector or scalar requested in Problems 38–41.

38. $2\|\mathbf{v}\| - 3\|\mathbf{w}\|$

39. $\|\mathbf{v}\|\mathbf{w}$

40. $\|2\mathbf{v} - 3\mathbf{w}\|$

41. $\|\mathbf{v} - \mathbf{w}\|(\mathbf{v} + \mathbf{w})$

Determine whether each vector in Problems 42–45 is parallel to $\mathbf{u} = 2\mathbf{i} - 3\mathbf{j} + 5\mathbf{k}$.

42. $\mathbf{v} = \langle 4, -6, 10 \rangle$

43. $\mathbf{v} = \langle -2, 6, -10 \rangle$

44. $\mathbf{v} = \langle 1, -\frac{3}{2}, 2 \rangle$

45. $\mathbf{v} = \langle -1, \frac{3}{2}, -\frac{5}{2} \rangle$

Ⓑ *The vertices A, B, and C of a triangle in $\mathbb{R}^3$ are given in Problems 46–49. Find the lengths of the sides of the triangle and determine whether it is a right triangle, an isosceles triangle, both, or neither.*

46. $A(3, -1, 0)$, $B(7, 1, 4)$, $C(1, 3, 4)$

47. $A(1, 1, 1)$, $B(3, 3, 2)$, $C(3, -3, 5)$

48. $A(1, 2, 3)$, $B(-3, 2, 4)$, $C(1, -4, 3)$

49. $A(2, 4, 3)$, $B(-3, 2, -4)$, $C(-6, 8, -10)$

50. Let $\mathbf{u} = \langle -1, 1, 2 \rangle$, $\mathbf{v} = \langle 0, 2, -3 \rangle$, and $\mathbf{w} = \langle 5, -1, 0 \rangle$. Find a vector $\mathbf{q}$ so that $2\mathbf{u} - \mathbf{v} + 5\mathbf{q} = 3\mathbf{w}$.

*In Problems 51 and 52, determine whether the given points are collinear (that is, lie on the same line). Note: For three points A, B, and C to be collinear, **AC** must be a multiple of **AB**.*

51. $(2, 3, 2)$, $(-1, 4, 0)$, $(-4, 5, -2)$

52. $(3, 0, 3)$, $(2, -3, 5)$, $(4, 3, 1)$

53. A 150-lb tennis official is sitting on a three-legged judge's stool. Suppose a coordinate system is chosen with the official at point $P(0, 0, 6)$ and the feet of the stool's legs at $A(0, -2, 0)$, $B(-\sqrt{3}, 1, 0)$, and $C(\sqrt{3}, 1, 0)$, as shown in Figure 9.23. Find the force exerted on each of the three legs.

Figure 9.23   Tennis judge

Ⓒ 54. Let $P(3, 2, -1)$, $Q(-2, 1, c)$, and $R(c, 1, 0)$ be points in $\mathbb{R}^3$. For what values of c (if any) is PQR a right triangle?

55. Find the point P that lies $\frac{2}{3}$ of the distance from the point $A(-1, 3, 9)$ to the midpoint of the line segment joining points $B(-2, 3, 7)$ and $C(4, 1, -3)$.

9.3   The Dot Product

IN THIS SECTION   definition and basic properties of the dot product, angle between vectors, direction cosines, projections, work as a dot product

DEFINITION AND BASIC PROPERTIES OF THE DOT PRODUCT

In the first two sections of this chapter, you have learned how to add vectors and multiply a vector by a scalar. We now turn our attention to the product of two vectors. The **dot (scalar) product** and the **cross (vector) product** are two important vector operations. We will examine the dot product in this section and the cross product in the next section. The dot product is also known as a *scalar product* because it is a product of vectors that gives a scalar (that is, real number) as a result. Sometimes the dot product is called the *inner product*.

Dot Product

> The **dot product** of vectors $\mathbf{v} = a_1\mathbf{i} + a_2\mathbf{j} + a_3\mathbf{k}$ and $\mathbf{w} = b_1\mathbf{i} + b_2\mathbf{j} + b_3\mathbf{k}$ is the scalar denoted by $\mathbf{v} \cdot \mathbf{w}$ and defined by
>
> $$\mathbf{v} \cdot \mathbf{w} = a_1 b_1 + a_2 b_2 + a_3 b_3$$

The dot product of two vectors $\mathbf{v} = a_1\mathbf{i} + a_2\mathbf{j}$ and $\mathbf{w} = b_1\mathbf{i} + b_2\mathbf{j}$ in a plane is given by a similar formula with $a_3 = b_3 = 0$, namely,

$$\mathbf{v} \cdot \mathbf{w} = a_1 b_1 + a_2 b_2$$

EXAMPLE 1   Dot product

Find the dot product of $\mathbf{v} = -3\mathbf{i} + 2\mathbf{j} + \mathbf{k}$ and $\mathbf{w} = 4\mathbf{i} - \mathbf{j} + 2\mathbf{k}$.

Solution $\mathbf{v} \cdot \mathbf{w} = -3(4) + 2(-1) + 1(2) = -12$ ∎

EXAMPLE 2   Dot product of vectors in component form

If $\mathbf{v} = \langle 4, -1, 3 \rangle$ and $\mathbf{w} = \langle -1, -2, 5 \rangle$, find the dot product, $\mathbf{v} \cdot \mathbf{w}$.

Solution $\mathbf{v} \cdot \mathbf{w} = 4(-1) + (-1)(-2) + 3(5) = 13$ ∎

Before we can apply the dot product to geometric and physical problems, we need to know how it behaves algebraically. A number of important general properties of the dot product are listed in the following theorem.

THEOREM 9.2   Properties of the dot product

If $\mathbf{u}$, $\mathbf{v}$, and $\mathbf{w}$ are vectors in $\mathbb{R}^2$ or $\mathbb{R}^3$ and c is a scalar, then

Magnitude of a vector	$\mathbf{v} \cdot \mathbf{v} = \|\mathbf{v}\|^2$
Zero product	$\mathbf{0} \cdot \mathbf{v} = 0$
Commutativity	$\mathbf{v} \cdot \mathbf{w} = \mathbf{w} \cdot \mathbf{v}$
Multiple of a dot product	$c(\mathbf{v} \cdot \mathbf{w}) = (c\mathbf{v}) \cdot \mathbf{w} = \mathbf{v} \cdot (c\mathbf{w})$
Distributivity	$\mathbf{u} \cdot (\mathbf{v} + \mathbf{w}) = \mathbf{u} \cdot \mathbf{v} + \mathbf{u} \cdot \mathbf{w}$

Proof   Let $\mathbf{u} = a_1\mathbf{i} + a_2\mathbf{j} + a_3\mathbf{k}$, $\mathbf{v} = b_1\mathbf{i} + b_2\mathbf{j} + b_3\mathbf{k}$, and $\mathbf{w} = c_1\mathbf{i} + c_2\mathbf{j} + c_3\mathbf{k}$.

Magnitude of a vector

$$\|\mathbf{v}\|^2 = \left(\sqrt{a_1^2 + a_2^2 + a_3^2} \right)^2 = a_1^2 + a_2^2 + a_3^2 = \mathbf{v} \cdot \mathbf{v}$$

Zero product, commutativity, and **multiple of a dot product** can be established in a similar fashion.

Distributivity

$$\mathbf{u} \cdot (\mathbf{v} + \mathbf{w})$$
$$= (a_1\mathbf{i} + a_2\mathbf{j} + a_3\mathbf{k}) \cdot [(b_1 + c_1)\mathbf{i} + (b_2 + c_2)\mathbf{j} + (b_3 + c_3)\mathbf{k}]$$
$$= a_1(b_1 + c_1) + a_2(b_2 + c_2) + a_3(b_3 + c_3)$$
$$= a_1b_1 + a_1c_1 + a_2b_2 + a_2c_2 + a_3b_3 + a_3c_3$$

and

$$\mathbf{u} \cdot \mathbf{v} + \mathbf{u} \cdot \mathbf{w} = (a_1b_1 + a_2b_2 + a_3b_3) + (a_1c_1 + a_2c_2 + a_3c_3)$$
$$= a_1b_1 + a_1c_1 + a_2b_2 + a_2c_2 + a_3b_3 + a_3c_3$$

Thus, $\mathbf{u} \cdot (\mathbf{v} + \mathbf{w}) = \mathbf{u} \cdot \mathbf{v} + \mathbf{u} \cdot \mathbf{w}$.   ❑

ANGLE BETWEEN VECTORS

The angle between two nonzero vectors $\mathbf{v}$ and $\mathbf{w}$ in $\mathbb{R}^2$ or $\mathbb{R}^3$ is defined to be the angle θ with $0 \leq \theta \leq \pi$ that is formed when the vectors are in standard position (initial points at the origin), as shown in Figure 9.24.

The angle between two vectors plays an important role in certain applications and may be computed by using the following formula involving the dot product.

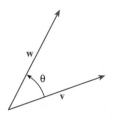

Figure 9.24 The angle between two vectors

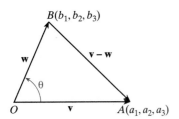

Figure 9.25 Finding an angle between two vectors

THEOREM 9.3   Angle between two vectors

If θ is the angle between the nonzero vectors **v** and **w**, then

$$\cos\theta = \frac{\mathbf{v}\cdot\mathbf{w}}{\|\mathbf{v}\|\|\mathbf{w}\|}$$

Proof   Suppose $\mathbf{v} = a_1\mathbf{i} + a_2\mathbf{j} + a_3\mathbf{k}$ and $\mathbf{w} = b_1\mathbf{i} + b_2\mathbf{j} + b_3\mathbf{k}$, and consider $\triangle AOB$ with vertices at the origin O and the points $A(a_1, a_2, a_3)$ and $B(b_1, b_2, b_3)$, as shown in Figure 9.25.

Note that sides $\overline{OA}$ and $\overline{OB}$ have lengths

$$\|\mathbf{v}\| = \sqrt{a_1^2 + a_2^2 + a_3^2} \quad \text{and} \quad \|\mathbf{w}\| = \sqrt{b_1^2 + b_2^2 + b_3^2}$$

respectively, and that side $\overline{AB}$ has length

$$\|\mathbf{v} - \mathbf{w}\| = \sqrt{(a_1 - b_1)^2 + (a_2 - b_2)^2 + (a_3 - b_3)^2}$$

Next, we apply the law of cosines to the sides $\overline{OA}$ and $\overline{OB}$:

$$\|\mathbf{v} - \mathbf{w}\|^2 = \|\mathbf{v}\|^2 + \|\mathbf{w}\|^2 - 2\|\mathbf{v}\|\|\mathbf{w}\|\cos\theta \qquad \textit{Law of cosines}$$

$$\cos\theta = \frac{\|\mathbf{v}\|^2 + \|\mathbf{w}\|^2 - \|\mathbf{v} - \mathbf{w}\|^2}{2\|\mathbf{v}\|\|\mathbf{w}\|} \qquad \textit{Solve for } \cos\theta.$$

$$= \frac{a_1^2 + a_2^2 + a_3^2 + b_1^2 + b_2^2 + b_3^2 - [(a_1 - b_1)^2 + (a_2 - b_2)^2 + (a_3 - b_3)^2]}{2\|\mathbf{v}\|\|\mathbf{w}\|}$$

$$= \frac{2a_1b_1 + 2a_2b_2 + 2a_3b_3}{2\|\mathbf{v}\|\|\mathbf{w}\|} = \frac{\mathbf{v}\cdot\mathbf{w}}{\|\mathbf{v}\|\|\mathbf{w}\|} \qquad \square$$

EXAMPLE 3   Angle between two given vectors

Let $\triangle ABC$ be the triangle with vertices $A(1, 1, 8)$, $B(4, -3, -4)$, and $C(-3, 1, 5)$. Find the angle formed at A (to two-place accuracy).

Solution

Draw $\triangle ABC$ and label the angle formed at A as α, as shown in Figure 9.26. The angle α is the angle between vectors **AB** and **AC**, where

$$\mathbf{AB} = (4 - 1)\mathbf{i} + (-3 - 1)\mathbf{j} + (-4 - 8)\mathbf{k} = 3\mathbf{i} - 4\mathbf{j} - 12\mathbf{k}$$
$$\mathbf{AC} = (-3 - 1)\mathbf{i} + (1 - 1)\mathbf{j} + (5 - 8)\mathbf{k} = -4\mathbf{i} - 3\mathbf{k}$$

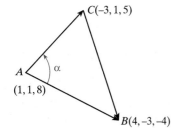

Figure 9.26 Find an angle of a triangle

Thus,

$$\cos\alpha = \frac{\mathbf{AB}\cdot\mathbf{AC}}{\|\mathbf{AB}\|\|\mathbf{AC}\|} = \frac{3(-4) + (-4)(0) + (-12)(-3)}{\sqrt{3^2 + (-4)^2 + (-12)^2}\sqrt{(-4)^2 + (-3)^2}}$$

$$= \frac{24}{\sqrt{169}\sqrt{25}} = \frac{24}{65}$$

and the required angle is $\alpha = \cos^{-1}\left(\frac{24}{65}\right) \approx 1.19$.   ■

The formula for the angle between vectors is often used in conjunction with the dot product. If we multiply both sides of the formula by $\|\mathbf{v}\|\|\mathbf{w}\|$, we obtain the following alternate form for the dot product formula.

Geometric Formula for the Dot Product

$$\mathbf{v}\cdot\mathbf{w} = \|\mathbf{v}\|\|\mathbf{w}\|\cos\theta$$

where θ is the angle ($0 \le \theta \le \pi$) between the vectors **v** and **w**.

Two vectors are said to be **perpendicular**, or **orthogonal**, if the angle between them is $\theta = \pi/2$. The following theorem provides a useful criterion for orthogonality.

THEOREM 9.4   The orthogonal vector theorem

Nonzero vectors $\mathbf{v}$ and $\mathbf{w}$ are **orthogonal** if and only if

$$\mathbf{v} \cdot \mathbf{w} = 0$$

Proof   If the vectors are orthogonal, then the angle between them is $\frac{\pi}{2}$; therefore,

$$\mathbf{v} \cdot \mathbf{w} = \|\mathbf{v}\| \|\mathbf{w}\| \cos \frac{\pi}{2} = 0$$

Conversely, if $\mathbf{v} \cdot \mathbf{w} = 0$, and $\mathbf{v}$ and $\mathbf{w}$ are nonzero vectors, then $\cos \theta = 0$, so that $\theta = \frac{\pi}{2}$ (since $0 \leq \theta \leq \pi$) and the vectors are orthogonal.   ❏

> **→ What This Says**   The orthogonal vector theorem allows you to show that two vectors are orthogonal by finding the dot product of the vectors and showing this dot product is 0. On the other hand, if you know the vectors are orthogonal, then it follows that the dot product of the vectors is 0. In particular, $\mathbf{i} \cdot \mathbf{j} = \mathbf{j} \cdot \mathbf{k} = \mathbf{k} \cdot \mathbf{i} = 0$.

EXAMPLE 4   Orthogonal vectors

Determine which (if any) pairs of the following vectors are orthogonal:

$$\mathbf{u} = 3\mathbf{i} + 7\mathbf{j} - 2\mathbf{k} \qquad \mathbf{v} = 5\mathbf{i} - 3\mathbf{j} - 3\mathbf{k} \qquad \mathbf{w} = \mathbf{j} - \mathbf{k}$$

Solution

$$\mathbf{u} \cdot \mathbf{v} = 3(5) + 7(-3) + (-2)(-3) = 0; \mathbf{u} \text{ and } \mathbf{v} \text{ are orthogonal.}$$
$$\mathbf{u} \cdot \mathbf{w} = 3(0) + 7(1) + (-2)(-1) = 9; \mathbf{u} \text{ and } \mathbf{w} \text{ are not orthogonal.}$$
$$\mathbf{v} \cdot \mathbf{w} = 5(0) + (-3)(1) + (-3)(-1) = 0; \mathbf{v} \text{ and } \mathbf{w} \text{ are orthogonal.}$$

■

DIRECTION COSINES

The direction of a vector in $\mathbb{R}^2$ positioned with its initial point at the origin may be specified by determining its angle of inclination; that is, the angle ϕ is the smallest positive angle the vector makes with the positive x-axis, as shown in Figure 9.27a. In $\mathbb{R}^3$, it is common practice to measure the direction of a nonzero vector $\mathbf{v}$ in terms of angles α, β, and γ between $\mathbf{v}$ and the positive coordinate axes as shown in Figure 9.27b. These angles are called the **direction angles** of $\mathbf{v}$, and their cosines are the **direction cosines** of $\mathbf{v}$.

a. In $\mathbb{R}^2$, the direction of $\mathbf{v}$ is given by the angle of inclination ϕ

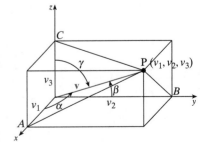

b. In $\mathbb{R}^3$, the direction of $\mathbf{v}$ is given by the angles $\alpha, \beta,$ and γ

Figure 9.27   The direction angles of a vector $\mathbf{v}$

The direction cosines of $\mathbf{v} = \langle v_1, v_2, v_3 \rangle$ can be computed by taking the dot product of $\mathbf{v}$ with the three standard basis vectors, $\mathbf{i} = \langle 1, 0, 0 \rangle$, $\mathbf{j} = \langle 0, 1, 0 \rangle$, and $\mathbf{k} = \langle 0, 0, 1 \rangle$. For example, to find $\cos \alpha$, note that

$$\mathbf{v} = \mathbf{v} \cdot \mathbf{i} = \|\mathbf{v}\| \|\mathbf{i}\| \cos \alpha, \quad \text{so that} \quad \cos \alpha = \frac{v_1}{\|\mathbf{v}\|}$$

and, similarly,

$$\cos \beta = \frac{v_2}{\|\mathbf{v}\|} \quad \text{and} \quad \cos \gamma = \frac{v_3}{\|\mathbf{v}\|}$$

It can be shown (Problem 53) that

$$\cos^2 \alpha + \cos^2 \beta + \cos^2 \gamma = 1$$

This identity can be used to check whether you have computed the direction cosines correctly.

EXAMPLE 5   Direction cosines and direction angles

Find the direction cosines and direction angles (to the nearest degree) of the vector $\mathbf{v} = -2\mathbf{i} + 3\mathbf{j} + 5\mathbf{k}$, and verify the formula $\cos^2 \alpha + \cos^2 \beta + \cos^2 \gamma = 1$.

Solution

We find $\|\mathbf{v}\| = \sqrt{(-2)^2 + 3^2 + 5^2} = \sqrt{38}$. Therefore,

$$\cos \alpha = \frac{v_1}{\|\mathbf{v}\|} = \frac{-2}{\sqrt{38}} \approx -0.3244428 \qquad \alpha \approx \cos^{-1}(-0.324428) \approx 109°$$

$$\cos \beta = \frac{v_2}{\|\mathbf{v}\|} = \frac{3}{\sqrt{38}} \approx 0.4866642 \qquad \beta \approx \cos^{-1}(0.4866642) \approx 61°$$

$$\cos \gamma = \frac{v_3}{\|\mathbf{v}\|} = \frac{5}{\sqrt{38}} \approx 0.8111071 \qquad \gamma \approx \cos^{-1}(0.8111071) \approx 36°$$

and

$$\cos^2 \alpha + \cos^2 \beta + \cos^2 \gamma = \left(\frac{-2}{\sqrt{38}} \right)^2 + \left(\frac{3}{\sqrt{38}} \right)^2 + \left(\frac{5}{\sqrt{38}} \right)^2 = 1$$

These direction angles are shown in Figure 9.28.

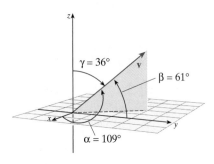

Figure 9.28 The vector $\mathbf{v} = -2\mathbf{i} + 3\mathbf{j} + 5\mathbf{k}$ forms the angles 109°, 61°, and 36° with the coordinate axes

PROJECTIONS

Let $\mathbf{v}$ and $\mathbf{w}$ be two vectors in $\mathbb{R}^3$ drawn so that they have a common initial point, as shown in Figure 9.29. If we drop a perpendicular from the tip of $\mathbf{v}$ to the line determined by $\mathbf{w}$, we determine a vector called the **vector projection of $\mathbf{v}$ onto $\mathbf{w}$**, which we have labeled $\mathbf{u}$ in Figure 9.29. To find a formula for the vector projection, note that $\mathbf{u} = t\mathbf{w}$ for some scalar t and that $\mathbf{v} - t\mathbf{w}$ is orthogonal to $\mathbf{w}$. Thus,

$$(\mathbf{v} - t\mathbf{w}) \cdot \mathbf{w} = 0$$
$$\mathbf{v} \cdot \mathbf{w} = t(\mathbf{w} \cdot \mathbf{w})$$
$$t = \frac{\mathbf{v} \cdot \mathbf{w}}{\mathbf{w} \cdot \mathbf{w}}$$

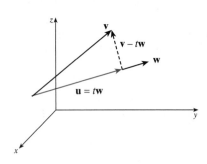

Figure 9.29 The vector $\mathbf{u}$ is the projection of $\mathbf{v}$ onto $\mathbf{w}$.

and the vector projection is

$$\mathbf{u} = \left(\frac{\mathbf{v} \cdot \mathbf{w}}{\mathbf{w} \cdot \mathbf{w}} \right) \mathbf{w}$$

We sometimes denote the vector projection of $\mathbf{v}$ onto $\mathbf{w}$ by $\text{proj}_{\mathbf{w}} \mathbf{v}$. Since $\mathbf{w} \cdot \mathbf{w} = \|\mathbf{w}\|^2$, we see that

$$\text{proj}_{\mathbf{w}} \mathbf{v} = \left(\frac{\mathbf{v} \cdot \mathbf{w}}{\|\mathbf{w}\|^2} \right) \mathbf{w}$$

Notice that $\text{proj}_{\mathbf{w}}\,\mathbf{v}$ can also be written

$$\text{proj}_{\mathbf{w}}\,\mathbf{v} = \left(\frac{\mathbf{v}\cdot\mathbf{w}}{\|\mathbf{w}\|}\right)\frac{\mathbf{w}}{\|\mathbf{w}\|}$$

where $\dfrac{\mathbf{w}}{\|\mathbf{w}\|}$ is the direction vector for $\mathbf{w}$. The scalar $\left(\dfrac{\mathbf{v}\cdot\mathbf{w}}{\|\mathbf{w}\|}\right)$ is called the **scalar projection** of $\mathbf{v}$ onto $\mathbf{w}$, or the **component** of $\mathbf{v}$ in the direction of $\mathbf{w}$, and is denoted by $\text{comp}_{\mathbf{w}}\,\mathbf{v}$. Thus,

$$\text{comp}_{\mathbf{w}}\,\mathbf{v} = \frac{\mathbf{v}\cdot\mathbf{w}}{\|\mathbf{w}\|} = \|\mathbf{v}\|\cos\theta \qquad \textit{Since } \mathbf{v}\cdot\mathbf{w} = \|\mathbf{v}\|\,\|\mathbf{w}\|\cos\theta$$

so $\text{comp}_{\mathbf{w}}\,\mathbf{v} > 0$ when $0 < \theta < \frac{\pi}{2}$ and $\text{comp}_{\mathbf{w}}\,\mathbf{v} < 0$ when $\frac{\pi}{2} < \theta < \pi$.
To summarize

WARNING Note that $\left(\dfrac{\mathbf{v}\cdot\mathbf{w}}{\mathbf{w}\cdot\mathbf{w}}\right)\mathbf{w}$ is not the same as $\left(\dfrac{\mathbf{v}}{\mathbf{w}}\right)\mathbf{w}$; you cannot "cancel" the vector $\mathbf{w}$. Remember, $\mathbf{v}\cdot\mathbf{w}$ and $\mathbf{w}\cdot\mathbf{w}$ are numbers, whereas $\mathbf{v}/\mathbf{w}$ is not defined.

Vector and Scalar Projections of $\mathbf{v}$ onto $\mathbf{w}$

If $\mathbf{v}$ and $\mathbf{w}$ are nonzero vectors,

then the **vector projection** of $\mathbf{v}$ onto $\mathbf{w}$ is the vector:

$$\text{proj}_{\mathbf{w}}\,\mathbf{v} = \left(\frac{\mathbf{v}\cdot\mathbf{w}}{\mathbf{w}\cdot\mathbf{w}}\right)\mathbf{w}$$

the **scalar projection** of $\mathbf{v}$ onto $\mathbf{w}$ is the number:

$$\text{comp}_{\mathbf{w}}\,\mathbf{v} = \frac{\mathbf{v}\cdot\mathbf{w}}{\|\mathbf{w}\|}$$

It is important to note that the $\mathbf{i}$, $\mathbf{j}$, and $\mathbf{k}$ components of any vector $\mathbf{v}$ are the scalar projections of $\mathbf{v}$ onto the appropriate basis vectors, as shown in Figure 9.30.

Figure 9.30   The $\mathbf{i}$, $\mathbf{j}$, and $\mathbf{k}$ components of $\mathbf{v} = a\mathbf{i} + b\mathbf{j} + c\mathbf{k}$ are scalar projections of $\mathbf{v}$ onto $\mathbf{i}$, $\mathbf{j}$, and $\mathbf{k}$.

EXAMPLE 6   Vector and scalar projections

Find the vector and scalar projections of $\mathbf{v} = 2\mathbf{i} + 3\mathbf{j} + 5\mathbf{k}$ onto $\mathbf{w} = 2\mathbf{i} - 2\mathbf{j} - \mathbf{k}$.

Solution

The vector projection of $\mathbf{v}$ onto $\mathbf{w}$ is

$$\text{proj}_{\mathbf{w}}\,\mathbf{v} = \left(\frac{\mathbf{v}\cdot\mathbf{w}}{\mathbf{w}\cdot\mathbf{w}}\right)\mathbf{w} = \left(\frac{2(2)+(3)(-2)+5(-1)}{2^2+(-2)^2+(-1)^2}\right)(2\mathbf{i}-2\mathbf{j}-\mathbf{k})$$

$$= -\tfrac{7}{9}(2\mathbf{i}-2\mathbf{j}-\mathbf{k}) = -\tfrac{14}{9}\mathbf{i}+\tfrac{14}{9}\mathbf{j}+\tfrac{7}{9}\mathbf{k}$$

To find the scalar projection, we use the scalar projection formula:

$$\text{comp}_{\mathbf{w}}\,\mathbf{v} = \frac{\mathbf{v}\cdot\mathbf{w}}{\|\mathbf{w}\|} = \frac{2\cdot 2 + 3\cdot(-2) + 5\cdot(-1)}{\sqrt{2^2+(-2)^2+(-1)^2}} = \frac{-7}{3} \qquad \blacksquare$$

WORK AS A DOT PRODUCT

In Section 6.5, we noted that when a constant force $\mathbf{F}$ acts in the direction of motion of an object moving from P to Q along a straight line, it produces work, W, given by

$$W = (\text{MAGNITUDE OF } \mathbf{F})(\text{DISTANCE MOVED}) = \|\mathbf{F}\|\|\mathbf{PQ}\|$$

where $\mathbf{PQ}$ is the displacement vector of the object's motion.

a. If the force $\mathbf{F}$ acts along the line of motion, then $W = \|\mathbf{F}\|\,\|\mathbf{PQ}\|$.

b. If $\mathbf{F}$ acts at a nonzero angle with the line of motion, then $W = \|\text{proj. of } \mathbf{F} \text{ onto } \mathbf{PQ}\|\,\|\mathbf{PQ}\|$.

Figure 9.31 The work performed

More generally, if the force $\mathbf{F}$ remains constant but points in a direction that makes an angle θ with the direction of motion (see Figure 9.31b), then the work W done by $\mathbf{F}$ in moving the object along the line from P to Q is given by the product of the *component* of $\mathbf{F}$ on the displacement vector $\mathbf{PQ}$ times the distance moved; that is,

$$W = (\text{SCALAR COMPONENT OF } \mathbf{F} \text{ ALONG } \mathbf{PQ})(\text{DISTANCE MOVED})$$
$$= (\text{comp}_{\mathbf{PQ}}\,\mathbf{F})(\|\mathbf{PQ}\|)$$
$$= (\|\mathbf{F}\|\cos\theta)\|\mathbf{PQ}\|$$
$$= \mathbf{F} \cdot \mathbf{PQ}$$

Work As a Dot Product

The work W done by a constant force $\mathbf{F}$ on an object moving along the line from point P to point Q is given by

$$W = \mathbf{F} \cdot \mathbf{PQ}$$

where $\mathbf{PQ}$ is the displacement vector of the object's motion.

Note that this formula also gives the work in the case where $\theta = 0$.

EXAMPLE 7 Work performed by a constant force

A boat sails north aided by a wind blowing in a direction of N30°E (30° east of north) with magnitude 500 lb. How much work is performed by the wind as the boat moves 100 ft?

Solution

The wind force is $\|\mathbf{F}\| = 500$ lb, acting in a direction $\theta = 30°$, as shown in Figure 9.32. The displacement vector is $\mathbf{PQ} = 100\mathbf{j}$, so $\|\mathbf{PQ}\| = 100$ ft. Thus, $\mathbf{F} = 500\cos 60° \,\mathbf{i} + 500\sin 60° \,\mathbf{j} = 250\mathbf{i} + 250\sqrt{3}\,\mathbf{j}$. The work performed is

$$W = \mathbf{F} \cdot \mathbf{PQ} = 100(250\sqrt{3}) = 25{,}000\sqrt{3}$$

Therefore, the work is approximately $43{,}300$ ft-lb.

Side View   Top View

Figure 9.32 Work performed on a sail

9.3   PROBLEM SET

A **1. WHAT DOES THIS SAY?**   Discuss how to find a dot product, and describe an application of dot product.

2. WHAT DOES THIS SAY?   What does it mean for vectors to be orthogonal? What is meant by the vector and scalar projections of a vector **v** onto another vector **w**.

Find the dot product **v** · **w** *in Problems 3–6.*

3. $\mathbf{v} = \langle 3, -2, 4 \rangle$; $\mathbf{w} = \langle 2, -1, -6 \rangle$
4. $\mathbf{v} = \langle 2, -6, 0 \rangle$; $\mathbf{w} = \langle 0, -3, 7 \rangle$
5. $\mathbf{v} = 2\mathbf{i} + 3\mathbf{j} - \mathbf{k}$; $\mathbf{w} = -3\mathbf{i} + 5\mathbf{j} + 4\mathbf{k}$
6. $\mathbf{v} = 3\mathbf{i} - \mathbf{j}$; $\mathbf{w} = 2\mathbf{i} + 5\mathbf{j}$

State whether the given pairs of vectors in Problems 7–10 are orthogonal.

7. $\mathbf{v} = \mathbf{i}$; $\mathbf{w} = \mathbf{k}$ 8. $\mathbf{v} = \mathbf{j}$; $\mathbf{w} = -\mathbf{k}$
9. $\mathbf{v} = 3\mathbf{i} - 2\mathbf{j}$; $\mathbf{w} = 6\mathbf{i} + 9\mathbf{j}$
10. $\mathbf{v} = 4\mathbf{i} - 5\mathbf{j} + \mathbf{k}$; $\mathbf{w} = 8\mathbf{i} + 10\mathbf{j} - 2\mathbf{k}$

Let $\mathbf{v} = 3\mathbf{i} - 2\mathbf{j} + \mathbf{k}$ *and* $\mathbf{w} = \mathbf{i} + \mathbf{j} - \mathbf{k}$. *Evaluate the expressions in Problems 11–14.*

11. $(\mathbf{v} + \mathbf{w}) \cdot (\mathbf{v} - \mathbf{w})$ 12. $(\mathbf{v} \cdot \mathbf{w})\mathbf{w}$

13. $(\|\mathbf{v}\|\mathbf{w}) \cdot (\|\mathbf{w}\|\mathbf{v})$ 14. $\dfrac{2\mathbf{v} + 3\mathbf{w}}{\|3\mathbf{v} + 2\mathbf{w}\|}$

Find the angle between the vectors given in Problems 15–18. Round to the nearest degree.

15. $\mathbf{v} = \mathbf{i} + \mathbf{j} + \mathbf{k}$; $\mathbf{w} = \mathbf{i} - \mathbf{j} + \mathbf{k}$
16. $\mathbf{v} = 2\mathbf{i} + \mathbf{k}$; $\mathbf{w} = \mathbf{j} - 3\mathbf{k}$
17. $\mathbf{v} = 2\mathbf{j} + \mathbf{k}$; $\mathbf{w} = \mathbf{i} - 2\mathbf{k}$
18. $\mathbf{v} = 4\mathbf{i} - \mathbf{j} + \mathbf{k}$; $\mathbf{w} = 2\mathbf{i} + 3\mathbf{j} + 5\mathbf{k}$

Find the scalar and vector projections of **v** *onto* **w** *in Problems 19–22.*

19. $\mathbf{v} = \mathbf{i} + \mathbf{j} + \mathbf{k}$; $\mathbf{w} = 2\mathbf{k}$ 20. $\mathbf{v} = 2\mathbf{i} - 3\mathbf{j}$; $\mathbf{w} = 2\mathbf{j} - 3\mathbf{k}$
21. $\mathbf{v} = \mathbf{i} + 2\mathbf{k}$; $\mathbf{w} = -3\mathbf{j}$ 22. $\mathbf{v} = \mathbf{i} + \mathbf{j} - 2\mathbf{k}$; $\mathbf{w} = \mathbf{i} + \mathbf{j} + \mathbf{k}$

23. Find two distinct unit vectors orthogonal to both $\mathbf{v} = \mathbf{i} + \mathbf{j} - \mathbf{k}$ and $\mathbf{w} = -\mathbf{i} + \mathbf{j} + \mathbf{k}$.

24. Find two distinct unit vectors orthogonal to both $\mathbf{v} = 2\mathbf{i} + \mathbf{j} + 2\mathbf{k}$ and $\mathbf{w} = -\mathbf{i} + 2\mathbf{j} - \mathbf{k}$.

25. Find a unit vector that points in the direction opposite to $\mathbf{v} = 2\mathbf{i} + 3\mathbf{j} - 2\mathbf{k}$.

26. Find a vector that points in the same direction as $\mathbf{v} = \mathbf{i} + 2\mathbf{j} - \mathbf{k}$ and has one-third its length.

27. Find x, y, and z that solve
$x(\mathbf{i} + \mathbf{j} + \mathbf{k}) + y(\mathbf{i} - \mathbf{j} + 2\mathbf{k}) + z(\mathbf{i} + \mathbf{k}) = 2\mathbf{i} + \mathbf{k}$.

28. Find x, y, and z that solve $x(\mathbf{i} - \mathbf{k}) + y(\mathbf{j} + \mathbf{k}) + z(\mathbf{i} - \mathbf{j}) = 5\mathbf{i} - \mathbf{k}$.

29. Find a number a that guarantees that the vectors $3\mathbf{i} - 2\mathbf{j} + \mathbf{k}$ and $2\mathbf{i} + a\mathbf{j} - 2a\mathbf{k}$ will be orthogonal.

30. Find x if the vectors $\mathbf{v} = 3\mathbf{i} - x\mathbf{j} + 2\mathbf{k}$ and $\mathbf{w} = x\mathbf{i} + \mathbf{j} - 2\mathbf{k}$ are to be orthogonal.

Find the direction cosines and the direction angles for the vectors given in Problems 31–34.

31. $\mathbf{v} = 2\mathbf{i} - 3\mathbf{j} - 5\mathbf{k}$ 32. $\mathbf{v} = 3\mathbf{i} - 2\mathbf{k}$
33. $\mathbf{v} = 5\mathbf{i} - 4\mathbf{j} + 3\mathbf{k}$ 34. $\mathbf{v} = \mathbf{j} - 5\mathbf{k}$

35. Find the cosine of the angle between the vectors $\mathbf{v} = \mathbf{i} - \mathbf{j} + 2\mathbf{k}$ and $\mathbf{w} = 2\mathbf{i} + \mathbf{j} - \mathbf{k}$. Then find the vector projection of **v** onto **w**.

36. Let $\mathbf{v} = 4\mathbf{i} - \mathbf{j} + \mathbf{k}$ and $\mathbf{w} = 2\mathbf{i} + 3\mathbf{j} - \mathbf{k}$. Find

   a. $\mathbf{v} \cdot \mathbf{w}$

   b. $\cos\theta$, where θ is the angle between **v** and **w**

   c. a scalar s such that **v** is orthogonal to $\mathbf{v} - s\mathbf{w}$

   d. a scalar s such that $s\mathbf{v} + \mathbf{w}$ is orthogonal to **w**

37. Let $\mathbf{v} = 2\mathbf{i} - 3\mathbf{j} + 6\mathbf{k}$ and $\mathbf{w} = 4\mathbf{i} + 3\mathbf{k}$. Find

   a. $\mathbf{v} \cdot \mathbf{w}$

   b. $\cos\theta$, where θ is the angle between **v** and **w**

   c. a scalar s such that **v** is orthogonal to $\mathbf{v} - s\mathbf{w}$

   d. a scalar t such that $\mathbf{v} + t\mathbf{w}$ is orthogonal to **w**

38. Find the scalar projection of the force $\mathbf{F} = 4\mathbf{i} - 2\mathbf{j} + 3\mathbf{k}$ in the direction of the vector $\mathbf{v} = \mathbf{i} - \mathbf{j} + 2\mathbf{k}$.

39. Find the work done by the constant force $\mathbf{F} = 2\mathbf{i} + 3\mathbf{j} + \mathbf{k}$ when it moves a particle along the line from $P(1, 0, -1)$ to $Q(3, 1, 2)$.

40. Find the work performed when a force $\mathbf{F} = \frac{6}{7}\mathbf{i} - \frac{2}{7}\mathbf{j} + \frac{6}{7}\mathbf{k}$ is applied to an object moving along the line from $P(-3, -5, 4)$ to $Q(4, 9, 11)$.

B **41.** Fred and his friend Sam are pulling a heavy log along flat horizontal ground by ropes attached to the front of the log. The ropes are 8 ft long. Fred holds his rope 2 ft above the log and 1 ft to the side, and Sam holds his end 1 ft above the log and 1 ft to the opposite side, as shown in Figure 9.33.

Side View   Top View

Figure 9.33   Problem 40

If Fred exerts a force of 30 lb and Sam exerts a force of 20 lb, what is the resultant force on the log?

42. Find the force required to keep a 5,000-lb van from rolling downhill if it is parked on a 10° slope (see Figure 9.34).

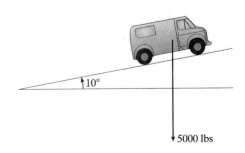

Figure 9.34   Problem 42

43. A block of ice is dragged 20 ft across a floor, using a force of 50 lb, as shown in Figure 9.35. Find the work done if the direction of the force is inclined θ to the horizontal, where

a. $\theta = \frac{\pi}{3}$ **b.** $\theta = \frac{\pi}{4}$

Figure 9.35   Problem 43

44. Suppose that the wind is blowing with a 1,000-lb magnitude force **F** in the direction of N60°W behind a boat's sail. How much work does the wind perform in moving the boat in a northerly direction a distance of 50 ft? Give your answer in foot-pounds.

Orthogonality of functions   *One of the big discoveries in modern mathematics was that it is very fruitful to extend some of the geometric notions of this chapter to functions—for example, taking the dot product of two functions or asking whether two functions are orthogonal, or attempting to project a function onto a second function (or onto a "plane" of functions). You are introduced to these important concepts in Problems 45 and 46.*

45. We say that two functions f and g are **orthogonal** on $[a, b]$ if

$$\int_a^b f(x)g(x)\,dx = 0$$

a. Show that the two functions x^2 and $x^3 - 5x$ are orthogonal on $[-b, b]$ for any positive b.

b. For distinct positive integers k and n, show that $\sin kx$ and $\sin nx$ are orthogonal on the interval $[-\pi, \pi]$. That is, the *family* of functions $\sin x, \sin 2x, \cdots$ are *mutually* orthogonal on $[-\pi, \pi]$. You may need the product-to-sum identity (trigonometric identity 28 from the *Student Mathematics Handbook*):

$$2 \sin \alpha \sin \beta = \cos(\alpha - \beta) - \cos(\alpha + \beta)$$

46. HISTORICAL QUEST   Jean Baptiste Joseph Fourier began his studies by training for the priesthood. In spite of this first calling, his interest in mathematics remained intense, and in 1789 he wrote, "Yesterday was my 21st birthday, and at that age Newton and Pascal had already acquired many claims to immortality." He did not take his religious vows but continued his mathematical research while teaching. Fourier was renowned as an outstanding teacher, but it was not until around 1804–1807 that he did his important mathematical work on the theory of heat. ∎

JEAN FOURIER
1768–1830

During that time, he expanded on an idea of Daniel Bernoulli by showing how many functions, both continuous and discontinuous, can be represented by trigonometric series of the general form

$$\tfrac{1}{2}a_0 + a_1 \cos x + a_2 \cos 2x + a_3 \cos 3x + \cdots$$
$$+ b_1 \sin x + b_2 \sin 2x + b_3 \sin 3x + \cdots$$

The study of such series, called **Fourier series**, is of fundamental importance in both theoretical and applied mathematics, and they are used for modeling phenomena in engineering and physics as well as in areas such as computer science and medical research.

The issue of whether a given function f can be represented by a Fourier series is a matter for a more advanced course. For this Quest, assume that $f(x)$ is such a function and that

$$f(x) = \tfrac{1}{2}a_0 + \sum_{k=1}^{\infty} a_k \cos kx + \sum_{k=1}^{\infty} b_k \sin kx$$

on the interval $[-\pi, \pi]$. In particular, if f is a continuous odd function (that is, $f(-x) = -f(x)$) on $[-\pi, \pi]$, it can be shown that $a_k = 0$ for $k = 0, 1, 2, \ldots$, and that

$$b_k = \frac{1}{\pi} \int_{-\pi}^{\pi} f(x) \sin kx \, dx$$
$$= \frac{2}{\pi} \int_0^{\pi} f(x) \sin kx \, dx \quad \text{for } k = 1, 2, \ldots$$

a. Using this result, show that the function $f(x) = x$ on $[-\pi, \pi]$ has the Fourier series representation

$$f(x) = 2 \sum_{k=1}^{\infty} \frac{(-1)^{k+1}}{k} \sin(kx)$$

b. Use the Fourier series in part **a** to evaluate the convergent alternating series

$$\sum_{k=1}^{\infty} \frac{(-1)^{k+1}}{2k-1} = 1 - \frac{1}{3} + \frac{1}{5} - \frac{1}{7} + \frac{1}{9} - \cdots$$

47. Let $\mathbf{u}_0 = a\mathbf{i} + b\mathbf{j} + c\mathbf{k}$ and let $\mathbf{u} = x\mathbf{i} + y\mathbf{j} + z\mathbf{k}$. Describe the set of points in $\mathbb{R}^3$ defined by

$$\|\mathbf{u}_0 - \mathbf{u}\| < r$$

where $r > 0$ and a, b, c are constants.

48. Find the angle (to the nearest degree) between the diagonal of a cube and a diagonal of one of its faces. Assume both diagonals have the same initial point.

49. Show that the vector $\mathbf{B} = \|\mathbf{v}\|\mathbf{u} + \|\mathbf{u}\|\mathbf{v}$ bisects the angle θ between the two nonzero vectors $\mathbf{u}$ and $\mathbf{v}$.

50. Use vector methods to prove that an angle inscribed in a semicircle must be a right angle.

51. a. Show that $(\mathbf{v} + \mathbf{w}) \cdot (\mathbf{v} + \mathbf{w}) = \|\mathbf{v}\|^2 + \|\mathbf{w}\|^2 + 2(\mathbf{v} \cdot \mathbf{w})$.

b. Use part **a** to prove the **triangle inequality:**

$$\|\mathbf{v} + \mathbf{w}\| \leq \|\mathbf{v}\| + \|\mathbf{w}\|$$

Hint: Note that

$$\|\mathbf{v} + \mathbf{w}\|^2 = (\mathbf{v} + \mathbf{w}) \cdot (\mathbf{v} + \mathbf{w})$$

52. The **Cauchy-Schwarz** inequality in $\mathbb{R}^3$ states that for any vectors $\mathbf{v}$ and $\mathbf{w}$

$$|\mathbf{v} \cdot \mathbf{w}| \leq \|\mathbf{v}\|\|\mathbf{w}\|$$

a. Prove the Cauchy-Schwarz inequality. *Hint*: Use the formula for the angle between vectors.

b. Show that equality $|\mathbf{v} \cdot \mathbf{w}| = \|\mathbf{v}\|\|\mathbf{w}\|$ occurs if and only if $\mathbf{v} = t\mathbf{w}$ for some scalar t.

c. Use the Cauchy-Schwarz inequality to prove the **triangle**

inequality:

$$\|\mathbf{v} + \mathbf{w}\| \leq \|\mathbf{v}\| + \|\mathbf{w}\|$$

53. Show that $\cos^2 \alpha + \cos^2 \alpha + \cos^2 \gamma = 1$, where α, β, and γ are the direction angles of a triangle $\triangle ABC$.

9.4   The Cross Product

IN THIS SECTION   definition and basic properties of the cross product; geometric interpretation of the cross product; applications of the cross product: area and torque; the scalar triple product

DEFINITION AND BASIC PROPERTIES OF THE CROSS PRODUCT

The dot product is not the only way to multiply vectors, In this section, we will study a second product involving vectors, called the *cross product*, *vector product*, or sometimes the *outer product*. Unlike the dot product, in which two vectors are multiplied to produce a scalar, the cross product produces a vector. Here is the definition.

Cross Product

> If $\mathbf{v} = a_1\mathbf{i} + a_2\mathbf{j} + a_3\mathbf{k}$ and $\mathbf{w} = b_1\mathbf{i} + b_2\mathbf{j} + b_3\mathbf{k}$, the **cross product** of the vector $\mathbf{v}$ with the vector $\mathbf{w}$, written $\mathbf{v} \times \mathbf{w}$, is the vector
>
> $$\mathbf{v} \times \mathbf{w} = (a_2b_3 - a_3b_2)\mathbf{i} + (a_3b_1 - a_1b_3)\mathbf{j} + (a_1b_2 - a_2b_1)\mathbf{k}$$

SMH Determinants are reviewed in Section 2.8 of the Student Mathematics Handbook.

The form of the cross product may seem odd to you, but you will soon discover that the cross product has a variety of useful properties.

A convenient way of remembering the cross product is the following determinant form:

$$\mathbf{v} \times \mathbf{w} = \begin{vmatrix} \mathbf{i} & \mathbf{j} & \mathbf{k} \\ a_1 & a_2 & a_3 \\ b_1 & b_2 & b_3 \end{vmatrix}$$

To verify the determinant formula for the cross product, we expand about the first row.

$$\mathbf{v} \times \mathbf{w} = \begin{vmatrix} \mathbf{i} & \mathbf{j} & \mathbf{k} \\ a_1 & a_2 & a_3 \\ b_1 & b_2 & b_3 \end{vmatrix} = \begin{vmatrix} a_2 & a_3 \\ b_2 & b_3 \end{vmatrix} \mathbf{i} - \begin{vmatrix} a_1 & a_3 \\ b_1 & b_3 \end{vmatrix} \mathbf{j} + \begin{vmatrix} a_1 & a_2 \\ b_1 & b_2 \end{vmatrix} \mathbf{k}$$

j is in row 1, column 2, so do not forget the negative sign here.

$$= (a_2b_3 - a_3b_2)\mathbf{i} - (a_1b_3 - a_3b_1)\mathbf{j} + (a_1b_2 - a_2b_1)\mathbf{k}$$
$$= (a_2b_3 - a_3b_2)\mathbf{i} + (a_3b_1 - a_1b_3)\mathbf{j} + (a_1b_2 - a_2b_1)\mathbf{k}$$

EXAMPLE 1   Cross product

Find $\mathbf{v} \times \mathbf{w}$, where $\mathbf{v} = 2\mathbf{i} - \mathbf{j} + 3\mathbf{k}$ and $\mathbf{w} = 7\mathbf{j} - 4\mathbf{k}$.

Solution

$$\mathbf{v} \times \mathbf{w} = \begin{vmatrix} \mathbf{i} & \mathbf{j} & \mathbf{k} \\ 2 & -1 & 3 \\ 0 & 7 & -4 \end{vmatrix}$$

Do not forget minus here (**j** is in the negative position).
↓

$$= [(-1)(-4) - 3(7)]\mathbf{i} - [2(-4) - 0(3)]\mathbf{j} + [2(7) - 0(-1)]\mathbf{k}$$
$$= -17\mathbf{i} + 8\mathbf{j} + 14\mathbf{k} \qquad\blacksquare$$

See Example 2.31 and Problem 66 of Problem Set 2 of the *Student Mathematics Handbook*.

TECHNOLOGY NOTE

Mathematical programs such as *Mathematica*, *Maple*, *Derive*, and *Matlab* all do vector operations, including dot and cross products. Many CAS programs and calculators use brackets to input vectors, with the components separated by commas, as in $[a, b, c]$.

The graphic in Figure 9.36 shows the dot product and the cross product of the vectors in Example 1.

Notice that the output shows a number, -19, for the dot product, and a vector, $[-17, 8, 14]$, for the cross product. In vector notation,

$$[-17, 8, 14] = -17\mathbf{i} + 8\mathbf{j} + 14\mathbf{k}$$

Figure 9.36 Example 1 using a calculator

Properties of determinants can also be used to establish properties of the cross product. For instance, the following computation shows that the cross product is not commutative. This property is sometimes called **anticommutativity**.

$$\mathbf{v} \times \mathbf{w} = \begin{vmatrix} \mathbf{i} & \mathbf{j} & \mathbf{k} \\ a_1 & a_2 & a_3 \\ b_1 & b_2 & b_3 \end{vmatrix} = -\begin{vmatrix} \mathbf{i} & \mathbf{j} & \mathbf{k} \\ b_1 & b_2 & b_3 \\ a_1 & a_2 & a_3 \end{vmatrix} = -(\mathbf{w} \times \mathbf{v})$$

This and other properties are listed in the following theorem.

THEOREM 9.5 **Properties of the cross product**

If $\mathbf{u}$, $\mathbf{v}$, $\mathbf{w}$, $\mathbf{a}$, $\mathbf{b}$, and $\mathbf{c}$ are vectors in $\mathbb{R}^3$ and s and t are scalars, then

Scalar distributivity $\qquad (s\mathbf{v}) \times (t\mathbf{w}) = st(\mathbf{v} \times \mathbf{w})$

Distributivity for cross product over addition
$$\mathbf{u} \times (\mathbf{v} + \mathbf{w}) = (\mathbf{u} \times \mathbf{v}) + (\mathbf{u} \times \mathbf{w})$$
$$(\mathbf{u} + \mathbf{v}) \times \mathbf{w} = (\mathbf{u} \times \mathbf{w}) + (\mathbf{v} \times \mathbf{w})$$

Anticommutativity $\qquad \mathbf{v} \times \mathbf{w} = -(\mathbf{w} \times \mathbf{v})$

Product of a multiple $\qquad \mathbf{v} \times s\mathbf{v} = \mathbf{0}$; in particular, $\mathbf{v} \times \mathbf{v} = \mathbf{0}$

Zero product $\qquad \mathbf{v} \times \mathbf{0} = \mathbf{0} \times \mathbf{v} = \mathbf{0}$

Lagrange's identity $\qquad \|\mathbf{v} \times \mathbf{w}\|^2 = \|\mathbf{v}\|^2\|\mathbf{w}\|^2 - (\mathbf{v} \cdot \mathbf{w})^2$

cab-bac formula $\qquad \mathbf{a} \times (\mathbf{b} \times \mathbf{c}) = (\mathbf{c} \cdot \mathbf{a})\mathbf{b} - (\mathbf{b} \cdot \mathbf{a})\mathbf{c}$

Proof Let $\mathbf{u} = a_1\mathbf{i} + a_2\mathbf{j} + a_3\mathbf{k}$, $\mathbf{v} = b_1\mathbf{i} + b_2\mathbf{j} + b_3\mathbf{k}$, and $\mathbf{w} = c_1\mathbf{i} + c_2\mathbf{j} + c_3\mathbf{k}$.

Scalar distributivity and *vector distributivity* are proved by using the definition of cross product and the corresponding properties of real numbers.

Anticommutativity was proved in the paragraph preceding this theorem.

Product of a multiple

$$\mathbf{v} \times s\mathbf{v} = \begin{vmatrix} \mathbf{i} & \mathbf{j} & \mathbf{k} \\ b_1 & b_2 & b_3 \\ sb_1 & sb_2 & sb_3 \end{vmatrix} = s\begin{vmatrix} \mathbf{i} & \mathbf{j} & \mathbf{k} \\ b_1 & b_2 & b_3 \\ b_1 & b_2 & b_3 \end{vmatrix} = \mathbf{0} \qquad \text{A determinant with two}$$
$$\qquad\qquad\qquad\qquad\qquad\qquad\qquad\qquad\qquad\text{identical rows is } 0.$$

Zero product is obvious.

Lagrange's identity

$$\|\mathbf{v} \times \mathbf{w}\|^2 = (b_2c_3 - b_3c_2)^2 + (b_3c_1 - b_1c_3)^2 + (b_1c_2 - b_2c_1)^2$$
$$= (b_1^2 + b_2^2 + b_3^2)(c_1^2 + c_2^2 + c_3^2) - (b_1c_1 + b_2c_2 + b_3c_3)^2$$
$$= \|\mathbf{v}\|^2\|\mathbf{w}\|^2 - (\mathbf{v} \cdot \mathbf{w})^2$$

You are asked to prove the *cab-bac formula* in Problem 55.   ❑

WARNING You are accustomed to dealing with products in which the order does not matter, but the order of the terms in the cross product $\mathbf{v} \times \mathbf{w}$ matters very much, since $\mathbf{w} \times \mathbf{v} = -(\mathbf{v} \times \mathbf{w})$.

Cross products involving the basic vectors $\mathbf{i}$, $\mathbf{j}$, and $\mathbf{k}$ occur so frequently in computations that it is useful to have the following list:

$$\begin{array}{lll} \mathbf{i} \times \mathbf{i} = \mathbf{0} & \mathbf{i} \times \mathbf{j} = \mathbf{k} & \mathbf{i} \times \mathbf{k} = -\mathbf{j} \\ \mathbf{j} \times \mathbf{i} = -\mathbf{k} & \mathbf{j} \times \mathbf{j} = \mathbf{0} & \mathbf{j} \times \mathbf{k} = \mathbf{i} \\ \mathbf{k} \times \mathbf{i} = \mathbf{j} & \mathbf{k} \times \mathbf{j} = -\mathbf{i} & \mathbf{k} \times \mathbf{k} = \mathbf{0} \end{array}$$

These nine rules can be established by using the definition of cross product, or by using the determinant form. For example,

$$\mathbf{j} \times \mathbf{i} = \begin{vmatrix} \mathbf{i} & \mathbf{j} & \mathbf{k} \\ 0 & 1 & 0 \\ 1 & 0 & 0 \end{vmatrix} = -\mathbf{k}$$

GEOMETRIC INTERPRETATION OF THE CROSS PRODUCT

The cross product is frequently used for geometric purposes, such as determining a vector that is orthogonal to two given vectors. The following theorem lists one of the most important geometric properties of the cross product. It says that for any two nonzero vectors $\mathbf{v}$ and $\mathbf{w}$, the cross product is a vector orthogonal to the plane determined by the vectors $\mathbf{v}$ and $\mathbf{w}$, as shown in Figure 9.37.

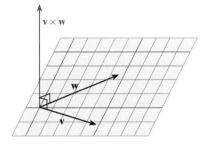

Figure 9.37 $\mathbf{v} \times \mathbf{w}$ is orthogonal to both $\mathbf{v}$ and $\mathbf{w}$

THEOREM 9.6 Orthogonality property of the cross product

If $\mathbf{v}$ and $\mathbf{w}$ are nonzero vectors in $\mathbb{R}^3$ that are not multiples of one another, then $\mathbf{v} \times \mathbf{w}$ is orthogonal to both $\mathbf{v}$ and $\mathbf{w}$.

Proof The theorem is illustrated in Figure 9.37. We will show that $(\mathbf{v} \times \mathbf{w})$ is orthogonal to $\mathbf{v}$ and leave the proof that $\mathbf{v} \times \mathbf{w}$ is orthogonal to $\mathbf{w}$ as an exercise. Let $\mathbf{v} = a_1\mathbf{i} + a_2\mathbf{j} + a_3\mathbf{k}$ and $\mathbf{w} = b_1\mathbf{i} + b_2\mathbf{j} + b_3\mathbf{k}$. Then

$$\mathbf{v} \times \mathbf{w} = \begin{vmatrix} \mathbf{i} & \mathbf{j} & \mathbf{k} \\ a_1 & a_2 & a_3 \\ b_1 & b_2 & b_3 \end{vmatrix} = (a_2b_3 - a_3b_2)\mathbf{i} - (a_1b_3 - a_3b_1)\mathbf{j} + (a_1b_2 - a_2b_1)\mathbf{k}$$

so that

$$\mathbf{v} \cdot (\mathbf{v} \times \mathbf{w}) = a_1(a_2b_3 - a_3b_2) - a_2(a_1b_3 - a_3b_1) + a_3(a_1b_2 - a_2b_1)$$
$$= a_1a_2b_3 - a_1a_3b_2 - a_1a_2b_3 + a_2a_3b_1 + a_1a_3b_2 - a_2a_3b_1 = 0$$

Thus, according to Theorem 9.4, $\mathbf{v} \times \mathbf{w}$ is orthogonal to $\mathbf{v}$.   ❏

EXAMPLE 2   A vector orthogonal to two given vectors

Find a nonzero vector that is orthogonal to both $\mathbf{v} = -2\mathbf{i} + 3\mathbf{j} - 7\mathbf{k}$ and $\mathbf{w} = 5\mathbf{i} + 9\mathbf{k}$.

Solution

The cross product $\mathbf{v} \times \mathbf{w}$ is orthogonal to both $\mathbf{v}$ and $\mathbf{w}$.

$$\mathbf{v} \times \mathbf{w} = \begin{vmatrix} \mathbf{i} & \mathbf{j} & \mathbf{k} \\ -2 & 3 & -7 \\ 5 & 0 & 9 \end{vmatrix} = (27 + 0)\mathbf{i} - (-18 + 35)\mathbf{j} + (0 - 15)\mathbf{k}$$

$$= 27\mathbf{i} - 17\mathbf{j} - 15\mathbf{k}$$ ■

In Section 9.3, we showed that the dot product satisfies $\mathbf{v} \cdot \mathbf{w} = \|\mathbf{v}\|\|\mathbf{w}\| \cos\theta$, where θ is the angle between $\mathbf{v}$ and $\mathbf{w}$. The following theorem establishes a similar result for the cross product.

THEOREM 9.7   Magnitude of the cross product

If $\mathbf{v}$ and $\mathbf{w}$ are nonzero vectors in $\mathbb{R}^3$ with θ the angle between $\mathbf{v}$ and $\mathbf{w}$ $(0 \leq \theta \leq \pi)$, then

$$\|\mathbf{v} \times \mathbf{w}\| = \|\mathbf{v}\|\|\mathbf{w}\| \sin\theta$$

Proof   We will use Lagrange's identity from Theorem 9.5.

$$\|\mathbf{v} \times \mathbf{w}\|^2 = \|\mathbf{v}\|^2\|\mathbf{w}\|^2 - [\|\mathbf{v}\|\|\mathbf{w}\| \cos\theta]^2 \qquad \text{Lagrange's identity}$$
$$= \|\mathbf{v}\|^2\|\mathbf{w}\|^2 - \|\mathbf{v}\|^2\|\mathbf{w}\|^2 \cos^2\theta$$
$$= \|\mathbf{v}\|^2\|\mathbf{w}\|^2(1 - \cos^2\theta)$$
$$= \|\mathbf{v}\|^2\|\mathbf{w}\|^2 \sin^2\theta$$

Thus,

$$\|\mathbf{v} \times \mathbf{w}\| = \|\mathbf{v}\|\|\mathbf{w}\| |\sin\theta| = \|\mathbf{v}\|\|\mathbf{w}\| \sin\theta \qquad \begin{array}{l} \text{Because } \sin\theta \geq 0 \\ \text{when } 0 \leq \theta \leq \pi \end{array}$$ ❏

From Theorem 9.7, we know the magnitude of the cross product $\mathbf{v} \times \mathbf{w}$, but what is its direction? It turns out that this direction can be described by the following rule:

The Right-Hand Rule

With **v** and **w** positioned so that their initial points coincide, place the little finger of your right hand along the vector **v** with the tip of your little finger pointing toward the tip of **v**, and curl your fingers toward **w**. Then your thumb points in the direction of **v** × **w**, as shown in Figure 9.38, so that

$$\mathbf{v} \times \mathbf{w} = \|\mathbf{v}\| \|\mathbf{w}\| \sin\theta \, \mathbf{n}$$

where **n** is a unit vector determined by the right-hand rule.

Figure 9.38 The right-hand rule: If you place the little finger of your right hand along **v** and curl your fingers toward **w** to cover the smaller angle between **v** and **w**, then your thumb points in the direction of **v** × **w**.

➡ **What This Says**   Theorem 9.7 provides a *coordinate-free way* of computing the cross product in the sense that it depends only on the geometry of the vectors involved and not on any particular choice of representation. This is important to physicists and engineers because it enables them to choose whichever coordinate system is most convenient to compute cross products that arise in applications.

APPLICATIONS OF THE CROSS PRODUCT: AREA AND TORQUE

The cross product can be used in a variety of ways in geometry and physics. For example, using Theorem 9.7, it can be shown that the parallelogram having **v** and **w** as adjacent sides has area $\|\mathbf{v} \times \mathbf{w}\|$ (see Figure 9.39). To see this, note that the height of the parallelogram is $h = \|\mathbf{w}\| \sin\theta$, so that the area is given by

$$\text{AREA} = (\text{BASE})(\text{HEIGHT}) = \|\mathbf{v}\|(\|\mathbf{w}\| \sin\theta) = \|\mathbf{v} \times \mathbf{w}\|$$

From this formula, it follows immediately that the area of a triangle in $\mathbb{R}^3$ is half the area of the corresponding parallelogram. For future reference, we let **v** be defined by the points A and B, and **w** by the points A and C, and summarize these formulas for area in the following box.

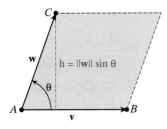

Figure 9.39 A geometric interpretation of $\|\mathbf{v} \times \mathbf{w}\|$

The parallelogram with adjacent sides $\overline{AB}$ and $\overline{AC}$ has

$$\text{AREA} = \|\mathbf{AB} \times \mathbf{AC}\|$$

Let A, B, and C be points in $\mathbb{R}^3$. Then the triangle with vertices A, B, and C has

$$\text{AREA} = \tfrac{1}{2}\|\mathbf{AB} \times \mathbf{AC}\|$$

EXAMPLE 3   Area of a triangle

Find the area of the triangle with vertices $P(-2, 4, 5)$, $Q(0, 7, -4)$, and $R(-1, 5, 0)$.

Solution

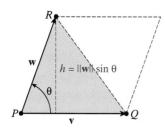

Draw this triangle as shown in Figure 9.40. Then $\triangle PQR$ has half the area of the parallelogram determined by the vectors **PQ** and **PR**; that is, the triangle has area
$$A = \tfrac{1}{2}\|\mathbf{PQ} \times \mathbf{PR}\|$$

First find
$$\mathbf{PQ} = (0+2)\mathbf{i} + (7-4)\mathbf{j} + (-4-5)\mathbf{k} = 2\mathbf{i} + 3\mathbf{j} - 9\mathbf{k}$$
$$\mathbf{PR} = (-1+2)\mathbf{i} + (5-4)\mathbf{j} + (0-5)\mathbf{k} = \mathbf{i} + \mathbf{j} - 5\mathbf{k}$$

Figure 9.40 Area of a triangle by cross product

and compute the cross product:
$$\mathbf{PQ} \times \mathbf{PR} = \begin{vmatrix} \mathbf{i} & \mathbf{j} & \mathbf{k} \\ 2 & 3 & -9 \\ 1 & 1 & -5 \end{vmatrix}$$
$$= (-15+9)\mathbf{i} - (-10+9)\mathbf{j} + (2-3)\mathbf{k}$$
$$= -6\mathbf{i} + \mathbf{j} - \mathbf{k}$$

Thus, the triangle has area
$$A = \tfrac{1}{2}\|\mathbf{PQ} \times \mathbf{PR}\|$$
$$= \tfrac{1}{2}\sqrt{(-6)^2 + 1^2 + (-1)^2}$$
$$= \tfrac{1}{2}\sqrt{38} \qquad\blacksquare$$

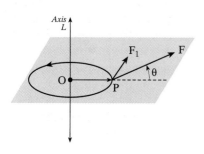

Figure 9.41 The torque of F about a point O

An important application of the cross product to physics involves computing the tendency of an object to rotate about an axis. Figure 9.41 shows an object at point P rotating about an axis L. A force **F** operates on the object to produce rotation and $\mathbf{F}_1$ is the component of **F** in the direction of motion (orthogonal to **OP**). Experience suggests that the tendency to rotate depends both on the magnitude of the component force $\mathbf{F}_1$ and on the distance $\|\mathbf{OP}\|$ of the rotating object from the axis of rotation. We have $\mathbf{F}_1 = \|\mathbf{F}\|\sin\theta$, where θ is the angle between **F** and **OP**, so the tendency to rotate can be measured by the product
$$\|\mathbf{OP}\|\|\mathbf{F}_1\| = \|\mathbf{OP}\|\|\mathbf{F}\|\sin\theta$$
$$= \|\mathbf{OP} \times \mathbf{F}\|$$

The cross product $\mathbf{OP} \times \mathbf{F}$ is called the **torque of F about O**. The magnitude of the torque measures the tendency to rotate, and the direction of the torque vector is the direction of the axis of rotation.

EXAMPLE 4   Torque on the hinge of a door

Figure 9.42 shows a half-open door that is 3 ft wide. A force of 30 lb is applied in a horizontal direction at the edge of the door. Find the torque of the force about the hinges on the door.

Solution

Figure 9.42 The torque of a force applied at the edge of a door about the hinges

We represent the force by $\mathbf{F} = -30\mathbf{i}$ (see Figure 9.42). Because the door is half open, it makes an angle of $\frac{\pi}{4}$ with the positive x-axis, and we can represent the "arm" **PQ** by the vector
$$\mathbf{PQ} = 3\left(\cos\tfrac{\pi}{4}\mathbf{i} + \sin\tfrac{\pi}{4}\mathbf{j}\right) = 3\left(\frac{\sqrt{2}}{2}\mathbf{i} + \frac{\sqrt{2}}{2}\mathbf{j}\right) = \frac{3\sqrt{2}}{2}\mathbf{i} + \frac{3\sqrt{2}}{2}\mathbf{j}$$

The torque can now be found:
$$\mathbf{T} = \mathbf{PQ} \times \mathbf{F} = \begin{vmatrix} \mathbf{i} & \mathbf{j} & \mathbf{k} \\ \frac{3\sqrt{2}}{2} & \frac{3\sqrt{2}}{2} & 0 \\ -30 & 0 & 0 \end{vmatrix} = 45\sqrt{2}\,\mathbf{k}$$

The magnitude of the torque ($45\sqrt{2}$ lb-ft) is a measure of the tendency of the door to rotate about its hinges. $\qquad\blacksquare$

THE SCALAR TRIPLE PRODUCT

The cross product can also be used to compute the volume of a parallelepiped in $\mathbb{R}^3$. Consider the parallelepiped determined by three nonzero vectors **u**, **v**, and **w** that do not all lie in the same plane, as shown in Figure 9.43.

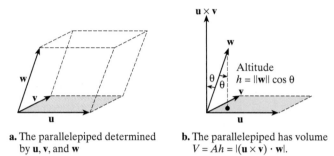

a. The parallelepiped determined by **u**, **v**, and **w**

b. The parallelepiped has volume $V = Ah = |(\mathbf{u} \times \mathbf{v}) \cdot \mathbf{w}|$.

Figure 9.43 Computing volume with the scalar triple product

It is known from solid geometry that this parallelepiped has volume $V = Ah$, where A is the area of the face determined by **u** and **v** and h is the altitude from the tip of **w** to this face (see Figure 9.43b).

The face determined by **u** and **v** is a parallelogram with area $A = \|\mathbf{u} \times \mathbf{v}\|$, and we know that the cross-product vector $\mathbf{u} \times \mathbf{v}$ is perpendicular to both **u** and **v** and hence to the face determined by **u** and **v**. From the geometric formula for the dot product, we have

$$(\mathbf{u} \times \mathbf{v}) \cdot \mathbf{w} = \|\mathbf{u} \times \mathbf{v}\| \|\mathbf{w}\| \cos \theta$$

where θ is the angle between $\mathbf{u} \times \mathbf{v}$ and **w**. Since the parallelepiped has altitude

$$h = |\|\mathbf{w}\| \cos \theta| = \left| \frac{(\mathbf{u} \times \mathbf{v}) \cdot \mathbf{w}}{\|\mathbf{u} \times \mathbf{v}\|} \right|$$

the volume is given by

$$V = Ah = \|\mathbf{u} \times \mathbf{v}\| \left| \frac{(\mathbf{u} \times \mathbf{v}) \cdot \mathbf{w}}{\|\mathbf{u} \times \mathbf{v}\|} \right| = |(\mathbf{u} \times \mathbf{v}) \cdot \mathbf{w}|$$

The combined operation $(\mathbf{u} \times \mathbf{v}) \cdot \mathbf{w}$ is called the **scalar triple product** of **u**, **v**, and **w**. These observations are summarized in the following box.

Volume Interpretation of the Scalar Triple Product

Let **u**, **v**, and **w** be nonzero vectors that do not all lie in the same plane. Then the parallelepiped determined by these vectors has volume

$$V = |(\mathbf{u} \times \mathbf{v}) \cdot \mathbf{w}|$$

EXAMPLE 5   Volume of a parallelepiped

Find the volume of the parallelepiped determined by the vectors

$$\mathbf{u} = \mathbf{i} - 2\mathbf{j} + 3\mathbf{k} \qquad \mathbf{v} = -4\mathbf{i} + 7\mathbf{j} - 11\mathbf{k} \qquad \mathbf{w} = 5\mathbf{i} + 9\mathbf{j} - \mathbf{k}$$

Solution

We first find the cross product

$$\mathbf{u} \times \mathbf{v} = \begin{vmatrix} \mathbf{i} & \mathbf{j} & \mathbf{k} \\ 1 & -2 & 3 \\ -4 & 7 & -11 \end{vmatrix}$$

$$= (22 - 21)\mathbf{i} - (-11 + 12)\mathbf{j} + (7 - 8)\mathbf{k}$$

$$= \mathbf{i} - \mathbf{j} - \mathbf{k}$$

Thus,

$$V = |(\mathbf{u} \times \mathbf{v}) \cdot \mathbf{w}| = |(\mathbf{i} - \mathbf{j} - \mathbf{k}) \cdot (5\mathbf{i} + 9\mathbf{j} - \mathbf{k})| = |5 - 9 + 1| = 3$$

The volume of the parallelepiped is 3 cubic units.   ∎

THEOREM 9.8   Determinant form for a scalar triple product

If $\mathbf{u} = a_1\mathbf{i} + a_2\mathbf{j} + a_3\mathbf{k}$, $\mathbf{v} = b_1\mathbf{i} + b_2\mathbf{j} + b_3\mathbf{k}$, and $\mathbf{w} = c_1\mathbf{i} + c_2\mathbf{j} + c_3\mathbf{k}$, then the scalar triple product can be found by evaluating the determinant

$$(\mathbf{u} \times \mathbf{v}) \cdot \mathbf{w} = \begin{vmatrix} a_1 & a_2 & a_3 \\ b_1 & b_2 & b_3 \\ c_1 & c_2 & c_3 \end{vmatrix} \begin{matrix} \leftarrow \text{ components of } \mathbf{u} \\ \leftarrow \text{ components of } \mathbf{v} \\ \leftarrow \text{ components of } \mathbf{w} \end{matrix}$$

Proof   The proof follows by expanding the determinant (see Problem 49).   ❏

We can use Theorem 9.8 to rework Example 5:

$$(\mathbf{u} \times \mathbf{v}) \cdot \mathbf{w} = \begin{vmatrix} 1 & -2 & 3 \\ -4 & 7 & -11 \\ 5 & 9 & -1 \end{vmatrix} = -3$$

See Problem 65 of Problem Set 2 of the Student Mathematics Handbook.
Thus, the volume is $|-3| = 3$, as computed directly in Example 5.

WARNING   In the problem set, we have included several exercises involving the scalar triple product and the triple vector product $\mathbf{u} \times \mathbf{v} \times \mathbf{w}$. In case you wonder why we neglect the product $(\mathbf{u} \cdot \mathbf{v}) \times \mathbf{w}$, notice that such a product makes no sense, because $\mathbf{u} \cdot \mathbf{v}$ is a scalar and the cross product is an operation involving only vectors. Thus, the product $\mathbf{u} \cdot \mathbf{v} \times \mathbf{w}$ must mean $\mathbf{u} \cdot (\mathbf{v} \times \mathbf{w})$.

9.4   PROBLEM SET

Ⓐ *Find $\mathbf{v} \times \mathbf{w}$ for the vectors given in Problems 1–10.*

1. $\mathbf{v} = \mathbf{i}$; $\mathbf{w} = \mathbf{j}$ 2. $\mathbf{v} = \mathbf{k}$; $\mathbf{w} = \mathbf{k}$

3. $\mathbf{v} = 3\mathbf{i} + 2\mathbf{k}$; $\mathbf{w} = 2\mathbf{i} + \mathbf{j}$ 4. $\mathbf{v} = \mathbf{i} - 3\mathbf{j}$; $\mathbf{w} = \mathbf{i} + 5\mathbf{k}$

5. $\mathbf{v} = 3\mathbf{i} - 2\mathbf{j} + 4\mathbf{k}$; $\mathbf{w} = \mathbf{i} + 4\mathbf{j} - 7\mathbf{k}$

6. $\mathbf{v} = 5\mathbf{i} - \mathbf{j} + 2\mathbf{k}$; $\mathbf{w} = 2\mathbf{i} + \mathbf{j} - 3\mathbf{k}$

7. $\mathbf{v} = 3\mathbf{i} - \mathbf{j} + 2\mathbf{k}$; $\mathbf{w} = 2\mathbf{i} + 3\mathbf{j} - 4\mathbf{k}$

8. $\mathbf{v} = -\mathbf{j} + 4\mathbf{k}$; $\mathbf{w} = 5\mathbf{i} + 6\mathbf{k}$

9. $\mathbf{v} = \mathbf{i} - 6\mathbf{j} + 10\mathbf{k}$; $\mathbf{w} = -\mathbf{i} + 5\mathbf{j} - 6\mathbf{k}$

10. $\mathbf{v} = \cos\theta\mathbf{i} + \sin\theta\mathbf{j}$; $\mathbf{w} = -\sin\theta\mathbf{i} + \cos\theta\mathbf{j}$ for any angle θ

Use the cross product to find $\sin\theta$ where θ is the angle between $\mathbf{v}$ and $\mathbf{w}$ in Problems 11–14.

11. $\mathbf{v} = \mathbf{i} + \mathbf{k}$; $\mathbf{w} = \mathbf{i} + \mathbf{j}$ 12. $\mathbf{v} = \mathbf{i} + \mathbf{j}$; $\mathbf{w} = \mathbf{i} + \mathbf{j} + \mathbf{k}$

13. $\mathbf{v} = \mathbf{j} + \mathbf{k}$; $\mathbf{w} = \mathbf{i} + \mathbf{k}$ 14. $\mathbf{v} = \mathbf{i} + \mathbf{j}$; $\mathbf{w} = \mathbf{j} + \mathbf{k}$

Find a unit vector that is orthogonal to both $\mathbf{v}$ and $\mathbf{w}$ in Problems 15–18.

15. $\mathbf{v} = 2\mathbf{i} + \mathbf{k}$; $\mathbf{w} = \mathbf{i} - \mathbf{j} - \mathbf{k}$

16. $\mathbf{v} = \mathbf{j} - 3\mathbf{k}$; $\mathbf{w} = -\mathbf{i} + \mathbf{j} + \mathbf{k}$

17. $\mathbf{v} = \mathbf{i} + \mathbf{j} + \mathbf{k}$; $\mathbf{w} = 3\mathbf{i} + 12\mathbf{j} - 4\mathbf{k}$

18. $\mathbf{v} = 2\mathbf{i} - 2\mathbf{j} + \mathbf{k}$; $\mathbf{w} = 4\mathbf{i} + 2\mathbf{j} - 3\mathbf{k}$

Find the area of the parallelogram determined by the vectors in Problems 19–22.

19. $\mathbf{v} = 3\mathbf{i} + 4\mathbf{j}$ and $\mathbf{w} = \mathbf{i} + \mathbf{j} - \mathbf{k}$

20. $\mathbf{v} = 2\mathbf{i} - \mathbf{j} + 2\mathbf{k}$ and $\mathbf{w} = 4\mathbf{i} - 3\mathbf{j}$

21. $\mathbf{v} = 4\mathbf{i} - \mathbf{j} + \mathbf{k}$ and $\mathbf{w} = 2\mathbf{i} + 3\mathbf{j} - \mathbf{k}$

22. $\mathbf{v} = 2\mathbf{i} + 3\mathbf{k}$ and $\mathbf{w} = 2\mathbf{j} - 3\mathbf{k}$

Find the area of △PQR in Problems 23–26.

23. $P(0, 1, 1)$, $Q(1, 1, 0)$, $R(1, 0, 1)$

24. $P(1, 0, 0)$, $Q(2, 1, -1)$, $R(0, 1, -2)$

25. $P(1, 2, 3)$, $Q(2, 3, 1)$, $R(3, 1, 2)$

26. $P(-1, -1, -1)$, $Q(1, -1, -1)$, $R(-1, 1, -1)$

Determine whether each product in Problems 27–29 is a scalar or a vector or does not exist. Explain your reasoning.

27. a. $\mathbf{u} \times (\mathbf{v} \cdot \mathbf{w})$ **b.** $\mathbf{u} \cdot (\mathbf{v} \times \mathbf{w})$

28. a. $\mathbf{u} \times (\mathbf{v} \times \mathbf{w})$ **b.** $\mathbf{u} \cdot (\mathbf{v} \cdot \mathbf{w})$

29. a. $(\mathbf{u} \times \mathbf{v}) \cdot (\mathbf{u} \times \mathbf{w})$ **b.** $(\mathbf{u} \times \mathbf{v}) \times (\mathbf{u} \times \mathbf{w})$

*In Problems 30–33, find the volume of the parallelepiped determined by vectors **u**, **v**, and **w**.*

30. $\mathbf{u} = \mathbf{i} + \mathbf{j}$; $\mathbf{v} = \mathbf{j} + 2\mathbf{k}$; $\mathbf{w} = 3\mathbf{k}$

31. $\mathbf{u} = \mathbf{j} + \mathbf{k}$; $\mathbf{v} = 2\mathbf{i} + \mathbf{j} + 2\mathbf{k}$; $\mathbf{w} = 5\mathbf{i}$

32. $\mathbf{u} = \mathbf{i} + \mathbf{j} + \mathbf{k}$; $\mathbf{v} = \mathbf{i} - \mathbf{j} - \mathbf{k}$; $\mathbf{w} = 2\mathbf{i} + 3\mathbf{k}$

33. $\mathbf{u} = 2\mathbf{i} + \mathbf{j} - \mathbf{k}$; $\mathbf{v} = 3\mathbf{i} + \mathbf{k}$; $\mathbf{w} = \mathbf{j} + \mathbf{k}$

B **34. WHAT DOES THIS SAY?**   Contrast dot and cross products of vectors, including a discussion of some of their properties.

35. WHAT DOES THIS SAY?   What is the right-hand rule?

36. Counterexample Problem

a. If $\mathbf{u} \times \mathbf{w} = \mathbf{v} \times \mathbf{w}$, does it follow that $\mathbf{u} = \mathbf{v}$?

b. If $\mathbf{u} \cdot \mathbf{w} = \mathbf{v} \cdot \mathbf{w}$, does it follow that $\mathbf{u} = \mathbf{v}$?

37. Find a number s that guarantees that the vectors $\mathbf{i}$, $\mathbf{i} + \mathbf{j} + \mathbf{k}$, and $\mathbf{i} + 2\mathbf{j} + s\mathbf{k}$ will all be coplanar (in the same plane).

38. Find a number t that guarantees the vectors $\mathbf{i} + \mathbf{j}$, $2\mathbf{i} - \mathbf{j} + \mathbf{k}$, and $\mathbf{i} + \mathbf{j} + t\mathbf{k}$ will all be coplanar (in the same plane).

39. Let $\mathbf{u} = \mathbf{i} + \mathbf{j}$, $\mathbf{v} = 2\mathbf{i} - \mathbf{j} + \mathbf{k}$, and $\mathbf{w} = 3\mathbf{i}$. Compute $(\mathbf{u} \times \mathbf{v}) \times \mathbf{w}$ and $\mathbf{u} \times (\mathbf{v} \times \mathbf{w})$. What does this say about the associativity of the cross product?

40. Show that $(a\mathbf{u}) \times (b\mathbf{v}) = ab(\mathbf{u} \times \mathbf{v})$ for any scalars a and b.

41. For a given vector $\mathbf{v}$ in $\mathbb{R}^3$, find all vectors $\mathbf{w}$ such that $\mathbf{v} \times \mathbf{w} = \mathbf{w}$.

42. What can be said about nonzero vectors $\mathbf{v}$ and $\mathbf{w}$ if $\mathbf{v} \cdot \mathbf{w} = 0$? What can be said if $\mathbf{v} \times \mathbf{w} = \mathbf{0}$?

43. One end of a 2-ft lever pivots about the origin in the yz-plane, as shown in Figure 9.44.

Figure 9.44   Finding the torque

If a vertical force of 40 lb is applied at the end of the lever, what is the torque of the lever about the pivot point (the origin) when the lever makes an angle of 30° with the xy-plane?

44. A 40-lb child sits on a seesaw, 3 ft from the fulcrum, as shown in Figure 9.45. What torque is exerted when the child is 2 ft above the horizontal?

Figure 9.45   Seesaw torque

45. Find the angle between the vector $2\mathbf{i} - \mathbf{j} + \mathbf{k}$ and the plane determined by the points $P(1, -2, 3)$, $Q(-1, 2, 3)$, and $R(1, 2, -3)$.

46. a. Show that the vectors $\mathbf{u}$, $\mathbf{v}$, and $\mathbf{w}$ are coplanar (all in the same plane) if

$$\mathbf{u} \cdot (\mathbf{v} \times \mathbf{w}) = 0 \qquad \text{or} \qquad (\mathbf{u} \times \mathbf{v}) \cdot \mathbf{w} = 0$$

b. Are the vectors $\mathbf{u} = \mathbf{i} + 3\mathbf{j} + \mathbf{k}$, $\mathbf{v} = 2\mathbf{i} - \mathbf{j} - \mathbf{k}$, and $\mathbf{w} = 7\mathbf{j} + 3\mathbf{k}$ coplanar?

47. Show that the triangle with vertices (x_1, y_1), (x_2, y_2), (x_3, y_3) has area $A = \frac{1}{2}|D|$, where

$$D = \begin{vmatrix} x_1 & y_1 & 1 \\ x_2 & y_2 & 1 \\ x_3 & y_3 & 1 \end{vmatrix}$$

48. Using the properties of determinants, show that

$$\mathbf{u} \cdot (\mathbf{v} \times \mathbf{w}) = (\mathbf{u} \times \mathbf{v}) \cdot \mathbf{w}$$

for any vectors $\mathbf{u}$, $\mathbf{v}$, and $\mathbf{w}$.

C **49.** Prove the determinant formula for evaluating a scalar triple product.

50. Let A, B, C, and D be four points that do not lie in the same plane; see Figure 9.46, for example. It can be shown that the volume of the tetrahedron with vertices A, B, C, and D satisfies

$$\begin{bmatrix} \text{VOLUME OF} \\ \text{TETRAHEDRON} \\ ABCD \end{bmatrix} = \frac{1}{3} \begin{pmatrix} \text{AREA OF} \\ \triangle ABC \end{pmatrix} \begin{pmatrix} \text{ALTITUDE} \\ \text{FROM } D \\ \text{TO } \triangle ABC \end{pmatrix}$$

Show that the volume is given by

$$V = \tfrac{1}{6} |(\mathbf{AB} \times \mathbf{AC}) \cdot \mathbf{AD}|$$

Figure 9.46   Problem 50

51. Show that if $\mathbf{u}$, $\mathbf{v}$, and $\mathbf{w}$ are vectors in $\mathbb{R}^3$ with $\mathbf{u} + \mathbf{v} + \mathbf{w} = \mathbf{0}$, then

$$\mathbf{u} \times \mathbf{v} = \mathbf{v} \times \mathbf{w} = \mathbf{w} \times \mathbf{u}$$

52. Suppose $\mathbf{u}$, $\mathbf{v}$, and $\mathbf{w}$ are nonzero vectors in $\mathbb{R}^3$ with $\mathbf{u} \times \mathbf{v} = \mathbf{w}$ and $\mathbf{u} \cdot \mathbf{v} = 0$. Show that

$$\mathbf{v} = s(\mathbf{w} \times \mathbf{u}) \quad \text{and} \quad \mathbf{u} = t(\mathbf{v} \times \mathbf{w})$$

for scalars s and t.

53. Show that

$$\tan \theta = \frac{\|\mathbf{v} \times \mathbf{w}\|}{\mathbf{v} \cdot \mathbf{w}}$$

where θ ($0 \le \theta < \frac{\pi}{2}$) is the angle between $\mathbf{v}$ and $\mathbf{w}$.

54. Let $\mathbf{u}$, $\mathbf{v}$, and $\mathbf{w}$ be nonzero vectors in $\mathbb{R}^3$ that do not all lie in the same plane. Show that

$$|\mathbf{u} \cdot (\mathbf{v} \times \mathbf{w})| = |\mathbf{v} \cdot (\mathbf{u} \times \mathbf{w})|$$

What other scalar triple products involving $\mathbf{u}$, $\mathbf{v}$, and $\mathbf{w}$ have the same absolute values?

55. Let $\mathbf{a}$, $\mathbf{b}$, and $\mathbf{c}$ be vectors in space. Prove the "cab - bac" formula.

$$\mathbf{a} \times (\mathbf{b} \times \mathbf{c}) = (\mathbf{c} \cdot \mathbf{a})\mathbf{b} - (\mathbf{b} \cdot \mathbf{a})\mathbf{c}$$

Establish the validity of the equations in Problems 56–59 for arbitrary vectors $\mathbf{u}$, $\mathbf{v}$, $\mathbf{w}$, *and* $\mathbf{z}$ *in* $\mathbb{R}^3$. *You may use the result of Problem 55.*

56. $(\mathbf{u} \times \mathbf{v}) \times (\mathbf{w} \times \mathbf{z}) = (\mathbf{u} \cdot \mathbf{w} \times \mathbf{z})\mathbf{v} - (\mathbf{v} \cdot \mathbf{w} \times \mathbf{z})\mathbf{u}$

57. $\mathbf{u} \times (\mathbf{v} \times \mathbf{w}) + \mathbf{v} \times (\mathbf{w} \times \mathbf{u}) + \mathbf{w} \times (\mathbf{u} \times \mathbf{v}) = \mathbf{0}$

58. $\mathbf{u} \times \mathbf{v} = (\mathbf{u} \cdot \mathbf{v} \times \mathbf{i})\mathbf{i} + (\mathbf{u} \cdot \mathbf{v} \times \mathbf{j})\mathbf{j} + (\mathbf{u} \cdot \mathbf{v} \times \mathbf{k})\mathbf{k}$

59. $\mathbf{u} \times [\mathbf{u} \times (\mathbf{u} \times \mathbf{v})] \cdot \mathbf{w} = -\|\mathbf{u}\|^2 \mathbf{u} \cdot \mathbf{v} \times \mathbf{w}$

60. Show that if vectors $\mathbf{OP}$, $\mathbf{OQ}$, $\mathbf{OR}$, and $\mathbf{OS}$ lie in the same plane, then

$$(\mathbf{OP} \times \mathbf{OQ}) \times (\mathbf{OR} \times \mathbf{OS}) = \mathbf{0}$$

9.5 Parametric Representation of Curves; Lines in $\mathbb{R}^3$

IN THIS SECTION parametric equations, parameterizing a curve, lines in $\mathbb{R}^3$

PARAMETRIC EQUATIONS

Until now, most of the curves we have considered have been graphs of functions of the form $y = f(x)$. This form of representation is limited by the requirements of the vertical line test, which precludes closed curves such as circles or any curve with a self-intersection. However, such restrictions are often not necessary for curves where the coordinates x and y of each point $P(x, y)$ on the curve are themselves functions of a third variable, called a *parameter*. We will begin by examining parametric representations in $\mathbb{R}^2$, and then extend our results to $\mathbb{R}^3$.

Parametric Representation for a Curve in $\mathbb{R}^2$

> Let f and g be continuous functions of t on an interval I; then the equations
>
> $$x = f(t) \quad \text{and} \quad y = g(t)$$
>
> are called **parametric equations** with **parameter** t. As t varies over the **parametric set** I, the points $(x, y) = (f(t), g(t))$ trace out a **parametric curve**.

Note: The letter "t" used for the parameter does not necessarily denote time, although in many applications, time is a suitable parameter. Indeed, any letter or symbol may be used to denote a parameter.

EXAMPLE 1 **Sketching the path of a parametric curve**

Sketch the path of the curve $x = t^2 - 9$, $y = \frac{1}{3}t$ for $-3 \le t \le 2$.

Solution

Values of x and y corresponding to various choices of the parameter t are shown in the following table:

t	x	y	
-3	0	-1	(Starting or **initial** point)
-2	-5	$-\frac{2}{3}$	
-1	-8	$-\frac{1}{3}$	
0	-9	0	
1	-8	$\frac{1}{3}$	
2	-5	$\frac{2}{3}$	(Ending or **terminal** point)

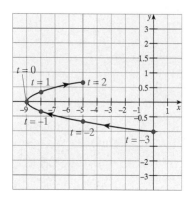

Figure 9.47 Graph of $x = t^2 - 9$, $y = \frac{1}{3}t$, for $-3 \le t \le 2$. Notice how the arrows show the orientation as t increases from -3 to 2.

The graph is shown in Figure 9.47. ∎

If you are using a computer or a graphing calculator, plotting points can be an effective way of sketching a parametric curve. Sometimes, however, we wish to eliminate the parameter to obtain a Cartesian equation. For instance, in Example 1, we have $y = \frac{1}{3}t$, so $t = 3y$, and by substituting into the equation $x = t^2 - 9$, we obtain

$$x = (3y)^2 - 9 = 9y^2 - 9$$

which is the Cartesian equation for a parabola that opens to the right. Because of the domain of the parameter t, we see that the parametric curve in Figure 9.47 is a subset of the set of points that satisfy the equation $x = 9y^2 - 9$.

Parameterizations are not unique. For example, the curve with parametric equations

$$x = 9(9t^2 - 1), \quad y = 3t \quad \text{for } -\tfrac{1}{3} \le t \le \tfrac{2}{9}$$

is the same as the curve in Figure 9.47.

EXAMPLE 2 Sketching the path of a parametric curve by eliminating the parameter

Describe the path $x = \sin \pi t$, $y = \cos 2\pi t$ for $0 \le t \le 0.5$.

Solution
Using a double angle identity, we find

$$\cos 2\pi t = 1 - 2 \sin^2 \pi t$$

so that

$$y = 1 - 2x^2$$

We recognize this as a Cartesian equation for a parabola. Because $y' = -4x$, we can find the critical number $x = 0$, which locates the vertex of the parabola at $(0, 1)$. The parabola is the curve shown in color as the dashed curve in Figure 9.48.

Because t is restricted to the interval $0 \le t \le 0.5$, the parametric representation involves only part of the right side of the parabola $y = 1 - 2x^2$. The curve is oriented from the point $(0, 1)$, where $t = 0$, to the point $(1, -1)$, where $t = 0.5$, and is the portion of the parabola shown in black in Figure 9.48. ∎

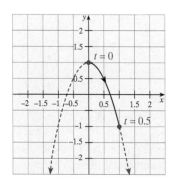

Figure 9.48 The parabolic arc $x = \sin \pi t$, $y = \cos 2\pi t$ for $0 \le t \le 0.5$

When it is difficult to eliminate the parameter from a given parametric representation, we can sometimes get a good picture of the parametric curve by plotting points.

EXAMPLE 3   Describing a spiraling path

Discuss the path of the curve described by the parametric equations

$$x = e^{-t} \cos t, \qquad y = e^{-t} \sin t \qquad \text{for } t \geq 0$$

Solution

We have no convenient way of eliminating the parameter so we write out a table of values (x, y) that correspond to various values of t. The curve is obtained by plotting these points in a Cartesian plane and passing a smooth curve through the plotted points, as shown in Figure 9.49.

t	x	y
0	1	0
$\frac{\pi}{4}$	0.32	0.32
1	0.20	0.31
$\frac{\pi}{2}$	0	0.21
2	−0.06	0.12
π	−0.04	0
$\frac{3\pi}{2}$	0	−0.01
2π	0.00	0

Figure 9.49   Graph of $x = e^{-t} \cos t$, $y = e^{-t} \sin t$ for $t \geq 0$

Note that for each value of t, the distance from $P(x, y)$ on the curve to the origin is

$$\sqrt{x^2 + y^2} = \sqrt{(e^{-t} \cos t)^2 + (e^{-t} \sin t)^2} = \sqrt{e^{-2t}(1)} = e^{-t}$$

Because e^{-t} decreases as t increases, it follows that P gets closer and closer to the origin as t increases. However, because $\cos t$ and $\sin t$ vary between -1 and $+1$, the approach is not direct but takes place along a spiral.   ■

PARAMETERIZING A CURVE

So far, our examples have dealt with sketching a parametric curve given the parametric equations. In general, this process may be tedious, but generally can be done easily using technology. However, the reverse process, finding a suitable set of parametric equations for a given curve, is an art for which there is no simple procedure. Indeed, a given curve can have many different parameterizations and there are curves for which no simple parameterization can be given. The following two examples illustrate various methods for parameterizing a given curve.

EXAMPLE 4   Parameterizing two curves

In each of the following cases, parameterize the given curve:

a. $y = 9x^2$ **b.** $r = 5 \cos^3 \theta$ in polar coordinates

Solution

a. The usual parameterization for a parabola is to let the parameter t be the variable that is squared: $x = t$, $y = 9t^2$. However, another parameterization is to let $t = 3x$ so that $x = \frac{1}{3}t$ and $y = t^2$.

b. In polar coordinates we have $x = r \cos \theta$, $y = r \sin \theta$, so we can parameterize x and y in terms of the parameter θ:

$$x = r \cos \theta \qquad\qquad y = r \sin \theta$$
$$= (5 \cos^3 \theta) \cos \theta \qquad\qquad = (5 \cos^3 \theta) \sin \theta$$
$$= 5 \cos^4 \theta$$

■

EXAMPLE 5   Modeling Problem: Finding parametric equations for a trochoid

A bicycle wheel has radius a and a reflector is attached at a point P on a spoke of the bicycle wheel at a fixed distance d from the center. Find parametric equations for the curve described by P as the wheel rolls along a straight line without slipping. Such a curve is called a **trochoid**, and is shown in Figure 9.50.

Figure 9.50   The path of a reflector placed 5 in. from the tire on the wheel of a bike with 30-in. wheel. This is an example of a trochoid.

Solution

Assume that the wheel rolls along the x-axis and that the center C of the wheel begins at $(0, a)$ on the y-axis. Further assume that P also starts on the y-axis, d units below C. Figure 9.51 shows the initial position of the wheel and its position after turning through an angle θ (radians).

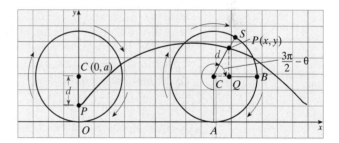

Figure 9.51   The path of a reflector on a bicycle

We begin by labeling some points: The point A is on the x-axis directly beneath C, whereas B is the point where the horizontal line through C meets the rim of the wheel. Finally, Q is the point on BC directly beneath P, and S is the point where the line through C and P intersects the rim. Let P have coordinates (x, y).

We need to find representations (in terms of a, d, and θ) for x and y.

$$x = |\overline{OA}| + |\overline{CQ}|$$
$$= a\theta + |\overline{CQ}|$$

Because the wheel rolls along the x-axis without slipping, $|\overline{OA}|$ is the same as the arc length from A to S, so $|\overline{OA}| = a\theta$.

$$y = |\overline{AC}| + |\overline{QP}|$$
$$= a + |\overline{QP}|$$

To complete our evaluation of x and y, we need to compute $\left|\overline{CQ}\right|$ and $\left|\overline{QP}\right|$. These are sides of $\triangle PCQ$. Note that $\angle PCQ = \frac{3\pi}{2} - \theta$; therefore, by the definitions of cosine and sine, we have

$$\cos\left(\frac{3\pi}{2} - \theta\right) = \frac{\left|\overline{CQ}\right|}{d} \quad \text{so} \quad \left|\overline{CQ}\right| = d\cos\left(\frac{3\pi}{2} - \theta\right) = -d\sin\theta$$

Similarly, $\left|\overline{QP}\right| = d\sin\left(\frac{3\pi}{2} - \theta\right) = -d\cos\theta$.

We can now substitute these values for $\left|\overline{CQ}\right|$ and $\left|\overline{QP}\right|$ into the equations we derived for x and y.

$$x = a\theta + \left|\overline{CQ}\right| = a\theta - d\sin\theta$$
$$y = a + \left|\overline{QP}\right| = a - d\cos\theta \qquad \blacksquare$$

The special case where P is on the rim of the wheel in Example 5 (when $d = a$) is a curve called a **cycloid**. There are several problems involving these and similar curves in the problem set.

If f, g, and h are continuous functions of a variable t on an interval I, then

$$x = f(t) \qquad y = g(t) \qquad z = h(t)$$

are parametric equations with parameter t, and as t varies over I, the points $(x, y, z) = (f(t), g(t), h(t))$ trace out a parametric curve in $\mathbb{R}^3$. The rest of this section is devoted to examining lines in $\mathbb{R}^3$, which we will represent in parametric form. More general curves in $\mathbb{R}^3$ will be studied in Chapter 10.

LINES IN $\mathbb{R}^3$

As in the plane, a line in $\mathbb{R}^3$ is completely determined once we know one of its points and its direction. We used the concept of slope to measure the direction of a line in the plane, but in space, it is more convenient to specify direction with vectors.

Suppose L is a line in $\mathbb{R}^3$ that is parallel to the vector $\mathbf{v} = A\mathbf{i} + B\mathbf{j} + C\mathbf{k}$ and also contains $Q(x_0, y_0, z_0)$, as shown in Figure 9.52. We say that the line has **direction numbers** A, B, and C and denote these direction numbers by $[A, B, C]$. The vector $\mathbf{v}$ is called a **direction vector** of the line L. If $P(x, y, z)$ is any point on L, then the vector $\mathbf{QP}$ is parallel to $\mathbf{v}$ and must satisfy the vector equation $\mathbf{QP} = t\mathbf{v}$ for some number t. If we introduce coordinates and use the standard representation, we can rewrite this vector equation as

$$(x - x_0)\mathbf{i} + (y - y_0)\mathbf{j} + (z - z_0)\mathbf{k} = t(A\mathbf{i} + B\mathbf{j} + C\mathbf{k})$$

By equating components on both sides of this equation, we find that the coordinates of P must satisfy the linear system

$$x - x_0 = tA \qquad y - y_0 = tB \qquad z - z_0 = tC$$

where t is a real number. Note that each point on L corresponds to exactly one value of the parameter t.

Figure 9.52 If L is parallel to **v** and contains Q, then P is on L whenever $QP = t\mathbf{v}$.

Parametric Form of a Line in $\mathbb{R}^3$

If L is a line that contains the point (x_0, y_0, z_0) and is parallel to the vector $\mathbf{v} = A\mathbf{i} + B\mathbf{j} + C\mathbf{k}$, then L has parametric form

$$x = x_0 + tA \qquad y = y_0 + tB \qquad z = z_0 + tC$$

Conversely, the set of all points (x, y, z) that satisfy such a set of equations is a line that passes through the point (x_0, y_0, z_0) and is parallel to a vector with direction numbers $[A, B, C]$.

EXAMPLE 6 Parametric equations of a line in $\mathbb{R}^3$

Find parametric equations for the line that contains the point $(3, 1, 4)$ and is parallel to the vector $\mathbf{v} = -\mathbf{i} + \mathbf{j} - 2\mathbf{k}$. Find where this line passes through the coordinate planes and sketch the line.

Solution

The direction numbers are $[-1, 1, -2]$ and $x_0 = 3$, $y_0 = 1$, $z_0 = 4$, so the line has the parametric form

$$x = 3 - t \qquad y = 1 + t \qquad z = 4 - 2t$$

This line will intersect the xy-plane when $z = 0$;

$$0 = 4 - 2t \quad \text{implies} \quad t = 2$$

If $t = 2$, then $x = 3 - 2 = 1$ and $y = 1 + 2 = 3$. This is the point $(1, 3, 0)$. Similarly, the line intersects the xz-plane at $(4, 0, 6)$ and the yz-plane at $(0, 4, -2)$. Plot these points and draw the line, as shown in Figure 9.53. ∎

In the special case where none of the direction numbers A, B, or C is 0, we can solve each of the parametric-form equations for t to obtain the following **symmetric equations** for a line.

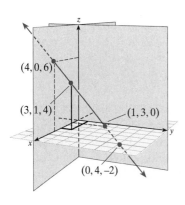

Figure 9.53 Graph of the line $x = 3 - t$, $y = 1 + t$, $z = 4 - 2t$

Symmetric Form of a Line in $\mathbb{R}^3$

If L is a line that contains the point (x_0, y_0, z_0) and is parallel to the vector $\mathbf{v} = A\mathbf{i} + B\mathbf{j} + C\mathbf{k}$ (A, B, and C nonzero numbers), then the point (x, y, z) is on L if and only if its coordinates satisfy

$$\frac{x - x_0}{A} = \frac{y - y_0}{B} = \frac{z - z_0}{C}$$

EXAMPLE 7 Symmetric form of the equation of a line in $\mathbb{R}^3$

Find symmetric equations for the line L through the points $P(-1, 3, 7)$ and $Q(4, 2, -1)$. Find the points of intersection with the coordinate planes and sketch the line.

Solution

The required line passes through P and is parallel to the vector

$$\mathbf{PQ} = [4 - (-1)]\mathbf{i} + [2 - 3]\mathbf{j} + [-1 - 7]\mathbf{k} = 5\mathbf{i} - \mathbf{j} - 8\mathbf{k}$$

Thus, the direction numbers of the line are $[5, -1, -8]$, and we can choose either P or Q as (x_0, y_0, z_0). Choosing P, we obtain

$$\frac{x + 1}{5} = \frac{y - 3}{-1} = \frac{z - 7}{-8}$$

Next, we find points of intersection with the coordinate planes. For the xy-plane $(z = 0)$, we have

$$\frac{x + 1}{5} = \frac{0 - 7}{-8} \quad \text{and} \quad \frac{y - 3}{-1} = \frac{0 - 7}{-8}$$

$$x = \frac{27}{8} \qquad\qquad y = \frac{17}{8}$$

The point of intersection of the line with the xy-plane is $\left(\frac{27}{8}, \frac{17}{8}, 0\right)$. Similarly, the other points of intersections are

$$xz\text{-plane:} \quad (14, 0, -17) \qquad\qquad yz\text{-plane:} \quad \left(0, \frac{14}{5}, \frac{27}{5}\right)$$

The graph is shown in Figure 9.54. ∎

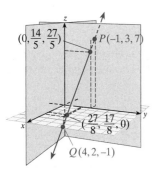

Figure 9.54 Graph of $\dfrac{x + 1}{5} = \dfrac{y - 3}{-1} = \dfrac{z - 7}{-8}$

If one or more of the direction numbers $[A, B, C]$ of a line L in $\mathbb{R}^3$ is zero, then L is parallel to at least one coordinate plane. For example, the line

$$x = 3 + 2t, \quad y = 5, \quad z = 4 - t$$

has direction numbers $[2, 0, -1]$ and lies in the plane $y = 5$, parallel to the xz-plane. The symmetric form for this line is

$$\frac{x - 3}{2} = \frac{z - 4}{-1}; \quad y = 5$$

Turning things around, the line with symmetric form

$$\frac{y + 1}{4} = \frac{z - 3}{-2}; \quad x = 7$$

lies in the plane $x = 7$, parallel to the yz-plane and has direction numbers $[0, 4, -2]$.

Recall that two lines in $\mathbb{R}^2$ must intersect if their slopes are different (because they cannot be parallel). However, two lines in $\mathbb{R}^3$ may have different direction numbers and still not intersect. In this case, the lines are said to be **skew.** The three different situations that can occur (ignoring the trivial case where the lines coincide) are shown in Figure 9.55.

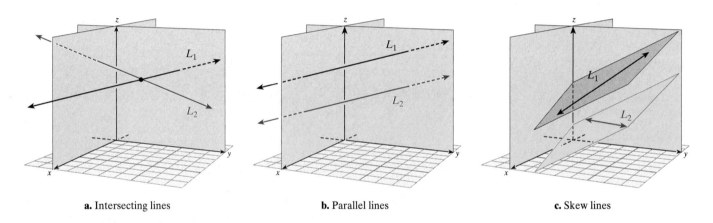

a. Intersecting lines   **b.** Parallel lines   **c.** Skew lines

Figure 9.55   Lines in space may intersect, be parallel, or skew.

EXAMPLE 8   Intersecting lines in $\mathbb{R}^3$

Show that the lines

$$L_1: \frac{x - 1}{2} = \frac{y + 1}{1} = \frac{z - 2}{4} \quad \text{and} \quad L_2: \frac{x + 2}{4} = \frac{y}{-3} = \frac{z - \frac{1}{2}}{-1}$$

intersect and find the point of intersection.

Solution

L_1 has direction numbers $[2, 1, 4]$ and L_2 has direction numbers $[4, -3, -1]$. Because there is no t for which $[2, 1, 4] = t[4, -3, -1]$, the lines are not parallel. Express the lines in parametric form:

$$L_1: x = 1 + 2s \qquad y = -1 + s \qquad z = 2 + 4s$$
$$L_2: x = -2 + 4t \qquad y = -3t \qquad z = \tfrac{1}{2} - t$$

At an intersection point we must have

$$1 + 2s = -2 + 4t \qquad \text{or} \qquad 2s - 4t = -3$$
$$-1 + s = -3t \qquad \text{or} \qquad s + 3t = 1$$
$$2 + 4s = \tfrac{1}{2} - t \qquad \text{or} \qquad 4s + t = -\tfrac{3}{2}$$

Solving the first two equations simultaneously, we find $s = -\frac{1}{2}, t = \frac{1}{2}$. This solution satisfies the third equation, namely,

$$4\left(-\tfrac{1}{2}\right) + \tfrac{1}{2} = -\tfrac{3}{2}$$

If the solution did not satisfy the third equation, the lines would be skew. To find the coordinates of the point of intersection, substitute $s = -\frac{1}{2}$ into the parametric-form equations for L_1 (or substitute $t = \frac{1}{2}$ into L_2) to obtain

$$x_0 = 1 + 2\left(-\tfrac{1}{2}\right) = 0$$
$$y_0 = -1 + \left(-\tfrac{1}{2}\right) = -\tfrac{3}{2}$$
$$z_0 = 2 + 4\left(-\tfrac{1}{2}\right) = 0$$

Thus, the lines intersect at $P\left(0, -\frac{3}{2}, 0\right)$. ∎

WARNING Notice that we used two parameters (s and t) for the lines in Example 8. This is because it is quite possible for two lines to intersect at "different times" (that is, different values of the parameter). In Example 8, we have $s = -\frac{1}{2}$ and $t = \frac{1}{2}$ at the point of intersection, rather than $s = t$.

EXAMPLE 9   Skew lines in $\mathbb{R}^3$

Determine whether the following lines intersect, are parallel, or are skew.

$$L_1 : \frac{x-1}{2} = \frac{y+1}{1} = \frac{z-2}{4} \quad \text{and} \quad L_2 : \frac{x+2}{4} = \frac{y}{-3} = \frac{z+1}{1}$$

Solution

Note that L_1 has direction numbers [2, 1, 4]; that is, L_1 is parallel to $(2\mathbf{i} + \mathbf{j} + 4\mathbf{k})$ and L_2 has direction numbers [4, −3, 1]. If we try to solve

$$\langle 2, 1, 4 \rangle = t\langle 4, -3, 1 \rangle$$

for t, we find no possible solution for any value of t. This implies that the lines are not parallel.

Next we determine whether the lines intersect or are skew. Note that $S(1, -1, 2)$ lies on L_1 and $T(-2, 0, -1)$ lies on L_2. The lines intersect if and only if there is a point P that lies on both lines. To determine this, we write the equations of the lines in parametric form:

$$L_1 : x = 1 + 2s \qquad y = -1 + s \qquad z = 2 + 4s$$
$$L_2 : x = -2 + 4t \qquad y = -3t \qquad z = -1 + t$$

The lines intersect if there are numbers s and t for which

$$
\begin{array}{lcl}
x = 1 + 2s = -2 + 4t & \text{or} & 2s - 4t = -3 \\
y = -1 + s = -3t & \text{or} & s + 3t = 1 \\
z = 2 + 4s = -1 + t & \text{or} & 4s - t = -3
\end{array}
$$

This is equivalent to the system of linear equations

$$\begin{cases} 2s - 4t = -3 \\ s + 3t = 1 \\ 4s - t = -3 \end{cases}$$

Any solution of this system must correspond to a point of intersection of L_1 and L_2, and if no solution exists, then L_1 and L_2 are skew. Because this is a system of three equations with two unknowns, we first solve the first two equations simultaneously to find $s = -\frac{1}{2}, t = \frac{1}{2}$. However, $(s, t) = \left(-\frac{1}{2}, \frac{1}{2}\right)$ does not satisfy the third equation, so it follows that L_1 and L_2 do not intersect and they must be skew. ∎

9.5 PROBLEM SET

A **1.** **WHAT DOES THIS SAY?**   What is a parameter?

2. **WHAT DOES THIS SAY?**   Contrast the parametric and symmetric forms of the equation of a line.

Find an explicit relationship between x and y in Problems 3–16 by eliminating the parameter. In each case, sketch the path described by the parametric equations over the prescribed interval.

3. $x = t + 1, y = t - 1, 0 \le t \le 2$

4. $x = -t, y = 3 - 2t, 0 \le t \le 1$

5. $x = 60t, y = 80t - 16t^2, 0 \le t \le 3$

6. $x = 30t, y = 60t - 9t^2, -1 \le t \le 2$

7. $x = t^3, y = t^2, t \ge 0$

8. $x = t^4, y = t^2, -1 \le t \le \sqrt{2}$

9. $x = 3\cos\theta, y = 3\sin\theta, 0 \le \theta \le 2\pi$

10. $x = 2\sin\theta, y = 2\cos\theta, 0 \le \theta \le 2\pi$

11. $x = 1 + \sin t, y = -2 + \cos t, 0 \le t \le 2\pi$

12. $x = 1 + \sin^2 t, y = -2 + \cos t, 0 \le t \le \pi$

13. $x = 4\tan 2t, y = 3\sec 2t, 0 \le t \le \pi$

14. $x = 4\sec 2t, y = 2\tan 2t, 0 \le t \le \pi$

15. $x = t^3, y = 3\ln t, t > 0$

16. $x = e^t, y = e^{-t}, (-\infty, \infty)$

Find the parametric and symmetric equations for the line(s) passing through the given points with the properties described in Problems 17–25.

17. $(1, -1, -2)$; parallel to $3\mathbf{i} - 2\mathbf{j} + 5\mathbf{k}$

18. $(1, 0, -1)$; parallel to $3\mathbf{i} + 4\mathbf{j}$

19. $(1, -1, 2)$; through $(2, 1, 3)$

20. $(2, 2, 3)$; through $(1, 3, -1)$

21. $(1, -3, 6)$; parallel to $\dfrac{x-5}{1} = \dfrac{y+2}{-3} = \dfrac{z}{-5}$

22. $(1, -1, 2)$; parallel to $\dfrac{x+3}{4} = \dfrac{y-2}{5} = \dfrac{z+5}{1}$

23. $(0, 4, -3)$; parallel to $\dfrac{x-1}{22} = \dfrac{y+2}{-6} = \dfrac{z-1}{10}$

24. $(1, 0, -4)$; parallel to $x = -2 + 3t, y = 4 + t, z = 2 + 2t$

25. $(-1, 1, 6)$; perpendicular to $3x + y - 2z = 5$

26. Find the parametric form of the equation of the line passing through $(3, -1, 0)$ parallel to both the xy- and yz-planes.

Find the points of intersection of each line in Problems 27–30 with each of the coordinate planes.

27. $\dfrac{x-4}{4} = \dfrac{y+3}{3} = \dfrac{z+2}{1}$

28. $\dfrac{x+1}{1} = \dfrac{y+2}{2} = \dfrac{z-6}{3}$

29. $x = 6 - 2t, y = 1 + t, z = 3t$

30. $x = 6 + 3t, y = 2 - t, z = 2t$

In Problems 31–36, tell whether the two lines intersect, are parallel, are skew, or coincide. If they intersect, give the point of intersection.

31. $\dfrac{x-4}{2} = \dfrac{y-6}{-3} = \dfrac{z+2}{5}; \dfrac{x}{4} = \dfrac{y+2}{-6} = \dfrac{z-3}{10}$

32. $x = 4 - 2t, y = 6t, z = 7 - 4t$;
    $x = 5 + t, y = 1 - 3t, z = -3 + 2t$

33. $x = 3 + 3t, y = 1 - 4t, z = -4 - 7t$;
    $x = 2 + 3t, y = 5 - 4t, z = 3 - 7t$

34. $x = 2 - 4t, y = 1 + t, z = \frac{1}{2} + 5t$;
    $x = 3t, y = -2 - t, z = 4 - 2t$

35. $\dfrac{x-3}{2} = \dfrac{y-1}{-1} = \dfrac{z-4}{1}; \dfrac{x+2}{3} = \dfrac{y-3}{-1} = \dfrac{z-2}{1}$

36. $\dfrac{x+1}{2} = \dfrac{y-3}{-1} = \dfrac{z-2}{1}; \dfrac{x+1}{2} = \dfrac{y+1}{3} = \dfrac{z-3}{-4}$

37. Find two unit vectors parallel to the line
$$\frac{x-3}{4} = \frac{y-1}{2} = \frac{z+1}{1}$$

38. Find two unit vectors parallel to the line
$$\frac{x-1}{2} = \frac{y+2}{4} = \frac{z+5}{1}$$

B *Find the parametric equations for each of the curves in Problems 39–44.*

39. A circle of radius 3, centered at the origin, oriented counterclockwise.

40. A circle of radius 2 centered at the origin, oriented clockwise.

41. The ellipse $\dfrac{x^2}{9} + \dfrac{y^2}{4} = 1$, oriented counterclockwise.

42. The parabola $y^2 = 4x + 9$, oriented from $(4, -5)$ to $(0, 3)$.

43. The hyperbola $\dfrac{x^2}{16} - \dfrac{y^2}{9} = 1$.

44. The ellipse $\dfrac{(x-2)^2}{3} + \dfrac{(y+3)^2}{5} = 1$.

45. Describe the path of the curve described by $x = \sin\pi t$, $y = \cos 2\pi t$ for $0 \le t < 1$.

46. Let $x = 4a\sin t, y = b\cos^2 t$. Express y as a function of x.

47. Show that the vector $\mathbf{v} = 3\mathbf{i} - 4\mathbf{j} + \mathbf{k}$ is orthogonal to the line that passes through the points $P(0, 0, 1)$ and $Q(2, 1, -1)$.

48. Show that the vector $\mathbf{v} = 7\mathbf{i} + 4\mathbf{j} + 3\mathbf{k}$ is orthogonal to the line passing through the points $P(-2, 2, 7)$ and $Q(3, -3, 2)$.

49. Find constants a and b so that the following lines coincide:
$$L_1 : \frac{x-a}{2} = \frac{y-1}{4} = \frac{z+2}{1}$$

and

$$L_2 : \frac{x-2}{-4} = \frac{y-b}{-8} = \frac{z+1}{-2}$$

50. Consider the lines
$$L_1 : x = -1 + 2t, \quad y = 3 - t, \quad z = 2 + 2t$$
$$L_2 : x = -2 - t, \quad y = 5 + 2t, \quad z = -2t$$

Show that L_1 and L_2 intersect and find the acute angle θ (to the nearest degree) between them.

51. Find an equation for the line L_1 that contains the point $P(-1, 3, 1)$ and is orthogonal to the line
$$L_1 : \quad x = 2 - t, \quad y = 1 - 2t, \quad z = 5 + t$$

52. What can be said about the lines

$$\frac{x - x_0}{a_1} = \frac{y - y_0}{b_1} = \frac{z - z_0}{c_1}$$

and

$$\frac{x - x_0}{a_2} = \frac{y - y_0}{b_2} = \frac{z - z_0}{c_2}$$

in the case where $a_1 a_2 + b_1 b_2 + c_1 c_2 = 0$?

53. Spy Problem Our friend, the internationally famous Spy, is on a mission to save his superior, Lord Newton Fleming, known as the "Flamer." Having escaped from the minions of his abductor, the archfiend Ernst Stavro Blohardt (Supplementary Problem 79 of Chapter 8), the Spy finds himself in the desert, where Blohardt has just taken off in a helicopter! Suppose the desert and the air space above are coordinatized (units in thousands of feet) and that two observation posts are located at points $A(7, 0, 0)$ and $B(0, 4, 0)$. At noon, Blohardt's helicopter is observed from A in the direction of the vector $\langle -4, 2, 5 \rangle$ and simultaneously from B in the direction of $\langle 3, -2, 5 \rangle$. One minute later, it is observed from A and B in the directions of $\langle 6, 7, 5.005 \rangle$ and $\langle 13, 3, 5.005 \rangle$, respectively. The Spy is notified and at 12:10 P.M. is flying at the point $Q(0, 0, 1)$ in an attack helicopter traveling at 150 mph. Assuming that Blohardt does not change his direction or rate, in what direction from Q should the Spy fly in order to intercept the villain?

54. Modeling Problem A circle of radius R rolls without slipping on the *outside* of a fixed circle of radius a. Assume the fixed circle is centered at the origin and that the moving point begins at $(a, 0)$. Let t be the angle measured from the positive x-axis to the ray from the origin to the center of the rolling circle.

Show that this *epicycloid* may be modeled by the parametric equations

$$x = (a + R) \cos t - R \cos \left(\frac{a + R}{R} \right) t,$$

$$y = (a + R) \sin t - R \sin \left(\frac{a + R}{R} \right) t$$

55. Modeling Problem A circle of radius R rolls without slipping on the inside of a fixed circle of radius a. Find parametric equations to model the curve traced out by a point P on the circumference of the rolling circle of radius R. Let t be the angle measured from the positive x-axis to the ray that passes through the center of the rolling circle, and assume that the point P begins on the x-axis (that is, P has coordinates $(a, 0)$ when $t = 0$). Show that this *hypocycloid* has parametric equations

$$x = (a - R) \cos t + R \cos \left(\frac{a - R}{R} \right) t,$$

$$y = (a - R) \sin t - R \sin \left(\frac{a - R}{R} \right) t$$

56. Journal Problem *The MATYC Journal** ■ A rectangle is "rolled" along a straight line by rotations of 90° about each vertex in turn. One vertex traces out a series of arches where each arch is the union of three circular arcs. Show that the area of the region bounded by the straight line and one arch equals the area of the rectangle plus twice the area of the circumscribed circle.

---

*Problem 148 by Frank Kocher, Vol. 14, 1980, p. 155.

9.6 Planes in $\mathbb{R}^3$

IN THIS SECTION forms for the equation of a plane in $\mathbb{R}^3$, vector methods for measuring distances in $\mathbb{R}^3$

FORMS FOR THE EQUATION OF A PLANE IN $\mathbb{R}^3$

Planes in space can also be characterized by vector methods. In particular, any plane is completely determined once we know one of its points and its orientation; that is, the "direction" it faces. A common way to specify the direction of a plane is by means of a vector **N** that is orthogonal to every vector in the plane, as shown in Figure 9.56. Such a vector is called a **normal** to the plane.

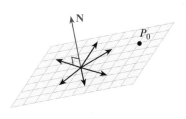

Figure 9.56 A plane may be described by specifying one of its points and a normal vector **N**.

EXAMPLE 1 Finding an equation for a plane

Find an equation for the plane that contains the point $Q(3, -7, 2)$ and is normal to the vector $\mathbf{N} = 2\mathbf{i} + \mathbf{j} - 3\mathbf{k}$.

Solution

The normal vector **N** is orthogonal to every vector in the plane. In particular, if $P(x, y, z)$ is any point in the plane, then **N** must be orthogonal to the vector

$$\mathbf{QP} = (x - 3)\mathbf{i} + (y + 7)\mathbf{j} + (z - 2)\mathbf{k}$$

Because the dot (or scalar) product of two orthogonal vectors is 0, we have

$$\mathbf{N} \cdot \mathbf{QP} = 2(x - 3) + (1)(y + 7) + (-3)(z - 2) = 0$$
$$2x - 6 + y + 7 - 3z + 6 = 0$$
$$2x + y - 3z + 7 = 0$$

Therefore, $2x + y - 3z + 7 = 0$ is the equation of the plane. ■

By generalizing the approach illustrated in Example 1, we can show that the plane that contains the point (x_0, y_0, z_0) and has normal vector $\mathbf{N} = A\mathbf{i} + B\mathbf{j} + C\mathbf{k}$ must have the Cartesian equation

$$A(x - x_0) + B(y - y_0) + C(z - z_0) = 0$$

This is called the **point-normal form** of the equation of a plane. By rearranging terms, we can rewrite this equation in the form $Ax + By + Cz + D = 0$. This is called the **standard form** of the equation of a plane. The numbers $[A, B, C]$ are called **attitude numbers** of the plane (see Figure 9.57).

Notice from Figure 9.57 that *attitude numbers of a plane are the same as direction numbers of a normal vector.* This means that you can find normal vectors to a plane *by inspecting the equation of the plane.*

Figure 9.57 The graph of a plane with attitude numbers $[A, B, C]$

EXAMPLE 2   Relationship between normal vectors and planes

Find normal vectors to the planes

a. $5x + 7y - 3z = 0$ **b.** $3x - 7z = 10$

Solution

a. A normal to the plane $5x + 7y - 3z = 0$ is $\mathbf{N} = 5\mathbf{i} + 7\mathbf{j} - 3\mathbf{k}$.
b. For $3x - 7z = 10$ a normal is $\mathbf{N} = 3\mathbf{i} - 7\mathbf{k}$. ■

Forms for the Equation of a Plane

An equation for the plane with normal $\mathbf{N} = A\mathbf{i} + B\mathbf{j} + C\mathbf{k}$ that contains the point (x_0, y_0, z_0) has the following forms:

     Point-normal form: $A(x - x_0) + B(y - y_0) + C(z - z_0) = 0$
     Standard form: $Ax + By + Cz + D = 0$

Conversely, a normal vector to the plane $Ax + By + Cz + D = 0$ is

$$\mathbf{N} = A\mathbf{i} + B\mathbf{j} + C\mathbf{k}$$

EXAMPLE 3   Equation of a plane containing three given points

Find the standard form equation of a plane containing $P(-1, 2, 1)$, $Q(0, -3, 2)$, and $R(1, 1, -4)$.

Solution

Because a normal $\mathbf{N}$ to the required plane is orthogonal to the vectors $\mathbf{PQ}$ and $\mathbf{PR}$, we find $\mathbf{N}$ by computing the cross product $\mathbf{N} = \mathbf{PR} \times \mathbf{PQ}$. The situation is illustrated in Figure 9.58.

$$\mathbf{PQ} = (0 + 1)\mathbf{i} + (-3 - 2)\mathbf{j} + (2 - 1)\mathbf{k} = \mathbf{i} - 5\mathbf{j} + \mathbf{k}$$
$$\mathbf{PR} = (1 + 1)\mathbf{i} + (1 - 2)\mathbf{j} + (-4 - 1)\mathbf{k} = 2\mathbf{i} - \mathbf{j} - 5\mathbf{k}$$

$$\mathbf{N} = \mathbf{PQ} \times \mathbf{PR} = \begin{vmatrix} \mathbf{i} & \mathbf{j} & \mathbf{k} \\ 1 & -5 & 1 \\ 2 & -1 & -5 \end{vmatrix}$$

$$= 26\mathbf{i} + 7\mathbf{j} + 9\mathbf{k}$$

Figure 9.58 A plane passing through three given points

We can now find the equation of the plane using this normal vector and any point in the plane. We will use the point P:

Attitude numbers of the plane—from $\mathbf{N} = 26\mathbf{i} + 7\mathbf{j} + 9\mathbf{k}$

$$26(x + 1) + 7(y - 2) + 9(z - 1) = 0$$

Point on the plane; we are using $P(-1, 2, 1)$.

Thus, the equation of the plane is

$$26x + 26 + 7y - 14 + 9z - 9 = 0$$
$$26x + 7y + 9z + 3 = 0$$ ∎

EXAMPLE 4   Equation of a line orthogonal to a given plane

Find an equation of the line that passes through the point $Q(2, -1, 3)$ and is orthogonal to the plane $3x - 7y + 5z + 55 = 0$. Where does the line intersect the plane?

Solution

By inspection of the equation of the plane we see that $\mathbf{N} = 3\mathbf{i} - 7\mathbf{j} + 5\mathbf{k}$ is a normal vector. Because the required line is also orthogonal to the plane, it must be parallel to $\mathbf{N}$ (see Figure 9.59). Thus, the line contains the point $Q(2, -1, 3)$ and has direction numbers $[3, -7, 5]$, so its parametric form is

$$x = 2 + 3t, \quad y = -1 - 7t, \quad \text{and} \quad z = 3 + 5t$$

Now, substitute into the equation of the plane:

$$3(2 + 3t) - 7(-1 - 7t) + 5(3 + 5t) = -55$$
$$6 + 9t + 7 + 49t + 15 + 25t = -55$$
$$t = -1$$

Then, the point of intersection is found by substituting $t = -1$ for x, y, and z:

$$x = 2 + 3(-1) = -1$$
$$y = -1 - 7(-1) = 6$$
$$z = 3 + 5(-1) = -2$$

The point of intersection is $(-1, 6, -2)$. ∎

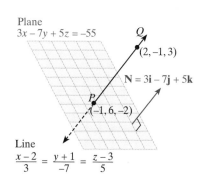

Plane
$3x - 7y + 5z = -55$

Q
$(2, -1, 3)$

$\mathbf{N} = 3\mathbf{i} - 7\mathbf{j} + 5\mathbf{k}$

P
$(-1, 6, -2)$

Line
$\dfrac{x-2}{3} = \dfrac{y+1}{-7} = \dfrac{z-3}{5}$

Figure 9.59 Intersection of a line and a plane

EXAMPLE 5   Equation of a line parallel to the intersection of two given planes

Find the equation of a line passing through $(-1, 2, 3)$ that is parallel to the line of intersection of the planes $3x - 2y + z = 4$ and $x + 2y + 3z = 5$.

Solution

By inspection, we see that normals to the given planes are $\mathbf{N}_1 = 3\mathbf{i} - 2\mathbf{j} + \mathbf{k}$ and $\mathbf{N}_2 = \mathbf{i} + 2\mathbf{j} + 3\mathbf{k}$. The desired line is perpendicular to both of these normals, so a vector parallel to the line is found by computing the cross product:

$$\mathbf{N}_1 \times \mathbf{N}_2 = \begin{vmatrix} \mathbf{i} & \mathbf{j} & \mathbf{k} \\ 3 & -2 & 1 \\ 1 & 2 & 3 \end{vmatrix} = (-6 - 2)\mathbf{i} - (9 - 1)\mathbf{j} + (6 + 2)\mathbf{k}$$

$$= -8\mathbf{i} - 8\mathbf{j} + 8\mathbf{k}$$
$$= -8(\mathbf{i} + \mathbf{j} - \mathbf{k})$$

Thus, the required line passes through $(-1, 2, 3)$ and is parallel to the vector $\langle 1, 1, -1 \rangle$ so it has parametric form

$$x = -1 + t, \quad y = 2 + t, \quad z = 3 - t$$ ∎

An equation for the line of intersection of two planes can be found by determining a point P in the intersection, and then proceeding as in Example 5 (using P instead of the point $(-1, 2, 3)$). In Example 6, we turn things around by finding the equation of a plane containing two given intersecting lines.

EXAMPLE 6   Equation of a plane containing two intersecting lines

Find the standard-form equation of the plane determined by the intersecting lines

$$\frac{x - 2}{3} = \frac{y + 5}{-2} = \frac{z + 1}{4} \quad \text{and} \quad \frac{x + 1}{2} = \frac{y}{-1} = \frac{z - 16}{5}$$

Solution

The given lines, respectively, have direction numbers $[3, -2, 4]$ and $[2, -1, 5]$. A vector orthogonal to both $\mathbf{v}_1$ and $\mathbf{v}_2$ and hence to the required planes determined by the intersecting lines is given by the cross product:

$$\mathbf{v}_1 \times \mathbf{v}_2 = \begin{vmatrix} \mathbf{i} & \mathbf{j} & \mathbf{k} \\ 3 & -2 & 4 \\ 2 & -1 & 5 \end{vmatrix} = (-10 + 4)\mathbf{i} - (15 - 8)\mathbf{j} + (-3 + 4)\mathbf{k}$$

$$= -6\mathbf{i} - 7\mathbf{j} + \mathbf{k}$$

The points $(2, -5, -1)$ and $(-1, 0, 16)$ are both in the required plane. We use $(2, -5, -1)$ to obtain

$$-6(x - 2) - 7(y + 5) + 1(z + 1) = 0$$
$$6x + 7y - z + 22 = 0$$

as the equation for the plane.   ■

VECTOR METHODS FOR MEASURING DISTANCES IN $\mathbb{R}^3$

Vector methods can also be used for measuring distances in $\mathbb{R}^3$, between points, lines, and planes. We begin by deriving the following formula for the distance from a point to a plane.

THEOREM 9.9   Distance from a point to a plane in $\mathbb{R}^3$

The distance from the point $P(x_0, y_0, z_0)$ to the plane $Ax + By + Cz + D = 0$ is given by

$$d = \frac{|\mathbf{QP} \cdot \mathbf{N}|}{\|\mathbf{N}\|} = \frac{|Ax_0 + By_0 + Cz_0 + D|}{\sqrt{A^2 + B^2 + C^2}}$$

where Q is any point in the given plane and $\mathbf{N}$ is a normal to the given plane.

Proof   The required distance is found by projecting the vector $\mathbf{QP}$ onto the normal $\mathbf{N}$. Thus, the distance from the point to the plane is given by

$$d = \|\mathbf{QP}\|\,|\cos\theta| = \frac{\|\mathbf{QP}\|\|\mathbf{N}\|\,|\cos\theta|}{\|\mathbf{N}\|} = \frac{|\mathbf{QP} \cdot \mathbf{N}|}{\|\mathbf{N}\|}$$

where θ is the angle between $\mathbf{QP}$ and $\mathbf{N}$ (see Figure 9.60). Suppose $Q(x_1, y_1, z_1)$ is any particular point in the plane, so that

$$\mathbf{QP} = (x_0 - x_1)\mathbf{i} + (y_0 - y_1)\mathbf{j} + (z_0 - z_1)\mathbf{k}$$

The dot product of $\mathbf{QP}$ with $\mathbf{N} = A\mathbf{i} + B\mathbf{j} + C\mathbf{k}$ is given by

$$\mathbf{QP} \cdot \mathbf{N} = (x_0 - x_1)A + (y_0 - y_1)B + (z_0 - z_1)C$$
$$= (Ax_0 + By_0 + Cz_0) - (Ax_1 + By_1 + Cz_1)$$
$$= Ax_0 + By_0 + Cz_0 - (-D) \qquad \text{Since } Ax_1 + By_1 + Cz_1 + D = 0$$

Because the normal vector has length $\|\mathbf{N}\| = \sqrt{A^2 + B^2 + C^2}$, we can substitute into the formula for d to obtain

$$d = \frac{|\mathbf{QP} \cdot \mathbf{N}|}{\|\mathbf{N}\|} = \frac{|Ax_0 + By_0 + Cz_0 + D|}{\sqrt{A^2 + B^2 + C^2}} \qquad \square$$

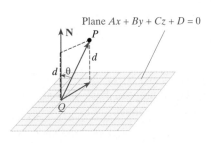

Figure 9.60 The distance from a point to a plane in $\mathbb{R}^3$

Figure 9.61 The sphere with center $C(-3, 1, 5)$ and radius 2 is tangent to the plane $6x - 2y + 3z = 9$.

EXAMPLE 7 **Finding an equation for a sphere tangent to a given plane**

Find an equation for the sphere with center $C(-3, 1, 5)$ that is tangent to the plane $6x - 2y + 3z = 9$.

Solution

The radius r of the sphere is the distance from the center C to the given plane, as shown in Figure 9.61; namely,

$$r = \left| \frac{6(-3) + (-2)(1) + 3(5) - 9}{\sqrt{6^2 + (-2)^2 + 3^2}} \right| = \left| \frac{-14}{7} \right| = 2$$

Therefore, an equation of the sphere is

$$(x + 3)^2 + (y - 1)^2 + (z - 5)^2 = 2^2 \qquad \blacksquare$$

Vector methods can also be used to derive a formula for the distance between two skew lines L_1 and L_2 in $\mathbb{R}^3$.

EXAMPLE 8 **Computing the distance between skew lines**

Find the distance between the skew lines

$$L_1 : x = 1 + 2s \qquad y = -1 + s \qquad z = 2 + 4s$$
$$L_2 : x = -2 + 4t \qquad y = -3t \qquad z = -1 + t$$

Solution

Recall from Section 9.5 that skew lines do not intersect and are not parallel, but do lie in parallel planes. Thus, the distance d between the lines L_1 and L_2 is the same as the distance between the parallel planes p_1 and p_2 that contain them (see Figure 9.62). Our strategy for finding the distance d is to find the distance from a point Q on p_1 to the plane p_2.

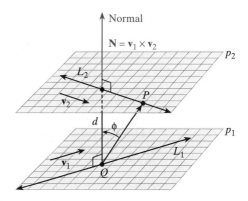

Figure 9.62 Distance between skew lines

First, find a point Q on p_1 by setting $s = 0$ in the parametric equations for L_1. Then $x = 1$, $y = -1$, and $z = 2$, so $Q(1, -1, 2)$ is such a point. Next, vectors $\mathbf{v}_1 = \langle 2, 1, 4 \rangle$ and $\mathbf{v}_2 = \langle 4, -3, 1 \rangle$ are parallel to L_1 and L_2, respectively, and the cross product $\mathbf{N} = \mathbf{v}_1 \times \mathbf{v}_2$ is normal to both p_1 and p_2. We find that

$$\mathbf{N} = \begin{vmatrix} \mathbf{i} & \mathbf{j} & \mathbf{k} \\ 2 & 1 & 4 \\ 4 & -3 & 1 \end{vmatrix} = 13\mathbf{i} + 14\mathbf{j} - 10\mathbf{k}$$

Setting $t = 0$, we see that the point $P(-2, 0, -1)$ is on p_2, so an equation for p_2 is

$$13(x + 2) + 14(y - 0) - 10(z + 1) = 0$$
$$13x + 14y - 10z + 16 = 0$$

The distance d between the lines L_1 and L_2 is the same as the distance from the point $Q(1, -1, 2)$ to the plane p_2. Thus

$$d = \frac{|13(1) + 14(-1) - 10(2) + 16|}{\sqrt{13^2 + 14^2 + (-10)^2}} = \frac{5}{\sqrt{465}} \approx 0.2319$$ ∎

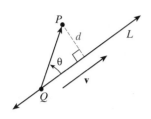

Figure 9.63 The distance from a point to a line in $\mathbb{R}^3$

In the final example of this section, we find the distance from a point P to a line L in $\mathbb{R}^3$. Let Q be a point on L and let $\mathbf{v}$ be a vector parallel to L. Then, as shown in Figure 9.63, the distance from P to L is given by

$$d = \|\mathbf{QP}\| \, |\sin \theta|$$

where θ is the angle between $\mathbf{v}$ and the vector $\mathbf{QP}$. This reminds us of the cross product $\mathbf{v} \times \mathbf{QP}$, and because

$$\|\mathbf{v} \times \mathbf{QP}\| = \|\mathbf{v}\| \|\mathbf{QP}\| \sin \theta$$

we have

$$d = \|\mathbf{QP}\| \sin \theta = \frac{\|\mathbf{v} \times \mathbf{QP}\|}{\|\mathbf{v}\|}$$

These observations are summarized in the following theorem.

THEOREM 9.10   Distance from a point to a line

The distance from point P to the line L is given by the formula

$$d = \frac{\|\mathbf{v} \times \mathbf{QP}\|}{\|\mathbf{v}\|}$$

where $\mathbf{v}$ is a vector parallel to L and Q is any point on L.

Proof   A sketch of the proof precedes the statement of the theorem.   ❏

EXAMPLE 9   Distance from a point to a line

Find the distance from the point $P(3, -8, 1)$ to the line

$$\frac{x - 3}{3} = \frac{y + 7}{-1} = \frac{z + 2}{5}$$

Solution

We need to find a point Q on the line. We see that $Q(3, -7, -2)$ is on the line and that $\mathbf{QP} = \langle 0, -1, 3 \rangle$. A vector parallel to L is $\mathbf{v} = \langle 3, -1, 5 \rangle$, so that

$$\mathbf{v} \times \mathbf{QP} = \begin{vmatrix} \mathbf{i} & \mathbf{j} & \mathbf{k} \\ 3 & -1 & 5 \\ 0 & -1 & 3 \end{vmatrix} = 2\mathbf{i} - 9\mathbf{j} - 3\mathbf{k}$$

Thus,

$$d = \frac{\|\mathbf{v} \times \mathbf{QP}\|}{\|\mathbf{v}\|} = \frac{\sqrt{(2)^2 + (-9)^2 + (-3)^2}}{\sqrt{3^2 + (-1)^2 + 5^2}} = \frac{\sqrt{94}}{\sqrt{35}} \approx 1.64$$ ∎

9.6 PROBLEM SET

A 1. **WHAT DOES THIS SAY?** Describe the relationship between normal vectors and planes.

2. **WHAT DOES THIS SAY?** Describe a procedure for finding the distance from a point to a plane.

3. **WHAT DOES THIS SAY?** Describe a procedure for finding the distance from a point to a line in $\mathbb{R}^3$.

Write each equation for a plane given in Problems 4–7 in standard form.

4. $4(x+1) - 2(y+1) + 6(z-2) = 0$
5. $5(x-2) - 3(y+2) + 4(z+3) = 0$
6. $-3(x-4) + 2(y+1) - 2(z+1) = 0$
7. $-2(x+1) + 4(y-3) - 8z = 0$

Find an equation for the plane that contains the point P and has the normal vector N given in Problems 8–13.

8. $P(-1, 3, 5); \mathbf{N} = 2\mathbf{i} + 4\mathbf{j} - 3\mathbf{k}$
9. $P(0, -7, 1); \mathbf{N} = -\mathbf{i} + \mathbf{k}$
10. $P(0, -3, 0); \mathbf{N} = -2\mathbf{j} + 3\mathbf{k}$
11. $P(1, 1, -1); \mathbf{N} = -\mathbf{i} - 2\mathbf{j} + 3\mathbf{k}$
12. $P(0, 0, 0); \mathbf{N} = \mathbf{k}$ 13. $P(0, 0, 0); \mathbf{N} = \mathbf{i}$
14. Find two unit vectors perpendicular to the plane $2x + 4y - 3z = 4$.
15. Find two unit vectors perpendicular to the plane $5x - 3y + 2z = 15$.

Find the distance between the point and the plane given in Problems 16–21.

16. $P(1, 0, -1); x + y - z = 1$
17. $P(0, 0, 0); 2x - 3y + 5z = 10$
18. $P(1, 1, -1); x - y + 2z = 4$
19. $P(2, 1, -2); 3x - 4y + z = -1$
20. $P(a, -a, 2a); 2ax - y + az = 4a, a \neq 0$
21. $P(a, 2a, 3a); 3x - 2y + z = -1/a, a \neq 0$

Find the distance from the point $(-1, 2, 1)$ to each plane given in Problems 22–25.

22. the plane through the points $(0, 0, 0)$, $(1, 2, 4)$, and $(-2, -1, 1)$
23. the plane through the points $(-1, 1, 1)$, $(4, 3, 7)$, and $(3, -1, 0)$
24. the plane through the point $(-3, 5, 1)$ with normal vector $3\mathbf{i} + \mathbf{j} + 5\mathbf{k}$
25. the plane through the point $(1, 0, 1)$ with normal vector $2\mathbf{i} - \mathbf{j} + 2\mathbf{k}$

Find the distance from the point P to the line L in Problems 26–31.

26. $P(4, 5); 3x - 4y + 8 = 0$
27. $P(9, -3); 3x - 4y + 8 = 0$
28. $P(4, -3); 12x + 5y - 2 = 0$
29. $P(1, 0, -1); \dfrac{x-2}{3} = \dfrac{y+1}{1} = \dfrac{z-1}{2}$
30. $P(1, 0, 1); \dfrac{x}{3} = \dfrac{y-1}{2} = \dfrac{z}{1}$
31. $P(1, -2, 2); \dfrac{x}{1} = \dfrac{2y}{1} = \dfrac{z}{-1}$

Find the distance between the lines given in Problems 32–35.

32. $\dfrac{x+1}{3} = \dfrac{y-2}{-2} = \dfrac{z-1}{1}$ and $\dfrac{x-2}{5} = \dfrac{y+1}{1} = \dfrac{z}{3}$
33. $x = 2 - t, \ y = 5 + 2t, \ z = 3t$ and $x = 2t, \ y = -1 - t, \ z = 1 + 2t$
34. $\dfrac{x+1}{1} = \dfrac{y-3}{2} = \dfrac{z+2}{3}$ and the line passing through $(1, 3, -2)$ and $(0, 1, -1)$
35. $x = -1 + t, \ y = -2t, \ z = 3$ and the line passing through $(0, -1, 2)$ and $(1, -2, 3)$
36. Find the equation of the sphere with center $C(-2, 3, 7)$ that is tangent to the plane $2x + 3y - 6z = 5$.

B 37. **a.** Show that the line

$$\frac{x-1}{3} = \frac{y}{-2} = \frac{z+1}{1}$$

   is parallel to the plane $x + 2y + z = 1$.

   b. Find the distance from the line to the plane in part **a**.

38. Three of the four vertices of a parallelogram $QRTS$ in $\mathbb{R}^3$ are $Q(-1, 3, 5)$, $R(6, -3, 2)$, and $S(2, 4, -3)$. (Note that the fourth vertex of the parallelogram may be located in one of several locations.) What is the area of the parallelogram?

39. Find an equation for the set of all points $P(x, y, z)$ such that the distance from P to the point $P_0(-1, 2, 4)$ is the same as the distance from P to the plane $2x - 5y + 3z = 7$. (Do not expand binomials.)

40. Find an equation for the set of all points $P(x, y, z)$ such that the distance from P to the line

$$\frac{x-1}{4} = \frac{y+1}{-1} = \frac{z}{3}$$

   is 5. (Do not expand.)

41. In Figure 9.64, $\mathbf{N}$ is normal to the plane P and L is a line that intersects p. Assume $\mathbf{N} = a\mathbf{i} + b\mathbf{j} + c\mathbf{k}$ and that L is given by $x = x_0 + At, \ y = y_0 + Bt, \ z = z_0 + Ct$

   a. Find $\cos\theta$ for the angle θ between L and the plane p.

   b. Find the angle between the plane $x + y + z = 10$ and the line

$$\frac{x-1}{2} = \frac{y+3}{3} = \frac{z-2}{-1}$$

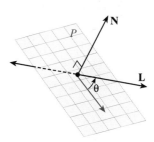

Figure 9.64 Problem 41

42. Find an equation for the line that passes through the point $P(0, 1, -1)$ and is parallel to the line of intersection of the planes $2x + y - 2z = 5$ and $3x - 6y - 2z = 7$.

43. Find a vector that is parallel to the line of intersection of the planes $2x + 3y = 0$ and $3x - y + z = 1$.

44. Find an equation for the line of intersection of the planes $2x - y + z = 8$ and $x + y - z = 5$.

45. Find the direction cosines of a vector determined by the line of intersection of the planes $x + y + z = 3$ and $2x + 3y - z = 4$.

46. Find an equation for the plane that passes through $P(1, -1, 2)$ and is normal to **PQ** where Q is $Q(2, 1, 3)$.

47. Find an equation for the line that passes through the point $(1, -5, 3)$ and is orthogonal to the plane $2x - 3y + z = 1$.

48. Find two unit vectors that are parallel to the line of intersection of the planes $x + y = 1$ and $x - 2z = 3$.

49. Find two unit vectors that are parallel to the line of intersection of the planes $x + y + z = 3$ and $x - y + z = 1$.

50. Find an equation for the plane that contains the point $(2, 1, -1)$ and is orthogonal to the line

$$\frac{x - 3}{3} = \frac{y + 1}{5} = \frac{z}{2}$$

51. Find a plane that passes through the point $(1, 2, -1)$ and is parallel to the plane $2x - y + 3z = 1$.

52. Show that the line

$$\frac{x - 1}{2} = \frac{y + 1}{3} = \frac{z - 2}{4}$$

is parallel to the plane $x - 2y + z = 6$.

53. Find the point where the line

$$\frac{x - 1}{2} = \frac{y + 1}{-1} = \frac{z}{3}$$

intersects the plane $3x + 2y - z = 5$.

54. The *angle between two planes* is defined to be the acute angle between their normal vectors. Find the angle between the planes $2x + y - 4z = 3$ and $x - y + z = 2$, rounded to the nearest degree.

55. Find an equation for the line that passes through the point $P(2, 3, 1)$ and is parallel to the line of intersection of the planes $x + 2y - 3z = 4$ and $x - 2y + z = 0$.

56. The planes $Ax + By + Cz = D_1$ and $Ax + By + Cz = D_2$ are parallel.

   a. Is the distance between the planes $|D_1 - D_2|$? If not, what is the correct formula?

   b. Find the distance between the parallel planes $x + y + 2z = 2$ and $x + y + 2z = 4$.

57. Show that the angle between the planes

$$A_1 x + B_1 y + C_1 z + D_1 = 0$$

and

$$A_2 x + B_2 y + C_2 z + D_2 = 0$$

is $\pi/2$ if and only if

$$A_1 A_2 + B_1 B_2 + C_1 C_2 = 0$$

58. Show that a plane with x-intercept a, y-intercept b, and z-intercept c has the equation

$$\frac{x}{a} + \frac{y}{b} + \frac{z}{c} = 1$$

assuming a, b, and c are all nonzero.

59. Suppose planes p_1 and p_2 intersect. If $\mathbf{v}_1$ and $\mathbf{w}_1$ are nonzero vectors on p_1 and $\mathbf{v}_2$ and $\mathbf{w}_2$ are nonzero vectors on plane p_2, then show that

$$(\mathbf{v}_1 \times \mathbf{w}_1) \times (\mathbf{v}_2 \times \mathbf{w}_2)$$

is parallel to the line of intersection of the planes.

60. Let L_1 and L_2 be skew lines that contain the points P_1 and P_2, respectively, and are parallel to the vectors $\mathbf{v}_1$ and $\mathbf{v}_2$. Show that the distance between the lines is given by

$$d = \left| \frac{(\mathbf{v}_1 \times \mathbf{v}_2) \cdot \mathbf{P}_1 \mathbf{P}_2}{\|\mathbf{v}_1 \times \mathbf{v}_2\|} \right|$$

9.7   Quadric Surfaces

IN THIS SECTION   methods for graphing quadric surfaces; a catalog of quadric surfaces:

You might want to review graphs whose equations are of the form

$$Ax^2 + Bxy + Cy^2 + Dx + Ey + F = 0$$

which are called **conic sections** (see Chapters 6 and 7 of the *Student Mathematics Handbook*). Conic sections are graphs in $\mathbb{R}^2$ that can be represented by a second-degree equation in two variables, and quadric surfaces are graphs in $\mathbb{R}^3$ which can be represented by a second-degree equation in three variables.

METHODS FOR GRAPHING QUADRIC SURFACES

Spheres, and elliptic, parabolic, and hyperbolic cylinders are examples of **quadric surfaces.** In general, such a surface is the graph of an equation of the form

$$Ax^2 + By^2 + Cz^2 + Dxy + Exz + Fyz + Gx + Hy + Iz + J = 0$$

where A, B, C, D, E, F, G, H, I, and J are constants. By using rotations and translations, it can be shown that any such equation can be expressed in one of the following standard forms:

$$z = Mx^2 + Ny^2 \quad \text{or} \quad Px^2 + Qy^2 + Rz^2 = S$$

We will concentrate on examining graphs of the standard forms.

The intersection of a surface with a plane is called the **trace** of the surface in that plane. The quadric surfaces are three-dimensional analogues of the conic sections in the sense that the traces of each quadric surface with respect to planes parallel to the coordinate planes are conic sections. The following three examples show how traces can be used to determine the appearance of a quadric surface.

EXAMPLE 1   Using traces to sketch the graph of an ellipsoid

Sketch the graph of the quadric surface given by

$$\frac{x^2}{4} + \frac{y^2}{5} + \frac{z^2}{9} = 1$$

Solution

Consider the trace of the given surface in the plane $z = 0$. We substitute $z = 0$ into the equation to obtain

$$\frac{x^2}{4} + \frac{y^2}{5} = 1$$

which we recognize as an ellipse. More generally, the trace in the plane $z = k$ is, by substitution,

$$\frac{x^2}{4} + \frac{y^2}{5} + \frac{k^2}{9} = 1 \quad \text{or} \quad \frac{x^2}{4} + \frac{y^2}{5} = 1 - \frac{k^2}{9}$$

which is also an ellipse for $k^2 \leq 9$, since the right-hand side is a constant. Similarly, the trace of the surface in a plane of the form $y = k$ is

$$\frac{x^2}{4} + \frac{k^2}{5} + \frac{z^2}{9} = 1 \quad \text{or} \quad \frac{x^2}{4} + \frac{z^2}{9} = 1 - \frac{k^2}{5}$$

which is an ellipse for $k^2 \leq 5$, as is the trace in the plane $x = k$ (for $k^2 \leq 4$):

$$\frac{k^2}{4} + \frac{y^2}{5} + \frac{z^2}{9} = 1 \quad \text{or} \quad \frac{y^2}{5} + \frac{z^2}{9} = 1 - \frac{k^2}{4}$$

Sketch these traces in $\mathbb{R}^3$ as shown in Figure 9.65. Since the traces in all three principal directions are ellipses, we call this surface an **ellipsoid**. ∎

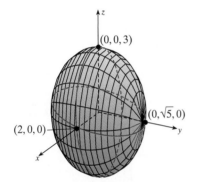

Figure 9.65 Graph of the ellipsoid
$$\frac{x^2}{4} + \frac{y^2}{5} + \frac{z^2}{9} = 1$$

EXAMPLE 2   Using traces to sketch the graph of a hyperboloid

Sketch the graph of the surface $9x^2 + 36y^2 - 4z^2 + 36 = 0$.

Solution

We begin by setting $z = k$, to obtain

$$9x^2 + 36y^2 = 4k^2 - 36$$

which is impossible for $k^2 < 9$, and represents an ellipse for $k^2 \geq 9$.

For $x = k$, we have

$$4z^2 - 36y^2 = 36 + 9k^2$$

and for $y = k$

$$4z^2 - 9x^2 = 36 + 36k^2$$

both of which represent hyperbolas for all k.

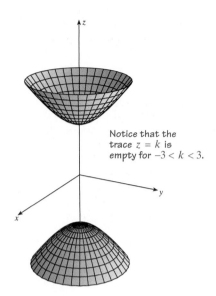

Notice that the trace $z = k$ is empty for $-3 < k < 3$.

Figure 9.66 The hyperboloid of two sheets $9x^2 + 36y^2 - 4z^2 + 36 = 0$

To summarize, the surface has hyperbolic traces for planes parallel to the yz-plane and the xz-plane and elliptical traces for planes parallel to the xy-plane, except for a "gap" corresponding to the planes of the form $z = k$ with $|k| < 3$. This surface is called a **hyperboloid of two sheets.** The graph is shown in Figure 9.66.

EXAMPLE 3   Using traces to sketch a hyperbolic paraboloid

Sketch the graph of the surface

$$z = \frac{y^2}{9} - \frac{x^2}{16}$$

Solution

For $z = 0$, we have

$$0 = \frac{y^2}{9} - \frac{x^2}{16}$$

$$y = \pm\frac{3}{4}x$$

which is a pair of intersecting lines, as shown in Figure 9.67a.
    For $z = k$,

$$k = \frac{y^2}{9} - \frac{x^2}{16}$$

and the corresponding trace is a hyperbola (along the y-axis if $k > 0$, and along the x-axis if $k < 0$). These traces are also shown in Figure 9.67a.

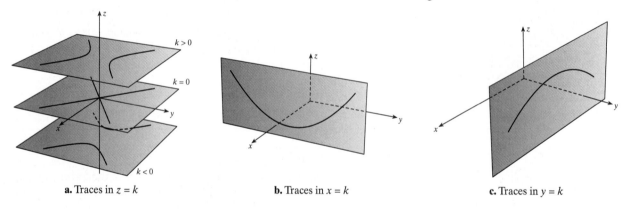

a. Traces in $z = k$ **b.** Traces in $x = k$ **c.** Traces in $y = k$

Figure 9.67   Traces in the surface $z = \dfrac{y^2}{9} - \dfrac{x^2}{16}$

If $x = k$, we have

$$z = \frac{y^2}{9} - \frac{k^2}{16}$$

and the trace is a parabola that opens up (Figure 9.67b), while if $y = k$, the corresponding equation

$$z = \frac{k^2}{9} - \frac{x^2}{16}$$

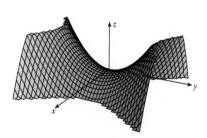

Figure 9.68 The hyperbolic paraboloid (saddle surface) $z = \dfrac{y^2}{9} - \dfrac{x^2}{16}$

is a parabola opening down (Figure 9.67c). This surface is called a **hyperbolic paraboloid**, but is more commonly referred to as a **saddle surface** because its shape near the origin resembles that of a saddle, as shown in Figure 9.68.   ∎

A CATALOG OF QUADRIC SURFACES

A catalog of quadric surfaces is shown in Table 9.1, where x, y, and z are variables and a, b, and c are constants.

TABLE 9.1 Quadric Surfaces

Surface	Description	Surface	Description
Elliptic cone 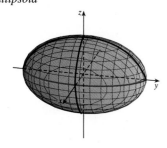	The trace in the xy-plane is a point; in planes parallel to the xy-plane, it is an ellipse. Traces in the xz- and yz-planes are intersecting lines; in planes parallel to these, they are hyperbolas: $$z^2 = \frac{x^2}{a^2} + \frac{y^2}{b^2}$$	*Elliptic paraboloid*	The trace in the xy-plane is a point; in planes parallel to the xy-plane, it is an ellipse. Traces in the xz- and yz-planes are parabolas: $$z = \frac{x^2}{a^2} + \frac{y^2}{b^2}$$
Hyperboloid of one sheet	The trace in the xy-plane is an ellipse; in the xz- and yz-planes, the traces are hyperbolas: $$\frac{x^2}{a^2} + \frac{y^2}{b^2} - \frac{z^2}{c^2} = 1$$	*Hyperboloid of two sheets*	There is no trace in the xy-plane. In planes parallel to the xy-plane that intersect the surface, the traces are ellipses. Traces in the xz- and yz-planes are hyperbolas: $$\frac{x^2}{a^2} + \frac{y^2}{b^2} - \frac{z^2}{c^2} = -1$$
Ellipsoid	The traces in the coordinate planes are ellipses: $$\frac{x^2}{a^2} + \frac{y^2}{b^2} + \frac{z^2}{c^2} = 1$$	*Hyperbolic paraboloid*, also called a *saddle surface*	The trace in the xy-plane is two intersecting lines; in planes parallel to the xy-plane, the traces are hyperbolas. Traces in the xz- and yz-planes are parabolas: $$z = \frac{y^2}{b^2} - \frac{x^2}{a^2}$$
Sphere	The sphere is a special kind of ellipsoid for which $a = b = c = r$: $$x^2 + y^2 + z^2 = r^2$$		

EXAMPLE 4 Identifying a quadric surface

Describe the given surfaces.

a. $7x^2 + 4y^2 - 28z^2 = 0$ **b.** $3x^2 + 5y^2 - 15z = 0$

Solution

a. We rewrite the equation in one of the two standard forms:

$$z^2 = \frac{x^2}{4} + \frac{y^2}{7}$$

We see this is an elliptic cone.

b. Rewrite as

$$z = \frac{x^2}{5} + \frac{y^2}{3}$$

We see that the surface is an elliptic paraboloid. ∎

9.7 PROBLEM SET

A 1. WHAT DOES THIS SAY? What is a quadric surface?

2. WHAT DOES THIS SAY? What is the trace of a surface?

In Problems 3–12, match the equation with its graph (A–L).

3. $x^2 = z^2 + y^2$

4. $z^2 = \frac{x^2}{4} + \frac{y^2}{9}$

5. $\frac{x^2}{2} - \frac{y^2}{4} + \frac{z^2}{9} = 1$

6. $\frac{x^2}{9} + \frac{y^2}{16} - \frac{z^2}{4} = 1$

7. $x^2 + y^2 + z^2 = 9$

8. $y = x^2 + z^2$

9. $x = \frac{y^2}{25} + \frac{z^2}{16}$

10. $y = \frac{z^2}{4} - \frac{x^2}{9}$

11. $y^2 + z^2 - x^2 = -1$

12. $\frac{x^2}{4} - \frac{y^2}{9} + \frac{z^2}{9} = -1$

E.

F.

A.

B.

G.

H.

C.

D.

I.

J.

K.

L.
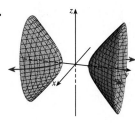

In Problems 13–20, identify the quadric surface and describe the traces in planes parallel to the coordinate planes. Sketch the graph.

13. $9x^2 + 4y^2 + z^2 = 1$

14. $\dfrac{x^2}{4} + y^2 + \dfrac{z^2}{9} = 1$

15. $\dfrac{x^2}{4} + \dfrac{y^2}{9} - z^2 = 1$

16. $\dfrac{x^2}{9} - y^2 - z^2 = 1$

17. $z = x^2 + \dfrac{y^2}{4}$

18. $z = \dfrac{x^2}{9} - \dfrac{y^2}{16}$

19. $x^2 + 2y^2 = 9z^2$

20. $z^2 = 1 + \dfrac{x^2}{9} + \dfrac{y^2}{4}$

The equations given in Problems 21–24 represent quadric surfaces whose axis of symmetry is either the x-axis or the y-axis. Identify and sketch the surface.

21. $y - x^2 - 9z^2 = 0$

22. $x = \dfrac{y^2}{4} - \dfrac{z^2}{9}$

23. $y^2 = x^2 + 3z^2$

24. $y^2 - x^2 + z^2 = 1$

Describe each of the translated quadric surfaces in Problems 25–30.

25. $\dfrac{(x-1)^2}{4} + (y+2)^2 + \dfrac{(z-3)^2}{9} = 1$

26. $z = (2x+1)^2 + \dfrac{(y-1)^2}{9} + 3$

27. $7x^2 + y^2 + 3z^2 - 9x + 4y + 7z = 1$

28. $z^2 = x^2 - y^2 + 2x + y - z + 5$

29. $y = x^2 - x + z^2 + z + 5$

30. $\dfrac{(x-1)^2}{2} - \dfrac{(y+1)^2}{4} - \dfrac{z^2}{9} = 1$

B Describe the curve of intersection of the quadric surfaces in Problems 31 and 32.

31. $\dfrac{x^2}{9} + \dfrac{y^2}{4} + \dfrac{z^2}{5} = 1$ and $z^2 = \dfrac{x^2}{3} + \dfrac{y^2}{2}$

32. $z = \dfrac{y^2}{9} - \dfrac{x^2}{16}$ and $\dfrac{x^2}{16} + \dfrac{y^2}{2} - \dfrac{z^2}{3} = 1$

33. Use the method of cross-sections (Section 6.2) to show that the ellipsoid
$$\dfrac{x^2}{a^2} + \dfrac{y^2}{b^2} + \dfrac{z^2}{c^2} = 1$$
has volume $V = \frac{4}{3}\pi abc$. *Hint:* You may assume that the ellipse $\dfrac{x^2}{A^2} + \dfrac{y^2}{B^2} = 1$ has area πAB.

34. Use the methods of Section 6.2 to find the volume bounded by the plane $z = 5$ and the paraboloid $9z = x^2 + y^2$.

35. What is the intersection of the surfaces $z = x^2 + y^2$ and $z = 1 - y^2$?

36. What is the intersection of the surfaces $z = x^2 + y^2$ and $z = y$?

37. Rotational forces cause the earth to flatten at the poles so its shape is really an ellipsoid rather than a true sphere. Assume that the equatorial radius of the earth is 6,378.2 km and that its polar radius is 6,356.5 km.

  a. Assuming that the earth is an ellipsoid with the general equation
  $$\dfrac{x^2}{a^2} + \dfrac{y^2}{a^2} + \dfrac{z^2}{b^2} = 1$$
  find a and b.

  b. Use the formula in Problem 33 to find the volume of the earth.

C **38.** The line
$$x = 1 - t, \qquad y = 2 + t, \qquad z = 3t$$
intersects the hyperboloid
$$\dfrac{z^2}{9} - \dfrac{x^2}{1} + \dfrac{y^2}{4} = 1$$
in two points P and Q. Find the distance between P and Q.

39. Find an equation for the surface S comprised of all points (x, y, z) the sum of whose distances from the points $A(1, 0, 0)$ and $B(-1, 0, 0)$ is 3. Is S a quadric surface?

40. Find an equation for the surface S comprised of all points (x, y, z) whose distance from the origin is the same as their distance from the plane $x + y + z = 5$. Is S a quadric surface?

CHAPTER 9 REVIEW

Proficiency Examination

CONCEPT PROBLEMS

1. What is a vector and what is a scalar?
2. Give both an algebraic and a geometric interpretation of multiplication of a vector by a scalar.
3. What is the parallelogram rule for vector sums?
4. State each of the following properties of vector operations:
   a. commutativity of vector addition
   b. associativity of vector addition
   c. identity for vector addition
   d. inverse property for vector addition
   e. magnitude of a vector for dot product
   f. commutativity for dot product
   g. multiple of a dot product
   h. distributivity for dot product over addition
   i. cross product of a zero vector
   j. anticommutativity for cross product
   k. distributivity for cross product over addition
5. If **u** and **v** are parallel vectors, what can be said about **u** × **v**?
6. How do you find the length of a vector? What is a unit vector?
7. State the triangle inequality.
8. What are the standard basis vectors in $\mathbb{R}^3$?
9. What is the standard-form equation of a sphere?
10. What is a cylinder?
11. What is the distance between two points in $\mathbb{R}^3$?
12. How do you find a unit vector **u** in the direction of a given vector **v**?
13. What is Lagrange's identity?
14. Define dot product.
15. What is the formula for the angle between two vectors?
16. What is meant by orthogonal vectors? What is the algebraic condition for orthogonality?
17. What is a vector projection? Give a formula for finding the vector projection of **v** onto **w**.
18. What is a scalar projection? Give a formula for finding the scalar projection of **v** onto **w**.
19. What is the vector formula for work?
20. Define cross product.
21. What is the right-hand rule for a coordinate system?
22. a. Give a geometric interpretation of cross product.
    b. What is the formula for the magnitude of a cross product?
23. What is the determinant form for the scalar triple product?
24. How do you find the volume of a parallelepiped?
25. a. What is the parametric form of a line in $\mathbb{R}^3$?
    b. What is the symmetric form of a line in $\mathbb{R}^3$?
26. What are the direction cosines of a vector in $\mathbb{R}^3$?
27. What is the normal vector of a plane in $\mathbb{R}^3$?
28. What is the point-normal form of a plane in $\mathbb{R}^3$?
29. What is the standard form of a plane in $\mathbb{R}^3$?
30. What is the formula for the distance from a point to a plane?
31. What is the formula for the distance from a point to a line?
32. What is a quadric surface?

PRACTICE PROBLEMS

33. Given $\mathbf{v} = 2\mathbf{i} - 3\mathbf{j} + \mathbf{k}$, $\mathbf{w} = 3\mathbf{i} - 2\mathbf{j}$. Find each of the following vectors or scalars.
    a. $2\mathbf{v} + 3\mathbf{w}$
    b. $\|\mathbf{v}\|^2 - \|\mathbf{w}\|^2$
    c. $\mathbf{v} \cdot \mathbf{w}$
    d. $\mathbf{v} \times \mathbf{w}$
    e. vector projection of **v** onto **w**
    f. scalar projection of **w** onto **v**
34. Given $\mathbf{u} = 2\mathbf{i} - 3\mathbf{j} + \mathbf{k}$, $\mathbf{v} = \mathbf{i} + \mathbf{j} - 2\mathbf{k}$, and $\mathbf{w} = 3\mathbf{i} + 5\mathbf{k}$. In each of the following cases, either perform the indicated computation or explain why it is not defined.
    a. $(\mathbf{u} \times \mathbf{v}) \cdot \mathbf{w}$
    b. $(\mathbf{u} \cdot \mathbf{v}) \times \mathbf{w}$
    c. $(\mathbf{u} \times \mathbf{v}) \times \mathbf{w}$
    d. $(\mathbf{u} \cdot \mathbf{v}) \cdot \mathbf{w}$

Find the equations for the lines and planes in Problems 35–38.

35. the line through the points $P(-1, 4, -3)$ and $Q(0, -2, 1)$
36. the plane that contains the point $P(1, 1, 3)$ and is normal to the vector $\mathbf{v} = 2\mathbf{i} + 3\mathbf{k}$
37. the line of intersection of the planes $2x + 3y + z = 2$ and $y - 3z = 5$
38. the plane that contains the points $P(0, 2, -1)$, $Q(1, -3, 5)$, and $R(3, 0, -2)$
39. Find the direction cosines and the direction angles of the vector $\mathbf{u} = -2\mathbf{i} + 3\mathbf{j} + \mathbf{k}$. Round to the nearest degree.
40. In each case, determine whether the lines intersect, are parallel, or are skew. If they intersect, find the point of intersection.
    a. $x = 2t - 3$, $y = 4 - t$, $z = 2t$ and $\dfrac{x+2}{3} = \dfrac{y-3}{5}$; $z = 3$
    b. $\dfrac{x-7}{5} = \dfrac{y-6}{4} = \dfrac{z-8}{5}$ and $\dfrac{x-8}{6} = \dfrac{y-6}{4} = \dfrac{z-9}{6}$
41. Let $\mathbf{u} = 2\mathbf{i} + \mathbf{j}$, $\mathbf{v} = \mathbf{i} - \mathbf{j} - \mathbf{k}$, and $\mathbf{w} = 3\mathbf{i} + \mathbf{k}$.
    a. Find the volume of the parallelepiped determined by these vectors.
    b. Find a positive number A that guarantees that the tetrahedron determined by $A\mathbf{u}$, $A\mathbf{v}$, and **w** has volume that is twice the volume of the original tetrahedron. *Hint:* See Problem 50, Section 9.4.
42. Find the distance from $P(-1, 1, 4)$ to $2x + 5y - z = 3$.
43. Find the distance between the skew lines $x = t$, $y = 2t$, $z = 3t - 1$ and $x = 1 - t$, $y = t + 2$, $z = t$.
44. Find the distance from the point $P(4, 5, 0)$ to the line

$$\frac{x-2}{3} = \frac{y}{5} = \frac{z+1}{-1}$$

45. An airplane flies at 200 mi/h parallel to the ground at an altitude of 10,000 ft. If the plane flies due south and the wind is blowing toward the northeast at 50 mi/h, what is the ground speed of the plane (that is, effective speed)?
46. A girl pulls a sled 50 ft on level ground with a rope inclined at an angle of 30° with the horizontal (the ground). If she applies 3 lb of tension to the rope, how much work is performed on the sled?

47. Sketch the graph of $x = 2^t$, $y = 2^{t+1}$ for $t > 0$ by
  a. using the parametric form   **b.** eliminating the parameter
48. Graph the line $\dfrac{x}{4} = -\dfrac{y}{9}$
  a. using the rectangular form   **b.** by parameterizing
49. Graph the line $\dfrac{x}{4} = -\dfrac{y}{9} = \dfrac{z}{1}$
  a. using the symmetric form   **b.** by parameterizing
50. Identify the graph of each given equation.
  a. $x^2 = z^2 + y^2$ **b.** $x^2 = z + y^2$
  c. $\dfrac{x^2}{2} - \dfrac{y^2}{4} - \dfrac{z^2}{9} = 1$ **d.** $\dfrac{x^2}{2} + \dfrac{y^2}{4} - \dfrac{z^2}{9} = 1$
  e. $\dfrac{x}{2} - \dfrac{y}{4} - \dfrac{z}{9} = 1$ **f.** $\dfrac{x}{2} = \dfrac{y}{4} = \dfrac{z}{9}$
  g. $x^2 + y^2 + z^2 = 9$ **h.** $y = x^2 + z^2$
  i. $y^2 + z^2 - x^2 = -1$ **j.** $x^2 - y^2 + z^2 = -1$

Supplementary Problems

1. A triangle in $\mathbb{R}^3$ has vertices $A(0, 2, -1)$, $B(1, 1, 3)$, and $C(1, 0, -4)$.
  a. Find the perimeter of the triangle.
  b. Find the area of the triangle.
  c. Find the three vertex angles of the triangle. (Round to the nearest degree.)
  d. Find a number p such that the points A, B, C, and $D(p, p, 0)$ form a tetrahedron of volume $\mathbf{v} = 100$ cubic units. *Hint*: See Problem 50, Section 9.4.

Find equations, in both parametric and symmetric forms, of the lines described in Problems 2–3. Find two additional points on each line.

2. passing through $A(1, -2, 3)$, $B(4, -1, 2)$
3. passing through $P(1, 4, 0)$ with direction numbers $[2, 0, 1]$
4. Find an equation for the line passing through $P(3, 4, -1)$ and parallel to the line of intersection of the planes $x + 2y + 2z + 5 = 0$ and $2x + y - 3z - 6 = 0$.

Find an equation for the plane satisfying the conditions given in Problems 5–12.

5. the xy-plane
6. the plane parallel to the xz-plane passing through $(4, 3, 7)$
7. the plane through $(1, -3, 4)$ with attitude numbers $[3, 4, -1]$
8. the plane through $(-1, 4, 5)$ and orthogonal to a line with direction numbers $[4, 4, -3]$
9. the plane through $(4, -3, 2)$ and parallel to the plane $5x - 2y + 3z - 10 = 0$
10. the plane containing
$$\dfrac{x-3}{4} = \dfrac{z-1}{2}; \ y = -2; \quad \text{and} \quad \dfrac{x-3}{3} = \dfrac{y+2}{1} = \dfrac{z-1}{-2}$$
11. the plane passing through $P(4, 1, 3)$, $Q(-4, 2, 1)$, and $R(1, 0, 2)$
12. the plane passing through $(4, -1, 2)$ and parallel to the lines
$$\dfrac{x+2}{3} = \dfrac{y-2}{-1} = \dfrac{z+1}{2} \quad \text{and} \quad \dfrac{x-2}{1} = \dfrac{y-3}{2} = \dfrac{z-4}{3}$$
13. If $\mathbf{u}$ and $\mathbf{v}$ are orthogonal unit vectors, show that $(\mathbf{u} \times \mathbf{v}) \times \mathbf{u} = \mathbf{v}$. What is $(\mathbf{u} \times \mathbf{v}) \times \mathbf{v}$?

14. Show that three planes whose normals $\mathbf{u}$, $\mathbf{v}$, $\mathbf{w}$ satisfy $\mathbf{u} \times \mathbf{v} \cdot \mathbf{w} \neq 0$ intersect in exactly one point.
15. Show that $\mathbf{u} \times (\mathbf{v} \times \mathbf{w}) = (\mathbf{u} \times \mathbf{v}) \times \mathbf{w}$ if and only if $\mathbf{v} \times (\mathbf{w} \times \mathbf{u}) = \mathbf{0}$.
16. Given the vectors $\mathbf{v} = 3\mathbf{i} - 2\mathbf{j} + \mathbf{k}$ and $\mathbf{w} = 4\mathbf{i} + \mathbf{j} - 3\mathbf{k}$, find $\|\mathbf{v}\|$, $\mathbf{v} - \mathbf{w}$, and $2\mathbf{v} + 3\mathbf{w}$.
17. Given $\mathbf{u} = \mathbf{i} - \mathbf{j} + \mathbf{k}$, $\mathbf{v} = 3\mathbf{i} - 2\mathbf{j} + 5\mathbf{k}$, and $\mathbf{w} = \mathbf{i} + \mathbf{j} - \mathbf{k}$, find $(\mathbf{u} - \mathbf{v}) \cdot \mathbf{w}$ and $(2\mathbf{u} + \mathbf{v}) \times (\mathbf{u} - \mathbf{w})$.
18. Given the vectors $\mathbf{v} = 4\mathbf{i} + 2\mathbf{j} + \mathbf{k}$, $\mathbf{w} = 2\mathbf{i} + \mathbf{j} - 5\mathbf{k}$. Find
  a. $5\mathbf{v} - 3\mathbf{w}$ **b.** $\|2\mathbf{v} - \mathbf{w}\|$
  c. vector projection of $\mathbf{v}$ onto $\mathbf{w}$
  d. scalar projection of $\mathbf{w}$ onto $\mathbf{v}$
19. Find the direction cosines for $\mathbf{v} = (2\mathbf{i} + \mathbf{j}) \times (\mathbf{i} + \mathbf{j} - 3\mathbf{k})$.
20. Find two unit vectors that are parallel to the line
$$\dfrac{x}{6} = \dfrac{y}{2} = \dfrac{z-1}{6}$$
21. Find the (acute) angle, rounded to the nearest degree, between the intersecting lines
$$\dfrac{x-1}{3} = \dfrac{y-3}{-1} = \dfrac{z+5}{2} \quad \text{and} \quad \dfrac{x-1}{2} = \dfrac{y-3}{-1} = \dfrac{z+5}{-2}$$
22. Find the area of the parallelogram determined by $3\mathbf{i} - 4\mathbf{j}$ and $-\mathbf{i} - \mathbf{j} + \mathbf{k}$.
23. Find the center and the radius of the sphere
$$4x^2 + 4y^2 + 4z^2 + 12y - 4z + 1 = 0$$
24. Find the points of intersection of the line $x = 6 + 3t$, $y = 10 - 2t$, $z = 5t$ with each of the coordinate planes.
25. Find the point of intersection of the planes $3x - y + 4z = 15$, $2x + y - 3z = 1$, and $x + 3y + 5z = 2$
26. Find the equation of the plane determined by the intersecting lines
$$\dfrac{x+3}{3} = \dfrac{y}{-2} = \dfrac{z-7}{6} \quad \text{and} \quad \dfrac{x+6}{1} = \dfrac{y+5}{-3} = \dfrac{z-1}{2}$$
27. Find an equation for the set of all points that are equidistant from the planes $3x - 4y + 12z = 6$ and $4x + 3z = 7$.
28. Graph the planes defined by the given equations.
  a. $x + 3y + 2z = 6$ **b.** $y + z = 5$ **c.** $x = 4$
29. Graph the curve defined by the functions $x = 2\cos t$, $y = 2\sin t$ for $0 \leq t < 2\pi$.
30. Vertices B and C of $\triangle ABC$ lie along the line
$$\dfrac{x+2}{2} = \dfrac{y-1}{1} = \dfrac{z}{4}$$
Find the area of the triangle given that A has coordinates $(1, -1, 2)$ and line segment $\overline{BC}$ has length 5.
31. Find the work done by the constant force $\mathbf{F} = 5\mathbf{i} + 4\mathbf{j} + \mathbf{k}$ in moving a particle along the line from $P(2, 1, -1)$ to $Q(4, 1, 2)$.
32. How much work does it take to move a container 25 m along a horizontal loading platform onto a truck using a constant force of 100 newtons at an angle of $\frac{\pi}{6}$ from the horizontal?

33. Find a formula for the surface area of the tetrahedron determined by vectors $\mathbf{u}$, $\mathbf{v}$, and $\mathbf{w}$. Assume the vectors do not all lie in the same plane.

34. Suppose $\mathbf{v}$ and $\mathbf{w}$ are nonzero vectors. Show that $\|\mathbf{v}\|\mathbf{w} + \|\mathbf{w}\|\mathbf{v}$ and $\|\mathbf{v}\|\mathbf{w} - \|\mathbf{w}\|\mathbf{v}$ are orthogonal vectors.

35. Let $\mathbf{v} = \cos\theta\,\mathbf{i} + \sin\theta\,\mathbf{j}$ and $\mathbf{w} = \cos\phi\,\mathbf{i} + \sin\phi\,\mathbf{j}$. Find $\mathbf{v} \times \mathbf{w}$. Interpret this cross product geometrically, and use it to derive a well-known trigonometric identity.

36. Find the three vertex angles (rounded to the nearest degree) of the triangle whose vertices are $(1, -2, 3)$, $(-1, 2, -3)$, and $(2, 1, -3)$.

37. Find a relationship between the numbers a_1, b_1, and c_1 so that the angle between the vectors $\mathbf{v} = a_1\mathbf{i} + b_1\mathbf{j} + c_1\mathbf{k}$ and $\mathbf{w}_1 = \mathbf{i} - 2\mathbf{j}$ is the same as the angle between $\mathbf{v}$ and $\mathbf{w}_2 = 2\mathbf{i} + \mathbf{k}$.

38. Find an equation of the plane that passes through $(a, 0, 0)$, $(0, a, 0)$, and $(0, 0, a)$.

39. Find the area of the triangle with vertices $A(0, -1, 2)$, $B(1, 2, -1)$, and $C(3, -1, 2)$.

40. Find an equation for the surface comprised of all points equidistant from $D(0, 0, 6)$ and the xy-plane. Is this a quadric surface?

41. Find the area of the triangle determined by the vectors $\mathbf{v} = \mathbf{i} - \mathbf{j} + \mathbf{k}$ and $\mathbf{w} = 2\mathbf{i} + \mathbf{j} - 2\mathbf{k}$.

42. Find an equation for the plane that passes through the origin and is parallel to the vectors $\mathbf{v} = \mathbf{i} - 2\mathbf{j} - 3\mathbf{k}$ and $\mathbf{w} = -\mathbf{i} + \mathbf{j} + 2\mathbf{k}$.

43. Find an equation for the plane that passes through the origin and whose normal vector is parallel to the line of intersection of the planes $2x - y + z = 4$ and $x + 3y - z = 2$.

44. Find a number A such that the angle between the planes $2Ax + 3y + z = 1$ and $x - Ay + 3z = 5$ is $\pi/2$.

45. The lines L_1 and L_2 are parallel to the vectors $\mathbf{v}_1 = \mathbf{i} - \mathbf{j}$ and $\mathbf{v}_2 = \mathbf{i} - \mathbf{j} + 2\mathbf{k}$, respectively. Find an equation for the line L that passes through the point $(-1, 2, 0)$ and is orthogonal to both L_1 and L_2.

46. A parallelepiped is determined by the vectors $\mathbf{u} = \mathbf{i} - \mathbf{j} + \mathbf{k}$, $\mathbf{v} = \mathbf{i} + 2\mathbf{j} - \mathbf{k}$, and $\mathbf{w} = 2\mathbf{i} + \mathbf{j} + \mathbf{k}$. Find the altitude from the tip of $\mathbf{w}$ to the side determined by $\mathbf{u}$ and $\mathbf{v}$.

47. In Chapter 10, we will show that $\mathbf{T} = \mathbf{i} + 2x\mathbf{j}$ is a vector in the direction of the tangent line at each point $P(x, x^2)$ on the parabola $y = x^2$. Find a unit vector normal to the parabola at the point $(3, 9)$.

48. For nonzero vectors $\mathbf{u}$, $\mathbf{v}$, and $\mathbf{w}$, show that the vector $(\mathbf{u} \times \mathbf{v}) \times (\mathbf{u} \times \mathbf{w})$ is parallel to $\mathbf{u}$.

49. Let $\mathbf{v} = a\mathbf{i} + b\mathbf{j} + c\mathbf{k}$ and $\mathbf{w} = A\mathbf{i} + B\mathbf{j} + C\mathbf{k}$, where a, b, c, A, B, and C are constants. Describe the set of vectors $\mathbf{v} + t\mathbf{w}$, where t is any scalar.

50. Use vectors to show that the sum of the squares of the lengths of the sides of a parallelogram equals the sum of the squares of the lengths of the diagonals.

51. The vectors $\mathbf{u}$, $\mathbf{v}$, and $\mathbf{w}$ are said to be *linearly independent* in $\mathbb{R}^3$ if the only solution to the equation $a\mathbf{u} + b\mathbf{v} + c\mathbf{w} = \mathbf{0}$ is $a = b = c = 0$. Otherwise, the vectors are *linearly dependent*. The simplest example of a set of linearly independent vectors is $\mathbf{i}$, $\mathbf{j}$, and $\mathbf{k}$. Determine whether the vectors $\mathbf{u} = -\mathbf{i} + 2\mathbf{k}$, $\mathbf{v} = 2\mathbf{i} - \mathbf{j} + 3\mathbf{k}$, $\mathbf{w} = \mathbf{i} + 3\mathbf{j} - 2\mathbf{k}$ are linearly independent or dependent.

52. Figure 9.69 shows a parallelogram $ABCD$. If M is the midpoint of side $\overline{AB}$, show that line $\overline{CM}$ intersects diagonal $\overline{BD}$ at a point P located one-third of the distance from B to D by completing the following steps.

Figure 9.69   Parallelogram $ABCD$

a. Let a and b be scalars such that $\mathbf{MP} = a\mathbf{MC}$ and $\mathbf{BP} = b\mathbf{BD}$. Show that

$$\tfrac{1}{2}\mathbf{AB} + b(\mathbf{AD} - \mathbf{AB}) = a\left(\mathbf{AD} + \tfrac{1}{2}\mathbf{AB}\right)$$

b. Two vectors $\mathbf{u}$ and $\mathbf{v}$ in $\mathbb{R}^2$ are *linearly independent* if the only solution to $a\mathbf{u} + b\mathbf{v} = \mathbf{0}$ is $a = b = 0$, and it can be shown that this is equivalent to saying that the two vectors are not parallel to each other. Use the fact that $\mathbf{AB}$ and $\mathbf{AD}$ are linearly independent to show that

$$\tfrac{1}{2} - b - \tfrac{1}{2}a = 0 \quad \text{and} \quad a - b = 0$$

Solve this system of equations to show that P has the required location.

53. Show that $\mathbf{u} = a_1\mathbf{i} + a_2\mathbf{j} + a_3\mathbf{k}$, $\mathbf{v} = b_1\mathbf{i} + b_2\mathbf{j} + b_3\mathbf{k}$ and $\mathbf{w} = c_1\mathbf{i} + c_2\mathbf{j} + c_3\mathbf{k}$ are linearly dependent (see Problem 51), if and only if

$$\begin{vmatrix} a_1 & a_2 & a_3 \\ b_1 & b_2 & b_3 \\ c_1 & c_2 & c_3 \end{vmatrix} = 0$$

54. Show that

$$\begin{vmatrix} \mathbf{u}_1 \cdot \mathbf{v}_1 & \mathbf{u}_1 \cdot \mathbf{v}_2 \\ \mathbf{u}_2 \cdot \mathbf{v}_1 & \mathbf{u}_2 \cdot \mathbf{v}_2 \end{vmatrix} = (\mathbf{u}_1 \times \mathbf{u}_2) \cdot (\mathbf{v}_1 \times \mathbf{v}_2)$$

55. Let $A(-2, 3, 7)$, $B(1, 5, -3)$, and $C(2, 8, -1)$ be the vertices of a triangle in $\mathbb{R}^3$. What are the coordinates of the point M where the medians of the triangle meet (the centroid)?

56. The medians of a triangle meet at a point (the centroid) located two-thirds of the distance from each vertex to the midpoint of the opposite side. Generalize this result by showing that the four lines that join each vertex of a tetrahedron to the centroid of the opposite face meet at a point located three-fourths of the distance from the vertex to the centroid.

57. Spy Problem   The Spy forces Blohardt's helicopter to land (see Problem 53, Section 9.5), but when he and his men approach, they discover no occupants—the helicopter is remote controlled! There is, however, a note attached to the control panel. There are map coordinates, a crude drawing, and these words: "From the snowman, pace off the distance to the woodpile, turn left, pace off an equal distance and drive a stake. From the snowman again, pace off the distance to the flagpole, turn right, pace off an equal distance and drive a second stake. Dig halfway between the stakes." A bloody letter "*F*" marks the spot on the map between the stakes.

"Flamer!" gasps the Spy. Gripped by trepidation, he arrives at the location indicated by the map coordinates. There is a woodpile and a flagpole, but the snowman has melted. How can the Spy know where to dig?

58. A triangle in $\mathbb{R}^3$ is determined by the vectors **v** and **w** as shown in Figure 9.70.

Figure 9.70   Area of a triangle

Show that the triangle has area

$$A = \tfrac{1}{2}\sqrt{\|\mathbf{v}\|^2 \|\mathbf{w}\|^2 - (\mathbf{v} \cdot \mathbf{w})^2}$$

59. Exploration Problem   In Figure 9.71, $\triangle ABC$ is equilateral and the points M, N, and O are located so that

$$\mathbf{AM} = \tfrac{1}{3}\mathbf{AB} \qquad \mathbf{BN} = \tfrac{1}{3}\mathbf{BC} \qquad \mathbf{CO} = \tfrac{1}{3}\mathbf{CA}$$

Figure 9.71   Problem 59

It can be shown that $\triangle PQR$ is also equilateral, and $\|\mathbf{PM}\| = \|\mathbf{QN}\| = \|\mathbf{RO}\|$ and $\|\mathbf{AP}\| = \|\mathbf{BQ}\| = \|\mathbf{CR}\|$.

a. Show that $\mathbf{PQ} = \tfrac{3}{7}\mathbf{AN}$, then show that $\|\mathbf{AN}\|^2 = \tfrac{7}{9}\|\mathbf{AB}\|^2$. *Hint*: Use the law of cosines.

b. Show that $\triangle PQR$ has area $\tfrac{1}{7}$ that of $\triangle ABC$.

c. Do you think the same result would hold if $\triangle ABC$ were not equilateral? Investigate your conjecture.

60. Gram-Schmidt orthogonalization process   Let **u**, **v**, and **w** be nonzero vectors in $\mathbb{R}^3$ that do not lie on the same plane. Define vectors α and β as follows:

$$\alpha = \mathbf{v} - \left[\frac{\mathbf{v} \cdot \mathbf{u}}{\|\mathbf{u}\|^2}\right]\mathbf{u} \quad \text{and} \quad \beta = \mathbf{w} - \left[\frac{\mathbf{w} \cdot \mathbf{u}}{\|\mathbf{u}\|^2}\right]\mathbf{u} - \left[\frac{\mathbf{w} \cdot \alpha}{\|\alpha\|^2}\right]\alpha$$

a. Show that **u**, α, β are mutually orthogonal (any pair is orthogonal).

b. If γ is any vector in $\mathbb{R}^3$, show that

$$\gamma = \left[\frac{\gamma \cdot \mathbf{u}}{\|\mathbf{u}\|^2}\right]\mathbf{u} + \left[\frac{\gamma \cdot \alpha}{\|\alpha\|^2}\right]\alpha + \left[\frac{\gamma \cdot \beta}{\|\beta\|^2}\right]\beta$$

61. Exploration Problem

a. In Figure 9.72, $ABCD$ is a rectangle, with M the midpoint of side $\overline{CD}$ and $\mathbf{AR}_k = \dfrac{1}{2k+1}\mathbf{AB}$. If P is the intersection of $\overline{AM}$ and $\overline{CR_k}$, and R_{k+1} is the foot of the perpendicular drawn from P to $\overline{AB}$, show that

$$\mathbf{AR}_{k+1} = \frac{1}{2k+3}\mathbf{AB}$$

Figure 9.72   Problem 61

b. It is easy to subdivide a line segment $\overline{AB}$ in half, or fourths, or eighths, etc. However, dividing it into thirds or fifths, ... is more difficult. Use the result obtained in part **a** to describe a procedure for subdividing $\overline{AB}$ into an odd number of equal parts.

62. Putnam Examination Problem   Find the equations of two straight lines, each of which cuts all four of the following lines:

$$\begin{aligned} L_1: \quad & x = 1, \quad y = 0 \\ L_2: \quad & y = 1, \quad z = 0 \\ L_3: \quad & z = 1, \quad x = 0 \\ L_4: \quad & x = y = -6z \end{aligned}$$

63. Putnam Examination Problem   Find the equation of the smallest sphere that is tangent to both the lines

$$\begin{aligned} L_1: \quad & x = t + 1, \ y = 2t + 4, \ z = -3t + 5 \text{ and} \\ L_2: \quad & x = 4t - 12, \ y = t + 8, \ z = t + 17 \end{aligned}$$

64. Putnam Examination Problem   The hands of an accurate clock have lengths 3 cm and 4 cm. Find the distance between the tips of the hands when the distance is increasing most rapidly.

Star Trek

This project is to be done in groups of three or four students. Each group will submit a single written report.

The starship *Enterprise* has been captured by the evil Romulans and is being held in orbit by a Romulan tractor beam. The orbit is elliptical with the planet Romulus at one focus of the ellipse. Repeated efforts to escape have been futile and have almost exhausted the fuel supplies. Morale is low and food reserves are dwindling.

In searching the ship's log, Lieutenant Commander Data discovers that the Enterprise had been captured long ago by a Romulan tractor beam and had escaped. The key to that escape was to fire the ship's thrusters at exactly the right position in the orbit. Captain Picard gives the command to feed the required information into the computer to find that position. But, alas, a Romulan virus has rendered the computer useless for this task. Everyone turns to you and asks for your help in solving the problem.

Here is what Data discovered. If F represents the focus of the ellipse and P is the position of the ship on the ellipse, then the vector **FP** can be written as a sum $\mathbf{T} + \mathbf{N}$ where $\mathbf{T}$ is tangent to the ellipse and $\mathbf{N}$ is normal to the ellipse (not necessarily unit vectors). The thrusters must be fired when the ratio $\|\mathbf{T}\|/\|\mathbf{N}\|$ is equal to the eccentricity of the ellipse.

Your mission is to save the starship from the evil Romulans.

TIME IS SAID to have only **one dimension**, and space to have **three dimensions**. ... The mathematical **quaternion** partakes of **both** of these elements; in technical language it may be said to be "time plus space," or "space plus time"; and in the sense it has, or at least involves a reference to, **four dimensions**.

——W. R. Hamilton
Graves' *Life of Hamilton*
(New York, 1882–1889)
Vol. 3, p. 635

Even though quaternions are not particularly important today, they preceded the modern concept of a vector. Our use of **i**, **j**, *and* **k** *derives from Hamilton's quaternions.*

*MAA Notes: 17 (1991) "Priming the Calculus Pump: Innovations and Resources," by Marcus S. Cohen, Edward D. Gaughan, R. Arthur Knoebel, Douglas S. Kurtz, and David J. Pengelley.

10

Vector-Valued Functions

PREVIEW

The marriage of calculus and vector methods forms what is called *vector calculus*. The key to using vector calculus is the concept of a *vector-valued function*. In this chapter, we introduce such functions and examine some of their properties. We will see that vector-valued functions behave much like the *scalar-valued functions* studied earlier in this text.

PERSPECTIVE

A car travels down a curved road at a constant speed of 55 mi/h. What additional information do we need about the car to determine whether it will stay on the road or skid off as it rounds a particular curve? Can we modify the road (say, by banking) so an average-sized car can travel at moderate speeds without skidding? A soldier fires a howitzer whose muzzle speed and angle of elevation are known. If the shell overshoots its target by 40 yd, how should the angle of elevation be changed to ensure a hit on the next shot? If a satellite is in orbit above the earth, how fast must it travel to remain stationary above a particular point on the equator? These and other similar questions can be answered using vector calculus.

10.1   Introduction to Vector Functions

IN THIS SECTION   vector-valued functions, operations with vector-valued functions, limits and continuity of vector functions

VECTOR-VALUED FUNCTIONS

In Section 9.5, we described a *plane curve* using parametric equations

$$x = f(t) \quad \text{and} \quad y = g(t)$$

where f and g are functions of t on some interval. We extend this definition to three dimensions. A **curve in** $\mathbb{R}^3$ is the set of all ordered triples $(f_1(t), f_2(t), f_3(t))$ satisfying the parametric equations $x = f_1(t)$, $y = f_2(t)$, $z = f_3(t)$, where f_1, f_2, and f_3 are functions of t on some domain D, which lies in $\mathbb{R}$.

The concept of a vector-valued function is fundamental to the ideas we plan to explore. Here is a definition.

Vector-Valued Function

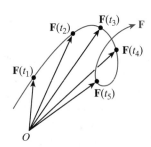

Figure 10.1 The graph of the vector function $\mathbf{F}(t)$ is traced out by the terminal point of $\mathbf{F}(t)$ as t varies over the domain D.

A **vector-valued function** (or, simply, a **vector function**) $\mathbf{F}$ of a real variable with *domain D* assigns to each number t in the set D a unique vector $\mathbf{F}(t)$. The set of all vectors $\mathbf{v}$ of the form $\mathbf{v} = \mathbf{F}(t)$ for t in D is the *range* of $\mathbf{F}$. That is,

$$\mathbf{F}(t) = f_1(t)\mathbf{i} + f_2(t)\mathbf{j} \qquad \text{Plane } (\mathbb{R}^2)$$

$$\mathbf{F}(t) = f_1(t)\mathbf{i} + f_2(t)\mathbf{j} + f_3(t)\mathbf{k} \qquad \text{Three-Space } (\mathbb{R}^3)$$

where f_1, f_2, and f_3 are real-valued (**scalar-valued**) functions of the real number t defined on the domain set D. In this context, f_1, f_2, and f_3 are called the **components** of $\mathbf{F}$. A vector function may also be denoted by $\mathbf{F}(t) = \langle f_1(t), f_2(t) \rangle$ or $\mathbf{F}(t) = \langle f_1(t), f_2(t), f_3(t) \rangle$.

Let $\mathbf{F}$ be a vector function, and suppose the initial point of the vector $\mathbf{F}(t)$ is at the origin. The graph of $\mathbf{F}$ is the curve traced out by the terminal point of the vector $\mathbf{F}(t)$ as t varies over the domain set D, as shown in Figure 10.1. In this context, $\mathbf{F}(t)$ is called the **position vector** at t for the point $P(f_1(t), f_2(t), f_3(t))$ on C.

EXAMPLE 1   Graph of a vector function

Sketch the graph of the vector function

$$\mathbf{F}(t) = (3 - t)\mathbf{i} + (2t)\mathbf{j} + (3t - 4)\mathbf{k}$$

for all t.

Solution

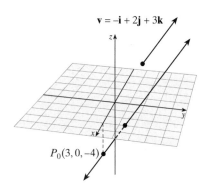

Figure 10.2 The graph of $\mathbf{F}(t) = (3 - t)\mathbf{i} + (2t)\mathbf{j} + (3t - 4)\mathbf{k}$

The graph is the collection of all points (x, y, z) with

$$x = 3 - t \qquad y = 2t \qquad z = -4 + 3t$$

for all t. We recognize these as the parametric equations for the line in $\mathbb{R}^3$ that contains the point $P_0(3, 0, -4)$ and is parallel to the vector

$$\mathbf{v} = -\mathbf{i} + 2\mathbf{j} + 3\mathbf{k}$$

as shown in Figure 10.2.   ∎

EXAMPLE 2   Graph of a circular helix

Sketch the graph of the vector function

$$\mathbf{F}(t) = (2 \sin t)\mathbf{i} - (2 \cos t)\mathbf{j} + (3t)\mathbf{k}$$

Solution

The graph of **F** is the collection of all points (x, y, z) in $\mathbb{R}^3$ whose coordinates satisfy

$$x = 2 \sin t \qquad y = -2 \cos t \qquad z = 3t \qquad \text{for all } t$$

The first two components satisfy

$$x^2 + y^2 = (2 \sin t)^2 + (-2 \cos t)^2 = 4(\sin^2 t + \cos^2 t) = 4$$

which means that the graph lies on the surface of the circular cylinder with radius 2 as shown in Figure 10.3.

Figure 10.3 The graph of $\mathbf{F}(t) = (2 \sin t)\mathbf{i} - (2 \cos t)\mathbf{j} + (3t)\mathbf{k}$

Next notice the axis of symmetry for the graph in Figure 10.3 is the z-axis. We also know that as t increases, the z-coordinate of the point $P(x, y, z)$ on the graph of **F** increases according to the formula $z = 3t$, which means that the point (x, y, z) on the graph rises in a spiral on the surface of the cylinder $x^2 + y^2 = 4$. The point on the graph of **F** that corresponds to $t = 0$ is $(0, -2, 0)$, and the points that correspond to $t = \frac{\pi}{2}$ and $t = \pi$ are $(2, 0, \frac{3\pi}{2})$ and $(0, 2, 3\pi)$, respectively. Thus, the graph spirals upward counterclockwise (as viewed from above). The graph, known as a **right circular helix**, is shown in Figure 10.3. ∎

A well-known example of a helix is the DNA (deoxyribonucleic acid) molecule, which has a structure consisting of two intertwined helixes, as shown in Figure 10.4. Some other computer-generated helixes are also shown.

a. DNA molecule

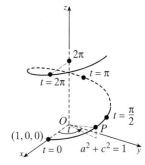

b. A computer-generated image of the helix in Example 2: $\mathbf{F}(t) = (2 \sin t)\mathbf{i} - (2 \cos t)\mathbf{j} + 3t\mathbf{k}$

c. A computer-generated image of the helix: $\mathbf{F}(t) = (\cos 2t)\mathbf{i} + (\sin 2t)\mathbf{j} + 0.2t\mathbf{k}$

Figure 10.4 Examples of helixes

Examples 1 and 2 illustrate how the graph of a vector function

$$\mathbf{F}(t) = f_1(t)\mathbf{i} + f_2(t)\mathbf{j} + f_3(t)\mathbf{k}$$

can be obtained by examining the parametric equations

$$x = f_1(t) \qquad y = f_2(t) \qquad z = f_3(t)$$

In Example 3, we turn things around and find a vector function whose graph is a given curve.

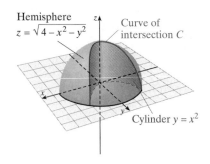

Hemisphere
$z = \sqrt{4 - x^2 - y^2}$

Curve of intersection C

Cylinder $y = x^2$

Figure 10.5 The curve of intersection of the hemisphere and the cylinder

EXAMPLE 3 **Find a vector function**

Find a vector function $\mathbf{F}$ whose graph is the curve of intersection of the hemisphere $z = \sqrt{4 - x^2 - y^2}$ and the parabolic cylinder $y = x^2$, as shown in Figure 10.5.

Solution

Finding a parametric representation is sometimes called **parameterizing the curve**. There are several ways this can be done, but one choice is to let $x = t$. Then $y = t^2$ (from the equation of the parabolic cylinder), and by substituting into the equation for the hemisphere, we find

$$z = \sqrt{4 - x^2 - y^2} = \sqrt{4 - (t)^2 - (t^2)^2} = \sqrt{4 - t^2 - t^4}$$

We can now state a formula for a vector-valued function of the given graph:

$$\mathbf{F}(t) = t\mathbf{i} + t^2\mathbf{j} + \sqrt{4 - t^2 - t^4}\,\mathbf{k} \qquad \blacksquare$$

OPERATIONS WITH VECTOR FUNCTIONS

It follows from the definition of vector operations that vector functions can be added, subtracted, multiplied by a scalar function, and multiplied together. We summarize these operations in the following box.

Vector Function Operations

Let $\mathbf{F}$ and $\mathbf{G}$ be vector functions of the real variable t, and let $f(t)$ be a scalar function. Then, $\mathbf{F} + \mathbf{G}$, $\mathbf{F} - \mathbf{G}$, $f\mathbf{F}$, and $\mathbf{F} \times \mathbf{G}$ are vector functions, and $\mathbf{F} \cdot \mathbf{G}$ is a scalar function. These operations are defined as follows:

Vector functions:

$$(\mathbf{F} + \mathbf{G})(t) = \mathbf{F}(t) + \mathbf{G}(t) \qquad (\mathbf{F} - \mathbf{G})(t) = \mathbf{F}(t) - \mathbf{G}(t)$$
$$(f\mathbf{F})(t) = f(t)\mathbf{F}(t) \qquad (\mathbf{F} \times \mathbf{G})(t) = \mathbf{F}(t) \times \mathbf{G}(t)$$

Scalar function:

$$(\mathbf{F} \cdot \mathbf{G})(t) = \mathbf{F}(t) \cdot \mathbf{G}(t)$$

These operations are defined on the intersection of the domain of the vector and scalar functions that occur in their definitions.

WARNING $(\mathbf{F} \times \mathbf{G})(t)$ is a vector and $(\mathbf{F} \cdot \mathbf{G})(t)$ is a scalar.

EXAMPLE 4 **Vector function operations**

Let $\mathbf{F}(t) = t^2\mathbf{i} + t\mathbf{j} - (\sin t)\mathbf{k}$ and $\mathbf{G}(t) = t\mathbf{i} + \dfrac{1}{t}\mathbf{j} + 5\mathbf{k}$. Find

a. $(\mathbf{F} + \mathbf{G})(t)$ **b.** $(e^t\mathbf{F})(t)$ **c.** $(\mathbf{F} \times \mathbf{G})(t)$ **d.** $(\mathbf{F} \cdot \mathbf{G})(t)$

Solution

a. $(\mathbf{F} + \mathbf{G})(t) = \mathbf{F}(t) + \mathbf{G}(t)$

$$= [t^2\mathbf{i} + t\mathbf{j} - (\sin t)\mathbf{k}] + \left[t\mathbf{i} + \frac{1}{t}\mathbf{j} + 5\mathbf{k}\right]$$

$$= (t^2 + t)\mathbf{i} + \left(t + \frac{1}{t}\right)\mathbf{j} + (5 - \sin t)\mathbf{k}$$

b. $(e^t \mathbf{F})(t) = e^t \mathbf{F}(t) = e^t t^2 \mathbf{i} + e^t t \mathbf{j} - (e^t \sin t)\mathbf{k}$

c. $(\mathbf{F} \times \mathbf{G})(t) = \mathbf{F}(t) \times \mathbf{G}(t)$

$$= [t^2 \mathbf{i} + t \mathbf{j} - (\sin t)\mathbf{k}] \times \left[t\mathbf{i} + \frac{1}{t}\mathbf{j} + 5\mathbf{k} \right]$$

$$= \begin{vmatrix} \mathbf{i} & \mathbf{j} & \mathbf{k} \\ t^2 & t & -\sin t \\ t & \dfrac{1}{t} & 5 \end{vmatrix}$$

$$= \left[5t + \frac{\sin t}{t} \right] \mathbf{i} - [5t^2 + t \sin t]\mathbf{j} + [t - t^2]\mathbf{k}$$

d. $(\mathbf{F} \cdot \mathbf{G})(t) = \mathbf{F}(t) \cdot \mathbf{G}(t)$

$$= [t^2 \mathbf{i} + t \mathbf{j} - (\sin t)\mathbf{k}] \cdot \left[t\mathbf{i} + \frac{1}{t}\mathbf{j} + 5\mathbf{k} \right]$$

$$= t^3 + 1 - 5 \sin t$$ ∎

TECHNOLOGY NOTE

Representing a vector function using technology is generally the same whether using a calculator or a software package. Square brackets are often used to denote vector functions. For Example 4, we denote $\mathbf{F}$ as the function $y1(x)$, and $\mathbf{G}$ as $y2(x)$:

$$y1(x) = [x^2, x, -\sin x]$$
$$y2(x) = [x, 1/x, 5]$$

A sample of these input values is shown in Figure 10.6a. Next, we work through each of the parts of Example 4, as shown in Figure 10.6b. Notice that if the answer is a vector, it is shown in brackets (as in parts **a**, **b**, and **c**). If the answer is a scalar (as in part **d**), the answer is not shown in brackets.

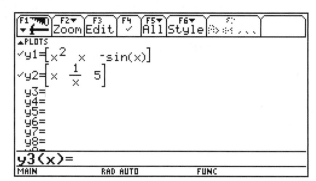

a. Input vector functions for Example 4

b. Technology solution for Example 4

Figure 10.6   Using technology with vector functions

LIMITS AND CONTINUITY OF VECTOR FUNCTIONS

Limit of a Vector Function

Suppose the components f_1, f_2, f_3 of the vector function

$$\mathbf{F}(t) = f_1(t)\mathbf{i} + f_2(t)\mathbf{j} + f_3(t)\mathbf{k}$$

all have finite limits as $t \to t_0$, where t_0 is any number or $\pm\infty$. Then the **limit** of $\mathbf{F}(t)$ as $t \to t_0$ is the vector

$$\lim_{t \to t_0} \mathbf{F}(t) = \left[\lim_{t \to t_0} f_1(t) \right] \mathbf{i} + \left[\lim_{t \to t_0} f_2(t) \right] \mathbf{j} + \left[\lim_{t \to t_0} f_3(t) \right] \mathbf{k}$$

EXAMPLE 5 **Limit of a vector function**

Find $\lim_{t \to 2} \mathbf{F}(t)$, where $\mathbf{F}(t) = (t^2 - 3)\mathbf{i} + e^t\mathbf{i} + (\sin \pi t)\mathbf{k}$.

Solution

$$\lim_{t \to 2} \mathbf{F}(t) = \left[\lim_{t \to 2}(t^2 - 3)\right]\mathbf{i} + \left[\lim_{t \to 2}(e^t)\right]\mathbf{j} + \left[\lim_{t \to 2}(\sin \pi t)\right]\mathbf{k}$$

$$= 1\mathbf{i} + e^2\mathbf{j} + (\sin 2\pi)\mathbf{k}$$

$$= \mathbf{i} + e^2\mathbf{j} \qquad \blacksquare$$

Vector limits behave like scalar limits. The following theorem contains some useful general properties of such limits.

THEOREM 10.1 **Rules for Vector Limits**

If the vector functions $\mathbf{F}$ and $\mathbf{G}$ are functions of a real variable t and $h(t)$ is a scalar function such that all three functions have finite limits as $t \to t_0$, then

Limit of a sum	$\lim_{t \to t_0}[\mathbf{F}(t) + \mathbf{G}(t)] = \lim_{t \to t_0}\mathbf{F}(t) + \lim_{t \to t_0}\mathbf{G}(t)$
Limit of a difference	$\lim_{t \to t_0}[\mathbf{F}(t) - \mathbf{G}(t)] = \lim_{t \to t_0}\mathbf{F}(t) - \lim_{t \to t_0}\mathbf{G}(t)$
Limit of a scalar multiple	$\lim_{t \to t_0}[h(t)\mathbf{F}(t)] = \left[\lim_{t \to t_0} h(t)\right]\left[\lim_{t \to t_0}\mathbf{F}(t)\right]$
Limit of a dot product	$\lim_{t \to t_0}[\mathbf{F}(t) \cdot \mathbf{G}(t)] = \left[\lim_{t \to t_0}\mathbf{F}(t)\right] \cdot \left[\lim_{t \to t_0}\mathbf{G}(t)\right]$
Limit of a cross product	$\lim_{t \to t_0}[\mathbf{F}(t) \times \mathbf{G}(t)] = \left[\lim_{t \to t_0}\mathbf{F}(t)\right] \times \left[\lim_{t \to t_0}\mathbf{G}(t)\right]$

These limit formulas are also valid as $t \to +\infty$ or as $t \to -\infty$, assuming all expressions have finite limits.

Proof We will establish the formula for the limit of a dot product and leave the rest of the proof as an exercise. Let

$$\mathbf{F}(t) = f_1(t)\mathbf{i} + f_2(t)\mathbf{j} + f_3(t)\mathbf{k} \quad \text{and} \quad \mathbf{G}(t) = g_1(t)\mathbf{i} + g_2(t)\mathbf{j} + g_3(t)\mathbf{k}$$

Apply the limit of a vector function along with the sum rule and product rule for scalar limits to write

$$\lim_{t \to t_0}[\mathbf{F}(t) \cdot \mathbf{G}(t)] = \lim_{t \to t_0}[f_1(t)g_1(t) + f_2(t)g_2(t) + f_3(t)g_3(t)]$$

$$= \left[\lim_{t \to t_0} f_1(t)\right]\left[\lim_{t \to t_0} g_1(t)\right] + \left[\lim_{t \to t_0} f_2(t)\right]\left[\lim_{t \to t_0} g_2(t)\right] + \left[\lim_{t \to t_0} f_3(t)\right]\left[\lim_{t \to t_0} g_3(t)\right]$$

$$= \left(\left[\lim_{t \to t_0} f_1(t)\right]\mathbf{i} + \left[\lim_{t \to t_0} f_2(t)\right]\mathbf{j} + \left[\lim_{t \to t_0} f_3(t)\right]\mathbf{k}\right) \cdot \left(\left[\lim_{t \to t_0} g_1(t)\right]\mathbf{i} + \left[\lim_{t \to t_0} g_2(t)\right]\mathbf{j} + \left[\lim_{t \to t_0} g_3(t)\right]\mathbf{k}\right)$$

$$= \left[\lim_{t \to t_0}\mathbf{F}(t)\right] \cdot \left[\lim_{t \to t_0}\mathbf{G}(t)\right] \qquad \blacksquare$$

EXAMPLE 6 **Limit of a cross product of vector functions**

Show that $\lim_{t \to 1}[\mathbf{F}(t) \times \mathbf{G}(t)] = \left[\lim_{t \to 1}\mathbf{F}(t)\right] \times \left[\lim_{t \to 1}\mathbf{G}(t)\right]$ for the vector functions

$$\mathbf{F}(t) = t\mathbf{i} + (1 - t)\mathbf{j} + t^2\mathbf{k} \quad \text{and} \quad \mathbf{G}(t) = e^t\mathbf{i} - (3 + e^t)\mathbf{k}$$

Solution

$$\mathbf{F}(t) \times \mathbf{G}(t) = \begin{vmatrix} \mathbf{i} & \mathbf{j} & \mathbf{k} \\ t & 1-t & t^2 \\ e^t & 0 & -(3+e^t) \end{vmatrix}$$

$$= [(1-t)(-3-e^t) - 0]\mathbf{i} - [-t(3+e^t) - t^2 e^t]\mathbf{j} + [0 - e^t(1-t)]\mathbf{k}$$

$$= (te^t + 3t - e^t - 3)\mathbf{i} + (t^2 e^t + te^t + 3t)\mathbf{j} + (te^t - e^t)\mathbf{k}$$

Thus, the limit of the cross product is

$$\lim_{t \to 1}[\mathbf{F}(t) \times \mathbf{G}(t)] = \lim_{t \to 1}[te^t + 3t - e^t - 3]\mathbf{i} + \lim_{t \to 1}[t^2 e^t + te^t + 3t]\mathbf{j} + \lim_{t \to 1}[te^t - e^t]\mathbf{k}$$

$$= (e + 3 - e - 3)\mathbf{i} + (e + e + 3)\mathbf{j} + (e - e)\mathbf{k} = (2e + 3)\mathbf{j}$$

Now we find the cross product of the limits.

$$\lim_{t \to 1} \mathbf{F}(t) = \left[\lim_{t \to 1} t\right]\mathbf{i} + \left[\lim_{t \to 1}(1-t)\right]\mathbf{j} + \left[\lim_{t \to 1} t^2\right]\mathbf{k} = \mathbf{i} + \mathbf{k}$$

$$\lim_{t \to 1} \mathbf{G}(t) = \left[\lim_{t \to 1} e^t\right]\mathbf{i} + \left[\lim_{t \to 1}(-3 - e^t)\right]\mathbf{k} = e\mathbf{i} + (-3 - e)\mathbf{k}$$

Thus,

$$\left[\lim_{t \to 1} \mathbf{F}(t)\right] \times \left[\lim_{t \to 1} \mathbf{G}(t)\right] = \begin{vmatrix} \mathbf{i} & \mathbf{j} & \mathbf{k} \\ 1 & 0 & 1 \\ e & 0 & -3-e \end{vmatrix}$$

$$= (0-0)\mathbf{i} - (-3 - e - e)\mathbf{j} + (0-0)\mathbf{k} = (3 + 2e)\mathbf{j}$$

We see $\lim_{t \to 1}[\mathbf{F}(t) \times \mathbf{G}(t)] = \left[\lim_{t \to 1} \mathbf{F}(t)\right] \times \left[\lim_{t \to 1} \mathbf{G}(t)\right]$. ∎

Continuity of a Vector Function

> A vector function $\mathbf{F}(t)$ is said to be **continuous** at t_0 if t_0 is in the domain of $\mathbf{F}$ and $\lim_{t \to t_0} \mathbf{F}(t) = \mathbf{F}(t_0)$.

> ➡ **What This Says** This is the same as requiring each component of $\mathbf{F}(t)$ to be continuous at t_0. That is,
>
> $$\mathbf{F}(t) = f_1(t)\mathbf{i} + f_2(t)\mathbf{j} + f_3(t)\mathbf{k}$$
>
> is continuous at t_0 when t_0 is in the domain of the component functions $f_1(t)$, $f_2(t)$, and $f_3(t)$ and
>
> $$\lim_{t \to t_0} f_1(t) = f_1(t_0) \qquad \lim_{t \to t_0} f_2(t) = f_2(t_0) \qquad \lim_{t \to t_0} f_3(t) = f_3(t_0)$$

The rules for vector limits listed in Theorem 10.1 can be used to derive general properties of vector function continuity.

EXAMPLE 7   Continuity of a vector function

For what values of t is $\mathbf{F}(t) = (\sin t)\mathbf{i} + (1-t)^{-1}\mathbf{j} + (\ln t)\mathbf{k}$ continuous?

Solution

The vector function $\mathbf{F}$ is continuous where its component functions

$$f_1(t) = \sin t \qquad f_2(t) = (1-t)^{-1} \qquad f_3(t) = \ln t$$

are continuous. The function f_1 is continuous for all t; f_2 is continuous whenever $1 - t \neq 0$ (that is, when $t \neq 1$); f_3 is continuous for $t > 0$. Thus, $\mathbf{F}$ is continuous when t is a positive number other than 1; that is, $t > 0, t \neq 1$. ∎

10.1 PROBLEM SET

Ⓐ *Find the domain of the vector functions given in Problems 1–8.*

1. $\mathbf{F}(t) = 2t\mathbf{i} - 3t\mathbf{j} + \dfrac{1}{t}\mathbf{k}$

2. $\mathbf{F}(t) = (1 - t)\mathbf{i} + \sqrt{t}\,\mathbf{j} - \dfrac{1}{t - 2}\mathbf{k}$

3. $\mathbf{F}(t) = (\sin t)\mathbf{i} + (\cos t)\mathbf{j} + (\tan t)\mathbf{k}$

4. $\mathbf{F}(t) = (\cos t)\mathbf{i} - (\cot t)\mathbf{j} + (\csc t)\mathbf{k}$

5. $h(t)\mathbf{F}(t)$, where $h(t) = \sin t$ and

$$\mathbf{F}(t) = \frac{1}{\cos t}\mathbf{i} + \frac{1}{\sin t}\mathbf{j} + \frac{1}{\tan t}\mathbf{k}$$

6. $\mathbf{F}(t) + \mathbf{G}(t)$, where $\mathbf{F}(t) = 3t\mathbf{j} + t^{-1}\mathbf{k}$ and $\mathbf{G}(t) = 5t\mathbf{i} + \sqrt{10 - t}\,\mathbf{j}$

7. $\mathbf{F}(t) - \mathbf{G}(t)$, where $\mathbf{F}(t) = (\ln t)\mathbf{i} + 3t\mathbf{j} - t^2\mathbf{k}$ and $\mathbf{G}(t) = \mathbf{i} + 5t\mathbf{j} - t^2\mathbf{k}$

8. $\mathbf{F}(t) \times \mathbf{G}(t)$ where $\mathbf{F}(t) = t^2\mathbf{i} - t\mathbf{j} + 2t\mathbf{k}$ and $\mathbf{G}(t) = \dfrac{1}{t + 2}\mathbf{i} + (t + 4)\mathbf{j} - \sqrt{-t}\,\mathbf{k}$

Describe the graph of the vector functions given in Problems 9–20 or sketch a graph in $\mathbb{R}^3$. A graph in $\mathbb{R}^2$ may help with your description.

9. $\mathbf{F}(t) = 2t\mathbf{i} + t^2\mathbf{j}$

10. $\mathbf{G}(t) = (1 - t)\mathbf{i} + \dfrac{1}{t}\mathbf{j}$

11. $\mathbf{G}(t) = (\sin t)\mathbf{i} - (\cos t)\mathbf{j}$

12. $\mathbf{F}(t) = (2\cos t)\mathbf{i} + (\sin t)\mathbf{j}$

13. $\mathbf{F}(t) = t\mathbf{i} - 4\mathbf{k}$

14. $\mathbf{G}(t) = e^t\mathbf{j} + t\mathbf{k}$

15. $\mathbf{F}(t) = (\cos t)\mathbf{i} + (\sin t)\mathbf{j} + t\mathbf{k}$

16. $\mathbf{F}(t) = e^t\mathbf{i} + e^t\mathbf{j} + e^{-t}\mathbf{k}$

17. $\mathbf{G}(t) = (1 - t)\mathbf{i} + t^2\mathbf{j} + t\mathbf{k}$

18. $\mathbf{F}(t) = \left(\dfrac{\sqrt{2}}{2}\sin t\right)\mathbf{i} + \left(\dfrac{\sqrt{2}}{2}\sin t\right)\mathbf{j} + (\cos^2 t)\mathbf{k}$

19. $\mathbf{F}(t) = t\mathbf{i} + (t^2 + 1)\mathbf{j} + t^2\mathbf{k}$

20. $\mathbf{G}(t) = (2\sin t)\mathbf{i} + (2\cos t)\mathbf{j} + 3\mathbf{k}$

Perform the operations indicated in Problems 21–32 with

$$\mathbf{F}(t) = 2t\mathbf{i} - 5\mathbf{j} + t^2\mathbf{k},$$

$$\mathbf{G}(t) = (1 - t)\mathbf{i} + \frac{1}{t}\mathbf{k},$$

$$\mathbf{H}(t) = (\sin t)\mathbf{i} + e^t\mathbf{j}$$

21. $2\mathbf{F}(t) - 3\mathbf{G}(t)$

22. $t^2\mathbf{F}(t) - 3\mathbf{H}(t)$

23. $\mathbf{F}(t) \cdot \mathbf{G}(t)$

24. $\mathbf{F}(t) \cdot \mathbf{H}(t)$

25. $\mathbf{G}(t) \cdot \mathbf{H}(t)$

26. $\mathbf{F}(t) \times \mathbf{G}(t)$

27. $\mathbf{F}(t) \times \mathbf{H}(t)$

28. $\mathbf{G}(t) \times \mathbf{H}(t)$

29. $2e^t\mathbf{F}(t) + t\mathbf{G}(t) + 10\mathbf{H}(t)$

30. $\mathbf{F}(t) \cdot [\mathbf{H}(t) \times \mathbf{G}(t)]$

31. $\mathbf{G}(t) \cdot [\mathbf{H}(t) \times \mathbf{F}(t)]$

32. $\mathbf{H}(t) \cdot [\mathbf{G}(t) \times \mathbf{F}(t)]$

Ⓑ 33. **WHAT DOES THIS SAY?** Discuss the concept of a vector-valued function.

34. **WHAT DOES THIS SAY?** Discuss the concept of the limit of a vector-valued function.

35. Show that the curve given by
$\mathbf{R}(t) = (2\sin t)\mathbf{i} + (2\sin t)\mathbf{j} + (\sqrt{8}\cos t)\mathbf{k}$
lies on a sphere centered at the origin.

36. Sketch the graph of the curve given by
$\mathbf{R}(t) = (2\cos t)\mathbf{i} + (\sin t)\mathbf{j} - 2t\mathbf{k}$.

Find a vector function $\mathbf{F}$ whose graph is the curve given in Problems 37–42.

37. $y = x^2; z = 2$

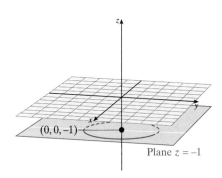

38. $x^2 + y^2 = 4; z = -1$

39. $x = 2t, y = 1 - t, z = \sin t$

40. $\dfrac{x - 2}{3} = \dfrac{y - 1}{2} = \dfrac{z}{4}$

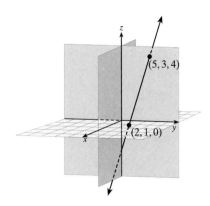

41. The curve of intersection of the hemisphere $z = \sqrt{9 - x^2 - y^2}$ and the parabolic cylinder $x = y^2$

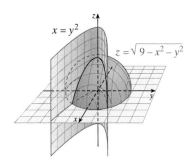

42. The line of intersection of the planes $2x + y + 3z = 6$ and $x - y - z = 1$

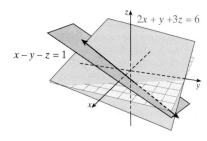

43. Which of the following curves lies on the surface $z = x^2 + y^2$?

a. $x = \sin t$, $y = \cos t$, $z = t$

b. $x = e^{-t} \cos t$, $y = e^{-t} \sin t$, $z = e^{-2t}$

c. $x = 1 + \dfrac{1}{t}$, $y = 1 - \dfrac{1}{t}$, $z = 2 + \dfrac{2}{t^2}$

44. Find parametric equations for the curve of intersection of the elliptical cylinder $x^2 + 3y^2 = 1$ and the parabolic cylinder $z = 2x^2 - 1$.

Find each limit indicated in Problems 45–52.

45. $\lim\limits_{t \to 1} \left[2t\mathbf{i} - 3\mathbf{j} + e^t\mathbf{k} \right]$

46. $\lim\limits_{t \to 1} [3t\mathbf{i} + e^{2t}\mathbf{j} + (\sin \pi t)\mathbf{k}]$

47. $\lim\limits_{t \to 0} \left[\dfrac{(\sin t)\mathbf{i} - t\mathbf{k}}{t^2 + t - 1} \right]$

48. $\lim\limits_{t \to 1} \left[\dfrac{t^3 - 1}{t - 1}\mathbf{i} + \dfrac{t^2 - 3t + 2}{t^2 + t - 2}\mathbf{j} + (t^2 + 1)e^{t-1}\mathbf{k} \right]$

49. $\lim\limits_{t \to 0} \left[\dfrac{te^t}{1 - e^t}\mathbf{i} + \dfrac{e^{t-1}}{\cos t}\mathbf{j} \right]$

50. $\lim\limits_{t \to 0} \left[\dfrac{\sin t}{t}\mathbf{i} + \dfrac{1 - \cos t}{t}\mathbf{j} + e^{1-t}\mathbf{k} \right]$

51. $\lim\limits_{t \to 0^+} \left[\dfrac{\sin 3t}{\sin 2t}\mathbf{i} + \dfrac{\ln(\sin t)}{\ln(\tan t)}\mathbf{j} + (t \ln t)\mathbf{k} \right]$

52. $\lim\limits_{t \to 2} \left[(2\mathbf{i} - t\mathbf{j} + e^t\mathbf{k}) \times (t^2\mathbf{i} + 4 \sin t\mathbf{j}) \right]$

Determine all values of t for which the vector function given in Problems 53–58 is continuous.

53. $\mathbf{F}(t) = t\mathbf{i} + 3\mathbf{j} - (1 - t)\mathbf{k}$ **54.** $\mathbf{G}(t) = t\mathbf{i} - \dfrac{1}{t}\mathbf{k}$

55. $\mathbf{G}(t) = \dfrac{\mathbf{i} + 2\mathbf{j}}{t^2 + t}$

56. $\mathbf{F}(t) = (e^t \sin t)\mathbf{i} + (e^t \cos t)\mathbf{k}$

57. $\mathbf{F}(t) = e^t \left[t\mathbf{i} + \dfrac{1}{t}\mathbf{j} + 3\mathbf{k} \right]$

58. $\mathbf{G}(t) = \dfrac{\mathbf{u}}{\|\mathbf{u}\|}$, where $\mathbf{u} = t\mathbf{i} + \sqrt{t}\,\mathbf{j}$

59. The graph of

$$\mathbf{R}(t) = t\mathbf{i} + \left(\frac{1-t}{t} \right)\mathbf{j} + \left(\frac{1-t^2}{t} \right)\mathbf{k}$$

lies in a plane. What is the equation of this plane?

60. How many revolutions are made by the circular helix

$$\mathbf{R}(t) = (2 \sin t)\mathbf{i} + (2 \cos t)\mathbf{j} + \tfrac{5}{8}t\mathbf{k}$$

in a vertical distance of 8 units?

61. Given the vector functions $\mathbf{F}(t) = t\mathbf{i} + t^2\mathbf{j} + t^3\mathbf{k}$ and $\mathbf{G}(t) = \tfrac{1}{t}\mathbf{i} - e^t\mathbf{j}$, directly verify each of the following limit formulas (that is, without using Theorem 10.1).

a. $\lim\limits_{t \to 0} e^t \mathbf{F}(t) = \left[\lim\limits_{t \to 0} e^t \right] \left[\lim\limits_{t \to 0} \mathbf{F}(t) \right]$

b. $\lim\limits_{t \to 1} \mathbf{F}(t) \cdot \mathbf{G}(t) = \left[\lim\limits_{t \to 1} \mathbf{F}(t) \right] \cdot \left[\lim\limits_{t \to 1} \mathbf{G}(t) \right]$

c. $\lim\limits_{t \to 1} [\mathbf{F}(t) \times \mathbf{G}(t)] = \left[\lim\limits_{t \to 1} \mathbf{F}(t) \right] \times \left[\lim\limits_{t \to 1} \mathbf{G}(t) \right]$

62. If $\mathbf{H}(t)$ is a vector function, we define the **difference operator** $\Delta\mathbf{H}$ by the formula

$$\Delta\mathbf{H} = \mathbf{H}(t + \Delta t) - \mathbf{H}(t)$$

where Δt is a change in the parameter t. (*Note:* Usually, $|\Delta t|$ is a small number.) If $\mathbf{F}(t)$ and $\mathbf{G}(t)$ are vector functions, show that $\Delta(\mathbf{F} \times \mathbf{G})(t) = \mathbf{F}(t + \Delta t) \times \Delta\mathbf{G}(t) + \Delta\mathbf{F}(t) \times \mathbf{G}(t)$. *Hint:* Observe that

$$\Delta(\mathbf{F} \times \mathbf{G})(t) = \mathbf{F}(t + \Delta t) \times \mathbf{G}(t + \Delta t)$$
$$- \mathbf{F}(t + \Delta t) \times \mathbf{G}(t) + \mathbf{F}(t + \Delta t) \times \mathbf{G}(t)$$
$$- \mathbf{F}(t) \times \mathbf{G}(t)$$

63. Prove the limit of a sum rule of Theorem 10.1. (The difference rule is proved similarly.)

64. a. Prove the limit of a scalar multiple rule of Theorem 10.1.

b. Prove the limit of a cross product rule of Theorem 10.1.

65. Counterexample Problem Let $\mathbf{F}(t)$ and $\mathbf{G}(t)$ be vector functions that are continuous at t_0, and let $h(t)$ be a scalar function continuous at t_0. In each of the following cases, either show that the given function is continuous at t_0 or provide a counterexample to show that it is not continuous.

a. $3\mathbf{F}(t) + 5\mathbf{G}(t)$ **b.** $\mathbf{F}(t) \cdot \mathbf{G}(t)$

c. $(h(t))^{-1}\mathbf{F}(t)$ **d.** $\mathbf{F}(t) \times \mathbf{G}(t)$

66. Counterexample Problem Find a vector function $\mathbf{F}(t) = \langle f_1(t), f_2(t), f_3(t) \rangle$ such that $\|\mathbf{F}(t)\|$ is continuous at $t = 0$ but $\mathbf{F}(t)$ is not continuous at $t = 0$.

10.2 Differentiation and Integration of Vector Functions

IN THIS SECTION vector derivatives, tangent vectors, properties of vector derivatives, modeling the motion of an object in $\mathbb{R}^3$, vector integrals

VECTOR DERIVATIVES

In Chapter 3, we defined the derivative of the (scalar) function f to be the limit as $\Delta x \to 0$ of the difference quotient $\Delta f / \Delta x$. The **difference quotient of a vector function F** is the vector expression

$$\frac{\Delta \mathbf{F}}{\Delta t} = \frac{\mathbf{F}(t + \Delta t) - \mathbf{F}(t)}{\Delta t}$$

and we define the derivative of $\mathbf{F}$ as follows.

Derivative of a Vector Function

The **derivative** of the vector function $\mathbf{F}$ is the vector function $\mathbf{F}'$ determined by the limit

$$\mathbf{F}'(t) = \lim_{\Delta t \to 0} \frac{\Delta \mathbf{F}}{\Delta t} = \lim_{\Delta t \to 0} \frac{\mathbf{F}(t + \Delta t) - \mathbf{F}(t)}{\Delta t}$$

wherever this limit exists. In the Leibniz notation, the derivative of $\mathbf{F}$ is denoted by $\dfrac{d\mathbf{F}}{dt}$. We say that the vector function $\mathbf{F}$ is **differentiable** at $t = t_0$ if $\mathbf{F}'(t)$ is defined at t_0.

The following theorem establishes a convenient method for computing the derivative of a vector function.

THEOREM 10.2 Derivative of a vector function

The vector function $\mathbf{F}(t) = f_1(t)\mathbf{i} + f_2(t)\mathbf{j} + f_3(t)\mathbf{k}$ is differentiable whenever the component functions f_1, f_2, and f_3 are each differentiable, and in this case

$$\mathbf{F}'(t) = f_1'(t)\mathbf{i} + f_2'(t)\mathbf{j} + f_3'(t)\mathbf{k}$$

Proof We use the definition of the derivative, along with rules of vector limits (Theorem 10.1), and the fact that the scalar derivatives $f_1'(t)$, $f_2'(t)$, and $f_3'(t)$ all exist.

$$\mathbf{F}'(t) = \lim_{\Delta t \to 0} \frac{\mathbf{F}(t + \Delta t) - \mathbf{F}(t)}{\Delta t}$$

$$= \lim_{\Delta t \to 0} \frac{[f_1(t + \Delta t)\mathbf{i} + f_2(t + \Delta t)\mathbf{j} + f_3(t + \Delta t)\mathbf{k}] - [f_1(t)\mathbf{i} + f_2(t)\mathbf{j} + f_3(t)\mathbf{k}]}{\Delta t}$$

$$= \left[\lim_{\Delta t \to 0} \frac{f_1(t + \Delta t) - f_1(t)}{\Delta t}\right]\mathbf{i} + \left[\lim_{\Delta t \to 0} \frac{f_2(t + \Delta t) - f_2(t)}{\Delta t}\right]\mathbf{j} + \left[\lim_{\Delta t \to 0} \frac{f_3(t + \Delta t) - f_3(t)}{\Delta t}\right]\mathbf{k}$$

$$= f_1'(t)\mathbf{i} + f_2'(t)\mathbf{j} + f_3'(t)\mathbf{k} \qquad \square$$

EXAMPLE 1 Differentiability of a vector function

For what values of t is $\mathbf{G}(t) = |t|\,\mathbf{i} + (\cos t)\mathbf{j} + (t - 5)\mathbf{k}$ differentiable?

Solution

The component functions $\cos t$ and $t - 5$ are differentiable for all t, but $|t|$ is not differentiable at $t = 0$. Thus, the vector function $\mathbf{G}$ is differentiable for all $t \neq 0$. ∎

EXAMPLE 2   Derivative of a vector function

Find the derivative of the vector function

$$\mathbf{F}(t) = e^t\mathbf{i} + (\sin t)\mathbf{j} + (t^3 + 5t)\mathbf{k}$$

Solution

Differentiating each component separately, we find that

$$\mathbf{F}'(t) = (e^t)'\mathbf{i} + (\sin t)'\mathbf{j} + (t^3 + 5t)'\mathbf{k} = e^t\mathbf{i} + (\cos t)\mathbf{j} + (3t^2 + 5)\mathbf{k} \qquad \blacksquare$$

TANGENT VECTORS

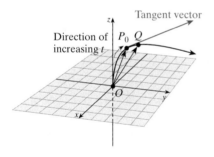

Figure 10.7 The difference quotient is a multiple of the secant line vector.

Recall that the scalar derivative $f'(x_0)$ gives the slope of the tangent line to the graph of f at the point where $x = x_0$ and thus provides a measure of the graph's direction at that point. Our first goal is to extend this interpretation by showing how the vector derivative can be used to find tangent vectors to curves in space. Let t be a number in the domain of the vector function $\mathbf{F}(t)$, and let P_0 be the point on the graph of $\mathbf{F}$ that corresponds to t_0, as shown in Figure 10.7. Then for any positive number Δt, the difference quotient

$$\frac{\Delta \mathbf{F}}{\Delta t} = \frac{\mathbf{F}(t_0 + \Delta t) - \mathbf{F}(t_0)}{\Delta t}$$

is a vector that points in the same direction as the secant vector

$$\mathbf{P_0Q} = \mathbf{F}(t_0 + \Delta t) - \mathbf{F}(t_0),$$

where Q is the point on the graph of $\mathbf{F}$ that corresponds to $t = t_0 + \Delta t$ (see Figure 10.7). If we choose $\Delta t < 0$, the difference quotient $\Delta \mathbf{F}/\Delta t$ points in the *opposite* direction to that of the secant vector $\mathbf{P_0Q}$.

Suppose the difference quotient $\Delta \mathbf{F}/\Delta t$ has a limit as $\Delta t \to 0$ and that

$$\lim_{\Delta t \to 0} \frac{\Delta \mathbf{F}}{\Delta t} \neq \mathbf{0}$$

Then, as $\Delta t \to 0$, the direction of the secant vector, $\mathbf{P_0Q}$, and hence that of the difference quotient $\Delta \mathbf{F}/\Delta t$, will approach the direction of the tangent vector at P_0, as shown in Figure 10.8. Thus, we expect the tangent vector at P_0 to be the limit vector

$$\lim_{\Delta t \to 0} \frac{\Delta \mathbf{F}}{\Delta t}$$

which we recognize as the vector derivative $\mathbf{F}'(t_0)$. These observations lead to the following interpretation.

Figure 10.8 As $\Delta t \to 0$, the vector $\mathbf{P_0Q}$ and hence the difference quotient $\Delta \mathbf{F}/\Delta t$ approaches the tangent vector at P_0.

Tangent Vector

> Suppose $\mathbf{F}(t)$ is differentiable at t_0 and that $\mathbf{F}'(t_0) \neq 0$. Then $\mathbf{F}'(t_0)$ is a **tangent vector** to the graph of $\mathbf{F}(t)$ at the point where $t = t_0$ and points in the direction of increasing t.

EXAMPLE 3   Finding a tangent vector

Find a tangent vector at the point P_0 where $t = 0.2$ on the graph of the vector function

$$\mathbf{F}(t) = e^{2t}\mathbf{i} + (t^2 - t)\mathbf{j} + (\ln t)\mathbf{k}$$

What is an equation for the tangent line at P_0?

Solution

The derivative of $\mathbf{F}(t)$ is

$$\mathbf{F}'(t) = 2e^{2t}\mathbf{i} + (2t - 1)\mathbf{j} + t^{-1}\mathbf{k}$$

so a tangent vector at the point where $t = 0.2$ is

$$\mathbf{F}'(0.2) = 2e^{0.4}\mathbf{i} - 0.6\mathbf{j} + 5\mathbf{k}$$

The tangent line to the graph of $\mathbf{F}(t)$ at P_0 is the line that passes through P_0 and is parallel to the vector $\mathbf{F}(0.2)$. Since

$$\mathbf{F}(0.2) = e^{0.4}\mathbf{i} - 0.16\mathbf{j} + (\ln 0.2)\mathbf{k}$$

the point of tangency is $(e^{0.4}, -0.16, \ln 0.2)$. Thus, the tangent line has parametric equations

$$x = e^{0.4} + 2e^{0.4}t, \qquad y = -0.16 - 0.6t, \qquad z = \ln 0.2 + 5t \qquad \blacksquare$$

If $\mathbf{F}'(t_0) \neq \mathbf{0}$ and we also require the derivative $\mathbf{F}'$ to be continuous at t_0, then the tangent vector at each point of the graph of $\mathbf{F}$ near P_0 will be close to the tangent vector at P_0. In this case, the graph is said to be *smooth* at P_0.

Smooth Curve

> The graph of the vector function defined by $\mathbf{F}(t)$ is **smooth** on any interval of t where $\mathbf{F}'$ is continuous and $\mathbf{F}'(t) \neq \mathbf{0}$. The graph is **piecewise smooth** on an interval that can be subdivided into a finite number of subintervals on which $\mathbf{F}$ is smooth.

EXAMPLE 4 Determining whether a curve is smooth

Determine whether the graph of the vector function

$$\mathbf{F}(t) = \langle t^2 + 1, \cos t, e^t + e^{-t} \rangle$$

is smooth for all t.

Solution

The derivative

$$\mathbf{F}'(t) = \langle 2t, -\sin t, e^t - e^{-t} \rangle$$

is continuous for all t, but $\mathbf{F}'(0) = \langle 0, 0, 1 - 1 \rangle = \mathbf{0}$. Thus, the graph is not smooth for all t, but it is smooth on any interval not containing $t = 0$. $\qquad \blacksquare$

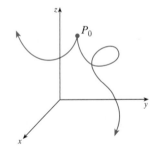

Figure 10.9 A curve that is not smooth

A graph will not be smooth on any interval containing a point where there is an abrupt change in direction. For example, the graph in Figure 10.9 is not smooth on any interval containing the point corresponding to the sharp corner P_0. In Example 4, such a point occurs when $t = 0$, namely $P_0(1, 1, 2)$. However, this curve is piecewise smooth because it consists of a finite number of smooth pieces.

PROPERTIES OF VECTOR DERIVATIVES

Higher-order derivatives of a vector function $\mathbf{F}$ are obtained by successively differentiating the components of $\mathbf{F}(t) = f_1(t)\mathbf{i} + f_2(t)\mathbf{j} + f_3(t)\mathbf{k}$. For instance, the **second derivative** of $\mathbf{F}$ is the vector function

$$\mathbf{F}''(t) = [\mathbf{F}'(t)]' = f_1''(t)\mathbf{i} + f_2''(t)\mathbf{j} + f_3''(t)\mathbf{k}$$

whereas the **third derivative** $\mathbf{F}'''(t)$ is the derivative of $\mathbf{F}''(t)$, and so on. In the Leibniz notation, the second vector derivative of $\mathbf{F}$ with respect to t is denoted by $\dfrac{d^2\mathbf{F}}{dt^2}$, and the third derivative, by $\dfrac{d^3\mathbf{F}}{dt^3}$.

EXAMPLE 5   Higher-order derivatives of a vector function

Find the second and third derivatives of the vector function

$$\mathbf{F}(t) = e^{2t}\mathbf{i} + (1 - t^2)\mathbf{j} + (\cos 2t)\mathbf{k}$$

Solution

$$\mathbf{F}'(t) = 2e^{2t}\mathbf{i} + (-2t)\mathbf{j} + (-2\sin 2t)\mathbf{k}$$
$$\mathbf{F}''(t) = 4e^{2t}\mathbf{i} - 2\mathbf{j} - (4\cos 2t)\mathbf{k}$$
$$\mathbf{F}'''(t) = 8e^{2t}\mathbf{i} + (8\sin 2t)\mathbf{k}$$

■

Several rules for computing derivatives of vector functions are listed in the following theorem.

THEOREM 10.3   Rules for differentiating vector functions

If the vector functions $\mathbf{F}$ and $\mathbf{G}$ and the scalar function h are differentiable at t, and if a and b are constants, then $a\mathbf{F} + b\mathbf{G}$, $h\mathbf{F}$, $\mathbf{F} \cdot \mathbf{G}$, and $\mathbf{F} \times \mathbf{G}$ are also differentiable at t, and

Linearity rule	$(a\mathbf{F} + b\mathbf{G})'(t) = a\mathbf{F}'(t) + b\mathbf{G}'(t)$
Scalar multiple rule	$(h\mathbf{F})'(t) = h'(t)\mathbf{F}(t) + h(t)\mathbf{F}'(t)$
Dot product rule	$(\mathbf{F} \cdot \mathbf{G})'(t) = (\mathbf{F}' \cdot \mathbf{G})(t) + (\mathbf{F} \cdot \mathbf{G}')(t)$
Cross product rule	$(\mathbf{F} \times \mathbf{G})'(t) = (\mathbf{F}' \times \mathbf{G})(t) + (\mathbf{F} \times \mathbf{G}')(t)$
Chain rule	$[\mathbf{F}(h(t))]' = h'(t)\mathbf{F}'(h(t))$

WARNING   The order of the factors is important in the cross product rule, because the cross product of vectors is not commutative.

Proof   We will prove the linearity rule and leave the rest of the proof as exercises. Note that if $\mathbf{F}$ and $\mathbf{G}$ are vector functions differentiable at t, and a and b are constants, then the linear combination $a\mathbf{F} + b\mathbf{G}$ has the difference quotient

$$\frac{\Delta(a\mathbf{F} + b\mathbf{G})}{\Delta t} = \frac{a\Delta\mathbf{F}}{\Delta t} + \frac{b\Delta\mathbf{G}}{\Delta t}$$

We can now find the derivative:

$$(a\mathbf{F} + b\mathbf{G})'(t) = \lim_{\Delta t \to 0} \left[\frac{\Delta(a\mathbf{F} + b\mathbf{G})}{\Delta t} \right]$$
$$= \lim_{\Delta t \to 0} \left[\frac{a\Delta\mathbf{F}}{\Delta t} + \frac{b\Delta\mathbf{G}}{\Delta t} \right]$$
$$= a \lim_{\Delta t \to 0} \frac{\Delta\mathbf{F}}{\Delta t} + b \lim_{\Delta t \to 0} \frac{\Delta\mathbf{G}}{\Delta t}$$
$$= a\mathbf{F}'(t) + b\mathbf{G}'(t)$$

❑

EXAMPLE 6   Derivative of a cross product

Let $\mathbf{F}(t) = \mathbf{i} + t\mathbf{j} + t^2\mathbf{k}$ and $\mathbf{G}(t) = t\mathbf{i} + e^t\mathbf{j} + 3\mathbf{k}$. Verify that

$$(\mathbf{F} \times \mathbf{G})'(t) = (\mathbf{F}' \times \mathbf{G})(t) + (\mathbf{F} \times \mathbf{G}')(t)$$

Solution

First find the derivative of the cross product:

$$(\mathbf{F} \times \mathbf{G})(t) = \begin{vmatrix} \mathbf{i} & \mathbf{j} & \mathbf{k} \\ 1 & t & t^2 \\ t & e^t & 3 \end{vmatrix} = (3t - t^2 e^t)\mathbf{i} - (3 - t^3)\mathbf{j} + (e^t - t^2)\mathbf{k}$$

so that $(\mathbf{F} \times \mathbf{G})'(t) = (3 - 2te^t - t^2 e^t)\mathbf{i} + (3t^2)\mathbf{j} + (e^t - 2t)\mathbf{k}$.

Next, find $(\mathbf{F}' \times \mathbf{G})(t) + (\mathbf{F} \times \mathbf{G}')(t)$ by first finding the derivatives of $\mathbf{F}$ and $\mathbf{G}$ and then the appropriate cross products.

$$\mathbf{F}'(t) = \mathbf{j} + 2t\mathbf{k} \quad \text{and} \quad \mathbf{G}'(t) = \mathbf{i} + e^t\mathbf{j}$$

$$(\mathbf{F}' \times \mathbf{G})(t) = \begin{vmatrix} \mathbf{i} & \mathbf{j} & \mathbf{k} \\ 0 & 1 & 2t \\ t & e^t & 3 \end{vmatrix} = (3 - 2te^t)\mathbf{i} - (-2t^2)\mathbf{j} + (-t)\mathbf{k}$$

$$(\mathbf{F} \times \mathbf{G}')(t) = \begin{vmatrix} \mathbf{i} & \mathbf{j} & \mathbf{k} \\ 1 & t & t^2 \\ 1 & e^t & 0 \end{vmatrix} = (-t^2 e^t)\mathbf{i} - (-t^2)\mathbf{j} + (e^t - t)\mathbf{k}$$

Finally, add these vector functions:

$$(\mathbf{F}' \times \mathbf{G})(t) + (\mathbf{F} \times \mathbf{G}')(t) = [(3 - 2te^t) - t^2 e^t]\mathbf{i} + [2t^2 + t^2]\mathbf{j} + [-t + e^t - t]\mathbf{k}$$
$$= (3 - 2te^t - t^2 e^t)\mathbf{i} + (3t^2)\mathbf{j} + (e^t - 2t)\mathbf{k}$$

Thus, $(\mathbf{F} \times \mathbf{G})' = (\mathbf{F}' \times \mathbf{G}) + (\mathbf{F} \times \mathbf{G}')$. ∎

EXAMPLE 7 Derivatives of vector function expressions

Let $\mathbf{F}(t) = \mathbf{i} + e^t\mathbf{j} + t^2\mathbf{k}$ and $\mathbf{G}(t) = 3t^2\mathbf{i} + e^{-t}\mathbf{j} - 2t\mathbf{k}$. Find

a. $\dfrac{d}{dt}[2\mathbf{F}(t) + t^3\mathbf{G}(t)]$ **b.** $\dfrac{d}{dt}[\mathbf{F}(t) \cdot \mathbf{G}(t)]$

Solution

Begin by finding $\dfrac{d\mathbf{F}}{dt}$ and $\dfrac{d\mathbf{G}}{dt}$:

$$\frac{d\mathbf{F}}{dt} = e^t\mathbf{j} + 2t\mathbf{k} \quad \text{and} \quad \frac{d\mathbf{G}}{dt} = 6t\mathbf{i} - e^{-t}\mathbf{j} - 2\mathbf{k}$$

a. $\dfrac{d}{dt}[2\mathbf{F}(t) + t^3\mathbf{G}(t)] = 2\dfrac{d\mathbf{F}}{dt} + \left[t^3\dfrac{d\mathbf{G}}{dt} + \dfrac{d}{dt}(t^3)\mathbf{G}(t)\right]$

$\qquad = 2(e^t\mathbf{j} + 2t\mathbf{k}) + t^3(6t\mathbf{i} - e^{-t}\mathbf{j} - 2\mathbf{k}) + 3t^2(3t^2\mathbf{i} + e^{-t}\mathbf{j} - 2t\mathbf{k})$

$\qquad = (6t^4 + 9t^4)\mathbf{i} + (2e^t - t^3 e^{-t} + 3t^2 e^{-t})\mathbf{j} + (4t - 2t^3 - 6t^3)\mathbf{k}$

$\qquad = 15t^4\mathbf{i} + [2e^t + e^{-t}t^2(3 - t)]\mathbf{j} - 4t(2t^2 - 1)\mathbf{k}$

b. $\dfrac{d}{dt}[\mathbf{F}(t) \cdot \mathbf{G}(t)] = \dfrac{d\mathbf{F}}{dt} \cdot \mathbf{G}(t) + \mathbf{F}(t) \cdot \dfrac{d\mathbf{G}}{dt}$

$\qquad = (e^t\mathbf{j} + 2t\mathbf{k}) \cdot (3t^2\mathbf{i} + e^{-t}\mathbf{j} - 2t\mathbf{k}) + (\mathbf{i} + e^t\mathbf{j} + t^2\mathbf{k}) \cdot (6t\mathbf{i} - e^{-t}\mathbf{j} - 2\mathbf{k})$

$\qquad = [0(3t^2) + e^t(e^{-t}) + 2t(-2t)] + [1(6t) + e^t(-e^{-t}) + t^2(-2)]$

$\qquad = (1 - 4t^2) + (6t - 1 - 2t^2)$

$\qquad = -6t^2 + 6t$ ∎

We will use Theorem 10.3 in both applied and theoretical problems. In the following theorem, we establish an important geometric property of vector functions.

THEOREM 10.4 Orthogonality of a function of constant length and its derivative

If the nonzero vector function $\mathbf{F}(t)$ is differentiable and has constant length, then $\mathbf{F}(t)$ is orthogonal to the derivative vector $\mathbf{F}'(t)$.

Proof We are given that $\|\mathbf{F}(t)\| = r$ for some constant r and all t. To prove that $\mathbf{F}$ and $\mathbf{F}'$ are orthogonal, we will show that $\mathbf{F}(t) \cdot \mathbf{F}'(t) = 0$ for all t. First, note that

$$r^2 = \|\mathbf{F}(t)\|^2 = \mathbf{F}(t) \cdot \mathbf{F}(t)$$

for all t. We take the derivative of both sides (remember that the derivative of r^2 is zero because it is a constant).

$$[r^2]' = [\mathbf{F}(t) \cdot \mathbf{F}(t)]'$$
$$0 = \mathbf{F}'(t) \cdot \mathbf{F}(t) + \mathbf{F}(t) \cdot \mathbf{F}'(t)$$
$$0 = 2\mathbf{F}(t) \cdot \mathbf{F}'(t)$$
$$0 = \mathbf{F}(t) \cdot \mathbf{F}'(t)$$

Thus, $\mathbf{F}$ and $\mathbf{F}'$ are orthogonal. ❑

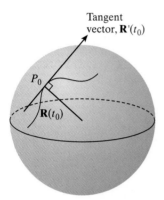

Figure 10.10 The tangent $\mathbf{R}'(t_0)$ at P_0 is orthogonal to the radius vector $\mathbf{R}(t_0)$ from the center of the sphere to P_0.

In plane geometry, it is shown that a tangent line to a circle is always perpendicular to the radial line from the center of a circle to the point of tangency. Theorem 10.4 can be used to extend this property to spheres in $\mathbb{R}^3$.

Suppose $\mathbf{R}(t) = \langle f(t), g(t), h(t) \rangle$ is a vector function whose graph lies entirely on the sphere $x^2 + y^2 + z^2 = r^2$ and let P_0 be the point on the graph that corresponds to $t = t_0$, as shown in Figure 10.10. Then $\mathbf{R}(t_0)$ is the vector from the center O to P_0, and $\mathbf{R}'(t_0)$ is a tangent vector to the graph of $\mathbf{R}(t)$ at P_0. Thus $\mathbf{R}'(t_0)$ is also tangent to the sphere at P_0, and since $\mathbf{R}(t)$ has constant length,

$$\|\mathbf{R}(t)\| = \sqrt{x^2(t) + y^2(t) + z^2(t)} = r \quad \text{for all } t$$

it follows from Theorem 10.4 that $\mathbf{R}$ is orthogonal to $\mathbf{R}'$. That is, the tangent vector $\mathbf{R}'(t_0)$ at P_0 is orthogonal to the vector $\mathbf{R}(t_0)$ from the center of the sphere to the point of tangency.

MODELING THE MOTION OF AN OBJECT IN $\mathbb{R}^3$

Recall from Chapter 3 that the derivative of an object's position with respect to time is the velocity, and the derivative of the velocity is the acceleration. We frequently know the acceleration and can use integration to find the velocity and the position. We will now express these concepts in terms of vector functions.

Vector Motion

An object that moves in such a way that its position at time t is given by the vector function $\mathbf{R}(t)$. Then

Velocity vector V is $\dfrac{d\mathbf{R}}{dt}$

Speed $\|\mathbf{V}\|$ is the magnitude of the velocity,

Direction of motion is the unit vector $\dfrac{\mathbf{V}}{\|\mathbf{V}\|}$

Acceleration vector is the derivative of the velocity:

$$\mathbf{A} = \frac{d\mathbf{V}}{dt} = \frac{d^2\mathbf{R}}{dt^2}$$

Figure 10.11 The velocity vector is tangent to the trajectory of the moving object

The graph of the position vector $\mathbf{R}(t)$ is called the **trajectory** of the moving object. According to the results obtained earlier in this section, the velocity $\mathbf{V}(t) = \mathbf{R}'(t)$ is a tangent vector to the trajectory at any point where $\mathbf{V}(t)$ exists and $\mathbf{V}(t) \neq \mathbf{0}$ (see Figure 10.11). When $\mathbf{V}(t) = \mathbf{0}$, the object is **stationary**. The *direction of motion* is given by the unit vector $\mathbf{V}/\|\mathbf{V}\|$, so the velocity satisfies

$$\mathbf{V} = \|\mathbf{V}\| \left(\frac{\mathbf{V}}{\|\mathbf{V}\|} \right) = (\text{SPEED})(\text{DIRECTION})$$

In practice, the position vector is often represented in the form

$$\mathbf{R}(t) = f_1(t)\mathbf{i} + f_2(t)\mathbf{j} + f_3(t)\mathbf{k}$$

and in this case, the velocity and the acceleration vectors are given by

$$\mathbf{V}(t) = \mathbf{R}'(t) = f_1'(t)\mathbf{i} + f_2'(t)\mathbf{j} + f_3'(t)\mathbf{k}$$

and

$$\mathbf{A}(t) = \mathbf{V}'(t) = \mathbf{R}''(t) = f_1''(t)\mathbf{i} + f_2''(t)\mathbf{j} + f_3''(t)\mathbf{k}$$

EXAMPLE 8   Speed and direction of a particle

A particle's position at time t is determined by the vector

$$\mathbf{R}(t) = (\cos t)\mathbf{i} + (\sin t)\mathbf{j} + t^3\mathbf{k}$$

Analyze the particle's motion. In particular, find the particle's velocity, speed, acceleration, and direction of motion at time $t = 2$.

Solution
$$\mathbf{V} = \frac{d\mathbf{R}}{dt} = (-\sin t)\mathbf{i} + (\cos t)\mathbf{j} + 3t^2\mathbf{k}$$

$$\mathbf{A} = \frac{d\mathbf{V}}{dt} = (-\cos t)\mathbf{i} - (\sin t)\mathbf{j} + 6t\mathbf{k}$$

The velocity at time $t = 2$ is $\mathbf{V}(2)$:

$$(-\sin 2)\mathbf{i} + (\cos 2)\mathbf{j} + 3(2)^2\mathbf{k} \approx -0.91\mathbf{i} - 0.42\mathbf{j} + 12\mathbf{k}$$

The acceleration at $t = 2$ is

$$(-\cos 2)\mathbf{i} - (\sin 2)\mathbf{j} + 6(2)\mathbf{k} \approx 0.42\mathbf{i} - 0.91\mathbf{j} + 12\mathbf{k}$$

The speed is

$$\|\mathbf{V}\| = \sqrt{(-\sin t)^2 + (\cos t)^2 + (3t^2)^2}$$
$$= \sqrt{1 + 9t^4}$$

At time $t = 2$, the speed is $\|\mathbf{V}(2)\| = \sqrt{1 + 144} \approx 12.04$. The direction of motion is

$$\frac{\mathbf{V}}{\|\mathbf{V}\|} = \frac{1}{\sqrt{145}}[(-\sin t)\mathbf{i} + (\cos t)\mathbf{j} + 3t^2\mathbf{k}]$$

At time $t = 2$, the direction is $\mathbf{V}(2)/\|\mathbf{V}(2)\|$:

$$\frac{1}{\sqrt{145}}[(-\sin 2)\mathbf{i} + (\cos 2)\mathbf{j} + 12\mathbf{k}] \approx -0.08\mathbf{i} - 0.03\mathbf{j} + \mathbf{k}$$

The geometry of this example is shown in Figure 10.12.   ■

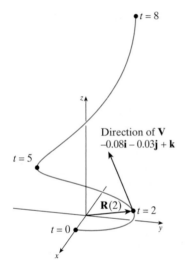

Figure 10.12 Graph of **R** and **V** at $t = 2$

VECTOR INTEGRALS

Like vector limits and derivatives, vector integration is performed in a componentwise fashion.

Vector Integrals

Let $\mathbf{F}(t) = f_1(t)\mathbf{i} + f_2(t)\mathbf{j} + f_3(t)\mathbf{k}$, where f_1, f_2, and f_3 are continuous on the closed interval $a \leq t \leq b$. Then the **indefinite integral** of $\mathbf{F}(t)$ is the vector function

$$\int \mathbf{F}(t)\, dt = \left[\int f_1(t)\, dt\right]\mathbf{i} + \left[\int f_2(t)\, dt\right]\mathbf{j} + \left[\int f_3(t)\, dt\right]\mathbf{k} + \mathbf{C}$$

where $\mathbf{C} = C_1\mathbf{i} + C_2\mathbf{j} + C_3\mathbf{k}$ is an arbitrary constant vector. The **definite integral** of $\mathbf{F}(t)$ on $a \leq t \leq b$ is the vector

$$\int_a^b \mathbf{F}(t)\, dt = \left[\int_a^b f_1(t)\, dt\right]\mathbf{i} + \left[\int_a^b f_2(t)\, dt\right]\mathbf{j} + \left[\int_a^b f_3(t)\, dt\right]\mathbf{k}$$

EXAMPLE 9   Integral of a vector function

Find $\int_0^\pi [t\mathbf{i} + 3\mathbf{j} - (\sin t)\mathbf{k}]\,dt$.

Solution

$$
\begin{aligned}
\int_0^\pi [t\mathbf{i} + 3\mathbf{j} - (\sin t)\mathbf{k}]\,dt &= \left[\int_0^\pi t\,dt\right]\mathbf{i} + \left[\int_0^\pi 3\,dt\right]\mathbf{j} - \left[\int_0^\pi \sin t\,dt\right]\mathbf{k} \\
&= \left(\frac{t^2}{2}\right)\Big|_0^\pi \mathbf{i} + (3t)|_0^\pi \,\mathbf{j} - (-\cos t)|_0^\pi \,\mathbf{k} \\
&= \tfrac{1}{2}\pi^2\mathbf{i} + 3\pi\mathbf{j} + (\cos \pi - \cos 0)]\mathbf{k} \\
&= \tfrac{1}{2}\pi^2\mathbf{i} + 3\pi\mathbf{j} - 2\mathbf{k}
\end{aligned}
$$

■

EXAMPLE 10   Position of a particle given its velocity

The velocity of a particle moving in space is

$$\mathbf{V}(t) = e^t\mathbf{i} + t^2\mathbf{j} + (\cos 2t)\mathbf{k}$$

Find the particle's position as a function of t if the position at time $t = 0$ is $\mathbf{R}(0) = 2\mathbf{i} + \mathbf{j} - \mathbf{k}$.

Solution

We need to solve the initial value problem that consists of

The differential equation $\mathbf{V}(t) = \dfrac{d\mathbf{R}}{dt} = e^t\mathbf{i} + t^2\mathbf{j} + (\cos 2t)\mathbf{k}$

The initial condition $\mathbf{R}(0) = 2\mathbf{i} + \mathbf{j} - \mathbf{k}$

We integrate both sides of the differential equation with respect to t:

$$
\int d\mathbf{R} = \left[\int e^t\,dt\right]\mathbf{i} + \left[\int t^2\,dt\right]\mathbf{j} + \left[\int \cos 2t\,dt\right]\mathbf{k}
$$

$$
\mathbf{R}(t) = (e^t + C_1)\mathbf{i} + \left(\tfrac{1}{3}t^3 + C_2\right)\mathbf{j} + \left(\tfrac{1}{2}\sin 2t + C_3\right)\mathbf{k}
$$

$$
= e^t\mathbf{i} + \tfrac{1}{3}t^3\mathbf{j} + \left(\tfrac{1}{2}\sin 2t\right)\mathbf{k} + \underbrace{C_1\mathbf{i} + C_2\mathbf{j} + C_3\mathbf{k}}_{\mathbf{C}}
$$

Now use the initial condition to find $\mathbf{C}$:

$$
\underbrace{2\mathbf{i} + \mathbf{j} - \mathbf{k}}_{\mathbf{R}(0)} = e^0\mathbf{i} + \tfrac{1}{3}(0)^3\mathbf{j} + \left(\tfrac{1}{2}\sin 0\right)\mathbf{k} + \mathbf{C} = \mathbf{i} + \mathbf{C}
$$

We solve three separate equations:

$$2\mathbf{i} = \mathbf{i} + C_1\mathbf{i} \qquad \mathbf{j} = C_2\mathbf{j} \qquad -\mathbf{k} = C_3\mathbf{k}$$

so

$$C_1 = 1, \qquad C_2 = 1, \qquad C_3 = -1$$

and

$$\mathbf{i} + \mathbf{j} - \mathbf{k} = \mathbf{C}$$

Thus, the particle's position at any time t is

$$
\begin{aligned}
\mathbf{R}(t) &= e^t\mathbf{i} + \tfrac{1}{3}t^3\mathbf{j} + \left(\tfrac{1}{2}\sin 2t\right)\mathbf{k} + \mathbf{i} + \mathbf{j} - \mathbf{k} \\
&= (e^t + 1)\mathbf{i} + \left(\tfrac{1}{3}t^3 + 1\right)\mathbf{j} + \left(\tfrac{1}{2}\sin 2t - 1\right)\mathbf{k}
\end{aligned}
$$

■

10.2   PROBLEM SET

Ⓐ *Find the vector derivative* **F**′ *in Problems 1–4.*

1. $\mathbf{F}(t) = t\mathbf{i} + t^2\mathbf{j} + (t + t^3)\mathbf{k}$
2. $\mathbf{F}(s) = (s\mathbf{i} + s^2\mathbf{j} + s^2\mathbf{k}) + (2s^2\mathbf{i} - s\mathbf{j} + 3\mathbf{k})$
3. $\mathbf{F}(s) = (\ln s)[s\mathbf{i} + 5\mathbf{j} - e^s\mathbf{k}]$
4. $\mathbf{F}(\theta) = (\cos\theta)[\mathbf{i} + (\tan\theta)\mathbf{j} + 3\mathbf{k}]$

Find **F**′ *and* **F**″ *for the vector functions given in Problems 5–8.*

5. $\mathbf{F}(t) = t^2\mathbf{i} + t^{-1}\mathbf{j} + e^{2t}\mathbf{k}$
6. $\mathbf{F}(s) = (1 - 2s^2)\mathbf{i} + (s\cos s)\mathbf{j} - s\mathbf{k}$
7. $\mathbf{F}(s) = (\sin s)\mathbf{i} + (\cos s)\mathbf{j} + s^2\mathbf{k}$
8. $\mathbf{F}(\theta) = (\sin^2\theta)\mathbf{i} + (\cos 2\theta)\mathbf{j} + \theta^2\mathbf{k}$

Differentiate the scalar functions in Problems 9–12.

9. $f(x) = [x\mathbf{i} + (x + 1)\mathbf{j}] \cdot [(2x)\mathbf{i} - (3x^2)\mathbf{j}]$
10. $f(x) = \langle\cos x, x, -x\rangle \cdot \langle\sec x, -x^2, 2x\rangle$
11. $g(x) = \|\langle\sin x, -2x, \cos x\rangle\|$
12. $f(x) = \|[x\mathbf{i} + x^2\mathbf{j} - 2\mathbf{k}] + [(1 - x)\mathbf{i} - e^x\mathbf{j}]\|$

In Problems 13–18, **R** *is the position vector for a particle in space at time t. Find the particle's velocity and acceleration vectors and then find the speed and direction of motion for the given value of t.*

13. $\mathbf{R}(t) = t\mathbf{i} + t^2\mathbf{j} + 2t\mathbf{k}$ at $t = 1$
14. $\mathbf{R}(t) = (1 - 2t)\mathbf{i} - t^2\mathbf{j} + e^t\mathbf{k}$ at $t = 0$
15. $\mathbf{R}(t) = (\cos t)\mathbf{i} + (\sin t)\mathbf{j} + 3t\mathbf{k}$ at $t = \frac{\pi}{4}$
16. $\mathbf{R}(t) = (2\cos t)\mathbf{i} + t^2\mathbf{j} + (2\sin t)\mathbf{k}$ at $t = \frac{\pi}{2}$
17. $\mathbf{R}(t) = e^t\mathbf{i} + e^{-t}\mathbf{j} + e^{2t}\mathbf{k}$ at $t = \ln 2$
18. $\mathbf{R}(t) = (\ln t)\mathbf{i} + \frac{1}{2}t^3\mathbf{j} - t\mathbf{k}$ at $t = 1$

Find the tangent vector to the graph of the given vector function **F** *at the points indicated in Problems 19–22.*

19. $\mathbf{F}(t) = t^2\mathbf{i} + 2t\mathbf{j} + (t^3 + t^2)\mathbf{k}$; $t = 0, t = 1, t = -1$
20. $\mathbf{F}(t) = \dfrac{t\mathbf{i} + t^2\mathbf{j} + t^3\mathbf{k}}{1 + 2t}$; $t = 0, t = 2$
21. $\mathbf{F}(t) = t^2\mathbf{i} + (\cos t)\mathbf{j} + (t^2\cos t)\mathbf{k}$; $t = 0, t = \frac{\pi}{2}$
22. $\mathbf{F}(t) = (e^t\sin\pi t)\mathbf{i} + (e^t\cos\pi t)\mathbf{j} + (\sin\pi t + \cos\pi t)\mathbf{k}$; $t = 0,$ $t = 1, t = 2$

In Problems 23–24, find parametric equations for the tangent line to the graph of the given vector function **F**(t) *at the point* P_0 *corresponding to* t_0.

23. $\mathbf{F}(t) = \langle t^{-3}, t^{-2}, t^{-1}\rangle$; $t_0 = -1$
24. $\mathbf{F}(t) = \langle\sin t, \cos t, 3t\rangle$; $t_0 = 0$

Find the indefinite vector integrals in Problems 25–30.

25. $\displaystyle\int \langle t, -e^{3t}, 3\rangle\, dt$ **26.** $\displaystyle\int \langle\cos t, \sin t, -2t\rangle\, dt$
27. $\displaystyle\int \langle\ln t, -t, 3\rangle\, dt$ **28.** $\displaystyle\int e^{-t}\langle 3, t, \sin t\rangle\, dt$
29. $\displaystyle\int \langle t\ln t, -\sin(1 - t), t\rangle\, dt$ **30.** $\displaystyle\int \langle\sinh t, -3, \cosh t\rangle\, dt$

Find the position vector **R**(t), *given the velocity* **V**(t) *and the initial position* **R**(0) *in Problems 31–34.*

31. $\mathbf{V}(t) = t^2\mathbf{i} - e^{2t}\mathbf{j} + \sqrt{t}\,\mathbf{k}$; $\mathbf{R}(0) = \mathbf{i} + 4\mathbf{j} - \mathbf{k}$
32. $\mathbf{V}(t) = t\mathbf{i} - \sqrt[3]{t}\mathbf{j} + e^t\mathbf{k}$; $\mathbf{R}(0) = \mathbf{i} - 2\mathbf{j} + \mathbf{k}$
33. $\mathbf{V}(t) = 2\sqrt{t}\mathbf{i} + (\cos t)\mathbf{j}$; $\mathbf{R}(0) = \mathbf{i} + \mathbf{j}$
34. $\mathbf{V}(t) = -3t\mathbf{i} + (\sin^2 t)\mathbf{j} + (\cos^2 t)\mathbf{k}$; $\mathbf{R}(0) = \mathbf{j}$

Find the position vector **R**(t) *and velocity vector* **V**(t), *given the acceleration* **A**(t) *and initial position and velocity vectors* **R**(0) *and* **V**(0), *respectively, in Problems 35–36.*

35. $\mathbf{A}(t) = (\cos t)\mathbf{i} - (t\sin t)\mathbf{k}$; $\mathbf{R}(0) = \mathbf{i} - 2\mathbf{j} + \mathbf{k}$; $\mathbf{V}(0) = 2\mathbf{i} + 3\mathbf{k}$
36. $\mathbf{A}(t) = t^2\mathbf{i} - 2\sqrt{t}\mathbf{j} + e^{3t}\mathbf{k}$; $\mathbf{R}(0) = 2\mathbf{i} + \mathbf{j} - \mathbf{k}$; $\mathbf{V}(0) = \mathbf{i} - \mathbf{j} - 2\mathbf{k}$
37. The velocity of a particle moving in space is $\mathbf{V}(t) = e^t\mathbf{i} + t^2\mathbf{j}$. Find the particle's position as a function of t if $\mathbf{R}(0) = \mathbf{i} - \mathbf{j}$.
38. The velocity of a moving particle is $\mathbf{V}(t) = (t\cos t)\mathbf{i} + e^{2t}\mathbf{k}$. Find the particle's position as a function of t if $\mathbf{R}(0) = \mathbf{i} + 2\mathbf{k}$.
39. The acceleration of a moving particle is $\mathbf{A}(t) = 24t^2\mathbf{i} + 4\mathbf{j}$. Find the particle's position as a function of t if $\mathbf{R}(0) = \mathbf{i} + 2\mathbf{j}$ and $\mathbf{V}(0) = \mathbf{0}$.

Ⓑ 40. WHAT DOES THIS SAY?   Describe the concept of a smooth curve.
41. WHAT DOES THIS SAY?   Compare or contrast the notions of speed and velocity of a moving object.

Let $\mathbf{v} = 2\mathbf{i} - \mathbf{j} + 5\mathbf{k}$ *and* $\mathbf{w} = \mathbf{i} + 2\mathbf{j} - 3\mathbf{k}$. *Find the derivatives in Problems 42–45.*

42. $\dfrac{d}{dt}(\mathbf{v} + t\mathbf{w})$ **43.** $\dfrac{d^2}{dt^2}(\mathbf{v} \cdot t^4\mathbf{w})$
44. $\dfrac{d^2}{dt^2}(t\|\mathbf{v}\| + t^2\|\mathbf{w}\|)$ **45.** $\dfrac{d}{dt}(t\mathbf{v} \times t^2\mathbf{w})$

In Problems 46–47, verify the indicated equations for the vector functions

$$\mathbf{F}(t) = (3 + t^2)\mathbf{i} - (\cos 3t)\mathbf{j} + t^{-1}\mathbf{k} \quad and$$
$$\mathbf{G}(t) = \sin(2 - t)\mathbf{i} - e^{2t}\mathbf{k}$$

46. $(3\mathbf{F} - 2\mathbf{G})'(t) = 3\mathbf{F}'(t) - 2\mathbf{G}'(t)$
47. $(\mathbf{F} \cdot \mathbf{G})'(t) = (\mathbf{F}' \cdot \mathbf{G})(t) + (\mathbf{F} \cdot \mathbf{G}')(t)$

In Problems 48–49, find a value of a that satisfies the given equation.

48. $\displaystyle\int_0^a \left[\left(t\sqrt{1 + t^2}\right)\mathbf{i} + \left(\dfrac{1}{1 + t^2}\right)\mathbf{j}\right] dt = \frac{1}{3}(2\sqrt{2} - 1)\mathbf{i} + \frac{\pi}{4}\mathbf{j}$
49. $\displaystyle\int_0^{2a} [(\cos t)\mathbf{i} + (\sin t)\mathbf{j} + (\sin t\cos t)\mathbf{k}]\, dt = \mathbf{i} + \mathbf{j} + \frac{1}{2}\mathbf{k}$

In Problems 50–51, show that **F**(t) *and* **F**″(t) *are parallel for all t if k is constant.*

50. $\mathbf{F}(t) = e^{kt}\,\mathbf{i} + e^{-kt}\,\mathbf{j}$ **51.** $\mathbf{F}(t) = (\cos kt)\mathbf{i} + (\sin kt)\mathbf{j}$

Determine whether the given vector function **F**(t) *in Problems 52–53 is smooth over the interval I.*

52. $\mathbf{F}(t) = \langle t^2 - t, \cos t, \sin t^2\rangle$ over $[-\pi, \pi]$
53. $\mathbf{F}(t) = \langle e^t - t, t^{3/2}, \cos t\rangle$ over $[-1, 2]$

Ⓒ 54. Let $\mathbf{F}(t) = u(t)\mathbf{i} + v(t)\mathbf{j} + w(t)\mathbf{k}$, where u, v, and w are differentiable scalar functions of t. Show that

$$\mathbf{F}(t) \cdot \mathbf{F}'(t) = \|\mathbf{F}(t)\| \,(\|\mathbf{F}(t)\|)'$$

55. Let $\mathbf{F}(t)$ be a differentiable vector function and let $h(t)$ be a differentiable scalar function of t. Show that

$$[h(t)\mathbf{F}(t)]' = h(t)\mathbf{F}'(t) + h'(t)\mathbf{F}(t)$$

56. If **F** and **G** are differentiable vector functions of t, prove

$$(\mathbf{F} \cdot \mathbf{G})'(t) = (\mathbf{F} \cdot \mathbf{G}')(t) + (\mathbf{F}' \cdot \mathbf{G})(t)$$

57. If $\mathbf{F}$ and $\mathbf{G}$ are differentiable vector functions of t, prove
$$(\mathbf{F} \times \mathbf{G})'(t) = (\mathbf{F}' \times \mathbf{G})(t) + (\mathbf{F} \times \mathbf{G}')(t)$$

58. If $\mathbf{F}$ is a differentiable vector function such that $\mathbf{F}(t) \neq \mathbf{0}$, show that
$$\frac{d}{dt}\left(\frac{\mathbf{F}(t)}{\|\mathbf{F}(t)\|}\right) = \frac{\mathbf{F}'(t)}{\|\mathbf{F}(t)\|} - \frac{[\mathbf{F}(t) \cdot \mathbf{F}'(t)]\mathbf{F}(t)}{\|\mathbf{F}(t)\|^3}$$

59. Show that $\dfrac{d}{dt}\|\mathbf{R}(t)\| = \dfrac{1}{\|\mathbf{R}\|}\mathbf{R} \cdot \dfrac{d\mathbf{R}}{dt}$.

60. Verify that
$$\frac{d}{dt}[\mathbf{F} \cdot (\mathbf{G} \times \mathbf{H})] = \mathbf{F}' \cdot (\mathbf{G} \times \mathbf{H}) + \mathbf{F} \cdot (\mathbf{G}' \times \mathbf{H}) + \mathbf{F} \cdot (\mathbf{G} \times \mathbf{H}')$$

61. Verify that $\dfrac{d}{dt}[\mathbf{F} \times (\mathbf{G} \times \mathbf{H})] = [(\mathbf{H} \cdot \mathbf{F})\mathbf{G}]' - [(\mathbf{G} \cdot \mathbf{F})\mathbf{H}]'$.

62. If $\mathbf{G} = \mathbf{F} \cdot (\mathbf{F}' \times \mathbf{F}'')$, what is $\mathbf{G}'$?

63. Counterexample Problem Is it true that whenever the vector functions $\mathbf{F}(t)$ and $\mathbf{G}(t)$ are smooth over the interval $[a, b]$, then $\mathbf{F}(t) + \mathbf{G}(t)$ is also smooth over the same interval? Either prove that this statement is generally true, or find a counterexample.

10.3 Modeling Ballistics and Planetary Motion

IN THIS SECTION modeling the motion of a projectile in a vacuum, Kepler's second law

MODELING THE MOTION OF A PROJECTILE IN A VACUUM

In general, it can be quite difficult to analyze the motion of a projectile, but the problem becomes manageable if we assume that the acceleration due to gravity is constant and that the projectile travels in a vacuum, so we will make this assumption. A model of the motion in the real world (with air resistance, for example) based on these assumptions is fairly realistic as long as the projectile is reasonably heavy, travels at relatively low speed, and stays close to the surface of the earth.

Figure 10.13 shows a projectile that is fired from a point P and travels in a vertical plane coordinatized so that P is directly above the origin O and the impact point I is on the x-axis at ground level. We let s_0 be the height of P above O, v_0 be the initial speed of the projectile (the **muzzle speed**), and α be the angle of elevation (the angle at which the projectile is launched).

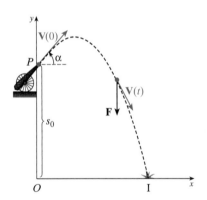

Figure 10.13 The motion of a projectile

Because the projectile is to travel in a vacuum, we will assume that the only force acting on the projectile at any given time is the force due to gravity. This force $\mathbf{F}$ is directed downward. The magnitude of this force is the projectile's weight. If m is the mass of the projectile and g is the "free-fall" acceleration due to gravity (g is approximately 32 ft/s^2 or 9.8 m/s^2), then the magnitude of $\mathbf{F}$ is mg. Also, according to **Newton's second law of motion**, the sum of all the forces acting on the projectile must equal $m\mathbf{A}(t)$, where $\mathbf{A}(t)$ is the acceleration of the projectile at time t. Because $\mathbf{F} = -mg\mathbf{j}$ is the only force acting on the projectile, it follows that

$$-mg\mathbf{j} = \mathbf{F} = m\mathbf{A}(t)$$
$$\mathbf{A}(t) = -g\mathbf{j} \qquad \text{for all } t \geq 0$$

The velocity $\mathbf{V}(t)$ of the projectile can now be obtained by integrating $\mathbf{A}(t)$. Specifically, we have

$$\mathbf{V}(t) = \int \mathbf{A}(t)\, dt = \int (-g\mathbf{j})\, dt = -gt\mathbf{j} + \mathbf{C}_1$$

Figure 10.14 Components of the initial velocity vector $\mathbf{V}(0)$

where $\mathbf{C}_1 = \mathbf{V}(0)$ is the velocity when $t = 0$ (that is, the *initial velocity*). Because the projectile is fired with initial speed $v_0 = \|\mathbf{V}(0)\|$ at an angle of elevation α, the initial velocity must be

$$\mathbf{C}_1 = \mathbf{V}(0) = (v_0 \cos\alpha)\mathbf{i} + (v_0 \sin\alpha)\mathbf{j}$$

(as shown in Figure 10.14),

and by substitution we find

$$\mathbf{V}(t) = -gt\mathbf{j} + (v_0 \cos \alpha)\mathbf{i} + (v_0 \sin \alpha)\mathbf{j}$$
$$= (v_0 \cos \alpha)\mathbf{i} + (v_0 \sin \alpha - gt)\mathbf{j}$$

Next, by integrating $\mathbf{V}$ with respect to t, we find that the position vector of the projectile is

$$\mathbf{R}(t) = \int \mathbf{V}(t)\,dt = \int [(v_0 \cos \alpha)\mathbf{i} + (v_0 \sin \alpha - gt)\mathbf{j}]\,dt$$
$$= (v_0 \cos \alpha)t\mathbf{i} + \left[(v_0 \sin \alpha)t - \tfrac{1}{2}gt^2\right]\mathbf{j} + \mathbf{C}_2$$

where $\mathbf{C}_2 = \mathbf{R}(0)$ is the position at time $t = 0$. Because the projectile begins its flight at P, the initial position is $\mathbf{R}(0) = s_0\mathbf{j}$, and by substituting this expression for $\mathbf{C}_2$, we find that the position at time t is

$$\mathbf{R}(t) = [(v_0 \cos \alpha)t]\mathbf{i} + \left[(v_0 \sin \alpha)t - \tfrac{1}{2}gt^2 + s_0\right]\mathbf{j}$$

Specifically, $\mathbf{R}(t)$ gives the position of the projectile at any time t after it is fired and before it hits the ground. If we let $x(t)$ and $y(t)$ denote the horizontal and vertical components of $\mathbf{R}(t)$, respectively, we can make the following statement.

Motion of a Projectile in a Vacuum

Consider a projectile that travels in a vacuum in a coordinate plane, with the x-axis along level ground. If the projectile is fired from a height of s_0 with initial speed v_0 and angle of elevation α, then at time t ($t \geq 0$) it will be at the point $(x(t), y(t))$, where

$$x(t) = (v_0 \cos \alpha)t \quad \text{and} \quad y(t) = -\tfrac{1}{2}gt^2 + (v_0 \sin \alpha)t + s_0$$

The parametric equations for the motion of a projectile in a vacuum provide useful general information about a projectile's motion. For instance, note that if $\alpha \neq 90°$, we can eliminate the parameter t by solving the first equation, obtaining

$$t = \frac{x}{v_0 \cos \alpha}$$

By substituting into the second equation, we find

$$y = -\frac{1}{2}g\left[\frac{x}{v_0 \cos \alpha}\right]^2 + (v_0 \sin \alpha)\left[\frac{x}{v_0 \cos \alpha}\right] + s_0$$

$$= \left[\frac{-g}{2(v_0 \cos \alpha)^2}\right]x^2 + (\tan \alpha)x + s_0$$

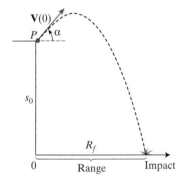

Figure 10.15 The motion of a projectile. The trajectory is part of a parabolic arc.

This is the Cartesian equation for the trajectory of the projectile, and because the equation has the general form $y = ax^2 + bx + c$, with $a < 0$, the trajectory must be part of a downward-opening parabola. That is, *if a projectile is not fired vertically (so that $\tan \alpha$ is defined), its trajectory will be part of a downward-opening parabolic arc,* as shown in Figure 10.15.

The *time of flight* T_f of a projectile is the elapsed time between launch and impact, and the *range* is the total horizontal distance R_f traveled by the projectile during its flight. Because $y = 0$ at impact, it follows that T_f must satisfy the quadratic equation

$$-\tfrac{1}{2}gT_f^2 + (v_0 \sin \alpha)T_f + s_0 = 0$$

Then, because the flight begins at $x = 0$ when $t = 0$ and ends at $x = R_f$ when $t = T_f$, the range R_f can be computed by evaluating $x(t)$ at $t = T_f$. That is,

$$R_f = (v_0 \cos \alpha)T_f$$

EXAMPLE 1   Modeling the path of a projectile

A boy standing at the edge of a cliff throws a ball upward at a 30° angle with an initial speed of 64 ft/s. Suppose that when the ball leaves the boy's hand, it is 48 ft above the ground at the base of the cliff.

a. What are the time of flight of the ball and its range?
b. What are the velocity of the ball and its speed at impact?
c. What is the highest point reached by the ball during its flight?

Solution

We know that $g = 32$ ft/s^2, and we are given $s_0 = 48$ ft, $v_0 = 64$ ft/s, and $\alpha = 30°$, so the ball's trajectory has parametric equations

$$x(t) = (v_0 \cos \alpha)t = (64 \cos 30°)t = 32\sqrt{3}\,t$$
$$y(t) = -\tfrac{1}{2}gt^2 + (v_0 \sin \alpha)t + s_0$$
$$= -\tfrac{1}{2}(32)t^2 + (64 \sin 30°)t + 48$$
$$= -16t^2 + 32t + 48$$

a. The ball hits the ground when $y = 0$, and by solving the equation $-16t^2 + 32t + 48 = 0$ for $t \geq 0$, we find that $t = 3$, so the time of flight is $T_f = 3$ seconds. The range is

$$x(3) = 32\sqrt{3}\,(3) \approx 166.27688$$

That is, the ball hits the ground about 166 ft from the base of the cliff.

b. We find that $x'(t) = 32\sqrt{3}$ and $y'(t) = -32t + 32$, so the velocity at time t is

$$\mathbf{V}(t) = 32\sqrt{3}\,\mathbf{i} + (-32t + 32)\mathbf{j}$$

Thus, at impact (when $t = 3$), the velocity is $\mathbf{V}(3) = 32\sqrt{3}\,\mathbf{i} - 64\mathbf{j}$, and its speed is

$$\|\mathbf{V}(3)\| = \sqrt{(32\sqrt{3})^2 + (-64)^2} \approx 84.664042$$

That is, the speed at impact is about 85 ft/s.

c. The ball attains its maximum height when the upward (vertical) component of its velocity $\mathbf{V}(t)$ is 0—that is, when $y'(t) = 0$. Solving the equation $-32t + 32 = 0$, we find that this occurs when $t = 1$. Therefore, the maximum height attained by the ball is

$$y_m = y(1) = -16(1)^2 + 32(1) + 48 = 64$$
$$x_m = x(1) = 32\sqrt{3}\,(1) \approx 55.425626$$

and the highest point reached by the ball has coordinates (rounded to the nearest foot) of $(55, 64)$.   ∎

In general, if the projectile is fired from ground level (so $s_0 = 0$), the time of flight T_f satisfies

$$-\tfrac{1}{2}gT_f^2 + (v_0 \sin \alpha)T_f = 0$$
$$T_f = \frac{2}{g}v_0 \sin \alpha$$

Thus, the range R_f is

$$R_f = x(T_f) = (v_0 \cos \alpha)\left(\frac{2v_0 \sin \alpha}{g}\right) = \frac{v_0^2}{g}(2 \sin \alpha \cos \alpha) = \frac{v_0^2}{g} \sin 2\alpha$$

Notice that for a given initial speed v_0, the range assumes its largest value when $\sin 2\alpha = 1$; that is, when $\alpha = 45° = \frac{\pi}{4}$. To summarize,

> **Time of Flight and Range of a Projectile Fired from Ground Level**
> A projectile fired from *ground level* has **time of flight** T_f and **range** R_f given by the equations
>
> $$T_f = \frac{2}{g} v_0 \sin \alpha \quad \text{and} \quad R_f = \frac{v_0^2}{g} \sin 2\alpha$$
>
> The maximal range is $R_m = \dfrac{v_0^2}{g}$, and it occurs when $\alpha = \frac{\pi}{4}$.

EXAMPLE 2   Time of flight and range for the motion of a projectile

A projectile is fired from ground level at an angle of 40° with muzzle speed 110 ft/s. Find the time of flight and the range.

Solution

With $g = 32$ ft/s^2, we see that the flight time T_f is

$$T_f = \frac{2}{g} v_0 \sin \alpha = \frac{2}{32}(110)(\sin 40°) \approx 4.4191648$$

The range R_f is

$$R_f = \frac{v_0^2}{g} \sin 2\alpha = \frac{(110)^2}{32}(\sin 80°) \approx 372.38043$$

That is, the projectile travels about 372 ft horizontally before it hits the ground, and the flight takes a little more than 4 seconds.   ■

KEPLER'S SECOND LAW

In the seventeenth century, the German astronomer Johannes Kepler (1571–1630) formulated three useful laws for describing planetary motion. The guest essay at the end of this chapter discusses the discovery of these laws.

> **Kepler's Laws**
>
> 1. The planets move about the sun in elliptical orbits, with the sun at one focus.
> 2. The radius vector joining a planet to the sun sweeps out equal areas in equal intervals of time.
> 3. The square of the time of one complete revolution of a planet about its orbit is proportional to the cube of the length of the semimajor axis of its orbit.

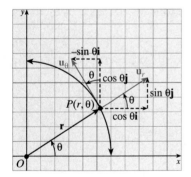

Figure 10.16 Description of the motion of a particle along a curve

Kepler's approach was empirical, but his three laws were proved eighty years later by Isaac Newton, in the *Principia Mathematica* (1687). We will prove the second law using vector methods. The other two laws can be established similarly.

We begin by introducing some useful notation involving polar coordinates. Let $\mathbf{u}_r$ and $\mathbf{u}_\theta$ denote unit vectors along the radial axis and orthogonal to that axis, respectively, as shown in Figure 10.16. Then, in terms of the unit Cartesian vectors $\mathbf{i}$ and $\mathbf{j}$, we have

$$\mathbf{u}_r = (\cos\theta)\mathbf{i} + (\sin\theta)\mathbf{j} \qquad \text{and} \qquad \mathbf{u}_\theta = (-\sin\theta)\mathbf{i} + (\cos\theta)\mathbf{j}$$

The derivatives $\dfrac{d\mathbf{u}_r}{d\theta}$ and $\dfrac{d\mathbf{u}_\theta}{d\theta}$ satisfy

$$\frac{d\mathbf{u}_r}{d\theta} = (-\sin\theta)\mathbf{i} + (\cos\theta)\mathbf{j} = \mathbf{u}_\theta$$

$$\frac{d\mathbf{u}_\theta}{d\theta} = (-\cos\theta)\mathbf{i} + (-\sin\theta)\mathbf{j} = -\mathbf{u}_r$$

Now, suppose the sun S is at the origin (pole) of a polar coordinate system, and consider the motion of a body B (planet, comet, artificial satellite) about S. The radial vector $\mathbf{R} = \mathbf{SB}$ can be expressed as

$$\mathbf{R} = r\mathbf{u}_r = (r\cos\theta)\mathbf{i} + (r\sin\theta)\mathbf{j}$$

where $r = \|\mathbf{R}\|$, and the velocity $\mathbf{V}$ satisfies

$$\mathbf{V} = \frac{d\mathbf{R}}{dt} = \frac{dr}{dt}\mathbf{u}_r + r\frac{d\mathbf{u}_r}{dt} \qquad \text{Derivative of a scalar multiple}$$

$$= \frac{dr}{dt}\mathbf{u}_r + r\frac{d\mathbf{u}_r}{d\theta}\frac{d\theta}{dt} \qquad \text{Chain rule}$$

$$= \frac{dr}{dt}\mathbf{u}_r + r\frac{d\theta}{dt}\mathbf{u}_\theta$$

A similar formula for acceleration can be derived (see Problem 41). We summarize these formulas in the following box.

Polar Formulas for Velocity and Acceleration

$$\mathbf{V}(t) = \frac{dr}{dt}\mathbf{u}_r + r\frac{d\theta}{dt}\mathbf{u}_\theta$$

$$\mathbf{A}(t) = \frac{d\mathbf{V}(t)}{dt} = \frac{d^2\mathbf{R}}{dt^2} = \left[\frac{d^2r}{dt^2} - r\left(\frac{d\theta}{dt}\right)^2\right]\mathbf{u}_r + \left[r\frac{d^2\theta}{dt^2} + 2\frac{dr}{dt}\frac{d\theta}{dt}\right]\mathbf{u}_\theta$$

EXAMPLE 3   Velocity of a moving body using polar unit vectors

The position vector of a moving body is $\mathbf{R}(t) = 2t\mathbf{i} - t^2\mathbf{j}$ for $t \geq 0$. Express $\mathbf{R}$ and the velocity vector $\mathbf{V}(t)$ in terms of $\mathbf{u}_r$ and $\mathbf{u}_\theta$.

Solution

We note that on the trajectory, $x = 2t$, $y = -t^2$, so $\mathbf{R}$ has length

$$r = \|\mathbf{R}(t)\| = \sqrt{(2t)^2 + (-t^2)^2} = \sqrt{4t^2 + t^4} = t\sqrt{t^2 + 4}$$

and $\mathbf{R} = r\mathbf{u}_r = t\sqrt{t^2 + 4}\,\mathbf{u}_r$. Because $\mathbf{V}(t) = \dfrac{dr}{dt}\mathbf{u}_r + r\dfrac{d\theta}{dt}\mathbf{u}_\theta$, we need $\dfrac{dr}{dt}$ and $\dfrac{d\theta}{dt}$. We find that

$$\frac{dr}{dt} = \sqrt{t^2 + 4} + t\left(\tfrac{1}{2}\right)(t^2 + 4)^{-1/2}(2t) = \frac{2t^2 + 4}{\sqrt{t^2 + 4}}$$

and because the polar angle satisfies $\theta = \tan^{-1}\left(\dfrac{y}{x}\right)$ for all points (x, y) on the curve (except where $x = 0$), we have

$$\theta = \tan^{-1}\left(\frac{-t^2}{2t}\right) = \tan^{-1}\left(-\frac{t}{2}\right)$$

$$\frac{d\theta}{dt} = \frac{1}{1 + \left(-\frac{t}{2}\right)^2}\left(-\frac{1}{2}\right) = \frac{-2}{t^2 + 4}$$

Thus,

$$\mathbf{V}(t) = \frac{2t^2 + 4}{\sqrt{t^2 + 4}}\mathbf{u}_r + t\sqrt{t^2 + 4}\left(\frac{-2}{t^2 + 4}\right)\mathbf{u}_\theta = \frac{(2t^2 + 4)\mathbf{u}_r - 2t\mathbf{u}_\theta}{\sqrt{t^2 + 4}} \qquad \blacksquare$$

We now have the tools we need to establish Kepler's second law.

THEOREM 10.5   Kepler's second law

The radius vector from the sun to a planet in its orbit sweeps out equal areas in equal intervals of time.

Time T Time T

Sun

Figure 10.17 The radial line sweeps out equal area in equal time

Proof   The situation described by Kepler's second law is illustrated in Figure 10.17. We will assume that the only force acting on a planet is the gravitational attraction of the sun. According to the universal law of gravitation, the force of attraction is given by

$$\mathbf{F} = -G\frac{mM}{r^2}\mathbf{u}_r$$

where G is a physical constant, and m and M are the masses of the planet and the sun, respectively. Because this is the only force acting on the planet, Newton's second law of motion tells us that $\mathbf{F} = m\mathbf{A}$, where $\mathbf{A}$ is the acceleration of the planet in its orbit. By equating these two expressions for the force $\mathbf{F}$ we find

$$m\mathbf{A} = -G\frac{mM}{r^2}\mathbf{u}_r$$

$$\mathbf{A} = \frac{-GM}{r^2}\mathbf{u}_r$$

This says that *the acceleration of a planet in its orbit has only a radial component*. Note how this conclusion depends on our assumption that the only force on the planet is the sun's gravitational attraction, which acts along radial lines from the sun to the planet.

This means that the $\mathbf{u}_\theta$ component of the planet's acceleration is 0, and by examining the polar formula for acceleration, we see that this condition is equivalent to the differential equation

$$r\frac{d^2\theta}{dt^2} + 2\frac{dr}{dt}\frac{d\theta}{dt} = 0$$

If we set $w = \dfrac{d\theta}{dt}$, we obtain $\dfrac{d^2\theta}{dt^2} = \dfrac{dw}{dt}$, so that

$$r\frac{dw}{dt} + 2w\frac{dr}{dt} = 0 \qquad \text{Separation of variables}$$

$$w^{-1}\frac{dw}{dt} = -2r^{-1}\frac{dr}{dt}$$

$$\int w^{-1}\,dw = \int (-2r^{-1})\,dr$$

$$\ln|w| + C_1 = -2\ln|r| + C_2$$

$$\ln w = \ln(Cr^{-2}) \qquad \text{Since } r > 0 \text{ and } w > 0$$

$$w = Cr^{-2}$$

$$\frac{d\theta}{dt} = Cr^{-2}$$

$$r^2\frac{d\theta}{dt} = C$$

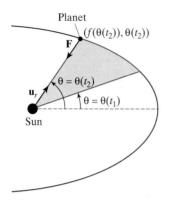

Now, let $[t_1, t_2]$ and $[t_3, t_4]$ be two time intervals of equal length, so $t_2 - t_1 = t_4 - t_3$. According to Theorem 6.1, the area swept out in the time period $[t_1, t_2]$ (see Figure 10.18) is

$$S_1 = \int_{t_1}^{t_2} \frac{1}{2} r^2 \, d\theta = \int_{t_1}^{t_2} \frac{1}{2} \left[r^2 \frac{d\theta}{dt} \right] dt = \int_{t_1}^{t_2} \frac{1}{2} C \, dt \qquad \text{Since } r^2 \frac{d\theta}{dt} = C$$

Thus, $S_1 = \frac{1}{2} C (t_2 - t_1)$. Similarly, the area swept out in time period $[t_3, t_4]$ is $S_2 = \frac{1}{2} C (t_4 - t_3)$, and we have

$$S_1 = \tfrac{1}{2} C (t_2 - t_1) = \tfrac{1}{2} C (t_4 - t_3) = S_2$$

so equal area is swept out in equal time, as claimed by Kepler.   ❑

Figure 10.18 The radial line sweeps out area at the rate $\dfrac{dA}{dt} = \dfrac{1}{2} r^2 \dfrac{d\theta}{dt}$.

An object is said to move in a **central force field** if it is subject to a single force that is always directed toward a particular point. Our derivation of Kepler's second law is based on the assumption that planetary motion occurs in a central force field. In particular, the formula

$$\mathbf{A} = \frac{-GM}{r^2} \mathbf{u}_r$$

shows that a planet's acceleration has only a radial component, and this turns out to be true for an object moving in a plane in any central force field.

Much of what we know about the motion of the planets is derived from Kepler's laws. Some of these data are listed in Table 10.1.

TABLE 10.1   Data on Planetary Orbits

Planet	Mean Orbit (in millions of miles)	Eccentricity	Period (in days)	Comparative Mass (Earth = 1.00)
Mercury	36.0	0.2056	88.0	0.06
Venus	67.3	0.0068	224.7	0.81
Earth	**93.0**	**0.0167**	**365.256**	**1.00**
Mars	141.7	0.0934	687.0	0.1074
Jupiter	484.0	0.0484	4,332.589	317.83
Saturn	887.0	0.0543	10,759.22	95.16
Uranus	1,787.0	0.0460	30,685.4	14.5
Neptune	2,797.0	0.0082	60,189.0	17.2
Pluto	3,675.0	0.2481	90,465.0	0.00215

10.3   PROBLEM SET

Ⓐ *In Problems 1–8, find the time of flight T_f (to the nearest tenth second) and the range R_f (to the nearest unit) of a projectile fired (in a vacuum) from ground level at the given angle α with the indicated initial speed v_0. Assume that $g = 32$ ft/s^2 or $g = 9.8$ m/s^2.*

1. $\alpha = 35°$, $v_0 = 128$ ft/s   **2.** $\alpha = 45°$, $v_0 = 80$ ft/s

3. $\alpha = 48.5°$, $v_0 = 850$ m/s   **4.** $\alpha = 43.5°$, $v_0 = 185$ m/s

5. $\alpha = 23.74°$, $v_0 = 23.3$ m/s   **6.** $\alpha = 31.04°$, $v_0 = 38.14$ m/s

7. $\alpha = 14.11°$, $v_0 = 100$ ft/s   **8.** $\alpha = 78.09°$, $v_0 = 88$ ft/s

In Problems 9–14, an object moves along the given curve in the plane (described in either polar or parametric form). Find its velocity and acceleration in terms of the unit polar vectors $\mathbf{u}_r$ and $\mathbf{u}_\theta$.

9. $x = 2t$, $y = t$ **10.** $x = \sin t$, $y = \cos t$

11. $r = \sin\theta$, $\theta = 2t$ **12.** $r = e^{-\theta}$, $\theta = 1 - t$

13. $r = 5(1 + \cos\theta)$, $\theta = 2t + 1$

14. $r = \dfrac{1}{1 - \cos\theta}$, $\theta = t$

B **15.** A shell fired from ground level at an angle of 45° hits the ground 2000 m away. What is the muzzle speed of the shell?

16. A gun at ground level has muzzle speed of 300 ft/s. What angle of elevation (to the nearest tenth of a degree) should be used to hit an object 1,500 ft away?

17. At what angle (to the nearest tenth of a degree) should a projectile be fired from ground level if its muzzle speed is 167.1 ft/s and the desired range is 600 ft?

18. A shell is fired at ground level with a muzzle speed of 280 ft/s and at an elevation of 45° from ground level.

   a. Find the maximum height attained by the shell.

   b. Find the time of flight and the range of the shell.

   c. Find the velocity and speed of the shell at impact.

19. Derek Jeter hits a baseball at a 30° angle with a speed of 144 ft/s. If the ball is 4 ft above ground level when it is hit, what is the maximum height reached by the ball? How far will it travel from home plate before it lands? If it just barely clears a 5-foot wall in the outfield (see Figure 10.19), how far (to the nearest foot) is the wall from home plate?

Figure 10.19 Home run distance

20. A baseball hit at a 24° angle from 3 ft above the ground just goes over the 9-ft fence 400 ft from home plate. About how fast was the ball traveling, and how long did it take the ball to reach the wall?

21. A nozzle discharges a stream of water with an initial velocity $v_0 = 50$ ft/s into the end of a horizontal pipe of inside diameter $d = 5$ ft (see Figure 10.20). What is the maximum horizontal distance that the stream can reach?

Figure 10.20 Problem 21

22. If a shotputter throws a shot from a height of 5 ft with an angle of 46° and initial speed of 25 ft/s, what is the horizontal distance of the throw?

23. **Modeling Problem** In 1974, Evel Knievel attempted a skycycle ride across the Snake River (see Figure 10.21), and at the time there was a great deal of hype about "will he make it?" If the angle of the launching ramp was 45° and if the horizontal distance the skycycle needed to travel was 4,700 ft, at what speed did Evel have to leave the ramp to make it across the Snake River Canyon?

Figure 10.21 Evel Knievel's jump

Assume that the opposite edges of the canyon are at the same height and neglect wind resistance.

24. Payton Manning, a quarterback for the Indianapolis Colts, throws a pass at a 45° angle from a height of 6.5 ft with a speed of 50 ft/s (see Figure 10.22). A receiver races straight downfield on a fly pattern at a constant speed of 32 ft/s and catches the ball at a height of 6 ft. If Manning and the receiver both line up on the 50 yard line at the start of the play, how far does Manning fade back (retreat) from the line of scrimmage before he releases the ball?

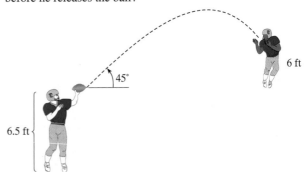

Figure 10.22 Downfield pass

25. A golf ball is hit from the tee to a green with an initial speed of 125 ft/s at an angle of elevation of 45°. How long will it take for the ball to hit the green?

26. **Modeling Problem** A 3-oz paddleball attached to a string is swung in a circular path with a 1-ft radius, as shown in Figure 10.23. If the string will break under a force of 2 lb, find the maximum speed the ball can attain without breaking the string. *Hint:* 3 oz is 3/16 lb.

Figure 10.23 Paddleball path

27. **Spy Problem** The Spy figures out where to dig (Supplementary Problem 57 of Chapter 9) and unearths a human-size box. With shaking hands, he opens the lid, and to his relief, finds not the body of Lord Fleming, but his favorite pipe and a ringing cell phone. He picks up the phone and hears the cold, calculating voice of Ernst Stavro Blohardt. "Do what you're told or the next box will be heavier!" he threatens. "You're going to help me eliminate my competition, starting with Scélérat." The Spy is still peeved with Scélérat for icing his friend, Siggy Leiter (Problem 62, Section 2.4), so he agrees. He's told that the Frenchman and his gang are holed up in a bunker on the side of a hill inclined at an angle of 15° to the horizontal, as shown in Figure 10.24. The Spy returns to his helicopter and flies toward the bunker traveling at 200 ft/s at an altitude of 10,000 ft. Just as the helicopter flies over the base of the hill, the Spy sights a ventilation hole in the top of the bunker at an angle of 20°. He decides to drop a canister of knockout gas into the bunker through the hole. How long should he wait before releasing the canister?

Figure 10.24   Spy and the bunker target

28. A child, running along level ground at the top of a 30-ft-high vertical cliff at a speed of 15 ft/s, throws a rock over the cliff into the sea below, as shown in Figure 10.25. Suppose the child's arm is 3 ft above the ground and her arm speed is 25 ft/s. If the rock is released 10 ft from the edge of the cliff at an angle of 30°, how long does it take for the rock to hit the water? How far from the base of the cliff does it hit?

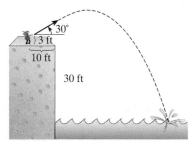

Figure 10.25   Tossing an object off a cliff

29. **Modeling Problem**   A calculus text (other than the one you are now reading) is hurled downward from the top of a 120-ft high building at an angle of 30° from the horizontal, as shown in Figure 10.26. Assume that the initial speed of the book is 8 ft/s. How far from the base of the building will the text land?

Figure 10.26   Path of an object thrown from a building

30. A particle moves along the polar path (r, θ), where $r(t) = 3 + 2\sin t$, $\theta(t) = t^3$. Find $\mathbf{V}(t)$ and $\mathbf{A}(t)$ in terms of $\mathbf{u}_r$ and $\mathbf{u}_\theta$.

In Problems 31–34, consider a particle moving on a circular path of radius a described by the equation

$$\mathbf{R}(t) = (a\cos\omega t)\mathbf{i} + (a\sin\omega t)\mathbf{j}, \text{ where } \omega = \frac{d\theta}{dt} \text{ is the constant}$$

angular velocity.

31. Find the velocity vector and show that it is orthogonal to $\mathbf{R}(t)$.
32. Find the speed of the particle.
33. Find the acceleration vector and show that its direction is always toward the center of the circle.
34. Find the magnitude of the acceleration vector.
35. A mortar is placed at the top of a hill sloping downward at a 20° angle, as shown in Figure 10.27.

Figure 10.27   Path of a projectile fired from a sloping surface

A shell is fired with muzzle speed of $v_0 = 100$ m/s at an angle of 40° relative to the horizontal.

a. How far down the slope does the shell land?
b. What is the impact speed of the shell?

36. Find two angles of elevation α_1 and α_2 (to the nearest degree) so that a shell fired at ground level with muzzle speed of 650 ft/s will hit a target 6,000 ft away (also at ground level).

37. A gun is fired with muzzle speed $v_0 = 550$ ft/s at an angle of $\alpha = 22.0°$. The shell overshoots the target by 50 ft. At what angle should a second shot be fired with the same muzzle speed to hit the target?

38. A gun is fired with muzzle speed $v_0 = 700$ ft/s at an angle of $\alpha = 25°$. It overshoots the target by 60 ft. The target moves away from the gun at a constant speed of 10 ft/s. If the gunner takes 30 seconds to reload, at what angle should a second shot be fired with the same muzzle speed to hit the target?

39. A shell is fired from ground level with muzzle speed of 750 ft/s at an angle of 25°. An enemy gun 20,000 ft away fires a shot 2 seconds later and the shells collide 50 ft above the ground at the same speed. What are the muzzle speed v_0 and angle of elevation α of the second gun?

C 40. Suppose a shell is fired with muzzle speed v_0 at an angle α from a height s_0 above level ground.

a. Show that the range R_f must satisfy

$$g(\sec^2\alpha)R_f^2 - 2v_0^2(\tan\alpha)R_f - 2v_0^2 s_0 = 0$$

b. Show that the maximum range R_m occurs at angle α_m, where $R_m\tan\alpha_m = \dfrac{v_0^2}{g}$.

c. Show that

$$R_{\max} = \frac{v_0}{g}\sqrt{v_0^2 + 2gs_0} \text{ and } \alpha_m = \tan^{-1}\left(\frac{v_0}{\sqrt{v_0^2 + 2gs_0}}\right)$$

41. Use the formula

$$\mathbf{V}(t) = \frac{dr}{dt}\mathbf{u}_r + r\frac{d\theta}{dt}\mathbf{u}_\theta$$

to derive the formula for polar acceleration:

$$\mathbf{A}(t) = \left[\frac{d^2r}{dt^2} - r\left(\frac{d\theta}{dt}\right)^2\right]\mathbf{u}_r + \left[r\frac{d^2\theta}{dt^2} + 2\frac{dr}{dt}\frac{d\theta}{dt}\right]\mathbf{u}_\theta$$

Hint: Begin by using the chain rule to show

$$\frac{d\mathbf{u}_r}{dt} = \mathbf{u}_\theta\frac{d\theta}{dt} \quad \text{and} \quad \frac{d\mathbf{u}_\theta}{dt} = -\mathbf{u}_r\frac{d\theta}{dt}$$

42. Use the results of Problem 41 to show that the force $\mathbf{F}(t)$ acting on an object moving with acceleration $\mathbf{A}(t)$ acts only in a radial direction (parallel to $\mathbf{u}_r$) if and only if

$$mr^2\frac{d\theta}{dt} = C$$

where C is a constant. Under what conditions will $\mathbf{F}(t)$ act only in a direction orthogonal to $\mathbf{u}_r$?

43. Modeling Problem A cannon is placed at the top of a downward-sloping hill as in Problem 35. A shell is fired with muzzle speed v_0 at an angle of elevation α with respect to the horizontal. Assume the hill is sloped at an angle β with the horizontal, as shown in Figure 10.28.

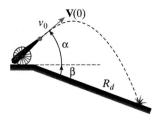

Figure 10.28   Projectile fired on a sloping hill

a. Derive a formula for the time of flight T_f of the shell and the downhill range R_d (measured along the slope).

b. What should α be in order to maximize the downhill range R_d?

10.4   Unit Tangent and Principal Unit Normal Vectors; Curvature

IN THIS SECTION   unit tangent and principal unit normal vectors, arc length as a parameter, curvature

UNIT TANGENT AND PRINCIPAL UNIT NORMAL VECTORS

Figure 10.29 Unit tangent vector $\mathbf{T}$ and vectors orthogonal to $\mathbf{T}$ on a given curve

In Section 10.2, we observed that if $\mathbf{R}(t)$ is a vector function whose graph is a smooth curve C, then the nonzero derivative $\mathbf{R}'(t)$ is tangent to C at the point P that corresponds to t. Since the curve C is smooth, we have $\mathbf{R}'(t) \neq \mathbf{0}$ and a **unit tangent vector** $\mathbf{T}$ to C can be obtained by dividing $\mathbf{R}'(t)$ by its length, that is,

$$\mathbf{T}(t) = \frac{\mathbf{R}'(t)}{\|\mathbf{R}'(t)\|}$$

Figure 10.29 shows a smooth curve C and the unit tangent $\mathbf{T}$ at a particular point P.

According to Theorem 10.4 (orthogonality of a function of constant length and its derivative), since $\mathbf{T}(t)$ has constant magnitude (namely $\|\mathbf{T}(t)\| = 1$), it follows that its derivative $\mathbf{T}'(t)$ is orthogonal to $\mathbf{T}(t)$ and hence is normal to the curve C at P. As indicated in Figure 10.29, there are infinitely many vectors that are orthogonal to $\mathbf{T}$ and thus deserve to be called normal vectors. To distinguish $\mathbf{T}'(t)$ from the others, we normalize it by dividing by its length and refer to it as the **principal unit normal** $\mathbf{N}(t)$, so

$$\mathbf{N}(t) = \frac{\mathbf{T}'(t)}{\|\mathbf{T}'(t)\|}$$

To summarize,

Unit Tangent and Principal Unit Normal Vectors

If $\mathbf{R}(t)$ is a vector function that defines a smooth graph, then at each point a **unit tangent vector** is

$$\mathbf{T}(t) = \frac{\mathbf{R}'(t)}{\|\mathbf{R}'(t)\|}$$

and the **principal unit normal** vector is

$$\mathbf{N}(t) = \frac{\mathbf{T}'(t)}{\|\mathbf{T}'(t)\|}$$

Note these are pairs of unit vectors that change direction depending on where you are on the curve.

EXAMPLE 1 Unit tangent and principal unit normal

Find the unit tangent vector $\mathbf{T}(t)$ and the principal unit normal vector $\mathbf{N}(t)$ at each point on the graph of the vector function

$$\mathbf{R}(t) = \langle 3\sin t, 4t, 3\cos t \rangle$$

Solution
We have $\mathbf{R}'(t) = \langle 3\cos t, 4, -3\sin t \rangle$, so

$$\begin{aligned}
\|\mathbf{R}'(t)\| &= \sqrt{(3\cos t)^2 + 4^2 + (-3\sin t)^2} \\
&= \sqrt{9\cos^2 t + 16 + 9\sin^2 t} \\
&= \sqrt{9 + 16} \qquad \text{Since } \sin^2 t + \cos^2 t = 1 \\
&= 5
\end{aligned}$$

Thus, the unit tangent vector is

$$\begin{aligned}
\mathbf{T}(t) = \frac{\mathbf{R}'(t)}{\|\mathbf{R}'(t)\|} &= \tfrac{1}{5}\langle 3\cos t, 4, -3\sin t \rangle \\
&= \langle \tfrac{3}{5}\cos t, \tfrac{4}{5}, -\tfrac{3}{5}\sin t \rangle
\end{aligned}$$

To find the principal unit normal $\mathbf{N}$, we first find $\mathbf{T}'(t)$ and then its magnitude:

$$\mathbf{T}'(t) = \langle -\tfrac{3}{5}\sin t, 0, -\tfrac{3}{5}\cos t \rangle$$

$$\begin{aligned}
\|\mathbf{T}'(t)\| &= \sqrt{\left(-\tfrac{3}{5}\sin t\right)^2 + 0 + \left(-\tfrac{3}{5}\cos t\right)^2} \\
&= \sqrt{\tfrac{9}{25}(\sin^2 t + \cos^2 t)} \\
&= \tfrac{3}{5}
\end{aligned}$$

So, the unit normal is

$$\begin{aligned}
\mathbf{N}(t) = \frac{\mathbf{T}'(t)}{\|\mathbf{T}'(t)\|} &= \frac{\langle -\tfrac{3}{5}\sin t, 0, -\tfrac{3}{5}\cos t \rangle}{\tfrac{3}{5}} \\
&= \langle -\sin t, 0, -\cos t \rangle
\end{aligned}$$

∎

Suppose an object moves along the graph C of a vector function $\mathbf{R}(t)$ in the direction of increasing t. Then at each point P on the trajectory C, the unit tangent $\mathbf{T}(t)$ points in the direction of motion, while the principal unit normal $\mathbf{N}(t)$ points in the direction the object is turning, as shown in Figure 10.30a. For a trajectory in a plane, this means that $\mathbf{N}(t)$ points toward the concave side of the trajectory (Figure 10.30b).

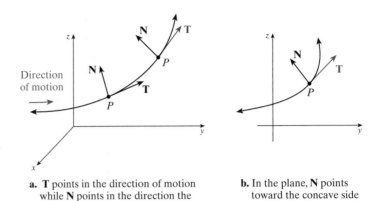

a. T points in the direction of motion while **N** points in the direction the moving object is turning

b. In the plane, **N** points toward the concave side of the trajectory

Figure 10.30   Unit tangent $\mathbf{T}$ and principal unit normal $\mathbf{N}$ at each point on the trajectory of a moving object

ARC LENGTH AS A PARAMETER

Recall from Section 6.4 that the arc length of the graph of $y = F(x)$ on the interval $[a, b]$ is given by the integral

$$s = \int_a^b \sqrt{1 + \left(\frac{dy}{dx}\right)^2} \, dx$$

If the graph is described by parametric equations $x = x(t)$ and $y = y(t)$, then by the chain rule $\frac{dy}{dt} = \frac{dy}{dx}\frac{dx}{dt}$, so we have

$$\frac{dy}{dx} = \frac{\dfrac{dy}{dt}}{\dfrac{dx}{dt}} \qquad \text{for } \frac{dx}{dt} \neq 0,$$

Thus, if the graph is traced out exactly once over the parameter interval $t_1 \leq t \leq t_2$, its arc length is given by

$$s = \int_a^b \sqrt{1 + \left(\frac{dy/dt}{dx/dt}\right)^2} \, dx$$

$$= \int_{t_1}^{t_2} \sqrt{\frac{\left(\dfrac{dx}{dt}\right)^2 + \left(\dfrac{dy}{dt}\right)^2}{\left(\dfrac{dx}{dt}\right)^2}} \left(\frac{dx}{dt}\right) dt$$

$$= \int_{t_1}^{t_2} \sqrt{\left(\frac{dx}{dt}\right)^2 + \left(\frac{dy}{dt}\right)^2} \, dt$$

Similarly, if $x = x(t)$, $y = y(t)$, $z = z(t)$ are parametric equations for a smooth curve C in $\mathbb{R}^3$ that is traced out exactly once over the parameter interval $t_1 \leq t \leq t_2$, then the arc length of C is given by the integral

$$s = \int_{t_1}^{t_2} \sqrt{\left(\frac{dx}{dt}\right)^2 + \left(\frac{dy}{dt}\right)^2 + \left(\frac{dz}{dt}\right)^2}\, dt$$

Often, however, we want a formula for arc length that does not depend on t or any other particular parameter used in its computation. It can be shown that the following definition of the arc length function is, indeed, independent of parameterization.

Arc Length Function

Let C be a piecewise-smooth curve that is the graph of the vector described parametrically by $\mathbf{R}(t) = x(t)\mathbf{i} + y(t)\mathbf{j} + z(t)\mathbf{k}$, and let $P_0 = P(t_0)$ be a particular point on C (called the **base point**). Then the length of C from the base point P_0 to the variable point $P(t)$ is given by the **arc length function** $s(t)$ defined by

$$s(t) = \int_{t_0}^{t} \sqrt{\left(\frac{dx}{du}\right)^2 + \left(\frac{dy}{du}\right)^2 + \left(\frac{dz}{du}\right)^2}\, du$$

➡ **What This Says**   The arc length function $s(t)$ measures the distance along the curve C from $P(t_0)$ to $P(t)$ if $t > t_0$ and the opposite of that distance if $t < t_0$ (see Figure 10.31). Since the length along C is a geometric property, it does not depend on the fact that we have used t as a parameter for describing $\mathbf{R}(t)$.

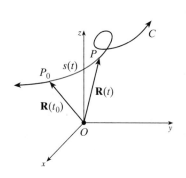

Figure 10.31 $s(t)$ is the length of the curve C from the point $P(t_0)$ to $P(t)$ if $t > t_0$.

This definition allows us to use arc length $s(t)$ as a parameter for describing a curve. While time is a natural parameter for studying motion in space, it is usually more fruitful to use the arc length parameter for describing a curve's geometric features.

EXAMPLE 2   Finding the arc length of a curve defined by parametric equations

Find the arc length of the curve defined by $\mathbf{R}(t) = \langle 12t, 5\cos t, 3 - 5\sin t \rangle$ from $t = 0$ to $t = 2$.

Solution

We find that

$$\frac{dx}{dt} = 12, \quad \frac{dy}{dt} = -5\sin t, \quad \frac{dz}{dt} = -5\cos t$$

Thus,

$$s = \int_0^2 \sqrt{\left(\frac{dx}{dt}\right)^2 + \left(\frac{dy}{dt}\right)^2 + \left(\frac{dz}{dt}\right)^2}\, dt$$

$$= \int_0^2 \sqrt{12^2 + (-5\sin t)^2 + (-5\cos t)^2}\, dt$$

$$= \int_0^2 13\, dt$$

$$= 26 \qquad\qquad\blacksquare$$

The formula for arc length can be used to obtain the following useful formula for the speed of an object moving in space.

THEOREM 10.6   Speed as the derivative of arc length

Suppose an object moves along a smooth curve C that is the graph of the position function $\mathbf{R}(t) = \langle x(t), y(t), z(t) \rangle$, where $\mathbf{R}'(t)$ is continuous on the interval $[t_1, t_2]$. Then the object has speed

$$\frac{ds}{dt} = \|\mathbf{V}(t)\| = \|\mathbf{R}'(t)\| \qquad \text{for } t_1 \le t \le t_2$$

Proof   Apply the second fundamental theorem of calculus to the arc length formula (where we use u as the variable of integration to distinguish it from t):

$$s = \int_{t_1}^{t} \sqrt{\left(\frac{dx}{du}\right)^2 + \left(\frac{dy}{du}\right)^2 + \left(\frac{dz}{du}\right)^2} \, du$$

$$\frac{ds}{dt} = \sqrt{\left(\frac{dx}{dt}\right)^2 + \left(\frac{dy}{dt}\right)^2 + \left(\frac{dz}{dt}\right)^2}$$

$$= \|\mathbf{R}'(t)\| \qquad \qquad \square$$

EXAMPLE 3   Speed and distance traveled by an object moving in space

The position vector of a moving object is $\mathbf{R}(t) = \langle e^t, \sqrt{2}t + 3, e^{-t} \rangle$. Find the speed of the object at time t and compute the distance the object travels between times $t = 0$ and $t = 1$.

Solution

By differentiating $\mathbf{R}(t)$ with respect to t, we find the velocity:

$$\mathbf{V}(t) = \mathbf{R}'(t) = \langle e^t, \sqrt{2}, -e^{-t} \rangle$$

Therefore, the speed at time t is

$$\frac{ds}{dt} = \|\mathbf{R}'(t)\| = \sqrt{(e^t)^2 + (\sqrt{2})^2 + (-e^{-t})^2}$$

$$= \sqrt{e^{2t} + 2 + e^{-2t}} = \sqrt{(e^t + e^{-t})^2} = e^t + e^{-t}$$

The distance traveled by the object between times $t = 0$ and $t = 1$ is the arc length and is given by

$$s = \int_0^1 (e^t + e^{-t}) \, dt = \left[e^t - e^{-t} \right]_0^1 = e - e^{-1} - 1 + 1 \approx 2.3504024 \qquad \blacksquare$$

To change a given parameterization from a given form, say $\mathbf{R}(t)$, to a form involving the parameter s, we first find $s(t)$ using the arc length formula (as in the proof of Theorem 10.6), and then solve for t as a function of s. The graph of $\mathbf{R}(t)$ can then be parameterized in terms of s by substituting $t = t(s)$. The general procedure is illustrated in the following example.

EXAMPLE 4   Using arc length to parameterize a curve

Express the helix $\mathbf{R}(t) = \langle \sin t, \cos t, 2t \rangle$ in terms of arc length measured from the point $P_0(0, 1, 0)$ in the direction of increasing t.

Solution

The point P_0 corresponds to $t = 0$, so arc length $s(t)$ is given by the integral

$$s(t) = \int_0^t \sqrt{\left(\frac{dx}{du}\right)^2 + \left(\frac{dy}{du}\right)^2 + \left(\frac{dz}{du}\right)^2}\, du$$

$$= \int_0^t \sqrt{(\cos u)^2 + (-\sin u)^2 + 2^2}\, du$$

$$= \int_0^t \sqrt{5}\, du$$

$$= \sqrt{5}\, t$$

Solving $s = \sqrt{5}\, t$ for t, we obtain $t = \dfrac{1}{\sqrt{5}} s$, and the required reparameterization is

$$x = \sin\left(\frac{s}{\sqrt{5}}\right), \quad y = \cos\left(\frac{s}{\sqrt{5}}\right), \quad z = \frac{2s}{\sqrt{5}} \qquad \blacksquare$$

Expressing a given position vector $\mathbf{R}$ in terms of the arc length parameter s often leads to complicated computation. However, once we have such a parameterization, the corresponding unit tangent vector $\mathbf{T}$ and principal unit normal vector $\mathbf{N}$ can be expressed in the relatively simple forms shown in the following theorem.

THEOREM 10.7 Formulas for T and N in terms of arc length *s*

If $\mathbf{R}(t)$ has a piecewise-smooth graph and is represented as $\mathbf{R}(s)$ in terms of the arc length parameter s, then the unit tangent vector $\mathbf{T}$ and the principal unit normal vector $\mathbf{N}$ satisfy

$$\mathbf{T} = \frac{d\mathbf{R}}{ds} \quad \text{and} \quad \mathbf{N} = \frac{1}{\kappa}\frac{d\mathbf{T}}{ds}$$

where $\kappa = \left\|\dfrac{d\mathbf{T}}{ds}\right\|$ is a scalar function of s.

Proof By the chain rule, we have $\dfrac{d\mathbf{R}}{dt} = \dfrac{d\mathbf{R}}{ds}\dfrac{ds}{dt}$, and since $\dfrac{ds}{dt} = \|\mathbf{R}'(t)\|$, it follows that

$$\frac{d\mathbf{R}}{ds} = \frac{\dfrac{d\mathbf{R}}{dt}}{\dfrac{ds}{dt}} = \frac{\mathbf{R}'(t)}{\|\mathbf{R}'(t)\|} = \mathbf{T}$$

Next, to obtain the formula for the principal unit normal vector

$$\mathbf{N} = \frac{\mathbf{T}'(t)}{\|\mathbf{T}'(t)\|}$$

in terms of the arc length parameter s, we first note that

$$\mathbf{T}'(t) = \frac{d\mathbf{T}}{dt} = \frac{d\mathbf{T}}{ds}\frac{ds}{dt} \qquad \text{Chain rule}$$

Because $\dfrac{ds}{dt} > 0$, it follows that the vector derivatives $\dfrac{d\mathbf{T}}{dt}$ and $\dfrac{d\mathbf{T}}{ds}$ point in the same direction, and $\mathbf{N}$ can be computed by finding $\dfrac{d\mathbf{T}}{ds}$ and dividing by its length; that is,

$$\mathbf{N} = \frac{\dfrac{d\mathbf{T}}{ds}}{\left\|\dfrac{d\mathbf{T}}{ds}\right\|} = \frac{1}{\kappa}\frac{d\mathbf{T}}{ds}$$

as claimed. □

CURVATURE

Let C be a smooth curve that is the graph of a vector function $\mathbf{R}(s)$ parameterized in terms of arc length s. Since the unit tangent $\mathbf{T}(s)$ is a unit vector, only its direction changes with s and the rate of change of that direction may be measured by the magnitude of the derivative $\dfrac{d\mathbf{T}}{ds}$. As shown in Figure 10.32a, if the curve is a line, then the direction remains constant so that $\dfrac{d\mathbf{T}}{ds} = \mathbf{0}$, but if the curve is bent sharply, the direction of $\mathbf{T}$ changes rapidly so $\dfrac{d\mathbf{T}}{ds}$ is relatively large in magnitude (Figure 10.32b). The magnitude of $\dfrac{d\mathbf{T}}{ds}$ is called the *curvature* of C.

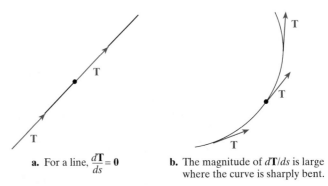

a. For a line, $\dfrac{d\mathbf{T}}{ds} = \mathbf{0}$ **b.** The magnitude of $d\mathbf{T}/ds$ is large where the curve is sharply bent.

Figure 10.32   The sharpness of a curve C is related to the magnitude of $\dfrac{d\mathbf{T}}{ds}$.

Curvature

> Suppose the smooth curve C is the graph of the vector function $\mathbf{R}(s)$, parameterized in terms of the arc length s. Then the **curvature** of C is the function
> $$\kappa(s) = \left\| \frac{d\mathbf{T}}{ds} \right\|$$
> where $\mathbf{T}(s)$ is the unit tangent vector.

First, notice that κ in Theorem 10.7 is the curvature of $\mathbf{R}(s)$. Since a vector function $\mathbf{R}$ is often represented in terms of a parameter other than arc length s, it may not be especially convenient to compute the curvature using the formula $\kappa = \left\| \dfrac{d\mathbf{T}}{ds} \right\|$. For instance, if the parameter is t, then according to the chain rule, we have

$$\frac{d\mathbf{T}}{dt} = \frac{d\mathbf{T}}{ds}\frac{ds}{dt}$$

so that

$$\left\| \frac{d\mathbf{T}}{ds} \right\| = \frac{\left\| \dfrac{dT}{dt} \right\|}{\dfrac{ds}{dt}}$$

(Remember, s increases in the same direction as t, so $\dfrac{ds}{dt} > 0$.) Since $\dfrac{ds}{dt} = \|\mathbf{R}'(t)\|$, it follows that

$$\kappa = \frac{\|\mathbf{T}'(t)\|}{\|\mathbf{R}'(t)\|}$$

which we will refer to as the **two derivatives form** of the curvature. This form is often used in practice to compute κ because it involves only the magnitudes of $\mathbf{R}'(t)$ and $\mathbf{T}'(t)$, both of which are usually easier to compute than $\left\|\dfrac{d\mathbf{T}}{ds}\right\|$. In Example 5, we illustrate the use of the two derivatives form by computing the curvature of a circle.

EXAMPLE 5   Finding the curvature of a circle

Show that the circle with radius r

$$\mathbf{R}(t) = \langle r \cos t, r \sin t \rangle$$

has curvature equal to the reciprocal of its radius; that is, $\kappa = \dfrac{1}{r}$.

Solution

Since $\mathbf{R}'(t) = \langle -r \sin t, r \cos t \rangle$, we have

$$\|\mathbf{R}'(t)\| = \sqrt{(-r \sin t)^2 + (r \cos t)^2} = r$$

Then

$$\mathbf{T}(t) = \frac{\mathbf{R}'(t)}{\|\mathbf{R}'(t)\|} = \frac{\langle -r \sin t, r \cos t \rangle}{r} = \langle -\sin t, \cos t \rangle$$

and

$$\mathbf{T}'(t) = \langle -\cos t, -\sin t \rangle$$

Since

$$\|\mathbf{T}'(t)\| = \sqrt{(-\cos t)^2 + (-\sin t)^2} = 1$$

the curvature is

$$\kappa = \frac{\|\mathbf{T}'(t)\|}{\|\mathbf{R}'(t)\|} = \frac{1}{r} \qquad\blacksquare$$

The definition of curvature, $\kappa = \dfrac{\|\mathbf{T}'(t)\|}{\|\mathbf{R}'(t)\|}$, can lead to difficult computations when $\|\mathbf{R}'(t)\|$ is not a constant. The formula for curvature derived in the following theorem may then be easier to apply.

THEOREM 10.8   The cross derivative formula for curvature

Suppose the smooth curve C is the graph of the vector function $\mathbf{R}(t)$. Then the curvature is given by

$$\kappa = \frac{\|\mathbf{R}' \times \mathbf{R}''\|}{\|\mathbf{R}'\|^3}$$

Proof   Since $\|\mathbf{R}'\| = \dfrac{ds}{dt}$, we have

$$\mathbf{R}' = \|\mathbf{R}'\|\mathbf{T} = \left(\frac{ds}{dt}\right)\mathbf{T}$$

$$\mathbf{R}'' = \frac{d}{dt}\left(\frac{ds}{dt}\right)\mathbf{T} + \frac{ds}{dt}\frac{d\mathbf{T}}{dt}$$

$$= \frac{d^2s}{dt^2}\mathbf{T} + \frac{ds}{dt}\mathbf{T}'$$

Therefore, we have

$$\mathbf{R}' \times \mathbf{R}'' = \left(\frac{ds}{dt}\mathbf{T}\right) \times \left(\frac{d^2s}{dt^2}\mathbf{T} + \frac{ds}{dt}\mathbf{T}'\right)$$

$$= \left(\frac{ds}{dt}\right)\left(\frac{d^2s}{dt^2}\right)(\mathbf{T} \times \mathbf{T}) + \left(\frac{ds}{dt}\right)^2(\mathbf{T} \times \mathbf{T}')$$

$$= \left(\frac{ds}{dt}\right)^2(\mathbf{T} \times \mathbf{T}') \qquad \text{Since } \mathbf{T} \times \mathbf{T} = \mathbf{0}$$

Since $\|\mathbf{T}\| = 1$, it follows from the orthogonality of a function of constant length and its derivative (Theorem 10.4) that $\mathbf{T} \cdot \mathbf{T}' = 0$, so the angle between $\mathbf{T}$ and $\mathbf{T}'$ is $\theta = \frac{\pi}{2}$. Thus,

$$\|\mathbf{R}' \times \mathbf{R}''\| = \left(\frac{ds}{dt}\right)^2 \|\mathbf{T} \times \mathbf{T}'\|$$

$$= \left(\frac{ds}{dt}\right)^2 \|\mathbf{T}\|\|\mathbf{T}'\| \sin\theta$$

$$= \left(\frac{ds}{dt}\right)^2 \|\mathbf{T}\|\|\mathbf{T}'\| \qquad \text{Since } \sin\frac{\pi}{2} = 1$$

$$= \left(\frac{ds}{dt}\right)^2 \|\mathbf{T}'\| \qquad \text{Since } \|\mathbf{T}\| = 1$$

$$= \|\mathbf{R}'\|^2 \|\mathbf{T}'\|$$

The curvature is given by

$$\kappa = \frac{\|\mathbf{T}'\|}{\|\mathbf{R}'\|}$$

$$= \frac{\dfrac{\|\mathbf{R}' \times \mathbf{R}''\|}{\|\mathbf{R}'\|^2}}{\|\mathbf{R}'\|}$$

$$= \frac{\|\mathbf{R}' \times \mathbf{R}''\|}{\|\mathbf{R}'\|^3} \qquad \square$$

EXAMPLE 6 Curvature of a helix

Given that a and b are both nonnegative, express the curvature κ of $\mathbf{R}(t) = (a\cos t)\mathbf{i} + (a\sin t)\mathbf{j} + bt\mathbf{k}$ in terms of a and b. The curve defined by $\mathbf{R}$ is shown in Figure 10.33.

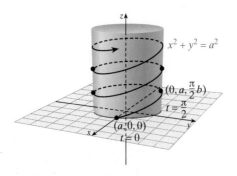

Figure 10.33 Graph of $\mathbf{R}(t) = (a\cos t)\mathbf{i} + (a\sin t)\mathbf{j} + bt\mathbf{k}$

Solution

$$\mathbf{R}'(t) = (-a \sin t)\mathbf{i} + (a \cos t)\mathbf{j} + b\mathbf{k}$$
$$\mathbf{R}''(t) = (-a \cos t)\mathbf{i} + (-a \sin t)\mathbf{j}$$

$$\mathbf{R}' \times \mathbf{R}'' = \begin{vmatrix} \mathbf{i} & \mathbf{j} & \mathbf{k} \\ -a \sin t & a \cos t & b \\ -a \cos t & -a \sin t & 0 \end{vmatrix} = (ab \sin t)\mathbf{i} - (ab \cos t)\mathbf{j} + a^2\mathbf{k}$$

We now need to find the magnitude of this vector, as well as that of $\mathbf{R}'$:

$$\|\mathbf{R}' \times \mathbf{R}''\| = \sqrt{(ab \sin t)^2 + (ab \cos t)^2 + a^4}$$

$$= \sqrt{a^2 b^2 (\sin^2 t + \cos^2 t) + a^4}$$

$$= \sqrt{a^2 b^2 + a^4}$$

$$= a\sqrt{a^2 + b^2} \qquad a > 0 \text{ is given}$$

and

$$\|\mathbf{R}'\| = \sqrt{a^2 \cos^2 t + a^2 \sin^2 t + b^2} = \sqrt{a^2 + b^2}$$

Then we have

$$\kappa = \frac{\|\mathbf{R}' \times \mathbf{R}''\|}{\|\mathbf{R}'\|^3} = \frac{a\sqrt{a^2 + b^2}}{(\sqrt{a^2 + b^2})^3} = \frac{a\sqrt{a^2 + b^2}}{(a^2 + b^2)\sqrt{a^2 + b^2}} = \frac{a}{a^2 + b^2} \qquad \blacksquare$$

In the special case where C is a curve in the plane with the equation $y = f(x)$, the curvature has the form given in the following theorem.

THEOREM 10.9   Curvature of a planar curve using $y = f(x)$

The graph C of the function $y = f(x)$ has curvature

$$\kappa = \frac{|f''(x)|}{\left(1 + [f'(x)]^2\right)^{3/2}}$$

wherever $f(x)$ and its derivatives $f'(x)$ and $f''(x)$ all exist.

Proof   Using x as a parameter, we can regard the curve C as the graph of the vector function $\mathbf{R}(x) = x\mathbf{i} + f(x)\mathbf{j}$. Then, $\mathbf{R}' = \mathbf{i} + f'(x)\mathbf{j}$ and $\mathbf{R}'' = f''(x)\mathbf{j}$, so

$$\mathbf{R}' \times \mathbf{R}'' = \begin{vmatrix} \mathbf{i} & \mathbf{j} & \mathbf{k} \\ 1 & f'(x) & 0 \\ 0 & f''(x) & 0 \end{vmatrix} = f''(x)\mathbf{k}$$

Therefore, since $\|\mathbf{R}'\| = \sqrt{1 + [f'(x)]^2}$, the curvature of C is given by

$$\kappa = \frac{\|\mathbf{R}' \times \mathbf{R}''\|}{\|\mathbf{R}'\|^3}$$

$$= \frac{|f''(x)|}{\left(\sqrt{1 + [f'(x)]^2}\right)^3}$$

$$= \frac{|f''(x)|}{\left(1 + [f'(x)]^2\right)^{3/2}} \qquad \square$$

EXAMPLE 7   Maximizing the curvature of a planar curve

Find the curvature of $y = x^{-1}$ for $x > 0$. For what value of x is the curvature the largest?

Solution

For $f(x) = x^{-1}$, we have $f'(x) = -x^{-2}$ and $f''(x) = 2x^{-3}$, so the curvature is

$$\kappa = \frac{|f''(x)|}{\left(1 + [f'(x)]^2\right)^{3/2}}$$

$$= \frac{2x^{-3}}{\left(1 + [-x^{-2}]^2\right)^{3/2}}$$

$$= \frac{2x^3}{(x^4 + 1)^{3/2}}$$

To find the largest value of κ, we compute the derivative of $\kappa(x)$ and solve the equation $\kappa'(x) = 0$:

$$\kappa'(x) = \frac{6x^2(1 - x^4)}{(x^4 + 1)^{5/2}}$$

Now $\kappa'(x) = 0$ when

$$6x^2(1 - x^4) = 0$$
$$6x^2(1 - x)(1 + x)(1 + x^2) = 0$$
$$x = 0, 1, -1 \quad \text{The 0 and } -1 \text{ are not in the domain.}$$

Since $\kappa'(x) > 0$ for $x < 1$ and $\kappa'(x) < 0$ for $x > 1$, it follows that the largest value of the curvature must occur at $x = 1$. ∎

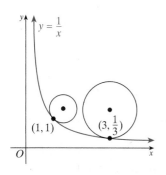

Figure 10.34 Osculating circles to the curve $y = x^{-1}$ at the points $(1, 1)$ and $\left(3, \frac{1}{3}\right)$

Suppose a curve C in the plane has curvature $\kappa \neq 0$ at a point P. Then the circle of radius $\rho = \frac{1}{\kappa}$ that is centered on the concave side of the curve and shares a common tangent with C at P is called the **osculating** (kissing) **circle** at P. The radius ρ of the osculating circle is called the **radius of curvature** and the center of the circle is called the **center of curvature**. The osculating circle and the curve C are tangent and have the same curvature at P, and in this sense, the osculating circle is the circle that best approximates the curve C at P.

Figure 10.34 shows the curve $y = x^{-1}$ for $x > 0$ whose curvature we found in Example 7, along with the osculating circle at the point $P(1, 1)$ that corresponds to $x = 1$. Note that the curvature at P is

$$\kappa(1) = \frac{2(1)^3}{\left(1^4 + 1\right)^{3/2}} = \frac{1}{\sqrt{2}}$$

so the radius of curvature is

$$\rho(1) = \frac{1}{\kappa(1)} = \sqrt{2}$$

In Problem 56, you are asked to show that the osculating circle at P has the equation

$$(x - 2)^2 + (y - 2)^2 = 2$$

We conclude this section by providing a summary of various formulas for computing curvature (Table 10.2).

TABLE 10.2   Curvature Formulas

Type	Given Information	Formula to Use	Where Derived
Arc length parameter	$\mathbf{R}(s)$	$\left\| \dfrac{d\mathbf{T}}{ds} \right\|$	Definition
Two derivatives form	$\mathbf{R}(t)$	$\dfrac{\|\mathbf{T}'(t)\|}{\|\mathbf{R}'(t)\|}$	Pages 666–667
Cross derivative form	$\mathbf{R}(t)$	$\dfrac{\|\mathbf{R}' \times \mathbf{R}''\|}{\|\mathbf{R}'\|^3}$	Theorem 10.8
Functional form	$y = f(x)$	$\dfrac{|f''(x)|}{(1 + [f'(x)]^2)^{3/2}}$	Theorem 10.9
Parametric form	$x = x(t),\ y = y(t)$	$\dfrac{|x'y'' - y'x''|}{[(x')^2 + (y')^2]^{3/2}}$	Problem 46
Polar form	$r = f(\theta)$	$\dfrac{|r^2 + 2r'^2 - rr''|}{(r^2 + r'^2)^{3/2}}$	Problem 47

10.4 PROBLEM SET

A *In Problems 1–8, find the unit tangent vector* $\mathbf{T}(t)$ *and the principal unit normal vector* $\mathbf{N}(t)$ *for the curve given by* $\mathbf{R}(t)$.

1. $\mathbf{R}(t) = t^2\mathbf{i} + t^3\mathbf{j},\ t \neq 0$ 2. $\mathbf{R}(t) = t^2\mathbf{i} + \sqrt{t}\,\mathbf{j},\ t > 0$
3. $\mathbf{R}(t) = (e^t \cos t)\mathbf{i} + (e^t \sin t)\mathbf{j}$
4. $\mathbf{R}(t) = (t \cos t)\mathbf{i} + (t \sin t)\mathbf{j}$
5. $\mathbf{R}(t) = (\cos t)\mathbf{i} + (\sin t)\mathbf{j} + t\mathbf{k}$
6. $\mathbf{R}(t) = (\sin t)\mathbf{i} - (\cos t)\mathbf{j} + t\mathbf{k}$
7. $\mathbf{R}(t) = (\ln t)\mathbf{i} + t^2\mathbf{k}$
8. $\mathbf{R}(t) = (e^{-t} \sin t)\mathbf{i} + e^{-t}\mathbf{j} + (e^{-t} \cos t)\mathbf{k}$

In Problems 9–14, find the length of the given curve over the given interval.

9. $\mathbf{R}(t) = 2t\mathbf{i} + t\mathbf{j}$, over $[0, 4]$
10. $\mathbf{R}(t) = t\mathbf{i} + 3t\mathbf{j}$, over $[0, 4]$
11. $\mathbf{R}(t) = 3t\mathbf{i} + (3 \cos t)\mathbf{j} + (3 \sin t)\mathbf{k}$, over $[0, \frac{\pi}{2}]$
12. $\mathbf{R}(t) = t\mathbf{i} + 2t\mathbf{j} + 3t\mathbf{k}$, over $[0, 2]$
13. $\mathbf{R}(t) = (4 \cos t)\mathbf{i} + (4 \sin t)\mathbf{j} + 5t\mathbf{k}$, over $[0, \pi]$
14. $\mathbf{R}(t) = (\cos^3 t)\mathbf{i} + (\cos^2 t)\mathbf{k}$, over $[0, \frac{\pi}{2}]$

Find the curvature of the plane curves at the points indicated in Problems 15–20.

15. $y = 4x - 2$ at $x = 2$ 16. $y = x - \frac{1}{9}x^2$ at $x = 3$
17. $y = x + \dfrac{1}{x}$ at $x = 1$ 18. $y = \sin x$ at $x = \frac{\pi}{2}$
19. $y = \ln x$ at $x = 1$ 20. $y = e^x$ at $x = 0$

Express the vector function $\mathbf{R}(t)$ *in Problems 21–24 in terms of the arc length parameter s measured from the point where* $t = 0$ *in the direction of increasing t.*

21. $\mathbf{R}(t) = \langle e^{-t}, 1 - e^{-t} \rangle$ 22. $\mathbf{R}(t) = \langle \sin t, \cos t \rangle$
23. $\mathbf{R}(t) = \langle 2 - 3t, 1 + t, -4t \rangle$
24. $\mathbf{R}(t) = \langle 3 \cos t + 3t \sin t, 2t^2, 3 \sin t - 3t \cos t \rangle$
25. Let $\mathbf{u}$ and $\mathbf{v}$ be constant, nonzero vectors. Show that the line given by $\mathbf{R}(t) = \mathbf{u} + t\mathbf{v}$ has curvature 0 at each point.

26. Find the unit tangent $\mathbf{T}$ for $\mathbf{R}(t) = (\cosh t)\mathbf{i} + (\sinh t)\mathbf{j}$ at the point where $t = 0$.
27. Find the unit tangent $\mathbf{T}$ for $\mathbf{R}(t) = [\ln(\sin t)]\mathbf{i} + [\ln(\cos t)]\mathbf{j}$ at the point where $t = \frac{\pi}{3}$.

B 28. A curve C in the plane is given parameterically by $x = 32t$, $y = 16t^2 - 4$.
    a. Sketch the graph of the curve.
    b. Find the unit tangent vector when $t = 3$.
    c. Find the radius of curvature of the point P on C where $t = 3$.

29. For the curve given by $\mathbf{R}(t) = (\sin t)\mathbf{i} + (\cos t)\mathbf{j} + t\mathbf{k}$,
    a. Find a unit tangent vector $\mathbf{T}$ at the point on the curve where $t = \pi$.
    b. Find the curvature when $t = \pi$.
    c. Find the length of the curve from $t = 0$ to $t = \pi$.

30. Let C be the curve given by
    $\mathbf{R}(t) = (t - \sin t)\mathbf{i} + (1 - \cos t)\mathbf{j} + \left(4 \sin \frac{1}{2}t\right)\mathbf{k}$.
    a. Find the unit tangent vector $\mathbf{T}(t)$ to C.
    b. Find $\dfrac{d\mathbf{T}}{ds}$ and the curvature $\kappa(t)$.

31. Find the point (or points) where the ellipse $9x^2 + 4y^2 = 36$ has maximum curvature.
32. Find the maximum curvature on the curve $y = e^{2x}$.
33. Find the radius of curvature at each relative extremum of the graph of $y = x^6 - 3x^2$.
34. Find the curvature of the graph defined by $x = t - \sin t$, $y = 1 - \cos t$. Sketch the curve for $0 \leq t \leq 2\pi$ and sketch the osculating circle at the point where $t = \frac{\pi}{2}$.

There are several different formulas for finding curvature, as summarized in Table 10.2. In Problems 35–42, find the curvature of each of the given curves using the indicated formula.

35. $\mathbf{R}(t) = t\mathbf{i} + t^2\mathbf{j} + t^3\mathbf{k}$; cross derivative form
36. $\mathbf{R}(t) = (t - \cos t)\mathbf{i} + (\sin t)\mathbf{j} + 3\mathbf{k}$; cross derivative form

37. $y = x^2$; functional form

38. $y = x^3$; functional form

39. $y = x^{-2}$; functional form

40. $y = \sin x$; functional form

41. the spiral $r = e^\theta$; polar form

42. the cardioid $r = 1 + \cos\theta$, for $0 \le \theta \le 2\pi$; polar form

43. A *pestus houseflyus* is observed to zip around a room in such a way that at time t its position with respect to the nose of an observer is given by the vector function

$$\mathbf{R}(t) = t\mathbf{u} + t^2\mathbf{v} + 2\left(\tfrac{2}{3}t\right)^{3/2}(\mathbf{u} \times \mathbf{v})$$

where $\mathbf{u}$ and $\mathbf{v}$ are unit vectors separated by an angle of $60°$. Compute the fly's speed and find how long it takes to move a distance of 20 units along its path (starting from the nose), as shown in Figure 10.35.

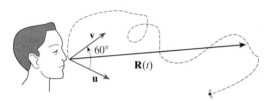

Figure 10.35   Path of a *pestus houseflyus*

44. WHAT DOES THIS SAY?   Describe what is meant by using arc length as a parameter.

45. WHAT DOES THIS SAY?   Discuss the various curvature formulas given in Table 10.2.

46. Let C be a smooth curve in $\mathbb{R}^2$ described by the parametric equations $x = x(t)$ and $y = y(t)$.

a. Show that the curvature of C is given by

$$\kappa = \frac{|x'y'' - y'x''|}{[(x')^2 + (y')^2]^{3/2}}$$

b. Find the curvature of the curve described by $x = 1 - \cos t$, $y = 2 + \sin t$.

47. Let f be a twice differentiable function in $\mathbb{R}^2$ described by the polar function $r = f(\theta)$. Show that the curvature is

$$\kappa(\theta) = \frac{|r^2 + 2r'^2 - rr''|}{(r^2 + r'^2)^{3/2}}$$

Hint: Use the formula in Problem 46 with $x = f(\theta)\cos\theta$, $y = f(\theta)\sin\theta$.

48. Use the formula in Problem 47 to find the curvature of the graphs with the given equations.

a. the cardioid $r = 1 + \cos\theta$

b. the spiral $r = \theta$

c. the circle $r = \sin\theta$

49. Show that the formula for polar curvature in Problem 47 reduces to $\kappa = \dfrac{2}{|r'|}$ at the pole (where $r = 0$).

50. Let $P(a, b)$ be a point on the graph C of the vector function $\mathbf{R}(t)$.

a. Describe a general procedure for finding an equation for the osculating circle to C at P.

b. Find an equation for the osculating circle at the point $P(32, 12)$ on the curve C defined by $x = 32t$, $y = 16t^2 - 4$.

51. The *Cornu spiral* is a parametric curve described by the equations

$$x = \int_0^t \cos\left(\frac{\pi u^2}{2}\right) du \qquad y = \int_0^t \sin\left(\frac{\pi u^2}{2}\right) du$$

It is used in optics to study amplitudes of light waves, and in the design of highway off ramps. It also has interesting mathematical properties.

a. Show that the arc length of the spiral measured from the origin $(0, 0)$ satisfies $s = t$.

b. Show that the curvature of the spiral satisfies $\kappa(s) = \pi |s|$. Thus, for $s \ge 0$, the curvature increases at a constant rate.

c. What happens to the curvature, κ, as $s \to +\infty$? That is, evaluate $\lim\limits_{s \to +\infty} \kappa(s)$.

52. If $\mathbf{T}$ and $\mathbf{N}$ are the unit tangent and normal vectors, respectively, on the trajectory of a moving body, then the cross product vector $\mathbf{B} = \mathbf{T} \times \mathbf{N}$ is called the unit **binormal** of the trajectory. Three planes determined by $\mathbf{T}$, $\mathbf{N}$, and $\mathbf{B}$ are shown in Figure 10.36.

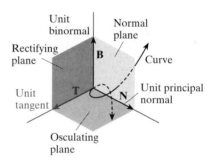

Figure 10.36   Three planes determined by $\mathbf{T}$, $\mathbf{N}$, and $\mathbf{B}$

a. Show that $\mathbf{T}$ is orthogonal to $\dfrac{d\mathbf{B}}{ds}$. *Hint*: Differentiate $\mathbf{B} \cdot \mathbf{T}$.

b. Show that $\mathbf{B}$ is orthogonal to $\dfrac{d\mathbf{B}}{ds}$. *Hint*: Differentiate $\mathbf{B} \cdot \mathbf{B}$.

c. Show that $\dfrac{d\mathbf{B}}{ds} = -\tau\mathbf{N}$ for some constant τ. *Note*: τ is called the **torsion** of the trajectory.

53. Prove the Frenet-Serret formulas:

$$\frac{d\mathbf{T}}{ds} = \kappa\mathbf{N} \qquad \frac{d\mathbf{N}}{ds} = -\kappa\mathbf{T} + \tau\mathbf{B} \qquad \frac{d\mathbf{B}}{ds} = -\tau\mathbf{N}$$

where κ is the curvature and $\tau = \tau(s)$ is a scalar function called the torsion, which provides a measure of the amount of twisting at each point on the trajectory.

54. Show that the torsion may be computed by the formula

$$\tau = \frac{[\mathbf{R}'(t) \times \mathbf{R}''(t)] \cdot \mathbf{R}'''(t)}{\|\mathbf{R}'(t) \times \mathbf{R}''(t)\|^2}$$

55. A highway has an exit ramp that begins at the origin and follows the curve $y = \frac{1}{32}x^{5/2}$ to the point $(4, 1)$. Then it follows the shape of the osculating circle at $(4, 1)$ until the point where $y = 3$. What is the total length of the exit ramp?

56. Show that the osculating circle at the point $(1, 1)$ on the curve $y = x^{-1}$ has the equation
$$(x - 2)^2 + (y - 2)^2 = 2$$

57. Find the curvature of the cycloid with parametric equations
$$x = t - \sin t, \quad y = 1 - \cos t$$

What are the largest and smallest values of the curvature $\kappa(t)$?

58. Counterexample Problem Let f be a function that is twice differentiable. Either prove that the largest curvature of the curve $y = f(x)$ occurs at a relative extremum of f or find a counterexample.

10.5 Tangential and Normal Components of Acceleration

IN THIS SECTION components of acceleration, applications

COMPONENTS OF ACCELERATION

When a body is caused to accelerate or brakes are applied, it is of interest to know how much of the acceleration acts in the direction of the body's motion, as indicated by the unit tangent vector $\mathbf{T}$ and how much in the direction of the principal unit normal $\mathbf{N}$. This question is answered by the following theorem.

THEOREM 10.10 Tangential and normal components of acceleration

An object moving along a smooth curve (with $\mathbf{T}' \neq \mathbf{0}$) has velocity $\mathbf{V}$ and acceleration $\mathbf{A}$, where

$$\mathbf{V} = \left(\frac{ds}{dt}\right)\mathbf{T} \quad \text{and} \quad \mathbf{A} = \left(\frac{d^2s}{dt^2}\right)\mathbf{T} + \kappa\left(\frac{ds}{dt}\right)^2\mathbf{N}$$

and s is the arc length along the trajectory.

Proof The formula for $\mathbf{V}$ follows from the fact that the velocity vector has magnitude $\|\mathbf{V}\| = \dfrac{ds}{dt}$ from Theorem 10.6 and points in the direction of the unit tangent. To establish the formula for $\mathbf{A}$, first note that

$$\frac{d\mathbf{T}}{ds} = \left\|\frac{d\mathbf{T}}{ds}\right\|\mathbf{N} = \kappa\mathbf{N}$$

since the derivative of $\mathbf{T}(s)$ points in a direction orthogonal to $\mathbf{T}$. Using this formula, we find

$$
\begin{aligned}
\mathbf{A} &= \frac{d\mathbf{V}}{dt} & &\text{Definition of } \mathbf{A}\\[4pt]
&= \frac{d}{dt}\left[\mathbf{T}\frac{ds}{dt}\right] & &\mathbf{V} = \left(\frac{ds}{dt}\right)\mathbf{T}\\[4pt]
&= \frac{d^2s}{dt^2}\mathbf{T} + \frac{ds}{dt}\frac{d\mathbf{T}}{dt} & &\text{Product rule}\\[4pt]
&= \frac{d^2s}{dt^2}\mathbf{T} + \frac{ds}{dt}\left[\frac{d\mathbf{T}}{ds}\frac{ds}{dt}\right] & &\text{Chain rule}\\[4pt]
&= \frac{d^2s}{dt^2}\mathbf{T} + \left(\frac{ds}{dt}\right)^2\frac{d\mathbf{T}}{ds} & &\\[4pt]
&= \frac{d^2s}{dt^2}\mathbf{T} + \left(\frac{ds}{dt}\right)^2(\kappa\mathbf{N}) & &\kappa\mathbf{N} = \frac{d\mathbf{T}}{ds} \qquad\square
\end{aligned}
$$

→ **What This Says** At each point on the trajectory of a moving object, the velocity **V** points in the direction of the unit tangent **T**, but the acceleration **A** may have both a tangential and normal component. The trajectory may twist and turn, but the acceleration is always in the plane determined by **T** and the unit normal **N**.

Tangential and Normal Components of Acceleration

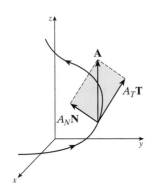

Figure 10.37 Components of acceleration $A = A_T\mathbf{T} + A_N\mathbf{N}$

The acceleration **A** of a moving object can be written as

$$\mathbf{A} = A_T\mathbf{T} + A_N\mathbf{N}$$

where

$$A_T = \frac{d^2s}{dt^2} \qquad \text{is the \textbf{tangential component.}}$$

$$A_N = \kappa \left(\frac{ds}{dt}\right)^2 \qquad \text{is the \textbf{normal component.}}$$

The tangential and normal components are shown in Figure 10.37.

The formulas $A_T = \dfrac{d^2s}{dt^2}$ and $A_N = \kappa \left(\dfrac{ds}{dt}\right)^2$ provide useful information about the way a moving object travels along its trajectory, but they are often not the most convenient computational forms. The following theorem provides alternative formulas for A_T and A_N that use only the derivatives $\mathbf{R}'$ and $\mathbf{R}''$ of the position vector $\mathbf{R}(t)$.

THEOREM 10.11 Formulas for the components of acceleration

Let $\mathbf{R}(t)$ be the position vector for an object moving along a smooth curve C. Then the tangential and normal components of the object's acceleration are given by

$$A_T = \frac{\mathbf{R}' \cdot \mathbf{R}''}{\|\mathbf{R}'\|} \qquad \text{and} \qquad A_N = \frac{\|\mathbf{R}' \times \mathbf{R}''\|}{\|\mathbf{R}'\|}$$

Proof Let θ be the angle between $\mathbf{R}'$ and $\mathbf{R}''$. Then, since the acceleration is given by $\mathbf{A} = \mathbf{R}''$, we have

$$A_T = \|\mathbf{A}\| \cos\theta = \frac{\|\mathbf{R}'\|\|\mathbf{A}\| \cos\theta}{\|\mathbf{R}'\|} = \frac{\mathbf{R}' \cdot \mathbf{R}''}{\|\mathbf{R}'\|}$$

and

$$A_N = \|\mathbf{A}\| \sin\theta = \frac{\|\mathbf{R}'\|\|\mathbf{A}\| \sin\theta}{\|\mathbf{R}'\|} = \frac{\|\mathbf{R}' \times \mathbf{R}''\|}{\|\mathbf{R}'\|} \qquad □$$

EXAMPLE 1 Finding tangential and normal components of acceleration

Find the tangential and normal components of the acceleration of an object that moves with position vector $\mathbf{R}(t) = \langle t^3, t^2, t \rangle$.

Solution

Since $\mathbf{R}'(t) = \mathbf{V}(t) = \langle 3t^2, 2t, 1 \rangle$ and $\mathbf{R}''(t) = \mathbf{A}(t) = \langle 6t, 2, 0 \rangle$, we have

$$\mathbf{R}' \cdot \mathbf{R}'' = \mathbf{V} \cdot \mathbf{A} = (3t^2)(6t) + (2t)(2) = 18t^3 + 4t$$

and

$$\mathbf{R}' \times \mathbf{R}'' = \mathbf{V} \times \mathbf{A} = \begin{vmatrix} \mathbf{i} & \mathbf{j} & \mathbf{k} \\ 3t^2 & 2t & 1 \\ 6t & 2 & 0 \end{vmatrix} = -2\mathbf{i} + 6t\mathbf{j} - 6t^2\mathbf{k}$$

Therefore, the components of acceleration are

$$A_T = \frac{\mathbf{R}' \cdot \mathbf{R}''}{\|\mathbf{R}'\|} = \frac{\mathbf{V} \cdot \mathbf{A}}{\|\mathbf{V}\|} = \frac{18t^3 + 4t}{\sqrt{(3t^2)^2 + (2t)^2 + 1}} = \frac{18t^3 + 4t}{\sqrt{9t^4 + 4t^2 + 1}}$$

$$A_N = \frac{\|\mathbf{R}' \times \mathbf{R}''\|}{\|\mathbf{R}'\|} = \frac{\|\mathbf{V} \times \mathbf{A}\|}{\|\mathbf{V}\|} = \frac{\sqrt{(-2)^2 + (6t)^2 + (-6t^2)^2}}{\sqrt{9t^4 + 4t^2 + 1}} = 2\sqrt{\frac{9t^4 + 9t^2 + 1}{9t^4 + 4t^2 + 1}}$$

■

Note in Figure 10.37 that $\|\mathbf{A}\|$ is the hypotenuse of a right triangle whose other two sides have lengths $\|A_T\mathbf{T}\| = |A_T|$ and $\|A_N\mathbf{N}\| = A_N$. Therefore, by the Pythagorean theorem, we have

$$\|\mathbf{A}\|^2 = A_T^2 + A_N^2$$

$$A_N = \sqrt{\|\mathbf{A}\|^2 - A_T^2}$$

This formula is used to compute A_N in the following example.

EXAMPLE 2   Finding components of acceleration on a helix

An object moves with position vector $\mathbf{R}(t) = \langle \cos t, \sin t, t \rangle$. Find the tangential and normal components of acceleration.

Solution

The trajectory is a helix, and we find

$$\mathbf{V} = \frac{d\mathbf{R}}{dt} = \langle -\sin t, \cos t, 1 \rangle \quad \text{and} \quad \mathbf{A} = \frac{d\mathbf{V}}{dt} = \langle -\cos t, -\sin t, 0 \rangle$$

$$\frac{ds}{dt} = \|\mathbf{V}\| = \sqrt{\sin^2 t + \cos^2 t + 1} = \sqrt{2} \quad \text{and} \quad A_T = \frac{d^2 s}{dt^2} = 0$$

$$A_N = \sqrt{\|\mathbf{A}\|^2 - A_T^2} = \sqrt{\left(\sqrt{\cos^2 t + \sin^2 t}\right)^2 - 0^2} = 1$$

The tangential and normal components of acceleration are 0 and 1, respectively. This means that the acceleration satisfies

$$\mathbf{A} = A_T\mathbf{T} + A_N\mathbf{N} = (0)\mathbf{T} + (1)\mathbf{N} = \mathbf{N}$$

That is, the acceleration vector is the principal unit normal $\mathbf{N}$, and the acceleration is always normal to the helical trajectory. ■

APPLICATIONS

Now that we know how to compute the tangential and normal components of acceleration, A_T and A_N, we will examine some applications. First, according to Newton's second law of motion, the total force acting on a moving object of mass m satisfies $\mathbf{F} = m\mathbf{A}$, where $\mathbf{A}$ is the acceleration of the object. Because $\mathbf{A} = A_T\mathbf{T} + A_N\mathbf{N}$, we have

$$\mathbf{F} = m\mathbf{A} = (mA_T)\mathbf{T} + (mA_N)\mathbf{N} = F_T\mathbf{T} + F_N\mathbf{N}$$

where

$$F_T = m\frac{d^2 s}{dt^2} \quad \text{and} \quad F_N = m\kappa \left(\frac{ds}{dt}\right)^2$$

Figure 10.38 Tendency to skid

For instance, experience leads us to expect a car to skid if it makes a sharp turn at moderate speed or even a gradual turn at high speed. Mathematically, a "sharp turn" occurs when the radius of curvature $\rho = 1/\kappa$ is small (that is, when κ is large), and "high speed" means that ds/dt is large. In either case,

$$F_N = m\kappa \left(\frac{ds}{dt}\right)^2$$

will be relatively large, and the car will stay on the road (its trajectory) only if (assuming no banking) there is a correspondingly large frictional force between the tires of the car and the surface of the road (see Figure 10.38).

When forces, masses, and accelerations are involved, it is customary to express the mass m in slugs of an object whose weight W has been given in pounds.* From the definition of weight,

$$m = \frac{W}{g}$$

where g is the acceleration of gravity ($g \approx 32$ ft/s^2).

EXAMPLE 3 **Modeling Application: Tendency of a vehicle to skid**

A car weighing 2,700 lb makes a turn on a flat road while traveling at 56 ft/s (about 38 mi/h). If the radius of the turn is 70 ft, how much frictional force is required to keep the car from skidding?

Solution

The force tending to push the car off the road is $F_N = m\kappa \left(\dfrac{ds}{dt}\right)^2$, where m is the mass of the car and κ is the curvature of the road. This is the force that must be compensated for by friction if the car is not to skid. We know that $\dfrac{ds}{dt} = 56$ ft/s, and because the car weighs $W = 2,700$ lb, its mass is $m = \dfrac{W}{g} = \dfrac{2,700}{32} \approx 84.38$ slugs. Because the turn radius is 70 ft, we have $\kappa = \frac{1}{70}$, so that

$$F_N = \left(\frac{2,700}{32} \text{ slugs}\right)\left(\frac{1}{70 \text{ ft}}\right)\left(56 \frac{\text{ft}}{\text{s}}\right)^2 = 3,780 \frac{\text{lb} \cdot \text{s}^2}{\text{ft}} \frac{1}{\text{ft}} \frac{\text{ft}^2}{\text{s}^2} = 3,780 \text{ lb} \qquad \blacksquare$$

There are certain important applications in which an object moves along its trajectory with constant speed ds/dt, and when this occurs, the acceleration $\mathbf{A}$ can have only a normal component, because $d^2s/dt^2 = 0$.

THEOREM 10.12 **Acceleration of an object with constant speed**

The acceleration of an object moving with constant speed is always orthogonal to the direction of motion.

Proof Notice that we really do not need to know anything about components of acceleration to prove this result. Saying that the object has constant speed means that $\|\mathbf{R}'(t)\|$ is constant, and by Theorem 10.4, we conclude that $\mathbf{R}'(t)$ is orthogonal to its derivative $\mathbf{R}''(t) = \mathbf{A}(t)$. But $\mathbf{R}'(t)$ points in the direction of the object's motion along its trajectory, which means that the acceleration $\mathbf{A}$ is orthogonal to the direction of motion. $\qquad \square$

---

*A slug is a unit of measurement defined as the unit of mass that receives an acceleration of 1 ft/s^2 when a force of 1 lb is applied to it. That is, $1 \text{ slug} = \dfrac{1 \text{ lb}}{1 \text{ ft/s}^2} = \dfrac{\text{lb s}^2}{\text{ft}}$

As an application of this result, note that when an object moves with constant speed v_0 along a circular path of radius R (so that $\kappa = 1/R$), its acceleration is directed toward the center of the path and has magnitude

$$A_N = \kappa \left(\frac{ds}{dt}\right)^2 = \frac{1}{R}v_0^2$$

(See Figure 10.39.)

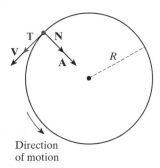

If the constant speed is v_0 and the path has radius R, the acceleration is

$$\mathbf{A} = \frac{v_0^2}{R}\mathbf{N}$$

Direction of motion

Figure 10.39   An object moving with constant speed on a circular path

EXAMPLE 4   Modeling Application: Period of a satellite

An artificial satellite travels at constant speed in a stable circular orbit 20,000 km above the earth's surface. How long does it take for the satellite to make one complete circuit of the earth?

Solution

Let m denote the satellite's mass and v denote its speed. We will assume that the earth is a sphere of radius 6,440 km, so the curvature of the path is $\kappa = 1/R$, where

$$R = \underbrace{6{,}440}_{\text{Radius of earth}} + \overbrace{20{,}000}^{\text{Height}} = 26{,}440 \text{ km}$$

Figure 10.40 A satellite in a stable orbit; magnitude of centripetal force $\mathbf{F}_c$ equals the magnitude of the force $\mathbf{F}_g$ due to gravity, but points in the opposite direction.

is the distance of the satellite from the center of the earth. The satellite maintains a stable orbit when the magnitude of the force $\mathbf{F}_c = mv^2/R$ produced by the rate of change of tangential velocity, called *centripetal* (center directed) *acceleration*, equals the magnitude of the force $\mathbf{F}_g$ due to gravity but is opposite in direction. (See Figure 10.40.)

According to Newton's law of universal gravitation, $\|\mathbf{F}_g\| = GmM/R^2$, where M is the mass of the earth and G is the *gravitational constant*. Thus, for stability we must have

$$\frac{mv^2}{R} = \frac{GmM}{R^2}$$

so $v = \sqrt{GM/R}$. Experiments indicate that $GM = 398{,}600 \text{ km}^3/\text{s}^2$, and by substituting $R = 26{,}440$, we obtain

$$v = \sqrt{\frac{GM}{R}} = \sqrt{\frac{398{,}600}{26{,}440}} \approx 3.88273653$$

For this example, we see that the speed of the satellite is approximately 3.883 km/s.

Finally, suppose T is the time required for the satellite to make one complete circuit of the earth (called the **period** of the satellite). In each period, the satellite travels a distance equal to the circumference of a circle of radius $R = 26{,}440$ km, and because it travels at v km/s, we must have $vT = 2\pi R$, so that

$$T = \frac{2\pi R}{v} \approx \frac{2\pi(26{,}440)}{3.88273653} \approx 42{,}786.16852 \text{ seconds}$$

or approximately 713 minutes (11 hr 53 min).   ∎

10.5 PROBLEM SET

Ⓐ *In Problems 1–8, $\mathbf{R}(t)$ is the position vector of a moving object. Find the tangential and normal components of the object's acceleration.*

1. $\mathbf{R}(t) = t\mathbf{i} + t^2\mathbf{j}$
2. $\mathbf{R}(t) = t\mathbf{i} + e^t\mathbf{j}$
3. $\mathbf{R}(t) = \langle t\sin t, t\cos t\rangle$
4. $\mathbf{R}(t) = \langle 3\cos t, 2\sin t\rangle$
5. $\mathbf{R}(t) = \langle t, t^2, t\rangle$
6. $\mathbf{R}(t) = \langle 4\cos t, 0, \sin t\rangle$
7. $\mathbf{R}(t) = (\sin t)\mathbf{i} + (\cos t)\mathbf{j} + (\sin t)\mathbf{k}$
8. $\mathbf{R}(t) = \left(\frac{5}{13}\cos t\right)\mathbf{i} + \frac{12}{13}(1 - \cos t)\mathbf{j} + \sin t\,\mathbf{k}$

In Problems 9–12, the velocity $\mathbf{V}$ and acceleration $\mathbf{A}$ of a moving object are given at a certain instant. Find A_T and A_N at this instant.

9. $\mathbf{V}_0 = \langle 1, -3\rangle$; $\mathbf{A}_0 = \langle 2, 5\rangle$
10. $\mathbf{V}_0 = -\mathbf{i} + 7\mathbf{j}$; $\mathbf{A}_0 = 4\mathbf{i} + \mathbf{j}$
11. $\mathbf{V}_0 = 2\mathbf{i} + 3\mathbf{j} - \mathbf{k}$; $\mathbf{A}_0 = -\mathbf{i} - 5\mathbf{j} + 2\mathbf{k}$
12. $\mathbf{V}_0 = \langle 5, -1, 2\rangle$; $\mathbf{A}_0 = \langle 1, 0, -7\rangle$

In Problems 13–16, the speed $\|\mathbf{V}\|$ of a moving object is given. Find A_T, the tangential component of acceleration, at the indicated time.

13. $\|\mathbf{V}\| = \sqrt{5t^2 + 3}$; $t = 1$
14. $\|\mathbf{V}\| = \sqrt{t^2 + t + 1}$; $t = 3$
15. $\|\mathbf{V}\| = \sqrt{\sin^2 t + \cos(2t)}$; $t = 0$
16. $\|\mathbf{V}\| = \sqrt{e^{-t} + t^4}$; $t = 0$

17. Find the maximum and minimum speeds of a particle whose position vector is

$$\mathbf{R}(t) = (4\sin 2t)\mathbf{i} - (3\cos 2t)\mathbf{j}$$

18. Where on the trajectory of

$$\mathbf{R}(t) = (2t^2 - 5t)\mathbf{i} + (5t + 2)\mathbf{j} + 4t^2\mathbf{k}$$

is the speed minimized? What is the minimum speed?

19. The position of an object at time t is (x, y), where $x = 1 + \cos 2t$, $y = \sin 2t$. Find the velocity, the acceleration, and the normal and tangential components of acceleration of the object at time t.

20. An object moves with constant angular velocity ω around the circle $x^2 + y^2 = r^2$ in the xy-plane. Find a parameterization for the circle and compute the tangential and normal components of acceleration for the moving object.

Ⓑ 21. Find the tangential and normal components of the acceleration of an object that moves along the parabolic path $y = 4x^2$ at the instant the speed is $ds/dt = 20$.

Modeling Problems In Problems 22–25, set up an appropriate model to answer the given question. Be sure to state your assumptions.

22. A pail attached to a rope 1 yd long is swung at the rate of 1 rev/s. Find the tangential and normal components of the pail's acceleration. Assume the rope is swung in a level plane.

23. A boy holds onto a pail of water weighing 2 lb and swings it in a vertical circle with a radius of 3 ft. If the pail travels at ω rpm, what is the force of the water on the bottom of the pail at the highest and lowest points of the swing? What is the *smallest* value of ω required to keep the water from spilling from the pail? Assume the pail is held by a handle so that its bottom is straight up when it is at its highest point.

24. A car weighing 2,700 lb moves along the elliptic path $900x^2 + 400y^2 = 1$, where x and y are measured in miles. If the car travels at the constant speed of 45 mi/h, how much frictional force is required to keep it from skidding as it turns the "corner" at $\left(\frac{1}{30}, 0\right)$? What about the corner $\left(0, \frac{1}{20}\right)$?

25. Curved sections of road such as expressway off ramps are often banked to protect against skidding. Consider a road with a circular curve of radius 150 feet, as shown in Figure 10.41, and assume the magnitude of the force of static friction $\mathbf{F}_s$ (the force resisting the tendency to skid) is proportional to the car's weight; that is $\|\mathbf{F}_s\| = \mu W$, where μ is a constant called the *coefficient of static friction*. For this problem, we assume $\mu = 0.47$.

Figure 10.41 A road with a circular arc

   a. First, suppose the road is not banked. What is the largest speed v_m that a 3,500 lb car can travel around the curve without skidding? Would the answer be different for a smaller car, say one weighing 2,000 lb?
   b. Now suppose the curve is banked at 17° and answer the questions in part **a**.
   c. Suppose the highway engineers want the maximum safe speed at 50 mi/h. At what angle should the road be banked?

26. What is the smallest radius that should be used for a circular highway exchange as shown in Figure 10.42 if the normal component of the acceleration of a car traveling at 45 mi/h is not to exceed 2.4 ft/s²?

Figure 10.42 Highway exchange

27. A Ferris wheel (see Figure 10.43) with radius 15 ft rotates in a vertical plane at ω rpm. What is the maximum value of ω for a Ferris wheel carrying a person of weight W? (That is, the largest ω so the passenger is not "thrown off.")

Figure 10.43 Ferris wheel

Use the formulas in Theorem 10.11 to find A_T and A_N for the given position vector $\mathbf{R}(t)$ in Problems 28–31.

28. $\mathbf{R}(t) = t^3\mathbf{i} + t^2\mathbf{j} + t\mathbf{k}$ **29.** $\mathbf{R}(t) = t\mathbf{i} + 2t\mathbf{j} + t^2\mathbf{k}$

30. $\mathbf{R}(t) = (\cos t)\mathbf{i} + (\sin t)\mathbf{j} + \mathbf{k}$

31. $\mathbf{R}(t) = (e^t \cos t)\mathbf{i} + (e^t \sin t)\mathbf{j} + e^t\mathbf{k}$

32. Modeling Problem   An amusement park ride consists of a large (25-ft radius), flat, horizontal wheel. Customers board the wheel while it is stationary and try to stay on as long as possible as it begins to rotate. The purpose of this problem is to discover how fast the wheel can rotate without losing its passengers.

    a. Suppose the wheel rotates at ω revolutions per minute and that a volunteer weighing W lb sits 15 feet from the center of the wheel, as shown Figure 10.44. Find $\mathbf{F}_N = m\mathbf{A}_N$, where $m = W/g$ is the volunteer's mass. This is the force tending to push the volunteer off the wheel.

Figure 10.44   Carnival fun ride

    b. If the frictional force of the volunteer in part **a** to stay on the wheel is $0.12W$, find the largest value of ω that will allow the volunteer to stay in place. Wait, we did not tell you the volunteer's weight. Does it matter?

33. Modeling Problem   A curve in a railroad track has the shape of the parabola $x = y^2/120$. If a train is loaded so that its scalar normal component of acceleration cannot exceed 30 units/s², what is its maximum possible speed as it rounds the curve at $(0, 0)$?

C 34. Let $\mathbf{T}$ be the unit tangent vector, $\mathbf{N}$ the principal unit normal, and $\mathbf{B}$ the unit binormal vector (see Problem 52, Section 10.4) to a given curve C. Show that

$$\frac{d\mathbf{B}}{ds} = \mathbf{T} \times \frac{d\mathbf{N}}{ds}$$

35. Suppose an object moves in the plane along the curve $y = f(x)$. Use the formulas obtained in Theorem 10.11 to

show that

$$A_T = \frac{f'(x)f''(x)}{\sqrt{1 + [f'(x)]^2}}; \quad A_N = \frac{|f''(x)|}{\sqrt{1 + [f'(x)]^2}}$$

36. If the graph of the function f has an inflection point at $x = a$ and $f''(a)$ exists, show that the graph of f has curvature 0 at $(a, f(a))$.

37. A projectile is fired from ground level with an angle of elevation α and muzzle speed v_0. Find formulas for the tangential and normal components of the projectile's acceleration at time t. What are A_T and A_N at the time the projectile is at its maximum height?

38. An object connected to a string of length r is spun counterclockwise in a circular path in a horizontal plane. Let ω be the constant angular velocity of the object.

    a. Show that the position vector of the object is
$\mathbf{R}(t) = (r \cos \omega t)\mathbf{i} + (r \sin \omega t)\mathbf{j}$.

    b. Find the normal component of the object's acceleration.

    c. If the angular velocity ω is doubled, what is the effect on the normal component of acceleration? What is the effect on $\mathbf{A}$ if ω is unchanged but the length of the string is doubled? What happens to A_N if ω is doubled and r is halved?

39. An artificial satellite travels at constant speed in a stable circular orbit R km above the earth's surface (as in Example 4).

    a. Show that the satellite's period is given by the formula

$$T = 2\pi \frac{(R + R_e)^{3/2}}{\sqrt{GM}}$$

    where R_e is the radius of the earth.

    b. A satellite is said to be in **geosynchronous orbit** if it completes one orbit every sidereal day (23 hours, 56 minutes). How high above the earth should the satellite be to achieve such an orbit? Assume the radius of the earth is approximately 6,440 miles.

    c. What is the speed of a satellite in geosynchronous orbit?

40. Modeling Problem   For Mars the following facts are known (all in relation to Earth):

diameter	0.533
length of day	1.029
mass	0.1074
gravity (g)	0.3776

Use these facts to determine how high above the surface of Mars a satellite must be to achieve Mars synchronous orbit. How fast would such a satellite be traveling?

CHAPTER 10 REVIEW

Proficiency Examination

CONCEPT PROBLEMS

    1. What is a vector-valued function?

    2. What are the components of a vector function?

    3. Describe what is meant by the graph of a vector function.

    4. Define the limit of a vector function.

    5. Define the derivative of a vector function.

    6. Define the integral of a vector function.

    7. What is a smooth curve?

8. State the following rules for differentiating vector functions:

    a. Linearity rule   **b.** Scalar multiple rule

    c. Dot product rule   **d.** Cross product rule

    e. Chain rule

9. State the theorem about the orthogonality of a vector function of constant length and its derivative.

10. What are the position, velocity, and acceleration vectors?

11. What is the speed of a particle moving on a curve C?

12. What are the formulas for the motion of a projectile in a vacuum?

13. What are the formulas for time of flight and range of a projectile?

14. State Kepler's laws.

15. What are the unit polar vectors, $\mathbf{u}_r$ and $\mathbf{u}_\theta$?

16. What are the unit tangent and principal unit normal vectors?

17. What is speed in terms of arc length?

18. Give a formula for the arc length of a curve in $\mathbb{R}^3$.

19. What is the curvature of a graph?

20. What is a formula for curvature in terms of the velocity and acceleration vectors?

21. What are the formulas for unit tangent and principal unit normal in terms of the arc length parameter?

22. What is the radius of curvature?

23. What are the tangential and normal components of acceleration?

PRACTICE PROBLEMS

24. Sketch the graph of $\mathbf{R}(t) = (3\cos t)\mathbf{i} + (3\sin t)\mathbf{j} + t\mathbf{k}$, and find the length of this curve from $t = 0$ to $t = 2\pi$.

25. If $\mathbf{F}(t) = \dfrac{t}{1+t}\mathbf{i} + \dfrac{\sin t}{t}\mathbf{j} + (\cos t)\mathbf{k}$, find $\mathbf{F}'(t)$ and $\mathbf{F}''(t)$.

26. Evaluate $\displaystyle\int_1^2 \langle 3t, 0, 3 \rangle \times \langle 0, \ln t, -t^2 \rangle \, dt$.

27. Find a vector function $\mathbf{F}$ such that $\mathbf{F}''(t) = \langle e^t, -t^2, 3 \rangle$ and $\mathbf{F}(0) = \langle 1, -2, 0 \rangle$, $\mathbf{F}'(0) = \langle 0, 0, 3 \rangle$.

28. Find the velocity $\mathbf{V}$, the speed $\dfrac{ds}{dt}$, and the acceleration $\mathbf{A}$ for the body with position vector $\mathbf{R}(t) = t\mathbf{i} + 2t\mathbf{j} + te^t\mathbf{k}$.

29. Find $\mathbf{T}$, $\mathbf{N}$, A_T and A_N (the tangential and normal components of acceleration), and κ (curvature) for an object with position vector $\mathbf{R}(t) = t^2\mathbf{i} + 3t\mathbf{j} - 3t\mathbf{k}$.

30. A projectile is fired from ground level with muzzle speed 50 ft/s at an angle of elevation of $\alpha = 30°$.

    a. What is the maximum height reached by the projectile?

    b. What are the time of flight and the range?

Supplementary Problems

Find the vector limits in Problems 1–6.

1. $\displaystyle\lim_{t\to 0}\left[t^2\mathbf{i} - 3te^t\mathbf{j} + \frac{\sin 2t}{t}\mathbf{k} \right]$

2. $\displaystyle\lim_{t\to 0}\left[\frac{\mathbf{i} + t\mathbf{j} - e^{-t}\mathbf{k}}{1-t} \right]$

3. $\displaystyle\lim_{t\to 0}\langle t, 0, 5 \rangle \cdot \langle \sin t, 3t, -(1-t) \rangle$

4. $\displaystyle\lim_{t\to \pi}\langle 1+t, -3, 0 \rangle \times \langle 0, t^2, \cos t \rangle$

5. $\displaystyle\lim_{t\to 0^+}\left[\left(1 + \frac{1}{t}\right)^t \mathbf{i} - \left(\frac{\sin t}{t}\right)\mathbf{j} - t\mathbf{k} \right]$

6. $\displaystyle\lim_{t\to \infty}\left[\left(\frac{1 - \cos t}{t}\right)\mathbf{i} + 4\mathbf{j} + \left(1 + \frac{3}{t}\right)^t \mathbf{k} \right]$

Find $\mathbf{F}'(t)$ and $\mathbf{F}''(t)$ for the vector functions in Problems 7–12.

7. $\mathbf{F}(t) = te^t\mathbf{i} + t^2\mathbf{j}$ **8.** $\mathbf{F}(t) = (t \ln 2t)\mathbf{i} + t^{3/2}\mathbf{k}$

9. $\mathbf{F}(t) = \langle 2t^{-1}, -2t, te^{-t} \rangle$

10. $\mathbf{F}(t) = \langle t, -(1-t), 0 \rangle \times \langle 0, t^2, e^{-t} \rangle$

11. $\mathbf{F}(t) = (t^2 + e^{at})\mathbf{i} + (te^{-at})\mathbf{j} + (e^{at+1})\mathbf{k}$, a constant

12. $\mathbf{F}(t) = (1-t)^{-1}\mathbf{i} + (\sin 2t)\mathbf{j} + (\cos^2 t)\mathbf{k}$

Sketch the graph of the vector functions in Problems 13–15.

13. $\mathbf{F}(t) = te^t\mathbf{i} + t^2\mathbf{j}$ **14.** $\mathbf{F}(t) = t^2\mathbf{i} - 3t\mathbf{j}$

15. $\mathbf{F}(t) = (1 - \cos t)\mathbf{i} + (\sin t)\mathbf{k}$

Sketch the graph of the vector functions in Problems 16–18.

16. $\mathbf{F}(t) = \langle 3\cos t, 3\sin t, t \rangle$ **17.** $\mathbf{F}(t) = \langle 2\sin t, 2\cos t, 5t \rangle$

18. $\mathbf{F}(t) = 2t^2\mathbf{i} + (1-t)\mathbf{j} + 3\mathbf{k}$

Find the indicated derivative in terms of $\mathbf{F}$ and $\mathbf{F}'$ in Problems 19–24.

19. $\dfrac{d}{dt}[\mathbf{F}(t) \cdot \mathbf{F}(t)]$ **20.** $\dfrac{d}{dt}[\|\mathbf{F}(t)\|\mathbf{F}(t)]$

21. $\dfrac{d}{dt}\left[\dfrac{\mathbf{F}(t)}{\|\mathbf{F}(t)\|} \right]$ **22.** $\dfrac{d}{dt}\|\mathbf{F}(t)\|$

23. $\dfrac{d}{dt}[\mathbf{F}(t) \times \mathbf{F}(t)]$ **24.** $\dfrac{d}{dt}\mathbf{F}(e^t)$

Evaluate the definite and indefinite vector integrals in Problems 25–30.

25. $\displaystyle\int_{-1}^1 (e^{-t}\mathbf{i} + t^3\mathbf{j} + 3\mathbf{k}) \, dt$

26. $\displaystyle\int_1^2 [(1-t)\mathbf{i} - t^{-1}\mathbf{j} + e^t\mathbf{k}] \, dt$

27. $\displaystyle\int [te^t\mathbf{i} - (\sin 2t)\mathbf{j} + t^2\mathbf{k}] \, dt$

28. $\displaystyle\int e^{2t}[2\mathbf{i} - t\mathbf{j} + (\sin t)\mathbf{k}] \, dt$

29. $\displaystyle\int t[e^t\mathbf{i} + (\ln t)\mathbf{j} + 3\mathbf{k}] \, dt$

30. $\displaystyle\int [e^t\mathbf{i} + 2\mathbf{j} - t\mathbf{k}] \cdot [e^{-t}\mathbf{i} - t\mathbf{j}] \, dt$

In Problems 31–34, $\mathbf{R}$ is the position vector of a moving body. Find the velocity $\mathbf{V}$, the speed ds/dt, and the acceleration $\mathbf{A}$.

31. $\mathbf{R}(t) = t\mathbf{i} + (3-t)\mathbf{j} + 2\mathbf{k}$

32. $\mathbf{R}(t) = (\sin 2t)\mathbf{i} + 2\mathbf{j} - (\cos 2t)\mathbf{k}$

33. $\mathbf{R}(t) = \langle t \sin t, te^{-t}, -(1-t) \rangle$

34. $\mathbf{R}(t) = \langle \ln t, e^t, -\tan t \rangle$

Find $\mathbf{T}$ and $\mathbf{N}$ for the vector functions given in Problems 35–38.

35. $\mathbf{R}(t) = t\mathbf{i} - t^2\mathbf{j}$

36. $\mathbf{R}(t) = (3\cos t)\mathbf{i} - (3\sin t)\mathbf{j}$

37. $\mathbf{R}(t) = \langle 4\cos t, -3t, 4\sin t \rangle$

38. $\mathbf{R}(t) = \langle e^t \sin t, e^t, e^t \cos t \rangle$

In Problems 39–42, find the tangential and normal components of acceleration, and the curvature of a moving object with position vector $\mathbf{R}(t)$.

39. $\mathbf{R}(t) = t^2\mathbf{i} + 2t\mathbf{j} + e^t\mathbf{k}$

40. $\mathbf{R}(t) = t^2\mathbf{i} - 2t\mathbf{j} + (t^2 - t)\mathbf{k}$

41. $\mathbf{R}(t) = (4 \sin t)\mathbf{i} + (4 \cos t)\mathbf{j} + 4t\mathbf{k}$

42. $\mathbf{R}(t) = (a \sin 3t)\mathbf{i} + (a + a \cos 3t)\mathbf{j} + (3a \sin t)\mathbf{k}$, for constant $a > 0$

A polar curve C is given by $r = f(\theta)$. If $f''(\theta)$ exists, then the curvature can be found by the formula

$$\kappa = \frac{|r^2 + 2r'^2 - rr''|}{(r^2 + r'^2)^{3/2}}$$

(see Problem 47, Section 10.4). Find the curvature at the given point on each of the polar curves given in Problems 43–48.

43. $r = 4 \cos \theta$, where $\theta = \frac{\pi}{3}$ **44.** $r = \theta^2$, where $\theta = 2$

45. $r = e^{-\theta}$, where $\theta = 1$

46. $r = 1 + \cos \theta$, where $\theta = \frac{\pi}{2}$

47. $r = 4 \cos 3\theta$, where $\theta = \frac{\pi}{6}$

48. $r = 1 - 2 \sin \theta$, where $\theta = \frac{\pi}{4}$

49. For what values of t is the following vector function continuous?

$$\mathbf{F}(t) = (2t - 1)\mathbf{i} + \left(\frac{t^2 - 1}{t - 1}\right)\mathbf{j} + 4\mathbf{k}$$

50. Find the length of the graph of the vector function

$$\mathbf{R}(t) = \left(\frac{t^2 - 2}{2}\right)\mathbf{i} + \frac{(2t + 1)^{3/2}}{3}\mathbf{j}$$

from $t = 0$ to $t = 6$.

51. Find a vector function $\mathbf{F}$ such that $\mathbf{F}(0) = \mathbf{F}'(0) = \mathbf{i}$, $\mathbf{F}''(0) = 2\mathbf{j} + \mathbf{k}$, and

$$\mathbf{F}'''(t) = (\cos t)\mathbf{i} + (\sin t)\mathbf{j} + \frac{t}{\pi}\mathbf{k}$$

52. Find parametric equations for the tangent line to the graph of

$$\mathbf{R}(t) = te^{-t}\mathbf{i} + t^2\mathbf{j} + te^{-2t}\mathbf{k}$$

at the highest point on the graph (that is, where z has the largest value).

53. Let $\mathbf{R}(t) = t\mathbf{i} + t\mathbf{j} + t^3\mathbf{k}$.

    a. Find parametric equations for the tangent line to the graph of $\mathbf{R}$ at the point where $x = 1$.

    b. The tangent line in part **a** intersects the graph of $\mathbf{R}(t)$ in a second point. Find the coordinates of this point.

54. An object moves in space with acceleration $\mathbf{A}(t) = \langle -t, 2, 2 - t \rangle$. When $t = 0$, it is known that the object is at the point $(1, 0, 0)$ and that it has velocity $\mathbf{V}(0) = \langle 2, -4, 0 \rangle$.

    a. Find the velocity $\mathbf{V}(t)$ and the position vector $\mathbf{R}(t)$.

    b. What are the speed and location of the object when $t = 1$?

    c. When is the object stationary and what is its position at that time?

55. The position vector for a curve is given in terms of arc length s by $\mathbf{R}(s) = \left\langle a \cos \frac{s}{a}, a \sin \frac{s}{a}, 2s \right\rangle$ for $0 \le s \le 2\pi a$, $a \ne 0$. Find the unit tangent vector $\mathbf{T}(s)$ and the principal unit normal $\mathbf{N}(s)$.

56. Find the radius of curvature of the curve given by $y = 1 + \sin x$ at the points where x is

a. $\dfrac{\pi}{6}$ **b.** $\dfrac{\pi}{4}$ **c.** $\dfrac{3\pi}{2}$

57. Find the radius of curvature of the ellipse given by $\mathbf{R}(t) = \langle a \cos t, b \sin t \rangle$, where $a > 0$, $b > 0$, $a \ne b$, and $0 \le t \le 2\pi$, at the points where $t = 0$ and $\pi/2$.

58. Modeling Problem A stunt pilot flying horizontally at an altitude of 4,000 ft with a speed of 180 mi/hr drops a weighted marker, attempting to hit a target on the ground below, as shown in Figure 10.45. How far away from the target (measured horizontally) should the pilot be when she releases the marker? You may neglect air resistance.

4000 ft

Figure 10.45   Problem 58

59. The position of an object moving in space is given by

$$\mathbf{R}(t) = (e^{-t} \cos t)\mathbf{i} + (e^{-t} \sin t)\mathbf{j} + e^{-t}\mathbf{k}$$

    a. Find the velocity, speed, and acceleration of the object at arbitrary time t.

    b. Determine the curvature of the trajectory at time t.

60. Find the point or points on the curve $y = e^{ax}$ ($a > 0$) where the curvature is maximized.

61. Modeling Problem A car weighing 3,000 lb travels at a constant speed of 60 mi/h on a flat road and then makes a circular turn on an interchange (see Figure 10.46). If the radius of the turn is 160 ft, what frictional force is needed to keep the car from skidding?

Figure 10.46   Problem 61

62. The position vector of an object in space is

$$\mathbf{R}(t) = (a \cos \omega t)\mathbf{i} + (a \sin \omega t)\mathbf{j} + \omega^2 t\mathbf{k}$$

Find $\omega \ne 0$ so that the sum of the object's tangential and normal components of acceleration equals half its speed.

63. A particle moves along a path given in parametric form where $r(t) = 1 + \cos at$ and $\theta(t) = e^{-at}$ (for positive constant a). Find the velocity and acceleration of the particle in terms of the unit polar vectors $\mathbf{u}_r$ and $\mathbf{u}_\theta$.

64. Basketball player Kobe Bryant attempts a shot while standing 20 ft from the basket, as shown in Figure 10.47. If Bryant shoots from a height of 7 ft and the ball reaches a maximum height of 12 ft before passing through the 10-ft-high basket, what is the initial speed of the ball?

12 ft

10 ft

7 ft

20 ft

Figure 10.47   Basketball shot

65. Sketch the graph of the vector function

$$\mathbf{R}(t) = \left(\frac{3t}{1+t^3}\right)\mathbf{i} + \left(\frac{3t^2}{1+t^3}\right)\mathbf{j}$$

then find parametric equations for the tangent line at the point where $t = 2$. This curve is called the *folium of Descartes*.

66. **Modeling Problem**   A fireman stands 5.5 m from the front of a burning building 15 m high. His fire hose discharges water at a speed of v_0 m/s from a height of 1.2 m at an angle of 62°, as shown in Figure 10.48.

A
D
B

15 m

F
62°
5.5 m

Figure 10.48   Problem 66

a. What is the smallest value of v_0 that will allow water from the hose to reach a window 11 m above the ground?

b. Suppose $v_0 = 50$ m/s. At what angle should the hose be held so the water can be directed onto a fire on the roof of the building (at a height of 15 m)?

67. A DNA molecule has the shape of a double helix (see Figure 10.4a). The radius of each helix is about 10^{-8} μm. Each helix rises about 3×10^{-8} μm during each complete turn and there are about 3×10^8 complete turns. Estimate the length of each helix.

68. Show that the tangential component of acceleration of a moving object is 0 if the object has constant speed. Is the converse statement also true? That is, if $A_T = 0$, can we conclude that the speed is constant?

69. The path of a particle P is an Archimedean spiral described in polar coordinates by $r = 10t$ and $\theta = 2\pi t$, where r is expressed in inches and t is in seconds. Determine the velocity of the particle (in terms of $\mathbf{u}_r$ and $\mathbf{u}_\theta$) when
a. $t = 0$ b. $t = 0.25$ sec

70. **Spy Problem**   After dropping the knockout gas into Scélérat's bunker (Problem 27, Section 10.3), the Spy lands and, as expected, enters a bunker full of sleeping thugs—but no Scélérat! He races through the back door of the bunker and finds himself in a ski resort. By the time he locates Scélérat, the French fiend is skiing down the mountain. The Spy grabs a pair of skis and runs over to a conveniently located ski jump. His keen mind quickly determines that the slope of the mountain is 17° and the angle at the lip of the jump is 10°, as shown in Figure 10.49.

Spy
S   10°

O   17°   V
Villain
17°

Figure 10.49   Spy scene

Suppose Scélérat is 150 ft down the mountainside skiing at 75 ft/s when the Spy launches himself from the end of the ski jump with a speed of 85 ft/s. How close will he be to Scélérat when he lands?

In Problems 71–72, find $\mathbf{F}'(x)$.

71. $\mathbf{F}(x) = \displaystyle\int_1^x [(\sin t)\mathbf{i} - (\cos 2t)\mathbf{j} + e^{-t}\mathbf{k}]\,dt$

72. $\mathbf{F}(x) = \displaystyle\int_1^{2x} [t^2\mathbf{i} + (\sec e^{-t})\mathbf{j} - (\tan e^{2t})\mathbf{k}]\,dt$

In Problems 73–76, solve the initial value problems for $\mathbf{F}$ as a vector function of t.

73. differential equation: $\mathbf{F}'(t) = t\mathbf{i} + t\mathbf{j} - t\mathbf{k}$;
    initial condition: $\mathbf{F}(0) = \mathbf{i} + 2\mathbf{j} - 3\mathbf{k}$

74. differential equation: $\mathbf{F}'(t) = \frac{3}{2}\sqrt{t+1}\,\mathbf{i} + (t+1)^{-1}\mathbf{j} + e^t\mathbf{k}$;
    initial condition: $\mathbf{F}(0) = \mathbf{j} + 2\mathbf{k}$

75. differential equation:

$$\mathbf{F}'(t) = (\sin 2t)\mathbf{i} + (e^t \cos t)\mathbf{j} - \left(\frac{3}{t+1}\right)\mathbf{k};$$

    initial condition: $\mathbf{F}(0) = \mathbf{i} - 3\mathbf{k}$

76. differential equation: $\dfrac{d^2\mathbf{F}}{dt^2} = -32\mathbf{j}$;
    initial conditions: $\mathbf{F} = 50\mathbf{j}$ and $\dfrac{d\mathbf{F}}{dt} = 5\mathbf{i} + 5\mathbf{k}$ at $t = 0$

77. **Putnam Examination Problem**   A shell strikes an airplane flying at a height h above the ground. It is known that the shell was fired from a gun on the ground with muzzle speed v_0, but the position of the gun and its angle of elevation are both unknown. Deduce that the gun is situated within a circle whose center lies directly below the airplane and whose radius is

$$\frac{v_0}{g}\sqrt{v_0^2 - 2gh}$$

Neglect resistance of the atmosphere.

78. **Putnam Examination Problem**   A coast artillery gun can fire at any angle of elevation between 0° and 90° in a fixed vertical plane. If air resistance is neglected and the muzzle speed is constant $(v = v_0)$, determine the set H of points in the plane that can be hit. Consider only those points above the horizontal.

79. **Putnam Examination Problem**   A particle moves on a circle with center O, starting from rest at a point P and coming to rest again at a point Q, without coming to rest at any intermediate point. Prove that the acceleration vector of the particle does not vanish at any point between P and Q, and that, at some point R between P and Q, the acceleration vector points in along the radius $\overline{RO}$.

The Stimulation of Science

Howard Eves was born in Paterson, New Jersey, in 1911 and is professor emeritus of mathematics at the University of Maine. This guest essay first appeared in Great Moments in Mathematics Before 1650.* Professor Eves reminds us, "It must be remembered that a moment in history is sometimes an inspired flash and sometimes an evolution extending over a long period of time." Howard Eves's lectures (from which this guest essay was taken) are so renowned that each college and university at which he taught awarded him, at one time or another, every available honor for distinguished teaching. Howard Eves is a prolific and successful textbook author with a real love for the history of mathematics. He now lives with his wife at the retirement retreat in Lubec, Maine.

THE DIE IS CAST; I have written my book; it will be read either in the present age or by posterity, it matters not which; it may well await a reader, since God has waited six thousand years for an interpreter of his words.

——Johannes Kepler
From James R. Newman, *The World of Mathematics, Volume I*, (New York: Simon and Schuster, 1956), p. 220.

The mighty Antaeus was the giant son of Neptune (god of the sea) and Ge (goddess of the earth), and his strength was invincible so long as he remained in contact with his Mother Earth. Strangers who came to his country were forced to wrestle to the death with him, and so it chanced one day that Hercules and Antaeus came to grips with one another. But Hercules, aware of the source of Antaeus' great strength, lifted and held the giant from the earth and crushed him in the air.

There is a parable here for mathematicians. For just as Antaeus was born of and nurtured by his Mother Earth, history has shown us that all significant and lasting mathematics is born of and nurtured by the real world. As in the case of Antaeus, so long as mathematics maintains its contact with the real world, it will remain powerful. But should it be lifted too long from the solid ground of its birth into the filmy air of pure abstraction, it runs the risk of weakening. It must of necessity return, at least occasionally, to the real world for renewed strength.

Such a rejuvenation of mathematics occurred in the seventeenth century, following discoveries made by two eminent mathematicians-scientists—Galileo Galilei (1564–1642) and Johannes Kepler (1571–1630). Galileo, through a sequence of experiments started before his 25th birthday, discovered a number of basic facts concerning the motion of bodies in the earth's gravitational field, and Kepler, by 1619, had deduced all three of his famous laws of planetary motion. These achievements proved to be so influential on the development of so much of subsequent mathematics that they must be ranked as two of the GREAT MOMENTS IN MATHEMATICS. Galileo's discoveries led to the creation of the modern science of dynamics and Kepler's to the creation of modern celestial mechanics; and each of these studies, in turn, required, for their development, the creation of a new mathematical tool—the calculus—capable of dealing with change, flux, and motion.

Galileo was born in Pisa in 1564 as the son of an impoverished Florentine nobleman. After a disinterested start as a medical student, Galileo obtained parental permission to change his studies to science and mathematics, fields in which he possessed a strong natural talent. While still a medical student at the University of Pisa, he made his historically famous observation that the great pendulous lamp in the cathedral there

---

oscillated to and fro with a period independent of the size of the arc of oscillation.[†]
Later, he showed that the period of a pendulum is also independent of the weight
of the pendulum's bob. When he was 25, he accepted an appointment as professor
of mathematics at the University of Pisa. It was during this appointment that he is
alleged to have performed experiments from the leaning tower of Pisa, showing that,
contrary to the teaching of Aristotle, heavy bodies do not fall faster than light ones. By
rolling balls down inclined planes, he arrived at the law that the distance a body falls
is proportional to the square of the time of falling, in accordance with the now familiar
formula $s = \frac{1}{2}gt^2$.

Unpleasant local controversies caused Galileo to resign his chair at Pisa in 1591,
and the following year he accepted a professorship in mathematics at the University of
Padua, where there reigned an atmosphere more friendly to scientific pursuits. Here at
Padua, for nearly 18 years, Galileo continued his experiments and his teaching, achieving a widespread fame. While at Padua, he heard of the discovery, in about 1607, of
the telescope by the Dutch lens-grinder Johann Lipersheim, and he set about making
instruments of his own, producing a telescope with a magnifying power of more than
30 diameters. With this telescope he observed sunspots (contradicting Aristotle's assertion that the sun is without blemish), saw mountains on the moon, and noticed the
phases of Venus, Saturn's rings, and the four bright satellites of Jupiter (all three of
these lending credence to the Copernican theory of the solar system). Galileo's discoveries roused the opposition of the Church, and finally, in the year 1633, he was
summoned to appear before the Inquisition, and there forced to recant his scientific
findings. Not many years later the great scientist became blind. He died, a prisoner in
his own home, in 1642, the year Isaac Newton was born.

Johannes Kepler was born near Stuttgart, Germany in 1571 and commenced his
studies at the University of Tübingen with the intention of becoming a Lutheran min-
ister. Like Galileo, he found his first choice of an occupation far less congenial than
his deep interest in science, particularly astronomy, and he accordingly changed his
plans. In 1594, when in his early twenties, he accepted a lectureship at the University
of Gräz in Austria. Five years later he became assistant to the famous Danish-Swedish
astronomer Tycho Brahe, who had moved to Prague to serve as the court astronomer
to Kaiser Rudolph II. Shortly after, in 1601, Brahe suddenly died, and Kepler inherited
both his master's position and his vast collection of very accurate data on the positions
of the planets as they moved about the sky. With amazing perseverance, Kepler set out
to find, from Brahe's enormous mass of observational data, just how the planets move
in space.

It has often been remarked that almost any problem can be solved if one but continuously worries over it and works at it a sufficiently long time. As Thomas Edison
said, genius is 1% inspiration and 99% perspiration. Perhaps nowhere in the history
of science is this more clearly demonstrated than in Kepler's incredible pertinacity in
solving the problem of the motion of the planets about the sun. Thoroughly convinced
of the Copernican theory that the planets revolve in orbits about the central sun, Kepler strenuously sought to determine the nature and position of those orbits and the
manner in which the planets travel in their orbits. With Brahe's great set of observational recordings at hand, the problem became this: to obtain a pattern of motion of
the planets that would exactly agree with Brahe's observations. So dependable were
Brahe's recordings that any solution that should differ from Brahe's observed positions
by even as little as a quarter of the moon's apparent diameter must be discarded as incorrect. Kepler had, then, first to guess with his *imagination* some plausible solution
and then, with painful *perseverance*, to endure the mountain of tedious calculations
needed to confirm or reject his guess. He made hundreds of fruitless attempts and performed reams and reams of calculations, laboring with undiminished zeal and patience
for many years. Finally he solved his problem, in the form of his three famous laws of

---

[†]This is only approximately true, the approximation being very close in the case of small amplitudes of
oscillation.

planetary motion, the first two around 1609 and the third one 10 years later in 1619:

I. The planets move about the sun in elliptical orbits with the sun at one focus.
II. The radius vector joining a planet to the sun sweeps over equal areas in equal intervals of time.
III. The square of the time of one complete revolution of a planet about its orbit is proportional to the cube of the orbit's semimajor axis.

The empirical discovery of these laws from Brahe's mass of data constitutes one of the most remarkable inductions ever made in science. With justifiable pride, Kepler prefaced his *Harmony of the Worlds* of 1619 with the following outburst:

> I am writing a book for my contemporaries or—it does not matter—for posterity. It may be that my book will wait for a hundred years for a reader. Has not God waited 6000 years for an observer?

Kepler's laws of planetary motion are landmarks in the history of astronomy and mathematics, for in the effort to justify them, Isaac Newton was led to create modern celestial mechanics. It is very interesting that 1800 years after the Greeks had developed the properties of the conic sections there should occur such an illuminating practical application of them. *One never knows when a piece of pure mathematics may receive an unexpected application.*

In order to compute the areas involved in his second law, Kepler had to resort to a crude form of the integral calculus, making him one of the precursors of that calculus. Also, in his *Stereometria doliorum vinorum* (*Solid Geometry of Wine Barrels*, 1615), he applied crude integration procedures to the finding of the volumes of 93 different solids obtained by revolving arcs of conic sections about axes in their planes. Among these solids were the torus and two that he called *the apple* and *the lemon*, these latter being obtained by revolving a major and a minor arc, respectively, of a circle about the arc's chord as an axis. Kepler's interest in these matters arose when he observed some of the poor methods in use by the wine gaugers of the time. (See Group Research Project, page 269.)

It was Kepler who introduced the word *focus* (Latin for "hearth") into the geometry of conic sections. He approximated the perimeter of an ellipse of semiaxes a and b by the formula $\pi(a + b)$. He also laid down a so-called *principle of continuity*, which postulates the existence in a plane of certain ideal points and an ideal line having many of the properties of ordinary points and lines, lying at infinity. Thus he explained that a straight line can be considered as closed at infinity and that a parabola may be regarded as the limiting case of either an ellipse or a hyperbola in which one of the foci has retreated to infinity. The ideas were extended by later geometers.

Kepler was a confirmed Pythagorean, with the result that his work is often a blend of the fancifully mystical and the carefully scientific. It is sad that his personal life was made almost unendurable by a multiplicity of worldly misfortunes. An infection from smallpox when he was four years old left his eyesight much impaired. In addition to his general lifelong weakness, he spent a joyless youth; his marriage was a constant source of unhappiness; his favorite child died of smallpox; his wife went mad and died; he was expelled from his lectureship at the University of Gräz when the city fell to the Catholics; his mother was charged and imprisoned for witchcraft, and for almost a year he desperately tried to save her from the torture chamber; he himself very narrowly escaped condemnation for heterodoxy; and his stipend was always in arrears. One report says that his second marriage was even less fortunate than his first, although he took the precaution to analyze carefully the merits and demerits of 11 women before choosing the wrong one. He was forced to augment his income by casting horoscopes, and he died of a fever in 1630 at the age of 59 while on a journey to try to collect some of his long overdue salary.

Mathematical Essays

1. Write a 500-word essay on the life and mathematics of Galileo Galilei.

2. HISTORICAL QUEST
   For this Quest, explain the remark in Galileo's *Discorsi e dimonstrazioni matematiche intorno à due nuove scienze* of 1638 that "neither is the number of squares less than the totality of all numbers, nor the latter greater than the former."

GALILEO GALILEI
1564–1642

3. HISTORICAL QUEST In 1638, Galileo published his ideas about dynamics in his book *Discorsi e dimonstrazioni matematiche intorno à due nuove scienze*. In this book, he considers the following problem:

   Suppose the larger circle of Figure 10.50 has made one revolution in rolling along the straight line from *A* to *B*, so that $|\overline{AB}|$ is equal to the circumference of the large circle. Then the small circle, fixed to the large one, has also made one revolution, so that $|\overline{CD}|$ is equal to the circumference of the small circle. It follows that the *two circles have equal circumferences*.

Figure 10.50 Aristotle's wheel

This paradox had been earlier described by Aristotle and is therefore sometimes referred to as *Aristotle's wheel*. Can you explain what is going on?

4. HISTORICAL QUEST In a previous Historical Quest (#60, Section 6.2), we outlined a procedure by Johannes Kepler for finding the volume of a torus.

   This Quest asks you to use Kepler's reasoning to find the area of an ellipse.

   a. Kepler divided a circle of radius *r* and circumference *C* into a large number of very thin sectors. By regarding each sector as a thin isosce-

JOHANNES KEPLER
1571–1630

les triangle of altitude *r* and base equal to the arc of the sector, he heuristically arrived at the formula $A = \frac{1}{2}rC$ for the area of a circle. Show how this was done.

   b. Use Kepler's reasoning to obtain a volume *V* of a sphere of radius *r* and surface area *S* as $V = \frac{1}{3}rS$.

   c. Use Kepler's reasoning to show that an ellipse of semimajor axis *a* and semiminor axis *b* has area $A = \pi ab$.

5. Prove Kepler's first law. You may search the literature or the internet for help with this proof.

6. Prove Kepler's third law. You may search the literature or the internet for help with this proof.

7. Where is a planet in its orbit when its speed is the greatest? Support your response.

8. Two hypothetical planets are moving about the sun in elliptical orbits having equal major axes. The minor axis of one, however, is half that of the other. How do the periods of the two planets compare? Support your response.

9. HISTORICAL QUEST The three laws of Kepler described in this section forever changed the way we view the universe, but it was not Kepler who correctly proved these laws. Isaac Newton (Historical Quest #1, Chapter 1 mathematical essays) proved these laws from the inverse-square law of gravitation.

   In Section 10.3 we proved Kepler's second law. Write a 500-word essay on the life and mathematics of Johannes Kepler, and include, as part of your essay, where Kepler made his mistake in his proof of his second law.

10. **Book Report** "Science is that body of knowledge that describes, defines, and where possible, explains the universe . . . we think of the history of science as a history of men. . . . [History] is the story of thousands of people who contributed to the knowledge and theories that constituted the science of their eras and made the 'great leaps' possible. Many of these people were women." So begins a history of women in science entitled *Hypatia's Heritage* by Margaret Alic (Boston: Beacon Press, 1986). Read this book and write a book report.

11. Make up a word problem involving the calculus of vector-valued functions. Send your problem and its solution to

    Strauss, Bradley, and Smith
    Prentice Hall Publishing Company
    1 Lake Street
    Upper Saddle River, NJ 07458

    The best ones submitted will appear in the next edition (along with credit to the problem poser).

CHAPTERS 1–10   Cumulative Review

Calculus is sometimes divided into *single-variable calculus* and *multivariable calculus*. The end of this chapter marks the transition between these major divisions of calculus, and as such it is a good time to take a look back at what we have done. There are three main ideas in single-variable calculus. The first is the notion of *limit*, which was introduced in Chapter 2. In Chapter 3 we introduced the idea of a *derivative*, the second major topic of calculus. In Chapter 5 we introduced one of the most important and revolutionary ideas in the history of mathematics—namely, that of a Riemann sum, and *integration*.

Definition of limit   The limit statement

$$\lim_{x \to c} f(x) = L$$

means that for each $\epsilon > 0$, there corresponds a number $\delta > 0$ with the property that

$$|f(x) - L| < \epsilon \quad \text{whenever} \quad 0 < |x - c| < \delta$$

The notion of a limit is fundamental for many of the ideas of calculus, primarily in the definitions of both the derivative and integral.

Definition of derivative   The derivative of f at x is given by

$$f'(x) = \lim_{\Delta x \to 0} \frac{f(x + \Delta x) - f(x)}{\Delta x}$$

provided this limit exists. If this limit exists the function f is said to be differentiable.

There are many applications of derivative, including related rates, curve sketching, and optimization.

Definition of integral   If f is defined on the closed interval $[a, b]$, we say f is integrable on $[a, b]$ if

$$I = \lim_{\|P\| \to 0} \sum_{k=1}^{n} f(x_k^*) \Delta x_k$$

exists, where $\|P\| = \max \Delta x_k, k = 1, 2, \ldots, n$. This limit is called the definite integral of f from a to b. The definite integral is denoted by

$$I = \int_a^b f(x)\, dx$$

Table 10.3 summarizes this relationship between modeling and the Riemann sum.

In the last two chapters we considered vectors in $\mathbb{R}^2$ and $\mathbb{R}^3$, as well as vector-valued functions. We summarize these ideas in Table 10.4.

TABLE 10.3   Modeling Formulas Involving Integration

Section	Situation	Riemann Sum Model	Outcome
5.3	**Introduction** Model a quantity in terms of one or more continuous functions on a closed interval $[a, b]$.	1. Partition interval. 2. Choose a number x_k^* arbitrarily from each subinterval of width Δx_k. 3. Form a sum of the type $$R_n = \sum_{k=1}^{n} f(x_k^*)\Delta x_k.$$	The approximations improve as the norm of the partitions goes to 0. The Riemann sums approach a limiting integral. Use this integral to define and calculate what we originally wanted to measure.
5.2	**Area** Find the area under a curve defined by $y = f(x)$, where $f(x) \geq 0$, above the x-axis and bounded by the lines $x = a$ and $x = b$.	Find the sum of n rectangles. $$\sum_{k=1}^{n} \underbrace{f(x_k^*)\Delta x_k}_{\text{Area of } k\text{th rectangle}}$$	$$\text{Area} = \int_a^b f(x)\,dx$$
5.3	**Displacement and distance traveled** Find the displacement and distance traveled for an object that is moving along a line having continuous velocity $v(t)$ for each time t between $t = a$ and $t = b$.	Find the sum of n distances. $$\sum_{k=1}^{n} \underbrace{\|v[a + (k-1)\Delta t]\|\, \Delta t}_{\substack{\text{Distance traveled over} \\ k\text{th time interval}}}$$	$$\text{Distance} = \int_a^b |v(t)|\, dt$$ $$\text{Displacement} = \int_a^b v(t)\, dt$$ $$= s(b) - s(a)$$
5.7	**Average value** Find the average value of a continuous function on the interval $[a, b]$.	Find the average of a weighted sum of n values. $$\frac{1}{b-a}\sum_{k=1}^{n} f(x_k^*)\Delta x, \text{ where}$$ $$\Delta x = \frac{b-a}{n}$$	$$\begin{array}{l}\text{Average} \\ \text{value}\end{array} \frac{1}{b-a}\int_a^b f(x)\, dx$$
6.1	**Area between curves** Find the area between the curves $y = f(x)$ and $y = g(x)$.	$$\sum_{k=1}^{n} \underbrace{[f(x_k^*) - g(x_k^*)]}_{\substack{\text{Top curve minus} \\ \text{bottom curve}}} \Delta x_k$$	$$\text{Area} = \int_a^b [f(x) - g(x)]\, dx$$
6.2	**Volume by cross section** Find the volume of a solid whose cross section perpendicular to the x-axis has area $A(x)$.	$$\sum_{k=1}^{n} A(x_k^*)\Delta x_k$$	$$\text{Volume} = \int_a^b A(x)\, dx$$
6.2	**Volumes of revolution** Find the volume of the region bounded by $y = f(x)$ and $y = g(x)$ revolved about the specified axis.		
	Disks: x-axis	$$\sum_{k=1}^{n} \pi[f(x_k^*)]^2 \Delta x_k$$	$$\text{Volume} = \pi \int_a^b [f(x)]^2\, dx$$
	Washers: x-axis	$$\sum_{k=1}^{n} \pi\left([f(x_k^*)]^2 - [g(x_k^*)]^2\right) \Delta x_k$$	$$V = \pi \int_a^b \left([f(x)]^2 - [g(x)]^2\right) dx$$
	Shells: y-axis	$$\sum_{k=1}^{n} 2\pi x_k^* f(x_k^*)\Delta x_k$$	$$\text{Volume} = 2\pi \int_a^b x f(x)\, dx$$
6.4	**Arc length** Find the length of a curve $y = f(x)$ on $[a, b]$.	$$\sum_{k=1}^{n} \sqrt{1 + [f'(x_k^*)]^2}\,\Delta x_k$$	$$\text{Length} = \int_a^b \sqrt{1 + [f'(x)]^2}\, dx$$
6.4	**Area of a surface of revolution** Find the area of a surface formed by rotating the curve $y = f(x)$ about the x-axis over $[a, b]$	$$\sum_{k=1}^{n} 2\pi f(x_k^*)\Delta s_k$$	$$\text{Area} = 2\pi \int_a^b f(x)\sqrt{1 + [f'(x)]^2}\, dx$$

TABLE 10.3   (continued)

Section	Situation	Riemann Sum Model	Outcome
6.5	**Mass** Find the mass of a thin plate with constant density ρ, bounded by $y = f(x)$ and $y = g(x)$.	$\displaystyle\sum_{k=1}^{n} \rho[f(x_k^*) - g(x_k^*)]\Delta x_k$	$\displaystyle m = \rho \int_a^b [f(x) - g(x)]\,dx$
6.5	**First moment** Find the moment of a thin plate with constant density ρ bounded by $y = f(x)$ and $y = g(x)$ (about the y-axis).	First moment about the y-axis $\displaystyle\sum_{k=2}^{n} \rho x_k^*[f(x_k^*) - g(x_k^*)]\Delta x_k$ First moment about the x-axis $\displaystyle\sum_{k=1}^{n} \frac{\rho}{2}\left\{[f(x_k^*)]^2 - [g(x_k^*)]^2\right\}\Delta x_k$	$\displaystyle M_y = \rho \int_a^b x[f(x) - g(x)]\,dx$ $\displaystyle M_x = \frac{\rho}{2}\int_a^b \left\{[f(x)]^2 - [g(x)]^2\right\}\,dx$
	Centroid Find the centroid of a homogeneous lamina. (ρ constant.)		$\displaystyle (\bar{x}, \bar{y}) = \left(\frac{M_y}{m}, \frac{M_x}{m}\right)$
6.5	**Work** Find the work done when an object moves along the x-axis from a to b when a variable force $F(x)$ is applied.	$\displaystyle\sum_{k=1}^{n} F(x_k^*)\Delta x_k$	$\displaystyle \text{Work} = \int_a^b F(x)\,dx$
6.5	**Hydrostatic force** Find the total force of a fluid against one side of a vertical plate, where h is the depth.	$\displaystyle\sum_{k=1}^{n} \rho(h_k^*)L(h_k^*)\Delta h_k$	$\displaystyle F = \int_a^b \rho h L(h)\,dh$

TABLE 10.4   Summary of Velocity, Acceleration, and Curvature

Position vector	$\mathbb{R}^2$ (Plane) $\mathbf{R}(t) = x(t)\mathbf{i} + y(t)\mathbf{j}$ $\mathbb{R}^3$ (Space) $\mathbf{R}(t) = x(t)\mathbf{i} + y(t)\mathbf{j} + z(t)\mathbf{k}$ The graph is called the **trajectory** of the object's motion.
Velocity vector	$\mathbf{V}(t) = \dfrac{d\mathbf{R}}{dt} = \mathbf{R}'(t)$ $\|\mathbf{V}\| = \dfrac{ds}{dt}$ is the object's **speed**. $\dfrac{\mathbf{V}}{\|\mathbf{V}\|}$ is the unit vector in the direction of $\mathbf{V}$ or *direction of motion*.
Unit tangent and principal unit normal vectors	$\mathbf{T} = \dfrac{\mathbf{V}}{\|\mathbf{V}\|} = \dfrac{\mathbf{R}'(t)}{\|\mathbf{R}'(t)\|} = \dfrac{d\mathbf{R}}{ds}$ is the **unit tangent vector** in the direction of *motion* $\mathbf{N} = \dfrac{\mathbf{T}'(t)}{\|\mathbf{T}'(t)\|} = \dfrac{1}{\kappa}\dfrac{d\mathbf{T}}{ds}$ is the **principal unit normal vector**.
Acceleration vector	$\mathbf{A}(t) = \dfrac{d^2\mathbf{R}}{dt^2} = \dfrac{d\mathbf{V}}{dt} = A_T\mathbf{T} + A_N\mathbf{N}$ where $A_T = \dfrac{d^2 s}{dt^2} = \dfrac{\mathbf{R}' \cdot \mathbf{R}''}{\|\mathbf{R}'\|}$ is the **tangential component**, and $A_N = \kappa\left(\dfrac{ds}{dt}\right)^2 = \dfrac{\|\mathbf{R}' \times \mathbf{R}''\|}{\|\mathbf{R}'\|} = \sqrt{\|\mathbf{A}\|^2 - A_T^2}$ is the **normal component**.
Curvature	$\kappa = \left\|\dfrac{d\mathbf{T}}{ds}\right\| = \dfrac{\|\mathbf{R}' \times \mathbf{R}''\|}{\|\mathbf{R}'\|^3}$; if $y = f(x)$ then $\kappa = \dfrac{|f''(x)|}{\left[1 + [f'(x)]^2\right]^{3/2}}$
Torsion	$\tau = \left\|\dfrac{d\mathbf{B}}{ds}\right\|$

Cumulative Review Problems for Chapters 1–10

1. **WHAT DOES THIS SAY?**   In your own words, outline a procedure for integrating a given function.

2. **WHAT DOES THIS SAY?**   In your own words, outline a procedure for deciding whether a given series converges or diverges.

3. **WHAT DOES THIS SAY?**   What is a vector? What is a vector function? In your own words, discuss what is meant by vector calculus.

Evaluate the limits in Problems 4–9, or explain why the limit does not exist.

4. $\lim\limits_{x\to 1} \dfrac{7x^3 - 5x - 2}{3x^2 + 5x - 8}$

5. $\lim\limits_{k\to +\infty} \dfrac{(2x^2 - 3x + 1)(1 - 5x)}{5x^3 + 4x - 9}$

6. $\lim\limits_{x\to 0} \dfrac{\sin 3x}{\tan 2x}$

7. $\lim\limits_{x\to +\infty} \left(1 - \dfrac{2}{x}\right)^{3x}$

8. $\lim\limits_{x\to 0^+} \left(\dfrac{2x + 3}{x} - \csc x\right)$

9. $\lim\limits_{x\to 0} \dfrac{\cos x - 1}{e^x - x - 1}$

Find the derivatives in Problems 10–15.

10. $y = 3x^4 - 7\sqrt{2x} + 4$

11. $y = \sin^3 x + 2\tan x$

12. $y = \ln \sqrt[3]{x} + xe^{-2x}$

13. $y = \sin^{-1} x + 2\tan^{-1} \dfrac{1}{x}$

14. $y = \dfrac{e^{-2x}(1 - x^2)}{\sqrt[3]{x^2 - x^3 + 3x^4}}$

15. $xy^3 + x^2 e^{-y} = 4$

Find the integrals in Problems 16–21.

16. $\displaystyle\int x \ln \sqrt[3]{x}\, dx$

17. $\displaystyle\int \sin^2 x \cos^3 x\, dx$

18. $\displaystyle\int x\sqrt{16 - x}\, dx$

19. $\displaystyle\int \dfrac{dx}{1 + \cos x}$

20. $\displaystyle\int \dfrac{dx}{x^2(x^2 + 5)}$

21. $\displaystyle\int \dfrac{dx}{\sqrt{2x - x^2}}$

22. A student graphs the function $f(x) = 2x^3 + 51x^2 - 360x$ using a calculator, and the graph is shown in Figure 10.51. Is this graph sufficient to get an idea of what the function looks like? Why or why not?

Figure 10.51   Calculator graph of $f(x) = 2x^3 + 51x^2 - 360x$

23. Let f be a continuous differentiable function with $f(0) = 0$ and $-2 \le f'(x) \le 4$ for $-3 \le x \le 3$.
    a. Can $f(2)$ be negative? Give a reason for your answer.
    b. Can $f(-2)$ be negative? Give a reason for your answer.
    c. Must f have a critical point between -3 and 3?

24. Find the area bounded by the curve $y = x\sqrt{x^2 + 8}$ and the x-axis on $[-2, 1]$.

25. Find the area bounded by the curves $y = \sin x$ and $y = \cos x$ and the lines $x = 0$ and $x = 1$.

26. Find the area of the region inside the circle $r = 4$ and to the right of the line $r = 2\sec\theta$.

27. Find the volume of the solid formed by revolving about the x-axis the region bounded by the curve

$$y = \dfrac{2}{\sqrt{3x - 2}}$$

28. Find the equation of the line through $(-1, 2, 5)$ and perpendicular to the plane $2x - 3y + z = 11$ in symmetric form.

29. Find the equation of the plane satisfying the given conditions.
    a. passing through $(5, 1, 2)$, $(3, 1, -2)$, and $(3, 2, 5)$
    b. passing through $(-2, -1, 4)$ and perpendicular to the line
    $$\dfrac{x}{2} = \dfrac{y + 1}{5} = \dfrac{3 - z}{2}$$

Test the series in Problems 30–35 for convergence.

30. $\displaystyle\sum_{k=0}^{\infty} \dfrac{k^3}{k^4 + 2}$

31. $\displaystyle\sum_{k=1}^{\infty} \dfrac{1}{k \cdot 4^k}$

32. $\displaystyle\sum_{k=0}^{\infty} \dfrac{3k^2 - 7k + 2}{(2k - 1)(k + 3)}$

33. $\displaystyle\sum_{k=1}^{\infty} \dfrac{k!}{2^k \cdot k}$

34. $\displaystyle\sum_{k=0}^{\infty} \dfrac{(-1)^{k+1} k}{k^2 + k - 1}$

35. $1 + \dfrac{1}{8} - \dfrac{1}{27} - \dfrac{1}{64} + \dfrac{1}{125} + \dfrac{1}{216} - - + + \cdots$

36. In each case, find the sum of the convergent series
    a. $\displaystyle\sum_{k=1}^{\infty} \dfrac{2^{k-1}}{5^{k+3}}$
    b. $\displaystyle\sum_{k=1}^{\infty} \dfrac{1}{(3k - 1)(3k + 2)}$

37. In each case, determine whether the given improper integral converges, and if it does, find its value.
    a. $\displaystyle\int_1^{\infty} x^2 e^{-x}\, dx$
    b. $\displaystyle\int_0^2 \dfrac{dx}{\sqrt{4 - x^2}}$

38. Find $\mathbf{v} \cdot \mathbf{w}$ and $\mathbf{v} \times \mathbf{w}$ for the vectors $\mathbf{v} = 3\mathbf{i} - 2\mathbf{j} + 5\mathbf{k}$ and $\mathbf{w} = \mathbf{i} - 3\mathbf{j} - \mathbf{k}$.

39. Find $\mathbf{F}'$ and $\mathbf{F}''$ for $\mathbf{F}(t) = 2t\mathbf{i} + e^{-3t}\mathbf{j} + t^4\mathbf{k}$.

40. Find $\displaystyle\int [e^t\mathbf{i} - \mathbf{j} - t\mathbf{k}] \cdot [e^{-t}\mathbf{i} + t\mathbf{j} - \mathbf{k}]\, dt$

41. Find $\mathbf{T}$ and $\mathbf{N}$ for $\mathbf{R}(t) = 2(\sin 2t)\mathbf{i} + (2 + 2\cos 2t)\mathbf{j} + 6t\mathbf{k}$.

42. Show that the alternating series

$$S = \sum_{k=1}^{\infty} \dfrac{(-1)^{k+1}}{\sqrt{k}}$$

converges and determine what N must be to guarantee that the partial sum

$$S_N = \sum_{k=1}^{N} \dfrac{(-1)^{k+1}}{\sqrt{k}}$$

approximates the entire sum S with four-decimal-place accuracy.

43. a. Find the Maclaurin series representation for
    $$f(x) = x^2 e^{-x^2}.$$
    b. Use the representation in **a** to approximate

    $$\int_0^1 x^2 e^{-x^2}\, dx$$

    with three-decimal-place accuracy.

44. Let $P(x) = 7 - 3(x-4) + 5(x-4)^2 - 2(x-4)^3 + 6(x-4)^4$ be the fourth-degree Taylor polynomial for the function f about 4.

   a. Find $f(4)$ and $f'''(4)$.

   b. Use the third-degree Taylor polynomial for f' about $x = 4$ to approximate $f'(4.2)$.

   c. Write a fifth-degree Taylor polynomial for

   $$F(x) = \int_4^x f(t)\, dt \text{ about } 4.$$

45. Solve the first-order linear differential equation

   $$\frac{dy}{dx} + 2y = x^2$$

   subject to the condition $y = 2$ when $x = 0$.

46. During the time period from $t = 0$ to $t = 6$ seconds, a particle moves along the path given by $x(t) = 3\cos(\pi t)$, $y(t) = 5\sin(\pi t)$.

   a. What is the position of the particle when $t = 1.5$?

   b. Graph the path of the particle from $t = 0$ to $t = 6$. Show direction.

   c. How many times does the particle pass through the point found in part **a**?

   d. Find the velocity vector for the particle at any time t.

   e. What is the distance traveled by the particle from $t = 1.25$ to $t = 1.75$.

47. Find an equation in terms of x and y for the line tangent to the curve given by

   $$x = t^2 - 2t - 1, \quad y = t^4 - 4t^2 + 2$$

   at the point where $t = 1$.

48. A particle moves in space with position vector

   $$\mathbf{R}(t) = (\sin t)\mathbf{i} - (\cos t)\mathbf{j} + \mathbf{k}$$

   a. Find the velocity and acceleration vectors for the particle and find its speed.

   b. Find a Cartesian equation for the particle's trajectory.

   c. Find the curvature κ and the tangential and normal components of acceleration for the particle's motion.

49. A mason lifts a 25-lb bucket of mortar from ground level to the top of a 50-ft building using a rope that weighs 0.25 lb/ft. Find the work done in lifting the bucket.

50. Modeling Problem At what speed must a satellite travel to maintain a circular orbit 1,000 mi above the surface of the earth? Assume the earth is a sphere of radius 4,000 mi and that GM in Newton's law of universal gravitation is approximately $9.56 \times 10^4 \text{ mi}^3/\text{s}^2$.

Appendices

Introduction to the Theory of Limits

In Section 2.1 we defined the limit of a function as follows:
The notation

$$\lim_{x \to c} f(x) = L$$

is read "the limit of $f(x)$ as x approaches c is L" and means that the function values $f(x)$ can be made arbitrarily close to L by choosing x sufficiently close to c but not equal to c.

This informal definition was valuable because it gave you an intuitive feeling for the limit of a function and allowed you to develop a working knowledge of this fundamental concept. For theoretical work, however, this definition will not suffice, because it gives no precise, quantifiable meaning to the terms "arbitrarily close to L" and "sufficiently close to c." The following definition, derived from the work of Cauchy and Weierstrass, gives precision to the limit definition and was also first stated in Section 2.1.

**Limit of a Function
(Formal definition)**

> The limit statement
>
> $$\lim_{x \to c} f(x) = L$$
>
> means that for each number $\epsilon > 0$, there corresponds a number $\delta > 0$ with the property that
>
> $$|f(x) - L| < \epsilon \qquad \text{whenever} \qquad 0 < |x - c| < \delta$$

Figure A.1 The epsilon-delta definition of limit

Behind the formal language is a fairly straightforward idea. In particular, to establish a specific limit, say $\lim_{x \to c} f(x) = L$, a number $\epsilon > 0$ is chosen first to establish a desired degree of proximity to L, and then a number $\delta > 0$ is found that determines how close x must be to c to ensure that $f(x)$ is within ϵ units of L.

The situation is illustrated in Figure A.1, which shows a function that satisfies the conditions of the definition. Notice that whenever x is within δ units of c (but not equal to c), the point $(x, f(x))$ on the graph of f must lie in the rectangle (shaded region) formed by the intersection of the horizontal band of width 2ϵ centered at L and the vertical band of width 2δ centered at c. The smaller the ϵ-interval around the proposed limit L, generally the smaller the δ-interval will need to be in order for $f(x)$ to lie in the ϵ-interval. If such a δ can be found no matter how small ϵ is, then L must be the limit.

THE BELIEVER/DOUBTER FORMAT

The limit process can be thought of as a "contest" between a "believer" who claims that $\lim\limits_{x \to c} f(x) = L$ and a "doubter" who disputes this claim. The contest begins with the doubter choosing a positive number ϵ and the believer countering with a positive number δ.

When will the believer win the argument and when will the doubter win? As you can see from Figure A.1, if the believer has the "correct limit" L, then no matter how the doubter chooses ϵ, the believer can find a δ so that the graph of $y = f(x)$ for $c - \delta < x < c + \delta$ will lie inside the heavily shaded rectangle in Figure A.1 (with the possible exception of $x = c$); that is, inside the rectangular region

$$R: \quad \text{width: } [c - \delta, c + \delta]; \quad \text{height: } [L - \epsilon, L + \epsilon]$$

On the other hand, if the believer tries to defend an incorrect value w (for "wrong") as the limit, then it will be possible for the doubter to choose an ϵ so that at least part of the curve $y = f(x)$ will lie outside the heavily shaded rectangle R, no matter what value of δ is chosen by the believer (see Figure A.2).

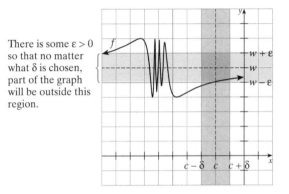

There is some $\epsilon > 0$ so that no matter what δ is chosen, part of the graph will be outside this region.

Figure A.2   False limit scenario

To avoid an endless chain of ϵ-δ challenges, the believer usually tries to settle the issue by producing a formula relating the choice of δ to the doubter's ϵ that will satisfy the requirement no matter what ϵ is chosen. The believer/doubter format is used in Example 1 to verify a given limit statement (believer "wins") and in Example 2 to show that a certain limit does not exist (doubter "wins").

EXAMPLE 1   Verifying a limit claim (believer wins)

Show that $\lim\limits_{x \to 2}(2x + 1) = 5$.

Solution

Let $f(x) = 2x + 1$. To verify the given limit statement, we begin by having the doubter choose a positive number ϵ. To help the believer, choose $\delta > 0$, and note the computation in the following box.

$$|(2x + 1) - 5| = |2x - 4| = 2|x - 2|$$
Thus, if $0 < |x - c| < \delta$ or $0 < |x - 2| < \delta$, then
$$|(2x + 1) - 5| < 2\delta$$

The believer *wants* $|f(x) - L| < \epsilon$, so we see that the believer should choose $2\delta = \epsilon$, or $\delta = \frac{\epsilon}{2}$.

With the information shown in this box, we can now make the following argument, which uses the formal definition of limit:

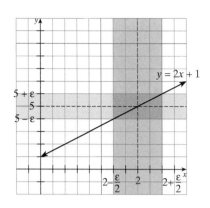

Figure A.3 Verifying $\lim\limits_{x \to 2}(2x + 1) = 5$:
Given $\epsilon > 0$, choose $\delta = \dfrac{\epsilon}{2}$.

Let $\epsilon > 0$ be given. The believer chooses $\delta = \frac{\epsilon}{2} > 0$ so that

$$|f(x) - L| = |(2x + 1) - 5| < 2\delta = 2\left(\frac{\epsilon}{2}\right) = \epsilon$$

whenever $0 < |x - 2| < \delta$, and the limit statement is verified. The graphical relationship between ϵ and δ is shown in Figure A.3. Notice that no matter what ϵ is chosen by the doubter, by choosing a number that is one-half of that ϵ, the believer will force the function to stay within the shaded portion of the graph, as shown in Figure A.3. ■

EXAMPLE 2 Disproving a limit claim (doubter wins)

Determine whether $\lim\limits_{x \to 2} \dfrac{2x^2 - 3x - 2}{x - 2} = 6$.

Solution

Let $f(x) = \dfrac{2x^2 - 3x - 2}{x - 2} = \dfrac{(2x + 1)(x - 2)}{x - 2} = 2x + 1$, $x \neq 2$. Once again, the doubter will choose a positive number ϵ and the believer must respond with a δ. As before, the believer does some preliminary work with $f(x) - L$:

$$\left| \frac{2x^2 - 3x - 2}{x - 2} - 6 \right| = \left| \frac{2x^2 - 3x - 2 - 6x + 12}{x - 2} \right|$$

$$= \left| \frac{2x^2 - 9x + 10}{x - 2} \right|$$

$$= \left| \frac{(2x - 5)(x - 2)}{x - 2} \right|$$

$$= |2x - 5|$$

The believer wants to write this expression in terms of $x - c = x - 2$. This example does not seem to "fall into place" as did Example 1. Let's analyze the situation shown in Figure A.4.

The doubter observes that if $\epsilon > 0$ is small enough, at least part of the line $y = 2x + 1$ for $0 < |x - 2| < \delta$ lies outside the shaded rectangle

$$R: \quad \text{width:} \quad [2 - \delta, 2 + \delta]; \quad \text{height:} \quad [6 - \epsilon, 6 + \epsilon]$$

regardless of the value of $\delta > 0$. For instance, suppose $\epsilon < 1$. Then, if δ is any positive number, we have

$$\left| f\left(2 - \frac{\delta}{2}\right) - 6 \right| = \left| \left[2\left(2 - \frac{\delta}{2}\right) + 1 \right] - 6 \right| = |-\delta - 1| > 1$$

so $|f(x) - 6| < \epsilon$ is not satisfied for all $0 < |x - 2| < \delta$ (in particular, not for $x = 2 - \frac{\delta}{2}$). Thus, $\lim\limits_{x \to 2} \dfrac{2x^2 - 3x - 2}{x - 2} = 6$ is false. ■

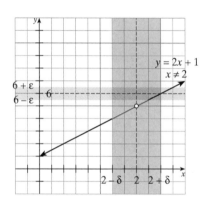

Figure A.4 Example of an incorrect limit statement

The believer/doubter format is a useful device for dramatizing the way certain choices are made in epsilon-delta arguments, but it is customary to be less "chatty" in formal mathematical proofs.

EXAMPLE 3 An epsilon-delta proof of a limit of a rational function

Show that $\lim\limits_{x \to 2} \dfrac{x^2 - 2x + 2}{x - 4} = -1$.

Solution

Let $f(x) = \dfrac{x^2 - 2x + 2}{x - 4}$. We have

$$|f(x) - L| = \left| \frac{x^2 - 2x + 2}{x - 4} - (-1) \right|$$

$$= \left| \frac{x^2 - x - 2}{x - 4} \right|$$

$$= \underbrace{|x - 2| \left| \frac{x + 1}{x - 4} \right|}$$

This must be less than the given ϵ
whenever x is near 2, but not equal to 2.

Certainly $|x - 2|$ is small if x is near 2, and the factor $\left| \dfrac{x + 1}{x - 4} \right|$ is not large (it is close to $\frac{3}{2}$). Note that if $|x - 2|$ is small it is reasonable to assume

$$|x - 2| < 1 \quad \text{so that} \quad 1 < x < 3$$

Let $g(x) = \dfrac{x + 1}{x - 4} = 1 + \dfrac{5}{x - 4}$; $g'(x) = -\dfrac{5}{(x - 4)^2} < 0$ on $(1, 3)$. Thus $g(x)$ is decreasing on $(1, 3)$ and

$$g(3) = \frac{4}{-1} < \frac{x + 1}{x - 4} < \frac{2}{-3} = g(1)$$

Hence, if $|x - 2| < 1$, then

$$-4 < \frac{x + 1}{x - 4} < 4 \qquad \text{Because } -\frac{2}{3} < 4$$

and

$$\left| \frac{x + 1}{x - 4} \right| < 4$$

Now let $\epsilon > 0$ be given. If simultaneously

$$|x - 2| < \frac{\epsilon}{4} \quad \text{and} \quad \left| \frac{x + 1}{x - 4} \right| < 4$$

then

$$|f(x) - L| = |x - 2| \left| \frac{x + 1}{x - 4} \right| < \frac{\epsilon}{4}(4) = \epsilon$$

Thus, we have only to take δ to be the smaller of the two numbers 1 and $\frac{\epsilon}{4}$ in order to guarantee that

$$|f(x) - L| < \epsilon$$

That is, given $\epsilon > 0$, choose δ to be the smaller of the numbers 1 and $\frac{\epsilon}{4}$. We write this as $\delta = \min\left(1, \frac{\epsilon}{4}\right)$. We can confirm this result by looking at the graph of $y = f(x)$ in Figure A.5). ∎

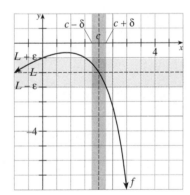

Figure A.5 $\displaystyle \lim_{x \to 2} \frac{x^2 - 2x + 2}{x - 4} = -1$

EXAMPLE 4   An epsilon-delta proof that a limit does not exist

Show that $\displaystyle \lim_{x \to 0} \frac{1}{x}$ does not exist.

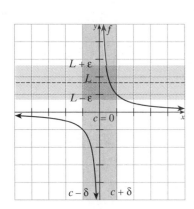

Figure A.6 $\lim\limits_{x\to 0}\dfrac{1}{x}$

Solution

Let $f(x)=\dfrac{1}{x}$ and L be any number. Suppose that $\lim\limits_{x\to 0}f(x)=L$. Look at the graph of f, as shown in Figure A.6. It would seem that no matter what value of ϵ is chosen, it would be impossible to find a corresponding δ. Consider the absolute value expression required by the definition of limit: If

$$|f(x)-L|<\epsilon, \qquad \text{or, for this example,} \qquad \left|\frac{1}{x}-L\right|<\epsilon$$

then

$$-\epsilon<\frac{1}{x}-L<\epsilon \qquad \begin{array}{l}\textit{Property of absolute value}\\ \textit{(Table 1.1, p. 3)}\end{array}$$

and

$$L-\epsilon<\frac{1}{x}<L+\epsilon$$

If $\epsilon=1$ (not a particularly small ϵ), then

$$\left|\frac{1}{x}\right|<|L|+1$$

$$|x|>\frac{1}{|L|+1}$$

and thus x is not close to zero which proves (since L was chosen arbitrarily) that $\lim\limits_{x\to 0}\dfrac{1}{x}$ does not exist. In other words, since $|x|$ can be chosen so that $0<|x|<\dfrac{1}{|L|+1}$, then $\dfrac{1}{|x|}$ will be very large, and it will be impossible to squeeze $\dfrac{1}{x}$ between $L-\epsilon$ and $L+\epsilon$ for any L. ∎

SELECTED THEOREMS WITH FORMAL PROOFS

Next, we will prove two theoretical results using the formal definition of the limit. These two theorems are useful tools in the development of calculus. The first states that the points on a graph that are on or above the x-axis cannot possibly "tend toward" a point *below* the axis, as shown in Figure A.7.

If $f(x)\geq 0$ for all x near c, then

$$\lim_{x\to c}f(x)\geq 0.$$

Figure A.7   Limit limitation theorem

THEOREM A.1   Limit limitation theorem

Suppose $\lim\limits_{x\to c}f(x)$ exists and $f(x)\geq 0$ throughout an open interval containing the number c, except possibly at c itself. Then

$$\lim_{x\to c}f(x)\geq 0$$

Proof Let $L = \lim\limits_{x \to c} f(x)$. To show that $L \geq 0$, assume the contrary; that is, assume $L < 0$. According to the definition of limit (with $\epsilon = -L$), there is a number $\delta > 0$ such that

$$|f(x) - L| < -L \qquad \text{whenever} \qquad 0 < |x - c| < \delta$$

Thus,

$$f(x) - L < -L$$
$$f(x) < 0$$

whenever $0 < |x - c| < \delta$, which contradicts the hypothesis that $f(x) \geq 0$ throughout an open interval containing c (with the possible exception of $x = c$). The contradiction forces us to reject the assumption that $L < 0$, so $L \geq 0$, as required. ❑

WARNING It may seem reasonable to conjecture that if $f(x) > 0$ throughout an open interval containing c, then $\lim\limits_{x \to c} f(x) > 0$. This is not necessarily true, and the most that can be said in this situation is that $\lim\limits_{x \to c} f(x) \geq 0$, if the limit exists. For example, if

$$f(x) = \begin{cases} x^2 & \text{for } x \neq 0 \\ 1 & \text{for } x = 0 \end{cases}$$

then $f(x) > 0$ for all x, but $\lim\limits_{x \to 0} f(x) = 0$.

   Useful information about the limit of a given function f can often be obtained by examining other functions that bound f from above and below. For example, in Section 2.1 we discovered

$$\lim_{x \to 0} \frac{\sin x}{x} = 1$$

by using a table. We justified the limit statement in Section 2.2 by using a geometric argument to show that

$$\cos x \leq \frac{\sin x}{x} \leq 1$$

for all x near 0 and then noting that since $\cos x$ and 1 both tend toward 1 as x approaches 0, the function

$$\frac{\sin x}{x}$$

which is "squeezed" between them, must converge to 1 as well. Theorem A.2 provides the theoretical basis for this method of proof.

THEOREM A.2 The squeeze rule

If $g(x) \leq f(x) \leq h(x)$ for all x in an open interval containing c (except possibly at c itself) and if

$$\lim_{x \to c} g(x) = \lim_{x \to c} h(x) = L$$

then $\lim\limits_{x \to c} f(x) = L$. (This is stated, without proof, in Section 2.2.)

Proof Let $\epsilon > 0$ be given. Since $\lim\limits_{x \to c} g(x) = L$ and $\lim\limits_{x \to c} h(x) = L$, there are positive numbers δ_1 and δ_2 such that

$$|g(x) - L| < \epsilon \quad \text{and} \quad |h(x) - L| < \epsilon$$

whenever

$$0 < |x - c| < \delta_1 \quad \text{and} \quad 0 < |x - c| < \delta_2$$

respectively. Let δ be the smaller of the numbers δ_1 and δ_2. Then, if x is a number that satisfies $0 < |x - c| < \delta$, we have

$$-\epsilon < g(x) - L \leq f(x) - L \leq h(x) - L < \epsilon$$

and it follows that $|f(x) - L| < \epsilon$. Thus, $\lim\limits_{x \to c} f(x) = L$, as claimed. ❑

The geometric interpretation of the squeeze rule is shown in Figure A.8. Notice that since $g(x) \le f(x) \le h(x)$, the graph of f is "squeezed" between those of g and h in the neighborhood of c. Thus, if the bounding graphs converge to a common point P as x approaches c, then the graph of f must also converge to P as well.

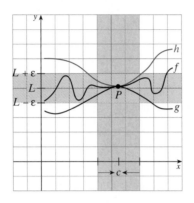

Figure A.8   The squeeze rule

A.1   PROBLEM SET

B *In Problems 1–6, use the believer/doubter format to prove or disprove the given limit statement.*

1. $\lim_{x \to 1}(2x - 5) = -3$

2. $\lim_{x \to -2}(3x + 7) = 1$

3. $\lim_{x \to 1}(3x + 1) = 5$

4. $\lim_{x \to 2}(x^2 - 2) = 5$

5. $\lim_{x \to 2}(x^2 + 2) = 6$

6. $\lim_{x \to 1}(x^2 - 3x + 2) = 0$

In Problems 7–12, use the formal definition of the limit to prove or disprove the given limit statement.

7. $\lim_{x \to 2}(x + 3) = 5$

8. $\lim_{t \to 0}(3t - 1) = 0$

9. $\lim_{x \to -2}(3x + 7) = 1$

10. $\lim_{x \to 1}(2x - 5) = -3$

11. $\lim_{x \to 2}(x^2 + 2) = 6$

12. $\lim_{x \to 2}\dfrac{1}{x} = \dfrac{1}{2}$

13. Prove that $f(x) = \begin{cases} \sin \dfrac{1}{x} & \text{if } x \ne 0 \\ 0 & \text{if } x = 0 \end{cases}$

is not continuous at $x = 0$. *Hint*: To show that $\lim_{x \to 0} f(x) \ne 0$, choose $\epsilon = 0.5$ and note that for any $\delta > 0$, there exists an x of the form $x = \dfrac{2}{\pi(2n + 1)}$ with n a natural number for which $0 < |x| < \delta$.

C *In Problems 14–19, construct a formal ϵ-δ proof to show that the given limit statement is valid for any number c.*

14. If $\lim_{x \to c} f(x)$ exists and k is a constant, then
$$\lim_{x \to c} kf(x) = k \lim_{x \to c} f(x).$$

15. If $\lim_{x \to c} f(x)$ and $\lim_{x \to c} g(x)$ both exist, then
$$\lim_{x \to c}[f(x) - g(x)] = \lim_{x \to c} f(x) - \lim_{x \to c} g(x).$$

16. If $\lim_{x \to c} f(x)$ and $\lim_{x \to c} g(x)$ both exist and a and b are constants, then $\lim_{x \to c}[af(x) + bg(x)] = a \lim_{x \to c} f(x) + b \lim_{x \to c} g(x).$

17. If $\lim_{x \to c} f(x) = 0$ and $\lim_{x \to c} g(x) = 0$, then $\lim_{x \to c} f(x)g(x) = 0$.

18. If $f(x) \ge g(x) \ge 0$ for all x and $\lim_{x \to c} f(x) = 0$, then $\lim_{x \to c} g(x) = 0$.

19. If $f(x) \ge g(x)$ for all x in an open interval containing the number c and $\lim_{x \to c} f(x)$ and $\lim_{x \to c} g(x)$ both exist, then $\lim_{x \to c} f(x) \ge \lim_{x \to c} g(x)$. *Hint*: Apply the limit limitation theorem (Theorem A.1) to the function $h(x) = f(x) - g(x)$.

Problems 20–23 lead to a proof of the product rule for limits.

20. If $\lim_{x \to c} f(x) = L$, show that $\lim_{x \to c} |f(x)| = |L|$. *Hint*: Note that $||f(x)| - |L|| \le |f(x) - L|$.

21. If $\lim_{x \to c} f(x) = L$, $L \ne 0$, show that there exists a $\delta > 0$ such that
$$\tfrac{1}{2}|L| < |f(x)| < \tfrac{3}{2}|L|$$
for all x for which $0 < |x - c| < \delta$. *Hint*: Use $\epsilon = \tfrac{1}{2}|L|$ in Problem 20.

22. If $\lim_{x \to c} f(x) = L$ and $L \ne 0$, show that $\lim_{x \to c}[f(x)]^2 = L^2$ by completing these steps:

a. Use Problem 21 to show that there exists a $\delta_1 > 0$ so that
$$|f(x) + L| < \tfrac{5}{2}|L|$$
whenever $0 < |x - c| < \delta_1$.

b. Given $\epsilon > 0$, show that there exists a $\delta_2 > 0$ such that
$$\left|[f(x)]^2 - L^2\right| < \tfrac{5}{2}|L|\epsilon$$
whenever $0 < |x - c| < \delta_2$.

c. Complete the proof that

$$\lim_{x \to c} [f(x)]^2 = L^2$$

23. Prove the product rule for limits: If $\lim_{x \to c} f(x) = L$ and $\lim_{x \to c} g(x) = M$, then $\lim_{x \to c} f(x)g(x) = LM$. *Hint*: Use the result of Problem 22 along with the identity

$$fg = \tfrac{1}{4}[(f+g)^2 - (f-g)^2]$$

24. Show that if f is continuous at c and $f(c) > 0$, then $f(x) > 0$ throughout an open interval containing c. *Hint*: Note that $\lim_{x \to c} f(x) = f(c)$ and use $\epsilon = \tfrac{1}{2}f(c)$ in the definition of limit.

25. Show that if f is continuous at L and $\lim_{x \to c} g(x) = L$, then $\lim_{x \to c} f[g(x)] = f(L)$ by completing the following steps:

a. Explain why there exists a $\delta_1 > 0$ such that $|f(w) - f(L)| < \epsilon$ whenever $|w - L| < \delta_1$.

b. Complete the proof by setting $w = g(x)$ and using part **a.**

26. For linear functions, the relation between ϵ and δ is clear; but for complicated functions, it is not. However, the correct graph can illustrate this relationship. For the following function, illustrate not only that f is continuous at $x = 3$, but determine graphically how you would pick δ to accommodate a given ϵ. (For example, $\delta = K^{-1}\epsilon$.)

$$f(x) = \frac{x^4 - 2x^3 + 3x^2 - 5x + 2}{x - 2}$$

APPENDIX B   Selected Proofs

CHAIN RULE (Chapter 3)

Suppose f is a differentiable function of u and u is a differentiable function of x. Then, $f[u(x)]$ is a differentiable function of x and

$$\frac{df}{dx} = \frac{df}{du}\frac{du}{dx}$$

Proof   Define an auxiliary function g by

$$g(t) = \frac{f[u(x) + t] - f[u(x)]}{t} - \frac{df}{du} \quad \text{if} \quad t \neq 0 \quad \text{and} \quad g(t) = 0 \quad \text{if} \quad t = 0$$

You can verify that g is continuous at $t = 0$. Notice that for $t = \Delta u$ and $t \neq 0$,

$$g(\Delta u) = \frac{f[u(x) + \Delta u] - f[u(x)]}{\Delta u} - \frac{df}{du}$$

$$g(\Delta u) + \frac{df}{du} = \frac{f[u(x) + \Delta u] - f[u(x)]}{\Delta u}$$

$$\left[g(\Delta u) + \frac{df}{du}\right]\Delta u = f[u(x) + \Delta u] - f[u(x)]$$

We now use the definition of derivative for f.

$$\frac{df}{dx} = \lim_{\Delta x \to 0} \frac{f[u(x + \Delta x)] - f[u(x)]}{\Delta x}$$

$$= \lim_{\Delta x \to 0} \frac{f[u(x) + \Delta u] - f[u(x)]}{\Delta x} \qquad \text{Where } \Delta u = u(x + \Delta x) - u(x)$$

$$= \lim_{\Delta x \to 0} \frac{\left[g(\Delta u) + \dfrac{df}{du}\right]\Delta u}{\Delta x} \qquad \text{Substitution}$$

$$= \lim_{\Delta x \to 0} \left[g(\Delta u) + \frac{df}{du}\right]\frac{\Delta u}{\Delta x}$$

$$= \lim_{\Delta x \to 0} \left[g(\Delta u) + \frac{df}{du}\right] \lim_{\Delta x \to 0} \frac{\Delta u}{\Delta x}$$

$$= \left[\lim_{\Delta x \to 0} g(\Delta u) + \lim_{\Delta x \to 0} \frac{df}{du}\right] \lim_{\Delta x \to 0} \frac{\Delta u}{\Delta x}$$

$$= \left[0 + \frac{df}{du}\right]\frac{du}{dx} \qquad \text{Since } g \text{ is continuous at } t = 0$$

$$= \frac{df}{du}\frac{du}{dx}$$

CAUCHY'S GENERALIZED MEAN VALUE THEOREM (Chapter 4)

Let f and g be functions that are continuous on the closed interval $[a, b]$ and differentiable on the open interval (a, b). If $g(b) \neq g(a)$ and $g'(x) \neq 0$ on (a, b), then

$$\frac{f(b) - f(a)}{g(b) - g(a)} = \frac{f'(c)}{g'(c)}$$

for at least one number c between a and b.

Proof   We begin by defining a special function, just as in the proof of the MVT as presented in Section 4.2. Specifically, let

$$F(x) = f(x) - f(a) - \frac{f(b) - f(a)}{g(b) - g(a)}[g(x) - g(a)]$$

for all x in the closed interval $[a, b]$. In the proof of the MVT in Section 4.2 we show that F satisfies the hypotheses of Rolle's theorem, which means that $F'(c) = 0$ for at least one number c in (a, b). For this number c, we have

$$0 = F'(c) = f'(c) - \frac{f(b) - f(a)}{g(b) - g(a)}g'(c)$$

and the result follows from this equation.

L'HÔPITAL'S RULE*(Chapter 4)

For any number a, let f and g be functions that are differentiable on an open interval (a, b), where $g'(x) \neq 0$. If $\lim_{x \to a^+} f(x) = 0$, $\lim_{x \to a^+} g(x) = 0$, and $\lim_{x \to a^+} \frac{f'(x)}{g'(x)}$ exists, then

$$\lim_{x \to a^+} \frac{f(x)}{g(x)} = \lim_{x \to a^+} \frac{f'(x)}{g'(x)}$$

Proof   First, define auxiliary functions F and G by

$$F(x) = f(x) \qquad \text{for } a < x \leq b \text{ and } F(a) = 0$$
$$G(x) = g(x) \qquad \text{for } a < x \leq b \text{ and } G(a) = 0$$

These definitions guarantee that $F(x) = f(x)$ and $G(x) = g(x)$ for $a < x \leq b$ and that $F(a) = G(a) = 0$. Thus, if w is any number between a and b, the functions F and G are continuous on the closed interval $[a, w]$ and differentiable on the open interval (a, w). According to the Cauchy generalized mean value theorem, there exists a number t between a and w for which

$$\frac{F(w) - F(a)}{G(w) - G(a)} = \frac{F'(t)}{G'(t)}$$

$$\frac{F(w)}{G(w)} = \frac{F'(t)}{G'(t)} \qquad \text{Because } F(a) = G(a) = 0$$

$$\frac{f(w)}{g(w)} = \frac{f'(t)}{g'(t)} \qquad \text{Because } F(x) = f(x) \text{ and } G(x) = g(x)$$

*This is a special case of l'Hôpital's rule. The other cases can be found in most advanced calculus textbooks.

$$\lim_{w \to a^+} \frac{f(w)}{g(w)} = \lim_{w \to a^+} \frac{f'(t)}{g'(t)}$$

$$\lim_{w \to a^+} \frac{f(w)}{g(w)} = \lim_{t \to a^+} \frac{f'(t)}{g'(t)}$$ Because t is "trapped" between a and w

$$\lim_{x \to a^+} \frac{f(x)}{g(x)} = \lim_{x \to a^+} \frac{f'(x)}{g'(x)}$$ ❏

LIMIT COMPARISON TEST (Chapter 8)

Suppose $a_k > 0$ and $b_k > 0$ for all sufficiently large k and that

$$\lim_{k \to \infty} \frac{a_k}{b_k} = L \qquad \text{where } L \text{ is finite and positive } (0 < L < \infty)$$

Then Σa_k and Σb_k either both converge or both diverge.

Proof   Assume that $\lim\limits_{k \to \infty} \dfrac{a_k}{b_k} = L$, where $L > 0$. Using $\epsilon = \dfrac{L}{2}$ in the definition of the limit of a sequence, we see that there exists a number N so that

$$\left| \frac{a_k}{b_k} - L \right| < \frac{L}{2} \qquad \text{whenever } k > N$$

$$-\frac{L}{2} < \frac{a_k}{b_k} - L < \frac{L}{2}$$

$$\frac{L}{2} < \frac{a_k}{b_k} < \frac{3L}{2}$$

$$\frac{L}{2} b_k < a_k < \frac{3L}{2} b_k \qquad b_k > 0$$

This is true for all $k > N$. Now we can complete the proof by using the direct comparison test. Suppose Σb_k converges. Then the series $\sum \dfrac{3L}{2} b_k$ also converges, and the inequality

$$a_k < \frac{3L}{2} b_k$$

tells us that the series Σa_k must also converge since it is dominated by a convergent series.

Similarly, if Σb_k diverges, the inequality

$$0 < \frac{L}{2} b_k < a_k$$

tells us that Σa_k dominates the divergent series $\sum \dfrac{L}{2} b_k$, and it follows that Σa_k also diverges.

Thus Σa_k and Σb_k either both converge or both diverge.   ❏

TAYLOR'S THEOREM (Chapter 8)

If f and all its derivatives exist in an open interval I containing c, then for each x in I

$$f(x) = f(c) + \frac{f'(c)}{1!}(x - c) + \frac{f''(c)}{2!}(x - c)^2 + \cdots + \frac{f^{(n)}(c)}{n!}(x - c)^n + R_n(x)$$

where the remainder function $R_n(x)$ is given by

$$R_n(x) = \frac{f^{(n+1)}(z_n)}{(n + 1)!}(x - c)^{n+1}$$

for some z_n that depends on x and lies between c and x.

Proof   We will prove Taylor's theorem by showing that if f and its first $n + 1$ derivatives are defined in an open interval I containing c, then for each fixed x in I

$$f(x) = f(c) + \frac{f'(c)}{1!}(x - c) + \frac{f''(c)}{2!}(x - c)^2 + \cdots + \frac{f^{(n)}(c)}{n!}(x - c)^n + \frac{f^{(n+1)}(z)}{(n+1)!}(x - c)^{n+1}$$

where z is some number between x and c. In our proof, we will apply Cauchy's generalized mean value theorem to the auxiliary functions F and G defined for all t in I as follows:

$$F(t) = f(x) - f(t) - \frac{f'(t)}{1!}(x - t) - \cdots - \frac{f^{(n)}(t)}{n!}(x - t)^n$$

$$G(t) = \frac{(x - t)^{n+1}}{(n+1)!}$$

Note that $F(x) = G(x) = 0$, and thus Cauchy's generalized mean value theorem tells us

$$\frac{F'(z)}{G'(z)} = \frac{F(x) - F(c)}{G(x) - G(c)} = \frac{F(c)}{G(c)}$$

for some number z between x and c. Rearranging the sides of this equation and finding the derivatives gives

$$\frac{F(c)}{G(c)} = \frac{F'(z)}{G'(z)}$$

$$= \frac{\dfrac{-f^{(n+1)}(z)}{n!}(x - z)^n}{\dfrac{-(x - z)^n}{n!}}$$

$$= f^{(n+1)}(z)$$

$$F(c) = f^{(n+1)}(z)G(c)$$

$$f(x) - f(c) - \frac{f'(c)}{1!}(x - c) - \cdots - \frac{f^{(n)}(c)}{n!}(x - c)^n = f^{(n+1)}(z)\left[\frac{(x - c)^{n+1}}{(n+1)!}\right]$$

Rearranging these terms, we obtain the required equation.

$$f(x) = f(c) + \frac{f'(c)}{1!}(x - c) + \cdots + \frac{f^{(n)}(c)}{n!}(x - c)^n + f^{(n+1)}(z)\left[\frac{(x - c)^{n+1}}{(n+1)!}\right]$$

❏

SUFFICIENT CONDITION FOR DIFFERENTIABILITY (Chapter 11)

If f is a function of x and y and f, f_x, and f_y are continuous in a disk D centered at (x_0, y_0), then f is differentiable at (x_0, y_0).

Proof   If (x, y) is a point in D, we have

$$f(x, y) - f(x_0, y_0) = f(x, y) - f(x_0, y) + f(x_0, y) - f(x_0, y_0)$$

The function $f(x, y)$ with y fixed satisfies the conditions of the mean value theorem, so that

$$f(x, y) - f(x_0, y) = f_x(x_1, y)(x - x_0)$$

for some number x_1 between x and x_0, and similarly, there is a number y_1 between y and y_0 such that

$$f(x_0, y) - f(x_0, y_0) = f_y(x_0, y_1)(y - y_0)$$

Substituting these expressions we obtain:

$$f(x, y) - f(x_0, y_0) = [f(x, y) - f(x_0, y)] + [f(x_0, y) - f(x_0, y_0)]$$
$$= [f_x(x_1, y)(x - x_0)] + [f_y(x_0, y_1)(y - y_0)]$$
$$= [f_x(x_1, y)(x - x_0)] + \underbrace{f_x(x_0, y_0)(x - x_0) - f_x(x_0, y_0)(x - x_0)}_{\text{This is zero}}$$
$$+ [f_y(x_0, y_1)(y - y_0)] + \underbrace{f_y(x_0, y_0)(y - y_0) - f_y(x_0, y_0)(y - y_0)}_{\text{This is also zero}}$$
$$= f_x(x_0, y_0)(x - x_0) + f_y(x_0, y_0)(y - y_0)$$
$$+ [f_x(x_1, y) - f_x(x_0, y_0)](x - x_0)$$
$$+ [f_y(x_0, y_1) - f_y(x_0, y_0)](y - y_0)$$

Let $\epsilon_1(x, y)$ and $\epsilon_2(x, y)$ be the functions

$$\epsilon_1(x, y) = f_x(x_1, y) - f_x(x_0, y_0) \quad \text{and} \quad \epsilon_2(x, y) = f_y(x_0, y_1) - f_y(x_0, y_0)$$

Then since x_1 is between x and x_0, and y_1 is between y and y_0, and the partial derivatives f_x and f_y are continuous at (x_0, y_0), we have

$$\lim_{(x,y)\to(x_0,y_0)} \epsilon_1(x, y) = \lim_{(x,y)\to(x_0,y_0)} [f_x(x_1, y) - f_x(x_0, y_0)] = 0$$

$$\lim_{(x,y)\to(x_0,y_0)} \epsilon_2(x, y) = \lim_{(x,y)\to(x_0,y_0)} [f_y(x_0, y_1) - f_y(x_0, y_0)] = 0$$

so that f is differentiable at (x_0, y_0), as required. ❏

CHANGE OF VARIABLES FORMULA FOR MULTIPLE INTEGRATION (Chapter 12)

Suppose f is a continuous function of a region D in $\mathbb{R}^2$ and let D^* be the image of the domain D under the change of variables $x = g(u, v)$, $y = h(u, v)$ where g and h are continuously differentiable on D^*. Then

$$\iint_D f(x, y)\,dy\,dx = \iint_{D^*} f[g(u, v), h(u, v)] \underbrace{\left|\frac{\partial(x, y)}{\partial(u, v)}\right|}_{\text{absolute value of Jacobian}} dv\,du$$

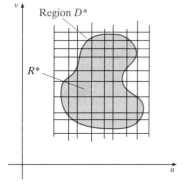

a. A rectangular grid in the uv-plane.

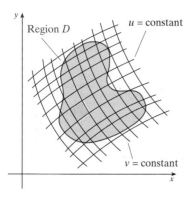

b. Corresponding grid in the xy-plane is curvilinear.

Figure B.1

Proof A proof of this theorem is found in advanced calculus texts, but we can provide a geometric argument that makes this formula plausible in the special case where $f(x, y) = 1$. In particular, we shall show that in order to find the area of a region D in the xy-plane using the change of variables $x = X(u, v)$ and $y = Y(u, v)$, it is reasonable to use the formula for area, A:

$$A = \iint_D dy\,dx = \iint_{D^*} \left|\frac{\partial(x, y)}{\partial(u, v)}\right| dv\,du$$

where D^* is the region in the uv-plane that corresponds to D.

Suppose the given change of variables has an inverse $u = u(x, y)$, $v = v(x, y)$ that transforms the region D in the xy-plane into a region D^* in the uv-plane. To find the area of D^* in the uv-coordinate system, it is natural to use a rectangular grid, with vertical lines $u = $ constant and horizontal lines $v = $ constant, as shown in Figure B.1. In the xy-plane, the equations $u = $ constant and $v = $ constant will be families of parallel curves, which provide a curvilinear grid for the region D (Figure B.1).

Next, let R^* be a typical rectangular cell in the uv-grid that covers D^*, and let R be the corresponding set in the xy-plane (that is, R^* is the image of R under the given change of variables). Then, as shown in Figure B.2, if R^* has vertices $A(\overline{u}, \overline{v})$,

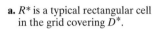

a. R^* is a typical rectangular cell in the grid covering D^*.

b. The image of R under the change of variables is a curvilinear rectangle R.

Figure B.2

$B(\overline{u} + \Delta u, \overline{v})$, $C(\overline{u}, \overline{v} + \Delta v)$, and $D(\overline{u} + \Delta u, \overline{v} + \Delta v)$, the set R will be the interior of a curvilinear rectangle with vertices

$$A[X(\overline{u}, \overline{v}), Y(\overline{u}, \overline{v})],$$
$$B[X(\overline{u} + \Delta u, \overline{v}), Y(\overline{u} + \Delta u, \overline{v})],$$
$$C[X(\overline{u}, \overline{v} + \Delta v), Y(\overline{u}, \overline{v} + \Delta v)]$$
$$D[X(\overline{u} + \Delta u, \overline{v} + \Delta v), Y(\overline{u} + \Delta u, \overline{v} + \Delta v)]$$

Note that the curved side of R joining A to B may be approximated by the secant vector

$$\mathbf{AB} = [X(\overline{u} + \Delta u, \overline{v}) - X(\overline{u}, \overline{v})]\mathbf{i} + [Y(\overline{u} + \Delta u, \overline{v}) - Y(\overline{u}, \overline{v})]\mathbf{j}$$

and by applying the mean value theorem, we find that

$$\mathbf{AB} = \left[\frac{\partial x}{\partial u}(a, \overline{v})\Delta u\right]\mathbf{i} + \left[\frac{\partial y}{\partial u}(b, \overline{v})\Delta u\right]\mathbf{j}$$

for some numbers a, b between $\overline{u}$ and $\overline{u} + \Delta u$. If Δu is very small, a and b are approximately the same as $\overline{u}$, and we can approximate $\mathbf{AB}$ by the vector

$$\boldsymbol{\alpha} = \left[\frac{\partial x}{\partial u}\Delta u\right]\mathbf{i} + \left[\frac{\partial y}{\partial u}\Delta u\right]\mathbf{j}$$

where the partials are evaluated at the point $(\overline{u}, \overline{v})$. Similarly, the curved side of R joining A and C may be approximated by the vector

$$\boldsymbol{\beta} = \left[\frac{\partial x}{\partial v}\Delta v\right]\mathbf{i} + \left[\frac{\partial y}{\partial v}\Delta v\right]\mathbf{j}$$

The area of the curvilinear rectangle R is approximately the same as that of the parallelogram determined by $\boldsymbol{\alpha}$ and $\boldsymbol{\beta}$; that is,

$$\|\boldsymbol{\alpha} \times \boldsymbol{\beta}\| = \begin{vmatrix} \mathbf{i} & \mathbf{j} & \mathbf{k} \\ \dfrac{\partial x}{\partial u}\Delta u & \dfrac{\partial y}{\partial u}\Delta u & 0 \\ \dfrac{\partial x}{\partial v}\Delta v & \dfrac{\partial y}{\partial v}\Delta v & 0 \end{vmatrix}$$

$$= \left\| \begin{vmatrix} \mathbf{i} & \mathbf{j} & \mathbf{k} \\ \dfrac{\partial x}{\partial u} & \dfrac{\partial y}{\partial u} & 0 \\ \dfrac{\partial x}{\partial v} & \dfrac{\partial y}{\partial v} & 0 \end{vmatrix} \Delta v \Delta u \right\|$$

$$= \left| \frac{\partial x}{\partial u} \frac{\partial y}{\partial v} - \frac{\partial y}{\partial u} \frac{\partial x}{\partial v} \right| \Delta v \Delta u$$

$$= \left| \frac{\partial(x, y)}{\partial(u, v)} \right| \Delta v \Delta u$$

By adding the contributions of all cells in the partition of D, we can approximate the area of D as follows:

$$\text{APPROXIMATE AREA OF } D = \sum \begin{array}{c} \text{APPROXIMATE AREA OF} \\ \text{CURVILINEAR RECTANGLES} \end{array}$$

$$= \sum \left| \frac{\partial(x, y)}{\partial(u, v)} \right| \Delta v \Delta u$$

Finally, using a limit to "smooth out" the approximation, we find

$$A = \iint_D dy\, dx = \lim \sum \left| \frac{\partial(x, y)}{\partial(u, v)} \right| \Delta v \Delta u = \iint_{D^*} \left| \frac{\partial(x, y)}{\partial(u, v)} \right| dv\, du \qquad \square$$

STOKES' THEOREM (Chapter 13)

Let S be an oriented surface with unit normal vector field $\mathbf{N}$, and assume that S is bounded by a piecewise smooth Jordan curve C whose orientation is compatible with that of S. If $\mathbf{F}$ is a vector field that is continuously differentiable on S, then

$$\int_C \mathbf{F} \cdot d\mathbf{R} = \iint_S (\text{curl } \mathbf{F} \cdot \mathbf{N})\, dS$$

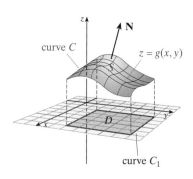

Figure B.3

Proof   The general proof requires the methods of advanced calculus. However, a proof for the case where S is a graph and $\mathbf{F}$, S and C are "well behaved" can be given. Let S be given by $z = g(x, y)$ where (x, y) is in a region D of the xy-plane. Assume g has continuous second-order partial derivatives. Let C_1 be the projection of C in the xy-plane, as shown in Figure B.3. Also let $\mathbf{F}(x, y, z) = f(x, y, z)\mathbf{i} + g(x, y, z)\mathbf{j} + h(x, y, z)\mathbf{k}$ where the partial derivatives of f, g, and h are continuous.

We will evaluate each side of Stokes's theorem separately and show that the results for each are the same. If $x = x(t)$, $y = y(t)$, and $z = z(t)$ for $a \le t \le b$ and $\mathbf{R}(t) = x(t)\mathbf{i} + y(t)\mathbf{j} + z(t)\mathbf{k}$, then

$$\int_C \mathbf{F} \cdot d\mathbf{R} = \int_a^b \left[f\frac{dx}{dt} + g\frac{dy}{dt} + h\frac{dz}{dt} \right] dt$$

$$= \int_a^b \left[f\frac{dx}{dt} + g\frac{dy}{dt} + h\left(\frac{\partial z}{\partial x}\frac{dx}{dt} + \frac{\partial z}{\partial y}\frac{dy}{dt} \right) \right] dt \qquad \text{Chain rule}$$

$$= \int_a^b \left[\left(f + h\frac{\partial z}{\partial x} \right)\frac{dx}{dt} + \left(g + h\frac{\partial z}{\partial y} \right)\frac{dy}{dt} \right] dt$$

$$= \int_{C_1} \left[\left(f + h\frac{\partial z}{\partial x} \right) dx + \left(g + h\frac{\partial z}{\partial y} \right) dy \right]$$

$$= \iint_D \left[\frac{\partial}{\partial x}\left(g + h\frac{\partial z}{\partial y} \right) - \frac{\partial}{\partial y}\left(f + h\frac{\partial z}{\partial x} \right) \right] dA \qquad \text{Green's theorem}$$

$$= \iint_D \left[\left(\frac{\partial g}{\partial x} + \frac{\partial g}{\partial z}\frac{\partial z}{\partial x} + \frac{\partial h}{\partial x}\frac{\partial z}{\partial y} + \frac{\partial h}{\partial z}\frac{\partial z}{\partial x}\frac{\partial z}{\partial y} + h\frac{\partial^2 z}{\partial y \partial x} \right) \right.$$

$$\left. - \left(\frac{\partial f}{\partial y} + \frac{\partial f}{\partial z}\frac{\partial z}{\partial y} + \frac{\partial h}{\partial y}\frac{\partial z}{\partial x} + \frac{\partial h}{\partial z}\frac{\partial z}{\partial y}\frac{\partial z}{\partial x} + h\frac{\partial^2 z}{\partial y \partial x} \right) \right] dA$$

$$= \iint_D \left[-\frac{\partial h}{\partial y}\frac{\partial z}{\partial x} + \frac{\partial g}{\partial z}\frac{\partial z}{\partial x} - \frac{\partial f}{\partial z}\frac{\partial z}{\partial y} + \frac{\partial h}{\partial x}\frac{\partial z}{\partial y} + \frac{\partial g}{\partial x} - \frac{\partial f}{\partial y} \right] dA$$

Next, we start over by evaluating the flux integral:

$$\iint\limits_{S} \text{curl } \mathbf{F} \cdot d\mathbf{S} = \iint\limits_{D} \left\langle -\left(\frac{\partial h}{\partial y} - \frac{\partial g}{\partial z}\right), \left(\frac{\partial f}{\partial z} - \frac{\partial h}{\partial x}\right), \left(\frac{\partial g}{\partial x} - \frac{\partial f}{\partial y}\right) \right\rangle \cdot \left\langle \frac{-\partial z}{\partial x}, \frac{-\partial z}{\partial y}, 1 \right\rangle dA$$

$$= \iint\limits_{D} \left[-\frac{\partial h}{\partial y}\frac{\partial z}{\partial x} + \frac{\partial g}{\partial z}\frac{\partial z}{\partial x} - \frac{\partial f}{\partial z}\frac{\partial z}{\partial y} + \frac{\partial h}{\partial x}\frac{\partial z}{\partial y} + \frac{\partial g}{\partial x} - \frac{\partial f}{\partial y} \right] dA$$

Since these results are the same, we have

$$\int_{C} \mathbf{F} \cdot d\mathbf{R} = \iint\limits_{S} (\text{curl } \mathbf{F} \cdot \mathbf{N}) \, dS \qquad \square$$

DIVERGENCE THEOREM (Chapter 13)

Let D be a region in space bounded by a smooth, orientable closed surface S. If $\mathbf{F}$ is a continuous vector field whose components have continuous partial derivatives in D, then

$$\iint\limits_{S} \mathbf{F} \cdot \mathbf{N} \, dS = \iiint\limits_{D} \text{div } \mathbf{F} \, dV$$

where $\mathbf{N}$ is an outward unit normal to the surface S.

Proof   Let $\mathbf{F}(x, y, z) = f(x, y, z)\mathbf{i} + g(x, y, z)\mathbf{j} + h(x, y, z)\mathbf{k}$. If we state the divergence theorem using this notation for $\mathbf{F}$, we have

$$\iint\limits_{S} [f(\mathbf{i} \cdot \mathbf{N}) + g(\mathbf{j} \cdot \mathbf{N}) + h(\mathbf{k} \cdot \mathbf{N})] \, dS = \iiint\limits_{D} \left(\frac{\partial f}{\partial x} + \frac{\partial g}{\partial y} + \frac{\partial h}{\partial z} \right) dV$$

$$\iint\limits_{S} f(\mathbf{i} \cdot \mathbf{N}) \, dS + \iint\limits_{S} g(\mathbf{j} \cdot \mathbf{N}) \, dS + \iint\limits_{S} h(\mathbf{k} \cdot \mathbf{N}) \, dS = \iiint\limits_{D} \frac{\partial f}{\partial x} \, dV + \iiint\limits_{D} \frac{\partial g}{\partial y} \, dV + \iiint\limits_{D} \frac{\partial h}{\partial z} \, dV$$

This result can be verified by proving

$$\iint\limits_{S} f(\mathbf{i} \cdot \mathbf{N}) \, dS = \iiint\limits_{D} \frac{\partial f}{\partial x} \, dV$$

$$\iint\limits_{S} g(\mathbf{j} \cdot \mathbf{N}) \, dS = \iiint\limits_{D} \frac{\partial g}{\partial y} \, dV$$

$$\iint\limits_{S} h(\mathbf{k} \cdot \mathbf{N}) \, dS = \iiint\limits_{D} \frac{\partial h}{\partial z} \, dV$$

Since the proof of each of these is virtually identical, we will show the verification for the last equation; the other two can be done in a similar fashion. We will evaluate this third integral by separately evaluating the left and right sides to show they are the same.

We will restrict our proof to a standard region; that is, a solid bounded above by a surface S_T with equation $z = u(x, y)$ and below by a surface S_B with equation $z = v(x, y)$. We assume that both S_T and S_B project onto the region R in the xy-plane. The lateral surface S_L of the region is the set of all (x, y, z) such that $v(x, y) \leq z \leq u(x, y)$ on the boundary of R, as shown in Figure B.4. The upward flux across the top surface S_T is

$$\iint_{S_T} h(\mathbf{k} \cdot \mathbf{N}_T) \, dS = \iint_R h\langle 0, 0, 1\rangle \cdot \left\langle -\frac{\partial u}{\partial x}, -\frac{\partial u}{\partial y}, 1 \right\rangle dA$$

$$= \iint_R h(x, y, z) \, dA$$

$$= \iint_R h(x, y, u(x, y)) \, dA$$

Similarly, the downward flux across the bottom surface S_B is

$$\iint_{S_B} h(\mathbf{k} \cdot \mathbf{N}_B) \, dS = \iint_R h\langle 0, 0, 1\rangle \cdot \left\langle \frac{\partial u}{\partial x}, \frac{\partial v}{\partial y}, -1 \right\rangle dA$$

$$= -\iint_R h(x, y, z) \, dA$$

$$= -\iint_R h(x, y, v(x, y)) \, dA$$

Because the outward unit normal $\mathbf{N}_L$ is horizontal on the lateral surface S_L, it is perpendicular to k, and

$$\iint_{S_L} h(\mathbf{k} \cdot \mathbf{N}_L) \, dS = 0$$

To find the flux integral over the entire closed surface S, add the separate flux integrals over the top, the bottom, and the sides:

$$\iint_S h(\mathbf{k} \cdot \mathbf{N}) \, dS = \iint_{S_T} h(\mathbf{k} \cdot \mathbf{N}_T) \, dS + \iint_{S_B} h(\mathbf{k} \cdot \mathbf{N}_B) \, dS + \iint_{S_L} h(\mathbf{k} \cdot \mathbf{N}_L) \, dS$$

$$= \iint_R h[x, y, u(x, y)] \, dA - \iint_R h[x, y, v(x, y)] \, dA + 0$$

$$= \iint_R \{h[x, y, u(x, y)] - h[x, y, v(x, y)]\} \, dA$$

Next, we start over by evaluating the triple integral $\iiint_D \dfrac{\partial h}{\partial z} \, dV$. Notice that we can describe the solid S as the set of all (x, y, z) for (x, y) in R and $v(x, y) \leq z \leq u(x, y)$. Thus,

$$\iiint_D \frac{\partial h}{\partial z} \, dV = \iint_R \left[\int_{v(x,y)}^{u(x,y)} \frac{\partial h}{\partial z}(x, y, z) \, dz \right] dA$$

$$= \iint_R \{h[x, y, u(x, y)] - h[x, y, v(x, y)]\} \, dA$$

$$= \iint_S h(\mathbf{k} \cdot \mathbf{N}) \, dS$$

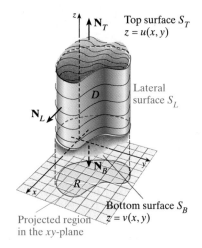

Figure B.4 A standard solid region in $\mathbb{R}^3$

By applying similar arguments to the integrals $\iiint\limits_D \frac{\partial f}{\partial x}\,dV$ and $\iiint\limits_D \frac{\partial g}{\partial y}\,dV$ and adding, we obtain

$$\iiint\limits_D \left[\frac{\partial f}{\partial x} + \frac{\partial g}{\partial y} + \frac{\partial h}{\partial z}\right] dV = \iint\limits_S [f(\mathbf{i}\cdot\mathbf{N}) + g(\mathbf{j}\cdot\mathbf{N}) + h(\mathbf{k}\cdot\mathbf{N})]\,dS$$

or

$$\iiint\limits_D \operatorname{div}\mathbf{F}\,dV = \iint\limits_S \mathbf{F}\cdot\mathbf{N}\,dS \qquad \square$$

APPENDIX C   Significant Digits

Throughout this book, various Technology Notes appear, and in the answers to the problems you will frequently find approximate (decimal) answers. Sometimes your answer may not exactly agree with the answer found in the back of the book. This does not necessarily mean that your answer is incorrect, particularly if your answer is very close to the given answer.

To use your calculator intelligently and efficiently, you should become familiar with its functions and practice doing problems with the same calculator, whenever possible. Read the Technology Notes provided throughout this text, and consult the owner's manual for your particular calculator when you have questions. In addition, there are *Technology Manuals* accompanying this text which are available for TI and HP graphic calculators, as well as for Mathematica and for Maple.

SIGNIFICANT DIGITS

Applications involving measurements can never be exact. It is particularly important that you pay attention to the accuracy of your measurements when you use a computer or calculator, because using available technology can give you a false sense of security about the accuracy in a particular problem. For example, if you measure a triangle and find the sides are approximately 1.2 and 3.4 and then find the ratio of $1.2/3.4 \approx 0.35294117$, it appears that the result is more accurate than the original measurements! Some discussion about accuracy and significant digits is necessary.

The digits known to be correct in a number obtained by a measurement are called **significant digits**. The digits 1, 2, 3, 4, 5, 6, 7, 8, and 9 are always significant, whereas the digit 0 may or may not be significant.

1. Zeros that come between two other digits are significant, as in 203 or 10.04.

2. If the zero's only function is to place the decimal point, it is not significant, as in

$$\underset{\text{Placeholders}}{0.00000}\,23 \qquad\text{or}\qquad \underset{\text{Placeholders}}{23,000}$$

If a zero does more than fix the decimal point, it is significant, as in

$$0.0023\,\underset{\substack{\uparrow\\ \text{This digit is significant.}}}{0} \qquad\text{or}\qquad 23,\underset{\text{These are significant.}}{000.01}$$

This second rule can, of course, result in certain ambiguities, such as 23,000 (measured to the exact unit). To avoid such confusion, we use scientific notation in this case:

$$2.3 \times 10^4 \quad \text{has two significant digits}$$

$$2.3000 \times 10^4 \quad \text{has five significant digits}$$

Numbers that come about by counting are considered to be exact and are correct to any number of significant digits.

When you compute an answer using a calculator, the answer may have 10 or more digits. In the Technology Notes we generally show the 10 or 12 digits that result from the numerical calculation but frequently the number in the answer section will have only 5 or 6 digits in the answer. It seems clear that if the first 3 or 4 nonzero digits of the answer coincide, you probably have the correct method of doing the problem.

However, you might ask why there are discrepancies and how many digits you should use when you write down your final answer. Roughly speaking, the significant digits in a number are the digits that have meaning. To clarify the concept, we must for the moment assume that we know the exact answer. We then assume that we have been able to compute an approximation to this exact answer.

Usually we do this by some sort of iterative process, in which the answers are getting closer and closer to the exact answer. In such a process, we hope that the number of significant digits in our approximate answer is increasing at each trial. If our approximate answer is, say, 6 digits long (some of those digits might even be zero), and the difference between our answer and the exact answer is 4 units or less in the last place, then the first 5 digits are significant.

For example, if the exact answer is 3.14159 and our approximate answer is 3.14162, then our answer has 6 significant digits but is correct to 5 significant digits. Note that saying that our answer is correct to 5 significant digits does not guarantee that all of those 5 digits exactly match the first 5 digits of the exact answer. In fact, if an exact answer is 6.001 and our computed answer is 5.997, then our answer is correct to 3 significant digits and not one of them matches the digits in the exact answer. Also note that it may be necessary for an approximation to have more digits than are actually significant for it to have a certain number of significant digits. For example, if the exact answer is 6.003 and our approximation is 5.998, then it has 3 significant digits, but only if we consider the total 4-digit number and do not strip off the last nonsignificant digit.

Again, suppose you know that all digits are significant in the number 3.456; then you know that the exact number is less than 3.4565 and at least 3.4555. Some people may say that the number 3.456 is correct to 3 decimal places. This is the same as saying that it has 4 significant digits. Why bother with significant digits? If you multiply (or divide) two numbers with the same number of significant digits, then the product will generally be at least twice as long, but will have roughly the same number of significant digits as the original factors. You can then dispense with the unneeded digits. In fact, to keep them would be misleading about the accuracy of the result.

Frequently we can make an educated guess of the number of significant digits in an answer. For example, if we compute an iterative approximation such as

$$2.3123, \quad 2.3125, \quad 2.3126, \quad 2.31261, \quad 2.31262, \ldots$$

we would generally conclude that the answer is 2.3126 to 5 significant digits. Of course, we may very well be wrong, and if we continued iterating, the answer might end up as 2.4.

ROUNDING AND RULES OF COMPUTATION USED IN THIS BOOK

In hand and calculator computations, rounding a number is done to reduce the number of digits displayed and make the number easier to comprehend. Furthermore, if you suspect that the digit in the last place is not significant, then you might be tempted to round and remove this last digit. This can lead to error. For example, if the computed value is 0.64 and the true value is known to be between 0.61 and 0.67, then the computed value has only 1 significant digit. However, if we round it to 0.6 and the true value is really 0.66, then 0.6 is not correct to even 1 significant digit. In the interest of making the text easier to read, we have used the following rounding procedure:

Rounding Procedure

To round off numbers,

1. Increase the last retained digit by 1 if the remainder is greater than or equal to 5; or
2. Retain the last digit unchanged if the remainder is less than 5.

Elaborate rules for computation of approximate data can be developed when needed (in chemistry, for example), but in this text we will use three simple rules:

Rules For Significant Digits

Addition-subtraction: Add or subtract in the usual fashion, and then round off the result so that the last digit retained is in the column farthest to the right in which both given numbers have significant digits.

Multiplication-division: Multiply or divide in the usual fashion, and then round off the results to the smaller number of significant digits found in either of the given numbers.

Counting numbers: Numbers used to count or whole numbers used as exponents are considered to be correct to any number of significant digits.

Rounding Rule   We use the following rounding procedure in problems requiring rounding by involving several steps: Round only once, at the end. *That is, do not work with rounded results, because round-off errors can accumulate. If the problem does not ask for an approximate answer, but requires a calculator result, we will show the calculator result and leave it to the reader to write the appropriate number of digits.*

CALCULATOR EXPERIMENTS

You should be aware that you are much better than your calculator at performing certain computations. For example, almost all calculators will fail to give the correct answer to

$$(10.0\ EE + 50.0) + 911.0 - (10.0\ EE + 50.0)$$

Calculators will return the value of 0, but you know at a glance that the answer is 911.0. We must reckon with this poor behavior on the part of calculators, which is called *loss of accuracy* due to *catastrophic cancellation*. In this case, it is easy to catch the error immediately, but what if the computation is so complicated (or hidden by other computations) that we do not see the error?

First, we want to point out that the order in which you perform computations can be very important. For example, most calculators will correctly conclude that

$$(10.0\ EE + 50.0) - (10.0\ EE + 50.0) + 911.0 = 911$$

There are other cases besides catastrophic cancellation where the order in which a computation is performed will substantially affect the result. For example, you may not be able to calculate

$$(10.0\ EE + 50.0) * (911.0\ EE + 73.0)/(20.0\ EE + 60.0)$$

but rearranging the factors as

$$((10.0\ EE + 50.0)/(20.0\ EE + 60.0)) * (911.0\ EE + 73.0)$$

should provide the correct answer of 4.555 EE 65. So, for what do we need to watch? If you subtract two numbers that are close to each other in magnitude, you may obtain an inaccurate result. When you have a sequence of multiplications and divisions in a string, try to arrange the factors so that the result of each intermediate calculation stays as close as possible to 1.0.

Second, since a calculator performs all computations with a finite number of digits, it is unable to do exact computations involving nonterminating decimals. This enables us to see how many digits the calculator actually uses when it computes a result. For example, the computation

$$(7.0/17.0) * (17.0)$$

should give the result 7.0, but on most calculators it does not. The size of the answer gives an indication of how many digits "Accuracy" the calculator uses internally. That is, the calculator may display decimal numbers that have 10 digits, but use 12 digits internally. If the answer to the above computation is something similar to 1.0 EE − 12, then the calculator is using 12 digits internally.

TRIGONOMETRIC EVALUATIONS

In many problems you will be asked to compute the values of trigonometric functions such as the sine, cosine, or tangent. In calculus, trigonometric arguments are usually assumed to be measured in radians. You must make sure the calculator is in radian mode. If it is in radian mode, then the sine of a small number will almost be equal to that number. For example,

$$\sin(0.00001) = 1 \, E - 5 \quad \text{(which is 0.00001)}$$

If not, then you are not using radian mode. Make sure you know how to put your calculator in radian mode.

GRAPHING BLUNDERS

When you are using the graphing features, you must always be careful to choose reasonable scales for the domain (horizontal scale) and range (vertical scale). If the scale is too large, you may not see important wiggles. If the scale is too small, you may not see important behavior elsewhere in the plane. Of course, knowing the techniques of graphing discussed in Chapter 4 will prevent you from making such blunders. Some calculators may have trouble with curves that jump suddenly at a point. An example of such a curve would be

$$y = \frac{e^x}{x}$$

which jumps at the origin. Try plotting this curve with your calculator using different horizontal and vertical scales, making sure that you understand how your calculator handles such graphs.

TECHNOLOGY NOTE

APPENDIX D   Short Table of Integrals

Each formula is numbered for easy reference. The numbers in this short table are not sequential because this short table is truncated from the table of integrals found in the *Student Mathematics Handbook*.

BASIC FORMULAS

1. Constant rule

$$\int 0 \, du = 0 + C$$

2. Power rule

$$\int u^n \, du = \frac{u^{n+1}}{n+1}; \quad n \neq -1$$

$$\int u^n \, du = \ln |u|; \quad n = -1$$

3. Natural Exponential rule

$$\int e^u \, du = e^u$$

4. Logarithmic rule

$$\int \ln |u| \, du = u \ln |u| - u$$

Trigonometric rules

5. $\displaystyle\int \sin u \, du = -\cos u$

6. $\displaystyle\int \cos u \, du = \sin u$

7. $\displaystyle\int \tan u \, du = -\ln |\cos u|$

8. $\displaystyle\int \cot u \, du = \ln |\sin u|$

9. $\displaystyle\int \sec u \, du = \ln |\sec u + \tan u|$

10. $\displaystyle\int \csc u \, du = -\ln |\csc u + \cot u|$

11. $\displaystyle\int \sec^2 u \, du = \tan u$

12. $\displaystyle\int \csc^2 u \, du = -\cot u$

13. $\displaystyle\int \sec u \tan u \, du = \sec u$

14. $\displaystyle\int \csc u \cot u \, du = -\csc u$

Exponential rule

15. $\displaystyle\int a^u \, du = \frac{a^u}{\ln a} \quad a > 0, a \neq 1$

Hyperbolic rules

16. $\displaystyle\int \cosh u \, du = \sinh u$

17. $\displaystyle\int \sinh u \, du = \cosh u$

18. $\displaystyle\int \tanh u \, du = \ln \cosh u$

19. $\displaystyle\int \coth u \, du = \ln |\sinh u|$

20. $\displaystyle\int \operatorname{sech} u \, du = \tan^{-1}(\sinh u)$
$$= 2 \tan^{-1}(e^u)$$

21. $\displaystyle\int \operatorname{csch} u \, du = \ln \left| \tanh \frac{u}{2} \right|$

Inverse rules

22. $\displaystyle\int \frac{du}{\sqrt{a^2 - u^2}} = \sin^{-1} \frac{u}{a}$

23. $\displaystyle\int \frac{du}{\sqrt{u^2 - a^2}} = \cosh^{-1} \frac{u}{a}$

24. $\displaystyle\int \frac{du}{a^2 + u^2} = \frac{1}{a} \tan^{-1} \frac{u}{a}$

25. $\displaystyle \int \frac{du}{a^2 - u^2} = \begin{cases} \dfrac{1}{a}\tanh^{-1}\dfrac{u}{a} & \text{if } \left|\dfrac{u}{a}\right| < 1 \\[4mm] \dfrac{1}{a}\coth^{-1}\dfrac{u}{a} & \text{if } \left|\dfrac{u}{a}\right| > 1 \end{cases}$

26. $\displaystyle \int \frac{du}{u\sqrt{u^2 - a^2}} = \frac{1}{a}\sec^{-1}\left|\frac{u}{a}\right|$

27. $\displaystyle \int \frac{du}{u\sqrt{a^2 - u^2}} = -\frac{1}{a}\operatorname{sech}^{-1}\left|\frac{u}{a}\right| = -\frac{1}{a}\ln\left|\frac{a + \sqrt{a^2 - u^2}}{u}\right|$

28. $\displaystyle \int \frac{du}{\sqrt{a^2 + u^2}} = \ln\left(u + \sqrt{a^2 + u^2}\right) = \sinh^{-1}\frac{u}{a}$

29. $\displaystyle \int \frac{du}{u\sqrt{a^2 + u^2}} = -\frac{1}{a}\ln\left|\frac{\sqrt{a^2 + u^2} + a}{u}\right| = -\frac{1}{a}\operatorname{csch}^{-1}\left|\frac{u}{a}\right|$

INTEGRALS INVOLVING $au + b$

30. $\displaystyle \int (au + b)^n\, du = \frac{(au + b)^{n+1}}{(n+1)a}$

31. $\displaystyle \int u(au + b)^n\, du = \frac{(au + b)^{n+2}}{(n+2)a^2} - \frac{b(au + b)^{n+1}}{(n+1)a^2}$

32. $\displaystyle \int u^2(au + b)^n\, du = \frac{(au + b)^{n+3}}{(n+3)a^3} - \frac{2b(au + b)^{n+2}}{(n+2)a^3} + \frac{b^2(au + b)^{n+1}}{(n+1)a^3}$

33. $\displaystyle \int u^m(au + b)^n\, du = \begin{cases} \dfrac{u^{m+1}(au + b)^n}{m+n+1} + \dfrac{nb}{m+n+1}\displaystyle\int u^m(au + b)^{n-1}\, du \\[5mm] \dfrac{u^m(au + b)^{n+1}}{(m+n+1)a} - \dfrac{mb}{(m+n+1)a}\displaystyle\int u^{m-1}(au + b)^n\, du \\[5mm] \dfrac{-u^{m+1}(au + b)^{n+1}}{(n+1)b} + \dfrac{m+n+2}{(n+1)b}\displaystyle\int u^m(au + b)^{n+1}\, du \end{cases}$

34. $\displaystyle \int \frac{du}{au + b} = \frac{1}{a}\ln|au + b|$

35. $\displaystyle \int \frac{u\, du}{au + b} = \frac{u}{a} - \frac{b}{a^2}\ln|au + b|$

36. $\displaystyle \int \frac{u^2\, du}{au + b} = \frac{(au + b)^2}{2a^3} - \frac{2b(au + b)}{a^3} + \frac{b^2}{a^3}\ln|au + b|$

37. $\displaystyle \int \frac{u^3\, du}{au + b} = \frac{(au + b)^3}{3a^4} - \frac{3b(au + b)^2}{2a^4} + \frac{3b^2(au + b)}{a^4} - \frac{b^3}{a^4}\ln|au + b|$

INTEGRALS INVOLVING $u^2 + a^2$

55. $\displaystyle \int \frac{du}{u^2 + a^2} = \frac{1}{a}\tan^{-1}\frac{u}{a}$

56. $\displaystyle \int \frac{u\, du}{u^2 + a^2} = \frac{1}{2}\ln(u^2 + a^2)$

57. $\displaystyle\int \frac{u^2\,du}{u^2+a^2} = u - a\tan^{-1}\frac{u}{a}$

58. $\displaystyle\int \frac{u^3\,du}{u^2+a^2} = \frac{u^2}{2} - \frac{a^2}{2}\ln(u^2+a^2)$

59. $\displaystyle\int \frac{du}{u(u^2+a^2)} = \frac{1}{2a^2}\ln\left(\frac{u^2}{u^2+a^2}\right)$

60. $\displaystyle\int \frac{du}{u^2(u^2+a^2)} = -\frac{1}{a^2u} - \frac{1}{a^3}\tan^{-1}\frac{u}{a}$

61. $\displaystyle\int \frac{du}{u^3(u^2+a^2)} = -\frac{1}{2a^2u^2} - \frac{1}{2a^4}\ln\left(\frac{u^2}{u^2+a^2}\right)$

INTEGRALS INVOLVING $u^2 - a^2, u^2 > a^2$

74. $\displaystyle\int \frac{du}{u^2-a^2} = \frac{1}{2a}\ln\left|\frac{u-a}{u+a}\right| \text{ or } -\frac{1}{a}\coth^{-1}\frac{u}{a}$

75. $\displaystyle\int \frac{u\,du}{u^2-a^2} = \frac{1}{2}\ln|u^2-a^2|$

76. $\displaystyle\int \frac{u^2\,du}{u^2-a^2} = u + \frac{a}{2}\ln\left|\frac{u-a}{u+a}\right|$

77. $\displaystyle\int \frac{u^3\,du}{u^2-a^2} = \frac{u^2}{2} + \frac{a^2}{2}\ln|u^2-a^2|$

78. $\displaystyle\int \frac{du}{u(u^2-a^2)} = \frac{1}{2a^2}\ln\left|\frac{u^2-a^2}{u^2}\right|$

79. $\displaystyle\int \frac{du}{u^2(u^2-a^2)} = \frac{1}{a^2u} + \frac{1}{2a^3}\ln\left|\frac{u-a}{u+a}\right|$

80. $\displaystyle\int \frac{du}{u^3(u^2-a^2)} = \frac{1}{2a^2u^2} - \frac{1}{2a^4}\ln\left|\frac{u^2}{u^2-a^2}\right|$

INTEGRALS INVOLVING $a^2 - u^2, u^2 < a^2$

93. $\displaystyle\int \frac{du}{a^2-u^2} = \frac{1}{2a}\ln\left|\frac{a+u}{a-u}\right| \text{ or } \frac{1}{a}\tanh^{-1}\frac{u}{a}$

94. $\displaystyle\int \frac{u\,du}{a^2-u^2} = -\frac{1}{2}\ln|a^2-u^2|$

95. $\displaystyle\int \frac{u^2\,du}{a^2-u^2} = -u + \frac{a}{2}\ln\left|\frac{a+u}{a-u}\right|$

96. $\displaystyle\int \frac{u^3\,du}{a^2-u^2} = -\frac{u^2}{2} - \frac{a^2}{2}\ln|a^2-u^2|$

97. $\displaystyle\int \frac{du}{u(a^2-u^2)} = \frac{1}{2a^2}\ln\left|\frac{u^2}{a^2-u^2}\right|$

98. $\displaystyle\int \frac{du}{u^2(a^2-u^2)} = -\frac{1}{a^2u} + \frac{1}{2a^3}\ln\left|\frac{a+u}{a-u}\right|$

99. $\displaystyle\int \frac{du}{u^3(a^2-u^2)} = -\frac{1}{2a^2u^2} + \frac{1}{2a^4}\ln\left|\frac{u^2}{a^2-u^2}\right|$

100. $\displaystyle\int \frac{du}{(a^2-u^2)^2} = \frac{u}{2a^2(a^2-u^2)} + \frac{1}{4a^3}\ln\left|\frac{a+u}{a-u}\right|$

101. $\displaystyle\int \frac{u\,du}{(a^2-u^2)^2} = \frac{1}{2(a^2-u^2)}$

102. $\displaystyle\int \frac{u^2\,du}{(a^2-u^2)^2} = \frac{u}{2(a^2-u^2)} - \frac{1}{4a}\ln\left|\frac{a+u}{a-u}\right|$

103. $\displaystyle\int \frac{u^3\,du}{(a^2-u^2)^2} = \frac{a^2}{2(a^2-u^2)} + \frac{1}{2}\ln\left|a^2-u^2\right|$

104. $\displaystyle\int \frac{du}{u(a^2-u^2)^2} = \frac{1}{2a^2(a^2-u^2)} + \frac{1}{2a^4}\ln\left|\frac{u^2}{a^2-u^2}\right|$

105. $\displaystyle\int \frac{du}{u^2(a^2-u^2)^2} = \frac{-1}{a^4u} + \frac{u}{2a^4(a^2-u^2)} + \frac{3}{4a^5}\ln\left|\frac{a+u}{a-u}\right|$

106. $\displaystyle\int \frac{du}{u^3(a^2-u^2)^2} = \frac{-1}{2a^4u^2} + \frac{1}{2a^4(a^2-u^2)} + \frac{1}{a^6}\ln\left|\frac{u^2}{a^2-u^2}\right|$

INTEGRALS INVOLVING $\sqrt{au+b}$

135. $\displaystyle\int \frac{du}{\sqrt{au+b}} = \frac{2\sqrt{au+b}}{a}$

136. $\displaystyle\int \frac{u\,du}{\sqrt{au+b}} = \frac{2(au-2b)}{3a^2}\sqrt{au+b}$

137. $\displaystyle\int \frac{u^2\,du}{\sqrt{au+b}} = \frac{2(3a^2u^2 - 4abu + 8b^2)}{15a^3}\sqrt{au+b}$

138. $\displaystyle\int \frac{du}{u\sqrt{au+b}} = \begin{cases} \dfrac{1}{\sqrt{b}}\ln\left|\dfrac{\sqrt{au+b}-\sqrt{b}}{\sqrt{au+b}+\sqrt{b}}\right| \\[3mm] \dfrac{2}{\sqrt{-b}}\tan^{-1}\sqrt{\dfrac{au+b}{-b}} \end{cases}$

139. $\displaystyle\int \frac{du}{u^2\sqrt{au+b}} = -\frac{\sqrt{au+b}}{bu} - \frac{a}{2b}\int \frac{du}{u\sqrt{au+b}}$

140. $\displaystyle\int \sqrt{au+b}\,du = \frac{2\sqrt{(au+b)^3}}{3a}$

141. $\displaystyle\int u\sqrt{au+b}\,du = \frac{2(3au-2b)}{15a^2}\sqrt{(au+b)^3}$

142. $\displaystyle\int u^2\sqrt{au+b}\,du = \frac{2(15a^2u^2 - 12abu + 8b^2)}{105a^3}\sqrt{(au+b)^3}$

INTEGRALS INVOLVING $\sqrt{u^2 + a^2}$

168. $\displaystyle\int \sqrt{u^2 + a^2}\, du = \frac{u\sqrt{u^2 + a^2}}{2} + \frac{a^2}{2}\ln\left(u + \sqrt{u^2 + a^2}\right)$

169. $\displaystyle\int u\sqrt{u^2 + a^2}\, du = \frac{(u^2 + a^2)^{3/2}}{3}$

170. $\displaystyle\int u^2\sqrt{u^2 + a^2}\, du = \frac{u(u^2 + a^2)^{3/2}}{4} - \frac{a^2 u\sqrt{u^2 + a^2}}{8} - \frac{a^4}{8}\ln\left(u + \sqrt{u^2 + a^2}\right)$

171. $\displaystyle\int u^3\sqrt{u^2 + a^2}\, du = \frac{(u^2 + a^2)^{5/2}}{5} - \frac{a^2(u^2 + a^2)^{3/2}}{3}$

172. $\displaystyle\int \frac{du}{\sqrt{u^2 + a^2}} = \ln\left(u + \sqrt{u^2 + a^2}\right) \text{ or } \sinh^{-1}\frac{u}{a}$

173. $\displaystyle\int \frac{u\, du}{\sqrt{u^2 + a^2}} = \sqrt{u^2 + a^2}$

174. $\displaystyle\int \frac{u^2\, du}{\sqrt{u^2 + a^2}} = \frac{u\sqrt{u^2 + a^2}}{2} - \frac{a^2}{2}\ln\left(u + \sqrt{u^2 + a^2}\right)$

175. $\displaystyle\int \frac{u^3\, du}{\sqrt{u^2 + a^2}} = \frac{(u^2 + a^2)^{3/2}}{3} - a^2\sqrt{u^2 + a^2}$

176. $\displaystyle\int \frac{du}{u\sqrt{u^2 + a^2}} = -\frac{1}{a}\ln\left|\frac{a + \sqrt{u^2 + a^2}}{u}\right|$

177. $\displaystyle\int \frac{du}{u^2\sqrt{u^2 + a^2}} = -\frac{\sqrt{u^2 + a^2}}{a^2 u}$

178. $\displaystyle\int \frac{du}{u^3\sqrt{u^2 + a^2}} = -\frac{\sqrt{u^2 + a^2}}{2a^2 u^2} + \frac{1}{2a^3}\ln\left|\frac{a + \sqrt{u^2 + a^2}}{u}\right|$

179. $\displaystyle\int \frac{\sqrt{u^2 + a^2}}{u}\, du = \sqrt{u^2 + a^2} - a\ln\left|\frac{a + \sqrt{u^2 + a^2}}{u}\right|$

180. $\displaystyle\int \frac{\sqrt{u^2 + a^2}}{u^2}\, du = -\frac{\sqrt{u^2 + a^2}}{u} + \ln\left(u + \sqrt{u^2 + a^2}\right)$

INTEGRALS INVOLVING $\sqrt{u^2 - a^2}$

196. $\displaystyle\int \frac{du}{\sqrt{u^2 - a^2}} = \ln\left|u + \sqrt{u^2 - a^2}\right|$

197. $\displaystyle\int \frac{u\, du}{\sqrt{u^2 - a^2}} = \sqrt{u^2 - a^2}$

198. $\displaystyle\int \frac{u^2\, du}{\sqrt{u^2 - a^2}} = \frac{u\sqrt{u^2 - a^2}}{2} + \frac{a^2}{2}\ln\left|u + \sqrt{u^2 - a^2}\right|$

199. $\displaystyle\int \frac{u^3\, du}{\sqrt{u^2 - a^2}} = \frac{(u^2 - a^2)^{3/2}}{3} + a^2\sqrt{u^2 - a^2}$

200. $\displaystyle\int \frac{du}{u\sqrt{u^2 - a^2}} = \frac{1}{a}\sec^{-1}\left|\frac{u}{a}\right|$

201. $\displaystyle\int \frac{du}{u^2\sqrt{u^2-a^2}} = \frac{\sqrt{u^2-a^2}}{a^2 u}$

202. $\displaystyle\int \frac{du}{u^3\sqrt{u^2-a^2}} = \frac{\sqrt{u^2-a^2}}{2a^2 u^2} + \frac{1}{2a^3}\sec^{-1}\left|\frac{u}{a}\right|$

203. $\displaystyle\int \sqrt{u^2-a^2}\,du = \frac{u\sqrt{u^2-a^2}}{2} - \frac{a^2}{2}\ln\left|u+\sqrt{u^2-a^2}\right|$

204. $\displaystyle\int u\sqrt{u^2-a^2}\,du = \frac{(u^2-a^2)^{3/2}}{3}$

205. $\displaystyle\int u^2\sqrt{u^2-a^2}\,du = \frac{u(u^2-a^2)^{3/2}}{4} + \frac{a^2 u\sqrt{u^2-a^2}}{8} - \frac{a^4}{8}\ln\left|u+\sqrt{u^2-a^2}\right|$

206. $\displaystyle\int u^3\sqrt{u^2-a^2}\,du = \frac{(u^2-a^2)^{5}/2}{5} + \frac{a^2(u^2-a^2)^{3/2}}{3}$

INTEGRALS INVOLVING $\sqrt{a^2-u^2}$

224. $\displaystyle\int \frac{du}{\sqrt{a^2-u^2}} = \sin^{-1}\frac{u}{a}$

225. $\displaystyle\int \frac{u\,du}{\sqrt{a^2-u^2}} = -\sqrt{a^2-u^2}$

226. $\displaystyle\int \frac{u^2\,du}{\sqrt{a^2-u^2}} = -\frac{u\sqrt{a^2-u^2}}{2} + \frac{a^2}{2}\sin^{-1}\frac{u}{a}$

227. $\displaystyle\int \frac{u^3\,du}{\sqrt{a^2-u^2}} = \frac{(a^2-u^2)^{3/2}}{3} - a^2\sqrt{a^2-u^2}$

228. $\displaystyle\int \frac{du}{u\sqrt{a^2-u^2}} = -\frac{1}{a}\ln\left|\frac{a+\sqrt{a^2-u^2}}{u}\right|$ or $-\frac{1}{a}\,\text{sech}^{-1}\left|\frac{u}{a}\right|$

229. $\displaystyle\int \frac{du}{u^2\sqrt{a^2-u^2}} = -\frac{\sqrt{a^2-u^2}}{a^2 u}$

230. $\displaystyle\int \frac{du}{u^3\sqrt{a^2-u^2}} = -\frac{\sqrt{a^2-u^2}}{2a^2 u^2} - \frac{1}{2a^3}\ln\left|\frac{a+\sqrt{a^2-u^2}}{u}\right|$

231. $\displaystyle\int \sqrt{a^2-u^2}\,du = \frac{u\sqrt{a^2-u^2}}{2} + \frac{a^2}{2}\sin^{-1}\frac{u}{a}$

232. $\displaystyle\int u\sqrt{a^2-u^2}\,du = -\frac{(a^2-u^2)^{3/2}}{3}$

233. $\displaystyle\int u^2\sqrt{a^2-u^2}\,du = -\frac{u(a^2-u^2)^{3/2}}{4} + \frac{a^2 u\sqrt{a^2-u^2}}{8} + \frac{a^4}{8}\sin^{-1}\frac{u}{a}$

234. $\displaystyle\int u^3\sqrt{a^2-u^2}\,du = \frac{(a^2-u^2)^{5/2}}{5} - \frac{a^2(a^2-u^2)^{3/2}}{3}$

235. $\displaystyle\int \frac{\sqrt{a^2-u^2}}{u}\,du = \sqrt{a^2-u^2} - a\ln\left|\frac{a+\sqrt{a^2-u^2}}{u}\right|$

INTEGRALS INVOLVING cos *au*

311. $\displaystyle\int \cos au\, du = \frac{\sin au}{a}$

312. $\displaystyle\int u \cos au\, du = \frac{\cos au}{a^2} + \frac{u \sin au}{a}$

313. $\displaystyle\int u^2 \cos au\, du = \frac{2u}{a^2} \cos au + \left(\frac{u^2}{a} - \frac{2}{a^3}\right) \sin au$

314. $\displaystyle\int u^3 \cos au\, du = \left(\frac{3u^2}{a^2} - \frac{6}{a^4}\right) \cos au + \left(\frac{u^3}{a} - \frac{6u}{a^3}\right) \sin au$

315. $\displaystyle\int u^n \cos au\, du = \frac{u^n \sin au}{a} - \frac{n}{a} \int u^{n-1} \sin au\, du$

316. $\displaystyle\int u^n \cos au\, du = \frac{u^n \sin au}{a} + \frac{nu^{n-1}}{a^2} \cos au - \frac{n(n-1)}{a^2} \int u^{n-2} \cos au\, du$

317. $\displaystyle\int \cos^2 au\, du = \frac{u}{2} + \frac{\sin 2au}{4a}$

INTEGRALS INVOLVING sin *au*

342. $\displaystyle\int \sin au\, du = -\frac{\cos au}{a}$

343. $\displaystyle\int u \sin au\, du = \frac{\sin au}{a^2} - \frac{u \cos au}{a}$

344. $\displaystyle\int u^2 \sin au\, du = \frac{2u}{a^2} \sin au + \left(\frac{2}{a^3} - \frac{u^2}{a}\right) \cos au$

345. $\displaystyle\int u^3 \sin au\, du = \left(\frac{3u^2}{a^2} - \frac{6}{a^4}\right) \sin au + \left(\frac{6u}{a^3} - \frac{u^3}{a}\right) \cos au$

346. $\displaystyle\int u^n \sin au\, du = -\frac{u^n \cos au}{a} + \frac{n}{a} \int u^{n-1} \cos au\, du$

347. $\displaystyle\int u^n \sin au\, du = -\frac{u^n \cos au}{a} + \frac{nu^{n-1} \sin au}{a^2} - \frac{n(n-1)}{a^2} \int u^{n-2} \sin au\, du$

348. $\displaystyle\int \sin^2 au\, du = \frac{u}{2} - \frac{\sin 2au}{4a}$

349. $\displaystyle\int u \sin^2 au\, du = \frac{u^2}{4} - \frac{u \sin 2au}{4a} - \frac{\cos 2au}{8a^2}$

350. $\displaystyle\int \sin^3 au\, du = -\frac{\cos au}{a} + \frac{\cos^3 au}{3a}$

351. $\displaystyle\int \sin^4 au\, du = \frac{3u}{8} - \frac{\sin 2au}{4a} + \frac{\sin 4au}{32a}$

INTEGRALS INVOLVING sin *au* and cos *au*

373. $\displaystyle\int \sin au \cos au \, du = \frac{\sin^2 au}{2a}$

374. $\displaystyle\int \sin pu \cos qu \, du = -\frac{\cos(p-q)u}{2(p-q)} - \frac{\cos(p+q)u}{2(p+q)}$

375. $\displaystyle\int \sin^n au \cos au \, du = \frac{\sin^{n+1} au}{(n+1)a}$

376. $\displaystyle\int \cos^n au \sin au \, du = -\frac{\cos^{n+1} au}{(n+1)a}$

377. $\displaystyle\int \sin^2 au \cos^2 au \, du = \frac{u}{8} - \frac{\sin 4au}{32a}$

378. $\displaystyle\int \frac{du}{\sin au \cos au} = \frac{1}{a} \ln |\tan au|$

379. $\displaystyle\int \frac{du}{\sin^2 au \cos au} = \frac{1}{a} \ln \left|\tan \left(\frac{\pi}{4} + \frac{au}{2}\right)\right| - \frac{1}{a \sin au}$

380. $\displaystyle\int \frac{du}{\sin au \cos^2 au} = \frac{1}{a} \ln \left|\tan \frac{au}{2}\right| + \frac{1}{a \cos au}$

INTEGRALS INVOLVING tan *au*

403. $\displaystyle\int \tan au \, du = -\frac{1}{a} \ln |\cos au| \text{ or } \frac{1}{a} \ln |\sec au|$

404. $\displaystyle\int \tan^2 au \, du = \frac{\tan au}{a} - u$

405. $\displaystyle\int \tan^3 au \, du = \frac{\tan^2 au}{2a} + \frac{1}{a} \ln |\cos au|$

406. $\displaystyle\int \tan^n au \, du = \frac{\tan^{n-1} au}{(n-1)a} - \int \tan^{n-2} au \, du$

407. $\displaystyle\int \tan^n au \sec^2 au \, du = \frac{\tan^{n+1} au}{(n+1)a}$

INTEGRALS INVOLVING cot *au*

414. $\displaystyle\int \cot au \, du = \frac{1}{a} \ln |\sin au|$

415. $\displaystyle\int \cot^2 au \, du = -\frac{\cot au}{a} - u$

416. $\displaystyle\int \cot^3 au \, du = -\frac{\cot^2 au}{2a} - \frac{1}{a} \ln |\sin au|$

417. $\displaystyle\int \cot^n au \, du = -\frac{\cot^{n-1} au}{(n-1)a} - \int \cot^{n-2} au \, du$

INTEGRALS INVOLVING sec *au*

425. $\displaystyle\int \sec au\, du = \frac{1}{a} \ln |\sec au + \tan au| = \frac{1}{a} \ln \left| \tan \left(\frac{au}{2} + \frac{\pi}{4} \right) \right|$

426. $\displaystyle\int \sec^2 au\, du = \frac{\tan au}{a}$

427. $\displaystyle\int \sec^3 au\, du = \frac{\sec au \tan au}{2a} + \frac{1}{2a} \ln |\sec au + \tan au|$

428. $\displaystyle\int \sec^n au\, du = \frac{\sec^{n-2} au \tan au}{a(n-1)} + \frac{n-2}{n-1} \int \sec^{n-2} au\, du$

INTEGRALS INVOLVING csc *au*

435. $\displaystyle\int \csc au\, du = \frac{1}{a} \ln |\csc au - \cot au| = \frac{1}{a} \ln \left| \tan \frac{au}{2} \right|$

436. $\displaystyle\int \csc^2 au\, du = -\frac{\cot au}{a}$

437. $\displaystyle\int \csc^3 au\, du = -\frac{\csc au \cot au}{2a} + \frac{1}{2a} \ln \left| \tan \frac{au}{2} \right|$

438. $\displaystyle\int \csc^n au\, du = -\frac{\csc^{n-2} au \cot au}{a(n-1)} + \frac{n-2}{n-1} \int \csc^{n-2} au\, du$

INTEGRALS INVOLVING INVERSE TRIGONOMETRIC FUNCTIONS

445. $\displaystyle\int \cos^{-1} \frac{u}{a}\, du = u \cos^{-1} \frac{u}{a} - \sqrt{a^2 - u^2}$

446. $\displaystyle\int u \cos^{-1} \frac{u}{a}\, du = \left(\frac{u^2}{2} - \frac{a^2}{4} \right) \cos^{-1} \frac{u}{a} - \frac{u\sqrt{a^2 - u^2}}{4}$

447. $\displaystyle\int u^2 \cos^{-1} \frac{u}{a}\, du = \frac{u^3}{3} \cos^{-1} \frac{u}{a} - \frac{(u^2 + 2a^2)\sqrt{a^2 - u^2}}{9}$

448. $\displaystyle\int \frac{\cos^{-1}(u/a)}{u}\, du = \frac{\pi}{2} \ln |u| - \int \frac{\sin^{-1}(u/a)}{u}\, du$

449. $\displaystyle\int \frac{\cos^{-1}(u/a)}{u^2}\, du = -\frac{\cos^{-1}(u/a)}{u} + \frac{1}{a} \ln \left| \frac{a + \sqrt{a^2 - u^2}}{u} \right|$

450. $\displaystyle\int \left(\cos^{-1} \frac{u}{a} \right)^2 du = u \left(\cos^{-1} \frac{u}{a} \right)^2 - 2u - 2\sqrt{a^2 - u^2} \cos^{-1} \frac{u}{a}$

451. $\displaystyle\int \sin^{-1} \frac{u}{a}\, du = u \sin^{-1} \frac{u}{a} + \sqrt{a^2 - u^2}$

452. $\displaystyle\int u \sin^{-1} \frac{u}{a}\, du = \left(\frac{u^2}{2} - \frac{a^2}{4} \right) \sin^{-1} \frac{u}{a} + \frac{u\sqrt{a^2 - u^2}}{4}$

453. $\displaystyle\int u^2 \sin^{-1} \frac{u}{a}\, du = \frac{u^3}{3} \sin^{-1} \frac{u}{a} + \frac{(u^2 + 2a^2)\sqrt{a^2 - u^2}}{9}$

454. $\displaystyle\int \frac{\sin^{-1}(u/a)}{u}\, du = \frac{u}{a} + \frac{(u/a)^3}{2 \cdot 3 \cdot 3} + \frac{1 \cdot 3(u/a)^5}{2 \cdot 4 \cdot 5 \cdot 5} + \frac{1 \cdot 3 \cdot 5(u/a)^7}{2 \cdot 4 \cdot 6 \cdot 7 \cdot 7} + \cdots$

455. $\displaystyle\int \frac{\sin^{-1}(u/a)}{u^2}\,du = -\frac{\sin^{-1}(u/a)}{u} - \frac{1}{a}\ln\left|\frac{a+\sqrt{a^2-u^2}}{u}\right|$

456. $\displaystyle\int \left(\sin^{-1}\frac{u}{a}\right)^2 du = u\left(\sin^{-1}\frac{u}{a}\right)^2 - 2u + 2\sqrt{a^2-u^2}\,\sin^{-1}\frac{u}{a}$

457. $\displaystyle\int \tan^{-1}\frac{u}{a}\,du = u\tan^{-1}\frac{u}{a} - \frac{a}{2}\ln\left(u^2+a^2\right)$

458. $\displaystyle\int u\tan^{-1}\frac{u}{a}\,du = \frac{1}{2}(u^2+a^2)\tan^{-1}\frac{u}{a} - \frac{au}{2}$

459. $\displaystyle\int u^2\tan^{-1}\frac{u}{a}\,du = \frac{u^3}{3}\tan^{-1}\frac{u}{a} - \frac{au^2}{6} + \frac{a^3}{6}\ln(u^2+a^2)$

INTEGRALS INVOLVING e^{au}

483. $\displaystyle\int e^{au}\,du = \frac{e^{au}}{a}$

484. $\displaystyle\int ue^{au}\,du = \frac{e^{au}}{a}\left(u - \frac{1}{a}\right)$

485. $\displaystyle\int u^2 e^{au}\,du = \frac{e^{au}}{a}\left(u^2 - \frac{2u}{a} + \frac{2}{a^2}\right)$

486. $\displaystyle\int u^n e^{au}\,du = \frac{u^n e^{au}}{a} - \frac{n}{a}\int u^{n-1}e^{au}\,du$

$$= \frac{e^{au}}{a}\left(u^n - \frac{nu^{n-1}}{a} + \frac{n(n-1)u^{n-2}}{a^2} - \cdots + \frac{(-1)^n n!}{a^n}\right)$$
if n = positive integer

487. $\displaystyle\int \frac{e^{au}}{u}\,du = \ln|u| + \frac{au}{1\cdot 1!} + \frac{(au)^2}{2\cdot 2!} + \frac{(au)^3}{3\cdot 3!} + \cdots$

488. $\displaystyle\int \frac{e^{au}}{u^n}\,du = \frac{-e^{au}}{(n-1)u^{n-1}} + \frac{a}{n-1}\int \frac{e^{au}}{u^{n-1}}\,du$

489. $\displaystyle\int \frac{du}{p+qe^{au}} = \frac{u}{p} - \frac{1}{ap}\ln\left|p+qe^{au}\right|$

490. $\displaystyle\int \frac{du}{(p+qe^{au})^2} = \frac{u}{p^2} + \frac{1}{ap(p+qe^{au})} - \frac{1}{ap^2}\ln\left|p+qe^{au}\right|$

491. $\displaystyle\int \frac{du}{pe^{au}+qe^{-au}} = \begin{cases} \dfrac{1}{a\sqrt{pq}}\tan^{-1}\left(\sqrt{\dfrac{p}{q}}\,e^{au}\right) \\[2em] \dfrac{1}{2a\sqrt{-pq}}\ln\left|\dfrac{e^{au}-\sqrt{-q/p}}{e^{au}+\sqrt{-q/p}}\right| \end{cases}$

492. $\displaystyle\int e^{au}\sin bu\,du = \frac{e^{au}(a\sin bu - b\cos bu)}{a^2+b^2}$

493. $\displaystyle\int e^{au}\cos bu\,du = \frac{e^{au}(a\cos bu + b\sin bu)}{a^2+b^2}$

494. $\displaystyle\int ue^{au}\sin bu\,du = \frac{ue^{au}(a\sin bu - b\cos bu)}{a^2+b^2}$

$$-\frac{e^{au}[(a^2-b^2)\sin bu - 2ab\cos bu]}{(a^2+b^2)^2}$$

495. $\int ue^{au} \cos bu \, du = \dfrac{ue^{au}(a \cos bu + b \sin bu)}{a^2 + b^2}$

$$- \dfrac{e^{au}[(a^2 - b^2) \cos bu + 2ab \sin bu]}{(a^2 + b^2)^2}$$

INTEGRALS INVOLVING $\ln |u|$

499. $\displaystyle\int \ln |u| \, du = u \ln |u| - u$

500. $\displaystyle\int (\ln |u|)^2 \, du = u(\ln |u|)^2 - 2u \ln |u| + 2u$

501. $\displaystyle\int (\ln |u|)^n \, du = u(\ln |u|)^n - n \int (\ln |u|)^{n-1} \, du$

502. $\displaystyle\int u \ln |u| \, du = \dfrac{u^2}{2}(\ln |u| - \dfrac{1}{2})$

503. $\displaystyle\int u^m \ln |u| \, du = \dfrac{u^{m+1}}{m+1}\left(\ln |u| - \dfrac{1}{m+1}\right)$

APPENDIX E   Trigonometric Formulas

 See Chapter 4 of the *Student Mathematics Handbook* for a review of trigonometry.

Definition of the Trigonometric Functions

Let θ be any angle in standard position and let $P(x, y)$ be any point on the terminal side of the angle a distance of r from the origin ($r \neq 0$). Then

$$\cos \theta = \frac{x}{r} \qquad \sin \theta = \frac{y}{r}$$

$$\tan \theta = \frac{y}{x} \qquad \sec \theta = \frac{r}{x}$$

$$\csc \theta = \frac{r}{y} \qquad \cot \theta = \frac{x}{y}$$

Definition of the Inverse Trigonometric Functions

Inverse Trigonometric Function	Domain	Range
$y = \arccos x$ or $y = \cos^{-1} x$	$-1 \leq x \leq 1$	$0 \leq y \leq \pi$
$y = \arcsin x$ or $y = \sin^{-1} x$	$-1 \leq x \leq 1$	$-\frac{\pi}{2} \leq y \leq \frac{\pi}{2}$
$y = \arctan x$ or $y = \tan^{-1} x$	All reals	$-\frac{\pi}{2} < y < \frac{\pi}{2}$
$y = \text{arccot}\, x$ or $y = \cot^{-1} x$	All reals	$0 < y < \pi$
$y = \text{arcsec}\, x$ or $y = \sec^{-1} x$	$|x| \geq 1$	$0 \leq y \leq \pi, y \neq \frac{\pi}{2}$
$y = \text{arccsc}\, x$ or $y = \csc^{-1} x$	$|x| \geq 1$	$-\frac{\pi}{2} \leq y \leq \frac{\pi}{2}, y \neq 0$

The principal values of the inverse trigonometric relations are those values defined by these inverse trigonometric functions. These are the values obtained when using a calculator.

Trigonometric Identities   Fundamental identities

Reciprocal identities

1. $\sec \theta = \dfrac{1}{\cos \theta}$ 　　　　　**2.** $\csc \theta = \dfrac{1}{\sin \theta}$ 　　　　　**3.** $\cot \theta = \dfrac{1}{\tan \theta}$

Ratio identities

4. $\tan \theta = \dfrac{\sin \theta}{\cos \theta}$ 　　　　　**5.** $\cot \theta = \dfrac{\cos \theta}{\sin \theta}$

Pythagorean identities

6. $\cos^2 \theta + \sin^2 \theta = 1$ 　　　**7.** $1 + \tan^2 \theta = \sec^2 \theta$ 　　　**8.** $\cot^2 \theta + 1 = \csc^2 \theta$

Cofunction identities

9. $\cos\left(\dfrac{\pi}{2} - \theta\right) = \sin \theta$ 　　　**10.** $\sin\left(\dfrac{\pi}{2} - \theta\right) = \cos \theta$ 　　　**11.** $\tan\left(\dfrac{\pi}{2} - \theta\right) = \cot \theta$

Opposite-angle identities

12. $\cos(-\theta) = \cos \theta$ 　　　**13.** $\sin(-\theta) = -\sin \theta$ 　　　**14.** $\tan(-\theta) = -\tan \theta$

Addition laws

15. $\cos(\alpha + \beta) = \cos \alpha \cos \beta - \sin \alpha \sin \beta$

16. $\cos(\alpha - \beta) = \cos \alpha \cos \beta + \sin \alpha \sin \beta$

17. $\sin(\alpha + \beta) = \sin \alpha \cos \beta + \cos \alpha \sin \beta$

18. $\sin(\alpha - \beta) = \sin \alpha \cos \beta - \cos \alpha \sin \beta$

19. $\tan(\alpha + \beta) = \dfrac{\tan \alpha + \tan \beta}{1 - \tan \alpha \tan \beta}$ 　　　**20.** $\tan(\alpha - \beta) = \dfrac{\tan \alpha - \tan \beta}{1 + \tan \alpha \tan \beta}$

Double-angle identities

21. $\begin{aligned}\cos 2\theta &= \cos^2 \theta - \sin^2 \theta \\ &= 2\cos^2 \theta - 1 \\ &= 1 - 2\sin^2 \theta\end{aligned}$ 　　　**22.** $\sin 2\theta = 2 \sin \theta \cos \theta$

23. $\tan 2\theta = \dfrac{2 \tan \theta}{1 - \tan^2 \theta}$

Half-angle identities

24. $\cos \dfrac{1}{2}\theta = \pm\sqrt{\dfrac{1 + \cos \theta}{2}}$ 　　　　　**25.** $\sin \dfrac{1}{2}\theta = \pm\sqrt{\dfrac{1 - \cos \theta}{2}}$

26. $\begin{aligned}\tan \dfrac{1}{2}\theta &= \dfrac{1 - \cos \theta}{\sin \theta} \\[2mm] &= \dfrac{\sin \theta}{1 + \cos \theta}\end{aligned}$

Product-to-sum identities

27. $2 \cos \alpha \cos \beta = \cos(\alpha - \beta) + \cos(\alpha + \beta)$

28. $2 \sin \alpha \sin \beta = \cos(\alpha - \beta) - \cos(\alpha + \beta)$

29. $2 \sin \alpha \cos \beta = \sin(\alpha + \beta) + \sin(\alpha - \beta)$

30. $2 \cos \alpha \sin \beta = \sin(\alpha + \beta) - \sin(\alpha - \beta)$

Sum-to-product identities

31. $\cos \alpha + \cos \beta = 2 \cos\left(\dfrac{\alpha + \beta}{2}\right) \cos\left(\dfrac{\alpha - \beta}{2}\right)$

32. $\cos \alpha - \cos \beta = -2 \sin\left(\dfrac{\alpha + \beta}{2}\right) \sin\left(\dfrac{\alpha - \beta}{2}\right)$

33. $\sin \alpha + \sin \beta = 2 \sin\left(\dfrac{\alpha + \beta}{2}\right) \cos\left(\dfrac{\alpha - \beta}{2}\right)$

34. $\sin \alpha - \sin \beta = 2 \sin \left(\dfrac{\alpha - \beta}{2} \right) \cos \left(\dfrac{\alpha + \beta}{2} \right)$

Hyperbolic identities

35. $\operatorname{sech} x = \dfrac{1}{\cosh x}$

36. $\operatorname{csch} x = \dfrac{1}{\sinh x}$

37. $\coth x = \dfrac{1}{\tanh x}$

38. $\tanh x = \dfrac{\sinh x}{\cosh x}$

39. $\coth x = \dfrac{\cosh x}{\sinh x}$

40. $\cosh^2 x - \sinh^2 x = 1$

41. $1 - \tanh^2 x = \operatorname{sech}^2 x$

42. $\coth^2 x - 1 = \operatorname{csch}^2 x$

43. $\sinh(-x) = -\sinh x$

44. $\cosh(-x) = \cosh x$

45. $\tanh(-x) = -\tanh x$

46. $\sinh(x \pm y) = \sinh x \cosh y \pm \cosh x \sinh y$

47. $\cosh(x \pm y) = \cosh x \cosh y \pm \sinh x \sinh y$

48. $\tanh(x \pm y) = \dfrac{\tanh x \pm \tanh y}{1 \pm \tanh x \tanh y}$

49. $\cosh 2x = \cosh^2 x + \sinh^2 x$
$$= 2\cosh^2 x - 1$$
$$= 1 + 2\sinh^2 x$$

50. $\sinh 2x = 2\sinh x \cosh x$

51. $\tanh 2x = \dfrac{2\tanh x}{1 + \tanh^2 x}$

52. $\cosh \dfrac{1}{2}x = \pm\sqrt{\dfrac{\cosh x + 1}{2}}$

53. $\sinh \dfrac{1}{2}x = \pm\sqrt{\dfrac{\cosh x - 1}{2}}$

54. $\tanh \dfrac{1}{2}x = \dfrac{\cosh x - 1}{\sinh x} = \dfrac{\sinh x}{\cosh x + 1}$

55. $\sinh^{-1} x = \ln(x + \sqrt{x^2 + 1})$

56. $\operatorname{csch}^{-1} x = \ln \left(\dfrac{1 + \sqrt{1 + x^2}}{x} \right)$ if $x > 0$

57. $\cosh^{-1} x = \ln(x + \sqrt{x^2 - 1})$

58. $\operatorname{sech}^{-1} x = \ln \left(\dfrac{1 + \sqrt{1 - x^2}}{x} \right)$ if $0 < x \le 1$

59. $\tanh^{-1} x = \dfrac{1}{2} \ln \left(\dfrac{1 + x}{1 - x} \right)$ if $-1 < x < 1$

60. $\coth^{-1} x = \dfrac{1}{2} \ln \left(\dfrac{x + 1}{x - 1} \right)$ if $x^2 > 1$

APPENDIX F   Answers to Selected Problems

Many problems in this book are labeled WHAT DOES THIS SAY? *These problems solicit answers in your own words or a statement for you to rephrase as given statement in your own words. For this reason, it seems inappropriate to include the answers to these questions.* COUNTEREXAMPLE PROBLEMS *generally ask for counterexamples for which answers may vary so these answers are also not given.*

We also believe that an answer section should function as a check on work done, so for that reason, when an answer has both an exact answer and an approximate solution (from technology), we usually show only the approximate solution in this appendix. The exact solution (which may be the more appropriate answer) can be **checked** *by using the given approximation.*

The Student Survival and Solutions Manual *offers some review, survival hints, and added explanations for selected problems.*

Chapter 1: Functions and Graphs

1.1 Preliminaries (Pages 10–13)

1. a. $(-3, 4)$ **b.** $3 \le x \le 5$ **c.** $-2 \le x < 1$ **d.** $(2, 7]$

3. a.

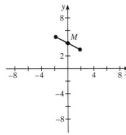

5. a.

$M = (0, 4); d = 2\sqrt{5}$

b.

$M = (1, 2); d = 2\sqrt{10}$

7. $x = 0, 1$ **9.** $y = 7, -2$ **11.** $x = \dfrac{b \pm \sqrt{b^2 + 12c}}{6}$

13. $x = 6, -10$ **15.** $w = -2, 5$ **17.** $\emptyset$

19. $x = \dfrac{7\pi}{6}, \dfrac{11\pi}{6}$ **21.** $x = \dfrac{3\pi}{4}, \dfrac{5\pi}{4}, \dfrac{\pi}{3}, \dfrac{5\pi}{3}$ **23.** $x = \dfrac{2\pi}{3}$

25. $\left(-\infty, -\dfrac{5}{3}\right)$ **27.** $\left(-\dfrac{5}{3}, 0\right)$ **29.** $(-8, -3]$

31. $[-1, 3]$ **33.** $[7.999, 8.001]$

35. $(x + 1)^2 + (y - 2)^2 = 9$ **37.** $x^2 + (y - 1.5)^2 = 0.0625$

39. Center $(1, -1); r = 1$ **41.** Center $(-1, 5); r = 1$

43. $\dfrac{\sqrt{2} - \sqrt{6}}{4} \approx -0.2588$ **45.** $2 - \sqrt{3} \approx 0.2679$

51. a. period 2π

b. period 2π

c. period π

53. period $\pi/2$

55. period 4π

57.

$A = 0, B = 60, C = \dfrac{\pi}{15}$, and $D = 7.5$

59. sun curve: $y = \cos \dfrac{\pi}{6} x$; moon curve: $y = 4\cos \dfrac{\pi}{6} x$; combined curve: $y = 5 \cos \dfrac{\pi}{6} x$

61. a. The apparent depth is 3.4 m.

   b. The angle of incidence is 59°.

1.2 Lines in the Plane (Pages 17–19)

3. $2x + y - 5 = 0$ **5.** $2y - 1 = 0$

7. $x + 2 = 0$ **9.** $8x - 7y - 56 = 0$

11. $3x + y - 5 = 0$ **13.** $3x + 4y - 1 = 0$

15. $4x + y + 3 = 0$

17. $m = -5/7$;

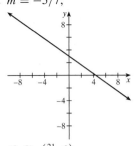

$(0, 3), \left(\frac{21}{5}, 0\right)$

19. $m = 6.001$;

$(1.50025, 0), (0, -9.003)$

21. $y = -\frac{3}{5}x - 3$; $m = -3/5$

$(-5, 0), (0, -3)$

23. $y = \frac{3}{5}x - 0.3$; $m = 3/5$

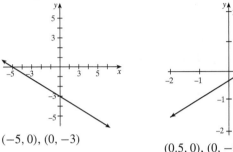

$(0.5, 0), (0, -0.3)$

25. $m = 3/2$

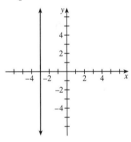

$(2, 0), (0, -3)$

27. $y = \frac{1}{5}x$; $m = 1/5$

$(0, 0)$

29. no slope

$(-3, 0)$

31. $y = 0$ and $y = 6$

33. $D(6, 6)$ or $E(2, 16)$; two parallelograms can be found

35. no values

37. $(4, -1)$

39. $\left(\frac{3}{4}, \frac{7}{2}\right)$

41. $\left(-\frac{64}{3}, \frac{100}{3}\right)$

43. $(\sqrt{2}, \sqrt{2}), (-\sqrt{2}, -\sqrt{2})$

45. $\left(-\frac{15}{8}, \frac{7\sqrt{15}}{8}\right), \left(-\frac{15}{8}, -\frac{7\sqrt{15}}{8}\right)$

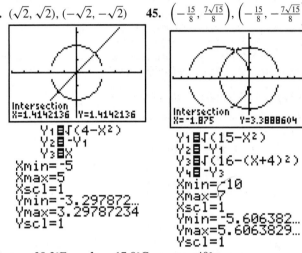

47. a. $-38.2°F$ **b.** $-17.8°C$ **c.** $-40°$

49. The Spy escapes since freedom is reached at the border after only 83.8 km.

51. $V(t) = -19,000t + 200,000$

The value in 4 years is $124,000.

55. For the 8th month, 216 gallons

57. a. $48   **b.** not linear

59. $x = \dfrac{a \pm \sqrt{a^2 - 4}}{2}$, $y = \dfrac{2}{a \pm \sqrt{a^2 - 4}}$

1.3 Functions and Graphs (Pages 31–33)

Let D represent the domain in Problems 1–11.

1. $D = (-\infty, \infty)$; $f(-2) = -1$; $f(1) = 5$; $f(0) = 3$

3. $D = (-\infty, \infty)$; $f(1) = 6$; $f(0) = -2$; $f(-2) = 0$

5. $D = (-\infty, -3) \cup (-3, \infty)$; $f(2) = 0$; $f(0) = -2$; $f(-3)$ is undefined

7. $D = (-\infty, -2] \cup [0, \infty)$; $f(-1)$ is undefined; $f\left(\frac{1}{2}\right) = \frac{\sqrt{5}}{2}$; $f(1) = \sqrt{3}$

9. $D = (-\infty, \infty)$; $f(-1) = \sin 3 \approx 0.1411$; $f\left(\frac{1}{2}\right) = 0$; $f(1) = \sin(-1) \approx -0.8415$

11. $D = (-\infty, \infty)$; $f(3) = 4$; $f(1) = 2$; $f(0) = 4$

13. 9   **15.** $10x + 5h$ **17.** -1 **19.** $\dfrac{-1}{x(x + h)}$

21. not equal   **23.** equal   **25.** not equal

27. even   **29.** neither   **31.** neither   **33.** even

35. $(f \circ g)(x) = 4x^2 + 1$; $(g \circ f)(x) = 2x^2 + 2$

37. $(f \circ g)(t) = |t|$; $(g \circ f)(t) = t$

39. $(f \circ g)(x) = \sin(2x + 3)$; $(g \circ f)(x) = 2\sin x + 3$

41. $u(x) = 2x^2 - 1$; $g(u) = u^4$

43. $u(x) = 2x + 3$; $g(u) = |u|$

45. $u(x) = \tan x$; $g(u) = u^2$

47. $u(x) = \sqrt{x}$; $g(u) = \sin u$

49. $u(x) = \dfrac{x + 1}{2 - x}$; $g(u) = \sin u$

51. $P(5, f(5))$; $Q(x_0, f(x_0))$

53. $-\frac{1}{3}, 2$ **55.** $15, -\frac{25}{2}, \frac{65}{3}, -\frac{1}{4}$

57. 0   **59.** ± 1

61. a. The cost is \$4,500.   **b.** The cost of the 20th unit is \$371.

63. a. $I = \dfrac{30}{t^2(6 - t)^2}$ **b.** $I(1) = \frac{6}{5}$ candles; $I(4) = \frac{15}{32}$ candles

65. a. $625t^2 + 25t + 900$ **b.** \$6,600   **c.** 4 hours

67. a.

b. $T = 273$

69. a. The maximum height of the cannonball is 280 ft.

     b. It hits the ground approximately 1,243 ft from the firing point.

     c. The basic shape of a parabola (standard quadratic function).

71. a. 19,400 people   **b.** 67 people

     c. The population will tend to 20,000 people in the long run.

73. Pythagorean theorem: $\triangle ABC$ with sides a, b, and c is a right triangle if and only if $c^2 = a^2 + b^2$. Proofs vary.

1.4 Inverse Functions; Inverse Trigonometric Functions (Pages 41–42)

3. These are inverse functions.

5. These are not inverse functions.

7. These are not inverse functions.

9. $\{(5, 4), (3, 6), (1, 7), (4, 2)\}$

11. $y = \frac{1}{2}x - \frac{3}{2}$ **13.** $y = \sqrt{x + 5}$

15. $y = (x - 5)^2$ **17.** $y = \dfrac{3x + 6}{2 - 3x}$

19. a. $\frac{\pi}{3}$ **b.** $-\frac{\pi}{3}$ **21. a.** $-\frac{\pi}{4}$ **b.** $\frac{5\pi}{6}$

23. a. $-\frac{\pi}{3}$ **b.** π **25.** $\frac{\sqrt{3}}{2}$ **27.** 3   **29.** $-\frac{2\sqrt{6}}{5} \approx -0.9798$

33. no inverse

35. no inverse

37. inverse exists

39. $\dfrac{2x}{x^2 + 1}$ **41.** $\dfrac{\sqrt{1 - x^2}}{x}$ **43.** 1

47. a. 0.588   **b.** 2.5536   **c.** 2.1997   **d.** -0.5746

49. $h = \dfrac{x \tan \beta \tan \alpha}{\tan \alpha - \tan \beta}$

Chapter 1 Review
Proficiency Examination (Page 42–43)

18. a. $6x + 8y - 37 = 0$ **b.** $3x + 10y - 41 = 0$

     c. $3x - 28y - 12 = 0$ **d.** $2x + 5y - 24 = 0$

     e. $4x - 3y + 8 = 0$

19. $y = -\frac{3}{2}x + 6$ **20.** $y - 3 = |x + 1|$

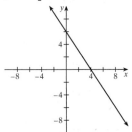

21. $y - 3 = -2(x - 1)^2$ **22.** $y = (x - 2)^2 - 14$

23. $y = 2\cos(x - 1)$ **24.** $y + 1 = \tan 2\left(x + \frac{3}{2}\right)$

25. $y = \sin^{-1}(2x)$

26. $y = \tan^{-1} x^2$

27. $-\frac{3}{2}, 1$

28. The two functions are not the same.

29. $f \circ g = \sin \sqrt{1 - x^2}$; $g \circ f = |\cos x|$

30. $V = \frac{2}{3} x(12 - x^2)$

Supplementary Problems (Pages 43–45)

1. $P = 30; A = 30$ **3.** $P \approx 27.1; A = 40$

5. $(x - 5)^2 + (y - 4)^2 = 16$ **7.** $5x - 3y + 3 = 0$

9. $A = -\frac{1}{4}$ and $B = \frac{3}{4}$

11. The medians meet at the point $\left(2, \frac{5}{3}\right)$.

13. 1,024 ft

15. a. 1.107149   **b.** 0.463647609   **c.** 1.899250945

17. $\frac{\pi^2}{4}$ **19.** $\frac{1}{2}(x^7 - 1)$ **21.** $\frac{7x + 5}{x - 1}$ **23.** $\frac{b - dx}{cx - a}$

27. a. 15 inches   **b.** 21 inches   **c.** 29 inches   **d.** 19 inches

29. a. The domain consists of all $x \neq 300$.

   b. x represents a percentage, so $0 \leq x \leq 100$ in order for $f(x) \geq 0$

   c. 120   **d.** 300

   e. The percentage of households should be 60%.

31. a. $V = \frac{256}{3}\pi; S = 64\pi$ **b.** $V = 30; S = 62$

   c. $V = 16\pi; S = 24\pi$ **d.** $V = 15\pi; A = 3\pi\sqrt{34}$

35. $c = -\frac{4}{5}; (-2, 0), (2, 0)$

37. $\theta = \frac{\pi}{4} - \tan^{-1}\frac{5}{12} \approx 0.3906$

39. a. $6 < p < 8$ **b.** $N(7) = 1$ is the maximum profit

41. The glass should be taken in on the 11th or 12th day.

X=3.5106383 Y=396.74966

Y1◻-3X²+21X+360
Xmin=0   Ymin=-100
Xmax=15 Ymax=500
Xscl=3   Yscl=100

43. $C(x) = 4x^2 + 1,000x^{-1}$

45. This is Putnam Examination Problem 4 in the morning session for 1959.

Chapter 2: Limits and Continuity

2.1 The Limit of a Function (Pages 59–61)

1. a. 0   **b.** 2   **c.** 6

3. a. 2   **b.** 7   **c.** 7.5

5. a. 6   **b.** 6   **c.** 6

7. 15   **9.** 10   **11.** 8

13. 2   **15.** -1 **19.** 1.00

21. 5.00   **23.** -0.17 **25.** does not exist

27. a. 0.00   **b.** 0.64

29. 8.00   **31.** -2.00 **33.** 0.17

35. 0.25   **37.** does not exist   **39.** 2.00

41. 1.00   **43.** does not exist

45. a. $-32t + 40$ **b.** 40 ft/s

   c. 3; impact velocity is -56 ft/s

   d. $t = 1.25$ seconds

47. 228

2.2 Algebraic Computation of Limits (Pages 68–69)

1. -9 **3.** -8 **5.** $-\frac{1}{2}$ **7.** 2   **9.** $\frac{\sqrt{3}}{9}$

11. 4   **13.** -1 **15.** $\frac{1}{9}$ **17.** $\frac{1}{2}$ **19.** $\frac{1}{2}$

21. 0   **23.** 2   **25.** $\frac{5}{2}$ **27.** 0   **29.** 1

31. 0   **35.** a **37.** -1 **39.** 0

41. the limit does not exist   **43.** the limit does not exist

49. the limit does not exist   **51.** the limit does not exist

53. the limit does not exist   **55.** 4   **57.** 8

2.3 Continuity (Pages 78–80)

1. Temperature is continuous, so TEMPERATURE $= f$(TIME) would be a continuous function. The domain could be midnight to midnight say, $0 \leq t < 24$.

3. The charges (range of the function) consist of rational numbers only (dollars and cents to the nearest cent), so the function CHARGE $= f$(MILEAGE) would be a step function (that is, not continuous). The domain would consist of the mileage from the beginning of the trip to its end.

5. No suspicious points and no points of discontinuity with a polynomial.

7. The denominator factors to $x(x - 1)$, so suspicious points would be $x = 0, 1$. There will be a hole discontinuity at $x = 0$ and a pole discontinuity at $x = 1$.

9. $x = 0$ is suspicious point and is a point of discontinuity.

11. $x = 1$ is a suspicious point; there are no points of discontinuity.

13. The sine and cosine are continuous on the reals, but the tangent is discontinuous at $x = \pi/2 + n\pi$, n an integer. Each of these values will have a pole type discontinuity.

15. 3   **17.** π **19.** no value

21. a. continuous   **b.** discontinuous on $[0, 1]$

23. discontinuous at $t = 0$ **25.** discontinuous at $x = 0$

39. $a = 1; b = -18/5$ **41.** $a = 1; b = \frac{1}{2}$

43. $a = 5; b = 5$ **45.** 0

47. a. 1.25872   **b.** 0.785398

2.4 Exponential and Logarithm Functions (Pages 89–91)

1.

3.

5. 31   **7.** 200.33681   **9.** 9,783.225896

11. 38,523.62544   **13.** 0   **15.** 2

17. -2 **19.** 3.5   **21.** $\frac{3}{10}$

23. 4   **25.** 0.23104906   **27.** -1.391662509

29. 729   **31.** 2, -1 **33.** 3

35. 2, $-\frac{5}{3}$ **37.** $-\frac{3}{2}$ **39.** 1

41. a. 0   **b.** e^{-1} **43. a.** e **b.** 2

45. $2e$ **47.** 27   **49.** 0.4

51. logarithmic   **53.** exponential

55. 3 m below surface   **57.** $0.7/r$ years to double

59. 6.9%

61. a. 793.7   **b.** 2 hr 40 min

63. a. $E^{1.5M+11.4}$ **b.** 1,000 times more energy

65. a.

b. It crosses the y-axis at $(0, 1)$. As $x \to +\infty$, $y \to +\infty$. As $x \to -\infty$, $y \to +\infty$.

c. The smallest value of $y = E$ is $E = 0.5$.

67. a. $10,285.33   **b.** $10,285.29

69. a. $\frac{4}{7}$ **b.** 59°   **c.** 70°

Chapter 2 Review
Proficiency Examination (Page 92)

19. $\frac{3}{2}$ **20.** $\frac{1}{4}$ **21.** $-\frac{1}{4}$

22. 0   **23.** $\frac{9}{5}$ **24.** -1

25. a. exponential   **b.** exponential
c. logarithmic   **d.** logarithmic

26. We have suspicious points where the denominators are 0 at $t = 0, -1$. There are pole discontinuities at each of these points.

27. Suspicious points $x = -2$ and $x = 1$ are also points of discontinuity (since the denominator is 0).

28. a. 11 years, 3 quarters   **b.** 11 years, 6 months
c. 11 years, 166 days

29. $A = -1$; $B = 1$

30. Let $f(x) = x + \sin x - \dfrac{1}{\sqrt{x + 3}}$. This function is continuous on $[0, \infty)$, and $f(0) = -\frac{1}{3}$, $f(\pi) = \pi - \dfrac{1}{\sqrt{\pi + 3}} \approx 2.93$. So by the root location theorem there must be some value c in $(0, \pi)$ where $f(c) = 0$.

Supplementary Problems (Pages 92–95)

1. a. 1.504077397   **b.** 16.44464677   **c.** π **d.** $\sqrt{2}$

3. -2 **5.** 16, -1 **7.** 1.46085

9. 5   **11.** 1   **13.** 0

15. $-\frac{1}{2}$ **17.** e^4 **19.** 5

21. $\frac{3}{2}$ **23.** 0.8415   **25.** 1

27. 3   **29.** 0   **31.** $\dfrac{1}{\sqrt{2x}}$

33. $\dfrac{-4}{x^2}$ **35.** e^x **37.** continuous on $[-5, 5]$

39. discontinuity is removable   **41.** not continuous anywhere

43. a. continuous on $[0, 5]$ **b.** not continuous at $x = -2$
c. not continuous at $x = -2$ **d.** continuous on $[-5, 5]$

45. a.

b. $\lim\limits_{x \to 3} [\![x]\!]$ does not exist

c. The limit exists for all nonintegral values.

47. does not exist   **49.** $a = 2$ and $b = 1$

51. a. 0.2973   **b.** 0.8232   **c.** 0.07955

53. The rock concert is one million times as intense as normal conversation.

67. a. The windchill for 20 mi/h is 3.75° and for 50 mi/h is $-7°$.
b. $v \approx 25.2$
c. at $v = 4$, $T \approx 91.4$; at $v = 45$, $T \approx 868$

71. This is a Problem A1 of the morning session of the 1956 Putnam Examination.

Chapter 3: Differentiation

3.1 An Introduction to the Derivative: Tangents
(Pages 107–110)

1. Some describe it as a five-step process:
(1) Find $f(x)$
(2) Find $f(x + \Delta x)$
(3) Find $f(x + \Delta x) - f(x)$
(4) Find $\dfrac{f(x + \Delta x) - f(x)}{\Delta x}$
(5) Find $\lim\limits_{\Delta x \to 0} \dfrac{f(x + \Delta x) - f(x)}{\Delta x}$

3. Answers vary; continuity does not imply differentiability, but differentiability implies continuity.

5.

7.

9.

11. a. 0   **b.** 0   **13. a.** 2   **b.** 2
15. a. $-\Delta x$ **b.** 0   **17.** 0; differentiable for all x
19. 3; differentiable for all x **21.** $6x$; differentiable for all x
23. $2x - 1$; differentiable for all x
25. $2s - 2$; differentiable for all real s
27. $\dfrac{\sqrt{5x}}{2x}$; differentiable for $x > 0$
29. $3x - y - 7 = 0$ **31.** $3s - 4y + 1 = 0$
33. $x + 25y - 7 = 0$ **35.** $x + 3y - 9 = 0$
37. $216x - 6y - 647 = 0$ **39.** 2
41. 0   **43. a.** -3.9 **b.** -4
45. The derivative is 0 when $x = \frac{1}{2}$; the graph has a horizontal tangent at $\left(\frac{1}{2}, -\frac{1}{4}\right)$.

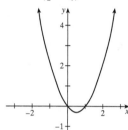

47. a. $-4x$ **b.** $y - 4 = 0$ **c.** $\left(\frac{2}{3}, \frac{28}{9}\right)$ **49.** yes
51. a.

b. f is not differentiable at $x = 1$, because the difference quotient has different limits from the left and right.
53. 4   **55.** $\frac{1}{2}$ **57.** $\frac{1}{4}$
59. The y-intercept is $\left(0, -Ac^2\right)$.   **61. a.** $2x$ **b.** $2x$
63. For $x < 0$, the tangent lines have negative slope and for $x > 0$ the tangent lines have positive slope. There is a corner at $x = 0$, and there is no tangent at $x = 0$, and consequently no derivative at that point.
65. Yes, the equation of the tangent line is $y = 1$.
67. The point $(6, 0)$ on the x-axis is the point through which the normal to the parabola at $(4, 4)$ passes. The required tangent is then the line through $(4, 4)$ perpendicular to this normal.
69. The equation of the normal line is $f'(c)y - f(c)f'(c) + x - c = 0$. If $f'(c) = 0$, then the normal line is vertical with equation $x = c$.

3.2 Techniques of Differentiation (Pages 117–118)

1. (11) 0   (12) 1   (13) 2   (14) 4   (15) 0   (16) -4
3. (23) $2x - 1$ (24) $-2t$ (25) $2s - 2$ (26) $-\frac{1}{2}x^{-2}$ (27) $\dfrac{\sqrt{5x}}{2x}$
5. a. $12x^3$ **b.** -1 **7. a.** $3x^2$ **b.** 1
9. $2t + 2t^{-3} - 20t^{-5}$ **11.** $-14x^{-3} + \frac{2}{3}x^{-1/3}$
13. $1 - x^{-2} + 14x^{-3}$ **15.** $-32x^3 - 12x^2 + 2$
17. $\dfrac{22}{(x+9)^2}$ **19.** $4x^3 + 12x^2 + 8x$

21. $f'(x) = 5x^4 - 15x^2 + 1$; $f''(x) = 20x^3 - 30x$; $f'''(x) = 60x^2 - 30$; $f^{(4)}(x) = 120x$
23. $f'(x) = 4x^{-3}$; $f''(x) = -12x^{-4}$; $f'''(x) = 48x^{-5}$; $f^{(4)}(x) = -240x^{-6}$
25. $\dfrac{d^2y}{dx^2} = 18x - 14$ **27.** $7x + y + 9 = 0$
29. $6x + y - 6 = 0$ **31.** $x - 6y + 5 = 0$
33. $(1, 0)$ and $\left(\frac{4}{3}, -\frac{1}{27}\right)$ **35.** $\left(\frac{29}{6}, -\frac{361}{12}\right)$
37. $(1, -2)$ **39.** no horizontal tangents
41. d. $-4x^{-3} + 9x^{-4}$ **43.** $2x - y - 2 = 0$
45. a. $4x + y - 1 = 0$ **b.** $x = 0$
47. $(0, 0)$ and $(4, 64)$
49. The equation is not satisfied.
51. This function satisfies the given equation.
53. $\pi \approx \frac{62,832}{20,000} \approx 3.1416$ **63.** $y = x + 3$

3.3 Derivatives of Trigonometric, Exponential, and Logarithmic Functions (Page 125)

1. $\cos x - \sin x$ **3.** $2t - \sin t$ **5.** $\sin 2t$
7. $-\sqrt{x}\,\sin x + \frac{1}{2}x^{-1/2}\cos x - x\csc^2 x + \cot x$
9. $-x^2 \sin x + 2x\cos x$ **11.** $\dfrac{x\cos x - \sin x}{x^2}$
13. $e^t \csc t(1 - \cot t)$ **15.** $x + 2x\ln x$
17. $e^{2x}(3\sin x - \cos x)$ **19.** $e^{-x}(\cos x - \sin x)$
21. $\dfrac{\sec^2 x - 2x\sec^2 x + 2\tan x}{(1 - 2x)^2}$
23. $\dfrac{t\cos t + 2\cos t - 2 - \sin t}{(t + 2)^2}$
25. $\dfrac{1}{\cos x - 1}$ **27.** $\dfrac{2\cos x - \sin x - 1}{(2 - \cos x)^2}$
29. $\dfrac{-2}{(\sin x - \cos x)^2}$ **31.** $-\sin x$
33. $-\sin\theta$ **35.** $2\sec^2\theta\tan\theta$
37. $\sec^3\theta + \sec\theta\tan^2\theta$ **39.** $-\sin x - \cos x$
41. $-2e^x \sin x$ **43.** $-\frac{1}{4}t^{-3/2}\ln t$
45. $4x - 2y - \pi + 2 = 0$
47. $\sqrt{3}x - 2y + \left(1 - \frac{\sqrt{3}\pi}{6}\right) = 0$
49. $2x - y = 0$ **51.** $x - y + 1 = 0$
53. a. yes   **b.** yes   **c.** no   **d.** no
55. $A = 0$, $B = -\frac{3}{2}$; $y = -\frac{3}{2}x\sin x$

3.4 Rates of Change: Modeling Rectilinear Motion (Pages 134–137)

1. A mathematical model is a mathematical framework whose results approximate the real-world situation. It involves abstractions, predictions, and then interpretations and comparisons with real world events. An excellent example of real world modeling from *Scientific American* (March 1991) is mentioned on page 129.
3. 1   **5.** -3 **7.** $\frac{13}{4}$
9. -1 **11.** $\frac{1}{2}$ **13.** -1 **15.** -6
17. a. $2t - 2$ **b.** 2   **c.** 2
   d. Because $a(t) > 0$, the object is always accelerating.
19. a. $3t^2 - 18t + 15$ **b.** $6t - 18$ **c.** 46
   d. decelerating on [0, 3]; accelerating on (3, 6]

21. a. $-2t^{-2} - 2t^{-3}$ **b.** $4t^{-3} + 6t^{-4}$

   c. $\frac{20}{9}$ **d.** accelerating on $[1, 3]$

23. a. $-3\sin t$ **b.** $-3\cos t$ **c.** 12

   d. decelerating on $\left[0, \frac{\pi}{2}\right)$ and on $\left(\frac{3\pi}{2}, 2\pi\right]$; accelerating on $\left(\frac{\pi}{2}, \frac{3\pi}{2}\right)$

25. quadratic model   **27.** exponential model

29. quadratic model   **31.** cubic model

33. logarithmic model

35. a. -6 **b.** The decline will be the same each year.

37. 136   **39. a.** 9.91 m/min   **b.** 9.961 m

41. a. 64 ft/s   **b.** 336 ft   **c.** $-32t + 64$ ft/s   **d.** -160 ft/s

43. 30 ft   **45.** 144 ft   **47.** $v_0 = 24$ ft/s; 126 ft

49. a. $200t + 50\ln t + 450$ newspapers per year.

   b. 1,530 newspapers per year   **c.** 1,635 newspapers

51. a. 0.2 ppm/yr   **b.** 0.15 ppm   **c.** 0.25 ppm

53. 91 thousand/hr

55. a. 20 persons per mo   **b.** 0.39% per mo

57. 7.5% per year   **59.** $\frac{dP}{dT} = -\frac{4\pi\mu^2 N}{9k}T^{-2}$

61. a. $v(t) = -7\sin t; a(t) = -7\cos t$ **b.** 2π **c.** 7

63. The Spy is on Mars.

3.5 The Chain Rule (Pages 143–145)

3. $6(3x - 2)$ **5.** $\frac{-8x}{(x^2 - 9)^3}$

7. $[\tan u + u\sec^2 u]\left[3 - \frac{6}{x^2}\right]$

9. a. $3u^2$ **b.** $2x$ **c.** $6x(x^2 + 1)^2$

11. a. $7u^6$ **b.** $-8 - 24x$ **c.** $-56(3x + 1)(12x^2 + 8x - 5)^6$

13. $25(5x - 2)^4$ **15.** $8(3x - 1)(3x^2 - 2x + 1)^3$

17. $4\cos(4\theta + 2)$ **19.** $(-2x + 3)e^{-x^2 + 3x}$

21. $e^{\sec x}\sec x\tan x$ **23.** $(2t + 1)\exp(t^2 + t + 5)$

25. $5x\cos 5x + \sin 5x$ **27.** $6(1 - 2x)^{-4}$

29. $(1 - 2x)e^{1-2x}$ **31.** $2x\cos 2x^2$

33. $3x^2(2 - 3x)(2 - 5x)$

35. $\frac{1}{2}\left(\frac{x^2 + 3}{x^2 - 5}\right)^{-1/2}\left[\frac{-16x}{(x^2 - 5)^2}\right]$

37. $\frac{1}{3}(x + \sqrt{2x})^{-2/3}\left(1 + \frac{1}{\sqrt{2x}}\right)$

39. $\frac{\cos x - \sin x}{\sin x + \cos x}$ **41.** $\frac{2}{9}$ **43.** 1, 7   **45.** $0, \frac{2}{3}$

47. a. 1   **b.** $\frac{3}{2}$ **c.** 1.5

49. a. $\frac{9}{40}$ **b.** $-\frac{3}{8}$ **c.** $\frac{3}{8}$

51. $-\frac{2}{3}$ **53.** 0.31 ppm/yr

55. decreasing by 6 lb/wk

57. a. increasing by 0.035 lux/s.   **b.** 7.15 m

61. $T'(2) \approx 5.23$; it is getting hotter

63. $I'(\theta) = \frac{2a\pi I_0}{\lambda\beta^3}\sin\beta(\beta\cos\beta - \sin\beta)\cos\frac{\theta}{\lambda}$

67. a. $\frac{3}{(3x - 1)^2 + 1}$ **b.** $\frac{-1}{x^2 + 1}$

71. $\frac{d}{dx}[f'(f(x))] = f''(f(x))f'(x)$ and

   $\frac{d}{dx}[f(f'(x))] = f'(f'(x))f''(x)$

3.6 Implicit Differentiation (Pages 155–157)

1. $-\frac{x}{y}$ **3.** $-\frac{y}{x}$

5. $\frac{-(2x + 3y)}{3x + 2y}$ **7.** $-\frac{y^2}{x^2}$

9. $\frac{1 - \cos(x + y)}{\cos(x + y) + 1}$ **11.** $\frac{2x - y\sin xy}{x\sin xy}$

13. $(2e^{2x} - x^{-1})y$

15. a. $-\frac{2x}{3y^2}$ **b.** $\frac{-2x}{3(12 - x^2)^{2/3}} = -\frac{2x}{3y^2}$

17. a. y^2 **b.** $\frac{1}{(x - 5)^2} = y^2$

19. $\frac{1}{\sqrt{-x^2 - x}}$ **21.** $\frac{x}{(x^2 + 2)\sqrt{x^2 + 1}}$

23. $\frac{6(\sin^{-1} 2x)^2}{\sqrt{1 - 4x^2}}$ **25.** $\frac{-1}{\sqrt{e^{-2x} - 1}}$

27. $\frac{-1}{x^2 + 1}$ **29.** ± 1

31. $\frac{1 - \sin^{-1} y - \dfrac{y}{1 + x^2}}{\dfrac{x}{\sqrt{1 - y^2}} + \tan^{-1} x}$

33. $2x - 3y + 13 = 0$ **35.** $(\pi + 1)x - y + \pi = 0$

37. $y = 0$ **39.** $y' = 0$

41. $y' = \frac{5}{4}$ **43.** $x - 1 = 0$

45. $y'' = -\frac{49}{100y^3}$

51. $\frac{dy}{dx} = y\left[\frac{10}{2x - 1} - \frac{1}{2(x - 9)} - \frac{2}{x + 3}\right]$

53. $\frac{dy}{dx} = y\left[6x - \frac{6x^2}{x^3 + 1} + \frac{8}{4x - 7}\right]$

55. $\frac{dy}{dx} = \frac{y\ln x}{x}$ **57.** $(3, -4), (-3, 4)$

61. c. The derivative does not exist.

63. $(2, 0), \left(-\frac{1}{4}, \frac{\sqrt{3}}{4}\right), \left(-\frac{1}{4}, -\frac{\sqrt{3}}{4}\right)$

65. 0.44 rad/s   **69.** $\frac{d^2y}{dx^2} = -\frac{ac}{b^2 y^3}$

73. The student was correct.

3.7 Related Rates and Applications (Pages 162–165)

1. -3 **3.** 1,000   **5.** 15

7. $\frac{4}{5}$ **9.** $\frac{30}{13}$ **11.** -3

13. $-10\sqrt{3}$ units/s   **15.** 0.637 ft/s

17. increasing at $34,000/yr

19. -30 lb/in.^{2}/s   **23.** -3π cm^3/s   **25.** 7.2 ft/s

27. decreasing at 7.5 ft/min

29. 200 ft/s toward the light.   **31.** 2.78 rad/s

33. Assume a sphere of radius r. The volume is changing at the rate of 60.3 cm^3/min.

35. 74.35 ft^3/min.

37. a. $\dfrac{dH}{dt} = \dfrac{3,125t - 6,250}{\sqrt{3,125t^2 - 12,500t + 62,500}}$

    b. $t = 2$ **c.** 224 mi

39. 1 rad/min   **41.** 3.927 mi/h

43. At $t = 2$ P.M., 8.875 knots; at $t = 5$ P.M., 10.417 knots

45. 0.001925 ft/min or 0.02310 in./min

47. a. $\theta = \cot^{-1}\frac{x}{150}$

    b. $d\theta/dt$ approaches 0.27 rad/s

    c. As v increases so will $d\theta/dt$ and it becomes more difficult to see the seals.

3.8 Linear Approximation and Differentials (Pages 173–176)

1. $6x^2\,dx$ **3.** $x^{-1/2}\,dx$

5. $(\cos x - x \sin x)\,dx$ **7.** $\dfrac{3x\sec^2 3x - \tan 3x}{2x^2}\,dx$

9. $\cot x\,dx$ **11.** $\dfrac{e^x}{x}(1 + x\ln x)\,dx$

13. $\dfrac{(x-3)(x^2\sec x\tan x + 2x\sec x) - x^2\sec x}{(x-3)^2}\,dx$

15. $\dfrac{x+13}{2(x+4)^{3/2}}\,dx$

19. 0.995; by calculator, 0.9949874371

21. 217.69; by calculator, 217.7155882 so we can see an error of approximately 0.0255882

23. 0.06 or 6%   **25.** 0.03 or 3%

27. 0.05 parts per million   **29.** reduced by 12,000 units

31. 28.37 in.^{3}

33. decrease of 2 beats every 3 minutes

35. $\pm 2\%$

37. S increases by 2% and the volume increases by 3%

39. 0.0525 ft

41. -6.93 (or about 7) particles/unit area

43. a. 472.7   **b.** 468.70

45. 1.2 units   **47.** -1.4142 **49.** 0.5671

51. b. $x \approx 1.367$ **53.** 0.5183

57. $\Delta x = -3$, $f(97) = 9.85$; by calculator 9.848857802; if $\Delta x = 16$, $f(97) \approx 9.89$

61. $x_n = x_{n+1} = x_{n+2} = \cdots$

Chapter 3 Review
Proficiency Examination (Page 177)

21. $\dfrac{dy}{dx} = 3x^2 + \dfrac{3}{2}x^{1/2} - 2\sin 2x$

22. $\dfrac{dy}{dx} = \dfrac{\sqrt{3x}}{2x} - \dfrac{6}{x^3}$

23. $\dfrac{dy}{dx} = -x[\cos(3 - x^2)][\sin(3 - x^2)]^{-1/2}$

24. $\dfrac{dy}{dx} = \dfrac{-y}{x + 3y^2}$ **25.** $y' = \frac{1}{2}xe^{-\sqrt{x}}(4 - \sqrt{x})$

26. $y' = \dfrac{\ln 1.5}{x(\ln 3x)^2}$ **27.** $y' = \dfrac{3}{\sqrt{1 - (3x+2)^2}}$

28. $y' = \dfrac{2}{1 + 4x^2}$ **29.** $\dfrac{dy}{dx} = 0$

30. $y' = y\left[\dfrac{2x}{(x^2-1)\ln(x^2-1)} - \dfrac{1}{3x} - \dfrac{9}{3x-1}\right]$

31. $y'' = 2(2x - 3)(40x^2 - 48x + 9)$

32. $\dfrac{dy}{dx} = 1 - 6x$ **33.** $14x - y - 6 = 0$

34. tangent line $y - \frac{1}{2} = \frac{\pi}{4}(x-1)$; normal line $y - \frac{1}{2} = -\frac{4}{\pi}(x-1)$

35. 2π ft^2/s

Supplementary Problems (Pages 177–180)

1. $y' = 4x^3 + 6x - 7$

3. $y' = \dfrac{-4x}{(x^2-1)^{1/2}(x^2-5)^{3/2}}$

5. $y' = \dfrac{4x - y}{x - 2}$

7. $y' = 10(x^3 + x)^9(3x^2 + 1)$

9. $y' = \dfrac{(x^3 + 1)^4(46x^3 + 1)}{3x^{2/3}}$

11. $y' = 8x^3(x^4 - 1)^9(2x^4 + 3)^6(17x^4 + 8)$

13. $y' = \dfrac{-\sin\sqrt{x}}{4\sqrt{x}\,\sqrt{\cos\sqrt{x}}}$

15. $y' = 5(x^{1/2} + x^{1/3})^4\left(\frac{1}{2}x^{-1/2} + \frac{1}{3}x^{-2/3}\right)$

17. $y' = (4x + 5)\exp(2x^2 + 5x - 3)$

19. $y' = 3^{2-x}(1 - x\ln 3)$ **21.** $y' = \dfrac{y(1 + xye^{xy})}{x(1 - xye^{xy})}$

23. $y' = e^{\sin x}(\cos x)$ **25.** $y' = \dfrac{e^{-x}}{x\ln 5}(1 - x\ln 3x)$

27. $y' = \dfrac{2x^2 + 2xy^2 - 1}{2y - 2x - 2y^2}$ **29.** $y' = [\cos(\sin x)]\cos x$

31. $y' = \dfrac{2\sqrt{xy} - \sqrt{y}}{\sqrt{x}}$ **33.** $y' = \dfrac{1 - y\cos xy}{x\cos xy - 1}$

35. $\dfrac{dy}{dx} = \dfrac{\sin^{-1}x - x(1-x^2)^{-1/2}}{(\sin^{-1}x)^2} + \dfrac{1}{x^2}\left(\dfrac{x}{1+x^2} - \tan^{-1}x\right)$

37. $\dfrac{d^2y}{dx^2} = 20x^3 - 60x^2 + 42x - 6$

39. $y'' = -\dfrac{2(3y^3 + 4x^2)}{9y^5}$

41. $y'' = -2[1 + y^2 - 4x^2y - 4x^2y^3]$

43. $33x - y - 32 = 0$ **45.** $2\pi x + 4y - \pi^2 = 0$

47. $x + y - 2 = 0$ **49.** $4x - 2y - 1 = 0$

51. $y - 1 = 0$ **53.** $4x - y - 3 = 0$

55. tangent line, $12x - y - 11 = 0$; normal line, $x + 12y - 13 = 0$

57. $\dfrac{dy}{dt} = (3x^2 - 7)(t\cos t + \sin t)$ or $(3t^2\sin^2 t - 7)(t\cos t + \sin t)$

59. $f'(x) = 4x^3 + 4x^{-5}$; $f''(x) = 12x^2 - 20x^{-6}$; $f'''(x) = 24x + 120x^{-7}$; $f^{(4)}(x) = 24 - 840x^{-8}$

61. $y' = \dfrac{-4x^3}{(x^4 - 2)^{4/3}(x^4 + 1)^{2/3}}$

63. $y' = \dfrac{x + 2y}{y - 2x}$; $y'' = \dfrac{5y^2 - 20xy - 5x^2}{(y - 2x)^3}$

65. a. $f'(0) = 0$ **b.** $f'(x) = 2x\sin\frac{1}{x} - \cos\frac{1}{x}$

    c. f' is not continuous at $x = 0$

67. 0.8634074099; calculator 0.8633955506

69. 0.012 rad/min

71. $\dfrac{d}{dx}f(x^2 + x) = (2x + 1)(x^4 + 2x^3 + 2x^2 + x)$

73. $4x\cos 2x^2 - 6x\sin 3x^2$

75. 75 mi/h

77. a. $\frac{4}{5}x + 3$ **b.** $4.00 **c.** $11.00 **d.** $11.40

81. 0.09 radians/s $\approx 5°$/s

83. a. -0.2 **b.** -0.4 **c.** 0.3

85. $\frac{d\theta}{dt} = 0.05$ **87.** $\frac{dx}{dt} = 20\pi$ mi/min

93. For the velocity, $\frac{dx}{dt} = \frac{12\pi x \sin\theta}{2\cos\theta - x}$; for acceleration,

$$\frac{d^2x}{dt^2} = \frac{144\pi^2 x}{(2\cos\theta - x)^3}\left[2\sin^2\theta\cos\theta + (2\cos\theta - x)(1 - \tfrac{1}{2}x\cos\theta)\right]$$

95. This is Putnam Problem 1, morning session in 1946.

97. This is Putnam Problem 6, morning session in 1946.

Chapter 4: Additional Applications of the Derivative

4.1 Extreme Values of a Continuous Function (Pages 193–195)

1. critical number $x = 5$ is not in the interval; maximum value is 26 and the minimum value is -34

3. critical numbers, $x = 0$, $x = 2$; maximum value is 0 and minimum value is -4

5. critical number $x = 0$; maximum value is 1 and minimum value is $-\frac{1}{8}$

7. critical numbers $x = 0$, $\frac{4}{5}$; maximum value is 0 and minimum value is -2

9. critical number, $t = 1$; maximum value is e^{-1} and minimum value is 0

11. critical number $x = 0$; maximum value is 1 and minimum value is 0

13. critical numbers 0, $\frac{\pi}{3}$; maximum value is 1.25 and minimum value is approximately 0.41067

15. a. Find the value of the function at the endpoints of an interval.

b. Find the critical points, that is, points at which the derivative of the function is zero or undefined.

c. Find the value of the function at each critical point.

d. State the absolute extrema.

17. The calculator does not seem to take the derivative at $x = 0$ and $x < 0$ into account. The derivative is not defined at $x = 0$, but it certainly is defined for $x < 0$. If you enter $(x^2)^\wedge(1/3)(5 - 2x)$ you will obtain the correct graph.

19. not defined at $t = -50$; the maximum value is 16 and the minimum value is 0.

21. the maximum value is 76 and the minimum value is -32

23. the maximum value is $\frac{5}{6}$ and the minimum value is $\frac{1}{6}$

25. the maximum value is 2π and the minimum value is $-\pi$

27. the maximum value is 0.2898 and the minimum value is -0.2898

29. the maximum value is 11 and the minimum value is -4

31. the smallest value is 3

33. the largest value is $\frac{9}{4}$

35. the largest value of g on [2, 3] is approximately -1.1

37. the maximum value is approximately 1.819 and the minimum value is 0

39. the maximum value is 6,496 and the minimum value is 0

41. the maximum value is 1.59 and the minimum value is -2.52

43. the maximum value is 1 and the minimum value is $-e^{-\pi}$

45. a. $f(x) = \begin{cases} x^2 & \text{for } -0.5 \le x \le 1 \\ -x + 3 & \text{for } 1 < x \le 1.5 \end{cases}$
The minimum is 0, but there is no maximum.

b. $f(x) = \begin{cases} -x^2 & \text{for } -1 < x \le 1 \\ x - 3 & \text{for } 1 < x \le 1.5 \end{cases}$
The maximum is 0, but there is no minimum.

c. $f(x) = \begin{cases} \sin x & \text{for } -\pi \le x \le \frac{\pi}{2} \\ 0.5 & \text{for } \frac{\pi}{2} < x \le 3 \end{cases}$
The maximum is 1 and the minimum is -1.

d. $f(x) = \begin{cases} (1+x)/2 & -1 \le x < 1 \\ 0 & 1 \le x \le 2 \\ (x-4)/2 & 2 < x \le 4 \end{cases}$
There is no maximum and there is no minimum.

47. a. No such function can be found because of the extreme value theorem (Theorem 4.1).

b. No such function can be found because of the extreme value theorem.

c. $f(x) = \sin x$ for [0, 2π].

d. No such function can be found because of the extreme value theorem.

49. $f(x) = x^{-1}$ on (0, 1) has no extremum.

51. The maximum velocity is 60 when $t = 0$.

53. The largest product occurs when $x = 3$ and $y = 6$.

55. The largest product occurs when $x = 21$ and $y = 63$.

57. The largest area occurs when the length of a side is one-fourth of the length of the perimeter.

59. a. The greatest difference occurs at $x = \frac{1}{2}$.

b. The greatest difference occurs at $x = 1/\sqrt{3}$.

c. The greatest difference occurs at $x = \left(\dfrac{1}{n}\right)^{1/(n-1)}$.

4.2 The Mean Value Theorem (Pages 199–201)

3. $c = 1$ **5.** $c \approx 1.5275$ **7.** $c \approx 1.0772$

9. $c = 9/4$ **11.** $c \approx 0.73$ **13.** $c \approx 0.6901$

15. $c \approx 0.54$ **17.** $c \approx 1.082$ **19.** $c \approx 0.5227$

21. does not apply **23.** applies **25.** does not apply

27. does not apply **29.** applies

31. $f(x) = 8x^3 - 6x + 10$ **33.** does not apply

35. applies **37.** $c = 2.5$ and $c = 6.25$

41. does not apply **43.** does not apply

47. does not apply **55.** does not apply

4.3 Using Derivatives to Sketch the Graph of a Function (Pages 214–217)

7. The black curve is the function and the one in color is the derivative.

9. **11.**

13. a. $x = 0, \quad x = -2$

b. increasing on $(-\infty, -2) \cup (0, +\infty)$; decreasing on $(-2, 0)$

c. critical points: $(0, 1)$, relative minimum; $(-2, 5)$, relative maximum

d. $x = -1$; concave down on $(-\infty, -1)$; concave up on $(-1, \infty)$

e.

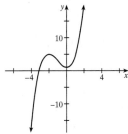

15. a. critical numbers: $x = \frac{5}{3}, x = -25$

b. increasing on $(-\infty, -25) \cup (\frac{5}{3}, +\infty)$; decreasing on $(-25, \frac{5}{3})$

c. critical points: $(\frac{5}{3}, -9{,}481)$, relative minimum; $(-25, 0)$, relative maximum

d. $x = -\frac{35}{3}$; concave down on $(-\infty, -\frac{35}{3})$; concave up on $(-\frac{35}{3}, \infty)$

e.

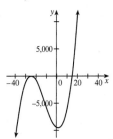

17. a. critical numbers: $x = -1, x = 3$

b. increasing on $(-1, 3)$; decreasing on $(-\infty, -1) \cup (3, +\infty)$

c. critical points: $(-1, -\frac{1}{2})$, relative minimum; $(3, \frac{1}{6})$, relative maximum

d. $x \approx -2.1, x \approx 0.3, x \approx 4.8$; concave down on $(-\infty, -2.1)$ and $(0.3, 4.8)$; concave up on $(-2.1, 0.3)$

e.

19. a. critical numbers: $t = -1, t = 3$

b. increasing on $(-\infty, -1) \cup (3, +\infty)$; decreasing on $(-1, 3)$

c. critical points: $(3, -32)$, relative minimum; $(-1, 0)$, relative maximum

d. $t = 1$; concave down on $(-\infty, 1)$; concave up on $(1, \infty)$

e.

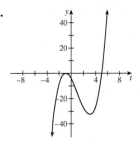

21. critical points: $(-3, -11)$, relative maximum; $(3, 13)$, relative minimum; increasing on $(-\infty, -3) \cup (3, +\infty)$; decreasing on $(-3, 0) \cup (0, 3)$; concave up on $(0, +\infty)$; concave down on $(-\infty, 0)$

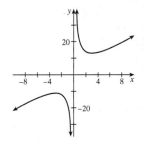

23. critical points: $(0, 26)$, relative maximum; $(-6.38, -852.22)$, relative minimum; $(1.88, -6.47)$, relative minimum; inflection points $(-4, -486), (1, 9)$; increasing on $(-6.38, 0) \cup (1.88, +\infty)$; decreasing on $(-\infty, -6.38) \cup (0, 1.88)$; concave up on $(-\infty, -4) \cup (1, +\infty)$; concave down on $(-4, 1)$

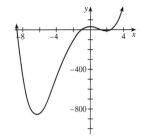

25. critical point: $(0, 0)$, relative minimum; increasing on $(0, +\infty)$; decreasing on $(-\infty, 0)$; concave up on $(-\infty, +\infty)$

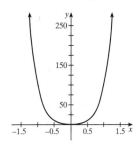

27. critical points: $(-1, -\frac{1}{2})$, relative minimum; $(1, \frac{1}{2})$, relative maximum; $(0, 0)$ is a point of inflection; points of inflection: $(\sqrt{3}, \frac{\sqrt{3}}{4}), (-\sqrt{3}, -\frac{\sqrt{3}}{4})$; increasing on $(-1, 1)$; decreasing on $(-\infty, -1) \cup (1, +\infty)$; concave up on $(-\sqrt{3}, 0) \cup (\sqrt{3}, +\infty)$; concave down on $(-\infty, -\sqrt{3}) \cup (0, \sqrt{3})$

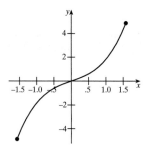

29. critical points: $(0, 0)$, relative minimum; $(0.67, 0.06)$, relative maximum; $(0.195, 0.021)$, point of inflection; $(1.138, 0.043)$, point of inflection; decreasing on $(-\infty, 0)$, $(0.67, +\infty)$; increasing on $(0, 0.67)$; concave up on $(-\infty, 0.195) \cup (1.138, +\infty)$; concave down on $(0.195, 1.138)$

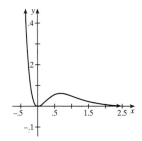

31. critical points $(1, 0)$, relative minimum; $(e, 1)$, point of inflection; decreasing on $(0, 1)$; increasing on $(1, +\infty)$; concave up for $(0, e)$; concave down for $(e, +\infty)$

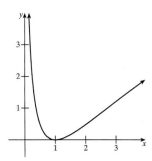

33. critical points: $(\frac{\pi}{12}, 1.13)$, relative maximum; $(\frac{5\pi}{12}, 0.44)$, relative minimum; inflection points: $(\frac{\pi}{4}, 0.79)$, $(\frac{3\pi}{4}, 2.36)$; increasing on $(0, \frac{\pi}{12}) \cup (\frac{5\pi}{12}, \pi)$; decreasing on $(\frac{\pi}{12}, \frac{5\pi}{12})$; concave up on $(\frac{\pi}{4}, \frac{3\pi}{4})$; concave down on $(0, \frac{\pi}{4}) \cup (\frac{3\pi}{4}, \pi)$

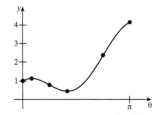

35. There are no critical points; the function is increasing for all x. Inflection point $(0, 0)$ concave down $(-1.57, 0)$; concave up $(0, 1.57)$

37. At $x = \frac{1}{2}$, relative maximum; at $x = 1$, relative minimum

39. At $x = 4$, relative minimum

41. $f(x) = \dfrac{x^2 - x + 5}{x + 4}$; $f'(x) = \dfrac{x^2 + 8x - 9}{(x + 4)^2}$; $f''(x) = \dfrac{50}{(x + 4)^3}$; $f''(1) = \frac{2}{5} > 0$; relative minimum $f''(-9) = -\frac{2}{5} < 0$; relative maximum

43. $f(x) = \sin x + \frac{1}{2} \cos 2x$; $f'(x) = \cos x - \sin 2x$; $f''(x) = -2 \cos 2x - \sin x$; $f''(\frac{\pi}{2}) = 1 > 0$; relative minimum; $f''(\frac{\pi}{6}) = -\frac{3}{2} < 0$; relative maximum

Answers for Problems 45–49 may vary.

45. **47.**

49. **51.**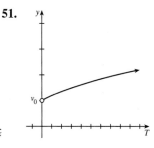

53. a. $-(2)^3 + 6(2)^2 + 13(2) = 42$; $-\frac{1}{3}(2)^3 + \frac{1}{2}(2)^2 + 25(2) = 49\frac{1}{3}$

b. $N(x) = -x^3 + 6x^2 + 13x - \frac{1}{3}(4-x)^3 + \frac{1}{2}(4-x)^2 + 25(4-x)$

c. $N'(x) = -2x^2 + 5x = 0$ when $x = 2.5$, so the optimum time for the break is 10:30 A.M. The optimum time is half the value for the diminishing return; $N''(2.5) < 0$.

55. Maximum deflection at $x = \dfrac{2\ell}{3}$

61. $f(x) = -3x^3 + 9x^2 - 1$

4.4 Curve Sketching with Asymptotes: Limits Involving Infinity (Pages 227–228)

5. 0 **7.** 3 **9.** 9

11. 1 **13.** 0 **15.** $-\frac{1}{2}$

17. $+\infty$ **19.** 1 **21.** $-\infty$

23. 0

25. asymptotes: $x = 7$, $y = -3$; graph rising on $(-\infty, 7) \cup (7, +\infty)$; concave up on $(-\infty, 7)$; concave down on $(7, +\infty)$; no critical points; no points of inflection;

27. asymptotes: $x = 3$, $y = 6$; graph falling on $(-\infty, 3) \cup (3, +\infty)$; concave up on $(3, +\infty)$; concave down on $(-\infty, 3)$; no critical points; no points of inflection;

29. asymptotes: $x = 2$, $y = 1$; graph falling on $(-\infty, 2) \cup (2, +\infty)$; concave up on $(-\sqrt[3]{4}, 0)$ or $(2, +\infty)$; concave down on $(-\infty, -\sqrt[3]{4})$ or $(0, 2)$; critical point is $(0, -\frac{1}{8})$; points of inflection $(0, -\frac{1}{8})$, $(-\sqrt[3]{4}, \frac{1}{4})$;

31. asymptotes: $x = -4$, $x = 1$, $y = 0$; graph falling on $(-\infty, -4) \cup (-4, 1) \cup (1, +\infty)$; concave up on $(-4, -1)$ or $(1, +\infty)$; concave down on $(-\infty, -4)$ or $(-1, 1)$; no critical points; point of inflection is $(-1, 5)$;

33. no asymptotes; graph rising on $(0, +\infty)$; graph falling on $(-\infty, 0)$; concave up on $(-\infty, +\infty)$; critical point is $(0, 0)$, which is a relative minimum; no points of inflection;

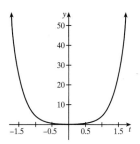

35. no asymptotes; graph rising on $(-3, 0) \cup (3, +\infty)$; graph falling on $(-\infty, -3) \cup (0, 3)$; concave up on $(-\infty, -\sqrt{3}) \cup (\sqrt{3}, +\infty)$; concave down on $(-\sqrt{3}, \sqrt{3})$; critical points are $(-3, 0)$, relative minimum; $(0, 81)$, relative maximum; $(3, 0)$, relative minimum; points of inflection are $(-\sqrt{3}, 36)$, $(\sqrt{3}, 36)$;

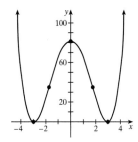

37. no asymptotes; graph rising on $(1, +\infty)$; graph falling on $(-\infty, 1)$; concave up on $(-\infty, -2) \cup (0, +\infty)$; concave down on $(-2, 0)$; critical points are $(1, -3)$, relative minimum; $(0, 0)$, vertical tangent; $(0, 0)$ and $(-2, 6\sqrt[3]{2})$ are points of inflection;

39. horizontal asymptote at $y = \frac{\pi}{2}$; relative minimum at $x = 0$;

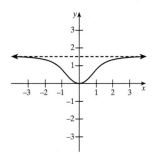

41. no asymptotes; graph rising on $(0, \frac{3\pi}{4}) \cup (\frac{7\pi}{4}, 2\pi)$; graph falling on $(\frac{3\pi}{4}, \frac{7\pi}{4})$; concave up on $(0, \frac{\pi}{4})$ or $(\frac{5\pi}{4}, 2\pi)$; concave down on $(\frac{\pi}{4}, \frac{5\pi}{4})$; critical points are $(\frac{3\pi}{4}, \sqrt{2})$, relative maximum; $(\frac{7\pi}{4}, -\sqrt{2})$, relative minimum; points of inflection: $(\frac{\pi}{4}, 0)$, $(\frac{5\pi}{4}, 0)$;

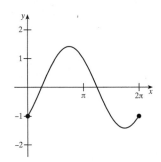

43. critical number at $x = \frac{\pi}{2}$; no asymptotes; graph rising on $(\frac{\pi}{2}, \pi)$; graph falling on $(0, \frac{\pi}{2})$; concave up on $(0, \pi)$; critical points are $(\frac{\pi}{2}, 0)$, relative minimum; $(0, 1)$, and $(\pi, 1)$ relative maxima; no points of inflection;

45.   **47.**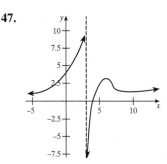

49. Frank is not correct.

51. a. $+\infty$ **b.** $+\infty$ **c.** $-\infty$

    d. $-\infty$ **e.** $+\infty$

53. $a = \frac{9}{5}, b = \frac{3}{5}$ **55.** ∞ **57.** True   **59.** False

4.5 l'Hôpital's Rule (Pages 236–237)

1. a. The limit is not an indeterminate form. The correct limit is $\frac{2}{\pi}$.

    b. The limit is not an indeterminate form. The correct limit is $\frac{2}{\pi}$.

3. $\frac{3}{2}$ **5.** 10   **7.** not defined

9. $\frac{1}{2}$ **11.** 0   **13.** $\frac{1}{2}$

15. 3   **17.** $\frac{3}{2}$ **19.** 2

21. $+\infty$ **23.** 2   **25.** $\frac{1}{2}$

27. 0   **29.** $-\infty$ **31.** 0

33. $e^{3/2}$ **35.** e^2 **37.** $+\infty$

39. $+\infty$ **41.** $+\infty$ **43.** $-\infty$

45. 1   **47.** 0   **49.** limit does not exist   **51.** $\frac{1}{120}$

53. horizontal asymptote, $y = 0$

55. horizontal asymptote, $y = 1$

61. b. 1   **63.** $a = -2; b = \frac{7}{3}$ **65.** If $C = \frac{1}{4}$, the limit is $\frac{5}{4}$.

67. $\displaystyle\lim_{\beta \to \alpha} \left[\frac{C}{\beta^2 - \alpha^2} (\sin \alpha t - \sin \beta t) \right] = \frac{-Ct}{2\alpha} \cos \alpha t$

4.6 Optimization in the Physical Sciences and Engineering (Pages 245–250)

3. The dimensions of the garden should be 8 ft by 8 ft.

7. The largest rectangle has sides of length $R/\sqrt{2}$ and $\sqrt{2}\, R$.

9. The dimensions are $h = \dfrac{R}{3}\sqrt{3}$ and $r = \dfrac{R}{3}\sqrt{6}$.

11. A circumscribed square of side $s = \dfrac{L}{\sqrt{2}} + \dfrac{L}{\sqrt{2}} = \sqrt{2}L$ yields a maximum area.

13. The minimum distance is 200 mi.

15. The minimal perimeter is obtained when the rectangle is a square.

17. $11,664$ in.$^3 = 6.75$ ft^3

19. a. When $s = 4$, Missy should row all the way.

    b. When $s = 6$, she should land at a point 4.5 km from point B and run the rest of the way (1.5 km).

21. $60.00

23. He has about 5 minutes 17 seconds to defuse the bomb.

25. $r \approx 3.84$ and $h \approx 7.67$ cm

27. b. When $p = 12$, the largest value of f is 6.

29. $x = \dfrac{Md}{\sqrt{4m^2 - M^2}}$

31. a. $T'(x) = -T\left[\dfrac{c}{(kx + c)x} + \ln p \right]$

    b. $x = \dfrac{-c \ln p + \sqrt{c^2(\ln p)^2 - 4kc \ln p}}{2k \ln p}$

    c.

33. a. The maximum height occurs when $x = \dfrac{mv^2}{32(m^2 + 1)}$

    b. The maximum height occurs when $y' = 0$ or when $m = v^2/(32x_0)$.

35.

As $\lambda \to +\infty$, $E(\lambda) \to m_0 c^2$.

37. 26 cm is a minimum.

39. The minimum value occurs when $T \approx 4°$.

41. 270 cm^3

43. b. $D'(\alpha) = 2 - 4\left[\dfrac{\cos \alpha}{1.33 \cos \beta} \right]$

    c. The rainbow angle is 0.742 (or about 42.5°).

4.7 Optimization in Business, Economics, and the Life Sciences (Pages 259–263)

1. $x = 26$ **3.** $x = 20$

5. The minimum average cost is 25.

7. a. $C(x) = 5x + \dfrac{x^2}{50}$; $R(x) = x\left(\dfrac{380 - x}{20}\right)$;

$P(x) = -0.07x^2 + 14x$

b. The maximum profit occurs when the price is \$14/item. The maximum profit is \$700.

9. Profit is maximized when $x = 60$.

11. a. 1.22 million people/yr   **b.** The percentage rate is 2%.

13. 208 years from now   **15.** 45 times per year

17. a. $x = 8$ **c.**

19. Since the optimum solution is over 100 years, you should will the book to your heirs so they can sell it in 117.19 years.

21. Sell the boards at a price of \$42, sell 39 per month, and have the maximum profit of \$507.

23. Lower the fare \$250.

25. Plant 80 total trees, have an average yield per tree of 320 oranges, and a maximum total crop of 25,600 oranges.

27. 62 vines

29. a. The most profitable time to conclude the project is 10 days from now.

b. Assume R is continuous over $[0, 10]$.

31. a. $R(x) = xp(x) = \dfrac{bx - x^2}{a}$ on $[0, b]$; R is increasing on $\left(0, \dfrac{b}{2}\right)$ and decreasing on $\left(\dfrac{b}{2}, b\right)$

b.

33. $v = \dfrac{kv_1}{k - 1}$

35. a. $x = r$ is a maximum

b.

c.

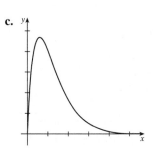

37. $v = 39$

39. a.

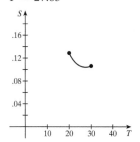

The largest survival percentage is 60.56%, and the smallest survival percentage is 22%.

b. $S'(T) = -(-0.06T + 1.67)(-0.03T^2 + 1.67T - 13.67)^{-2}$; $T \approx 27.83$

c.

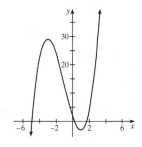

The largest hatching occurs when $T \approx 23.58$ and the smallest when $T = 30$.

41. a. $C = Sx + \dfrac{pQ}{nx}$; $x = \sqrt{\dfrac{pQ}{nS}}$

51. $\theta \approx 0.9553$; this is about $55°$.

Chapter 4 Review
Proficiency Examination (Page 263)

18. 2   **19.** $-\frac{1}{2}$ **20.** 0   **21.** e^6

22. Relative maximum at $(-3, 29)$; relative minimum at $(1, -3)$; inflection point at $(-1, 13)$;

23. inflection points at approximately $\left(-\frac{27}{2}, -96.43\right)$ and $(0, 0)$; relative maximum at $\left(\frac{27}{4}, 38.27\right)$;

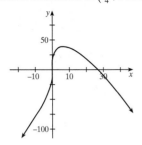

24. relative maximum at $\left(0, \frac{1}{4}\right)$

25. relative minimum at $(-1, -2e)$; relative maximum at $(3, 6e^{-3})$;

26. $(0, 0)$ is a point of infection; the graph is concave up for $x < 0$ and down for $x > 0$.

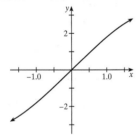

27. relative maximum where $x = \pi$; inflection points at $x = \frac{\pi}{3}$ and $x = \frac{5\pi}{3}$

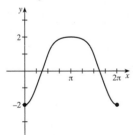

28. absolute maximum at $(0.4, 5.005)$; absolute minimum at $(1, 4)$

29. The dimensions of the box are 19 in. × 19 in. × 10 in.
30. hire 7 people

Supplementary Problems (Pages 263–268)

1.

3.

5.

7.

9.

11.

13.

15.

17.

19.

21. c   **23.** b
25. maximum $f(0) = 12$; minimum $f(2) = -4$
27. minimum $f(0) = 0$; maximum $f(\sqrt{3}/2) \approx 0.68$
29. 0   **31.** $-\frac{1}{2}$ **33.** 0
35. This limit does not exist.
37. 1   **39.** 9   **41.** 0
43. e^4 **45.** 1   **47.** 0
49. $+\infty$ **51.** f is the function and g is the derivative
55. $A = \frac{1}{2}, B = 0, C = -\frac{3}{2}, D = 0$
57. The maximum profit of $108,900$ is reached when 165 units are rented at $740 each.

59. maximum yield of 6,125 for 35 trees per acre

61. 28,072 ft of pipe laid on the shore gives the minimum cost.

63. $c = 4ab$

65. The price is 90 and the maximum profit is 1,100.

73. $x = \sqrt{ab}$ leads to a relative minimum; $x = -\sqrt{ab}$ leads to a relative maximum.

75. a. The highest value is 165,000 when $N = 20$; this is in the year 2002.

　b. The model predicts that the highest point occurs when $x \approx 11.21$ or in the year 1991. This model predicts it will reach 0 when $x \approx 15.25$, or in the year 1995.

　c. This model seems to do a better job of fitting the data.

　d. Linear model: $y = 5,122x - 16,387$; exponential model: $y = 347e^{0.4888x}$; logarithmic model: $y = -21,712\ln(12 - x) + 49,700$

77. a. $f'(x) > 0$ on $\left(-\frac{3}{2}, -\frac{1}{3}\right) \cup (1, 2)$ **b.** $f'(x) < 0$ on $\left(-\frac{1}{3}, 1\right)$

　c. $f''(x) > 0$ on $\left(\frac{1}{3}, +\infty\right)$ **d.** $f''(x) < 0$ on $\left(-\infty, \frac{1}{3}\right)$

　e. $f'(x) = 0$ at $x = -\frac{1}{3}, 1$ **f.** $f'(x)$ exists everywhere

　g. $f''(x) = 0$ at $x = -\frac{1}{3}$

79. a. $f'(x) > 0$ on $(-2, -0.876) \cup (0.876, 2)$

　b. $f'(x) < 0$ on $(-0.876, 0) \cup (0, 0.876)$

　c. $f''(x) > 0$ on $(0, 2)$ **d.** $f''(x) < 0$ on $(-2, 0)$

　e. $f'(x) = 0$ at $x \approx \pm 0.876$

　f. $f'(x)$ does not exist at $x = 0$ **g.** $f''(x) \neq 0$

81. Distance is minimized in the neighborhood of $x = 0.460355$.

83. a. Cost is minimized in the neighborhood of 27 mi/hr.

　b. The total cost is minimized at $x = 55$ mi/hr.

85. This is Putnam Problem 11 of the afternoon session of 1938.

87. This is Putnam Problem 1 of the morning session of 1987.

89. This is Putnam Problem 15 of the afternoon session of 1940.

Chapter 5: Integration
5.1 Antidifferention (Pages 281–282)

1. $2x + C$　　　　　　　　**3.** $x^2 + 3x + C$

5. $t^4 + t^3 + C$　　　　　**7.** $\frac{1}{2}\ln|x| + C$

9. $2u^3 - 3\sin u + C$　　**11.** $\tan\theta + C$

13. $-2\cos\theta + C$　　　**15.** $5\sin^{-1} y + C$

17. $\frac{2}{5}u^{5/2} - \frac{2}{3}u^{3/2} - \frac{1}{9}u^{-9} + C$

19. $\frac{1}{3}x^3 + \frac{2}{5}x^{5/2} + C$　　　**21.** $-t^{-1} + \frac{1}{2}t^{-2} - \frac{1}{3}t^{-3} + C$

23. $\frac{4}{5}x^5 + \frac{20}{3}x^3 + 25x + C$　**25.** $-x^{-1} - \frac{3}{2}x^{-2} + \frac{1}{3}x^{-3} + C$

27. $x + \ln|x| + 2x^{-1} + C$　　**29.** $x - \sin^{-1} x + C$

31. $F(x) = \frac{1}{3}x^3 + \frac{3}{2}x^2$　　　**33.** $F(x) = \frac{1}{2}x^2 + 4x^{3/2} + 9x - 40$

35. $F(x) = \ln|x| - x^{-1} - 1$　**37.** $F(x) = \frac{1}{2}x^2 + e^x + 1$

39. a. $F(x) = 2\sqrt{x} - 4x + 2$

　b.

　c. $C_0 = -\frac{9}{4}$

41. $249

43. The population in 8 months will be 10,128.

45. $k = 20$ ft/s^2

47. The acceleration of the plane at liftoff is 4.3 ft/s^2

49. $k = -32$ ft/s^2　　**51.** 28.8　　　**53.** 21

55. $e^2 - 3$　　　　　　**57.** 1

59. a. $\frac{1}{2}\sec^{-1} y + C$　　　　**b.** $\frac{1}{2}\tan^{-1}\sqrt{y^2 - 1}$

　c. Technology does not use "$+C$," and to reconcile the inverse trigonometric functions, use a reference triangle where $\theta = \sec^{-1} y$, so $y = \sec\theta$. Then $\tan\theta = \sqrt{y^2 - 1}$ so that $\tan^{-1}(y^2 - 1) = \sec^{-1} y$.

61. $A = \frac{1}{2}(d - c)[m(d + c) + 2b]$

5.2 Area as the Limit of a Sum (Pages 288–290)

1. 6　　　　**3.** 120　　**5.** 225　　**7.** 9,800

9. $\frac{1}{2}$　　　**11.** 3　　　　**13. a.** 3.5　　**b.** 3.25

15. a. 2.71875　　**b.** 2.58796

17. 0.795　　　　　**19.** 1.269　　　**21.** 1.2033　　**23.** 18

25. 68　　　　　　　**27.** 2　　　　　**29.** false　　**31.** true

33. true　　**37. a.** $\frac{14}{3}$　　**c.** The statement is true.

39. 21.3646　　　　　**41.** 0.6018907　　　**43.** 0.5038795

45. a. 2　　**c.** The statement is true.

49. $A = \lim\limits_{n \to +\infty}\left[\sum\limits_{k=1}^{n} f\left(a + \dfrac{(2k+1)(b-a)}{2n}\right)\left(\dfrac{b-a}{n}\right)\right]$

5.3 Riemann Sums and the Definite Integral (Pages 301–302)

1. 2.25　　　　　　　**3.** 10.75　　　　　**5.** −0.875

7. 1.18　　　　　　　**9.** 1.94　　　　　　**11.** 28.875

13. 1.896　　　　　　**15.** 0.556　　　　　**17.** $-\frac{1}{3}$

19. $\frac{3}{2}$　　　　　　　**21.** $-\frac{1}{2}$　　　　　　**23.** $\frac{1}{2}$

25. $\int_{-2}^{4} f(x)\,dx = 1$; $\int_{-2}^{4} g(x)\,dx = 1$　**27.** 10

29.

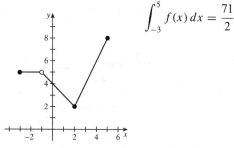

$$\int_{-3}^{5} f(x)\,dx = \frac{71}{2}$$

35. $2.3; \ \|P\| = 1.1$

5.4 The Fundamental Theorems of Calculus (Pages 308–309)

1. 140 **3.** $16 + 8a$ **5.** $\frac{15}{4}a$

7. $\frac{3}{8}c$ **9.** 18 **11.** $\frac{5}{8} + \pi^2$

13. $\dfrac{2^{2a+1} - 1}{2a + 1}$ **15.** 15 **17.** $\frac{272}{15}$

19. 2 **21.** $\dfrac{a\pi}{2}$ **23.** 5

25. $2 - e$ **27.** $\frac{5}{2}$ **29.** 4

31. $\frac{8}{3}$ **33.** 1 **35.** $e - \frac{3}{2}$

37. $3\ln 2 - \frac{1}{2}$ **39.** $(x - 1)\sqrt{x + 1}$ **41.** $\dfrac{\sin t}{t}$

43. $\dfrac{-1}{\sqrt{1 + 3x^2}}$ **55.** $\frac{1}{8}(8 + 3\pi^2) \approx 4.7011$

57. a. relative minimum   **b.** $g(1) = 0$ **c.** $x = 0.75$

d.

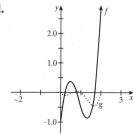

61. $8x^7 - 12x^3$ **63.** $y = \frac{1}{2}(x - 1)$

65. $F'(x) = f(v)\dfrac{dv}{dx} - f(u)\dfrac{du}{dx}$

5.5 Integration by Substitution (Pages 315–316)

1. a. 32 **b.** $2\sqrt{3} - 2$ **3. a.** 0 **b.** 0

5. a. $\frac{128}{5}$ **b.** $\frac{128}{5}$

7. a. $\frac{2}{9}\sqrt{2}\,x^{9/2} + C$ **b.** $\frac{1}{9}(2x^3 - 5)^{3/2} + C$

9. $\frac{1}{10}(2x + 3)^5 + C$

11. $-\ln\left|\cos(x^2 + 5x + 3)\right| + C$ or $\ln\left|\sec(x^2 + 5x + 3)\right| + C$

13. $\frac{1}{3}x^3 - \frac{1}{3}\sin 3x + C$ **15.** $\cos(4 - x) + C$

17. $\frac{1}{6}(t^{3/2} + 5)^4 + C$ **19.** $-\frac{1}{2}\cos(3 + x^2) + C$

21. $\frac{1}{4}\ln(2x^2 + 3) + C$ **23.** $\frac{1}{6}(2x^2 + 1)^{3/2} + C$

25. $\frac{2}{3}e^{x^{3/2}} + C$ **27.** $\frac{1}{3}(x^2 + 4)^{3/2} + C$

29. $\frac{1}{2}(\ln x)^2 + C$ **31.** $2\ln(\sqrt{x} + 7) + C$

33. $\ln(e^t + 1) + C$ **35.** $\frac{5}{6}\ln 3$

37. 0 **39.** $e - e^{1/2}$

41. $\frac{1}{2}\ln 2$ **43.** 1.80

45. a. We take 1 Frdor as the variable so the note from the students reads "Because of illness I cannot lecture between Easter and Michaelmas."

b. The Dirichlet function is defined as a function f so that $f(x)$ equals a determined constant c (usually 1) when the variable x takes a rational value, and equals another constant d (usually 0) when this variable is irrational. This famous function is one which is discontinuous everywhere.

47. $\frac{16}{3}$ **49.** $\sqrt{10} - \sqrt{2}$ **51.** 0 **53.** 0

55. a. true **b.** true **c.** false   **57.** $F(x) = -\frac{1}{3}\ln\left|1 - 3x^2\right| + 5$

59. The amount of water at 4 seconds is 2 ft³. The depth at that time is about $\frac{1}{4}$ in.

61. a. $L(t) = 0.03\sqrt{36 + 16t - t^2} + 3.82$

b. The highest level is 4.12 ppm.

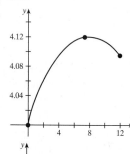

c. It is the same as 11:00 A.M. $(t = 4)$ at 7:00 P.M. (when $t = 12$).

5.6 Introduction to Differential Equations (Pages 325–328)

9. $x^2 + y^2 = 8$ **11.** $y = \tan\left(x - \frac{3\pi}{4}\right)$ **13.** $x^{3/2} - y^{3/2} = 7$

15.

17.

19.

21. $y = Be^{(3/2)x^2}$

23. $2(1 - x^2)^{3/2} + 3y^2 = C$

25. $x^{3/2} + 3y^{1/2} = C$

27. $\cos x + \sin y = C$

29. $-\frac{1}{3}(1 - y^2)^{3/2} = \frac{1}{2}(\ln x)^2 + C$ **31.** $xy = C$ **33.** $y = Cx$

35.

Orthogonal trajectories

37.

Orthogonal trajectories

39.   **41.**

Orthogonal trajectories

Orthogonal trajectories

43. $\frac{dQ}{dt} = kQ$ **45.** $\frac{dT}{dt} = c(T - T_m)$ **47.** $\frac{dQ}{dt} = kQ(B - Q)$

49. Conjecture: The orthogonal trajectories are circles.
51. There was still 78% of the ^{14}C still present.
53. 1326 A.D.   **55.** 28 min
57. a. 98.5 ft/s **b.** $s = 3,956.067; h \approx 352$ ft
59. a. 33,000 **b.** 42,000
61. 10 years from now   **63.** 2 hr and 53 min
65. It will take about 65 hr for 43% to be left. Neptunium-139 has a half-life of 53.70 hours.
67. 7.182%; the percentage lost is always the same.

5.7 The Mean Value Theorem for Integrals; Average Value (Pages 332–333)

1. 1.55 is in the interval [1, 2].
3. $\sqrt{5} \approx 2.24$ is in the interval.
5. The mean value theorem does not apply.
7. 0.0807 is in the interval.
9. −0.187 is in the interval.
11. $A = 25$ **13.** $A = 38/3$

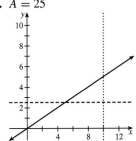

15. $A \approx 1.839$ **17.** $\frac{3}{2}$ **19.** 0
21. 0.3729 **23.** $\frac{4}{3}$
25. −10 **27.** −25.375
29. 2.36

33. Avg temperature = 49.4 °F
35. a. 1,987.24
   b. The average population is reached at just under 14 minutes.
37. $f(x) = (x + 1)\cos x + \sin x$

5.8 Numerical Integration: The Trapezoidal Rule and Simpson's Rule (Pages 339–342)

1. Trapezoidal rule; 2.34375; Simpson's rule, 2.33333; exact value, $\frac{7}{3}$
3. a. 0.7825   **b.** 0.785

5. a. 2.037871   **b.** 2.048596
7. a. 0.584   **b.** 0.594
11. $A \approx 0.775$; the exact answer is between $0.775 - 0.05$ and $0.775 + 0.05$
13. $A \approx 0.455$; the exact answer is between $0.455 - 0.0005$ and $0.455 + 0.0005$
15. $A \approx 3.25$; the exact answer is between $3.25 - 0.01$ and $3.25 + 0.01$
17. $A \approx 0.44$; the exact answer is between $0.44 - 0.01$ and $0.44 + 0.01$
19. a. $n = 164$ **b.** $n = 18$ **21. a.** $n = 184$ **b.** $n = 22$
23. a. $n = 82$ **b.** $n = 8$ **25. a.** 3.1 **b.** 3.1
27. 430   **29.** 613 ft^2 **31.** 79.17
33. a. Simpson error is 0. **b.** $f^{(4)}(x)$ is unbounded near $x = 2$
35. The inscribed circles have diameter $13\frac{1}{3}$ and $8\frac{1}{3}$.
39. $\frac{32}{3}$ **41.** Error is zero.

5.9 An Alternative Approach: The Logarithm as an Integral (Pages 345–346)

3. $E \approx 0.00104$; the number of subintervals should be 18

Chapter 5 Review
Proficiency Examination (Pages 346–347)
22. $-\frac{3}{5}$ **23.** $x^5\sqrt{\cos(2x + 1)}$
24. $\frac{1}{2}\tan^{-1}(2x) + C$ **25.** $-\frac{1}{2}e^{-x^2} + C$ **26.** $\frac{17}{3}$
27. $-\frac{35}{3}$ **28.** $\frac{1}{2}$ **29.** 0   **30.** 36
31. $\int_1^2 e^{x^2}\, dx$ does not have a closed form, so we must use approximate integration or technology to find
$$\int_1^2 e^{x^2}\, dx \approx 14.99.$$
32. 0
33. The percent is approximately 1.33×10^{-184}, which exceeds the accuracy of most calculators and measuring devices.
34.   **e.** $y = -x - 1$
35. a. $n \geq 26$ **b.** $n \geq 4$

Supplementary Problems (Pages 347–350)

1. 25   **3.** 6   **5.** 1,710

7. $\dfrac{6 - 2\sqrt{2}}{3}$ **9.** $\sqrt{3} - 1$ **11.** $-\frac{3}{4}(1 - \sqrt[3]{9})$

13. $2x^{5/2} - \frac{4}{3}x^{3/2} + 2x^{1/2} + C$ **15.** $3\sin^{-1}x + \sqrt{1 - x^2} + C$

17. $-\ln|\sin x + \cos x| + C$ **19.** $\frac{1}{3}(x - 1)^3 + C$ **21.** $x + C$

23. $\dfrac{x^3(x^2 + 1)^{3/2}}{3} + C$ **25.** $-\frac{1}{15}(1 - 5x^2)^{3/2} + C$

27. $\frac{1}{3}(1 + x^2)^{3/2} - (1 + x^2)^{1/2} + C$

29. $\frac{2}{5}(1 + \ln 2x)^{5/2} - \frac{2}{3}(1 + \ln 2)(1 + \ln 2x)^{3/2} + C$ **31.** 0

33.   **35.**

37. $\ln 4$ **39.** $\frac{1}{4}(e^8 - 1)$

41. $f(t) = -\frac{1}{16}\sin 4t + \frac{1}{4}\cos 2t + \frac{5}{4}t + \frac{5}{4} - \frac{5}{8}\pi$

43. $y = 1 - \dfrac{1}{x - C}$ **45.** $\tan y + \cot x = C$

47. $y = Ce^{(1/3)x^3 + x}$ **49.** $y = \tan^{-1}(C - \ln|\cos x|)$

51. 0.5274   **53.** 1.9541   **55.** 0.9091   **57.** $n = 6$

59. Number of primes less than x

x	$\pi(x)$	Gauss
10^3	168	178
10^4	1,229	1,246
10^5	9,592	9,628
10^6	78,498	78,628
10^{10}	455,052,511	455,055,614

61. 1.01297   **63. b.** ± 8 m/s

65. The revenue for the five years is $7,666.67.

67. The tree was 2.33 ft tall when it was transplanted.

69. $y = \frac{1}{3}(x^2 + 5)^{3/2} + 1$ **71.** 4.45 ppm   **73.** 126 people

75. The average price was $1.32 per pound.

77. about 2 min, 45 seconds

81. a. 51.7%

     b. The time for 90% to disintegrate is about $17\frac{1}{2}$ years.

83. 35 years   **85.** 1:44 P.M.

87. This is Putnam Problem 1 from the morning session in 1958.

Cumulative Review for Chapters 1–5 (Pages 353–354)

5. $\frac{7}{5}$ **7.** $\frac{1}{2}$ **9.** 0   **11.** 1   **13.** 1

15. $6(x^2 + 1)^2(3x - 4)(2x - 1)^2$ **17.** $\dfrac{\cos x + x\sin x}{(x + \cos x)^2}$

19. $6\sec^2 3x \tan 3x$ **21.** $\dfrac{10x + 3}{5x^2 + 3x - 2}$

23. 5   **25.** $\sqrt{10} - 3$ **27.** $\ln(e^x + 2) + C$ **29.** 1.812

31.

33. $y = \dfrac{9}{2(4 - x^3)^{3/2}} + C$

35. $y = 5 \pm 2e^{-2x}$

37. a. $\frac{9}{4}\pi - 8$ **b.** 3

     c. $4x + y - 20 - \frac{9}{4}\pi = 0$

     d. $x = 0$ and $x = 7$

39. 0.1 rad/s

41. a. $y = -\frac{4}{\pi}x + 2$ **b.** $y = -2x + \pi$

     c. $x \approx 0.690$; mean value theorem

43. a. relative maximum at $(0, k)$; relative minimum at $(4, k - 32)$

     b. $0 < k < 32$ **c.** $k = \frac{27}{4}$

45. $y = \dfrac{3}{\cos 3x} + C$

Chapter 6: Additional Applications of the Integral

6.1 Area Between Two Curves (Pages 361–362)

1. vertical strip; $\frac{125}{48}$ **3.** vertical strip; 2

5. horizontal strip; $\frac{125}{6}$

7. 1   **9.** $\frac{1}{12}$ **11.** $\frac{8}{3}$

13. $\frac{324}{5}$ **15.** $\frac{9}{2}$ **17.** $\frac{253}{12}$

19. $\frac{5}{2}$ **21.** $\frac{323}{12}$ **23.** $\frac{131}{4}$

25. $2\ln 1.6 - \sin^{-1} 0.6$ **27.** $k \approx 0.34$ **29.** π

31. c. The V values are approximately 72.5299, 196.539, 344.337, 502.655, 660.972, 808.77, 932.78, 1,005.31

6.2 Volume (Pages 372–375)

The given volumes are all in cubic units.

1. 9   **3.** $\frac{1}{30}$ **5.** $36\sqrt{3}$

7. $\frac{\sqrt{3}}{2}$ **9.** $\frac{64\pi}{15}$ **11.** $\frac{\pi}{8}\left(1 - \frac{\pi}{4}\right)$

13. $\frac{\pi}{2}$ **15.** π **17.** 2,555

19. $\frac{\pi}{2}$ **21.** $\frac{\pi}{12}$ **23.** $\frac{\pi}{2}$ **25.** $\frac{\pi}{2}$

27. a. $\pi \displaystyle\int_0^4 (4 - x)^2\,dx$ **b.** $2\pi \displaystyle\int_0^4 x(4 - x)\,dx$

     c. $\pi \displaystyle\int_0^4 [24 - 10x + x^2]\,dx$ **d.** $2\pi \displaystyle\int_0^4 (x + 2)(4 - x)\,dx$

29. a. $2\pi \displaystyle\int_0^2 y\sqrt{4 - y^2}\,dy$ **b.** $\pi \displaystyle\int_0^2 (4 - y^2)\,dy$

     c. $2\pi \displaystyle\int_0^2 (y + 1)\sqrt{4 - y^2}\,dy$

     d. $\pi \displaystyle\int_0^2 \left[4 - y^2 + 4\sqrt{4 - y^2}\right]\,dy$

31. a. $\pi \displaystyle\int_0^1 \left(e^{-x}\right)^2\,dx$ **b.** $2\pi \displaystyle\int_0^1 xe^{-x}\,dx$

     c. $\pi \displaystyle\int_0^1 \left(e^{-2x} + 2e^{-x}\right)\,dx$ **d.** $2\pi \displaystyle\int_0^1 (x + 2)e^{-x}\,dx$

33. a. $2\pi \displaystyle\int_0^{\pi/2} y(1 - \sin y)\,dy$ **b.** $\pi \displaystyle\int_0^{\pi/2} [1^2 - \sin^2 y]\,dy$

     c. $2\pi \displaystyle\int_0^{\pi/2} (y + 1)(1 - \sin y)\,dy$

     d. $\pi \displaystyle\int_0^{\pi/2} [5 - \sin^2 y - 4\sin y]\,dy$

35. a. $V = \pi \displaystyle\int_0^1 \left[\left(\sqrt{x}\right)^2 - \left(x^2\right)^2\right]\,dx$

b. $V = \pi \int_0^1 \left[\left(\sqrt{y}\right)^2 - \left(y^2\right)^2 \right] dy$

37. a. $V = \pi \int_0^2 [(x^2 + 1)^2 - 1^2] dx$

    b. $V = \pi \int_1^5 \left[(2)^2 - \left(\sqrt{y-1}\right)^2 \right] dy$

39. a. $V = \pi \int_{1.138}^{3.566} \left[(\ln x)^2 - \left(0.1x^2\right)^2 \right] dx$

    b. $V = \pi \int_{0.130}^{1.271} \left[\left(\sqrt{10y}\right)^2 - (e^y)^2 \right] dy$

41. $\frac{2\pi}{35}$ **43.** $\frac{\pi}{3}$ **45.** $\frac{4}{3}\pi r^3$

47. $36\sqrt{3}$ **49.** 18π **51.** $\frac{32\sqrt{3}}{3}$

53. $90{,}000{,}000 \text{ ft}^3$ **55. a.** $\pi \ln 4$ **b.** $\frac{32\pi}{3}$ **c.** $\pi(8 + \ln 4)$

57. 63.187 ft^3 **59.** 20.13 **61.** $\frac{4}{3}\pi r^3$

6.3 Polar Forms and Area (Pages 384–385)

3. a. lemniscate   **b.** circle

    c. rose (3 petals)   **d.** none (spiral)

    e. cardioid   **f.** line

    g. lemniscate   **h.** limaçon

5. a. rose (4 petals)   **b.** circle

    c. limaçon   **d.** cardioid

    e. line   **f.** line

    g. rose (5 petals)   **h.** line

7.   **9.**

11.   **13.**

15.   **17.**

19.   **21.**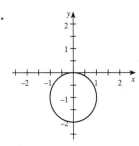

23. $P_1\left(2, \frac{\pi}{6}\right)$ and $P_2\left(2, \frac{5\pi}{6}\right)$ **25.** $P_1(0,0),\ P_2(2,0),\ P_3(2,\pi)$

27. $P_1(0,0),\ P_2\left(1, \frac{\pi}{4}\right)$

29. There are no intersection points.

31. $\frac{\pi}{24} + \frac{\sqrt{3}}{16} \approx 0.2392$ **33.** $\frac{\sqrt{3}}{4} \approx 0.4330$

35. $\frac{1}{8}(\pi + 2) \approx 0.6427$ **37.** $\frac{16\pi^3}{5} \approx 99.2201$

39.   **41.**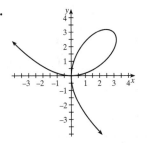

43. $\dfrac{\pi a^2}{4}$ **45.** $a^2\left(2 - \frac{\pi}{4}\right)$ **47.** 4π

49. $4\pi + 12\sqrt{3} \approx 33.3510$ **51.** 0.0674

53. The maximum value of x is $9\sqrt{3}/4 \approx 3.8971$.

55.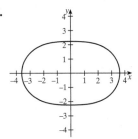

59. a. $-\cot\theta$ **b.** $\dfrac{1 - \cos\theta}{\sin\theta}$ **c.** $\frac{1}{3}$

6.4 Arc Length and Surface Area (Pages 393–395)

1. $3\sqrt{10}$ **3.** $2\sqrt{5}$ **5.** $\frac{10\sqrt{5}}{3} - \frac{2}{3}$

7. $\frac{331}{120}$ **9.** $\frac{123}{32}$ **11.** $\frac{14}{3}$

13. $12\pi\sqrt{5}$ **15.** 25.28 **17.** π

19. $\frac{\sqrt{10}}{3}\left[e^{3\pi/2} - 1\right]$ **21.** $\frac{1}{3}\left(5^{3/2}\right) - \frac{8}{3}$

23. 25π **25.** $\frac{2\pi}{3}\left(3 - \sqrt{3}\right)$

27. It appears the arc length is approximately 3.82.

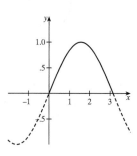

Y = SIN(X)

# OF TERMS	ARC LENGTH OVER [0, π]
2	3.724192
4	3.790091
8	3.812529
16	3.818275
32	3.819717
64	3.820077
128	3.820168
256	3.82019
512	3.820196
THIS IS THE FINAL ESTIMATE	

29. a. $\frac{1,505\pi}{18}$ **b.** 127.4

31. $3\pi\sqrt{10}$ **33.** $\pi\left(\frac{28\sqrt{3}}{5} - \frac{16}{15}\right)$

35. 6   **37.** 8.6322   **39.** 141.6974

45. $L = C\left[b^{2n} - a^{2n}\right] - D\left[b^{2(1-n)} - a^{2(1-n)}\right]$

47. $f(x) = \pm \ln|\cos x|$

6.5 Physical Applications: Work, Liquid Force, and Centroids (Pages 405–407)

5. 12,750 ft-lb   **7.** 6,500 ft-lb

9. $\frac{10}{3}$ ft-lb   **11.** 2,000 ft-lb

13. 4 ergs   **15.** 192 lb   **17.** 59.7 lb

19. $F = 64.5 \int_0^{1/24} 2\left(x + \frac{1}{24}\right)\sqrt{\left(\frac{1}{24}\right)^2 - x^2}\, dx$

21. $F = 51.2 \int_{-2}^0 2(x+3)\sqrt{4 - x^2}\, dx$

23. $\left(0, -\frac{18}{5}\right)$ **25.** $\left(\frac{1}{\ln 2}, \frac{1}{4\ln 2}\right)$ **27.** $\left(\frac{7}{3}, \frac{1}{2}\ln 2\right)$

29. about $y = -1$, $V = \frac{200\pi}{3}$ **31.** $V = \frac{8\pi}{3}(4 + 3\pi)$

33. a. 4.284 ft-lb   **b.** 12,603 ft-lb

35. The amount of work is approximately 345,800 ft-lb.

37. 24,881 ft-lb   **39.** 2,932.8 lb   **41.** 152,381 mi-lb

43. a. -7.5 ergs   **b.** -37.5 ergs

45. The total force on the bottom is about 85,169 lb.

47. $I_x = \frac{64}{15}$, $I_y = \frac{256}{7}$ **49.** $\rho = \frac{2}{3}\sqrt{6}$

53. $V = \pi(2sL^2 + L^3)$

6.6 Applications to Business, Economics, and Life Sciences (Pages 414–417)

5. a. 0.75   **b.** 0   **7. a.** 75   **b.** 432

9. a. $468.00   **b.**

11. a. $12.46   **b.**

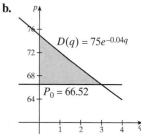

13. $6.25   **15.** $2.84   **17.** $5.33   **19.** $42.67

21. a. $FV = \$10,171.84$ **b.** $PV = \$8,001.46$

23. a. The machine will be profitable for 9 years.

     b. $12,150; in geometric terms, the net earnings is presented by the area of the region between the curves $y = R'(x)$ and $y = C'(x)$ from $x = 0$ to $x = 9$.

25. a. The second plan is more profitable for the first 18 years.

     b. $7,776; in geometric terms, the net extra profit generated by the second plan is the area of the region between the curves $y = P_2(x)$ and $y = P_1(x)$ from $x = 0$ to $x = 18$.

27. a. $x = 20$ is a maximum   **b.** $400

29. a. The revenue function is $R(q) = qp(q) = \frac{1}{4}q(10-q)^2$; the marginal revenue function is $R'(q) = \frac{1}{4}(10-q)(10-3q)$.

     b. $q = 2$

     c. The consumer's surplus is $8.67.

31. $\frac{8}{3}$

33. The number of people entering the fair during the prescribed time period is 1,220 people.

35. The total future revenue is $176,256.

37. $4,081,077

39. a. The second plan will be profitable for 12 years.

     b. $100,800   **c.**

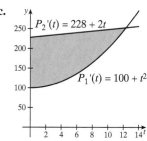

41. a. The second plan will be more profitable for about 24.5 years.

     b. The excess profit is about $3.2 million.

43. a. The machine will be profitable for 8 years.

     b. The net earnings are $7,168.

     c.

45. a. $\frac{184}{3}$ **b.**

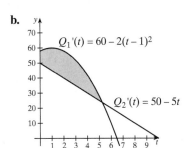

$Q_1'(t) = 60 - 2(t-1)^2$

$Q_2'(t) = 50 - 5t$

47. There would be approximately $208,128$ people living in the ring.

49. 0.04π or $0.126 \text{ cm}^3/\text{s}$.

51. The population will increase by about 98 people.

53. The demand is approximately 515.48 billion barrels.

55. There are about 240 spies still alive.

57. There should be approximately $4,207$ active members.

59. a. $f(t_k)e^{-rt_{k-1}}$

b. $\displaystyle\sum_{k=1}^{n} f(t_k)e^{-rt_{k-1}}\Delta_n t$, where $\Delta_n t = \dfrac{N}{n}$

c. $PV = \displaystyle\lim_{n\to\infty} \sum_{k=1}^{n} f(t_k)e^{-rt_{k-1}}\Delta_n t = \int_0^N f(t)e^{-rt}\,dt$

Chapter 6 Review
Proficiency Examination (Pages 417–418)

20. The definite integrals could represent the following:
  A. Disks revolved about the y-axis.
  B. Disks revolved about the x-axis.
  C. Slices taken perpendicular to the x-axis.
  D. Slices taken perpendicular to the y-axis.
  E. Mass of a lamina with density π.
  F. Washers taken along the x-axis.
  G. Washers taken along the y-axis.

  a. All but E are formulas for volumes of solids.
  b. A, B, F, G **c.** F, G **d.** C, D **e.** B, F **f.** A, G

21. $8\sqrt{2}$ **22.** $\frac{1}{2}$ **23.** $\frac{256}{3}$

24. a. $\frac{32}{3}$ **b.** $\frac{256\pi}{5}$ **c.** $\frac{128\pi}{3}$

25. $a^2\left(\frac{\pi}{2}-1\right) \approx 0.5708a^2$

26. $\frac{1}{27}\left(13^{3/2}-8\right) \approx 1.4397$

27. $\frac{\pi}{6}\left(5\sqrt{5}-1\right) \approx 5.33$

28. $M = \frac{5}{12}$; $M_y = \frac{13}{60}$; $M_x = \frac{3}{140}$; the centroid is $\left(\frac{13}{25}, \frac{9}{175}\right)$

29. $F = 128 \displaystyle\int_0^2 (5-x)(1+\sqrt{1-x^2})\,dx$

30. The amount in the account will be $\$7,377.37$.

Supplementary Problems (Pages 418–423)

1. circle   **3.** line

5. circle   **7.** parabola

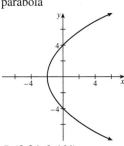

9. $P_1(0, 0), P_2(2, 0)$ **11.** $P_1(2.24, 0.464)$

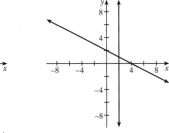

13. $\frac{9}{4}$ **15.** $\frac{4}{15}$ **17.** 8π

19. $\frac{1}{2}$ **21.** $a^2\left(\dfrac{\sqrt{3}}{2}+\dfrac{\pi}{3}\right)$ **23.** $\frac{81\pi}{4}$

25. $\frac{\pi}{2}(\sqrt{3}-\sqrt{2})$ **27.** $\frac{\pi}{240}$ **29.** 225 in.-lb

31. $V_1 = \frac{16}{15}$; $V_2 = 2$; S_2 has greater volume.

33. $S = \pi r\sqrt{r^2+h^2}$ **35.** $30,264$ ft-lb

37. $2,954.67$ ft-lb   **39.** 44.108 ft-lb

41. The spring should be stretched about 14 in.

43. 5.625×10^{-5} **45.** $\left(\frac{7}{3}, \frac{1}{3}\right)$

47. $\$7,191.64$ **49.** $309,030$ people

51. a. 3 units   **b.** $\$31.50$

53. $17,500$ ft-lb   **55.** 135 lb   **57.** $F(t) = \frac{1}{2}t + \frac{2}{3}$

59. The total work is $6,176\pi$.

61. The desired distances are 4.88 cm, 11.12 cm from the right, 10.34 cm from the top, and 5.66 cm from the bottom.

63. a. $\frac{35\pi}{3}$ ft-lb   **b.** $\frac{70}{9}\pi$ ft-lb

65. The force is approximately $1,321$ ft-lb.   **67.** $S = 2\pi\sqrt{2}$

69. $W = (4,000)^2 P\left(\dfrac{1}{4,000}-\dfrac{1}{t}\right)$ mi-lb   **71.** 344π

73. a. $h \approx -0.644391$ **b.** $h \approx 0.865202$

75. $W \approx 1.728 \times 10^{12}$ ft-lb

77. 4.7999 **79.** $\frac{\pi}{4}$ **81.** 1.3833

83. This is Putnam Examination Problem 1 of the morning session in 1939.

85. This is Putnam Examination Problem 1 of the morning session in 1993.

Chapter 7: Methods of Integration

7.1 Review of Substitution and Integration by Table
(Pages 433–435)

1. $2(x^2+5x)^{1/2}+C$ **3.** $\ln|\ln x|+C$

5. $\frac{1}{4}\tan^{-1}\dfrac{x^2}{2}+C$ **7.** $-\frac{1}{5}(1+\cot x)^5+C$

9. $-\frac{1}{4}(x^4-2x^2+3)^{-1}+C$ **11.** $\ln(x^2+x+1)+C$

13. $\dfrac{\sqrt{x^2-a^2}}{a^2x}+C$ **15.** $\frac{1}{2}x^2\left(\ln x-\frac{1}{2}\right)+C$

17. $a^{-1}e^{ax}(x - a^{-1}) + C$

19. $\frac{1}{2}x\sqrt{x^2 + 1} - \frac{1}{2}\ln\left|x + \sqrt{x^2 + 1}\right| + C$

21. $\frac{1}{4}\sqrt{4x^2 + 1} + C$

23. $\dfrac{-4\sin 5x - 5\cos 5x}{41e^{4x}} + C$

25. $b^{-1}\ln|1 + bx| + C$

27. $\frac{1}{20}(x + 1)^4(4x - 1) + C$

29. $\frac{1}{4}e^{4x}\left(x - \frac{1}{4}\right) + C$

31. $x - \frac{1}{2}\ln|1 + e^{2x}| + C$

33. $\frac{1}{8}(4x^4 + 1)^{1/2} + C$

35. $\sec\dfrac{x}{2}\tan\dfrac{x}{2} + \ln\left|\sec\dfrac{x}{2} + \tan\dfrac{x}{2}\right| + C$

37. $-\dfrac{1}{3(3x + 1)} + C$

39. $-\sin x + \ln\left|\tan\left(\dfrac{x}{2} + \dfrac{\pi}{4}\right)\right| + C$

41. $\frac{1}{2}x + \frac{1}{4}\sin 2x + C$

43. $-\frac{1}{5}\cos^5 x + \frac{1}{7}\cos^7 x + C$

45. If m is odd, let $u = \cos x$. If n is odd, let $u = \sin x$. If both m and n are even, use the identities shown in Problems 40 and 41 until one exponent is odd.

47. $4\left[\dfrac{x^{1/2}}{2} - x^{1/4} + \ln(x^{1/4} + 1)\right] + C$

49. $-6(2 + \tan^3 t)^{-1} + C$

51. $\dfrac{1}{e^x + e^{-x}} + C$

53. 94.68   55. 40.3053   57. 0.7207   59. 3.8097

7.2 Integration by Parts (Pages 439–441)

1. $-\frac{1}{2}xe^{-2x} - \frac{1}{4}e^{-2x} + C$

3. $\frac{1}{2}x^2\ln x - \frac{1}{4}x^2 + C$

5. $x\sin^{-1}x + \sqrt{1 - x^2} + C$

7. $\frac{4}{25}e^{-3x}\sin 4x - \frac{3}{25}e^{-3x}\cos 4x + C$

9. $\frac{1}{3}x^3\ln x - \frac{1}{9}x^3 + C$

11. $\dfrac{x}{2}\left[\sin(\ln x) - \cos(\ln x)\right] + C$

13. $x\ln(x^2 + 1) - 2x + 2\tan^{-1}x + C$

15. $\dfrac{-xe^{-x}}{x - 1} + e^{-x} + C$

17. $\frac{32}{3}\ln 2 - \frac{28}{9}$

19. $e - 2$

21. $\frac{1}{4}(e^{2\pi} - 1)$

25. $\frac{1}{2}\cos^2 x - (\cos^2 x)\ln(\cos x) + C$

27. $-(2 + \cos x)\ln(2 + \cos x) + (2 + \cos x) + C$

29. $\frac{x}{2} + \frac{1}{4}\sin 2x + C$

31. $\frac{1}{4}x^2 + \frac{1}{4}x\sin 2x + \frac{1}{8}\cos 2x + C$

33. $\dfrac{x^{n+1}}{n + 1}\ln x - \dfrac{x^{n+1}}{(n + 1)^2} + C$

35. 177 units

37. 3.73218

39. a. 2.2565   b. 13.1775   41. (0.68, 1.27)

43. $2y^{1/2} = \frac{2}{3}x^{3/2}(\ln x - \frac{2}{3}) + C$

45. $y = \exp(1 - \cos x - x\sin x)$

47. 65 miles   49. 0.07   51. $1.558nRT$

7.3 Trigonometric Methods (Page 447)

5. $\sin x - \frac{1}{3}\sin^3 x + C$

7. $\frac{1}{3}\sin^3 x - \frac{1}{5}\sin^5 x + C$

9. $-\frac{2}{3}(\cos t)^{3/2} + C$

11. $-e^{\cos x} + C$

13. $\frac{1}{8}x - \frac{1}{32}\sin 4x + C$

15. $-\frac{1}{2}\ln|\cos 2\theta| + C$

17. $\frac{1}{6}\tan^6 x + \frac{1}{4}\tan^4 x + C$

19. $2\tan x - x + C$

21. $\frac{1}{2}\sec u\tan u - \frac{1}{2}\ln|\sec u + \tan u| + C$

23. $\frac{3}{4}(\tan x)^{4/3} + C$ 25. $\frac{1}{4}\sin^2 x^2 + C$

27. $\frac{1}{4}\sec^3 t\tan t - \frac{5}{8}\sec t\tan t + \frac{3}{8}\ln|\sec t + \tan t| + C$

29. $-\frac{1}{3}\csc^3 x + C$

31. $-\csc x + C$

33. $2\sin^{-1}\dfrac{t}{2} + \frac{1}{2}t\sqrt{4 - t^2} + C$

35. $\sqrt{4 + x^2} + \ln\left|\sqrt{4 + x^2} + x\right| + C$

37. $\ln\left|x + \sqrt{x^2 - 7}\right| + C$

39. $\sin^{-1}\dfrac{x}{\sqrt{5}} + C$

41. $-\dfrac{\sqrt{4 - x^2}}{4x} + C$

43. $\sqrt{x^2 - 4} - 2\sec^{-1}\dfrac{x}{2} + C$

45. $\frac{1}{3}\ln\left|\dfrac{x + 4}{\sqrt{9 - (x + 1)^2}}\right| + C$

47. $\ln\left|\sqrt{x^2 - 2x + 6} + x - 1\right| + C$

49. $\frac{1}{4}\tan^4 u + C$ 51. $\frac{1}{2}$ 53. 15.5031

55. $\pm 2\sqrt{2}\sin\frac{x}{2} + C$

57. $-\frac{1}{3}\cos\left(\frac{3}{2}x\right) - \frac{1}{5}\cos\left(\frac{5}{2}x\right) + C$

59. $\frac{1}{8}\sin 4x - \frac{1}{8}\sin 2x - \frac{1}{40}\sin 10x + C$

61. $-\ln\left|\sqrt{9 - x^2} - 1\right| + C$

7.4 Method of Partial Fractions (Pages 455–457)

1. $\dfrac{-1}{3x} + \dfrac{1}{3(x - 3)}$

3. $3 - \dfrac{1}{x}$

5. $\dfrac{4}{x} + \dfrac{-8}{2x + 1}$

7. $\dfrac{3}{x} + \dfrac{-1}{x^2} + \dfrac{1}{x + 1} + \dfrac{-2}{(x + 1)^2}$

9. $\dfrac{-4}{9x^2} + \dfrac{17}{27x} - \dfrac{13}{9(x + 3)^2} + \dfrac{10}{27(x + 3)}$

11. $\dfrac{1}{4(1 - x)} + \dfrac{1}{4(1 + x)} + \dfrac{1}{2(1 + x^2)}$

13. $\dfrac{1}{x} - \dfrac{1}{3(x + 1)} - \dfrac{1}{3(2x - 1)}$

15. $\frac{1}{3}\ln\left|\dfrac{x - 3}{x}\right| + C$

17. $3x - \ln|x| + C$

19. $4\ln|x| - 4\ln|2x + 1| + C$

21. $-9\ln|x| - x^{-1} + 6\ln|x + 1| + 5\ln|x - 1| + C$

23. $x - \frac{5}{3}\ln|x + 2| + \frac{2}{3}\ln|x - 1| + C$

25. $x + \frac{1}{2}\ln|x - 1| - \frac{1}{2}\ln|x + 1| - \tan^{-1}x + C$

27. $\ln|x + 1| + (x + 1)^{-1} + C$

29. $-\frac{1}{2}\ln|x| + \frac{1}{3}\ln|x + 1| + \frac{1}{6}\ln|x - 2| + C$

31. $\ln\left|\dfrac{x + 2}{x + 1}\right| - \dfrac{2}{x + 2} + C$

33. $3\ln|x - 1| + 2\ln|x + 3| + C$

35. $\ln|x^3 - x^2 + 4x - 4| + C$

39. $\frac{1}{7}\ln|e^x - 3| - \frac{1}{7}\ln(2e^x + 1) + C$

41. $\dfrac{1}{1 + \cos x} + C$

43. $\ln|\tan x + 4| + C$

45. $3x^{1/3} + 6x^{1/6} + 6\ln|x^{1/6} - 1| + C$

47. $-\frac{1}{5}\ln\left|\tan\dfrac{x}{2} - 3\right| + \frac{1}{5}\ln\left|3\tan\dfrac{x}{2} + 1\right| + C$

49. $-\ln|\sin x + \cos x| + C$

51. $-\ln(1 - \sin x) + C$

53. $\dfrac{1}{\tan\frac{x}{2} - 2} + C$

55. $\frac{1}{2}\ln|\ln x - 3| - \frac{1}{2}\ln|\ln x - 1| + C$

57. 0.6931   **59. a.** 1.0888   **b.** 7.6402

61. $\frac{1}{2}\ln|x^2 - 9| + C$ **65.** $\frac{1}{b}\ln\left|\dfrac{x}{ax+b}\right| + C$

7.5 Summary of Integration Techniques (Pages 460–461)

1. $\dfrac{1}{2x^2(x-1)^2} + C$ **3.** $\frac{1}{4}\ln\left|\sec 2x^2 + \tan 2x^2\right| + C$

5. $\ln|\sin e^x| + C$ **7.** $-\ln|\cos(\ln x)| + C$

9. $3\ln|\sec t + \tan t| + 2\ln|\sec t| + C$

11. $\frac{1}{2}\tan^{-1} e^{2t} + C$ **13.** $x + \frac{1}{2}\ln(x^2 + 9) - \frac{8}{3}\tan^{-1}\frac{x}{3} + C$

15. $-x - 2\ln|e^{-x} - 1| + C$

17. $\frac{2}{3}\sin^{-1} t^3 + C$ **19.** $x - \frac{1}{2}\ln(e^{2x} + 1) + C$

21. $\tan^{-1}(x+1) + C$ **23.** $\dfrac{2}{\sqrt{3}}\tan^{-1}\dfrac{1}{\sqrt{3}}(2x+1) + C$

25. $x\tan^{-1} x - \frac{1}{2}\ln(x^2 + 1) + C$

27. $\frac{1}{2}e^{-x}(\sin x - \cos x) + C$

29. $x\cos^{-1}(-x) + \sqrt{1 - x^2} + C$

31. $-\cos x + \frac{1}{3}\cos^3 x + C$ **33.** $\frac{1}{5}\cos^5 x - \frac{1}{3}\cos^3 x + C$

35. $\frac{1}{16}x - \frac{1}{64}\sin 4x + \frac{1}{48}\sin^3 2x + C$

37. $-\frac{1}{9}\cos^9 x + \frac{2}{7}\cos^7 x - \frac{1}{5}\cos^5 x + C$

39. $\frac{1}{8}\tan^8 x + \frac{1}{6}\tan^6 x + C$

41. $\sqrt{1 - x^2} - \ln\left|\dfrac{1 + \sqrt{1-x^2}}{x}\right| + C$

43. $\sqrt{2x^2 - 1} + \dfrac{3}{\sqrt{2}}\ln\left|\sqrt{2}\,x + \sqrt{2x^2 - 1}\right| + C$

45. $\ln|x| - \ln\left|1 + \sqrt{x^2 + 1}\right| + C$

47. $-2\sqrt{4x - x^2 - 2} + 5\sin^{-1}\dfrac{x-2}{\sqrt{2}} + C$

49. $\ln(\sin x + \sqrt{1 + \sin^2 x}) + C$

51. $-\dfrac{\cos^5 x}{5} + \dfrac{2\cos^3 x}{3} - \cos x + C$

53. $\frac{1}{3}\tan^3 x - \tan x + x + C$

55. π **57.** 1.8101   **59.** 0.1353   **61.** 113.4327

63. $\ln\left(\sqrt{1 + e^{2x}} + e^x\right) + C$

65. $3\ln|x| - \ln|x^2 + x + 1| + \dfrac{4}{\sqrt{3}}\tan^{-1}\dfrac{\sqrt{3}}{3}(2x+1) + C$

67. $5\ln|x - 7| + 2(x + 2)^{-1} + C$

69. $3\ln|x + 1| - 2(x + 1)^{-1} + C$

71. $-\frac{1}{2}\ln|x + 1| + 2\ln|x + 2| - \frac{3}{2}\ln|x + 3| + C$

73. 0.09   **77.** 2.9579   **79.** 4   **81.** 101.7876

83. $(0.71, 0.12)$ **87.** $-\frac{1}{16}\sin^3 4x\cos 4x + \frac{3}{8}x - \frac{3}{64}\sin 8x + C$

7.6 First-Order Differential Equations (Pages 469–472)

1. $\frac{1}{5}x^2 + Cx^{-3}$ **3.** $-5x^{-3} + Cx^{-2}$

5. $\frac{1}{3}x^{-2}e^{x^3} + Cx^{-2}$ **7.** $\left(\dfrac{1}{2}x + \dfrac{1}{2x}\right)\tan^{-1} x - \dfrac{1}{2} + \dfrac{C}{x}$

9. $y = -\cos x\ln|\cos x| + C\cos x$

11. $y = x^2 - 1$ for $x > -1$ **13.** $y = 2x^3 - 6x^2$

15. $2X^2 + Y^2 = C$ **17.** $Y = CX$

19. The GDP in the year 2010 will be about \$17,828 billion.

21. The predicted number of divorces in 2005 is about 2,378,000.

23. a. $Q(t) = 30 - 20e^{-t/15}$ **b.** 4 minutes, 19 seconds

25. In the "long run" (as $t \to +\infty$), the concentration will be $\lim\limits_{t\to+\infty}\dfrac{\alpha}{\beta}(1 - e^{-\beta t}) = \dfrac{\alpha}{\beta}$; the "half-way point" is reached when $t = \dfrac{\ln 2}{\beta}$.

27. 46.8 million   **29.** $M(t) = \dfrac{k}{r}(1 - e^{-rt})$

31. It will take about an hour to drain.

33. a. $y = -\dfrac{k}{24}(x^4 - 2Lx^3 + L^3 x)$

b. maximum deflection occurs where $y_m \approx -0.0130 kL^4$

c. The maximum deflection is $y_m \approx -0.0054 kL^4$; the maximum deflection in the cantilevered case is less than that in part **b.**

35. a. $S_1(t) = 200(1 - e^{-t/100})$

b. $S_2(t) = 200 - 2te^{-t/100} - 200e^{-t/100}$

c. The maximum excess is $S(100) \approx 73.58$ lb.

37. $P(t) = P_\infty \exp\left[-\left(\ln\dfrac{P_\infty}{P_0}\right)e^{-kt}\right]$

39. $x = \dfrac{1}{2}(y + 1)e^y + \dfrac{Ce^y}{y + 1}$

41. a. $I(t) = \frac{3}{2}(1 - e^{-2t})$ **b.** $I(t) = e^{-2t}(1 - \cos t)$

43. 9.872 days   **45.** $N = \left(\dfrac{1}{m} + Ce^{pt}\right)S$

47. a. $v(t) = \dfrac{-mg}{k} + \left(\dfrac{mg}{k} + v_0\right)e^{-kt/m}$;

$s(t) = \dfrac{-mg}{k}t + \dfrac{m}{k}\left(\dfrac{mg}{k} + v_0\right)(1 - e^{-kt/m})$

b. The object reaches its maximum height when $v(t) = 0$. The maximum height is

$$s_{max} = \dfrac{mv_0}{k} - \dfrac{m^2 g}{k^2}\ln\left(1 + \dfrac{kv_0}{mg}\right).$$

c. The maximum height is 82.98 ft; the object hits the ground when $t \approx 5.51$ seconds.

d. 9.38 seconds

7.7 Improper Integrals (Pages 479–481)

3. $\frac{1}{2}$ **5.** diverges   **7.** 10   **9.** diverges

11. $\frac{1}{10}$ **13.** $\frac{5}{2}$ **15.** $\frac{1}{9}$ **17.** diverges

19. $\dfrac{2}{e}$ **21.** 0   **23.** diverges   **25.** 2

27. diverges   **29.** diverges   **31.** 0   **33.** $\frac{5}{4}$ **35.** 2

37. $\frac{3}{2}\left[(1 - e^{-1})^{2/3} - (1 - e)^{2/3}\right]$

39. -1 **41.** 1   **43.** 2   **45.** diverges   **47.** $\frac{1}{4}$

49. There will be 100,000 millirads.

51.

$$\int_{-\infty}^{+\infty} f(x)\,dx = 4$$

53. $(1 - p)^{-1}$ if $p < 1$, and diverges if $p \geq 1$

55. both diverge   **57.** 8   **59.** $\frac{3}{40}$

65. b. $\dfrac{s^2 - 4}{(s^2 + 4)^2}$

67. b. $\mathcal{L}(y) = \dfrac{-\frac{2}{13}s}{s^2 + 4} + \dfrac{\frac{6}{13}}{s^2 + 4} + \dfrac{\frac{28}{13}}{s + 3}$

   c. $y = -\frac{2}{13} \cos 2t + \frac{3}{13} \sin 2t + \frac{28}{13} e^{-3t}$

7.8 Hyperbolic and Inverse Hyperbolic Functions (Pages 486–487)

1. 3.6269   **3.** -0.7616 **5.** 1.1995   **7.** 0.9624

9. 1.6667   **11.** 0.6481   **13.** $3 \cosh 3x$

15. $(4x + 3) \sinh(2x^2 + 3x)$ **17.** $-x^{-2} \cosh x^{-1}$

19. $\dfrac{3x^2}{\sqrt{1 + x^6}}$ **21.** $\sec x$ **23.** $\sec x$

25. $\dfrac{x - \sqrt{1 + x^2} \sinh^{-1} x}{x^2 \sqrt{1 + x^2}}$ **27.** $\cosh^{-1} x$

29. $\left[e^x - 2e^{2x} \sinh^{-1} x - \dfrac{e^x}{\sqrt{1 + x^2}} \right] \sinh(e^x - e^{2x} \sinh^{-1} x)$

31. $-\cosh x^{-1} + C$ **33.** $\ln |\sinh x| + C$

35. $\dfrac{1}{3} \cosh^{-1} \dfrac{3t}{4} + C$ **37.** $\sinh^{-1}(\sin x) + C$

39. $\frac{1}{3} \tanh^{-1} x^3 + C$ **41.** -0.2027 **43.** 1.0311

45. 0.3808   **47.** 1

49. The curve is rising for all x. The curve is concave up on $(-\infty, 0)$ and concave down on $(0, +\infty)$.

53. b. $y = \cosh 2x + \sinh 2x$ **55.** $2.3504a$ **57.** 17.68

Chapter 7 Review
Proficiency Examination (Pages 487–488)

12. a. 0.5493   **b.** $\frac{4}{3}$ **c.** 0.5493

13. $2\sqrt{x^2 + 1} + 3 \sinh^{-1} x + C$

14. $-\dfrac{x}{2} \cos 2x + \dfrac{1}{4} \sin 2x + C$

15. $-\frac{1}{2} \cosh(1 - 2x) + C$ **16.** $\sin^{-1} \dfrac{x}{2} + C$

17. $\frac{1}{4} \ln(x^2 + 1) + \frac{1}{2} \tan^{-1} x + \frac{1}{2} \ln |x - 1| + C$

18. $\dfrac{x^2}{2} + \dfrac{1}{2} \ln |x^2 - 1| + C$

19. 1.9089   **20.** 0.1144   **21.** -0.2027

22. 0.6095   **23.** $\frac{1}{4}$ **24.** 2

25. diverges   **26.** $\frac{1}{2}$

27. $\dfrac{1}{\sqrt{\tanh^{-1} 2x}} \left[\dfrac{1}{1 - 4x^2} \right]$ **28.** 4.7999

29. $y = \dfrac{x + 1}{e^x} (\ln |x + 1| + 1)$ **30.** 485.2 lb

Supplementary Problems (Pages 488–490)

1. $(1 - x^2)^{-1}$ **3.** e^{-2x}

5. $x \cosh x + \sinh x + (e^x - e^{-x}) \cosh(e^x + e^{-x})$

7. $-\dfrac{x}{2} \sqrt{4 - x^2} + 2 \sin^{-1} \dfrac{x}{2} + C$

9. $-\ln |x| - x^{-1} + \ln |x - 2| + C$

11. $4x^{1/4} - 4 \ln(1 + x^{1/4}) + C$

13. $\dfrac{x^3}{3} \tan^{-1} x - \dfrac{x^2}{6} + \dfrac{1}{6} \ln(1 + x^2) + C$

15. $\dfrac{1}{2} e^x \sqrt{4 - e^{2x}} + 2 \sin^{-1} \left(\dfrac{e^x}{2} \right) + C$

17. $-\frac{1}{5}(1 + x^{-2})^{5/2} + \frac{1}{3}(1 + x^{-2})^{3/2} + C$

19. $-2\sqrt{1 - \sin x} + C$

21. $\dfrac{x}{2} [\sin(\ln x) - \cos(\ln x)] + C$

23. $\ln(e^{2x} + 4e^x + 1) - x + C$

25. $\dfrac{x^3}{3} \cot^{-1} x + \dfrac{x^2}{6} - \dfrac{1}{6} \ln(1 + x^2) + C$

27. $\frac{1}{3} \ln |x^3 + 6x + 1| + C$

29. $2 \cos \sqrt{x + 2} + 2\sqrt{x + 2} \sin \sqrt{x + 2} + C$

31. $\frac{1}{4} \ln(x^4 + 4x^2 + 3) + C$

33. $\sqrt{5 - x^2} - \sqrt{5} \ln \left| \dfrac{\sqrt{5} + \sqrt{5 - x^2}}{x} \right| + C$

35. $\frac{1}{3} \sqrt{x^2 + 4} \, (x^2 - 8) + C$ **37.** $\dfrac{\sec^3 x}{3} + C$

39. $\frac{3}{2} x^{2/3} - 2x^{1/2} + 3x^{1/3} - 6x^{1/6} + 6 \ln |x^{1/6} + 1| + C$

41. $-\ln |1 + \cos x - \sin x| + C$

43. $-2e^{x/2} - 3e^{x/3} - 6e^{x/6} - 6 \ln |e^{x/6} - 1| + C$

45. $y = \dfrac{1}{x + 1} [x + C]$

47. $y = \sin x + \frac{1}{3} \cos x \tan^3 x + C \cos x$

49. 2   **51.** 0.8101   **53.** $\dfrac{n!}{a^{n+1}}$

57. -1 **59.** $T = T_0 + Be^{kt}$ **63.** $(0.65, 1)$

65. 2.848   **67.** 1.2876   **69.** 4.739

71. 23.38   **73.** $A_1 = A_2 = \ln |a|$

75. $3 |a| + \dfrac{b^2}{8 |a|} \ln 2$ **81.** 0.5199   **83.** $\dfrac{3\pi}{16}$

85. a. $v(t) = \left(\dfrac{W - B}{k} \right) \left(1 - e^{-(kg/W)t} \right)$

   b. $S(t) = \dfrac{W - B}{k} \left[t + \dfrac{W}{kg} e^{-(kg/W)t} - \dfrac{W}{kg} \right]$

   c. The critical depth is 269.07 m.

87. about 5.6 years

89. This is Putnam Examination Problem 3 in the morning session of 1980.

Chapter 8: Infinite Series
8.1 Sequences and Their Limits (Pages 503–505)

3. 0, 2, 0, 2, 0   **5.** $1, \frac{1}{2}, \frac{1}{3}, \frac{1}{4}, \frac{1}{5}$

7. $\frac{4}{3}, \frac{7}{4}, \frac{10}{5}, \frac{13}{6}, \frac{16}{7}$ **9.** $256, 16, 4, 2, \sqrt{2}$

11. 1, 3, 13, 183, 33673   **13.** 5

15. -7 **17.** 0   **19.** $\frac{1}{2}$

21. 4   **23.** 0   **25.** 1

27. 1   **29.** 1   **31.** $\frac{1}{2}$

33. 1   **35.** 1   **47.** $a_4 = 6.25\%, a_n = 100(\frac{1}{2})^n \%$

49. $N = 100$ **51.** $N = 1,001$

8.2 Introduction to Infinite Series; Geometric Series (Pages 511–514)

3. 5   **5.** 3   **7.** diverges

9. $\frac{5}{27}$ **11.** 4.51665   **13.** $-\frac{2}{45}$

15. $\frac{1}{3}$ **17.** $\frac{16}{63}$ **19.** $2(2 + \sqrt{2})$

21. $\frac{1}{2}(4 + 3\sqrt{2})$ **23.** converges to 1   **25.** converges to 1

27. diverges   **29.** The series converges to 1   **31.** $\frac{1}{99}$

33. $\frac{52}{37}$ **35. a.** $A = 1, B = 1$ **b.** $\frac{1}{2}$

37. 0   **39.** 9.14   **41.** 4.858   **43.** $\frac{1}{a}$

45. 2   **47.** 1,500   **49.** 3 ft

51. The total depreciation is $10,000.

53. The patient will have about 30.8 units of the drug.

55. In the long run there will be 33 members on the board.

59. 4 miles   **63. a.** false   **b.** true   **65. d.** $\frac{3}{4}$ **67.** 3

8.3 The Integral Test; p-Series (Pages 520–521)

3. $p = 3$; converges   **5.** $p = \frac{1}{3}$; diverges

7. converges   **9.** diverges   **11.** converges

13. diverges   **15.** diverges

17. diverges   **19.** converges   **21.** diverges

23. converges   **25.** diverges   **27.** diverges

29. converges   **31.** diverges   **33.** converges

35. converges   **37.** diverges   **39.** diverges

41. diverges   **43.** diverges   **45.** diverges

47. converges   **49.** converges   **51.** diverges

53. diverges

55. converges if $p > 1$ and diverges if $p \leq 1$

57. converges if $p > 1$ and diverges if $p \leq 1$

59. converges if $p > 1$ and diverges if $p \leq 1$

61. false

8.4 Comparison Tests (Pages 526–527)

1. converges if $|r| < 1$ and diverges if $|r| \geq 1$

3. geometric; converges   **5.** geometric; diverges

7. p-series; diverges   **9.** p-series; converges

11. geometric; diverges   **13.** converges

15. diverges   **17.** diverges   **19.** converges

21. converges   **23.** converges   **25.** diverges

27. converges   **29.** converges   **31.** converges

33. converges   **35.** diverges   **37.** converges

39. converges   **41.** converges   **43.** diverges

45. converges   **47.** converges   **49.** converges

51. diverges   **53.** converges   **55.** converges

8.5 The Ratio Test and the Root Test (Pages 532–533)

3. converges   **5.** diverges   **7.** converges

9. converges   **11.** diverges   **13.** converges

15. converges   **17.** converges   **19.** diverges

21. converges   **23.** converges   **25.** converges

27. diverges   **29.** converges   **31.** diverges

33. converges   **35.** converges   **37.** diverges

39. converges   **41.** converges   **43.** converges

45. converges for $0 \leq x < 1$

47. converges for $0 \leq x \leq 0.5$

49. converges for $x \geq 0$

51. converges for $0 \leq x < a^{-1}$

53. The integral also converges.

8.6 Alternating Series; Absolute and Conditional Convergence (Pages 542–544)

5. converges conditionally   **7.** diverges

9. converges absolutely   **11.** converges conditionally

13. converges absolutely   **15.** diverges

17. converges absolutely   **19.** converges conditionally

21. converges conditionally   **23.** converges conditionally

25. diverges   **27.** converges conditionally

29. converges conditionally   **31.** converges absolutely

33. a. $\frac{5}{8}$ **b.** $S_7 \approx 0.632$ **35. a.** 0.0823   **b.** 0.083

37. a. -0.1664 **b.** $S_5 \approx -0.167$

39. The interval of convergence is $\left[-\frac{1}{2}, \frac{1}{2}\right)$.

41. The interval of convergence is $[-1, 1]$.

43. The interval of convergence is $(-1, 1)$ for all $p > 0$.

45. $E_{max} = \frac{1}{6^2} \approx 0.0278$ **47.** $E_{max} = \frac{7}{2^7} \approx 0.0547$

49. converges absolutely

8.7 Power Series (Pages 552–553)

1. $(-1, 1)$ **3.** $(-1, 1)$ **5.** $\left(\frac{8}{3}, \frac{10}{3}\right)$

7. $\left(-\frac{13}{3}, -\frac{5}{3}\right)$ **9.** $x = 1$ **11.** $(-1, 3)$

13. $\left[1, \frac{5}{3}\right)$ **15.** $(-7, 7)$ **17.** $x = 0$

19. $[-1, 1]$ **21.** $(-\infty, \infty)$ **23.** $x = 0$

25. $(-\infty, \infty)$ **27.** $[-1, 1]$ **29.** $R = 1$

31. $R = e$ **33.** $R = \frac{1}{|a|}$

35. $f'(x) = \sum_{k=1}^{\infty} \frac{kx^{k-1}}{2^k}$ **37.** $f'(x) = \sum_{k=1}^{\infty} k(k+2)x^{k-1}$

39. $F(x) = \sum_{k=1}^{\infty} \frac{x^k}{k(2)^{k-1}}$ **41.** $F(x) = \sum_{k=0}^{\infty} \frac{(k+2)x^{k+1}}{(k+1)}$

47. $|x| > 1$ **49.** $(-1, 1)$ **51.** $R = \sqrt{2}$

8.8 Taylor and Maclaurin Series (Pages 564–566)

3. $e^{2x} = \sum_{k=0}^{\infty} \frac{(2x)^k}{k!}$ **5.** $e^{x^2} = \sum_{k=0}^{\infty} \frac{x^{2k}}{k!}$

7. $\sin x^2 = \sum_{k=0}^{\infty} \frac{(-1)^k x^{4k+2}}{(2k+1)!}$

9. $\sin ax = \sum_{k=0}^{\infty} \frac{(-1)^k (ax)^{2k+1}}{(2k+1)!}$

11. $\cos 2x^2 = \sum_{k=0}^{\infty} \frac{(-1)^k (2x^2)^{2k}}{(2k)!}$

13. $x^2 \cos x = \sum_{k=0}^{\infty} \frac{(-1)^k x^{2k+2}}{(2k)!}$

15. $x^2 + 2x + 1$ is its own Maclaurin series.

17. $xe^x = \sum_{k=0}^{\infty} \frac{x^{k+1}}{k!}$

19. $e^x + \sin x = 1 + 2x + \frac{x^2}{2!} + \frac{x^4}{4!} + \frac{2x^5}{5!} + \frac{x^6}{6!}$
$+ \frac{x^8}{8!} + \frac{2x^9}{9!} + \cdots$

21. $\frac{1}{1+4x} = \sum_{k=0}^{\infty} (-4)^k x^k$ **23.** $\frac{1}{a+x} = \sum_{k=0}^{\infty} \frac{(-1)^k x^k}{a^{k+1}}$

25. $\ln(3+x) = \ln 3 + \sum_{k=0}^{\infty} \frac{(-1)^k x^{k+1}}{(k+1)3^{k+1}}$

27. $\tan^{-1}(2x) = \displaystyle\sum_{k=0}^{\infty} \frac{(-1)^k 2^{2k+1} x^{2k+1}}{2k+1}$

29. $e^{-x^2} = \displaystyle\sum_{k=0}^{\infty} \frac{(-1)^k x^{2k}}{k!}$

31. $e^x \approx e + e(x-1) + \dfrac{1}{2!}e(x-1)^2 + \dfrac{1}{3!}e(x-1)^3$

33. $\cos x \approx \cos\dfrac{\pi}{3} - \left(x - \dfrac{\pi}{3}\right)\sin\dfrac{\pi}{3} - \dfrac{\left(x - \dfrac{\pi}{3}\right)^2}{2!}\cos\dfrac{\pi}{3}$

$\qquad + \dfrac{\left(x - \dfrac{\pi}{3}\right)^3}{3!}\sin\dfrac{\pi}{3}$

35. $\tan x \approx 0 + 1 \cdot x + \dfrac{0x^2}{2!} + \dfrac{2x^3}{3!}$

37. $f(x) = (x-2)^3 + 4(x-2)^2 + 5(x-2) - 3$

39. $f(x) \approx -\dfrac{1}{3} + \dfrac{1}{9}(x-5) - \dfrac{1}{27}(x-5)^2 + \dfrac{1}{81}(x-5)^3$

41. $f(x) \approx 1 - \dfrac{2}{3}(x-2) + \dfrac{4}{9}(x-2)^2 - \dfrac{8}{27}(x-2)^3$

43. $f(x) = 1 + \dfrac{1}{2}x - \dfrac{1}{8}x^2 + \dfrac{1}{16}x^3 - \dfrac{5}{128}x^4 + \cdots$
Interval of convergence is $[-1, 1]$.

45. $f(x) = 1 + \dfrac{2}{3}x - \dfrac{1}{9}x^2 + \dfrac{4}{81}x^3 - \cdots$ Interval of convergence is $[-1, 1]$.

47. $f(x) = x + \dfrac{1}{2}x^3 + \dfrac{3}{8}x^5 + \dfrac{5}{16}x^7 + \cdots$ Interval of convergence is $(-1, 1)$.

51. $x + \dfrac{1}{3!}x^3 + \dfrac{1}{5!}x^5 + \cdots = \displaystyle\sum_{k=0}^{\infty} \frac{x^{2k+1}}{(2k+1)!}$

53. $n = 4$ **55.** $f(x) = \displaystyle\sum_{k=0}^{\infty}\left[1 - \frac{1}{2^{k+1}}\right]x^k$

57. $f(x) = \displaystyle\sum_{k=0}^{\infty} \frac{(-1)^k 2^{2k} x^{2k+1}}{(2k+1)!}$

59. $f(x) = \dfrac{1}{2}\displaystyle\sum_{k=0}^{\infty}(-1)^k\left[\frac{1 + 2^{2k}}{(2k)!}\right]x^{2k}$

61. $f(x) = \displaystyle\sum_{k=0}^{\infty}[(-1)^k 2^{k+1} + 2^{k+1} + 1]\frac{x^{k+1}}{k+1}$

63. 1   **69.** 0.499514

Chapter 8 Review
Proficiency Examination (Pages 566–567)

31. This is the definition of e.   **32. a.** 0 **b.** converges

33. diverges   **34.** diverges   **35.** diverges

36. diverges   **37.** converges absolutely

38. The interval of convergence is $(-1, 1)$.

39. $\sin 2x = \displaystyle\sum_{k=0}^{\infty} \frac{(-1)^k (2x)^{2k+1}}{(2k+1)!}$

40. $f(x) = -\dfrac{2}{5}\displaystyle\sum_{k=0}^{\infty}\left(\frac{2}{5}\right)^k\left(x - \frac{1}{2}\right)^k$

Supplementary Problems (Pages 567–569)

1. does not exist   **3.** converges to e^{-2}

5. converges to 1   **7.** converges to $\dfrac{5}{3}$

9. converges to 0   **11.** converges to 0

13. converges to e^4 **15.** converges

17. $\dfrac{1}{3}$ **19.** $\dfrac{e}{3-e}$ **21.** $\dfrac{1}{2}$

23. $\dfrac{15 - e}{9(6 - e)}$ **25.** $\dfrac{-1}{\ln 2}$ **27.** $\dfrac{1}{4}$

29. diverges   **31.** diverges   **33.** converges

35. converges   **37.** converges   **39.** converges

41. diverges   **43.** diverges   **45.** converges

47. converges absolutely   **49.** converges absolutely

51. diverges   **53.** converges absolutely

55. converges absolutely

57. $(0, 2)$ **59.** $(-1, 1)$ **61.** $(-\infty, \infty)$

63. $[-3, -1)$ **65.** $x = 0$ **67.** $(-\infty, \infty)$

69. $f(x) = \displaystyle\sum_{k=0}^{\infty}(-1)^k \frac{x^{2k+4}}{(2k+1)!}$

71. $f(x) = x + x^2 + \displaystyle\sum_{k=1}^{\infty} \frac{(-1)^k x^{2k+1}}{2k+1}$

73. $f(x) = \displaystyle\sum_{k=0}^{\infty}\left[2 + \frac{5(-1)^{k+1}}{2}\left(\frac{3}{2}\right)^k\right]x^k$

75. 0.000022   **77.** $\dfrac{20}{9}$ ft   **79.** $p = 0.95$

81. $a_k = (-1)^k(4k+1)\left[\dfrac{1 \cdot 3 \cdot 5 \cdots (2k-1)}{2 \cdot 4 \cdot 6 \cdots (2k)}\right]^3$

There is an interesting discussion of this identity, as well as biographical information on Ramanujan, in the article "Ramanujan the Phenomenon," by S.G. Gindikin, *Quantum*, pp. 4–9, April 1998.

85. $\displaystyle\int \sqrt{x^3 + 1}\, dx = x + \dfrac{1}{8}x^4 - \dfrac{1}{56}x^7 + \dfrac{1}{160}x^{10} - \cdots$

89. 0.1986693   **91.** $R_3(x) < 10^{-12}$ **93.** $|x| < 0.187$

95. a. $t \approx \dfrac{\lambda}{\ln 2}\left[\dfrac{S(t) - S(0)}{R(t)}\right]$

b. $t \approx 5.61 \times 10^7$ years (56 million years)

97. This is Putnam Problem 3 in the morning session of 1951.

Cumulative Review for Chapters 6–8 (Page 571)

5. $\dfrac{1}{2}x \ln x - \dfrac{1}{2}x + C$ **7.** $\dfrac{1}{2}(\tan^{-1} x)^2 + C$

9. $2\sin^{-1}\left(\dfrac{x}{2}\right) + \dfrac{1}{2}x\sqrt{4 - x^2} + C$

11. a. $\dfrac{3}{x - 1} + \dfrac{2}{(x - 1)^2} - \dfrac{5}{2x + 3}$

b. $3\ln|x - 1| - \dfrac{2}{x - 1} - \dfrac{5}{2}\ln|2x + 3| + C$ **c.** diverges

13. converges   **15.** diverges   **17.** converges

19. -6 **21.** $-1 \le x \le 1$ **23.** 16,937 ft-lb

25. 9.7828

27. a. $y = x\tan^{-1} x + Cx$ **b.** $y = 1 + 2\cos x$

29. a. $x^2 - \dfrac{x^6}{3!} + \dfrac{x^{10}}{5!} - \cdots = \displaystyle\sum_{k=0}^{\infty} \frac{(-1)^k x^{4k+2}}{(2k+1)!}$

b. $\displaystyle\sum_{k=0}^{\infty} \frac{1}{k!}\left(\frac{1}{2}\right)^k x^k$ **c.** $\dfrac{3}{2} - \dfrac{19}{10}\displaystyle\sum_{k=0}^{\infty}\left(-\frac{2}{5}x\right)^k$

Chapter 9: Vectors in the Plane and in Space

9.1 Introduction to Vectors (Pages 580–582)

1.   **3.**

5. $\langle 4, 3 \rangle$; $\|\mathbf{PQ}\| = \sqrt{4^2 + 3^2} = 5$ **7.** $\langle -5, 0 \rangle$; $\|\mathbf{PQ}\| = 5$

9. $\mathbf{PQ} = 2\mathbf{i}$; $\|\mathbf{PQ}\| = 2$ **11.** $\mathbf{PQ} = 4\mathbf{i} + 2\mathbf{j}$; $\|\mathbf{PQ}\| = 2\sqrt{5}$

13. $\dfrac{1}{\sqrt{2}}(\mathbf{i} + \mathbf{j})$ **15.** $\frac{3}{5}\mathbf{i} - \frac{4}{5}\mathbf{j}$

17. $s = 6$; $t = 24$ **19.** $s = -5$; $t = -1$

21. $17\mathbf{i} - 18\mathbf{j}$ **23.** $35\mathbf{i} - 35\mathbf{j}$

25. $(3, 2)$ **27.** $(2, 4), (-4, -2)$

29. $\dfrac{\sqrt{3}}{2}\mathbf{i} + \dfrac{1}{2}\mathbf{j}$ **31.** $\dfrac{4}{\sqrt{17}}\mathbf{i} - \dfrac{1}{\sqrt{17}}\mathbf{j}$

33. $\dfrac{5}{\sqrt{26}}\mathbf{i} + \dfrac{1}{\sqrt{26}}\mathbf{j}$ **35.** $(3, 10)$

37. a. $6\mathbf{i} + 3\mathbf{j}$ **b.** $10\mathbf{i} + 5\mathbf{j}$

39. $\|\mathbf{v}\| = \sqrt{\cos^2\theta + \sin^2\theta} = 1$ **41.** Not necessarily equal.

43. a. This is the set of points on the circle with center (x_0, y_0) and radius 1.

   b. $\|\mathbf{u} - \mathbf{u}_0\| \le r$ is the set of points on or interior to the circle with center (x_0, y_0) and radius r.

45. $a = -2t$, $b = t$, and $c = t$, for any t

47. $-6\mathbf{i} + 3\mathbf{j}$ **49.** 5.22 mi/hr; N53.6°W

9.2 Coordinates and Vectors in $\mathbb{R}^3$ (Pages 587–588)

1. a. $\langle 2, 2, 4 \rangle$ **b.** $\langle 6, -8, -2 \rangle$ **c.** $\langle 10, -\frac{15}{2}, \frac{5}{2} \rangle$ **d.** $\langle 2, 9, 11 \rangle$

3. a. $\langle 1, -3, 8 \rangle$ **b.** $\langle 1, -1, 2 \rangle$ **c.** $\langle \frac{5}{2}, -5, \frac{25}{2} \rangle$ **d.** $\langle 2, -7, 19 \rangle$

5. $d = \sqrt{149}$

7. $d = \sqrt{382}$

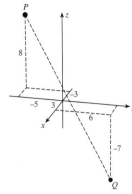

9. $x^2 + y^2 + z^2 = 1$

11. $x^2 + (y - 4)^2 + (z + 5)^2 = 9$

13. $C(0, 1, -1,)$; $r = 2$

15. $C(3, -1, 1)$; $r = 1$

17. $\mathbf{PQ} = -2\mathbf{i} + 2\mathbf{j} + \mathbf{k}$; $\|\mathbf{PQ}\| = 3$

19. $\mathbf{PQ} = -4\mathbf{i} - 4\mathbf{j} - 4\mathbf{k}$; $\|\mathbf{PQ}\| = 4\sqrt{3}$

21. $-7\mathbf{i} - 16\mathbf{k}$

23. $13\mathbf{i} - 4\mathbf{j} + 19\mathbf{k}$

25. $\dfrac{1}{\sqrt{14}} \langle 3, -2, 1 \rangle$

27. $\dfrac{1}{5\sqrt{2}} \langle -5, 3, 4 \rangle$

29.

31.

33. $(2x + 1)^2 + (2y - 5)^2 + 4z^2 = 46$

35. $\sqrt{3}$ **37.** 66   **39.** $6\mathbf{i} + 12\mathbf{j} - 3\mathbf{k}$

41. $\sqrt{46}\,(3\mathbf{i} + 2\mathbf{j} + \mathbf{k})$ **43.** no   **45.** yes

47. $\triangle ABC$ is right, but not isosceles

49. $\triangle ABC$ is neither right nor isosceles

51. collinear

53. $\mathbf{F}_1 = \frac{25}{3} \langle 0, -2, -6 \rangle$, $\mathbf{F}_2 = \frac{25}{3} \langle -\sqrt{3}, 1, -6 \rangle$, $\mathbf{F}_3 = \frac{25}{3} \langle \sqrt{3}, 1, -6 \rangle$

55. $P\left(\frac{1}{3}, \frac{7}{3}, \frac{13}{3}\right)$

9.3 The Dot Product (Pages 595–597)

3. -16 **5.** 5   **7.** orthogonal

9. orthogonal   **11.** 11   **13.** 0

15. $71°$ **17.** $114°$ **19.** $1; \mathbf{k}$

21. $0; \mathbf{0}$

23. $\mathbf{u}_1 = \langle \frac{\sqrt{2}}{2}, 0, \frac{\sqrt{2}}{2} \rangle, \mathbf{u}_2 = \langle -\frac{\sqrt{2}}{2}, 0, -\frac{\sqrt{2}}{2} \rangle$

25. $\mathbf{u} = -\dfrac{1}{\sqrt{17}}(2\mathbf{i} + 3\mathbf{j} - 2\mathbf{k})$

27. $x = -1, y = -1,$ and $z = 4$ **29.** $a = \frac{3}{2}$

31. $\cos\alpha = \dfrac{2}{\sqrt{38}}, \alpha \approx 1.24$ or $71°$; $\cos\beta = \dfrac{-3}{\sqrt{38}}, \beta \approx 2.08$ or $119°$; $\cos\gamma = \dfrac{-5}{\sqrt{38}}, \gamma \approx 2.52$ or $144°$

33. $\cos\alpha = \dfrac{1}{\sqrt{2}}, \alpha \approx 0.79$ or $45°$; $\cos\beta = \dfrac{-4}{5\sqrt{2}}, \beta \approx 2.17$ or $124°$; $\cos\gamma = \dfrac{3}{5\sqrt{2}}, \gamma \approx 1.13$ or $65°$

35. $\cos\theta = -\frac{1}{6}; -\frac{1}{6}\mathbf{w}$

37. a. 26   **b.** $\frac{26}{35}$ **c.** $\frac{49}{26}$ **d.** $\frac{-26}{25}$

39. 10

41. The force has magnitude $\|\mathbf{F}\| \approx 49.53$ lbs and points in the direction of the unit vector $\langle 0.979, -0.025, 0.202 \rangle$

43. a. 500 ft-lb   **b.** $500\sqrt{2}$ ft-lb

47. The region inside a sphere with center (a, b, c) and radius r.

9.4 The Cross Product (Pages 604–606)

1. $\mathbf{k}$ **3.** $-2\mathbf{i} + 4\mathbf{j} + 3\mathbf{k}$

5. $-2\mathbf{i} + 25\mathbf{j} + 14\mathbf{k}$ **7.** $-2\mathbf{i} + 16\mathbf{j} + 11\mathbf{k}$

9. $-14\mathbf{i} - 4\mathbf{j} - \mathbf{k}$ **11.** $\dfrac{\sqrt{3}}{2}$ **13.** $\dfrac{\sqrt{3}}{2}$

15. $\dfrac{1}{\sqrt{14}}\mathbf{i} + \dfrac{3}{\sqrt{14}}\mathbf{j} - \dfrac{2}{\sqrt{14}}\mathbf{k}$

17. $\dfrac{-16}{\sqrt{386}}\mathbf{i} + \dfrac{7}{\sqrt{386}}\mathbf{j} + \dfrac{9}{\sqrt{386}}\mathbf{k}$

19. $\sqrt{26}$ **21.** $2\sqrt{59}$ **23.** $\frac{1}{2}\sqrt{3}$

25. $\frac{3}{2}\sqrt{3}$ **27. a.** does not exist   **b.** scalar

29. a. scalar   **b.** vector

31. 5   **33.** 8   **37.** $s = 2$

39. Cross product is not an associative operation.

41. $\mathbf{w} = 0$ **43.** $T = -40\sqrt{3}\mathbf{i}$ **45.** $40°$

9.5 Parametric Representation of Curves; Lines in $\mathbb{R}^3$ (Pages 614–615)

3. $y = x - 2; 1 \le x \le 3$ **5.** $y = \frac{4}{3}x - \frac{1}{225}x^2;$ $0 \le x \le 180$

7. $y = x^{2/3}; x \ge 0$

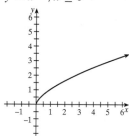

9. $x^2 + y^2 = 9; -3 \le x \le 3$

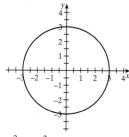

11. $(x - 1)^2 + (y + 2)^2 = 1;$ $0 \le x \le 2$ **13.** $\dfrac{y^2}{9} - \dfrac{x^2}{16} = 1;$ $-\infty < x < \infty$

15. $y = \ln x; x > 0$

17. $x = 1 + 3t, y = -1 - 2t, z = -2 + 5t;$ $\dfrac{x - 1}{3} = \dfrac{y + 1}{-2} = \dfrac{z + 2}{5}$

19. $x = 1 + t, y = -1 + 2t, z = 2 + t;$ $\dfrac{x - 1}{1} = \dfrac{y + 1}{2} = \dfrac{z - 2}{1}$

21. $x = 1 + t, y = -3 - 3t, z = 6 - 5t;$ $\dfrac{x - 1}{1} = \dfrac{y + 3}{-3} = \dfrac{z - 6}{-5}$

23. $x = 11t, y = 4 - 3t, z = -3 + 5t;$ $\dfrac{x}{11} = \dfrac{y - 4}{-3} = \dfrac{z + 3}{5}$

25. $x = -1 + 3t, y = 1 + t, z = 6 - 2t;$ $\dfrac{x + 1}{3} = \dfrac{y - 1}{1} = \dfrac{z - 6}{-2}$

27. $(0, -6, -3), (8, 0, -1), (12, 3, 0)$

29. $(0, 4, 9), (8, 0, -3), (6, 1, 0)$

31. parallel   **33.** parallel   **35.** $(1, 2, 3)$

37. $\pm\dfrac{1}{\sqrt{21}}(4\mathbf{i} + 2\mathbf{j} + \mathbf{k})$

39. $x = \sqrt{3}\cos t, y = \sqrt{3}\sin t; 0 \le t \le 2\pi$

41. $x = 3\cos t, y = 2\sin t; 0 \le t \le 2\pi$

43. $x = \dfrac{4}{\cos t}, y = 3\tan t, 0 \le t \le 2\pi, t \ne \frac{\pi}{2}, t \ne \frac{3\pi}{2}$

45. parabolic arc

49. $a = 0, b = 5$; other choices will work

51. $x = -1 - 23t, y = 3 + 2t, z = 1 - 19t$

53. The Spy should travel from $Q(0, 0, 1)$ in the direction of the vector $\mathbf{v} = 674.35\mathbf{i} + 337.68\mathbf{j} + 4.34\mathbf{k}$

9.6 Planes in $\mathbb{R}^3$ (Pages 621–622)

5. $5x - 3y + 4z - 4 = 0$ **7.** $x - 2y + 4z + 7 = 0$

9. $x - z + 1 = 0$ **11.** $x + 2y - 3z - 6 = 0$

13. $x = 0$ **15.** $\pm \dfrac{1}{\sqrt{38}}(5\mathbf{i} - 3\mathbf{j} + 2\mathbf{k})$

17. $\dfrac{5\sqrt{38}}{19}$ **19.** $\dfrac{\sqrt{26}}{26}$ **21.** $\dfrac{2a^2 + 1}{|a|\sqrt{14}}$

23. $\dfrac{29}{\sqrt{1{,}265}}$ **25.** 2 **27.** $\dfrac{47}{5}$

29. $\dfrac{4\sqrt{3}}{\sqrt{14}}$ **31.** $\dfrac{\sqrt{65}}{3}$ **33.** $\dfrac{65}{\sqrt{122}}$

35. $\dfrac{2}{\sqrt{6}}$ **37. b.** $\dfrac{1}{\sqrt{6}}$

39. $(x + 1)^2 + (y - 2)^2 + (z - 4)^2 = \frac{1}{38}(2x - 5y + 3z - 7)^2$

41. a. $\cos\theta = \dfrac{aA + bB + cC}{\sqrt{a^2 + b^2 + c^2}\,\sqrt{A^2 + B^2 + C^2}}$ **b.** $\theta \approx 52°$

43. $3\mathbf{i} - 2\mathbf{j} - 11\mathbf{k}$

45. $\cos\alpha = \dfrac{-4}{\sqrt{26}}$, $\alpha \approx 2.47$ or $142°$; $\cos\beta = \dfrac{3}{\sqrt{26}}$, $\beta \approx 0.94$ or $54°$; $\cos\gamma = \dfrac{1}{\sqrt{26}}$, $\gamma \approx 1.37$ or $79°$

47. $\dfrac{x - 1}{2} = \dfrac{y + 5}{-3} = \dfrac{z - 3}{1}$

49. The unit vectors are $\mathbf{N} = \pm\dfrac{1}{\sqrt{2}}(\mathbf{i} - \mathbf{k})$

51. $2x - y + 3z + 3 = 0$ **53.** $P(9, -5, 12)$

55. $\dfrac{x - 2}{1} = \dfrac{y - 3}{1} = \dfrac{z - 1}{1}$

9.7 Quadric Surfaces (Pages 626–627)

3. circular cone; B **5.** hyperboloid of one sheet; E

7. sphere; A **9.** paraboloid, G

11. hyperboloid of two sheets; I

13. ellipsoid: xy-plane, ellipse; yz-plane, ellipse; xz-plane, ellipse

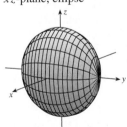

15. hyperboloid of one sheet: xy-plane, ellipse; yz-plane, hyperbola; xz-plane, hyperbola

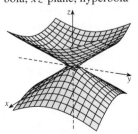

17. elliptic paraboloid: parallel to xy-plane, $z > 0$, ellipse; yz-plane, parabola; xz-plane, parabola

19. elliptic cone: parallel to xy-plane, $z \neq 0$, ellipse; parallel to yz-plane, $x \neq 0$, hyperbola; parallel to xz-plane, $z \neq 0$; hyperbola

21. elliptic paraboloid: xy-plane, parabola; yz-plane, parabola; xz-plane, ellipse

23. elliptic cone: xy-plane, hyperbola; yz-plane, hyperbola; xz-plane, ellipse

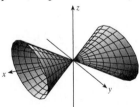

25. ellipsoid centered at $(1, -2, 3)$ with semi-axes 2, 1, and 3.

27. ellipsoid centered at $\left(\frac{9}{14}, -2, -\frac{7}{6}\right)$ with semi-axes approximately 1.31, 3.46, and 2.00.

29. elliptic paraboloid with vertex $\left(\frac{1}{2}, \frac{9}{2}, -\frac{1}{2}\right)$ and the y-axis as the axis of symmetry.

31. $\dfrac{8}{45}x^2 + \dfrac{7}{20}y^2 = 1$; this is an ellipse in the xy-plane

35. $x^2 + 2y^2 = 1$; this is the equation of an ellipse.

37. a. The equatorial radius is $a = 6{,}278.2$ and the polar radius is $b = 6{,}356.5$.

  b. The volume is approximately 1.08319×10^{12} km^3.

39. $20x^2 + 36y^2 + 36z^2 = 45$ is an ellipsoid.

Chapter 9 Review
Proficiency Examination (Page 628–629)

33. a. $13\mathbf{i} - 12\mathbf{j} + 2\mathbf{k}$ **b.** 1
   c. 12   **d.** $2\mathbf{i} + 3\mathbf{j} + 5\mathbf{k}$
   e. $\frac{12}{13}(3\mathbf{i} - 2\mathbf{j})$ **f.** $\dfrac{6\sqrt{14}}{7}$

34. a. 40   **b.** not possible
   c. $25\mathbf{i} - 10\mathbf{j} - 15\mathbf{k}$ **d.** not possible

35. $\dfrac{x}{1} = \dfrac{y+2}{-6} = \dfrac{z-1}{4}$ **36.** $2x + 3z - 11 = 0$

37. $x = -\frac{13}{2} - 10t,\ y = 5 + 6t,\ z = 2t$

38. $17x + 19y + 13z - 25 = 0$

39. $\cos\alpha = \dfrac{-2}{\sqrt{14}},\ \alpha \approx 2.13$ or $122°$; $\cos\beta = \dfrac{3}{\sqrt{14}},\ \beta \approx 0.64$ or $37°$; $\cos\gamma = \dfrac{1}{\sqrt{14}},\ \gamma \approx 1.30$ or $74°$

40. a. skew   **b.** intersect at $(2, 2, 3)$

41. a. 6   **b.** $\sqrt{2}$

42. $\dfrac{2\sqrt{30}}{15}$ **43.** $\dfrac{6}{\sqrt{26}}$ **44.** $\dfrac{5\sqrt{6}}{\sqrt{35}}$

45. 168.4 mph   **46.** 130 ft-lb

47. a.

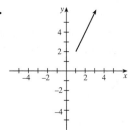

   b. $\dfrac{y}{x} = \dfrac{2^{t+1}}{2^t} = 2$ or $y = 2x,\ x \geq 1$

48. a.

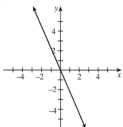

   b. $\dfrac{x}{4} = t$, so $x = 4t$, $\dfrac{y}{-9} = t$, so $y = -9t$

49. a.

   b. $x = 4t,\ y = -9t,\ z = t$

50. a. elliptic cone   **b.** hyperbolic paraboloid
   c. hyperboloid of two sheets
   d. hyperboloid of one sheet   **e.** plane   **f.** line
   g. sphere   **h.** elliptic (circular) paraboloid
   i. hyperboloid of two sheets
   j. hyperboloid of two sheets

Supplementary Problems (Pages 629–631)

1. a. $8\sqrt{2} + \sqrt{14}$ **b.** $\frac{1}{2}\sqrt{171}$

---

     c. $A \approx 125°,\ B \approx 26°,\ C \approx 30°$ **d.** $p = \dfrac{205}{6}$

3. $\dfrac{x-1}{2} = \dfrac{z}{1}$ and $y = 4$; $x = 1 + 2t,\ y = 4,\ z = t$; answers vary for the two points

5. $z = 0$ **7.** $3x + 4y - z + 13 = 0$

9. $5x - 2y + 3z - 32 = 0$ **11.** $3x + 2y - 11z + 19 = 0$

17. $(\mathbf{u} - \mathbf{v}) \cdot \mathbf{w} = 3$; $(2\mathbf{u} + \mathbf{v}) \times (\mathbf{u} - \mathbf{w}) = 6\mathbf{i} - 10\mathbf{j} - 10\mathbf{k}$

19. $\cos\alpha = \dfrac{-3}{\sqrt{46}}$; $\cos\beta = \dfrac{6}{\sqrt{46}}$; $\cos\gamma = \dfrac{1}{\sqrt{46}}$

21. $\theta = 74°$

23. The center is $\left(0, -\frac{3}{2}, \frac{1}{2}\right)$ and the radius is $\frac{3}{2}$.

25. $P(3, -2, 1)$

27. $37X + 20Y - 21Z = 61$ and $67X - 20Y + 99Z = 121$

29. $x = 2\cos t,\ y = 2\sin t$ is a circle.

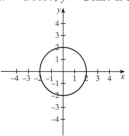

31. 13

33. $\frac{1}{2}\left(\|\mathbf{u} \times \mathbf{v}\| + \|\mathbf{u} \times \mathbf{w}\| + \|\mathbf{v} \times \mathbf{w}\| + \|(\mathbf{u} - \mathbf{v}) \times (\mathbf{w} - \mathbf{v})\|\right)$

37. $a_1 + 2b_1 + c_1 = 0$ **39.** $\dfrac{9\sqrt{2}}{2}$ **41.** $\dfrac{\sqrt{26}}{2}$

43. $2x - 3y - 7z = 0$

45. $\dfrac{x+1}{1} = \dfrac{y-2}{1}$ and $z = 0$ **47.** $\mathbf{N} = \dfrac{\pm 1}{\sqrt{37}}(-6\mathbf{i} + \mathbf{j})$

49. Let P have coordinates (a, b, c); then $\mathbf{r} = \mathbf{v} + \mathbf{w}t$ is the set of position vectors in the line $x = a + At$, $y = b + Bt$, and $z = c + Ct$.

51. linearly independent

55. $M\left(\frac{1}{3}, \frac{16}{3}, 1\right)$

57. Let S denote the position where the snowman had been, and let W and F denote the positions of the woodpile and flag-pole. Also, for any vector $\mathbf{v}$, let $\mathbf{v}_L$ and $\mathbf{v}_R$ denote the vectors obtained by rotating $\mathbf{v}$ through $90°$ counterclockwise and clockwise, respectively. (Note that $\mathbf{v}_R = -\mathbf{v}_L$.) Then, the two stakes are driven at points A and B, where $\mathbf{SA} = \mathbf{SW} + \mathbf{SW}_L$ and $\mathbf{SB} = \mathbf{SF} + \mathbf{SF}_R$. Find the digging point from W by pacing off half the distance from W to F, turning right, and pacing off an equal distance.

59. c. The result is still true.

61. Let M_1 be the midpoint of $\overline{AB}$. To divide $\overline{AB}$ into thirds, drop the perpendicular from the intersection of $\overline{AM}$ and $\overline{CM_1}$ to $\overline{AB}$. The point R_1 at the foot of the perpendicular satisfies $\mathbf{AR}_1 = \frac{1}{3}\mathbf{AB}$. Then drop the perpendicular from the intersection of $\overline{AM}$ and $\overline{CR_1}$ to obtain R_2, which divides $\overline{AB}$ into fifths. The process may be continued as far as desired.

63. This is Putnam Problem 5 of the afternoon session of 1959.

Chapter 10: Vector-Valued Functions

10.1 Introduction to Vector Functions (Pages 640–641)

1. $t \neq 0$ **3.** $t \neq \dfrac{(2n+1)\pi}{2}$, n an integer

5. $t \neq \dfrac{n\pi}{2}$, n an integer   **7.** $t > 0$

9. parabolic cylinder; in $\mathbb{R}^2$, the graph is a parabola in the xy-plane

11. cylinder; in $\mathbb{R}^2$, the graph is a circle in the xy-plane

13. plane; in $\mathbb{R}^2$, the graph is a line in the xz plane parallel
   to and four units below the x-axis

15. a circular helix; in $\mathbb{R}^2$, the graph is a circle in the xy-plane

17. the curve is in the intersection of the parabolic cylinder
   $y = (1 - x)^2$ with the plane $x + z = 1$; in $\mathbb{R}^2$, the graph
   is a parabola in the xy-plane

19. the curve is the intersection of the cylinder $y = x^2 + 1$ and the
   plane $y = z + 1$; in $\mathbb{R}^2$, the graph is a parabola in the xy-plane

21. $(7t - 3)\mathbf{i} - 10\mathbf{j} + \left(2t^2 - \dfrac{3}{t}\right)\mathbf{k}$

23. $3t - 2t^2$ **25.** $(1 - t)\sin t$

27. $-t^2 e^t \mathbf{i} + (t^2 \sin t)\mathbf{j} + (2te^t + 5\sin t)\mathbf{k}$

29. $(4te^t - t^2 + t + 10\sin t)\mathbf{i} + (2e^t t^2 + 1)\mathbf{k}$

31. $t^2 e^t - t^3 e^t - 2e^t - \dfrac{5}{t}\sin t$

37. $\mathbf{F}(t) = t\mathbf{i} + t^2\mathbf{j} + 2\mathbf{k}$

39. $\mathbf{F}(t) = 2t\mathbf{i} + (1 - t)\mathbf{j} + (\sin t)\mathbf{k}$

41. $\mathbf{F}(t) = t^2\mathbf{i} + t\mathbf{j} + \sqrt{9 - t^2 - t^4}\,\mathbf{k}$

43. a. no   **b.** yes   **c.** yes

45. $2\mathbf{i} - 3\mathbf{j} + e\mathbf{k}$ **47.** 0 **49.** $-\mathbf{i} + e^{-1}\mathbf{j}$ **51.** $\frac{3}{2}\mathbf{i} + \mathbf{j}$

53. continuous for all t **55.** continuous for $t \neq 0$, $t \neq -1$

57. continuous for all $t \neq 0$ **59.** $x - y + z = 1$

65. a. The function is continuous at t_0.

   b. The function is continuous at t_0.

   c. The function is discontinuous at $t_0 = 0$ if $h(t) = t$.

   d. The function is continuous at t_0.

**10.2 Differentiation and Integration of Vector Functions
(Pages 650–651)**

1. $\mathbf{F}'(t) = \mathbf{i} + 2t\mathbf{j} + (1 + 3t^2)\mathbf{k}$

3. $\mathbf{F}'(s) = (1 + \ln s)\mathbf{i} + 5s^{-1}\mathbf{j} - e^s(\ln s + s^{-1})\mathbf{k}$

5. $\mathbf{F}'(t) = 2t\mathbf{i} - t^{-2}\mathbf{j} + 2e^{2t}\mathbf{k}$; $\mathbf{F}''(t) = 2\mathbf{i} + 2t^{-3}\mathbf{j} + 4e^{2t}\mathbf{k}$

7. $\mathbf{F}'(s) = (\cos s)\mathbf{i} - (\sin s)\mathbf{j} + 2s\mathbf{k}$;
   $\mathbf{F}''(s) = (-\sin s)\mathbf{i} - (\cos s)\mathbf{j} + 2\mathbf{k}$

9. $f'(x) = -9x^2 - 2x$ **11.** $g'(x) = \dfrac{4x}{\sqrt{1 + 4x^2}}$

13. $\mathbf{V}(1) = \mathbf{i} + 2\mathbf{j} + 2\mathbf{k}$; $\mathbf{A}(1) = 2\mathbf{j}$; speed $= 3$; direction of motion
   is $\frac{1}{3}\mathbf{i} + \frac{2}{3}\mathbf{j} + \frac{2}{3}\mathbf{k}$

15. $\mathbf{V}\left(\dfrac{\pi}{4}\right) = -\dfrac{\sqrt{2}}{2}\mathbf{i} + \dfrac{\sqrt{2}}{2}\mathbf{j} + 3\mathbf{k}$; $\mathbf{A}\left(\dfrac{\pi}{4}\right) = -\dfrac{\sqrt{2}}{2}\mathbf{i} - \dfrac{\sqrt{2}}{2}\mathbf{j}$;
   speed $= \sqrt{10}$; direction of motion $-\dfrac{1}{2\sqrt{5}}\mathbf{i} + \dfrac{1}{2\sqrt{5}}\mathbf{j} + \dfrac{3}{\sqrt{10}}\mathbf{k}$

17. $\mathbf{V}(\ln 2) = 2\mathbf{i} - \frac{1}{2}\mathbf{j} + 8\mathbf{k}$; $\mathbf{A}(\ln 2) = 2\mathbf{i} + \frac{1}{2}\mathbf{j} + 16\mathbf{k}$;
   speed $= \dfrac{\sqrt{273}}{2}$; direction of motion $\dfrac{1}{\sqrt{273}}(4\mathbf{i} - \mathbf{j} + 16\mathbf{k})$

19. $\mathbf{F}'(0) = 2\mathbf{j}$; $\mathbf{F}'(1) = 2\mathbf{i} + 2\mathbf{j} + 5\mathbf{k}$; $\mathbf{F}'(-1) = -2\mathbf{i} + 2\mathbf{j} + \mathbf{k}$

21. $\mathbf{F}'(0) = 0$; $\mathbf{F}'(\frac{\pi}{2}) = \pi\mathbf{i} - \mathbf{j} - \frac{\pi^2}{4}\mathbf{k}$

23. $x = -1 - 3t$, $y = 1 + 2t$, $z = -1 - t$

25. $\dfrac{t^2}{2}\mathbf{i} - \dfrac{e^{3t}}{3}\mathbf{j} + 3t\mathbf{k} + \mathbf{C}$ **27.** $(t\ln t - t)\mathbf{i} - \frac{1}{2}t^2\mathbf{j} + 3t\mathbf{k} + \mathbf{C}$

29. $\dfrac{t^2}{2}\left(\ln t - \dfrac{1}{2}\right)\mathbf{i} - \cos(1 - t)\mathbf{j} + \dfrac{t^2}{2}\mathbf{k} + \mathbf{C}$

31. $\mathbf{R}(t) = \left(\frac{1}{3}t^3 + 1\right)\mathbf{i} + \left(-\frac{1}{2}e^{2t} + \frac{9}{2}\right)\mathbf{j} + \left(\frac{2}{3}t^{3/2} - 1\right)\mathbf{k}$

33. $\mathbf{R}(t) = \left(\frac{4}{3}t^{3/2} + 1\right)\mathbf{i} + (\sin t + 1)\mathbf{j}$

35. $\mathbf{V}(t) = \langle 2 + \sin t, 0, 3 + t\cos t - \sin t \rangle$;
   $\mathbf{R}(t) = \langle 2 + 2t - \cos t, -2, -1 + 3t + 2\cos t + t\sin t \rangle$

37. $\mathbf{R}(t) = e^t\mathbf{i} + (\frac{1}{3}t^3 - 1)\mathbf{j}$ **39.** $\mathbf{R}(t) = \langle 2t^4 + 1, 2t^2 + 2 \rangle$

43. $-180t^2$ **45.** $3t^2(-7\mathbf{i} + 11\mathbf{j} + 5\mathbf{k})$ **49.** $a = \frac{\pi}{4}$

53. not smooth

61. $[\mathbf{F} \times (\mathbf{G} \times \mathbf{H})]' = [(\mathbf{H} \cdot \mathbf{F})\mathbf{G}]' - [(\mathbf{G} \cdot \mathbf{F})\mathbf{H}]'$ **63.** false

**10.3 Modeling Ballistics and Planetary Motion
(Pages 657–660)**

1. 4.6 sec; 481 ft   **3.** 129.9 sec; 73,175 m   **5.** 1.9 sec; 41 m

7. 1.5 sec; 148 ft   **9.** $\mathbf{V}(t) = \sqrt{5}\,\mathbf{u}_r$; $\mathbf{A}(t) = 0$

11. $\mathbf{V} = (2\cos 2t)\mathbf{u}_r + (2\sin 2t)\mathbf{u}_\theta$;
   $\mathbf{A} = (-8\sin 2t)\mathbf{u}_r + (8\cos 2t)\mathbf{u}_\theta$

13. $\mathbf{V} = -10\sin(2t + 1)\mathbf{u}_r + 10[1 + \cos(2t + 1)]\mathbf{u}_\theta$;
   $\mathbf{A} = [-40\cos(2t + 1) - 20]\mathbf{u}_r - 40\sin(2t + 1)\mathbf{u}_\theta$

15. 140 m/s   **17.** 21.7°

19. The maximum height is 85 ft; the ball will land at a distance
   of 568 ft; the distance to the fence is 560 ft

21. 52.3 ft   **23.** 387.8 ft/s   **25.** 5.5 sec

27. The Spy should wait one minute before releasing the canister.

29. 18.13 ft   **31.** $\mathbf{V} = (-a\omega\sin\omega t)\mathbf{i} + (a\omega\cos\omega t)\mathbf{j}$; $\mathbf{R} \cdot \mathbf{V} = 0$

33. $-a\omega^2\mathbf{R}$

35. a. 1,533.33 m   **b.** 142.41 m/s

37. 21.8°   **39.** 37.2°; 471.77 ft/s

43. a. $R_f = v_0\cos\alpha\sec\beta\,T_f$ **b.** $\alpha = \frac{\pi}{4} - \frac{1}{2}\beta$

**10.4 Unit Tangent and Principal Unit Normal Vectors;
Curvature (Pages 671–673)**

1. $\mathbf{T}(t) = \dfrac{2}{\sqrt{4 + 9t^2}}\mathbf{i} + \dfrac{3t}{\sqrt{4 + 9t^2}}\mathbf{j}$;
   $\mathbf{N}(t) = \dfrac{-3t}{\sqrt{4 + 9t^2}}\mathbf{i} + \dfrac{2}{\sqrt{4 + 9t^2}}\mathbf{j}$

3. $\mathbf{T}(t) = \dfrac{\sqrt{2}}{2}[(\cos t - \sin t)\mathbf{i} + (\cos t + \sin t)\mathbf{j}]$;
   $\mathbf{N}(t) = -\dfrac{\sqrt{2}}{2}[(\sin t + \cos t)\mathbf{i} + (\sin t - \cos t)\mathbf{j}]$

5. $\mathbf{T}(t) = \dfrac{1}{\sqrt{2}}(-\sin t\mathbf{i} + \cos t\mathbf{j} + \mathbf{k})$; $\mathbf{N}(t) = -\cos t\mathbf{i} - \sin t\mathbf{j}$

7. $\mathbf{T}(t) = \dfrac{1}{\sqrt{1 + 4t^4}}(\mathbf{i} + 2t^2\mathbf{k})$; $\mathbf{N}(t) = \dfrac{1}{\sqrt{1 + 4t^4}}(-2t^2\mathbf{i} + \mathbf{k})$

9. $4\sqrt{5}$ **11.** $\dfrac{3\sqrt{2}\pi}{2}$ **13.** $\sqrt{41}\pi$

15. $\kappa = 0$ **17.** $\kappa = 2$ **19.** $\kappa = 2^{-3/2}$

21. $\mathbf{R}(s) = \left\langle \dfrac{\sqrt{2} - s}{\sqrt{2}}, \dfrac{s}{\sqrt{2}} \right\rangle$

23. $\mathbf{R}(s) = \left\langle 2 - \dfrac{3s}{\sqrt{26}}, 1 + \dfrac{s}{\sqrt{26}}, \dfrac{-4s}{\sqrt{26}} \right\rangle$

27. $\mathbf{T} = \dfrac{1}{\sqrt{10}}(\mathbf{i} - 3\mathbf{j})$

29. a. $\mathbf{T}(\pi) = \dfrac{\sqrt{2}}{2}(-\mathbf{i} + \mathbf{k})$ **b.** $\frac{1}{2}$ **c.** $\sqrt{2}\pi$

31. The points at which a maximum occurs are $P(0, 3)$, and $Q(0, -3)$.

33. $\rho(0) = \frac{1}{6}$; $\rho(1) = \rho(-1) = \frac{1}{24}$

35. $\kappa = \dfrac{2\sqrt{9t^4 + 9t^2 + 1}}{(1 + 4t^2 + 9t^4)^{3/2}}$ **37.** $\kappa = \dfrac{2}{(1 + 4x^2)^{3/2}}$

39. $\kappa = \dfrac{6x^5}{(x^6 + 4)^{3/2}}$ **41.** $\dfrac{1}{\sqrt{2}\, e^\theta}$

43. The fly's speed at time t is $4t^2 + 4t + 1$; 4 sec

51. c. $\lim\limits_{s \to +\infty} \kappa(s) = +\infty$, so the spiral keeps winding tighter and tighter.

55. The total length of the ramp is about 7 units.

57. $\kappa = \dfrac{1}{2^{3/2}} \dfrac{1}{\sqrt{1 - \cos t}}$; the minimum is $\frac{1}{4}$ and there is no maximum value.

10.5 Tangential and Normal Components of Acceleration (Pages 678–679)

1. $A_T = \dfrac{4t}{\sqrt{1 + 4t^2}}$; $A_N = \dfrac{2}{\sqrt{1 + 4t^2}}$

3. $A_T = \dfrac{t}{\sqrt{1 + t^2}}$; $A_N = \dfrac{t^2 + 2}{\sqrt{t^2 + 1}}$

5. $A_T = \dfrac{4t}{\sqrt{2 + 4t^2}}$; $A_N = \dfrac{2}{\sqrt{1 + 2t^2}}$

7. $A_T = -\dfrac{\sin t \cos t}{\sqrt{1 + \cos^2 t}}$; $A_N = \sqrt{\dfrac{2}{1 + \cos^2 t}}$

9. $A_T = \dfrac{-13}{\sqrt{10}}$; $A_N = \dfrac{11}{\sqrt{10}}$

11. $A_T = \dfrac{-19}{\sqrt{14}}$; $A_N = \sqrt{\dfrac{59}{14}}$

13. $\dfrac{5}{\sqrt{8}}$ **15.** 0

17. The maximum speed is 8 at $t = 0$, and the minimum speed is 6 at $t = \pi/4$.

19. $\mathbf{V}(t) = (-2\sin 2t)\mathbf{i} + (2\cos 2t)\mathbf{j}$;
$\mathbf{A}(t) = (-4\cos 2t)\mathbf{i} - (4\sin 2t)\mathbf{j}$; $A_T = 0$; $A_N = 4$

21. $A_T\left(\dfrac{\sqrt{399}}{8}\right) \approx 7.98999$; $A_N\left(\dfrac{\sqrt{399}}{8}\right) = 0.4$

23. $\omega = 0.52$ rev/s

25. a. the maximum safe speed is 47 ft/s (about 32 mi/h)
  b. 65.94 ft/s (about 45 mi/h)   **c.** 23.1°

27. 14 rev/min

29. $A_T = \dfrac{4t}{\sqrt{5 + 4t^2}}$; $A_N = \dfrac{2\sqrt{5}}{\sqrt{5 + 4t^2}}$

31. $A_T = \sqrt{3}\, e^t$; $A_N = \sqrt{2}\, e^t$ **33.** 42.43 units/s

37. $A_T = \dfrac{\mathbf{V} \cdot \mathbf{A}}{\|\mathbf{V}\|} = 0$ and $A_N = \dfrac{\|\mathbf{V} \times \mathbf{A}\|}{\|\mathbf{V}\|} = g$

39. b. About 36,000 km   **c.** About 200 km/h

Chapter 10 Review
Proficiency Examination (Page 679–680)

24.

$s = 2\pi\sqrt{10}$

25. $\mathbf{F}' = \dfrac{1}{(1 + t)^2}\mathbf{i} + \dfrac{t\cos t - \sin t}{t^2}\mathbf{j} + (-\sin t)\mathbf{k}$;
$\mathbf{F}'' = -\dfrac{2}{(1 + t)^3}\mathbf{i} + \dfrac{-t^2\sin t - 2t\cos t + 2\sin t}{t^3}\mathbf{j} - (\cos t)\mathbf{k}$

26. $(3 - 6\ln 2)\mathbf{i} + \frac{45}{4}\mathbf{j} + \left(6\ln 2 - \frac{9}{4}\right)\mathbf{k}$

27. $(e^t - t)\mathbf{i} - \left(\dfrac{t^4}{12} + 2\right)\mathbf{j} + \left(\dfrac{3t^2}{2} + 3t\right)\mathbf{k}$

28. $\mathbf{V} = \mathbf{i} + 2\mathbf{j} + e^t(t + 1)\mathbf{k}$; speed is $\sqrt{5 + e^{2t}(t + 1)^2}$;
$\mathbf{A} = e^t(t + 2)\mathbf{k}$

29. $\mathbf{T} = \dfrac{2t\mathbf{i} + 3\mathbf{j} - 3\mathbf{k}}{\sqrt{4t^2 + 18}}$; $\mathbf{N} = \dfrac{3\mathbf{i} - t\mathbf{j} + t\mathbf{k}}{(2t^2 + 9)^{1/2}}$; $A_T = \dfrac{2\sqrt{2}\, t}{\sqrt{2t^2 + 9}}$;
$A_N = \dfrac{6}{\sqrt{2t^2 + 9}}$

30. a. The maximum height is 9.77 ft.
  b. $\frac{25}{16}$ sec; maximum range is 67.7 ft

Supplementary Problems (Pages 680–683)

1. $2\mathbf{k}$ **3.** -5 **5.** $\mathbf{i} - \mathbf{j}$

7. $\mathbf{F}'(t) = (1 + t)e^t\mathbf{i} + 2t\mathbf{j}$; $\mathbf{F}''(t) = (2 + t)e^t\mathbf{i} + 2\mathbf{j}$

9. $\mathbf{F}'(t) = -2t^{-2}\mathbf{i} - 2\mathbf{j} + (1 - t)e^{-t}\mathbf{k}$; $\mathbf{F}''(t) = 4t^{-3}\mathbf{i} + (t - 2)e^{-t}\mathbf{k}$

11. $\mathbf{F}'(t) = (2t + ae^{at})\mathbf{i} + (1 - at)e^{-at}\mathbf{j} + ae^{at+1}\mathbf{k}$;
$\mathbf{F}''(t) = (2 + a^2 e^{at})\mathbf{i} + (a^2 t - 2a)e^{-at}\mathbf{j} + a^2 e^{at+1}\mathbf{k}$

13.

15.

17. The graph is a circular helix of radius 2, traversed clockwise. It begins (t = 0) at (0, 2, 0) and rises 10π units with each revolution.

19. $\dfrac{d}{dt}[\mathbf{F}(t) \cdot \mathbf{F}(t)] = 2\mathbf{F} \cdot \mathbf{F}'$

21. $\dfrac{d}{dt}\left[\dfrac{\mathbf{F}(t)}{\|\mathbf{F}(t)\|}\right] = \dfrac{\|\mathbf{F}\|^2 \mathbf{F}' - (\mathbf{F} \cdot \mathbf{F}')\mathbf{F}}{\|\mathbf{F}\|^3}$

23. $\dfrac{d}{dt}[\mathbf{F}(t) \times \mathbf{F}(t)] = 0$ **25.** $(e - e^{-1})\mathbf{i} + 6\mathbf{k}$

27. $(te^t - e^t)\mathbf{i} + \dfrac{\cos 2t}{2}\mathbf{j} + \dfrac{t^3}{3}\mathbf{k} + \mathbf{C}$

29. $(te^t - e^t)\mathbf{i} + \dfrac{1}{4}t^2(2\ln t - 1)\mathbf{j} + \dfrac{3t^2}{2}\mathbf{k} + \mathbf{C}$

31. $\mathbf{V}(t) = \mathbf{i} - \mathbf{j}; \dfrac{ds}{dt} = \sqrt{2}; \mathbf{A}(t) = \mathbf{0}$

33. $\mathbf{V}(t) = (t\cos t + \sin t)\mathbf{i} + (1 - t)e^{-t}\mathbf{j} + \mathbf{k};$

$\dfrac{ds}{dt} = \sqrt{t^2\cos^2 t + t\sin 2t + \sin^2 t + (1-t)^2 e^{-2t} + 1};$

$\mathbf{A}(t) = (2\cos t - t\sin t)\mathbf{i} + e^{-t}(t-2)\mathbf{j}$

35. $\mathbf{T}(t) = \dfrac{\mathbf{i} - 2t\mathbf{j}}{\sqrt{4t^2 + 1}}; \mathbf{N}(t) = \dfrac{-2t\mathbf{i} - \mathbf{j}}{\sqrt{4t^2 + 1}}$

37. $\mathbf{T}(t) = \tfrac{1}{5}(-4\sin t\,\mathbf{i} - 3\mathbf{j} + 4\cos t\,\mathbf{k});$
$\mathbf{N}(t) = (-\cos t)\mathbf{i} + (-\sin t)\mathbf{j}$

39. $A_T = \dfrac{4t + e^{2t}}{\sqrt{4 + 4t^2 + e^{2t}}}$; $A_N = 2\sqrt{\dfrac{t^2 e^{2t} - 2t e^{2t} + 2e^{2t} + 4}{4 + 4t^2 + e^{2t}}};$

$\kappa = \dfrac{2\sqrt{t^2 e^{2t} - 2t e^{2t} + 2e^{2t} + 4}}{(4t^2 + 4 + e^{2t})^{3/2}}$

41. $A_T = 0; A_N = 4; \kappa = \tfrac{1}{8}$

43. $\kappa = \tfrac{1}{2}$ **45.** $\kappa \approx 1.9221$ **47.** $\kappa = \tfrac{1}{6}$

49. $\mathbf{F}(t)$ is continuous for $t \neq 1$.

51. $\mathbf{F}(t) = (2t - \sin t + 1)\mathbf{i} + (\dfrac{3t^2}{2} + \cos t - 1)\mathbf{j} + \left(\dfrac{t^4}{24\pi} + \dfrac{t^2}{2}\right)\mathbf{k}$

53. a. $x = 1 + s, y = 1 + s, z = 1 + 3s$
b. $(-2, -2, -8)$

55. $\mathbf{T}(s) = \dfrac{1}{\sqrt{5}}\left[\left(-\sin\dfrac{s}{a}\right)\mathbf{i} + \left(\cos\dfrac{s}{a}\right)\mathbf{j} + 2\mathbf{k}\right];$

$\mathbf{N} = -\left(\cos\dfrac{s}{a}\right)\mathbf{i} - \left(\sin\dfrac{s}{a}\right)\mathbf{j}$

57. At $t = 0, \rho = \dfrac{b^2}{a}$; at $t = \dfrac{\pi}{2}, \rho = \dfrac{a^2}{b}$

59. a. $\mathbf{V}(t) = e^{-t}(-\cos t - \sin t)\mathbf{i} + e^{-t}(-\sin t + \cos t)\mathbf{j} - e^{-t}\mathbf{k};$
$\mathbf{A}(t) = e^{-t}(2\sin t)\mathbf{i} + e^{-t}(-2\cos t)\mathbf{j} + e^{-t}\mathbf{k}$

b. $\kappa = \dfrac{\sqrt{2}}{3}e^t$

61. 4,537.5 lb

63. $\mathbf{V}(t) = -a\sin at\,\mathbf{u}_r - ae^{-at}(1 + \cos at)\mathbf{u}_\theta;$
$\mathbf{A}(t) = [-a^2\cos at - a^2 e^{-2at}(1 + \cos at)]\mathbf{u}_r$
$\quad + [a^2 e^{-at}(1 + \cos at) + 2a^2 e^{-at}\sin at]\mathbf{u}_\theta$

65.

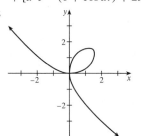

$x = \tfrac{2}{3} - \tfrac{5}{9}t, y = \tfrac{1}{3} - \tfrac{4}{9}t$

67. 20.888 μm

69. a. At $t = 0, \mathbf{V}(0) = 10\mathbf{u}_r$
b. At $t = 0.25, \mathbf{V}(0.25) = 10\mathbf{u}_r + 5\pi\,\mathbf{u}_\theta$

71. $\sin x\mathbf{i} - \cos 2x\mathbf{j} + e^{-x}\mathbf{k}$

73. $\mathbf{F}(t) = (\dfrac{t^2}{2} + 1)\mathbf{i} + (\dfrac{t^2}{2} + 2)\mathbf{j} - (\dfrac{t^2}{2} + 3)\mathbf{k}$

75. $\mathbf{F}(t) = \left[\left(\dfrac{1}{2}\right)(3 - \cos 2t)\right]\mathbf{i}$

$\quad + \left[\left(\dfrac{e^t}{2}\right)(\sin t + \cos t) - \dfrac{1}{2}\right]\mathbf{j} - [3\ln|t + 1| + 3]\mathbf{k}$

77. This is Putnam Problem 6ii of the morning session of 1939.

79. This is Putnam Problem 6 of the afternoon session of 1946.

Cumulative Review for Chapters 1-10 (Pages 691–692)

5. -2 **7.** e^{-6} **9.** -1

11. $y' = 3\sin^2 x\cos x + 2\sec^2 x$ **13.** $y' = \dfrac{1}{\sqrt{1-x^2}} - \dfrac{2}{x^2 + 1}$

15. $y' = \dfrac{-(y^3 + 2xe^{-y})}{3xy^2 - x^2 e^{-y}}$ **17.** $\tfrac{1}{3}\sin^3 x - \tfrac{1}{5}\sin^5 x + C$

19. $-\cot x + \csc x + C$ **21.** $\sin^{-1}(x - 1) + C$

23. a. yes   **b.** yes   **c.** no

25. 0.4467   **27.** 5.8069

29. a. $2x + 7y - z - 15 = 0$ **b.** $2x + 5y - 2z + 17 = 0$

31. converges by the ratio test

33. diverges by the ratio test

35. converges by the absolute convergence test

37. a. converges; $5e^{-1}$ **b.** converges; $\tfrac{\pi}{2}$

39. $\mathbf{F}'(t) = 2\mathbf{i} - 3e^{-3t}\mathbf{j} + 4t^3\mathbf{k}; \mathbf{F}''(t) = 9e^{-3t}\mathbf{j} + 12t^2\mathbf{k}$

41. $\mathbf{T} = \dfrac{1}{\sqrt{13}}[2(\cos 2t)\mathbf{i} - 2(\sin 2t)\mathbf{j} + 3\mathbf{k}];$
$\mathbf{N} = -\sin 2t\,\mathbf{i} - \cos 2t\,\mathbf{j}$

43. a. $\displaystyle\sum_{k=0}^{\infty} \dfrac{(-1)^k x^{2k+2}}{k!}$ **b.** 0.189

45. $y = \dfrac{x^2}{2} - \dfrac{x}{2} + \dfrac{1}{4} + \dfrac{7}{4}e^{-2x}$ **47.** $x = -2$

49. 1,562.5 ft-lb

Appendix A (Pages A-7–A-8)

1. $\delta = \tfrac{\epsilon}{2}$ **3.** false   **5.** $\delta = \min(1, \tfrac{\epsilon}{5})$

7. $\delta = \epsilon$ **9.** $\delta = \tfrac{\epsilon}{3}$ **11.** $\delta = \min(1, \tfrac{\epsilon}{5})$

APPENDIX G   Credits

This page constitutes an extension of the copyright page.

CHAPTER 1

12   Problems 59 and 60 from *Introductory Oceanography*, 5th ed., by H. V. Thurman, p. 253. Reprinted with permission of Merrill, an imprint of Macmillan Publishing Company. Copyright ©1988 Merrill Publishing Company, Columbus, Ohio.

12   Problem 61 adapted from R. A. Serway, *Physics*, 3rd ed., Philadelphia: Saunders, 1992, p. 1007.

12   Journal problem (#62) by Murray Klamkin reprinted from *The Mathematics Student Journal*, Vol. 28, 1980, issue 3, p. 2.

18–19   Babylonian system of equations in Problem 58 and 59 taken from *A History of Mathematics*, 2nd ed., by Carl B. Boyer, revised by Uta C. Merzbach, John Wiley and Sons, Inc., New York, 1968.

19   Journal problem (#64) reprinted from *Ontario Secondary School Mathematics Bulletin*, Vol. 18, 1982, issue 2. p. 7.

33   Journal problem (#72) reprinted from *The Mathematics Student Journal*, Vol. 28, 1980, issue 3, p. 2.

45   Putnam examination problem (#45) 1959, reprinted by permission from The Mathematical Association of America.

45   Putnam examination problem (#46) 1960, reprinted by permission from The Mathematical Association of America.

46–47   Guest essay, "Calculus Was Inevitable," by John L. Troutman.

48   Historical Quest (#9) information from *Ethnomathematics* by Marcia Ascher, (Pacific Grove: Brooks/Cole, 1991) pp. 128–129 and 188–189.

CHAPTER 2

78   Graph (#36) from Michael D. La Grega, Philip L. Buckingham, and Jeffery C. Evans, *Hazardous Waste Management*, New York: McGraw-Hill, 1994, pp. 565–566.

79   Modeling problem (#37) from E. Batschelet, *Introduction to Mathematics for Life Scientists*, 2nd ed., New York: Springer-Verlag, 1976, p. 280.

95   Modeling problem (#67) from William Bosch and L. G. Cobb, "Windchill," *UMAP Module No. 658*, (1984), pp. 244–247.

95   Putnam examination problem (#71) 1956, reprinted by permission from The Mathematical Association of America.

95   Putnam examination problem (#72) 1986, reprinted by permission from The Mathematical Association of America.

CHAPTER 3

129   Global modeling illustration from *Scientific American*, March 1991.

157   Journal problem (#73) by Bruce W. King from *The Pi Mu Epsilon Journal*, Volume 7 (1981), p. 346.

174   Modeling problem (#30) by John A. Helms, "Environmental Control of Net Photosynthesis in Naturally Grown Pinus Ponderosa Nets," *Ecology* (Winter 1972) p. 92.

174   Modeling problems (#35, #36) from *Introduction to Mathematics for Life Scientists*, 2nd ed. New York: Springer-Verlag (1976), pp. 102–103.

178   Problem 65 from 1982 BC AP Examination.

180   Putnam examination problem (#95) 1946, reprinted by permission from The Mathematical Association of America.

180   Putnam examination problem (#96) 1939, reprinted by permission from The Mathematical Association of America.

180   Putnam examination problem (#97) 1946, reprinted by permission from The Mathematical Association of America.

181   Chaos example is by Jack Wadhams of Golden West College.

181   Photo courtesy of Gregory Sams/Photo Researchers, Inc.

CHAPTER 4

183   Energy expended by a pigeon by Edward Batschelet, *Introduction to Mathematics for Life Scientists*, 2nd ed., New York: Springer-Verlag, 1979, pp. 276–277.

209   MAL cartoon, © 1974, by permission from the estate of Malcolm Hancock.

216   Problem 54 from "The Mechanics of Bird Migration," by C. J. Pennycuick, *Ibis III*, pp. 525–556.

216   Journal problem (#57) from *Mathematics Magazine*, Volume 55 (1982), p. 300.

228   Journal problem (#54) from *Parabola*, Vol. 20, Issue 1, 1984.

228   Historical Quest (#55) from "Mathematics in India in the Middle Ages," by Chandra B. Sharma, *Mathematical Spectrum*, Volume 14(1), pp. 6–8, 1982.

237   Historical Quest (#58) from *A Source Book in Mathematics, 1200–1800* by D. J. Struik, Cambridge, MA: University Press, 1969, pp. 313–316.

245   Journal problem (#6) from *Parabola*, Vol. 19, Issue 1, 1983, p. 22.

248   Modeling problem (#31) by Paul J. Campbell, "Calculus Optimization in Information Technology," *UMAP Module 1991: Tool for Teaching*, Lexington, MA, CUPM Inc., 1992, pp. 175–199.

248   Modeling problem (#32) by John C. Lewis and Peter P. Gillis, "Packing Factors in Diatomic Crystals," *American Journal of Physics*, Vol. 61, No. 5 (1993), pp. 434–438.

249–250   Modeling problem (#42) by Thomas O'Neil, "A Mathematical Model of a Universal Joint," *UMAP Modules 1982: Tools for Teaching*. Lexington, MA: Consortium for Mathematics and Its Applications, Inc., 1983, pp. 393–405.

250   Modeling problem (#43) is based on an article by Steve Janke, "Somewhere Within the Rainbow," *UMAP Modules 1992: Tools for Teaching*. Lexington, MA: Consortium for Mathematics and Its Applications, Inc., 1993.

256   Concentration of drugs in the bloodstream application adapted from E. Heinz, "Probleme bei der Diffusion kleiner Substanzmengen innerhalb des menschlichen Körpers," *Biochem.*, Vol. 319 (1949), pp. 482–492.

257   Modeling vascular branching is adapted from *Introduction to Mathematics for Life Scientists*, 2nd ed., by Edward Batschelet, New York: Springer-Verlag, 1976, pp. 278–280.

261   Modeling problem (#35) from "A Blood Cell Population Model, Dynamical Diseases, and Chaos," by W. B. Gearhart and M. Martelli, *UMAP Modules 1990: Tools for Teaching*, Arlington, MA: Consortium for Mathematics and Its Applications (CUPM) Inc., 1991.

261   Modeling problem (#37) adapted from "Flight of Birds in Relation to Energetics and Wind Directions," by V. A. Tucker and K. Schmidt-Koening, *The Auk*, Vol. 88, 1971, pp. 97–107.

261   Modeling problem (#39) adapted from P. L. Shaffer and H. J. Gold, "A Simulation Model of Population Dynamics of the Codling Moth Cydia Pomonella," *Ecological Modeling*, Vol. 30, 1985, pp. 247–274.

262   Problem #47 adapted from "Price Elasticity of Demand: Gambling, Heroin, Marijuana, Prostitution, and Fish," by Yves Nievergelt, *UMAP Modules 1987: Tools for Teaching*. Arlington, MA, CUPM Inc., 1988, pp. 153–181.

263   Photo of a beehive, courtesy of Scott Camazine/Photo Researchers, Inc.

264   Journal problem (#51) from the *Mathematics Teacher*, December 1990, p. 718.

265   Journal problem (#63) from the *AMATYC Review*, Spring 1995, pp. 67–68.

265   Journal problem (#64) from *Quantum*, 1997, Vol. 7, No. 4, pp. 50–53.

266   Modeling problem (#75) adapted from "Modeling the AIDS Epidemic," by Allyn Jackson, *Notices of the American Mathematical Society*, Vol. 36., No. 8 (October 1989), pp. 981–983.

267   Historical Quest (#84) quotation about Euler from *Elements of Algebra, 5th Ed*, by Leonhard Euler, London: Longman, Orme, and Co., 1840.

268   Putnam examination problem (#85) 1938, reprinted by permission from The Mathematical Association of America.

268   Putnam examination problem (#86) 1941, reprinted by permission from The Mathematical Association of America.

268   Putnam examination problem (#87) 1987, reprinted by permission from The Mathematical Association of America.

268   Putnam examination problem (#88) 1985, reprinted by permission from The Mathematical Association of America.

268   Putnam examination problem (#89) 1940, reprinted by permission from The Mathematical Association of America.

268   Putnam examination problem (#90) 1961, reprinted by permission from The Mathematical Association of America.

269   Photograph of a wine cellar showing wooden cask, courtesy of Peter Menzel/Stock Boston.

269   Group research project is from Elgin Johnston of Iowa State University. This group research project comes from research done at Iowa State University as part of a National Science Foundation grant.

CHAPTER 5

285   Software output from *Converge* by John R. Mowbray, JEMware, 567 South King Street, Suite 178, Honolulu, HI 96813.

308   Journal problem (#52) from *FOCUS*, February 1995, p. 15.

315   Historical Quest (#45) quotation from "A Short Account of the History of Mathematics," as quoted in *Mathematical Circles Adieu* by Howard Eves, Boston: Prindle, Weber & Schmidt, Inc., 1977.

316   Journal problem (#60) by Murray Klamkin from the *College Mathematics Journal*, Sept. 1989, p. 343. Problem by Murray Klamkin.

327   Shroud of Turin, courtesy of Gianni Tortoli/Science Source/Photo Researches, Inc.

341   Historical Quest (#34) Early Asian Calculus from *Mathematics*, by David Bergamini, p. 108. Reprinted by permission of Time, Incorporated, 1963.

349   Modeling problem (#76) adapted from "Flight Speeds of Birds in Relation to Energies and Wind Directions," by V. A. Tucker and K. Schmidt-Koenig, *The Auk*, Vol. 88 (1971), pp. 97–107.

350   Putnam examination problem (#86) 1951, reprinted by permission from The Mathematical Association of America.

350   Putnam examination problem (#87) 1958, reprinted by permission from The Mathematical Association of America.

351–352 Guest essay, "Kinematics of Jogging," by Ralph Boas.

CHAPTER 6

374   Photograph of the great pyramid of Cheops, courtesy of Paul W. Liebhardt.

385   Journal problem (#60) from *School Science and Mathematics*, Problem 3949 by V. C. Bailey, Vol. 83, 1983, p. 356.

395   *Blondie* cartoon reprinted with special permission of King Features Syndicate.

418   Journal problem (#20) *Mathematics Teacher*, December 1990, p. 695.

422   Putnam examination problem (#83) 1939, reprinted by permission from The Mathematical Association of America.

422   Putnam examination problem (#84) 1938, reprinted by permission from The Mathematical Association of America.

423 Putnam examination problem (#85) 1993, reprinted by permission from The Mathematical Association of America.

424 Photograph of Harry Houdini, courtesy of AP/Wide World Photos.

424 Group research project is from *MAA Notes*, Vol. 17, 1991, "Priming the Calculus Pump: Innovations and Resources," by Marcus S. Cohen, Edward D. Gaughan, R. Arthur Knoebel, Douglas S. Kurtz, and David J. Pengelley.

CHAPTER 7

456 Historical Quest (#61) from "Pólya, Problem Solving, and Education," by Alan H. Schoenfeld, *Mathematics Magazine*, Vol. 60, No. 5, December 1987, p. 290.

480 Journal problem (#56) by Peter Lindstrom from the *College Mathematics Journal*, Vol. 24, No. 4, September 1993, p. 343.

489 Modeling problem (#60) adapted from "The Learning Function," by L. L. Thurstone, *J. of General Psychology 3* (1930), pp. 469–493.

490 Putnam examination problem (#88) 1968, reprinted by permission from The Mathematical Association of America.

490 Putnam examination problem (#89) 1980, reprinted by permission from The Mathematical Association of America.

490 Putnam examination problem (#90) 1985, reprinted by permission from The Mathematical Association of America.

491 Photograph of buoy, courtesy of G. Walts/The Image Works.

491 Group research project is adapted from a computer project used at the U.S. Coast Guard Academy.

CHAPTER 8

543 Journal problem (#54) by Michael Brozinsky, *School Science and Mathematics*, Vol. 82, 1982, p. 175.

565 Historical Quest (#68) from *The Historical Development of the Calculus* by C. H. Edwards, Jr., Springer-Verlag, 1979, p. 292.

566 Journal problem (#75) by Robert C. Gebhardt, *The Pi Mu Epsilon Journal*, Vol. 8, 1984, Problem 586.

568 Ramanujan's series (#81) from *Quantum*, March/April, 1998.

569 Modeling problem (#95) adapted, "How Old Is the Earth?" by Paul J. Campbell, *UMAP Modules 1992: Tools for Teaching*, Lexington, MA: Consortium for Mathematics and Its Applications, Inc., 1993, pp. 105–137.

569 Putnam examination problem (#96) 1984, reprinted by permission from The Mathematical Association of America.

569 Putnam examination problem (#97) 1951, reprinted by permission from The Mathematical Association of America.

569 Putnam examination problem (#98) 1982, reprinted by permission from The Mathematical Association of America.

570 Photograph of a tightrope walker, courtesy of Paul Conklin, PhotoEdit.

570 Group research project is from David Pengelley of New Mexico State University.

CHAPTER 9

609 Photograph of a bicycle, courtesy of David Young-Wolff, PhotoEdit.

615 Journal problem (#56) by Frank Kocher, *The MATYC Journal*, Problem 148, Vol. 14, 1980, p. 155.

631 Putnam examination problem (#62) 1939, reprinted by permission from The Mathematical Association of America.

631 Putnam examination problem (#63) 1959, reprinted by permission from The Mathematical Association of America.

631 Putnam examination problem (#64) 1983, reprinted by permission from The Mathematical Association of America.

632 Photograph of the Starship Enterprise from *Star Trek: The Next Generation* © Paramount Pictures, courtesy of CorbisSygma.

632 Group research project is from *MAA Notes*, Vol. 17, 1991, "Priming the Calculus Pump: Innovations and Resources," by Marcus S. Cohen, Edward D. Gaughan, R. Arthur Knoebel, Douglas S. Kurtz, and David J. Pengelley.

CHAPTER 10

635 Drawing of double helix of DNA, courtesy of *Scientific American*, October 1985, p. 60.

678 Photograph of a Ferris wheel, courtesy of Marc D. Longwood.

682 Putnam examination problem (#77) 1939, reprinted by permission from The Mathematical Association of America.

683 Putnam examination problem (#78) 1947, reprinted by permission from The Mathematical Association of America.

683 Putnam examination problem (#79) 1946, reprinted by permission from The Mathematical Association of America.

684 Quotation by Johann Kepler from *The World of Mathematics*, Vol. I, James R. Newman, New York: Simon & Schuster, 1956, p. 220.

684–686 Guest essay, "The Simulation of Science," by Howard Eves. This article is reprinted by permission from *Great Moments in Mathematics Before 1650*, published by The Mathematical Association of America, 1983, p. 194.

Index

DIFFERENTIATION FORMULAS

PROCEDURAL RULES

Constant multiple rule $\quad (cf)' = cf'$

Sum rule $\quad (f + g)' = f' + g'$

Difference rule $\quad (f - g)' = f' - g'$

Linearity rule $\quad (af + bg)' = af' + bg'$

Product rule $\quad (fg)' = fg' + f'g$

Quotient rule $\quad \left(\dfrac{f}{g}\right)' = \dfrac{gf' - fg'}{g^2}$

Chain rule $\quad \dfrac{dy}{dx} = \dfrac{dy}{du}\dfrac{du}{dx}$

BASIC FORMULAS

Extended power rule $\quad \dfrac{d}{dx}u^n = nu^{n-1}\dfrac{du}{dx}$

Trigonometric rules

$$\frac{d}{dx}\sin u = \cos u \frac{du}{dx} \qquad\qquad \frac{d}{dx}\cos u = -\sin u \frac{du}{dx}$$

$$\frac{d}{dx}\tan u = \sec^2 u \frac{du}{dx} \qquad\qquad \frac{d}{dx}\cot u = -\csc^2 u \frac{du}{dx}$$

$$\frac{d}{dx}\sec u = \sec u \tan u \frac{du}{dx} \qquad\qquad \frac{d}{dx}\csc u = -\csc u \cot u \frac{du}{dx}$$

Inverse trigonometric rules

$$\frac{d}{dx}\sin^{-1} u = \frac{1}{\sqrt{1-u^2}}\frac{du}{dx} \qquad\qquad \frac{d}{dx}\cos^{-1} u = \frac{-1}{\sqrt{1-u^2}}\frac{du}{dx}$$

$$\frac{d}{dx}\tan^{-1} u = \frac{1}{1+u^2}\frac{du}{dx} \qquad\qquad \frac{d}{dx}\cot^{-1} u = \frac{-1}{1+u^2}\frac{du}{dx}$$

$$\frac{d}{dx}\sec^{-1} u = \frac{1}{|u|\sqrt{u^2-1}}\frac{du}{dx} \qquad\qquad \frac{d}{dx}\csc^{-1} u = \frac{-1}{|u|\sqrt{u^2-1}}\frac{du}{dx}$$

Logarithmic rules

$$\frac{d}{dx}\ln u = \frac{1}{u}\frac{du}{dx} \qquad\qquad \frac{d}{dx}\log_b u = \frac{\log_b e}{u}\frac{du}{dx} = \frac{1}{u\ln b}\frac{du}{dx}$$

Exponential rules

$$\frac{d}{dx}e^u = e^u \frac{du}{dx} \qquad\qquad \frac{d}{dx}b^u = b^u \ln b \frac{du}{dx}$$

Hyperbolic rules

$$\frac{d}{dx}\sinh u = \cosh u \frac{du}{dx} \qquad\qquad \frac{d}{dx}\cosh u = \sinh u \frac{du}{dx}$$

$$\frac{d}{dx}\tanh u = \operatorname{sech}^2 u \frac{du}{dx} \qquad\qquad \frac{d}{dx}\coth u = -\operatorname{csch}^2 u \frac{du}{dx}$$

$$\frac{d}{dx}\operatorname{sech} u = -\operatorname{sech} u \tanh u \frac{du}{dx} \qquad\qquad \frac{d}{dx}\operatorname{csch} u = -\operatorname{csch} u \coth u \frac{du}{dx}$$

Inverse hyperbolic rules

$$\frac{d}{dx}\sinh^{-1} u = \frac{1}{\sqrt{u^2+1}}\frac{du}{dx} \qquad\qquad \frac{d}{dx}\cosh^{-1} u = \frac{1}{\sqrt{u^2-1}}\frac{du}{dx}$$

$$\frac{d}{dx}\tanh^{-1} u = \frac{1}{1-u^2}\frac{du}{dx} \qquad\qquad \frac{d}{dx}\coth^{-1} u = \frac{1}{1-u^2}\frac{du}{dx}$$

$$\frac{d}{dx}\operatorname{sech}^{-1} u = \frac{-1}{u\sqrt{1-u^2}}\frac{du}{dx} \qquad\qquad \frac{d}{dx}\operatorname{csch}^{-1} u = \frac{-1}{|u|\sqrt{1+u^2}}\frac{du}{dx}$$

INTEGRATION FORMULAS

PROCEDURAL RULES   **Constant multiple rule**

$$\int cf(u)\,du = c\int f(u)\,du$$

Sum rule

$$\int [f(u) + g(u)]\,du = \int f(u)\,du + \int g(u)\,du$$

Difference rule

$$\int [f(u) - g(u)]\,du = \int f(u)\,du - \int g(u)\,du$$

Linearity rule

$$\int [af(u) + bg(u)]\,du = a\int f(u)\,du + b\int g(u)\,du$$

BASIC FORMULAS   **Constant rule**

$$\int 0\,du = C$$

Power rules

$$\int u^n\,du = \frac{u^{n+1}}{n+1} + C; \qquad n \neq -1$$

$$\int u^{-1}\,du = \ln|u| + C$$

Exponential rules

$$\int e^u\,du = e^u + C$$

$$\int a^u\,du = \frac{a^u}{\ln a} + C \qquad a > 0, a \neq 1$$

Logarithmic rule

$$\int \ln u\,du = u\ln u - u + C, u > 0$$

Trigonometric rules

$$\int \sin u\,du = -\cos u + C \qquad\qquad \int \cos u\,du = \sin u + C$$

$$\int \tan u\,du = -\ln|\cos u| + C \qquad\qquad \int \cot u\,du = \ln|\sin u| + C$$

$$= \ln|\sec u| + C$$

$$\int \sec u\,du = \ln|\sec u + \tan u| + C \qquad\qquad \int \csc u\,du = -\ln|\csc u + \cot u| + C$$

$$\int \sec^2 u\,du = \tan u + C \qquad\qquad \int \csc^2 u\,du = -\cot u + C$$

$$\int \sec u \tan u\,du = \sec u + C \qquad\qquad \int \csc u \cot u\,du = -\csc u + C$$

Cosine squared formula

$$\int \cos^2 u\,du = \tfrac{1}{2}u + \tfrac{1}{4}\sin 2u + C$$

Sine squared formula

$$\int \sin^2 u\,du = \tfrac{1}{2}u - \tfrac{1}{4}\sin 2u + C$$

Hyperbolic rules

$$\int \sinh u\,du = \cosh u + C \qquad\qquad \int \cosh u\,du = \sinh u + C$$

$$\int \tanh u\,du = \ln(\cosh u) + C \qquad\qquad \int \coth u\,du = \ln|\sinh u| + C$$

Inverse rules

$$\int \frac{du}{\sqrt{a^2 - u^2}} = \sin^{-1}\frac{u}{a} + C \qquad\qquad \int \frac{du}{\sqrt{u^2 - a^2}} = \cosh^{-1}\frac{u}{a} + C$$

$$\int \frac{du}{a^2 + u^2} = \frac{1}{a}\tan^{-1}\frac{u}{a} + C \qquad\qquad \int \frac{du}{a^2 - u^2} = \begin{cases} \dfrac{1}{a}\tanh^{-1}\dfrac{u}{a} + C & \text{if } \left|\dfrac{u}{a}\right| < 1 \\[2ex] \dfrac{1}{a}\coth^{-1}\dfrac{u}{a} + C & \text{if } \left|\dfrac{u}{a}\right| > 1 \end{cases}$$

$$\int \frac{du}{u\sqrt{u^2 - a^2}} = \frac{1}{a}\sec^{-1}\frac{|u|}{a} + C$$

$$\int \frac{du}{u\sqrt{a^2 - u^2}} = -\frac{1}{a}\operatorname{sech}^{-1}\left|\frac{u}{a}\right| + C = -\frac{1}{a}\ln\left|\frac{a + \sqrt{a^2 - u^2}}{u}\right| + C$$

$$\int \frac{du}{\sqrt{a^2 + u^2}} = \sinh^{-1}\frac{u}{a} + C = \ln\left(\sqrt{a^2 + u^2} + u\right) + C$$

$$\int \frac{du}{u\sqrt{a^2 + u^2}} = -\frac{1}{a}\operatorname{csch}^{-1}\left|\frac{u}{a}\right| + C = -\frac{1}{a}\ln\left|\frac{\sqrt{a^2 + u^2} + a}{u}\right| + C$$